T0180466

Mathematical Methods

Sadri Hassani

Mathematical Methods

For Students of Physics and Related Fields

 Springer

Sadri Hassani
Illinois State University
Normal, IL
USA
hassani@entropy.phy.ilstu.edu

ISBN: 978-1-4939-3712-7   e-ISBN: 978-0-387-09504-2 (eBook)
DOI 10.1007/978-0-387-09504-2

springer.com

To my wife, Sarah,
and to my children,
Dane Arash and Daisy Bita

Preface to the Second Edition

In this new edition, which is a substantially revised version of the old one, I have added five new chapters: Vectors in Relativity (Chapter 8), Tensor Analysis (Chapter 17), Integral Transforms (Chapter 29), Calculus of Variations (Chapter 30), and Probability Theory (Chapter 32). The discussion of vectors in Part II, especially the introduction of the inner product, offered the opportunity to present the special theory of relativity, which unfortunately, in most undergraduate physics curricula receives little attention. While the main motivation for this chapter was *vectors*, I grabbed the opportunity to develop the Lorentz transformation and Minkowski distance, the bedrocks of the special theory of relativity, from first principles.

The short section, *Vectors and Indices*, at the end of Chapter 8 of the first edition, was too short to demonstrate the importance of what the indices are really used for, tensors. So, I expanded that short section into a somewhat comprehensive discussion of tensors. Chapter 17, **Tensor Analysis**, takes a fresh look at vector transformations introduced in the earlier discussion of vectors, and shows the necessity of classifying them into the covariant and contravariant categories. It then introduces tensors based on—and as a generalization of—the transformation properties of covariant and contravariant vectors. In light of these transformation properties, the Kronecker delta, introduced earlier in the book, takes on a new look, and a natural and extremely useful generalization of it is introduced leading to the Levi-Civita symbol. A discussion of connections and metrics motivates a four-dimensional treatment of Maxwell's equations and a manifest unification of electric and magnetic fields. The chapter ends with Riemann curvature tensor and its place in Einstein's general relativity.

The Fourier series treatment alone does not do justice to the many applications in which aperiodic functions are to be represented. Fourier transform is a powerful tool to represent functions in such a way that the solution to many (partial) differential equations can be obtained elegantly and succinctly. Chapter 29, **Integral Transforms**, shows the power of Fourier transform in many illustrations including the calculation of Green's functions for Laplace, heat, and wave differential operators. Laplace transforms, which are useful in solving initial-value problems, are also included.

The Dirac delta function, about which there is a comprehensive discussion in the book, allows a very smooth transition from multivariable calculus to the **Calculus of Variations**, the subject of Chapter 30. This chapter takes an intuitive approach to the subject: replace the sum by an integral and the Kronecker delta by the Dirac delta function, and you get from multivariable calculus to the calculus of variations! Well, the transition may not be as simple as this, but the heart of the intuitive approach is. Once the transition is made and the master Euler-Lagrange equation is derived, many examples, including some with constraint (which use the Lagrange multiplier technique), and some from electromagnetism and mechanics are presented.

Probability Theory is essential for quantum mechanics and thermodynamics. This is the subject of Chapter 32. Starting with the basic notion of the probability space, whose prerequisite is an understanding of elementary set theory, which is also included, the notion of random variables and its connection to probability is introduced, average and variance are defined, and binomial, Poisson, and normal distributions are discussed in some detail.

Aside from the above major changes, I have also incorporated some other important changes including the rearrangement of some chapters, adding new sections and subsections to some existing chapters (for instance, the dynamics of fluids in Chapter 15), correcting all the mistakes, both typographic and conceptual, to which I have been directed by many readers of the first edition, and adding more problems at the end of each chapter. Stylistically, I thought splitting the sometimes very long chapters into smaller ones and collecting the related chapters into Parts make the reading of the text smoother. I hope I was not wrong!

I would like to thank the many instructors, students, and general readers who communicated to me comments, suggestions, and errors they found in the book. Among those, I especially thank Dan Holland for the many discussions we have had about the book, Rafael Benguria and Gebhard Grübl for pointing out some important historical and conceptual mistakes, and Ali Erdem and Thomas Ferguson for reading multiple chapters of the book, catching many mistakes, and suggesting ways to improve the presentation of the material. Jerome Brozek meticulously and diligently read most of the book and found numerous errors. Although a lawyer by profession, Mr. Brozek, as a hobby, has a keen interest in mathematical physics. I thank him for this interest and for putting it to use on my book. Last but not least, I want to thank my family, especially my wife Sarah for her unwavering support.

                                                                                                S.H.

Normal, IL
January, 2008

Preface

Innocent light-minded men, who think that astronomy can be learnt by looking at the stars without knowledge of mathematics will, in the next life, be birds.

—Plato, Timaeos

This book is intended to help bridge the wide gap separating the level of mathematical sophistication expected of students of introductory physics from that expected of students of advanced courses of undergraduate physics and engineering. While nothing beyond simple calculus is required for introductory physics courses taken by physics, engineering, and chemistry majors, the next level of courses—both in physics and engineering—already demands a readiness for such intricate and sophisticated concepts as divergence, curl, and Stokes' theorem. It is the aim of this book to make the transition between these two levels of exposure as smooth as possible.

Level and Pedagogy

I believe that the best pedagogy to teach mathematics to beginning students of physics and engineering (even mathematics, although some of my mathematical colleagues may disagree with me) is to introduce and use the concepts in a *multitude of applied settings*. This method is not unlike teaching a language to a child: it is by *repeated* usage—by the parents or the teacher—of the same word in different circumstances that a child learns the meaning of the word, and by repeated active (and sometimes wrong) usage of words that the child learns to use them in a sentence.

And what better place to use the language of mathematics than in Nature itself in the context of physics. I start with the familiar notion of, say, a derivative or an integral, but interpret it entirely in terms of physical ideas. Thus, a derivative is a means by which one obtains velocity from position vectors or acceleration from velocity vectors, and integral is a means by which one obtains the gravitational or electric field of a large number of charged or massive particles. If concepts (e.g., infinite series) do not succumb easily to physical interpretation, then I immediately subjugate the physical situation to the mathematical concepts (e.g., multipole expansion of electric potential).

Because of my belief in this pedagogy, I have kept formalism to a bare minimum. After all, a child needs no knowledge of the formalism of his or her language (i.e., grammar) to be able to read and write. Similarly, a novice in physics or engineering needs to see a lot of examples in which mathematics is used to be able to "speak the language." And I have spared no effort to provide these examples throughout the book. Of course, formalism, at some stage, becomes important. Just as grammar is taught at a higher stage of a child's education (say, in high school), mathematical formalism is to be taught at a higher stage of education of physics and engineering students (possibly in advanced undergraduate or graduate classes).

Features

The unique features of this book, which set it apart from the existing text-books, are

- the inseparable treatments of physical and mathematical concepts,

- the large number of original illustrative examples,

- the accessibility of the book to sophomores and juniors in physics and engineering programs, and

- the large number of historical notes on people and ideas.

All mathematical concepts in the book are either introduced as a natural tool for expressing some physical concept or, upon their introduction, immediately used in a physical setting. Thus, for example, differential equations are not treated as some mathematical equalities seeking solutions, but rather as a statement about the laws of Nature (e.g., the second law of motion) whose solutions describe the behavior of a physical system.

Almost all examples and problems in this book come directly from physical situations in mechanics, electromagnetism, and, to a lesser extent, quantum mechanics and thermodynamics. Although the examples are drawn from physics, they are conceptually at such an introductory level that students of engineering and chemistry will have no difficulty benefiting from the mathematical discussion involved in them.

Most mathematical-methods books are written for readers with a higher level of sophistication than a sophomore or junior physics or engineering student. This book is directly and precisely targeted at sophomores and juniors, and seven years of teaching it to such an audience have proved both the need for such a book and the adequacy of its level.

My experience with sophomores and juniors has shown that peppering the mathematical topics with a bit of history makes the subject more enticing. It also gives a little boost to the motivation of many students, which at times can run very low. The history of ideas removes the myth that all mathematical concepts are clear cut, and come into being as a finished and polished product. It reveals to the students that ideas, just like artistic masterpieces, are molded into perfection in the hands of many generations of mathematicians and physicists.

Use of Computer Algebra

As soon as one applies the mathematical concepts to real-world situations, one encounters the impossibility of finding a solution in "closed form." One is thus forced to use approximations and numerical methods of calculation. Computer algebra is especially suited for many of the examples and problems in this book.

Because of the variety of the computer algebra softwares available on the market, and the diversity in the preference of one software over another among instructors, I have left any discussion of computers out of this book. Instead, all computer and numerical chapters, examples, and problems are collected in *Mathematical Methods Using Mathematica*®, a relatively self-contained companion volume that uses *Mathematica*®.

By separating the computer-intensive topics from the text, I have made it possible for the instructor to use his or her judgment in deciding how much and in what format the use of computers should enter his or her pedagogy. The usage of *Mathematica*® in the accompanying companion volume is only a reflection of my limited familiarity with the broader field of symbolic manipulations on the computers. Instructors using other symbolic algebra programs such as Maple® and Macsyma® may generate their own examples or translate the *Mathematica*® commands of the companion volume into their favorite language.

Acknowledgments

I would like to thank all my PHY 217 students at Illinois State University who gave me a considerable amount of feedback. I am grateful to Thomas von Foerster, Executive Editor of Mathematics, Physics and Engineering at Springer-Verlag New York, Inc., for being very patient and supportive of the project as soon as he took over its editorship. Finally, I thank my wife, Sarah, my son, Dane, and my daughter, Daisy, for their understanding and support.

Unless otherwise indicated, all biographical sketches have been taken from the following sources:

Kline, M. *Mathematical Thought: From Ancient to Modern Times*, Vols. 1–3, Oxford University Press, New York, 1972.

History of Mathematics archive at *www-groups.dcs.st-and.ac.uk:80*.

Simmons, G. *Calculus Gems*, McGraw-Hill, New York, 1992.

Gamow, G. *The Great Physicists: From Galileo to Einstein*, Dover, New York, 1961.

Although extreme care was taken to correct all the misprints, it is very unlikely that I have been able to catch all of them. I shall be most grateful to those readers kind enough to bring to my attention any remaining mistakes, typographical or otherwise. Please feel free to contact me.

Sadri Hassani

Department of Physics, Illinois State University, Normal, Illinois

Note to the Reader

> "Why," said the Dodo, "the best way to explain it is to do it."
>
> —Lewis Carroll

Probably the best advice I can give you is, if you want to learn mathematics and physics, "Just do it!" As a first step, read the material in a chapter carefully, tracing the logical steps leading to important results. As a (very important) second step, *make sure you can reproduce* these logical steps, as well as all the relevant examples in the chapter, *with the book closed.* No amount of following other people's logic—whether in a book or in a lecture—can help you learn as much as a single logical step that you have taken yourself. Finally, do as many problems at the end of each chapter as your devotion and dedication to this subject allows!

Whether you are a physics or an engineering student, almost all the material you learn in this book will become handy in the rest of your academic training. Eventually, you are going to take courses in mechanics, electromagnetic theory, strength of materials, heat and thermodynamics, quantum mechanics, etc. A solid background of the mathematical methods at the level of presentation of this book will go a long way toward your deeper understanding of these subjects.

As you strive to grasp the (sometimes) difficult concepts, glance at the historical notes to appreciate the efforts of the past mathematicians and physicists as they struggled through a maze of uncharted territories in search of the correct "path," a path that demands courage, perseverance, self-sacrifice, and devotion.

At the end of most chapters, you will find a short list of references that you may want to consult for further reading. In addition to these specific references, as a general companion, I frequently refer to my more advanced book, *Mathematical Physics: A Modern Introduction to Its Foundations*, Springer-Verlag, 1999, which is abbreviated as [Has 99]. There are many other excellent books on the market; however, my own ignorance of their content and the parallelism in the pedagogy of my two books are the only reasons for singling out [Has 99].

Contents

Preface to Second Edition   vii

Preface   ix

Note to the Reader   xiii

I   Coordinates and Calculus   1

1   Coordinate Systems and Vectors   **3**
   1.1   Vectors in a Plane and in Space   3
       1.1.1   Dot Product   5
       1.1.2   Vector or Cross Product   7
   1.2   Coordinate Systems   11
   1.3   Vectors in Different Coordinate Systems   16
       1.3.1   Fields and Potentials   21
       1.3.2   Cross Product   28
   1.4   Relations Among Unit Vectors   31
   1.5   Problems   37

2   Differentiation   **43**
   2.1   The Derivative   44
   2.2   Partial Derivatives   47
       2.2.1   Definition, Notation, and Basic Properties   47
       2.2.2   Differentials   53
       2.2.3   Chain Rule   55
       2.2.4   Homogeneous Functions   57
   2.3   Elements of Length, Area, and Volume   59
       2.3.1   Elements in a Cartesian Coordinate System   60
       2.3.2   Elements in a Spherical Coordinate System   62
       2.3.3   Elements in a Cylindrical Coordinate System   65
   2.4   Problems   68

3 Integration: Formalism   **77**

3.1 "∫" Means "∫um" 77

3.2 Properties of Integral 81

    3.2.1 Change of Dummy Variable 82

    3.2.2 Linearity 82

    3.2.3 Interchange of Limits 82

    3.2.4 Partition of Range of Integration 82

    3.2.5 Transformation of Integration Variable 83

    3.2.6 Small Region of Integration 83

    3.2.7 Integral and Absolute Value 84

    3.2.8 Symmetric Range of Integration 84

    3.2.9 Differentiating an Integral 85

    3.2.10 Fundamental Theorem of Calculus 87

3.3 Guidelines for Calculating Integrals 91

    3.3.1 Reduction to Single Integrals 92

    3.3.2 Components of Integrals of Vector Functions 95

3.4 Problems 98

4 Integration: Applications   **101**

4.1 Single Integrals 101

    4.1.1 An Example from Mechanics 101

    4.1.2 Examples from Electrostatics and Gravity 104

    4.1.3 Examples from Magnetostatics 109

4.2 Applications: Double Integrals 115

    4.2.1 Cartesian Coordinates 115

    4.2.2 Cylindrical Coordinates 118

    4.2.3 Spherical Coordinates 120

4.3 Applications: Triple Integrals 122

4.4 Problems 128

5 Dirac Delta Function   **139**

5.1 One-Variable Case 139

    5.1.1 Linear Densities of Points 143

    5.1.2 Properties of the Delta Function 145

    5.1.3 The Step Function 152

5.2 Two-Variable Case 154

5.3 Three-Variable Case 159

5.4 Problems 166

II Algebra of Vectors   **171**

6 Planar and Spatial Vectors   **173**

6.1 Vectors in a Plane Revisited 174

    6.1.1 Transformation of Components 176

    6.1.2 Inner Product 182

        6.1.3   Orthogonal Transformation 190
    6.2 Vectors in Space 192
        6.2.1   Transformation of Vectors 194
        6.2.2   Inner Product 198
    6.3 Determinant 202
    6.4 The Jacobian 207
    6.5 Problems 211

7   Finite-Dimensional Vector Spaces   215
    7.1 Linear Transformations 216
    7.2 Inner Product 218
    7.3 The Determinant 222
    7.4 Eigenvectors and Eigenvalues 224
    7.5 Orthogonal Polynomials 227
    7.6 Systems of Linear Equations 230
    7.7 Problems 234

8   Vectors in Relativity   237
    8.1 Proper and Coordinate Time 239
    8.2 Spacetime Distance 240
    8.3 Lorentz Transformation 243
    8.4 Four-Velocity and Four-Momentum 247
        8.4.1   Relativistic Collisions 250
        8.4.2   Second Law of Motion 253
    8.5 Problems 254

III   Infinite Series   257

9   Infinite Series   259
    9.1 Infinite Sequences 259
    9.2 Summations 262
        9.2.1   Mathematical Induction 265
    9.3 Infinite Series 266
        9.3.1   Tests for Convergence 267
        9.3.2   Operations on Series 273
    9.4 Sequences and Series of Functions 274
        9.4.1   Properties of Uniformly Convergent Series 277
    9.5 Problems 279

10 Application of Common Series   283
    10.1 Power Series 283
        10.1.1 Taylor Series 286
    10.2 Series for Some Familiar Functions 287
    10.3 Helmholtz Coil 291
    10.4 Indeterminate Forms and L'Hôpital's Rule 294

10.5 Multipole Expansion 297
10.6 Fourier Series 299
10.7 Multivariable Taylor Series 305
10.8 Application to Differential Equations 307
10.9 Problems 311

11 Integrals and Series as Functions   **317**
11.1 Integrals as Functions 317
     11.1.1 Gamma Function 318
     11.1.2 The Beta Function 320
     11.1.3 The Error Function 322
     11.1.4 Elliptic Functions 322
11.2 Power Series as Functions 327
     11.2.1 Hypergeometric Functions 328
     11.2.2 Confluent Hypergeometric Functions 332
     11.2.3 Bessel Functions 333
11.3 Problems 336

IV   Analysis of Vectors   **341**

12 Vectors and Derivatives   **343**
12.1 Solid Angle 344
     12.1.1 Ordinary Angle Revisited 344
     12.1.2 Solid Angle 347
12.2 Time Derivative of Vectors 350
     12.2.1 Equations of Motion in a Central Force Field 352
12.3 The Gradient 355
     12.3.1 Gradient and Extremum Problems 359
12.4 Problems 362

13 Flux and Divergence   **365**
13.1 Flux of a Vector Field 365
     13.1.1 Flux Through an Arbitrary Surface 370
13.2 Flux Density = Divergence 371
     13.2.1 Flux Density 371
     13.2.2 Divergence Theorem 374
     13.2.3 Continuity Equation 378
13.3 Problems 383

14 Line Integral and Curl   **387**
14.1 The Line Integral 387
14.2 Curl of a Vector Field and Stokes' Theorem 391
14.3 Conservative Vector Fields 398
14.4 Problems 404

15 Applied Vector Analysis   **407**
  15.1 Double Del Operations 407
  15.2 Magnetic Multipoles 409
  15.3 Laplacian 411
      15.3.1 A Primer of Fluid Dynamics 413
  15.4 Maxwell's Equations 415
      15.4.1 Maxwell's Contribution 416
      15.4.2 Electromagnetic Waves in Empty Space 417
  15.5 Problems 420

16 Curvilinear Vector Analysis   **423**
  16.1 Elements of Length 423
  16.2 The Gradient 425
  16.3 The Divergence 427
  16.4 The Curl 431
      16.4.1 The Laplacian 435
  16.5 Problems 436

17 Tensor Analysis   **439**
  17.1 Vectors and Indices 439
      17.1.1 Transformation Properties of Vectors 441
      17.1.2 Covariant and Contravariant Vectors 445
  17.2 From Vectors to Tensors 447
      17.2.1 Algebraic Properties of Tensors 450
      17.2.2 Numerical Tensors 452
  17.3 Metric Tensor 454
      17.3.1 Index Raising and Lowering 457
      17.3.2 Tensors and Electrodynamics 459
  17.4 Differentiation of Tensors 462
      17.4.1 Covariant Differential and Affine Connection 462
      17.4.2 Covariant Derivative 464
      17.4.3 Metric Connection 465
  17.5 Riemann Curvature Tensor 468
  17.6 Problems 471

V   Complex Analysis   **475**

18 Complex Arithmetic   **477**
  18.1 Cartesian Form of Complex Numbers 477
  18.2 Polar Form of Complex Numbers 482
  18.3 Fourier Series Revisited 488
  18.4 A Representation of Delta Function 491
  18.5 Problems 493

19 Complex Derivative and Integral   **497**
   19.1 Complex Functions 497
      19.1.1 Derivatives of Complex Functions 499
      19.1.2 Integration of Complex Functions 503
      19.1.3 Cauchy Integral Formula 508
      19.1.4 Derivatives as Integrals 509
   19.2 Problems 511

20 Complex Series   **515**
   20.1 Power Series 516
   20.2 Taylor and Laurent Series 518
   20.3 Problems 522

21 Calculus of Residues   **525**
   21.1 The Residue 525
   21.2 Integrals of Rational Functions 529
   21.3 Products of Rational and Trigonometric Functions 532
   21.4 Functions of Trigonometric Functions 534
   21.5 Problems 536

VI   Differential Equations   **539**

22 From PDEs to ODEs   **541**
   22.1 Separation of Variables 542
   22.2 Separation in Cartesian Coordinates 544
   22.3 Separation in Cylindrical Coordinates 547
   22.4 Separation in Spherical Coordinates 548
   22.5 Problems 550

23 First-Order Differential Equations   **551**
   23.1 Normal Form of a FODE 551
   23.2 Integrating Factors 553
   23.3 First-Order Linear Differential Equations 556
   23.4 Problems 561

24 Second-Order Linear Differential Equations   **563**
   24.1 Linearity, Superposition, and Uniqueness 564
   24.2 The Wronskian 566
   24.3 A Second Solution to the HSOLDE 567
   24.4 The General Solution to an ISOLDE 569
   24.5 Sturm–Liouville Theory 570
      24.5.1 Adjoint Differential Operators 571
      24.5.2 Sturm–Liouville System 574
   24.6 SOLDEs with Constant Coefficients 575
      24.6.1 The Homogeneous Case 576
      24.6.2 Central Force Problem 579

24.6.3 The Inhomogeneous Case 583
24.7 Problems 587

25 Laplace's Equation: Cartesian Coordinates   **591**
25.1 Uniqueness of Solutions 592
25.2 Cartesian Coordinates 594
25.3 Problems 603

26 Laplace's Equation: Spherical Coordinates   **607**
26.1 Frobenius Method 608
26.2 Legendre Polynomials 610
26.3 Second Solution of the Legendre DE 617
26.4 Complete Solution 619
26.5 Properties of Legendre Polynomials 622
    26.5.1 Parity 622
    26.5.2 Recurrence Relation 622
    26.5.3 Orthogonality 624
    26.5.4 Rodrigues Formula 626
26.6 Expansions in Legendre Polynomials 628
26.7 Physical Examples 631
26.8 Problems 635

27 Laplace's Equation: Cylindrical Coordinates   **639**
27.1 The ODEs 639
27.2 Solutions of the Bessel DE 642
27.3 Second Solution of the Bessel DE 645
27.4 Properties of the Bessel Functions 646
    27.4.1 Negative Integer Order 646
    27.4.2 Recurrence Relations 646
    27.4.3 Orthogonality 647
    27.4.4 Generating Function 649
27.5 Expansions in Bessel Functions 653
27.6 Physical Examples 654
27.7 Problems 657

28 Other PDEs of Mathematical Physics   **661**
28.1 The Heat Equation 661
    28.1.1 Heat-Conducting Rod 662
    28.1.2 Heat Conduction in a Rectangular Plate 663
    28.1.3 Heat Conduction in a Circular Plate 664
28.2 The Schrödinger Equation 666
    28.2.1 Quantum Harmonic Oscillator 667
    28.2.2 Quantum Particle in a Box 675
    28.2.3 Hydrogen Atom 677
28.3 The Wave Equation 680
    28.3.1 Guided Waves 682

28.3.2 Vibrating Membrane 686
28.4 Problems 687

VII   Special Topics   691

29 Integral Transforms   693
29.1 The Fourier Transform 693
      29.1.1 Properties of Fourier Transform 696
      29.1.2 Sine and Cosine Transforms 697
      29.1.3 Examples of Fourier Transform 698
      29.1.4 Application to Differential Equations 702
29.2 Fourier Transform and Green's Functions 705
      29.2.1 Green's Function for the Laplacian 708
      29.2.2 Green's Function for the Heat Equation 709
      29.2.3 Green's Function for the Wave Equation 711
29.3 The Laplace Transform 712
      29.3.1 Properties of Laplace Transform 713
      29.3.2 Derivative and Integral of the Laplace Transform 717
      29.3.3 Laplace Transform and Differential Equations 718
      29.3.4 Inverse of Laplace Transform 721
29.4 Problems 723

30 Calculus of Variations   727
30.1 Variational Problem 728
      30.1.1 Euler-Lagrange Equation 729
      30.1.2 Beltrami identity 731
      30.1.3 Several Dependent Variables 734
      30.1.4 Several Independent Variables 734
      30.1.5 Second Variation 735
      30.1.6 Variational Problems with Constraints 738
30.2 Lagrangian Dynamics 740
      30.2.1 From Newton to Lagrange 740
      30.2.2 Lagrangian Densities 744
30.3 Hamiltonian Dynamics 747
30.4 Problems 750

31 Nonlinear Dynamics and Chaos   753
31.1 Systems Obeying Iterated Maps 754
      31.1.1 Stable and Unstable Fixed Points 755
      31.1.2 Bifurcation 757
      31.1.3 Onset of Chaos 761
31.2 Systems Obeying DEs 763
      31.2.1 The Phase Space 764
      31.2.2 Autonomous Systems 766
      31.2.3 Onset of Chaos 770

31.3 Universality of Chaos 773
    31.3.1 Feigenbaum Numbers 773
    31.3.2 Fractal Dimension 775
31.4 Problems 778

32 Probability Theory   **781**
32.1 Basic Concepts 781
    32.1.1 A Set Theory Primer 782
    32.1.2 Sample Space and Probability 784
    32.1.3 Conditional and Marginal Probabilities 786
    32.1.4 Average and Standard Deviation 789
    32.1.5 Counting: Permutations and Combinations 791
32.2 Binomial Probability Distribution 792
32.3 Poisson Distribution 797
32.4 Continuous Random Variable 801
    32.4.1 Transformation of Variables 804
    32.4.2 Normal Distribution 806
32.5 Problems 809

Bibliography   **815**

Index   **817**

Part I

Coordinates and Calculus

Chapter 1

Coordinate Systems and Vectors

Coordinates and vectors—in one form or another—are two of the most fundamental concepts in any discussion of mathematics as applied to physical problems. So, it is beneficial to start our study with these two concepts. Both vectors and coordinates have generalizations that cover a wide variety of physical situations including not only ordinary three-dimensional space with its ordinary vectors, but also the four-dimensional spacetime of relativity with its so-called *four vectors*, and even the infinite-dimensional spaces used in quantum physics with their vectors of infinite components. Our aim in this chapter is to review the *ordinary space* and how it is used to describe physical phenomena. To facilitate this discussion, we first give an outline of some of the properties of vectors.

1.1 Vectors in a Plane and in Space

We start with the most common definition of a vector as a directed line segment without regard to where the vector is located. In other words, a vector is a directed line segment whose only important attributes are its direction and its length. As long as we do not change these two attributes, the vector is not affected. Thus, we are allowed to move a vector parallel to itself without changing the vector. Examples of vectors[1] are position $\mathbf{r}$, displacement $\Delta\mathbf{r}$, velocity $\mathbf{v}$, momentum $\mathbf{p}$, electric field $\mathbf{E}$, and magnetic field $\mathbf{B}$. The vector that has no length is called the **zero vector** and is denoted by $\mathbf{0}$.

general properties of vectors

Vectors would be useless unless we could perform some kind of operation on them. The most basic operation is changing the length of a vector. This is accomplished by multiplying the vector by a real positive number. For example, $3.2\mathbf{r}$ is a vector in the same direction as $\mathbf{r}$ but 3.2 times longer. We

---

[1] Vectors will be denoted by Roman letters printed in boldface type.

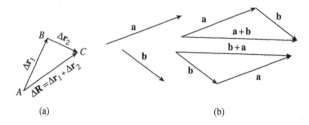

(a)   (b)

Figure 1.1: Illustration of the commutative law of addition of vectors.

operations on
vectors

can flip the direction of a vector by multiplying it by -1. That is, $(-1) \times \mathbf{r} = -\mathbf{r}$ is a vector having the same length as $\mathbf{r}$ but pointing in the opposite direction. We can combine these two operations and think of multiplying a vector by any real (positive or negative) number. The result is another vector *lying along the same line* as the original vector. Thus, $-0.732\mathbf{r}$ is a vector that is 0.732 times as long as $\mathbf{r}$ and points in the opposite direction. The zero *vector* is obtained every time one multiplies any vector by the *number* zero.

Another operation is the addition of two vectors. This operation, with which we assume the reader to have some familiarity, is inspired by the obvious addition law for displacements. In Figure 1.1(a), a displacement, $\Delta\mathbf{r}_1$ from A to B is added to the displacement $\Delta\mathbf{r}_2$ from B to C to give $\Delta\mathbf{R}$ their resultant, or their sum, i.e., the displacement from A to C: $\Delta\mathbf{r}_1 + \Delta\mathbf{r}_2 = \Delta\mathbf{R}$. Figure 1.1(b) shows that addition of vectors is commutative: $\mathbf{a} + \mathbf{b} = \mathbf{b} + \mathbf{a}$. It is also associative, $\mathbf{a} + (\mathbf{b} + \mathbf{c}) = (\mathbf{a} + \mathbf{b}) + \mathbf{c}$, i.e., the order in which you add vectors is irrelevant. It is clear that $\mathbf{a} + \mathbf{0} = \mathbf{0} + \mathbf{a} = \mathbf{a}$ for any vector $\mathbf{a}$.

Example 1.1.1. The parametric equation of a line through two given points can be obtained in vector form by noting that any point in space defines a vector whose components are the coordinates of the given point.[2] If the components of the points P and Q in Figure 1.2 are, respectively, (p_x, p_y, p_z) and (q_x, q_y, q_z), then we can define vectors $\mathbf{p}$ and $\mathbf{q}$ with those components. An arbitrary point X with components (x, y, z) will lie on the line PQ if and only if the vector $\mathbf{x} = (x, y, z)$ has its tip on that line. This will happen if and only if the vector joining P and X, namely $\mathbf{x} - \mathbf{p}$, is proportional to the vector joining P and Q, namely $\mathbf{q} - \mathbf{p}$. Thus,

vector form of the
parametric
equation of a line

for some real number t, we must have

$$\mathbf{x} - \mathbf{p} = t(\mathbf{q} - \mathbf{p}) \qquad \text{or} \qquad \mathbf{x} = t(\mathbf{q} - \mathbf{p}) + \mathbf{p}.$$

This is the vector form of the equation of a line. We can write it in component form by noting that the equality of vectors implies the equality of corresponding components. Thus,

$$x = (q_x - p_x)t + p_x,$$
$$y = (q_y - p_y)t + p_y,$$
$$z = (q_z - p_z)t + p_z,$$

which is the usual parametric equation for a line.   ∎

---

[2]We shall discuss components and coordinates in greater detail later in this chapter. For now, the knowledge gained in calculus is sufficient for our discussion.

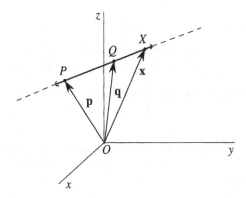

Figure 1.2: The parametric equation of a line in space can be obtained easily using vectors.

There are some special vectors that are extremely useful in describing physical quantities. These are the **unit vectors**. If one divides a vector by its length, one gets a unit vector in the direction of the original vector. Unit vectors are generally denoted by the symbol $\hat{\mathbf{e}}$ with a subscript which designates its direction. Thus, if we divided the vector $\mathbf{a}$ by its length $|\mathbf{a}|$ we get the unit vector $\hat{\mathbf{e}}_a$ in the direction of $\mathbf{a}$. Turning this definition around, we have

use of unit vectors

> **Box 1.1.1.** *If we know the magnitude $|\mathbf{a}|$ of a vector quantity as well as its direction $\hat{\mathbf{e}}_a$, we can construct the vector:* $\mathbf{a} = |\mathbf{a}|\hat{\mathbf{e}}_a$.

This construction will be used often in the sequel.

The most commonly used unit vectors are those in the direction of coordinate axes. Thus $\hat{\mathbf{e}}_x$, $\hat{\mathbf{e}}_y$, and $\hat{\mathbf{e}}_z$ are the unit vectors pointing in the positive directions of the x-, y-, and z-axes, respectively.[3] We shall introduce unit vectors in other coordinate systems when we discuss those coordinate systems later in this chapter.

unit vectors along the x-, y-, and z-axes

1.1.1 Dot Product

The reader is no doubt familiar with the concept of **dot product** whereby two vectors are "multiplied" and the result is a number. The dot product of $\mathbf{a}$ and $\mathbf{b}$ is defined by

dot product defined

$$\mathbf{a} \cdot \mathbf{b} \equiv |\mathbf{a}|\,|\mathbf{b}|\cos\theta, \tag{1.1}$$

where $|\mathbf{a}|$ is the length of $\mathbf{a}$, $|\mathbf{b}|$ is the length of $\mathbf{b}$, and θ is the angle between the two vectors. This definition is motivated by many physical situations.

---

[3]These unit vectors are usually denoted by $\mathbf{i}$, $\mathbf{j}$, and $\mathbf{k}$, a notation that can be confusing when other non-Cartesian coordinates are used. We shall not use this notation, but adhere to the more suggestive notation introduced above.

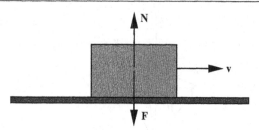

Figure 1.3: No work is done by a force orthogonal to displacement. If such a work were not zero, it would have to be positive or negative; but no consistent rule exists to assign a sign to the work.

The prime example is work which is defined as the scalar product of force and displacement. The presence of $\cos\theta$ ensures the requirement that the work done by a force perpendicular to the displacement is zero. If this requirement were not met, we would have the precarious situation of Figure 1.3 in which the two vertical forces add up to zero but the total work done by them is not zero! This is because it would be impossible to assign a "sign" to the work done by forces being displaced perpendicular to themselves, and make the rule of such an assignment in such a way that the work of **F** in the figure cancels that of **N**. (The reader is urged to try to come up with a rule—e.g., assigning a positive sign to the work if the velocity points to the right of the observer and a negative sign if it points to the observer's left—and see that it will not work, no matter how elaborate it may be!) The only logical definition of work is that which includes a $\cos\theta$ factor.

properties of dot product

The dot product is clearly commutative, $\mathbf{a} \cdot \mathbf{b} = \mathbf{b} \cdot \mathbf{a}$. Moreover, it distributes over vector addition

$$(\mathbf{a} + \mathbf{b}) \cdot \mathbf{c} = \mathbf{a} \cdot \mathbf{c} + \mathbf{b} \cdot \mathbf{c}.$$

To see this, note that Equation (1.1) can be interpreted as the product of the length of **a** with the projection of **b** along **a**. Now Figure 1.4 demonstrates[4] that the projection of $\mathbf{a} + \mathbf{b}$ along **c** is the sum of the projections of **a** and **b** along **c** (see Problem 1.2 for details). The third property of the inner product is that $\mathbf{a} \cdot \mathbf{a}$ is always a *positive* number unless **a** is the zero vector in which

properties defining the dot (inner) product

case $\mathbf{a} \cdot \mathbf{a} = 0$. In mathematics, the collection of these three properties—commutativity, positivity, and distribution over addition—*defines* a dot (or inner) product on a vector space.

The definition of the dot product leads directly to $\mathbf{a} \cdot \mathbf{a} = |\mathbf{a}|^2$ or

$$|\mathbf{a}| = \sqrt{\mathbf{a} \cdot \mathbf{a}}, \tag{1.2}$$

which is useful in calculating the length of sums or differences of vectors.

---

[4]Figure 1.4 appears to prove the distributive property only for vectors lying in the same plane. However, the argument will be valid even if the three vectors are not coplanar. Instead of dropping perpendicular *lines* from the tips of **a** and **b**, one drops perpendicular *planes*.

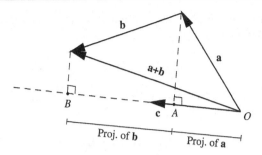

Figure 1.4: The distributive property of the dot product is clearly demonstrated if we interpret the dot product as the length of one vector times the projection of the other vector on the first.

One can use the distributive property of the dot product to show that if (a_x, a_y, a_z) and (b_x, b_y, b_z) represent the components of $\mathbf{a}$ and $\mathbf{b}$ along the axes x, y, and z, then

$$\mathbf{a} \cdot \mathbf{b} = a_x b_x + a_y b_y + a_z b_z. \tag{1.3}$$

dot product in terms of components

From the definition of the dot product, we can draw an important conclusion. If we divide both sides of $\mathbf{a} \cdot \mathbf{b} = |\mathbf{a}|\,|\mathbf{b}|\cos\theta$ by $|\mathbf{a}|$, we get

$$\frac{\mathbf{a} \cdot \mathbf{b}}{|\mathbf{a}|} = |\mathbf{b}|\cos\theta \quad \text{or} \quad \left(\frac{\mathbf{a}}{|\mathbf{a}|}\right) \cdot \mathbf{b} = |\mathbf{b}|\cos\theta \Rightarrow \hat{\mathbf{e}}_a \cdot \mathbf{b} = |\mathbf{b}|\cos\theta.$$

Noting that $|\mathbf{b}|\cos\theta$ is simply the projection of $\mathbf{b}$ along $\mathbf{a}$, we conclude

a useful relation to be used frequently in the sequel

> **Box 1.1.2.** *To find the perpendicular projection of a vector* $\mathbf{b}$ *along another vector* $\mathbf{a}$, *take the dot product of* $\mathbf{b}$ *with* $\hat{\mathbf{e}}_a$, *the unit vector along* $\mathbf{a}$.

Sometimes "component" is used for perpendicular projection. This is not entirely correct. For any set of three *mutually perpendicular* unit vectors in space, Box 1.1.2 can be used to find the *components* of a vector along the three unit vectors. Only if the unit vectors are mutually perpendicular do components and projections coincide.

1.1.2 Vector or Cross Product

Given two space vectors, $\mathbf{a}$ and $\mathbf{b}$, we can find a third *space* vector $\mathbf{c}$, called the **cross product** of $\mathbf{a}$ and $\mathbf{b}$, and denoted by $\mathbf{c} = \mathbf{a} \times \mathbf{b}$. The magnitude of $\mathbf{c}$ is defined by $|\mathbf{c}| = |\mathbf{a}|\,|\mathbf{b}|\sin\theta$ where θ is the angle between $\mathbf{a}$ and $\mathbf{b}$. The direction of $\mathbf{c}$ is given by the **right-hand rule**: If $\mathbf{a}$ is turned to $\mathbf{b}$ (note the order in which $\mathbf{a}$ and $\mathbf{b}$ appear here) through the angle between $\mathbf{a}$ and $\mathbf{b}$,

cross product of two space vectors

right-hand rule explained

a (right-handed) screw that is perpendicular to **a** and **b** will advance in the direction of **a** × **b**. This definition implies that

$$\mathbf{a} \times \mathbf{b} = -\mathbf{b} \times \mathbf{a}.$$

cross product is antisymmetric

This property is described by saying that the cross product is **antisymmetric**. The definition also implies that

$$\mathbf{a} \cdot (\mathbf{a} \times \mathbf{b}) = \mathbf{b} \cdot (\mathbf{a} \times \mathbf{b}) = 0.$$

That is, **a** × **b** is perpendicular to both **a** and **b**.[5]

The vector product has the following properties:

$$\mathbf{a} \times (\alpha\mathbf{b}) = (\alpha\mathbf{a}) \times \mathbf{b} = \alpha(\mathbf{a} \times \mathbf{b}), \qquad \mathbf{a} \times \mathbf{b} = -\mathbf{b} \times \mathbf{a},$$
$$\mathbf{a} \times (\mathbf{b}+\mathbf{c}) = \mathbf{a} \times \mathbf{b} + \mathbf{a} \times \mathbf{c}, \qquad \mathbf{a} \times \mathbf{a} = 0. \qquad (1.4)$$

cross product in terms of components

Using these properties, we can write the vector product of two vectors in terms of their components. We are interested in a more general result valid in other coordinate systems as well. So, rather than using x, y, and z as subscripts for unit vectors, we use the numbers 1, 2, and 3. In that case, our results can also be used for spherical and cylindrical coordinates which we shall discuss shortly.

$$\begin{aligned}\mathbf{a} \times \mathbf{b} &= (\alpha_1\hat{\mathbf{e}}_1 + \alpha_2\hat{\mathbf{e}}_2 + \alpha_3\hat{\mathbf{e}}_3) \times (\beta_1\hat{\mathbf{e}}_1 + \beta_2\hat{\mathbf{e}}_2 + \beta_3\hat{\mathbf{e}}_3)\\ &= \alpha_1\beta_1\hat{\mathbf{e}}_1 \times \hat{\mathbf{e}}_1 + \alpha_1\beta_2\hat{\mathbf{e}}_1 \times \hat{\mathbf{e}}_2 + \alpha_1\beta_3\hat{\mathbf{e}}_1 \times \hat{\mathbf{e}}_3\\ &\quad + \alpha_2\beta_1\hat{\mathbf{e}}_2 \times \hat{\mathbf{e}}_1 + \alpha_2\beta_2\hat{\mathbf{e}}_2 \times \hat{\mathbf{e}}_2 + \alpha_2\beta_3\hat{\mathbf{e}}_2 \times \hat{\mathbf{e}}_3\\ &\quad + \alpha_3\beta_1\hat{\mathbf{e}}_3 \times \hat{\mathbf{e}}_1 + \alpha_3\beta_2\hat{\mathbf{e}}_3 \times \hat{\mathbf{e}}_2 + \alpha_3\beta_3\hat{\mathbf{e}}_3 \times \hat{\mathbf{e}}_3.\end{aligned}$$

But, by the last property of Equation (1.4), we have

$$\hat{\mathbf{e}}_1 \times \hat{\mathbf{e}}_1 = \hat{\mathbf{e}}_2 \times \hat{\mathbf{e}}_2 = \hat{\mathbf{e}}_3 \times \hat{\mathbf{e}}_3 = 0.$$

right-handed set of unit vectors

Also, if we assume that $\hat{\mathbf{e}}_1$, $\hat{\mathbf{e}}_2$, and $\hat{\mathbf{e}}_3$ form a so-called **right-handed set**, i.e., if

$$\hat{\mathbf{e}}_1 \times \hat{\mathbf{e}}_2 = -\hat{\mathbf{e}}_2 \times \hat{\mathbf{e}}_1 = \hat{\mathbf{e}}_3,$$
$$\hat{\mathbf{e}}_1 \times \hat{\mathbf{e}}_3 = -\hat{\mathbf{e}}_3 \times \hat{\mathbf{e}}_1 = -\hat{\mathbf{e}}_2, \qquad (1.5)$$
$$\hat{\mathbf{e}}_2 \times \hat{\mathbf{e}}_3 = -\hat{\mathbf{e}}_3 \times \hat{\mathbf{e}}_2 = \hat{\mathbf{e}}_1,$$

then we obtain

$$\mathbf{a} \times \mathbf{b} = (\alpha_2\beta_3 - \alpha_3\beta_2)\hat{\mathbf{e}}_1 + (\alpha_3\beta_1 - \alpha_1\beta_3)\hat{\mathbf{e}}_2 + (\alpha_1\beta_2 - \alpha_2\beta_1)\hat{\mathbf{e}}_3$$

---

[5]This fact makes it clear why **a** × **b** is not defined in the plane. Although it is possible to define **a** × **b** for vectors **a** and **b** lying in a plane, **a** × **b** *will not lie in that plane* (it will be perpendicular to that plane). For the vector product, **a** and **b** (although lying in a plane) must be considered as space vectors.

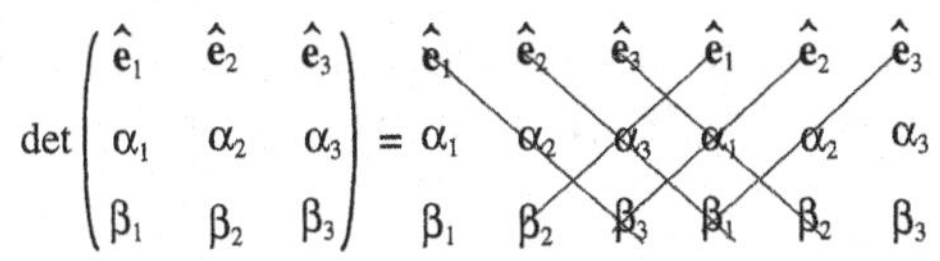

$$\det \begin{pmatrix} \hat{e}_1 & \hat{e}_2 & \hat{e}_3 \\ \alpha_1 & \alpha_2 & \alpha_3 \\ \beta_1 & \beta_2 & \beta_3 \end{pmatrix} = \alpha_1 \quad \begin{matrix} \hat{e}_1 & \hat{e}_2 & \hat{e}_3 & \hat{e}_1 & \hat{e}_2 & \hat{e}_3 \\ \alpha_2 & \alpha_3 & \alpha_1 & \alpha_2 & \alpha_3 \\ \beta_1 & \beta_2 & \beta_3 & \beta_1 & \beta_2 & \beta_3 \end{matrix}$$

Figure 1.5: A 3×3 determinant is obtained by writing the entries twice as shown, multiplying all terms on each slanted line and adding the results. The lines from upper left to lower right bear a positive sign, and those from upper right to lower left a negative sign.

which can be nicely written in a determinant form[6]

$$\mathbf{a} \times \mathbf{b} = \det \begin{pmatrix} \hat{e}_1 & \hat{e}_2 & \hat{e}_3 \\ \alpha_1 & \alpha_2 & \alpha_3 \\ \beta_1 & \beta_2 & \beta_3 \end{pmatrix}. \qquad (1.6)$$

cross product in terms of the determinant of components

Figure 1.5 explains the rule for "expanding" a determinant.

Example 1.1.2. From the definition of the vector product and Figure 1.6(a), we note that

$$|\mathbf{a} \times \mathbf{b}| = \text{area of the parallelogram defined by } \mathbf{a} \text{ and } \mathbf{b}.$$

area of a parallelogram in terms of cross product of its two sides

So we can use Equation (1.6) to find the area of a parallelogram defined by two vectors directly in terms of their components. For instance, the area defined by $\mathbf{a} = (1, 1, -2)$ and $\mathbf{b} = (2, 0, 3)$ can be found by calculating their vector product

$$\mathbf{a} \times \mathbf{b} = \det \begin{pmatrix} \hat{e}_1 & \hat{e}_2 & \hat{e}_3 \\ 1 & 1 & -2 \\ 2 & 0 & 3 \end{pmatrix} = 3\hat{e}_1 - 7\hat{e}_2 - 2\hat{e}_3,$$

and then computing its length

$$|\mathbf{a} \times \mathbf{b}| = \sqrt{3^2 + (-7)^2 + (-2)^2} = \sqrt{62}. \qquad \blacksquare$$

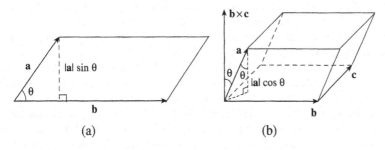

(a) (b)

Figure 1.6: (a) The area of a parallelogram is the absolute value of the cross product of the two vectors describing its sides. (b) The volume of a parallelepiped can be obtained by mixing the dot and the cross products.

---

[6]No knowledge of determinants is necessary at this point. The reader may consider (1.6) to be a mnemonic device useful for remembering the components of $\mathbf{a} \times \mathbf{b}$.

Example 1.1.3. The volume of a parallelepiped defined by three non-coplanar vectors, $\mathbf{a}$, $\mathbf{b}$, and $\mathbf{c}$, is given by $|\mathbf{a} \cdot (\mathbf{b} \times \mathbf{c})|$. This can be seen from Figure 1.6(b), where it is clear that

volume of a parallelepiped as a combination of dot and cross products

$$\text{volume} = (\text{area of base})(\text{altitude}) = |\mathbf{b} \times \mathbf{c}|(|\mathbf{a}|\cos\theta) = |(\mathbf{b} \times \mathbf{c}) \cdot \mathbf{a}|.$$

The absolute value is taken to ensure the positivity of the area. In terms of components we have

$$\text{volume} = |(\mathbf{b} \times \mathbf{c})_1\alpha_1 + (\mathbf{b} \times \mathbf{c})_2\alpha_2 + (\mathbf{b} \times \mathbf{c})_3\alpha_3|$$
$$= |(\beta_2\gamma_3 - \beta_3\gamma_2)\alpha_1 + (\beta_3\gamma_1 - \beta_1\gamma_3)\alpha_2 + (\beta_1\gamma_2 - \beta_2\gamma_1)\alpha_3|,$$

volume of a parallelepiped as the determinant of the components of its side vectors

which can be written in determinant form as

$$\text{volume} = |\mathbf{a} \cdot (\mathbf{b} \times \mathbf{c})| = \left| \det \begin{pmatrix} \alpha_1 & \alpha_2 & \alpha_3 \\ \beta_1 & \beta_2 & \beta_3 \\ \gamma_1 & \gamma_2 & \gamma_3 \end{pmatrix} \right|.$$

Note how we have put the absolute value sign around the determinant of the matrix, so that the area comes out positive. ∎

Historical Notes

The concept of vectors as directed line segments that could represent velocities, forces, or accelerations has a very long history. Aristotle knew that the effect of two forces acting on an object could be described by a single force using what is now called *the parallelogram law*. However, the real development of the concept took an unexpected turn in the nineteenth century.

With the advent of *complex numbers* and the realization by Gauss, Wessel, and especially Argand, that they could be represented by points in a plane, mathematicians discovered that complex numbers could be used to study vectors in a plane. A complex number is represented by a pair[7] of real numbers—called the real and imaginary parts of the complex number—which could be considered as the two components of a planar vector.

This connection between vectors in a plane and complex numbers was well established by 1830. Vectors are, however, useful only if they are treated as objects in *space*. After all, velocities, forces, and accelerations are mostly three-dimensional objects. So, the two-dimensional complex numbers had to be generalized to three dimensions. This meant inventing ways of adding, subtracting, multiplying, and dividing objects such as (x, y, z).

The invention of a spatial analogue of the planar complex numbers is due to **William R. Hamilton**. Next to Newton, Hamilton is the greatest of all English mathematicians, and like Newton he was even greater as a physicist than as a mathematician. At the age of five Hamilton could read Latin, Greek, and Hebrew. At eight he added Italian and French; at ten he could read Arabic and Sanskrit, and at fourteen, Persian. A contact with a lightning calculator inspired him to study mathematics. In 1822 at the age of seventeen and a year before he entered Trinity College in Dublin, he prepared a paper on caustics which was read before the Royal Irish Academy in 1824 but not published. Hamilton was advised to rework and expand it. In 1827 he submitted to the Academy a revision which initiated the science of geometrical optics and introduced new techniques in analytical mechanics.

William R.
Hamilton
1805–1865

---

[7]See Chapter 18.

In 1827, while still an undergraduate, he was appointed Professor of Astronomy at Trinity College in which capacity he had to manage the astronomical observations and teach science. He did not do much of the former, but he was a fine lecturer.

Hamilton had very good intuition, and knew how to use analogy to reason from the known to the unknown. Although he lacked great flashes of insight, he worked very hard and very long on special problems to see what generalizations they would lead to. He was patient and systematic in working on specific problems and was willing to go through detailed and laborious calculations to check or prove a point.

After mastering and clarifying the concept of complex numbers and their relation to planar vectors (see Problem 18.11 for the connection between complex multiplication on the one hand, and dot and cross products on the other), Hamilton was able to think more clearly about the three-dimensional generalization. His efforts led unfortunately to frustration because the vectors (a) required four components, and (b) defied commutativity! Both features were revolutionary and set the standard for algebra. He called these new numbers **quaternions**.

In retrospect, one can see that the new three-dimensional complex numbers *had* to contain four components. Each "number," when acting on a vector, rotates the latter about an axis and stretches (or contracts) it. Two angles are required to specify the axis of rotation, one angle to specify the amount of rotation, and a fourth number to specify the amount of stretch (or contraction).

Hamilton announced the invention of quaternions in 1843 at a meeting of the Royal Irish Academy, and spent the rest of his life developing the subject.

1.2   Coordinate Systems

Coordinates are "functions" that specify **points** of a space. The smallest number of these functions necessary to specify a point is called the **dimension** of that space. For instance, a point of a plane is specified by two numbers, and as the point moves in the plane the two numbers change, i.e., the coordinates are functions of the position of the point. If we designate the point as P, we may write the coordinate functions of P as $(f(P), g(P))$.[8] Each pair of such functions is called a **coordinate system**.

coordinate systems as functions.

There are two coordinate systems used for a plane, **Cartesian**, denoted by $(x(P), y(P))$, and **polar**, denoted by $(r(P), \theta(P))$. As shown in Figure 1.7,

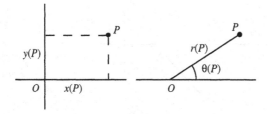

Figure 1.7: Cartesian and polar coordinates of a point P in two dimensions.

---

[8]Think of f (or g) as a rule by which a unique number is assigned to each point P.

the "function" x is defined as giving the distance from P to the vertical axis, while θ is the function which gives the angle that the line OP makes with a given fiducial (usually horizontal) line. The origin O and the fiducial line are completely arbitrary. Similarly, the functions r and y give distances from the origin and to the horizontal axis, respectively.

Box 1.2.1. *In practice, one drops the argument P and writes (x, y) and (r, θ).*

the three common
coordinate
systems:
Cartesian,
cylindrical and
spherical

We can generalize the above concepts to three dimensions. There are three coordinate functions now. So for a point P in space we write

$$(f(P), g(P), h(P)),$$

where f, g, and h are functions on the three-dimensional space. There are three widely used coordinate systems, **Cartesian** $(x(P), y(P), z(P))$, **cylindrical** $(\rho(P), \varphi(P), z(P))$, and **spherical** $(r(P), \theta(P), \varphi(P))$. $\varphi(P)$ is called the **azimuth** or the **azimuthal angle** of P, while $\theta(P)$ is called its **polar angle**. To find the spherical coordinates of P, one chooses an arbitrary point as the origin O and an arbitrary line through O called the **polar axis**. One measures $\overline{OP}$ and calls it $r(P)$; $\theta(P)$ is the angle between $\overline{OP}$ and the polar axis. To find the third coordinate, we construct the plane through O and perpendicular to the polar axis, drop a projection from P to the plane meeting the latter at H, draw an arbitrary fiducial line through O in this plane, and measure the angle between this line and $\overline{OH}$. This angle is $\varphi(P)$. Cartesian and cylindrical coordinate systems can be described similarly. The three coordinate systems are shown in Figure 1.8. As indicated in the figure, the polar axis is usually taken to be the z-axis, and the fiducial line from which $\varphi(P)$ is measured is chosen to be the x-axis. Although there are other coordinate systems, the three mentioned above are by far the most widely used.

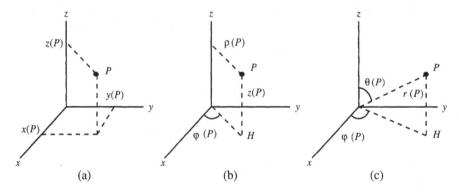

Figure 1.8: (a) Cartesian, (b) cylindrical, and (c) spherical coordinates of a point P in three dimensions.

Which one of the three systems of coordinates to use in a given physical problem is dictated mainly by the geometry of that problem. As a rule, spherical coordinates are best suited for spheres and spherically symmetric problems. Spherical symmetry describes situations in which quantities of interest are functions only of the distance from a fixed point and not on the orientation of that distance. Similarly, cylindrical coordinates ease calculations when cylinders or cylindrical symmetries are involved. Finally, Cartesian coordinates are used in rectangular geometries.

Of the three coordinate systems, Cartesian is the most complete in the following sense: *A point in space can have only* one triplet *as its coordinates.* This property is not shared by the other two systems. For example, a point P located on the z-axis of a cylindrical coordinate system does not have a well-defined $\varphi(P)$. In practice, such imperfections are not of dire consequence and we shall ignore them.

limitations of non-Cartesian coordinates

Once we have three coordinate systems to work with, we need to know how to translate from one to another. First we give the transformation rule from spherical to cylindrical. It is clear from Figure 1.9 that

$$\rho = r\sin\theta, \quad \varphi_{\mathrm{cyl}} = \varphi_{\mathrm{sph}}, \quad z = r\cos\theta. \tag{1.7}$$

transformation from spherical to cylindrical coordinates

Thus, given (r, θ, φ) of a point P, we can obtain (ρ, φ, z) of the same point by substituting in the RHS.

Next we give the transformation rule from cylindrical to Cartesian. Again Figure 1.9 gives the result:

$$x = \rho\cos\varphi, \quad y = \rho\sin\varphi, \quad z_{\mathrm{car}} = z_{\mathrm{cyl}}. \tag{1.8}$$

transformation from cylindrical to Cartesian coordinates

We can combine (1.7) and (1.8) to connect Cartesian and spherical coordinates:

$$x = r\sin\theta\cos\varphi, \quad y = r\sin\theta\sin\varphi, \quad z = r\cos\theta. \tag{1.9}$$

transformation from spherical to Cartesian coordinates

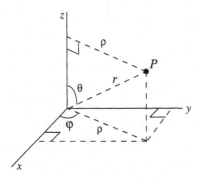

Figure 1.9: The relation between the cylindrical and spherical coordinates of a point P can be obtained using this diagram.

The transformations given are in their standard form. We can turn them around and give the inverse transformations. For instance, squaring the first and third equations of (1.7) and adding gives $\rho^2 + z^2 = r^2$ or $r = \sqrt{\rho^2 + z^2}$. Similarly, dividing the first and third equation yields $\tan\theta = \rho/z$, which implies that $\theta = \tan^{-1}(\rho/z)$, or equivalently,

$$\frac{z}{r} = \cos\theta \;\Rightarrow\; \theta = \cos^{-1}\left(\frac{z}{r}\right) = \cos^{-1}\left(\frac{z}{\sqrt{\rho^2 + z^2}}\right).$$

transformation from cylindrical to spherical coordinates

Thus, the inverse of (1.7) is

$$r = \sqrt{\rho^2 + z^2}, \quad \theta = \tan^{-1}\left(\frac{\rho}{z}\right) = \cos^{-1}\left(\frac{z}{\sqrt{\rho^2 + z^2}}\right), \quad \varphi_{\text{sph}} = \varphi_{\text{cyl}}.$$

$$(1.10)$$

Similarly, the inverse of (1.8) is

$$\rho = \sqrt{x^2 + y^2},$$

$$\varphi = \tan^{-1}\left(\frac{y}{x}\right) = \cos^{-1}\left(\frac{x}{\sqrt{x^2 + y^2}}\right) = \sin^{-1}\left(\frac{y}{\sqrt{x^2 + y^2}}\right), \quad (1.11)$$

$$z_{\text{cyl}} = z_{\text{car}},$$

transformation from Cartesian to spherical coordinates

and that of (1.9) is

$$r = \sqrt{x^2 + y^2 + z^2},$$

$$\theta = \tan^{-1}\left(\frac{\sqrt{x^2 + y^2}}{z}\right) = \cos^{-1}\left(\frac{z}{\sqrt{x^2 + y^2 + z^2}}\right)$$

$$= \sin^{-1}\left(\frac{\sqrt{x^2 + y^2}}{\sqrt{x^2 + y^2 + z^2}}\right), \quad (1.12)$$

$$\varphi = \tan^{-1}\left(\frac{y}{x}\right) = \cos^{-1}\left(\frac{x}{\sqrt{x^2 + y^2}}\right) = \sin^{-1}\left(\frac{y}{\sqrt{x^2 + y^2}}\right).$$

An important question concerns the *range* of these quantities. In other words: *In what range should we allow these quantities to vary in order to cover the whole space?* For Cartesian coordinates all three variables vary between $-\infty$ and $+\infty$. Thus,

range of coordinate variables

$$-\infty < x < +\infty, \quad -\infty < y < +\infty, \quad -\infty < z < +\infty.$$

The ranges of cylindrical coordinates are

$$0 \leq \rho < \infty, \quad 0 \leq \varphi \leq 2\pi, \quad -\infty < z < \infty.$$

Note that ρ, being a distance, cannot have negative values.[9] Similarly, the ranges of spherical coordinates are

$$0 \le r < \infty, \quad 0 \le \theta \le \pi, \quad 0 \le \varphi \le 2\pi.$$

Again, r is never negative for similar reasons as above. Also note that the range of θ excludes values larger than π. This is because the range of φ takes care of points where θ "appears" to have been increased by π.

Historical Notes

One of the greatest achievements in the development of mathematics since Euclid was the introduction of coordinates. Two men take credit for this development: Fermat and Descartes. These two great French mathematicians were interested in the unification of geometry and algebra, which resulted in the creation of a most fruitful branch of mathematics now called **analytic geometry**. Fermat and Descartes who were heavily involved in physics, were keenly aware of both the need for quantitative methods and the capacity of algebra to deliver that method.

Fermat's interest in the unification of geometry and algebra arose because of his involvement in optics. His interest in the attainment of maxima and minima—thus his contribution to calculus—stemmed from the investigation of the passage of light rays through media of different indices of refraction, which resulted in *Fermat's principle* in optics and the law of refraction. With the introduction of coordinates, Fermat was able to quantify the study of optics and set a trend to which all physicists of posterity would adhere. It is safe to say that without analytic geometry the progress of science, and in particular physics, would have been next to impossible.

Pierre de Fermat
1601–1665

Born into a family of tradespeople, **Pierre de Fermat** was trained as a lawyer and made his living in this profession becoming a councillor of the parliament of the city of Toulouse. Although mathematics was but a hobby for him and he could devote only spare time to it, he made great contributions to number theory, to calculus, and, together with **Pascal**, initiated work on probability theory.

The coordinate system introduced by Fermat was not a convenient one. For one thing, the coordinate axes were not at right angles to one another. Furthermore, the use of negative coordinates was not considered. Nevertheless, he was able to translate geometric curves into algebraic equations.

René Descartes was a great philosopher, a founder of modern biology, and a superb physicist and mathematician. His interest in mathematics stemmed from his desire to understand nature. He wrote:

> ...I have resolved to quit only abstract geometry, that is to say, the consideration of questions which *serve only to exercise the mind*, and this, in order to study another kind of geometry, which has for its object the explanation of the phenomena of nature.

René Descartes
1596–1650

His father, a relatively wealthy lawyer, sent him to a Jesuit school at the age of eight where, due to his delicate health, he was allowed to spend the mornings in bed, during which time he worked. He followed this habit during his entire life. At twenty he graduated from the University of Poitier as a lawyer and went to Paris where he studied mathematics with a Jesuit priest. After one year he decided to

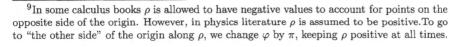

[9] In some calculus books ρ is allowed to have negative values to account for points on the opposite side of the origin. However, in physics literature ρ is assumed to be positive. To go to "the other side" of the origin along ρ, we change φ by π, keeping ρ positive at all times.

join the army of Prince Maurice of Orange in 1617. During the next nine years he vacillated between various armies while studying mathematics.

He eventually returned to Paris, where he devoted his efforts to the study of optical instruments motivated by the newly discovered power of the telescope. In 1628 he moved to Holland to a quieter and freer intellectual environment. There he lived for the next twenty years and wrote his famous works. In 1649 Queen Christina of Sweden persuaded Descartes to go to Stockholm as her private tutor. However the Queen had an uncompromising desire to draw curves and tangents at 5 a.m., causing Descartes to break the lifelong habit of getting up at 11 o'clock! After only a few months in the cold northern climate, walking to the palace for the 5 o'clock appointment with the queen, he died of pneumonia in 1650.

Descartes described his algebraic approach to geometry in his monumental work *La Géométrie*. It is in this work that he solves geometrical problems using algebra by introducing coordinates. These coordinates, as in Fermat's case, were not lengths along perpendicular axes. Nevertheless they paved the way for the later generations of scientists such as **Newton** to build on Descartes' and Fermat's ideas and improve on them.

Newton uses polar coordinates for the first time

Throughout the seventeenth century, mathematicians used one axis with the y values drawn at an oblique or right angle onto that axis. Newton, however, in a book called *The Method of Fluxions and Infinite Series* written in 1671, and translated much later into English in 1736, describes a coordinate system in which points are located in reference to a fixed point and a fixed line through that point. This was the first introduction of essentially the *polar coordinates* we use today.

1.3 Vectors in Different Coordinate Systems

Many physical situations require the study of vectors in different coordinate systems. For example, the study of the solar system is best done in spherical coordinates because of the nature of the gravitational force. Similarly calculation of electromagnetic fields in a cylindrical cavity will be easier if we use cylindrical coordinates. This requires not only writing functions in terms of these coordinate variables, but also expressing vectors in terms of unit vectors suitable for these coordinate systems. It turns out that, for the three coordinate systems described above, the most natural construction of such vectors renders them mutually perpendicular.

orthonormal basis

Any set of three (two) mutually perpendicular unit vectors in space (in the plane) is called an **orthonormal basis**.[10] Basis vectors have the property that any vector can be written in terms of them.

Let us start with the plane in which the coordinate system could be Cartesian or polar. In general, we construct an orthonormal basis *at a point* and note that

---

[10]The word "orthonormal" comes from *orthogonal* meaning "perpendicular," and *normal* meaning "of unit length."

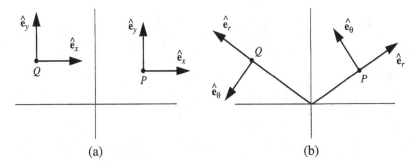

Figure 1.10: The unit vectors in (a) Cartesian coordinates and (b) polar coordinates. The unit vectors at P and Q are the same for Cartesian coordinates, but different in polar coordinates.

Box 1.3.1. *The orthonormal basis, generally speaking, depends on the point at which it is constructed.*

The vectors of a basis are constructed as follows. To find the unit vector corresponding to a coordinate at a point P, hold the other coordinate fixed and *increase* the coordinate in question. The initial direction of motion of P is the direction of the unit vector sought. Thus, we obtain the Cartesian unit vectors at point P of Figure 1.10(a): $\hat{\mathbf{e}}_x$ is obtained by holding y fixed and letting x vary in the increasing direction; and $\hat{\mathbf{e}}_y$ is obtained by holding x fixed at P and letting y increase. In each case, the unit vectors show the initial direction of the motion of P. It should be clear that one obtains the same set of unit vectors regardless of the location of P. However, the reader should take note that this is true only for coordinates that are defined in terms of axes whose directions are fixed, such as Cartesian coordinates.

general rule for constructing a basis at a point

If we use polar coordinates for P, then holding θ fixed at P gives the direction of $\hat{\mathbf{e}}_r$ as shown in Figure 1.10(b), because for fixed θ, that is the direction of increase for r. Similarly, if r is fixed at P, the initial direction of motion of P when θ is increased is that of $\hat{\mathbf{e}}_\theta$ shown in the figure. If we choose another point such as Q shown in the figure, then a new set of unit vectors will be obtained which are different form those of P. This is because polar coordinates are not defined in terms of any fixed axes.

Since $\{\hat{\mathbf{e}}_x, \hat{\mathbf{e}}_y\}$ and $\{\hat{\mathbf{e}}_r, \hat{\mathbf{e}}_\theta\}$ form a basis in the plane, any vector $\mathbf{a}$ in the plane can be expressed in terms of either basis as shown in Figure 1.11. Thus, we can write

$$\mathbf{a} = a_{x_P}\hat{\mathbf{e}}_{x_P} + a_{y_P}\hat{\mathbf{e}}_{y_P} = a_{r_P}\hat{\mathbf{e}}_{r_P} + a_{\theta_P}\hat{\mathbf{e}}_{\theta_P} = a_{r_Q}\hat{\mathbf{e}}_{r_Q} + a_{\theta_Q}\hat{\mathbf{e}}_{\theta_Q}, \quad (1.13)$$

where the coordinates are subscripted to emphasize their dependence on the points at which the unit vectors are erected. In the case of Cartesian coordinates, this, of course, is not necessary because the unit vectors happen to be independent of the point. In the case of polar coordinates, although this

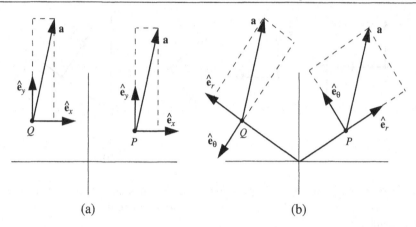

(a)   (b)

Figure 1.11: (a) The vector **a** has the same components along unit vectors at P and Q in Cartesian coordinates. (b) The vector **a** has different components along unit vectors at different points for a polar coordinate system.

dependence exists, we normally do not write the points as subscripts, being aware of this dependence every time we use polar coordinates.

angle brackets denote vector components

So far we have used parentheses to designate the (components of) a vector. Since, parentheses—as a universal notation—are used for coordinates of points, we shall write components of a vector in angle brackets. So Equation (1.13) can also be written as

$$\mathbf{a} = \langle a_x, a_y \rangle_P = \langle a_r, a_\theta \rangle_P = \langle a_r, a_\theta \rangle_Q,$$

where again the subscript indicating the point at which the unit vectors are defined is normally deleted. However, we need to keep in mind that although $\langle a_x, a_y \rangle$ is independent of the point in question, $\langle a_r, a_\theta \rangle$ is very much point-dependent. Caution should be exercised when using this notation as to the location of the unit vectors.

The unit vectors in the coordinate systems of space are defined the same way. We follow the rule given before:

> **Box 1.3.2. (*Rule for Finding Coordinate Unit Vectors*).** *To find the unit vector corresponding to a coordinate at a point P, hold the other coordinates fixed and increase the coordinate in question. The initial direction of motion of P is the direction of the unit vector sought.*

It should be clear that the Cartesian basis $\{\hat{\mathbf{e}}_x, \hat{\mathbf{e}}_y, \hat{\mathbf{e}}_z\}$ is the same for all points, and usually they are drawn at the origin along the three axes. An arbitrary vector **a** can be written as

$$\mathbf{a} = a_x\hat{\mathbf{e}}_x + a_y\hat{\mathbf{e}}_y + a_z\hat{\mathbf{e}}_z \qquad \text{or} \qquad \mathbf{a} = \langle a_x, a_y, a_z \rangle, \qquad (1.14)$$

where we used angle brackets to denote *components* of the vector, reserving the parentheses for *coordinates* of points in space.

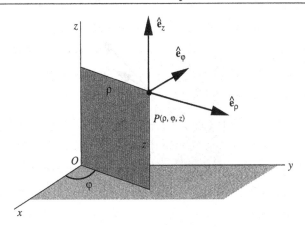

Figure 1.12: Unit vectors of cylindrical coordinates.

The unit vectors at a point P in the other coordinate systems are obtained similarly. In cylindrical coordinates, $\hat{\mathbf{e}}_\rho$ lies along and points in the direction of increasing ρ at P; $\hat{\mathbf{e}}_\varphi$ is perpendicular to the plane formed by P and the z-axis and points in the direction of increasing φ; $\hat{\mathbf{e}}_z$ points in the direction of positive z (see Figure 1.12). We note that only $\hat{\mathbf{e}}_z$ is independent of the point at which the unit vectors are defined because z is a fixed axis in cylindrical coordinates. Given any vector $\mathbf{a}$, we can write it as

$$\mathbf{a} = a_\rho \hat{\mathbf{e}}_\rho + a_\varphi \hat{\mathbf{e}}_\varphi + a_z \hat{\mathbf{e}}_z \quad \text{or} \quad \mathbf{a} = \langle a_\rho, a_\varphi, a_z \rangle. \tag{1.15}$$

The unit vectors in spherical coordinates are defined similarly: $\hat{\mathbf{e}}_r$ is taken along r and points in the direction of increasing r; this direction is called **radial**; $\hat{\mathbf{e}}_\theta$ is taken to lie in the plane formed by P and the z-axis, is perpendicular to r, and points in the direction of increasing θ; $\hat{\mathbf{e}}_\varphi$ is as in the cylindrical case (Figure 1.13). An arbitrary vector in space can be expressed in terms of the spherical unit vectors at P:

radial direction

$$\mathbf{a} = a_r \hat{\mathbf{e}}_r + a_\theta \hat{\mathbf{e}}_\theta + a_\varphi \hat{\mathbf{e}}_\varphi \quad \text{or} \quad \mathbf{a} = \langle a_r, a_\theta, a_\varphi \rangle. \tag{1.16}$$

It should be emphasized that

Box 1.3.3. *The cylindrical and spherical unit vectors $\hat{\mathbf{e}}_\rho$, $\hat{\mathbf{e}}_r$, $\hat{\mathbf{e}}_\theta$, and $\hat{\mathbf{e}}_\varphi$ are dependent on the position of P.*

Once an origin O is designated, every point P in space will define a vector, called a **position vector** and denoted by $\mathbf{r}$. This is simply the vector drawn from O to P. In Cartesian coordinates this vector has *components* $\langle x, y, z \rangle$, thus one can write

position vector

$$\mathbf{r} = x\hat{\mathbf{e}}_x + y\hat{\mathbf{e}}_y + z\hat{\mathbf{e}}_z. \tag{1.17}$$

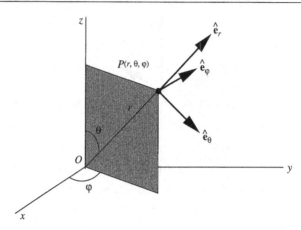

Figure 1.13: Unit vectors of spherical coordinates. Note that the intersection of the shaded plane with the xy-plane is a line along the cylindrical coordinate ρ.

difference between coordinates and components explained

But (x, y, z) are also the *coordinates* of the point P. This can be a source of confusion when other coordinate systems are used. For example, in spherical coordinates, the *components* of the vector $\mathbf{r}$ at P are $\langle r, 0, 0 \rangle$ because $\mathbf{r}$ has only a component along $\hat{\mathbf{e}}_r$ and none along $\hat{\mathbf{e}}_\theta$ or $\hat{\mathbf{e}}_\varphi$. One writes[11]

$$\mathbf{r} = r\hat{\mathbf{e}}_r. \tag{1.18}$$

However, the *coordinates* of P are still (r, θ, φ)! Similarly, the coordinates of P are (ρ, φ, z) in a cylindrical system, while

$$\mathbf{r} = \rho\,\hat{\mathbf{e}}_\rho + z\hat{\mathbf{e}}_z, \tag{1.19}$$

because $\mathbf{r}$ lies in the ρz-plane and has no component along $\hat{\mathbf{e}}_\varphi$. Therefore,

Box 1.3.4. *Make a clear distinction between the* **components** *of the vector* $\mathbf{r}$ *and the* **coordinates** *of the point* P.

A common symptom of confusing components with coordinates is as follows. Point P_1 has position vector $\mathbf{r}_1$ with spherical components $\langle r_1, 0, 0 \rangle$ at P_1. The position vector of a second point P_2 is $\mathbf{r}_2$ with spherical components $\langle r_2, 0, 0 \rangle$ at P_2. It is easy to fall into the trap of thinking that $\mathbf{r}_1 - \mathbf{r}_2$ has spherical components $\langle r_1 - r_2, 0, 0 \rangle$! This is, of course, not true, because the spherical unit vectors at P_1 are completely different from those at P_2, and, therefore, contrary to the Cartesian case, we cannot simply subtract components.

---

[11]We should really label everything with P. But, as usual, we assume this labeling to be implied.

One of the great advantages of vectors is their ability to express results independent of any specific coordinate systems. Physical laws are always coordinate-independent. For example, when we write $\mathbf{F} = m\mathbf{a}$ both $\mathbf{F}$ and $\mathbf{a}$ could be expressed in terms of Cartesian, spherical, cylindrical, or any other convenient coordinate system. This independence allows us the freedom to choose the coordinate systems most convenient for the problem at hand. For example, it is extremely difficult to solve the planetary motions in Cartesian coordinates, while the use of spherical coordinates facilitates the solution of the problem tremendously.

Physical laws ought to be coordinate independent!

Example 1.3.1. We can express the *coordinates* of the **center of mass** (CM) of a collection of particles in terms of their position vectors.[12] Thus, if $\bar{\mathbf{r}}$ denotes the position vector of the CM of the collection of N mass points, $m_1, m_2, \ldots, m_N$ with respective position vectors $\mathbf{r}_1, \mathbf{r}_2, \ldots, \mathbf{r}_N$ relative to an origin O, then[13]

center of mass

$$\bar{\mathbf{r}} = \frac{m_1\mathbf{r}_1 + m_2\mathbf{r}_2 + \cdots + m_N\mathbf{r}_N}{m_1 + m_2 + \cdots + m_N} = \frac{\sum_{k=1}^{N} m_k\mathbf{r}_k}{M}, \qquad (1.20)$$

where $M = \sum_{k=1}^{N} m_k$ is the total mass of the system. One can also think of Equation (1.20) as a *vector* equation. To find the component equations in a coordinate system, one needs to pick a *fixed* point (say the origin), a set of unit vectors at that point (usually the unit vectors along the axes of some coordinate system), and substitute the *components* of $\mathbf{r}_k$ along those unit vectors to find the components of $\bar{\mathbf{r}}$ along the unit vectors. ∎

1.3.1 Fields and Potentials

The distributive property of the dot product and the fact that the unit vectors of the bases in all coordinate systems are mutually perpendicular can be used to derive the following:

$$\begin{aligned}
\mathbf{a} \cdot \mathbf{b} &= a_x b_x + a_y b_y + a_z b_z \quad \text{(Cartesian)}, \\
\mathbf{a} \cdot \mathbf{b} &= a_\rho b_\rho + a_\varphi b_\varphi + a_z b_z \quad \text{(cylindrical)}, \\
\mathbf{a} \cdot \mathbf{b} &= a_r b_r + a_\theta b_\theta + a_\varphi b_\varphi \quad \text{(spherical)}.
\end{aligned} \qquad (1.21)$$

dot product in terms of components in the three coordinate systems

The first of these equations is the same as (1.3).

It is important to keep in mind that the components are to be expressed in the same set of unit vectors. This typically means setting up mutually perpendicular unit vectors (an orthonormal basis) *at a single point* and resolving all vectors along those unit vectors.

The dot product, in various forms and guises, has many applications in physics. As pointed out earlier, it was introduced in the definition of work, but soon spread to many other concepts of physics. One of the simplest—and most important—applications is its use in writing the laws of physics in a coordinate-independent way.

---

[12]This implies that the equation is most useful only when Cartesian coordinates are used, because only for these coordinates do the components of the position vector of a point coincide with the coordinates of that point.

[13]We assume that the reader is familiar with the symbol $\sum$ simply as a summation symbol. We shall discuss its properties and ways of manipulating it in Chapter 9.

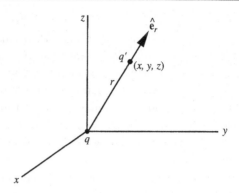

Figure 1.14: The diagram illustrating the electrical force when one charge is at the origin.

Example 1.3.2. A point charge q is situated at the origin. A second charge q' is located at (x, y, z) as shown in Figure 1.14. We want to express the electric force on q' in Cartesian, spherical, and cylindrical coordinate systems.

We know that the electric force, as given by **Coulomb's law**, lies along the line joining the two charges and is either attractive or repulsive according to the signs of q and q'. All of this information can be summarized in the formula

Coulomb's law

$$\mathbf{F}_{q'} = \frac{k_e q q'}{r^2} \hat{\mathbf{e}}_r \tag{1.22}$$

where $k_e = 1/(4\pi\epsilon_0) \approx 9 \times 10^9$ in SI units. Note that if q and q' are unlike, $qq' < 0$ and $\mathbf{F}_{q'}$ is opposite to $\hat{\mathbf{e}}_r$, i.e., it is *attractive*. On the other hand, if q and q' are of the same sign, $qq' > 0$ and $\mathbf{F}_{q'}$ is in the same direction as $\hat{\mathbf{e}}_r$, i.e., *repulsive*.

Equation (1.22) expresses $\mathbf{F}_{q'}$ in spherical coordinates. Thus, its components in terms of unit vectors at q' are $\left\langle k_e q q'/r^2, 0, 0 \right\rangle$. To get the components in the other coordinate systems, we rewrite (1.22). Noting that $\hat{\mathbf{e}}_r = \mathbf{r}/r$, we write

$$\mathbf{F}_{q'} = \frac{k_e q q'}{r^2} \frac{\mathbf{r}}{r} = \frac{k_e q q'}{r^3} \mathbf{r}. \tag{1.23}$$

For Cartesian coordinates we use (1.12) to obtain $r^3 = (x^2 + y^2 + z^2)^{3/2}$. Substituting this and (1.17) in (1.23) yields

$$\mathbf{F}_{q'} = \frac{k_e q q'}{(x^2 + y^2 + z^2)^{3/2}} (x \hat{\mathbf{e}}_x + y \hat{\mathbf{e}}_y + z \hat{\mathbf{e}}_z).$$

Therefore, the components of $\mathbf{F}_{q'}$ in Cartesian coordinates are

$$\left\langle \frac{k_e q q' x}{(x^2 + y^2 + z^2)^{3/2}}, \frac{k_e q q' y}{(x^2 + y^2 + z^2)^{3/2}}, \frac{k_e q q' z}{(x^2 + y^2 + z^2)^{3/2}} \right\rangle.$$

Finally, using (1.10) and (1.19) in (1.23), we obtain

$$\mathbf{F}_{q'} = \frac{k_e q q'}{(\rho^2 + z^2)^{3/2}} (\rho \, \hat{\mathbf{e}}_\rho + z \hat{\mathbf{e}}_z).$$

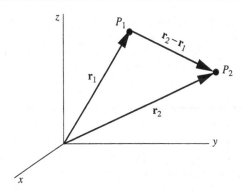

Figure 1.15: The displacement vector between P_1 and P_2 is the difference between their position vectors.

Thus the components of $\mathbf{F}_{q'}$ along the cylindrical unit vectors constructed at the location of q' are

$$\left\langle \frac{k_e q q' \rho}{(\rho^2 + z^2)^{3/2}}, 0, \frac{k_e q q' z}{(\rho^2 + z^2)^{3/2}} \right\rangle.$$ ∎

Since $\mathbf{r}$ gives the position of a point in space, one can use it to write the distance between two points P_1 and P_2 with position vectors $\mathbf{r}_1$ and $\mathbf{r}_2$. Figure 1.15 shows that $\mathbf{r}_2 - \mathbf{r}_1$ is the displacement vector from P_1 to P_2. The importance of this vector stems from the fact that many physical quantities are functions of distances between point particles, and $\mathbf{r}_2 - \mathbf{r}_1$ is a concise way of expressing this distance. The following example illustrates this.

Historical Notes

During the second half of the eighteenth century many physicists were engaged in a quantitative study of electricity and magnetism. **Charles Augustin de Coulomb**, who developed the so-called *torsion balance* for measuring weak forces, is credited with the discovery of the law governing the force between electrical charges.

Coulomb was an army engineer in the West Indies. After spending nine years there, due to his poor health, he returned to France about the same time that the French Revolution began, at which time he retired to the country to do scientific research.

Charles Coulomb
1736–1806

Beside his experiments on electricity, Coulomb worked on applied mechanics, structural analysis, the fracture of beams and columns, the thrust of arches, and the thrust of the soil.

At about the same time that Coulomb discovered the law of electricity, there lived in England a very reclusive character named **Henry Cavendish**. He was born into the nobility, had no close friends, was afraid of women, and disinterested in music or arts of any kind. His life revolved around experiments in physics and chemistry that he carried out in a private laboratory located in his large mansion.

During his long life he published only a handful of relatively unimportant papers. But after his death about one million pounds sterling were found in his bank account and twenty bundles of notes in his laboratory. These notes remained in the possession of his relatives for a long time, but when they were published one

Henry Cavendish
1731–1810

hundred years later, it became clear that Henry Cavendish was one of the greatest experimental physicists ever. He discovered all the laws of electric and magnetic interactions at the same time as Coulomb, and his work in chemistry matches that of Lavoisier. Furthermore, he used a torsion balance to measure the universal gravitational constant for the first time, and as a result was able to arrive at the exact mass of the Earth.

Example 1.3.3. COULOMB'S LAW FOR TWO ARBITRARY CHARGES

Suppose there are point charges q_1 at P_1 and q_2 at P_2. Let us write the force exerted on q_2 by q_1. The magnitude of the force is

$$F_{21} = \frac{k_e q_1 q_2}{d^2},$$

where $d = \overline{P_1 P_2}$ is the distance between the two charges. We use d because the usual notation r has special meaning for us: it is one of the coordinates in spherical systems. If we multiply this magnitude by the unit vector describing the direction of the force, we obtain the full force vector (see Box 1.1.1). But, assuming repulsion for the moment, this unit vector is

$$\frac{\mathbf{r}_2 - \mathbf{r}_1}{|\mathbf{r}_2 - \mathbf{r}_1|} \equiv \hat{\mathbf{e}}_{21}.$$

Also, since $d = |\mathbf{r}_2 - \mathbf{r}_1|$, we have

$$\mathbf{F}_{21} = \frac{k_e q_1 q_2}{d^2} \hat{\mathbf{e}}_{21} = \frac{k_e q_1 q_2}{|\mathbf{r}_2 - \mathbf{r}_1|^2} \frac{\mathbf{r}_2 - \mathbf{r}_1}{|\mathbf{r}_2 - \mathbf{r}_1|}$$

Coulomb's law when charges are arbitrarily located

or

$$\mathbf{F}_{21} = \frac{k_e q_1 q_2}{|\mathbf{r}_2 - \mathbf{r}_1|^3} (\mathbf{r}_2 - \mathbf{r}_1). \tag{1.24}$$

Although we assumed repulsion, we see that (1.24) includes attraction as well. Indeed, if $q_1 q_2 < 0$, $\mathbf{F}_{21}$ is opposite to $\mathbf{r}_2 - \mathbf{r}_1$, i.e., $\mathbf{F}_{21}$ is directed from P_2 to P_1. Since $\mathbf{F}_{21}$ is the force *on* q_2 *by* q_1, this is an *attraction*. We also note that Newton's third law is included in (1.24):

$$\mathbf{F}_{12} = \frac{k_e q_2 q_1}{|\mathbf{r}_1 - \mathbf{r}_2|^3} (\mathbf{r}_1 - \mathbf{r}_2) = -\mathbf{F}_{21}$$

because $\mathbf{r}_2 - \mathbf{r}_1 = -(\mathbf{r}_1 - \mathbf{r}_2)$ and $|\mathbf{r}_2 - \mathbf{r}_1| = |\mathbf{r}_1 - \mathbf{r}_2|$.

vector form of gravitational force

We can also write the gravitational force immediately

$$\mathbf{F}_{21} = -\frac{G m_1 m_2}{|\mathbf{r}_2 - \mathbf{r}_1|^3} (\mathbf{r}_2 - \mathbf{r}_1), \tag{1.25}$$

where m_1 and m_2 are point masses and the minus sign is introduced to ensure attraction. ∎

Now that we have expressions for electric and gravitational forces, we can obtain the electric field of a point charge and the gravitational field of a point mass. First recall that the **electric field** at a point P is defined to be the force on a test charge q located at P divided by q. Thus if we have a charge q_1, at P_1 with position vector $\mathbf{r}_1$ and we are interested in its fields at P with position vector $\mathbf{r}$, we introduce a charge q at $\mathbf{r}$ and calculate the force on q from Equation (1.24):

$$\mathbf{F}_q = \frac{k_e q_1 q}{|\mathbf{r} - \mathbf{r}_1|^3}(\mathbf{r} - \mathbf{r}_1).$$

Dividing by q gives

electric field of a
point charge

$$\mathbf{E}_1 = \frac{k_e q_1}{|\mathbf{r} - \mathbf{r}_1|^3}(\mathbf{r} - \mathbf{r}_1), \qquad (1.26)$$

where we have given the field the same index as the charge producing it.

The calculation of the gravitational field follows similarly. The result is

$$\mathbf{g}_1 = -\frac{G m_1}{|\mathbf{r} - \mathbf{r}_1|^3}(\mathbf{r} - \mathbf{r}_1). \qquad (1.27)$$

In (1.26) and (1.27), P is called the **field point** and P_1 the **source point**. Note that in both expressions, the field position vector comes first.

field point and
source point

If there are several point charges (or masses) producing an electric (gravitational) field, we simply add the contributions from each source. The principle behind this procedure is called the **superposition principle**. It is a principle that "seems" intuitively obvious, but upon further reflection its validity becomes surprising. Suppose a charge q_1 produces a field $\mathbf{E}_1$ around itself. Now we introduce a second charge q_2 which, far away and isolated from any other charges, produced a field $\mathbf{E}_2$ around itself. It is not at all obvious that once we move these charges together, the individual fields should not change. After all, this is not what happens to human beings! We act completely differently when we are alone than when we are in the company of others. The presence of others drastically changes our individual behaviors. Nevertheless, charges and masses, unfettered by any social chains, retain their individuality and produce fields as if no other charges were present.

superposition
principle explained

It is important to keep in mind that the superposition principle applies only to point sources. For example, a charged conducting sphere will not produce the same field when another charge is introduced nearby, because the presence of the new charge alters the charge distribution of the sphere and indeed does change the sphere's field. However each individual point charge (electron) on the sphere, whatever location on the sphere it happens to end up in, will retain its individual electric field.[14]

Going back to the electric field, we can write

$$\mathbf{E} = \mathbf{E}_1 + \mathbf{E}_2 + \cdots + \mathbf{E}_n$$

for n point charges $q_1, q_2, \ldots, q_n$ (see Figure 1.16). Substituting from (1.26), with appropriate indices, we obtain

$$\mathbf{E} = \frac{k_e q_1}{|\mathbf{r} - \mathbf{r}_1|^3}(\mathbf{r} - \mathbf{r}_1) + \frac{k_e q_2}{|\mathbf{r} - \mathbf{r}_2|^3}(\mathbf{r} - \mathbf{r}_2) + \cdots + \frac{k_e q_n}{|\mathbf{r} - \mathbf{r}_n|^3}(\mathbf{r} - \mathbf{r}_n)$$

or, using the summation symbol, we obtain

---

[14]The superposition principle, which in the case of electrostatics and gravity is needed to calculate the fields of large sources consisting of many point sources, becomes a vital pillar upon which quantum theory is built and by which many of the strange phenomena of quantum physics are explained.

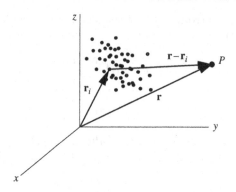

Figure 1.16: The electrostatic field of N point charges is the sum of the electric fields of the individual charges.

Box 1.3.5. *The electric field of n point charges $q_1, q_2, \ldots, q_n$, located at position vectors $\mathbf{r}_1, \mathbf{r}_2, \ldots, \mathbf{r}_n$ is $\mathbf{E} = \sum_{i=1}^{n} \frac{k_e q_i}{|\mathbf{r}-\mathbf{r}_i|^3}(\mathbf{r} - \mathbf{r}_i)$, and the analogous expression for the gravitational field of n point masses $m_1, m_2, \ldots, m_n$ is $\mathbf{g} = -\sum_{i=1}^{n} \frac{G m_i}{|\mathbf{r}-\mathbf{r}_i|^3}(\mathbf{r} - \mathbf{r}_i)$.*

Historical Notes

The concept of force has a fascinating history which started in the works of Galileo around the beginning of the seventeenth century, mathematically formulated and precisely defined by Sir Isaac Newton in the second half of the seventeenth century, revised and redefined in the form of **fields** by Michael Faraday and James Maxwell in the mid nineteenth century, and finally brought to its modern quantum field theoretical form by Dirac, Heisenberg, Feynman, Schwinger, and others by the mid twentieth century.

notion of field elaborated

Newton, in his theory of gravity, thought of gravitational force as "action-at-a-distance," an agent which affects something that is "there" because of the influence of something that is "here." This kind of interpretation of force had both philosophical and physical drawbacks. It is hard to accept a ghostlike influence on a distant object. Is there an agent that "carries" this influence? What is this agent, if any? Does the influence travel infinitely fast? If we remove the Sun from the Solar System would the Earth and other planets "feel" the absence of the Sun immediately?

These questions, plus others, prompted physicists to come up with the idea of a field. According to this interpretation, the Sun, by its mere presence, creates around itself an invisible three dimensional "sheet" such that, if any object is placed in this sheet, it feels the gravitational force. The reason that planets feel the force of gravity of the Sun is because they happen to be located in the gravitational field of the Sun. The reason that an apple falls to the Earth is because it is in the gravitational field of the Earth and not due to some kind of action-at-a-distance ghost.

Therefore, according to this concept, the force acts on an object *here*, because there exists a field right *here*. And force becomes a local concept. The field concept removes the difficulties associated with action-at-a-distance. The "agent" that transmits the influence from the source to the object, is the field. If the Sun is stolen from the solar system, the Earth will not feel the absence of the Sun immediately. It will receive the information of such cosmic burglary after a certain time-lapse corresponding to the time required for the disturbance to travel from the Sun to the Earth. We can liken such a disturbance (disappearance of the Sun) to a disturbance in the smooth water of a quiet pond (by dropping a stone into it). Clearly, the disturbance travels from the source (where the stone was dropped) to any other point with a finite speed, the speed of the water waves.

The concept of a field was actually introduced first in the context of electricity and magnetism by Michael Faraday as a means of "visualizing" electromagnetic effects to replace certain mathematical ideas for which he had little talent. However, in the hands of James Maxwell, fields were molded into a physical entity having an existence of their own in the form of electromagnetic waves to be produced in 1887 by Hertz and used in 1901 by Marconi in the development of radio.

A concept related to that of fields is **potential** which is closely tied to the work done by the fields on a charge (in the case of electrostatics) or a mass (in the case of gravity). It can be shown[15] that the gravitational potential $\Phi(\mathbf{r})$ at $\mathbf{r}$, of n point masses, is given by

$$\Phi(\mathbf{r}) = -\sum_{i=1}^{n} \frac{Gm_i}{|\mathbf{r} - \mathbf{r}_i|} \qquad (1.28)$$

and that of n point charges by

$$\Phi(\mathbf{r}) = \sum_{i=1}^{n} \frac{k_e q_i}{|\mathbf{r} - \mathbf{r}_i|}. \qquad (1.29)$$

Note that in both cases, the potential goes to zero as $\mathbf{r}$ goes to infinity. This has to do with the choice of the location of the zero of potential, which we have chosen to be the point at infinity in Equations (1.28) and (1.29).

Example 1.3.4. The electric charges q_1, q_2, q_3, and q_4 are located at Cartesian $(a, 0, 0)$, $(0, a, 0)$, $(-a, 0, 0)$, and $(0, -a, 0)$, respectively. We want to find the electric field and the electrostatic potential at an arbitrary point on the z-axis. We note that

$$\mathbf{r}_1 = a\hat{\mathbf{e}}_x, \quad \mathbf{r}_2 = a\hat{\mathbf{e}}_y, \quad \mathbf{r}_3 = -a\hat{\mathbf{e}}_x, \quad \mathbf{r}_4 = -a\hat{\mathbf{e}}_y, \quad \mathbf{r} = z\hat{\mathbf{e}}_z,$$

so that

$$\mathbf{r} - \mathbf{r}_1 = -a\hat{\mathbf{e}}_x + z\hat{\mathbf{e}}_z, \qquad \mathbf{r} - \mathbf{r}_2 = -a\hat{\mathbf{e}}_y + z\hat{\mathbf{e}}_z,$$
$$\mathbf{r} - \mathbf{r}_3 = a\hat{\mathbf{e}}_x + z\hat{\mathbf{e}}_z, \qquad \mathbf{r} - \mathbf{r}_4 = a\hat{\mathbf{e}}_y + z\hat{\mathbf{e}}_z,$$

---

[15]See Chapter 14 for details.

and $|\mathbf{r} - \mathbf{r}_i|^3 = (a^2 + z^2)^{3/2}$ for all i. The electric field can now be calculated using Box 1.3.5:

$$
\begin{aligned}
\mathbf{E} &= \frac{k_e q_1}{(a^2 + z^2)^{3/2}}(-a\hat{\mathbf{e}}_x + z\hat{\mathbf{e}}_z) + \frac{k_e q_2}{(a^2 + z^2)^{3/2}}(-a\hat{\mathbf{e}}_y + z\hat{\mathbf{e}}_z) \\
&\quad + \frac{k_e q_3}{(a^2 + z^2)^{3/2}}(a\hat{\mathbf{e}}_x + z\hat{\mathbf{e}}_z) + \frac{k_e q_4}{(a^2 + z^2)^{3/2}}(a\hat{\mathbf{e}}_y + z\hat{\mathbf{e}}_z) \\
&= \frac{k_e}{(a^2 + z^2)^{3/2}} \left[(-aq_1 + aq_3)\hat{\mathbf{e}}_x + (-aq_2 + aq_4)\hat{\mathbf{e}}_y + (q_1 + q_2 + q_3 + q_4)z\hat{\mathbf{e}}_z \right].
\end{aligned}
$$

It is interesting to note that if the sum of all charges is zero, the z-component of the electric field vanishes at all points on the z-axis. Furthermore, if, in addition, $q_1 = q_3$ and $q_2 = q_4$, there will be no electric field at any point on the z-axis.

The potential is obtained similarly:

$$
\begin{aligned}
\Phi &= \frac{k_e q_1}{(a^2 + z^2)^{1/2}} + \frac{k_e q_2}{(a^2 + z^2)^{1/2}} + \frac{k_e q_3}{(a^2 + z^2)^{1/2}} + \frac{k_e q_4}{(a^2 + z^2)^{1/2}} \\
&= \frac{k_e(q_1 + q_2 + q_3 + q_4)}{\sqrt{a^2 + z^2}}.
\end{aligned}
$$

So, the potential is zero at all points of the z-axis, as long as the total charge is zero. ∎

1.3.2   Cross Product

The unit vectors in the three coordinate systems are not only mutually perpendicular, but in the order in which they are given, they also form a right-handed set [see Equation (1.5)]. Therefore, we can use Equation (1.6) and write

$$
\mathbf{a} \times \mathbf{b} = \det \underbrace{\begin{pmatrix} \hat{\mathbf{e}}_x & \hat{\mathbf{e}}_y & \hat{\mathbf{e}}_z \\ a_x & a_y & a_z \\ b_x & b_y & b_z \end{pmatrix}}_{\text{in Cartesian CS}} = \det \underbrace{\begin{pmatrix} \hat{\mathbf{e}}_\rho & \hat{\mathbf{e}}_\varphi & \hat{\mathbf{e}}_z \\ a_\rho & a_\varphi & a_z \\ b_\rho & b_\varphi & b_z \end{pmatrix}}_{\text{in cylindrical CS}} = \det \underbrace{\begin{pmatrix} \hat{\mathbf{e}}_r & \hat{\mathbf{e}}_\theta & \hat{\mathbf{e}}_\varphi \\ a_r & a_\theta & a_\varphi \\ b_r & b_\theta & b_\varphi \end{pmatrix}}_{\text{in spherical CS}}
$$

$$\tag{1.30}$$

Two important prototypes of the concept of cross product are **angular momentum** and **torque**. A particle moving with instantaneous linear momentum $\mathbf{p}$ relative to an origin O has *instantaneous* angular momentum $\mathbf{L} = \mathbf{r} \times \mathbf{p}$ if its instantaneous position vector with respect to O is $\mathbf{r}$. In Figure 1.17 we have shown $\mathbf{r}$, $\mathbf{p}$, and $\mathbf{r} \times \mathbf{p}$. Similarly, if the instantaneous force on the above particle is $\mathbf{F}$, then the instantaneous torque acting on it is $\mathbf{T} = \mathbf{r} \times \mathbf{F}$.

If there are more than one particle we simply add the contribution of individual particles. Thus, the total angular momentum $\mathbf{L}$ of N particles and the total torque $\mathbf{T}$ acting on them are

angular momentum and torque as examples of cross products

$$
\mathbf{L} = \sum_{k=1}^{N} \mathbf{r}_k \times \mathbf{p}_k \qquad \text{and} \qquad \mathbf{T} = \sum_{k=1}^{N} \mathbf{r}_k \times \mathbf{F}_k, \tag{1.31}
$$

where $\mathbf{r}_k$ is the position of the kth particle, $\mathbf{p}_k$ its instantaneous momentum, and $\mathbf{F}_k$ the instantaneous force acting on it.

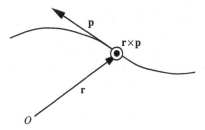

Figure 1.17: Angular momentum of a moving particle with respect to the origin O. The circle with a dot in its middle represents a vector pointing out of the page. It is assumed that $\mathbf{r}$ and $\mathbf{p}$ lie in the page.

Example 1.3.5. In this example, we show that the torque on a collection of three particles is caused by *external* forces only. The torques due to the internal forces add up to zero. The generalization to an arbitrary number of particles will be done in Example 9.2.1 when we learn how to manipulate summation symbols.

For $N = 3$, the second formula in Equation (1.31) reduces to

$$\mathbf{T} = \mathbf{r}_1 \times \mathbf{F}_1 + \mathbf{r}_2 \times \mathbf{F}_2 + \mathbf{r}_3 \times \mathbf{F}_3.$$

Each force can be divided into an external part and an internal part, the latter being the force caused by the presence of the other particles. So, we have

$$\mathbf{F}_1 = \mathbf{F}_1^{(\text{ext})} + \mathbf{F}_{12} + \mathbf{F}_{13},$$
$$\mathbf{F}_2 = \mathbf{F}_2^{(\text{ext})} + \mathbf{F}_{21} + \mathbf{F}_{23},$$
$$\mathbf{F}_3 = \mathbf{F}_3^{(\text{ext})} + \mathbf{F}_{31} + \mathbf{F}_{32},$$

where $\mathbf{F}_{12}$ is the force on particle 1 exerted by particle 2, etc. Substituting in the above expression for the torque, we get

$$\begin{aligned}
\mathbf{T} &= \mathbf{r}_1 \times \mathbf{F}_1^{(\text{ext})} + \mathbf{r}_2 \times \mathbf{F}_2^{(\text{ext})} + \mathbf{r}_3 \times \mathbf{F}_3^{(\text{ext})} \\
&\quad + \mathbf{r}_1 \times \mathbf{F}_{12} + \mathbf{r}_1 \times \mathbf{F}_{13} + \mathbf{r}_2 \times \mathbf{F}_{21} + \mathbf{r}_2 \times \mathbf{F}_{23} + \mathbf{r}_3 \times \mathbf{F}_{31} + \mathbf{r}_3 \times \mathbf{F}_{32} \\
&= \mathbf{T}^{(\text{ext})} + (\mathbf{r}_1 - \mathbf{r}_2) \times \mathbf{F}_{12} + (\mathbf{r}_1 - \mathbf{r}_3) \times \mathbf{F}_{13} + (\mathbf{r}_2 - \mathbf{r}_3) \times \mathbf{F}_{23},
\end{aligned}$$

where we used the third law of motion: $\mathbf{F}_{12} = -\mathbf{F}_{21}$, etc. Now we note that the internal force between two particles, 1 and 2 say, is along the line joining them, i.e., along $\mathbf{r}_1 - \mathbf{r}_2$. It follows that all the cross products in the last line of the equation above vanish and we get $\mathbf{T} = \mathbf{T}^{(\text{ext})}$. ∎

We have already seen that multiplying *a vector by a number* gives another vector. A physical example of this is electric force which is obtained by multiplying electric field by electric charge. In fact we divided the electric force by charge to get the electric field. Historically, it was the law of the force which was discovered first and *then* the concept of electric field was defined. We have also seen that one can get a new vector by cross-multiplying *two vectors*. The rule of this kind of multiplication is, however, more complicated. It turns out

from electric field to electric force

that the magnetic force is related to the magnetic field via such a cross multiplication. What is worse is that the magnetic field is also related to its source
(electric charges *in motion*) via such a product. Little wonder that magnetic
phenomena are mathematically so much more complicated than their electric
counterparts. That is why in the study of magnetism, one first introduces
the concept of magnetic field and how it is related to the motion of charges
producing it, and *then* the force of this field on moving charges.

Example 1.3.6. A charge q, located instantaneously at the origin, is moving
with velocity $\mathbf{v}$ relative to P [see Figure 1.18(a)]. Assuming that $|\mathbf{v}|$ is much smaller
than the speed of light, the *instantaneous* magnetic field at P due to q is given by

magnetic field of a
moving charge or
Biot–Savart law

$\mathbf{B} = \dfrac{k_m q\,\mathbf{v} \times \hat{\mathbf{e}}_r}{r^2}$, or, using $\hat{\mathbf{e}}_r = \mathbf{r}/r$, by $\mathbf{B} = \dfrac{k_m q\,\mathbf{v} \times \mathbf{r}}{r^3}$. This is a simple version of
a more general formula known as the **Biot–Savart law**. In the above relations, k_m
is the analog of k_e in the electric case.

If we are interested in the magnetic field when q is located at a point other
than the origin, we replace $\mathbf{r}$ with the vector from the instantaneous location of the
moving charge to P. This is shown in Figure 1.18(b), where the vector from q_1 to
P is to replace $\mathbf{r}$ in the above equation. More specifically, we have

$$\mathbf{B}_1 = \frac{k_m q_1 \mathbf{v}_1 \times (\mathbf{r} - \mathbf{r}_1)}{|\mathbf{r} - \mathbf{r}_1|^3}. \tag{1.32}$$

If there are N charges, the total magnetic field will be

$$\mathbf{B} = \sum_{k=1}^{N} \frac{k_m q_k \mathbf{v}_k \times (\mathbf{r} - \mathbf{r}_k)}{|\mathbf{r} - \mathbf{r}_k|^3}, \tag{1.33}$$

where we have used the superposition principle.   ∎

When a charge q moves with velocity $\mathbf{v}$ in a magnetic field $\mathbf{B}$, it experiences

magnetic force on
a moving charge.

a force given by

$$\mathbf{F} = q\mathbf{v} \times \mathbf{B}. \tag{1.34}$$

It is instructive to write the magnetic force exerted by a charge q_1 moving
with velocity $\mathbf{v}_1$ on a second charge q_2 moving with velocity $\mathbf{v}_2$. We leave this
as an exercise for the reader.

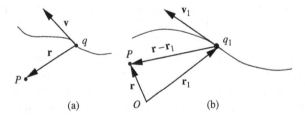

Figure 1.18: The (instantaneous) magnetic field at P of a moving point charge (a)
when P is at the origin, and (b) when P is different from the origin. The field points
out of the page for the configuration shown.

Example 1.3.7. A charge q moves with constant speed v (assumed to be small compared to the speed of light) on a straight line. We want to calculate the magnetic field produced by the charge at a point P located at a distance ρ from the line as a function of time. Cylindrical coordinates are most suitable for this problem because of the existence of a natural axis. Choose the path of the charge to be the z-axis. Also assume that P lies in the xy-plane, and that q was at the origin at $t = 0$. Then $\mathbf{v} = v\hat{\mathbf{e}}_z$, $\mathbf{r} = \rho\hat{\mathbf{e}}_\rho$, $\mathbf{r}_1 = vt\hat{\mathbf{e}}_z$, $\mathbf{r} - \mathbf{r}_1 = \rho\hat{\mathbf{e}}_\rho - vt\hat{\mathbf{e}}_z$. So

$$|\mathbf{r} - \mathbf{r}_1| = \sqrt{(\rho\hat{\mathbf{e}}_\rho - vt\hat{\mathbf{e}}_z) \cdot (\rho\hat{\mathbf{e}}_\rho - vt\hat{\mathbf{e}}_z)} = \sqrt{\rho^2 + v^2 t^2}$$

and $\mathbf{v} \times (\mathbf{r} - \mathbf{r}_1) = v\hat{\mathbf{e}}_z \times (\rho\hat{\mathbf{e}}_\rho - vt\hat{\mathbf{e}}_z) = \rho v\hat{\mathbf{e}}_\varphi$. Therefore, the magnetic field is

$$\mathbf{B} = \frac{k_m q \mathbf{v} \times (\mathbf{r} - \mathbf{r}_1)}{|\mathbf{r} - \mathbf{r}_1|^3} = \frac{k_m q \rho v}{(\rho^2 + v^2 t^2)^{3/2}} \hat{\mathbf{e}}_\varphi.$$

Readers familiar with the relation between magnetic fields and currents in long wires will note that the magnetic field above obeys the right-hand rule. ∎

1.4 Relations Among Unit Vectors

We have seen that, depending on the nature of problems encountered in physics, one coordinate system may be more useful than others. We have also seen that the *coordinates* can be transformed back and forth using functional relations that connect them. Since many physical quantities are vectors, transformation and expression of components in bases of various coordinate systems also become important. The key to this transformation is writing one set of unit vectors in terms of others. In the derivation of these relations, we shall make heavy use of Box 1.1.2.

First we write the cylindrical unit vectors in terms of Cartesian unit vectors. Since $\{\hat{\mathbf{e}}_x, \hat{\mathbf{e}}_y, \hat{\mathbf{e}}_z\}$ form a basis, any vector can be written in terms of them. In particular, $\hat{\mathbf{e}}_\rho$ can be expressed as

$$\hat{\mathbf{e}}_\rho = a_1 \hat{\mathbf{e}}_x + b_1 \hat{\mathbf{e}}_y + c_1 \hat{\mathbf{e}}_z \tag{1.35}$$

with a_1, b_1, and c_1 to be determined. Next we recall that

Box 1.4.1. *The dot product of two unit vectors is the cosine of the angle between them.*

Furthermore, Figure 1.12 shows that the angle between $\hat{\mathbf{e}}_\rho$ and $\hat{\mathbf{e}}_x$ is φ, and that between $\hat{\mathbf{e}}_\rho$ and $\hat{\mathbf{e}}_y$ is $\pi/2 - \varphi$. So, by dotting both sides of Equation (1.35) by $\hat{\mathbf{e}}_x$, $\hat{\mathbf{e}}_y$, and $\hat{\mathbf{e}}_z$ in succession, we obtain

$$\underbrace{\hat{\mathbf{e}}_x \cdot \hat{\mathbf{e}}_\rho}_{=\cos\varphi} = a_1 + 0 + 0 = a_1 \Rightarrow a_1 = \cos\varphi,$$

$$\underbrace{\hat{\mathbf{e}}_y \cdot \hat{\mathbf{e}}_\rho}_{=\sin\varphi} = 0 + b_1 + 0 = b_1 \Rightarrow b_1 = \sin\varphi,$$

$$\underbrace{\hat{\mathbf{e}}_z \cdot \hat{\mathbf{e}}_\rho}_{=0} = 0 + 0 + c_1 = c_1 \Rightarrow c_1 = 0.$$

Therefore,

$$\hat{e}_\rho = \hat{e}_x \cos\varphi + \hat{e}_y \sin\varphi.$$

With the first and third cylindrical unit vectors $\hat{e}_\rho$ and $\hat{e}_z$ at our disposal,[16] we can determine the second, using Equation (1.5):

$$\hat{e}_\varphi = \hat{e}_z \times \hat{e}_\rho = \det \begin{pmatrix} \hat{e}_x & \hat{e}_y & \hat{e}_z \\ 0 & 0 & 1 \\ \cos\varphi & \sin\varphi & 0 \end{pmatrix} = -\hat{e}_x \sin\varphi + \hat{e}_y \cos\varphi.$$

Thus,

cylindrical unit
vectors in terms of
Cartesian unit
vectors

$$\begin{aligned} \hat{e}_\rho &= \hat{e}_x \cos\varphi + \hat{e}_y \sin\varphi, \\ \hat{e}_\varphi &= -\hat{e}_x \sin\varphi + \hat{e}_y \cos\varphi, \\ \hat{e}_z &= \hat{e}_z. \end{aligned} \qquad (1.36)$$

This equation can easily be inverted to find the Cartesian unit vectors in terms of the cylindrical unit vectors. For example, the coefficients in

$$\hat{e}_x = a_2 \hat{e}_\rho + b_2 \hat{e}_\varphi + c_2 \hat{e}_z$$

can be obtained by dotting both sides of it with $\hat{e}_\rho$, $\hat{e}_\varphi$, and $\hat{e}_z$, respectively,

$$\begin{aligned} \hat{e}_\rho \cdot \hat{e}_x &= a_2 + 0 + 0 \;\Rightarrow\; \cos\varphi = a_2, \\ \hat{e}_\varphi \cdot \hat{e}_x &= 0 + b_2 + 0 \;\Rightarrow\; -\sin\varphi = b_2, \\ \hat{e}_z \cdot \hat{e}_x &= 0 + 0 + c_2 \;\Rightarrow\; 0 = c_2, \end{aligned}$$

where we have used $\hat{e}_\rho \cdot \hat{e}_x = \cos\varphi$, and $\hat{e}_\varphi \cdot \hat{e}_x = -\sin\varphi$—obtained by dotting the first and second equations of (1.36) with $\hat{e}_x$—as well as $\hat{e}_z \cdot \hat{e}_x = 0$. Similarly, one can obtain $\hat{e}_y$ in terms of the cylindrical unit vectors. The entire result is

Cartesian unit
vectors in terms of
cylindrical unit
vectors

$$\begin{aligned} \hat{e}_x &= \hat{e}_\rho \cos\varphi - \hat{e}_\varphi \sin\varphi \\ \hat{e}_y &= \hat{e}_\rho \sin\varphi + \hat{e}_\varphi \cos\varphi \\ \hat{e}_z &= \hat{e}_z \end{aligned} \qquad (1.37)$$

Now we express the spherical unit vectors in terms of the cylindrical ones. This is easily done for $\hat{e}_r$, because it has only $\hat{e}_\rho$ and $\hat{e}_z$ components (why?). Thus, with

$$\hat{e}_r = a_3 \hat{e}_\rho + b_3 \hat{e}_z,$$

we obtain

$$\begin{aligned} \hat{e}_\rho \cdot \hat{e}_r &= a_3 + 0 \;\Rightarrow\; a_3 = \sin\theta, \\ \hat{e}_z \cdot \hat{e}_r &= 0 + b_3 \;\Rightarrow\; b_3 = \cos\theta, \end{aligned}$$

---

[16]Remember that $\hat{e}_z$ is a unit vector in both coordinate systems. So, one can say that the *cylindrical* $\hat{e}_z$ has components $\langle 0,0,1 \rangle$ in the *Cartesian* basis $\{\hat{e}_x, \hat{e}_y, \hat{e}_z\}$.

where in the last step of each line, we used the fact that the angle between $\hat{\mathbf{e}}_r$ and $\hat{\mathbf{e}}_z$ is θ and that between $\hat{\mathbf{e}}_r$ and $\hat{\mathbf{e}}_\rho$ is $\pi/2 - \theta$ (see Figure 1.13). With a_3 and b_3 so determined, we can write

$$\hat{\mathbf{e}}_r = \hat{\mathbf{e}}_\rho \sin\theta + \hat{\mathbf{e}}_z \cos\theta.$$

Having two spherical unit vectors $\hat{\mathbf{e}}_r$ and $\hat{\mathbf{e}}_\varphi$ at our disposal,[17] we can determine the third one, using (1.5) and (1.30):

$$\hat{\mathbf{e}}_\theta = \hat{\mathbf{e}}_\varphi \times \hat{\mathbf{e}}_r = \det \begin{pmatrix} \hat{\mathbf{e}}_\rho & \hat{\mathbf{e}}_\varphi & \hat{\mathbf{e}}_z \\ 0 & 1 & 0 \\ \sin\theta & 0 & \cos\theta \end{pmatrix} = \hat{\mathbf{e}}_\rho \cos\theta - \hat{\mathbf{e}}_z \sin\theta.$$

Thus,

$$\begin{aligned} \hat{\mathbf{e}}_r &= \hat{\mathbf{e}}_\rho \sin\theta + \hat{\mathbf{e}}_z \cos\theta, \\ \hat{\mathbf{e}}_\theta &= \hat{\mathbf{e}}_\rho \cos\theta - \hat{\mathbf{e}}_z \sin\theta, \\ \hat{\mathbf{e}}_\varphi &= \hat{\mathbf{e}}_\varphi. \end{aligned} \tag{1.38}$$

spherical unit vectors in terms of cylindrical unit vectors

The inverse relations can be obtained as before. We leave the details of the calculation as an exercise for the reader.

Combining Equations (1.36) and (1.38), we can express spherical unit vectors in terms of the Cartesian unit vectors:

$$\begin{aligned} \hat{\mathbf{e}}_r &= \hat{\mathbf{e}}_x \sin\theta \cos\varphi + \hat{\mathbf{e}}_y \sin\theta \sin\varphi + \hat{\mathbf{e}}_z \cos\theta, \\ \hat{\mathbf{e}}_\theta &= \hat{\mathbf{e}}_x \cos\theta \cos\varphi + \hat{\mathbf{e}}_y \cos\theta \sin\varphi - \hat{\mathbf{e}}_z \sin\theta, \\ \hat{\mathbf{e}}_\varphi &= -\hat{\mathbf{e}}_x \sin\varphi + \hat{\mathbf{e}}_y \cos\varphi. \end{aligned} \tag{1.39}$$

spherical unit vectors in terms of Cartesian unit vectors

Equations (1.39) and (1.36) are very useful when calculating vector quantities in spherical and cylindrical coordinates as we shall see in many examples to follow. These equations also allow us to express a unit vector in one of the three coordinate systems in terms of the unit vectors of any other coordinate system.

Example 1.4.1. P_1 and P_2 have Cartesian coordinates $(1, 1, 1)$ and $(-1, 2, -1)$, respectively. A vector $\mathbf{a}$ has spherical components $\langle 0, 2, 0 \rangle$ at P_1. We want to find the spherical components of $\mathbf{a}$ at P_2. These are given by $\mathbf{a} \cdot \hat{\mathbf{e}}_{r_2}$, $\mathbf{a} \cdot \hat{\mathbf{e}}_{\theta_2}$, and $\mathbf{a} \cdot \hat{\mathbf{e}}_{\varphi_2}$. In order to calculate these dot products, it is most convenient to express all vectors in Cartesian form. So, using Equation (1.39), we have

$$\mathbf{a} = 2\hat{\mathbf{e}}_{\theta_1} = 2\left(\hat{\mathbf{e}}_x \cos\theta_1 \cos\varphi_1 + \hat{\mathbf{e}}_y \cos\theta_1 \sin\varphi_1 - \hat{\mathbf{e}}_z \sin\theta_1\right),$$

where $(r_1, \theta_1, \varphi_1)$ are coordinates of P_1. We can calculate these from the Cartesian coordinates of P_1:

$$r_1 = \sqrt{1^2 + 1^2 + 1^2} = \sqrt{3}, \quad \cos\theta_1 = \frac{z_1}{r_1} = \frac{1}{\sqrt{3}}, \quad \tan\varphi_1 = \frac{y_1}{x_1} = 1 \Rightarrow \varphi_1 = \frac{\pi}{4}.$$

---

[17]Recall that $\hat{\mathbf{e}}_\varphi$ is both a cylindrical and a spherical unit vector.

Therefore,

$$\mathbf{a} = 2\left(\hat{\mathbf{e}}_x \frac{1}{\sqrt{3}}\frac{1}{\sqrt{2}} + \hat{\mathbf{e}}_y \frac{1}{\sqrt{3}}\frac{1}{\sqrt{2}} - \hat{\mathbf{e}}_z \sqrt{\frac{2}{3}}\right) = \frac{2}{\sqrt{6}}\hat{\mathbf{e}}_x + \frac{2}{\sqrt{6}}\hat{\mathbf{e}}_y - \frac{4}{\sqrt{6}}\hat{\mathbf{e}}_z.$$

Now we need to express $\hat{\mathbf{e}}_{r_2}$, $\hat{\mathbf{e}}_{\theta_2}$, and $\hat{\mathbf{e}}_{\varphi_2}$ in terms of Cartesian unit vectors. Once again we use Equation (1.39) for which we need the spherical coordinates of P_2:

$$r_2 = \sqrt{(-1)^2 + 2^2 + (-1)^2} = \sqrt{6}, \quad \cos\theta_2 = \frac{z_2}{r_2} = -\frac{1}{\sqrt{6}}, \quad \tan\varphi_2 = \frac{y_2}{x_2} = -2.$$

Similarly, Equations (1.11) and (1.12) yield

$$\sin\theta_2 = +\sqrt{\frac{5}{6}}, \qquad \cos\varphi_2 = -\frac{1}{\sqrt{5}}, \qquad \sin\varphi_2 = +\frac{2}{\sqrt{5}}.$$

Then

$$\hat{\mathbf{e}}_{r_2} = \hat{\mathbf{e}}_x \sin\theta_2 \cos\varphi_2 + \hat{\mathbf{e}}_y \sin\theta_2 \sin\varphi_2 + \hat{\mathbf{e}}_z \cos\theta_2$$

$$= \hat{\mathbf{e}}_x \sqrt{\frac{5}{6}}\left(-\frac{1}{\sqrt{5}}\right) + \hat{\mathbf{e}}_y \sqrt{\frac{5}{6}}\frac{2}{\sqrt{5}} - \hat{\mathbf{e}}_z \frac{1}{\sqrt{6}}$$

$$= -\frac{1}{\sqrt{6}}\hat{\mathbf{e}}_x + \frac{2}{\sqrt{6}}\hat{\mathbf{e}}_y - \frac{1}{\sqrt{6}}\hat{\mathbf{e}}_z,$$

$$\hat{\mathbf{e}}_{\theta_2} = \hat{\mathbf{e}}_x \cos\theta_2 \cos\varphi_2 + \hat{\mathbf{e}}_y \cos\theta_2 \sin\varphi_2 - \hat{\mathbf{e}}_z \sin\theta_2$$

$$= \hat{\mathbf{e}}_x\left(-\frac{1}{\sqrt{6}}\right)\left(-\frac{1}{\sqrt{5}}\right) + \hat{\mathbf{e}}_y\left(-\frac{1}{\sqrt{6}}\right)\frac{2}{\sqrt{5}} - \hat{\mathbf{e}}_z \sqrt{\frac{5}{6}}$$

$$= \frac{1}{\sqrt{30}}\hat{\mathbf{e}}_x - \frac{2}{\sqrt{30}}\hat{\mathbf{e}}_y - \frac{5}{\sqrt{30}}\hat{\mathbf{e}}_z,$$

$$\hat{\mathbf{e}}_{\varphi_2} = -\hat{\mathbf{e}}_x \sin\varphi_2 + \hat{\mathbf{e}}_y \cos\varphi_2 = -\frac{2}{\sqrt{5}}\hat{\mathbf{e}}_x - \frac{1}{\sqrt{5}}\hat{\mathbf{e}}_y.$$

We can now take the dot products required for the components:

$$r \text{ comp} = \mathbf{a} \cdot \hat{\mathbf{e}}_{r_2} = \left(\frac{2}{\sqrt{6}}\hat{\mathbf{e}}_x + \frac{2}{\sqrt{6}}\hat{\mathbf{e}}_y - \frac{4}{\sqrt{6}}\hat{\mathbf{e}}_z\right) \cdot \left(-\frac{1}{\sqrt{6}}\hat{\mathbf{e}}_x + \frac{2}{\sqrt{6}}\hat{\mathbf{e}}_y - \frac{1}{\sqrt{6}}\hat{\mathbf{e}}_z\right)$$

$$= -\frac{2}{6} + \frac{4}{6} + \frac{4}{6} = 1,$$

$$\theta \text{ comp} = \mathbf{a} \cdot \hat{\mathbf{e}}_{\theta_2} = \left(\frac{2}{\sqrt{6}}\hat{\mathbf{e}}_x + \frac{2}{\sqrt{6}}\hat{\mathbf{e}}_y - \frac{4}{\sqrt{6}}\hat{\mathbf{e}}_z\right) \cdot \left(\frac{1}{\sqrt{30}}\hat{\mathbf{e}}_x - \frac{2}{\sqrt{30}}\hat{\mathbf{e}}_y - \frac{5}{\sqrt{30}}\hat{\mathbf{e}}_z\right)$$

$$= \frac{2}{6\sqrt{5}} - \frac{4}{6\sqrt{5}} + \frac{20}{6\sqrt{5}} = \frac{3}{\sqrt{5}},$$

$$\varphi \text{ comp} = \mathbf{a} \cdot \hat{\mathbf{e}}_{\varphi_2} = \left(\frac{2}{\sqrt{6}}\hat{\mathbf{e}}_x + \frac{2}{\sqrt{6}}\hat{\mathbf{e}}_y - \frac{4}{\sqrt{6}}\hat{\mathbf{e}}_z\right) \cdot \left(-\frac{2}{\sqrt{5}}\hat{\mathbf{e}}_x - \frac{1}{\sqrt{5}}\hat{\mathbf{e}}_y\right)$$

$$= -\frac{4}{\sqrt{30}} - \frac{2}{\sqrt{30}} = -\frac{6}{\sqrt{30}} = -\sqrt{\frac{6}{5}}.$$

It now follows that

$$\mathbf{a} = \hat{\mathbf{e}}_{r_2} + \frac{3}{\sqrt{5}}\hat{\mathbf{e}}_{\theta_2} - \sqrt{\frac{6}{5}}\hat{\mathbf{e}}_{\varphi_2}.$$

As a check, we note that

$$|\mathbf{a}| = \sqrt{1^2 + \left(\frac{3}{\sqrt{5}}\right)^2 + \left(-\sqrt{\frac{6}{5}}\right)^2} = \sqrt{\frac{5+9+6}{5}} = \sqrt{4} = 2,$$

which agrees with the length of $\mathbf{a}$.   ∎

Example 1.4.2. Points P_1 and P_2 have spherical coordinates $(r_1, \theta_1, \varphi_1)$ and $(r_2, \theta_2, \varphi_2)$, respectively. We want to find: (a) the angle between their position vectors $\mathbf{r}_1$ and $\mathbf{r}_2$ in terms of their coordinates; (b) the spherical components of $\mathbf{r}_2$ at P_1; and (c) the spherical components of $\mathbf{r}_1$ at P_2. Once again, we shall express all vectors in terms of Cartesian unit vectors when evaluating dot products.

(a) The cosine of the angle—call it γ_{12}—between the position vectors is simply $\hat{\mathbf{e}}_{r_1} \cdot \hat{\mathbf{e}}_{r_2}$. We can readily find this by using Equation (1.39):

$$\begin{aligned}
\cos\gamma_{12} = \hat{\mathbf{e}}_{r_1} \cdot \hat{\mathbf{e}}_{r_2} &= (\hat{\mathbf{e}}_x \sin\theta_1 \cos\varphi_1 + \hat{\mathbf{e}}_y \sin\theta_1 \sin\varphi_1 + \hat{\mathbf{e}}_z \cos\theta_1) \\
&\quad \cdot (\hat{\mathbf{e}}_x \sin\theta_2 \cos\varphi_2 + \hat{\mathbf{e}}_y \sin\theta_2 \sin\varphi_2 + \hat{\mathbf{e}}_z \cos\theta_2) \\
&= \sin\theta_1 \cos\varphi_1 \sin\theta_2 \cos\varphi_2 + \sin\theta_1 \sin\varphi_1 \sin\theta_2 \sin\varphi_2 + \cos\theta_1 \cos\theta_2 \\
&= \sin\theta_1 \sin\theta_2 (\cos\varphi_1 \cos\varphi_2 + \sin\varphi_1 \sin\varphi_2) + \cos\theta_1 \cos\theta_2 \\
&= \sin\theta_1 \sin\theta_2 \cos(\varphi_1 - \varphi_2) + \cos\theta_1 \cos\theta_2.
\end{aligned}$$

(b) To find the spherical components of $\mathbf{r}_2$ at P_1, we need to take the dot product of $\mathbf{r}_2$ with the spherical unit vectors at P_1:

$$\begin{aligned}
r \text{ comp} = \mathbf{r}_2 \cdot \hat{\mathbf{e}}_{r_1} &= r_2 \hat{\mathbf{e}}_{r_2} \cdot \hat{\mathbf{e}}_{r_1} \\
&= r_2 \left[\sin\theta_1 \sin\theta_2 \cos(\varphi_1 - \varphi_2) + \cos\theta_1 \cos\theta_2\right],
\end{aligned}$$

$$\begin{aligned}
\theta \text{ comp} = \mathbf{r}_2 \cdot \hat{\mathbf{e}}_{\theta_1} &= r_2 \hat{\mathbf{e}}_{r_2} \cdot \hat{\mathbf{e}}_{\theta_1} \\
&= r_2 (\hat{\mathbf{e}}_x \sin\theta_2 \cos\varphi_2 + \hat{\mathbf{e}}_y \sin\theta_2 \sin\varphi_2 + \hat{\mathbf{e}}_z \cos\theta_2) \\
&\quad \cdot (\hat{\mathbf{e}}_x \cos\theta_1 \cos\varphi_1 + \hat{\mathbf{e}}_y \cos\theta_1 \sin\varphi_1 - \hat{\mathbf{e}}_z \sin\theta_1) \\
&= r_2(\sin\theta_2 \cos\varphi_2 \cos\theta_1 \cos\varphi_1 + \sin\theta_2 \sin\varphi_2 \cos\theta_1 \sin\varphi_1 - \cos\theta_2 \sin\theta_1) \\
&= r_2[\sin\theta_2 \cos\theta_1 \cos(\varphi_1 - \varphi_2) - \cos\theta_2 \sin\theta_1],
\end{aligned}$$

$$\begin{aligned}
\varphi \text{ comp} = \mathbf{r}_2 \cdot \hat{\mathbf{e}}_{\varphi_1} &= r_2 \hat{\mathbf{e}}_{r_2} \cdot \hat{\mathbf{e}}_{\varphi_1} \\
&= r_2 (\hat{\mathbf{e}}_x \sin\theta_2 \cos\varphi_2 + \hat{\mathbf{e}}_y \sin\theta_2 \sin\varphi_2 + \hat{\mathbf{e}}_z \cos\theta_2) \cdot (-\hat{\mathbf{e}}_x \sin\varphi_1 + \hat{\mathbf{e}}_y \cos\varphi_1) \\
&= r_2 (-\sin\theta_2 \cos\varphi_2 \sin\varphi_1 + \sin\theta_2 \sin\varphi_2 \cos\varphi_1) = r_2 \sin\theta_2 \sin(\varphi_2 - \varphi_1).
\end{aligned}$$

(c) The spherical components of $\mathbf{r}_1$ at P_2 can be found similarly. In fact, switching the indices "1" and "2" in the expressions of part (b) gives the desired formulas.   ∎

Example 1.4.3. To illustrate further the conversion of vectors from one coordinate system to another, consider a charge q that is located at the *cylindrical* coordinates $(a, \pi/3, -a)$. We want to find the *spherical* components of the electrostatic field $\mathbf{E}$ of this charge at a point P with *Cartesian* coordinates (a, a, a).

The most straightforward way of doing this is to convert all coordinates to Cartesian, find the field, and then take the dot products with appropriate unit vectors. The Cartesian coordinates of the charge are

$$x_q = \rho_q \cos\varphi_q = a\cos\left(\frac{\pi}{3}\right) = \tfrac{1}{2}a,$$

$$y_q = \rho_q \sin\varphi_q = a\sin\left(\frac{\pi}{3}\right) = \frac{\sqrt{3}a}{2} = 0.866a,$$

$$z_q = -a.$$

Thus,

$$\mathbf{r} - \mathbf{r}_q = (a - \tfrac{1}{2}a)\hat{\mathbf{e}}_x + (a - 0.866a)\hat{\mathbf{e}}_y + (a - (-a))\hat{\mathbf{e}}_z = 0.5a\hat{\mathbf{e}}_x + 0.134a\hat{\mathbf{e}}_y + 2a\hat{\mathbf{e}}_z$$

and

$$|\mathbf{r} - \mathbf{r}_q| = \sqrt{(0.5a)^2 + (0.134a)^2 + (2a)^2} = 2.066a,$$

and the electric field at P can be written in terms of Cartesian unit vectors at P:

$$\mathbf{E} = \frac{k_e q}{|\mathbf{r} - \mathbf{r}_q|^3}(\mathbf{r} - \mathbf{r}_q) = k_e q \frac{0.5a\hat{\mathbf{e}}_x + 0.134a\hat{\mathbf{e}}_y + 2a\hat{\mathbf{e}}_z}{(2.066a)^3}$$

$$= k_e q \frac{0.5\hat{\mathbf{e}}_x + 0.134\hat{\mathbf{e}}_y + 2\hat{\mathbf{e}}_z}{8.818a^2} = \frac{k_e q}{a^2}(0.0567\hat{\mathbf{e}}_x + 0.0152\hat{\mathbf{e}}_y + 0.2268\hat{\mathbf{e}}_z).$$

To find the *spherical* components of the field at P, we first express the spherical unit vectors at P in terms of Cartesian unit vectors. For this, we need the spherical coordinates of P:

$$r = \sqrt{a^2 + a^2 + a^2} = \sqrt{3}\,a = 1.732a,$$

$$\cos\theta = \frac{z}{r} = \frac{a}{\sqrt{3}\,a} = \frac{1}{\sqrt{3}} = 0.577 \;\Rightarrow\; \theta = 0.955,$$

$$\tan\varphi = \frac{y}{x} = \frac{a}{a} = 1 \;\Rightarrow\; \varphi = \frac{\pi}{4} = 0.785.$$

It now follows that

$$\hat{\mathbf{e}}_r = \hat{\mathbf{e}}_x \sin\theta \cos\varphi + \hat{\mathbf{e}}_y \sin\theta \sin\varphi + \hat{\mathbf{e}}_z \cos\theta = 0.577\hat{\mathbf{e}}_x + 0.577\hat{\mathbf{e}}_y + 0.577\hat{\mathbf{e}}_z,$$

$$\hat{\mathbf{e}}_\theta = \hat{\mathbf{e}}_x \cos\theta \cos\varphi + \hat{\mathbf{e}}_y \cos\theta \sin\varphi - \hat{\mathbf{e}}_z \sin\theta = 0.408\hat{\mathbf{e}}_x + 0.408\hat{\mathbf{e}}_y - 0.816\hat{\mathbf{e}}_z,$$

$$\hat{\mathbf{e}}_\varphi = -\hat{\mathbf{e}}_x \sin\varphi + \hat{\mathbf{e}}_y \cos\varphi = -0.707\hat{\mathbf{e}}_x + 0.707\hat{\mathbf{e}}_y.$$

Now we take the dot product of $\mathbf{E}$ with these unit vectors to find its spherical components at P. The reader may first easily check that

$$\hat{\mathbf{e}}_r \cdot \hat{\mathbf{e}}_x = 0.577, \qquad \hat{\mathbf{e}}_r \cdot \hat{\mathbf{e}}_y = 0.577, \qquad \hat{\mathbf{e}}_r \cdot \hat{\mathbf{e}}_z = 0.577,$$

$$\hat{\mathbf{e}}_\theta \cdot \hat{\mathbf{e}}_x = 0.408, \qquad \hat{\mathbf{e}}_\theta \cdot \hat{\mathbf{e}}_y = 0.408, \qquad \hat{\mathbf{e}}_\theta \cdot \hat{\mathbf{e}}_z = -0.816,$$

$$\hat{\mathbf{e}}_\varphi \cdot \hat{\mathbf{e}}_x = -0.707, \qquad \hat{\mathbf{e}}_\varphi \cdot \hat{\mathbf{e}}_y = 0.707, \qquad \hat{\mathbf{e}}_\varphi \cdot \hat{\mathbf{e}}_z = 0.$$

We can now finally calculate the field components:

$$E_r = \mathbf{E} \cdot \hat{\mathbf{e}}_r = \frac{k_e q}{a^2}(0.0567\hat{\mathbf{e}}_r \cdot \hat{\mathbf{e}}_x + 0.0152\hat{\mathbf{e}}_r \cdot \hat{\mathbf{e}}_y + 0.2268\hat{\mathbf{e}}_r \cdot \hat{\mathbf{e}}_z)$$

$$= \frac{k_e q}{a^2}(0.0567 \times 0.577 + 0.0152 \times 0.577 + 0.2268 \times 0.577) = 0.1724\frac{k_e q}{a^2},$$

$$E_\theta = \mathbf{E} \cdot \hat{\mathbf{e}}_\theta = \frac{k_e q}{a^2}(0.0567\hat{\mathbf{e}}_\theta \cdot \hat{\mathbf{e}}_x + 0.0152\hat{\mathbf{e}}_\theta \cdot \hat{\mathbf{e}}_y + 0.2268\hat{\mathbf{e}}_\theta \cdot \hat{\mathbf{e}}_z)$$

$$= \frac{k_e q}{a^2}(0.0567 \times 0.408 + 0.0152 \times 0.408 - 0.2268 \times 0.816) = -0.1558\frac{k_e q}{a^2},$$

$$E_\varphi = \mathbf{E} \cdot \hat{\mathbf{e}}_\varphi = \frac{k_e q}{a^2}(0.0567\hat{\mathbf{e}}_\varphi \cdot \hat{\mathbf{e}}_x + 0.0152\hat{\mathbf{e}}_\varphi \cdot \hat{\mathbf{e}}_y + 0.2268\hat{\mathbf{e}}_\varphi \cdot \hat{\mathbf{e}}_z)$$

$$= \frac{k_e q}{a^2}(-0.0567 \times 0.707 + 0.0152 \times 0.707) = -0.0294\frac{k_e q}{a^2}.$$

The choice of Cartesian coordinates was the most straightforward one, but one can choose any other coordinate system to calculate the field and find the components in any other set of unit vectors. The reader is urged to try the other choices. ∎

1.5 Problems

1.1. Find the equation of a line that passes through the following pairs of points:

    (a) $(1, 0, 1)$ and $(-1, 1, 0)$.   (b) $(2, 2, -1)$ and $(-2, -1, 1)$.

    (c) $(1, 1, 1)$ and $(-1, 1, -1)$.   (d) $(1, 1, 1)$ and $(-2, 2, 0)$.

    (e) $(0, 2, -1)$ and $(3, -1, 1)$.   (f) $(0, 1, 0)$ and $(-1, 0, -1)$.

1.2. Use Figure 1.4 and the interpretation of the $\mathbf{a} \cdot \mathbf{b}$ as the product of the length of $\mathbf{a}$ with the projection of $\mathbf{b}$ along $\mathbf{a}$ to show that

$$(\mathbf{a} + \mathbf{b}) \cdot \mathbf{c} = \mathbf{a} \cdot \mathbf{c} + \mathbf{b} \cdot \mathbf{c}.$$

1.3. Take the dot product of $\mathbf{a} = \mathbf{b} - \mathbf{c}$ with itself and prove the law of cosines by interpreting the result geometrically. Note that the three vectors form a triangle.

1.4. Find the angle between $\mathbf{a} = 2\hat{\mathbf{e}}_x + 3\hat{\mathbf{e}}_y + \hat{\mathbf{e}}_z$ and $\mathbf{b} = \hat{\mathbf{e}}_x - 6\hat{\mathbf{e}}_y + 2\hat{\mathbf{e}}_z$.

1.5. Find the angle between $\mathbf{a} = 9\hat{\mathbf{e}}_x + \hat{\mathbf{e}}_y - 6\hat{\mathbf{e}}_z$ and $\mathbf{b} = 4\hat{\mathbf{e}}_x - 6\hat{\mathbf{e}}_y + 5\hat{\mathbf{e}}_z$.

1.6. Show that $\mathbf{a} = \hat{\mathbf{e}}_x \cos\alpha + \hat{\mathbf{e}}_y \sin\alpha$ and $\mathbf{b} = \hat{\mathbf{e}}_x \cos\beta + \hat{\mathbf{e}}_y \sin\beta$ are unit vectors in the xy-plane making angles α and β with the x-axis. Then take their dot product and obtain a formula for $\cos(\alpha - \beta)$. Now use $\sin x = \cos(\pi/2 - x)$ to find the formula for $\sin(\alpha - \beta)$.

1.7. Vectors $\mathbf{a}$ and $\mathbf{b}$ are the sides of a parallelogram, $\mathbf{c}$ and $\mathbf{d}$ are its diagonals, and θ is the angle between $\mathbf{a}$ and $\mathbf{b}$. Show that

$$|\mathbf{c}|^2 + |\mathbf{d}|^2 = 2(|\mathbf{a}|^2 + |\mathbf{b}|^2)$$

and that

$$|\mathbf{c}|^2 - |\mathbf{d}|^2 = 4|\mathbf{a}|\,|\mathbf{b}| \cos\theta.$$

1.8. Given $\mathbf{a}$, $\mathbf{b}$, and $\mathbf{c}$—vectors from the origin to the points A, B, and C—show that the vector $(\mathbf{a} \times \mathbf{b}) + (\mathbf{b} \times \mathbf{c}) + (\mathbf{c} \times \mathbf{a})$ is perpendicular to the plane ABC.

1.9. Show that the vectors $\mathbf{a} = 2\hat{\mathbf{e}}_x - \hat{\mathbf{e}}_y + \hat{\mathbf{e}}_z$, $\mathbf{b} = \hat{\mathbf{e}}_x - 3\hat{\mathbf{e}}_y - 5\hat{\mathbf{e}}_z$, and $\mathbf{c} = 3\hat{\mathbf{e}}_x - 4\hat{\mathbf{e}}_y - 4\hat{\mathbf{e}}_z$ form the sides of a right triangle.

1.10. (a) Find the vector form of the equation of the plane defined by the three points P, Q, and R with coordinates (p_1, p_2, p_3), (q_1, q_2, q_3), and (r_1, r_2, r_3), respectively. Hint: The position vector of a point $X = (x, y, z)$ in the plane is perpendicular to the cross product of $\overrightarrow{PQ}$ and $\overrightarrow{PR}$.
(b) Determine an equation for the plane passing through the points $(2, -1, 1)$, $(3, 2, -1)$, and $(-1, 3, 2)$.

1.11. Derive the law of sines for a triangle using vectors.

1.12. Using vectors, show that the diagonals of a rhombus are orthogonal.

1.13. Show that a necessary and sufficient condition for three vectors to be in the same plane is that the dot product of one with the cross product of the other two be zero.

1.14. Show that two nonzero vectors have the same direction if and only if their cross product vanishes.

1.15. Show the following vector identities by writing each vector in terms of Cartesian unit vectors and showing that each component of the LHS is equal to the corresponding component of the RHS.

(a) $\mathbf{a} \cdot (\mathbf{b} \times \mathbf{c}) = \mathbf{c} \cdot (\mathbf{a} \times \mathbf{b}) = \mathbf{b} \cdot (\mathbf{c} \times \mathbf{a})$.

(b) $\mathbf{a} \times (\mathbf{b} \times \mathbf{c}) = \mathbf{b}(\mathbf{a} \cdot \mathbf{c}) - \mathbf{c}(\mathbf{a} \cdot \mathbf{b})$, this is called the **bac cab rule**.

(c) $(\mathbf{a} \times \mathbf{b}) \cdot (\mathbf{c} \times \mathbf{d}) = (\mathbf{a} \cdot \mathbf{c})(\mathbf{b} \cdot \mathbf{d}) - (\mathbf{a} \cdot \mathbf{d})(\mathbf{b} \cdot \mathbf{c})$.

(d) $(\mathbf{a} \times \mathbf{b}) \times (\mathbf{c} \times \mathbf{d}) = \mathbf{b}[\mathbf{a} \cdot (\mathbf{c} \times \mathbf{d})] - \mathbf{a}[\mathbf{b} \cdot (\mathbf{c} \times \mathbf{d})]$.

(e) $(\mathbf{a} \times \mathbf{b}) \times (\mathbf{c} \times \mathbf{d}) = \mathbf{c}[\mathbf{a} \cdot (\mathbf{b} \times \mathbf{d})] - \mathbf{d}[\mathbf{a} \cdot (\mathbf{b} \times \mathbf{c})]$.

(f) $(\mathbf{a} \times \mathbf{b}) \cdot (\mathbf{a} \times \mathbf{b}) = |\mathbf{a}|^2 |\mathbf{b}|^2 - (\mathbf{a} \cdot \mathbf{b})^2$.

1.16. Convert the following triplets from the given coordinate system to the other two. All angles are in radians.
Cartesian: $(1, 2, 1)$, $(0, 0, 1)$, $(1, -1, 0)$, $(0, 1, 0)$, $(1, 1, 1)$, $(2, 2, 2)$, $(0, 0, 5)$, $(1, 1, 0)$, $(1, 0, 0)$.
Spherical: $(2, \pi/3, \pi/4)$, $(5, 0, \pi/3)$, $(3, \pi/3, 3\pi/4)$, $(1, 1, 0)$, $(1, 0, 0)$, $(5, 0, \clubsuit)$, $(3, \pi, \heartsuit)$, $(0, \spadesuit, \diamondsuit)$.
Cylindrical: $(0, \clubsuit, 4), (2, \pi, 0), (0, 217, -18), (1, 3\pi/4, -2), (1, 2, 3), (1, 0, 0)$.

1.17. Derive the second and third relations in Equation (1.21).

1.18. Points P and P' have spherical coordinates (r, θ, φ) and (r', θ', φ'), cylindrical coordinates (ρ, φ, z) and (ρ', φ', z'), and Cartesian coordinates (x, y, z) and (x', y', z'), respectively. Write $|\mathbf{r} - \mathbf{r}'|$ in all three coordinate systems. Hint: Use Equation (1.2) with $\mathbf{a} = \mathbf{r} - \mathbf{r}'$ and $\mathbf{r}$ and $\mathbf{r}'$ written in terms of appropriate unit vectors.

1.19. Show that Equation (1.24) is independent of where we choose the origin to be. Hint: Pick a different origin O' whose position vector relative to O is $\mathbf{R}$ and write the equation in terms of position vectors relative to O' and show that the final result is the same as in Equation (1.24).

1.20. Three point charges are located at the corners of an equilateral triangle of sides a with the origin at the center of the triangle as shown in Figure 1.19.
(a) Find the general expression for the electric field and electric potential at $(0, 0, z)$.
(b) Find a relation between q and Q such that the z-component of the field vanishes for all values of z. What are $\mathbf{E}$ and Φ for such charges?
(c) Calculate $\mathbf{E}$ and Φ for $z = a$.

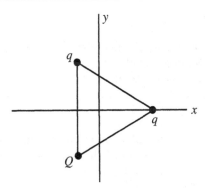

Figure 1.19:

1.21. A point charge Q and two point charges q are located in the xy-plane at the corners of an equilateral triangle of side a as shown in Figure 1.20.
(a) Find the potential and the Cartesian components of the electrostatic field at $(0, 0, z)$.
(b) Show that it is impossible for **E** to be along the z-axis.
(c) Calculate **E** for $z = a$ and find Q in terms of q such that E_z vanishes for this value of z.
(d) What is the value of Φ at $z = a$ for the charges found in (c)?

1.22. Three point charges each of magnitude Q and one point charge q are located at the corners of a square of side $2a$. Using an appropriate coordinate system.
(a) Find the electric field and potential at point P located on the diagonal from Q to q (and beyond) a distance $2\sqrt{2}\,a$ from the center.
(b) Find a relation, if it exists, between q and Q such that the field vanishes at P.

1.23. A charge q is located at the spherical coordinates $(a, \pi/4, \pi/3)$. Find the electrostatic potential and the *Cartesian* components of the electrostatic field of this charge at a point P with spherical coordinates $(a, \pi/6, \pi/4)$. Write the field components as numerical multiples of $k_e q/a^2$, and the potential as a

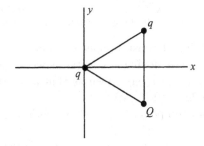

Figure 1.20:

numerical multiple of $k_e q/a$.

1.24. A charge q is located at the cylindrical coordinates $(a, \pi/4, 2a)$. Find the *Cartesian* components of the electrostatic field of this charge at a point P with cylindrical coordinates $(2a, \pi/6, a)$. Write your answers as a numerical multiple of $k_e q/a^2$. Find the electrostatic potential at P and express it as a numerical multiple of $k_e q/a$.

1.25. A charge q is located at the cylindrical coordinates $(a, \pi/3, -a)$.
(a) Find the *Cartesian* components of the electrostatic field $\mathbf{E}$ of this charge at a point P with cylindrical coordinates $(a, \pi/4, 2a)$. Write your answers as a numerical multiple of $k_e q/a^2$.
(b) Write $\mathbf{E}$ in terms of the *cylindrical* unit vectors at P.
(c) Find the electrostatic potential at P as a numerical multiple of $k_e q/a$.

1.26. Two charges q and $-2q$ are located at the cylindrical coordinates $(a, \pi/4, a)$ and $(a, 2\pi/3, -a)$, respectively.
(a) Find the *Cartesian* components of the electrostatic field at a point P with spherical coordinates $(3a, \pi/6, \pi/4)$. Write your answers as a numerical multiple of $k_e q/a^2$.
(b) Find the electrostatic potential at P. Write your answer as a numerical multiple of $k_e q/a$.

1.27. Two charges $3q$ and $-q$ are located at the spherical coordinates $(a, \pi/3, \pi/6)$ and $(2a, \pi/6, \pi/4)$, respectively.
(a) Find the *cylindrical* components of the electrostatic field at a point P with spherical coordinates $(3a, \pi/4, \pi/4)$. Write your answers as a numerical multiple of $k_e q/a^2$.
(b) Find the electrostatic potential at P. Write your answer as a numerical multiple of $k_e q/a$.

1.28. A charge q is located at the spherical coordinates $(a, \pi/3, \pi/6)$. Find the Cartesian components of the electrostatic field of this charge at a point P with cylindrical coordinates $(a, 2\pi/3, 2a)$. Write your answers as a numerical multiple of $k_e q/a^2$. Also find the electrostatic potential at P.

1.29. Four charges are located at Cartesian coordinates as follows: q at $(2a, 0, 0)$, $-2q$ at $(0, 2a, 0)$, $\dfrac{4\sqrt{2}}{5\sqrt{5}}q$ at $(-a, 0, 0)$, and $-\dfrac{2\sqrt{2}}{5\sqrt{5}}q$ at $(0, -a, 0)$. Find the Cartesian components of the electrostatic field at $(0, 0, a)$.

1.30. Charge q is moving at constant speed v along the positive x-axis. Two other charges $-q$ and $2q$ are moving at constant speeds v and $2v$ along positive y and negative z axes, respectively. Assume that at $t = 0$, q is at the origin, $-q$ is at $(0, a, 0)$, and $2q$ at $(0, 0, -a)$.
(a) Find the Cartesian components of the magnetic field at a point (x, y, z) for $t > 0$.
(a) Find the cylindrical components of the magnetic field at a point (ρ, φ, z) for $t > 0$.

(a) Find the spherical components of the magnetic field at a point (r, θ, φ) for $t > 0$.

1.31. A charge q is moving at *constant speed* v along a curve parametrized by

$$x' = 6as, \quad y' = 3as^2, z' = -2as^3$$

(a) Find the Cartesian components of the magnetic field at a point (x, y, z) as a function of s.
(a) Find the cylindrical components of the magnetic field at a point (ρ, φ, z) as a function of s.
(a) Find the spherical components of the magnetic field at a point (r, θ, φ) as a function of s.

1.32. Points P_1 and P_2 have Cartesian coordinates $(1, 1, 1)$ and $(1, 1, 0)$, respectively.
(a) Find the spherical coordinates of P_1 and P_2.
(b) Write down the components of $\mathbf{r}_1$, the position vector of P_1, in terms of spherical unit vectors at P_1.
(c) Write down the components of $\mathbf{r}_2$, the position vector of P_2, in terms of spherical unit vectors at P_1.

1.33. Points P_1 and P_2 have Cartesian coordinates $(2, 2, 0)$ and $(1, 0, 1)$, respectively.
(a) Find the spherical coordinates of P_1.
(b) Express $\hat{\mathbf{e}}_{r_1}$, $\hat{\mathbf{e}}_{\theta_1}$, and $\hat{\mathbf{e}}_{\varphi_1}$, the spherical unit vectors at P_1, in terms of the Cartesian unit vectors.
(c) Find the components of the position vector of P_2 along the spherical unit vectors at P_1.
(d) From its components in (c) find the length of $\mathbf{r}_2$, and show that it agrees with the length as calculated from its Cartesian components.

1.34. Points P_1 and P_2 have spherical coordinates

$$P_1 : (a, \pi/4, \pi/3) \quad \text{and} \quad P_2 : (a, \pi/3, \pi/4).$$

(a) Find the angle between their position vectors $\mathbf{r}_1$ and $\mathbf{r}_2$.
(b) Find the spherical components of $\mathbf{r}_2 - \mathbf{r}_1$ at P_1.
(c) Find the spherical components of $\mathbf{r}_2 - \mathbf{r}_1$ at P_2.

1.35. Point P_1 has Cartesian coordinates $(1, 1, 0)$, point P_2 has cylindrical coordinates $(1, 1, 0)$, and point P_3 has spherical coordinates $(1, 1, 0)$ where all angles are in radians. Express $\mathbf{r}_3 - \mathbf{r}_1$ in terms of the spherical unit vectors at P_2.

1.36. Points P_1 and P_2 have Cartesian coordinates $(1, 1, 1)$ and $(1, 2, 1)$, and position vectors $\mathbf{r}_1$ and $\mathbf{r}_2$, respectively.
(a) Find the spherical coordinates of P_1 and P_2.
(b) Find the components of $\mathbf{r}_1$, in terms of spherical unit vectors at P_1.

(c) Find the components of $\mathbf{r}_2$, in terms of spherical unit vectors at P_2.
(d) Find the components of $\mathbf{r}_1$, in terms of spherical unit vectors at P_2.
(e) Find the components of $\mathbf{r}_2$, in terms of spherical unit vectors at P_1.

1.37. Points P_1 and P_2 have Cartesian coordinates

$$(x_1, y_1, z_1) \quad \text{and} \quad (x_2, y_2, z_2).$$

(a) Find the angle between their position vectors $\mathbf{r}_1$ and $\mathbf{r}_2$ in terms of their coordinates.
(b) Find the Cartesian components of $\mathbf{r}_2 - \mathbf{r}_1$ at P_1.
(c) Find the Cartesian components of $\mathbf{r}_2 - \mathbf{r}_1$ at P_2.

1.38. Points P_1 and P_2 have cylindrical coordinates

$$(\rho_1, \varphi_1, z_1) \quad \text{and} \quad (\rho_2, \varphi_2, z_2)$$

(a) Find the angle between their position vectors $\mathbf{r}_1$ and $\mathbf{r}_2$ in terms of their coordinates.
(b) Find the cylindrical components of $\mathbf{r}_2 - \mathbf{r}_1$ at P_1.
(c) Find the cylindrical components of $\mathbf{r}_2 - \mathbf{r}_1$ at P_2.

1.39. Write the Cartesian unit vectors in terms of spherical unit vectors with coefficients written in spherical coordinates.

1.40. Write the spherical unit vectors in terms of Cartesian unit vectors with coefficients written in Cartesian coordinates.

1.41. In Example 1.4.3, calculate the electric field using cylindrical coordinates, then find the components in terms of (a) Cartesian and (b) spherical unit vectors.

1.42. In Example 1.4.3, calculate the electric field using spherical coordinates, then find the components in terms of (a) Cartesian and (b) cylindrical unit vectors.

Chapter 2

Differentiation

Physics deals with both the large and the small. Its domain of study includes the interior of the nucleus of an atom as well as the exterior of a galaxy. It is, therefore, natural for the scope of physical theories to switch between *global*, or large-scale, and *local*, or small-scale regimes. Such an interplay between the local and the global has existed ever since **Newton** and others discovered the mathematical translation of this interplay: Derivatives are defined as local objects while integrals encompass global properties. This chapter is devoted to the concept of differentiation, which we shall consider as a natural tool with which many physical concepts are expressed most concisely and conveniently.

All physical quantities reside in space and change with time. Even a static quantity—once scrutinized—will reveal noticeable attributes of change, validating the old adage "The only thing that doesn't change is the change itself." Thus, static, or time-independent, quantities are so only as approximations to the true physical quantity which is dynamic.

Take the temperature of the surface of the Earth. As we move about on the globe, we notice the variation of this quantity with location—poles as opposed to the equator—and with time—winter versus summer. A specification of temperature requires that of location and time. We thus speak of *local* and *instantaneous* temperature. This is an example of the fact that, generally speaking, *all physical quantities are functions of space and time.*

Locality and instantaneity have both a mathematical and a physical (or operational) interpretation. Mathematically, they correspond to a point in space and an instant of time with no extension or spread whatsoever. Physically, or operationally, many quantities require an extension in space and an interval in time to be defined. Thus, a local weatherman's morning statement "Today's high will be 45" limits the location to the size of a city, and the time to at most a.m. or p.m. This is admittedly a rough localization, suitable for a weatherman's forecast. Nevertheless, even the most precise statements in physics embody a space extension as well as a time interval whose "sizes" are determined by the physical system under investigation. If we are studying heat conduction by a metal bar several inches long, then "local" temperature takes a completely different meaning from the weatherman's "local" temperature. In the latter case, a city is as local as one gets, while in the former, variations over a centimeter are significant.

2.1   The Derivative

A prime example of an instantaneously defined quantity is **velocity**. To find the velocity of a moving particle at time t_0, determine its position $\mathbf{r}_0$ at time t_0, determine also its position $\mathbf{r}$ at time t with t close to t_0, divide $\mathbf{r} - \mathbf{r}_0$ by $t - t_0$, and make $t - t_0$ as small as possible. This defines the **derivative** of $\mathbf{r}$ with respect to t which we call velocity $\mathbf{v}$:

$$\mathbf{v}(t_0) = \lim_{t \to t_0} \frac{\mathbf{r} - \mathbf{r}_0}{t - t_0} \equiv \left. \frac{d\mathbf{r}}{dt} \right|_{t=t_0} \equiv \dot{\mathbf{r}}(t_0).$$

Acceleration is defined similarly:

$$\mathbf{a}(t_0) = \lim_{t \to t_0} \frac{\mathbf{v} - \mathbf{v}_0}{t - t_0} \equiv \left. \frac{d\mathbf{v}}{dt} \right|_{t=t_0} \equiv \left. \frac{d^2\mathbf{r}}{dt^2} \right|_{t=t_0} \equiv \ddot{\mathbf{r}}(t_0).$$

Velocity and acceleration are examples of derivatives which are generally called **rate of change**. In the rate of change, one is interested in the way a quantity (dependent variable) changes as another quantity (independent variable) is allowed to vary. In the majority of rates of change, the independent variable is either time or one of the space coordinates.

The second type of derivative is simply the ratio of two infinitesimal physical quantities. In general, whenever a physical quantity Q is defined as the ratio of two other physical quantities R and S, one must define Q in a small neighborhood (small volume, area, length, or time interval). One, therefore, writes

$$Q = \lim_{\Delta S \to 0} \frac{\Delta R}{\Delta S} \equiv \frac{dR}{dS}, \tag{2.1}$$

where ΔR and ΔS are both local small quantities. Being physical quantities, both R and S, and therefore ΔR and ΔS are, in general, functions of position and time. Hence, their ratio, Q, is also a function of position and time. The last sentence requires further elaboration.

In physics, we deal with two completely different, yet subtly related, objects: particles and fields. The former is no doubt familiar to the reader. Examples of the latter are the gravitational, electric, and magnetic fields, as well as the less familiar velocity field of a fluid such as water in a river or air in the atmosphere. Suppose we want to specify the "state" of the two types of objects at a particular time t. For a particle, this means determining its position and momentum or velocity[1] at t. Imagine the particle carrying with

---

[1] It is a fundamental result of classical mechanics that such a specification completely determines the subsequent motion of the particle and, therefore, any other property of the particle will be specified by the initial position and momentum.

it a vector representing its velocity. Then a snapshot of the particle at time t depicts its location as well as its velocity, and thus, a complete specification of the particle. A large collection of such snapshots specifies the motion of the particle. Since each snapshot represents an instant of time, and since the collection of snapshots specifies the motion, we conclude that, for particles, the only independent variable is time.[2] A problem involving a classical particle is solved once we find its position as a function of *time alone*.

How do we specify the "state" of a fluid? A fluid is an *extended* object, different parts of which behave differently. Attaching a vector to *different points* of the fluid to represent the velocity at that point, and taking snapshots at different times, we can get an idea of how the fluid behaves. This is done constantly (without the arrows, of course) by weather satellites whose snapshots are sometimes shown on our TV screens and reveal, for example, the turbulence developed by a hurricane. A complete determination of the fluid, therefore, entails a specification of the velocity vector at different points of the fluid for different times. A vector which varies from point to point is called a **vector field**. A problem involving a classical fluid is, therefore, solved once we find its velocity field as a function of *position and time*. The concept of a field can be abstracted from the physical reality of the fluid.[3] It then becomes a legitimate physical entity whose specification requires a position, a time, and a direction (if the field happens to be a vector field), just like the specification of the velocity field of a fluid.

vector field

The reason for going into so much detail in the last two paragraphs is to prevent a possible confusion. In the case of velocity and acceleration, one divides two quantities and the limit of the ratio turns out to be a *function of the denominator*, and one might get the impression that in (2.1), Q is a function of S. This is not the case, as, in general, all three quantities, R, S, and Q are functions of other (independent) variables, for instance, the three coordinates specifying position and time.

Velocity and acceleration are examples of the first interpretation of derivative, the rate of change. There are many situations in which the second interpretation of derivative is applicable. One important example is the **density** of a physical quantity R:

density: an example of the second interpretation of derivative

$$\rho_R = \lim_{\Delta V \to 0} \frac{\Delta R}{\Delta V} \equiv \frac{dR}{dV}, \tag{2.2}$$

where ΔR is the amount of the quantity R in the small volume ΔV. Examples of densities are mass density ρ_m, electric charge density ρ_q, number density

---

[2]This is true only in a classical picture of particles. A quantum mechanical picture disallows a complete determination of the position and momentum of a particle.

[3]Historically, this abstraction was very hard to achieve in the case of electromagnetism, where, for a long time a hypothetical "fluid" called æther was assumed to support the electromagnetic field. It was Einstein who suggested getting rid of the fluid altogether, and attaching physical reality and significance to the field itself.

ρ_n, energy density ρ_E, and momentum density ρ_p. Sometimes it is convenient to define surface and linear densities:

$$\sigma_R = \lim_{\Delta a \to 0} \frac{\Delta R}{\Delta a} \equiv \frac{dR}{da},$$

$$\lambda_R = \lim_{\Delta l \to 0} \frac{\Delta R}{\Delta l} \equiv \frac{dR}{dl}, \tag{2.3}$$

where ΔR is the amount of R on the small area Δa or along the small length Δl. The most frequently encountered surface density is that of electric charge which is commonly found on the surface of a conductor.

Another example of Equation (2.1) is **pressure** defined as

pressure: another
example of the
second
interpretation of
derivative

$$P = \lim_{\Delta a \to 0} \frac{\Delta F_\perp}{\Delta a} \equiv \frac{dF_\perp}{da}, \tag{2.4}$$

where $\Delta F_\perp$ is the force perpendicular to the surface Δa. This discussion makes it clear that *The most natural setting for the concept of derivative is the ratio of two physical quantities which are defined locally.* Equations (2.2) and (2.3) are hardly interpreted as the rate of change of density with respect to volume, area, or length!

Historical Notes

Descartes said that he "neither admits nor hopes for any principles in Physics other than those which are in Geometry or in abstract mathematics." And Nature couldn't agree more! The start of modern physics coincides with the start of modern mathematics. Calculus was, in large parts, motivated by the need for a quantitative analysis of physical problems. Calculation of instantaneous velocities and accelerations, determination of tangents to lens surfaces, evaluation of the angle corresponding to the maximum range of a projectile, and calculation of the lengths of curves such as the orbits of planets around the Sun were only a few of the physical motivations that instigated the intense activities of the seventeenth-century physicists and mathematicians alike.

The problems mentioned above were tackled by at least a dozen of the greatest mathematicians of the seventeenth century and many other minor ones. All of these efforts climaxed in the monumental achievements of **Newton** and **Leibniz**. Newton, in particular, noted the generality of the concept of rate of change—a concept he used for calculating instantaneous velocities—and bestowed a universal character upon the notion of derivative.

Of the several methods advanced to find the tangent to a curve, Fermat's is the closest to the modern treatment. He approximates the increment of the tangent *line* with the increment of the function describing the curve and takes the ratio of the two increments to find the angle of the tangent line. Fermat, however, ignores the question of limits as the increments go to zero, a procedure necessary for finding the slope of tangents. Descartes method, on the other hand, is purely algebraic and is not plagued by the question of the limits. However, his method worked only for polynomials.

Another great name associated with the development of calculus is **Isaac Barrow** who used elaborate geometrical methods to find tangents. He was the first to point out the connection between integration and differentiation. Barrow was a professor of mathematics at Cambridge University. Well versed in both Greek and Arabic (he was once nominated for a chair of Greek at Cambridge in 1655 but was denied the chair due to his loyalist views), he was able to translate some of Euclid's works and to improve the translations of other works of Euclid as well as Archimedes.

Isaac Barrow
1630–1677

After spending some time in eastern Europe, he returned to England and accepted the Greek chair denied him before. To supplement his income, he taught geometry at Gresham College, London. However, he soon gave up his geometry chair to serve as the first Lucasian professor of mathematics at Cambridge from 1663 to 1669, at which time Barrow resigned his chair of mathematics in favor of his student Isaac Newton and turned to theological studies.

His chief work *Lectiones Geometricae* is one of the great contributions to calculus. In it he used geometrical methods, "freed from the loathsome burdens of calculations," as he put it.

2.2 Partial Derivatives

All physical quantities are real functions of space and time. This means that given the three coordinates of a point in space, and an instant of time, we can associate a real number with them which happens to be the value of the physical quantity at that point and time.[4] Thus, $Q(x, y, z, t)$ is the value of the physical quantity Q at time t at a point whose Cartesian coordinates are (x, y, z). Similarly, we write $Q(r, \theta, \varphi, t)$ and $Q(\rho, \varphi, z, t)$ for spherical and cylindrical coordinates, respectively. Thus, ultimately, the physical quantities are functions of four real variables. However, there are many circumstances in which the quantity may be a function of less or more variables. An example of the former is all **static** phenomena in which the quantity is assumed—really approximated—to be independent of time. Then the quantity is a function of only three variables.[5] Physical quantities that depend on more than four variables are numerous in physics: In the mechanics of many particles, all quantities of interest depend, in general, on the coordinates of *all* particles, and in thermodynamics one encounters a multitude of *thermodynamical variables* upon which many quantities of interest depend.

2.2.1 Definition, Notation, and Basic Properties

We consider real functions $f(x_1, x_2, \ldots, x_n)$ of many variables. Generalizing the notation that denotes the set of real numbers by $\mathbb{R}$, the set of points in a plane by $\mathbb{R}^2$, and those in space by $\mathbb{R}^3$, we consider the n-tuples $(x_1, x_2, \ldots, x_n)$ as *points* in a (hyper)space $\mathbb{R}^n$. Similarly, just as the triplet (x, y, z) can be identified with the position vector $\mathbf{r}$, we abbreviate the n-tuple $(x_1, x_2, \ldots, x_n)$ by $\mathbf{r}$. Constant n-tuples will be denoted by the same letter

---

[4]This statement is not strictly true. There are many physical quantities which require more than *one* real number for their specification. A vector is a prime example which requires three real numbers to be specified. Thus, a vector field, which we discussed earlier, is really a collection of *three* real functions.

[5]If the natural setting of the problem is a surface or a line, then the number of variables is further reduced to two or one.

used for components but in boldface type. For example $(a_1, a_2, \ldots, a_n) \equiv \mathbf{a}$ and $(b_1, b_2, \ldots, b_n) \equiv \mathbf{b}$. This suggests using $\mathbf{x}$ in place of $\mathbf{r}$, and we shall do so once in a while.

Being independent, we can vary any one of the variables of a function at will while keeping the others constant. The concept of derivative is now applied to such a variation. The result is **partial derivative**. To be more precise, the partial derivative of $f(\mathbf{r})$ with respect to the independent variable

partial derivative
defined

x_k at $(a_1, a_2, \ldots, a_n)$ is denoted[6] by $\frac{\partial f}{\partial x_k}(\mathbf{a})$ and is defined as follows:

$$\frac{\partial f}{\partial x_k}(\mathbf{a}) \equiv \lim_{\epsilon \to 0} \frac{f(a_1, \ldots, a_k + \epsilon, \ldots, a_n) - f(a_1, \ldots, a_k, \ldots, a_n)}{\epsilon}. \quad (2.5)$$

One usually leaves out the a's and simply writes $\frac{\partial f}{\partial x_k}$, keeping in mind that the result has to be evaluated at some specific "point" of $\mathbb{R}^n$. As the definition suggests, the partial derivative with respect to x_k is obtained by the usual rules of differentiation with the proviso that all the other variables are assumed to be constants.

A useful strategy is to turn Equation (2.5) around and write the increment in f in terms of the partial derivative. This possibility is the result of the meaning of the limit: The closer ϵ gets to zero the better the ratio approximates the partial derivative. Thus we can leave out $\lim_{\epsilon \to 0}$ and approximate the two sides. After multiplying both sides by ϵ, we obtain

$$\Delta_k f \equiv f(a_1, \ldots, a_k + \epsilon, \ldots, a_n) - f(a_1, \ldots, a_k, \ldots, a_n) \approx \epsilon \frac{\partial f}{\partial x_k},$$

where the subscript k on the LHS indicates the independent variable being varied. Sometimes we use the notation $\Delta_k f(\mathbf{a})$ to emphasize the point at which the increment of the function—due to an increment in the kth argument—is being evaluated. Most of the time, however, for notational convenience, we shall leave out the arguments, it being understood that all quantities are to be evaluated at some specific "point." Since ϵ is an increment in x_k, it is natural to denote it as Δx_k, and write the above equation as

$$\Delta_k f \equiv f(a_1, \ldots, a_k + \Delta x_k, \ldots, a_n) - f(a_1, \ldots, a_k, \ldots, a_n) \approx \frac{\partial f}{\partial x_k} \Delta x_k.$$

If two independent variables, say x_k and x_j, are varied we still can find the increment in f:

$$\begin{aligned}
\Delta_{k,j} f &\equiv f(a_1, \ldots, a_k + \Delta x_k, \ldots, a_j + \Delta x_j, \ldots, a_n) \\
&\quad - f(a_1, \ldots, a_k, \ldots, a_j, \ldots, a_n) \\
&= f(a_1, \ldots, a_k + \Delta x_k, \ldots, a_j + \Delta x_j, \ldots, a_n) \\
&\quad - f(a_1, \ldots, a_k, \ldots, a_j + \Delta x_j, \ldots, a_n) \\
&\quad + f(a_1, \ldots, a_k, \ldots, a_j + \Delta x_j, \ldots, a_n) \\
&\quad - f(a_1, \ldots, a_k, \ldots, a_j, \ldots, a_n),
\end{aligned}$$

---

[6]This notation may be confusing because of the a's and the x's. A better notation will be introduced shortly.

where we have added and subtracted the same term on the RHS of this equation. Now we use the definition of the change in a function at a point to write

$$\Delta_{k,j}f = \Delta_k f(a_1, \ldots, a_k, \ldots, a_j + \Delta x_j, \ldots, a_n)$$
$$+ \Delta_j f(a_1, \ldots, a_k, \ldots, a_j, \ldots, a_n)$$
$$\approx \Delta x_k \frac{\partial f}{\partial x_k}(a_1, \ldots, a_k, \ldots, a_j + \Delta x_j, \ldots, a_n)$$
$$+ \Delta x_j \frac{\partial f}{\partial x_j}(a_1, \ldots, a_k, \ldots, a_j, \ldots, a_n).$$

The first term on the RHS expresses the change in the function due to a change in x_k, and the second expresses the change in the function due to a change in x_j. As their arguments show, the derivatives in the last two lines are not evaluated at the same point. However, the difference between these arguments is small—of order Δx_j—which, when multiplied by the small Δx's in front of them, will be even smaller. In the limit that Δx_j and Δx_k go to zero, we can ignore this subtle difference and write

$$\Delta_{k,j}f \approx \frac{\partial f}{\partial x_k}\Delta x_k + \frac{\partial f}{\partial x_j}\Delta x_j. \tag{2.6}$$

This shows that the total change is simply the sum of the change due to x_j and x_k.

> **Box 2.2.1.** *In general, the change in f due to a change in all the independent variables is $\Delta f \approx \sum_{i=1}^{n} \frac{\partial f}{\partial x_i}\Delta x_i$.*

Some of the Δx's may be zero of course. For example, if all of the Δx's are zero except Δx_j and Δx_k, then the equation in the Box above reduces to (2.6). The following example describes a situation which occurs frequently in thermodynamics.

Example 2.2.1. Suppose a physical quantity Q is a function of other physical quantities U, V, and W. We write this as $Q = f(U, V, W)$ with the intention that U, V, and W are the independent variables. It is possible, however, to solve for one of the independent variables in terms of Q and the rest of the independent variables.[7] It is therefore legitimate to seek the partial derivative of any one of the four quantities with respect to any other one. Because of the multitude of thermodynamic variables, it may become confusing as to which variables are kept constant. Therefore, it is common in thermodynamics to use the variables held constant as subscripts of the partial derivative. Thus,

$$\left(\frac{\partial Q}{\partial V}\right)_{U,W}, \quad \left(\frac{\partial V}{\partial Q}\right)_{U,W}, \quad \left(\frac{\partial U}{\partial V}\right)_{Q,W}, \tag{2.7}$$

an example that is useful for thermodynamics

---

[7]That this can be done under very mild assumptions regarding the function f is the content of the celebrated **implicit function theorem** proved in higher analysis.

are typical examples of partial derivatives, and in priciple, one can solve for V in terms of Q, U, and W and differentiate the resulting funtion with respect to Q to find the second term in Equation (2.7). Similarly, one can solve for U in terms of Q, V, and W and differentiate the resulting funtion with respect to V to find the last term. However, Box 2.2.1 allows us to bypass this (sometimes impossible) task and evaluate derivatives by directly differentiating the given function. Let's see how.

The first term is obvious:

$$\left(\frac{\partial Q}{\partial V}\right)_{U,W} = \left(\frac{\partial f}{\partial V}\right)_{U,W}.$$

The key to the evaluation of the other two is Box 2.2.1 as applied to Q. We thus write

$$\Delta Q \approx \left(\frac{\partial f}{\partial U}\right)_{V,W} \Delta U + \left(\frac{\partial f}{\partial V}\right)_{U,W} \Delta V + \left(\frac{\partial f}{\partial W}\right)_{U,V} \Delta W. \qquad (2.8)$$

If U and W are kept constant, then $\Delta U = 0 = \Delta W$, and we have

$$\Delta Q \approx \left(\frac{\partial f}{\partial V}\right)_{U,W} \Delta V \Rightarrow 1 \approx \left(\frac{\partial f}{\partial V}\right)_{U,W} \frac{\Delta V}{\Delta Q}.$$

In the limit that ΔQ goes to zero, the ratio of the Δ's becomes the corresponding partial derivative and the approximation becomes equality, leading to the relation

$$1 = \left(\frac{\partial f}{\partial V}\right)_{U,W} \left(\frac{\partial V}{\partial Q}\right)_{U,W}.$$

Changing f to Q,[8] and solving for the partial derivative, we obtain

$$\left(\frac{\partial V}{\partial Q}\right)_{U,W} = \frac{1}{\left(\frac{\partial Q}{\partial V}\right)_{U,W}} \qquad (2.9)$$

which is a result we should have expected. This equation shows that we don't have to solve for V in terms of the other three variables to find its derivative with respect to Q. Just differentiate $f(U,V,W)$ with respect to V and take its reciprocal!

The last partial derivative is obtained by setting ΔQ and ΔW equal to zero in (2.8). The result is

$$0 \approx \left(\frac{\partial f}{\partial U}\right)_{V,W} \Delta U + \left(\frac{\partial f}{\partial V}\right)_{U,W} \Delta V \Rightarrow \frac{\Delta U}{\Delta V} \approx -\frac{\left(\frac{\partial f}{\partial V}\right)_{U,W}}{\left(\frac{\partial f}{\partial U}\right)_{V,W}}.$$

Once again, taking the limit as $\Delta V \to 0$, noting that the LHS becomes a partial derivative, subscripting this partial with the variables held constant, and substituting Q for f,[9] we obtain

$$\left(\frac{\partial U}{\partial V}\right)_{Q,W} = -\frac{\left(\frac{\partial Q}{\partial V}\right)_{U,W}}{\left(\frac{\partial Q}{\partial U}\right)_{V,W}}. \qquad (2.10)$$

---

[8] Recall that if $y = f(x)$, then dy/dx and df/dx represent the same quantity.

[9] This is an abuse of notation because Q is held *constant* and the derivative of any constant is always zero, while the derivative of f is well defined. This abuse of notation is so common in thermodynamics that we shall adopt it here as well.

Thus, by differentiating $f(U, V, W)$ with respect to V and U and taking their ratios, we obtain the derivative of U with respect to V; no need to solve for U in terms of the other three variables!

Equation (2.10) is ususlly written in a more symmetric way. The numerator of the fraction on the RHS can be replaced using Equation (2.9). Then, the result can be written as

$$\left(\frac{\partial U}{\partial V}\right)_{Q,W} \left(\frac{\partial V}{\partial Q}\right)_{U,W} \left(\frac{\partial Q}{\partial U}\right)_{V,W} = -1. \tag{2.11}$$

A simpler version of this result, in which the fourth variable W is absent, is commonly used in thermodynamics. ∎

an important relation used often in thermodynamics

A word of caution about notation is in order. We chose the set of variables $(x_1, x_2, \ldots, x_n)$ as arguments of the function f, and then denoted the derivative by $\partial f/\partial x_k$. We could have chosen any other set of symbols such as $(y_1, y_2, \ldots, y_n)$, or $(t_1, t_2, \ldots, t_n)$ as the arguments. Then we would have had to write $\partial f/\partial y_k$, or $\partial f/\partial t_k$ for partial derivatives. This freedom of choice can become confusing because, little effort is made in the literature to distinguish between the "free" general arguments and the specific point at which the derivative is to be evaluated. For example, the symbol $(\partial f/\partial x)(y, x)$ can be interpreted in two ways: It can be the derivative of a function of two variables with respect to its first argument, subsequently evaluated at the point with coordinates (y, x), or it could be the derivative with respect to the second argument, in a seemingly strange world in which y is used as the first argument! The longstanding usage of x as the first partner of a doublet by no means reserves the first slot for x at all times. Therefore, the confusion above is indeed a legitimate one.

confusion surrounding the expression $(\partial f/\partial x)(y, x)$ and a notation that resolves the confusion

We started the discussion by distinguishing between the free arguments $(x_1, x_2, \ldots, x_n)$ and the specific point $(a_1, a_2, \ldots, a_n)$. However, making this distinction every time we write down a partial derivative can become very clumsy. Nevertheless, the reader should always keep in mind this distinction and write it down explicitly whenever necessary. To minimize the confusion, we leave out all symbols but keep only the position of the variable in the array. Specifically,

> **Box 2.2.2.** *We write* $\partial_k f$ *for the derivative of* f *with respect to its kth argument. This derivative is a* function*: We can evaluate it at* $(a_1, a_2, \ldots, a_n)$, *for which we write* $\partial_k f(a_1, a_2, \ldots, a_n) \equiv \partial_k f(\mathbf{a})$.

This notation avoids any reference to the "free" arguments. One can choose *any* symbol for the free arguments; the final answer is independent of this choice:

$$\partial_k f(a_1, a_2, \ldots, a_n) = \left.\frac{\partial f(t_1, t_2, \ldots, t_n)}{\partial t_k}\right|_{\mathbf{t}=\mathbf{a}} = \left.\frac{\partial f(y_1, y_2, \ldots, y_n)}{\partial y_k}\right|_{\mathbf{y}=\mathbf{a}}$$

$$= \left.\frac{\partial f(\heartsuit_1, \heartsuit_2, \ldots, \heartsuit_n)}{\partial \heartsuit_k}\right|_{(\heartsuit_1=a_1, \ldots, \heartsuit_n=a_n)}$$

because the only thing that matters is the index k which tells us with respect to what variable we are differentiating.

Example 2.2.2. Consider the function $f(x, y, z) = e^{xy/z}$. We write it first as $f(x_1, x_2, x_3) = e^{x_1 x_2/x_3}$. Then

$$\partial_1 f(x_1, x_2, x_3) = (x_2/x_3)e^{x_1 x_2/x_3},$$

$$\partial_2 f(x_1, x_2, x_3) = (x_1/x_3)e^{x_1 x_2/x_3},$$

$$\partial_3 f(x_1, x_2, x_3) = -(x_1 x_2/x_3^2)e^{x_1 x_2/x_3}.$$

Now that the functional form of all partial derivatives are derived, we can evaluate them at any point we want. For example,

$$\partial_2 f(1, 2, 3) = \tfrac{1}{3}e^{2/3}, \qquad \partial_3 f(1, 1, 1) = -e,$$

$$\partial_1 f(t, u, v) = (u/v)e^{tu/v}, \qquad \partial_3 f(z, x, y) = -(zx/y^2)e^{xz/y}. \qquad \blacksquare$$

Higher-order derivatives are defined just as in the single-variable case, except that now **mixed derivatives** are also possible. Thus,

$$\partial_1(\partial_1 f) \equiv \partial_1^2 f \equiv \frac{\partial^2 f}{\partial x_1{}^2}, \quad \partial_1(\partial_5 f) \equiv \frac{\partial^2 f}{\partial x_1 \partial x_5}, \quad \partial_j(\partial_k f) \equiv \frac{\partial^2 f}{\partial x_j \partial x_k},$$

order of differentiation in a mixed derivative is immaterial

are all legitimate derivatives. An important property of mixed derivatives is that—for well-behaved functions—the *order of differentiation is immaterial*.

Example 2.2.3. Functions which can be written as the product of single-variable functions are important in the solution of partial differential equations. Suppose that $F(x, y, z) = f(x)g(y)h(z)$. Then $\partial_1 F(x, y, z) = f'(x)g(y)h(z)$ and the function

$$\frac{\partial_1 F}{F}(x, y, z) = \frac{f'(x)g(y)h(z)}{f(x)g(y)h(z)} = \frac{f'(x)}{f(x)}$$

is seen to be independent of y and z. One can show similarly that

$$\frac{\partial_2 F}{F}(x, y, z) = \frac{g'(y)}{g(y)}, \qquad \frac{\partial_3 F}{F}(x, y, z) = \frac{h'(z)}{h(z)},$$

each one depending on only one variable. $\qquad \blacksquare$

Example 2.2.4. It is sometimes necessary to find the most general function, one of whose partial derivatives is given. This can be done by antidifferentiating (indefinite integral) with respect to the variable of the partial derivative, treating the rest of the variables constant. The usual "constant" of integration is replaced by a function of the undifferentiating variables. For example, suppose $\partial_3 f(z, x, y) = ye^{x^2 y^2/z}$. Since the third variable is y, and the partial derivative is with respect to the third variable, we need to integrate with respect to y, keeping x and z constant. This gives

$$f(z, x, y) = \frac{z}{2x^2}e^{x^2 y^2/z} + g(x, z) \;\Rightarrow\; f(x, y, z) = \frac{x}{2y^2}e^{y^2 z^2/x} + g(y, x),$$

where g, the "constant" of integration, is an arbitrary function of the first two variables. $\qquad \blacksquare$

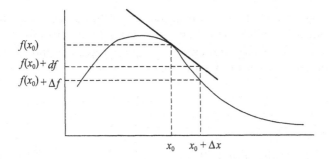

Figure 2.1: The tangent line at x_0 approximates the curve in a small neighborhood of x_0. If confined in this neighborhood, i.e., if Δx—which is equal to dx—is small, Δf and df are approximately equal. However, df is defined regardless of the size of Δx.

2.2.2 Differentials

We now introduce the notion of **differentials**. Recall from calculus that, in the case of one variable, the differential of a function is related to a linear approximation of that function (see Figure 2.1). Basically, the tangent line at a point x_0 is considered as the linear approximation to the curve representing the function f in the neighborhood of x_0. The increment in the value of the function *representing the tangent line*—denoted by $df(x_0)$—when x_0 changes to $x_0 + \Delta x$, is given by

differentials

$$df(x_0) \equiv \left(\frac{df}{dx}\right)_{x_0} \Delta x \equiv \left(\frac{df}{dx}\right)_{x_0} dx,$$

where, as a matter of notation, Δx has been replaced by dx, because by definition, the differential of an *independent variable* is nothing but its increment. The above equation is *not* an approximation: dx can be any number, large or small, and $df(x_0)$ will be correspondingly large or small. The approximation starts when we try to replace Δf with df: The smaller the $\Delta x = dx$, the better the approximation $\Delta f(x_0) \approx df(x_0)$. The generalization of this idea to two variables involves approximating the surface representing the function $f(x, y)$ by its tangent plane. For more variables, no visualizable geometric interpretation is possible, but the basic idea is to replace the Δ's with d's and the approximation with equality in Box 2.2.1. The result is

$$df = \frac{\partial f}{\partial x_1}dx_1 + \frac{\partial f}{\partial x_2}dx_2 + \cdots + \frac{\partial f}{\partial x_n}dx_n = \sum_{i=1}^{n} \frac{\partial f}{\partial x_i}dx_i. \qquad (2.12)$$

We note that dx_i's in Equation (2.12) determine the independent variables on which f depends, and the coefficient of dx_i is $\partial f/\partial x_i$. This observation is the basis of transforming functions in such a way that the resulting functions depend on variables which are physically more useful. To be specific, suppose a function f exists which depends on (x, y, z), but from a physical perspective, a function which depends on the derivative of f with respect to its second argument, and not on the second argument itself, is more valuable. This function can be obtained by a **Legendre transformation** on f, obtained by subtracting from f the product of the second argument and the derivative of f with respect to that argument. So, define a new function g by

$$g \equiv f - y\partial_2 f \equiv f - yh \qquad \text{where} \quad h = \partial_2 f.$$

Then, we get

$$dg = df - h\,dy - y\,dh = \partial_1 f\,dx + \partial_2 f\,dy + \partial_3 f\,dz - h\,dy - y\,dh$$
$$= \partial_1 f\,dx + \partial_3 f\,dz - y\,dh.$$

The differentials on the RHS of the last line indicate that the "natural" independent variables for g are x, z, and h, and that

$$\frac{\partial g}{\partial x} = \partial_1 f, \qquad \frac{\partial g}{\partial z} = \partial_3 f, \qquad \frac{\partial g}{\partial h} = -y.$$

Legendre transformation is used frequently in thermodynamics and mechanics.

Example 2.2.5. The internal energy U of a thermodynamical system is a function of entropy S, volume V, and number of moles N. These variables are called the

natural variables of U, and we write $U(S, V, N)$. Temperature T, pressure P, and chemical potential μ, are defined as follows:

$$T = \left(\frac{\partial U}{\partial S}\right)_{V,N}, \qquad P = -\left(\frac{\partial U}{\partial V}\right)_{S,N}, \qquad \mu = \left(\frac{\partial U}{\partial N}\right)_{S,V},$$

where, as is common in thermodynamics, we have indicated the variables that are held constant as subscripts. Entropy is a hard quantity to measure: If we were to measure $\partial U/\partial S$, we would have to find the ratio of the change of U to that of S; not an easy task! On the other hand, T is easy to measure, and thus it is desirable to Legendre transform U to obtain a function which has T as a natural variable.

The **Helmholtz free energy** F is defined as $F = U - ST$. We note that since

$$dU = \frac{\partial U}{\partial S}\,dS + \frac{\partial U}{\partial V}\,dV + \frac{\partial U}{\partial N}\,dN = T\,dS - P\,dV + \mu\,dN,$$

we have

$$dF = dU - SdT - TdS = \underbrace{TdS - PdV + \mu dN}_{=dU} - SdT - TdS$$
$$= -SdT - PdV + \mu dN$$

and, therefore

$$\left(\frac{\partial F}{\partial T}\right)_{V,N} = -S, \qquad \left(\frac{\partial F}{\partial V}\right)_{T,N} = -P, \qquad \left(\frac{\partial F}{\partial N}\right)_{T,V} = \mu.$$

Helmholtz free energy is by far the most frequently used thermodynamic function, because all its "natural" variables, namely, T, V, and N, are easily measurable quantities. ∎

2.2.3 Chain Rule

In many cases of physical interest, the "independent" variables x_i may in turn depend on one or more variables. Let us denote these new independent variables by $(t_1, t_2, \ldots, t_m)$ and the functional dependence of x_i by g_i, so that

$$x_i = g_i(t_1, t_2, \ldots, t_m) \equiv g_i(\mathbf{t}), \quad i = 1, 2, \ldots, n. \tag{2.13}$$

As the t's vary, so will the x's and consequently the function f. Therefore, f becomes dependent on the t's and we can talk about partial derivatives of f with respect to one of the t's. To find such a partial derivative, we go back to Box 2.2.1 and substitute for Δx_i in terms of Δt's. From (2.13), we have

$$\Delta x_i \approx \frac{\partial g_i}{\partial t_1}\Delta t_1 + \frac{\partial g_i}{\partial t_2}\Delta t_2 + \cdots + \frac{\partial g_i}{\partial t_m}\Delta t_m = \sum_{j=1}^{m} \frac{\partial g_i}{\partial t_j}\Delta t_j, \;\; i = 1, 2, \ldots, n.$$

Substituting this in the equation of Box 2.2.1 yields

$$\Delta f \approx \frac{\partial f}{\partial x_1}\sum_{j=1}^{m} \frac{\partial g_1}{\partial t_j}\Delta t_j + \frac{\partial f}{\partial x_2}\sum_{j=1}^{m} \frac{\partial g_2}{\partial t_j}\Delta t_j + \cdots + \frac{\partial f}{\partial x_n}\sum_{j=1}^{m} \frac{\partial g_n}{\partial t_j}\Delta t_j$$

$$= \sum_{i=1}^{n} \frac{\partial f}{\partial x_i}\sum_{j=1}^{m} \frac{\partial g_i}{\partial t_j}\Delta t_j. \tag{2.14}$$

Now suppose that we keep all of the t's constant except for one, say t_7. Then $\Delta t_j = 0$ for all j except $j = 7$ and the sum over j will have only one nonzero term, i.e., the seventh term. In such a case, Equation (2.14) becomes

$$\Delta f \approx \frac{\partial f}{\partial x_1}\frac{\partial g_1}{\partial t_7}\Delta t_7 + \frac{\partial f}{\partial x_2}\frac{\partial g_2}{\partial t_7}\Delta t_7 + \cdots + \frac{\partial f}{\partial x_n}\frac{\partial g_n}{\partial t_7}\Delta t_7 = \sum_{i=1}^{n} \frac{\partial f}{\partial x_i}\frac{\partial g_i}{\partial t_7}\Delta t_7.$$

Dividing both sides by Δt_7, taking limit, and replacing the approximation by equality, we obtain

$$\frac{\partial f}{\partial t_7} = \frac{\partial f}{\partial x_1}\frac{\partial g_1}{\partial t_7} + \frac{\partial f}{\partial x_2}\frac{\partial g_2}{\partial t_7} + \cdots + \frac{\partial f}{\partial x_n}\frac{\partial g_n}{\partial t_7} = \sum_{i=1}^{n} \frac{\partial f}{\partial x_i}\frac{\partial g_i}{\partial t_7}.$$

Instead of t_7, we could have used any other one of the t's, say t_{19}, or t_{217}.   the chain rule

Theorem 2.2.6. (***The Chain Rule***). *Let $f(\mathbf{x})$ be a function of the x_i and $x_i = g_i(\mathbf{t})$. Let $h(\mathbf{t}) = f(g_1(\mathbf{t}), g_2(\mathbf{t}), \ldots, g_n(\mathbf{t}))$ be a function of the t_k, called the **composite** of f and the g_i. If t_p is any one of these t's, then*

$$\partial_p h(\mathbf{t}) = \sum_{i=1}^{n} \partial_i f(\mathbf{g}(\mathbf{t}))\partial_p g_i(\mathbf{t}), \tag{2.15}$$

where $\mathbf{g} = (g_1, g_2, \ldots, g_n)$, and $\mathbf{g}(\mathbf{t}) \equiv (g_1(\mathbf{t}), g_2(\mathbf{t}), \ldots, g_n(\mathbf{t}))$.

In words, the chain rule states that to evaluate the partial derivative of h with respect to its pth argument (of which there are m) at $\mathbf{t}$, multiply the ith partial of f evaluated at $\mathbf{g}(\mathbf{t})$ by the pth partial of g_i evaluated at $\mathbf{t}$ and sum over i.

Sometimes the chain rule is written in the following less precise form:

$$\frac{\partial f}{\partial t_p} = \sum_{i=1}^{n} \frac{\partial f}{\partial x_i} \frac{\partial g_i}{\partial t_p} = \sum_{i=1}^{n} \frac{\partial f}{\partial x_i} \frac{\partial x_i}{\partial t_p}, \tag{2.16}$$

where in the last line we have substituted x_i for g_i.

Example 2.2.7. Suppose F is a function of three variables given by

$$F(x, y, z) = f\left(\frac{x^2 y}{az^2}\right),$$

where f is some given function, and a is a constant. Let us calculate all partial derivatives of F at $(a, 2a, a)$ assuming that $f'(2) = a$. Denote the single variable of f by u, so that F is obtained by substituting $x^2 y/(az^2)$ for u in $f(u)$. The chain rule gives

$$\partial_1 F(x, y, z) = f'(u)\partial_1 u = f'(u)\frac{2xy}{az^2},$$

$$\partial_2 F(x, y, z) = f'(u)\partial_2 u = f'(u)\frac{x^2}{az^2},$$

$$\partial_3 F(x, y, z) = f'(u)\partial_3 u = -2f'(u)\frac{x^2 y}{az^3}.$$

If $x = a$, $y = 2a$, and $z = a$, then $u = a^2(2a)/a^3 = 2$, and

$$\partial_1 F(a, 2a, a) = f'(2)\frac{2a(2a)}{a^3} = 4.$$

Similarly, $\partial_2 F(a, 2a, a) = 1$ and $\partial_3 F(a, 2a, a) = -4$.

In the notation of Theorem 2.2.6, there are three t's: $t_1 = x$, $t_2 = y$, $t_3 = z$, and only one g: $g(t_1, t_2, t_3) = t_1^2 t_2/(at_3^2)$. Then F becomes the composite function of f and g. ∎

Example 2.2.8. One of the important occasions of the use of the chain rule is in the transformation of derivatives from Cartesian to spherical coordinates. A good example of such a transformation occurs in quantum mechanics where an expression such as $x\partial f/\partial y - y\partial f/\partial x$ turns out to be related to angular momentum, and it is most conveniently expressed in spherical coordinates. In this example we go through the detailed exercise of converting that expression into spherical coordinates.

We start with the transformations

$$x = r\sin\theta\cos\varphi, \qquad y = r\sin\theta\sin\varphi, \qquad z = r\cos\theta,$$

and their inverse

$$r = \sqrt{x^2 + y^2 + z^2}, \qquad \cos\theta = \frac{z}{\sqrt{x^2 + y^2 + z^2}}, \qquad \tan\varphi = \frac{y}{x}. \tag{2.17}$$

We shall need the derivatives of spherical coordinates with respect to x and y written in terms of spherical coordinates. We easily find these by differentiating both sides of the equations in (2.17):

how derivatives are transformed under a coordinate transformation

$$\frac{\partial r}{\partial x} = \frac{\partial}{\partial x}\left(\sqrt{x^2+y^2+z^2}\right) = \frac{1}{2}\frac{1}{\sqrt{x^2+y^2+z^2}}(2x) = \frac{x}{r} = \sin\theta\cos\varphi,$$

$$-\sin\theta\frac{\partial\theta}{\partial x} = z\left[-\frac{1}{2}\frac{1}{r^3}2x\right] = -\frac{xz}{r^3} = -\frac{\sin\theta\cos\varphi\cos\theta}{r} \Rightarrow \frac{\partial\theta}{\partial x} = \frac{\cos\theta\cos\varphi}{r},$$

$$\sec^2\varphi\frac{\partial\varphi}{\partial x} = -\frac{y}{x^2} = -\frac{\sin\varphi}{r\sin\theta\cos^2\varphi} \Rightarrow \frac{\partial\varphi}{\partial x} = -\frac{\sin\varphi}{r\sin\theta}.$$

Similarly,

$$\frac{\partial r}{\partial y} = \sin\theta\sin\varphi, \quad \frac{\partial\theta}{\partial y} = \frac{\cos\theta\sin\varphi}{r}, \quad \frac{\partial\varphi}{\partial y} = \frac{\cos\varphi}{r\sin\theta}.$$

Therefore, using the chain rule as given in Equation (2.16), we get

$$\frac{\partial f}{\partial x} = \frac{\partial f}{\partial r}\frac{\partial r}{\partial x} + \frac{\partial f}{\partial\theta}\frac{\partial\theta}{\partial x} + \frac{\partial f}{\partial\varphi}\frac{\partial\varphi}{\partial x}$$

$$= \sin\theta\cos\varphi\frac{\partial f}{\partial r} + \frac{\cos\theta\cos\varphi}{r}\frac{\partial f}{\partial\theta} - \frac{\sin\varphi}{r\sin\theta}\frac{\partial f}{\partial\varphi},$$

$$\frac{\partial f}{\partial y} = \frac{\partial f}{\partial r}\frac{\partial r}{\partial y} + \frac{\partial f}{\partial\theta}\frac{\partial\theta}{\partial y} + \frac{\partial f}{\partial\varphi}\frac{\partial\varphi}{\partial y}$$

$$= \sin\theta\sin\varphi\frac{\partial f}{\partial r} + \frac{\cos\theta\sin\varphi}{r}\frac{\partial f}{\partial\theta} + \frac{\cos\varphi}{r\sin\theta}\frac{\partial f}{\partial\varphi}.$$

If we multiply the first of the last two equations by $y = r\sin\theta\sin\varphi$ and subtract it from the second equation multiplied by $x = r\sin\theta\cos\varphi$, the terms involving derivatives with respect to r and θ cancel while the terms with φ derivatives add to give

$$x\frac{\partial f}{\partial y} - y\frac{\partial f}{\partial x} = \frac{\partial f}{\partial\varphi}.$$

Details are left as an exercise for the reader. ∎

There is a multitude of examples in thermodynamics, for which a mastery of the techniques of partial differentiation is essential. A property that is used often in thermodynamics is homogeneity of functions which we derive below.

2.2.4 Homogeneous Functions

A function is called homogeneous of degree q if multiplying all of its arguments by a parameter λ results in the multiplication of the function itself by λ^q. More precisely,

Box 2.2.3. *We say that $f(x_1, x_2, \ldots, x_n)$ is homogeneous of degree q if $f(\lambda x_1, \lambda x_2, \ldots, \lambda x_n) = \lambda^q f(x_1, x_2, \ldots, x_n)$.*

extensive and
intensive functions

Two cases merit special consideration. When $q = 1$, the function changes at exactly the same rate as its arguments: Doubling all its arguments doubles the function and so on. Such a function is called **extensive**. When $q = 0$, the function is called **intensive**, and it *will not change* if *all* its arguments are changed by exactly the same factor.

In many cases, we want a relation between f and its partial derivatives. We shall find this relation by differentiating both sides of Box 2.2.3 with respect to λ. To avoid any confusion, let us evaluate both sides at the point $(b_1, b_2, \ldots, b_n)$ after differentiation. Differentiation of the RHS is easy:

$$RHS = q\lambda^{q-1} f(b_1, b_2, \ldots, b_n).$$

For the LHS, we first let $y_i = \lambda x_i$ for all $i = 1, 2, \ldots, n$—so that we have a single variable (one symbol) in the ith place—and note that

$$LHS = \frac{\partial f}{\partial \lambda} = \sum_{i=1}^{n} \frac{\partial f}{\partial y_i} \frac{\partial y_i}{\partial \lambda} = \sum_{i=1}^{n} [\partial_i f(y_1, y_2, \ldots, y_n)] x_i,$$

where we have used the fact that $\frac{\partial y_i}{\partial \lambda} = x_i$—by the definition of y_i. Evaluating the result at $x_i = b_i$, we obtain

$$LHS = \sum_{i=1}^{n} b_i \partial_i f(\lambda b_1, \lambda b_2, \ldots, \lambda b_n).$$

Equating the LHS and the RHS, we obtain the important result

$$q\lambda^{q-1} f(b_1, b_2, \ldots, b_n) = \sum_{i=1}^{n} b_i \partial_i f(\lambda b_1, \lambda b_2, \ldots, \lambda b_n)$$

This relation holds for all values of λ, in particular we can substitute $\lambda = 1$ to obtain $q f(b_1, b_2, \ldots, b_n) = \sum_{i=1}^{n} b_i \partial_i f(b_1, b_2, \ldots, b_n)$. Keep in mind that the b's, although fixed, are completely arbitrary. In particular, one can substitute x's for them and arrive at the *functional* relation

relation between a
homogeneous
function and the
sum of its partial
derivatives

$$q f(x_1, x_2, \ldots, x_n) = \sum_{i=1}^{n} x_i \partial_i f(x_1, x_2, \ldots, x_n). \tag{2.18}$$

This is the relation we were looking for.

Another important result, which the reader is asked to derive in Problem 2.17, is

> **Box 2.2.4.** *If f is homogeneous of degree q, then $\partial_i f$ is homogeneous of degree $q - 1$.*

Example 2.2.9. We have already seen that the natural variables of the internal energy U of a thermodynamical system are entropy S, volume V, and number of moles N. Based on physical intuition, we expect the total internal energy, entropy, volume and number of moles of the *combined system* to be doubled when two identical systems are brought together. We conclude that the internal energy function increases by the same factor as its arguments. A thermodynamic quantity that has this property is called an **extensive variable**. It follows that U is an extensive variable and a homogeneous function of degree one.

Now consider temperature T, pressure P, and chemical potential μ, which are all partial derivatives of U with respect to its natural variables. From Problem 2.17, we conclude that these quantities are homogeneous of degree zero. It follows that, if we bring two identical systems together, temperature, pressure, and the chemical potential will not change, a result expected on physical grounds. Such a thermodynamic quantity is called an **intensive variable**. ∎

extensive and intensive variables of thermodynamics and their relation to homogeneous functions

2.3   Elements of Length, Area, and Volume

We mentioned earlier the significance of the second interpretation of the derivative in conjunction with density. This interpretation is often used in reverse order, i.e., in writing the infinitesimal (element) of the physical quantity as a product of density and the element of volume (or area, or length). These elements appear inside integrals and will be integrated over (see the next chapter). As a concrete example, let us consider the mass element which can be expressed as

volume distribution: $dm(\mathbf{r'}) = \rho(\mathbf{r'})\, dV(\mathbf{r'})$

surface distribution: $dm(\mathbf{r'}) = \sigma(\mathbf{r'})\, da(\mathbf{r'})$

linear distribution: $dm(\mathbf{r'}) = \lambda(\mathbf{r'})\, dl(\mathbf{r'})$

where $\mathbf{r'}$ denotes the *coordinates* of the location of the element of mass.

The relations above reduce the problem to that of writing the elements of volume, area, and length. Most of the time, the evaluation of the integral simplifies considerably if we choose the correct coordinate system. Therefore, we need these elements in all three coordinate systems.

Basic to the calculation of all elements are elements of length in the direction of unit vectors in any of the three coordinate systems. First we define

> **Box 2.3.1.** *The **primary curve** along any given coordinate is the curve obtained when that coordinate is allowed to vary while the other two coordinates are held fixed.*

The **primary length elements** are infinitesimal lengths along the primary curves. By construction, they are also infinitesimal lengths along unit vectors. To find a primary length element at point P' with position vector $\mathbf{r'}$ along

primary length elements

a given unit vector, one keeps the other two coordinates fixed and allows the given coordinate to change by an infinitesimal amount.[10] This procedure displaces P' an infinitesimal distance. The length of this displacement, written in terms of the coordinates of P', is the primary length element along the given unit vector. Once the three primary length elements are found, we can calculate area and volume elements by multiplying appropriate length elements.

A notion related to the primary length is

> **Box 2.3.2.** *A **primary surface** perpendicular to a primary length is obtained when the coordinate determining the primary length is held fixed and the other two coordinates are allowed to vary arbitrarily.*

primary area
elements

The **primary element of area** at a point on a primary surface is, by definition, the product of the two primary length elements whose coordinates define that surface.

Integrating over a primary surface of a coordinate system is facilitated if all boundaries of the surface can be described by $q_i = c_i$ where q_i is either of the two coordinates that vary on the surface and c_i is a constant. For example, the third primary surface in Cartesian coordinates is a plane parallel to the xy-plane. A problem involving integration on this plane becomes simplest if the boundaries of the region of integration are of the form, $x = c_1$ and $y = c_2$, i.e., if the region of integration is a rectangle.

Finally, by taking the product of all three primary length elements, we obtain the volume element in the given coordinate system.

2.3.1   Elements in a Cartesian Coordinate System

Consider the point P' with coordinates (x', y', z') as shown in Figure 2.2. To find the primary length along $\hat{\mathbf{e}}_{x'} = \hat{\mathbf{e}}_x$,[11] keep y' and z' fixed and let x' change to $x' + dx'$. Then P' will be displaced by dx' along $\hat{\mathbf{e}}_x$. Thus, the first primary length element—denoted by dl_1—is simply dx'. Similarly, we have $dl_2 = dy'$, and $dl_3 = dz'$. A general infinitesimal displacement, which is a *vector*, can be written as

general Cartesian
infinitesimal
displacement

$$\vec{dl} = \hat{\mathbf{e}}_x \, dl_1 + \hat{\mathbf{e}}_y \, dl_2 + \hat{\mathbf{e}}_z \, dl_3 = \hat{\mathbf{e}}_x \, dx' + \hat{\mathbf{e}}_y \, dy' + \hat{\mathbf{e}}_z \, dz' \equiv d\mathbf{r}'. \qquad (2.19)$$

Figure 2.2 shows that $\vec{dl}$ represents the displacement vector from P', with position vector $\mathbf{r}'$, to a *neighboring* point P'', with position vector $\mathbf{r}''$. But this displacement is simply the *increment in the position vector of P'*. That is why $d\mathbf{r}'$ is also used for $\vec{dl}$. Note that this vectorial infinitesimal displacement

---

[10]Usually an infinitesimal amount is expressed by a differential. Thus, an increment in x is simply dx.

[11]Recall that this equality holds in Cartesian—and *only* in Cartesian—coordinates, where the unit vectors are independent of the coordinates of P'.

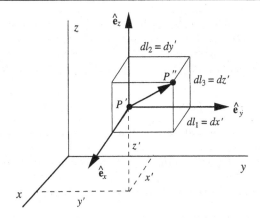

Figure 2.2: Elements of length, area, and volume in Cartesian coordinates.

includes the primary length elements as special cases: When a coordinate is held fixed, the corresponding differential will be zero. Thus, setting $dy' = 0 = dz'$, i.e., holding y' and z' fixed, we recover the first primary length element. The *length* of $d\vec{l}$ is also of interest:

$$dl \equiv |d\vec{l}| = \sqrt{dl_1^2 + dl_2^2 + dl_3^2}$$
$$= \sqrt{(dx')^2 + (dy')^2 + (dz')^2} \equiv \sqrt{dx'^2 + dy'^2 + dz'^2}. \qquad (2.20)$$

In one-dimensional problems involving curves, one is either given, or has to find, the **parametric equation** of a curve γ whereby the coordinates (x', y', z') of a point on γ are expressed as functions of a parameter, usually denoted by t. This is concisely written as

parametric equation of a curve

$$\gamma(t) = (x', y', z') \equiv \big(f(t), g(t), h(t)\big),$$

so that the "curve function" γ takes a real number t and gives three real numbers $f(t)$, $g(t)$, and $h(t)$ which are the coordinates x', y', and z' of a point on the curve in space. Usually one considers an interval[12] (a, b) for the real variable t. Then $\big(f(a), g(a), h(a)\big)$ is the initial point of the curve and $\big(f(b), g(b), h(b)\big)$ its final point. The parameter t and the functions f, g, and h are not unique. For example, the three functions

$$f_1(t) = a\cos t, \quad g_1(t) = a\sin t, \quad h_1(t) = 0, \qquad 0 \le t \le \pi,$$

describe a semicircle in the xy-plane. However,

$$f_2(t) = a\cos\left(t^3\right), \quad g_2(t) = a\sin\left(t^3\right), \quad h_2(t) = 0, \qquad 0 \le t \le \pi^{1/3},$$

also describe the same semicircle. This arbitrariness is useful, because it allows us to choose f, g, and h so that calculations become simple.

---

[12]Do not confuse this with the coordinates of a point in the plane. The notation (a, b) here means all the real numbers between a and b excluding a and b themselves.

For "flat" curves [lying in the xy-plane and given by an equation $y = f(x)$], one obvious parameterization—which may not be the most convenient one—is $x = t$, $y = f(t)$.

Let us assume that we have chosen the three functions and they are of the form

$$x' = f(t), \quad y' = g(t), \quad z' = h(t).$$

Then the primary lengths can be written as

$$dx' = f'(t)\, dt, \quad dy' = g'(t)\, dt, \quad dz' = h'(t)\, dt,$$

the infinitesimal element of displacement along a curve

and the element of displacement *along the curve* becomes

$$d\mathbf{r}'(t) = d\vec{l}(t) = \hat{\mathbf{e}}_x\, f'(t)\, dt + \hat{\mathbf{e}}_y\, g'(t)\, dt + \hat{\mathbf{e}}_z\, h'(t)\, dt,$$

$$|d\mathbf{r}'(t)| = dl(t) = \sqrt{[f'(t)\, dt]^2 + [g'(t)\, dt]^2 + [h'(t)\, dt]^2}$$

$$= \sqrt{[f'(t)]^2 + [g'(t)]^2 + [h'(t)]^2}\ \ dt, \tag{2.21}$$

where a prime on a function denotes its derivative with respect to its argument.[13]

primary surfaces of Cartesian coordinates are planes

The first primary surface at P' is obtained by holding x' constant and letting the other two coordinates vary arbitrarily. It is clear that the resulting surface is a plane passing through P' and parallel to the yz-plane. It is also clear that the first primary length element, dx' is perpendicular to the first primary surface. The first primary element of area, denoted by da_1, is simply $dy'\, dz'$. The second and third primary surfaces are the xz-plane and the xy-plane, respectively. These planes are perpendicular to their corresponding length elements. The primary elements of area are obtained similarly. We

primary elements of area in Cartesian coordinates

thus have

$$da_1 = dy'\, dz', \quad da_2 = dx'\, dz', \quad da_3 = dx'\, dy'. \tag{2.22}$$

element of volume in Cartesian coordinates

Finally, the volume element is

$$dV = dl_1\, dl_2\, dl_3 = dx'\, dy'\, dz'. \tag{2.23}$$

2.3.2   Elements in a Spherical Coordinate System

The point P' in Figure 2.3 now has coordinates (r', θ', φ'). To find the primary length along $\hat{\mathbf{e}}_{r'}$, keep θ' and φ' fixed and let r' change to $r' + dr'$. Then P' will be displaced by dr' along $\hat{\mathbf{e}}_{r'}$. Thus, the first primary length element, dl_1, is simply dr'. To find the primary length along $\hat{\mathbf{e}}_{\theta'}$, keep r' and φ' fixed, i.e.,

---

[13]The use of primes to represent both the derivative and the coordinates of the element of the source (such as dm) is unfortunately confusing. However, this practice is so widespread that any alteration to it would result in more confusion. The context of any given problem is usually clear enough to resolve such confusion.

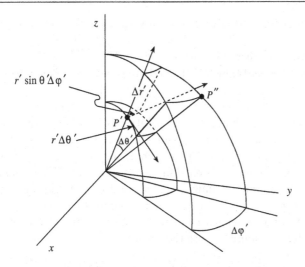

Figure 2.3: Elements of length, area, and volume in spherical coordinates. We have used "Δ" instead of "d."

confine yourself to the plane passing through P' and the polar—or z—axis, and let θ' change to $\theta' + d\theta'$. Then P' will be displaced by[14] $r'\,d\theta'$ along $\hat{\mathbf{e}}_{\theta'}$. The primary length along $\hat{\mathbf{e}}_{\varphi'}$ is obtained by keeping r' and θ' fixed, i.e., confining oneself to a plane passing through P' and perpendicular to the z-axis,[15] and letting φ' change to $\varphi' + d\varphi'$. Then P' will be displaced along a circle of radius $r'\sin\theta'$ by an angle $d\varphi'$. This can be seen by noting that P' lies in the xy-plane and that its distance from the z-axis is given by

$$x'^2 + y'^2 = (r'\sin\theta'\cos\varphi')^2 + (r'\sin\theta'\sin\varphi')^2 = r'^2\sin^2\theta'$$

and that the RHS, which is the square of the radius of the circle, is a constant. The displacement of P' is therefore $r'\sin\theta'\,d\varphi'$ along $\hat{\mathbf{e}}_{\varphi'}$. A general infinitesimal (vector) displacement can, therefore, be written as

general spherical infinitesimal displacement

$$dr' = d\vec{l} = \hat{\mathbf{e}}_{r'}\,dl_1 + \hat{\mathbf{e}}_{\theta'}\,dl_2 + \hat{\mathbf{e}}_{\varphi'}\,dl_3$$
$$= \hat{\mathbf{e}}_{r'}\,dr' + \hat{\mathbf{e}}_{\theta'}\,r'\,d\theta' + \hat{\mathbf{e}}_{\varphi'}\,r'\sin\theta'\,d\varphi'. \tag{2.24}$$

Note again that this vectorial infinitesimal displacement includes the primary length elements as special cases. Thus, setting $d\theta' = 0 = d\varphi'$, i.e., holding θ' and φ' fixed, we recover the first primary length element. The length of dr' (or $d\vec{l}$) is

do not confuse $|dr'|$ with dr', they are not equal.

$$|dr'| = dl = \sqrt{(dr')^2 + (r'\,d\theta')^2 + (r'\sin\theta'\,d\varphi')^2}$$
$$= \sqrt{dr'^2 + r'^2\,d\theta'^2 + r'^2\sin^2\theta'\,d\varphi'^2}. \tag{2.25}$$

---

[14]Since r' is held fixed, P' is confined to move on a circle of radius r', describing an infinitesimal arc subtended by the angle $d\theta'$.

[15]Fixing r' and θ' fixes $z' = r'\cos\theta'$ which describes a plane parallel to the xy-plane, i.e., a plane perpendicular to the z-axis.

If we know the parametric equation of a curve in spherical coordinates, i.e., if the coordinates r', θ', and φ' of a point on the curve can be expressed as functions of the parameter t, then we can find the differentials in terms of dt and substitute in Equation (2.25) to find an expression analogous to Equation (2.21). We leave this as an exercise for the reader.

primary surfaces
of spherical
coordinates
consist of a
sphere, a cone,
and a plane.

The first primary surface at P' is obtained by holding r' constant and letting the other two coordinates vary arbitrarily. It is clear that the resulting surface is a sphere of radius r' passing through P'. It is also clear that the first primary length element dr' is perpendicular to the first primary surface. It is not hard to convince oneself that the second and third primary surfaces are, respectively, a cone of (half) angle θ', and a plane containing the z-axis and making an angle of φ' with the x-axis. These surfaces are perpendicular to their corresponding length elements. The primary elements of area are obtained easily. We simply quote the results:

primary elements
of area in spherical
coordinates

$$da_1 = (r'\,d\theta')(r'\sin\theta'\,d\varphi') = r'^2 \sin\theta'\,d\theta'\,d\varphi',$$
$$da_2 = (dr')(r'\sin\theta'\,d\varphi') = r'\sin\theta'\,dr'\,d\varphi',$$
$$da_3 = (dr')(r'\,d\theta') = r'\,dr'\,d\theta'.$$

(2.26)

element of volume
in spherical
coordinates

Finally, the volume element is

$$dV = (dr')(r'\,d\theta')(r'\sin\theta'\,d\varphi') = r'^2 \sin\theta'\,dr'\,d\theta'\,d\varphi'.$$

(2.27)

Table 2.1 gathers together all the primary curves and surfaces for the three coordinate systems used frequently in this book. The reader is advised to remember that

> **Box 2.3.3.** *All the differentials of Table 2.1 carry a prime to emphasize that they are evaluated at P', the location of infinitesimal elements.*

Coordinate system	Primary curves	Primary surfaces
Cartesian	1st: Straight line (x-axis) 2nd: Straight line (y-axis) 3rd: Straight line (z-axis)	yz-plane xz-plane xy-plane
Cylindrical	1st: Rays perp. to z-axis 2nd: Circle centered on z-axis 3rd: Straight line (z-axis)	Cylinder with axis z Half-plane from z-axis Plane perp. z-axis
Spherical	1st: Rays from origin 2nd: Half-circle 3rd: Circle centered on polar axis	Sphere Cone of half angle θ Half-plane from z-axis

Table 2.1: Primary curves and surfaces of the three common coordinate systems.

2.3.3 Elements in a Cylindrical Coordinate System

The coordinates of P' are now (ρ', φ', z') as shown in Figure 2.4. To find the primary length along $\hat{\mathbf{e}}_{\rho'}$, keep φ' and z' fixed and let ρ' change to $\rho' + d\rho'$. Then P' will be displaced by $d\rho'$ along $\hat{\mathbf{e}}_{\rho'}$. Thus, the first primary length element dl_1 is simply $d\rho'$. To find the primary length along $\hat{\mathbf{e}}_{\varphi'}$, keep ρ' and z' fixed, i.e., confine yourself to a circle of radius ρ' in the plane passing through P' and perpendicular to the z-axis, and let φ' change to $\varphi' + d\varphi'$. Then P' will be displaced by $\rho' \, d\varphi'$ along $\hat{\mathbf{e}}_{\varphi'}$. The primary length along $\hat{\mathbf{e}}_{z'} = \hat{\mathbf{e}}_z$ is[16] obtained by keeping ρ' and φ' fixed, and letting z' change to $z' + dz'$. Then **general cylindrical** P' will be displaced by dz'. A general infinitesimal (vector) displacement can, **infinitesimal** therefore, be written as **displacement**

$$dr' = d\vec{l} = \hat{\mathbf{e}}_{\rho'} \, dl_1 + \hat{\mathbf{e}}_{\varphi'} \, dl_2 + \hat{\mathbf{e}}_{z'} \, dl_3$$
$$= \hat{\mathbf{e}}_{\rho'} \, d\rho' + \hat{\mathbf{e}}_{\varphi'} \, \rho' \, d\varphi' + \hat{\mathbf{e}}_z \, dz'. \tag{2.28}$$

Note again that this infinitesimal displacement includes the primary length elements as special cases. The length of this vector is

$$|dr'| = dl = \sqrt{(d\rho')^2 + (\rho' \, d\varphi')^2 + (dz')^2}$$
$$= \sqrt{d\rho'^2 + \rho'^2 d\varphi'^2 + dz'^2}. \tag{2.29}$$

If we know the parametric equation of a curve in cylindrical coordinates, i.e., if the coordinates ρ', φ', and z' of a point on the curve can be expressed as

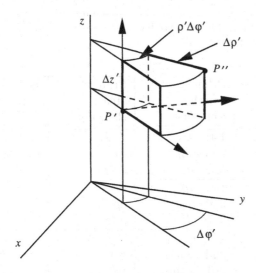

Figure 2.4: Elements of length, area, and volume in cylindrical coordinates. We have used "Δ" instead of "d."

---

[16]This is the only unit vector in "curvilinear coordinates" which is independent of the position of P'.

functions of the parameter t, then we can find the differentials in terms of dt and substitute in Equation (2.29) to find an expression analogous to Equation (2.21). We leave this as an exercise for the reader.

The first primary surface at P' is obtained by holding ρ' constant and letting the other two coordinates vary arbitrarily. It is clear that the resulting surface is a cylinder of radius ρ' passing through P'. It is also clear that the first primary length element $d\rho'$ is perpendicular to the first primary surface. The second and third primary surfaces are, respectively, a plane containing the z-axis and making an angle of φ' with the x-axis, and a plane perpendicular to the z-axis and cutting it at z'. These surfaces are perpendicular to their corresponding length elements. The primary elements of area are again obtained easily, and we merely quote the results

primary surfaces of cylindrical coordinates consist of a cylinder and two planes.

primary elements of area in cylindrical coordinates

$$da_1 = (\rho'\, d\varphi')(dz') = \rho'\, d\varphi'\, dz',$$
$$da_2 = d\rho'\, dz',$$
$$da_3 = (d\rho')(\rho'\, d\varphi') = \rho'\, d\rho'\, d\varphi'. \tag{2.30}$$

element of volume in cylindrical coordinates

Finally, the volume element is

$$dV = (d\rho')(\rho'\, d\varphi')(dz') = \rho'\, d\rho'\, d\varphi'\, dz'. \tag{2.31}$$

Table 2.2 gathers together all the elements of primary length, surface, and volume for the three commonly used coordinate systems.

Example 2.3.1. EXAMPLES OF ELEMENTS IN VARIOUS COORDINATE SYSTEMS
(a) The element of length in the φ direction at a point with spherical coordinates (a,γ,φ) is $a\sin\gamma\, d\varphi$. Note that this element is independent of φ, and for a fixed a, it has the largest value when $\gamma = \pi/2$, corresponding to the equatorial plane.
(b) The element of area for a cone of half-angle α is $r\sin\alpha\, dr\, d\varphi$, because for a cone, θ is a constant (in this case, α).

Coordinate system	Primary length elements	Primary area elements	Volume element
Cartesian (x,y,z)	1st: dx 2nd: dy 3rd: dz	$dy\, dz$ $dx\, dz$ $dx\, dy$	$dx\, dy\, dz$
Cylindrical (ρ,φ,z)	1st: $d\rho$ 2nd: $\rho\, d\varphi$ 3rd: dz	$\rho\, d\varphi\, dz$ $d\rho\, dz$ $\rho\, d\rho\, d\varphi$	$\rho\, d\rho\, d\varphi\, dz$
Spherical (r,θ,φ)	1st: dr 2nd: $r\, d\theta$ 3rd: $r\sin\theta\, d\varphi$	$r^2\sin\theta\, d\theta\, d\varphi$ $r\sin\theta\, dr\, d\varphi$ $r\, dr\, d\theta$	$r^2\sin\theta\, dr\, d\theta\, d\varphi$

Table 2.2: Primary length and area as well as volume elements in the three common coordinate systems. In almost all applications of the next chapter each of these variables carries a prime.

(c) The element of area of a cylinder of radius a is $a\,d\varphi\,dz$.

(d) The element of area of a sphere of radius a is $a^2\sin\theta\,d\theta\,d\varphi$. Note that the largest element of area (for given $d\theta$ and $d\varphi$) is at the equator and the smallest (zero) at the two poles.

(e) The element of area of a half-plane containing the z-axis and making an angle α with the x-axis is $d\rho\,dz$, independent of the angle α. ∎

Example 2.3.2. Suppose Cartesian coordinates of the plane are related to two other variables u and v via the formulas

finding unit vectors without use of geometry!

$$x = f(u,v), \qquad y = g(u,v).$$

We want to consider u and v as coordinates and find the unit vectors corresponding to them using our knowledge of differentiation gained in this chapter without any resort to geometric arguments.

In general, the unit vector in the direction of any coordinate variable at a point P is obtained by increasing the coordinate slightly (keeping other coordinate variables constant), calculating the displacement vector described by the motion of P, and dividing this vector by its length. So, consider changing u while v is kept constant. Call the displacement obtained $\Delta\vec{l}_1$. Then

$$\Delta\vec{l}_1 = \hat{e}_x\Delta x + \hat{e}_y\Delta y = \hat{e}_x\frac{\partial f}{\partial u}\Delta u + \hat{e}_y\frac{\partial g}{\partial u}\Delta u$$

and

$$|\Delta\vec{l}_1| = \sqrt{(\Delta x)^2 + (\Delta y)^2} = \sqrt{\left(\frac{\partial f}{\partial u}\right)^2 + \left(\frac{\partial g}{\partial u}\right)^2}\,\Delta u.$$

Therefore,

$$\hat{e}_u = \lim_{\Delta u\to 0}\frac{\Delta\vec{l}_1}{|\Delta\vec{l}_1|} = \frac{\hat{e}_x\dfrac{\partial f}{\partial u} + \hat{e}_y\dfrac{\partial g}{\partial u}}{\sqrt{\left(\dfrac{\partial f}{\partial u}\right)^2 + \left(\dfrac{\partial g}{\partial u}\right)^2}} = \frac{\hat{e}_x\dfrac{\partial x}{\partial u} + \hat{e}_y\dfrac{\partial y}{\partial u}}{\sqrt{\left(\dfrac{\partial x}{\partial u}\right)^2 + \left(\dfrac{\partial y}{\partial u}\right)^2}}.$$

For $\hat{e}_v$, we keep u fixed and vary v. Calling the resulting displacement $\Delta\vec{l}_2$, we easily obtain

$$\hat{e}_v = \lim_{\Delta v\to 0}\frac{\Delta\vec{l}_2}{|\Delta\vec{l}_2|} = \frac{\hat{e}_x\dfrac{\partial f}{\partial v} + \hat{e}_y\dfrac{\partial g}{\partial v}}{\sqrt{\left(\dfrac{\partial f}{\partial v}\right)^2 + \left(\dfrac{\partial g}{\partial v}\right)^2}} == \frac{\hat{e}_x\dfrac{\partial x}{\partial v} + \hat{e}_y\dfrac{\partial y}{\partial v}}{\sqrt{\left(\dfrac{\partial x}{\partial v}\right)^2 + \left(\dfrac{\partial y}{\partial v}\right)^2}}.$$

Note that for general f and g, $\hat{e}_u$ and $\hat{e}_v$ are not perpendicular.

The result can easily be generalized to three variables. In fact, if

$$x = f(u,v,w), \qquad y = g(u,v,w), \qquad z = h(u,v,w),$$

then, a similar calculation as above will yield

$$\hat{e}_u = \frac{\hat{e}_x \frac{\partial f}{\partial u} + \hat{e}_y \frac{\partial g}{\partial u} + \hat{e}_z \frac{\partial h}{\partial u}}{\sqrt{\left(\frac{\partial f}{\partial u}\right)^2 + \left(\frac{\partial g}{\partial u}\right)^2 + \left(\frac{\partial h}{\partial u}\right)^2}} = \frac{\hat{e}_x \frac{\partial x}{\partial u} + \hat{e}_y \frac{\partial y}{\partial u} + \hat{e}_z \frac{\partial z}{\partial u}}{\sqrt{\left(\frac{\partial x}{\partial u}\right)^2 + \left(\frac{\partial y}{\partial u}\right)^2 + \left(\frac{\partial z}{\partial u}\right)^2}},$$

$$\hat{e}_v = \frac{\hat{e}_x \frac{\partial f}{\partial v} + \hat{e}_y \frac{\partial g}{\partial v} + \hat{e}_z \frac{\partial h}{\partial v}}{\sqrt{\left(\frac{\partial f}{\partial v}\right)^2 + \left(\frac{\partial g}{\partial v}\right)^2 + \left(\frac{\partial h}{\partial v}\right)^2}} = \frac{\hat{e}_x \frac{\partial x}{\partial v} + \hat{e}_y \frac{\partial y}{\partial v} + \hat{e}_z \frac{\partial z}{\partial v}}{\sqrt{\left(\frac{\partial x}{\partial v}\right)^2 + \left(\frac{\partial y}{\partial v}\right)^2 + \left(\frac{\partial z}{\partial v}\right)^2}},$$

$$\hat{e}_w = \frac{\hat{e}_x \frac{\partial f}{\partial w} + \hat{e}_y \frac{\partial g}{\partial w} + \hat{e}_z \frac{\partial h}{\partial w}}{\sqrt{\left(\frac{\partial f}{\partial w}\right)^2 + \left(\frac{\partial g}{\partial w}\right)^2 + \left(\frac{\partial h}{\partial w}\right)^2}} = \frac{\hat{e}_x \frac{\partial x}{\partial w} + \hat{e}_y \frac{\partial y}{\partial w} + \hat{e}_z \frac{\partial z}{\partial w}}{\sqrt{\left(\frac{\partial x}{\partial w}\right)^2 + \left(\frac{\partial y}{\partial w}\right)^2 + \left(\frac{\partial z}{\partial w}\right)^2}}.$$

2.4 Problems

2.1. Find the partial derivatives of the following functions at the given points with respect to the given variables. In the following $\mathbf{r} = (x, y, z)$ and $\mathbf{r}' = (x', y', z')$:

$$e^{xyz} \quad \text{with respect to } x \quad \text{at} \quad (1, 0, -1),$$
$$\cos(xy/z) \quad \text{with respect to } z \quad \text{at} \quad (\pi, 1, 1),$$
$$x^2 y + y^2 z + z^2 x \quad \text{with respect to } y \quad \text{at} \quad (1, -1, 2),$$
$$\ln\left(\frac{ax + by + cz}{x^2 + y^2 + z^2}\right) \quad \text{with respect to } x \quad \text{at} \quad (a, b, c),$$
$$r \equiv \sqrt{x^2 + y^2 + z^2} \quad \text{with respect to } x \quad \text{at} \quad (x, y, z),$$
$$|\mathbf{r} - \mathbf{r}'| \quad \text{with respect to } y \quad \text{at} \quad (x, y, z, x', y', z'),$$
$$\frac{1}{|\mathbf{r} - \mathbf{r}'|} \quad \text{with respect to } z' \quad \text{at} \quad (x, y, z, x', y', z').$$

2.2. The Earth has a radius of 6400 km. The thickness of the atmosphere is about 50 km. Starting with the volume of a sphere and using differentials, estimate the volume of the atmosphere. Hint: Find the change in the volume of a sphere when its radius changes by a "small" amount.

2.3. The gravitational potential (potential energy per unit mass) at a distance r from the center of the Earth (assumed to be the origin of a Cartesian coordinate system) is given by

$$\Phi = -\frac{GM}{r}, \qquad r = \sqrt{x^2 + y^2 + z^2},$$

where $G = 6.67 \times 10^{-11}$ N·m^2/kg^2 and $M = 6 \times 10^{24}$ kg. Using differentials, find the energy needed to raise a 10-kg object from the point with coordinates (4000 km, 4000 km, 3000 km) to a point with coordinates (4020 km, 4050 km, 3010 km).

2.4. Show that the function $f(x \pm ct)$ satisfies the one-dimensional **wave equation**:

one-dimensional wave equation

$$\frac{\partial^2 f}{\partial x^2} - \frac{1}{c^2}\frac{\partial^2 f}{\partial t^2} = 0.$$

Hint: Let $y = x \pm ct$ and use the chain rule.

2.5. Assume that $f'' + kf = 0$ and $g'' - kg = 0$. Show that $F(x,y) \equiv f(x)g(y)$ satisfies the two-dimensional **Laplace's equation**:

two-dimensional Laplace's equation

$$\frac{\partial^2 F}{\partial x^2} + \frac{\partial^2 F}{\partial y^2} = 0.$$

2.6. Suppose that $f'' - \alpha f = 0$, $g'' - \beta g = 0$, and $h'' - \gamma h = 0$. Write an equation relating α, β, and γ such that the function

$$F(x,y,z) \equiv f(x)g(y)h(z)$$

satisfies the three-dimensional **Laplace's equation**:

three-dimensional Laplace's equation

$$\frac{\partial^2 F}{\partial x^2} + \frac{\partial^2 F}{\partial y^2} + \frac{\partial^2 F}{\partial z^2} = 0.$$

2.7. Suppose that $f'' - \alpha f = 0$, $g'' - \beta g = 0$, $h'' - \gamma h = 0$, and $u' - \omega u = 0$. Write an equation relating α, β, γ, and ω such that the function

$$F(x,y,z,t) \equiv f(x)g(y)h(z)u(t)$$

satisfies the **heat equation**:

heat equation

$$\frac{\partial^2 F}{\partial x^2} + \frac{\partial^2 F}{\partial y^2} + \frac{\partial^2 F}{\partial z^2} = a\frac{\partial F}{\partial t}.$$

where a is a constant.

2.8. Suppose that $f'' + k_x^2 f = 0$, $g'' + k_y^2 g = 0$, $h'' + k_z^2 h = 0$, and $u'' + \omega^2 u = 0$.
(a) Write an equation relating k_x, k_y, k_z, and ω such that the function

$$F(x,y,z,t) \equiv f(x)g(y)h(z)u(t)$$

satisfies the three-dimensional **wave equation**:

three-dimensional wave equation

$$\frac{\partial^2 F}{\partial x^2} + \frac{\partial^2 F}{\partial y^2} + \frac{\partial^2 F}{\partial z^2} - \frac{1}{c^2}\frac{\partial^2 F}{\partial t^2} = 0.$$

(b) If ω is considered as *angular frequency*, and c as the speed of the wave, what is the magnitude of the vector $\mathbf{k} \equiv \langle k_x, k_y, k_z \rangle$?

2.9. Consider the function $F(x,y,z) \equiv f\left(\dfrac{x^2 y + y^2 x}{a^2 z}\right)$ in which a is a constant. Assuming that $f'(2) = a$, find the unit vector $\hat{\mathbf{e}}_v$ in the direction of

$$\mathbf{v} = \hat{\mathbf{e}}_x \partial_1 F(a,a,a) + \hat{\mathbf{e}}_y \partial_2 F(a,a,a) + \hat{\mathbf{e}}_z \partial_3 F(a,a,a).$$

2.10. Consider the function $F(x, y, z) \equiv f\left(\dfrac{x^3y - y^3z + 2z^3x}{a^4}\right)$ in which a is a constant. Assuming that $f'(17) = a$, find the unit vector $\hat{\mathbf{e}}_v$ in the direction of

$$\mathbf{v} = \hat{\mathbf{e}}_x \partial_1 F(a, -a, 2a) + \hat{\mathbf{e}}_y \partial_2 F(a, -a, 2a) + \hat{\mathbf{e}}_z \partial_3 F(a, -a, 2a).$$

2.11. Given that $f(x, y, z) = e^{-k\sqrt{x^2+y^2+z^2}} / \sqrt{x^2 + y^2 + z^2}$, where k is a constant, find the radial component (component along $\hat{\mathbf{e}}_r$) of the vector

$$\mathbf{V} = \hat{\mathbf{e}}_x \partial_1 f(x, y, z) + \hat{\mathbf{e}}_y \partial_2 f(x, y, z) + \hat{\mathbf{e}}_z \partial_3 f(x, y, z).$$

2.12. Given that

$$\partial_1 f(x, y, z) = \partial_2 f(z, x, y) = \partial_3 f(y, z, x) = \frac{2kx}{y} - \frac{kz^2}{x^2},$$

where k is a constant, find the function $f(x, y, z)$. Note the order of the variables in each pair of parentheses.

2.13. Given that $f(x, y, z) = x^2y \sin(yz/x)$, find

$$\partial_2 f(1, 1, \pi/2), \quad \partial_1 f(2, \pi, 1), \quad \partial_3 f(4, \pi, 1), \quad \partial_1 f(y, z, x), \quad \partial_1 f(t, u, v).$$

2.14. Derive the analogue of Equation (2.11) assuming this time that Q is held constant in all derivatives instead of W.

2.15. Which of the following functions are homogeneous?

$$e^{xy/z^2}, \quad xyz \sin\frac{xy}{az}, \quad \frac{x^2y^2}{z} \cos\frac{xz}{y^2}, \quad x^2 + y^2 - z^2, \quad ax + y(z - x),$$

where a is a constant. For those functions that are homogeneous, find their degree and verify that they satisfy Equation (2.18).

2.16. Suppose f and g are homogeneous functions of degrees q and p, respectively. What can you say about the homogeneity of $f \pm g$, fg, and f/g. If they are homogeneous, find their degree, and verify that they satisfy Equation (2.18).

2.17. If f is homogeneous of degree q, show that $\partial_i f$ is homogeneous of degree $q - 1$. Hint: Use the definition of homogeneity and differentiate with respect to x_i.

2.18. A function $f(x, y, z)$ of Cartesian coordinates can also be thought of as a function of cylindrical coordinates ρ, φ, z, because the latter are functions of the former via the relations $\rho = \sqrt{x^2 + y^2}$ and $\tan\varphi = y/x$.
(a) Using the chain rule for differentiation, find $\partial f/\partial x$ and $\partial f/\partial y$ in terms of $\partial f/\partial \rho$ and $\partial f/\partial \varphi$. Express your answers entirely in terms of cylindrical coordinates.

(b) Show that the vector $\hat{e}_x \dfrac{\partial f}{\partial x} + \hat{e}_y \dfrac{\partial f}{\partial y} + \hat{e}_z \dfrac{\partial f}{\partial z}$, when written entirely in terms of cylindrical coordinates and cylindrical unit vectors, becomes

$$\hat{e}_\rho \frac{\partial f}{\partial \rho} + \hat{e}_\varphi \frac{1}{\rho} \frac{\partial f}{\partial \varphi} + \hat{e}_z \frac{\partial f}{\partial z}.$$

2.19. In each of the following, the partial derivative of a function is given. Find the most general function with such a derivative.

(a) $\partial_2 f(x, y, z) = xy^2 z$.　　　　(b) $\partial_1 f(x, y, z) = xy^2 z$.

(c) $\partial_1 h(z, x, y) = \dfrac{e^x \sin z}{x}$.　　　(d) $\partial_1 g(z, x, y) = e^x y^2$.

(e) $\partial_2 g(z, x, y) = e^x y^2$.　　　　(f) $\partial_2 h(x, y, z) = \dfrac{e^x \sin z}{xy}$.

(g) $\partial_3 f(x, y, z) = xy^2 z$.　　　　(h) $\partial_3 g(z, x, y) = e^x y^2$.

(i) $\partial_3 h(y, x, z) = \dfrac{e^x \sin z}{x}$

2.20. Finish the calculation of Example 2.2.8.

2.21. Find $y\partial f/\partial z - z\partial f/\partial y$ and $z\partial f/\partial x - x\partial f/\partial z$ in terms of spherical coordinates. Warning! These will not be as nice-looking as the expression calculated in Example 2.2.8.

2.22. Given that $f'(1) = 2$, find

$$\frac{\partial f}{\partial x}\hat{e}_x + \frac{\partial f}{\partial y}\hat{e}_y + \frac{\partial f}{\partial z}\hat{e}_z$$

for $f(xyz)$ at the Cartesian point $(-1, 2, -1/2)$.

2.23. Given that $f'(3) = -1$, find the **radial component** of the vector

$$\frac{\partial f}{\partial x}\hat{e}_x + \frac{\partial f}{\partial y}\hat{e}_y + \frac{\partial f}{\partial z}\hat{e}_z$$

for $f(\sqrt{x^2 + y^2 + z^2})$ at the Cartesian point $(2, 1, -2)$.

2.24. Show that the function $F(\mathbf{k} \cdot \mathbf{r} - \omega t)$ satisfies the three-dimensional wave equation:

$$\frac{\partial^2 F}{\partial x^2} + \frac{\partial^2 F}{\partial y^2} + \frac{\partial^2 F}{\partial z^2} - \frac{1}{c^2}\frac{\partial^2 F}{\partial t^2} = 0$$

if $\mathbf{k} \equiv \langle k_x, k_y, k_z \rangle$ is a *constant vector*, ω is a constant, and a certain relation exists between $k = |\mathbf{k}|$ and ω. Find this relation.

2.25. In electromagnetic radiation theory one encounters an equation of the form

$$t = \frac{1}{\sqrt{[x - f(t)]^2 + [y - g(t)]^2 + [z - h(t)]^2}}$$

and one is interested in the partial derivative of t with respect to x, y, and z. Note the hybrid role that t plays here as both a dependent and an independent variable. Show that

$$\frac{\partial t}{\partial x} = \frac{x - f(t)}{[x - f(t)]f'(t) + [y - g(t)]g'(t) + [z - h(t)]h'(t) - F^{3/2}},$$

where $F(x, y, z, t) \equiv [x - f(t)]^2 + [y - g(t)]^2 + [z - h(t)]^2$. Find similar expressions for partial derivatives of t with respect to y and z.

2.26. Consider the function $f(|\mathbf{r} - \mathbf{r}'|)$ with $\mathbf{r} = x\hat{\mathbf{e}}_x + y\hat{\mathbf{e}}_y + z\hat{\mathbf{e}}_z$ and $\mathbf{r}' = x'\hat{\mathbf{e}}_x + y'\hat{\mathbf{e}}_y + z'\hat{\mathbf{e}}_z$ being the position vectors of P and P'.
(a) Find a general expression for the vector

$$\mathbf{V} = \frac{\partial f}{\partial x}\hat{\mathbf{e}}_x + \frac{\partial f}{\partial y}\hat{\mathbf{e}}_y + \frac{\partial f}{\partial z}\hat{\mathbf{e}}_z$$

in terms of $\mathbf{r}$ and $\mathbf{r}'$.
(b) If $f'(3) = 3$ and the coordinates of P and P' are $(1, -1, 0)$, and $(0, 1, 2)$, respectively, find the numerical value of $\mathbf{V}$.

2.27. Find an expression in cylindrical and spherical coordinates analogous to Equation (2.21).

2.28. A function $f(x, y)$ of Cartesian coordinates can also be thought of as a function of some other coordinates u and v defined by

$$x = u\sin v, \qquad y = u\cos v.$$

(a) Applying the procedure of Example 2.3.2, find the unit vectors $\hat{\mathbf{e}}_u$ and $\hat{\mathbf{e}}_v$.
(b) Find u and v as functions of x and y.
(c) Calculate $\hat{\mathbf{e}}_x$ and $\hat{\mathbf{e}}_y$ in terms of $\hat{\mathbf{e}}_u$ and $\hat{\mathbf{e}}_v$.
(d) Write the vector

$$\mathbf{A} = \hat{\mathbf{e}}_x \frac{\partial f}{\partial x} + \hat{\mathbf{e}}_y \frac{\partial f}{\partial y}$$

entirely in the (u, v) coordinate system.

2.29. Find the cylindrical unit vectors in terms of Cartesian unit vectors using the procedure of Example 2.3.2.

2.30. Find the spherical unit vectors in terms of Cartesian unit vectors using the procedure of Example 2.3.2.

2.31. In the first part of Example 2.3.2, assume that $f(u, v) = uf_1(v)$ and $g(u, v) = ug_1(v)$ where f_1 and g_1 are functions of only one variable.
(a) Find a relation between f_1 and g_1 to make $\hat{\mathbf{e}}_u$ and $\hat{\mathbf{e}}_v$ perpendicular.
(b) Can you recover the polar coordinates as a special case of (a)?

2.32. The **elliptic coordinates** (u, θ) are given by

$$x = a \cosh u \cos \theta,$$
$$y = a \sinh u \sin \theta,$$

where a is a constant.
(a) What are the curves of constant u?
(b) What are the curves of constant θ?
(c) Find $\hat{\mathbf{e}}_u$ and $\hat{\mathbf{e}}_\theta$ in terms of the Cartesian unit vectors, and examine their orthogonality.

2.33. The **parabolic coordinates** (u, v) are given by

$$x = a(u^2 - v^2),$$
$$y = 2auv,$$

where a is a constant.
(a) What are the curves of constant u?
(b) What are the curves of constant v?
(c) Find $\hat{\mathbf{e}}_u$ and $\hat{\mathbf{e}}_v$ in terms of the Cartesian unit vectors, and examine their orthogonality.

2.34. The **two-dimensional bipolar coordinates** (u, v) are given by

$$x = \frac{a \sinh u}{\cosh u + \cos v},$$
$$y = \frac{a \sin v}{\cosh u + \cos v},$$

where a is a constant.
(a) What are the curves of constant u?
(b) What are the curves of constant v?
(c) Find $\hat{\mathbf{e}}_u$ and $\hat{\mathbf{e}}_v$ in terms of the Cartesian unit vectors, and examine their orthogonality.

2.35. The **elliptic cylindrical coordinates** (u, θ, z) are given by

$$x = a \cosh u \cos \theta,$$
$$y = a \sinh u \sin \theta,$$
$$z = z,$$

where a is a constant.
(a) What are the surfaces of constant u?
(b) What are the surfaces of constant θ?
(c) Find $\hat{\mathbf{e}}_u$, $\hat{\mathbf{e}}_\theta$, and $\hat{\mathbf{e}}_z$ in terms of the Cartesian unit vectors and examine their orthogonality.

2.36. The **prolate spheroidal coordinates** (u, θ, φ) are given by

$$x = a \sinh u \sin \theta \cos \varphi,$$
$$y = a \sinh u \sin \theta \sin \varphi,$$
$$z = a \cosh u \cos \theta,$$

where a is a constant.
(a) What are the surfaces of constant u?
(b) What are the surfaces of constant θ?
(c) Find $\hat{\mathbf{e}}_u$, $\hat{\mathbf{e}}_\theta$, and $\hat{\mathbf{e}}_\varphi$ in terms of the Cartesian unit vectors, and examine their mutual orthogonality.

2.37. The **toroidal coordinates** (u, θ, φ) are given by

$$x = \frac{a \sinh u \cos \varphi}{\cosh u - \cos \theta},$$
$$y = \frac{a \sinh u \sin \varphi}{\cosh \theta - \cos \theta},$$
$$z = \frac{a \sin u}{\cosh u - \cos \theta},$$

where a is a constant.
(a) What are the surfaces of constant u?
(b) What are the surfaces of constant θ?
(c) Find $\hat{\mathbf{e}}_u$, $\hat{\mathbf{e}}_\theta$, and $\hat{\mathbf{e}}_\varphi$ in terms of the Cartesian unit vectors, and examine their mutual orthogonality.

2.38. The **paraboloidal coordinates** (u, v, φ) are given by

$$x = 2auv \cos \varphi,$$
$$y = 2auv \sin \varphi,$$
$$z = a(u^2 - v^2),$$

where a is a constant.
(a) What are the surfaces of constant u?
(b) What are the surfaces of constant v?
(c) Find $\hat{\mathbf{e}}_u$, $\hat{\mathbf{e}}_v$, and $\hat{\mathbf{e}}_\varphi$ in terms of the Cartesian unit vectors, and examine their mutual orthogonality.

2.39. The **three-dimensional bipolar coordinates** (u, θ, φ) are given by

$$x = \frac{a \sin \theta \cos \varphi}{\cosh u - \cos \theta},$$
$$y = \frac{a \sin \theta \sin \varphi}{\cosh u - \cos \theta},$$
$$z = \frac{a \sinh u}{\cosh u - \cos \theta},$$

where a is a constant.
(a) What are the surfaces of constant u?
(b) What are the surfaces of constant θ?
(c) Find $\hat{\mathbf{e}}_u$, $\hat{\mathbf{e}}_\theta$, and $\hat{\mathbf{e}}_\varphi$ in terms of the Cartesian unit vectors, and examine their mutual orthogonality.

2.40. A coordinate system (R, Θ, ϕ) in space is defined by

$$x = R \cos \Theta \cos \phi + b \cos \phi,$$
$$y = R \cos \Theta \sin \phi + b \sin \phi,$$
$$z = R \sin \Theta,$$

where b is a constant, and $0 < R < b$.

1. Express the unit vectors $\hat{\mathbf{e}}_R$, $\hat{\mathbf{e}}_\Theta$, and $\hat{\mathbf{e}}_\phi$ in terms of Cartesian unit vectors with coefficients being functions of (R, Θ, ϕ).

2. Are unit vectors mutually perpendicular?

Chapter 3

Integration: Formalism

It is not an exaggeration to say that the most important concept, whose mastery ensures a much greater understanding of all undergraduate physics, is the concept of integral. Generally speaking, physical laws are given in local form while their application to the real world requires a departure from locality. For instance, Coulomb's law in electrostatics and the universal law of gravity are both given in terms of point particles. These are mathematical points and the laws assume that. In real physical situations, however, we never deal with a mathematical point. Usually, we *approximate* the objects under consideration as points, as in the case of the gravitational force between the Earth and the Sun. Whether such an approximation is good depends on the properties of the objects and the parameters of the law. In the example of gravity, on the sizes of the Earth and the Sun as compared to the distance between them. On the other hand, the precise motion of a satellite circling the earth requires more than approximating the Earth as a point; all the bumps and grooves of the Earth's surface will affect the satellite's motion.

Physical laws are given for mathematical points but applied to extended objects.

This chapter is devoted to a thorough discussion of integrals *from a physical standpoint*, i.e., the meaning and the use of the concept of integration rather than the technique and the art of evaluating integrals.

3.1   "$\int$" Means "$\int$um"

One of the first difficulties we have to overcome is the preconception instilled in all of us from calculus that integral is "area under a curve." This preconception is so strong that in some introductory physics books the authors translate physical concepts, in which integral plays a natural role, into the unphysical and unnatural notion of area under a curve. It is true that calculation of the area under a curve employs the concept of integration, but it does so only because the calculation happens to be the limit of a sum, and such limits find their natural habitat in many physical situations.

Integral is not just area under a curve!

Take the gravitational force, for example. As a fundamental physical law, it is given for point masses, but when we want to calculate the force between the Earth and the Moon, we cannot apply the law directly because the Earth and the Moon cannot be considered as points with the Moon being only 60 Earth radii away. This problem was recognized by Newton who found its solution in integration. Inherent in the concept of integration is the superposition principle whereby, as mentioned in Chapter 1, different parts of a system are assumed to act independently. Thus a natural procedure is to divide the big Earth and the big Moon into small pieces, write down the gravitational force between these small pieces, invoke the superposition principle, and add the contribution of these pieces to get the total force. Now, nothing is more natural than this process, and no example is a more illustrative example of integration than such a calculation.

calculation of
fields of
continuous
distribution of
sources as the
natural setting for
the concept of
integral

In order to define and elucidate the concept of integration,[1] let us reconsider the gravitational field of Box 1.3.5. Instead of a known collection of *point* masses, let us calculate the gravitational field at a point P of a **continuous distribution** of mass such as that distributed in the volume of the Earth. The point P is called the **field point**.[2] We divide the large mass into N pieces, denoting the mass of the ith piece, located around the point P_i, by Δm_i as shown in Figure 3.1. To be able to even write the field equation for the ith piece of mass, we have to make sure that the size of Δm_i is small enough. We thus write

$$\mathbf{g}_i \approx -\frac{G\Delta m_i}{|\mathbf{r} - \mathbf{r}_i|^3}(\mathbf{r} - \mathbf{r}_i).$$

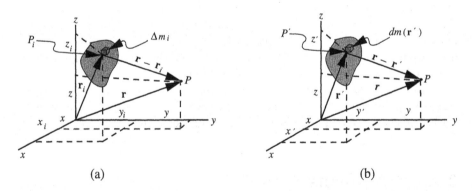

(a)   (b)

Figure 3.1: The mass distribution giving rise to a gravitational field. (a) The mass is divided into discrete pieces labeled 1 through N with the ith piece singled out. (b) The mass is divided into infinitesimal pieces with the piece located at $\mathbf{r}'$ singled out.

---

[1]The discussion that follows may seem specific to one example, but in reality, it is much more general. Instead of the gravitational law one can substitute any other local law, and instead of mass, the appropriate physical quantity must be used. The examples that follow throughout this chapter will clarify any vague points.

[2]The same terminology applies to electrostatic fields as well.

The smaller the size, the better this expression approximates the field due to Δm_i. Invoking the superposition principle, we write

$$\mathbf{g}(\mathbf{r}) \approx \sum_{i=1}^{N} \mathbf{g}_i = -\sum_{i=1}^{N} \frac{G\Delta m_i}{|\mathbf{r}-\mathbf{r}_i|^3}(\mathbf{r}-\mathbf{r}_i) = -\sum_{i=1}^{N} \frac{G\Delta m(\mathbf{r}_i)}{|\mathbf{r}-\mathbf{r}_i|^3}(\mathbf{r}-\mathbf{r}_i), \quad (3.1)$$

where in the last equality we have replaced Δm_i with $\Delta m(\mathbf{r}_i)$. Aside from a change in notation, this replacement emphasizes the dependence of the small piece of mass on its "location." The quotation marks around the last word need some elaboration. In any practical slicing of the gravitating object, such as the Earth, each piece still has some nonzero size. This makes it impossible to define the distance between Δm_i and the point P. We can define this distance to be that of the "center" of Δm_i from P, but then the difficulty shifts to defining the center of the piece. Fortunately, it turns out that, as long as we ultimately make the size of all Δm_i's indefinitely small, any point—such as P_i shown in the figure—in Δm_i can be chosen to define its distance from P. We are thus led to taking the limit of Equation (3.1) as the size of all pieces tends to zero, and, necessarily, the number of pieces tends to infinity. If such a limit exists, we call it the **integral** of the gravitational field and denote it as follows:[3]

integral as the limit of a sum

$$\mathbf{g}(\mathbf{r}) = -\lim_{\substack{\Delta m \to 0 \\ N \to \infty}} \sum_{i=1}^{N} \frac{G\Delta m(\mathbf{r}_i)}{|\mathbf{r}-\mathbf{r}_i|^3}(\mathbf{r}-\mathbf{r}_i) \equiv -\iint_{\Omega} \frac{G\,dm(\mathbf{r}')}{|\mathbf{r}-\mathbf{r}'|^3}(\mathbf{r}-\mathbf{r}'). \quad (3.2)$$

An identical procedure leads to a similar formula for the electrostatic field and potential:

$$\mathbf{E} = \iint_{\Omega} \frac{k_e dq(\mathbf{r}')}{|\mathbf{r}-\mathbf{r}'|^3}(\mathbf{r}-\mathbf{r}'), \qquad \Phi = \iint_{\Omega} \frac{k_e dq(\mathbf{r}')}{|\mathbf{r}-\mathbf{r}'|}. \quad (3.3)$$

Equations (3.2) and (3.3) will be used frequently in the sequel as we try to illustrate their use in various physical situations. Note that *Equations (3.2) and (3.3) are independent of any coordinate systems as all physical laws should be.*

In the symbolic representation of integral on the RHS, Ω, called the **region of integration**,[4] is the region—for example, the volume of the Earth—in which the mass distribution resides, and $dm(\mathbf{r}')$ is called an **element of mass** located[5] at point P' whose position vector is $\mathbf{r}'$. P' is called the **source point** because it is the location of the source of the gravitational field, i.e., the mass element at that point. We also call it the **integration point**. The

region of integration

integration point, integration variables, and integrand defined

---

[3]We shall use the symbol $\iint_{\Omega}$ (or simply $\int_{\Omega}$) to indicate general integration without regard to the dimensionality (single, double, or triple) of the integral.

[4]When the region of integration is one dimensional, such as an interval (a,b) on the real line, one uses $\int_a^b$ instead of $\int_{(a,b)}$.

[5]Whenever $\mathbf{r}'$ is used as an argument of a quantity, it will refer to the *coordinates* of a point *not* the components of its position vector.

coordinates of $\mathbf{r}'$ upon which the mass element depends—and in terms of which it will eventually be expressed—are called the **integration variables**, and whatever multiplies the products of the differentials of these variables is called the **integrand**.

It is not hard to abstract the concept of integration from the specific example of gravity. Instead of the specific form of the integral in Equations (3.2) and (3.3), we use $f(\mathbf{r}, \mathbf{r}')$, and instead of the element of mass, we use the element of some other quantity which we generically designate as $dQ(\mathbf{r}')$. We thus write

$$h(\mathbf{r}) = \lim_{\substack{\Delta Q \to 0 \\ N \to \infty}} \sum_{i=1}^{N} f(\mathbf{r}, \mathbf{r}_i) \Delta Q(\mathbf{r}_i) \equiv \int_{\Omega} f(\mathbf{r}, \mathbf{r}') \, dQ(\mathbf{r}'), \qquad (3.4)$$

integration
parameters

where $h(\mathbf{r})$, the result of integration, will be a function of $\mathbf{r}$, the position vector of P whose coordinates are called the **parameters of integration**. Although we have used $\mathbf{r}$ and $\mathbf{r}'$, the concept of integration does not require the parameters and integration variables to be position vectors. They could be any collection of parameters and variables. Nevertheless, we continue to use the *terminology* of position vectors and call such collections the coordinates of **points**.

An immediate—and important—consequence of the definition of integral is that if the region of integration Ω is small, then, for practical calculations, we do not need to subdivide it into many pieces. In fact, if Ω is small enough, only one piece may be a good approximation to the integral. We thus write

$$\int_{\Delta\Omega} f(\mathbf{r}, \mathbf{r}') \, dQ(\mathbf{r}') \approx f(\mathbf{r}, \mathbf{r}_M) \, \Delta Q, \qquad (3.5)$$

where it is understood that $\Delta\Omega$ is a small region around point M whose "position vector" is $\mathbf{r}_M$.

Another immediate and important consequence of the definition of integral is that if Ω is divided into two regions Ω_1 and Ω_2, then

$$\int_{\Omega} f(\mathbf{r}, \mathbf{r}') \, dQ(\mathbf{r}') = \int_{\Omega_1} f(\mathbf{r}, \mathbf{r}') \, dQ(\mathbf{r}') + \int_{\Omega_2} f(\mathbf{r}, \mathbf{r}') \, dQ(\mathbf{r}') \qquad (3.6)$$

In order to be able to evaluate integrals, one has to express both $dQ(\mathbf{r}')$ and $f(\mathbf{r}, \mathbf{r}')$ in terms of a suitable set of coordinates. $f(\mathbf{r}, \mathbf{r}')$ poses no problem, and in most cases it involves a mere substitution. The element of Q, on the other hand, is often related, via density, to the element of volume (or area, or length) whose expression is more involved. Section 2.3 dealt with the construction of elements of length, area, and volume in the three coordinate systems.

Historical Notes

Integral calculus, in its geometric form, was known to the ancient Greeks. For example, **Euclid**, by adding pieces to the area of a square inscribed in a circle, constructing newer polygons of larger numbers of sides, and continuing the process until the circle is "exhausted" by regular polygons, proved the theorem: Circles are to one another as the squares on the diameters. In essence, Euclid thinks of a circle as the limiting case of a regular polygon and proves the above theorem for polygons. Then he uses the argument of "exhaustion" to get to the result. Although mathematicians of antiquity made frequent use of the method of exhaustion, no one did it with the mastery of **Archimedes**.

Archimedes is arguably believed to be the greatest mathematician of antiquity. The son of an astronomer, he was born in Syracuse, a Greek settlement in Sicily. As a young man he went to Alexandria to study mathematics, and although he went back to Syracuse to spend the rest of his life there, he never lost contact with Alexandria.

Archimedes
287–212 B.C.

Archimedes possessed a lofty intellect, great breadth of interest—both theoretical and practical—and excellent mechanical skills. He is credited with finding the areas and volumes of many geometric figures using the method of exhaustion, the calculation of π, a new scheme of presenting large numbers in verbal language, finding the centers of gravity of many solids and plane figures, and founding the science of hydrostatics.

His great achievements in mathematics—he is ranked with Newton and Gauss as one of the three greatest mathematicians of all time—did not overshadow his practical inventions. He invented the first planetarium and a pump (Archimedean screw). He showed how to use levers to move great weights, and used compound pulleys to launch a galley of the king of Syracuse. Taking advantage of the focusing power of a parabolic mirror, so the story goes, he concentrated the Sun's rays on the Roman ships besieging Syracuse and burned them!

Perhaps the most famous story about Archimedes is his discovery of the method of testing the debasement of a crown of gold. The king of Syracuse had ordered the crown. Upon delivery, he suspected that it was filled with baser metal and sent it to Archimedes to test it for purity. Archimedes pondered about the problem for some time, until one day, as he was taking a bath, he observed that his body was partly buoyed up by the water and suddenly grasped the principle—now called **Archimedes' principle**—by which he could solve the problem. He was so excited about the discovery that he forgetfully ran out into the street naked shouting "Eureka!" ("I have found it!").

3.2   Properties of Integral

Now that we have developed the formalism of integration, we should look at some applications in which integrals are evaluated. As we shall see, *all integral evaluations eventually reduce to integrals involving only one variable.* Thus, it is important to have a thorough understanding of the properties of such integrals. Some of these properties are familiar, others may be less familiar or completely new. We gather all these properties here for the sake of completeness.

3.2.1   Change of Dummy Variable

The symbol used as the variable of integration in the integral is completely irrelevant. Thus, we have

Feel free to use
any symbol you
like for the
integration
variable!

$$\int_{t_1}^{t_2} g(t)\, dt = \int_{t_1}^{t_2} g(x)\, dx = \int_{t_1}^{t_2} g(s)\, ds$$

$$= \int_{t_1}^{t_2} g(t')\, dt' = \int_{t_1}^{t_2} g(\bigstar)\, d\bigstar.$$

Note how the limits of integration remain the same in all integrals. The fact that these limits use the same symbol as the first dummy variable should not confuse the reader. What is important is that they are fixed real numbers.

3.2.2   Linearity

For arbitrary constant real numbers a and b, we have

$$\int_{c_1}^{c_2} [af(t) + bg(t)]\, dt = a \int_{c_1}^{c_2} f(t)\, dt + b \int_{c_1}^{c_2} g(t)\, dt.$$

3.2.3   Interchange of Limits

Interchanging the limits of integration introduces a minus sign:

$$\int_{c}^{d} f(t)\, dt = - \int_{d}^{c} f(t)\, dt. \tag{3.7}$$

This relation implies that $\int_{s}^{s} f(t)\, dt = 0$. (Show this implication!)

3.2.4   Partition of Range of Integration

If q is a real number between the two limits, i.e., if $p < q < r$, then

$$\int_{p}^{r} f(t)\, dt = \int_{p}^{q} f(t)\, dt + \int_{q}^{r} f(t)\, dt. \tag{3.8}$$

piecewise
continuous
functions

which is a special case of Equation (3.6). This property is used to evaluate **piecewise continuous** functions, i.e., functions that have a finite number of discontinuities in the interval of integration. For instance, suppose $f(t)$ is defined to be

$$f(t) = \begin{cases} f_1(t) & \text{if } p < t < q_1, \\ f_2(t) & \text{if } q_1 < t < q_2, \\ f_3(t) & \text{if } q_2 < t < r, \end{cases}$$

where $f_1(t)$, $f_2(t)$, and $f_3(t)$ are, in general, totally unrelated (continuous) functions. Then one divides the interval of integration into three natural parts and writes

$$\int_{p}^{r} f(t)\, dt = \int_{p}^{q_1} f_1(t)\, dt + \int_{q_1}^{q_2} f_2(t)\, dt + \int_{q_2}^{r} f_3(t)\, dt.$$

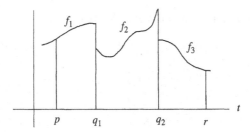

Figure 3.2: The integral is defined as long as there is only a finite number of discontinuities (jumps) in the function.

This is illustrated in Figure 3.2.

3.2.5   Transformation of Integration Variable

When evaluating an integral it is sometimes convenient to use a new variable
of integration of which the old one is a function. Call the new integration
variable y and assume that $t = h(y)$. Then we have

$$\int_a^b f(t)\,dt = \int_p^q f(h(y))\,h'(y)\,dy,\tag{3.9}$$

where p and q are the solutions to the two equations

$$a = h(p),\quad b = h(q).$$

Transformation of integration variable accompanies a change in the limits of integration.

Each of these two equations must have a *unique* solution, otherwise, the transformation of the integration variable will not be a valid procedure. This condition puts restrictions on the type of function h can be. Note that we have
essentially substituted $h(y)$ for t in the original integral including the differential $h'(y)\,dy$ for dt. It is vital to remember to **change the limits of
integration** when transforming variables.

3.2.6   Small Region of Integration

When the region of integration is small, in the sense that the *integrand does
not change much over the range of integration*, then the integral can be approximated by the product of integrand and the size of the range.[6] We thus
can write

$$\int_a^b f(t)\,dt \approx (b-a)f(t_0),\tag{3.10}$$

When is the region of integration small?

where t_0 is a number between a and b, mostly taken to be the midpoint of
the interval (a,b).

---

[6]This is simply a restatement of Equation (3.5) for the case of one variable.

3.2.7   Integral and Absolute Value

A useful property of integrals that we shall be using sometimes is

$$\left| \int_a^b f(t)\, dt \right| \le \int_a^b |f(t)|\, dt. \tag{3.11}$$

This should be clear once we realize that an integral is the limit of a sum and the absolute value of a sum is always less than or equal to the sum of the absolute values.

3.2.8   Symmetric Range of Integration

By a symmetric range of integration, we mean a range that has 0—the origin—as its midpoint. For certain functions, partitioning such a range into two equal pieces can simplify the evaluation of the integral considerably. So, let us write

$$\int_{-T}^{+T} f(t)\, dt = \int_{-T}^{0} f(t)\, dt + \int_0^{+T} f(t)\, dt.$$

For the first integral, make a change of variable $t = -y$ to obtain

$$h(y) = -y \;\Rightarrow\; h'(y)\, dy = (-1)\, dy = -dy.$$

The limits of integration in y are determined by

$$h(-T) = y_{\text{lower}},\; h(0) = y_{\text{upper}} \;\Rightarrow\; y_{\text{lower}} = +T,\; y_{\text{upper}} = 0.$$

We therefore have

$$\int_{-T}^{0} f(t)\, dt = \int_{+T}^{0} f(-y)(-dy) = \int_0^{+T} f(-y)\, dy = \int_0^{+T} f(-t)\, dt,$$

where we have used the properties in Subsections 3.2.3 and 3.2.1. Combining our results and using the second property, we get

$$\int_{-T}^{+T} f(t)\, dt = \int_0^{+T} f(-t)\, dt + \int_0^{+T} f(t)\, dt$$

$$= \int_0^{+T} [f(t) + f(-t)]\, dt. \tag{3.12}$$

even and odd
functions defined

A real-valued function f is called **even** if $f(-x) = f(x)$, and **odd** if $f(-x) = -f(x)$. Thus, from Equation (3.12), we obtain

$$\int_{-T}^{+T} f(t)\, dt = \int_0^{+T} [f(t) + f(-t)]\, dt = 2 \int_0^{+T} f(t)\, dt \tag{3.13}$$

when f is even, and

$$\int_{-T}^{+T} f(t)\, dt = \int_0^{+T} [f(t) + f(-t)]\, dt = 0 \tag{3.14}$$

when it is odd.

3.2.9   Differentiating an Integral

We have seen that an integral can have an integrand which depends on a set of parameters, and that the result of integration will depend on these parameters. Thus, we can think of the integral as a function of those parameters, and in particular, we may want to know its derivative with respect to one of the parameters. Using the definition of integral as the limit of a sum, and the fact that the derivative of a sum is the sum of derivatives, it is easy to show that

$$\frac{\partial}{\partial x_i} \int_a^b f(x_1, x_2, \ldots, x_n, t)\, dt = \int_a^b \frac{\partial}{\partial x_i} f(x_1, x_2, \ldots, x_n, t)\, dt, \qquad (3.15)$$

where we have represented the list of parameters as $(x_1, x_2, \ldots, x_n)$. We can write exactly the same relation for the integral of Equation (3.4). Assuming that $\mathbf{r} = (x_1, x_2, \ldots, x_n)$, we have

$$\frac{\partial}{\partial x_i} h(\mathbf{r}) = \frac{\partial}{\partial x_i} \iint_\Omega f(\mathbf{r}, \mathbf{r}')\, dQ(\mathbf{r}') = \iint_\Omega \frac{\partial}{\partial x_i} f(\mathbf{r}, \mathbf{r}')\, dQ(\mathbf{r}'). \qquad (3.16)$$

In both cases the region of integration is assumed to be independent of x_i.

Restricting ourselves to single integrals,[7] we now consider the case where the limits of integration depend on some parameters. First, consider an integral of the form

$$\int_u^v f(t)\, dt$$

and treat the result as a function of the limits. So, let us write

$$F(u, v) \equiv \int_u^v f(t)\, dt \;\Rightarrow\; F(v, u) = -F(u, v)$$

and evaluate the partial derivative of F with respect to its arguments:

$$\frac{\partial F}{\partial u} \equiv \partial_1 F(u, v) = \lim_{\epsilon \to 0} \frac{F(u + \epsilon, v) - F(u, v)}{\epsilon}$$

$$= \lim_{\epsilon \to 0} \frac{\int_{u+\epsilon}^v f(t)\, dt - \int_u^v f(t)\, dt}{\epsilon} = -\lim_{\epsilon \to 0} \frac{\int_u^v f(t)\, dt + \int_v^{u+\epsilon} f(t)\, dt}{\epsilon}$$

$$= -\lim_{\epsilon \to 0} \frac{\int_u^{u+\epsilon} f(t)\, dt}{\epsilon} = -\lim_{\epsilon \to 0} \frac{\epsilon f(u_0)}{\epsilon} = -\lim_{\epsilon \to 0} f(u_0) = -f(u).$$

The last equality follows from the fact that as $\epsilon \to 0$, u_0, lying between u and $u + \epsilon$, will be squeezed to u. Note that the derivative above is independent of the second variable. To find the other derivative, we use the result obtained above and simply note that

$$\frac{\partial F(u, v)}{\partial v} = -\frac{\partial F(v, u)}{\partial v} = -\partial_1 F(v, u) = -\big(-f(v)\big) = f(v).$$

[7]Since all multiple integrals are reducible to single integrals, this restriction is not severe.

Putting these two results together, we can write

$$\frac{\partial}{\partial v} \int_u^v f(t)\, dt = f(v), \qquad \frac{\partial}{\partial u} \int_u^v f(t)\, dt = -f(u). \qquad (3.17)$$

In words,

> **Box 3.2.1.** *The derivative of an integral with respect to its upper (lower) limit equals the integrand (minus the integrand) evaluated at the upper (lower) limit.*

By evaluation, we mean replacing the *variable of integration*. If the integrand has parameters, they are to be left alone.

By combining Equations (3.15) and (3.17) we can derive the most general equation. So, assume that both u and v are functions of $(x_1, x_2, \ldots, x_n)$, and write

$$G(x_1, x_2, \ldots, x_n, u, v) \equiv \int_{u(x_1,x_2,\ldots,x_n)}^{v(x_1,x_2,\ldots,x_n)} f(x_1, x_2, \ldots, x_n, t)\, dt.$$

Then, using the chain rule. we get

$$D_i G \equiv \frac{\partial G}{\partial u}\frac{\partial u}{\partial x_i} + \frac{\partial G}{\partial v}\frac{\partial v}{\partial x_i} + \partial_i G,$$

total derivative   where $D_i G$ stands for the "total" derivative with respect to x_i. This means that the dependence of u and v on x_i is taken into account. In contrast, $\partial_i G$ is evaluated assuming that u and v are constants. We note that

$$\frac{\partial G}{\partial u} = \frac{\partial}{\partial u} \int_u^v f(x_1, x_2, \ldots, x_n, t)\, dt = -f(x_1, x_2, \ldots, x_n, u),$$

$$\frac{\partial G}{\partial v} = \frac{\partial}{\partial v} \int_u^v f(x_1, x_2, \ldots, x_n, t)\, dt = f(x_1, x_2, \ldots, x_n, v),$$

$$\frac{\partial G}{\partial x_i} = \frac{\partial}{\partial x_i} \int_u^v f(x_1, x_2, \ldots, x_n, t)\, dt = \int_u^v \frac{\partial}{\partial x_i} f(x_1, x_2, \ldots, x_n, t)\, dt,$$

where the partial derivative in the last equation treats u and v as constants. It follows that

> **Box 3.2.2.** *The most general formula for the derivative of an integral is*
>
> $$\frac{\partial}{\partial x_i} \int_{u(\mathbf{r})}^{v(\mathbf{r})} f(\mathbf{r}, t)\, dt = \frac{\partial v}{\partial x_i} f(\mathbf{r}, v) - \frac{\partial u}{\partial x_i} f(\mathbf{r}, u) + \int_{u(\mathbf{r})}^{v(\mathbf{r})} \frac{\partial}{\partial x_i} f(\mathbf{r}, t)\, dt,$$
>
> *where* $\mathbf{r} = (x_1, x_2, \ldots, x_n)$.

As indicated in Equation (2.16), it is common to ignore the difference between D_i and ∂_i; and the formula in Box 3.2.2 reflects this.

3.2.10 Fundamental Theorem of Calculus

A special case of Box 3.2.2 is extremely useful. Consider a function g of a single variable x. We want to find a function called the **primitive**, or **antiderivative**, or indefinite integral[8] whose derivative is g. This can be easily done using integrals. In fact using Box 3.2.2, we have

$$G(x) \equiv \int_a^x g(s)\,ds \;\Rightarrow\; \frac{dG}{dx} = \frac{d}{dx}\int_a^x g(s)\,ds = g(x), \qquad (3.18)$$

primitive (antiderivative) of a function

where a is an arbitrary constant. We can add an arbitrary constant to the RHS of the above equation and still get a primitive. Adding such a constant, evaluating both sides at $x = a$, and noting that the integral vanishes, we find that the constant must be $G(a)$. We, therefore, obtain

$$G(x) - G(a) = \int_a^x g(s)\,ds. \qquad (3.19)$$

Now suppose that $F(x)$ is *any* function whose derivative is $g(x)$. Then, from Equation (3.18), we see that

$$\frac{d}{dx}[G(x) - F(x)] = \frac{dG}{dx} - \frac{dF}{dx} = g(x) - g(x) = 0.$$

Therefore, $G(x) - F(x)$ must be a constant C. It now follows from (3.19) that

$$F(x) - F(a) = G(x) - C - [G(a) - C] = G(x) - G(a) = \int_a^x g(s)\,ds,$$

and we have

fundamental theorem of calculus

> **Box 3.2.3. (*Fundamental Theorem of Calculus*).** *Let $F(x)$ be any primitive of $g(x)$ defined in the interval (a, b), i.e., any function whose derivative is $g(x)$ in that interval. Then,*
>
> $$F(b) - F(a) = \int_a^b g(s)\,ds. \qquad (3.20)$$

The founders of calculus such as Barrow, Newton, and Leibniz thought of an integral as a sum. At the beginning no connection between integration and differentiation was established, and to obtain the result of an integral one had to go through the painstaking process of adding the terms of a (infinite) sum. It was later, that the founders of calculus realized (but did not prove) that the process of summation and taking limits was intimately connected

Connection between integrals and antiderivatives was not apparent at the time integration was introduced. It was discovered later.

---

[8]We would like to emphasize the concept of integral as the limit of a sum. Therefore, we think it is better to reserve the word "integral" for such sums and will avoid using the phrase "indefinite integral."

to the process of (anti) differentiation. In this respect, Equation (3.20) is indeed a *fundamental* result, because it eliminates the cumbersome labor of summation.

Another useful result is

$$G(x) - G(a) = \int_a^x g(s)\, ds = \int_a^x \frac{dG}{ds}(s)\, ds = \int_a^x dG. \qquad (3.21)$$

In words, *the integral of the differential of a physical quantity is equal to the quantity evaluated at the upper limit minus the quantity evaluated at the lower limit.*

Example 3.2.1. The properties mentioned above can be very useful in evaluating some integrals. Consider the integral $\int_{-\infty}^{\infty} e^{-t^2}\, dt$ whose value is known to be $\sqrt{\pi}$ (see Example 3.3.1). We want to use this information to obtain the integral $\int_{-\infty}^{\infty} t^2 e^{-t^2}\, dt$. First, we note that

$$\int_{-\infty}^{\infty} e^{-xt^2}\, dt = \sqrt{\frac{\pi}{x}}.$$

This can be shown readily by changing the variable of integration to $u = \sqrt{x}\, t$ and using the result of Example 3.3.1. Next, we differentiate both sides with respect to x and use Box 3.2.2 with $u = -\infty$ and $v = \infty$. We then get

$$LHS = \frac{\partial}{\partial x} \int_{-\infty}^{\infty} e^{-xt^2}\, dt = \int_{-\infty}^{\infty} \frac{\partial}{\partial x} e^{-xt^2}\, dt = \int_{-\infty}^{\infty} (-t^2) e^{-xt^2}\, dt$$

using derivative of integral to obtain new integral formulas from known integral formulas

for the LHS, and $\dfrac{\partial}{\partial x} \sqrt{\dfrac{\pi}{x}} = -\dfrac{1}{2} \dfrac{\sqrt{\pi}}{x^{3/2}}$ for the RHS. So

$$\int_{-\infty}^{\infty} t^2 e^{-xt^2}\, dt = \sqrt{\pi}\, \tfrac{1}{2} x^{-3/2} \qquad (3.22)$$

or, setting $x = 1$, $\int_{-\infty}^{\infty} t^2 e^{-t^2}\, dt = \frac{\sqrt{\pi}}{2}$.

We can obtain more general results. Differentiating both sides of Equation (3.22), we obtain

$$\int_{-\infty}^{\infty} t^4 e^{-xt^2}\, dt = \sqrt{\pi}\, \tfrac{1}{2} \tfrac{3}{2} x^{-5/2} = \sqrt{\pi}\, \frac{1 \cdot 3}{2^2} x^{-5/2}.$$

Continuing the process n times, we obtain

$$\int_{-\infty}^{\infty} t^{2n} e^{-xt^2}\, dt = \sqrt{\pi}\, \frac{1 \cdot 3 \cdot 5 \cdots (2n-1)}{2^n} x^{-(2n+1)/2}. \qquad (3.23)$$

In particular, if $x = 1$, we have

$$\int_{-\infty}^{\infty} t^{2n} e^{-t^2}\, dt = \sqrt{\pi}\, \frac{1 \cdot 3 \cdot 5 \cdots (2n-1)}{2^n}. \qquad \blacksquare$$

Example 3.2.2. Integrals involving only trigonometric functions are easy to evaluate:

$$\int_a^b \sin t\, dt = -\cos t \Big|_a^b = \cos a - \cos b,$$

$$\int_a^b \cos t\, dt = \sin t \Big|_a^b = \sin b - \sin a.$$

However, integrals of the form $I \equiv \int_a^b t^n \sin t \, dt$, in which n is a positive integer, are not as easy to evaluate although they occur frequently in applications. One can of course evaluate these integrals using integration by parts. But that is a tedious process. A more direct method of evaluation is to use the ideas developed above.

A pair of slightly more complicated trigonometric integrals which will be useful for our purposes is

$$\int_a^b \sin st \, dt = -\frac{1}{s} \cos st \Big|_a^b = \frac{\cos sa - \cos sb}{s},$$

$$\int_a^b \cos st \, dt = \frac{1}{s} \sin st \Big|_a^b = \frac{\sin sb - \sin sa}{s}. \tag{3.24}$$

If we differentiate both sides with respect to s, we can obtain the integrals we are after.[9] This is because each differentiation introduces one power of t in the integrand. For example if we are interested in I with $n = 1$, then we can differentiate the *second* equation in (3.24):

$$LHS = \frac{d}{ds} \int_a^b \cos st \, dt = \int_a^b \frac{\partial}{\partial s} (\cos st) \, dt = -\int_a^b t \sin st \, dt.$$

On the other hand,

$$RHS = \frac{\partial}{\partial s} \left(\frac{\sin sb - \sin sa}{s} \right)$$

$$= -\frac{\sin sb - \sin sa}{s^2} + \frac{b \cos sb - a \cos sa}{s}.$$

Setting $s = 1$ in these equations yields

$$\int_a^b t \sin t \, dt = \sin b - \sin a - b \cos b + a \cos a. \tag{3.25}$$

We can also find the primitive of functions of the form $x^n \sin x$. All we need to do is change b to x as suggested by Equation (3.18). For example, the primitive (indefinite integral) of $x \sin x$ is obtained by substituting x for b in Equation (3.25):

$$\int x \sin x \, dx = \sin x - \sin a - x \cos x + a \cos a = \sin x - x \cos x + C$$

because $-\sin a + a \cos a$ is simply a constant.   ■

Historical Notes

After a lull of almost two millennia, the subject of "exhaustion," like any other form of human intellectual activity, was picked up after the Renaissance. Johannes Kepler is reportedly the first one to begin work on finding areas, volumes, and centers of gravity. He is said to have been attracted to such problems because he noted the inaccuracy of methods used by wine dealers to find the volumes of their kegs.

Some of the results he obtained were the relations between areas and perimeters. For example, by considering the area of a circle to be covered by an infinite number of triangles, each with a vertex at the center, he shows that the area of a circle is $\frac{1}{2}$

---

[9]We can set $s = 1$ at the end if need be.

its radius times its circumference. Similarly, he regarded the volume of a sphere as the sum of a large number of small cones with vertices at the center. Since he knew the volume of each cone to be $\frac{1}{3}$ its height times the area of its base, he concluded that the volume of a sphere should be $\frac{1}{3}$ its radius times the surface area.

Galileo used the same technique as Kepler to treat the uniformly accelerated motion and essentially arrived at the formula $x = \frac{1}{2}at^2$. They both regarded an area as the sum of infinitely many lines, and a volume as the sum of infinitely many planes, without questioning the validity of manipulating infinities. Galileo regarded a line as an *indivisible* element of area, and a plane as an indivisible element of volume.

Influenced by the idea of "indivisibles," **Bonaventura Cavalieri**, a pupil of Galileo and professor in a lyceum in Bologna, took up the study of calculus upon Galileo's recommendation. He developed the ideas of Galileo and others on indivisibles into a geometrical method and in 1635 published a book on the subject called *Geometry Advanced by a thus far Unknown Method, Indivisible of Continua*.

Cavalieri joined the religious order Jesuati in Milan in 1615 while he was still a boy. In 1616 he transferred to the Jesuati monastery in Pisa. His interest in mathematics was stimulated by Euclid's works and after meeting Galileo, considered himself a disciple of the astronomer. The meeting with Galileo was set up by Cardinal Federico Borromeo who saw clearly the genius in Cavalieri while he was at the monastery in Milan.

Bonaventura
Cavalieri
1598–1647

Cavalieri was largely responsible for introducing logarithms as a computational tool in Italy. The tables of logarithms which he published included logarithms of trigonometric functions for use by astronomers. Cavalieri also wrote on conic sections, trigonometry, optics, and astronomy. He showed by his methods of indivisibles that, in the modern notation,

$$\int_0^a x^n \, dx = \frac{a^{n+1}}{n+1}$$

for positive integral values of n up to 9.

The next important step in the development of integral calculus began when the seventeenth-century mathematicians generalized the Greek method of exhaustion. Whereas this method requires different rectilinear approximation for different geometrical figures, the new generation of mathematicians approximated the area under any curve by a large number of rectangles of equal width (much like it is done today), summed up the areas, and neglected the "small corrections" in the sum. Using essentially this kind of summation technique, Fermat showed the above integral formula for all rational n except -1 before 1636.

Before Newton and Leibniz, the man who did most to replace the geometrical techniques with analytical ones in calculus was **John Wallis**. Although he did not begin to learn mathematics until he was about twenty, he became professor of geometry at Oxford and the ablest British mathematician of the century, next to Newton. One of Wallis's notable results, obtained while he was trying to find the area of a circle analytically, was a new formula for π. He calculated the area bounded by the axes and the curves $y = (1 - x^2)^n$ for $n = 0, 1, 2, \ldots$. Then by interpolation and further complicated reasoning he related the area of a unit circle $y = (1 - x^2)^{1/2}$ to the previous areas and showed that

John Wallis
1616–1703

$$\frac{\pi}{2} = \frac{2.2.4.4.6.6.8.8 \ldots}{1.3.3.5.5.7.7.9 \ldots}$$

3.3   Guidelines for Calculating Integrals

The number of situations in which integrals are used is unlimited, and we shall see many examples of such usage in this chapter and throughout the book. Before embarking on specific examples, let us summarize some guidelines which will be helpful in applying integrals in physical problems:

- Make sure you understand what physical quantity you are trying to calculate. Instead of searching randomly for formulas, think about the problem and let *it* determine the formulas.

  Let the problem determine formulas!

- Determine which coordinate system is most suited for the problem. Then place the origin and orient the axes in such a way that the problem takes the simplest form. Usually spherical coordinates are suited for regions of integration which are symmetric about a single point. If there is a natural "axis" associated with the problem, then cylindrical coordinates are useful, and if the region of integration is in the shape of a rectangular box, Cartesian coordinates may be most suitable. If there is no obvious symmetry, then any one of the systems is just as good (or just as bad).

  Choose coordinates, origin, and orientation of axes wisely!

- Write down the *local* formula first, i.e., confine the problem to a small region and write the formula, for instance, in terms of $dQ(\mathbf{r}')$, $dm(\mathbf{r}')$, etc., then put the formula inside the integral. Do this in a coordinate-independent way first. All physical laws are written with no reference to a particular coordinate system, anyway.

  Write the local formula, then put it inside the integral.

- Now express the formula in terms of the coordinates you have chosen. When dealing with vector quantities, pay particular attention to unit vectors *whose directions depend on the integration point.* They cannot in general be taken out of the integral sign (see Section 3.3.2 for details).

- Determine the limits of integration. In a typical situation, if you have chosen a good coordinate system, placed the origin properly, and oriented the axes nicely, then the limits of integration should be easy to write.

- Never take anything out of the integral unless you are absolutely sure that it is independent of the integration variables. This is easily said, but most often also easily forgotten.

  Never take anything out of the integral unless....

- Once you have evaluated the integrals and found the physical quantity you are after, try to express your result in a coordinate-free language. This is not, in general, easy, but in special circumstances you can immediately guess the coordinate-free form of the result.

- As a general rule—valid in all physical calculations—check your final answer for correct dimensions. The dimension of the LHS must match that of the RHS.

  Always check the dimension of your final result!

3.3.1   Reduction to Single Integrals

Most integrals encountered in physics are multidimensional. Thus, it is important to know how to evaluate multiple integrals. Let us concentrate on triple integration, and for definiteness, let us assume that the integration variables are actual coordinates in a Cartesian coordinate system.[10] The most general integral, namely Equation (3.4), will then be rewritten as

$$\iint_\Omega f(\mathbf{r}, \mathbf{r}')\, dQ(\mathbf{r}') \equiv \iint_\Omega f(\mathbf{r}, x', y', z')\, dQ(x', y', z')$$

$$= \iiint_V f(\mathbf{r}, x', y', z')\rho_Q(x', y', z')\, dx'\, dy'\, dz',$$

where we have reexpressed dQ in terms of some density. The region of integration V may have to be divided into a number of other more easily integrable regions. However, in most applications, *by a good choice of the order of integration*, one can avoid such division. Let us assume that by integrating the z' variable first, we will not need to divide the region. The z' integration is a single integral and is done by keeping x' and y' constant. To find the upper limit of this integral, we pick **an arbitrary point**[11] in the region, fix its first two coordinates, move "up" until we hit the boundary of V at a point. The third coordinate of **this boundary point**, when expressed in terms of x' and y', will be the upper limit of the z' integration. The lower limit is obtained similarly. In most cases, V is bounded by a given upper surface of the form $z = g(x, y)$, and a lower surface of the form $z = h(x, y)$ as shown in Figure 3.3.

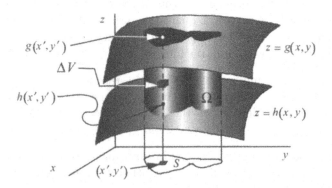

Figure 3.3: The limits of the first integration of a triple integral are defined by two surfaces.

---

[10]Recall that the integration variables, although considered as "coordinates of a point," need not be an actual geometric point in space. They could, for instance, be a set of thermodynamical variables describing a thermodynamical system.

[11]A common mistake at this stage is to pick a *special* point. To make sure that you have picked an arbitrary point, go through the following process using the point chosen, then pick a different point, go through the process and see if you obtain the same result for the upper and lower limits of the integral.

Thus, since the first two coordinates of the boundary points are x' and y', the upper limit will be $g(x', y')$ and the lower limit will be $h(x', y')$. We thus write the integral as

$$\iint_\Omega f(\mathbf{r}, \mathbf{r}') \, dQ(\mathbf{r}') = \iint_S dx' \, dy' \int_{h(x',y')}^{g(x',y')} f(\mathbf{r}, x', y', z') \rho_Q(x', y', z') \, dz',$$

where S is the projection of V on the xy-plane. For S to be useful, it must have the following property: Every point of the upper and lower boundaries of V has one and only one image in S, and no two points of the upper (or lower) boundary project onto the same point in S. If this property is not fulfilled, then we must choose another coordinate as our first integration variable, or, if this does not work, divide the region of integration into pieces for each one of which this property holds.

Let us assume that the property holds for S, and that we can do the integral in z'. The result of this integration is a complete elimination of the z'-coordinate and the reduction of the triple integral down to a double integral. To be more specific, assume that the primitive of the integrand, as a function of z', is $F(\mathbf{r}, x', y', z')$, i.e., that

$$\frac{\partial F}{\partial z'} = f(\mathbf{r}, x', y', z') \rho_Q(x', y', z').$$

Then, the z' integration yields

$$\int_{h(x',y')}^{g(x',y')} f(\mathbf{r}, x', y', z') \rho_Q(x', y', z') \, dz'$$
$$= F(\mathbf{r}, x', y', g(x', y')) - F(\mathbf{r}, x', y', h(x', y')) \equiv G(\mathbf{r}, x', y'),$$

where the last line defines G. The triple integration has now been reduced to a double integral, and we have

$$\iint_\Omega f(\mathbf{r}, \mathbf{r}') \, dQ(\mathbf{r}') = \iint_S dx' \, dy' \, G(\mathbf{r}, x', y').$$

We follow the same procedure as above to do the double integral. Once again, the region of integration S may have to be divided into a number of other more easily integrable regions. However, let us assume that by integrating the x' variable first, we will not need to divide the region. The x' integration is again a single integral and is done by keeping y' constant. To find the upper limit of this integral, we pick *an arbitrary point* in S, fix its second coordinate, and move "to the right" until we hit the boundary of S at a point. The first coordinate of *this boundary point*, when expressed in terms of y', will be the upper limit of the x' integration. The lower limit is obtained by "moving to the left." Once again, in most cases, S is bounded by a given upper curve of the form $x = v(y)$, and a lower curve of the form $x = u(y)$ (see Figure 3.4). Thus, since the second coordinate of both boundary points is y',

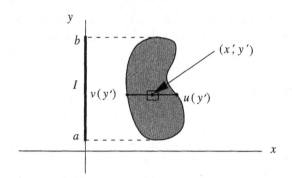

Figure 3.4: The limits of the second integration of a triple integral are defined by two curves.

the upper limit will be $v(y')$ and the lower limit will be $u(y')$. We thus write the integral as

$$\iint_{\Omega} f(\mathbf{r}, \mathbf{r}') \, dQ(\mathbf{r}') = \int_{I} dy' \int_{u(y')}^{v(y')} dx' \, G(\mathbf{r}, x', y'),$$

where I is the projection of S on the y-axis. For I to be useful, it must have the same property as S, namely: Every point of the right and left boundaries of S has one and only one image in I, and no two points of the right (or left) boundary project onto the same point in I. If this property is not fulfilled, then we must choose y' as our first integration variable, or, if this does not work, divide the region of integration into pieces for each one of which this property holds. Assuming that I satisfies this property, and that the primitive of the integrand, as a function of x', is $W(\mathbf{r}, x', y')$, i.e, that

$$\frac{\partial W}{\partial x'} = G(\mathbf{r}, x', y'),$$

we get

$$\int_{u(y')}^{v(y')} G(\mathbf{r}, x', y') \, dx' = W(\mathbf{r}, v(y'), y') - W(\mathbf{r}, u(y'), y') \equiv H(\mathbf{r}, y'),$$

where the last line defines H. The triple integration has now been reduced to a single integral, and we have

$$\iint_{\Omega} f(\mathbf{r}, \mathbf{r}') \, dQ(\mathbf{r}') = \int_{I} H(\mathbf{r}, y') \, dy' = \int_{a}^{b} H(\mathbf{r}, y') \, dy',$$

where a and b are the end points of the interval I.

Sometimes the inverse of the foregoing operation is useful whereby a single integral is turned into a multiple integral. This happens when the integrand is given in terms of an integral. To be specific, suppose in the integral

$$I \equiv \int_a^b g(x)\, dx,$$

$g(x)$ is given by

$$g(x) = \int_u^v h(x,t)\, dt,$$

where u and v could be functions of x. Then, the original integral can be written as

$$I = \int_a^b \left\{ \int_u^v h(x,t)\, dt \right\} dx = \int_a^b \int_u^v h(x,t)\, dt\, dx.$$

Example 3.3.1. A historical example of this inverse operation is the evaluation of the integral

$$I \equiv \int_0^\infty e^{-x^2}\, dx.$$

integral of e^{-x^2} over the positive real line

As the reader attempting to solve this integral will soon find out, it is impossible to find a primitive of the integrand. However, with

$$I^2 = \underbrace{\int_0^\infty e^{-x^2}\, dx}_{=I} \underbrace{\int_0^\infty e^{-y^2}\, dy}_{=I} = \int_0^\infty \int_0^\infty \underbrace{e^{-x^2} e^{-y^2}}_{=e^{-(x^2+y^2)}}\, dx\, dy$$

we end up with an integration over the first quadrant of the xy-plane which opens up the possibility of using other coordinate systems. In polar coordinates, the integrand becomes e^{-r^2} and the Cartesian element of area $dx\, dy$ becomes the element of area in polar coordinates, namely $r\, dr\, d\theta$. The limits of integration correspond to the first quadrant, with the range of θ being from 0 to $\pi/2$ and that of r being from 0 to infinity. This leads to

$$I^2 = \int_0^{\pi/2} \int_0^\infty e^{-r^2} r\, dr\, d\theta = \underbrace{\int_0^{\pi/2} d\theta}_{=\pi/2} \underbrace{\int_0^\infty e^{-r^2} r\, dr}_{=-\frac{1}{2}e^{-r^2}\big|_0^\infty}.$$

This shows that $I^2 = \pi/4$ and, therefore, $I = \sqrt{\pi}/2$. The reader may verify that

$$\int_{-\infty}^\infty e^{-x^2}\, dx = \sqrt{\pi} \tag{3.26}$$

by either invoking the evenness of the integrand or starting from scratch as done above. ∎

3.3.2 Components of Integrals of Vector Functions

Many calculations involve an integrand which is a vector and whose integration also leads to a vector. Let us write this as

finding the components of the vector resulting from integration of another vector

$$\mathbf{F}(\mathbf{r}) = \iint_\Omega \mathbf{A}(\mathbf{r},\mathbf{r}')\, dQ(\mathbf{r}')$$

$$= \iint_\Omega [A_1(\mathbf{r},\mathbf{r}')\hat{\mathbf{e}}_1(\mathbf{r}') + A_2(\mathbf{r},\mathbf{r}')\hat{\mathbf{e}}_2(\mathbf{r}') + A_3(\mathbf{r},\mathbf{r}')\hat{\mathbf{e}}_3(\mathbf{r}')]\, dQ(\mathbf{r}'),$$

where A_1, A_2, and A_3 are the components of the vector $\mathbf{A}$ along the *mutually perpendicular* unit vectors $\hat{\mathbf{e}}_1$, $\hat{\mathbf{e}}_2$, and $\hat{\mathbf{e}}_3$, respectively.[12] Note that these unit vectors are, in general, functions of the variables of integration, and that

> **Box 3.3.1.** *The geometry of the distribution of the source determines the most convenient variables of integration (coordinate variables).*

To find the component of $\mathbf{F}(\mathbf{r})$ along any unit vector $\hat{\mathbf{e}}_a$, one simply takes the dot product of $\mathbf{F}(\mathbf{r})$ with $\hat{\mathbf{e}}_a$. Thus,

$$F_a(\mathbf{r}) \equiv \hat{\mathbf{e}}_a \cdot \mathbf{F}(\mathbf{r}) = \hat{\mathbf{e}}_a \cdot \iint_\Omega \mathbf{A}(\mathbf{r}, \mathbf{r}') \, dQ(\mathbf{r}') = \iint_\Omega [\hat{\mathbf{e}}_a \cdot \mathbf{A}(\mathbf{r}, \mathbf{r}')] \, dQ(\mathbf{r}')$$

$$\equiv \iint_\Omega [A_1(\mathbf{r}, \mathbf{r}') f_1(\mathbf{r}') + A_2(\mathbf{r}, \mathbf{r}') f_2(\mathbf{r}') + A_3(\mathbf{r}, \mathbf{r}') f_3(\mathbf{r}')] \, dQ(\mathbf{r}'), \quad (3.27)$$

where $f_1(\mathbf{r}') \equiv \hat{\mathbf{e}}_a \cdot \hat{\mathbf{e}}_1$, $f_2(\mathbf{r}') \equiv \hat{\mathbf{e}}_a \cdot \hat{\mathbf{e}}_2$, and $f_3(\mathbf{r}') \equiv \hat{\mathbf{e}}_a \cdot \hat{\mathbf{e}}_3$. Once these dot products are expressed in terms of the variables of integration, the integral becomes an ordinary integral which, in principle, can be performed using the guidelines above.

> **Box 3.3.2.** *In practice, $\hat{\mathbf{e}}_a$ is one of the unit vectors of some convenient coordinate system* which need not be the same as the coordinate system used for integration.

For example, one may be interested in the Cartesian components of the gravitational field of a spherical distribution of mass. In that case, one uses spherical coordinates for integration and the unit vectors inside the integral, and $\hat{\mathbf{e}}_x$, $\hat{\mathbf{e}}_y$, or $\hat{\mathbf{e}}_z$ for $\hat{\mathbf{e}}_a$. We shall illustrate this point extensively with numerous examples scattered throughout this chapter.

Historical Notes

By the time Newton entered the scene, an immense amount of knowledge of calculus had accumulated. In his book *Lectiones Geometricae*, Barrow, for example, shows a method of finding tangents, theorems on the differentiation of products and quotients of functions, change of variables in a definite integral, and even differentiation of implicit functions. So, why, one may wonder, is the word "calculus" so much attached to Newton and Leibniz? The answer is in these two men's recognition of the generality of the methods of calculus, and, more importantly, their emphasis on the newly discovered *analytic geometry*.

Isaac Newton was born in the hamlet of Woolsthorpe, England, two months after his father's death. His mother, in need of help for the management of the family farm, wanted Isaac to pursue a farming career. However, Isaac's uncle persuaded him to enter Trinity College, Cambridge University. Newton took the entrance exam

---

[12]These unit vectors are usually those of a convenient coordinate system.

and was accepted to the College in 1661 with a deficiency in Euclidean geometry. Apparently receiving very little stimulation from his teachers, except possibly Barrow, he studied Descartes's *Géométrie*, as well as the works of Copernicus, Kepler, Galileo, Wallis, and Barrow, by himself.

Upon his graduation, Newton had to leave Cambridge due to the widespread plague in the London area to spend the next eighteen months, during 1665 and 1666, in the quiet of his family farm at Woolsthorpe. These eighteen months were the most productive of his (as well as any other scientist's) life. In his own words:

Isaac Newton
1642–1727

> In the beginning of 1665 I found the ...rule for reducing any dignity
> [power] of binomial to a series.[13] The same year, in May, I found the method
> of tangents ... and in November the direct method of Fluxions [the elements
> of what is now called differential calculus], and the next year in January had
> the theory of Colours, and in May following I had entrance into the inverse
> method of Fluxions [integral calculus], and in the same year I began to think
> of gravity extending to the orb of the Moon ... and ... compared the force
> requisite to keep the Moon in her orb with the force of gravity at the surface
> of the Earth.

Newton spent the rest of his scientific life developing and refining the ideas conceived at his family farm. At the age of 26 he became the second Lucasian professor of mathematics at Cambridge replacing Isaac Barrow who stepped aside in favor of Newton. At 30 he was elected a Fellow of the Royal Society, the highest scientific honor in England.

Newton often worked until early morning, kept forgetting to eat his meals, and when he appeared, once in a while, in the dining hall of the college, his shoes were down at the heels, stockings untied, and his hair scarcely combed. Being always absorbed in his thoughts, he was very naive and impractical concerning daily routines. It is said that once he made a hole in the door of his house for his cat to come in and out. When the cat had kittens, he added some smaller holes in the door!

Newton did not have a pleasant personality, and was often involved in controversy with his colleagues. He quarreled bitterly with *Robert* Hooke (founder of the theory of elasticity and the discoverer of Hooke's law) concerning his theory of color as well as priority in the discovery of the universal law of gravitation. He was also involved in another priority squabble with the German mathematician Gottfried Leibniz over the development of calculus. With Christian Huygens, the Dutch physicist, he got into an argument over the theory of light. Astronomer John Flamsteed, who was hardly on speaking terms with Newton, described him as "insidious, ambitious, excessively covetous of praise, and impatient of contradictions ... a good man at the bottom but, through his nature, suspicious."

De Morgan says that "a morbid fear of opposition from others ruled his whole life." Because of this fear of criticism, Newton hesitated to publish his works. When in 1672 he did publish his theory of light and his philosophy of science, he was criticized by his contemporaries. Newton decided not to publish in the future, a decision that had to be abandoned frequently.

His theory of gravity, although germinated in 1665 under the influence of works by Hooke and Huygens, was not published until much later, partly because of his fear of criticism. Another reason for this hesitance in publishing this result was his

---

[13]Newton is talking about the binomial theorem here.

lack of proof that the gravitational attraction of a solid sphere acts as if the sphere's map were concentrated at the center. So, when his friend Edmund Halley urged him in 1684 to publish his results, he refused. However, in 1685 he showed that a sphere whose density varies only with distance to the center does in fact attract particles as though its mass were concentrated at the center, and agreed to write up his work. Halley then assisted Newton editorially and paid for the publication. The first edition of *Philosophiae Naturalis Principia Mathematica* appeared in 1687, and the *Newtonian* age began.

3.4   Problems

3.1. Use Equation (3.7) to show that $\int_a^a f(t)\,dt = 0$.

3.2. In Equation (3.8), it was assumed that $p < q < r$. Show that the equation holds even if q is not between p and r.

3.3. For each of the following integrals make the given change of variables:

(a) $\int_0^8 t\,dt$, $t = y^3$.   (b) $\int_0^1 \frac{dt}{1+t^2}$, $t = \tan y$, $0 \le y \le \pi/2$.

(c) $\int_0^1 \frac{dt}{1+t}$, $t = \ln y$.   (d) $\int_1^\infty \frac{t\,dt}{1+t^3}$, $t = \frac{1}{y}$.

3.4. By a suitable change of variables, show the following integral identities:

(a) $\int_{-\infty}^\infty \frac{dt}{(a^2+t^2)^{3/2}} = \frac{2}{a^2}\int_0^{\pi/2}\cos t\,dt$.   (b) $\int_0^\infty \frac{dt}{(1+t)^2} = \int_0^1 dt$.

3.5. If

$$g(x) = \int_{x^2-1}^{\sin(\pi x)} \left\{\cos[\pi(t+x)]\right\} e^{-t^4 \sin^2[(\pi/2)\ln(tx^2+1)]}\,dt,$$

find $g'(1)$.

3.6. Suppose that $F(x) = \int_0^{\cos x} e^{xt^2}\,dt$, $G(x) = \int_0^{\cos x} t^2 e^{xt^2}\,dt$, and $H(x) = G(x) - F'(x)$. Find $H(x)$ in terms of elementary functions. Show that $H(\pi/4) = e^{\pi/8}/\sqrt{2}$.

3.7. Suppose that $F(x) = \int_0^{\sin x}\ln(\cos^2 x + t^2 + 1)\,dt$, $G(x) = \int_0^{\sin x}(\cos^2 x + t^2 + 1)^{-1}dt$, and $H(x) = F'(x) + 2\sin x \cos x G(x)$. Find $H(x)$ in terms of elementary functions. Show that $H(\pi/3) = \ln 2/2$.

3.8. Evaluate the *derivative* of the following integrals with respect to x at the given values of x:

(a) $\int_0^x e^{-t^2}\,dt$ at $x = 1$.   (b) $\int_{-3}^x \cos t\,dt$ at $x = \pi$.

(c) $\int_{-\infty}^{\sqrt{\cos(x/3)}} e^{-t^2}\,dt$ at $x = \pi$.   (d) $\int_0^{x^2}\cos(\sqrt{s})\,ds$ at $x = \pi$.

3.9. Find the numerical value of the *derivative* of the following two integrals at $x = 1$:

(a) $\int_0^{\ln x} e^{-x(t^2-2)} dt$.   (b) $\int_a^{x^2+a-1} \sin\left[\frac{\pi x e^{-t^2}}{2e^{-(x^2+a-1)^2}}\right] dt$.

3.10. Write the derivatives with respect to x of the following integrals in terms of other integrals. *Do not try to evaluate the integrals.*

(a) $\int_a^b \ln(1 + sx) \, ds$.   (b) $\int_a^b \frac{dt}{t^2+x^2}$.   (c) $\int_0^1 \sqrt{x^2 + a^2 - 2ax\cos t} \, dt$.

3.11. Differentiate $\int_{-\infty}^{\infty} dt/(z+t^2) = \pi/\sqrt{z}$ with respect to z to show that

(a) $\int_{-\infty}^{\infty} \frac{dt}{(1+t^2)^2} = \frac{\pi}{2}$.   (b) $\int_{-\infty}^{\infty} \frac{dt}{(1+t^2)^3} = \frac{3\pi}{8}$.

3.12. Using the method of Example 3.2.2, find the following integrals:

(a) $\int_a^b t^2 \sin t \, dt$.   (b) $\int_a^b t^3 \sin t \, dt$.   (c) $\int_a^b t^4 \sin t \, dt$.

(d) $\int_a^b t^2 \cos t \, dt$.   (e) $\int_a^b t^3 \cos t \, dt$.   (f) $\int_a^b t^4 \cos t \, dt$.

In each case calculate the primitive of the integrand and verify your answer by differentiating the primitive.

3.13. Find the integral

$$\Gamma(n+1) = \int_0^{\infty} t^n e^{-t} dt$$

by first evaluating the integral

$$\int_0^{\infty} e^{-xt} dt$$

and then differentiating the result n times, and setting $x = 1$ at the end. Can you see why $\Gamma(n+1)$ is called the **factorial function?**

3.14. Sketch each of the following integrands to decide whether the approximation to the integral is good or not.

(a) $\int_{-0.1}^{0.1} \frac{dt}{10+t^2} \approx 0.02$.   (b) $\int_{-0.1}^{0.1} \frac{dt}{0.001+t^2} \approx 200$.

(c) $\int_{-0.1}^{0.1} \cos(5\pi x) \, dx \approx 0.2$.   (d) $\int_{-0.1}^{0.1} \cos\frac{\pi x}{10} \, dx \approx 0.2$.

(e) $\int_{-0.1}^{0.1} e^{-100t^2} dt \approx 0.2$.   (f) $\int_{-0.1}^{0.1} e^{-t^2/100} dt \approx 0.2$.

3.15. Show that if a function is even (odd), then its derivative is odd (even).

3.16. Use the result of Example 3.3.1 to show that

$$\int_{-\infty}^{\infty} e^{-xt^2} dt = \sqrt{\frac{\pi}{x}}.$$

3.17. By differentiating the electrostatic potential

$$\Phi(\mathbf{r}) = \int_{\Omega} \frac{k_e \, dq(\mathbf{r}')}{|\mathbf{r} - \mathbf{r}'|}$$

with respect to x, y, and z, and assuming that Ω is independent of x, y, and z, show that the electric field

$$\mathbf{E}(\mathbf{r}) = \int_{\Omega} \frac{k_e \, dq(\mathbf{r}')}{|\mathbf{r} - \mathbf{r}'|^3} (\mathbf{r} - \mathbf{r}')$$

can be written as

$$\mathbf{E} = -\frac{\partial \Phi}{\partial x} \hat{\mathbf{e}}_x - \frac{\partial \Phi}{\partial y} \hat{\mathbf{e}}_y - \frac{\partial \Phi}{\partial z} \hat{\mathbf{e}}_z.$$

Chapter 4

Integration: Applications

The preceding chapter introduced integration and dealt with its formal aspects. It also gave some general guidelines concerning the calculation and manipulation of integrals, in particular how to reduce the process of multiple integration to a number of single integrations. In this chapter, we apply the formalism of the previous chapter to concrete examples.

4.1 Single Integrals

This section is devoted to the simple but important case of single integrals with examples from mechanics, electrostatics and gravity, and magnetostatics. Generally, we encounter problems which are defined and set up in a single dimension leading to integrals that have a single variable to be integrated.

4.1.1 An Example from Mechanics

In our discussion of primitive, Equation (3.18) clearly shows that integration can be interpreted as the inverse of differentiation. Thus, if we know the functional form of the derivative of a quantity, we should be able to express the quantity in terms of an integral.

Velocity is the derivative of displacement. So, we seek to write displacement in terms of an integral of velocity. This is easily done as follows:[1]

$$\frac{d\mathbf{r}}{dt} = \mathbf{v}(t) \;\Rightarrow\; \frac{d\mathbf{r}}{ds} = \mathbf{v}(s) \;\Rightarrow\; d\mathbf{r} = \mathbf{v}(s)\,ds.$$

Integrating both sides from 0 to t, we get

$$\int_0^t d\mathbf{r} = \int_0^t \mathbf{v}(s)\,ds \;\Rightarrow\; \mathbf{r}(t) - \mathbf{r}_0 = \int_0^t \mathbf{v}(s)\,ds, \qquad (4.1)$$

where $\mathbf{r}_0 = \mathbf{r}(0)$, and we used Equation (3.21).

---

[1] As cautioned below, we change t to s because we anticipate using t as the upper limit of the integral.

There is an alternative derivation of the last formula which relies directly on the definition of integral. Since the velocity of the particle is changing, we cannot find the displacement by simple multiplication with time. However, if we divide the time interval (from 0 to t) into N small subintervals, and concentrate on the motion of the particle in each subinterval, then each displacement can be approximated by the product of velocity and the small time-interval, and the total displacement $\mathbf{r}(t) - \mathbf{r}_0$ will be simply the sum of all such displacements. This is summarized as

$$\mathbf{r}(t) - \mathbf{r}_0 \approx \sum_{i=1}^{N} \mathbf{v}(s_i)\, \Delta s_i$$

which, in the limit of larger and larger N, gives

$$\mathbf{r}(t) - \mathbf{r}_0 = \int_0^t \mathbf{v}(s)\, ds.$$

Notice how careful we have been to avoid using the same variable for integration as well as the limit of integration. This is a practice the reader **important caution!** should constantly keep in mind. As a rule

> **Box 4.1.1.** (*Caution!*). *Never use the same symbol for the variable of an integral and its limits, or of an integral and of another integral of which the first integral is the integrand.*

The following example is a good illustration of the significance of the concept of an integral and the rule in the Box above.

Example 4.1.1. In mechanics, Newton's second law places special importance on acceleration,[2] and a knowledge of acceleration is normally sufficient to solve a mechanical problem, i.e., find displacement as a function of time. A particular example of this situation is when acceleration is known as a function of time, in which case we can immediately find the velocity in exact analogy with Equation (4.1). We thus have

$$\mathbf{v}(t) - \mathbf{v}_0 = \int_0^t \mathbf{a}(s)\, ds \;\Rightarrow\; \mathbf{v}(t) = \mathbf{v}_0 + \int_0^t \mathbf{a}(s)\, ds.$$

Notice how the argument of $\mathbf{v}$ is the same as the upper limit of integration. Now that we have velocity, we can substitute it in Equation (4.1) to find the displacement. This gives

$$\mathbf{r}(t) - \mathbf{r}_0 = \int_0^t \left\{ \mathbf{v}_0 + \int_0^s \mathbf{a}(u)\, du \right\} ds$$

or

$$\mathbf{r}(t) = \mathbf{r}_0 + \mathbf{v}_0 t + \int_0^t ds \int_0^s \mathbf{a}(u)\, du.$$

---

[2]Because the second law of motion connects acceleration and the cause of motion, force.

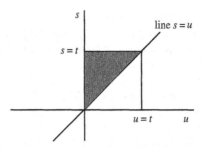

Figure 4.1: The region of integration for calculating position as a double integral.

In the double integral, it is understood that the u-integration is to be done first, followed by the s-integration. As the last double integral suggests, the region of integration, in the us-plane, is a right triangle bounded by the vertical axis (the s-axis, or $u = 0$), the line $u = s$, and the horizontal line $s = t$ as shown in Figure 4.1. It is convenient, in this case, to change the order of integration. The lower limit of the s-integral—the first integration—is u and the upper limit is t. Once this integral is done, the u-integral goes from 0 to t, as can easily be verified. We, therefore, have

given a definite double integral, one can reconstruct the region of integration in a plane.

$$\mathbf{r}(t) = \mathbf{r}_0 + \mathbf{v}_0 t + \int_0^t du \int_u^t \mathbf{a}(u)\,ds = \mathbf{r}_0 + \mathbf{v}_0 t + \int_0^t \mathbf{a}(u)\,du \int_u^t ds \qquad (4.2)$$

$$= \mathbf{r}_0 + \mathbf{v}_0 t + \int_0^t \mathbf{a}(u)(t - u)\,du = \mathbf{r}_0 + \mathbf{v}_0 t + t\int_0^t \mathbf{a}(u)\,du - \int_0^t u\,\mathbf{a}(u)\,du.$$

It is instructive for the reader to show that the first derivative of this expression gives the velocity and the second derivative the acceleration. ∎

Historical Notes

Two men are credited with the invention of calculus, **Newton** and Leibniz. Of course, as we have seen, the "invention" of calculus was a long process involving many generations of mathematicians. Nevertheless, Newton and Leibniz made great contributions to the subject and gave it a prominent role in the subsequent evolution of mathematical thought.

Gottfried Wilhelm Leibniz studied law and, after defending a thesis in logic, received a Bachelor of Philosophy degree. He wrote a second thesis on a universal method of reasoning in 1666 which completed his work for a doctorate in philosophy at the University of Altdorf and qualified him for a professorship. During the years 1670 and 1671, Leibniz wrote his first papers on mechanics and produced his calculating machine.

Leibniz was also involved in the politics of his time. In March, 1672, he went to Paris on a political mission as an ambassador of the Elector of Mainz. While in Paris, he made contact with notable mathematicians and scientists including **Huygens**. This stirred up his interest in mathematics, a subject that he knew nothing about prior to 1672. In 1673 he went to London and met other scientists and mathematicians including the secretary of the Royal Society of London.

While making his living as a diplomat, he delved further into mathematics and read **Descartes** and **Pascal**. In 1676 Leibniz was appointed librarian and councilor to the Elector of Hanover. Twenty-four years later the Elector of Brandenburg invited

Gottfried Wilhelm Leibniz 1646–1716

Leibniz to work for him in Berlin. While involved in many political maneuvers, including the succession of George Ludwig of Hanover to the English throne, Leibniz worked in many fields and his side activities encompassed an enormous range. He died in 1716, undeservedly neglected.

In addition to being a diplomat, Leibniz was a philosopher, lawyer, historian, and pioneer geologist. He did important work in logic, mathematics, optics, mechanics, hydrostatics, nautical science, and calculating machines. Although law was his profession, his contributions to mathematics and philosophy are among the best. He tried endlessly to reconcile the Catholic and Protestant faiths. He founded the Berlin Academy in 1700. He criticized the universities for being "monkish" and charged that they possessed learning but no judgment and were absorbed in trifles. Instead he urged that true knowledge—mathematics, physics, chemistry, anatomy, botany, zoology, history, and geography be pursued. He favored the German language over Latin because Latin was tied to the older, useless thought. Men mask their ignorance, he said, by using the Latin language to impress people.

His numerous mathematical notes on differentiation and integration is full of novel ideas. His notations were quite ingenious: He introduced the notation dy/dx for the derivative and $\int$ for the integral. He recognized the operations of integration and differentiation as the inverse of one another.

4.1.2   Examples from Electrostatics and Gravity

In electrostatics or magnetostatics, one is sometimes interested in calculating the electric or magnetic field of a *linear* charge or current distribution. In electrostatics, one can imagine sprinkling electric charges on a thin piece of string and asking for the electric field of the charge distribution. In magnetostatics, one flows an electric current through a thin wire and asks for the resulting magnetic field. In general, the string or the wire, being a curve in space, has a **parametric equation** given, in Cartesian coordinates say, by $\big(f(t), g(t), h(t)\big)$, where f, g, and h are known functions of the parameter t. The problems of gravity are entirely analogous to those of electrostatics. The master equation of electrostatocs is Equation (3.3) which we reproduce here for convenience:

$$\mathbf{E} = \iint_{\Omega} \frac{k_e \, dq(\mathbf{r}')}{|\mathbf{r} - \mathbf{r}'|^3}(\mathbf{r} - \mathbf{r}'), \qquad \Phi = \iint_{\Omega} \frac{k_e \, dq(\mathbf{r}')}{|\mathbf{r} - \mathbf{r}'|}. \qquad (4.3)$$

Cartesian Coordinates

Let us assume that Cartesian coordinates are suitable for the problem, and we want to calculate the electrostatic field at a point P with coordinates (x, y, z) as shown in Figure 4.2. We reduce the integrals in Equation (4.3) to single integrals by calculating their various parts entirely in terms of t. First we note that the source point P' lies on the curve, and therefore, its coordinates (x', y', z') are functions of t. Since we are using Cartesian coordinates, the *components* of the position vector of P' are the same as the source point's *coordinates*. Therefore, $\mathbf{r}' = x'\hat{\mathbf{e}}_x + y'\hat{\mathbf{e}}_y + z'\hat{\mathbf{e}}_z = \langle x', y', z' \rangle$.

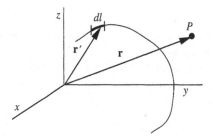

Figure 4.2: Electrostatic field of a general linear charge distribution.

The element of charge

$$dq(\mathbf{r}') = \lambda(\mathbf{r}')\, dl(\mathbf{r}') = \lambda(\mathbf{r}')\sqrt{(dx')^2 + (dy')^2 + (dz')^2} \qquad (4.4)$$

turns into a function of t (times dt) after the substitutions:

$$x' = f(t), \qquad y' = g(t), \qquad z' = h(t),$$
$$dx' = f'(t)dt, \quad dy' = g'(t)dt, \quad dz' = h'(t)dt.$$

Similarly,

$$\mathbf{r} - \mathbf{r}' = x\hat{\mathbf{e}}_x + y\hat{\mathbf{e}}_y + z\hat{\mathbf{e}}_z - x'\hat{\mathbf{e}}_x - y'\hat{\mathbf{e}}_y - z'\hat{\mathbf{e}}_z$$
$$= (x - x')\,\hat{\mathbf{e}}_x + (y - y')\,\hat{\mathbf{e}}_y + (z - z')\,\hat{\mathbf{e}}_z \qquad (4.5)$$

and

$$|\mathbf{r} - \mathbf{r}'| = \sqrt{(x - x')^2 + (y - y')^2 + (z - z')^2},$$
$$|\mathbf{r} - \mathbf{r}'|^3 = \left[(x - x')^2 + (y - y')^2 + (y - y')^2\right]^{3/2}. \qquad (4.6)$$

Substituting all the above in Equation (4.3) yields an integral in t for $\mathbf{E}$ and another integral in t for Φ. The limits of these integrals are determined from the parametric equation of the curve describing the linear charge distribution.

As a general rule, in order to find the components of the field along a unit vector, we use Box 1.1.2, i.e., we take the dot product of the field with that unit vector. This involves taking the dot product of the *integrand* with the unit vector. In the case of Cartesian unit vectors, this procedure simply picks out the integral multiplying one of the unit vectors. For other coordinate systems, this is not the case, as we shall see shortly.

Box 4.1.2. *Although the geometry of the source (charge distribution) may dictate a particular coordinate system, the components of the field can be calculated in any coordinate system desired.*

Thus, by multiplying the integrand by $\hat{\mathbf{e}}_\rho$, $\hat{\mathbf{e}}_\varphi$, and $\hat{\mathbf{e}}_z$ and expressing the dot products $\hat{\mathbf{e}}_\rho \cdot \hat{\mathbf{e}}_x$, $\hat{\mathbf{e}}_\varphi \cdot \hat{\mathbf{e}}_x$, etc., in terms of Cartesian coordinates, we can obtain E_ρ, E_φ, and E_z as integrals over t. A similar derivation gives the electric potential Φ as an integral over t. Although a formula can be obtained for the components of the electric field for a general curve (see Problem 4.3), it is best to learn the formalism by an example.

Example 4.1.2. The simplest example of the general discussion above is a thin rod of length L that is uniformly charged with constant linear density λ. We want to find the electric field and the electrostatic potential at an arbitrary point P in space, as shown in Figure 4.3(a).

As discussed at the beginning of this section, it pays to choose one's coordinates wisely. Clearly, the rod defines an axis naturally. So, let us choose our z-axis to lie along the rod. Once this is done, we are free to move the origin up and down, and orient the x- and y-axes. Let us use this freedom to put the field (or observation) point P on the x-axis. We then have a situation depicted in Figure 4.3(b).

To continue, we need the parametric equation of the rod. Clearly, the x' and y' parts have the (unique) "parameterization" $x' = 0$ and $y' = 0$. There are many ways to parameterize the z' part of the curve. However, in situations involving only one coordinate, it is most natural to set that coordinate equal to the parameter t. So, we choose the following simple parameterization:

$$x' = 0, \ y' = 0, \ z' = t, \ a \le t \le a + L \equiv b.$$

Substituting this and $\mathbf{r} = x\hat{\mathbf{e}}_x$ in Equations (4.5) and (4.6) yields

$$\mathbf{r} - \mathbf{r}' = x\hat{\mathbf{e}}_x - t\hat{\mathbf{e}}_z,$$

as well as $|\mathbf{r} - \mathbf{r}'| = \sqrt{x^2 + t^2}$ and $|\mathbf{r} - \mathbf{r}'|^3 = (x^2 + t^2)^{3/2}$.

Putting all this in Equation (4.3) yields

$$\mathbf{E}(x, y, z) = \int_a^b \frac{k_e \lambda \, dt}{(x^2 + t^2)^{3/2}} (x\hat{\mathbf{e}}_x - t\hat{\mathbf{e}}_z) \, dt \tag{4.7}$$

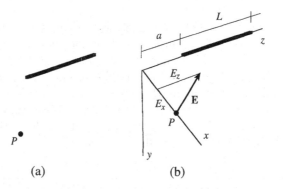

(a)   (b)

Figure 4.3: Electrostatic field of a uniformly charged rod of length L. (a) The point P and the rod, and (b) a convenient Cartesian coordinate system for the calculation of the field. The figure assumes a negative λ.

To find the components of the field in any coordinate system, dot-multiply Equation (4.7) by the unit vectors of that coordinate system. For Cartesian components, $E_x = \mathbf{E} \cdot \hat{\mathbf{e}}_x$, which picks the term multiplying $\hat{\mathbf{e}}_x$ in (4.7); $E_y = \mathbf{E} \cdot \hat{\mathbf{e}}_y$, which is zero; $E_z = \mathbf{E} \cdot \hat{\mathbf{e}}_z$, which picks the term multiplying $\hat{\mathbf{e}}_z$ in (4.7). Thus,

$$E_x = k_e \lambda x \int_a^b \frac{dt}{(x^2 + t^2)^{3/2}} = \frac{k_e \lambda}{x} \left(\frac{b}{\sqrt{x^2 + b^2}} - \frac{a}{\sqrt{x^2 + a^2}} \right),$$

$$E_y = 0, \tag{4.8}$$

$$E_z = -k_e \lambda \int_a^b \frac{t \, dt}{(x^2 + t^2)^{3/2}} = -k_e \lambda \left(\frac{1}{\sqrt{x^2 + a^2}} - \frac{1}{\sqrt{x^2 + b^2}} \right).$$

It is instructive to consider special cases of these formulas, such as when $a = -L/2$ and $b = +L/2$ (especially when L is large compared to x), which may be more familiar to the reader. We leave such considerations as exercises.

The electrostatic potential can be obtained similarly. From Equation (4.3), we get

$$\Phi(x, y, z) = \int_a^b \frac{k_e \lambda}{(x^2 + t^2)^{1/2}} \, dt = k_e \, \lambda \ln(t + \sqrt{x^2 + t^2}) \Big|_a^b$$

$$= k_e \lambda \ln \left(\frac{b + \sqrt{x^2 + b^2}}{a + \sqrt{x^2 + a^2}} \right). \qquad \blacksquare$$

Cylindrical Coordinates

For cylindrical coordinates the *components* of the position vector of P' are **not** the same as the *coordinates* of P'. In fact, $\mathbf{r}' = \rho' \hat{\mathbf{e}}_{\rho'} + z' \hat{\mathbf{e}}_z$.

caution! coordinates and components are not the same.

Various parts of the "master" equation (4.3) [or (3.3)] can be calculated as before—this time, of course, in cylindrical coordinates—and the results substituted in it to arrive at the expression for $\mathbf{E}$ entirely in terms of t. Thus

$$dq(\mathbf{r}') = \lambda(\mathbf{r}') \, dl(\mathbf{r}') = \lambda(\mathbf{r}') \sqrt{(d\rho')^2 + \rho'^2 (d\varphi')^2 + (dz')^2}, \tag{4.9}$$

where use has been made of Equation (2.29). Similarly, we have

$$\mathbf{r} - \mathbf{r}' = \rho \hat{\mathbf{e}}_\rho + z \hat{\mathbf{e}}_z - \rho' \hat{\mathbf{e}}_{\rho'} - z' \hat{\mathbf{e}}_z = \rho \hat{\mathbf{e}}_\rho - \rho' \hat{\mathbf{e}}_{\rho'} + (z - z') \hat{\mathbf{e}}_z \tag{4.10}$$

which leads to the absolute value

$$|\mathbf{r} - \mathbf{r}'| = \sqrt{(\mathbf{r} - \mathbf{r}') \cdot (\mathbf{r} - \mathbf{r}')}$$

$$= \sqrt{[\rho \hat{\mathbf{e}}_\rho - \rho' \hat{\mathbf{e}}_{\rho'} + (z - z') \hat{\mathbf{e}}_z] \cdot [\rho \hat{\mathbf{e}}_\rho - \rho' \hat{\mathbf{e}}_{\rho'} + (z - z') \hat{\mathbf{e}}_z]}.$$

Carrying out the dot product and keeping in mind that $\hat{\mathbf{e}}_\rho$ and $\hat{\mathbf{e}}_{\rho'}$ are *neither the same nor perpendicular* to each other, but make the two different angles φ and φ' with the x-axis, we obtain

$$|\mathbf{r} - \mathbf{r}'| = \sqrt{\rho^2 + \rho'^2 - 2\rho\rho' \cos(\varphi - \varphi') + (z - z')^2},$$

$$|\mathbf{r} - \mathbf{r}'|^3 = \left\{ \rho^2 + \rho'^2 - 2\rho\rho' \cos(\varphi - \varphi') + (z - z')^2 \right\}^{3/2}.$$

Putting everything together, we obtain

$$\mathbf{E} = \iint_\Omega \frac{k_e \lambda(\mathbf{r}') \sqrt{(d\rho')^2 + \rho'^2 (d\varphi')^2 + (dz')^2}}{\left\{\rho^2 + \rho'^2 - 2\rho\rho' \cos(\varphi - \varphi') + (z - z')^2\right\}^{3/2}}$$
$$\times \left(\rho \hat{\mathbf{e}}_\rho - \rho' \hat{\mathbf{e}}_{\rho'} + (z - z')\hat{\mathbf{e}}_z\right). \qquad (4.11)$$

To find components in any coordinate system, use Box 1.1.2 and take the dot product of Equation (4.11) with the appropriate unit vectors. The electrostatic potential is derived in a similar way.

Example 4.1.3. Let us reconsider the example of a rod. Obviously we should choose our z-axis along the rod. We further move the origin so that P ends up in the xy-plane (see Figure 4.4). This will reduce $\mathbf{r}$ to $\rho \hat{\mathbf{e}}_\rho$. The simplest parameterization of the rod is

$$\rho' = 0, \ z' = t, \ a \le t \le a + L \equiv b.$$

We note that φ' is undefined. This poses no problem because, as will be seen below, it will drop out of the equations. Putting these in Equation (4.11) we obtain

$$\mathbf{E} = k_e \lambda \iint_\Omega \frac{\sqrt{(0)^2 + (0)(d\varphi')^2 + (dz')^2}}{\left\{\rho^2 + (0)^2 - 2\rho(0) \cos(\varphi - \varphi') + (0 - z')^2\right\}^{3/2}}$$
$$\times \left[\rho \hat{\mathbf{e}}_\rho + (0)\hat{\mathbf{e}}_\rho + (0 - z')\hat{\mathbf{e}}_z\right] \qquad (4.12)$$
$$= k_e \lambda \int_a^b \frac{dt}{(\rho^2 + t^2)^{3/2}} \left(\rho \hat{\mathbf{e}}_\rho - t\hat{\mathbf{e}}_z\right)$$

To find the components of the electric field, take the dot product of one of the unit vectors of a coordinate system and Equation (4.12). For the ρ component, we have

$$E_\rho = \mathbf{E} \cdot \hat{\mathbf{e}}_\rho = k_e \lambda \int_a^b \frac{dt}{(\rho^2 + t^2)^{3/2}} \left(\rho \hat{\mathbf{e}}_\rho - t\hat{\mathbf{e}}_z\right) \cdot \hat{\mathbf{e}}_\rho$$
$$= k_e \lambda \int_a^b \frac{dt}{(\rho^2 + t^2)^{3/2}} \left(\rho \underbrace{\hat{\mathbf{e}}_\rho \cdot \hat{\mathbf{e}}_\rho}_{=1} - t \underbrace{\hat{\mathbf{e}}_z \cdot \hat{\mathbf{e}}_\rho}_{=0}\right) \qquad (4.13)$$
$$= k_e \lambda \rho \int_a^b \frac{dt}{(\rho^2 + t^2)^{3/2}} = \frac{k_e \lambda}{\rho} \left\{\frac{b}{\sqrt{\rho^2 + b^2}} - \frac{a}{\sqrt{\rho^2 + a^2}}\right\};$$

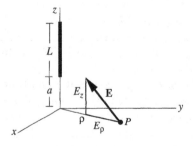

Figure 4.4: Electrostatic field of a uniformly charged rod of length L in cylindrical coordinates. The figure assumes a negative λ.

for the φ component, we obtin

$$E_\varphi = \mathbf{E} \cdot \hat{\mathbf{e}}_\varphi = k_e\lambda \int_a^b \frac{dt}{(\rho^2 + t^2)^{3/2}} (\rho\hat{\mathbf{e}}_\rho - t\hat{\mathbf{e}}_z) \cdot \hat{\mathbf{e}}_\varphi$$

$$= k_e\lambda \int_a^b \frac{dt}{(\rho^2 + t^2)^{3/2}} (\rho\underbrace{\hat{\mathbf{e}}_\rho \cdot \hat{\mathbf{e}}_\varphi}_{=0} - t\underbrace{\hat{\mathbf{e}}_z \cdot \hat{\mathbf{e}}_\varphi}_{=0}) = 0. \quad (4.14)$$

Note how the dependence on φ has completely disappeared because of the azimuthal symmetry of the rod. Finally the z component is

$$E_z = \mathbf{E} \cdot \hat{\mathbf{e}}_z = k_e\lambda \int_a^b \frac{dt}{(\rho^2 + t^2)^{3/2}} (\rho\hat{\mathbf{e}}_\rho - t\hat{\mathbf{e}}_z) \cdot \hat{\mathbf{e}}_z$$

$$= k_e\lambda \int_a^b \frac{dt}{(\rho^2 + t^2)^{3/2}} (\rho\underbrace{\hat{\mathbf{e}}_\rho \cdot \hat{\mathbf{e}}_z}_{=0} - t\underbrace{\hat{\mathbf{e}}_z \cdot \hat{\mathbf{e}}_z}_{=1}) \quad (4.15)$$

$$= -k_e\lambda \int_a^b \frac{t\,dt}{(\rho^2 + t^2)^{3/2}} = k_e\lambda \left\{ \frac{1}{\sqrt{\rho^2 + b^2}} - \frac{1}{\sqrt{\rho^2 + a^2}} \right\}$$

The electrostatic potential Φ can be calculated similarly.

We can also find the components in Cartesian coordinates by dot-multiplying Equation (4.12) with Cartesian unit vectors. For example,

$$E_x = \mathbf{E} \cdot \hat{\mathbf{e}}_x = k_e\lambda \int_a^b \frac{dt}{(\rho^2 + t^2)^{3/2}} (\rho\hat{\mathbf{e}}_\rho - t\hat{\mathbf{e}}_z) \cdot \hat{\mathbf{e}}_x$$

$$= k_e\lambda \int_a^b \frac{dt}{(\rho^2 + t^2)^{3/2}} (\rho\underbrace{\hat{\mathbf{e}}_\rho \cdot \hat{\mathbf{e}}_x}_{=\cos\varphi} - t\underbrace{\hat{\mathbf{e}}_z \cdot \hat{\mathbf{e}}_x}_{=0}),$$

$$= k_e\lambda\rho\cos\varphi \int_a^b \frac{dt}{(\rho^2 + t^2)^{3/2}} = \frac{k_e\lambda\cos\varphi}{\rho} \left\{ \frac{b}{\sqrt{\rho^2 + b^2}} - \frac{a}{\sqrt{\rho^2 + a^2}} \right\}$$

E_y will be the same except that instead of $\cos\varphi$ it will have $\sin\varphi$, and E_z will be identical to the E_z of Equation (4.15). When $\varphi = 0$, we recover the result of Example 4.1.2, because $\rho = x$ when $\varphi = 0$. ∎

All the foregoing derivations in electrostatics can be applied almost verbatim to the theory of gravitation. The only difference is the appearance of G instead of k_e and the interpretation of λ as linear *mass* density.

4.1.3   Examples from Magnetostatics

Probably the most realistic physical application of single integrals appears in the calculation of magnetic fields of currents in (thin) wires. Before looking at examples, let us briefly review magnetism.

We already mentioned in Chapter 1 that the magnetic field of N (slowly) moving point charges is given by[3]

$$\mathbf{B} = \sum_{k=1}^N \frac{k_m q_k \mathbf{v}_k \times (\mathbf{r} - \mathbf{r}_k)}{|\mathbf{r} - \mathbf{r}_k|^3}. \quad (4.16)$$

---

[3] "Slow" compared to the speed of light which is 3×10^8 m/s.

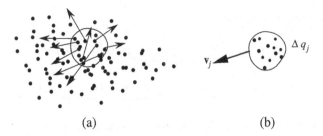

(a)   (b)

Figure 4.5: Magnetic field of a moving charge distribution. (a) All charges in motion with a "sample" singled out. The vectors show the velocities of some of the charges in the sample. (b) The sample is described by a charge Δq_j and an *average* velocity $\mathbf{v}_j$.

In a typical situation, N is of the order of 10^{25} or more. So, instead of adding all the terms individually, we lump together those that are close to one another, i.e., in a small region, and subsequently describe the situation by a **current density** (see Figure 4.5). This boils down to writing Equation (4.16) as

$$\mathbf{B} \approx \sum_{j=1}^{M} \frac{k_m \Delta q_j \mathbf{v}_j \times (\mathbf{r} - \mathbf{r}_j)}{|\mathbf{r} - \mathbf{r}_j|^3},$$

where Δq_j is the amount of charge in the jth region, $\mathbf{v}_j$ is the *average* velocity of all charges in the jth region, and $\mathbf{r}_j$ is the position vector of the "center" of the jth region. We can rewrite the equation above as

$$\mathbf{B} \approx \sum_{j=1}^{M} \frac{k_m \left[\Delta q(\mathbf{r}_j) \mathbf{v}(\mathbf{r}_j)\right] \times (\mathbf{r} - \mathbf{r}_j)}{|\mathbf{r} - \mathbf{r}_j|^3}.$$

Biot–Savart law   In the limit that $M \to \infty$ and $\Delta q \to 0$, we obtain

---

Box 4.1.3. *The magnetic field of a moving charge distribution is given by*

$$\mathbf{B}(\mathbf{r}) = k_m \iint_\Omega \frac{dq(\mathbf{r}') \mathbf{v}(\mathbf{r}') \times (\mathbf{r} - \mathbf{r}')}{|\mathbf{r} - \mathbf{r}'|^3}. \tag{4.17}$$

*This is the most general form of the **Biot–Savart law**.*

The product of the element of charge and velocity appearing in the equation is related to the various forms of current we may encounter. These are described below:

volume current density: $dq(\mathbf{r}')\mathbf{v}(\mathbf{r}') = \underbrace{\rho(\mathbf{r}')\mathbf{v}(\mathbf{r}')}_{\equiv \mathbf{J}(\mathbf{r}')} \, dV(\mathbf{r}') \equiv \mathbf{J}(\mathbf{r}') \, dV(\mathbf{r}'),$

surface current density: $dq(\mathbf{r}')\mathbf{v}(\mathbf{r}') = \underbrace{\sigma(\mathbf{r}')\mathbf{v}(\mathbf{r}')}_{\equiv \mathbf{j}(\mathbf{r}')}\, da(\mathbf{r}') \equiv \mathbf{j}(\mathbf{r}')\, da(\mathbf{r}'),$

linear current density: $dq(\mathbf{r}')\mathbf{v}(\mathbf{r}') = \underbrace{\lambda(\mathbf{r}')\mathbf{v}(\mathbf{r}')}_{\equiv \mathbf{I}(\mathbf{r}')}\, dl(\mathbf{r}') \equiv \mathbf{I}(\mathbf{r}')\, dl(\mathbf{r}').$

The volume current density $\mathbf{J}(\mathbf{r}')$ describes a situation in which charges are free to move in all directions. The surface current density $\mathbf{j}(\mathbf{r}')$ is used when charges are confined to a surface. The most familiar current density is the linear current density which is usually rewritten as

$$\mathbf{I}(\mathbf{r}')\, dl(\mathbf{r}') = I\vec{dl}(\mathbf{r}') = I d\mathbf{r}'.$$

This follows from the fact that $\mathbf{I}(\mathbf{r}')$ is in the same direction as the velocity (at $\mathbf{r}'$) which, since charges are confined to a curve (the wire), has the same direction as the (infinitesimal) tangent displacement along the wire, namely $d\mathbf{r}'$.

We are particularly interested in the linear case as shown in Figure 4.6. Thus, assuming that the current I is constant—it has to be due to charge conservation—we obtain

Biot–Savart law for circuits

> **Box 4.1.4.** *The general expression for the magnetic field of a circuit is given by*
>
> $$\mathbf{B}(\mathbf{r}) = k_m I \oint \frac{d\mathbf{r}' \times (\mathbf{r} - \mathbf{r}')}{|\,\mathbf{r} - \mathbf{r}'|^3}, \qquad (4.18)$$
>
> *where the circle on the integral sign implies a closed loop.*

This equation is independent of any coordinate systems. We now specialize to Cartesian and cylindrical systems.

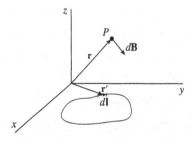

Figure 4.6: A general current filament described parametrically and used to calculate the magnetic field in Cartesian coordinates.

Cartesian Coordinates

To obtain the magnetic field we substitute

$$\mathbf{r} = x\hat{\mathbf{e}}_x + y\hat{\mathbf{e}}_y + z\hat{\mathbf{e}}_z,$$
$$\mathbf{r}' = x'\hat{\mathbf{e}}_x + y'\hat{\mathbf{e}}_y + z'\hat{\mathbf{e}}_z,$$
$$d\mathbf{r}' = \hat{\mathbf{e}}_x dx' + \hat{\mathbf{e}}_y dy' + \hat{\mathbf{e}}_z dz',$$
$$\mathbf{r} - \mathbf{r}' = (x - x')\hat{\mathbf{e}}_x + (y - y')\hat{\mathbf{e}}_y + (z - z')\hat{\mathbf{e}}_z,$$
$$|\mathbf{r} - \mathbf{r}'|^3 = \left[(x - x')^2 + (y - y')^2 + (z - z')^2 \right]^{3/2}$$

in Equation (4.18). For the cross product, we need to expand the determinant

$$d\mathbf{r}' \times (\mathbf{r} - \mathbf{r}') = \det \begin{pmatrix} \hat{\mathbf{e}}_x & \hat{\mathbf{e}}_y & \hat{\mathbf{e}}_z \\ dx' & dy' & dz' \\ x - x' & y - y' & z - z' \end{pmatrix},$$

using Figure 1.5.

Cylindrical Coordinates

The cylindrical coordinates can be handled in exact analogy with the Cartesian case. Using Equations (1.19) and (2.28), we have

$$\mathbf{r} = \rho\hat{\mathbf{e}}_\rho + z\hat{\mathbf{e}}_z, \qquad \mathbf{r}' = \rho'\hat{\mathbf{e}}_{\rho'} + z'\hat{\mathbf{e}}_z,$$
$$\mathbf{r} - \mathbf{r}' = \rho\hat{\mathbf{e}}_\rho - \rho'\hat{\mathbf{e}}_{\rho'} + (z - z')\hat{\mathbf{e}}_z, \tag{4.19}$$
$$d\mathbf{r}' = \hat{\mathbf{e}}_{\rho'} d\rho' + \hat{\mathbf{e}}_{\varphi'} \rho' d\varphi' + \hat{\mathbf{e}}_z dz',$$
$$|\mathbf{r} - \mathbf{r}'|^3 = \left\{ \rho^2 + \rho'^2 - 2\rho\rho' \cos(\varphi - \varphi') + (z - z')^2 \right\}^{3/2}.$$

The cross product cannot be done using determinants because not everything is written in terms of the three *mutually perpendicular* unit vectors: $\hat{\mathbf{e}}_\rho$ is different from $\hat{\mathbf{e}}_{\rho'}$ but not perpendicular to it. In fact, this difference is the cause for the appearance of the cosine term in the last equation of (4.19). To find the cross product, we simply multiply the two terms and use the following relations, most of which should be familiar, and the unfamiliar ones can be obtained using Figure 4.7:

$$\hat{\mathbf{e}}_{\rho'} \times \hat{\mathbf{e}}_\rho = \hat{\mathbf{e}}_z \sin(\varphi - \varphi'), \qquad \hat{\mathbf{e}}_{\rho'} \times \hat{\mathbf{e}}_z = -\hat{\mathbf{e}}_{\varphi'},$$
$$\hat{\mathbf{e}}_{\varphi'} \times \hat{\mathbf{e}}_\rho = -\hat{\mathbf{e}}_z \cos(\varphi - \varphi'), \qquad \hat{\mathbf{e}}_{\varphi'} \times \hat{\mathbf{e}}_{\rho'} = -\hat{\mathbf{e}}_z,$$
$$\hat{\mathbf{e}}_{\varphi'} \times \hat{\mathbf{e}}_z = \hat{\mathbf{e}}_{\rho'}, \qquad \hat{\mathbf{e}}_z \times \hat{\mathbf{e}}_\rho = \hat{\mathbf{e}}_\varphi, \tag{4.20}$$
$$\hat{\mathbf{e}}_z \times \hat{\mathbf{e}}_{\rho'} = \hat{\mathbf{e}}_{\varphi'}.$$

The cross product can be written as

$$\begin{aligned} d\mathbf{r}' \times (\mathbf{r} - \mathbf{r}') = &\hat{\mathbf{e}}_z \left[\rho'^2 d\varphi' + \rho \sin(\varphi - \varphi') \, d\rho' - \rho\rho' \cos(\varphi - \varphi') \, d\varphi' \right] \\ &- \hat{\mathbf{e}}_{\varphi'} \left[(z - z') \, d\rho' + \rho' dz' \right] \\ &+ \hat{\mathbf{e}}_\varphi \rho \, dz' + \hat{\mathbf{e}}_{\rho'} \rho' (z - z') \, d\varphi'. \end{aligned} \tag{4.21}$$

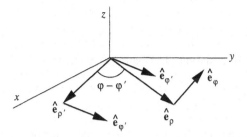

Figure 4.7: The orientation of some of the cylindrical unit vectors drawn for the calculation of cross products.

To find the components of the magnetic field, we substitute this in Equation (4.18), take the dot product of cylindrical unit vectors with the integrand, and use

$$\hat{\mathbf{e}}_\rho \cdot \hat{\mathbf{e}}_{\rho'} = \cos(\varphi' - \varphi), \qquad \hat{\mathbf{e}}_{\rho'} \cdot \hat{\mathbf{e}}_\varphi = \sin(\varphi' - \varphi),$$
$$\hat{\mathbf{e}}_\rho \cdot \hat{\mathbf{e}}_{\varphi'} = -\sin(\varphi' - \varphi), \quad \hat{\mathbf{e}}_\varphi \cdot \hat{\mathbf{e}}_{\varphi'} = \cos(\varphi' - \varphi), \qquad (4.22)$$

as well as the other more obvious dot products of unit vectors.

We can derive a general expression for the components of the electric field in terms of the parametric functions of a general curve (see Problem 4.6). However, a simple example will also illustrate the general procedure without entangling the formulas with complicated expressions.

Example 4.1.4. A simple application of the foregoing general formalism is to calculate the magnetic field of a circular loop of radius a. The choice of the axes and origin of Figure 4.8 yields the following parameterization of the loop:

$$\rho' = a, \, d\rho' = 0; \quad \varphi' = t, \, d\varphi' = dt; \quad z' = 0, \, dz' = 0, \qquad 0 \le t \le 2\pi.$$

Furthermore, because of the azimuthal symmetry of the current distribution, the final answer will be independent of φ. Thus, we can set that equal to zero. Inserting this information in Equations (4.19) and (4.21) gives

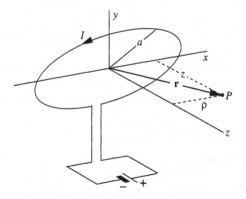

Figure 4.8: The geometry of the circular loop of current.

$$|\mathbf{r} - \mathbf{r}'|^3 = \left[\rho^2 + a^2 - 2\rho a \cos(t) + z^2\right]^{3/2}$$

$$d\mathbf{r}' \times (\mathbf{r} - \mathbf{r}') = \hat{\mathbf{e}}_z\left[a^2 dt - \rho a \cos(t)\, dt\right] + \hat{\mathbf{e}}_{\rho'} az\, dt.$$

The magnetic field of Equation (4.18) can now be written as

$$\mathbf{B} = k_m I \int_0^{2\pi} \frac{\hat{\mathbf{e}}_z\left[a^2 - \rho a \cos(t)\right] + \hat{\mathbf{e}}_{\rho'} az}{\left[\rho^2 + a^2 - 2\rho a \cos(t) + z^2\right]^{3/2}} \, dt \qquad (4.23)$$

Finally, to find the cylindrical components, dot-multiply (4.23) with the cylindrical unit vectors and use Equation (4.22) with $\varphi = 0$ (and $\varphi' = t$):

$$B_\rho = \mathbf{B} \cdot \hat{\mathbf{e}}_\rho = k_m I a \int_0^{2\pi} \frac{\left[\hat{\mathbf{e}}_z\left(a - \rho \cos t\right) + \hat{\mathbf{e}}_{\rho'} z\right] \cdot \hat{\mathbf{e}}_\rho}{\left(\rho^2 + a^2 - 2\rho a \cos t + z^2\right)^{3/2}} \, dt$$

$$= k_m I a \int_0^{2\pi} \frac{\overbrace{\hat{\mathbf{e}}_z \cdot \hat{\mathbf{e}}_\rho}^{=0}\left(a - \rho \cos t\right) + \overbrace{\hat{\mathbf{e}}_{\rho'} \cdot \hat{\mathbf{e}}_\rho}^{=\cos t} z}{\left(\rho^2 + a^2 - 2\rho a \cos t + z^2\right)^{3/2}} \, dt \qquad (4.24)$$

$$= k_m I a z \int_0^{2\pi} \frac{\cos t \, dt}{(\rho^2 + a^2 - 2\rho a \cos t + z^2)^{3/2}},$$

Similarly,

$$B_\varphi = \mathbf{B} \cdot \hat{\mathbf{e}}_\varphi = k_m I a \int_0^{2\pi} \frac{\overbrace{\hat{\mathbf{e}}_z \cdot \hat{\mathbf{e}}_\varphi}^{=0}\left(a - \rho \cos t\right) + \overbrace{\hat{\mathbf{e}}_{\rho'} \cdot \hat{\mathbf{e}}_\varphi}^{=\sin t} z}{\left(\rho^2 + a^2 - 2\rho a \cos t + z^2\right)^{3/2}} \, dt$$

$$= k_m I a z \int_0^{2\pi} \frac{\sin t \, dt}{(\rho^2 + a^2 - 2\rho a \cos t + z^2)^{3/2}} \qquad (4.25)$$

$$= -\frac{k_m I z}{\rho}(\rho^2 + a^2 - 2a\rho \cos t + z^2)^{-1/2} \Big|_0^{2\pi} = 0,$$

and

$$B_z = \mathbf{B} \cdot \hat{\mathbf{e}}_z = k_m I a \int_0^{2\pi} \frac{\overbrace{\hat{\mathbf{e}}_z \cdot \hat{\mathbf{e}}_z}^{=1}\left(a - \rho \cos t\right) + \overbrace{\hat{\mathbf{e}}_{\rho'} \cdot \hat{\mathbf{e}}_z}^{=0} z}{\left(\rho^2 + a^2 - 2\rho a \cos t + z^2\right)^{3/2}} \, dt$$

$$= k_m I a \int_0^{2\pi} \frac{(a - \rho \cos t) \, dt}{(\rho^2 + a^2 - 2\rho a \cos t + z^2)^{3/2}}. \qquad (4.26)$$

Once again the azimuthal symmetry prohibits a φ-component for the field. These integrals cannot be evaluated analytically, but if we specialize to the case where P is on the z-axis (i.e., when $\rho = 0$), the integrals become trivial. In fact, we have

$$B_\rho = k_m I a z \int_0^{2\pi} \frac{\cos t \, dt}{(a^2 + z^2)^{3/2}} = 0,$$

$$B_\varphi = 0,$$

$$B_z = -k_m I a \int_0^{2\pi} \frac{-a \, dt}{(a^2 + z^2)^{3/2}} = \frac{2\pi k_m I a^2}{(a^2 + z^2)^{3/2}}.$$

■

After graduating from the college of Louis-le-Grand in Paris and subsequently spending some time in the army, **Jean-Baptiste Biot** entered the Ecole Polytechnique in Paris where **Monge** (a noted mathematician of the time and an expert in differential geometry) realized his potential. Because of his political views and his participation in an attempted insurrection by the royalists against the Convention, Biot was captured by government forces and taken prisoner. Had it not been for Monge's intervention and plead for his release, Biot's promising career might have ended.

Jean-Baptiste Biot
1774–1862

Biot became Professor of Mathematics at the Ecole Centrale at Beauvais in 1797, and three years later joined the faculty of the Collège de France as Professor of Mathematical Physics an appointment which was due to the influence of **Laplace**.

Biot studied a wide range of mathematical topics, mostly on the applied mathematics side. He made advances in astronomy, elasticity, heat, and optics while, in pure mathematics, he also did important work in geometry. He collaborated with Arago on the refractive properties of gases.

Biot's most notable contribution was done in collaboration with **Felix Savart** (1791–1841), who was an acoustics expert and developed the Savart disk, a device which produced a sound wave of known frequency, using a rotating cog wheel as a measuring device.

Biot and Savart jointly discovered that the intensity of the magnetic field set up by a current flowing through a wire varies inversely with the distance from the wire. This is a special case of what is now known as *Biot–Savart's Law* and is fundamental to modern electromagnetic theory.

For his work on the polarization of light passing through chemical solutions Biot was awarded the Rumford Medal of the Royal Society. He tried twice for the post of Secretary to the Académie des Sciences but lost out in 1822 to Fourier for this post. When Fourier died he applied again only to lose to Arago.

4.2   Applications: Double Integrals

Whenever areas are sources of physical quantities such as fields, or interactions take place on areas, such as pressure applied on a surface, double integrals are used. We can be as general as in the previous section and consider a general surface given by a parametric equation in two variables (instead of one used for curves). However, since the geometry of surfaces is much more complicated, and much less illuminating, we shall confine our discussion to very simple geometries which require trivial and obvious parameterization. More specifically, we restrict ourselves to *primary surfaces* of the three coordinate systems.

4.2.1   Cartesian Coordinates

Since we are restricting ourselves to primary surfaces, our choice for Cartesian coordinates is narrowed down to planes, and if we want the boundaries of the plane to be simple in Cartesian coordinates, we are limited to just a rectangle.

Example 4.2.1. We start with an example from electrostatics. A rectangular flat surface of sides a and b is charged uniformly with surface charge density σ, and we are interested in the electric field at a general point P in space. This is given by

$$\mathbf{E} = \int_{\Omega} \frac{k_e dq(\mathbf{r}')}{|\mathbf{r} - \mathbf{r}'|^3} (\mathbf{r} - \mathbf{r}')$$

with $\mathbf{r} = x\hat{\mathbf{e}}_x + y\hat{\mathbf{e}}_y + z\hat{\mathbf{e}}_z = \langle x, y, z \rangle$ and $\mathbf{r}' = x'\hat{\mathbf{e}}_x + y'\hat{\mathbf{e}}_y = \langle x', y', 0 \rangle$, where we have chosen the plane of the rectangle to be the xy-plane. If we choose the center of the rectangle to be our origin, our z-axis perpendicular to the plane of the rectangle, and our x-and y-axes parallel to the sides as shown in Figure 4.9, then the element of area coincides with the third primary element, and we can write

$$dq(\mathbf{r}') = \sigma(\mathbf{r}')\, da(\mathbf{r}') = \sigma\, dx'\, dy'.$$

We also have

$$\mathbf{r} - \mathbf{r}' = (x - x')\hat{\mathbf{e}}_x + (y - y')\hat{\mathbf{e}}_y + z\hat{\mathbf{e}}_z = \langle x - x', y - y', z \rangle,$$

$$|\mathbf{r} - \mathbf{r}'| = \sqrt{(x - x')^2 + (y - y')^2 + z^2},$$

$$|\mathbf{r} - \mathbf{r}'|^3 = \left\{ (x - x')^2 + (y - y')^2 + z^2 \right\}^{3/2}.$$

Inserting all these relations in the expression for $\mathbf{E}$, we obtain

$$\mathbf{E} = \int_{\Omega} \frac{k_e \sigma\, dx'\, dy'}{\left\{ (x - x')^2 + (y - y')^2 + z^2 \right\}^{3/2}} \left[(x - x')\hat{\mathbf{e}}_x + (y - y')\hat{\mathbf{e}}_y + z\hat{\mathbf{e}}_z \right]$$

electric field of a uniformly charged rectangle

with components

$$E_x = k_e \sigma \int_{\Omega} \frac{(x - x')\, dx'\, dy'}{\left\{ (x - x')^2 + (y - y')^2 + z^2 \right\}^{3/2}},$$

$$E_y = k_e \sigma \int_{\Omega} \frac{(y - y')\, dx'\, dy'}{\left\{ (x - x')^2 + (y - y')^2 + z^2 \right\}^{3/2}},$$

$$E_z = k_e \sigma z \int_{\Omega} \frac{dx'\, dy'}{\left\{ (x - x')^2 + (y - y')^2 + z^2 \right\}^{3/2}},$$

where everything independent of the variables of integration, x' and y', is taken out of the integrals.

We have already discussed a general procedure for evaluating multiple integrals by reducing them to lower-dimensional integrals. We follow the same procedure here: The y' integration has the lower limit $-b/2$ and the upper limit $+b/2$, both independent of x'.[4] Similarly, the x' integration has $-a/2$ and $a/2$ as its limits. This means that the components can be written as

$$E_x = k_e \sigma \int_{-a/2}^{a/2} (x - x')\, dx' \int_{-b/2}^{b/2} \frac{dy'}{\left\{ (x - x')^2 + (y - y')^2 + z^2 \right\}^{3/2}},$$

$$E_y = k_e \sigma \int_{-a/2}^{a/2} dx' \int_{-b/2}^{b/2} \frac{(y - y')\, dy'}{\left\{ (x - x')^2 + (y - y')^2 + z^2 \right\}^{3/2}},$$

---

[4]The independence of the limits is one reason that Cartesian coordinates are useful for rectangular regions of integration. If we had chosen cylindrical coordinates, then the limits of integration, the lines $y' = -b/2$ and $y' = b/2$, would have had to be written in cylindrical coordinates, giving, for the upper limit, for example, $\rho' \sin \varphi' = b/2$ or $\rho' = b/(2 \sin \varphi')$. Thus a ρ' integration with limits dependent on φ' would have been involved.

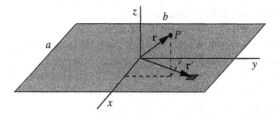

Figure 4.9: Electrostatic field of a flat rectangular charge distribution.

$$E_z = k_e \sigma z \int_{-a/2}^{a/2} dx' \int_{-b/2}^{b/2} \frac{dy'}{\{(x - x')^2 + (y - y')^2 + z^2\}^{3/2}}.$$

Note that the x' integration cannot be done until after the y' integration, because the latter has an x'-dependent integrand. ∎

Having exhausted the (simple) possibilities for electrostatics (and gravity, since the two are almost identical), we now turn to magnetostatics.

Example 4.2.2. Approximate the belt of a Van de Graff machine with an isolated moving rectangle having sides a and b, and velocity $\mathbf{v}$ along the side b as shown in Figure 4.10. Furthermore, assume that the charges are uniformly distributed on the belt with surface charge density σ. We want to find the magnetic field of the belt at a general point P in space. Let us choose the positive y-direction to be that of the velocity. Then, Equation (4.17) becomes

$$\mathbf{B}(\mathbf{r}) = k_m \iint_\Omega \frac{\sigma da \mathbf{v} \times (\mathbf{r} - \mathbf{r}')}{|\mathbf{r} - \mathbf{r}'|^3}.$$

The geometry of this example is identical to that of Example 4.2.1. Therefore, we can immediately write the integral for $\mathbf{B}$:

magnetic field of a charged rectangular moving belt

$$\mathbf{B}(\mathbf{r}) = k_m \iint_\Omega \frac{\sigma dx' dy' v \hat{\mathbf{e}}_y \times [(x - x')\hat{\mathbf{e}}_x + (y - y')\hat{\mathbf{e}}_y + z\hat{\mathbf{e}}_z]}{\{(x - x')^2 + (y - y')^2 + z^2\}^{3/2}},$$

from which the components of the magnetic field are easily calculated:

$$B_x = k_m \sigma v z \int_{-a/2}^{a/2} dx' \int_{-b/2}^{b/2} \frac{dy'}{\{(x - x')^2 + (y - y')^2 + z^2\}^{3/2}},$$

$$B_y = 0, \tag{4.27}$$

$$B_z = -k_m \sigma v \int_{-a/2}^{a/2} (x - x') dx' \int_{-b/2}^{b/2} \frac{dy'}{\{(x - x')^2 + (y - y')^2 + z^2\}^{3/2}}. \quad ∎$$

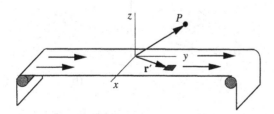

Figure 4.10: A rectangular distribution of moving charges whose magnetic field can be calculated using Cartesian coordinates.

4.2.2   Cylindrical Coordinates

The cylindrical system has two types of primary surface: planes and cylinders. Although we considered planes in the previous subsection, we shall reconsider them here because the third primary surface, that perpendicular to the z-axis, gives us the possibility of solving planar problems with nonrectangular regions of integration. Let us start with such a problem.

Example 4.2.3. In this example we want to calculate the gravitational field of a uniform surface mass distribution of density σ_m which is a segment of a planar annular region with inner radius a and outer radius b, and whose sides make an angle of α as shown in Figure 4.11(a). Let us choose our origin to coincide with the center of the annular region, our x-axis to be along one of the sides, and the xy-plane to be the plane of the mass distribution [see Figure 4.11(b)].

Recall that in cylindrical coordinates, the *components* of the position vector of P' are not the same as the source point's *coordinates*. In fact, we have

$$\mathbf{r}' = \rho'\hat{\mathbf{e}}_{\rho'}, \qquad \mathbf{r} = \rho\hat{\mathbf{e}}_\rho + z\hat{\mathbf{e}}_z,$$
$$\mathbf{r} - \mathbf{r}' = \rho\hat{\mathbf{e}}_\rho + z\hat{\mathbf{e}}_z - \rho'\hat{\mathbf{e}}_{\rho'},$$
$$|\mathbf{r} - \mathbf{r}'|^3 = (\rho^2 + \rho'^2 - 2\rho\rho'\cos\varphi' + z^2)^{3/2},$$

where in the last line we have made the simplification that the field point is in the xz-plane, so that $\varphi = 0$; otherwise, we would have $\cos(\varphi - \varphi')$ instead of $\cos\varphi'$. The element of mass is given by

$$dm(\mathbf{r}') = \sigma_m\,da(\mathbf{r}') = \sigma_m(d\rho')(\rho'\,d\varphi') = \sigma_m\rho'\,d\rho'\,d\varphi'.$$

Thus, the gravitational field is

$$\mathbf{g} = -\iint_\Omega \frac{G\,dm(\mathbf{r}')}{|\mathbf{r} - \mathbf{r}'|^3}(\mathbf{r} - \mathbf{r}'),$$
$$= -G\sigma_m \int_a^b \rho'\,d\rho' \int_0^\alpha \frac{d\varphi'(\rho\hat{\mathbf{e}}_\rho + z\hat{\mathbf{e}}_z - \rho'\hat{\mathbf{e}}_{\rho'})}{(\rho^2 + \rho'^2 - 2\rho\rho'\cos\varphi' + z^2)^{3/2}}. \tag{4.28}$$

To find the components, we take the dot product of this integral with the cylindrical

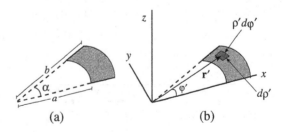

(a)   (b)

Figure 4.11: The annular region whose gravitational field is being calculated. The position vector of the source point and the lengths of the sides of the element of area are also shown.

unit vectors. The result will then be

$$g_\rho = -G\sigma_m \int_a^b \rho' \, d\rho' \int_0^\alpha \frac{(\rho - \rho' \cos\varphi') \, d\varphi'}{(\rho^2 + \rho'^2 - 2\rho\rho' \cos\varphi' + z^2)^{3/2}},$$

$$g_\varphi = G\sigma_m \int_a^b \rho' \, d\rho' \int_0^\alpha \frac{\rho' \sin\varphi' \, d\varphi'}{(\rho^2 + \rho'^2 - 2\rho\rho' \cos\varphi' + z^2)^{3/2}}, \tag{4.29}$$

$$g_z = -G\sigma_m z \int_a^b \rho' \, d\rho' \int_0^\alpha \frac{d\varphi'}{(\rho^2 + \rho'^2 - 2\rho\rho' \cos\varphi' + z^2)^{3/2}}.$$

Let us look at some special cases of this. For a complete annular region, we simply replace α with 2π:

$$g_\rho = -G\sigma_m \int_a^b \rho' \, d\rho' \int_0^{2\pi} \frac{(\rho - \rho' \cos\varphi') \, d\varphi'}{(\rho^2 + \rho'^2 - 2\rho\rho' \cos\varphi' + z^2)^{3/2}},$$

$$g_\varphi = G\sigma_m \int_a^b \rho' \, d\rho' \int_0^{2\pi} \frac{\rho' \sin\varphi' \, d\varphi'}{(\rho^2 + \rho'^2 - 2\rho\rho' \cos\varphi' + z^2)^{3/2}} = 0, \tag{4.30}$$

$$g_z = -G\sigma_m z \int_a^b \rho' \, d\rho' \int_0^{2\pi} \frac{d\varphi'}{(\rho^2 + \rho'^2 - 2\rho\rho' \cos\varphi' + z^2)^{3/2}}.$$

As expected, the φ-component has disappeared.

We can further simplify the geometry by locating the field point on the z-axis. Then, $\rho = 0$ and we have

$$g_\rho = G\sigma_m \int_a^b \frac{\rho'^2 \, d\rho'}{(\rho'^2 + z^2)^{3/2}} \int_0^{2\pi} \cos\varphi' \, d\varphi' = 0,$$

$$g_\varphi = 0,$$

$$g_z = -G\sigma_m z \int_a^b \frac{\rho' \, d\rho'}{(\rho'^2 + z^2)^{3/2}} \int_0^{2\pi} d\varphi' = -2\pi G\sigma_m z \int_a^b \frac{\rho' \, d\rho'}{(\rho'^2 + z^2)^{3/2}}$$

$$= -2\pi G\sigma_m z \left\{ \frac{1}{\sqrt{a^2 + z^2}} - \frac{1}{\sqrt{b^2 + z^2}} \right\}.$$

If we take the limit $a \to 0$ and $b \to \infty$, we obtain

$$\mathbf{g} = -2\pi G\sigma_m \frac{z}{\sqrt{z^2}} \hat{\mathbf{e}}_z = -2\pi G\sigma_m \frac{z}{|z|} \hat{\mathbf{e}}_z,$$

where we have used Box 4.2.1 (see below). Now note that $z/|z| = \pm 1$ depending on the sign of z. When $z > 0$, we get $z/|z| \hat{\mathbf{e}}_z = \hat{\mathbf{e}}_z$ which is the unit normal to the surface. When $z < 0$, we get $z/|z| \hat{\mathbf{e}}_z = -\hat{\mathbf{e}}_z$ which is again the unit normal to (the other side of) the surface. Denoting the unit normal by $\hat{\mathbf{e}}_n$, we can write

$$\mathbf{g} = -2\pi G\sigma_m \hat{\mathbf{e}}_n.$$

The electrostatic analogue of this is obtained by substituting $-k_e = -1/4\pi\epsilon_0$ for G. This yields

$$\mathbf{E} = \frac{\sigma_q}{2\epsilon_0} \hat{\mathbf{e}}_n$$

which is the field of an infinite sheet of charge with which the reader is familiar. Note that while $\mathbf{g}$ always points toward the sheet (opposite to $\hat{\mathbf{e}}_n$, because σ_m is always positive), the direction of $\mathbf{E}$ is determined by the sign of σ_q. ∎

4.2.3   Spherical Coordinates

One of the primary surfaces of a spherical coordinate system is a sphere, and since there are a lot of spherical objects around, it is useful to gain experience in calculations involving spheres.

In the following, we shall be taking square roots of functions. Care needs to be taken when doing so:

> **Box 4.2.1.** *For any real-valued quantity A, $\sqrt{A^2} \equiv |A|$, i.e., the square root of the square of a quantity is the **absolute value** of that quantity.*

Failure to keep this in mind will result in incorrect conclusions, as we shall see below.

Example 4.2.4. In this example we are interested in the gravitational field at a general point P of a spherical cap, i.e., a segment of a spherical shell of radius a and uniform surface density σ such that the cone defined by the segment and the center of the sphere has a half-angle α (see Figure 4.12). It is clear that the choice of axes and origin resulting in the greatest simplification is as shown in Figure 4.12. Notice that P is taken to lie in the xz-plane, so that $\varphi = 0$. We can immediately write

$$\mathbf{g} = -G \iint_{\Omega} \frac{dm(\mathbf{r}')}{|\mathbf{r} - \mathbf{r}'|^3}(\mathbf{r} - \mathbf{r}') \tag{4.31}$$

with

$$\mathbf{r}' = a\hat{\mathbf{e}}_{r'}, \qquad \mathbf{r} = r\hat{\mathbf{e}}_r, \qquad \mathbf{r} - \mathbf{r}' = r\hat{\mathbf{e}}_r - a\hat{\mathbf{e}}_{r'},$$

$$|\mathbf{r} - \mathbf{r}'|^3 = \left\{ r^2 + a^2 - 2ra\overbrace{(\sin\theta\sin\theta'\cos\varphi' + \cos\theta\cos\theta')}^{\hat{\mathbf{e}}_r\cdot\hat{\mathbf{e}}_{r'}} \right\}^{3/2}, \tag{4.32}$$

$$dm(\mathbf{r}') = \sigma da_1 = \sigma a^2 \sin\theta'\, d\theta'\, d\varphi'.$$

By inserting these relations in (4.31) and dotting the result with unit vectors, we obtain the three components of $\mathbf{g}$ in various coordinate systems. In spherical coordinates these are

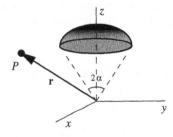

Figure 4.12: A spherical cap whose gravitational field can be calculated using spherical coordinates.

$$g_r = -G\sigma a^2 \iint_\Omega \frac{\sin\theta' \{r - a(\sin\theta \sin\theta' \cos\varphi' + \cos\theta \cos\theta')\}\, d\theta'\, d\varphi'}{\{r^2 + a^2 - 2ra(\sin\theta \sin\theta' \cos\varphi' + \cos\theta \cos\theta')\}^{3/2}},$$

$$g_\theta = G\sigma a^3 \iint_\Omega \frac{\sin\theta'(\cos\theta \sin\theta' \cos\varphi' - \sin\theta \cos\theta')\, d\theta'\, d\varphi'}{\{r^2 + a^2 - 2ra(\sin\theta \sin\theta' \cos\varphi' + \cos\theta \cos\theta')\}^{3/2}}, \qquad (4.33)$$

$$g_\varphi = G\sigma a^3 \iint_\Omega \frac{\sin^2\theta' \sin\varphi'\, d\theta'\, d\varphi'}{\{r^2 + a^2 - 2ra(\sin\theta \sin\theta' \cos\varphi' + \cos\theta \cos\theta')\}^{3/2}} = 0.$$

The region of integration Ω is one in which θ' varies from 0 to α, and φ' from 0 to 2π. The last integral vanishes because of the φ' integration. The vanishing of the φ-component is simply the result of the azimuthal symmetry.

The result above is not interesting, but if we move P to the polar axis, so that $\theta = 0$, then the equations simplify considerably, and we get

$$g_r = -G\sigma a^2 \int_0^\alpha \frac{\sin\theta'(r - a\cos\theta')\, d\theta'}{(r^2 + a^2 - 2ra\cos\theta')^{3/2}} \int_0^{2\pi} d\varphi'$$

$$= -2\pi G\sigma a^2 \int_0^\alpha \frac{\sin\theta'(r - a\cos\theta')\, d\theta'}{(r^2 + a^2 - 2ra\cos\theta')^{3/2}},$$

$$g_\theta = G\sigma a^3 \int_0^\alpha \frac{\sin^2\theta'\, d\theta'}{(r^2 + a^2 - 2ra\cos\theta')^{3/2}} \int_0^{2\pi} \cos\varphi'\, d\varphi' = 0,$$

$$g_\varphi = 0.$$

The most interesting result is obtained when $\alpha = \pi$, i.e., when we have a complete spherical shell. Then using

$$\int_0^\pi \frac{\sin\theta'(r - a\cos\theta')\, d\theta'}{(r^2 + a^2 - 2ra\cos\theta')^{3/2}} = \frac{1}{r^2}\left(1 - \frac{\sqrt{(a-r)^2}}{a - r}\right) \equiv \frac{1}{r^2}\left(1 - \frac{|a - r|}{a - r}\right),$$

which can be looked up in a good integral table, we obtain

$$g_r = -\frac{2\pi G\sigma a^2}{r^2}\left(1 - \frac{|a - r|}{a - r}\right), \qquad g_\theta = 0, \qquad g_\varphi = 0.$$

For points inside the shell, $r < a$; therefore $\dfrac{|a - r|}{a - r} = \dfrac{a - r}{a - r} = 1$, and the field vanishes. Thus, the *gravitational field inside a spherical shell is zero.* On the other hand, for points outside, $r > a$, and $\dfrac{|a - r|}{a - r} = \dfrac{r - a}{a - r} = -1$, leading to

> gravitational field inside a spherical shell is zero.

$$g_r = -\frac{4\pi G\sigma a^2}{r^2} = -\frac{GM}{r^2},$$

where $M = 4\pi a^2 \sigma$ is the total mass of the shell. This is identical to the gravitational field of a point charge of mass M located at the center of the shell. Now, if we have a number of concentric shells, then, at a point outside the outermost one, the field must be that of a point charge at the common center having a mass equal to the total mass of all the shells. Note that each shell can have a different uniform density than others. In particular, if we have a *solid sphere*, with a density which is a function of r alone, the same result holds. A density which is a function of r alone is called a **spherical mass distribution**. We thus have the famous result:

> concept of spherical mass distribution elaborated

> **Box 4.2.2.** *When gravitationally attracting objects outside it, a spherical mass distribution acts as if all its mass were concentrated into a point at its center.*

Newton spent approximately twenty years convincing himself of this result.

Because of the similarity between gravity and electrostatics, the conclusion above can be applied to the electrostatic field as well. Thus, in particular, the electrostatic field inside any uniformly charged shell is zero. ∎

moment of inertia

We take the final example of this section from mechanics and calculate the **moment of inertia** of the foregoing shell about the polar axis. Recall that the moment of inertia of a mass distribution about an axis is defined as

$$I = \iint_\Omega R^2 \, dm, \tag{4.34}$$

where R is the distance from the integration point—location of dm—to the reference axis.

Example 4.2.5. The moment of inertia of the spherical shell segment is obtained easily. All we need to note is that $R = a \sin \theta'$. Then Equation (4.34) gives

$$I = \iint_\Omega (a \sin \theta')^2 \sigma a^2 \sin \theta' \, d\theta' \, d\varphi' = a^4 \sigma \int_0^\alpha \sin^3 \theta' \, d\theta' \int_0^{2\pi} d\varphi'$$

$$= 2\pi a^4 \sigma \left(\tfrac{1}{3} \cos^3 \theta' - \cos \theta' \right)\big|_0^\alpha = \frac{2\pi a^4 \sigma}{3} (\cos^3 \alpha - 3 \cos \alpha + 2).$$

We can express this in terms of total mass if we note that the area is given by

$$A = \iint_\Omega a^2 \sin \theta' \, d\theta' \, d\varphi' = 2\pi a^2 \int_0^\alpha \sin \theta' \, d\theta' = 2\pi a^2 (1 - \cos \alpha)$$

so that

$$\sigma = \frac{M}{A} = \frac{M}{2\pi a^2 (1 - \cos \alpha)}.$$

Therefore,

$$I = \tfrac{1}{3} M a^2 \frac{\cos^3 \alpha - 3 \cos \alpha + 2}{1 - \cos \alpha},$$

which reduces to $I = \tfrac{2}{3} M a^2$ for a complete spherical shell (with $\alpha = \pi$). ∎

4.3   Applications: Triple Integrals

To illustrate the difficulty of calculations when appropriate coordinate systems are not chosen, in the following example we calculate the gravitational field of a uniform hemisphere at a point P on its axis in Cartesian coordinates.

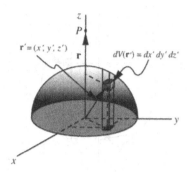

Figure 4.13: Calculating the gravitational field of a hemisphere in the "unnatural" Cartesian coordinates.

Example 4.3.1. The geometry of the problem is shown in Figure 4.13. The location of P and the choice of axes indicate that

$$\mathbf{r} = z\hat{\mathbf{e}}_z, \qquad \mathbf{r}' = x'\hat{\mathbf{e}}_x + y'\hat{\mathbf{e}}_y + z'\hat{\mathbf{e}}_z,$$
$$|\mathbf{r} - \mathbf{r}'|^3 = \left\{x'^2 + y'^2 + (z - z')^2\right\}^{3/2},$$
$$dm(\mathbf{r}') = \rho_m dV(\mathbf{r}') = \rho_m dx' dy' dz',$$

where ρ_m is the uniform mass density. Thus,

$$\mathbf{g} = -\int_\Omega \frac{G\, dm(\mathbf{r}')}{|\mathbf{r} - \mathbf{r}'|^3}(\mathbf{r} - \mathbf{r}') = G\rho_m \int_\Omega \frac{dx'\, dy'\, dz'\, \left\{x'\hat{\mathbf{e}}_x + y'\hat{\mathbf{e}}_y + (z' - z)\hat{\mathbf{e}}_z\right\}}{\left\{x'^2 + y'^2 + (z - z')^2\right\}^{3/2}}$$

with components

$$g_x = G\rho_m \int_\Omega \frac{x'\, dx'\, dy'\, dz'}{\left\{x'^2 + y'^2 + (z - z')^2\right\}^{3/2}},$$
$$g_y = G\rho_m \int_\Omega \frac{y'\, dx'\, dy'\, dz'}{\left\{x'^2 + y'^2 + (z - z')^2\right\}^{3/2}},$$
$$g_z = G\rho_m \int_\Omega \frac{(z' - z)\, dx'\, dy'\, dz'}{\left\{x'^2 + y'^2 + (z - z')^2\right\}^{3/2}}.$$

The limits of integrals associated with Ω can be done as discussed in Section 3.3. In Figure 4.13, we have chosen the first integral to be along the z-axis. Then the lower limit will be the xy-plane, or $z' = 0$, and the upper limit, the surface of the hemisphere. A general point P' in Ω with coordinates (x', y', z') will hit the hemisphere at $z' = \sqrt{a^2 - x'^2 - y'^2}$. So, this will be the upper limit of the z' integration. Concentrating on the x-component for a moment, we thus write

$$g_x = G\rho_m \iint_S x'\, dx'\, dy' \int_0^{\sqrt{a^2 - x'^2 - y'^2}} \frac{dz'}{\left\{x'^2 + y'^2 + (z - z')^2\right\}^{3/2}},$$

where S is the projection of the hemispherical surface on the xy-plane. To do the remaining integrations, we refer to Figure 4.14, where the projections of the hemisphere and the point P' are shown. It is clear that the y' integration has

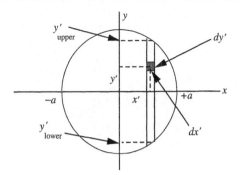

Figure 4.14: The projection of Ω, a hemisphere, in the xy-plane.

the lower semicircle as the lower limit and the upper semicircle as the upper limit. Finally the x' integration has lower and upper limits of $-a$ and $+a$, respectively. We, therefore, have

$$g_x = G\rho_m \int_{-a}^{+a} x'\, dx' \int_{-\sqrt{a^2-x'^2}}^{+\sqrt{a^2-x'^2}} dy' \int_0^{\sqrt{a^2-x'^2-y'^2}} \frac{dz'}{\{x'^2 + y'^2 + (z-z')^2\}^{3/2}}.$$

Instead of looking up the integrals in an integral table, we note that the integrand of the x' integration is an odd function. This is because it is the product of x', which is odd, and another function, in the form of a double integral whose integrand and limits are even functions of x'. Since the interval of integration is symmetric, the x' integration vanishes. A similar argument shows that the y' integration vanishes as well. This is as expected intuitively: We expect the field to be along the z-axis. Therefore, $g_x = 0$, $g_y = 0$, and

$$g_z = G\rho_m \int_{-a}^{+a} dx' \int_{-\sqrt{a^2-x'^2}}^{+\sqrt{a^2-x'^2}} dy' \int_0^{\sqrt{a^2-x'^2-y'^2}} \frac{(z'-z)\, dz'}{\{x'^2 + y'^2 + (z-z')^2\}^{3/2}}$$

$$= G\rho_m \int_{-a}^{+a} dx' \int_{-\sqrt{a^2-x'^2}}^{+\sqrt{a^2-x'^2}} \frac{dy'}{\sqrt{x'^2 + y'^2 + z^2}}$$

$$- G\rho_m \int_{-a}^{+a} dx' \int_{-\sqrt{a^2-x'^2}}^{+\sqrt{a^2-x'^2}} \frac{dy'}{\sqrt{a^2 + z^2 + x'^2 + y'^2 - 2z\sqrt{a^2 - x'^2 - y'^2}}}.$$

The y' integration in the first integral can be done, but the remaining x' integration will be complicated. The second y' integral cannot even be performed in closed form. This difficulty is a result of our poor choice of coordinates whereby the boundary of the region of integration does not turn out to be a "natural" surface. ∎

The example of the hemisphere in Cartesian coordinates indicates the difficulty encountered when the boundaries of the integration region do not match the primary surfaces of the coordinate system. In the next example, we calculate the gravitational field of the hemisphere in spherical coordinates.

Example 4.3.2. The spherical coordinate system makes the problem so manageable that we can consider a more general mass distribution. We will calculate

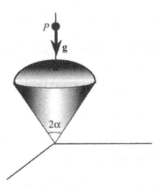

Figure 4.15: The gravitational field of a solid cone with a spherically curved top.

the gravitational field of a cone-shaped segment of a solid sphere of half-angle α as shown in Figure 4.15. We are interested in the field at a point P on the axis of the cone as shown. Since $\hat{\mathbf{e}}_\theta$ and $\hat{\mathbf{e}}_\varphi$ cannot be defined at P (why?), we expect, from physical intuition, that the only surviving component of the gravitational field is radial. This component is obtained by dotting the vector field with $\hat{\mathbf{e}}_r$:

$$g_r = \hat{\mathbf{e}}_r \cdot \mathbf{g} = \hat{\mathbf{e}}_r \cdot \left\{ -\int_\Omega \frac{G\,dm(\mathbf{r}')}{|\mathbf{r}-\mathbf{r}'|^3}(\mathbf{r}-\mathbf{r}') \right\} = -G \int_\Omega \frac{dm(\mathbf{r}')}{|\mathbf{r}-\mathbf{r}'|^{3/2}}(r - r'\cos\theta')$$

$$= -G\rho_m \int_\Omega \frac{r'^2 \sin\theta'\,dr'\,d\theta'\,d\varphi'}{(r^2+r'^2-2rr'\cos\theta')^{3/2}}(r - r'\cos\theta')$$

$$= -G\rho_m \int_0^{2\pi} d\varphi' \int_0^a r'^2\,dr' \int_0^\alpha \frac{\sin\theta'\,d\theta'}{(r^2+r'^2-2rr'\cos\theta')^{3/2}}(r - r'\cos\theta').$$

To do the integrations, we use the technique of differentiating inside the integral and note that

$$\frac{r - r'\cos\theta'}{(r^2+r'^2-2rr'\cos\theta')^{3/2}} = -\frac{\partial}{\partial r}\frac{1}{\sqrt{r^2+r'^2-2rr'\cos\theta'}}.$$

Therefore, the integral becomes

$$g_r = 2\pi G\rho_m \int_0^a r'^2\,dr' \int_0^\alpha \sin\theta'\,d\theta' \frac{\partial}{\partial r}\frac{1}{\sqrt{r^2+r'^2-2rr'\cos\theta'}}$$

$$= 2\pi G\rho_m \frac{\partial}{\partial r} \int_0^a r'^2\,dr' \int_0^\alpha \frac{\sin\theta'\,d\theta'}{\sqrt{r^2+r'^2-2rr'\cos\theta'}}$$

$$= 2\pi G\rho_m \frac{\partial}{\partial r} \int_0^a r'^2\,dr' \left(\frac{1}{rr'}\sqrt{r^2+r'^2-2rr'\cos\theta'} \;\Big|_0^\alpha \right)$$

$$= 2\pi G\rho_m \frac{\partial}{\partial r} \left\{ \frac{1}{r} \int_0^a r'\,dr' \left(\sqrt{r^2+r'^2-2rr'\cos\alpha} - \sqrt{(r-r')^2} \right) \right\}$$

$$= 2\pi G\rho_m \frac{\partial}{\partial r} \left\{ \frac{1}{r} \int_0^a r'\,dr' \left(\sqrt{r^2+r'^2-2rr'\cos\alpha} - |r-r'| \right) \right\}. \qquad (4.35)$$

The integral involving the absolute value can be done easily. However, we have to be careful about the relative size of r, a, and r'. We therefore consider two cases: $r \geq a$ and $r \leq a$. Keeping in mind that $r' \leq a$, the first case yields

$$\int_0^a r'|r - r'|\,dr' = \int_0^a r'(r - r')\,dr' = \frac{ra^2}{2} - \frac{a^3}{3}, \qquad r \geq a.$$

For the second case, we have to split the interval of integration in two, and write the absolute value accordingly:

$$\int_0^a r'|r - r'|\,dr' = \int_0^r r'(r - r')\,dr' + \int_r^a r'(r' - r)\,dr'$$

$$= \frac{r^3}{3} + \frac{a^3}{3} - \frac{ra^2}{2}, \qquad r \leq a.$$

Substituting these in Equation (4.35), we get

$$g_r = 2\pi G \rho_m \frac{\partial}{\partial r} \begin{cases} \dfrac{a^3}{3r} - \dfrac{a^2}{2} + \dfrac{1}{r}\displaystyle\int_0^a r'\sqrt{r^2 + r'^2 - 2rr'\cos\alpha}\,dr' & \text{if } r \geq a, \\[4mm] \dfrac{a^2}{2} - \dfrac{r^2}{3} - \dfrac{a^3}{3r} + \dfrac{1}{r}\displaystyle\int_0^a r'\sqrt{r^2 + r'^2 - 2rr'\cos\alpha}\,dr' & \text{if } r \leq a. \end{cases}$$

The remaining integral can also be performed with the result

$$\frac{1}{r}\int_0^a r'\sqrt{r^2 + r'^2 - 2rr'\cos\alpha}\,dr' = -\frac{r^2}{12}(1 - 3\cos 2\alpha)$$

$$+ \sqrt{r^2 + a^2 - 2ra\cos\alpha}\left(\frac{a^2}{3r} + \frac{r}{12} - \frac{a\cos\alpha}{6} - \frac{r\cos 2\alpha}{4}\right)$$

$$+ \frac{r^2\cos\alpha\sin^2\alpha}{2}\ln\left(\frac{a - r\cos\alpha + \sqrt{r^2 + a^2 - 2ra\cos\alpha}}{r - r\cos\alpha}\right).$$

The special case of $\alpha = \pi$, i.e., a full sphere, is very important, because historically it motivated the rapid development of integral calculus. For this case, we have

$$\frac{1}{r}\int_0^a r'\sqrt{r^2 + r'^2 - 2rr'\cos\alpha}\,dr' \xrightarrow{\alpha=\pi} \frac{r^2}{6} + (a+r)\left(\frac{a^2}{3r} - \frac{r}{6} + \frac{a}{6}\right),$$

whereby the radial component of the field becomes

$$g_r = 2\pi G \rho_m \frac{\partial}{\partial r} \begin{cases} \dfrac{a^3}{3r} - \dfrac{a^2}{2} + \dfrac{r^2}{6} + (a+r)\left(\dfrac{a^2}{3r} - \dfrac{r}{6} + \dfrac{a}{6}\right) & \text{if } r \geq a, \\[4mm] \dfrac{a^2}{2} - \dfrac{r^2}{6} - \dfrac{a^3}{3r} + (a+r)\left(\dfrac{a^2}{3r} - \dfrac{r}{6} + \dfrac{a}{6}\right) & \text{if } r \leq a, \end{cases}$$

$$= 2\pi G \rho_m \begin{cases} -\dfrac{2a^3}{3r^2} & \text{if } r \geq a \\[4mm] -\dfrac{2r}{3} & \text{if } r \leq a \end{cases} = \begin{cases} -\dfrac{GM}{r^2} & \text{if } r \geq a, \\[4mm] -\dfrac{4\pi G \rho_m r}{3} & \text{if } r \leq a. \end{cases}$$

The first result is the well-known fact that the field outside a uniform sphere is the same as the field of a point charge with the same mass concentrated at the center of the original sphere. The second result, usually obtained in electrostatics by using Gauss's law, *would not have been obtained* if we had not used absolute values when extracting a square root.   ∎

Example 4.3.3. A uniformly charged hollow cylinder of length L and volume charge density ρ_q has an inner radius a and an outer radius b (see Figure 4.16). The cylinder is rotating with constant angular speed ω about its axis. We want to find the magnetic field produced by this motion of charges. We note that the problem has an azimuthal symmetry, so we do not lose generality if we choose our coordinates so that our field point P lies in the xz-plane. This is equivalent to setting $\varphi = 0$.

We use cylindrical coordinates in Equation (4.17) to find the magnetic field. For a general field point, we have

$$\mathbf{r} = \rho\,\hat{\mathbf{e}}_\rho + z\hat{\mathbf{e}}_z, \qquad \mathbf{r}' = \rho'\hat{\mathbf{e}}_{\rho'} + z'\hat{\mathbf{e}}_z,$$
$$\mathbf{r} - \mathbf{r}' = \rho\,\hat{\mathbf{e}}_\rho - \rho'\hat{\mathbf{e}}_{\rho'} + (z - z')\hat{\mathbf{e}}_z,$$
$$|\mathbf{r} - \mathbf{r}'|^3 = \left\{\rho^2 + \rho'^2 - 2\rho\rho'\cos\varphi' + (z - z')^2\right\}^{3/2},$$
$$dq(\mathbf{r}') = \rho_q\,dV(\mathbf{r}') = \rho_q\rho'd\rho'd\varphi'dz', \qquad \mathbf{v}(\mathbf{r}') = \rho'\omega\hat{\mathbf{e}}_{\varphi'},$$

so that

$$\mathbf{v}(\mathbf{r}') \times (\mathbf{r} - \mathbf{r}') = \omega\rho\rho'\underbrace{\hat{\mathbf{e}}_{\varphi'} \times \hat{\mathbf{e}}_\rho}_{-\hat{\mathbf{e}}_z\cos\varphi'} - \omega\rho'^2\underbrace{\hat{\mathbf{e}}_{\varphi'} \times \hat{\mathbf{e}}_{\rho'}}_{-\hat{\mathbf{e}}_z} + \omega\rho'(z - z')\underbrace{\hat{\mathbf{e}}_{\varphi'} \times \hat{\mathbf{e}}_z}_{\hat{\mathbf{e}}_{\rho'}}$$
$$= \omega\rho'(z - z')\hat{\mathbf{e}}_{\rho'} + \omega(\rho'^2 - \rho\rho'\cos\varphi')\hat{\mathbf{e}}_z.$$

Substituting all these results in Equation (4.17), we obtain

$$\mathbf{B} = \int_\Omega \frac{\omega k_m(\rho_q\rho'd\rho'd\varphi'dz')\left[\rho'(z - z')\hat{\mathbf{e}}_{\rho'} + (\rho'^2 - \rho\rho'\cos\varphi')\hat{\mathbf{e}}_z\right]}{\left\{\rho^2 + \rho'^2 - 2\rho\rho'\cos\varphi' + (z - z')^2\right\}^{3/2}}.$$

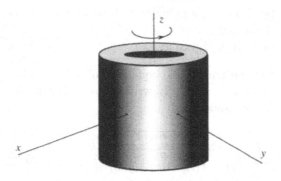

Figure 4.16: The charged rotating hollow cylinder produces a magnetic field due to the motion of charges.

The cylindrical components are obtained by dotting this equation with the cylindrical unit vectors at P:

$$B_\rho = \mathbf{B} \cdot \hat{\mathbf{e}}_\rho = \omega k_m \rho_q \int_0^{2\pi} \int_a^b \int_{-L/2}^{L/2} \frac{\rho'^2 (z - z') \cos \varphi' \, d\rho' d\varphi' dz'}{\left\{ \rho^2 + \rho'^2 - 2\rho\rho' \cos \varphi' + (z - z')^2 \right\}^{3/2}},$$

$$B_\varphi = \mathbf{B} \cdot \hat{\mathbf{e}}_\varphi = \omega k_m \rho_q \int_0^{2\pi} \int_a^b \int_{-L/2}^{L/2} \frac{\rho'^2 (z - z') \sin \varphi' \, d\rho' d\varphi' dz'}{\left\{ \rho^2 + \rho'^2 - 2\rho\rho' \cos \varphi' + (z - z')^2 \right\}^{3/2}} = 0,$$

$$B_z = \mathbf{B} \cdot \hat{\mathbf{e}}_z = \omega k_m \rho_q \int_0^{2\pi} \int_a^b \int_{-L/2}^{L/2} \frac{\left(\rho'^3 - \rho\rho'^2 \cos \varphi' \right) d\rho' d\varphi' dz'}{\left\{ \rho^2 + \rho'^2 - 2\rho\rho' \cos \varphi' + (z - z')^2 \right\}^{3/2}}.$$

The middle equation gives zero as a result of the φ' integration. It turns out that the z' and ρ' integrations of the remaining integrals can be performed in closed form. However, the results are very complicated and will not be reproduced here. Furthermore, the φ' integration has no closed form and must be done numerically.

We can also obtain the components of $\mathbf{B}$ in other coordinate systems by dotting $\mathbf{B}$ into the corresponding unit vectors. The reader may check, for example, that in Cartesian coordinates, $B_x = \mathbf{B} \cdot \hat{\mathbf{e}}_x$ is the same as B_ρ above and B_y is the same as B_φ, i.e., $B_y = 0$. This is due to the particular choice of our coordinate system ($\varphi = 0$). ∎

4.4   Problems

4.1. Differentiate Equation (4.2) to find the velocity and acceleration and compare with the expected results.

4.2. By choosing a coordinate system properly, write down the simplest parametric equation for the following curves. In each case specify the range of the parameter you use:
(a) a rectangle of sides a and b, lying in the xy-plane with center at the origin and sides parallel to the axes;
(b) an ellipse with semi-major and semi-minor axes a and b;
(c) a helix wrapped around a cylinder with an elliptical cross section of the type described in (b); and
(d) a helix wrapped around a cone.

4.3. Assume that the parametric equations of a linear charge density are $x' = f(t), y' = g(t), z' = h(t)$. By writing everything in Equation (4.3) in Cartesian coordinates, show that

$$\mathbf{E} = \int_a^b \frac{k_e \Lambda(t) \sqrt{[f'(t)]^2 + [g'(t)]^2 + [h'(t)]^2}}{\left\{ [x - f(t)]^2 + [y - g(t)]^2 + [z - h(t)]^2 \right\}^{3/2}}$$
$$\times \left([x - f(t)] \hat{\mathbf{e}}_x + [y - g(t)] \hat{\mathbf{e}}_y + [z - h(t)] \hat{\mathbf{e}}_z \right) dt. \qquad (4.36)$$

and that

$$E_x = \int_a^b \frac{k_e \Lambda(t) \sqrt{[f'(t)]^2 + [g'(t)]^2 + [h'(t)]^2}}{\left\{ [x - f(t)]^2 + [y - g(t)]^2 + [z - h(t)]^2 \right\}^{3/2}} [x - f(t)] \, dt$$

$$E_y = \int_a^b \frac{k_e \Lambda(t) \sqrt{[f'(t)]^2 + [g'(t)]^2 + [h'(t)]^2}}{\left\{ [x - f(t)]^2 + [y - g(t)]^2 + [z - h(t)]^2 \right\}^{3/2}} [y - g(t)] \, dt \quad (4.37)$$

$$E_z = \int_a^b \frac{k_e \Lambda(t) \sqrt{[f'(t)]^2 + [g'(t)]^2 + [h'(t)]^2}}{\left\{ [x - f(t)]^2 + [y - g(t)]^2 + [z - h(t)]^2 \right\}^{3/2}} [z - h(t)] \, dt$$

and

$$\Phi(x, y, z) = \int_a^b \frac{k_e \Lambda(t) \sqrt{[f'(t)]^2 + [g'(t)]^2 + [h'(t)]^2}}{\left\{ [x - f(t)]^2 + [y - g(t)]^2 + [z - h(t)]^2 \right\}^{1/2}} \, dt$$

How is $\Lambda(t)$ related to $\lambda(\mathbf{r}')$?

4.4. (a) Show that

$$\hat{e}_\rho \cdot \hat{e}_x = \frac{x}{\sqrt{x^2 + y^2}}, \qquad \hat{e}_\rho \cdot \hat{e}_y = \frac{y}{\sqrt{x^2 + y^2}}.$$

(b) Similarly, express $\hat{e}_\varphi \cdot \hat{e}_x$ and $\hat{e}_\varphi \cdot \hat{e}_y$ in Cartesian coordinates.
(c) Use (a), (b), and Equation (4.36) to find the general expressions for E_ρ and E_φ as integrals in Cartesian coordinates similar to the integrals of Equation (4.37).

4.5. (a) Find the nine dot products of all Cartesian and spherical unit vectors and express the results in terms of Cartesian coordinates.
(b) Use (a) and Equation (4.36) to find general expressions for E_r, E_θ, and E_φ as integrals in Cartesian coordinates similar to the integrals of Equation (4.37).

4.6. Assume that the parametric equations of a linear charge density are $\rho' = f(t), \varphi' = g(t), z' = h(t)$. By writing everything in Equation (4.3) in cylindrical coordinates, show that Equation (4.11) holds and that

$$E_\rho = \int_a^b \frac{k_e \Lambda(t) \sqrt{[f'(t)]^2 + [f(t)]^2 [g'(t)]^2 + [h'(t)]^2}}{\left\{ \rho^2 + [f(t)]^2 - 2\rho f(t) \cos(\varphi - g(t)) + [z - h(t)]^2 \right\}^{3/2}}$$
$$\times \left[\rho - f(t) \cos(g(t) - \varphi) \right] dt \qquad (4.38)$$

$$E_\varphi = -\int_a^b \frac{k_e \Lambda(t) \sqrt{[f'(t)]^2 + [f(t)]^2 [g'(t)]^2 + [h'(t)]^2}}{\left\{ \rho^2 + [f(t)]^2 - 2\rho f(t) \cos(\varphi - g(t)) + [z - h(t)]^2 \right\}^{3/2}} \qquad (4.39)$$
$$\times f(t) \sin(g(t) - \varphi) \, dt$$

$$E_z = \int_a^b \frac{k_e\Lambda(t)\sqrt{[f'(t)]^2 + [f(t)]^2[g'(t)]^2 + [h'(t)]^2}}{\left\{\rho^2 + [f(t)]^2 - 2\rho f(t)\cos(\varphi - g(t)) + [z - h(t)]^2\right\}^{3/2}} \qquad (4.40)$$
$$\times \left[z - h(t)\right] dt$$

and

$$\Phi = \int_a^b \frac{k_e\Lambda(t)\sqrt{[f'(t)]^2 + f^2(t)[g'(t)]^2 + [h'(t)]^2}}{\left\{\rho^2 + f^2(t) - 2\rho f(t)\cos(\varphi - g(t)) + [z - h(t)]^2\right\}^{1/2}} dt$$

How is $\Lambda(t)$ related to $\lambda(\mathbf{r}')$?

4.7. Use (4.11) to calculate Cartesian and spherical components of the electric field in terms of integrals in cylindrical variables similar to (4.38).

4.8. Use the cylindrical coordinates for the integration variables of Example 4.1.3, but calculate the Cartesian components of $\mathbf{E}$.

4.9. A uniformly charged infinitely thin circular ring of radius a has total charge Q. Place the ring in the xy-plane with its center at the origin. Use cylindrical coordinates.
(a) Find the electrostatic potential at P with cylindrical coordinates (ρ, φ, z) in terms of a single integral.
(b) Find the analytic form of the potential if P is on the z-axis (evaluate the integral).
(c) Find the potential at a point in the xy-plane a distance $2a$ from the origin. Give your answer as a number times k_eQ/a.

4.10. Write a general formula for $\Phi(\mathbf{r})$ of a charged curve in spherical coordinates.

4.11. A straight-line segment of length $2L$ is placed on the z-axis with its midpoint at the origin. The segment has a linear charge density given by

$$\lambda(x, y, z) = \frac{Q}{|z| + a},$$

where Q and a are constants with $a > 0$. Find the electrostatic potential of this charge distribution at a point on the x-axis in Cartesian coordinates.

4.12. Same as the previous problem, except that

$$\lambda(x, y, z) = \frac{aQ}{z^2 + a^2}.$$

Look up the integral in an integral table.
(a) Does anything peculiar happen at $x = \pm a$? Based on the integration result? Based on physical intuition? Look at the result carefully and reconcile any conflict.
(b) What is the potential when $L \to \infty$?

4.13. A segment of the parabola $y = x^2/a$—with a a constant—extending from $x = 0$ to $x = L$ has a linear charge density given by

$$\lambda(x, y, z) = \frac{\lambda_0}{\sqrt{1 + (2x/a)^2}},$$

where λ_0 is a constant. Find the potential and the electric field at the point $(0, 0, z)$. What are Φ and $\mathbf{E}$ at $(0, 0, a/2)$? Simplify your results as much as possible.

4.14. A circular ring of radius a is uniformly charged with linear density λ.
(a) Find an expression for each of the three components of the electric field at an arbitrary point in space in terms of an integral in an appropriate coordinate system. Evaluate the integrals whenever possible.
(b) Find the components of the field at the point P shown in Figure 4.17. Express your answers as a numerical multiple of $k_e\lambda/a$.
(c) Find the electrostatic potential at the point P shown in Figure 4.17. Express your answer as a numerical multiple of $k_e\lambda$.
For (b) and (c) you will need to evaluate certain integrals numerically.

4.15. Consider a uniform linear charge distribution in the form of an ellipse with linear charge density λ. The semi-major and semi-minor axes of the ellipse are a and b, respectively. Use Cartesian coordinates and the parametric equation of the ellipse.
(a) Write down the integrals that give the electric field and the electric potential at an arbitrary point P in space.
(b) Specialize to the case where P lies on the axis that is perpendicular to the plane of the ellipse and passes through its center.
(c) Specialize (a) to the case where P lies on the line containing the minor axis.

4.16. Consider a uniform linear charge distribution in the form of an ellipse with linear charge density λ located in the xy-plane. The semi-major and semi-minor axes of the ellipse are $2a$ and a, respectively.
(a) Write the Cartesian parameterization of the ellipse in terms of trigonometric functions.

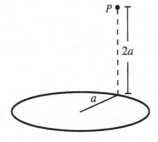

Figure 4.17: The figure for Problems 4.14 and 4.21.

(b) Write the integral that gives the Cartesian components of the electric field at an arbitrary point (x, y, z) in space.

(c) Specialize to the point $(a, 2a, 2a)$, and write your answer as a numerical multiple of $k_e \lambda / a$.

4.17. Consider a uniform linear charge distribution, with linear charge density λ, in the form of an elliptical helix whose parametric equation is given by

$$x' = a \cos t, \quad y' = b \sin t, \quad z' = ct$$

Use Cartesian coordinates.

(a) Write down the integrals that give the electric field and the electric potential at an arbitrary point P in space.

(b) Verify that when $c = 0$, you get the field and potential of an ellipse (see Problem 4.15).

(c) Verify that when $c = 0 = b$, you get the field and potential of a straight line segment.

(d) Verify that when $c = 0 = b$ and $a \to \infty$, you get the field of an infinite straight line.

4.18. Find the three components of the electric field and the potential of Example 4.1.2 when $a = -L/2$ and $b = L/2$. Approximate the three components of the electric field for the case where $L \gg x$.

4.19. Derive all relations in Equations (4.20) and (4.21).

4.20. Figure 4.18 shows a hyperbola $y = \sqrt{x^2 + a^2}$. Only the segment between $x = 0$ and $x = a$ is charged uniformly with *linear* density λ.

(a) Write the expression for $\mathbf{E}$ as an integral in Cartesian coordinates.

(b) Find the three components of $\mathbf{E}$ as integrals over x'.

(c) Making the substitution $x' = au$, write each component as a numerical multiple of $k_e \lambda / a$.

4.21. A circular ring of radius a is uniformly charged with linear density λ. The ring rotates with angular speed ω about the axis perpendicular to the plane of the ring, passing through its center.

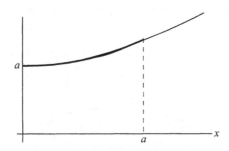

Figure 4.18: The segment of the hypebola that is charged.

(a) Find an expression for each of the three components of the magnetic field at an arbitrary point in space in terms of an integral in an appropriate coordinate system. Evaluate the integrals whenever possible.

(b) Find the components of the field at the point P shown in Figure 4.17. Express your answers as a numerical multiple of $k_m \lambda \omega$. (You will need to evaluate some integrals numerically!)

4.22. An elliptical conducting ring of semi-major axis a and semi-minor axis b carries a current I.

(a) Find an expression for each of the three Cartesian components of the magnetic field at an arbitrary point in space in terms of an integral in the Cartesian coordinate system.

(b) Find an integral expression for the components of the field at a point on the line perpendicular to the ellipse that passes through its center.

4.23. Perform the integrals for E_x, E_y, and E_z of Example 4.2.1 when the field point is on the z-axis. Hint: You can get E_x and E_y without doing the integrals.

4.24. Assume that the parametric equations of a current loop are $x' = f(t), y' = g(t), z' = h(t)$. By writing everything in Equation (4.18) in Cartesian coordinates, show that

$$B_x(\mathbf{r}) = k_m I \int_a^b \frac{g'(t)\left[z - h(t)\right] - h'(t)\left[y - g(t)\right]}{\left\{\left[x - f(t)\right]^2 + \left[y - g(t)\right]^2 + \left[z - h(t)\right]^2\right\}^{3/2}} dt,$$

$$B_y(\mathbf{r}) = k_m I \int_a^b \frac{h'(t)\left[x - f(t)\right] - f'(t)\left[z - h(t)\right]}{\left\{\left[x - f(t)\right]^2 + \left[y - g(t)\right]^2 + \left[z - h(t)\right]^2\right\}^{3/2}} dt,$$

$$B_z(\mathbf{r}) = k_m I \int_a^b \frac{f'(t)\left[y - g(t)\right] - g'(t)\left[x - f(t)\right]}{\left\{\left[x - f(t)\right]^2 + \left[y - g(t)\right]^2 + \left[z - h(t)\right]^2\right\}^{3/2}} dt,$$

where a and b are the initial and final values of the parameter t.

4.25. By writing everything in Equation (4.18) in cylindrical coordinates, show that

$$B_\rho = k_m I \oint \frac{N_1 \, d\rho' + \rho' N_2 \, d\varphi' + \rho' \sin(\varphi' - \varphi) \, dz'}{\left\{\rho^2 + \rho'^2 - 2\rho\rho' \cos(\varphi - \varphi') + (z - z')^2\right\}^{3/2}}$$

$$B_\varphi = k_m I \oint \frac{\rho' N_1 \, d\varphi' - N_2 \, d\rho' + \left[\rho - \rho' \cos(\varphi' - \varphi)\right] dz'}{\left\{\rho^2 + \rho'^2 - 2\rho\rho' \cos(\varphi - \varphi') + (z - z')^2\right\}^{3/2}}$$

$$B_z = -k_m I \oint \frac{\rho \sin(\varphi' - \varphi) \, d\rho' + \left[\rho\rho' \cos(\varphi' - \varphi) - \rho'^2\right] d\varphi'}{\left\{\rho^2 + \rho'^2 - 2\rho\rho' \cos(\varphi - \varphi') + (z - z')^2\right\}^{3/2}}$$

where

$$N_1 \equiv (z - z') \sin(\varphi' - \varphi), \quad N_2 \equiv (z - z') \cos(\varphi' - \varphi)$$

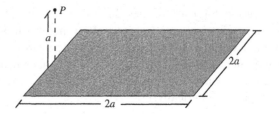

Figure 4.19: The figure for Problem 4.28.

4.26. Derive Equation (4.27).

4.27. Derive Equation (4.29) from Equation (4.28).

4.28. A square of side $2a$ is uniformly charged with surface density σ.
(a) Find the electrostatic potential at an arbitrary point in space. Do one of the integrals and express your answer in terms of a single integral in an appropriate coordinate system.
(b) Find the potential at a point a distance a directly above the midpoint of one of the sides as shown in Figure 4.19. Express your answer as a numerical multiple of $k_e \sigma a$.

4.29. The area in the xy-plane shown in Figure 4.21 is uniformly charged with surface charge density σ. The equation of the parabolic boundary is $y = x^2/a$. Assume that the observation point (field point) P is on the z-axis at $z = a$.
(a) Derive the Cartesian components of the electric field at P as double integrals.
(b) Do the y' integration first and then the x' integration to find the components of the electric field. Write your answers as a numerical multiples of $k_e \sigma$. You will need to evaluate certain integral(s) numerically.

4.30. Using cylindrical coordinates, find the electrostatic field of a uniformly charged circular disk of charge density σ and radius a:
(a) at an arbitrary point in space;
(b) at an arbitrary point on the perpendicular axis of the disk; and

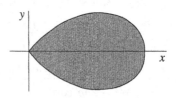

Figure 4.20: The region of the xy-plane that is charged.

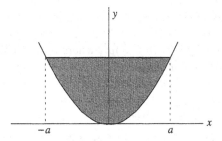

Figure 4.21: The shaded region is uniformly charged.

(c) at an arbitrary point in the plane of the disk.
(d) For (b), consider the case of infinite radius and compare your result with the infinite rectangle discussed in introductory physics books and Example 4.2.3.

4.31. Figure 4.20 shows a region of the xy-plane that is uniformly charged with surface charge density σ. The *boundary* of the region is given in a polar/cylindrical coordinate system by $\rho = a\cos(2\varphi)$ with $-\pi/4 \leq \varphi \leq \pi/4$. We are interested in the electrostatic potential at a point P on the z-axis with $z = a$.
(a) Write the position vector of P and P' (a typical source point) in cylindrical coordinates. Now evaluate $|\mathbf{r} - \mathbf{r}'|$.
(b) Write the expression for $dq(\mathbf{r}')$ in cylindrical coordinates.
(c) Write the expression for the potential Φ as a double integral in cylindrical coordinates.
(d) Perform one of the integrations, and wrtie your final answer as a single integral.
(e) Find the value of the potential as a numerical multiple of $k_e\sigma a$.

4.32. A cylindrical shell of radius a and length L is uniformly charged with surface charge density σ. Using an appropriate coordinate system and axis orientation:
(a) Find the electric field at an arbitrary point in space.
(b) Now let the length go to infinity and find a closed-form expression for the field in (a). You will have to look up the integral in an integral table.
(c) Find the expression of the field for a point outside and a point inside the cylinder.

4.33. A uniformly charged disk of radius a and surface charge density σ is inthe xy-plane with its center at the origin and is rotating about its perpendicular axis with angular frequency ω.
(a) Find the cylindrical components of the magnetic field produced at a point $P = (\rho, 0, z)$ as double integrals in cylindrical coordinates.
(b) Now assume that P is on the z-axis and find the components of $\mathbf{B}$ by performing all the integrals involved.

4.34. An electrically charged disk of radius a is rotating about its perpendicular axis with angular frequency ω. Its surface charge density is given in cylindrical coordinates by $\sigma = (\sigma_0/a^2)\rho^2$, where σ_0 is a constant.
(a) Find the *Cartesian* components of the magnetic field produced at an arbitrary point $P = (\rho, 0, z)$ as double integrals in cylindrical coordinates.
(b) Now assume that P is on the z-axis and find the components of **B** by performing all the integrals involved.

4.35. Express the components of **g** of Example 4.2.4 in Cartesian and cylindrical coordinates in terms of integrals similar to Equation (4.33).

4.36. A conic surface of (maximum) radius a and half-angle α is uniformly charged with surface density σ.
(a) Find the three components of the electric field at a point on the cone's axis a distance r from its vertex. Express your answers in terms of single integrals in an appropriate coordinate system.
(b) Find the components of the field at $r = a/\sqrt{3}$ when $\alpha = \pi/6$. By evaluating integrals numerically if necessary, express your answer as a numerical multiple of $k_e\sigma$.

4.37. A cone with half-angle α, the distance of whose vertex from its circular rim is L, is rotating with angular speed ω about its axis. Electric charge is distributed uniformly on the cone with surface charge density σ. Use the coordinate system appropriate for this geometry.
(a) Express the components of the magnetic field produced at an arbitrary point in space in terms of double integrals. Evaluate those components whose integrals are easily done.
(b) Move the field point to the axis of the cone, and write the components of the field in terms of single integrals. Evaluate the remaining components whose integrals are easily done.
(c) Now assume that $\alpha = \pi/3$, and express the magnitude of the field on the axis at a distance L from the vertex of the cone as a number times $k_m\omega\sigma L$.

4.38. A uniformly charged solid cylinder of length L, radius a, and total charge q is rotated about its axis with angular speed ω. Find the magnetic field at a point on this axis.

4.39. Use cylindrical coordinates to calculate the gravitational field of the hemisphere of Example 4.3.1 at a point on the z-axis.
(a) Show that

$$g_z = 2\pi G\rho_m\left\{\sqrt{z^2 + a^2} - |z| - \frac{(a^2 + z^2)^{3/2} - |a - z|(a^2 + z^2 + az)}{3z^2}\right\}$$

with the other components being zero.
(b) Simplify this expression for points outside ($z < 0$ and $z > a$), and inside ($0 < z < a$).
(c) Using the result of (b), find the gravitational field of a hemisphere whose flat side points up.
(d) Add the results of (b) and (c) to find the field of a full sphere.

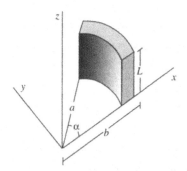

Figure 4.22: The segment of a cylinder with uniform charge density used in Problem 4.41.

4.40. Find the moment of inertia of a uniform solid cone of mass M and half-angle α cut out of a solid sphere of radius a. What is the moment of inertia of a whole solid sphere?

4.41. A solid cylinder of length L has a cross section which is in the shape of a segment of an annular ring with outer radius b and inner radius a. It is subtended by an angle α and is uniformly charged with total charge q (Figure 4.22). Find the electric field at:
(a) an arbitrary point in space; and
(b) a point on the axis of the ring.
(c) What is the answer to (b) if we have a complete ring?
(d) What is the answer to (a) if we have a complete ring that is infinitely long? Consider the three regions: $\rho \leq a$, $a \leq \rho \leq b$, and $\rho \geq b$.

4.42. Find the moment of inertia of the (incomplete) cylinder of the previous problem about the perpendicular axis passing through the common center of the inner and outer radii. Assume that the total mass is M. From this result obtain the moment of inertia of a hollow as well as a solid cylinder.

Chapter 5

Dirac Delta Function

Paul Adrian Maurice Dirac, one of the most inventive mathematical physicists of all time, co-founder of quantum theory, inventor of relativistic quantum mechanics in the form of an equation which bears his name, predictor of the existence of anti-matter, clarifier of the concept of spin, and contributor to the unraveling of the mathematical difficulties associated with the quantization of the general theory of relativity, came across the subject matter of this chapter in his study of quantum mechanical scattering. In order to appreciate the usefulness of this function, we shall start with an intuitive approach drawn from electrostatics.

5.1 One-Variable Case

Consider a straight linear charge distribution of length L with uniform charge density as shown in Figure 5.1(a). If the total charge of the line segment is q, then the linear charge density will be $\lambda = q/L$. We are interested in the graph of the function describing the linear density in the interval $(-\infty, +\infty)$. Assuming that the midpoint of the segment is x_0 and its length L, we can easily draw the graph of the function. This is shown in Figure 5.1(b). The graph

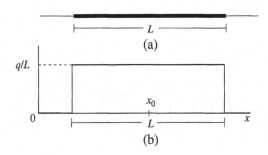

Figure 5.1: (a) The charged line segment and (b) its linear density function.

is that of a function that is zero for values less than $x_0 - L/2$, q/L for values between $x_0 - L/2$ and $x_0 + L/2$, and zero again for values greater than $x_0 + L/2$. Let us call this function $\lambda(x)$. Then, we can write

$$\lambda(x) = \begin{cases} 0 & \text{if } x < x_0 - L/2, \\ q/L & \text{if } x_0 - L/2 < x < x_0 + L/2, \\ 0 & \text{if } x > x_0 + L/2. \end{cases}$$

Now suppose that we squeeze the segment on both sides so that the length shrinks to $L/2$ without changing the position of the midpoint and the amount of charge. The new function describing the linear charge density will now be

$$\lambda(x, x_0) = q \begin{cases} 0 & \text{if } x < x_0 - L/4, \\ 2/L & \text{if } x_0 - L/4 < x < x_0 + L/4, \\ 0 & \text{if } x > x_0 + L/4. \end{cases}$$

We have "factored out" q for later convenience. We have also introduced a second argument to emphasize the dependence of the function on the midpoint. Instead of one-half, we can shrink the segment to any fraction, still keeping both the amount of charge and the midpoint unchanged. Shrinking the size to L/n and renaming the function $\lambda_n(x, x_0)$ to reflect its dependence on n, gives

$$\lambda_n(x, x_0) = q \begin{cases} 0 & \text{if } x < x_0 - L/2n, \\ n/L & \text{if } x_0 - L/2n < x < x_0 + L/2n, \\ 0 & \text{if } x > x_0 + L/2n, \end{cases}$$

which is depicted in Figure 5.2 for $n = 10$ as well as some smaller values of n. The important property of $\lambda_n(x, x_0)$ is that its height increases at the same time that its width decreases.

Instead of a charge distribution that abruptly changes from zero to some finite value and just as abruptly drops to zero, let us consider a distribution

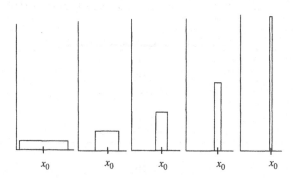

Figure 5.2: The linear density function as the length shrinks.

that smoothly rises to a maximum value and just as smoothly falls to zero. There are many functions describing such a distribution. For example,

$$\lambda_n(x, x_0) = q\sqrt{\frac{n}{\pi}}e^{-n(x-x_0)^2}$$

has a peak of $q\sqrt{n/\pi}$ at $x = x_0$ and drops to smaller and smaller values as we get farther and farther away from x_0 in either direction. This function is plotted for various values of n in Figure 5.3. It is clear from the figure that the "width" of the graph of $\lambda_n(x, x_0)$ gets smaller as $n \to \infty$.

In both cases $\lambda_n(x, x_0)$ is a true linear (charge) density in the sense that its integral gives the total charge. This is evident in the first case because of the way the function was defined. In the second case, *once we integrate* $\lambda_n(x, x_0)$ *from* $-\infty$ *to* $+\infty$, we also obtain the total charge q. The region of integration extends over all real numbers in the second case because at every point of the real line we have some nonzero charge. Furthermore, we can extend the interval of integration over all the real numbers *even for the first case*, because the function vanishes outside the interval $(x_0 - L/2n, x_0 + L/2n)$ and no extra contribution to the integral arises. We thus write

$$\int_{-\infty}^{+\infty} \lambda_n(x, x_0)\, dx = q$$

for all such functions. It is convenient to divide by q and define new functions $\delta_n(x, x_0)$ by

$$\delta_n(x, x_0) \equiv \frac{\lambda_n(x, x_0)}{q}$$

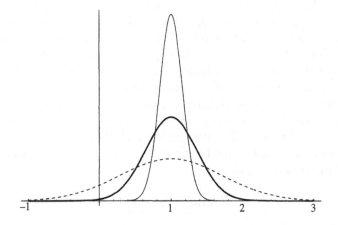

Figure 5.3: The Gaussian bell-shaped curve approaches the Dirac delta function as the width of the curve approaches zero. The value of n is 1 for the dashed curve, 4 for the heavy curve, and 20 for the light curve.

so that

$$\delta_n(x, x_0) = \begin{cases} 0 & \text{if} \quad x < x_0 - L/2n, \\ n/L & \text{if} \quad x_0 - L/2n < x < x_0 + L/2n, \\ 0 & \text{if} \quad x > x_0 + L/2n, \end{cases}$$

in the first case, and

$$\delta_n(x, x_0) = \sqrt{\frac{n}{\pi}} e^{-n(x-x_0)^2} \qquad (5.1)$$

in the second case. Both these functions have the property that

$$\int_{-\infty}^{+\infty} \delta_n(x, x_0)\, dx = 1, \qquad (5.2)$$

Dirac delta function defined

i.e., their integral over all the real numbers is one, and, in particular, independent of n.

Box 5.1.1. *The **Dirac delta function** $\delta(x, x_0)$ is defined as*

$$\delta(x, x_0) \equiv \lim_{n \to \infty} \delta_n(x, x_0) \qquad (5.3)$$

and has the following property:

$$\int_{-\infty}^{+\infty} \delta(x, x_0)\, dx = 1. \qquad (5.4)$$

Equation (5.4) follows from the fact that the integral in (5.2) is independent of n. The Dirac delta function has infinite height and zero width at x_0, but these two undefined quantities compensate for one another to give a *finite area* under the "graph" of the function. The Dirac delta function is not a well-behaved mathematical function as defined in elementary textbooks because at the only point that it is nonzero, it is infinite! Nevertheless, this function has been investigated rigorously in higher mathematics. For us, the Dirac delta function is a convenient way of describing densities.

Although we have separated the arguments of the Dirac delta function by a comma, the function depends only on the *difference* between the two arguments. This becomes clear if we think of the Dirac delta function as the limit of the exponential because the latter is a function of $x - x_0$. We therefore have the important relation

$$\delta(x, x_0) = \delta(x - x_0). \qquad (5.5)$$

In particular, since the delta function becomes infinite at $x = x_0$, we have

$$\delta(x, x_0)\Big|_{x=x_0} = \delta(x - x_0)\Big|_{x=x_0} = \delta(0) = \infty. \qquad (5.6)$$

One can think of the last equality as an identity satisfied by the Dirac delta function:

> **Box 5.1.2.** *The Dirac delta function is zero everywhere except at the point which makes its argument zero, in which case the Dirac delta function is infinite.*

Since the Dirac delta function is zero almost everywhere, we can shrink the region of integration to a smaller interval. In fact,

$$\int_a^b \delta(x - x_0)\, dx = 1$$

as long as x_0 lies in the interval (a, b). If x_0 is outside the interval, then the integral will be zero because the delta function would always be zero in the region of integration. We summarize these results:

> **Box 5.1.3.** *The Dirac delta function satisfies the following relation*
>
> $$\int_a^b \delta(x - x_0)\, dx = \begin{cases} 1 & \text{if } a < x_0 < b, \\ 0 & \text{otherwise.} \end{cases} \qquad (5.7)$$
>
> *Equation (5.4) is a special case of this, because $-\infty < x_0 < +\infty$ for any value of x_0.*

5.1.1   Linear Densities of Points

Any function $\lambda(x)$ whose integral over all real numbers is one is called a **linear density function**. The δ_n's defined above are such functions. If we multiply a linear density function by a physical quantity Q, the result will be a linear density for Q. In fact, this was how we arrived at δ_n. Thus, $Q_0\lambda(x)$ is a Q-linear density with total magnitude Q_0. Similarly, if M represents a mass, then $M\lambda(x)$ is a linear mass density with total mass M. Conversely, if $f(x)$ describes the linear density of a physical quantity with total magnitude Q, then $\lambda(x) \equiv f(x)/Q$ is a linear density function.

linear density function

Because of Equation (5.4) the Dirac delta function is a linear density function. What kind of a distribution does it describe? To be specific, consider $m\delta(x, x_0)$ with m designating mass. This function is zero everywhere except at x_0. Thus, if it is to be a mass distribution, it has to be a *point mass* located at x_0. Keep in mind that $m\delta(x, x_0)$ is a linear mass *density*, so that its *integral* is the total mass m. The linear "density" of a point mass is infinite because

δ function and densities of point charges and point masses

its length is zero, and this is precisely what $m\delta(x, x_0)$ describes. In fact, the linear density of a point physical quantity of magnitude Q located at x_0 can be written as $Q\delta(x, x_0) = Q\delta(x - x_0)$, or generalizing,

> **Box 5.1.4.** *The linear density $\lambda(x)$ of N point physical quantities $Q_1, Q_2, \ldots, Q_N$ located at $x_1, x_2, \ldots, x_N$, respectively, can be written as*
>
> $$\lambda(x) = \sum_{k=1}^{N} Q_k \delta(x - x_k). \qquad (5.8)$$

We see that with the help of the Dirac delta function we can express discrete charge distributions (collection of point charges) in terms of functions. This is the most useful property of the Dirac delta function.

Example 5.1.1. Three charges $-q$, $2q$, and $-q$ are located along the x-axis at $-a$, the origin, and $+a$, respectively. How do we write the linear charge density for such a charge distribution? We use Equation (5.8) with Q replaced by q:

$$\lambda(x) = \sum_{k=1}^{3} q_k \delta(x - x_k) = -q\delta(x - (-a)) + 2q\delta(x - 0) - q\delta(x - a)$$

$$= -q\delta(x + a) + 2q\delta(x) - q\delta(x - a).$$

Note that the Dirac delta functions ensure that no electric charge is present anywhere except at $x = a$, $x = -a$, and $x = 0$. ∎

Example 5.1.2. A more interesting example of a linear charge distribution using the Dirac delta function is that of an infinite array of point charges equally spaced on a straight line having equal magnitudes and alternating in sign. This is a one-dimensional model of ionic crystals.

density of one-dimensional ionic crystal

Let us assume that the magnitude of each charge is $\pm q$, the spacing between it and the neighboring charge is a, and that the charges start at $-\infty$ and extend to $+\infty$ with one positive charge at the origin as shown in Figure 5.4. Then it is easy to write the density of this distribution. It is

$$\lambda(x) = \sum_{k=-\infty}^{+\infty} (-1)^k q\delta(x - ka).$$

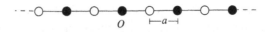

Figure 5.4: A one-dimensional ionic crystal. The black circles represent positive charges and the white circles the negative charges.

Note that for odd k the charge is negative and for even k it is positive. This is because we placed a positive charge at the origin. Had we chosen the origin to be the site of a negative charge, the above arrangement would have shifted by one spacing.

■

5.1.2   Properties of the Delta Function

From a mathematical point of view, the most important property, which is sometimes used to *define* the Dirac delta function, occurs when it multiplies a "smooth"[1] function in an integrand. First look at an integral with a $\delta_n(x-x_0)$ inside. If the function $f(x)$ multiplying $\delta_n(x - x_0)$ is smooth and n is large enough, the product $f(x)\delta_n(x - x_0)$ practically vanishes outside a narrow interval in which $\delta_n(x - x_0)$ is appreciably different from zero. For example, if $n = 10^7$, $x = x_0 + 0.001$, and we use the exponential function of Equation (5.1), then $\delta_n(x - x_0) = 0.08$, so that $f(x)\delta_n(x - x_0)$ drops to about 8% of the value it has at x_0, assuming that f does not change appreciably in the small interval of width 0.002 around x_0. For larger values of n this drop is even sharper. In fact, no matter what function we choose, there is always a large enough n such that the product $f(x)\delta_n(x - x_0)$ will drop to as small a value as we please in as short an interval as we please. Therefore, we can approximate the integral over all real numbers to an integral over that small interval. Let the interval be $(x_0 - \epsilon, x_0 + \epsilon)$. Then, we have

$$\int_{-\infty}^{+\infty} f(x)\delta_n(x - x_0)\, dx \approx \int_{x_0-\epsilon}^{x_0+\epsilon} f(x)\delta_n(x - x_0)\, dx$$

$$\approx f(x_0) \int_{x_0-\epsilon}^{x_0+\epsilon} \delta_n(x - x_0)\, dx$$

$$\approx f(x_0) \int_{-\infty}^{+\infty} \delta_n(x - x_0) = f(x_0).$$

The approximation in the second line follows from the fact that $f(x)$ is almost constant in the small interval $(x_0 - \epsilon, x_0 + \epsilon)$. The third approximation is a result of the smallness of δ_n outside the interval, and the equality follows because δ_n is a linear density function. The approximation above reaches equality once the limit of $n \to \infty$ is taken in which case δ_n becomes the Dirac delta function. Thus, we have the important relation

$$\int_{-\infty}^{+\infty} f(x)\delta(x - x_0)\, dx = f(x_0). \tag{5.9}$$

---

[1]In the present context, a smooth function is one that does not change abruptly when its argument changes by a small amount.

integral of product
of $\delta(x - x_0)$ and
$f(x)$ is simply
$f(x_0)$

This is equivalent to the following statement:

Box 5.1.5. *The Dirac delta function satisfies*

$$\int_a^b f(x)\delta(x - x_0)\, dx = \begin{cases} f(x_0) & \text{if } a < x_0 < b, \\ 0 & \text{otherwise.} \end{cases} \tag{5.10}$$

*In words, the result of integration is **the value of** f **at the root of the argument of the delta function**, provided this root is inside the range of integration.*

Example 5.1.3. In this example we illustrate some of the properties of the Dirac delta function. For instance $\int_1^\infty f(t)\delta(t)\, dt = 0$ because the root of the argument of the Dirac delta function (the point that makes the argument of the Dirac delta function zero)—namely $t = 0$—is outside the range of integration. The integral $\int_{-\infty}^{+\infty} x\delta(x)\, dx$ is zero because the function x vanishes at the point $x = 0$ (the root of the argument of the delta function). Also,

$$\int_{-\infty}^{+3} \cos y\,\delta(y - \pi)\, dy = 0$$

because π—which makes the argument of the delta function vanish—lies outside the range of integration. However,

$$\int_{-\infty}^{+3.2} \cos y\,\delta(y - \pi)\, dy = \cos\pi = -1$$

because now π lies inside the range of integration.

The reader is urged to check the following results:

$$\int_{-\infty}^{+\infty} \cos y\,\delta(y - \pi)\, dy = -1, \qquad \int_{-\infty}^{+\infty} \sin z\,\delta(z)\, dz = 0,$$

$$\int_0^{+\infty} \cos y\,\delta(y + \pi)\, dy = 0, \qquad \int_{-\infty}^{+\infty} \cos\frac{y}{2}\,\delta(y - \pi)\, dy = 0,$$

$$\int_{-1}^1 e^t \delta(t)\, dt = 1, \qquad \int_{-\infty}^{+\infty} x f(x)\delta(x)\, dx = 0,$$

$$\int_{-\infty}^{2.7} \ln t\,\delta(t - e)\, dt = 0, \qquad \int_{-\infty}^{2.8} \ln t\,\delta(t - e)\, dt = 1.$$

∎

As noted earlier, the Dirac delta function is not an ordinary over-the-counter function. Nevertheless, it is possible to study it, along with many other "weird" functions called **distributions**, in a mathematically rigorous and systematic way. It turns out that, in all physical applications, distributions occur *inside an integral*, and once they do, Equation (5.10) tells us how to manipulate such integrals. The result of integration is always well defined because it is simply the value of a "good" function at a point, say x_0. In fact,

distributions

the result of integration is so nice that one can even define the derivative of the Dirac delta function by differentiating (5.10) with respect to x_0. We leave the details as an exercise and simply quote the result:

$$\int_{-\infty}^{+\infty} f(x)\delta'(x - x_0)\, dx = -f'(x_0) \tag{5.11}$$

Higher order derivatives of the Dirac delta function can be obtained similarly. In fact, we have

derivatives of Dirac delta function

Box 5.1.6. *The nth derivative of the Dirac delta function satisfies*

$$\int_a^b f(x)\delta^{(n)}(x - x_0)(x - x_0)\, dx = \begin{cases} (-1)^n f^{(n)}(x_0) & \text{if } a < x_0 < b, \\ 0 & \text{otherwise,} \end{cases} \tag{5.12}$$

where the superscript (n) indicates the nth derivatives.

In many applications the argument of the Dirac delta function is not of the simple form $(x - x_0)$, but may itself be a function $g(x)$ whose derivative is assumed to be continuous in (a, b). Since by Equation (5.6) the delta function vanishes except when its argument is zero, in such a case, one has to concentrate on the roots of $g(x)$, i.e., values c for which $g(c) = 0$. For simplicity, first assume that there is only one root c of g in the interval (a, b) and that $g'(c) > 0$. Then, since the Dirac delta function is zero everywhere in the interval (a, b), except at $x = c$, we can shrink the region of integration to $(c - \epsilon, c + \epsilon)$, and write

what happens when the argument of δ is itself a function?

$$\int_a^b \delta\left(g(x)\right)\, dx = \int_{c-\epsilon}^{c+\epsilon} \delta\left(g(x)\right)\, dx.$$

Now make the change of variable $y = g(x)$, $dy = g'(x)\, dx$ with the appropriate transformation of limits of integration to get

$$\int_a^b \delta\left(g(x)\right)\, dx = \int_{g(c-\epsilon)}^{g(c+\epsilon)} \delta(y)\frac{dy}{g'(x)}.$$

With $g(c) = 0$ and $g'(c) > 0$, we conclude that g is increasing in the interval $(c - \epsilon, c + \epsilon)$, that $g(c - \epsilon) < 0$, and that $g(c + \epsilon) > 0$. We can therefore write

$$\int_a^b \delta\left(g(x)\right)\, dx = \int_{g(c-\epsilon)}^{g(c+\epsilon)} \delta(y)\frac{dy}{g'(x)} = \left.\frac{1}{g'(x)}\right|_{y=0} = \frac{1}{g'(c)} > 0,$$

because zero is in the region of integration and $y = 0$ is equivalent to $x = c$ there.

When $g'(c) < 0$, g will be decreasing in the interval $(c - \epsilon, c + \epsilon)$, and $g(c - \epsilon) > 0$ and $g(c + \epsilon) < 0$. Thus, flipping the limits of integration so that the smaller number corresponds to the lower limit, we obtain

$$\int_a^b \delta\left(g(x)\right) dx = -\int_{g(c+\epsilon)}^{g(c-\epsilon)} \delta(y) \frac{dy}{g'(x)} = -\frac{1}{g'(x)}\Big|_{y=0} = -\frac{1}{g'(c)} > 0.$$

We summarize the two results as

$$\int_a^b \delta\left(g(x)\right) dx = \frac{1}{|g'(c)|}.$$

If there are two roots of g in the interval, say c_1 and c_2 with $c_2 > c_1$, we break up (a, b):

$$\int_a^b \delta\left(g(x)\right) dx = \overbrace{\int_a^{c_1-\epsilon} \delta\left(g(x)\right) dx}^{0} + \int_{c_1-\epsilon}^{c_1+\epsilon} \delta\left(g(x)\right) dx$$

$$+ \overbrace{\int_{c_1+\epsilon}^{c_2-\epsilon} \delta\left(g(x)\right) dx}^{0} + \int_{c_2-\epsilon}^{c_2+\epsilon} \delta\left(g(x)\right) dx$$

$$+ \overbrace{\int_{c_2+\epsilon}^b \delta\left(g(x)\right) dx}^{0} = \frac{1}{|g'(c_1)|} + \frac{1}{|g'(c_2)|},$$

where in the last line we used the result obtained in the previous paragraph. It should be clear that if g has n roots $c_1, c_2, \ldots, c_n$ in (a, b), there will be a summation of n terms in the last line of the above equation. In fact, we can summarize the result of the foregoing discussion as

> **Box 5.1.7.** *If $g(x)$ has the roots $c_1, c_2, \ldots, c_n$, and $g'(c_k) \neq 0$ for all k between 1 and n, then*
>
> $$\int_a^b \delta(g(x)) dx = \begin{cases} \sum_{k=1}^n 1/|g'(c_k)| & \text{if } a < c_k < b, \\ 0 & \text{otherwise.} \end{cases} \tag{5.13}$$

When the delta function is multiplied by a smooth function $f(x)$, a similar argument as above—which is left to the reader—can be used to show that

$$\int_a^b f(x)\delta(g(x)) dx = \begin{cases} \sum_{k=1}^n f(c_k)/|g'(c_k)| & \text{if } a < c_k < b, \\ 0 & \text{otherwise,} \end{cases} \tag{5.14}$$

provided $g'(c_k) \neq 0$. These results are sometimes written as an identity among the delta functions.

a *very* important relation

> **Box 5.1.8.** *The Dirac delta function satisfies the following relation:*
>
> $$\delta(g(x)) = \sum_{k=1}^{n} \frac{\delta(x - c_k)}{|g'(c_k)|}, \quad g'(c_k) \neq 0, \tag{5.15}$$
>
> *where $\{c_k\}_{k=1}^{n}$ are all the roots of the equation $g(x) = 0$.*

The formula analogous to Equation (5.14) involving the derivative of the Dirac delta function is

$$\int_a^b f(x)\delta'(g(x))\,dx = \begin{cases} -\sum_{k=1}^{n} f'(c_k)/|g'(c_k)| & \text{if } a < c_k < b, \\ 0 & \text{otherwise.} \end{cases} \tag{5.16}$$

Example 5.1.4. As a concrete example, let us evaluate the integral

$$I \equiv \int_{-\infty}^{+\infty} f(t)\delta(t^2 - a^2)\,dt,$$

where f is a smooth function and a is a real constant. We can identify $g(t)$ as $t^2 - a^2$ with roots $c_1 = -a$, $c_2 = a$ and derivative $g'(t) = 2t$. Therefore, Equation (5.15) reduces to

$$\begin{aligned}
\delta(t^2 - a^2) &= \frac{\delta(t - c_1)}{|g'(c_1)|} + \frac{\delta(t - c_2)}{|g'(c_2)|} = \frac{\delta(t - (-a))}{|-2a|} + \frac{\delta(t - a)}{|2a|} \\
&= \frac{1}{|2a|}\left\{\delta(t + a) + \delta(t - a)\right\}.
\end{aligned}$$

Substituting in the integral, we obtain

$$\begin{aligned}
I &= \frac{1}{2|a|} \int_{-\infty}^{+\infty} f(t)\left\{\delta(t + a) + \delta(t - a)\right\} \\
&= \frac{1}{2|a|}\left\{\int_{-\infty}^{+\infty} f(t)\delta(t + a) + \int_{-\infty}^{+\infty} f(t)\delta(t - a)\right\} \\
&= \frac{1}{2|a|}\left\{f(-a) + f(a)\right\}.
\end{aligned}$$

Note that the integral vanishes—as expected—if f is odd. ∎

Example 5.1.5. We illustrate further the foregoing general discussions with some more concrete examples. To evaluate the integral

$$\int_1^{\infty} \sin t \; \delta(t^2 - \pi^2/4)\,dt,$$

we note that $g(t) = t^2 - \pi^2/4$ which has two roots $c_1 = \pi/2$ and $c_2 = -\pi/2$ with only the positive root lying in the range of integration. Moreover, $g'(t) = 2t$. Thus,

$$\int_1^\infty \sin t \, \delta(t^2 - \pi^2/4) \, dt = \frac{f(c_1)}{|g'(c_1)|} = \frac{\sin(c_1)}{|2c_1|} = \frac{\sin(\pi/2)}{\pi} = \frac{1}{\pi}.$$

On the other hand,

$$\int_{-\infty}^\infty \sin t \, \delta(t^2 - \pi^2/4) \, dt = 0$$

because the second root c_2 is also included in the range of integration and its contribution cancels that of c_1.

To evaluate the integral

$$\int_0^\infty \ln z \, \delta(z^2 - 4) \, dz$$

we note that $g(z) = z^2 - 4$ which has two roots $c_1 = 2$ and $c_2 = -2$ with only the positive root lying in the range of integration. Thus, with $g'(z) = 2z$, we have

$$\int_0^\infty \ln z \, \delta(z^2 - 4) \, dz = \frac{f(c_1)}{|g'(c_1)|} = \frac{\ln(c_1)}{|2c_1|} = \frac{\ln 2}{4} = 0.1733.$$

The integral

$$\int_{-\infty}^{+\infty} f(y) \, \delta(y^2 + a^2) \, dy$$

is zero because there is no point in the range of integration at which the argument of the Dirac delta function vanishes. In other words, $g(y) = y^2 + a^2$ has no real roots at all.

To evaluate the integral

$$\int_{-\pi/2}^{+\pi/2} (t+1)^2 \, \delta(\sin \pi t) \, dt$$

we note that $g(t) = \sin \pi t$ which has three roots $c_1 = -1$, $c_2 = 0$, and $c_3 = +1$ in the range of integration. Thus, with $g'(t) = \pi \cos \pi t$, we have

$$\int_{-\pi/2}^{+\pi/2} (t+1)^2 \, \delta(\sin \pi t) \, dt = \sum_{k=1}^3 \frac{f(c_k)}{|g'(c_k)|} = \sum_{k=1}^3 \frac{(c_k + 1)^2}{|\pi \cos(c_k \pi)|}$$

$$= \frac{(-1+1)^2}{|\pi \cos(-\pi)|} + \frac{(0+1)^2}{|\pi \cos(0)|} + \frac{(1+1)^2}{|\pi \cos(\pi)|} = \frac{5}{\pi}.$$

Some other concrete examples are:

$$\int_{-\infty}^{+\infty} \sin |t| \, \delta(t^2 - \pi^2/4) \, dt = 2/\pi, \qquad \int_{-\infty}^{+\infty} \cos x \, \delta(x^2 - \pi^2) \, dx = -1/\pi,$$

$$\int_0^\infty \ln z \, \delta(z^2 - 1) \, dz = 0, \qquad \int_{-\infty}^{+3} \cos y \, \delta(y^2 + \pi^2) \, dy = 0,$$

$$\int_{-\pi}^{+\pi} (t+1)^2 \, \delta(\sin \pi t) \, dt = 35/\pi, \qquad \int_{-\infty}^{+\infty} f(t) \, \delta(e^t - 1) \, dt = f(0),$$

$$\int_0^\infty \ln x \, \delta(10x^2 + 3x - 1) \, dx = -0.23, \qquad \int_{-\infty}^{+\infty} f(t) \, \delta(e^t) \, dt = 0.$$

The reader is urged to derive all the above relations.   ∎

Historical Notes

"Physical laws should have mathematical beauty." This statement was Dirac's response to the question of his philosophy of physics, posed to him in Moscow in 1955. He wrote it on a blackboard that is still preserved today.

Paul Adrien Maurice Dirac (1902–1984), was born in 1902 in Bristol, England, of a Swiss, French-speaking father and an English mother. His father, a taciturn man who refused to receive friends at home, enforced young Paul's silence by requiring that only French be spoken at the dinner table. Perhaps this explains Dirac's later disinclination toward collaboration and his general tendency to be a loner in most aspects of his life. The fundamental nature of his work made the involvement of students difficult, so perhaps Dirac's personality was well-suited to his extraordinary accomplishments.

Dirac went to Merchant Venturer's School, the public school where his father taught French, and while there displayed great mathematical abilities. Upon graduation, he followed in his older brother's footsteps and went to Bristol University to study electrical engineering. He was 19 when he graduated from Bristol University in 1921. Unable to find a suitable engineering position due to the economic recession that gripped post-World War I England, Dirac accepted a fellowship to study mathematics at Bristol University. This fellowship, together with a grant from the Department of Scientific and Industrial Research, made it possible for Dirac to go to Cambridge as a research student in 1923. At Cambridge Dirac was exposed to the experimental activities of the Cavendish Laboratory, and he became a member of the intellectual circle over which Rutherford and Fowler presided. He took his PhD in 1926 and was elected in 1927 as a fellow. His appointment as university lecturer came in 1929. He assumed the Lucasian professorship following Joseph Larmor in 1932 and retired from it in 1969. Two years later he accepted a position at Florida State University where he lived out his remaining years. The FSU library now carries his name.

In the late 1920s the relentless march of ideas and discoveries had carried physics to a generally accepted relativistic theory of the electron. Dirac, however, was dissatisfied with the prevailing ideas and, somewhat in isolation, sought for a better formulation. By 1928 he succeeded in finding an equation, the *Dirac equation*, that accorded with his own ideas and also fitted most of the established principles of the time. Ultimately, this equation, and the physical theory behind it, proved to be one of the great intellectual achievements of the period. It was particularly remarkable for the internal beauty of its mathematical structure, which not only clarified previously mysterious phenomena such as **spin** and the **Fermi–Dirac** statistics associated with it, but also predicted the existence of an electron-like particle of negative energy, the antielectron, or *positron*, and, more recently, it has come to play a role of great importance in modern mathematics, particularly in the interrelations between topology, geometry, and analysis. **Heisenberg** characterized the discovery of antimatter by Dirac as "the most decisive discovery in connection with the properties or the nature of elementary particles .... This discovery of particles and antiparticles by Dirac ...changed our whole outlook on atomic physics completely." One of the interesting implications of his work that predicted the positron was the prediction of a *magnetic monopole*. Dirac won the Nobel Prize in 1933 for this work.

Dirac is not only one of the chief authors of quantum mechanics, but he is also the creator of *quantum electrodynamics* and one of the principal architects of

"The amount of theoretical ground one has to cover before being able to solve problems of real practical value is rather large, but this circumstance is an inevitable consequence of the fundamental part played by transformation theory and is likely to become more pronounced in the theoretical physics of the future."
P.A.M. Dirac (1930)

Paul Adrien
Maurice Dirac
1902–1984

quantum field theory. While studying the scattering theory of quantum particles, he invented the (Dirac) *delta function*; in his attempt at quantizing the general theory of relativity, he founded *constrained Hamiltonian dynamics*, which is one of the most active areas of theoretical physics research today. One of his greatest contributions is the invention of the *bra* $\langle |$ and *ket* $| \rangle$ notation used in quantum theory.

While at Cambridge, Dirac did not accept many research students. Those who worked with him generally thought that he was a good supervisor, but one who did not spend much time with his students. A student needed to be extremely independent to work under Dirac. One such student was **Dennis Sciama**, who later became the supervisor of **Stephen Hawking**, the current holder of the Lucasian chair.

Salam and **Wigner** in their Preface to the Festschrift that honors Dirac on his seventieth birthday and commemorates his contributions to quantum mechanics succinctly assessed the man:

> Dirac is one of the chief creators of quantum mechanics.... Posterity will rate Dirac as one of the greatest physicists of all time. The present generation values him as one of its greatest teachers.... On those privileged to know him, Dirac has left his mark ... by his human greatness. He is modest, affectionate, and sets the highest possible standards of personal and scientific integrity. He is a legend in his own lifetime and rightly so.

(Taken from Schweber, S. S. "Some chapters for a history of quantum field theory: 1938–1952," in *Relativity, Groups, and Topology II*, vol. 2, B. S. DeWitt and R. Stora, eds., North-Holland, Amsterdam, 1984.)

5.1.3   The Step Function

The **step function** θ is defined as

$$\theta(x) = \begin{cases} 1 & \text{if } x > 0 \\ 0 & \text{if } x < 0 \end{cases} \tag{5.17}$$

The θ function (as it is often called) is useful in writing functions that have discontinuities or cusps. For instance, absolute values can be written in terms of the step function:

$$|x| = x\theta(x) - x\theta(-x) \quad \text{or} \quad |x - y| = (x - y)[\theta(x - y) - \theta(y - x)]$$

A piecewise continuous function such as

$$g(x) = \begin{cases} g_1(x) & \text{if } 0 < x < 1 \\ g_2(x) & \text{if } x > 1 \end{cases} \tag{5.18}$$

can be written as

$$g(x) = g_1(x)\theta(x)\theta(1 - x) + g_2(x)\theta(x - 1)$$

Because θ is constant everywhere except at 0, its derivative is zero everywhere except at 0. The discontinuity at 0 makes the derivative infinite there:

$$\theta'(0) = \lim_{\epsilon \to 0} \frac{\theta(\epsilon) - \theta(-\epsilon)}{2\epsilon} = \lim_{\epsilon \to 0} \frac{1 - 0}{2\epsilon} \to \infty$$

This strongly suggests the identification of the derivative of the step function as the Dirac delta function. In fact, noting that

$$\theta(x - x_0) = \begin{cases} 1 & \text{if } x > x_0 \\ 0 & \text{if } x < x_0, \end{cases} \tag{5.19}$$

and the fact that $\theta'(x - x_0)$ is zero everywhere except at x_0, for any well-behaved function $f(x)$ we obtain

$$\int_{-\infty}^{\infty} f(x)\theta'(x - x_0)\, dx = \int_{x_0-\epsilon}^{x_0+\epsilon} f(x)\theta'(x - x_0)\, dx \approx f(x_0) \int_{x_0-\epsilon}^{x_0+\epsilon} \theta'(x - x_0)\, dx$$

$$= f(x_0)\, \theta(x - x_0)|_{x_0-\epsilon}^{x_0+\epsilon} = f(x_0)[\underbrace{\theta(\epsilon)}_{=1} - \underbrace{\theta(-\epsilon)}_{=0}] = f(x_0)$$

We thus have another important representation of the Dirac delta function:

$$\delta(x - x_0) = \theta'(x - x_0) \tag{5.20}$$

Example 5.1.6. For positive a, $\tanh(ax)$ goes to 1 as $x \to \infty$ and to -1 as $x \to -\infty$ and it makes a smooth transition from one of these asymptotic values to the other. This transition gets steeper and steeper for larger and larger values of a. This suggests the following relation:

$$\theta(x - x_0) = \tfrac{1}{2} \lim_{a \to \infty} \{1 + \tanh[a(x - x_0)]\}$$

Let $\theta_a(x - x_0)$ stand for the function on the right-hand side for any finite positive a. Then

$$\theta_a'(x - x_0) = \frac{1}{2} \frac{d}{dx} \{1 + \tanh[a(x - x_0)]\} = \frac{a \operatorname{sech}^2[a(x - x_0)]}{2}$$

and

$$\int_{-\infty}^{\infty} \theta_a'(x - x_0)\, dx = \theta_a(x - x_0)|_{-\infty}^{\infty} = \tfrac{1}{2} \{1 + \tanh[a(x - x_0)]\}|_{-\infty}^{\infty} = 1$$

for any value of $a > 0$, in particular for $a \to \infty$. Thus, we get yet another representation of the Dirac delta function:

$$\delta(x - x_0) = \lim_{a \to \infty} \theta_a'(x - x_0) = \lim_{a \to \infty} \frac{a \operatorname{sech}^2[a(x - x_0)]}{2} \qquad \blacksquare$$

5.2   Two-Variable Case

We can generalize the discussion of the previous section to the case of many variables. For example, in two dimensions using Cartesian coordinates, we can define the functions δ_n as

$$\delta_n(x - x_0, y - y_0) = Ce^{-n\left[(x-x_0)^2 + (y-y_0)^2\right]} = Ce^{-n(x-x_0)^2}e^{-n(y-y_0)^2}, \quad (5.21)$$

where C is a constant to be determined in such a way as to make the integral of δ_n over the entire xy-plane equal to one. A simple calculation will show that $C = n/\pi$. This constant is simply the product of two "one-dimensional constants": one for the exponential in x and the other for the exponential in y. This is as expected, because $\delta_n(x - x_0, y - y_0)$ is *defined* to be the product of two one-dimensional δ_n's. Such a simplicity is the result of the coordinate systems we have used and does not prevail in other—non-Cartesian—coordinate systems, for which the constant C must be evaluated separately.

It should be clear from (5.21) that as n increases, the height of δ_n at (x_0, y_0) increases while its width decreases (see Figure 5.5). What may not be clear is that this reciprocal behavior takes place in such a way as to keep the volume under the surface equal to one. We can define—as we did in the one dimensional case—a **surface density function** as a function whose integral over the entire plane is one. For any n, then, δ_n will be a surface density function.

surface density
function

The passage to the **two-dimensional Dirac delta function** is now clear:

two-dimensional
Dirac delta
function

$$\delta(x - x_0, y - y_0) \equiv \lim_{n \to \infty} \delta_n(x - x_0, y - y_0). \quad (5.22)$$

The two-dimensional Dirac delta function above is zero everywhere except at (x_0, y_0) where it is infinite. Thus for the Dirac delta function not to be zero *both of its arguments must be zero*. It is convenient to define points P and P_0

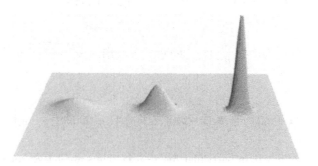

Figure 5.5: As n gets larger and larger, the two-dimensional Gaussian exponential approaches the two-dimensional Dirac delta function. For the left bump, $n = 400$; for the middle, $n = 1000$; and for the right spike $n = 4000$.

with respective Cartesian coordinates (x, y) and (x_0, y_0), and position vectors $\mathbf{r} = \langle x, y \rangle$, $\mathbf{r}_0 = \langle x_0, y_0 \rangle$, and write

$$\delta(x - x_0, y - y_0) \equiv \delta(\mathbf{r} - \mathbf{r}_0) = \begin{cases} \delta(\vec{0}) \equiv \delta(0,0) = \infty & \text{if} \quad \mathbf{r} = \mathbf{r}_0, \\ 0 & \text{otherwise.} \end{cases} \quad (5.23)$$

This means

> **Box 5.2.1.** *The two-dimensional Dirac delta function is zero everywhere except at the point which makes both of its arguments zero, in which case the two-dimensional Dirac delta function is infinite.*

We noted above that in Cartesian coordinates—and only in Cartesian coordinates—the product of two one-dimensional δ_n's gave rise to a two-dimensional δ_n which subsequently yielded the two-dimensional Dirac delta function. Thus *only in Cartesian coordinates* can we conclude that

$$\delta(\mathbf{r} - \mathbf{r}_0) = \delta(x - x_0, y - y_0) = \delta(x - x_0)\,\delta(y - y_0). \quad (5.24)$$

We shall see that in polar coordinates, the two-dimensional delta function is not merely the product of two one-dimensional delta functions, but some other factor is also present.

The density property of the two-dimensional Dirac delta function survives the $n \to \infty$ process because the integral of δ_n is independent of n. On the other hand, the delta function is zero everywhere except at the point which makes both of its arguments zero. Therefore, for any two-dimensional region Ω, we have

$$\iint_\Omega \delta(\mathbf{r} - \mathbf{r}_0)\, da(\mathbf{r}) = \begin{cases} 1 & \text{if} \quad P_0 \text{ is in } \Omega, \\ 0 & \text{otherwise.} \end{cases} \quad (5.25)$$

Equation (5.25) is written independently of coordinates, and as such, the vector arguments are to be interpreted as *coordinates* not components. We can use this equation in polar coordinates to write the two-dimensional Dirac delta function as a product of two one-dimensional delta functions. First write[2]

$$\delta(\mathbf{r} - \mathbf{r}_0) = C\delta(\rho - \rho_0)\delta(\varphi - \varphi_0).$$

---

[2]We use ρ and φ instead of the more common r and θ because we have reserved the latter for the three-dimensional spherical coordinates. There is no danger of confusing the pair (ρ, φ) with the corresponding pair in *cylindrical* coordinates because the two pairs are identical.

Now substitute this in Equation (5.25) with Ω being the entire plane, and note that $da = \rho\,d\rho\,d\varphi$:

$$1 = \iint_\Omega C\delta(\rho - \rho_0)\delta(\varphi - \varphi_0)\,\rho\,d\rho\,d\varphi$$

$$= C\int_0^\infty \delta(\rho - \rho_0)\rho\,d\rho \underbrace{\int_0^{2\pi}\delta(\varphi - \varphi_0)\,d\varphi}_{=1}$$

$$= C\rho_0 \;\Rightarrow\; C = \frac{1}{\rho_0}.$$

In the above derivation, we have used properties of the one-dimensional delta function as applied to $\delta(\rho - \rho_0)$ and $\delta(\varphi - \varphi_0)$.

> **Box 5.2.2.** *The two-dimensional Dirac delta function can be written in polar coordinates as*
>
> $$\delta(\mathbf{r} - \mathbf{r}_0) = \frac{1}{\rho_0}\delta(\rho - \rho_0)\delta(\varphi - \varphi_0) = \frac{1}{\rho}\delta(\rho - \rho_0)\delta(\varphi - \varphi_0). \qquad (5.26)$$

The last equality follows because the Dirac delta function in ρ forces ρ and ρ_0 to be equal.

A collection of point physical quantities $Q_1, Q_2, \ldots, Q_n$ located on a surface can be described by a surface density $\sigma_Q(\mathbf{r})$ using the two-dimensional Dirac delta function:

surface density and two-dimensional delta function

$$\sigma_Q(\mathbf{r}) = \sum_{k=1}^n Q_k\,\delta(\mathbf{r} - \mathbf{r}_k), \qquad (5.27)$$

where $\mathbf{r}_k$ is the position vector of Q_k. This equation can be rewritten as

$$\sigma_Q(x,y) = \sum_{k=1}^n Q_k\,\delta(x - x_k)\delta(y - y_k)$$

in Cartesian coordinates, and as

$$\sigma_Q(\rho,\varphi) = \sum_{k=1}^n \frac{Q_k}{\rho_k}\,\delta(\rho - \rho_k)\delta(\varphi - \varphi_k) = \frac{1}{\rho}\sum_{k=1}^n Q_k\,\delta(\rho - \rho_k)\delta(\varphi - \varphi_k)$$

in polar coordinates.

Example 5.2.1. With an appropriate choice of the origin and the axes of a Cartesian coordinate system, the surface charge density for four charges q_1, q_2, q_3, q_4 located at the four corners of a square of sides $2a$ can be written as

$$\sigma_q(x,y) = \sum_{k=1}^4 q_k\delta(x - x_k)\delta(y - y_k)$$

$$= q_1\delta(x - a)\delta(y - a) + q_2\delta(x + a)\delta(y - a)$$

$$+ q_3\delta(x + a)\delta(y + a) + q_4\delta(x - a)\delta(y + a).$$

If polar coordinates are used, the surface charge density becomes

$$\sigma_q(\rho,\varphi) = \sum_{k=1}^{4} \frac{q_k}{\rho_k}\,\delta(\rho-\rho_k)\delta(\varphi-\varphi_k)$$

$$= \frac{\delta(\rho-\sqrt{2}\,a)}{\sqrt{2}\,a}\Big\{ q_1\delta(\varphi-\pi/4) + q_2\delta(\varphi-3\pi/4)$$

$$+ q_3\delta(\varphi-5\pi/4) + q_4\delta(\varphi-7\pi/4) \Big\}.$$

The reader is urged to study these two equations carefully and make sure to understand the details of their derivation.   ∎

A more interesting example is the two-dimensional ionic crystal.

Example 5.2.2. Suppose positive and negative charges $\pm q$ are arranged on an infinite square grid in such a way that the *nearest neighbors* of each charge have charges of opposite sign, i.e., charges alternate both horizontally and vertically (see Figure 5.6). Assume that the distance between each charge and its nearest neighbor is a, and that we place our Cartesian origin at the location of a positive charge. Then the surface charge density can be written as

two-dimensional ionic crystal

$$\sigma_q(x,y) = q\sum_{i=-\infty}^{\infty}\sum_{j=-\infty}^{\infty} (-1)^{i+j}\delta(x-ia)\delta(y-ja).$$

For a finite $2M \times 2N$ grid one substitutes the first infinity with M and the second one with N. Similarly, one can consider rectangular units of sides a and b for the grid. Then one should change the second argument of the delta function (or the argument of the delta function corresponding to y) to $y - jb$.   ∎

With an extra dimension at our disposal, we can invent many new varieties of distribution of point physical quantities that were not possible in one dimension. For example, we can put the points on a *curve* in the xy-plane. It is instructive to find the *surface* density of such a collection of points. The following example examines this problem.

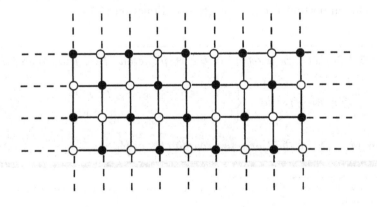

Figure 5.6: A two-dimensional ionic crystal.

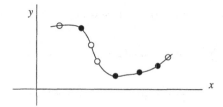

Figure 5.7: Point charges located on a curve in the xy-plane.

Example 5.2.3. For concreteness, we consider n point charges located at n points $\{P_k\}_{k=1}^n$ with P_k having Cartesian coordinates (x_k, y_k). These points are assumed to be on a curve with the Cartesian equation $y = f(x)$ as shown in Figure 5.7. The surface charge density in Cartesian coordinates becomes

$$\sigma_q(x,y) = \sum_{k=1}^n q_k \delta(x - x_k)\delta(y - y_k) = \sum_{k=1}^n q_k \delta(x - x_k)\delta\Big(y - f(x_k)\Big).$$

If the curve is given as $\rho = g(\varphi)$, then polar coordinates are more appropriate, and the surface charge density will be[3]

$$\sigma_q(\rho,\varphi) = \sum_{k=1}^n \frac{q_k}{\rho_k}\, \delta(\rho - \rho_k)\delta(\varphi - \varphi_k) = \sum_{k=1}^n \frac{q_k}{g(\varphi_k)}\, \delta(\rho - g(\varphi_k))\,\delta(\varphi - \varphi_k).$$

For instance, if the charges are located on a circle of radius a each separated from its nearest neighbor by an angle α, with the first charge on the x-axis, then

$$\sigma_q(\rho,\varphi) = \frac{\delta(\rho - a)}{a}\sum_{k=1}^n q_k\delta(\varphi - (k-1)\alpha),$$

where we have used the fact that $g(\varphi) = a$ for a circle of radius a.   ∎

All the properties of the delta function can be generalized to two dimensions. One important property is given in Equation (5.10).

Box 5.2.3. *Let Ω be a region in the xy-plane and P_0 a point there; then*

$$\iint_\Omega f(\mathbf{r})\delta(\mathbf{r} - \mathbf{r}_0)\, da = \begin{cases} f(\mathbf{r}_0) \equiv f(x_0, y_0) & \text{if } P_0 \text{ is in } \Omega, \\ 0 & \text{otherwise,} \end{cases} \tag{5.28}$$

where (x_0, y_0) are the Cartesian coordinates of P_0.

---

[3]Because of the two delta functions, one can substitute ρ for ρ_k and φ for φ_k in the denominators.

Differentiating both sides with respect to the first argument x_0, we easily obtain the analog of Equation (5.12):

$$\iint_\Omega f(\mathbf{r}) \partial_1 \delta(\mathbf{r} - \mathbf{r}_0) \, da = \begin{cases} -\partial_1 f(\mathbf{r}_0) \equiv -\partial_1 f(x_0, y_0) & \text{if } P_0 \text{ is in } \Omega, \\ 0 & \text{otherwise,} \end{cases}$$

with a similar relation for differentiation with respect to the second argument. We can combine the two relations into a single relation:

Box 5.2.4. *The derivative of the Dirac delta function in two dimensions satisfies*

$$\iint_\Omega f(\mathbf{r}) \partial_i \delta(\mathbf{r} - \mathbf{r}_0) \, da = \begin{cases} -\partial_i f(\mathbf{r}_0) \equiv -\partial_i f(x_0, y_0) & \text{if } P_0 \text{ is in } \Omega, \\ 0 & \text{otherwise,} \end{cases}$$

where i can be 1 or 2, $\partial_1 = \partial_x$ and $\partial_2 = \partial_y$.

5.3 Three-Variable Case

Once the generalization to two variables is realized, the three—and more—variable cases become trivial. In fact, we had such generalizations in mind when we wrote most of the formulas in the last section: All that is needed is to change da to dV and keep in mind that the vectors $\mathbf{r}$ and $\mathbf{r}_0$ have three components, and points in space have three coordinates. Nevertheless, we shall summarize the most important properties of the three-dimensional Dirac delta function.

First we note that

$$\delta(\mathbf{r} - \mathbf{r}_0) = \begin{cases} \delta(\vec{0}) \equiv \delta(0,0,0) = \infty & \text{if } \mathbf{r} = \mathbf{r}_0, \\ 0 & \text{otherwise.} \end{cases} \tag{5.29}$$

This means

Box 5.3.1. *The three-dimensional Dirac delta function is zero everywhere except at the point which makes all three of its arguments zero in which case it is infinite.*

In Cartesian coordinates, we have

$$\begin{aligned} \delta(\mathbf{r} - \mathbf{r}_0) &= \delta(x - x_0, y - y_0, z - z_0) \\ &= \delta(x - x_0)\, \delta(y - y_0)\, \delta(z - z_0). \end{aligned} \tag{5.30}$$

3D Dirac delta function in Cartesian coordinates

An argument similar to the two-dimensional case can be used to show that

---

Box 5.3.2. *In cylindrical coordinates*

$$\delta(\mathbf{r} - \mathbf{r}_0) = \frac{1}{\rho_0}\delta(\rho - \rho_0)\delta(\varphi - \varphi_0)\delta(z - z_0), \qquad (5.31)$$

where $\mathbf{r}$ *and* $\mathbf{r}_0$ *on the LHS are to be understood as cylindrical coordinates, not cylindrical position vectors. The corresponding formula for the spherical coordinate system is*

$$\delta(\mathbf{r} - \mathbf{r}_0) = \frac{1}{r_0^2 \sin\theta_0}\delta(r - r_0)\delta(\theta - \theta_0)\delta(\varphi - \varphi_0), \qquad (5.32)$$

with $\mathbf{r}$ *and* $\mathbf{r}_0$ *representing the coordinates* (r, θ, φ) *and* $(r_0, \theta_0, \varphi_0)$, *respectively.*

The density property of the three-dimensional Dirac delta function is given by

$$\iiint_\Omega \delta(\mathbf{r} - \mathbf{r}_0)\, dV(\mathbf{r}) = \begin{cases} 1 & \text{if } P_0 \text{ is in } \Omega, \\ 0 & \text{otherwise}, \end{cases} \qquad (5.33)$$

where Ω is a region of space and P_0 is the point with Cartesian coordinates (x_0, y_0, z_0), spherical coordinates $(r_0, \theta_0, \varphi_0)$, and cylindrical coordinates (ρ_0, φ_0, z_0). Similarly,

---

Box 5.3.3. *If* Ω *is a region of space, then for a "good" function* $f(\mathbf{r})$,

$$\iiint_\Omega f(\mathbf{r})\delta(\mathbf{r} - \mathbf{r}_0)\, dV(\mathbf{r}) = \begin{cases} f(\mathbf{r}_0) & \text{if } P_0 \text{ is in } \Omega, \\ 0 & \text{otherwise}. \end{cases}$$

Thus integration reduces to the evaluation of the function f *at the coordinates of* P_0.

The density property allows us to write the distribution of discrete physical quantities in terms of the three-dimensional Dirac delta function. In general,

$$\rho_Q(\mathbf{r}) = \sum_{k=1}^{n} Q_k \delta(\mathbf{r} - \mathbf{r}_k) \qquad (5.34)$$

which can be rewritten as

$$\rho_Q(x, y, z) = \sum_{k=1}^{n} Q_k \delta(x - x_k)\delta(y - y_k)\delta(z - z_k)$$

in Cartesian coordinates, as

$$\rho_Q(\rho, \varphi, z) = \sum_{k=1}^{n} \frac{Q_k}{\rho_k} \delta(\rho - \rho_k)\delta(\varphi - \varphi_k)\delta(z - z_k)$$

in cylindrical coordinates, and as

$$\rho_Q(r, \theta, \varphi) = \sum_{k=1}^{n} \frac{Q_k}{r_k^2 \sin\theta_k} \delta(r - r_k)\delta(\theta - \theta_k)\delta(\varphi - \varphi_k)$$

in the spherical coordinate system. In fact, the linear and surface distributions of a physical quantity involving the Dirac delta function are special cases of the volume distribution. For instance, a collection of point quantities in the xy-plane can be described by the *volume* density

$$\rho_Q(x, y, z) = \sum_{k=1}^{n} Q_k \delta(x - x_k)\delta(y - y_k)\delta(z)$$

$$= \delta(z) \sum_{k=1}^{n} Q_k \delta(x - x_k)\delta(y - y_k).$$

The delta function outside the sum restricts the z-coordinates of point quantities to zero, and thus their location, to the xy-plane. Similarly,

$$\rho_Q(r, \theta, \varphi) = \frac{\delta(r - a)}{a^2} \sum_{k=1}^{n} \frac{Q_k}{\sin\theta_k} \delta(\theta - \theta_k)\delta(\varphi - \varphi_k)$$

describes a distribution of n point quantities on a sphere of radius a.

Example 5.3.1. Let us calculate the electrostatic field of the one-dimensional infinite ionic crystal in Cartesian coordinates. Assume that the charges are located on the z-axis (Figure 5.8). We treat this as a *three-dimensional* charge distribution with density

$$\rho_q(x, y, z) = q \sum_{k=-\infty}^{\infty} (-1)^k \, \delta(x)\delta(y)\delta(z - ka). \qquad (5.35)$$

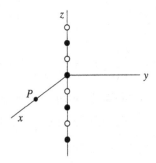

Figure 5.8: The geometry for the calculation of the electrostatic field of the one-dimensional ionic crystal.

The first two delta functions restrict the charges to the z-axis and the third locates them. This density is to be substituted in the equation for the electric field in Cartesian coordinates. Let us concentrate on the x-component

$$E_x(x, y, z) = k_e \iiint_\Omega \frac{\rho_q(x', y', z')(x - x') \, dx' \, dy' \, dz'}{\{(x - x')^2 + (y - y')^2 + (z - z')^2\}^{3/2}}.$$

We can always take Ω to be the entire space because the delta function will restrict the integration to the region of charges automatically. We can also choose our coordinate system so that the field point lies in the xz-plane, i.e., $y = 0$. Note that we have to prime all the arguments of ρ_q before we substitute it in the integral. Having done this, we obtain

$$E_x(x, y, z) = k_e q \sum_{k=-\infty}^{\infty} (-1)^k \iiint_\Omega \frac{(x - x')\delta(x')\delta(y')\delta(z' - ka) \, dx' \, dy' \, dz'}{\{(x - x')^2 + y'^2 + (z - z')^2\}^{3/2}}.$$

Using Box 5.3.3, noting that

$$f(x', y', z') = \frac{(x - x')}{\{(x - x')^2 + y'^2 + (z - z')^2\}^{3/2}},$$

and that the result of integration is the evaluation of f at $x' = 0 = y'$, $z' = ka$, we obtain

$$E_x(x, y, z) = k_e q \sum_{k=-\infty}^{\infty} (-1)^k \frac{x}{\{x^2 + (z - ka)^2\}^{3/2}}$$

$$= k_e q \sum_{k=-\infty}^{-1} (-1)^k \frac{x}{\{x^2 + (z - ka)^2\}^{3/2}}$$

$$+ k_e q \frac{x}{(x^2 + z^2)^{3/2}} + k_e q \sum_{k=+1}^{\infty} (-1)^k \frac{x}{\{x^2 + (z - ka)^2\}^{3/2}},$$

where we have broken up the summation into three pieces, a permissible act as long as the series converges. We can combine the first and third terms by changing k to $-k$ in the first and noting that

$$(-1)^{-k} = \frac{1}{(-1)^k} = (-1)^k.$$

Doing so, we get

$$E_x(x, 0, z) = k_e q \frac{x}{(x^2 + z^2)^{3/2}}$$

$$+ k_e q \sum_{k=1}^{\infty} (-1)^k \left(\frac{x}{\{x^2 + (z + ka)^2\}^{3/2}} + \frac{x}{\{x^2 + (z - ka)^2\}^{3/2}} \right).$$

The other components of the field can be found similarly:

$$E_y(x, 0, z) = 0,$$

$$E_z(x, 0, z) = k_e q \frac{z}{(x^2 + z^2)^{3/2}} \tag{5.36}$$

$$+ k_e q \sum_{k=1}^{\infty} (-1)^k \left(\frac{z + ka}{\{x^2 + (z + ka)^2\}^{3/2}} + \frac{z - ka}{\{x^2 + (z - ka)^2\}^{3/2}} \right).$$

Let us further simplify the problem by positioning the field point on the x-axis, i.e., setting $z = 0$. This reduces the above expressions to

$$E_x(x,0,0) = k_e q \frac{x}{|x|^3} + 2k_e q \sum_{k=1}^{\infty} (-1)^k \frac{x}{(x^2 + k^2 a^2)^{3/2}},$$

$$E_y(x,0,0) = 0,$$

$$E_z(x,0,0) = k_e q \sum_{k=1}^{\infty} (-1)^k \left(\frac{ka}{\{x^2 + (ka)^2\}^{3/2}} + \frac{-ka}{\{x^2 + (-ka)^2\}^{3/2}} \right) = 0.$$

At a distance a from the origin on the x-axis, the field strength is

$$E_x(x,0,0) = \frac{k_e q}{a^2} \left\{ 1 + 2 \underbrace{\sum_{k=1}^{\infty} \frac{(-1)^k}{(1+k^2)^{3/2}}}_{=-0.286269} \right\} = 0.42746 \frac{k_e q}{a^2},$$

$$E_y(x,0,0) = 0,$$

$$E_z(x,0,0) = 0,$$

where the numerical value for the sum—accurate to six decimal places—is obtained by adding its first 150 terms.

Another useful quantity is the electrostatic potential which for an arbitrary charge distribution is given by

$$\Phi(\mathbf{r}) = k_e \iint_{\Omega} \frac{dq(\mathbf{r}')}{|\mathbf{r} - \mathbf{r}'|}. \tag{5.37}$$

For the one-dimensional crystal, with the volume charge density of Equation (5.35), the electrostatic potential at an arbitrary point (x, y, z) in space becomes

$$\Phi(x,y,z) = k_e \iint_{\Omega} \frac{\rho_q(x',y',z')\, dx'\, dy'\, dz'}{\sqrt{(x-x')^2 + (y-y')^2 + (z-z')^2}}$$

$$= k_e q \sum_{k=-\infty}^{\infty} (-1)^k \iint_{\Omega} \frac{\delta(x')\delta(y')\delta(z'-ka)\, dx'\, dy'\, dz'}{\sqrt{(x-x')^2 + (y-y')^2 + (z-z')^2}}$$

$$= k_e q \sum_{k=-\infty}^{\infty} \frac{(-1)^k}{\sqrt{x^2 + y^2 + (z-ka)^2}}.$$

If we are interested in the potential at a specific point such as $(x,0,0)$, the expression simplifies to

$$\Phi(x,0,0) = k_e q \sum_{k=-\infty}^{\infty} \frac{(-1)^k}{\sqrt{x^2 + k^2 a^2}}$$

$$= k_e q \frac{1}{\sqrt{x^2}} + 2k_e q \sum_{k=1}^{\infty} \frac{(-1)^k}{\sqrt{x^2 + k^2 a^2}}$$

$$= \frac{k_e q}{|x|} + 2k_e q \sum_{k=1}^{\infty} \frac{(-1)^k}{\sqrt{x^2 + k^2 a^2}}.$$

For $x = a$, this further simplifies to

$$\Phi(a,0,0) = \frac{k_e q}{a}\left\{1 + 2\underbrace{\sum_{k=1}^{\infty}\frac{(-1)^k}{\sqrt{1+k^2}}}_{=-0.4409}\right\} = 0.1182\frac{k_e q}{a}.$$

We note that the potential is positive, because the field point is closest to the *positive* charge at the origin. To obtain the numerical value of the sum accurate to only four decimal places, we have to add at least 40,000 terms! This sum is, therefore, much less convergent than the sum encountered in the evaluation of E_x above. ∎

An important physical quantity for real crystals is the potential *energy* U of the crystal. Physically, this is the amount of energy required to assemble the charges in their final configuration. A positive potential energy corresponds to *positive* energy stored in the system, i.e., a tendency for the system to provide energy to the outside, once disrupted slightly from its equilibrium position. A negative potential energy is a sign of the stability of the system, i.e., the tendency for the system to restore its original configuration if disrupted slightly from its equilibrium position.[4] It is shown in electrostatics that the potential energy of a system located within the region Ω is

$$U = \frac{1}{2}\iiint_\Omega dq(\mathbf{r})\Phi(\mathbf{r}). \tag{5.38}$$

electrostatic potential energy of a one-dimensional crystal

Example 5.3.2. Let us calculate the electrostatic potential energy of the one-dimensional crystal. Let us assume that there are a total of $2N+1$ charges stretching from $z = -Na$ to $z = +Na$ with a positive charge at the origin. Eventually we shall let N go to infinity, but, in order not to deal explicitly with infinities, we assume that N is finite but large. Substituting in (5.38) the element of charge in terms of volume density, and electrostatic potential found in the previous example, we find

$$U = \frac{1}{2}\iiint_\Omega \rho_q(x,y,z)\Phi(x,y,z)\,dx\,dy\,dz$$

$$= \frac{1}{2}\iiint_\Omega\left\{q\sum_{j=-N}^{N}(-1)^j\,\delta(x)\delta(y)\delta(z-ja)\right\}$$

$$\times\left\{k_e q\sum_{k=-N}^{N}\frac{(-1)^k}{\sqrt{x^2+y^2+(z-ka)^2}}\right\}dx\,dy\,dz$$

$$= \frac{k_e q^2}{2}\sum_{j=-N}^{N}\sum_{\substack{k=-N\\k\neq j}}^{N}\frac{(-1)^{j+k}}{\sqrt{(ja-ka)^2}}.$$

[4]A system that has negative potential energy requires some positive energy (such as kinetic energy of a projectile) to reach a state of zero potential energy corresponding to dissociation of its parts and their removal to infinity (where potential energy is zero).

The restriction $k \neq j$ is necessary, because the $k = j$ terms correspond to the interaction energy of each charge with itself, and should be excluded. Continuing with the calculation, we write

$$U = \frac{k_e q^2}{2a} \sum_{j=-N}^{N} \sum_{\substack{k=-N \\ k \neq j}}^{N} \frac{(-1)^{j+k}}{|j-k|}$$

$$= \frac{k_e q^2}{2a} \sum_{j=-N}^{N} \left\{ \sum_{k=-N}^{j-1} \frac{(-1)^{j+k}}{j-k} + \sum_{k=j+1}^{N} \frac{(-1)^{j+k}}{k-j} \right\}.$$

In the first inner sum, let $j - k = m$, and in the second let $k - j = m$. These substitutions change the limits of the sums, and we get

$$U = \frac{k_e q^2}{2a} \sum_{j=-N}^{N} \left\{ \sum_{m=N+j}^{1} \frac{(-1)^{2j-m}}{m} + \sum_{m=1}^{N-j} \frac{(-1)^{2j+m}}{m} \right\}$$

$$= \frac{k_e q^2}{2a} \sum_{j=-N}^{N} \left\{ \sum_{m=1}^{N+j} \frac{(-1)^m}{m} + \sum_{m=1}^{N-j} \frac{(-1)^m}{m} \right\}.$$

To evaluate the inner sums, denoted by S, we now assume that N is very large—compared to j—so that $N - j \approx N \approx N + j$. Then the inner sum yields[5]

$$S = \sum_{m=1}^{N+j} \frac{(-1)^m}{m} + \sum_{m=1}^{N-j} \frac{(-1)^m}{m} \approx \sum_{m=1}^{N} \frac{(-1)^m}{m} + \sum_{m=1}^{N} \frac{(-1)^m}{m}$$

$$= 2 \sum_{m=1}^{N} \frac{(-1)^m}{m} \approx -2 \sum_{m=1}^{\infty} \frac{(-1)^{m+1}}{m} = -2 \ln 2.$$

Substituting S in the expression for U, we get

$$U \approx \frac{k_e q^2}{2a} \sum_{j=-N}^{N} (-2 \ln 2) = -\frac{k_e q^2}{a} \ln 2 \sum_{j=-N}^{N} 1 = -(2N+1)\frac{k_e q^2}{a} \ln 2.$$

The negative sign indicates that the one-dimensional salt crystal is stable. A useful quantity used in solid-state physics is ionization energy per molecule which is defined to be the potential energy divided by the number of molecules. Noting that the number of molecules is *half* the number of particles, we obtain

$$u \equiv U/N = -\frac{2N+1}{N} \frac{k_e q^2}{a} \ln 2 \approx -\frac{k_e q^2}{a} 2 \ln 2 \equiv -\alpha \frac{k_e q^2}{a}.$$

A real three-dimensional salt crystal has exactly the same expression. However, the constant α, called the **Madelung constant** has the value of 1.747565 instead of $2 \ln 2 = 1.386294$. (See Problem 5.17 for an alternative way of calculating the potential energy of the one-dimensional ionic crystal.) ∎

Madelung constant

---

[5]We are really cheating here! The sum over j indicates that j can assume values close to N, and therefore, the approximation is not valid for such j's. However, a careful analysis, in which one breaks up the sum over j and separates large and small values of j, shows that the original approximation is valid as long as N is large enough.

5.4   Problems

5.1. Plot the distribution on the real line of each of the following electric linear charge densities:

(a) $\lambda(x) = \delta(x - 2)$.   (b) $\lambda(x) = -\delta(x + 1)$.

(c) $\lambda(x) = 5\delta(x) - 3\delta(x + 3)$.   (d) $\lambda(x) = \delta(x + 1) + 3\delta(x - 1)$.

5.2. Evaluate the following integrals:

(a) $\displaystyle\int_0^\infty e^x \sin\frac{\pi x}{2}\delta(x^2 - 1)\,dx$.   (b) $\displaystyle\int_{-2}^2 e^x \sin\frac{\pi x}{2}\delta(x^2 - 1)\,dx$.

(c) $\displaystyle\int_0^\infty e^x \sin\frac{\pi x}{2}\delta(x^3 + 1)\,dx$.   (d) $\displaystyle\int_{-\infty}^\infty \sin\left(\frac{\pi e^x}{2}\right)\delta(x^4 + 1)\,dx$.

(e) $\displaystyle\int_0^\infty \sin^{-1}(1/x)\delta(x^4 - 1)\,dx$.   (f) $\displaystyle\int_{-\infty}^\infty \cos(\pi x)\delta(6x^2 - x - 1)\,dx$.

(g) $\displaystyle\int_{-0.1}^\infty \sin\left(\frac{\pi e^x}{2}\right)\delta(x^2 + x)\,dx$.   (h) $\displaystyle\int_{-\infty}^\infty e^x \sin\frac{\pi x}{2}\delta(e^x \sin\frac{\pi x}{2})\,dx$.

(i) $\displaystyle\int_0^5 e^{\sin x}\delta(\cos x)\,dx$.   (j) $\displaystyle\int_0^\infty \sin^{-1}\left(\frac{1}{x}\right)\delta(x^4 - 4)\,dx$.

(k) $\displaystyle\int_{-\infty}^\infty e^x \sin\frac{\pi x}{3}\delta(4x^2 - 1)\,dx$.   (l) $\displaystyle\int_{-\infty}^\infty \ln(1 + x)\sin\frac{\pi x}{2}\delta(x^3 - 1)\,dx$.

(m) $\displaystyle\int_{-\infty}^\infty \sin\frac{\pi e^x}{2}\delta(x^3 + 1)\,dx$.

5.3. Show that

$$\int_{-\infty}^{+\infty} f(x)\delta'(x - x_0)\,dx = -f'(x_0)$$

and

$$\int_{-\infty}^{+\infty} f(x)\delta'(g(x))\,dx = -\int_{-\infty}^{+\infty} f'(x)\delta(g(x))\,dx.$$

5.4. Evaluate the following integrals:

(a) $\displaystyle\int_0^\infty e^x \sin\frac{\pi x}{2}\delta'(x^2 - 1)\,dx$.   (b) $\displaystyle\int_{-2}^2 e^x \sin\frac{\pi x}{2}\delta'(x^2 - 1)\,dx$.

(c) $\displaystyle\int_0^\infty e^x \sin\frac{\pi x}{2}\delta'(x^3 + 1)\,dx$.   (d) $\displaystyle\int_{-\infty}^\infty \sin\left(\frac{\pi e^x}{2}\right)\delta'(x^4 + 1)\,dx$.

(e) $\displaystyle\int_0^\infty \sin^{-1}(1/x)\delta'(x^4 - 1)\,dx$.   (f) $\displaystyle\int_{-\infty}^\infty \cos(\pi x)\delta'(6x^2 - x - 1)\,dx$.

(g) $\int_{-0.1}^{\infty} \sin\left(\frac{\pi e^x}{2}\right) \delta'(x^2 + x)\,dx.$ (h) $\int_{-\infty}^{\infty} e^x \sin\frac{\pi x}{2} \delta'(e^x \sin\frac{\pi x}{2})\,dx.$

(i) $\int_{0}^{5} e^{\sin x} \delta'(\cos x)\,dx.$ (j) $\int_{0}^{\infty} \sin^{-1}\left(\frac{1}{x}\right) \delta'(x^4 - 4)\,dx.$

(k) $\int_{-\infty}^{\infty} e^x \sin\frac{\pi x}{3} \delta'(4x^2 - 1)\,dx.$ (l) $\int_{-\infty}^{\infty} \ln(1 + x) \sin\frac{\pi x}{2} \delta'(x^3 - 1)\,dx.$

(m) $\int_{-\infty}^{\infty} \sin\frac{\pi e^x}{2} \delta'(x^3 + 1)\,dx.$

5.5. Use integration by parts (or differentiation with respect to x_0) to show that

$$\int_{-\infty}^{+\infty} f(x)\delta''(x - x_0)\,dx = f''(x_0)$$

and

$$\int_{-\infty}^{+\infty} f(x)\delta'''(x - x_0)\,dx = -f'''(x_0)$$

and, in general,

$$\int_{-\infty}^{+\infty} f(x)\delta^{(n)}(x - x_0)\,dx = (-1)^n f^{(n)}(x_0)$$

where $\delta^{(n)}$ and $f^{(n)}$ represent the nth derivatives.

5.6. Derive Equation (5.16). Hint: Use the result of Problem 5.3.

5.7. Six point charges of equal strength q are equally spaced on a circle of radius a. What is the volume charge density describing such a distribution in cylindrical coordinates?

5.8. Convince yourself that

$$\sigma_q(x, y) = q \sum_{i=-\infty}^{\infty} \sum_{j=-\infty}^{\infty} (-1)^{i+j} \delta(x - ia)\delta(y - ja)$$

indeed describes a two-dimensional ionic crystal. Pay particular attention to the power of (-1).

5.9. Derive Equations (5.31) and (5.32).

5.10. Plot (or describe) the distribution in space of each of the following volume charge densities:

$$\rho_q(x, y, z) = \delta(x)\delta(y)\left\{2\delta(z) - 3\delta(z + 3)\right\},$$
$$\rho_q(x, y, z) = 5\delta(x + 1)\delta(y - 1)\left\{\delta(z - 1) - \delta(z + 1)\right\},$$

$$\rho_q(\rho,\varphi,z) = -2\delta(\rho-3)\delta(\varphi-\pi)\delta(z),$$

$$\rho_q(\rho,\varphi,z) = 2\delta(\varphi-\pi/4)\delta(z)\left\{\sum_{k=1}^{10}(-1)^{k+1}\delta(\rho-0.5k)\right\},$$

$$\rho_q(r,\theta,\varphi) = 2\delta(\varphi-\pi/4)\delta(r-2)\left\{\sum_{k=1}^{10}(-1)^{k+1}\delta\left(\theta-\frac{\pi}{20}k\right)\right\},$$

$$\rho_q(r,\theta,\varphi) = 2\delta(\theta-\pi/4)\delta(r-2)\left\{\sum_{k=1}^{20}(-1)^{k+1}\delta\left(\varphi-\frac{\pi}{10}k\right)\right\}.$$

5.11. Derive Equation (5.36).

5.12. Plot $\theta(t)\theta(1-t)$, $\theta(t)-\theta(-t)$, and $\theta(t^2+1)$ for $-\infty < t < +\infty$.

5.13. Write $\theta(t^2-1)$ as a product of two step functions.

5.14. For the two-dimensional ionic crystal shown in Figure 5.6:
(a) write the *volume* charge density describing the distribution (charges are in the xy-plane);
(b) calculate the electrostatic field at $(0,0,a)$; and
(c) calculate the electrostatic potential at an arbitrary point in space with coordinates (x,y,z).
(d) Show that the ionization energy is of the form $-\alpha k_e q^2/a$ with α given in terms of a sum.
(e) Numerically evaluate α.

5.15. For the three-dimensional ionic crystal:
(a) write the *volume* charge density describing the distribution; and
(b) calculate the electrostatic potential at an arbitrary point in space with coordinates (x,y,z).
(c) Show that the ionization energy is of the form $-\alpha k_e q^2/a$ with α given in terms of a sum.
(d) Numerically evaluate α.

5.16. Two electric charges $+q$ and $-q$ are located at P_1 and P_2 with position vectors $\mathbf{r}_1$ and $\mathbf{r}_2$.
(a) Write the volume charge density describing these charges.
(b) Use (a) to find their dipole moment defined by $\iiint \mathbf{r}'dq(\mathbf{r}')$.

5.17. The electric charge density of the one-dimensional ionic crystal can be written as $\rho(\mathbf{r}) = \sum_{i=-N}^{N} q_i\delta(\mathbf{r}-\mathbf{r}_i)$.
(a) Substitute this in Equation (5.38) and get

$$U = \frac{1}{2}\sum_{i=-N}^{N} q_i\Phi(\mathbf{r}_i)$$

(b) Assuming that N is very large (infinite), convince yourself that all products $q_i\Phi(\mathbf{r}_i)$ in the sum are equal (in particular the sign of the charge does not matter). Therefore, $U = \frac{1}{2}(2N+1)q_0\Phi(\mathbf{r}_0)$, where the subscript denotes the zeroth charge.

(c) Show that $\Phi(\mathbf{r}_0) = \sum_{j=-N}^{N} k_e q_j/|\mathbf{r}_j - \mathbf{r}_0|$.

(d) Place the origin at the location of the zeroth charge, and assume that the this charge is positive. Then, $\mathbf{r}_0 = 0$, $\mathbf{r}_j = ja\hat{\mathbf{e}}_z$, and $q_j = -(-1)^j q$. Now show that

$$U = -(N + \tfrac{1}{2})q^2 k_e \sum_{j=-N}^{N} \frac{(-1)^j}{|aj|}$$

(e) By breaking up the sum into two parts show that

$$U = -(2N+1)\frac{q^2 k_e}{a} \sum_{j=1}^{N} \frac{(-1)^j}{j}$$

5.18. $2N$ charges of equal sign and magnitude q are arranged equally spaced on a circle of radius a located in the xy-plane. Assume that the charge numbered $2N$ is at $(a,0,0)$.

(a) Write the volume charge density of such a distribution in cylindrical coordinates.

(b) Starting with an integral expression for the electric field, find the cylindrical components of the field at an arbitrary point P in space in terms of a sum. The coordinates of P are (ρ, φ, z). Simplify your answer as much as possible.

(c) Now let P have coordinates $(2a,0,0)$. Show that all components of the field are of the form $(k_e q/a^2)\alpha$. Express the α for each component in terms of a sum. What do you expect the value of α to be? Can you find that value?

(d) For $N = 3$, i.e., six charges, calculate the numerical value of α in part (c) for all components.

5.19. $2N+1$ charges of equal sign and magnitude q are arranged on the x-axis of a Cartesian coordinate system as shown in Figure 5.9, with the zeroth charge at the origin. The numbers below the axis are labels of the charges.

(a) From the pattern of the figure, determine the location of the kth charge for $-N \le k \le N$.

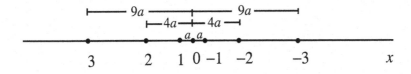

Figure 5.9: The charges and their distances on the x-axis.

(b) Write a volume charge density in terms of the Dirac delta function describing such a charge distribution.

(c) Calculate the components of the electric field at a general point P with coordinates (x, y, z).

(d) Now let P have coordinates $(a, a, 0)$. Show that all components of the field are of the form $(k_e q/a^2)\alpha$ where α is a numerical factor. Find this factor for each component.

5.20. $2N$ positive and negative charges of equal magnitude are arranged equally spaced and alternating in sign on a circle of radius a.

(a) Write the expression of the volume charge density describing this charge distribution.

(b) Find the ionization energy in the form $-\alpha k_e q^2/a$ with α given in terms of a sum. Simplify this sum as much as possible.

Part II

Algebra of Vectors

Chapter 6

Planar and Spatial Vectors

The preceding chapters made heavy use of vectors in the plane and in space. The enormous utility of the concept of vectors has prompted mathematicians and physicists to generalize this concept to include other objects that at first glance have no resemblance whatsoever with the planar and spatial vectors. In this chapter, we shall study this generalization in its limited form, i.e., only in an *algebraic* context. Although the *analysis* of vectors is discussed in Chapters 12 through 17, it is confined to vectors in space. The analysis of generalized vectors is the subject of differential geometry and functional analysis that are beyond the scope of this book.[1]

There are many mathematical objects used in physics that allow for the two operations of addition and multiplication by a number. The collection of such objects is called a **vector space**. Thus, a vector space is a bunch of "things" having the property that when you add two "things" you get a third one, and if you multiply a "thing" by a number you get another one of those "things." Furthermore, the operation of multiplication by a number and addition of "things" is distributive, and a vector space always has a "thing" that we call the **zero vector**.

vector spaces defined

Using the two operations of multiplication by a number and addition, we can form a sum,

$$\alpha_1 \mathbf{a}_1 + \alpha_2 \mathbf{a}_2 + \cdots + \alpha_n \mathbf{a}_n, \qquad (6.1)$$

where $\alpha_1, \alpha_2, \ldots, \alpha_n$, are real numbers and $\mathbf{a}_1, \mathbf{a}_2, \ldots, \mathbf{a}_n$ are vectors. The sum in Equation (6.1) is called a **linear combination** of the n vectors and $\alpha_1, \alpha_2, \ldots, \alpha_n$ are called the **coefficients** of the linear combinations.

linear combination

coefficients

---

[1]Hassani, S. *Mathematical Physics: A Modern Introduction to Its Foundations*, Springer-Verlag, 1999, discusses differential geometry and functional analysis in some detail.

> **Box 6.0.1.** *If we can find some set of real numbers,* $\alpha_1, \alpha_2, \ldots, \alpha_n$ *(not all of which are zero), such that the sum in (6.1) is zero, we say that the vectors are **linearly dependent**. If no such set of real numbers can be found, then the vectors are called **linearly independent**.*

6.1   Vectors in a Plane Revisited

Before elaborating further on the generalization of vectors and their spaces, it is instructive to revisit the familiar vectors in a plane from a point of view suitable for generalization. We first discuss the notion of linear independence as applied to vectors in the plane.

The two vectors $\hat{\mathbf{e}}_x$ and $\hat{\mathbf{e}}_y$ (sometimes denoted as $\mathbf{i}$ and $\mathbf{j}$) are linearly independent because $\alpha\hat{\mathbf{e}}_x + \beta\hat{\mathbf{e}}_y = 0$ can be satisfied only if both α and β are zero. If one of them, say α, were different from zero, one could divide the equation by α and get

$$\hat{\mathbf{e}}_x = -\frac{\beta}{\alpha}\hat{\mathbf{e}}_y$$

which is impossible because $\hat{\mathbf{e}}_x$ and $\hat{\mathbf{e}}_y$ cannot lie along the same line.

Example 6.1.1. The arrows in the plane are not the only kinds of vectors dealt with in physics. For instance, consider the set of all linear functions, or polynomials of degree one (or less), i.e., functions of the form $\alpha_0 + \alpha_1 t$ where α_0 and α_1 are real numbers and t is an *arbitrary* variable. Let us call this set $\mathcal{P}_1[t]$, where $\mathcal{P}$ stands for "polynomial," 1 signifies the degree of these polynomials, and t is just the variable used. We can add two such polynomials and get a third one of the same form. We can multiply any such polynomial by a real number and get another polynomial. In fact, $\mathcal{P}_1[t]$ has all the properties of the vectors in a plane. We say that $\mathcal{P}_1[t]$ and the vectors in a plane are **isomorphic** which literally means they have the "same shape."

polynomials as vectors?

It is important to emphasize that two polynomials are equal if and only if all their coefficients are equal. In particular, a polynomial is equal to zero only if it is so *for all values of* t, i.e., only if its coefficients vanish. This immediately leads to the fact that the two polynomials 1 and t are linearly independent because if $\alpha + \beta t = 0$ (for all values of t), then $\alpha = \beta = 0$ (try $t = 0$ and $t = 1$). ∎

proof of the fact that any three vectors in the plane are linearly dependent

It is easy to show that any three vectors in the plane are linearly dependent. Figure 6.1 shows three arbitrary vectors drawn in a plane. From the tip of one of the vectors ($\mathbf{a}_3$ in the figure), a line is drawn parallel to one of the other two vectors such that it meets the third vector (or its extension) at point D. The vectors $\overrightarrow{OD}$ and $\overrightarrow{DC}$ are proportional to $\mathbf{a}_1$ and $\mathbf{a}_2$, respectively, and their sum is equal to $\mathbf{a}_3$. So we can write

$$\mathbf{a}_3 = \overrightarrow{OD} + \overrightarrow{DC} = \alpha\mathbf{a}_1 + \beta\mathbf{a}_2 \quad \Rightarrow \quad \alpha\mathbf{a}_1 + \beta\mathbf{a}_2 - \mathbf{a}_3 = 0$$

and $\mathbf{a}_1$, $\mathbf{a}_2$, and $\mathbf{a}_3$ are linearly dependent. Clearly we cannot do the same with two arbitrary vectors. Thus

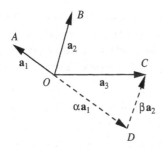

Figure 6.1: Any three vectors $\mathbf{a}_1$, $\mathbf{a}_2$, and $\mathbf{a}_3$ in the plane are linearly dependent.

Box 6.1.1. *The maximum number of linearly independent vectors in a plane is two. Any vector in a plane can be written as a linear combination of only two non-collinear (not lying along the same line) vectors.*

We also say that any two non-collinear vectors **span** the plane.

Suppose that we can write a vector $\mathbf{a}$ as a linear combination of n vectors

$$\mathbf{a} = \alpha_1 \mathbf{a}_1 + \alpha_2 \mathbf{a}_2 + \cdots + \alpha_n \mathbf{a}_n.$$

We want to see under what conditions the coefficients are unique. Suppose that we can also write

$$\mathbf{a} = \beta_1 \mathbf{a}_1 + \beta_2 \mathbf{a}_2 + \cdots + \beta_n \mathbf{a}_n,$$

where the β's are different from the α's. Then, subtracting these two linear combinations, we get

$$\mathbf{0} = (\alpha_1 - \beta_1)\mathbf{a}_1 + (\alpha_2 - \beta_2)\mathbf{a}_2 + \cdots + (\alpha_n - \beta_n)\mathbf{a}_n.$$

This is possible only if the vectors are linearly dependent. Therefore, if we want the coefficients to be unique, the vectors have to be linearly independent. In particular, we can have at most two such vectors in the plane. Thus, choosing any two linearly independent vectors $\mathbf{a}_1$ and $\mathbf{a}_2$ in the plane, we can expand any other vector *uniquely* as a linear combination of $\mathbf{a}_1$ and $\mathbf{a}_2$. This brings us to the notion of a basis.

basis defined

Box 6.1.2. *Vectors that span the plane and are linearly independent are called a **basis** for the plane.*

The foregoing argument showed that any two non-collinear vectors form a basis for the plane.

With the notion of a basis comes the concept of components of a vector. Given a basis, there is a unique way in which a particular vector can be written in terms of the vectors in the basis. The unique coefficients of the basis vectors are called the **components** of the particular vector in that basis. Another concept associated with the basis is **dimension** which is defined to be the number of vectors in a basis. It follows that the plane has two dimensions.

Example 6.1.2. The components of $\mathbf{a}_3$ in the basis $\{\mathbf{a}_1, \mathbf{a}_2\}$ of Figure 6.1 are (α, β).[2] Given any basis $\{\mathbf{a}_1, \mathbf{a}_2\}$ of the plane, it is readily seen that the components of $\mathbf{a}_1$ are $(1, 0)$ and those of $\mathbf{a}_2$ are $(0, 1)$. ∎

Example 6.1.3. The polynomials $\{1, t\}$ form a basis for $\mathcal{P}_1[t]$, because they are linearly independent and they span $\mathcal{P}_1[t]$. Therefore $\mathcal{P}_1[t]$ is a two-dimensional vector space. The components of $\mathbf{f} = \alpha_0 + \alpha_1 t$ are (α_0, α_1) in this basis. How do we determine the components of $\mathbf{f}$ in another basis $\{\mathbf{a}_1, \mathbf{a}_2\}$ with $\mathbf{a}_1 = 1 + t$ and $\mathbf{a}_2 = 1 - t$? Since $\{\mathbf{a}_1, \mathbf{a}_2\}$ is a basis, we can write

$$\mathbf{f} = x_1 \mathbf{a}_1 + x_2 \mathbf{a}_2 = x_1(1 + t) + x_2(1 - t) = (x_1 + x_2) + (x_1 - x_2)t$$

or

$$\alpha_0 + \alpha_1 t = (x_1 + x_2) + (x_1 - x_2)t \;\Rightarrow\; (\alpha_0 - x_1 - x_2) \cdot 1 + (\alpha_1 - x_1 + x_2)t = 0.$$

The linear independence of 1 and t now tells us that the coefficients of 1 and t should vanish. This leads to two equations in two unknowns:

$$x_1 + x_2 = \alpha_0,$$
$$x_1 - x_2 = \alpha_1.$$

The solution of these equations are easily found to be

$$x_1 = \tfrac{1}{2}(\alpha_0 + \alpha_1), \qquad x_2 = \tfrac{1}{2}(\alpha_0 - \alpha_1).$$

Thus, the components of $\mathbf{f}$ are $(\tfrac{1}{2}(\alpha_0 + \alpha_1), \tfrac{1}{2}(\alpha_0 - \alpha_1))$ in the new basis. ∎

6.1.1 Transformation of Components

There are infinitely many bases in a plane, because there are infinitely many pairs of vectors that are linearly independent. Therefore, there are infinitely many sets of components for any given vector, and it is desirable to be able to find a relation between any two such sets. Such a relation employs the machinery of matrices.

Consider a vector $\mathbf{a}$ with components (α_1, α_2) in the basis $\{\mathbf{a}_1, \mathbf{a}_2\}$ and components (α_1', α_2') in the basis $\{\mathbf{a}_1', \mathbf{a}_2'\}$. We can write

$$\mathbf{a} = \alpha_1 \mathbf{a}_1 + \alpha_2 \mathbf{a}_2 \qquad \text{and} \qquad \mathbf{a} = \alpha_1' \mathbf{a}_1' + \alpha_2' \mathbf{a}_2'. \tag{6.2}$$

Since $\{\mathbf{a}_1', \mathbf{a}_2'\}$ form a basis, any vector, in particular, $\mathbf{a}_1$ or $\mathbf{a}_2$, can be written in terms of them:

$$\mathbf{a}_1 = a_{11} \mathbf{a}_1' + a_{21} \mathbf{a}_2',$$
$$\mathbf{a}_2 = a_{12} \mathbf{a}_1' + a_{22} \mathbf{a}_2', \tag{6.3}$$

---

[2]Since in this chapter we are dealing primarily with components (and not coordinates), we shall use parentheses—instead of angle brackets—to list the components.

where (a_{11}, a_{21}) and (a_{12}, a_{22}) are, respectively, components of $\mathbf{a}_1$ and $\mathbf{a}_2$ in the basis $\{\mathbf{a}'_1, \mathbf{a}'_2\}$. Combining Equations (6.2) and (6.3), we obtain

$$\alpha_1(a_{11}\mathbf{a}'_1 + a_{21}\mathbf{a}'_2) + \alpha_2(a_{12}\mathbf{a}'_1 + a_{22}\mathbf{a}'_2) = \alpha'_1\mathbf{a}'_1 + \alpha'_2\mathbf{a}'_2$$

or

$$(\alpha'_1 - a_{11}\alpha_1 - a_{12}\alpha_2)\mathbf{a}'_1 + (\alpha'_2 - a_{21}\alpha_1 - a_{22}\alpha_2)\mathbf{a}'_2 = 0.$$

The linear independence of $\mathbf{a}'_1$ and $\mathbf{a}'_2$ gives

$$\alpha'_1 = a_{11}\alpha_1 + a_{12}\alpha_2,$$
$$\alpha'_2 = a_{21}\alpha_1 + a_{22}\alpha_2. \tag{6.4}$$

These equations can be written concisely as[3]

$$\begin{pmatrix} \alpha'_1 \\ \alpha'_2 \end{pmatrix} = \begin{pmatrix} a_{11} & a_{12} \\ a_{21} & a_{22} \end{pmatrix} \begin{pmatrix} \alpha_1 \\ \alpha_2 \end{pmatrix} \qquad \text{or} \qquad \mathsf{a}' = \mathsf{A}\mathsf{a}, \tag{6.5}$$

where we have introduced the **matrices**

matrix and column vector

$$\mathsf{a} \equiv \begin{pmatrix} \alpha_1 \\ \alpha_2 \end{pmatrix}, \qquad \mathsf{a}' \equiv \begin{pmatrix} \alpha'_1 \\ \alpha'_2 \end{pmatrix}, \qquad \mathsf{A} \equiv \begin{pmatrix} a_{11} & a_{12} \\ a_{21} & a_{22} \end{pmatrix}. \tag{6.6}$$

The matrices a and a' are called **column vectors** or 2×1 matrices because they each have two rows and one column. Similarly, A is called a 2×2 matrix.

Let us now choose a third basis, $\{\mathbf{a}''_1, \mathbf{a}''_2\}$, and write $\mathbf{a} = \alpha''_1\mathbf{a}''_1 + \alpha''_2\mathbf{a}''_2$. If (a'_{11}, a'_{21}) and (a'_{12}, a'_{22}) are, respectively, the components of $\mathbf{a}'_1$ and $\mathbf{a}'_2$ in this third basis, then

$$\mathbf{a}'_1 = a'_{11}\mathbf{a}''_1 + a'_{21}\mathbf{a}''_2,$$
$$\mathbf{a}'_2 = a'_{12}\mathbf{a}''_1 + a'_{22}\mathbf{a}''_2.$$

Substituting these in the second equation of (6.2) and equating the result to $\mathbf{a} = \alpha''_1\mathbf{a}''_1 + \alpha''_2\mathbf{a}''_2$ yields

$$\alpha''_1 = a'_{11}\alpha'_1 + a'_{12}\alpha'_2,$$
$$\alpha''_2 = a'_{21}\alpha'_1 + a'_{22}\alpha'_2. \tag{6.7}$$

We can write Equation (6.7) in matrix form:

$$\begin{pmatrix} \alpha''_1 \\ \alpha''_2 \end{pmatrix} = \begin{pmatrix} a'_{11} & a'_{12} \\ a'_{21} & a'_{22} \end{pmatrix} \begin{pmatrix} \alpha'_1 \\ \alpha'_2 \end{pmatrix} \qquad \text{or} \qquad \mathsf{a}'' = \mathsf{A}'\mathsf{a}', \tag{6.8}$$

where

$$\mathsf{a}'' \equiv \begin{pmatrix} \alpha''_1 \\ \alpha''_2 \end{pmatrix}, \qquad \mathsf{A}' \equiv \begin{pmatrix} a'_{11} & a'_{12} \\ a'_{21} & a'_{22} \end{pmatrix}, \tag{6.9}$$

and a' is as defined before.

---

[3] At this point, think of Equation (6.5) as a short-hand way of writing Equation (6.4). Further significance of this notation will become clear after Box 6.1.3.

We can also discover how a'' and a are related by substituting (6.4) in (6.7). This leads to the equation

two
transformations in
a row suggest the
rule of matrix
multiplication.

$$\alpha_1'' = (a_{11}'a_{11} + a_{12}'a_{21})\alpha_1 + (a_{11}'a_{12} + a_{12}'a_{22})\alpha_2,$$
$$\alpha_2'' = (a_{21}'a_{11} + a_{22}'a_{21})\alpha_1 + (a_{21}'a_{12} + a_{22}'a_{22})\alpha_2,$$

which, in matrix form, becomes

$$a'' = A''a \quad \text{where} \quad A'' \equiv \begin{pmatrix} a_{11}'a_{11} + a_{12}'a_{21} & a_{11}'a_{12} + a_{12}'a_{22} \\ a_{21}'a_{11} + a_{22}'a_{21} & a_{21}'a_{12} + a_{22}'a_{22} \end{pmatrix}. \quad (6.10)$$

On the other hand, the matrix equations (6.8) and (6.5) yield $a'' = A'(Aa)$, which is consistent with Equation (6.10) only if **matrix multiplication** is defined so that $A'' = A'A$, i.e.,

$$\begin{pmatrix} a_{11}' & a_{12}' \\ a_{21}' & a_{22}' \end{pmatrix} \begin{pmatrix} a_{11} & a_{12} \\ a_{21} & a_{22} \end{pmatrix} = \begin{pmatrix} a_{11}'a_{11} + a_{12}'a_{21} & a_{11}'a_{12} + a_{12}'a_{22} \\ a_{21}'a_{11} + a_{22}'a_{21} & a_{21}'a_{12} + a_{22}'a_{22} \end{pmatrix}. \quad (6.11)$$

All discussions and all the equations obtained so far are based on fixing a vector and looking at its components in different bases. However, there is another, more physical, way of interpreting these equations. Consider (6.5).

active and passive
transformations
distinguished

Here the column vector on the RHS represents the components of a vector a in the basis $\{a_1, a_2\}$. Applying the matrix A to this column vector yields a new column vector given on the LHS, *which can be interpreted as the components of a new vector a' in the same basis.* So, in essence we have changed the vector a into a new vector a' via the transformation A. The first interpretation mentioned above is called a **passive transformation** (a is "passively" unchanged as basis vectors are altered); the second interpretation is called **active transformation** (a is actively changed into a'). We shall have occasion to employ both interpretations. However, the active transformation is more direct and we shall use that more often. The reader may convince himself or herself that passive transformation in one "direction" is completely equivalent to active transformation in the "opposite" direction. A good example to keep in mind is the rotation of axes (passive rotation) versus the rotation of a vector (active rotation) in the plane as shown in Figure 6.2.

Equation (6.11) defines the "product" of two matrices in a prescribed manner. To find the entry in the first row and first column of the product, multiply the entries of the first row of the first matrix by the corresponding entries of the first column of the second matrix and add the terms thus obtained. To find the entry in the first row and second column of the product, multiply the entries of the first row of the first matrix by the corresponding entries of the second column of the second matrix and add the terms. Other entries are found similarly. This leads us to the following rule which applies to all matrices, not just those that are 2×2:

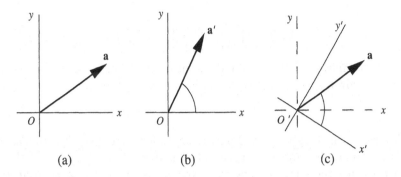

Figure 6.2: (a) A vector **a** in a coordinate system Oxy can be (b) actively transformed to a *new vector* **a′** in the same coordinate system, or (c) passively transformed to a *new coordinate system* $O'x'y'$. Note that the relation of **a′** to Oxy is identical to the relation of **a** to $O'x'y'$.

Box 6.1.3. (*Matrix Multiplication Rule*). *To obtain the entry in the ith row and jth column of the product of two matrices, multiply the entries of the ith row of the matrix on the left by the corresponding entries of the jth column of the matrix on the right and add the products thus obtained.*

For this rule to make sense, the number of entries in a row of the matrix on the left must equal the number of entries in a column of the matrix on the right.

We identified a column vector as a 2×1 matrix. With this identification, the RHS of Equation (6.5) can be interpreted as the product of two matrices, a 2×2 matrix and a 2×1 matrix, resulting in a 2×1 matrix, the column vector on the LHS.

Matrices were obtained in a natural way in the discussion of basis changes, and the natural operation ensued was that of multiplication. Once a mathematical entity is created in this manner, a full mathematical structure becomes irresistibly enticing. For example, such operations as addition, subtraction, division, inversion, etc., also demand our attention. We now consider such operations.

First, we need to define the equality of matrices: Two matrices are equal if they have the same number of rows and columns, and their corresponding elements are equal. Addition of two matrices is defined if they have the same number of rows and columns in which case the sum is defined to be the sum of corresponding elements. A 2×2 matrix can be added to another 2×2 matrix, but a column vector cannot. Thus if

$$A = \begin{pmatrix} a_{11} & a_{12} \\ a_{21} & a_{22} \end{pmatrix} \quad \text{and} \quad B = \begin{pmatrix} b_{11} & b_{12} \\ b_{21} & b_{22} \end{pmatrix},$$

(margin notes)

for the product rule to make sense, number of columns of the left matrix must equal number of rows of the right matrix.

matrices forming a mathematical structure with operations other than multiplication then

$$A + B = \begin{pmatrix} a_{11} + b_{11} & a_{12} + b_{12} \\ a_{21} + b_{21} & a_{22} + b_{22} \end{pmatrix}.$$

From the definition of the sum and the product of matrices, it is clear that addition is always commutative but product need not be:

$$A + B = B + A \qquad \text{but} \qquad AB \neq BA. \qquad (6.12)$$

2 × 2 matrices
form a vector
space

zero matrix

We can turn the set of 2 × 2 matrices into a vector space by defining the product of a number and a matrix as a new matrix whose elements are the old elements times the number. The zero "vector" is simply the **zero matrix**—the 2 × 2 matrix all of whose elements are zero. The reader may verify that all the usual operations of vectors apply to this set.[4] If you multiply a matrix by the number 0, you get the zero matrix.

Example 6.1.4. Suppose

$$A = \begin{pmatrix} 1 & -1 \\ 2 & 3 \end{pmatrix} \qquad \text{and} \qquad B = \begin{pmatrix} -1 & 0 \\ 1 & 2 \end{pmatrix}.$$

Then

$$A + B = \begin{pmatrix} 1-1 & -1+0 \\ 2+1 & 3+2 \end{pmatrix} = \begin{pmatrix} 0 & -1 \\ 3 & 5 \end{pmatrix} = B + A$$

and

$$AB = \begin{pmatrix} 1 & -1 \\ 2 & 3 \end{pmatrix} \begin{pmatrix} -1 & 0 \\ 1 & 2 \end{pmatrix} = \begin{pmatrix} 1 \cdot (-1) + (-1) \cdot 1 & 1 \cdot 0 + (-1) \cdot 2 \\ 2 \cdot (-1) + 3 \cdot 1 & 2 \cdot 0 + 3 \cdot 2 \end{pmatrix}$$
$$= \begin{pmatrix} -2 & -2 \\ 1 & 6 \end{pmatrix},$$

while

$$BA = \begin{pmatrix} -1 & 0 \\ 1 & 2 \end{pmatrix} \begin{pmatrix} 1 & -1 \\ 2 & 3 \end{pmatrix} = \begin{pmatrix} (-1) \cdot 1 + 0 \cdot 2 & (-1) \cdot (-1) + 0 \cdot 3 \\ 1 \cdot 1 + 2 \cdot 2 & 1 \cdot (-1) + 2 \cdot 3 \end{pmatrix} = \begin{pmatrix} -1 & 1 \\ 5 & 5 \end{pmatrix}.$$

Clearly, $AB \neq BA$. ∎

The 2 × 2 matrix

$$1 \equiv \begin{pmatrix} 1 & 0 \\ 0 & 1 \end{pmatrix}$$

identity matrix or
unit matrix

inverse of a matrix

is called the 2 × 2 **identity matrix** or **unit matrix,** and has the property that when it multiplies any other matrix (on the right or on the left), the latter does not get affected. The unit matrix is used to define the **inverse** of a matrix A as a matrix B that multiplies A on either side and gives the unit matrix. The inversion of a matrix is a much more complicated process than that of ordinary numbers, and we shall discuss it in greater length later. At this point, suffice it to say that, contrary to numbers, not all nonzero matrices have an inverse. For example, the reader can easily verify that the nonzero matrix $\begin{pmatrix} 1 & 0 \\ 0 & 0 \end{pmatrix}$ cannot have an inverse.

---

[4]Note that the extra operation of multiplication of a matrix by another matrix is *not* part of the requirement for the set to be a vector space.

We have introduced 2×2 and column (or 2×1) matrices. To complete the picture, we also introduce a **row vector**, or a 1×2 matrix. The rule of matrix multiplication allows the multiplication of a 2×2 matrix and a column vector, as long as the latter is to the right of the former: You cannot multiply a 2×1 matrix situated to the left of a 2×2 matrix. Similarly, you cannot multiply two 2×1 matrices. However, the product of a row vector (a 1×2 matrix) and a column vector (a 2×1 matrix) is defined—as long as the latter is to the right of the former—and the result is a 1×1 matrix, i.e., a number. This is because we have only one row to the left of a single column. What about the product of a row vector and a 2×2 matrix? As long as the matrix is to the right of the row vector, the product is defined and the result is a row vector.

row vector

Example 6.1.5. With A and B as defined in Example 6.1.4 and

$$x = \begin{pmatrix} 1 \\ -1 \end{pmatrix}, \qquad y = \begin{pmatrix} -1 & 2 \end{pmatrix},$$

we have

$$Ax = \begin{pmatrix} 1 & -1 \\ 2 & 3 \end{pmatrix} \begin{pmatrix} 1 \\ -1 \end{pmatrix} = \begin{pmatrix} 2 \\ -1 \end{pmatrix},$$

$$yB = \begin{pmatrix} -1 & 2 \end{pmatrix} \begin{pmatrix} -1 & 0 \\ 1 & 2 \end{pmatrix} = \begin{pmatrix} 3 & 4 \end{pmatrix},$$

$$yx = \begin{pmatrix} -1 & 2 \end{pmatrix} \begin{pmatrix} 1 \\ -1 \end{pmatrix} = \begin{pmatrix} -3 \end{pmatrix} = -3,$$

$$yAx = \begin{pmatrix} -1 & 2 \end{pmatrix} \begin{pmatrix} 1 & -1 \\ 2 & 3 \end{pmatrix} \begin{pmatrix} 1 \\ -1 \end{pmatrix} = \begin{pmatrix} -1 & 2 \end{pmatrix} \begin{pmatrix} 2 \\ -1 \end{pmatrix} = -4,$$

$$yBAx = \underbrace{\begin{pmatrix} 3 & 4 \end{pmatrix}}_{=2} \begin{pmatrix} 2 \\ -1 \end{pmatrix} = \begin{pmatrix} -1 & 2 \end{pmatrix} \underbrace{\begin{pmatrix} -1 & 1 \\ 5 & 5 \end{pmatrix}}_{=BA} \begin{pmatrix} 1 \\ -1 \end{pmatrix} = \begin{pmatrix} -1 & 2 \end{pmatrix} \begin{pmatrix} -2 \\ 0 \end{pmatrix} = 2,$$

$$yABx = \begin{pmatrix} -1 & 2 \end{pmatrix} \begin{pmatrix} -2 & -2 \\ 1 & 6 \end{pmatrix} \begin{pmatrix} 1 \\ -1 \end{pmatrix} = \begin{pmatrix} 4 & 14 \end{pmatrix} \begin{pmatrix} 1 \\ -1 \end{pmatrix} = -10.$$

In the manipulations above, we have used the associativity of matrix multiplication and multiplied matrices in different orders without, of course, commuting them. Products such as Ay, By, yy, and xx are not defined; therefore, we have not considered them here.   ∎

There is a new operation on matrices which does not exist for ordinary numbers. This is called transposition and is defined as follows:

transpose of a matrix

> **Box 6.1.4.** *The **transpose** of a matrix is a new matrix whose rows are the columns of the old matrix and whose columns are the rows of the old matrix. The transpose of A is denoted by A^t or $\tilde{A}$.*

Therefore

$$A = \begin{pmatrix} a_{11} & a_{12} \\ a_{21} & a_{22} \end{pmatrix} \Rightarrow A^t = \widetilde{A} = \begin{pmatrix} a_{11} & a_{21} \\ a_{12} & a_{22} \end{pmatrix}.$$

symmetric matrix   If $A^t = A$, we say that A is **symmetric**.

Example 6.1.6. With A, B, x, and y as defined in Example 6.1.5, we have

$$A^t = \begin{pmatrix} 1 & 2 \\ -1 & 3 \end{pmatrix}, \qquad \widetilde{B} = \begin{pmatrix} -1 & 1 \\ 0 & 2 \end{pmatrix}, \qquad x^t = \begin{pmatrix} 1 & -1 \end{pmatrix}, \qquad \widetilde{y} = \begin{pmatrix} -1 \\ 2 \end{pmatrix}.$$

Note that although xx and yy are not defined, all the combinations $\widetilde{x}x$, $y\widetilde{y}$, $\widetilde{y}y$, and $x\widetilde{x}$ are defined: In the first two cases one gets a number, and in the last two cases a 2×2 matrix. ∎

properties of   It should be clear from the definition of the transpose that
transposition

$$(A + B)^t = A^t + B^t, \qquad (AB)^t = B^t A^t, \qquad (A^t)^t = A. \qquad (6.13)$$

Of the three relations, the middle one is the least obvious, but the reader can verify it directly by choosing appropriate *general* matrices and carrying through the multiplications on both sides of the relation.

6.1.2 Inner Product

From our discussion of Chapter 1, we know that if **a** and **b** are vectors in the plane having components (a_x, a_y) and (b_x, b_y) along the x- and y-axes, then their dot product is

$$\mathbf{a} \cdot \mathbf{b} = a_x b_x + a_y b_y. \qquad (6.14)$$

We want to generalize this dot product so that it applies to *arbitrary* bases. This generalization is called the **inner product**.

Recall that any two non-collinear vectors $\{\mathbf{a}_1, \mathbf{a}_2\}$ in the plane form a basis and any vector can be written as a linear combination of them with the unique coefficients being the components of the vector in the basis. In particular, the components of $\mathbf{a}_1$ are $(1, 0)$ and those of $\mathbf{a}_2$ are $(0, 1)$. If we were to define the dot product in terms of components, we would have to modify

equation (6.14)   Equation (6.14) because that equation would give zero for $\mathbf{a}_1 \cdot \mathbf{a}_2$ which would
will not work for   be inconsistent with (1.1). How should we modify (6.14)? Since we want to
arbitrary bases!   deal with components, a natural setting would be the language of matrices. If a and b are the column vectors $\begin{pmatrix} a_x \\ a_y \end{pmatrix}$ and $\begin{pmatrix} b_x \\ b_y \end{pmatrix}$, respectively, then we can

inner (dot)   rewrite Equation (6.14) as
product in terms
of row and column
vectors

$$\mathbf{a} \cdot \mathbf{b} = a^t b = \begin{pmatrix} a_x & a_y \end{pmatrix} \begin{pmatrix} b_x \\ b_y \end{pmatrix} = a_x b_x + a_y b_y. \qquad (6.15)$$

It is this matrix relation that we want to generalize so that the result is the true dot product of vectors no matter what basis we choose in which to express our vectors.

Besides the failure of Equation (6.15) for general bases, the demand for generalization stems from another source: There are other kinds of "vectors" that are not just arrows in the plane. For instance, the polynomials $\mathcal{P}_1[t]$ of degree one that we introduced in Example 6.1.1 are such vectors. How do we define inner products for these vectors? We cannot use Equation (1.1) because neither the length of a polynomial nor the angle between two polynomials is defined. In fact, both the length and the angle are defined only *after* an inner product has been introduced. Furthermore, there is no guarantee that Equation (6.15) will make sense.

Let's see how far we can go using the general properties of the inner product discussed at the beginning of Section 1.1.1. Write **a** and **b** as a linear combination of the basis vectors $\{\mathbf{a}_1, \mathbf{a}_2\}$:

$$\mathbf{a} = \alpha_1\mathbf{a}_1 + \alpha_2\mathbf{a}_2, \qquad \mathbf{b} = \beta_1\mathbf{a}_1 + \beta_2\mathbf{a}_2$$

Take the dot-product of these vectors and write it in terms of the dot-products of the basis vectors:

$$\begin{aligned}
\mathbf{a} \cdot \mathbf{b} &= \left(\alpha_1\mathbf{a}_1 + \alpha_2\mathbf{a}_2\right) \cdot \left(\beta_1\mathbf{a}_1 + \beta_2\mathbf{a}_2\right) \\
&= \alpha_1\beta_1\mathbf{a}_1 \cdot \mathbf{a}_1 + \alpha_1\beta_2\mathbf{a}_1 \cdot \mathbf{a}_2 \\
&\quad + \alpha_2\beta_1\mathbf{a}_2 \cdot \mathbf{a}_1 + \alpha_2\beta_2\mathbf{a}_2 \cdot \mathbf{a}_2
\end{aligned}$$

Define a matrix with elements

$$g_{11} = \mathbf{a}_1 \cdot \mathbf{a}_1, \quad g_{12} = \mathbf{a}_1 \cdot \mathbf{a}_2 = \mathbf{a}_2 \cdot \mathbf{a}_1 = g_{21}, \quad g_{22} = \mathbf{a}_2 \cdot \mathbf{a}_2$$

Then, representing **a** and **b** as column vectors $\mathbf{a} = \left(\begin{smallmatrix}\alpha_1\\\alpha_2\end{smallmatrix}\right)$ and $\mathbf{b} = \left(\begin{smallmatrix}\beta_1\\\beta_2\end{smallmatrix}\right)$, the dot product can be generalized to

> a symmetric matrix G is needed to generalize the inner product.

$$\mathbf{a} \cdot \mathbf{b} = \mathbf{a}^t \mathsf{G}\mathbf{b} = \begin{pmatrix}\alpha_1 & \alpha_2\end{pmatrix}\begin{pmatrix}g_{11} & g_{12} \\ g_{21} & g_{22}\end{pmatrix}\begin{pmatrix}\beta_1 \\ \beta_2\end{pmatrix}, \tag{6.16}$$

where G is a *symmetric* matrix.

Example 6.1.7. In this example, we shall define an inner product for the vectors in $\mathcal{P}_1[t]$ that happens to be useful in physical applications. The idea is to find a rule that takes two "vectors" in $\mathcal{P}_1[t]$ and gives a real number. Since the vectors in $\mathcal{P}_1[t]$ are functions (albeit a very special kind), one natural way of getting numbers out of functions is by integrating them. It turns out that this is indeed the most useful way of defining the inner product for such polynomials. So, let (a, b) be an interval on the real line and let $\mathbf{f} = \alpha_0 + \alpha_1 t$ and $\mathbf{g} = \beta_0 + \beta_1 t$ be two "vectors" in $\mathcal{P}_1[t]$. We define

> inner (dot) product of two polynomials

$$\mathbf{f} \cdot \mathbf{g} \equiv \int_a^b f(t)g(t)\,dt. \tag{6.17}$$

One can show that Equation (6.17) exhibits all the properties expected of an inner product (as outlined in Section 1.1.1). For instance, $\mathbf{f} \cdot \mathbf{f}$ is always positive because the integrand $[f(t)]^2$ is always positive. Furthermore, $\mathbf{f} \cdot \mathbf{g} = \mathbf{g} \cdot \mathbf{f}$, and, as the reader may check,

$$\mathbf{f} \cdot (\mathbf{g} + \mathbf{h}) = \mathbf{f} \cdot \mathbf{g} + \mathbf{f} \cdot \mathbf{h}.$$

These all indicate that we are on the right track.

We also note that the inner product depends on the interval chosen on the real line. For different (a, b), we get a different inner product. The choice is usually dictated by the physical application. We shall choose $a = 0, b = 1$, although this may not be a physically suitable choice. With such a choice and with $\{\mathbf{f}_1 = 1, \mathbf{f}_2 = t\}$ as a basis, we obtain

$$g_{11} = \mathbf{f}_1 \cdot \mathbf{f}_1 = \int_0^1 f_1(t) f_1(t)\, dt = \int_0^1 dt = 1,$$

$$g_{12} = \mathbf{f}_1 \cdot \mathbf{f}_2 = \int_0^1 f_1(t) f_2(t)\, dt = \int_0^1 t\, dt = \tfrac{1}{2} = g_{21},$$

$$g_{22} = \mathbf{f}_2 \cdot \mathbf{f}_2 = \int_0^1 f_2(t) f_2(t)\, dt = \int_0^1 t^2\, dt = \tfrac{1}{3}.$$

So the inner product matrix is

$$\mathsf{G} = \begin{pmatrix} 1 & \tfrac{1}{2} \\ \tfrac{1}{2} & \tfrac{1}{3} \end{pmatrix}. \qquad \blacksquare$$

We started with Equation (1.1) as the definition of the inner product. This definition assumed a knowledge of lengths and angles. These are notions with which we become intuitively familiar very early in our mental development. However, such notions are not intuitively obvious for two polynomials. That is why the concepts of lengths and angles for objects such as polynomials come *after* introducing the notion of inner product. Of course, we want these notions to agree with the intuitive notions of lengths and angles, i.e., we want them to be related to the inner product in precisely the same manner as given in Equation (1.1). If we let $\mathbf{b} = \mathbf{a}$ in that equation, we get $\mathbf{a} \cdot \mathbf{a} = |\mathbf{a}|^2$. This becomes our definition for length:

the notion of length comes after *that of the inner product!*

Box 6.1.5. *Given any inner product on a set of objects that we can call "vectors," we define the* **length of a vector** *a as* $|\mathbf{a}| \equiv +\sqrt{\mathbf{a} \cdot \mathbf{a}}$.

Once the notion of length is established for a general set of vectors, we can define the angle between two vectors $\mathbf{a}$ and $\mathbf{b}$ as

$$\cos\theta \equiv \frac{\mathbf{a} \cdot \mathbf{b}}{|\mathbf{a}|\,|\mathbf{b}|} = \frac{\mathbf{a} \cdot \mathbf{b}}{\sqrt{\mathbf{a} \cdot \mathbf{a}}\,\sqrt{\mathbf{b} \cdot \mathbf{b}}}. \qquad (6.18)$$

This equation and the one in Box 6.1.5 clearly show that lengths and angles are given entirely in terms of inner products. For these concepts to be valid, we must ensure that however we define the inner product, it will have the property that $\mathbf{a} \cdot \mathbf{a} > 0$ for a nonzero vector. It turns out that most inner products encountered in applications have this property. Nevertheless, there *are* cases (very important ones) for which $\mathbf{a} \cdot \mathbf{a} \leq 0$. In such cases, the concepts of length and angles, as we know them, break down, and we have to be content with "dot products" that may produce nonpositive numbers when a nonzero vector is "dotted" with itself.

Even if $\mathbf{a} \cdot \mathbf{a} > 0$, there is no a priori guarantee that the cosine obtained in Equation (6.18) will lie between -1 and $+1$, as it should. However, there is a famous inequality in mathematics called the **Schwarz inequality**, which establishes this fact for those inner products which satisfy $\mathbf{a} \cdot \mathbf{a} > 0$. We shall come back to this later in this chapter.

in some important physical situations the "length" of a nonzero vector can be zero—even negative!

Example 6.1.8. The lengths of the basis vectors $\{\mathbf{f}_1 = 1, \mathbf{f}_2 = t\}$ of $\mathcal{P}_1[t]$ can be found easily using the results of Example 6.1.7:

$$|\mathbf{f}_1| = \sqrt{\mathbf{f}_1 \cdot \mathbf{f}_1} = +\sqrt{1} = 1$$
$$|\mathbf{f}_2| = \sqrt{\mathbf{f}_2 \cdot \mathbf{f}_2} = +\sqrt{\tfrac{1}{3}}.$$

We can also find the "angle" between the two polynomials

$$\cos \theta = \frac{\mathbf{f}_1 \cdot \mathbf{f}_2}{|\mathbf{f}_1| \, |\mathbf{f}_2|} = \frac{\tfrac{1}{2}}{1 \cdot (1/\sqrt{3})} = \frac{\sqrt{3}}{2} \;\Rightarrow\; \theta = \frac{\pi}{6}.$$

∎

The matrix G, called the **inner product matrix** or **metric matrix**, completely determines the inner product of vectors when they are written as linear combinations of $\mathbf{a}_1$ and $\mathbf{a}_2$. For example, consider a vector $\mathbf{a}$ with components (α_1, α_2) in the basis $\{\mathbf{a}_1, \mathbf{a}_2\}$. Figure 6.3 shows $\mathbf{a}$ as the sum of $\overrightarrow{OA}$ (which is the same as $\alpha_1 \mathbf{a}_1$) and $\overrightarrow{OA'}$ (which is the same as $\alpha_2 \mathbf{a}_2$). Using the law of cosines for the triangle OAP, we get

G is the inner product matrix or the metric matrix.

$$|\mathbf{a}|^2 = \overline{OP}^2 = \overline{OA}^2 + \overline{AP}^2 - 2\,\overline{OA}\,\overline{AP}\cos\varphi$$
$$= \alpha_1^2 |\mathbf{a}_1|^2 + \alpha_2^2 |\mathbf{a}_2|^2 + 2\alpha_1\alpha_2 |\mathbf{a}_1| |\mathbf{a}_2| \cos\theta_{12}.$$

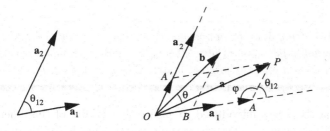

Figure 6.3: The length of **a** is the same whether we use the law of cosine or the inner product matrix G.

On the other hand, using Equation (6.16), we obtain

$$|\mathbf{a}|^2 = \mathbf{a} \cdot \mathbf{a} = \begin{pmatrix} \alpha_1 & \alpha_2 \end{pmatrix} \begin{pmatrix} g_{11} & g_{12} \\ g_{21} & g_{22} \end{pmatrix} \begin{pmatrix} \alpha_1 \\ \alpha_2 \end{pmatrix}$$

$$= \begin{pmatrix} \alpha_1 & \alpha_2 \end{pmatrix} \begin{pmatrix} g_{11}\alpha_1 + g_{12}\alpha_2 \\ g_{21}\alpha_1 + g_{22}\alpha_2 \end{pmatrix} = g_{11}\alpha_1^2 + 2g_{12}\alpha_1\alpha_2 + g_{22}\alpha_2^2$$

$$= \mathbf{a}_1 \cdot \mathbf{a}_1 \alpha_1^2 + 2\alpha_1\alpha_2\mathbf{a}_1 \cdot \mathbf{a}_2 + \mathbf{a}_2 \cdot \mathbf{a}_2\alpha_2^2$$

$$= |\mathbf{a}_1|^2\alpha_1^2 + 2\alpha_1\alpha_2|\mathbf{a}_1|\,|\mathbf{a}_2|\cos\theta_{12} + |\mathbf{a}_2|^2\alpha_2^2$$

and the two expressions agree. In fact, we can show this agreement very generally:

$$\mathbf{a} \cdot \mathbf{b} = (\alpha_1\mathbf{a}_1 + \alpha_2\mathbf{a}_2) \cdot (\beta_1\mathbf{a}_1 + \beta_2\mathbf{a}_2)$$

$$= \alpha_1\beta_1\mathbf{a}_1 \cdot \mathbf{a}_1 + \alpha_1\beta_2\mathbf{a}_1 \cdot \mathbf{a}_2 + \alpha_2\beta_1\mathbf{a}_2 \cdot \mathbf{a}_1 + \alpha_2\beta_2\mathbf{a}_2 \cdot \mathbf{a}_2$$

$$= \alpha_1\beta_1 g_{11} + \alpha_1\beta_2 g_{12} + \alpha_2\beta_1 g_{21} + \alpha_2\beta_2 g_{22}$$

$$= \begin{pmatrix} \alpha_1 & \alpha_2 \end{pmatrix} \begin{pmatrix} g_{11} & g_{12} \\ g_{21} & g_{22} \end{pmatrix} \begin{pmatrix} \beta_1 \\ \beta_2 \end{pmatrix} = \tilde{\mathsf{a}}\mathsf{G}\mathsf{b},$$

where we used the distributive property of the inner product.

It should now be clear to the reader that the matrix G contains all the information needed to evaluate the inner product of any pair of vectors. Suppose now that instead of $\{\mathbf{a}_1, \mathbf{a}_2\}$ we choose $\{\hat{\mathbf{e}}_1, \hat{\mathbf{e}}_2\}$ where $\hat{\mathbf{e}}_1$ and $\hat{\mathbf{e}}_2$ are *unit vectors* and *perpendicular* to one another. Then, the matrix G will have elements

$$g_{11} = \hat{\mathbf{e}}_1 \cdot \hat{\mathbf{e}}_1 = 1, \qquad g_{12} = g_{21} = \hat{\mathbf{e}}_1 \cdot \hat{\mathbf{e}}_2 = 0, \qquad g_{22} = \hat{\mathbf{e}}_2 \cdot \hat{\mathbf{e}}_2 = 1,$$

i.e., G is the unit matrix. In that case, we obtain

$$\tilde{\mathsf{a}}\mathsf{G}\mathsf{b} = \begin{pmatrix} \alpha_1 & \alpha_2 \end{pmatrix} \begin{pmatrix} 1 & 0 \\ 0 & 1 \end{pmatrix} \begin{pmatrix} \beta_1 \\ \beta_2 \end{pmatrix} = \alpha_1\beta_1 + \alpha_2\beta_2$$

which is the usual expression of the dot product of two vectors in terms of their components. A basis whose vectors have *unit length* and are *mutually perpendicular* to one another is called an **orthonormal** basis. Thus,

orthonormal basis

> **Box 6.1.6.** *Only in an orthonormal basis is the dot (inner) product of two vectors equal to the sum of the products of their corresponding components. In such a basis the inner product matrix G is the unit matrix.*

The matrix G was introduced to ensure the validity of the inner product in an arbitrary basis. This poses some restriction on G; for example, we saw that it had to be symmetric, i.e., $g_{12} = g_{21}$ because of the symmetry of the dot product. Another restriction—if we want the dot product of a basis vector with itself to be positive—is that $g_{11} > 0$ and $g_{22} > 0$, in which case the inner product is called **positive definite** (or **Riemannian**). It turns out, however, that such a restriction constrains G too much to be useful in physical applications. Although, in most of this book, we shall adhere to the usual positive definite or Euclidean inner product, the reader should be aware that non-Euclidean inner products also have important applications in physics.

positive definite, or Riemannian inner product

> **Box 6.1.7.** *Regardless of the nature of* G, *we call two vectors* **a** *and* **b** G-*orthogonal if* $\mathbf{a} \cdot \mathbf{b} \equiv \tilde{\mathbf{a}}\mathsf{G}\mathbf{b} = 0$.

Every point in the plane can be thought of as the tip of a vector whose tail is the origin. With this interpretation, we can express the (G-dependent) distance between two points in terms of vectors. Let $\mathbf{r}_1$ be the vector to point P_1 and $\mathbf{r}_2$ the vector to point P_2. Then the "length" of the **displacement** vector $\Delta\mathbf{r} \equiv \mathbf{r}_1 - \mathbf{r}_2$ is the "distance" between P_1 and P_2:

$$\overline{P_1 P_2}^2 = \Delta\mathbf{r} \cdot \Delta\mathbf{r} = (\mathbf{r}_1 - \mathbf{r}_2) \cdot (\mathbf{r}_1 - \mathbf{r}_2) = (\widetilde{\Delta\mathbf{r}})\mathsf{G}(\Delta\mathbf{r}). \qquad (6.19)$$

Keep in mind that only in the positive definite (Euclidean) case is $\overline{P_1 P_2}^2$ nonnegative. There are physical situations in which the square of the length of the displacement vectors can be zero or even negative. We shall encounter one such example when we discuss the special theory of relativity.

The simplicity of G in orthonormal bases makes them very much in demand. So, it is important to know whether it is always possible to construct orthonormal vectors out of general basis vectors. The construction should involve linear combinations only. In other words, given a basis $\{\mathbf{a}_1, \mathbf{a}_2\}$, we want to know if there are linear combinations of $\mathbf{a}_1$ and $\mathbf{a}_2$ which are orthonormal. We assume that the inner product is positive definite, so that the inner product of every nonzero vector with itself is positive. First we divide $\mathbf{a}_1$ by its length to get

$$\hat{\mathbf{e}}_1 \equiv \frac{\mathbf{a}_1}{|\mathbf{a}_1|} = \frac{\mathbf{a}_1}{\sqrt{\mathbf{a}_1 \cdot \mathbf{a}_1}}.$$

To obtain the second orthonormal vector, we refer to Figure 6.4 which shows that if we take away from $\mathbf{a}_2$ its projection on $\mathbf{a}_1$, the remaining vector will be orthogonal to $\mathbf{a}_1$. So consider

$$\mathbf{a}_2' = \mathbf{a}_2 - \underbrace{(\mathbf{a}_2 \cdot \hat{\mathbf{e}}_1)\hat{\mathbf{e}}_1}_{\substack{\text{projection of} \\ \mathbf{a}_2 \text{ on } \mathbf{a}_1}}$$

and note that

$$\hat{\mathbf{e}}_1 \cdot \mathbf{a}_2' = \hat{\mathbf{e}}_1 \cdot \mathbf{a}_2 - (\mathbf{a}_2 \cdot \hat{\mathbf{e}}_1) \underbrace{\hat{\mathbf{e}}_1 \cdot \hat{\mathbf{e}}_1}_{=1} = 0.$$

Gram–Schmidt process for the plane

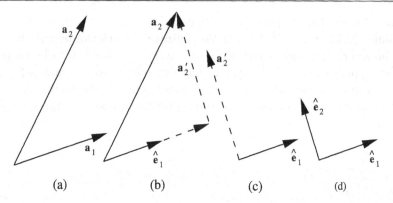

Figure 6.4: The illustration of the Gram–Schmidt process for two linearly independent vectors in the plane.

This suggests defining $\hat{\mathbf{e}}_2$ as

$$\hat{\mathbf{e}}_2 \equiv \frac{\mathbf{a}_2'}{|\mathbf{a}_2'|} = \frac{\mathbf{a}_2'}{\sqrt{\mathbf{a}_2' \cdot \mathbf{a}_2'}}.$$

The reader should note that in the construction of $\{\hat{\mathbf{e}}_1, \hat{\mathbf{e}}_2\}$, we have added vectors and multiplied them by numbers, i.e., we have taken a linear combination of $\mathbf{a}_1$ and $\mathbf{a}_2$. This process, and its generalization to arbitrary number of vectors, is called the **Gram–Schmidt process**, and shows that by appropriately taking linear combinations, it is always possible to find orthonormal vectors out of any linearly independent set of vectors.

Example 6.1.9. The basis $\{1, t\}$ introduced for $\mathcal{P}_1[t]$ is not orthonormal when the inner product is integration over the interval $(0, 1)$ as in Example 6.1.7. Let us use the Gram–Schmidt process to find an orthonormal basis. We note that the first basis vector already has a unit length; so we let $\hat{\mathbf{e}}_1 = \mathbf{f}_1 = 1$. To find the second vector, we first construct

$$\mathbf{f}_2' = \mathbf{f}_2 - (\mathbf{f}_2 \cdot \hat{\mathbf{e}}_1)\hat{\mathbf{e}}_1 = t - (\tfrac{1}{2})1 = t - \tfrac{1}{2}$$

with

$$|\mathbf{f}_2'|^2 = \mathbf{f}_2' \cdot \mathbf{f}_2' = \int_0^1 (t - \tfrac{1}{2})^2 dt = \tfrac{1}{12}.$$

Then the second vector will be

$$\hat{\mathbf{e}}_2 = \frac{\mathbf{f}_2'}{|\mathbf{f}_2'|} = \frac{t - \tfrac{1}{2}}{\sqrt{\tfrac{1}{12}}} = \sqrt{12}(t - \tfrac{1}{2}) = \sqrt{3}(2t - 1).$$

The reader may verify directly that $\{\hat{\mathbf{e}}_1, \hat{\mathbf{e}}_2\}$ is an orthonormal basis.   ∎

Example 6.1.10. Consider the vectors

$$\mathbf{a}_1 = \hat{\mathbf{e}}_x + \hat{\mathbf{e}}_y \qquad \text{and} \qquad \mathbf{a}_2 = 2\hat{\mathbf{e}}_x + \hat{\mathbf{e}}_y.$$

The inner product matrix elements in the basis $\{a_1, a_2\}$ are

$$g_{11} = a_1 \cdot a_1 = (\hat{e}_x + \hat{e}_y) \cdot (\hat{e}_x + \hat{e}_y) = 2, \qquad g_{12} = (\hat{e}_x + \hat{e}_y) \cdot (2\hat{e}_x + \hat{e}_y) = 3,$$
$$g_{21} = a_2 \cdot a_1 = g_{12} = 3, \qquad\qquad g_{22} = (2\hat{e}_x + \hat{e}_y) \cdot (2\hat{e}_x + \hat{e}_y) = 5.$$

or, in matrix form, $G = \begin{pmatrix} 2 & 3 \\ 3 & 5 \end{pmatrix}$.

Now consider vectors b and c, whose components in $\{a_1, a_2\}$ are, respectively, $(1, 1)$ and $(-3, 2)$. We can compute the scalar product of b and c in terms of these components using Equation (6.16):

$$b \cdot c = \tilde{b}Gc = \begin{pmatrix} 1 & 1 \end{pmatrix} \begin{pmatrix} 2 & 3 \\ 3 & 5 \end{pmatrix} \begin{pmatrix} -3 \\ 2 \end{pmatrix} = \begin{pmatrix} 1 & 1 \end{pmatrix} \begin{pmatrix} 0 \\ 1 \end{pmatrix} = 1.$$

We can also write b and c in terms of $\hat{e}_x$ and $\hat{e}_y$ and use the usual definition of the inner product (in terms of components) to find $b \cdot c$. Since b has the components $(1, 1)$ in $\{a_1, a_2\}$, it can be written as

$$b = a_1 + a_2 = (\hat{e}_x + \hat{e}_y) + (2\hat{e}_x + \hat{e}_y) = 3\hat{e}_x + 2\hat{e}_y.$$

Similarly,

$$c = -3a_1 + 2a_2 = -3(\hat{e}_x + \hat{e}_y) + 2(2\hat{e}_x + \hat{e}_y) = \hat{e}_x - \hat{e}_y.$$

Thus, in $\{\hat{e}_x, \hat{e}_y\}$, b has components $(3, 2)$, and c has components $(1, -1)$. Then

$$b \cdot c = b_x c_x + b_y c_y = 3 \cdot 1 + 2 \cdot (-1) = 1$$

which agrees with the previous result obtained above. ∎

Example 6.1.11. Consider two vectors f and g in $\mathcal{P}_1[t]$ with

$$f \equiv f(t) = \alpha_0 + \alpha_1 t, \qquad g \equiv g(t) = \beta_0 + \beta_1 t.$$

We want to find the inner product of these two vectors. First, we use the basis $\{1, t\}$ and its corresponding G matrix found in Example 6.1.7:

$$f \cdot g = f^t G g = \begin{pmatrix} \alpha_0 & \alpha_1 \end{pmatrix} \begin{pmatrix} 1 & \frac{1}{2} \\ \frac{1}{2} & \frac{1}{3} \end{pmatrix} \begin{pmatrix} \beta_0 \\ \beta_1 \end{pmatrix} = \alpha_0 \beta_0 + \tfrac{1}{2}(\alpha_0 \beta_1 + \alpha_1 \beta_0) + \tfrac{1}{3}\alpha_1 \beta_1.$$

Next, we use the orthonormal basis found in Example 6.1.9. In this basis G is the identity matrix and the inner product is the usual one in terms of components. However, the components of f and g need to be found in $\{\hat{e}_1, \hat{e}_2\}$. The reader may check that

$$f = \alpha_0 + \alpha_1 t = \alpha_0 \hat{e}_1 + \alpha_1 \left(\frac{1}{2\sqrt{3}}\hat{e}_2 + \frac{1}{2}\hat{e}_1 \right) = (\alpha_0 + \tfrac{1}{2}\alpha_1)\hat{e}_1 + \frac{\alpha_1}{2\sqrt{3}}\hat{e}_2,$$

$$g = \beta_0 + \beta_1 t = (\beta_0 + \tfrac{1}{2}\beta_1)\hat{e}_1 + \frac{\beta_1}{2\sqrt{3}}\hat{e}_2.$$

It then follows that

$$f \cdot g = (\alpha_0 + \tfrac{1}{2}\alpha_1)(\beta_0 + \tfrac{1}{2}\beta_1) + \left(\frac{\alpha_1}{2\sqrt{3}} \right)\left(\frac{\beta_1}{2\sqrt{3}} \right) = \alpha_0 \beta_0 + \tfrac{1}{2}(\alpha_0 \beta_1 + \alpha_1 \beta_0) + \tfrac{1}{3}\alpha_1 \beta_1.$$

Finally, we take the dot product of the two vectors using the definition of this dot product:

$$\mathbf{f} \cdot \mathbf{g} = \int_0^1 (\alpha_0 + \alpha_1 t)(\beta_0 + \beta_1 t)\, dt$$

$$= \alpha_0 \beta_0 \int_0^1 dt + (\alpha_0 \beta_1 + \alpha_1 \beta_0) \int_0^1 t\, dt + \alpha_1 \beta_1 \int_0^1 t^2\, dt$$

$$= \alpha_0 \beta_0 + \tfrac{1}{2}(\alpha_0 \beta_1 + \alpha_1 \beta_0) + \tfrac{1}{3}\alpha_1 \beta_1.$$

All three ways of calculating the inner product agree, as they should. ∎

6.1.3 Orthogonal Transformation

Now that we have defined inner products, we may combine it with the concept of transformation. More specifically, we seek transformations that leave the inner product—which we shall assume to be positive definite (Euclidean)—unchanged. Under such transformations, the length of a vector and the angle between two vectors will not change. That is why such transformations are called **rigid transformations**. We choose an orthonormal basis, so that $G = 1$, and denote the transformed vectors by a prime: $a' = Aa$, $b' = Ab$. Then the invariance of the inner product yields

$$\widetilde{a'}b' = \widetilde{a}b \;\Rightarrow\; \widetilde{(Aa)}Ab = \widetilde{a}\widetilde{A}Ab = \widetilde{a}b.$$

rigid
transformations

This will hold for arbitrary a and b only if

$$\widetilde{A}A = 1. \tag{6.20}$$

orthogonal
matrices

Matrices that satisfy this relation are called **orthogonal**. We now investigate conditions under which Equation (6.20) holds by writing out the matrices:

$$\begin{pmatrix} a_{11} & a_{21} \\ a_{12} & a_{22} \end{pmatrix} \begin{pmatrix} a_{11} & a_{12} \\ a_{21} & a_{22} \end{pmatrix} = \begin{pmatrix} a_{11}^2 + a_{21}^2 & a_{11}a_{12} + a_{21}a_{22} \\ a_{12}a_{11} + a_{22}a_{21} & a_{12}^2 + a_{22}^2 \end{pmatrix} = \begin{pmatrix} 1 & 0 \\ 0 & 1 \end{pmatrix}$$

which is equivalent to the following three equations:

$$a_{11}^2 + a_{21}^2 = 1, \qquad a_{11}a_{12} + a_{21}a_{22} = 0, \qquad a_{12}^2 + a_{22}^2 = 1. \tag{6.21}$$

Squaring the second equation and substituting from the first and third, we get

$$a_{11}^2 a_{12}^2 = a_{21}^2 a_{22}^2 \;\Rightarrow\; (1 - a_{21}^2)a_{12}^2 = a_{21}^2(1 - a_{12}^2) \;\Rightarrow\; a_{21}^2 = a_{12}^2.$$

The first and third equations of (6.21) now yield

$$a_{22}^2 = a_{11}^2 \qquad \text{and} \qquad a_{12}^2 = a_{21}^2 = 1 - a_{11}^2.$$

Therefore, all parameters are given in terms of a_{11}. Now the first equation of (6.21) indicates that $-1 \leq a_{11} \leq 1$. It follows that a_{11} can be thought of as a sine or a cosine of some angle, say θ. Let us choose cosine. Then $a_{22} = \pm \cos \theta$. If we choose the plus sign for cosine, then the middle equation of (6.21) shows that $a_{12} = -a_{21} = \pm \sin \theta$, and if we choose the minus sign, $a_{12} = a_{21} = \pm \sin \theta$. Let us choose the plus sign for cosine. Then, we obtain two possibilities for A:

2 × 2 orthogonal matrices are described in terms of a single parameter.

$$\mathsf{A} = \begin{pmatrix} \cos \theta & -\sin \theta \\ \sin \theta & \cos \theta \end{pmatrix} \quad \text{or} \quad \mathsf{A} = \begin{pmatrix} \cos \theta & \sin \theta \\ -\sin \theta & \cos \theta \end{pmatrix}.$$

The difference is in the sign of the angle θ.

Writing (x, y) for the components of a vector in the plane [instead of (α_1, α_2)], and (x', y') for the transformed vector, and using the first choice for A, we have

$$\begin{pmatrix} x' \\ y' \end{pmatrix} = \begin{pmatrix} \cos \theta & -\sin \theta \\ \sin \theta & \cos \theta \end{pmatrix} \begin{pmatrix} x \\ y \end{pmatrix}$$

or

$$x' = x \cos \theta - y \sin \theta, \tag{6.22}$$
$$y' = x \sin \theta + y \cos \theta. \tag{6.23}$$

This is how the coordinates of a point in the plane transform under a *counterclockwise rotation* of angle θ. Had we chosen the second form of A, we would have obtained a clockwise rotation of the coordinates. Notice how we chose the signs of sines and cosines to ensure that when $\theta = 0$, the rotation is the unit matrix, i.e., no rotation at all. Although rotations are part of orthogonal transformations, the converse is not true: There are orthogonal transformations that do not correspond to a rotation. For example, the matrix

$$\mathsf{A} = \begin{pmatrix} \cos \theta & \sin \theta \\ \sin \theta & -\cos \theta \end{pmatrix} \tag{6.24}$$

is orthogonal (as the reader can verify), but it does not correspond to a rotation because at $\theta = 0$ it does not give the identity matrix.

In general, the inner product of the transformed (primed) vectors will be

$$\widetilde{\mathsf{a}}'\mathsf{G}\mathsf{b}' = \widetilde{(\mathsf{A}\mathsf{a})}\mathsf{G}\mathsf{A}\mathsf{b} = \widetilde{\mathsf{a}}\widetilde{\mathsf{A}}\mathsf{G}\mathsf{A}\mathsf{b}.$$

For A to preserve the inner product, i.e., for $\widetilde{\mathsf{a}}'\mathsf{G}\mathsf{b}'$ to be equal to $\widetilde{\mathsf{a}}\mathsf{G}\mathsf{b}$, we need to have

G-orthogonal matrices

$$\widetilde{\mathsf{A}}\mathsf{G}\mathsf{A} = \mathsf{G}. \tag{6.25}$$

A matrix that satisfies Equation (6.25) is called **G-orthogonal**.

Historical Notes

Matrices entered mathematics slowly and somewhat reluctantly. The related notion of *determinant*, which is a number associated with an array of numbers, was introduced as early as the middle of the eighteenth century in the study of a system of linear equations. However, the recognition that the array itself could be treated as a mathematical object, obeying certain rules of manipulation, came much later.

Logically, the idea of a matrix precedes that of a determinant as Arthur Cayley has pointed out; however, the order was reversed historically. In fact, many of the properties of matrices were known as a result of their connection to determinants. Because the uses of matrices were well established, it occurred to Cayley to introduce them as distinct entities. He says, "I certainly did not get the notion of a matrix in any way through quaternions; it was either directly from that of a determinant or as a convenient way of expression of" a system of two equations in two unknowns. Because Cayley was the first to single out the matrix itself and was the first to publish a series of articles on them, he is generally credited with being the creator of the theory of matrices.

Arthur Cayley
1821–1895

Arthur Cayley's father, Henry Cayley, although from a family who had lived for many generations in Yorkshire, England, lived in St. Petersburg, Russia. It was in St. Petersburg that Arthur spent the first eight years of his childhood before his parents returned to England and settled near London. Arthur showed great skill in numerical calculations at school and, after he moved to King's College in 1835, his aptitude for advanced mathematics became apparent. His mathematics teacher advised that Arthur be encouraged to pursue his studies in this area rather than follow his father's wishes to enter the family business as a merchant.

In 1838 Arthur began his studies at Trinity College, Cambridge, from where he graduated in 1842. While still an undergraduate he had three papers published in the newly founded *Cambridge Mathematical Journal*. For four years he taught at Cambridge having won a Fellowship and, during this period, he published 28 papers.

A Cambridge Fellowship had a limited tenure so Cayley had to find a profession. He chose law and was admitted to the bar in 1849. He spent 14 years as a lawyer but Cayley, although very skilled in conveyancing (his legal speciality), always considered it as a means to make money so that he could pursue mathematics. During this period he met *Sylvester* who was also in the legal profession. Both worked at the courts of Lincoln's Inn in London and discussed deep mathematical questions during their working day. During these 14 years as a lawyer Cayley published about 250 mathematical papers!

In 1863 Cayley was appointed to the newly created Sadleirian professorship of mathematics at Cambridge. Except for the year 1882, spent at the Johns Hopkins University at the invitation of Sylvester, he remained at Cambridge until his death in 1895.

6.2 Vectors in Space

The ideas developed so far can be easily generalized to vectors in space. For example, a linear combination of vectors in space is again a vector in space. We can also find a basis for space. In fact, any three non-coplanar (not lying in the same plane) vectors constitute a basis. To see this, let $\{\mathbf{a}_1, \mathbf{a}_2, \mathbf{a}_3\}$ be three such vectors drawn from a common point[5] and assume that $\mathbf{b}$ is any fourth vector in space. If $\mathbf{b}$ is along any of the $\mathbf{a}$'s, we are done, because then $\mathbf{b}$ is a multiple of that vector, i.e., a linear combination of the three vectors

---

[5]If the vectors are not originally drawn from the same point, we can transport them parallel to themselves to a common point.

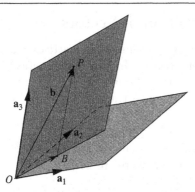

Figure 6.5: Any vector in space can be written as a linear combination of three non-coplanar vectors.

(with two coefficients being zero). So assume that $\mathbf{b}$ is not along any of the $\mathbf{a}$'s. The plane formed by $\mathbf{b}$ and $\mathbf{a}_3$ intersects the plane of $\mathbf{a}_1$ and $\mathbf{a}_2$ along a certain line common to both (see Figure 6.5). Draw a line from the tip of $\mathbf{b}$ parallel to $\mathbf{a}_3$. This line will resolve $\mathbf{b}$ into a vector $\overrightarrow{OB}$ in the plane of $\mathbf{a}_1$ and $\mathbf{a}_2$ and a vector $\overrightarrow{BP}$ parallel to $\mathbf{a}_3$. So, we write $\mathbf{b} = \overrightarrow{OB} + \alpha_3\mathbf{a}_3$. Furthermore, since $\overrightarrow{OB}$ is in the plane of $\mathbf{a}_1$ and $\mathbf{a}_2$, it can be written as a linear combination of these two vectors: $\overrightarrow{OB} = \alpha_1\mathbf{a}_1 + \alpha_2\mathbf{a}_2$. Putting all of this together, we get

$$\mathbf{b} = \alpha_1\mathbf{a}_1 + \alpha_2\mathbf{a}_2 + \alpha_3\mathbf{a}_3.$$

This shows that

Box 6.2.1. *The maximum number of linearly independent vectors in space is three. Any three non-coplanar vectors form a basis for the space.*

It follows that the space is a three-dimensional vector space.

In the previous section we introduced $\mathcal{P}_1[t]$, the set of polynomials of first degree, and showed that they could be treated as vectors. We even defined an inner product for these vectors, and from that, we calculated the length of a vector and the angle between two vectors. This process can be generalized to three dimensions. Let $\mathcal{P}_2[t]$ be the set of polynomials of degree 2 (or less) in the variable t. One can easily show that such a set, a typical element of which looks like $\alpha_0 + \alpha_1 t + \alpha_2 t^2$, has all the properties of arrows in space. We shall use $\mathcal{P}_2[t]$ as a prototype of vectors that are not directed line segments. Clearly, $\{1, t, t^2\}$ form a basis for $\mathcal{P}_2[t]$; therefore, $\mathcal{P}_2[t]$ is a three-dimensional vector space.

polynomials of degree 2 or less form a 3-dimensional vector space.

6.2.1 Transformation of Vectors

In the case of the plane, the machinery of matrices connected the components of a vector in different bases. In the same context, we contrasted active versus passive transformation. From now on, we want to concentrate on active transformations, i.e., we consider transformations that alter the vectors rather that the axes.

Consider a vector $\mathbf{a}$ with components $(\alpha_1, \alpha_2, \alpha_3)$ in the basis $B \equiv \{\mathbf{a}_1, \mathbf{a}_2, \mathbf{a}_3\}$. If we transform this vector, it will acquire new components, $(\alpha_1', \alpha_2', \alpha_3')$, in the same basis B. We can therefore write

$$\mathbf{a} = \alpha_1\mathbf{a}_1 + \alpha_2\mathbf{a}_2 + \alpha_3\mathbf{a}_3 \quad \text{and} \quad \mathbf{a}' = \alpha_1'\mathbf{a}_1 + \alpha_2'\mathbf{a}_2 + \alpha_3'\mathbf{a}_3, \quad (6.26)$$

where $\mathbf{a}'$ is the transform of $\mathbf{a}$. Now suppose that we transform *both* $\mathbf{a}$ and the basis vectors *in exactly the same manner*. Then the components of the transformed $\mathbf{a}$ will be the same in the new basis as the original $\mathbf{a}$ was in the old basis:

$$\mathbf{a}' = \alpha_1\mathbf{a}_1' + \alpha_2\mathbf{a}_2' + \alpha_3\mathbf{a}_3'. \quad (6.27)$$

Since B is a basis, any vector, in particular, the transformed basis vectors can be written in terms of them:

$$\begin{aligned}
\mathbf{a}_1' &= a_{11}\mathbf{a}_1 + a_{21}\mathbf{a}_2 + a_{31}\mathbf{a}_3, \\
\mathbf{a}_2' &= a_{12}\mathbf{a}_1 + a_{22}\mathbf{a}_2 + a_{32}\mathbf{a}_3, \\
\mathbf{a}_3' &= a_{13}\mathbf{a}_1 + a_{23}\mathbf{a}_2 + a_{33}\mathbf{a}_3.
\end{aligned} \quad (6.28)$$

Now substitute Equation (6.28) in the RHS of (6.27), and the second equation of (6.26) in the LHS of (6.27) and rearrange terms to obtain

$$(\alpha_1' - a_{11}\alpha_1 - a_{12}\alpha_2 - a_{13}\alpha_3)\mathbf{a}_1 + (\alpha_2' - a_{21}\alpha_1 - a_{22}\alpha_2 - a_{23}\alpha_3)\mathbf{a}_2$$
$$+ (\alpha_3' - a_{31}\alpha_1 - a_{32}\alpha_2 - a_{33}\alpha_3)\mathbf{a}_3 = 0.$$

The linear independence of $\mathbf{a}_1$, $\mathbf{a}_2$, and $\mathbf{a}_3$ gives

$$\begin{aligned}
\alpha_1' &= a_{11}\alpha_1 + a_{12}\alpha_2 + a_{13}\alpha_3, \\
\alpha_2' &= a_{21}\alpha_1 + a_{22}\alpha_2 + a_{23}\alpha_3, \\
\alpha_3' &= a_{31}\alpha_1 + a_{32}\alpha_2 + a_{33}\alpha_3,
\end{aligned} \quad (6.29)$$

which, with the introduction of 3×1 (column), and 3×3 matrices, can be written concisely as

$$\begin{pmatrix} \alpha_1' \\ \alpha_2' \\ \alpha_3' \end{pmatrix} = \begin{pmatrix} a_{11} & a_{12} & a_{13} \\ a_{21} & a_{22} & a_{23} \\ a_{31} & a_{32} & a_{33} \end{pmatrix} \begin{pmatrix} \alpha_1 \\ \alpha_2 \\ \alpha_3 \end{pmatrix} \quad \text{or} \quad \mathsf{a}' = \mathsf{A}\mathsf{a}. \quad (6.30)$$

To know how a general vector transforms, we only need the transformation matrix, namely the 3×3 matrix in Equation (6.30). This, in turn, is obtained completely from the transformation of basis vectors as given in Equation (6.28). The reader should note, however, that the coefficients in each line of (6.28) appear as a *column* in the transformation matrix. Thus,

Box 6.2.2. *To find the **transformation matrix**, apply the transformation to the basis vectors, and write the transformed basis vectors in terms of the old basis vectors. The "horizontal" coefficients become the columns of the transformation matrix.*

Let us apply a transformation to $\mathbf{a}'$ and to $\{\mathbf{a}'_1, \mathbf{a}'_2, \mathbf{a}'_3\}$. We could denote the new vectors by a second prime; but, then it would give the impression that it is the *same* transformation as the earlier one. This is not the case. Therefore, we use a new symbol "˘" to emphasize that the second transformations is of a completely different nature, and denote the new transformed vectors by $\breve{\mathbf{a}}'$ and $\{\breve{\mathbf{a}}'_1, \breve{\mathbf{a}}'_2, \breve{\mathbf{a}}'_3\}$. In the basis $\{\mathbf{a}_1, \mathbf{a}_2, \mathbf{a}_3\}$, $\breve{\mathbf{a}}'$ can be written as

$$\breve{\mathbf{a}}' = \alpha''_1 \mathbf{a}_1 + \alpha''_2 \mathbf{a}_2 + \alpha''_3 \mathbf{a}_3, \tag{6.31}$$

while the application of the new transformation to the second equation of (6.26) gives

$$\breve{\mathbf{a}}' = \alpha'_1 \breve{\mathbf{a}}_1 + \alpha'_2 \breve{\mathbf{a}}_2 + \alpha'_3 \breve{\mathbf{a}}_3.$$

The vectors on the RHS can be written as a linear combination of $\{\mathbf{a}_1, \mathbf{a}_2, \mathbf{a}_3\}$:

$$\begin{aligned}
\breve{\mathbf{a}}_1 &= a'_{11} \mathbf{a}_1 + a'_{21} \mathbf{a}_2 + a'_{31} \mathbf{a}_3, \\
\breve{\mathbf{a}}_2 &= a'_{12} \mathbf{a}_1 + a'_{22} \mathbf{a}_2 + a'_{32} \mathbf{a}_3, \\
\breve{\mathbf{a}}_3 &= a'_{13} \mathbf{a}_1 + a'_{23} \mathbf{a}_2 + a'_{33} \mathbf{a}_3.
\end{aligned} \tag{6.32}$$

Using the by-now-familiar procedure, we can relate the coefficients as follows:

$$\begin{pmatrix} \alpha''_1 \\ \alpha''_2 \\ \alpha''_3 \end{pmatrix} = \begin{pmatrix} a'_{11} & a'_{12} & a'_{13} \\ a'_{21} & a'_{22} & a'_{23} \\ a'_{31} & a'_{32} & a'_{33} \end{pmatrix} \begin{pmatrix} \alpha'_1 \\ \alpha'_2 \\ \alpha'_3 \end{pmatrix} \qquad \text{or} \qquad \mathsf{a}'' = \mathsf{A}' \mathsf{a}'. \tag{6.33}$$

We can also find how a'' and a are related in two ways. The first way applies "˘" to both sides of Equations (6.27) and (6.28), substitutes (6.32) in the transformed (6.28), and the result of this substitution in (6.27). This will give $\breve{\mathbf{a}}'$ as a linear combination of $\mathbf{a}_1$, $\mathbf{a}_2$, and $\mathbf{a}_3$. Equating this with Equation (6.31) will give us a matrix relation between the a'' and a. Second, we can substitute the matrix relation of Equation (6.30) in that of (6.33) and obtain a relation between the a'' and a via the product of two matrices. Comparison of these two relations will give us the rules of multiplication for 3×3 matrices which, except for the number of elements involved, is identical to the multiplication rule for the 2×2 matrices. Similarly, the multiplication by a row or a column vector, etc., is exactly as before.

There is a new kind of matrix associated with the space that we could not consider in our discussion of the plane. Let $B = \{\mathbf{a}_1, \mathbf{a}_2, \mathbf{a}_3\}$ be a basis for the space, and take any two of the vectors in B, say $\mathbf{a}_1$ and $\mathbf{a}_2$. These two vectors form a plane any vector of which has only two components: If **a** is in this plane, it can be written as

$$\mathbf{a} = \alpha_1 \mathbf{a}_1 + \alpha_2 \mathbf{a}_2.$$

Now suppose we apply the same transformation to both **a** and $\{\mathbf{a}_1, \mathbf{a}_2\}$. Then, on the one hand, $\mathbf{a}' = \alpha_1 \mathbf{a}_1' + \alpha_2 \mathbf{a}_2'$, and on the other hand, $\mathbf{a}' = \alpha_1' \mathbf{a}_1 + \alpha_2' \mathbf{a}_2 + \alpha_3' \mathbf{a}_3$, because the transformed **a**, in general, comes out of the plane of $\mathbf{a}_1$ and $\mathbf{a}_2$. Therefore,

$$\alpha_1 \mathbf{a}_1' + \alpha_2 \mathbf{a}_2' = \alpha_1' \mathbf{a}_1 + \alpha_2' \mathbf{a}_2 + \alpha_3' \mathbf{a}_3. \tag{6.34}$$

But we also have

$$\mathbf{a}_1' = a_{11} \mathbf{a}_1 + a_{21} \mathbf{a}_2 + a_{31} \mathbf{a}_3,$$
$$\mathbf{a}_2' = a_{12} \mathbf{a}_1 + a_{22} \mathbf{a}_2 + a_{32} \mathbf{a}_3.$$

Substituting these in Equation (6.34) yields

$$(\alpha_1' - a_{11}\alpha_1 - a_{12}\alpha_2)\mathbf{a}_1 + (\alpha_2' - a_{21}\alpha_1 - a_{22}\alpha_2)\mathbf{a}_2 + (\alpha_3' - a_{31}\alpha_1 - a_{32}\alpha_2)\mathbf{a}_3 = 0.$$

Linear independence of the vectors in B now gives

$$\begin{aligned} \alpha_1' &= a_{11}\alpha_1 + a_{12}\alpha_2, \\ \alpha_2' &= a_{21}\alpha_1 + a_{22}\alpha_2, \\ \alpha_3' &= a_{31}\alpha_1 + a_{32}\alpha_2, \end{aligned} \tag{6.35}$$

which can be written in matrix form as

$$\begin{pmatrix} \alpha_1' \\ \alpha_2' \\ \alpha_3' \end{pmatrix} = \begin{pmatrix} a_{11} & a_{12} \\ a_{21} & a_{22} \\ a_{31} & a_{32} \end{pmatrix} \begin{pmatrix} \alpha_1 \\ \alpha_2 \end{pmatrix} \qquad \text{or} \qquad \mathsf{a}' = \mathsf{A}\mathsf{a}. \tag{6.36}$$

The matrix A is now a 3×2 matrix. It relates two-component column vectors to three-component column vectors.

Example 6.2.1. Another way to illustrate the preceding discussion is to use first degree polynomials. Let us multiply all polynomials of $\mathcal{P}_1[t]$ by a fixed first degree polynomial, say $1 + t$. This will transform vectors of $\mathcal{P}_1[t]$ into vectors of $\mathcal{P}_2[t]$. In particular, it will transform the basis $\{1, t\}$ into vectors in $\mathcal{P}_2[t]$ which can be expressed as a linear combination of the basis vectors $\{1, t, t^2\}$ of $\mathcal{P}_2[t]$. Let $\mathbf{f}_1 = 1$, $\mathbf{f}_2 = t$, and $\mathbf{f}_3 = t^2$, and note that

$$\mathbf{f}_1' = 1 \cdot (1 + t) = 1 + t = 1 \cdot \mathbf{f}_1 + 1 \cdot \mathbf{f}_2 + 0 \cdot \mathbf{f}_3,$$
$$\mathbf{f}_2' = t(1 + t) = t + t^2 = 0 \cdot \mathbf{f}_1 + 1 \cdot \mathbf{f}_2 + 1 \cdot \mathbf{f}_3.$$

According to Box 6.2.2, the transformation matrix is

$$\begin{pmatrix} 1 & 0 \\ 1 & 1 \\ 0 & 1 \end{pmatrix}$$

from which we can find the transform of a general vector $\mathbf{f} = \alpha_0 + \alpha_1 t$ in $\mathcal{P}_1[t]$. If the transformed vector is written as $\mathbf{f}' = \alpha_0' + \alpha_1' t + \alpha_2' t^2$, then

$$\begin{pmatrix} \alpha_0' \\ \alpha_1' \\ \alpha_2' \end{pmatrix} = \begin{pmatrix} 1 & 0 \\ 1 & 1 \\ 0 & 1 \end{pmatrix} \begin{pmatrix} \alpha_0 \\ \alpha_1 \end{pmatrix}.$$

This can be verified directly by multiplying $\mathbf{f} = \alpha_0 + \alpha_1 t$ by $1 + t$. ∎

In the discussion above, we started with the plane (with two dimensions) and transformed to space (with three dimensions). Example 6.2.1 illustrated this transformation for $\mathcal{P}_1[t]$ and $\mathcal{P}_2[t]$. We can also start with three dimensions and end up in two dimensions. The result will be a matrix relation of the form

$$\begin{pmatrix} \alpha_1 \\ \alpha_2 \end{pmatrix} = \begin{pmatrix} b_{11} & b_{12} & b_{13} \\ b_{21} & b_{22} & b_{23} \end{pmatrix} \begin{pmatrix} \alpha_1' \\ \alpha_2' \\ \alpha_3' \end{pmatrix} \qquad \text{or} \qquad \mathbf{a} = \mathsf{B}\mathbf{a}' \qquad (6.37)$$

with B a 2×3 matrix. The following example illustrates this point.

Example 6.2.2. Let us start with $\mathcal{P}_2[t]$ and as transformation, consider *differentiation* which acts on the basis $\{1, t, t^2\}$. It is clear that the resulting vectors will belong to $\mathcal{P}_1[t]$, because they will be linear combinations of 1 and t. With $\mathbf{f}_1 = 1$, $\mathbf{f}_2 = t$, and $\mathbf{f}_3 = t^2$, and using a prime to denote the transformed vector, we can write

differentiation is a (linear) transformation.

$$\mathbf{f}_1' = \frac{d}{dt}(1) = 0 = 0 \cdot \mathbf{f}_1 + 0 \cdot \mathbf{f}_2,$$

$$\mathbf{f}_2' = \frac{d}{dt}(t) = 1 = 1 \cdot \mathbf{f}_1 + 0 \cdot \mathbf{f}_2,$$

$$\mathbf{f}_3' = \frac{d}{dt}(t^2) = 2t = 0 \cdot \mathbf{f}_1 + 2 \cdot \mathbf{f}_2,$$

giving rise to the transformation matrix

$$\begin{pmatrix} 0 & 1 & 0 \\ 0 & 0 & 2 \end{pmatrix}.$$

The reader may verify that the coefficients (α_0', α_1') in $\mathcal{P}_1[t]$ of the derivative of an arbitrary polynomial $f(t) = \alpha_0 + \alpha_1 t + \alpha_2 t^2$ are given by

$$\begin{pmatrix} \alpha_0' \\ \alpha_1' \end{pmatrix} = \begin{pmatrix} 0 & 1 & 0 \\ 0 & 0 & 2 \end{pmatrix} \begin{pmatrix} \alpha_0 \\ \alpha_1 \\ \alpha_2 \end{pmatrix}$$

which can also be obtained directly by differentiating $f(t)$. ∎

The point of this discussion is that if you have a collection of vectors with various numbers of components, then it is possible to construct matrices that relate the two sets of vectors. These matrices have different numbers of rows and columns. The mathematics of these new matrices, their notion of equality, their addition, subtraction, multiplication, transposition, etc., is exactly the same as before

Example 6.2.3. Suppose

$$A = \begin{pmatrix} 1 & -1 \\ -1 & 2 \\ 0 & 1 \end{pmatrix} \quad \text{and} \quad B = \begin{pmatrix} -1 & 0 & 1 \\ 1 & 2 & -2 \end{pmatrix}.$$

Then $A + B$ is not defined, but

$$A^t + B = \begin{pmatrix} 1 & -1 & 0 \\ -1 & 2 & 1 \end{pmatrix} + \begin{pmatrix} -1 & 0 & 1 \\ 1 & 2 & -2 \end{pmatrix} = \begin{pmatrix} 0 & -1 & 1 \\ 0 & 4 & -1 \end{pmatrix}$$

and

$$A + B^t = \begin{pmatrix} 1 & -1 \\ -1 & 2 \\ 0 & 1 \end{pmatrix} + \begin{pmatrix} -1 & 1 \\ 0 & 2 \\ 1 & -2 \end{pmatrix} = \begin{pmatrix} 0 & 0 \\ -1 & 4 \\ 1 & -1 \end{pmatrix} = (A^t + B)^t.$$

As for multiplication, we have

$$AB = \begin{pmatrix} 1 & -1 \\ -1 & 2 \\ 0 & 1 \end{pmatrix} \begin{pmatrix} -1 & 0 & 1 \\ 1 & 2 & -2 \end{pmatrix} = \begin{pmatrix} -2 & -2 & 3 \\ 3 & 4 & -5 \\ 1 & 2 & -2 \end{pmatrix}$$

and

$$BA = \begin{pmatrix} -1 & 0 & 1 \\ 1 & 2 & -2 \end{pmatrix} \begin{pmatrix} 1 & -1 \\ -1 & 2 \\ 0 & 1 \end{pmatrix} = \begin{pmatrix} -1 & 2 \\ -1 & 1 \end{pmatrix},$$

where the element in the ith row and jth column of the product is obtained by multiplying the ith row of the left factor by the jth row of the right factor term-by-term and adding the products (see Box 6.1.3). ∎

The 3×3 matrix

$$1 \equiv \begin{pmatrix} 1 & 0 & 0 \\ 0 & 1 & 0 \\ 0 & 0 & 1 \end{pmatrix}$$

is the 3×3 identity matrix (or unit matrix), and has the property that when it multiplies any other 3×3 matrix on either side, the latter does not get affected. Similarly, when this identity matrix multiplies a three-column vector on the left or a three-row vector on the right, it does not affect them. As in the case of the plane, the unit matrix is used to define the **inverse** of a matrix A as a matrix B that multiplies A on either side and gives the unit matrix.

6.2.2   Inner Product

As in the case of two dimensions, the usual rule of the dot product of space vectors in terms of their components along $\hat{\mathbf{e}}_x$, $\hat{\mathbf{e}}_y$, and $\hat{\mathbf{e}}_z$ does not apply in the general case. For that, we need an inner product matrix G. As in the plane, this is a matrix whose elements are dot products of the basis vectors. If $B = \{\mathbf{a}_1, \mathbf{a}_2, \mathbf{a}_3\}$ is a basis for space, then G is a 3×3 *symmetric* matrix

$$G = \begin{pmatrix} g_{11} & g_{12} & g_{13} \\ g_{21} & g_{22} & g_{23} \\ g_{31} & g_{32} & g_{33} \end{pmatrix}, \qquad g_{ij} = g_{ji} = \mathbf{a}_i \cdot \mathbf{a}_j, \quad i, j = 1, 2, 3. \qquad (6.38)$$

Example 6.2.4. Let us find the inner product matrix for the basis $\{1, t, t^2\}$ of $\mathcal{P}_2[t]$ when the inner product integration is from 0 to 1. Because of the symmetry of the matrix and the fact that we have already calculated the 2×2 submatrix of G, we need to find g_{13}, g_{23}, and g_{33}. Let $\mathbf{f}_1 = f_1(t) = 1$, $\mathbf{f}_2 = f_2(t) = t$, and $\mathbf{f}_3 = f_3(t) = t^2$; then

$$g_{13} = \mathbf{f}_1 \cdot \mathbf{f}_3 = \int_0^1 f_1(t) f_3(t)\, dt = \int_0^1 t^2 dt = \tfrac{1}{3},$$

$$g_{23} = \mathbf{f}_2 \cdot \mathbf{f}_3 = \int_0^1 f_2(t) f_3(t)\, dt = \int_0^1 t^3 dt = \tfrac{1}{4},$$

$$g_{33} = \mathbf{f}_3 \cdot \mathbf{f}_3 = \int_0^1 f_3(t) f_3(t)\, dt = \int_0^1 t^4 dt = \tfrac{1}{5}.$$

It follows that

$$\mathsf{G} = \begin{pmatrix} 1 & \tfrac{1}{2} & \tfrac{1}{3} \\ \tfrac{1}{2} & \tfrac{1}{3} & \tfrac{1}{4} \\ \tfrac{1}{3} & \tfrac{1}{4} & \tfrac{1}{5} \end{pmatrix}.$$

This matrix can be used to find the dot product of any two vectors in terms of their components in the basis $\{1, t, t^2\}$ of $\mathcal{P}_2[t]$. ■

If $\mathbf{a}$ and $\mathbf{b}$ have components $(\alpha_1, \alpha_2, \alpha_3)$ and $(\beta_1, \beta_2, \beta_3)$ in B, then their inner product is given by

$$\widetilde{\mathsf{a}}\mathsf{G}\mathsf{b} = \begin{pmatrix} \alpha_1 & \alpha_2 & \alpha_3 \end{pmatrix} \begin{pmatrix} g_{11} & g_{12} & g_{13} \\ g_{21} & g_{22} & g_{23} \\ g_{31} & g_{32} & g_{33} \end{pmatrix} \begin{pmatrix} \beta_1 \\ \beta_2 \\ \beta_3 \end{pmatrix}. \tag{6.39}$$

If this expression is zero, we say that $\mathbf{a}$ and $\mathbf{b}$ are G-orthogonal. For an orthonormal basis, the inner product matrix G becomes the unit matrix[6] and we recover the usual inner product of space vectors in terms of components.

G-orthogonal vectors in space

As discussed in the case of the plane, every point in space can be thought of as the tip of a vector whose tail is the origin. Then, we can express the (G-dependent) distance between two points in terms of vectors. Let $\mathbf{r}_1$ be the vector to point P_1 and $\mathbf{r}_2$ the vector to point P_2. Then the length of the displacement vector is the "distance" between P_1 and P_2:

$$\Delta \mathbf{r} \cdot \Delta \mathbf{r} = (\mathbf{r}_1 - \mathbf{r}_2) \cdot (\mathbf{r}_1 - \mathbf{r}_2) = (\widetilde{\Delta \mathbf{r}})\mathsf{G}(\Delta \mathbf{r}). \tag{6.40}$$

Recall that only in the positive definite case is $\overline{P_1 P_2}^2$ nonnegative.

As in the case of the plane, it is convenient to construct orthonormal basis vectors in space. This can be done by the Gram–Schmidt process. Suppose $B = \{\mathbf{a}_1, \mathbf{a}_2, \mathbf{a}_3\}$ is a basis for space as shown in Figure 6.6. Again, to avoid complications, we assume that the inner product is positive definite, so that the inner product of every nonzero vector with itself is positive. We know how to construct two orthonormal vectors out of $\{\mathbf{a}_1, \mathbf{a}_2\}$; we did that in

Gram–Schmidt process for vectors in space

---

[6] Only if the inner product is positive definite.

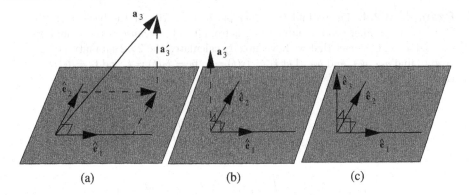

Figure 6.6: The Gram–Schmidt process for three linearly independent vectors in space.

our discussion of the plane. Call these new orthonormal vectors $\{\hat{\mathbf{e}}_1, \hat{\mathbf{e}}_2\}$ and construct the vector $\mathbf{a}_3'$,

$$\mathbf{a}_3' = \mathbf{a}_3 - (\mathbf{a}_3 \cdot \hat{\mathbf{e}}_1)\hat{\mathbf{e}}_1 - (\mathbf{a}_3 \cdot \hat{\mathbf{e}}_2)\hat{\mathbf{e}}_2$$

which is obtained from $\mathbf{a}_3$ by taking away its projections along $\hat{\mathbf{e}}_1$ and $\hat{\mathbf{e}}_2$. Now note that

$$\hat{\mathbf{e}}_1 \cdot \mathbf{a}_3' = \hat{\mathbf{e}}_1 \cdot \mathbf{a}_3 - (\mathbf{a}_3 \cdot \hat{\mathbf{e}}_1)\underbrace{\hat{\mathbf{e}}_1 \cdot \hat{\mathbf{e}}_1}_{=1} - (\mathbf{a}_3 \cdot \hat{\mathbf{e}}_2)\underbrace{\hat{\mathbf{e}}_2 \cdot \hat{\mathbf{e}}_1}_{=0} = 0,$$

$$\hat{\mathbf{e}}_2 \cdot \mathbf{a}_3' = \hat{\mathbf{e}}_2 \cdot \mathbf{a}_3 - (\mathbf{a}_3 \cdot \hat{\mathbf{e}}_1)\underbrace{\hat{\mathbf{e}}_2 \cdot \hat{\mathbf{e}}_1}_{=0} - (\mathbf{a}_3 \cdot \hat{\mathbf{e}}_2)\underbrace{\hat{\mathbf{e}}_2 \cdot \hat{\mathbf{e}}_2}_{=1} = 0,$$

i.e., $\mathbf{a}_3'$ is orthogonal to both $\hat{\mathbf{e}}_1$ and $\hat{\mathbf{e}}_2$. This suggests defining $\hat{\mathbf{e}}_3$ as

$$\hat{\mathbf{e}}_3 \equiv \frac{\mathbf{a}_3'}{|\mathbf{a}_3'|} = \frac{\mathbf{a}_3'}{\sqrt{\mathbf{a}_3' \cdot \mathbf{a}_3'}}.$$

The reader should note that in the construction of $\{\hat{\mathbf{e}}_1, \hat{\mathbf{e}}_2, \hat{\mathbf{e}}_3\}$, we have simply taken the linear combination of $\mathbf{a}_1$, $\mathbf{a}_2$, and $\mathbf{a}_3$.

Transformations that leave the inner products unchanged can be obtained in exactly the same way as for the plane. For A to preserve the inner product, we need to have

G-orthogonal
matrices

$$\widetilde{\mathsf{A}}\mathsf{G}\mathsf{A} = \mathsf{G}, \tag{6.41}$$

i.e., it has to be G-orthogonal. If G is the identity matrix, then A can be thought of as a *rigid rotation* and is simply called *orthogonal*; it satisfies

$$\widetilde{\mathsf{A}}\mathsf{A} = 1. \tag{6.42}$$

If we write A as

$$\begin{pmatrix} a_{11} & a_{12} & a_{13} \\ a_{21} & a_{22} & a_{23} \\ a_{31} & a_{32} & a_{33} \end{pmatrix},$$

then Equation (6.42) can be written as

$$\begin{pmatrix} a_{11} & a_{21} & a_{31} \\ a_{12} & a_{22} & a_{32} \\ a_{13} & a_{23} & a_{33} \end{pmatrix} \begin{pmatrix} a_{11} & a_{12} & a_{13} \\ a_{21} & a_{22} & a_{23} \\ a_{31} & a_{32} & a_{33} \end{pmatrix} = \begin{pmatrix} 1 & 0 & 0 \\ 0 & 1 & 0 \\ 0 & 0 & 1 \end{pmatrix}. \qquad (6.43)$$

It is clear from Equation (6.43) that the columns of the matrix A, considered as vectors, have unit length and are orthogonal to other columns in the usual positive definite inner product.[7] This is why A is called orthogonal.

The product on the LHS of Equation (6.43) is a 3×3 matrix whose elements must equal the corresponding elements of the unit matrix on the RHS. For example,

$$a_{11}^2 + a_{21}^2 + a_{31}^2 = 1. \qquad (6.44)$$

Similarly, the equality of the elements located in the first row and second column on both sides gives

$$a_{11}a_{12} + a_{21}a_{22} + a_{31}a_{32} = 0$$

and so on. Thus we obtain nine equations. However, simple inspection of these equations reveals that only six of them are independent. Therefore, we can only solve for the nine unknowns in terms of three of them (see Section 7.6). It does not matter which three matrix elements we choose. If we choose a_{11}, a_{21}, and a_{31}, for example, then Equation (6.44) reveals that these parameters can be sines and cosines. What this means physically is that *Three parameters are required to specify a rigid rotation of the axes.*

There are many ways to specify these three parameters. One of the most useful and convenient ways is by using **Euler angles** ψ, φ, and θ (see Figure 6.7). Example 6.2.5 below shows that in terms of these angles, the matrix A can be written as

> orthogonal matrices in space are determined by three parameters such as the Euler angles.

> Euler angles

$$A = \begin{pmatrix} \cos\psi\cos\varphi - \sin\psi\cos\theta\sin\varphi & -\cos\psi\sin\varphi - \sin\psi\cos\theta\cos\varphi & \sin\psi\sin\theta \\ \sin\psi\cos\varphi + \cos\psi\cos\theta\sin\varphi & -\sin\psi\sin\varphi + \cos\psi\cos\theta\cos\varphi & -\cos\psi\sin\theta \\ \sin\theta\sin\varphi & \sin\theta\cos\varphi & \cos\theta \end{pmatrix}.$$

It is straightforward to verify that $A^t A = 1$. Euler angles are useful in describing the rotational motion of a rigid body in mechanics.

Example 6.2.5. From Figure 6.7 it should be clear that the primed basis is obtained from the basis $\{\hat{e}_1, \hat{e}_2, \hat{e}_3\}$ by the following three operations.
(a) Rotate the coordinate system about the $\hat{e}_3$-axis through angle φ. This corresponds to a rotation in the $\hat{e}_1\hat{e}_2$-plane, leaving the $\hat{e}_3$-axis unchanged. We saw in the previous section how the 2×2 part of the matrix looked like. The complete 3×3 matrix corresponding to such a rotation is

> a general orthogonal matrix in space can be written as the product of three successive rotations.

$$A_1 = \begin{pmatrix} \cos\varphi & -\sin\varphi & 0 \\ \sin\varphi & \cos\varphi & 0 \\ 0 & 0 & 1 \end{pmatrix}. \qquad (6.45)$$

---

[7]This holds for 2×2 orthogonal matrices as well.

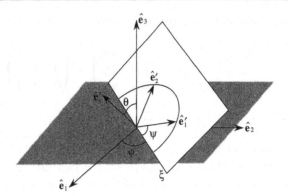

Figure 6.7: The Euler angles and the rotations about three axes making up a general rotation in space.

It is clear that this matrix leaves the third (z) component of a column vector unchanged while rotating the first two $(x$ and $y)$ components by φ.

(b) Rotate the new coordinate system around the new $\hat{\mathbf{e}}_1$-axis (the ξ-axis in the figure) through an angle θ. The corresponding matrix is

$$A_2 = \begin{pmatrix} 1 & 0 & 0 \\ 0 & \cos\theta & -\sin\theta \\ 0 & \sin\theta & \cos\theta \end{pmatrix}. \tag{6.46}$$

(c) Rotate the system about the new $\hat{\mathbf{e}}_3$-axis (the $\hat{\mathbf{e}}_3'$-axis in the figure) through an angle ψ. The corresponding matrix is

$$A_3 = \begin{pmatrix} \cos\psi & -\sin\psi & 0 \\ \sin\psi & \cos\psi & 0 \\ 0 & 0 & 1 \end{pmatrix}. \tag{6.47}$$

It is easily verified that $A = A_3A_2A_1$, i.e., the rotation A has the same effect as that of A_1, A_2, and A_3 performed in succession. ∎

6.3 Determinant

from matrices to systems of linear equations to determinants

Matrices have found application in many diverse fields of pure and applied mathematics. One such application is in the solution of linear equations. Consider the first set of equations in which we introduced matrices, Equations (6.4) and (6.5). The first of these equations associates a pair of numbers (α_1', α_2') to a given pair (α_1, α_2), i.e., if we *know* the latter pair, Equation (6.4) gives the former. What if we treat (α_1, α_2) as unknown? Under what conditions can we find these unknowns in terms of the *known* pair (α_1', α_2')? Let us use a more suggestive notation and write Equation (6.4) as

$$a_{11}x + a_{12}y = b_1,$$
$$a_{21}x + a_{22}y = b_2. \tag{6.48}$$

We want to investigate conditions under which a pair (x, y) exists which satisfies Equation (6.48). Let us assume that none of the a_{ij}'s is zero. The case in which one of them is zero is included in the final conclusion we are about to draw. Multiply the first equation of (6.48) by a_{22} and the second by a_{12} and subtract the resulting two equations. This yields $(a_{11}a_{22} - a_{12}a_{21})x = a_{22}b_1 - a_{12}b_2$, which has a solution for x of the form

$$x = \frac{a_{22}b_1 - a_{12}b_2}{a_{11}a_{22} - a_{12}a_{21}} \equiv \frac{a_{22}b_1 - a_{12}b_2}{\det A} \qquad (6.49)$$

if $a_{11}a_{22} - a_{12}a_{21} \neq 0$. In the last equality we have defined the **determinant** of A:

determinant of a
2×2 matrix

$$A = \begin{pmatrix} a_{11} & a_{12} \\ a_{21} & a_{22} \end{pmatrix} \Rightarrow \det A \equiv a_{11}a_{22} - a_{12}a_{21}. \qquad (6.50)$$

We can also find y. Multiply the first equation of (6.48) by a_{21} and the second by a_{11} and subtract the resulting two equations. This yields

$$(a_{11}a_{22} - a_{12}a_{21})y = a_{11}b_2 - a_{21}b_1$$

which has a solution for y of the form

$$y = \frac{a_{11}b_2 - a_{21}b_1}{\det A}. \qquad (6.51)$$

We can combine Equations (6.49) and (6.51) into a single *matrix* equation:

$$\begin{pmatrix} x \\ y \end{pmatrix} = \frac{1}{\det A} \begin{pmatrix} a_{22} & -a_{12} \\ -a_{21} & a_{11} \end{pmatrix} \begin{pmatrix} b_1 \\ b_2 \end{pmatrix}. \qquad (6.52)$$

This is the inverse of the matrix form of Equation (6.48). Indeed if we had written that equation in the form $Ax = b$, and *if* A *had an inverse*, say B, then we could have multiplied both sides of the equation by B and obtained

$$\underbrace{BA}_{=1} x = Bb \Rightarrow x = Bb.$$

This is precisely what we have in Equation (6.52)! Is the matrix multiplying the column vector b the inverse of A? Let us find out

$$\frac{1}{\det A} \begin{pmatrix} a_{22} & -a_{12} \\ -a_{21} & a_{11} \end{pmatrix} \begin{pmatrix} a_{11} & a_{12} \\ a_{21} & a_{22} \end{pmatrix}$$

$$= \frac{1}{\det A} \begin{pmatrix} a_{22}a_{11} - a_{12}a_{21} & 0 \\ 0 & -a_{21}a_{12} + a_{11}a_{22} \end{pmatrix} = \begin{pmatrix} 1 & 0 \\ 0 & 1 \end{pmatrix}.$$

So, it is indeed the inverse of A. We denote this inverse by A^{-1}.

Theorem 6.3.1. *A matrix* $A = \begin{pmatrix} a_{11} & a_{12} \\ a_{21} & a_{22} \end{pmatrix}$ *has an **inverse** if and only if its determinant, defined by* $\det A \equiv a_{11}a_{22} - a_{12}a_{21}$, *is not zero, in which case*

$$A^{-1} = \frac{1}{\det A} \begin{pmatrix} a_{22} & -a_{12} \\ -a_{21} & a_{11} \end{pmatrix}.$$

The reader may verify that, not only $A^{-1}A = 1$, but also $AA^{-1} = 1$.

Equation (6.48) gives the components b_1 and b_2 of a new vector obtained from an old vector with components x and y when the matrix A acts on the latter. We want to see what conditions A must satisfy for it to transform vectors in a basis into vectors of a new basis. Let $B = \{\mathbf{a}_1, \mathbf{a}_2\}$ be the old basis. The components of $\mathbf{a}_1$ in B are $x = 1$ and $y = 0$; so by (6.48), $\mathbf{a}_1'$, the vector obtained from $\mathbf{a}_1$ by the action of A, has components $b_1 = a_{11}$ and $b_2 = a_{21}$. The components of $\mathbf{a}_2$ in B are $x = 0$ and $y = 1$; so $\mathbf{a}_2'$, the vector obtained from $\mathbf{a}_2$ by the action of A, has components $c_1 = a_{12}$ and $c_2 = a_{22}$. The vectors (b_1, b_2) and (c_1, c_2) form a basis if and only if they are linearly independent, i.e.,

$$(b_1, b_2) = k(c_1, c_2) = (kc_1, kc_2) \Rightarrow b_1 = kc_1, \quad b_2 = kc_2,$$

does not hold for any constant k. This is equivalent to saying that

$$\frac{b_1}{c_1} \neq \frac{b_2}{c_2} \quad \text{or} \quad b_1 c_2 - b_2 c_1 \neq 0.$$

Expressing the b's and c's in terms of a_{ij}'s, we recognize the last relation as a condition on the determinant of A. Using Theorem 6.3.1, we thus have

> **Box 6.3.1.** *A transformation (or a matrix) transforms a basis into another basis if and only if it is invertible.*

Let us now consider three equations in three unknowns:

$$
\begin{aligned}
a_{11}x + a_{12}y + a_{13}z &= b_1, \\
a_{21}x + a_{22}y + a_{23}z &= b_2, \\
a_{31}x + a_{32}y + a_{33}z &= b_3,
\end{aligned}
\tag{6.53}
$$

which can also be written in matrix form as

$$
\begin{pmatrix} a_{11} & a_{12} & a_{13} \\ a_{21} & a_{22} & a_{23} \\ a_{31} & a_{32} & a_{33} \end{pmatrix} \begin{pmatrix} x \\ y \\ z \end{pmatrix} = \begin{pmatrix} b_1 \\ b_2 \\ b_3 \end{pmatrix} \Rightarrow A x = b.
\tag{6.54}
$$

We eliminate z from the set of equations by multiplying the first equation of (6.53) by a_{23} and the second by a_{13} and subtracting. This will give one equation in x and y. Similarly, multiplying the first equation by a_{33} and the third by a_{13} and subtracting gives another equation in x and y. These two equations are

from three equations in three unknowns to two equations in two unknowns, and from the determinant of a 2×2 matrix to that of a 3×3 matrix

$$
\underbrace{(a_{11}a_{23} - a_{21}a_{13})}_{\equiv a_{11}} x + \underbrace{(a_{12}a_{23} - a_{22}a_{13})}_{\equiv a_{12}} y = \underbrace{a_{23}b_1 - a_{13}b_2}_{\equiv b_1},
$$

$$
\underbrace{(a_{11}a_{33} - a_{31}a_{13})}_{\equiv a_{21}} x + \underbrace{(a_{12}a_{33} - a_{32}a_{13})}_{\equiv a_{22}} y = \underbrace{a_{33}b_1 - a_{13}b_3}_{\equiv b_2}.
\tag{6.55}
$$

Thus, we have reduced the three equations in three unknowns to two equations in two unknowns. We know how to find the solution for this set of equations. These solutions are given in Equations (6.49) and (6.51). In order for this equation to have a solution, the determinant of the coefficients must not vanish. Let us calculate this determinant:

$$\mathbf{a}_{11}\mathbf{a}_{22} - \mathbf{a}_{12}\mathbf{a}_{21} = (a_{11}a_{23} - a_{21}a_{13})(a_{12}a_{33} - a_{32}a_{13})$$
$$- (a_{12}a_{23} - a_{22}a_{13})(a_{11}a_{33} - a_{31}a_{13})$$
$$= a_{11}a_{23}a_{12}a_{33} - a_{11}a_{23}a_{32}a_{13} - a_{21}a_{13}a_{12}a_{33} + a_{21}a_{13}a_{32}a_{13}$$
$$- a_{12}a_{23}a_{11}a_{33} + a_{12}a_{23}a_{31}a_{13} + a_{22}a_{13}a_{11}a_{33} - a_{22}a_{13}a_{31}a_{13}$$
$$= a_{13}[a_{11}(a_{22}a_{33} - a_{23}a_{32}) - a_{12}(a_{21}a_{33} - a_{31}a_{23})$$
$$+ a_{13}(a_{21}a_{32} - a_{22}a_{31})]$$
$$= a_{13}\left[a_{11}\det\begin{pmatrix} a_{22} & a_{23} \\ a_{32} & a_{33} \end{pmatrix} - a_{12}\det\begin{pmatrix} a_{21} & a_{23} \\ a_{31} & a_{33} \end{pmatrix} + a_{13}\det\begin{pmatrix} a_{21} & a_{22} \\ a_{31} & a_{32} \end{pmatrix}\right].$$

If the original set of equations is to have a solution, the expression in the square brackets must not vanish. We call this expression the determinant of the 3×3 matrix A. We can give a cookbook recipe for calculating the determinant; but first we need the following definition:

> **Box 6.3.2.** *The **cofactor** A_{ij} of an element a_{ij} of a matrix A is defined as the product of $(-1)^{i+j}$ (i.e., $+1$ if $i+j$ is even and -1 if $i+j$ is odd) and the determinant of the smaller matrix (2×2, if A is a 3×3 matrix) obtained from A when its ith row and jth column are deleted.*

The following recipe applies to any (square) matrix, not just to 3×3 matrices:

> **Box 6.3.3.** *The **determinant** of A is obtained by multiplying each element of a row (or a column) by its cofactor and adding the products.*

If $\det A \neq 0$, then Equation (6.49) gives

$$x = \frac{\mathbf{a}_{22}\mathbf{b}_1 - \mathbf{a}_{12}\mathbf{b}_2}{a_{13}\det A}.$$

The numerator is

$$\mathbf{a}_{22}\mathbf{b}_1 - \mathbf{a}_{12}\mathbf{b}_2 = (a_{12}a_{33} - a_{32}a_{13})(a_{23}b_1 - a_{13}b_2)$$
$$- (a_{12}a_{23} - a_{22}a_{13})(a_{33}b_1 - a_{13}b_3)$$
$$= a_{13}\Big[\underbrace{(a_{22}a_{33} - a_{32}a_{23})}_{\equiv C_{11}} b_1 + \underbrace{(a_{32}a_{13} - a_{12}a_{33})}_{\equiv C_{12}} b_2 + \underbrace{(a_{12}a_{23} - a_{22}a_{13})}_{\equiv C_{13}} b_3\Big]$$
$$= a_{13}(C_{11}b_1 + C_{12}b_2 + C_{13}b_3).$$

Therefore,

$$x = \frac{C_{11}b_1 + C_{12}b_2 + C_{13}b_3}{\det A}. \tag{6.56}$$

Similarly, using Equation (6.51), we find

$$y = \frac{\mathbf{a}_{11}\mathbf{b}_2 - \mathbf{a}_{21}\mathbf{b}_1}{a_{13}\det A}$$

with

$$\mathbf{a}_{11}\mathbf{b}_2 - \mathbf{a}_{21}\mathbf{b}_1 = (a_{11}a_{23} - a_{21}a_{13})(a_{33}b_1 - a_{13}b_3)$$
$$- (a_{11}a_{33} - a_{31}a_{13})(a_{23}b_1 - a_{13}b_2)$$
$$= a_{13}\Big[\underbrace{(a_{31}a_{23} - a_{21}a_{33})}_{\equiv C_{21}} b_1 + \underbrace{(a_{11}a_{33} - a_{31}a_{13})}_{\equiv C_{22}} b_2 + \underbrace{(a_{21}a_{13} - a_{11}a_{23})}_{\equiv C_{23}} b_3\Big]$$
$$= a_{13}(C_{21}b_1 + C_{22}b_2 + C_{23}b_3),$$

so that

$$y = \frac{C_{21}b_1 + C_{22}b_2 + C_{23}b_3}{\det A}. \tag{6.57}$$

With x and y thus determined, we can substitute them in any of the three original equations and find z. Let us use the first equation; then

$$z = \frac{b_1 - a_{11}x - a_{12}y}{a_{13}}$$
$$= \frac{b_1 - a_{11}\dfrac{C_{11}b_1 + C_{12}b_2 + C_{13}b_3}{\det A} - a_{12}\dfrac{C_{21}b_1 + C_{22}b_2 + C_{23}b_3}{\det A}}{a_{13}}$$
$$= \frac{b_1(\det A - a_{11}C_{11} - a_{12}C_{21}) - b_2(a_{11}C_{12} + a_{12}C_{22}) - b_3(a_{11}C_{13} + a_{12}C_{23})}{a_{13}\det A}.$$

The numerator N can be calculated:

$$N = b_1[a_{11}(a_{22}a_{33} - a_{23}a_{32}) - a_{12}(a_{21}a_{33} - a_{31}a_{23}) + a_{13}(a_{21}a_{32} - a_{22}a_{31})$$
$$- a_{11}(a_{22}a_{33} - a_{32}a_{23}) - a_{12}(a_{31}a_{23} - a_{21}a_{33})]$$
$$- b_2[a_{11}(a_{32}a_{13} - a_{12}a_{33}) + a_{12}(a_{11}a_{33} - a_{31}a_{13})]$$
$$- b_3[a_{11}(a_{12}a_{23} - a_{22}a_{13}) + a_{12}(a_{21}a_{13} - a_{11}a_{23})]$$
$$= a_{13}\Big[\underbrace{(a_{21}a_{32} - a_{22}a_{31})}_{\equiv C_{31}} b_1 + \underbrace{(a_{12}a_{31} - a_{11}a_{32})}_{\equiv C_{32}} b_2 + \underbrace{(a_{11}a_{22} - a_{12}a_{21})}_{\equiv C_{33}} b_3\Big]$$
$$= a_{13}(C_{31}b_1 + C_{32}b_2 + C_{33}b_3).$$

It now follows that

$$z = \frac{C_{31}b_1 + C_{32}b_2 + C_{33}b_3}{\det A}. \tag{6.58}$$

We can put Equations (6.56), (6.57), and (6.58) in matrix form:

$$\begin{pmatrix} x \\ y \\ z \end{pmatrix} = \frac{1}{\det A} \begin{pmatrix} C_{11} & C_{12} & C_{13} \\ C_{21} & C_{22} & C_{23} \\ C_{31} & C_{32} & C_{33} \end{pmatrix} \begin{pmatrix} b_1 \\ b_2 \\ b_3 \end{pmatrix} \;\Rightarrow\; \mathbf{x} = \frac{1}{\det A}\mathbf{C}\mathbf{b}. \tag{6.59}$$

This is the inverse of Equation (6.54). The reader may verify that multiplying
A on either side of $C/\det A$ yields the identity matrix, so that $C/\det A$ is indeed
the inverse of A. The rule for calculating this inverse is as follows. Construct
a matrix out of the cofactors and denote it by **A**:

$$\mathbf{A} \equiv \begin{pmatrix} A_{11} & A_{12} & A_{13} \\ A_{21} & A_{22} & A_{23} \\ A_{31} & A_{32} & A_{33} \end{pmatrix} \tag{6.60}$$

and note that

$$\begin{pmatrix} C_{11} & C_{12} & C_{13} \\ C_{21} & C_{22} & C_{23} \\ C_{31} & C_{32} & C_{33} \end{pmatrix} = \tilde{\mathbf{A}}$$

so, we obtain the important result

inverse of a 3×3 matrix

$$A^{-1} = \frac{1}{\det A}\tilde{\mathbf{A}} = \frac{1}{\det A}\begin{pmatrix} A_{11} & A_{21} & A_{31} \\ A_{12} & A_{22} & A_{32} \\ A_{13} & A_{23} & A_{33} \end{pmatrix}. \tag{6.61}$$

Equation (6.61), although derived for a 3×3 matrix, applies to all matrices,
including a 2×2 one whose inverse was given in Theorem 6.3.1, as the reader
is asked to verify.

As in the case of 2×2 matrices, a transformation in space that takes a
basis onto another basis is invertible.

6.4   The Jacobian

With the machinery of determinants at our disposal, we can formalize the
geometric construction of area and volume elements in Chapter 2 to a procedure which can be used for *all* coordinate transformations. We start with
two dimensions and consider the coordinate transformation

$$x = f(u, v), \qquad y = g(u, v). \tag{6.62}$$

Our goal is to write the element of area in the (u, v) coordinate system. This
is the area formed by infinitesimal elements in the direction of u and v, i.e.,
elements in the direction of the primary curves of the (u, v) coordinate system.
For an arbitrary change du and dv in u and v, the Cartesian coordinates
change as follows:

$$dx = \frac{\partial f}{\partial u}\,du + \frac{\partial f}{\partial v}\,dv,$$
$$dy = \frac{\partial g}{\partial u}\,du + \frac{\partial g}{\partial v}\,dv.$$

The element in the direction of the first primary curve is obtained by holding v constant and letting u vary. This corresponds to setting $dv = 0$ in the above equations. It follows that the first primary (vector) length element is

$$\vec{dl_1} = \hat{\mathbf{e}}_x\, dx_1 + \hat{\mathbf{e}}_y\, dy_1 = \hat{\mathbf{e}}_x \frac{\partial f}{\partial u}\, du + \hat{\mathbf{e}}_y \frac{\partial g}{\partial u}\, du. \tag{6.63}$$

Similarly, the second primary (vector) length element, obtained by fixing u and letting v vary, is

$$\vec{dl_2} = \hat{\mathbf{e}}_x\, dx_2 + \hat{\mathbf{e}}_y\, dy_2 = \hat{\mathbf{e}}_x \frac{\partial f}{\partial v}\, dv + \hat{\mathbf{e}}_y \frac{\partial g}{\partial v}\, dv. \tag{6.64}$$

When we derived the elements of area and volume in the three coordinate systems in Chapter 2, we used the fact that the set of unit vectors in each system were mutually perpendicular. Therefore, the area and volume elements were obtained by mere multiplication of length elements. We are not assuming that $\hat{\mathbf{e}}_u$ and $\hat{\mathbf{e}}_v$ are perpendicular. Thus, we cannot simply multiply the lengths to get the area. However, we can use the result of Example 1.1.2 which gives the area of a parallelogram formed by two non-collinear vectors. Writing the cross product in terms of the determinant, we have

$$\vec{dl_1} \times \vec{dl_2} = \det \begin{pmatrix} \hat{\mathbf{e}}_x & \hat{\mathbf{e}}_y & \hat{\mathbf{e}}_z \\ \frac{\partial f}{\partial u}\, du & \frac{\partial g}{\partial u}\, du & 0 \\ \frac{\partial f}{\partial v}\, dv & \frac{\partial g}{\partial v}\, dv & 0 \end{pmatrix} = \hat{\mathbf{e}}_z \det \begin{pmatrix} \frac{\partial f}{\partial u} & \frac{\partial g}{\partial u} \\ \frac{\partial f}{\partial v} & \frac{\partial g}{\partial v} \end{pmatrix} du\, dv$$

Jacobian matrix and Jacobian

and the area is simply the absolute value of this cross product:

$$da = \left| \det \begin{pmatrix} \frac{\partial f}{\partial u} & \frac{\partial g}{\partial u} \\ \frac{\partial f}{\partial v} & \frac{\partial g}{\partial v} \end{pmatrix} \right| du\, dv \equiv \begin{vmatrix} \frac{\partial x}{\partial u} & \frac{\partial y}{\partial u} \\ \frac{\partial x}{\partial v} & \frac{\partial y}{\partial v} \end{vmatrix} du\, dv, \tag{6.65}$$

where we substituted x and y for f and g and introduced a new notation for the (absolute value of the) determinant. The matrix whose determinant multiplies $du\, dv$ is called the **Jacobian matrix**, and the absolute value of its determinant, the **Jacobian**.

Example 6.4.1. Let us apply Equation (6.65) to polar coordinates. The transformation is

$$x = f(r,\theta) = r \cos\theta, \qquad y = g(r,\theta) = r \sin\theta.$$

This gives

$$\frac{\partial x}{\partial r} = \frac{\partial f}{\partial r} = \cos\theta, \qquad\qquad \frac{\partial x}{\partial \theta} = \frac{\partial f}{\partial \theta} = -r \sin\theta,$$

$$\frac{\partial y}{\partial r} = \frac{\partial g}{\partial r} = \sin\theta, \qquad\qquad \frac{\partial y}{\partial \theta} = \frac{\partial g}{\partial \theta} = r \cos\theta,$$

and

$$da = \begin{vmatrix} \frac{\partial x}{\partial r} & \frac{\partial y}{\partial r} \\ \frac{\partial x}{\partial \theta} & \frac{\partial y}{\partial \theta} \end{vmatrix} dr\, d\theta = \begin{vmatrix} \cos\theta & \sin\theta \\ -r\sin\theta & r\cos\theta \end{vmatrix} dr\, d\theta$$

$$= (r\cos^2\theta + r\sin^2\theta)\, dr\, d\theta = r\, dr\, d\theta,$$

which is the familiar element of area in polar coordinates.   ∎

The procedure discussed above for two dimensions can be generalized to three dimensions using the result of Example 1.1.3 which gives the volume of a parallelepiped formed by three non-coplanar vectors. Suppose the coordinate transformations are of the form

$$x = f(u, v, w), \qquad y = g(u, v, w), \qquad z = h(u, v, w).$$

Then

$$dx = \frac{\partial f}{\partial u}\, du + \frac{\partial f}{\partial v}\, dv + \frac{\partial f}{\partial w}\, dw,$$

$$dy = \frac{\partial g}{\partial u}\, du + \frac{\partial g}{\partial v}\, dv + \frac{\partial g}{\partial w}\, dw,$$

$$dz = \frac{\partial h}{\partial u}\, du + \frac{\partial h}{\partial v}\, dv + \frac{\partial h}{\partial w}\, dw.$$

The first primary element of length is obtained by fixing v and w and allowing u to vary; similarly for the second and third primary elements of length. We therefore have

$$\vec{dl_1} = \hat{e}_x\, dx_1 + \hat{e}_y\, dy_1 + \hat{e}_z\, dz_1 = \hat{e}_x\frac{\partial f}{\partial u}\, du + \hat{e}_y\frac{\partial g}{\partial u}\, du + \hat{e}_z\frac{\partial h}{\partial u}\, du,$$

$$\vec{dl_2} = \hat{e}_x\, dx_2 + \hat{e}_y\, dy_2 + \hat{e}_z\, dz_2 = \hat{e}_x\frac{\partial f}{\partial v}\, dv + \hat{e}_y\frac{\partial g}{\partial v}\, dv + \hat{e}_z\frac{\partial h}{\partial v}\, dv,$$

$$\vec{dl_3} = \hat{e}_x\, dx_3 + \hat{e}_y\, dy_3 + \hat{e}_z\, dz_3 = \hat{e}_x\frac{\partial f}{\partial w}\, dw + \hat{e}_y\frac{\partial g}{\partial w}\, dw + \hat{e}_z\frac{\partial h}{\partial w}\, dw.$$

Example 1.1.3 now yields

$$dV = \left| \vec{dl_1} \cdot (\vec{dl_2} \times \vec{dl_3}) \right| = \left| \det \begin{pmatrix} \frac{\partial f}{\partial u}\, du & \frac{\partial g}{\partial u}\, du & \frac{\partial h}{\partial u}\, du \\ \frac{\partial f}{\partial v}\, dv & \frac{\partial g}{\partial v}\, dv & \frac{\partial h}{\partial v}\, dv \\ \frac{\partial f}{\partial w}\, dw & \frac{\partial g}{\partial w}\, dw & \frac{\partial h}{\partial w}\, dw \end{pmatrix} \right|.$$

We summarize the foregoing argument in

Theorem 6.4.2. *For the coordinates u, v, and w, related to the Cartesian coordinates by $x = f(u, v, w)$, $y = g(u, v, w)$, and $z = h(u, v, w)$, the volume element is given by*

$$dV = \begin{vmatrix} \dfrac{\partial x}{\partial u} & \dfrac{\partial y}{\partial u} & \dfrac{\partial z}{\partial u} \\[6pt] \dfrac{\partial x}{\partial v} & \dfrac{\partial y}{\partial v} & \dfrac{\partial z}{\partial v} \\[6pt] \dfrac{\partial x}{\partial w} & \dfrac{\partial y}{\partial w} & \dfrac{\partial z}{\partial w} \end{vmatrix} du\, dv\, dw. \tag{6.66}$$

Jacobian defined

The (absolute value of the) determinant multiplying $du\, dv\, dw$ is called the **Jacobian** of the coordinate transformation.

Historical Notes

Determinants were mathematical objects created in the process of solving a system of linear equations. As early as 1693 Leibniz used a systematic set of indices for the coefficients of a system of three equations in two unknowns. By eliminating the two unknowns from the set of three equations, he obtained an expression involving the coefficients that "determined" whether a solution existed for the set of equations.

The solution of simultaneous linear equations in two, three, and four unknowns by the method of determinants was created by **Maclaurin** around 1729. Though not as good in notation, his rule is the one we use today and which **Cramer** used in connection with his study of the conic sections. In 1764, **Bezout** systematized the process of determining the signs of the terms of a determinant for n equations in n unknowns and showed that the vanishing of the determinant is a necessary condition for nonzero solutions to exist.

Vandermonde was the first to give a connected and logical exposition of the theory of determinants detached from any system of linear equations, although he used his theory mostly as applied to such systems. He also gave a rule for expanding a determinant by using second-order minors and their complementary minors. In the sense that he concentrated on determinants, he is aptly considered the founder of the theory.

One of the consistent workers in determinant theory over a period of over fifty years was **James Joseph Sylvester**.

In 1833 he became a student at St. John's College, Cambridge, and took the difficult tripos examination in the same year along with two other famous mathematicians, Gregory and **Green** (the creator of the important *Green's functions*). Sylvester came second, Green who was 20 years older than the other two came fourth with Duncan Gregory fifth. (The first-place winner did little work of importance after graduating.)

James Joseph
Sylvester
1814–1897

At this time it was necessary for a student to sign a religious oath to the Church of England before graduating and Sylvester, being Jewish, refused to take the oath, so could not graduate. For the same reason he was not eligible for a Smith's prize nor for a Fellowship.

From 1838 Sylvester started to teach physics at the University of London, one of the few places which did not bar him because of his religion. Three years later he was appointed to a chair in the University of Virginia but he resigned after a few months. A student who had been reading a newspaper in one of Sylvester's lectures insulted him and Sylvester struck him with a sword stick. The student collapsed in shock and Sylvester believed (wrongly) that he had killed him. He fled to New York boarding the first available ship back to England.

On his return, Sylvester worked as an actuary and lawyer but gave private mathematics lessons. His pupils included **Florence Nightingale**. By good fortune

Cayley was also a lawyer, and both worked at the courts of Lincoln's Inn in London. Cayley and Sylvester discussed mathematics as they walked around the courts and, although very different in temperament, they became life-long friends.

Sylvester tried hard to return to mathematics as a profession, and he applied unsuccessfully for a lectureship in geometry at Gresham College, London, in 1854. Another failed application was for the chair in mathematics at the Royal Military Academy at Woolwich, but, after the successful applicant died within a few months of being appointed, Sylvester became professor of mathematics at Woolwich. Being at a military academy, Sylvester had to retire at age 55. At first it looked as though he might give up mathematics since he had published his only book at this time, and it was on poetry. Apparently Sylvester was proud of this work, entitled *The Laws of Verse*, since after this he sometimes signed himself "J. J. Sylvester, author of The Laws of Verse."

In 1877 Sylvester accepted a chair at the Johns Hopkins University and founded in 1878 the *American Journal of Mathematics*, the first mathematical journal in the USA.

In 1883 Sylvester, although 68 years old at this time, was appointed to the Savilian chair of geometry at Oxford. However he only liked to lecture on his own research and this was not well liked at Oxford where students wanted only to do well in examinations. In 1892, at the age of 78, Oxford appointed a deputy professor in his place and Sylvester, by this time partially blind and suffering from loss of memory, returned to London where he spent his last years at the Athenaeum Club.

Sylvester did important work on matrix and determinant theory, a topic in which he became interested during the walks with Cayley while they were at the courts of Lincoln's Inn. In particular he used matrix theory to study higher-dimensional geometry. He also devised an improved method of determining conditions under which a system of *polynomial* equations has a solution.

The formula for the derivative of a determinant when the elements are functions of a variable was first given in 1841 by **Jacobi** who had earlier used them in the change of variables in a multiple integral. In this context the determinant is called the *Jacobian* of the transformation (as discussed in the current section of this book).

6.5 Problems

6.1. What vector is obtained when the vector $\mathbf{a}_2$ of a basis $\{\mathbf{a}_1, \mathbf{a}_2\}$ is actively transformed with the matrix $\left(\begin{smallmatrix} 0 & 1 \\ 0 & 0 \end{smallmatrix} \right)$.

6.2. Show that the nonzero matrix $A = \left(\begin{smallmatrix} 1 & 0 \\ 0 & 0 \end{smallmatrix} \right)$ cannot have an inverse. Hint: Suppose that $B = \left(\begin{smallmatrix} a & b \\ c & d \end{smallmatrix} \right)$ is the inverse of A. Calculate AB and BA, set them equal to the unit matrix and show that no solution exists for a, b, c, and d.

6.3. Let $A = \left(\begin{smallmatrix} a_1 & b_1 \\ c_1 & d_1 \end{smallmatrix} \right)$ and $B = \left(\begin{smallmatrix} a_2 & b_2 \\ c_2 & d_2 \end{smallmatrix} \right)$ be arbitrary matrices. Find AB, A^t, and B^t and show that $(AB)^t = B^t A^t$.

6.4. Find the angle between $1 + t$ and $1 - t$ when the inner product is integration over the interval $(0, 1)$.

6.5. Instead of $(0, 1)$, choose $(-1, 1)$ as the interval of integration for $\mathcal{P}_1[t]$. From the basis $\{1, t\}$, construct an orthonormal basis using the Gram–Schmidt process.

6.6. Take the interval of the integration to be $(-1, +1)$, and find the inner product matrix for the basis $\{1, t\}$ of $\mathcal{P}_1[t]$.

6.7. Find the angle between two vectors $\mathbf{a}$ and $\mathbf{b}$, whose components in an orthonormal basis are, respectively, $(1, 2)$ and $(2, -3)$. Use the Gram–Schmidt process to find the orthonormal vectors obtained from $\mathbf{a}$ and $\mathbf{b}$.

6.8. Use the Gram–Schmidt process to find an orthonormal basis in three dimensions from each of the following:

      (a) $(-1, 1, 1), (1, -1, 1), (1, 1, -1)$ (b) $(1, 2, 2), (0, 0, 1), (0, 1, 0)$

6.9. (a) Find the inner product matrix associated with the basis vectors $\mathbf{a}_1 = \hat{\mathbf{e}}_x + \hat{\mathbf{e}}_y$, $\mathbf{a}_2 = \hat{\mathbf{e}}_x + \hat{\mathbf{e}}_z$, and $\mathbf{a}_3 = \hat{\mathbf{e}}_y + \hat{\mathbf{e}}_z$.
(b) Calculate the inner product of two vectors $\mathbf{a}$ and $\mathbf{b}$, whose components in the basis above are, respectively, $(1, -1, 2)$ and $(0, 2, 3)$.
(c) Use the Gram–Schmidt process to find three orthonormal vectors out of the basis of (a).

6.10. Use Gram–Schmidt process to find orthonormal vectors out of the three vectors $(2, -1, 3)$, $(-1, 1, -2)$, and $(3, 1, 2)$. What do you get as the last vector? What can you say about the linear independence of the original vectors?

6.11. What is the angle between the second and fourth vectors in the standard basis of $\mathcal{P}_3[t]$ when the interval of integration of the inner product is $(0, 1)$? Between the first and fourth vectors?

6.12. Calculate the inner product matrix for the standard basis of $\mathcal{P}_3[t]$ when the interval of integration of the inner product is $(-1, +1)$. Now find the angle between all vectors in that basis.

6.13. The inner product matrix in a basis $\{\mathbf{a}_1, \mathbf{a}_2\}$ is given by

$$G = \begin{pmatrix} 2 & -1 \\ -1 & 3 \end{pmatrix}.$$

(a) Calculate the cosine of the angle between $\mathbf{a}_1$ and $\mathbf{a}_2$.
(b) Suppose that $\mathbf{a} = -\mathbf{a}_1 + \mathbf{a}_2$ and $\mathbf{b} = 2\mathbf{a}_1 - \mathbf{a}_2$. Calculate $|\mathbf{a}|$, $|\mathbf{b}|$, $\mathbf{a} \cdot \mathbf{b}$, and the cosine of the angle between $\mathbf{a}$ and $\mathbf{b}$.

6.14. Let $\mathbf{a}_1 = 1 + t$ and $\mathbf{a}_2 = 1 - t$ be a basis of $\mathcal{P}_1[t]$. Define the inner product as the integral of products of polynomials over the interval $(0, a)$ with $a > 0$.
(a) Determine a such that $\mathbf{a}_1$ and $\mathbf{a}_2$ are orthogonal.
(b) Given this value of a, calculate $|\mathbf{a}_1|$ and $|\mathbf{a}_2|$.
(c) Find two orthogonal polynomials $\{\hat{\mathbf{e}}_1, \hat{\mathbf{e}}_2\}$ of unit length that form a basis for $\mathcal{P}_1[t]$.
(d) Write the polynomial $\mathbf{b} = 3 - 2t$ as a linear combination of $\hat{\mathbf{e}}_1$ and $\hat{\mathbf{e}}_2$.
(e) Calculate $\mathbf{b} \cdot \mathbf{b}$ using the definition of the inner product.
(f) Calculate $\mathbf{b} \cdot \mathbf{b}$ by squaring (and then adding) the components in $\{\hat{\mathbf{e}}_1, \hat{\mathbf{e}}_2\}$.

6.15. Show that the matrix C defined in Equations (6.56)–(6.59) is indeed the transpose of the matrix **A** of cofactors of A.

6.16. Show directly that the matrix given in Equation (6.61) is indeed the inverse of the matrix A.

6.17. From the transformation rules (1.8) and (1.9) giving the Cartesian coordinates as functions of cylindrical and spherical coordinates, and using the Jacobian (6.66), find the volume elements in cylindrical and spherical coordinates

6.18. The **elliptic coordinates** are given by

$$x = a \cosh u \cos \theta$$
$$y = a \sinh u \sin \theta.$$

Using the Jacobian for two variables (6.65), find the element of area for the elliptic coordinate system.

6.19. The **elliptic cylindrical coordinates** are given by

$$x = a \cosh u \cos \theta$$
$$y = a \sinh u \sin \theta$$
$$z = z$$

Using the Jacobian for three variables (6.66), find the element of volume for the elliptic cylindrical coordinate system.

6.20. The **prolate spheroidal coordinates** are given by

$$x = a \sinh u \sin \theta \cos \varphi$$
$$y = a \sinh u \sin \theta \sin \varphi$$
$$z = a \cosh u \cos \theta$$

Using the Jacobian for three variables (6.66), find the element of volume for the prolate spheroidal coordinate system.

6.21. The **toroidal coordinates** are given by

$$x = \frac{a \sinh \theta \cos \varphi}{\cosh \theta - \cos u}$$
$$y = \frac{a \sinh \theta \sin \varphi}{\cosh \theta - \cos u}$$
$$z = \frac{a \sin u}{\cosh \theta - \cos u}$$

Using the Jacobian for three variables (6.66), find the element of volume for the toroidal coordinate system.

6.22. A coordinate system (R, Θ, ϕ) in space is defined by

$$x = R \cos \Theta \cos \phi + b \cos \phi$$
$$y = R \cos \Theta \sin \phi + b \sin \phi$$
$$z = R \sin \Theta$$

where b is a constant, and $0 < R < b$. Using the Jacobian for three variables (6.66), find the element of volume for this coordinate system.

Chapter 7

Finite-Dimensional Vector Spaces

Human visual perception of dimension is limited to two and three, the plane and space. However, his mental perception, and his ability to abstract, recognizes no bounds. If this abstraction were a mere useless mental exercise, we would not bother to add this chapter to the book. It is an intriguing coincidence that Nature plays along with the tune of human mental abstraction in the most harmonious way. This harmony was revealed to Hermann Minkowski in 1908 when he convinced physicists and mathematicians alike, that the most natural setting for the newly discovered special theory of relativity was a four-dimensional space. Eight years later, Einstein used this concept to formulate his general theory of relativity which is the only viable theory of gravity for the large-scale structure of space and time. In 1921, Kaluza, in a most beautiful idea, unified the electromagnetic interaction with gravity using a five-dimensional spacetime. Today string theory, one of the most promising candidates for the unification of all forces of nature, uses 11-dimensional spacetime; and the language of quantum mechanics—a theory that describes atomic, molecular, and solid-state physics, as well as all of chemistry—is best spoken in an infinite-dimensional space, called Hilbert space.

The key to this multidimensional abstraction is Descartes' ingenious idea of translating Euclid's geometry into the language of coordinates whereby the abstract Euclidean point in a plane is given the two coordinates (x, y), and that in space, the three coordinates (x, y, z), where x, y, and z are real numbers. Once this crucial step is taken, the generalization to multidimensional spaces becomes a matter of adding more and more coordinates to the list: (x, y, z, w) is a point in a four-dimensional space, and (x, y, z, w, u) describes a point in a five-dimensional space. In the spirit of this chapter, we want to identify points with vectors as in the plane and space, in which we drew a directed line segment from the origin to the point in question. In general, an n-dimensional **Cartesian vector x** is

$$\mathbf{x} = (x_1, x_2, \ldots, x_n) \tag{7.1}$$

in which x_j is called the jth **component** of the vector. These have all the properties expected of vectors: You can add them

$$\mathbf{x} + \mathbf{y} = (x_1, x_2, \ldots, x_n) + (y_1, y_2, \ldots, y_n) \equiv (x_1 + y_1, x_2 + y_2, \ldots, x_n + y_n),$$

you can multiply a vector by a number

$$\alpha \mathbf{x} = \alpha(x_1, x_2, \ldots, x_n) \equiv (\alpha x_1, \alpha x_2, \ldots, \alpha x_n),$$

and the **zero vector** is $\mathbf{0} = (0, 0, \ldots, 0)$. Two vectors are equal if and only if their corresponding components are equal. Sometimes, it will be convenient to denote these vectors as columns rather than rows.

The set of real numbers, or the set of points on a line, is denoted by $\mathbb{R}$. It is common to denote the set of points in a plane—or, in the language of Cartesian coordinates, the set of pairs of real numbers (x, y)—by $\mathbb{R}^2$, and the set of points in space by $\mathbb{R}^3$. Generalizing this notation, we denote the set of points in the n-dimensional Cartesian space by $\mathbb{R}^n$. We now have an infinite collection of "spaces" of various dimensions, starting with the one-dimensional real line $\mathbb{R}^1 = \mathbb{R}$, moving on to the two-dimensional plane $\mathbb{R}^2$, and the three-dimensional space $\mathbb{R}^3$, and continuing to all the abstract spaces $\mathbb{R}^n$ with $n \geq 4$. The concepts of linear combination, linear independence, and basis are exactly the same as before. The vectors

$$\hat{\mathbf{e}}_1 \equiv (1, 0, \ldots, 0), \quad \hat{\mathbf{e}}_2 \equiv (0, 1, \ldots, 0), \quad \ldots \quad \hat{\mathbf{e}}_n \equiv (0, 0, \ldots, 1) \tag{7.2}$$

form a basis for $\mathbb{R}^n$, called the **standard basis**.

7.1 Linear Transformations

formal definition
of a linear
transformation or
a linear operator

A **linear transformation** or a **linear operator** is a correspondence that takes a vector in one space and produces a vector in another space in such a way that the operation of summation of vectors and multiplication of vectors by numbers is preserved. If we denote the linear transformation by **T**, then in mathematical symbolism, the above statement becomes

$$\mathbf{T}(\alpha \mathbf{x} + \beta \mathbf{y}) = \alpha \mathbf{T}(\mathbf{x}) + \beta \mathbf{T}(\mathbf{y}). \tag{7.3}$$

Matrices are prototypes of linear transformations. In fact, we saw earlier that it was possible to transform vectors in the plane to vectors in space and vice versa via 3×2 or 2×3 matrices. We did not attempt to verify Equation (7.3) for those transformations, but the reader can easily do so.

In fact, denoting vectors of $\mathbb{R}^n$ and $\mathbb{R}^m$ by column vectors, we can immediately generalize Equations (6.36) and (6.37) to

$$\begin{pmatrix} \alpha_1' \\ \alpha_2' \\ \vdots \\ \alpha_m' \end{pmatrix} = \begin{pmatrix} a_{11} & a_{12} & \dots & a_{1n} \\ a_{21} & a_{22} & \dots & a_{2n} \\ \vdots & \vdots & & \vdots \\ a_{m1} & a_{m2} & \dots & a_{mn} \end{pmatrix} \begin{pmatrix} \alpha_1 \\ \alpha_2 \\ \vdots \\ \alpha_n \end{pmatrix} \qquad \text{or} \qquad \mathsf{a}' = \mathsf{A}\mathsf{a}, \qquad (7.4)$$

where A is an $m \times n$ matrix—i.e., it has m rows and n columns—whose elements a_{ij} are real numbers. The reader may verify that Equation (7.4) is a linear transformation that maps vectors of $\mathbb{R}^n$ to those of $\mathbb{R}^m$.

Other linear operators of importance are various differential operators, i.e., derivatives of various order. For example, it is easily verified that d/dx is a linear operator acting on the space of differentiable functions.[1] This is because

$$\frac{d}{dx}(\alpha f + \beta g) = \alpha \frac{df}{dx} + \beta \frac{dg}{dx}$$

derivative is a linear operator

for α and β real constants. Similarly d^2/dx^2 and derivative of higher orders, as well as partial derivatives of various kinds and orders, are all linear operators. In fact, even when these derivatives are multiplied by functions (on the left), they are still linear. In particular, the *second-order linear differential operator*

$$\mathsf{L} \equiv p_2(x)\frac{d^2}{dx^2} + p_1(x)\frac{d}{dx} + p_0(x)$$

second-order linear differential operator

is indeed a linear operator.

If a linear transformation T maps vectors of $\mathbb{R}^n$ to vectors of $\mathbb{R}^m$, and S maps vectors of $\mathbb{R}^m$ to vectors of $\mathbb{R}^k$, then we can "compose" or "multiply" the two transformations to obtain a linear transformation ST which maps vectors of $\mathbb{R}^n$ to vectors of $\mathbb{R}^k$. In terms of matrices, T is represented by an $m \times n$ matrix T, S is represented by a $k \times m$ matrix S, and ST is represented by an $k \times n$ matrix which is the product of S and T with S to the left of T. The product of matrices is as outlined in Box 6.1.3.

Box 7.1.1. *If A is a $k \times m$ matrix, and B is an $m \times n$ matrix, then AB is a $k \times n$ matrix whose entries are given by Box 6.1.3.*

The product BA is not defined unless $k = n$, in which case BA will be an $m \times m$ matrix.

Using polynomials, we can generate multidimensional vector spaces by adding increasing powers of t. Then, the collection $\mathcal{P}_n[t]$ of polynomials of degree n and less becomes an $(n+1)$-dimensional vector space. A convenient basis for this vector space is $\{1, t, t^2, \dots, t^n\}$ which we call the **standard**

polynomials can generate multidimensional vector spaces!

---

[1]The reader may want to check that the collection of differentiable functions is indeed a vector space with the "zero function" being the zero vector.

basis of $\mathcal{P}_n[t]$. The reader may verify that the operation of differentiation (of any order) is a linear transformation on $\mathcal{P}_n[t]$ which can be represented by matrices as done in Example 6.2.2.

Example 7.1.1. Let us find the matrix that represents the operation of second differentiation on $\mathcal{P}_3[t]$ using the standard basis of $\mathcal{P}_3[t]$. Recall that we only need to apply the second derivative to the basis vectors $\mathbf{f}_1 = 1$, $\mathbf{f}_2 = t$, $\mathbf{f}_3 = t^2$, and $\mathbf{f}_4 = t^3$. We use a prime to denote the transformed vector:

$$\mathbf{f}_1' = \frac{d^2}{dt^2}(1) = 0 = 0 \cdot \mathbf{f}_1 + 0 \cdot \mathbf{f}_2,$$

$$\mathbf{f}_2' = \frac{d^2}{dt^2}(t) = 0 = 0 \cdot \mathbf{f}_1 + 0 \cdot \mathbf{f}_2,$$

$$\mathbf{f}_3' = \frac{d^2}{dt^2}(t^2) = 2 = 2 \cdot \mathbf{f}_1 + 0 \cdot \mathbf{f}_2,$$

$$\mathbf{f}_4' = \frac{d^2}{dt^2}(t^3) = 6t = 0 \cdot \mathbf{f}_1 + 6 \cdot \mathbf{f}_2,$$

where we have anticipated the fact that double differentiation of $\mathcal{P}_3[t]$ results in $\mathcal{P}_1[t]$. Following the rule of Box 6.2.2, we can write the transformation matrix as

$$\begin{pmatrix} 0 & 0 & 2 & 0 \\ 0 & 0 & 0 & 6 \end{pmatrix}.$$

We may verify that the coefficients in $\mathcal{P}_1[t]$ of the second derivative of an arbitrary polynomial $f(t) = \alpha_0 + \alpha_1 t + \alpha_2 t^2 + \alpha_3 t^3$ can be obtained by the product of the matrix of second derivative and the 4×1 column vector representing $f(t)$. In fact,

$$\begin{pmatrix} 0 & 0 & 2 & 0 \\ 0 & 0 & 0 & 6 \end{pmatrix} \begin{pmatrix} \alpha_0 \\ \alpha_1 \\ \alpha_2 \\ \alpha_3 \end{pmatrix} = \begin{pmatrix} 2\alpha_2 \\ 6\alpha_3 \end{pmatrix}.$$

These are the two coefficients of the resulting polynomial in $\mathcal{P}_1[t]$. The polynomial itself is $2\alpha_2 + 6\alpha_3 t$ which is indeed the derivative of the third degree polynomial $f(t)$. ∎

7.2   Inner Product

Since the concepts of length and angle are not familiar for $\mathbb{R}^n$, we need to define the inner product first and then deduce those concepts. We can generalize the usual inner product of $\mathbb{R}^2$ and $\mathbb{R}^3$ in terms of components of vectors. Let

$$\mathbf{a} = (a_1, a_2, \ldots, a_n) \qquad \text{and} \qquad \mathbf{b} = (b_1, b_2, \ldots, b_n).$$

inner product in $\mathbb{R}^n$ defined in terms of components in the standard basis

Then

$$\mathbf{a} \cdot \mathbf{b} = a_1 b_1 + a_2 b_2 + \cdots + a_n b_n \qquad (7.5)$$

is the immediate generalization of the dot product to $\mathbb{R}^n$.

This, of course, is not the most general inner product. For that, we need an inner product matrix G. As in the case of the plane and space, this is simply a symmetric $n \times n$ matrix whose elements determine the dot products of the vectors of the basis in which we are working.

$$\mathsf{G} = \begin{pmatrix} g_{11} & g_{12} & \cdots & g_{1n} \\ g_{21} & g_{22} & \cdots & g_{2n} \\ \vdots & \vdots & & \vdots \\ g_{n1} & g_{n2} & \cdots & g_{nn} \end{pmatrix}, \qquad g_{ij} = g_{ji}, \quad i,j = 1, 2, \ldots, n. \qquad (7.6)$$

Example 7.2.1. Let us find the inner product matrix for the basis $\{1, t, t^2, t^3\}$ of $\mathcal{P}_3[t]$. As usual, we assume that the interval of integration for the inner product is $(0, 1)$. Because of the symmetry of the matrix and the fact that we have already calculated the 3×3 submatrix of G, we need to find g_{14}, g_{24}, g_{34}, and g_{44}. Once again, let $\mathbf{f}_1 = f_1(t) = 1$, $\mathbf{f}_2 = f_2(t) = t$, $\mathbf{f}_3 = f_3(t) = t^2$, and $\mathbf{f}_4 = f_4(t) = t^3$; then

$$g_{14} = \mathbf{f}_1 \cdot \mathbf{f}_4 = \int_0^1 f_1(t) f_4(t) \, dt = \int_0^1 t^3 dt = \tfrac{1}{4},$$

$$g_{24} = \mathbf{f}_2 \cdot \mathbf{f}_4 = \int_0^1 f_2(t) f_4(t) \, dt = \int_0^1 t^4 dt = \tfrac{1}{5}.$$

Similarly, $g_{34} = \tfrac{1}{6}$ and $g_{44} = \tfrac{1}{7}$. It follows that

$$\mathsf{G} = \begin{pmatrix} 1 & \tfrac{1}{2} & \tfrac{1}{3} & \tfrac{1}{4} \\ \tfrac{1}{2} & \tfrac{1}{3} & \tfrac{1}{4} & \tfrac{1}{5} \\ \tfrac{1}{3} & \tfrac{1}{4} & \tfrac{1}{5} & \tfrac{1}{6} \\ \tfrac{1}{4} & \tfrac{1}{5} & \tfrac{1}{6} & \tfrac{1}{7} \end{pmatrix}.$$

This matrix can be used to find the dot product of any two vectors in terms of their components in the basis $\{1, t, t^2, t^3\}$ of $\mathcal{P}_3[t]$. ∎

If $\mathbf{a}$ and $\mathbf{b}$ have components $(a_1, a_2, \ldots, a_n)$ and $(b_1, b_2, \ldots, b_n)$, then their inner product is given by

inner product in $\mathbb{R}^n$ defined in terms of the metric matrix and components in a general basis

$$\tilde{\mathsf{a}}\mathsf{G}\mathsf{b} = \begin{pmatrix} a_1 & a_2 & \cdots & a_n \end{pmatrix} \begin{pmatrix} g_{11} & g_{12} & \cdots & g_{1n} \\ g_{21} & g_{22} & \cdots & g_{2n} \\ \vdots & \vdots & & \vdots \\ g_{n1} & g_{n2} & \cdots & g_{nn} \end{pmatrix} \begin{pmatrix} b_1 \\ b_2 \\ \vdots \\ b_n \end{pmatrix}. \qquad (7.7)$$

As usual, if this expression is zero, we say that $\mathbf{a}$ and $\mathbf{b}$ are G-orthogonal. For an orthonormal basis, the inner product matrix G becomes the unit matrix[2] and we recover the usual inner product of vectors in terms of components.

With a *positive definite* inner product at hand, we can define the length of a vector as the (positive) square root of the inner product of the vector with itself. Can we define the angle as well? We can always define

length of a vector defined in terms of inner product

---

[2]Only if the inner product is positive definite.

angle defined in
terms of inner
product

$$\cos\theta \equiv \frac{\mathbf{a}\cdot\mathbf{b}}{|\mathbf{a}|\,|\mathbf{b}|} = \frac{\mathbf{a}\cdot\mathbf{b}}{\sqrt{\mathbf{a}\cdot\mathbf{a}}\,\sqrt{\mathbf{b}\cdot\mathbf{b}}}.$$

But how do we know that the ratio on the RHS is less than one? After all, a true cosine must have this property! It is an amazing fact of nature that *any* positive definite inner product has precisely this property. To show this, let $\mathbf{a}$ and $\mathbf{b}$ be two vectors in *any* vector space on which an inner product is defined. Denote the unit vector in the $\mathbf{a}$ direction by $\hat{\mathbf{e}}_a$, and construct the vector

$$\mathbf{b}' = \mathbf{b} - \underbrace{(\mathbf{b}\cdot\hat{\mathbf{e}}_a)}_{\text{a number}}\hat{\mathbf{e}}_a \qquad (7.8)$$

which is easily seen to be perpendicular to $\hat{\mathbf{e}}_a$ (and therefore to $\mathbf{a}$). If the inner product is positive definite, then

derivation of the
Schwarz inequality

$$\mathbf{b}'\cdot\mathbf{b}' \geq 0 \;\Rightarrow\; [\mathbf{b}-(\mathbf{b}\cdot\hat{\mathbf{e}}_a)\hat{\mathbf{e}}_a]\cdot[\mathbf{b}-(\mathbf{b}\cdot\hat{\mathbf{e}}_a)\hat{\mathbf{e}}_a] \geq 0$$

or

$$\underbrace{\mathbf{b}\cdot\mathbf{b}}_{=|\mathbf{b}|^2}-2\underbrace{\mathbf{b}\cdot[(\mathbf{b}\cdot\hat{\mathbf{e}}_a)\hat{\mathbf{e}}_a]}_{=(\mathbf{b}\cdot\hat{\mathbf{e}}_a)^2}+(\mathbf{b}\cdot\hat{\mathbf{e}}_a)^2\underbrace{\hat{\mathbf{e}}_a\cdot\hat{\mathbf{e}}_a}_{=1} \geq 0.$$

It follows that

$$|\mathbf{b}|^2 - (\mathbf{b}\cdot\hat{\mathbf{e}}_a)^2 \geq 0 \;\Rightarrow\; |\mathbf{b}|^2 \geq \left[\mathbf{b}\cdot\left(\frac{\mathbf{a}}{|\mathbf{a}|}\right)\right]^2$$

and

$$|\mathbf{b}|^2 \geq \left(\frac{\mathbf{b}\cdot\mathbf{a}}{|\mathbf{a}|}\right)^2 \;\Rightarrow\; |\mathbf{b}|^2|\mathbf{a}|^2 \geq (\mathbf{b}\cdot\mathbf{a})^2.$$

This is the desired inequality.

> **Box 7.2.1. (*Schwarz Inequality*).** *If $\mathbf{a}$ and $\mathbf{b}$ are two nonzero vectors of a vector space for which a positive definite inner product is defined, then*
>
> $$|\mathbf{a}|\,|\mathbf{b}| \geq |\mathbf{a}\cdot\mathbf{b}|.$$
>
> *The equality holds only if $\mathbf{b}$ is a multiple of $\mathbf{a}$.*

The last statement follows from the fact that $\mathbf{b}'\cdot\mathbf{b}' = 0$ only if $\mathbf{b}' = 0$ when the inner product is positive definite [see Equation (7.8)].

Schwarz inequality
holds in all inner
product spaces
regardless of their
dimensionality.

    The Schwarz inequality holds not only for finite-dimensional vector spaces such as $\mathbb{R}^n$ or $\mathcal{P}_n[t]$, but also for infinite-dimensional vector spaces. It is one of the most important inequalities in mathematical physics. One of its consequences is that we can actually define the angle between two nonzero vectors in $\mathbb{R}^n$ or $\mathcal{P}_n[t]$ (or any other vector space, finite or infinite, for which a positive definite inner product exists).

Example 7.2.2. What is the angle between the third and fourth vectors in the standard basis of $\mathcal{P}_3[t]$ when the interval of integration of the inner product is $(0, 1)$? All the inner products are calculated in Example 7.2.1. Therefore,

$$\cos\theta = \frac{\mathbf{f}_3 \cdot \mathbf{f}_4}{\sqrt{\mathbf{f}_3 \cdot \mathbf{f}_3}\sqrt{\mathbf{f}_4 \cdot \mathbf{f}_4}} = \frac{g_{34}}{\sqrt{g_{33}}\sqrt{g_{44}}} = \frac{1/6}{\sqrt{1/5}\sqrt{1/7}} = \frac{\sqrt{35}}{6}$$

or $\theta = 9.594°$. ■

As in the case of the plane and space, it is convenient to construct orthonormal basis vectors in $\mathbb{R}^n$. This is done by the Gram–Schmidt process which can easily be generalized. Suppose $B = \{\mathbf{a}_1, \mathbf{a}_2, \ldots, \mathbf{a}_n\}$ is a basis for $\mathbb{R}^n$. Again, to avoid complications, we assume that the inner product is Euclidean so that the inner product of every nonzero vector with itself is positive. We know how to construct three orthonormal vectors out of $\{\mathbf{a}_1, \mathbf{a}_2, \mathbf{a}_3\}$, we did that in our discussion of the space vectors. Call these new orthonormal vectors $\{\hat{\mathbf{e}}_1, \hat{\mathbf{e}}_2, \hat{\mathbf{e}}_3\}$. Now construct the vector $\mathbf{a}_4'$,

Gram–Schmidt process

$$\mathbf{a}_4' = \mathbf{a}_4 - (\mathbf{a}_4 \cdot \hat{\mathbf{e}}_1)\hat{\mathbf{e}}_1 - (\mathbf{a}_4 \cdot \hat{\mathbf{e}}_2)\hat{\mathbf{e}}_2 - (\mathbf{a}_4 \cdot \hat{\mathbf{e}}_3)\hat{\mathbf{e}}_3$$

which is obtained from $\mathbf{a}_4$ by taking away its projections along $\hat{\mathbf{e}}_1$, $\hat{\mathbf{e}}_2$, and $\hat{\mathbf{e}}_3$. Now note that

$$\hat{\mathbf{e}}_1 \cdot \mathbf{a}_4' = \hat{\mathbf{e}}_1 \cdot \mathbf{a}_4 - (\mathbf{a}_4 \cdot \hat{\mathbf{e}}_1)\underbrace{\hat{\mathbf{e}}_1 \cdot \hat{\mathbf{e}}_1}_{=1} - (\mathbf{a}_4 \cdot \hat{\mathbf{e}}_2)\underbrace{\hat{\mathbf{e}}_2 \cdot \hat{\mathbf{e}}_1}_{=0} - (\mathbf{a}_4 \cdot \hat{\mathbf{e}}_3)\underbrace{\hat{\mathbf{e}}_3 \cdot \hat{\mathbf{e}}_1}_{=0} = 0.$$

Similarly, $\hat{\mathbf{e}}_2 \cdot \mathbf{a}_4' = 0$ and $\hat{\mathbf{e}}_3 \cdot \mathbf{a}_4' = 0$; i.e., $\mathbf{a}_4'$ is orthogonal to $\hat{\mathbf{e}}_1$, $\hat{\mathbf{e}}_2$, and $\hat{\mathbf{e}}_3$. This suggests defining $\hat{\mathbf{e}}_4$ as

$$\hat{\mathbf{e}}_4 \equiv \frac{\mathbf{a}_4'}{|\mathbf{a}_4'|} = \frac{\mathbf{a}_4'}{\sqrt{\mathbf{a}_4' \cdot \mathbf{a}_4'}}.$$

This process can continue until we come up with n orthonormal vectors. This will happen only if the n vectors with which we started are linearly independent.

> **Box 7.2.2.** *If $\{\mathbf{a}_1, \mathbf{a}_2, \ldots, \mathbf{a}_n\}$ are linearly independent vectors of $\mathbb{R}^n$, then we can construct a set of n orthonormal vectors out of them by the Gram–Schmidt process.*

An orthonormal basis will be denoted by $\{\hat{\mathbf{e}}_1, \hat{\mathbf{e}}_2, \ldots, \hat{\mathbf{e}}_n\}$, where, as usual, the symbol $\hat{\mathbf{e}}$ stands for unit vectors. We can abbreviate the orthonormal property of these vectors by writing

$$\hat{\mathbf{e}}_i \cdot \hat{\mathbf{e}}_j = \begin{cases} 1 & \text{if } i = j, \\ 0 & \text{if } i \neq j. \end{cases}$$

Kronecker delta
and its use in
discussing
orthonormal
vectors

There is a symbol that shortens the above statement even further. It is called the **Kronecker delta** and denoted by δ_{ij}. It is defined by

$$\delta_{ij} = \begin{cases} 1 & \text{if } i = j, \\ 0 & \text{if } i \neq j. \end{cases} \tag{7.9}$$

Therefore, the orthonormality condition can be expressed as

$$\hat{\mathbf{e}}_i \cdot \hat{\mathbf{e}}_j = \delta_{ij}. \tag{7.10}$$

We shall see many examples of the use of the Kronecker delta in the sequel.

Transformations that leave the inner products unchanged can be obtained in exactly the same way as for the plane and the space. For A to preserve the inner product, we need to have

G-orthogonal
matrix

$$\widetilde{\mathsf{A}}\mathsf{G}\mathsf{A} = \mathsf{G}, \tag{7.11}$$

i.e., it has to be G-orthogonal. If G is the identity matrix, then A can be thought of as an n-dimensional rigid rotation and is simply called *orthogonal*; it satisfies

$$\widetilde{\mathsf{A}}\mathsf{A} = 1 \tag{7.12}$$

or

$$\begin{pmatrix} a_{11} & a_{12} & \cdots & a_{1n} \\ a_{21} & a_{22} & \cdots & a_{2n} \\ \vdots & \vdots & & \vdots \\ a_{n1} & a_{n2} & \cdots & a_{nn} \end{pmatrix} \begin{pmatrix} a_{11} & a_{21} & \cdots & a_{n1} \\ a_{12} & a_{22} & \cdots & a_{n2} \\ \vdots & \vdots & & \vdots \\ a_{1n} & a_{2n} & \cdots & a_{nn} \end{pmatrix} = \begin{pmatrix} 1 & 0 & \cdots & 0 \\ 0 & 1 & \cdots & 0 \\ \vdots & \vdots & & \vdots \\ 0 & 0 & \cdots & 1 \end{pmatrix}.$$

It should be clear from this that the columns of the matrix A, considered as vectors, have unit length and are orthogonal to other columns in the usual Euclidean inner product.

7.3 The Determinant

The determinant of an $n \times n$ matrix is obtained in terms of cofactors in exactly the same way as in the case of 3×3 matrices. The cofactors are themselves determinants of $(n-1) \times (n-1)$ matrices which can be expanded in terms of cofactors of their elements which are determinants of $(n-2) \times (n-2)$ matrices, etc. Continuing this process, we finally end up with determinants of 2×2 matrices. The determinant is also related to the inverse of a matrix [see Equations (6.60) and (6.61)]:

Theorem 7.3.1. *The matrix* A *has an inverse if and only if* $\det \mathsf{A} \neq 0$ *in which case*

$$\mathsf{A}^{-1} = \frac{1}{\det \mathsf{A}} \widetilde{\mathsf{A}} = \frac{1}{\det \mathsf{A}} \begin{pmatrix} A_{11} & A_{21} & \cdots & A_{n1} \\ A_{12} & A_{22} & \cdots & A_{n2} \\ \vdots & \vdots & & \vdots \\ A_{1n} & A_{2n} & \cdots & A_{nn} \end{pmatrix}, \tag{7.13}$$

where A_{ij} *is the cofactor of* a_{ij} *as defined in Box 6.3.2.*

Calculation of the determinant becomes extremely cumbersome when the dimension of the matrix increases beyond 4 or 5. However, there are certain properties of the determinant which may sometimes facilitate its calculation. The determinant has the following properties:

some properties of the determinant

1. To obtain the determinant of an $n \times n$ matrix, multiply each element of one row (or one column) by its cofactor and then add the results.

2. The determinant of the unit matrix is 1.

3. The determinant of a matrix is equal to the determinant of its transpose: $\det A = \det A^t$.

4. If two rows (or two columns) of a matrix are proportional (in particular, equal), the determinant of the matrix is zero.

5. If a row or column—treated as a vector in $\mathbb{R}^n$—of a matrix is multiplied by a constant, the determinant of the matrix will be multiplied by the same constant.

6. If two rows (or two columns) of a matrix are interchanged, the determinant changes sign.

7. The determinant will not change if we add to one row (or one column) a multiple of another row (or another column). The addition of rows or columns and their multiplication by numbers are to be understood as operations in $\mathbb{R}^n$.

An important relation, which we state without proof,[3] is

$$\det(AB) = \det A \det B. \qquad (7.14)$$

the determinant of a product is the product of determinants.

This, in combination with $\det 1 = 1$ and $AA^{-1} = 1$, gives

$$\det(AA^{-1}) = \det 1 \;\Rightarrow\; \det A \det(A^{-1}) = 1 \;\Rightarrow\; \det(A^{-1}) = \frac{1}{\det A}. \quad (7.15)$$

In words, the determinant of the inverse of a matrix is the inverse of its determinant.

determinant of inverse is inverse determinant

Recall that an orthogonal matrix A satisfies $AA^t = 1$. The third property of the determinant given above and (7.14) can be used to obtain

$$\det(AA^t) = \det 1 \;\Rightarrow\; (\det A)^2 = 1 \;\Rightarrow\; \det A = \pm 1. \qquad (7.16)$$

So

Box 7.3.1. *The determinant of an orthogonal matrix is either $+1$ or -1.*

---

[3]See Hassani, S. *Mathematical Physics: A Modern Introduction to Its Foundations*, Springer-Verlag, 1999, Chapters 3 and 25.

7.4   Eigenvectors and Eigenvalues

One of the most important applications of the determinant is in finding certain vectors that are not affected by transformations. As an example, consider rotation which is a linear transformation of space onto itself (or a transformation from $\mathbb{R}^3$ to $\mathbb{R}^3$). A general rotation in space is very complicated (see Example 6.2.5 and the discussion immediately preceding it), but if we can find an axis which is unaffected by the operation, then the process becomes a simple rotation about this axis.

When we say that a vector is unaffected, we mean that its direction (and not necessarily its magnitude) is unchanged. We use $n \times n$ matrices to represent transformations of $\mathbb{R}^n$. If x is a (column) vector in $\mathbb{R}^n$ whose direction is not affected by the transformation T, then we can write

$$\mathsf{T}x = \lambda x \qquad \text{or} \qquad (\mathsf{T} - \lambda 1)x = 0, \qquad (7.17)$$

where λ is a real number and we introduced the unit matrix to give meaning to the subtraction of λ from T. In Equation (7.17), x is called the **eigenvector** and λ the **eigenvalue** of the linear transformation. Since the zero vector trivially satisfies (7.17), we demand that *eigenvectors always be nonzero*. Equation (7.17) itself is called an **eigenvalue equation**; its solution involves calculating both the eigenvalues and the eigenvectors. It is clear from (7.17) that a multiple of an eigenvector is also an eigenvector (see Problem 7.6). Therefore, an eigenvalue equation (7.17) has no unique solution. By convention, we normalize eigenvectors so that their length is unity.

eigenvector, eigenvalue, and eigenvalue equation

To find the solution to (7.17), we note that the matrix $(\mathsf{T} - \lambda 1)$ must have no inverse, because if it did, then we could multiply both sides of the equation by $(\mathsf{T} - \lambda 1)^{-1}$ and obtain

$$\underbrace{(\mathsf{T} - \lambda 1)^{-1}(\mathsf{T} - \lambda 1)}_{=1}x = \underbrace{(\mathsf{T} - \lambda 1)^{-1}0}_{=0} \Rightarrow x = 0$$

a necessary condition for an eigenvalue equation to have nontrivial solutions is that the determinant of $\mathsf{T} - \lambda 1$ be zero.

which is not an acceptable solution. So, we must demand that the matrix $(\mathsf{T} - \lambda 1)$ have no inverse. This will happen only if the determinant of this matrix vanishes. So, the problem is reduced to finding those λ's which make the determinant of the matrix vanish. In other words, the eigenvalues are the solutions of the equation

$$\det(\mathsf{T} - \lambda 1) = 0. \qquad (7.18)$$

Once the eigenvalues are determined, we substitute them one by one in the matrix equation (7.17) and find the corresponding eigenvectors by solving the resulting n linear equations in n unknowns. The best way to explain this is through an example.

Example 7.4.1. Let T be a linear transformation of space (or $\mathbb{R}^3$) represented by the matrix

$$\mathsf{T} = \begin{pmatrix} 1 & 0 & 0 \\ 0 & 1 & 2 \\ 0 & 2 & 1 \end{pmatrix}.$$

The eigenvalue equation is

$$(T - \lambda 1)x = 0 \quad \text{or} \quad \left[\begin{pmatrix} 1 & 0 & 0 \\ 0 & 1 & 2 \\ 0 & 2 & 1 \end{pmatrix} - \lambda \begin{pmatrix} 1 & 0 & 0 \\ 0 & 1 & 0 \\ 0 & 0 & 1 \end{pmatrix} \right] \begin{pmatrix} x_1 \\ x_2 \\ x_3 \end{pmatrix} = \begin{pmatrix} 0 \\ 0 \\ 0 \end{pmatrix}.$$

This can also be written as

$$\begin{pmatrix} 1 - \lambda & 0 & 0 \\ 0 & 1 - \lambda & 2 \\ 0 & 2 & 1 - \lambda \end{pmatrix} \begin{pmatrix} x_1 \\ x_2 \\ x_3 \end{pmatrix} = \begin{pmatrix} 0 \\ 0 \\ 0 \end{pmatrix} \tag{7.19}$$

whose nontrivial solution is obtained by setting the determinant of the matrix equal to zero:

$$\det \begin{pmatrix} 1 - \lambda & 0 & 0 \\ 0 & 1 - \lambda & 2 \\ 0 & 2 & 1 - \lambda \end{pmatrix} = 0$$

or

$$(1 - \lambda) \det \begin{pmatrix} 1 - \lambda & 2 \\ 2 & 1 - \lambda \end{pmatrix} = (1 - \lambda) \left[(1 - \lambda)^2 - 4 \right] = 0.$$

This equation has the solutions

$$1 - \lambda = 0 \quad \text{or} \quad (1 - \lambda)^2 = 4 \implies 1 - \lambda = \pm 2.$$

It follows that there are three eigenvalues: $\lambda_1 = 1$, $\lambda_2 = -1$, and $\lambda_3 = 3$. We now find the eigenvectors corresponding to each eigenvalue.

Substituting $\lambda_1 = 1$ for λ in Equation (7.19) yields

$$\begin{pmatrix} 0 & 0 & 0 \\ 0 & 0 & 2 \\ 0 & 2 & 0 \end{pmatrix} \begin{pmatrix} x_1 \\ x_2 \\ x_3 \end{pmatrix} = \begin{pmatrix} 0 \\ 0 \\ 0 \end{pmatrix} \quad \text{or} \quad \begin{pmatrix} 0 \\ 2x_3 \\ 2x_2 \end{pmatrix} = \begin{pmatrix} 0 \\ 0 \\ 0 \end{pmatrix}.$$

It follows that $x_2 = 0 = x_3$. Therefore, the first eigenvector is

$$a_1 = \begin{pmatrix} x_1 \\ 0 \\ 0 \end{pmatrix} = x_1 \begin{pmatrix} 1 \\ 0 \\ 0 \end{pmatrix}$$

with x_1 an arbitrary real number. This arbitrariness comes from the fact that a multiple of an eigenvector is also an eigenvector. We choose $x_1 = 1$ to normalize the eigenvector to unit length. Denoting this eigenvector by e_1, we have

$$e_1 = \begin{pmatrix} 1 \\ 0 \\ 0 \end{pmatrix}.$$

To find the second eigenvector, we substitute $\lambda_2 = -1$ for λ in Equation (7.19). This gives

$$\begin{pmatrix} 2 & 0 & 0 \\ 0 & 2 & 2 \\ 0 & 2 & 2 \end{pmatrix} \begin{pmatrix} x_1 \\ x_2 \\ x_3 \end{pmatrix} = \begin{pmatrix} 0 \\ 0 \\ 0 \end{pmatrix} \quad \text{or} \quad \begin{pmatrix} 2x_1 \\ 2x_2 + 2x_3 \\ 2x_2 + 2x_3 \end{pmatrix} = \begin{pmatrix} 0 \\ 0 \\ 0 \end{pmatrix}.$$

It follows that $x_1 = 0$, and $2x_2 + 2x_3 = 0$ or $x_3 = -x_2$. Therefore, the second eigenvector is

$$\mathsf{a}_2 = \begin{pmatrix} 0 \\ x_2 \\ -x_2 \end{pmatrix} = x_2 \begin{pmatrix} 0 \\ 1 \\ -1 \end{pmatrix}$$

with x_2 arbitrary. To normalize the eigenvector, we divide it by its length.[4] This amounts to choosing $x_2 = 1/\sqrt{2}$ (see Problem 7.7). We thus have

$$\mathsf{e}_2 = \frac{1}{\sqrt{2}} \begin{pmatrix} 0 \\ 1 \\ -1 \end{pmatrix}.$$

For the third eigenvector, we substitute $\lambda_3 = 3$ in Equation (7.19) to obtain

$$\begin{pmatrix} -2 & 0 & 0 \\ 0 & -2 & 2 \\ 0 & 2 & -2 \end{pmatrix} \begin{pmatrix} x_1 \\ x_2 \\ x_3 \end{pmatrix} = \begin{pmatrix} 0 \\ 0 \\ 0 \end{pmatrix} \quad \text{or} \quad \begin{pmatrix} -2x_1 \\ -2x_2 + 2x_3 \\ 2x_2 - 2x_3 \end{pmatrix} = \begin{pmatrix} 0 \\ 0 \\ 0 \end{pmatrix}$$

or $x_1 = 0$, and $x_3 = x_2$. Therefore, the third eigenvector is

$$\mathsf{a}_3 = \begin{pmatrix} 0 \\ x_2 \\ x_2 \end{pmatrix} = x_2 \begin{pmatrix} 0 \\ 1 \\ 1 \end{pmatrix}$$

with x_2 arbitrary. To normalize the eigenvector, we divide it by its length and get

$$\mathsf{e}_3 = \frac{1}{\sqrt{2}} \begin{pmatrix} 0 \\ 1 \\ 1 \end{pmatrix}.$$

∎

The unit eigenvectors $\hat{\mathsf{e}}_1$, $\hat{\mathsf{e}}_2$, and $\hat{\mathsf{e}}_3$ of the preceding example are mutually perpendicular as the reader may easily verify. This is no accident! The matrix of that example happens to be *symmetric*, and for such matrices, we have the following general property:

> **Box 7.4.1.** *Eigenvectors of a symmetric matrix corresponding to different eigenvalues are orthogonal.*

To show this, let x and y be eigenvectors of a symmetric matrix T corresponding to eigenvalues λ and λ', respectively:

$$\mathsf{T}\mathsf{x} = \lambda\mathsf{x}, \qquad \mathsf{T}\mathsf{y} = \lambda'\mathsf{y}.$$

Multiply both sides of the first equation by $\tilde{\mathsf{y}}$ and the second by $\tilde{\mathsf{x}}$ to get

$$\tilde{\mathsf{y}}\,\mathsf{T}\mathsf{x} = \lambda\tilde{\mathsf{y}}\mathsf{x}, \qquad \tilde{\mathsf{x}}\,\mathsf{T}\mathsf{y} = \lambda'\,\tilde{\mathsf{x}}\mathsf{y}. \tag{7.20}$$

---

[4] Here we are assuming that the inner product for the calculation of length is the usual Euclidean one.

Now take the transpose of both sides of the first equation in (7.20). This gives[5]

$$(\tilde{y}\,\mathsf{T}\mathsf{x})^t = \lambda(\tilde{y}\,\mathsf{x})^t \;\Rightarrow\; \tilde{\mathsf{x}}\,\tilde{\mathsf{T}}\,\tilde{\tilde{y}} = \lambda\tilde{\mathsf{x}}\tilde{\tilde{y}}.$$

But double transposing y gives back y. Furthermore, $\tilde{\mathsf{T}} = \mathsf{T}$, because T is symmetric. So,

$$\tilde{\mathsf{x}}\,\mathsf{T}\mathsf{y} = \lambda\tilde{\mathsf{x}}\mathsf{y}.$$

Subtracting both sides of this equation from those of the second equation in (7.20), we obtain

$$0 = (\lambda - \lambda')\tilde{\mathsf{x}}\mathsf{y}.$$

By assumption, $\lambda \neq \lambda'$; so, we must have $\tilde{\mathsf{x}}\mathsf{y} = 0$, i.e., that x and y are orthogonal.

7.5 Orthogonal Polynomials

The last section generalized the two- and three-dimensional "arrows" and polynomials to higher dimensions in which many of the original properties of vectors—such as the inner product—were retained. In this section, we want to make two more generalizations which are necessary for many physical applications. The first is the introduction of a **weight function** in the definition of inner product. A weight function is a function that is positive definite[6] in the interval (a, b) of integration of the inner product. More specifically, let $\mathbf{p} = p(t)$ and $\mathbf{q} = q(t)$ be polynomials in $\mathcal{P}_n[t]$. We define their inner product as

$$\mathbf{p} \cdot \mathbf{q} = \int_a^b p(t)q(t)w(t)\,dt, \qquad (7.21)$$

where $w(t)$ is a function that is never zero or negative for $a < t < b$, and its form is usually dictated by the physical application. The reader may verify that Equation (7.21) defines a *positive definite* inner product.

The second generalization is to consider the collection of all polynomials of arbitrary degree. In other words, instead of confining ourselves to $\mathcal{P}_n[t]$ for some fixed n, we shall allow all polynomials without any restriction on their degree. Clearly, such a collection is indeed a vector space; however it does not have a finite basis. We denote this infinite-dimensional space by $\mathcal{P}^w_{(a,b)}[t]$, in which notation both the weight function and the interval of integration are included.

Given any basis for $\mathcal{P}^w_{(a,b)}[t]$, we can apply the Gram–Schmidt process on it to turn it into an orthonormal basis. Due to historical reasons, the normality is not a desirable property for the basis vectors. So, one seeks polynomials that are orthogonal, but not necessarily of unit length. Instead of normalizing the vectors, one *standardizes* them. Standardization is a rule—dictated by tradition—that fixes some of the coefficients of the polynomials. The procedure for finding these orthogonal polynomials is to start from the constant

weight function

tradition and history
"standardize" polynomials instead of "normalizing" them.

---

[5]Recall that ˜ and ^{t} mean the same thing.
[6]This just means that the function is positive and never zero.

polynomial (of degree zero) and standardize it to get the first polynomial. Next apply the standardization to the polynomial of degree one (with two unknown coefficients), and make sure that it is perpendicular to the first polynomial, where the inner product is defined by (7.21). These two requirements (standardization and perpendicularity) provide two equations and two unknowns which can be solved to find the coefficients of the second polynomial. The next polynomial has degree two with three unknown coefficients. Standardization and orthogonality to the first two polynomials provide three equations in three unknowns, the solution of which equations determines the third polynomial. This process can be continued indefinitely determining the coefficients of orthogonal polynomials up to any desired degree.

Example 7.5.1. The procedure above is best illustrated by a concrete example. The **Legendre polynomial** of degree n, denoted by $P_n(t)$, is characterized by the standardization $P_n(1) = 1$. We denote the collection of these polynomials by $\mathcal{P}^1_{(-1,1)}[t]$, indicating that the interval of integration for them is from -1 to $+1$ and that the weight function is unity. Because of standardization, we must choose $P_0(t) = 1$. The first degree polynomial is generally written as $P_1(t) = \alpha_0 + \alpha_1 t$. Standardization gives $\alpha_0 + \alpha_1 = 1$. Orthogonality to $P_0(t)$ gives

$$0 = \int_{-1}^{1} P_0(t) P_1(t) w(t) \, dt = \int_{-1}^{1} 1 \cdot (\alpha_0 + \alpha_1 t) \cdot 1 \, dt = 2\alpha_0.$$

So, $\alpha_0 = 0$ and $\alpha_1 = 1$. Therefore, $P_1(t) = t$.

For $P_2(t) = \alpha_0 + \alpha_1 t + \alpha_2 t^2$ we have (reader please verify!)

$$\begin{aligned}
\alpha_0 + \alpha_1 + \alpha_2 &= 1 && \text{(by standardization)}, \\
2\alpha_0 + 0 \cdot \alpha_1 + \tfrac{2}{3}\alpha_2 &= 0 && \text{(by orthogonality to } P_0\text{)}, \\
0 \cdot \alpha_0 + \tfrac{2}{3} \cdot \alpha_1 + 0 \cdot \alpha_2 &= 0 && \text{(by orthogonality to } P_1\text{)}.
\end{aligned}$$

The solution to these equations is $\alpha_0 = -\frac{1}{2}$, $\alpha_1 = 0$, and $\alpha_2 = -\frac{3}{2}$, so that $P_2(t) = \frac{1}{2}(3t^2 - 1)$. Other Legendre polynomials can be found analogously. ∎

By their very construction, orthogonal polynomials, which are denoted by $F_n(t)$, satisfy the following orthogonality condition:

$$\int_a^b F_n(t) F_m(t) w(t) \, dt = \begin{cases} 0 & \text{if} \quad m \neq n, \\ h_n & \text{if} \quad m = n, \end{cases} \tag{7.22}$$

where h_n is just a positive number (depending on n, of course) which is different for different types of F_n.[7] As before, let us treat these polynomials as vectors and write $\mathbf{F}_n$ for $F_n(t)$. Then using the Kronecker delta of (7.9), Equation (7.22) can be written as

$$\mathbf{F}_n \cdot \mathbf{F}_m = \begin{cases} 0 & \text{if} \quad m \neq n \\ h_n & \text{if} \quad m = n \end{cases} = h_n \delta_{mn}.$$

---

[7]There are many different *types* of orthogonal polynomials, distinguished from each other by different intervals, and different $w(t)$. Different symbols—such as $P_n(t)$, $H_n(t)$, $T_n(t)$, etc., are used for different types. We have used $F_n(t)$ to represent any one of these types in our general discussion.

In particular, $\mathbf{F}_n \cdot \mathbf{F}_n = h_n$ or $|\mathbf{F}_n|^2 = h_n$. So, the "length" of $\mathbf{F}_n$ is $\sqrt{h_n}$.

Now consider the set of all *functions* defined in the interval (a, b) any two of which give a finite result when integrated as in Equation (7.22). The reader may easily verify that this set is indeed a *vector space*. If $\mathbf{f} = f(t)$ and $\mathbf{g} = g(t)$ are two vectors in this space, then we define their inner product as

the set of all *functions* (not just polynomials) is also a vector space.

$$\mathbf{f} \cdot \mathbf{g} = \int_a^b f(t)g(t)w(t)\, dt. \tag{7.23}$$

It is clear that the $\mathbf{F}_n$ belong to this space. Furthermore, it can be shown that they form a convenient basis for the vector space. In fact, any function of the space can be written as a (infinite) linear combination of the orthogonal polynomials

$$\mathbf{f} = \sum_{n=0}^{\infty} a_n \mathbf{F}_n,$$

whose coefficients can be determined by taking the inner product of both sides with $\mathbf{F}_m$:

$$\mathbf{f} \cdot \mathbf{F}_m = \left(\sum_{n=0}^{\infty} a_n \mathbf{F}_n \right) \cdot \mathbf{F}_m = \sum_{n=0}^{\infty} a_n \mathbf{F}_n \cdot \mathbf{F}_m = a_m \mathbf{F}_m \cdot \mathbf{F}_m = h_m a_m$$

because in the last infinite sum all the terms are zero except one. We can solve this equation for a_m to obtain $a_m = \mathbf{f} \cdot \mathbf{F}_m / h_m$. Thus,

$$\mathbf{f} = \sum_{n=0}^{\infty} a_n \mathbf{F}_n \qquad \text{where} \quad a_n = \frac{\mathbf{f} \cdot \mathbf{F}_n}{h_n}. \tag{7.24}$$

In terms of functions and polynomials, we have the important result:

expansion of functions in terms of orthogonal polynomials

Theorem 7.5.2. *A function $f(t)$, defined in the interval (a, b), can be represented as an infinite sum in orthogonal polynomials given by*

$$f(t) = \sum_{n=0}^{\infty} a_n F_n(t), \qquad \text{where} \quad a_n = \frac{1}{h_n} \int_a^b f(t) F_n(t) w(t)\, dt. \tag{7.25}$$

There are a number of so-called **classical orthogonal polynomials** used in mathematical physics a number of whose properties we simply cite here. We have already mentioned Legendre polynomials for which the interval is $(-1, +1)$ and $w(x) = 1$.[8] For Legendre polynomials, $h_n = 2/(2n+2)$, i.e.,

classical orthogonal polynomials

$$\int_{-1}^{1} P_n(t) P_m(t)\, dt = \begin{cases} 0 & \text{if} \quad m \neq n \\ \\ \dfrac{2}{2n+1} & \text{if} \quad m = n \end{cases} = \frac{2}{2n+1} \delta_{mn}. \tag{7.26}$$

If the interval is $(-\infty, \infty)$ and $w(t) = e^{-t^2}$, then the resulting polynomials, denoted by $H_n(t)$, are called **Hermite polynomials**. For Hermite polynomials, we have

Hermite polynomials

---

[8] A detailed discussion of Legendre polynomials and their origin can be found in Chapter 26.

$$\int_{-\infty}^{\infty} H_n(t)H_m(t)e^{-t^2}\,dt = \begin{cases} 0 & \text{if } m \neq n \\ \\ \sqrt{\pi}\,2^n n! & \text{if } m = n \end{cases} = \sqrt{\pi}\,2^n n!\,\delta_{mn}. \quad (7.27)$$

If the interval is $(0, \infty)$ and $w(t) = t^m e^{-t}$ with m a positive integer,[9] then the resulting polynomials, denoted by $L_n^m(t)$, are called **Laguerre poly-nomials**. For Laguerre polynomials, we have

Laguerre polynomials

$$\int_0^{\infty} L_n^m(t)L_k^m(t)t^m e^{-t}\,dt = \begin{cases} 0 & \text{if } k \neq n \\ \\ \sqrt{\pi}\,(n+m)!/n! & \text{if } k = n \end{cases}$$

$$= \sqrt{\pi}\,\frac{(n+m)!}{n!}\delta_{kn}. \quad (7.28)$$

There are other (classical) orthogonal polynomials which we shall not investigate here.[10]

7.6   Systems of Linear Equations

Our discussion of determinants in Section 6.3 started with a system of two linear equations in two unknowns and led to the result that if the determinant of the matrix of coefficients is nonzero, then the inverse of this matrix exists, and the unknowns can be found conveniently using this inverse [see Equation (6.52) and Theorem 6.3.1]. This was further generalized to the case of three linear equations in three unknowns and stated in Equation (6.59). A system of n linear equations in n unknowns can be handled in the same way. We write such a system as

$$\begin{pmatrix} x_1 \\ x_2 \\ \vdots \\ x_n \end{pmatrix} = \begin{pmatrix} a_{11} & a_{12} & \cdots & a_{1n} \\ a_{21} & a_{22} & \cdots & a_{2n} \\ \vdots & \vdots & & \vdots \\ a_{n1} & a_{n2} & \cdots & a_{nn} \end{pmatrix} \begin{pmatrix} b_1 \\ b_2 \\ \vdots \\ b_n \end{pmatrix} \Rightarrow \mathsf{x} = \mathsf{Ab} \quad (7.29)$$

and note that, if $\det \mathsf{A} \neq 0$, we can calculate A^{-1} according to Box 7.3.1, and multiply both sides of (7.29) by this inverse and obtain $\mathsf{x} = \mathsf{A}^{-1}\mathsf{b}$. The case of the vanishing determinant is best treated in the context of a system of equations for which the number of unknowns is not equal to the number of equations.

The process that led to Equations (6.49) and (6.51) is called elimination,

m linear equations in n unknowns

and can be extended to m linear equations in n unknowns of the form

---

[9]Actually m need not be an integer. However, the space and scope of this book does not permit us to consider the general case.

[10]The interested reader may find Hassani, S. *Mathematical Physics: A Modern Introduction to Its Foundations*, Springer-Verlag, 1999, Chapter 7, a useful reference for all orthogonal polynomials including many derivations and proofs that we have skipped here.

$$a_{11}x_1 + a_{12}x_2 + \cdots + a_{1n}x_n = b_1,$$
$$a_{21}x_1 + a_{22}x_2 + \cdots + a_{2n}x_n = b_2,$$
$$\vdots \qquad (7.30)$$
$$a_{m1}x_1 + a_{m2}x_2 + \cdots + a_{mn}x_n = b_m.$$

We will now describe a general process known as **Gauss elimination**, for finding all solutions of the given system of linear equations. The idea is to replace the given system by a simpler system, which is equivalent to the original system in the sense that it has precisely the same solutions. For example, the degenerate equation

$$0 \cdot x_1 + 0 \cdot x_2 + \cdots + 0 \cdot x_n = b_j$$

is equivalent to $0 = b_j$, which cannot be satisfied unless b_j is zero.

In a more compact notation, we write only the ith equation, indicating its form by a sample term $a_{ij}x_j$ and the statement that the equation is to be summed over j from 1 to n by writing[11]

$$\sum_{j=1}^{n} a_{ij}x_j = b_i \qquad \text{for} \quad i = 1, 2, \ldots, m. \qquad (7.31)$$

We distinguish two cases:

1. Every $a_{i1} = 0$, i.e., all coefficients of the unknown x_1 vanish. Then, trivially, the system (7.31) is equivalent to a smaller system of m equations in the $n-1$ unknowns $x_2, \ldots, x_n$ with x_1 arbitrary for any solution of the smaller system.

2. Some $a_{i1} \neq 0$. By interchanging the first equation with another if necessary, we get an equivalent system with $a_{11} \neq 0$. Dividing the first equation by a_{11}, we then get an equivalent system in which $a_{11} = 1$. Then subtracting a_{i1} times the new first equation from each ith equation for $i = 2, \ldots, m$, we get an equivalent system of the form

$$x_1 + a'_{12}x_2 + a'_{13}x_3 + \cdots + a'_{1n}x_n = b'_1,$$
$$a'_{22}x_2 + a'_{23}x_3 + \cdots + a'_{2n}x_n = b'_2,$$
$$\vdots \qquad (7.32)$$
$$a'_{m2}x_2 + a'_{m3}x_3 + \cdots + a'_{mn}x_n = b'_m.$$

Now we apply the same procedure to the system of equations in (7.32) involving only x_2 through x_n so that x_2 will appear only in the first of these equations. If case 2 always arises, the given system is said to be **compatible**. If case 1 arises once in a while, then we may get degenerate equations of the form $0 = d_k$. If all d_k turn out to be zero, these can be ignored; if one $d_k \neq 0$, the original system (7.30) is **incompatible** (has no solutions). We summarize these findings as

Gauss elimination

compatible and incompatible systems of linear equations

---

[11] The reader may find an adequate discussion of summations and "dummy" indices in Section 9.2.

Theorem 7.6.1. *Any system (7.30) of m linear equations in n unknowns can be reduced to an equivalent system of r linear equations whose ith equation has the form*

$$x_i + c_{i,i+1}x_{i+1} + c_{i,i+2}x_{i+2} + \cdots + c_{in}x_n = d_i \qquad (7.33)$$

plus $m - r$ equations of the form $0 = d_k$.

Written out in full, Equation (7.33) looks like

$$x_1 + c_{12}x_2 + c_{13}x_3 + c_{14}x_4 + \cdots + c_{1n}x_n = d_1,$$
$$x_2 + c_{23}x_3 + c_{24}x_4 \cdots + c_{2n}x_n = d_2,$$
$$x_3 + c_{34}x_4 \cdots + c_{3n}x_n = d_3, \qquad (7.34)$$
$$\vdots$$
$$x_r + \cdots + c_{rn}x_n = d_r \qquad (r \leq m),$$

echelon form of a system of linear equations

which is said to be in **echelon** form.

Solutions of any system of the echelon form (7.34) are easily described. Consider the succession of the unknowns starting with x_n and going down to x_1. If a given x_i appears as the first variable in an equation of (7.34), then it can be written in terms of all preceding unknowns:[12]

$$x_i = d_i - c_{i,i+1}x_{i+1} - c_{i,i+2}x_{i+2} - \cdots - c_{in}x_n. \qquad (7.35)$$

If x_i does not appear as the first variable in an equation of (7.34), then it can be chosen arbitrarily. We thus have

> **Box 7.6.1.** *In the compatible case of Theorem 7.6.1, the set of all solutions of Equation (7.30) are determined as follows. The $m - r$ unknowns x_k not occurring in (7.34) can be chosen arbitrarily (they are free parameters). For any choice of these x_k's, the remaining x_i can be computed by substituting in (7.35).*

Example 7.6.2. Consider the following four linear equations in three unknowns (so $m = 4$ and $n = 3$):

$$-x_2 + 2x_3 = 1,$$
$$x_1 + x_2 - 3x_3 = 0,$$
$$-x_1 + x_2 + x_3 = -2, \qquad (7.36)$$
$$x_1 + 2x_2 - x_3 = -1.$$

The coefficient of x_1 in the first equation is zero. So, we switch this equation with one of the other equations, say the second. Then we multiply the new first equation by the negative of the coefficient of x_1 in each remaining equation and add the result

---

[12]If $r = n$, then the last equation of (7.34) will be $x_n = d_n$, and (if the set of equations is compatible) all unknowns will be determined.

to that equation to eliminate x_1. Thus, we add the new first equation to the third equation of (7.36), and subtract the new first equation from the last equation of (7.36). The result is

$$
\begin{aligned}
x_1 + x_2 - 3x_3 &= 0, \\
-x_2 + 2x_3 &= 1, \\
2x_2 - 2x_3 &= -2, \\
x_2 + 2x_3 &= -1.
\end{aligned}
\tag{7.37}
$$

To eliminate x_2 from the last two equations, multiply the second equation of (7.37) by 2 (or 1 for the last) and add it to the third (or last) equation. This will yield

$$
\begin{aligned}
x_1 + x_2 - 3x_3 &= 0, \\
-x_2 + 2x_3 &= 1, \\
4x_3 &= 0, \\
4x_3 &= 0.
\end{aligned}
\tag{7.38}
$$

Multiply the second equation by -1, divide the third equation in (7.38) by 4, and finally subtract the result from the last equation. The final result is the following echelon form:

$$
\begin{aligned}
x_1 + x_2 - 3x_3 &= 0, \\
x_2 - 2x_3 &= -1, \\
x_3 &= 0, \\
0 &= 0,
\end{aligned}
\tag{7.39}
$$

which corresponds to Equation (7.34) with $r = n = 3$. Thus, we have one equation of the form $0 = d_k$ for which d_k is zero. So, the system has a solution. To find this solution, start with the third equation of (7.39) which gives $x_3 = 0$. Substitute in the equation above it to get $x_2 = -1$, and these values in the first equation to obtain $x_1 = 1$. ■

Example 7.6.3. As another example, consider the following:

$$
\begin{aligned}
x_1 + x_2 + x_3 &= 0, \\
2x_1 - x_2 + x_3 &= -2, \\
-x_1 + 2x_2 + x_3 &= -1, \\
x_1 - 2x_2 + x_3 &= 2.
\end{aligned}
\tag{7.40}
$$

Multiply the first equation successively by -2, 1, and -1 and add it to the second, third and fourth equations. The result will be

$$
\begin{aligned}
x_1 + x_2 + x_3 &= 0, \\
-3x_2 - x_3 &= -2, \\
3x_2 + 2x_3 &= -1, \\
-3x_2 + 0 \cdot x_3 &= 2.
\end{aligned}
$$

Now divide the second equation by -3,

$$x_1 + x_2 + x_3 = 0,$$
$$x_2 + \tfrac{1}{3}x_3 = \tfrac{2}{3},$$
$$3x_2 + 2x_3 = -1, \tag{7.41}$$
$$-3x_2 + 0 \cdot x_3 = 2.$$

Multiply the second equation of (7.41) successively by -3 and $+3$ and add it to the third and last equations. This will yield

$$x_1 + x_2 + x_3 = 0,$$
$$x_2 + \tfrac{1}{3}x_3 = \tfrac{2}{3},$$
$$x_3 = -3, \tag{7.42}$$
$$x_3 = 4.$$

Subtract the third equation from the last to get

$$x_1 + x_2 + x_3 = 0,$$
$$x_2 + \tfrac{1}{3}x_3 = \tfrac{2}{3},$$
$$x_3 = -3, \tag{7.43}$$
$$0 = 7.$$

In this case, we have an equation of the form $0 = d_k$ for which $d_k = 7$. So, the system is incompatible, i.e., it has no solution. ∎

homogeneous system of linear equations

A system of linear equations (7.30) is **homogeneous** if the constants b_i on the RHS are all zero. Such a system always has a trivial solution with all the unknowns equal to zero. There may be no further solutions, but if the number of variables exceeds the number of equations, the last equation of (7.32) will always contain more than one variable at least one of which can be chosen at will. Furthermore, the inconsistent equations $0 = d_k$ can never arise for such homogeneous equations. Hence,

> **Box 7.6.2.** *A system of m homogeneous linear equations in n unknowns, with $n > m$, always has a solution in which not all the unknowns are zero.*

7.7 Problems

7.1. Show that Equation (7.4) is a linear transformation.

7.2. Verify that the operation of differentiation of any order is a linear transformation on $\mathcal{P}_n[t]$.

7.3. Show that

$$\mathsf{L} \equiv p_2(x)\frac{d^2}{dx^2} + p_1(x)\frac{d}{dx} + p_0(x)$$

is a linear operator on the space of differentiable functions.

7.4. Show that the coefficients in $\mathcal{P}_1[t]$ of the second derivative of an arbitrary polynomial $f(t) = \alpha_0 + \alpha_1 t + \alpha_2 t^2 + \alpha_3 t^3$ can be obtained by the product of the matrix of the second derivative obtained in Example 7.1.1, and the 4×1 column vector representing $f(t)$.

7.5. Express the element in the ith row and jth column of a unit matrix in terms of the Kronecker delta.

7.6. Suppose x is an eigenvector of T with eigenvalue λ. Show that, for any constant α, αx is also an eigenvector of T with the same eigenvalue.

7.7. Find the length of a_2 of Example 7.4.1 in terms of x_2. Now show that $a_2/|a_2| = \hat{e}_2$.

7.8. Show that the rotation of the plane affects *all* vectors in the plane. Hint: Try to find an eigenvector of the 2×2 rotation matrix (6.24).

7.9. Find the eigenvalues and normalized (unit length) eigenvectors of the following matrices. In cases where the matrix is symmetric, verify directly that its eigenvectors corresponding to different eigenvalues are orthogonal.

(a) $\begin{pmatrix} 1 & 2 \\ 2 & -2 \end{pmatrix}$.   (b) $\begin{pmatrix} 2 & 4 \\ 5 & 3 \end{pmatrix}$.   (c) $\begin{pmatrix} 3 & 2 \\ 2 & 3 \end{pmatrix}$.

(d) $\begin{pmatrix} 1 & 1 & 0 \\ 1 & 0 & 1 \\ 0 & 1 & 1 \end{pmatrix}$.   (e) $\begin{pmatrix} 2 & 0 & 0 \\ 0 & 1 & 1 \\ 0 & 1 & 1 \end{pmatrix}$.   (f) $\begin{pmatrix} 1 & 1 & 1 \\ 1 & 1 & 1 \\ 1 & 1 & 1 \end{pmatrix}$.

7.10. Show that Box 7.4.1 is not necessarily true for a general inner product with matrix G. However, if G and T commute (i.e., if GT $=$ TG), then Box 7.4.1 holds. Hint: Follow the argument after Box 7.4.1 and see how far you can proceed.

7.11. Show that the inner product defined in Equation (7.21) is indeed a *positive definite* inner product.

7.12. Find the fourth Legendre polynomial using the results of Example 7.5.1.

7.13. Find the first three Hermite polynomials using the standardization (or normalization) Equation (7.27).

7.14. The volume element of a four-dimensional Euclidean space with Cartesian coordinates x, y, z, and w is $dxdydzdw$. In any other coordinate system, it is given by a 4-dimensional generalization of the Jacobian (6.66)
(a) Write this Jacobian for a general transformation to coordinates s, t, u, and v where x, y, z, and w are functions of these new coordinates.
(b) Now consider the 4-dimensional spherical coordinates:

$$x = r \sin\mu \sin\theta \cos\varphi$$
$$y = r \sin\mu \sin\theta \sin\varphi$$
$$z = r \sin\mu \cos\theta$$
$$w = r \cos\mu$$

and calculate the 4-dimensional Jacobian to find the volume element of a 4-dimensional sphere.

(c) With $0 \le \varphi \le 2\pi$, $0 \le \theta \le \pi$, $0 \le \mu \le \pi$, find the volume of a 4-sphere of radius a.

7.15. Determine the r of Equation (7.34) for each of the following systems of linear equations and whether or not the system is compatible. If the system is compatible, find a solution for it.

(a)
$$2x - y - 4z = 1,$$
$$x + 2y + 2z = 0,$$
$$-x - y + 6z = 3.$$

(b)
$$x + y + z = -1,$$
$$2x - y + 2z = -5,$$
$$3x + 3y + z = 1.$$

(c)
$$x + y + z = 2,$$
$$2x - y + 2z = -2,$$
$$3x + y - z = 4.$$

(d)
$$2x + y - 2z = 2,$$
$$3x - y - 4z = -1,$$
$$3x + 4y - 2z = 7.$$

(e)
$$3x + 2y = 7,$$
$$x + y + z = 6,$$
$$5x + 4y + 2z = 19,$$
$$x - 2y = -5.$$

(f)
$$x + 5y - z = 2,$$
$$2x + y + 3z = -1,$$
$$-x + 3y + 2z = -3,$$
$$3x + 2y - z = 4.$$

Chapter 8

Vectors in Relativity

One of the most rewarding applications of vectors is to relativity. The **special theory of relativity** (STR) was a direct consequence of Maxwell's equations, which summarize the entire theory of electromagnetism (see Section 15.4). These equations predict mathematically that there must exist electromagnetic (EM) waves which travel at the speed of light in *empty space*. This speed c is found in terms of purely electric and magnetic measurements:

$$c = \frac{1}{\sqrt{\mu_0 \epsilon_0}} = \frac{1}{\sqrt{(4\pi \times 10^{-7})(8.854 \times 10^{-12})}} = 2.998 \times 10^8 \text{ m/s},$$

where $\epsilon_0 = 1/4\pi k_e$ and $\mu_0 = 4\pi k_m$, with k_e and k_m the electric and magnetic constants introduced in Chapter 1.

Imagine two laboratories on two spaceships, S_1 and S_2, with S_1 behind (and moving towards) S_2 at $0.9c$ relative to S_2. The physicists on S_1 perform electric and magnetic experiments, measure ϵ_0 and μ_0, and conclude that EM waves travel at 300,000 km/s in empty space. The physicists on S_2 also perform electric and magnetic experiments, measure ϵ_0 and μ_0, and also conclude that EM waves travel at 300,000 km/s in empty space. Now a physicist on S_1 takes a flashlight and sends a beam of light in the forward direction in empty space. The consequence of Maxwell's equations is that the physicists on S_2, although seeing S_1 moving towards them at $0.9c$ and the light beam moving away from S_1 at c, conclude that the speed of the light beam is c and not $1.9c$, as expected from the Newtonian *law of addition of velocities*.

To appreciate the strange consequence of Maxwell's equations, consider the following example: A train moving at 30 m/s and a passenger throwing a ball in the forward direction with a speed of 20 m/s. A ground observer measures the speed of the ball to be $30 + 20 = 50$ m/s: velocities add. Here is another familiar example: A car moves at 75 mph on a highway on which your car is moving at 50 mph. The speed of the fast car *relative* to you is 25 mph. You speed up to 70 mph. Then the other car appears to have "slowed down," because, now you measure its speed relative to you to be only 5 mph. Go to

law of addition
of velocities

outer space, let someone in your spaceship fire a bullet moving at 500 mph. Increase your speed to 450 mph, the bullet appears to be moving at 50 mph away from you. Increase your speed by another 100 mph. You catch up with the bullet, and if you decrease your speed by 50 mph, the bullet appears *stationary* relative to you.

Now shoot a beam of light forward, and once the beam leaves your flashlight, accelerate your spaceship to a speed of 299,000 km/s. Measure the speed of the light beam. It is still 300,000 km/s, and *not* 1000 km/s, as intuitively expected! Maxwell's equations defy intuition, and the (STR), which is entirely based on these equations is extremely counter-intuitive. Let us summarize these observations:

> **Box 8.0.1. (Principle of Relativity)** *Every time you detect an electromagnetic wave, it moves at the rate of 300,000 km per second in vacuum, regardless of the motion of its source or its detector. Speed of light in vacuum is a universal constant.*

the
Michelson–Morley
clock

An immediate consequence of the principle of relativity is the fact that time is observer-dependent. As Einstein said "Time is something that is measured by clocks." So, let us look at the effect of motion on clocks. The clock best suited for this investigation is the "arm" of the Michelson–Morley apparatus shown in Figure 8.1. It consists of a source S of light, or electromagnetic waves, and a mirror M. The distance between S and M is L. Therefore, it takes light $\Delta\tau_{\text{tick}} \equiv 2L/c$ to go from S to M and back. If we place a light sensitive "ticker" at S, the clock will tick every $\Delta\tau_{\text{tick}}$ second. We call such a clock a Michelson–Morley clock, or an **MM clock**, and $\Delta\tau_{\text{tick}}$ the **proper** tick of the MM clock. $\Delta\tau_{\text{tick}}$ is the tick measured by an observer for whom the clock is at rest, or for whom the beginning and the end of a tick occur *at the same location.*

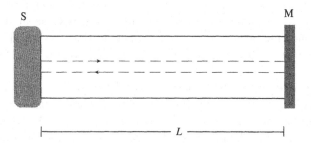

Figure 8.1: A Michelson–Morley clock. A "tick" of this clock occurs when the light signal makes a round trip along the length L.

8.1   Proper and Coordinate Time

An MM clock is placed on a train and observed by two observers, O (on the ground) and O' (on the train) moving to the right of O. Consider three events: The emission of a light beam at S, its reflection at M, and its reception at S. These three events constitute one tick. Let us denote them by E_1, E_2, and E_3, respectively. How does O' see the ticking of the clock? The clock is sitting right beside her, and she observes the whole process of ticking as the light going straight up and coming straight down. She concludes that her clock's ticks are $\Delta\tau_{\text{tick}}$ long.

Now, let us see how O perceives the succession of these three events. Since the clock is moving to the right, the light signal that leaves S will reach M only after M has moved to the right. Thus, to O, the events E_1 and E_2 are separated not only by a vertical distance, but also by a horizontal distance (see Figure 8.2). Since the *speed of light is the same for all observers*, O concludes that it takes light more than $2L/c$ to travel $\overline{E_1E_2}$ and $\overline{E_2E_3}$. Therefore, he concludes that the *clock on the train must tick slower!*

moving clocks slow down.

We can quantify the above statement by referring to the triangle E_1AE_2 of Figure 8.2. Pythagoras' theorem implies

$$\left(\overline{E_1E_2}\right)^2 = \left(\overline{E_1A}\right)^2 + \left(\overline{AE_2}\right)^2.$$

Let the speed of the train be v and the light beam's travel time from S to M be δt according to O. Then $\overline{E_1A} = v\delta t$ and $\overline{E_1E_2} = c\delta t$ with c the (universal) speed of light. Putting all of this in the above equation gives

$$(c\delta t)^2 = (v\delta t)^2 + L^2 \;\Rightarrow\; c^2(\delta t)^2 = v^2(\delta t)^2 + L^2, \tag{8.1}$$

or

$$(\delta t)^2 - \frac{v^2}{c^2}(\delta t)^2 = \frac{L^2}{c^2}, \;\Rightarrow\; (\delta t)^2\left(1 - \frac{v^2}{c^2}\right) = \frac{L^2}{c^2}.$$

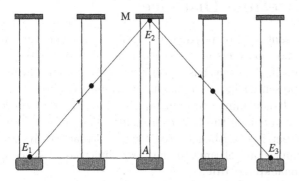

Figure 8.2: A moving Michelson–Morley clock. The path of light (represented by a black dot) is not a vertical line but a slanted one due to the motion of M.

This yields

$$(\delta t)^2 = \frac{L^2/c^2}{1 - v^2/c^2} \;\Rightarrow\; \delta t = \frac{L/c}{\sqrt{1 - (v/c)^2}}.$$

Let us denote by Δt_{tick} the duration of the light's round trip as seen by O. Then

$$\Delta t_{\text{tick}} = \frac{2L/c}{\sqrt{1 - (v/c)^2}} = \frac{\Delta \tau_{\text{tick}}}{\sqrt{1 - (v/c)^2}}. \tag{8.2}$$

motion does not affect transverse lengths.

In deriving this equation, we have tacitly assumed that *motion does not affect transverse lengths*. Thus the length of the MM clock does not change because it is perpendicular to the direction of motion. To see this, consider the distance between two wheels of a train, and suppose that this distance shrinks[1] due to its motion as seen by a ground observer. This means that the wheels will fall *between the rails*. On the other hand, the engineer of the train sees the rail moving and concludes that the distance between the rails shrink; i.e., that the wheels fall *outside the rails*. This contradicts the previous conclusion. Thus, the length perpendicular to the direction of motion must not change.

Although Equation (8.2) is derived for a single tick, it really applies to all time intervals, because any such interval is a multiple of a single tick. We now rewrite Equation (8.2) without the subscript "tick," realizing that $\Delta \tau$ is the proper time between *any two events*, i.e., the time interval between the two events measured by a clock that is present at both events:

relation between proper time and coordinate time

$$\Delta t = \frac{\Delta \tau}{\sqrt{1 - (v/c)^2}}. \tag{8.3}$$

$\Delta \tau$ can also be defined as the time measured by an observer for whom the two events occur at the same *spatial* point. Δt, called the **coordinate time**, is the time measured by another observer, moving relative to the first one with speed v, for whom the two events occur at two different spatial points.

8.2   Spacetime Distance

The most elegant way of relating an event's space and time properties as described by two observers is to use geometry. We start with the description of the event itself. An event has a position and an instant of time. Therefore, it can be represented by a set of four coordinates: three for position and one for time. It is common to multiply the time t by c (to make a distance out of it) and put it as the first coordinate. Thus in Cartesian coordinate

spacetime introduced

system, an event is described by (ct, x, y, z). Geometrically, we have added the extra "dimension" of time to the three-dimensional space to create the four-dimensional **spacetime**.

At the heart of any geometry is the distance between two nearby points, and how it is written in terms of the coordinates of the points. Euclidean

---

[1] The same argument applies to the case where the distance expands.

geometry started without coordinates, with the notion of the distance between two points being "evident." In fact, we *use* the properties of Euclidean distance (such as the Pythagoras' theorem involving three distances corresponding to the three sides of a right triangle) to show that the distance between two points whose Cartesian coordinates differ by $(\Delta x, \Delta y, \Delta z)$ is $\sqrt{(\Delta x)^2 + (\Delta y)^2 + (\Delta z)^2}$.

geometry and distance formula

In the case of the spacetime geometry, we have *started* with coordinates. Now we have to find a distance formula in terms of the difference between coordinates of two events. We get some clues from Euclidean distance as expressed in terms of coordinates. The first clue is that distance is observer-independent: If observer O uses his Cartesian coordinate system to label point P_1 by (x_1, y_1, z_1) and P_2 by (x_2, y_2, z_2), and finds

$$(\overline{P_1 P_2})_O \equiv \Delta r = \sqrt{(x_2 - x_1)^2 + (y_2 - y_1)^2 + (z_2 - z_1)^2},$$

and if observer O' uses her Cartesian coordinate system to label point P_1 by (x_1', y_1', z_1') and P_2 by (x_2', y_2', z_2'), and finds

$$(\overline{P_1 P_2})_{O'} \equiv \Delta r' = \sqrt{(x_2' - x_1')^2 + (y_2' - y_1')^2 + (z_2' - z_1')^2},$$

then $\Delta r' = \Delta r$. The second clue is that if P_1 and P_2 lie along a single axis of an observer, then the distance is the (absolute value of the) difference between the coordinates of P_1 and P_2.

Now consider two *events* E_1 and E_2, which occur at the same *spatial location* according to O', with E_2 happening after E_1. This means that O' (his clock) is present at both events, i.e., that E_1 and E_2 lie along the time axis of O', and that O' is measuring the *proper time* interval between the two events: $\Delta \tau = t_2' - t_1'$. By the second clue above, $c\Delta \tau = c(t_2' - t_1')$ is the distance we are looking for (again we multiply by c to make a distance out of it). We introduce the notation $\Delta s \equiv c\Delta \tau$ and call Δs the **spacetime distance** or the **invariant interval** between the two events.

Another observer O assigns spacetime coordinates (ct_1, x_1, y_1, z_1) to E_1 and (ct_2, x_2, y_2, z_2) to E_2. Now the spatial separation between E_1 and E_2 according to O is

$$\sqrt{(x_2 - x_1)^2 + (y_2 - y_1)^2 + (z_2 - z_1)^2},$$

and since O' is at E_1 when it happens and at E_2 when *it* happens, this equation is precisely the distance that O' travels in time $t_2 - t_1$ with respect to O. Therefore, the speed of O' relative to O is

$$v = \frac{\sqrt{(x_2 - x_1)^2 + (y_2 - y_1)^2 + (z_2 - z_1)^2}}{t_2 - t_1},$$

or

$$v^2 = \frac{(x_2 - x_1)^2 + (y_2 - y_1)^2 + (z_2 - z_1)^2}{(t_2 - t_1)^2}.$$

Up to this point, we have not used any physics (except for the definition of speed). Now comes the crucial final step. Equation (8.3) (which is a direct result of Box 8.0.1) can now be used to find the expression of Δs in terms of coordinate differences. Equation (8.3) implies that

$$c\Delta t = \frac{c\Delta\tau}{\sqrt{1-(v/c)^2}} = \frac{\Delta s}{\sqrt{1-(v/c)^2}},$$

or

$$\Delta s = c\Delta t\sqrt{1-(v/c)^2} = c(t_2-t_1)\sqrt{1-v^2/c^2}$$
$$= \sqrt{c^2(t_2-t_1)^2 - v^2(t_2-t_1)^2}.$$

spacetime distance   Substituting the expression for v^2 above, we get

$$\Delta s = \sqrt{c^2(t_2-t_1)^2 - (x_2-x_1)^2 - (y_2-y_1)^2 - (z_2-z_1)^2}.$$

We rewrite this important formula as

$$(\Delta s)^2 = (c\Delta\tau)^2 = c^2(\Delta t)^2 - (\Delta x)^2 - (\Delta y)^2 - (\Delta z)^2. \qquad (8.4)$$

Let's emphasize the significance of this equation: If observer O uses his Cartesian coordinate system to label event E_1 by (ct_1, x_1, y_1, z_1) and E_2 by (ct_2, x_2, y_2, z_2), and finds

$$(\Delta s)^2 = c^2(t_2-t_1)^2 - (x_2-x_1)^2 - (y_2-y_1)^2 - (z_2-z_1)^2,$$

and if observer O' uses her Cartesian coordinate system to label event E_1 by $(ct_1', x_1', y_1', z_1')$ and E_2 by $(ct_2', x_2', y_2,' z_2)$, and finds

$$(\Delta s')^2 = c^2(t_2'-t_1')^2 - (x_2'-x_1')^2 - (y_2'-y_1')^2 - (z_2'-z_1')^2,$$

then $(\Delta s')^2 = (\Delta s)^2$. Thus, although events are coordinatized differently by different observers, the *spacetime distance* between two events is universal. In contrast to Newtonian physics, neither the time interval nor the spatial distance between two events is universal in relativity.

Example 8.2.1. Observer O spots a light beam (event E_1) at (x_1, y_1, z_1) at time t_1. A little later he finds the beam (event E_2) at (x_2, y_2, z_2) at time t_2. What is the spacetime interval for this light beam (i.e., for the two events E_1 and E_2)?

zero spacetime
distance for two
different events

Since light travels from (x_1, y_1, z_1) to (x_2, y_2, z_2) with speed c, we have

$$\sqrt{(x_2-x_1)^2 + (y_2-y_1)^2 + (z_2-z_1)^2} = c(t_2-t_1).$$

Therefore,

$$(\Delta s)^2 = c^2(t_2-t_1)^2 - (x_2-x_1)^2 - (y_2-y_1)^2 - (z_2-z_1)^2 = 0,$$

which holds for any light signal, as the two events above are quite general. Thus the spacetime distance between two *different* events which can be connected by a light signal is zero. This is in contrast to the Euclidean case where two different points *always* have a nonzero distance between them. ∎

8.3   Lorentz Transformation

Because of the intuitiveness of the concept of distance in Euclidean geometry, it is not essential to know how the coordinates of a point in one coordinate system (CS) are related to the coordinates of that same point in another CS. This transformation was found long after the maturity of the Euclidean geometry [see Section 6.1.3 and especially Equation (6.22) for a discussion of the two-dimensional version of coordinate transformation], and it was based entirely on the expression for the distance between two points in terms of the coordinates of those points.

In spacetime geometry such a transformation is indispensable due to the counter-intuitive properties of the invariant interval (see Example 8.2.1 above). And while in Euclidean geometry, one can picture different coordinate systems and how they relate to one another (see Figure 6.7, for example), spacetime geometry does not readily allow such a direct pictorial representation without some preliminary *algebraic* discussion.

Let $\mathbf{r}_1 = (ct_1, x_1, y_1, z_1)$ and $\mathbf{r}_2 = (ct_2, x_2, y_2, z_2)$ be the spacetime "position vectors" of two events E_1 and E_2 relative to a coordinate system O. Construct the difference

$$\Delta\mathbf{r} = \mathbf{r}_2 - \mathbf{r}_1 = (ct_2 - ct_1, x_2 - x_1, y_2 - y_1, z_2 - z_1),$$

and define the square of the "length" of this vector to be $(\Delta s)^2$. In fact, this is generalized for any four-dimensional vector. But first, let's introduce a notation.

A spacetime vector has the form $\mathbf{a} = (a_0, a_1, a_2, a_3)$, which is usually called a **four-vector** or a **4-vector**.[2] It is also denoted by $(a_0, \vec{a})$ where $\vec{a} \equiv (a_1, a_2, a_3)$ is the *space part* (or the *3-vector part*) of the 4-vector. A primary example of a four-vector is $\mathbf{r} = (ct, x, y, z) \equiv (ct, \vec{r})$. The generalization mentioned above defines the square of the length of $\mathbf{a}$ (or the inner product of $\mathbf{a}$ with itself) as

four-vectors introduced

$$\mathbf{a} \cdot \mathbf{a} \equiv a_0^2 - a_1^2 - a_2^2 - a_3^2 \equiv a_0^2 - \vec{a} \cdot \vec{a} = a_0^2 - |\vec{a}|^2. \qquad (8.5)$$

Then it is easy (see Problem 8.1) to show that the inner product of any *two* vectors must be given by

$$\mathbf{a} \cdot \mathbf{b} \equiv a_0 b_0 - a_1 b_1 - a_2 b_2 - a_3 b_3 = a_0 b_0 - \vec{a} \cdot \vec{b}. \qquad (8.6)$$

In matrix form this can be written as

$$\mathbf{a} \cdot \mathbf{b} = \begin{pmatrix} a_0 & a_1 & a_2 & a_3 \end{pmatrix} \begin{pmatrix} 1 & 0 & 0 & 0 \\ 0 & -1 & 0 & 0 \\ 0 & 0 & -1 & 0 \\ 0 & 0 & 0 & -1 \end{pmatrix} \begin{pmatrix} b_0 \\ b_1 \\ b_2 \\ b_3 \end{pmatrix}, \qquad (8.7)$$

---

[2]Note that the first component of $\mathbf{a}$ has zero as an index, and is called the time component. This is common in relativity.

or

$$\mathbf{a} \cdot \mathbf{b} = \widetilde{\mathbf{a}} \eta \mathbf{b} \qquad \text{where} \qquad \eta = \begin{pmatrix} 1 & 0 & 0 & 0 \\ 0 & -1 & 0 & 0 \\ 0 & 0 & -1 & 0 \\ 0 & 0 & 0 & -1 \end{pmatrix}, \tag{8.8}$$

and $\widetilde{\mathbf{a}}$ and $\mathbf{b}$ are the row and column vectors in Equation (8.7).

general Lorentz transformation   A linear transformation that leaves the inner product of Equation (8.8)—and therefore the spacetime length Δs—invariant is called a **Lorentz transformation**. By Equation (7.11), such a transformation Λ—which is a 4×4 matrix—satisfies

$$\widetilde{\Lambda} \eta \Lambda = \eta. \tag{8.9}$$

The study of the general structure of Lorentz transformations is beyond the scope of this book. Here we shall confine ourselves to the Lorentz transformations in two dimensions, in which the third and fourth components of vectors are ignored. This means that vectors are of the form $\mathbf{a} = (a_0, a_1)$, $\mathbf{b} = (b_0, b_1)$, the inner product is of the form $\mathbf{a} \cdot \mathbf{b} \equiv a_0 b_0 - a_1 b_1$, and the matrix η reduces to

$$\eta = \begin{pmatrix} 1 & 0 \\ 0 & -1 \end{pmatrix}.$$

In addition, the Lorentz transformations become 2×2 matrices.

Let $\Lambda = \begin{pmatrix} a_{11} & a_{12} \\ a_{21} & a_{22} \end{pmatrix}$ be a two-dimensional Lorentz transformation that acts on 2-vectors in O to give the corresponding 2-vectors in O'. Then Λ must satisfy Equation (8.9) or

$$\begin{pmatrix} a_{11} & a_{21} \\ a_{12} & a_{22} \end{pmatrix} \begin{pmatrix} 1 & 0 \\ 0 & -1 \end{pmatrix} \begin{pmatrix} a_{11} & a_{12} \\ a_{21} & a_{22} \end{pmatrix} = \begin{pmatrix} 1 & 0 \\ 0 & -1 \end{pmatrix}, \tag{8.10}$$

which is equivalent to the following three equations [see (6.21) for a guide]:

$$a_{11}^2 - a_{21}^2 = 1, \qquad a_{11} a_{12} - a_{21} a_{22} = 0, \qquad a_{12}^2 - a_{22}^2 = -1. \tag{8.11}$$

As in the case of rotations (see Section 6.1.3), we can conclude that

$$a_{22}^2 = a_{11}^2, \qquad a_{12}^2 = a_{21}^2, \qquad a_{12}^2 = a_{11}^2 - 1. \tag{8.12}$$

So, all parameters are once again given in terms of a_{11}.

To determine a_{11}, consider the 2-vector $(c\Delta t, \Delta x)$, the difference between the time and position of two events in O. This 2-vector is represented by $(c\Delta t', \Delta x')$ in O', and, by the definition of the Lorentz transformations,

$$\begin{pmatrix} c\Delta t' \\ \Delta x' \end{pmatrix} = \begin{pmatrix} a_{11} & a_{12} \\ a_{21} & a_{22} \end{pmatrix} \begin{pmatrix} c\Delta t \\ \Delta x \end{pmatrix}. \tag{8.13}$$

Now suppose that $\Delta x = 0$, i.e., that the two events occur at the same location. Then O is measuring the *proper time*, so that $\Delta t = \Delta \tau$. From Equation (8.13), we also have $c\Delta t' = a_{11}c\Delta t$ or $\Delta t' = a_{11}\Delta\tau$. Comparison with Equation (8.3) yields

$$a_{11} = \frac{1}{\sqrt{1 - (v/c)^2}}.$$

Introducing the two symbols $\beta \equiv v/c$ and $\gamma = 1/\sqrt{1 - (v/c)^2}$, we obtain

$$a_{11} = \frac{1}{\sqrt{1 - \beta^2}} \equiv \gamma. \tag{8.14}$$

The rest of the matrix elements can now be found. The first equation in (8.12) gives $a_{22} = \pm\gamma$. To choose the correct sign for a_{22}, note that if O and O' are not moving relative to one another, the coordinates do not change. Therefore Λ must be the unit matrix. So, $a_{22} = 1$ when $v = 0$. This can happen only if $a_{22} = +\gamma$. The second equation in (8.12) now gives $a_{12} = a_{21}$; and the third equation yields

$$a_{12}^2 = \gamma^2 - 1 = \frac{1}{1 - \beta^2} - 1 = \frac{\beta^2}{1 - \beta^2} = \beta^2\gamma^2 \ \Rightarrow\ a_{12} = \pm\beta\gamma.$$

The ambiguity in the sign comes from the choice we have for the direction of motion. We absorb this choice of sign in β, and write

$$\Lambda = \begin{pmatrix} \gamma & \gamma\beta \\ \gamma\beta & \gamma \end{pmatrix}. \tag{8.15}$$

For the important case of spacetime "position" vector (ct, x), this yields

$$ct' = \gamma\,(ct + \beta x),$$
$$x' = \gamma(x + \beta ct). \tag{8.16}$$

Lorentz transformation in two spacetime dimensions

β is positive (negative) when observer O—who uses (ct, x) for events—travels in the positive (negative) direction of O'—who uses primed coordinates. Equation (8.16) displays the celebrated Lorentz transformations in two spacetime dimensions.

Example 8.3.1. Emmy (observer O) is riding a train and she is standing in the middle of one of the cars of length L at the two ends of which are two firecrackers that explode simultaneously. Karl (observer O') is standing on the platform watching Emmy go by with speed β. Time zero *for both* coincides with the moment that Emmy passes by Karl. Suppose that the simultaneous explosion of the two forecrackers (according to Emmy) also takes place at $t = 0$. We want to see how all this appears to Karl.

Assume that Emmy and Karl are located at their respective origins. Let the front firecracker be labeled as 1 and the back as 2. Then the front and back firecrackers have coordinates $(0, L/2)$ and $(0, -L/2)$, respectively, in Emmy's RF. Karl, on the other hand, measures the coordinates of the firecrackers as

$$ct_1' = \gamma\beta L/2, \quad x_1' = \gamma L/2 \quad ct_2' = \gamma(-\beta L/2), \quad x_2' = \gamma(-L/2)$$

from Equation (8.16). This shows that, for Karl, the back firecracker occurs first. In fact, it occurs *before* Emmy reaches him (at time $t' = 0$). The time difference between the two events is

$$\Delta t' = t_1' - t_2' = \gamma\beta L/c.$$

Take L to be 30 m. Then, for the time difference to be a mere one second, we must have

$$30\gamma\beta = 3 \times 10^8 \quad \text{or} \quad \frac{\beta}{\sqrt{1 - \beta^2}} = 10^7,$$

giving $\beta = 0.999999999999995$, awfully close to the speed of light!

On the other hand, if L is a typical interstellar distance of say 10 light years, then

$$\gamma\beta = \frac{\Delta t'}{10}$$

with $\Delta t'$ measured in years. For a time difference of one hour, we have $\gamma\beta = 1.14 \times 10^{-5}$, yielding $\beta = 1.14 \times 10^{-5}$, or $v = 3425$ m/s, an easily attainable speed. ∎

Example 8.3.2. Observer O moves in the positive space direction of observer O' at speed v (or $\beta = v/c$). A particle moves at speed β_p in the positive space direction of O. What is β_p', the speed of the particle relative to O'?

The definition of speed is distance between two events divided by time interval between those events: spotting of the particle at a point in space and an instant in time (first event), and spotting the particle at a nearby point a little later (second event). For example, observer O assigns the coordinates (ct, x) to the first event and $(ct + c\Delta t, x + \Delta x)$ to the second event, and concludes that the (dimensionless) speed of the particle is $\beta_p = \Delta x/(c\Delta t)$.

Similarly, observer O' assigns the coordinates (ct', x') to the first event and $(ct' + c\Delta t', x' + \Delta x')$ to the second event, and concludes that the speed of the particle is $\beta_p' = \Delta x'/(c\Delta t')$, where $\Delta x'$ and $c\Delta t'$ are related to Δx and $c\Delta t$ via the Lorentz transformation. Using Equation (8.16), we find

$$\beta_p' = \frac{\Delta x'}{c\Delta t'} = \frac{\gamma(\Delta x + \beta c\Delta t)}{\gamma(c\Delta t + \beta\Delta x)},$$

relativistic law of addition of velocities

dividing the numerator and denominator by $c\Delta t$, we get

$$\beta_p' = \frac{\beta_p + \beta}{1 + \beta\beta_p}, \tag{8.17}$$

which is called the **relativistic law of addition of velocities**.

One can show that if $0 < \beta_p < 1$ and $0 < \beta < 1$, then $0 < \beta_p' < 1$. So, it is impossible to add two velocities close to light speed and get a velocity larger than light speed. Furthermore, if the particle happens to be a photon (or a light beam), then $\beta_p = 1$ and

$$\beta_p' = \frac{1 + \beta}{1 + \beta} = 1,$$

verifying the universality of the speed of light, the starting point of relativity theory! ∎

In many situations, an observer in *three dimensions* moves along the x-axis. Then, the y and z coordinates of events—being perpendicular to the direction of motion—do not change. This suggests a slightly more general Lorentz transformation than (8.16):

$$
\begin{aligned}
ct' &= \gamma \left(ct + \beta x\right), \\
x' &= \gamma(x + \beta ct), \\
y' &= y, \\
z' &= z.
\end{aligned}
\tag{8.18}
$$

If an object moves in the xy-plane of an observer O with a velocity whose components are (v_x, v_y), then the same object moves in the $x'y'$-plane of another observer O' with a velocity whose components are

$$
\begin{aligned}
v_{x'} &= \frac{dx'}{dt'} = \frac{\gamma(dx + \beta c\, dt)}{\gamma \left(dt + \beta dx/c\right)} = \frac{v_x + \beta c}{1 + \beta v_x/c}, \\
v_{y'} &= \frac{dy'}{dt'} = \frac{dy}{\gamma \left(dt + \beta dx/c\right)} = \frac{v_y}{\gamma(1 + \beta v_x/c)},
\end{aligned}
\tag{8.19}
$$

where β is the velocity of O relative to O'. In particular, if the object is light and the angle it makes with the x-axis is α, then $v_x = c\cos\alpha$, $v_y = c\sin\alpha$, $v_{x'} = c\cos\alpha'$ and $v_{y'} = c\sin\alpha'$, and the equations above yield

$$
\begin{aligned}
\cos\alpha' &= \frac{\cos\alpha + \beta}{1 + \beta\cos\alpha}, \\
\sin\alpha' &= \frac{\sin\alpha}{\gamma(1 + \beta\cos\alpha)}.
\end{aligned}
\tag{8.20}
$$

Now suppose that an observer O carries an EM radiation source which radiates uniformly in all directions. If β is very close to 1, then (8.20) implies that $\cos\alpha' \to 1$ (and of course, $\sin\alpha' \to 0$), *regardless of* α. Thus, an ultrarelativistic source of EM wave radiates (almost) only in the forward direction.

an ultrarelativistic source radiates only in the forward direction.

8.4   Four-Velocity and Four-Momentum

In Newtonian mechanics velocity is defined as the derivative of the position vector with respect to time. In terms of (Cartesian) coordinates, an observer O locates the object in motion by assigning it the coordinates (x, y, z), and differentiates these coordinates with respect to (the universal) time t to get the velocity of the object: $\vec{v} = (\dot{x}, \dot{y}, \dot{z})$.

In relativity, the "position vector" is $\mathbf{r} = (ct, x, y, z) \equiv (ct, \vec{r})$, and there is no universal time. However, each moving object has a proper time (measured by a clock carried by the object), which *is* universal in the sense that all observers measure it to be the same [see Equation (8.4) and the comments after it]. Therefore, it is natural to define the dimensionless **four-velocity** as

$$\mathbf{u} \equiv \frac{d\mathbf{r}}{ds} = \frac{1}{c}\frac{d\mathbf{r}}{d\tau} = \left(\frac{dt}{d\tau}, \frac{1}{c}\frac{dx}{d\tau}, \frac{1}{c}\frac{dy}{d\tau}, \frac{1}{c}\frac{dz}{d\tau} \right) = \gamma \left(1, \frac{\dot{x}}{c}, \frac{\dot{y}}{c}, \frac{\dot{z}}{c} \right) = \gamma \left(1, \vec{v}/c \right),$$

$$(8.21)$$

where a dot represents differentiation with respect to the *coordinate* time t, and we used $dt = \gamma d\tau$ [see Equation (8.3)].

An interesting property of the four-velocity is that its spacetime length is one:

4-velocity has constant length

$$\mathbf{u} \cdot \mathbf{u} = u_0^2 - u_1^2 - u_2^2 - u_3^2 = \gamma^2 \left[1 - (\vec{v}/c) \cdot (\vec{v}/c) \right] = \gamma^2 \left(1 - v^2/c^2 \right) = 1, \quad (8.22)$$

from the definition of γ in (8.14). The four-velocity of an object in the object's rest frame is $(1, 0, 0, 0)$, i.e., it is a unit vector in the *time direction*. If we define the **four-acceleration** as the rate of change of the four-velocity with respect to proper time, then the inner product of the 4-velocity and the 4-acceleration of any object is zero, i.e., because of (8.22), the 4-acceleration is η-orthogonal to the 4-velocity. Summarizing these two properties of the 4-velocity, we get

4-velocity is perpendicular to 4-acceleration

$$\mathbf{u} \cdot \mathbf{u} = 1, \qquad \mathbf{u} \cdot \mathbf{a} = 0. \qquad (8.23)$$

Example 8.4.1. A particle is moving in the two-dimentional spacetime of an inertial frame on a path given parametrically as

$$t(\sigma) = b \sinh(\sigma), \quad x(\sigma) = cb \cosh(\sigma),$$

where σ is a dimensionless parameter. The differential of the particle's proper time is

$$(cd\tau)^2 = (cdt)^2 - (dx)^2 = (cb)^2 \cosh^2(\sigma) (d\sigma)^2 - (cb)^2 \sinh^2(\sigma) (d\sigma)^2$$

$$= (cb)^2 (d\sigma)^2 \;\Rightarrow\; d\sigma = \frac{1}{b} d\tau,$$

and $\sigma = \tau/b$. Thus, as a function of the proper time, the path becomes

$$t(\tau) = b \sinh(\tau/b), \quad x(\tau) = cb \cosh(\tau/b).$$

The components of the (dimensionless) 4-velocity are

$$u_0 = \frac{dt}{d\tau} = \cosh(\tau/b), \quad u_1 = \frac{dx}{cd\tau} = \sinh(\tau/b),$$

which satisfy $u_0^2 - u_1^2 = 1$ as they should.

The acceleration of the particle has components

$$a_0 = \frac{du_0}{d\tau} = \frac{1}{b} \sinh(\tau/b), \quad a_1 = \frac{du_1}{d\tau} = \frac{1}{b} \cosh(\tau/b).$$

It is easily verified that $\mathbf{a} \cdot \mathbf{u} = 0$ and that

$$\mathbf{a} \cdot \mathbf{a} = a_0^2 - a_1^2 = - \left(\frac{1}{b} \right)^2.$$

So, the particle has a uniform acceleration of $1/b$. The negative sign in the last equation is due to the fact that the magnitude of the acceleration has to be defined as $-\mathbf{a} \cdot \mathbf{a} = \vec{a}^2 - a_0^2$, with the space part appearing as positive (so that when a_0 is absent, we get back the Newtonian acceleration). ∎

The (kinematic) 4-velocity leads to the (dynamic) 4-momentum: just multiply **u** by mc—the c is to give dimension to the 4-velocity. In a reference frame in which an object of mass m moves with velocity $\vec{v}$, the **4-momentum** **p** is given by

$$\mathbf{p} \equiv (p_0, p_1, p_2, p_3) \equiv (p_0, \vec{p}) = mc\mathbf{u} = \gamma mc\,(1, \vec{v}/c) = (\gamma mc, \gamma m\vec{v})\,. \quad (8.24)$$

4-momentum defined

The space part of the 4-momentum is

$$\vec{p} = \gamma m\vec{v} = \frac{m\vec{v}}{\sqrt{1 - (v/c)^2}}, \quad (8.25)$$

relativistic momentum

and gives ordinary Newtonian momentum when $|\vec{v}| \ll c$, because in that limit, $\gamma \approx 1$. Therefore, we call $\vec{p}$ the **relativistic momentum**.

What about p_0? How are we to interpret that? If we set $\gamma \approx 1$, we get $p_0 \approx mc$ which does not correspond to any Newtonian quantity.[3] However, if we make the next best approximation to γ (see Example 10.2.1 and Problem 10.8), i.e.,

$$\gamma = \frac{1}{\sqrt{1 - (v/c)^2}} \approx 1 + \tfrac{1}{2}(v/c)^2,$$

then

$$p_0 = mc\gamma \approx mc\left(1 + \tfrac{1}{2}v^2/c^2\right) \;\Rightarrow\; p_0 c \approx mc^2 + \tfrac{1}{2}mv^2.$$

The second term gives us the clue that $p_0 c$ must be the **relativistic energy** E. So we write

relativistic energy

$$\mathbf{p} \equiv (p_0, \vec{p}) = (E/c, \vec{p}) = (\gamma mc, \gamma m\vec{v}), \quad E = \gamma mc^2 = \frac{mc^2}{\sqrt{1 - (v/c)^2}}. \quad (8.26)$$

An important special case of this is the 4-momentum **p** of a particle in its rest frame:

$$\mathbf{p} = (mc, \vec{0}) = (mc, 0, 0, 0). \quad (8.27)$$

The definition of the relativistic energy allows objects to have *rest energy*: when $v = 0$, we get

$$E = mc^2, \quad (8.28)$$

the most famous equation in physics!

which states the equivalence of mass and energy and allows their conversion into one another.

The invariance of the length of a 4-vector tells us that $\mathbf{p} \cdot \mathbf{p}$ is a quantity that is independent of observers. From Equation (8.26), we get

$$\mathbf{p} \cdot \mathbf{p} = (E/c)^2 - |\vec{p}|^2 = \gamma^2 m^2 c^2 - \gamma^2 m^2 v^2 = \gamma^2 m^2 c^2 (1 - v^2/c^2) = m^2 c^2,$$

which we rewrite for future reference

$$\mathbf{p} \cdot \mathbf{p} = m^2 c^2 \quad \text{or} \quad E^2 - |\vec{p}|^2 c^2 = m^2 c^4. \quad (8.29)$$

---

[3]One may interpret mc as the momentum of an object moving at the speed of light. However, while objects moving at light speed are possible in Newtonian physics, relativity does not allow a massive object to go at the speed of light [see (8.25)].

Thus, although different observers measure different values for the energy and 3-momentum of an object, when they subtract the square of their value of momentum (times c) from their corresponding value of energy squared, all get the same numerical value, namely the square of the mass of the object (time c^4).

Equation (8.29) allows particles with *zero mass* to have energy and momentum. For such particles,

$$E^2 - |\vec{p}|^2 c^2 = 0 \quad \text{or} \quad E = |\vec{p}|c. \tag{8.30}$$

Since $\vec{p}/E = \vec{v}/c^2$ [see Equation (8.26)], we conclude from (8.29) and (8.30) that

Box 8.4.1. *A particle is massless if and only if it moves at light speed.*

photon is massless!

The particle (quantum) of electromagnetic waves is **photon**. It travels at the speed of light (obviously!). Therefore, it must be massless.

Example 8.4.2. A particle has 4-momentum **p** relative to an observer O' whose 4-velocity is $\mathbf{u}'$. In the rest frame of this observer $\mathbf{u}' = (1,0,0,0)$, and if $\mathbf{p} = (E'/c, \vec{p}')$ in this frame, then

$$\mathbf{p} \cdot \mathbf{u}' = E'/c.$$

Now consider another observer O with respect to whom the 4-momentum of the particle is $\mathbf{p} = (E/c, \vec{p})$ and the 4-velocity of O' is $\mathbf{u}' = (\gamma, \gamma \vec{v}/c)$. In the frame of O,

$$\mathbf{p} \cdot \mathbf{u}' = \gamma E/c - \gamma \vec{p} \cdot \vec{v}/c.$$

The invariance of the inner product now gives

$$E' = \gamma(E - \vec{p} \cdot \vec{v}). \tag{8.31}$$

In the special case in which the particle is at rest with respect to O, $\vec{p} = 0$ and $E = mc^2$. This leads to

$$E' = \gamma mc^2 = \frac{mc^2}{\sqrt{1 - (v/c)^2}},$$

which is the expected expression for the relativistic energy of a particle moving with velocity $\vec{v}$ relative to O'. ∎

8.4.1 Relativistic Collisions

Conservation of energy and momentum in relativistic collisions is stated succinctly in terms of the total four-momenta before and after: $\mathbf{p}_{\text{tot}}^{\text{bef}} = \mathbf{p}_{\text{tot}}^{\text{aft}}$, where in each case, $\mathbf{p}_{\text{tot}}$ is the sum of the 4-momenta of all particles involved.

As a first example, consider two particles that collide and form a single third particle. Let the masses of the first two particles be m_1 and m_2. We can immediately find the mass M of the third particle. Before doing so, we set

$c = 1$ to avoid the cluttering of calculations. This is common in high energy physics, in which energy, momentum, and mass are all measured in the same unit (usually electron volt, eV). If desired, we can easily restore the factors of c at the end by a simple dimensional analysis. With this convention, Equation (8.29) becomes $\mathbf{p} \cdot \mathbf{p} = m^2$.

The conservation of 4-momentum in the present situation is $\mathbf{p}_1 + \mathbf{p}_2 = \mathbf{P}$, where $\mathbf{P}$ is the four-momentum of the final particle. Since this is a vector equation, all components must equal. In particulare, separating the time and the space parts, we get

$$p_{01} + p_{02} = P_0, \quad \text{or} \quad E_1 + E_2 = E,$$
$$\vec{p}_1 + \vec{p}_2 = \vec{P}, \tag{8.32}$$

which are the conservation of energy and momentum.

Squaring both sides of $\mathbf{p}_1 + \mathbf{p}_2 = \mathbf{P}$ gives

$$(\mathbf{p}_1 + \mathbf{p}_2) \cdot (\mathbf{p}_1 + \mathbf{p}_2) = \mathbf{P} \cdot \mathbf{P},$$

or

$$\mathbf{p}_1 \cdot \mathbf{p}_1 + \mathbf{p}_2 \cdot \mathbf{p}_2 + 2\mathbf{p}_1 \cdot \mathbf{p}_2 = \mathbf{P} \cdot \mathbf{P},$$

or

$$m_1^2 + m_2^2 + 2\mathbf{p}_1 \cdot \mathbf{p}_2 = M^2. \tag{8.33}$$

Because of the invariance of the dot product, this equation holds *in any inertial frame.*

Let us evaluate (8.33) in the rest frame of the second particle, where $\mathbf{p}_2 = (m_2, \vec{0})$ by (8.27), and the energy of the first particle is assumed to be E_1. Then

$$\mathbf{p}_1 \cdot \mathbf{p}_2 = (E_1, \vec{p}_1) \cdot (m_2, \vec{0}) = E_1 m_2,$$

and Equation (8.33) immediately gives the mass of the final particle:

$$M^2 = m_1^2 + m_2^2 + 2m_2 E_1, \quad \text{or} \quad M^2 = m_1^2 + m_2^2 + 2m_2 E_1/c^2, \tag{8.34}$$

where the second equation restores the necessary powers of c. Note how the initial energy E_1 on the right-hand side has turned into (part of) the final mass M on the left-hand side. This is how large accelerators create new particles out of the energy of collision.

We can also find the momentum of the final particle from the second equation in (8.32). This easily gives $\vec{P} = \vec{p}_1$, indicating that, in the rest frame of particle 2, the final particle moves in the initial direction of particle 1. The magnitude of $\vec{P}$ can be calculated in terms of energies and masses:

$$|\vec{P}| = |\vec{p}_1| = \sqrt{E_1^2 - m_1^2}. \tag{8.35}$$

The first equation in (8.32) gives the energy of the final particle

$$E = E_1 + m_2. \tag{8.36}$$

Combining Equations (8.35) and (8.36), we can obtain the speed of the final particle:

$$V = \frac{|\vec{P}|}{E} = \frac{\sqrt{E_1^2 - m_1^2}}{E_1 + m_2}. \tag{8.37}$$

A more common collision has two particles initially and two finally. So the conservation of 4-momentum becomes $\mathbf{p}_1 + \mathbf{p}_2 = \mathbf{p}_3 + \mathbf{p}_4$. Separating the time and the space parts yields the conservation of energy and momentum:

$$E_1 + E_2 = E_3 + E_4,$$
$$\vec{p}_1 + \vec{p}_2 = \vec{p}_3 + \vec{p}_4. \tag{8.38}$$

Squaring both sides of $\mathbf{p}_1 + \mathbf{p}_2 = \mathbf{p}_3 + \mathbf{p}_4$ gives

$$m_1^2 + m_2^2 + 2\mathbf{p}_1 \cdot \mathbf{p}_2 = m_3^2 + m_4^2 + 2\mathbf{p}_3 \cdot \mathbf{p}_4, \tag{8.39}$$

which holds in any inertial frame. Evaluating this equation in the rest frame of the second particle, yields

$$m_1^2 + m_2^2 + 2m_2 E_1 = m_3^2 + m_4^2 + 2(E_3 E_4 - \vec{p}_3 \cdot \vec{p}_4). \tag{8.40}$$

In this frame, Equation (8.38) becomes $E_1 + m_2 = E_3 + E_4$ and $\vec{p}_1 = \vec{p}_3 + \vec{p}_4$. Solving for E_4 and $\vec{p}_4$ from these equations and substituting the results in (8.40) yields (after some algebra and using $E_3^2 - |\vec{p}_3|^2 = m_3^2$)

$$m_1^2 + m_2^2 + 2m_2 E_1 = m_4^2 - m_3^2 + 2E_3(E_1 + m_2) - 2\vec{p}_1 \cdot \vec{p}_3,$$

or

$$m_1^2 + m_2^2 + 2m_2 E_1 = m_4^2 - m_3^2 + 2E_3(E_1 + m_2) - 2|\vec{p}_1||\vec{p}_3| \cos\theta_{13}, \tag{8.41}$$

where θ_{13} is the *scattering angle* of the third particle. Once the energy E_1 of the initial incident particle is known, Equation (8.41) gives the scattering angle as a function of the energy of the third particle ($|\vec{p}_1|$ and $|\vec{p}_3|$ are related to E_1 and E_3, respectively).

Compton scattering

Example 8.4.3. The particle nature of light, which had been proposed by Einstein in his explanation of the photoelectric effect, was demonstrated by Compton in what is now called the **Compton scattering**. In this scattering, a photon of energy E is scattered off a stationary electron of mass m_e. The scattered photon is detected at an angle θ from the direction of the incident photon. What is the change in the wavelength of the photon as a function of θ?

In (8.41), let 1 denote the incident photon, 2 the stationary electron, 3 the scattered photon, and 4 the scattered electron. Let E' denote the energy of the scatterd photon, then, with $m_1 = m_3 = 0$, Equation (8.41) becomes

$$m_e^2 + 2m_e E = m_e^2 + 2E'(E + m_e) - 2EE' \cos\theta,$$

or

$$m_e E = E'(E + m_e) - EE' \cos\theta \;\; \Rightarrow \;\; m_e(E - E') = EE'(1 - \cos\theta).$$

Restoring the factors of c and noting that $E = hc/\lambda$, we obtain

$$m_e c^2 \left(\frac{hc}{\lambda} - \frac{hc}{\lambda'} \right) = \left(\frac{hc}{\lambda} \right) \left(\frac{hc}{\lambda'} \right) (1 - \cos\theta),$$

which can be simplified to

$$\Delta\lambda \equiv \lambda' - \lambda = \frac{h}{m_e c}(1 - \cos\theta) \equiv \lambda_c(1 - \cos\theta), \qquad (8.42)$$

where $\lambda_c = h/m_e c$ is called the **Compton wavelength** of the electron. By measuring the difference between the wavelengths of scattered and incident photons, Compton could verify Equation (8.42) and demonstrate that light had particle property. ∎

8.4.2   Second Law of Motion

The Newtonian mechanics defines force as the rate of change of momentum. We generalize this to relativity and define

$$\mathbf{f} = \frac{d\mathbf{p}}{d\tau} = m\frac{d\mathbf{u}}{d\tau} = m\mathbf{a}, \qquad (8.43)$$

where τ is the proper time of the moving object with mass m, four-velocity $\mathbf{u}$, and four-momentum $\mathbf{p}$. Let us explore the meaning of the components of $\mathbf{f}$.

In a particular inertial frame, we assume that Newton's second law holds:

$$\frac{d\vec{p}}{dt} = \vec{F}, \qquad (8.44)$$

where $\vec{p}$ is the space part of the 4-momentum. The space part of $\mathbf{f}$ can now be written as

$$\vec{f} = \frac{d\vec{p}}{d\tau} = \frac{d\vec{p}}{dt}\frac{dt}{d\tau} = \gamma\vec{F}. \qquad (8.45)$$

The time part of $\mathbf{f}$ is a little trickier. First note that

$$f_0 = \frac{dp_0}{d\tau} = \frac{1}{c}\frac{dE}{d\tau}.$$

Next differentiate (8.29) with respect to τ to obtain $E(dE/d\tau) = c^2\vec{p}\cdot(d\vec{p}/d\tau)$. Finally use $\vec{p}/E = \vec{v}/c^2$ to arrive at

$$f_0 = \frac{1}{c}\frac{dE}{d\tau} = \frac{1}{c}\frac{c^2\vec{p}}{E}\cdot\frac{d\vec{p}}{d\tau} = \frac{1}{c}\gamma\vec{v}\cdot\vec{F} = \gamma\vec{\beta}\cdot\vec{F},$$

where $\vec{\beta} \equiv \vec{v}/c$. Thus,

$$\mathbf{f} \equiv (f_0, \vec{f}) = (\gamma\vec{\beta}\cdot\vec{F}, \gamma\vec{F}). \qquad (8.46)$$

The fact that $f_0 = \gamma\vec{\beta}\cdot\vec{F}$ could also be obtained by using $\mathbf{f}\cdot\mathbf{u} = 0$, which is a result of Equation (8.43) and the orthogonality of the 4-velocity and 4-acceleration (see Problem 8.13).

Example 8.4.4. Let a constant force act on a particle of mass m in some inertial frame. What is the speed of the particle at time t if it starts from rest?

Equation (8.44) can be trivially integrated to give $\vec{p} = \vec{F}t$. Since the force is constant, the motion takes place in one dimension. So, we can ignore the vector sign and (remembering that $\beta = v/c$) write

$$m\gamma v = Ft, \quad \text{or} \quad m\gamma\beta = \frac{Ft}{c}, \quad \text{or} \quad \frac{\beta}{\sqrt{1-\beta^2}} = \frac{Ft}{mc}.$$

Squaring both sides and solving for β gives

$$\beta = \frac{Ft/mc}{\sqrt{1+(Ft/mc)^2}}, \quad \text{or} \quad v = \frac{Ft/m}{\sqrt{1+(Ft/mc)^2}}. \tag{8.47}$$

Note that for large t (i.e., when $Ft >> mc$), $\beta \approx 1$ or $v \approx c$. However, the particle can never attain the speed of light no matter how long we wait. On the other hand, if $Ft << mc$, then $v = (F/m)t$, which is the Newtonian speed of a particle moving with constant acceleration.

It is interesting to consider a particle having a constant acceleration of 10 m/s^2 (approximately Earth's gravitational acceleration). How long does it take to attain a speed of $0.999c$? Over 21 years! (See Problem 8.14). On the other hand, Newtonian mechanics requires under one year to achieve the same speed! ∎

8.5 Problems

8.1. Show that Equation (8.6) follows from Equation (8.5). Hint: Consider the three vectors **a**, **b**, and **c** = **a** + **c**.

8.2. Multiply the matrices in Equation (8.10) to obtain the three equations of (8.11). Solve these equations to find all matrix elements in terms of a_{11}.

8.3. In Example 8.3.1, Emmy receives the two signals from the explosions at the same time.
(a) Show that this time is $L/(2c)$ according to Emmy, and $\gamma L/(2c)$ according to Karl.
(b) Let T_1' and T_2' denote the times that Karl receives the signal from the front and back firecrackers, respectively. Show that

$$T_1' = \frac{L}{2c}\sqrt{\frac{1+\beta}{1-\beta}}, \quad T_2' = \frac{L}{2c}\sqrt{\frac{1-\beta}{1+\beta}}.$$

(c) How is $\Delta T' \equiv T_1' - T_2'$ related to $\Delta t'$ calculated in Example 8.3.1? Discuss your answer.

8.4. Show that the relativistic law of addition of velocities (8.17) prohibits the sum of two large velocities to be larger than the speed of light. Hint: Multiply both sides of $\beta_p < 1$ by $1 - \beta$.

8.5. Show that the 4-acceleration is η-orthogonal to the 4-velocity.

8.6. Provide the details of the proof of the statement: *a particle is massless if and only if it moves at light speed.*

8.7. Apply (8.31) to a photon moving in the x-direction and use $|\vec{p}| = E/c$ to show that

$$E' = \sqrt{\frac{1-\beta}{1+\beta}}\, E.$$

Now use $E = hc/\lambda$ to find a formula for the **relativistic Doppler shift**.

8.8. Two identical particles of mass m approach each other along a straight line with speed $v = \beta c$ as measured in the lab frame. Show that the energy of one particle as measured in the rest frame of the other is

$$\frac{1+\beta^2}{1-\beta^2}mc^2.$$

8.9. A particle of mass m and relativistic energy $4mc^2$ collides with another stationary particle of mass $2m$ and sticks to it. What is the mass of the resulting composite particle.

8.10. An electron of kinetic energy 1 GeV (10^9 eV) strikes a positron (anti-electron) at rest and the two particles annihilate each other and produce two photons, one moving in the forward direction (the direction that electron had before collision) and the other in the backward direction. What are the energies of the two photons. The mass (times c^2) of electron and positron are the same and equal to 0.511 MeV (10^6 eV).

8.11. A particle of mass m and energy E collides with an identical particle at rest. The collision results in the formation of a single particle. Show that the mass and the speed of the formed particle are, respectively, $\sqrt{2m(E+m)}$ and $\sqrt{(E-m)/(E+m)}$, assuming that $c = 1$.

8.12. A photon of energy E is absorbed by a stationary nucleus of mass m. The collision results in an excitation of the nucleus. Show that the mass and the speed of the excited nucleus are, respectively, $\sqrt{m(2E+m)}$ and $E/(E+m)$, assuming that $c = 1$.

8.13. Use Equations (8.21), (8.43), (8.45), and the orthogonality of the 4-velocity and 4-acceleration to show that $f_0 = \gamma \vec{\beta} \cdot \vec{F}$.

8.14. How long does it take a particle to attain a speed of $0.999c$, if its acceleration is 10 m/s^2? What is the answer based on Newtonian mechanics? How do the answers change if the ultimate speed of the particle is $0.99999c$?

Part III

Infinite Series

Chapter 9

Infinite Series

Physics is an exact science of approximation. Although this statement sounds like an oximoron, it does summarize the nature of physics. All the laws we deal with in physics are mathematical laws, and as such, they are exact. However, once we try to apply them to Nature, they become only approximations. Therefore, methods of approximation play a central role in physics. One such method is infinite series which we study in this chapter.

9.1 Infinite Sequences

An **infinite sequence** is an association between the set of natural numbers (often zero is also included) and the real numbers, so that for every natural number k there is a real number s_k. Instead of the *association*, one calls the collection of real numbers the infinite sequence. Two common notations for a sequence are an indicated list, and enclosure in a pair of braces, as given below:

$$\{s_1, s_2, \ldots, s_k, \ldots\} \equiv \{s_k\}_{k=1}^{\infty}.$$

Instead of k, one can use any other symbol usually used for natural numbers such as i, j, n, m, etc. We call s_n the *nth term* of the sequence.

In practice, elements of a sequence are given by a rule or formula. The following are examples of sequences:

$$\left\{\frac{1}{2}, \frac{1}{4}, \frac{1}{8}, \ldots\right\} = \left\{\frac{1}{2^n}\right\}_{n=1}^{\infty}, \quad \left\{2, \left(\frac{3}{2}\right)^2, \left(\frac{4}{3}\right)^3, \ldots\right\} = \left\{\left(\frac{k+1}{k}\right)^k\right\}_{k=1}^{\infty},$$

$$\left\{1, \frac{1}{2^3}, \frac{1}{3^3}, \ldots\right\} = \left\{\frac{1}{j^3}\right\}_{j=1}^{\infty}, \quad \left\{\frac{1}{2}, -\frac{2}{3}, \frac{3}{4}, \ldots\right\} = \left\{(-1)^{m+1}\frac{m}{m+1}\right\}_{m=1}^{\infty}.$$

$$(9.1)$$

An important sequence is the **sequence of partial sums** in which each term is a sum. Examples of such sequences are the following:

infinite sequence

sequence of partial sums

$$\left\{1, 1+\frac{1}{2}, 1+\frac{1}{2}+\frac{1}{4}, \cdots\right\}, \qquad \left\{1, 1+\frac{1}{2}, 1+\frac{1}{2}+\frac{1}{3}, \cdots\right\},$$

$$\left\{1, 1+\frac{1}{2^3}, 1+\frac{1}{2^3}+\frac{1}{3^3}, \cdots\right\}, \qquad \left\{1, 1+1, 1+1+\frac{1}{2!}, \cdots\right\}.$$

The nth term of the sequences above are, respectively,

$$s_n = 1 + \frac{1}{2} + \frac{1}{4} + \cdots + \frac{1}{2^n} = \sum_{k=0}^{n} \frac{1}{2^k},$$

$$s_n = 1 + \frac{1}{2} + \frac{1}{3} + \cdots + \frac{1}{n} = \sum_{i=1}^{n} \frac{1}{i},$$

$$s_n = 1 + \frac{1}{2^3} + \frac{1}{3^3} + \cdots + \frac{1}{n^3} = \sum_{j=1}^{n} \frac{1}{j^3},$$

$$s_n = 1 + 1 + \frac{1}{2!} + \cdots + \frac{1}{n!} = \sum_{k=0}^{n} \frac{1}{k!},$$

convention:
$0! = 1.$

so that the sequences can be written, respectively, as

$$\left\{\sum_{k=0}^{n} \frac{1}{2^k}\right\}_{n=0}^{\infty}, \quad \left\{\sum_{i=1}^{n} \frac{1}{i}\right\}_{n=1}^{\infty}, \quad \left\{\sum_{j=1}^{n} \frac{1}{j^3}\right\}_{n=1}^{\infty}, \quad \left\{\sum_{k=0}^{n} \frac{1}{k!}\right\}_{n=0}^{\infty}. \tag{9.2}$$

convergence and
limit of a sequence

In the last sequence, we have used the usual definition, $0! \equiv 1$. A sequence is said to **converge** to the number s or to have **limit** s if for every positive (usually very small) real number ϵ there exists a (usually large) natural number N such that $|s_n - s| < \epsilon$ whenever $n > N$. We then write

$$\lim_{n\to\infty} s_n = \lim_{\nu\to\infty} s_\nu = \lim_{\clubsuit\to\infty} s_\clubsuit = \lim_{\heartsuit\to\infty} s_\heartsuit = s. \tag{9.3}$$

Note the freedom of choice in using the symbol of the limit. A sequence that does not converge is said to **diverge**. The first three sequences in Equation (9.1) are convergent and their limits are

$$\lim_{n\to\infty}\left(\frac{1}{2^n}\right) = 0, \qquad \lim_{n\to\infty}\left(\frac{n+1}{n}\right)^n = e, \qquad \lim_{n\to\infty}\left(\frac{1}{n^3}\right) = 0.$$

The last sequence diverges because there is no *single* number to which the terms get closer and closer.

There are many ways that a sequence can converge to its limit. For instance, the terms s_n may steadily increase toward s after some large integer

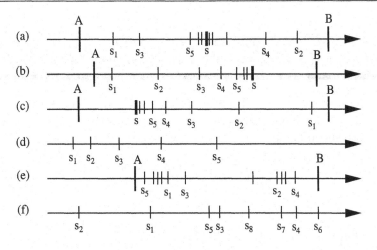

Figure 9.1: Types of sequences and modes of their convergence: (a) convergent, (b) convergent monotone increasing, (c) convergent monotone decreasing, (d) divergent monotone increasing, (e) divergent bounded, (f) divergent unbounded.

N, so that for all $n \geq N$, $s_n \leq s_{n+1} \leq s_{n+2} \leq s_{n+3} \leq \cdots$. [1] In this case we say that the sequence is **monotone increasing**. If the terms s_n steadily decrease toward s after some large integer N, the sequence is called **monotone decreasing**. A sequence may bounce back and forth on either side of its limit, getting closer and closer to it. A sequence is called **bounded** if there exist two numbers A and B such that

$$A \leq s_n \leq B \qquad \text{for all } n.$$

monotone increasing, monotone decreasing, and bounded sequences

A sequence may be bounded but divergent. Various forms of convergence and divergence are depicted in Figure 9.1.

A sequence may have an *upper* and/or a *lower limit*. The upper limit is a number $\bar{s}$ such that there are infinitely many n's with the property that s_n is very close to $\bar{s}$ if n is large enough, and there is no other number larger than $\bar{s}$ with the same property. Similarly, the lower limit is a number $\underline{s}$ such that there are infinitely many n's with the property that s_n is very close to $\underline{s}$ if n is large enough, and there is no other number smaller than $\underline{s}$ with the same property. The last sequence of Equation (9.1) has an upper limit of 1 and a lower limit of -1. It is intuitively obvious that a sequence converges if and only if its upper and lower limits are finite and equal. For instance, the sequence $\{(-1)^n/n\}_{n=1}^{\infty}$ converges to the single limit 0 after bouncing left and right of it infinitely many times.

One can decide whether a sequence converges or not without knowing its limit:

Cauchy criterion

---

[1] We often use the loose phrase: "For large enough n, ...". The precise statement would be: There exists an N such that for all $n \geq N$, ...

Box 9.1.1. (*Cauchy Criterion*). *The sequence* $\{s_n\}_{n=1}^{\infty}$ *converges if the difference* $s_n - s_m$ *approaches zero as* both m *and* n *approach infinity.*

We can add, subtract, multiply, and divide two convergent sequences term by term and obtain a new sequence. The limit of the new sequence is obtained by the corresponding operation of the limits. Thus, if

$$\lim_{n \to \infty} x_n = x, \quad \lim_{n \to \infty} y_n = y,$$

then

$$\lim_{n \to \infty} (x_n \pm y_n) = x \pm y, \quad \lim_{n \to \infty} (x_n \cdot y_n) = x \cdot y, \quad \lim_{n \to \infty} \frac{x_n}{y_n} = \frac{x}{y},$$

provided, of course, that $y \neq 0$ when it is in the denominator.

9.2   Summations

We have been using summation signs on a number of occasions, and we shall be making heavy use of them in this chapter as well. It is appropriate at this point to study some of the properties associated with such sums. Every summation has a **dummy index** which has a **lower limit**, usually written under the **summation symbol** $\sum$, and an **upper limit**, usually written above it. The limits are always fixed, but the dummy index can be any symbol one wishes to use except the symbols used in the expression being summed. Therefore, all the following sums are identical:

dummy
summation index
can be any symbol
you want it to be!

$$\sum_{i=1}^{N} a_i x^i, \quad \sum_{k=1}^{N} a_k x^k, \quad \sum_{\alpha=1}^{N} a_\alpha x^\alpha, \quad \sum_{\clubsuit=1}^{N} a_\clubsuit x^\clubsuit, \quad \sum_{\aleph=1}^{N} a_\aleph x^\aleph. \tag{9.4}$$

It is not a good idea, however, to use a or x as the dummy index for the summation above!

When adding or subtracting sums of equal length, it is better to use the same symbol for the dummy index of the sum:

$$\sum_{i=1}^{N} a_i + \sum_{\heartsuit=1}^{N} b_\heartsuit = \sum_{i=1}^{N} (a_i + b_i) = \sum_{\heartsuit=1}^{N} (a_\heartsuit + b_\heartsuit) = \sum_{k=1}^{N} (a_k + b_k).$$

However,

Box 9.2.1. *When multiplying two sums (not necessarily of equal length), it is essential to choose two different dummy indices for the two sums.*

Thus, to multiply $\sum_{i=1}^{N} a_i$ by $\sum_{i=1}^{M} b_i$, one writes

$$\sum_{i=1}^{N} a_i \sum_{j=1}^{M} b_j = \sum_{i=1}^{N} \sum_{j=1}^{M} a_i b_j.$$

Failure to obey this simple rule can lead to catastrophe. For example, one may end up with $\sum_{i=1}^{N} a_i \sum_{i=1}^{M} b_i = \sum_{i=1}^{N} \sum_{i=1}^{M} a_i b_i$, which is a sum of terms of the form $a_1 b_1 + a_2 b_2 + \cdots$, excluding terms such as $a_1 b_2$ or $a_3 b_5$, etc.

The freedom of choice for the symbol of dummy index can be used to manipulate sums and get results very quickly. As an example, suppose that $\{a_{ij}\}$ is a set of (doubly indexed) numbers which are *symmetric* under interchange of their indices, i.e., $a_{ij} = a_{ji}$. Similarly, suppose that b_{ij} are *antisymmetric* under interchange of their indices, i.e., $b_{ij} = -b_{ji}$. Furthermore, assume that i and j have the lower limit of 1 and the upper limit of n. What is $\sum_{i=1}^{n} \sum_{j=1}^{n} a_{ij} b_{ij}$? Call this sum S. Since the choice of the dummy symbol is irrelevant, we have

$$S = \sum_{i=1}^{n} \sum_{j=1}^{n} a_{ij} b_{ij} = \sum_{\alpha=1}^{n} \sum_{\beta=1}^{n} a_{\alpha\beta} b_{\alpha\beta} = -\sum_{\alpha=1}^{n} \sum_{\beta=1}^{n} a_{\beta\alpha} b_{\beta\alpha}, \qquad (9.5)$$

where we used the symmetry of a_{ij} and the antisymmetry of b_{ij}. Since the order of summation is irrelevant, we can write S as $S = -\sum_{\beta=1}^{n} \sum_{\alpha=1}^{n} a_{\beta\alpha} b_{\beta\alpha}$. Once again, change the dummy symbols: Choose i for β and j for α. Then Equation (9.5) becomes

$$S = -\sum_{i=1}^{n} \sum_{j=1}^{n} a_{ij} b_{ij} = -S \implies 2S = 0 \implies S = 0.$$

As another illustration, suppose we want to multiply $\sum_{i=0}^{M} a_i t^i$ and $\sum_{i=0}^{N} b_i t^i$, and express the coefficient of a typical power of t in the product in terms of a_i and b_i. Call the product P. Then

$$P = \sum_{i=0}^{M} a_i t^i \sum_{j=0}^{N} b_j t^j = \sum_{i=0}^{M} \sum_{j=0}^{N} a_i b_j t^{i+j}.$$

We need to use a single symbol for the power of t in the double sum. So, let $\alpha = i + j$. Our goal is to write $P = \sum c_\alpha t^\alpha$, find c_α in terms of a_i and b_i, and determine the lower and upper limits of the summation on α. The latter is easy: α has a lower limit of 0 (when both i and j are zero), and an upper limit of $M + N$.

For the second dummy index we choose one of the original indices, say i. The limits of i cannot be the original limits, because i is now mixed up with α and j through $j = \alpha - i$. Because of the original bounds of i and j, we have $0 \leq i \leq M$ as well as

$$0 \leq \alpha - i \leq N \quad \text{or} \quad -\alpha \leq -i \leq N - \alpha \quad \text{or} \quad \alpha \geq i \geq \alpha - N.$$

Since i is greater than both 0 and $\alpha - N$, it must be greater than the maximum of the two: $i \geq \max(0, \alpha - N)$. This means that the lower limit of the i-summation is $\max(0, \alpha - N)$. Similarly, since i is smaller than both M and α, it must be smaller than the minimum of the two: $i \leq \min(M, \alpha)$, making the upper limit of the i-summation $\min(M, \alpha)$. We therefore have

$$P = \sum_{\alpha=0}^{M+N} \sum_{i=\max(0,\alpha-N)}^{\min(M,\alpha)} a_i b_{\alpha-i} t^\alpha = \sum_{\alpha=0}^{M+N} \underbrace{\left(\sum_{i=\max(0,\alpha-N)}^{\min(M,\alpha)} a_i b_{\alpha-i} \right)}_{\equiv c_\alpha} t^\alpha. \quad (9.6)$$

Example 9.2.1. As further practice in working with the summation symbol, we show that the torque on a collection of particles is caused by *external* forces only. The torques due to the internal forces add up to zero. We have already illustrated this for three particles in Example 1.3.5. Here, we generalize the result to any number of particles.

We use the second formula in Equation (1.31) and separate the forces

$$\mathbf{T} = \sum_{k=1}^{N} \mathbf{r}_k \times \mathbf{F}_k = \sum_{k=1}^{N} \mathbf{r}_k \times \left(\mathbf{F}_k^{(\text{ext})} + \sum_{i \neq k} \mathbf{F}_{ki} \right)$$

$$= \overbrace{\sum_{k=1}^{N} \mathbf{r}_k \times \mathbf{F}_k^{(\text{ext})}}^{\mathbf{T}^{(\text{ext})}} + \overbrace{\sum_{k=1}^{N} \sum_{i \neq k} \mathbf{r}_k \times \mathbf{F}_{ki}}^{\mathbf{T}^{(\text{int})}}.$$

We need to show that the double sum is zero. To do so, we break the inner sum into two parts, $i > k$ and $i < k$. This yields

$$\mathbf{T}^{(\text{int})} \equiv \sum_{\substack{i,k=1 \\ i \neq k}}^{N} \mathbf{r}_k \times \mathbf{F}_{ki} = \sum_{\substack{i,k=1 \\ i > k}}^{N} \mathbf{r}_k \times \mathbf{F}_{ki} + \sum_{\substack{i,k=1 \\ i < k}}^{N} \mathbf{r}_k \times \mathbf{F}_{ki}$$

$$= \sum_{\substack{i,k=1 \\ i > k}}^{N} \mathbf{r}_k \times \mathbf{F}_{ki} - \sum_{\substack{i,k=1 \\ i < k}}^{N} \mathbf{r}_k \times \mathbf{F}_{ik},$$

because, by the third law of motion, $\mathbf{F}_{ik} = -\mathbf{F}_{ki}$. Now, in the second sum, change the dummy indices twice:

$$\mathbf{T}^{(\text{int})} = \sum_{\substack{i,k=1 \\ i > k}}^{N} \mathbf{r}_k \times \mathbf{F}_{ki} - \sum_{\substack{\alpha,\beta=1 \\ \alpha > \beta}}^{N} \mathbf{r}_\alpha \times \mathbf{F}_{\beta\alpha}$$

$$= \sum_{\substack{i,k=1 \\ i > k}}^{N} \mathbf{r}_k \times \mathbf{F}_{ki} - \sum_{\substack{i,k=1 \\ i > k}}^{N} \mathbf{r}_i \times \mathbf{F}_{ki} = \sum_{\substack{i,k=1 \\ i > k}}^{N} (\mathbf{r}_k - \mathbf{r}_i) \times \mathbf{F}_{ki}.$$

As in Example 1.3.5, we assume that $\mathbf{F}_{ki}$ and $\mathbf{r}_k - \mathbf{r}_i$ lie along the same line in which case the cross products in the sum are all zero. ∎

In the sequel, we shall have many occasions to use summations and manipulate them in ways similar to above. The reader is urged to go through such manipulations with great care and diligence. The skill of summation techniques is acquired only through such diligent pursuit.

9.2.1 Mathematical Induction

Many a time it is desirable to make a mathematical statement that is true for all natural numbers. For example, we may want to establish a formula involving an integer parameter that will hold for all positive integers. One encounters this situation when, after experimenting with the first few positive integers, one recognizes a pattern and discovers a formula, and wants to make sure that the formula holds for all natural numbers. For this purpose, one uses **mathematical induction**. The essence of mathematical induction is stated in

induction principle

> **Box 9.2.2.** (*Mathematical Induction*). *Suppose that there is associated with a natural number (positive integer) n a statement S_n. Then S_n is true for every positive integer provided the following two conditions hold:*
>
> 1. *S_1 is true.*
>
> 2. *If S_m is true for some given positive integer m, then S_{m+1} is also true.*

We illustrate the use of mathematical induction by proving the **binomial theorem**:

binomial theorem

$$(a+b)^m = \sum_{k=0}^{m} \binom{m}{k} a^{m-k} b^k = \sum_{k=0}^{m} \frac{m!}{k!(m-k)!} a^{m-k} b^k$$

$$= a^m + m a^{m-1} b + \frac{m(m-1)}{2!} a^{m-2} b^2 + \cdots + m a b^{m-1} + b^m, \quad (9.7)$$

where we have used the shorthand notation

$$\binom{m}{k} \equiv \frac{m!}{k!(m-k)!}. \quad (9.8)$$

The mathematical statement S_m is Equation (9.7). We note that S_1 is trivially true: $(a+b)^1 = a + b$. Now we assume that S_m is true and show that S_{m+1} is also true. This means starting with Equation (9.7) and showing that

$$(a+b)^{m+1} = \sum_{k=0}^{m+1} \binom{m+1}{k} a^{m+1-k} b^k.$$

Then the induction principle ensures that the statement (equation) holds for all positive integers.

Multiply both sides of Equation (9.7) by $a + b$ to obtain

$$(a+b)^{m+1} = \sum_{k=0}^{m} \binom{m}{k} a^{m-k+1} b^k + \sum_{k=0}^{m} \binom{m}{k} a^{m-k} b^{k+1}.$$

Now separate the $k = 0$ term from the first sum and the $k = m$ term from the second sum:

$$(a+b)^{m+1} = a^{m+1} + \sum_{k=1}^{m} \binom{m}{k} a^{m-k+1} b^k + \underbrace{\sum_{k=0}^{m-1} \binom{m}{k} a^{m-k} b^{k+1}}_{\text{let } k = j-1 \text{ in this sum}} + b^{m+1}$$

$$= a^{m+1} + \sum_{k=1}^{m} \binom{m}{k} a^{m-k+1} b^k + \sum_{j=1}^{m} \binom{m}{j-1} a^{m-j+1} b^j + b^{m+1}.$$

The second sum in the last line involves j. Since this is a dummy index, we can substitute any symbol we please. The choice k is especially useful because then we can unite the two summations. This gives

$$(a+b)^{m+1} = a^{m+1} + \sum_{k=1}^{m} \left\{ \binom{m}{k} + \binom{m}{k-1} \right\} a^{m-k+1} b^k + b^{m+1}.$$

If we now use

$$\binom{m+1}{k} = \binom{m}{k} + \binom{m}{k-1}$$

which the reader can easily verify, we finally obtain

$$(a+b)^{m+1} = a^{m+1} + \sum_{k=1}^{m} \binom{m+1}{k} a^{m-k+1} b^k + b^{m+1}$$

$$= \sum_{k=0}^{m+1} \binom{m+1}{k} a^{m-k+1} b^k.$$

inductive definitions

Mathematical induction is also used in *defining* quantities involving integers. Such definitions are called **inductive definitions**. For example, inductive definition is used in defining powers: $a^1 = a$ and $a^m = a^{m-1} a$.

9.3 Infinite Series

An infinite series is an indicated sum of the members of a sequence $\{a_k\}_{k=1}^{\infty}$. This sum is written as

$$a_1 + a_2 + a_3 + \cdots \equiv \sum_{k=1}^{\infty} a_k \equiv \sum_{j=1}^{\infty} a_j \equiv \sum_{n=1}^{\infty} a_n \equiv \sum_{\clubsuit=1}^{\infty} a_\clubsuit,$$

where we have exploited the freedom of choice in using the dummy index as emphasized in the previous section.

> **Box 9.3.1.** *Associated with an infinite series is the **sequence of partial sums** $\{S_n\}_{n=1}^{\infty}$ with $S_n = a_1 + a_2 + \cdots + a_n = \sum_{k=1}^{n} a_k$. A series is **convergent** (divergent) if its associated sequence of partial sums converges (diverges).*

For a convergent series the nth member of the sequence of partial sums will be a good approximation to the series if n is large enough. This is a simple but important property of the series that is very useful in practice. It should be clear that the convergence property of a series is not affected by changing a *finite* number of terms in the series. Convergent series can be added or multiplied by a constant to obtain new convergent series. In other words, if $\sum_{n=1}^{\infty} a_n = A$ and $\sum_{n=1}^{\infty} b_n = B$, then

$$\sum_{n=1}^{\infty}(a_n \pm b_n) = A \pm B, \qquad r\sum_{n=1}^{\infty} a_n = rA,$$

for any real number r.

9.3.1 Tests for Convergence

When adding, subtracting, or multiplying finite sums, no problem occurs because these operations are all well defined for a finite number of terms. However, when adding an infinite number of terms, no operation on the infinite sum will be defined unless the series converges. It is therefore important to have criteria to test whether a series converges or not. We list various tests which are helpful in determining whether an infinite series is convergent or not.

The nth Term Test

If $\lim_{n \to \infty} a_n \neq 0$, then $\sum_{n=1}^{\infty} a_n$ diverges. This is easily shown by looking at the difference $S_n - S_{n-1}$ and noting that it is simply a_n, and that if the series converges, then this difference must approach zero by the Cauchy criterion. Thus none of the following series converges:

if the infinite series is to converge, its nth term must approach zero. But that by itself is not enough for convergence!

$$\sum_{n=1}^{\infty}\frac{n}{n+1}, \quad \sum_{k=1}^{\infty}(-1)^k\frac{k-1}{5k-1}, \quad \sum_{j=1}^{\infty}(-1)^j, \quad \sum_{m=1}^{\infty}\frac{m^2-10}{8m^2+1}.$$

On the other hand, the series

$$\sum_{n=1}^{\infty}\frac{n}{n^2+1}, \quad \sum_{k=1}^{\infty}(-1)^k\frac{1}{k}, \quad \sum_{j=1}^{\infty}\frac{1}{j}, \quad \sum_{m=1}^{\infty}\frac{1}{m^2},$$

may or may not converge: The approach of a_n to zero does not guarantee the convergence of the series. In fact, the first and third of the series above diverge while the second and last converge.

> **Box 9.3.2.** *Do not confuse the convergence of an infinite series with the convergence of its nth term. If the nth term converges to anything but zero, the series will not converge!*

Absolute Convergence

absolute
convergence

If $\sum_{n=1}^{\infty} |a_n|$ converges, so does $\sum_{n=1}^{\infty} a_n$. The series is then said to be *absolutely convergent*. For example, the series $\sum_{k=1}^{\infty} (-1)^k / 2^k$ converges because $\sum_{k=1}^{\infty} 1/2^k$ converges. However, although the series $\sum_{k=1}^{\infty} 1/k$ can be shown to diverge, $\sum_{k=1}^{\infty} (-1)^k / k$ is known to converge.

Comparison Test

If $|a_n| \leq b_n$ for large enough values of n and $\sum_{n=1}^{\infty} b_n$ converges, then $\sum_{n=1}^{\infty} a_n$ is absolutely convergent and $\sum_{n=1}^{\infty} a_n \leq \sum_{n=1}^{\infty} b_n$. On the other hand, if $a_n \geq b_n \geq 0$ for large values of n and $\sum_{n=1}^{\infty} b_n$ diverges, then so does $\sum_{n=1}^{\infty} a_n$.

Integral Test

This is probably the most powerful test of convergence for infinite series. Assume that $\lim_{n\to\infty} a_n = 0$, so that the series is at least a candidate for convergence. Now find a function f which expresses a_n, i.e., such that $f(n) = a_n$, and assume that $f(n)$ decreases monotonically for large values of n. Then

Theorem 9.3.1. *The series $\sum_{n=1}^{\infty} a_n$ converges if and only if the integral $\int_c^{\infty} f(t)\, dt$ exists and is finite for some real number $c > 1$.*

To see this, refer to Figure 9.2 and suppose that c lies between two consecutive positive integers m and $m+1$. Since the convergence or divergence of a series is not affected by the removal of a finite number of terms of the series, we are allowed to consider either the series $\sum_{k=m}^{\infty} a_k$ or $\sum_{k=m+1}^{\infty} a_k$. Figure 9.2(a) compares the area under the curve $f(t)$ with the shaded area which is the sum of the areas of an infinite number of rectangles each of height $f(k) = a_k$ for some positive integer k larger than (or equal to) $m+1$. The width of all rectangles is unity. The shaded area A is therefore

$$A = \sum_{k=m+1}^{\infty} \underbrace{f(k)}_{=a_k} \Delta t = \sum_{k=m+1}^{\infty} a_k \cdot 1 = \sum_{k=m+1}^{\infty} a_k.$$

It is clear from Figure 9.2(a) that

$$A < \int_c^{\infty} f(t)\, dt \;\Rightarrow\; \sum_{k=m+1}^{\infty} a_k < \int_c^{\infty} f(t)\, dt.$$

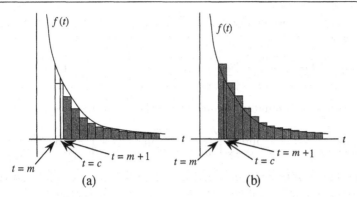

Figure 9.2: The area under the curve (a) bounds, and (b) is bounded by, the infinite sum obtained from the series by removing a finite number of terms. This finite number of terms is the first m terms for (a) and the first $m-1$ terms for (b).

Similarly, Figure 9.2(b) shows that $\sum_{k=m}^{\infty} a_k$ is larger than the area under the curve. We thus can write

$$\sum_{k=m+1}^{\infty} a_k < \int_c^{\infty} f(t)\, dt < \sum_{k=m}^{\infty} a_k.$$

Hence, if the integral is finite $\sum_{k=m+1}^{\infty} a_k$ (being smaller than the integral) is also finite and the series converges. If, on the other hand, the integral is infinite then $\sum_{k=m}^{\infty} a_k$ (being larger than the integral) diverges.

The integral test leads directly to the observation that the **Riemann zeta function**, also called the **harmonic series of order** p defined by

$$\zeta(p) \equiv \sum_{k=1}^{\infty} \frac{1}{k^p} = 1 + \frac{1}{2^p} + \frac{1}{3^p} + \cdots \qquad (9.9)$$

Riemann zeta function or harmonic series of order p

converges for $p > 1$ and diverges for $p \leq 1$. In particular,

$$\zeta(1) = \sum_{k=1}^{\infty} \frac{1}{k} = 1 + \frac{1}{2} + \frac{1}{3} + \cdots,$$

called simply the *harmonic* series, diverges.

harmonic series

Ratio Test

Consider the series $\sum_{n=1}^{\infty} a_n$. If $a_n \neq 0$ for large enough n and

$$\lim_{n \to \infty} \left| \frac{a_{n+1}}{a_n} \right| = R,$$

then the series is absolutely convergent if $R < 1$ and is divergent if $R > 1$.

The terms that we choose for the ratio test need not be consecutive. To see this, note that

$$\lim_{n\to\infty}\left|\frac{a_{n+2}}{a_n}\right| = \lim_{n\to\infty}\left|\frac{a_{n+2}}{a_{n+1}}\right| \cdot \lim_{n\to\infty}\left|\frac{a_{n+1}}{a_n}\right| = \left(\lim_{n\to\infty}\left|\frac{a_{n+1}}{a_n}\right|\right)^2.$$

In going to the last equality, we have used the following:

$$\lim_{n\to\infty}\left|\frac{a_{n+2}}{a_{n+1}}\right| = \lim_{(m-1)\to\infty}\left|\frac{a_{m+1}}{a_m}\right| = \lim_{m\to\infty}\left|\frac{a_{m+1}}{a_m}\right| = \lim_{n\to\infty}\left|\frac{a_{n+1}}{a_n}\right|,$$

where we have substituted $m = n + 1$ and used Equation (9.3) and the fact that $m \to \infty$ if and only if $(m - 1) \to \infty$. It now follows that

$$\lim_{n\to\infty}\left|\frac{a_{n+1}}{a_n}\right| = \sqrt{\lim_{n\to\infty}\left|\frac{a_{n+2}}{a_n}\right|}$$

and the LHS will be less than or greater than one if the term inside the square root sign is. In fact, one can generalize the above argument and state that the series is convergent (divergent) if

$$\lim_{n\to\infty}\left|\frac{a_{n+j}}{a_n}\right| = \left(\lim_{n\to\infty}\left|\frac{a_{n+1}}{a_n}\right|\right)^j \tag{9.10}$$

is less than (greater than) one for any finite j.

The Riemann zeta function can sharpen the ratio test of convergence to allow for certain cases in which the ratio is one. Instead of taking the complete limit, we *approximate* the ratio of consecutive terms for the Riemann zeta function to first order in $1/n$. This yields

$$\frac{a_{n+1}}{a_n} = \left(\frac{n}{n+1}\right)^p = \left(\frac{n+1}{n}\right)^{-p} = \left(1 + \frac{1}{n}\right)^{-p} \approx 1 - \frac{p}{n},$$

where we used the binomial expansion formula, to which we shall come back [see Equation (10.15)]. We know that such a ratio leads to a convergent series if $p > 1$ and to a divergent series if $p \leq 1$. Therefore, we obtain

Theorem 9.3.2. (*Generalized Ratio Test*). *If the ratio of consecutive terms of a series satisfies* $\left|\dfrac{a_{n+1}}{a_n}\right| \to 1 - \dfrac{p}{n}$, *then the series converges if* $p > 1$ *and diverges if* $p \leq 1$.

Alternating Series Test

An alternating series

$$a_1 - a_2 + a_3 - a_4 + \cdots = \sum_{j=1}^{\infty}(-1)^{j+1}a_j, \qquad a_j > 0,$$

converges if $\lim_{j\to\infty} a_j = 0$, and if there exists a positive integer N such that $a_k > a_{k+1}$ for all $k > N$.

Example 9.3.3. A useful series is the **geometric series**:   geometric series

$$b + bu + bu^2 + bu^3 + \cdots = \sum_{k=0}^{\infty} bu^k.$$

We claim that this series converges to $b/(1-u)$ if $|u| < 1$, and diverges if $|u| \geq 1$. To show this, let S_n represent the sum of the first n terms, so that $\{S_n\}_{n=0}^{\infty}$ is the *sequence* of partial sums. We calculate S_n as follows. First note that

$$S_n = \sum_{k=0}^{n} bu^k \Rightarrow uS_n = \sum_{k=0}^{n} bu^{k+1}.$$

Next separate the zeroth term from the rest of S_n and rewrite it as

$$S_n = b + \sum_{k=1}^{n} bu^k = b + \sum_{m=0}^{n-1} bu^{m+1} = b + \sum_{k=0}^{n-1} bu^{k+1},$$

where in the second equality, we changed k to $m = k-1$ and in the last equality we changed the dummy index back to k. Subtracting uS_n from S_n, we obtain

$$S_n - uS_n = (1-u)S_n = b + \sum_{k=0}^{n-1} bu^{k+1} - \sum_{k=0}^{n} bu^{k+1}$$

$$= b + \sum_{k=0}^{n-1} bu^{k+1} - \left(\sum_{k=0}^{n-1} bu^{k+1} + bu^{n+1} \right) = b - bu^{n+1}$$

or

$$S_n = \frac{b - bu^{n+1}}{1-u}.$$

It is now clear that $u^{n+1} \to 0$ for $n \to \infty$ only if $|u| < 1$. For $|u| > 1$, the series clearly diverges. For $|u| = 1$ the partial sum is either $S_n = nb$ (when $u = 1$), which diverges for any nonzero b, or $S_n = b\sum_{n=0}^{\infty}(-1)^n$, which bounces back and forth between $+b$ and $-b$, and never converges. So the series diverges for $|u| \geq 1$

For example, if $b = 0.3$ and $u = 0.1$, then the series gives

$$0.3 + 0.3 \times 0.1 + 0.3 \times 0.01 + \cdots = 0.33333\cdots = \frac{0.3}{1 - 0.1} = \frac{1}{3}.$$

For $b = 1$ the series gives

$$1 + u + u^2 + \cdots = \frac{1}{1-u} = (1-u)^{-1}, \tag{9.11}$$

which can be thought of as the binomial expansion when the power is -1. As we shall see in Section 10.1, there is a generalization of binomial expansion for any real power. ∎

The result of Example 9.3.3 is important enough to be summarized:

Box 9.3.3. *The series $b + bu + bu^2 + bu^3 + \cdots = \sum_{k=0}^{\infty} bu^k$ is called the **geometric series**. It converges to $b/(1-u)$ if $|u| < 1$, and diverges if $|u| \geq 1$.*

Example 9.3.4. Another example of a series used often is

$$1 + 1 + \frac{1}{2!} + \frac{1}{3!} + \cdots = \sum_{k=0}^{\infty} \frac{1}{k!}.$$

The ratio test shows only that the series converges, but the comparison test gives us more information. In fact, since $1/n! \leq 1/2^{n-1}$ for $n \geq 1$, we conclude that

$$1 + \frac{1}{2!} + \frac{1}{3!} + \cdots \leq 1 + \frac{1}{2} + \frac{1}{2^2} + \frac{1}{2^3} + \cdots.$$

But the RHS is the geometric series with $u = 1/2$ which is known to converge to 2. We thus obtain the upper bound to our series:

$$2 \leq \sum_{k=0}^{\infty} \frac{1}{k!} \leq 3.$$

It is well known that the series converges to $e = 2.718281828 \cdots$.   ∎

Example 9.3.5. If one alternates the sign of the terms in the harmonic series, one obtains the series

$$1 - \frac{1}{2} + \frac{1}{3} - \frac{1}{4} + \cdots$$

conditional
convergence

which is convergent by the alternating series test. In fact, we shall show in Example 9.4.4 that the series converges to ln 2. Note that the series is not *absolutely* convergent. A convergent series that does not converge absolutely is called **conditionally convergent**.   ∎

Historical Notes

The invention of calculus motivated several other areas of investigation in mathematics. One of these areas was infinite series. For example, it was not always possible to find a closed formula for the integral of a function. So, it was common to expand the integrand in powers of the variable and integrate the resulting infinite series. No question was asked as to the legitimacy of the operations performed. In fact, **Newton**, **Leibniz**, and **Euler** regarded infinite series as an extension of the algebra of polynomials, and they did not realize that new problems would arise if a finite sum were extended to an infinite series. However the apparent difficulties that did arise caused them occasionally to bring up the question of convergence and divergence.

Some mathematicians of the seventeenth century had observed the difference between convergence and divergence. In 1668 *Lord Brouncker*, while studying the relation between $\ln x$ and the area under $y = 1/x$, demonstrated the convergence of the series for ln 2 and $\ln(\frac{5}{4})$ by comparison with a geometric series. Newton and **James Gregory**, who made much use of the numerical values of series to calculate logarithmic and other function tables and to evaluate integrals, were aware that the sum of a series can be finite or infinite. The terms "convergent" and "divergent" were actually used by Gregory in 1668, but he did not develop the ideas.

Leibniz, too, felt some concern about convergence and noted in a letter of October 25, 1713 to **John Bernoulli** what is now a theorem that we call the *alternating series test*. Maclaurin used series as a regular method for integration. He recognized that the terms of a convergent series must continually decrease and become less than any given quantity no matter how small.

D'Alembert also distinguished convergent from divergent series. In his article "Série" in the *Encyclopédie* he describes a convergent series as that which approaches a finite value and consequently has terms that keep diminishing. In this same volume, d'Alembert gave a test for the absolute convergence of the series $\sum_{k=1}^{\infty} a_k$, namely, if for all $k > N$, the ratio $|a_{k+1}/a_k| < r$ where r is a positive number independent of k and less than 1, the series converges absolutely.

Edward Waring (1734–1798), Lucasian professor of mathematics at Cambridge University, held advanced views on convergence. He showed that the harmonic series of order p converges if $p > 1$ and diverges if $p < 1$. He also gave the well-known test for convergence and divergence, now known as the *ratio test*.

9.3.2 Operations on Series

It has already been mentioned that *convergent* series can be added, subtracted, and multiplied by a constant. There are other important operations one can perform on convergent series. These operations may be "obvious" for finite sums, but they have to be justified for infinite series. In fact, performing such obvious operations on *divergent* series leads to contradictory results.

One such operation is **grouping**:

grouping of convergent series

> **Box 9.3.4.** *One can group the terms of a finite sum or a convergent infinite series in any way one desires, and the sum will not change.*

The operation of grouping is essentially putting parentheses around a collection of terms of the series (or the sum), adding the terms inside each parentheses first, and then adding the results. This is simply the *associative property of addition*. It turns out that this associative property of addition does not apply to divergent infinite series.[2] For example, $\sum_{m=0}^{\infty}(-1)^m$ gives an infinite number of zeros if every $+1$ is grouped with one -1. On the other hand, the same series can be grouped such that the first $+1$ is set aside and the rest of the terms are paired. The result would then be a $+1$ with an infinite number of zeros. If a series is divergent and not bounded, so that the sum is infinite, then any grouping of terms gives infinity.

Another operation is the **rearrangement** of terms of a series. This is the *commutative property of addition*:

warning! rearranging terms is not, in general, allowed!

> **Box 9.3.5.** *If a series is absolutely convergent then the rearrangement of terms does not change either the nature of convergence or the limit of the series. A conditionally convergent series does not share this property.*

---

[2]Caution is to be exercised not to move the terms around, as this will, in general, affect the sum as explained in the property of rearrangement described below.

To see the importance of absolute convergence, consider the alternating series $\sum_{k=1}^{\infty}(-1)^{k+1}/k$—which converges conditionally to $\ln 2$—and rearrange terms as follows:

$$\sum_{k=1}^{\infty}\frac{(-1)^{k+1}}{k} = 1 + \tfrac{1}{3} + \tfrac{1}{5} + \cdots - \tfrac{1}{2}\left(1 + \tfrac{1}{2} + \tfrac{1}{3} + \cdots\right)$$

$$= 1 + \tfrac{1}{2} + \tfrac{1}{3} + \tfrac{1}{4} + \tfrac{1}{5} + \cdots - \tfrac{1}{2} - \tfrac{1}{4} - \tfrac{1}{6} - \cdots$$
$$- \tfrac{1}{2}\left(1 + \tfrac{1}{2} + \tfrac{1}{3} + \cdots\right)$$
$$= 1 + \tfrac{1}{2} + \tfrac{1}{3} + \tfrac{1}{4} + \tfrac{1}{5} + \cdots - \left(1 + \tfrac{1}{2} + \tfrac{1}{3} + \cdots\right) = 0,$$

where in the second line, terms with even denominators have been added and subtracted with the positive ones interspersed among terms with odd denominators.

multiplication of two series

The third operation is **multiplication of two series**. As for rearrangement,

> **Box 9.3.6.** *Multiplication is defined only for absolutely convergent series: If the two series $\sum_{k=1}^{\infty} a_k$ and $\sum_{j=1}^{\infty} b_j$ are absolutely convergent, then their product $\left(\sum_{k=1}^{\infty} a_k\right) \cdot \left(\sum_{j=1}^{\infty} b_j\right) \equiv \sum_{k=1}^{\infty}\sum_{j=1}^{\infty} a_k b_j \equiv \sum_{i=1}^{\infty} c_i$ is also absolutely convergent.*

The last series is a rearrangement of the terms $a_k b_j$ into a single term c_i. This rearrangement makes it necessary for the original series to be absolutely convergent.

9.4   Sequences and Series of Functions

The infinite series of the last section are useful when we want to approximate a number, such as e or $\ln 2$ by a (large) sum of other (rational, decimal) numbers. Physics, however, deals with functions as well as numbers. It is therefore useful to know how to approximate functions in terms of "elementary" functions. In this section we shall investigate the possibility of expressing a given function in terms of a series of functions. Since functions give numbers once their arguments are assigned a value, many of the ideas developed in the preceding two sections will be employed.

sequence of functions

Suppose for each natural number n there is a *function $f_n(x)$*. Then, the set $\{f_n(x)\}_{n=1}^{\infty}$ is called a **sequence of functions**. Just as in the case of sequences of numbers, we need to address the question of the convergence of the sequence of functions. This reduces to the question of convergence of ordinary numbers once we substitute values for x. Variation of $f_n(x)$ with x opens up the possibility of convergence for some values of x and divergence for others. For instance, the sequence $\{x^n\}_{n=1}^{\infty}$ converges for $-1 < x < 1$ and diverges for all other values of x.

More interesting than sequences of functions are **series of functions**: series of functions

$$f_1(x) + f_2(x) + f_3(x) + \cdots = \sum_{k=1}^{\infty} f_k(x).$$

The nth partial sum of such a series is

$$S_n(x) = f_1(x) + f_2(x) + \cdots + f_n(x) = \sum_{j=1}^{n} f_j(x).$$

The convergence of a series of functions $\sum_{k=1}^{\infty} f_k(x)$ depends on x. For example, the series may converge for $x = 0.35$. This means that the series of *numbers* $\sum_{k=1}^{\infty} f_k(0.35)$ converges, i.e., there exists a real number s such that for every ϵ there exists an N with the property that $|\sum_{k=1}^{n} f_k(0.35) - s| < \epsilon$ whenever $n > N$. It should be clear that an N that works for one value of x—here 0.35—and ϵ, may not work for other values of x and ϵ. Thus, N depends on x and ϵ, and this dependence is denoted by $N(x, \epsilon)$.

We can imagine making a table with one column consisting of the values of x and a second column consisting of the corresponding limits of the series of numbers whose terms are f_n evaluated at the value of x. The table then defines a real-valued function, say $S(x)$, which is called the *limit of the series of functions*, and one writes

$$S(x) = \lim_{n \to \infty} S_n(x) \equiv \sum_{k=1}^{\infty} f_k(x). \tag{9.12}$$

We have already seen examples of series of functions: the geometric series $\sum_{n=0}^{\infty} u^n$ —convergent for $|u| < 1$—in which the terms are functions of u with $f_n(u) = u^n$, and the Riemann zeta function (or harmonic series of degree p)—convergent for $|p| > 1$—in which the terms were functions of p with $f_n(p) = 1/n^p$.

In general, the sum in Equation (9.12) may converge only for a limited range of values of x. To find this range, we impose the ratio test on the terms of the series. This yields

$$r(x) \equiv \lim_{k \to \infty} \left| \frac{f_{k+1}(x)}{f_k(x)} \right| < 1, \tag{9.13}$$

which is an inequality in x that can be solved to find the values of x for which the series converges.

Example 9.4.1. As an example of the application of Equation (9.13), let us find the values of x for which the series

$$\sum_{k=1}^{\infty} \frac{[\ln(x+1)]^k}{k}$$

converges. The ratio in (9.13) is

$$r(x) = \lim_{k \to \infty} \left| \frac{[\ln(x+1)]^{k+1}/(k+1)}{[\ln(x+1)]^k/k} \right| = \lim_{k \to \infty} \left| \frac{[\ln(x+1)]^{k+1}}{[\ln(x+1)]^k} \cdot \frac{k}{k+1} \right|$$

$$= |\ln(x+1)| \lim_{k \to \infty} \left| \frac{k}{k+1} \right| = |\ln(x+1)|.$$

So, the condition for convergence is

$$|\ln(x+1)| < 1 \; \Rightarrow \; -1 < \ln(x+1) < 1$$

or

$$e^{-1} < x+1 < e \; \Rightarrow \; e^{-1} - 1 < x < e - 1$$

and the series converges for $-0.632 < x < 1.718$.

Let us now check the convergence of the series for the two end points. The left end point corresponds to $\ln(x+1) = -1$ for which the series becomes $\sum_{n=1}^{\infty} (-1)^n/n$ which is convergent (see Example 9.3.5). On the other hand, for the right end point, $\ln(x+1) = 1$, and the series becomes $\sum_{n=1}^{\infty} 1/n$ which is the divergent harmonic series. Thus, the interval of convergence is $-0.632 \le x < 1.718$. ∎

uniform convergence

An important notion is **uniform convergence**:

> **Box 9.4.1.** *If, for a given ϵ, it is possible to find an N such that $|S_n(x) - S(x)| < \epsilon$ whenever $n > N$ for all values of x in some interval (a, b)—so that N is independent of x—then the series is said to converge **uniformly** on (a, b).*

Clearly, for uniform convergence to have any meaning, there must exist a *range* of values of x for which the series converges uniformly because a series may converge for all values of x on the real line without converging uniformly for any interval of the real line. A pictorial representation of uniform convergence is shown in Figure 9.3. Basically, we say that a series is uniformly convergent if the graphs of partial sums $S_n(x)$, after a certain large N, all lie within a (narrow) strip of width ϵ containing the graph of the limit function $f(x)$.

test for uniform convergence

There is a useful test for the uniform convergence which works for a large number of familiar series and goes by the name of the **Weierstrass M-test**: Let $\sum_{k=1}^{\infty} f_k(x)$ be a series of functions all defined in an interval[3] (a, b). If there is a convergent series of positive *numbers* $\sum_{k=1}^{\infty} M_k$, such that $|f_k(x)| \le M_k$ for all x in (a, b), then $\sum_{k=1}^{\infty} f_k(x)$ converges absolutely for each such x, and is uniformly convergent in (a, b).

Example 9.4.2. Consider the series $\sum_{n=1}^{\infty} x^n/n^p$, which is a generalization of the geometric series (for which $p = 0$). We want to see for what values of p and in what

---

[3]Instead of an interval, one may use the union of many intervals. In fact, the statement is true even when the interval (a, b) is replaced with a *general subset* of the real line.

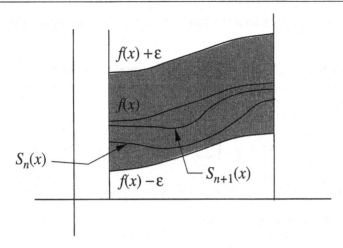

Figure 9.3: Uniform convergence.

interval of x is the series convergent. One way to get the answer is to apply the ratio test:

$$\lim_{n\to\infty} \left| \frac{a_{n+1}}{a_n} \right| = \lim_{n\to\infty} \left| \frac{x^{n+1}}{(n+1)^p} \cdot \frac{n^p}{x^n} \right| = |x| \lim_{n\to\infty} \left(\frac{n}{n+1} \right)^p = |x|.$$

It follows that, regardless of the value of p, the series converges for $|x| < 1$, and diverges for $|x| > 1$. For $x = 1$, the series becomes $\sum_{n=1}^{\infty} 1/n^p$ which converges for $p > 1$ and diverges for $p \le 1$ as pointed out in the integral test of convergence. Finally if $x = -1$, the alternating series test of convergence tells us that the series converges for all $p > 0$. What about the uniformity of convergence? We note that for $M_n = 1/n^p$, and for $|x| \le 1$, we have

$$\left| \frac{x^n}{n^p} \right| \le \frac{1}{n^p} \equiv M_n$$

and the series of M_n converges as long as $p > 1$. Thus, for $p > 1$, the series $\sum_{n=1}^{\infty} x^n/n^p$ is *uniformly* convergent.   ■

9.4.1   Properties of Uniformly Convergent Series

The importance of uniformly convergent series lies in the nice properties such series possess. For instance, if $u_i(x)$ is continuous in the interval $a \le x \le b$, and if the series $\sum_{i=1}^{\infty} u_i(x)$ is *uniformly* convergent in that interval, then the *function* defined by $f(x) = \sum_{i=1}^{\infty} u_i(x)$ is also continuous in the interval. This statement is equivalent to saying that for x and x_0 in the interval (a,b), one has

$$\lim_{x\to x_0} \left[\lim_{n\to\infty} S_n(x) \right] = \lim_{n\to\infty} \left[\lim_{x\to x_0} S_n(x) \right].$$

Accordingly, *uniform convergence permits the interchange of the two limit processes*.

Another property, which is extremely useful in physical applications, is the fact that

you can integrate a uniformly convergent series term by term.

Theorem 9.4.3. *If $f(x) = \sum_{i=1}^{\infty} u_i(x)$ is uniformly convergent, and each $u_i(x)$ is continuous for $a \le x \le b$, then the series can be integrated term by term, i.e.,*

$$\int_a^b f(x)\,dx = \int_a^b \left(\sum_{i=1}^{\infty} u_i(x) \right) dx = \sum_{i=1}^{\infty} \int_a^b u_i(x)\,dx,$$

i.e., integration and summation can be interchanged.

Example 9.4.4. Consider the geometric series $\frac{1}{1-t} = \sum_{i=0}^{\infty} t^i$, which, by Example 9.4.2, converges uniformly for $-1 < t < 1$. Changing t to $-t$ does not change either the interval or the nature of convergence of the series. We thus have

$$\frac{1}{1+t} = \sum_{i=0}^{\infty} (-t)^i = \sum_{i=0}^{\infty} (-1)^i t^i. \tag{9.14}$$

Because of the uniform convergence of the series, we can integrate both sides from 0 to x with $-1 < x < 1$ to obtain

$$\int_0^x \frac{dt}{1+t} = \ln(1+x) = \sum_{i=0}^{\infty} (-1)^i \int_0^x t^i\,dt = \sum_{i=0}^{\infty} (-1)^i \frac{x^{i+1}}{i+1}.$$

With $x = 1$, we obtain the result alluded to in Example 9.3.5.

Note that the integral of a series may be convergent for a bigger range of values of its argument than the original series. Here, the original series was divergent (for $t = 1$) while its integral converges (for $x = 1$). ∎

The property stated in Theorem 9.4.3 is a useful tool for the expansion of physical quantities in terms of some more "elementary" quantities. For example, one can expand the electric potential—usually given in terms of an integral—as a sum of the potentials of a single charge, a dipole, a quadrupole, etc. (see Section 10.5). In many physical situations only the first few terms of the series expansion will be of importance. Thus, for instance, in atomic transitions, it is only the dipole term that participates significantly.

One can also differentiate a uniformly convergent series. To be specific,

you can differentiate a uniformly convergent series term by term.

Theorem 9.4.5. *Suppose that $u'_n(x) = du_n/dx$ is continuous for $a \le x \le b$, that the series $\sum_{n=1}^{\infty} u_n(x)$ converges to $f(x)$ for $a \le x \le b$, and that the series $\sum_{n=1}^{\infty} u'_n(x)$ converges uniformly for $a \le x \le b$. Then*

$$f'(x) = \frac{d}{dx} \sum_{n=1}^{\infty} u_n(x) = \sum_{n=1}^{\infty} u'_n(x), \qquad a \le x \le b,$$

i.e., one can change the order of differentiation and summation.

Other operations defined on uniformly convergent series are addition, subtraction, and multiplication by a continuous function: If $\sum_{i=1}^{\infty} u_i(x)$ and $\sum_{i=1}^{\infty} v_i(x)$ are uniformly convergent for $a \le x \le b$ and $h(x)$ is continuous in the same interval, then the series

$$\sum_{i=1}^{\infty} [u_i(x) \pm v_i(x)], \quad \sum_{i=1}^{\infty} h(x) u_i(x),$$

are also uniformly convergent for $a \le x \le b$.

Historical Notes

The mathematicians of the seventeenth and eighteenth centuries used series indiscriminately. By the beginning of the nineteenth century some absurd results from manipulating infinite series stirred up some interest in questioning the validity of operations performed on them. Around 1810 a number of mathematicians began the exact handling of infinite series.

In his 1811 paper and his *Analytical Theory of Heat,* **Fourier** gave a satisfactory definition of convergence, though in general he worked freely with divergent series. His definition of convergence was essentially in terms of the sequence of partial sums. Moreover, he recognized that the convergence of a series of functions of the variable x may be achieved only in an interval of x values. Although Fourier stressed that a necessary condition for convergence is that the terms of the series approach zero, he was fooled by the series $\sum_{k=0}^{\infty} (-1)^k$, and thought that its sum was $\frac{1}{2}$ [substitute $t = 1$ on both sides of (9.14)].

The first important and strictly rigorous investigation of convergence was made by **Gauss** in his 1812 paper *Disquisitiones Generales Circa Seriem Infinitam* wherein he studied the *hypergeometric series* (see Section 11.2.1). Though Gauss is often mentioned as one of the first to recognize the need for restricting the series to their interval of convergence, he avoided any decisive position. He was so much concerned to solve concrete problems by numerical calculations that he used a divergent expansion of the gamma function. When he did investigate the convergence of the hypergeometric series, he remarked that he did so to please those who favored the rigor of the ancient geometers.

Cauchy's work on the convergence of series is the first extensive treatment of the subject. In his *Cours d'Analyse* Cauchy clearly defines the *sequence of partial sums* and gives a rigorous definition of the convergence and divergence of the series in terms of this sequence. It is also in this work that he gives what is now called the *Cauchy criterion* for convergence of a *sequence* (see Box 9.1.1). He proves this to be a necessary condition, but merely remarks that if the condition holds, the convergence of the series is assured. He lacked the knowledge of the properties of real numbers to provide a proof. Cauchy then goes on to state and prove many of the results that we have outlined in our discussion of the tests for convergence.

9.5   Problems

9.1. Show that

(a) $\sum_{k=1}^{n} k z^{k-1} = \sum_{k=0}^{n-1} (k+1) z^k$.   (b) $x^2 \sum_{k=0}^{n} a_k x^k = \sum_{k=2}^{n+2} a_{k-2} x^k$.

9.2. Use some small values of M and N (say $M = 2, N = 3$) and verify the validity of Equation (9.6).

9.3. Use Equation (9.8) to show that

$$\binom{m+1}{k} = \binom{m}{k} + \binom{m}{k-1}.$$

9.4. Use mathematical induction to prove the following relations:

(a) $\frac{d}{dx}(x^n) = nx^{n-1}$.   (b) $\sum_{k=0}^{n} x^k = \frac{x^{n+1}-1}{x-1}$.

9.5. Use the integral test to show that the harmonic series of order p is convergent for $p > 1$ and divergent for $p \leq 1$.

9.6. Test the following series for convergence or divergence:

(a) $\sum_{n=1}^{\infty} \frac{(-1)^n n}{n^2+1}$.   (b) $\sum_{n=1}^{\infty} \frac{(-1)^n \sin^2 n\alpha}{n+1}$.   (c) $\sum_{n=1}^{\infty} \frac{\ln n}{n^p}$.

(d) $\sum_{n=1}^{\infty} \frac{n+1}{3n^2+3n}$.   (e) $\sum_{n=1}^{\infty} \frac{n+1}{3n^2+5n-10}$.   (f) $\sum_{n=2}^{\infty} \frac{1}{n\ln n}$.

where α is some real number. For (c), consider the three cases $p > 1$, $p < 1$, and $p = 1$.

9.7. Prove convergence or divergence by the comparison test:

$$\sum_{n=1}^{\infty} \frac{\sin n}{n^2}, \quad \sum_{n=2}^{\infty} \frac{1}{n^3-1}, \quad \sum_{n=1}^{\infty} \frac{n+5}{n^2-3n-5}, \quad \sum_{n=2}^{\infty} \frac{1}{\sqrt{n}\ln n}.$$

9.8. Prove convergence or divergence by the integral test:

$$\sum_{n=1}^{\infty} \frac{1}{n^2+1}, \quad \sum_{n=1}^{\infty} \frac{n}{n^2+1}, \quad \sum_{n=2}^{\infty} \frac{1}{n\ln^2 n}, \quad \sum_{n=2}^{\infty} \frac{1}{n\ln n\ln\ln n}.$$

9.9. Prove convergence or divergence by the ratio test:

$$\sum_{n=1}^{\infty} \frac{2^n+1}{3^n+n}, \quad \sum_{n=1}^{\infty} \frac{(-1)^n}{n!}, \quad \sum_{n=1}^{\infty} \frac{5^n}{n!}.$$

9.10. Use the ratio test to find the range of values of x for which the following series converge. Make sure to investigate the end points.

(a) $\sum_{n=1}^{\infty} \frac{(\ln x)^n}{n+1}$.   (b) $\sum_{n=1}^{\infty} \frac{4^n \sin^n x}{(n+1)5^n}$.   (c) $\sum_{n=1}^{\infty} \frac{x^n}{\sqrt{n}}$.

(d) $\sum_{n=1}^{\infty} \frac{(\ln x)^n}{n!}$.   (e) $\sum_{n=1}^{\infty} \frac{x^n}{3^n n!}$.   (f) $\sum_{n=3}^{\infty} \frac{n^2}{(x-2)^n}$.

(g) $\sum_{n=1}^{\infty} nx^n$.   (h) $\sum_{n=1}^{\infty} n!x^n$.   (i) $\sum_{n=1}^{\infty} \frac{n^3}{(\ln x)^n}$.

(j) $\sum_{n=0}^{\infty} \frac{nx^n}{n^2+1}$.   (k) $\sum_{n=1}^{\infty} \frac{(x^2+1)^n}{n^3}$.   (l) $\sum_{n=1}^{\infty} \frac{n^2}{(x+1)^n}$.

(m) $\sum_{n=0}^{\infty} \left(\frac{x^2+1}{3}\right)^n$.   (n) $\sum_{n=1}^{\infty} \left(\frac{x^2}{\sqrt{n}}\right)^n$.   (o) $\sum_{n=0}^{\infty} \frac{(x-2)^n}{n^2+1}$.

(p) $\sum_{n=0}^{\infty} \left(\frac{x}{2}\right)^n$.   (q) $\sum_{n=1}^{\infty} \left(\frac{x}{n}\right)^n$.   (r) $\sum_{n=0}^{\infty} \frac{x^n}{n^2+1}$.[6bp]

9.11. Write the first four terms of the following series:

$$\sum_{n=1}^{\infty} \frac{n!}{2 \cdot 4 \cdots 2n}, \quad \sum_{n=1}^{\infty} \frac{(-1)^n}{\ln(n+1)}, \quad \sum_{n=1}^{\infty} \frac{1}{\sqrt[10]{n^9}}, \quad \sum_{n=1}^{\infty} \frac{1}{\sqrt[9]{n^{10}}}.$$

Test for convergence or divergence of these series.

Chapter 10

Application of Common Series

The preceding chapter concerned itself with the formal properties of infinite sequences and series, especially the sequences and series of functions. One of the useful properties of the infinite series of functions is that they can be approximated by *finite sums*. In this approximation, two important features of the series play crucial roles: the simplicity of the functions used in the series and the convergence of the series. This chapter deals with some of the series of functions most commonly used in mathematical physics.

10.1 Power Series

One of the most common series of functions is the **power series** where the nth term of the series is $c_n(x - a)^n$ with c_n a real number. To be specific, a power series in powers of $(x - a)$ is of the form

$$\sum_{n=0}^{\infty} c_n(x - a)^n = c_0 + c_1(x - a) + c_2(x - a)^2 + \cdots. \qquad (10.1)$$

An important special case is when $a = 0$, so that we have

$$\sum_{n=0}^{\infty} c_n x^n = c_0 + c_1 x + c_2 x^2 + \cdots. \qquad (10.2)$$

Sometimes negative powers are also included, but by power series we usually mean Equation (10.1).

We note that Equation (10.1) converges for $x = a$. The question is whether it converges for any other values of x, and if so, what these values are. It turns out that:

radius of convergence of a power series

Theorem 10.1.1. *Every power series* $\sum_{n=0}^{\infty} c_n(x-a)^n$ *has a **radius of convergence** r^* such that the series converges* absolutely and uniformly *when* $|x-a| < r^*$ *and diverges for* $|x-a| > r^*$. *If* $r^* \neq 0$ *and* r_1 *is a number such that* $0 < r_1 < r^*$, *then the series converges* absolutely and uniformly *for* $|x-a| \leq r_1$.

The number r^* can be 0 (in which case the series converges only for $x = a$), a finite positive number, or ∞ (in which case the series converges for all x).

The radius of convergence can be evaluated by using the ratio test. Consider the ratio

$$r(x) = \lim_{n \to \infty} \left| \frac{c_{n+1}(x-a)^{n+1}}{c_n(x-a)^n} \right| = |x-a| \lim_{n \to \infty} \left| \frac{c_{n+1}}{c_n} \right|$$

and note that the series converges if $r(x) < 1$, or

$$|x-a| < \lim_{n \to \infty} \left| \frac{c_n}{c_{n+1}} \right|.$$

The RHS is naturally defined to be the radius of convergence

$$r^* = \lim_{n \to \infty} \left| \frac{c_n}{c_{n+1}} \right| \qquad \text{if the limit exists.} \tag{10.3}$$

It can be shown that the radius of convergence can also be found from the following formula:

$$r^* = \lim_{n \to \infty} \frac{1}{\sqrt[n]{|c_n|}} \qquad \text{if the limit exists.} \tag{10.4}$$

Example 10.1.2. Consider the exponential function e^x which, as we shall see, has a power series expansion

$$e^x \equiv \sum_{n=0}^{\infty} c_n x^n = \sum_{n=0}^{\infty} \frac{x^n}{n!}.$$

By the ratio test, we have

$$r(x) = \lim_{n \to \infty} \left| \frac{x^{n+1}/(n+1)!}{x^n/n!} \right| = \lim_{n \to \infty} |x| \left| \frac{n!}{(n+1)!} \right| = |x| \lim_{n \to \infty} \left| \frac{1}{n+1} \right| = 0$$

for all values of x. So, regardless of x, the series representation of e^x converges, i.e., the radius of convergence is infinite. We can also use Equation (10.3) to calculate the radius of convergence

$$r^* = \lim_{n \to \infty} \left| \frac{c_n}{c_{n+1}} \right| = \lim_{n \to \infty} \left| \frac{1/n!}{1/(n+1)!} \right| = \lim_{n \to \infty} |n+1| = \infty. \qquad \blacksquare$$

Example 10.1.3. Let us find the interval of convergence of

$$\sum_{k=0}^{\infty} \frac{(-1)^k x^k}{4^k(k+1)}.$$

The ratio test gives

$$r(x) = \lim_{k \to \infty} \left| \frac{f_{k+1}(x)}{f_k(x)} \right| = \lim_{k \to \infty} \left| \frac{(-1)^{k+1} x^{k+1} / [4^{k+1}(k+2)]}{(-1)^k x^k / [4^k(k+1)]} \right|$$

$$= \lim_{k \to \infty} \left| \frac{x(k+1)}{4(k+2)} \right| = \left| \frac{x}{4} \right| \underbrace{\lim_{k \to \infty} \left| \frac{k+1}{k+2} \right|}_{=1} = \frac{|x|}{4}.$$

So, the series converges if $r(x) < 1$, i.e., if $|x| < 4$, or $-4 < x < 4$.

What about the end points? For $x = 4$, the series becomes

$$\sum_{k=0}^{\infty} \frac{(-1)^k}{k+1}$$

which is the alternating series and it converges. On the other hand, if $x = -4$, the series becomes

$$\sum_{k=0}^{\infty} \frac{(-1)^k (-4)^k}{4^k (k+1)} = \sum_{k=0}^{\infty} \frac{(-1)^k (-1)^k}{k+1} = \sum_{k=0}^{\infty} \frac{1}{k+1},$$

which is the divergent harmonic series. So, the interval of convergence of the series is $-4 < x \le 4$, and its radius of convergence is $r^* = 4$.   ■

Because of the uniform convergence of power series, we can perform all the common operations used for ordinary functions on the power series. We list all these properties in the following:

▶ **Continuity.** A power series represents a continuous function within its radius of convergence; i.e., if r^* is the radius of convergence, then the series [a convergent power series represents a continuous function]

$$f(x) = \sum_{n=0}^{\infty} c_n (x-a)^n \qquad \text{for} \quad a - r^* < x < a + r^* \qquad (10.5)$$

is continuous.

▶ **Integration.** The power series (10.5) can be integrated term by term within its radius of convergence; i.e., for $a - r^* < p < q < a + r^*$, [a convergent power series can be integrated term by term]

$$\int_p^q f(t)\, dt = \sum_{n=0}^{\infty} c_n \int_p^q (t-a)^n dt = \sum_{n=0}^{\infty} c_n \frac{(q-a)^{n+1} - (p-a)^{n+1}}{n+1}.$$

$$(10.6)$$

▶ **Differentiation.** The power series (10.5) can be differentiated term by term within its radius of convergence; that is, [a convergent power series can be differentiated term by term]

$$f'(x) = \sum_{n=1}^{\infty} n c_n (x-a)^{n-1}, \quad a - r^* < x < a + r^*. \qquad (10.7)$$

▶ **Zero Power Series.** If a power series has nonzero radius of convergence and has a sum which is identically zero, then every coefficient of the series must be zero. This leads to the following [if two power series are equal, so are their corresponding coefficients]

Theorem 10.1.4. *If two power series $\sum_{n=0}^{\infty} c_n(x-a)^n$ and $\sum_{n=0}^{\infty} b_n$ $(x-a)^n$ have nonzero convergence radii and have equal sums whenever both series converge, then the two series are identical, i.e.,*

$$c_n = b_n, \quad n = 0, 1, 2, \ldots.$$

This property is very effectively used to find solutions of differential equations in terms of infinite power series.

10.1.1   Taylor Series

A power series whose coefficients are derivatives of the function representing the sum is called **Taylor series**. More precisely, let

$$f(x) = \sum_{n=0}^{\infty} c_n(x-a)^n, \qquad a - r^* < x < a + r^*. \tag{10.8}$$

This series is called the Taylor series of $f(x)$ at $x = a$ if the coefficients c_n are given by the rule:

$$c_0 = f(a), \quad c_1 = \frac{f'(a)}{1!}, \quad c_2 = \frac{f''(a)}{2!}, \ldots, \quad c_k = \frac{f^{(k)}(a)}{k!},$$

Taylor series   so that

$$f(x) = f(a) + \frac{f'(a)}{1!}(x-a) + \cdots + \frac{f^{(k)}(a)}{k!}(x-a)^k + \cdots$$

$$= \sum_{k=0}^{\infty} \frac{f^{(k)}(a)}{k!}(x-a)^k \qquad \text{where} \quad f^{(0)}(a) \equiv f(a), \ 0! \equiv 1. \tag{10.9}$$

From Theorem 10.1.4 and the equality of (10.8) and (10.9), we conclude that every power series with nonzero convergence radius is the Taylor series of the function denoting its sum, and conversely every infinitely differentiable function can be represented by a Taylor series within the interval of convergence of the series.

Taylor series and   An alternative way of writing the Taylor series which suggests approxima-
approximating   tion is to let $\Delta x = x - a$. Then Equation (10.9) becomes
functions

$$f(a + \Delta x) = f(a) + \frac{f'(a)}{1!}\Delta x + \cdots = \sum_{k=0}^{\infty} \frac{f^{(k)}(a)}{k!}(\Delta x)^k.$$

Since a is an arbitrary real number, we can replace it with x which is more suggestive of the generality of this formula:

$$f(x + \Delta x) = f(x) + \frac{f'(x)}{1!}\Delta x + \cdots = \sum_{k=0}^{\infty} \frac{f^{(k)}(x)}{k!}(\Delta x)^k. \tag{10.10}$$

With Δx interpreted as the increment in x, Equation (10.10) states that the function at the incremented value of x is $f(x)$ plus a "correction" involving all powers of Δx. The smaller the increment, the smaller the number of terms of the correction we need to keep to achieve a given accuracy.

A convenient value for a is 0, in which case the series is called **Maclaurin series**:

$$f(x) = f(0) + \frac{f'(0)}{1!}x + \cdots = \sum_{k=0}^{\infty} \frac{f^{(k)}(0)}{k!}x^k. \qquad (10.11)$$

10.2 Series for Some Familiar Functions

In this subsection, we give the Maclaurin series representation of a few familiar functions. These representations are so useful that the reader is urged to commit them to memory.

The Exponential Function

For e^x, the derivatives of all orders are e^x implying that $f^{(n)}(0) = 1$ for all n. Therefore,

Maclaurin series of exponential function

$$e^x = 1 + \frac{x}{1!} + \frac{x^2}{2!} + \cdots = \sum_{n=0}^{\infty} \frac{x^n}{n!}. \qquad (10.12)$$

This series converges uniformly for all x as we saw in Example 10.1.2.

The Trigonometric Functions

The sine function has the following derivatives:

Maclaurin series of trigonometric function

$$f'(x) = \cos x, \ f''(x) = -\sin x, \ f'''(x) = -\cos x, \ f^{(\text{iv})}(x) = \sin x, \ \ldots.$$

This can be summarized as

$$f^{(n)}(x) = \begin{cases} (-1)^{n/2} \sin x & \text{if } n \text{ is even,} \\ (-1)^{(n-1)/2} \cos x & \text{if } n \text{ is odd.} \end{cases}$$

Evaluating at $x = 0$ for the Maclaurin series yields

$$f^{(n)}(0) = \begin{cases} 0 & \text{if } n \text{ is even,} \\ (-1)^{(n-1)/2} & \text{if } n \text{ is odd,} \end{cases}$$

so that

$$\sin x = 0 + x - 0 - \frac{x^3}{3!} + 0 + \frac{x^5}{5!} - \cdots = \sum_{k=0}^{\infty} (-1)^k \frac{x^{2k+1}}{(2k+1)!}. \qquad (10.13)$$

The combination $2k+1$ ensures that only odd terms are included even though there is no restriction on the sum over k. The radius of convergence is

$$r^* = \lim_{k \to \infty} \left| \frac{(-1)^k/(2k+1)!}{(-1)^{k+1}/(2k+3)!} \right| = \lim_{k \to \infty} \frac{(2k+3)!}{(2k+1)!} = \infty.$$

Thus the Taylor series representation of the sine function is convergent for all x.

The Maclaurin series representation of the cosine function can be obtained similarly. We leave the details to the reader, and simply quote the result:

$$\cos x = \sum_{k=0}^{\infty} (-1)^k \frac{x^{2k}}{(2k)!}, \qquad -\infty < x < \infty. \qquad (10.14)$$

The Binomial Function

Another useful function which is used extensively in physics is the binomial function with arbitrary exponent, i.e., $(1+x)^\alpha$ with α an arbitrary real number. It is easy to find the nth derivative of this function:

$$f^{(n)}(x) = \alpha(\alpha - 1)(\alpha - 2) \cdots (\alpha - n + 1)(1+x)^{\alpha - n}, \qquad n \geq 1.$$

Evaluating this at $x = 0$ gives

$$c_n = \frac{f^{(n)}(0)}{n!} = \frac{\alpha(\alpha - 1) \cdots (\alpha - n + 1)}{n!}, \qquad n \geq 1.$$

From this, we can immediately find the radius of convergence:

$$r^* = \lim_{n \to \infty} \left| \frac{c_n}{c_{n+1}} \right| = \lim_{n \to \infty} \left| \frac{\alpha(\alpha - 1) \cdots (\alpha - n + 1)}{n!} \cdot \frac{(n+1)!}{\alpha(\alpha - 1) \cdots (\alpha - n)} \right|$$

$$= \lim_{n \to \infty} \left| \frac{n+1}{\alpha - n} \right| = 1.$$

Maclaurin series of binomial function

Thus, the series is convergent for $-1 < x < 1$, and we can write

$$(1+x)^\alpha = 1 + \sum_{n=1}^{\infty} \frac{\alpha(\alpha - 1) \cdots (\alpha - n + 1)}{n!} x^n, \qquad -1 < x < 1. \qquad (10.15)$$

Example 10.2.1. Because of the frequent occurrence of the square root, we work through the calculation of (10.15) for $\alpha = \pm \frac{1}{2}$. For $\alpha = +\frac{1}{2}$, we have

$$\sqrt{1+x} = (1+x)^{1/2} = 1 + \sum_{n=1}^{\infty} \frac{\frac{1}{2}(\frac{1}{2} - 1) \cdots (\frac{1}{2} - n + 1)}{n!} x^n$$

$$= 1 + \tfrac{1}{2}x + \sum_{n=2}^{\infty} (-1)^{n-1} \frac{1 \cdot 3 \cdot 5 \cdots (2n - 3)}{2^n n!} x^n.$$

Now let $n = m + 1$ and rewrite the sum as

$$\sqrt{1+x} = 1 + \tfrac{1}{2}x + \sum_{m=1}^{\infty} (-1)^m \frac{1 \cdot 3 \cdot 5 \cdots (2m - 1)}{2^{m+1}(m + 1)!} x^{m+1}$$

$$= 1 + \tfrac{1}{2}x - \tfrac{1}{8}x^2 + \tfrac{3}{48}x^3 - \cdots. \qquad (10.16)$$

The case of $\alpha = -\frac{1}{2}$ can be handled in exactly the same way. We simply quote the result

$$\frac{1}{\sqrt{1+x}} = 1 + \sum_{m=1}^{\infty} (-1)^m \frac{1 \cdot 3 \cdot 5 \cdots (2m-1)}{2^m \, m!} x^m$$

$$= 1 - \tfrac{1}{2}x + \tfrac{3}{8}x^2 - \tfrac{15}{48}x^3 \cdots, \tag{10.17}$$

and urge the reader to fill in the details.   ∎

It is important to note the limitations of the power series representation of a function: Although $(1+x)^\alpha$ is defined for all positive[1] values of x, the power series *representation* of it is good only for a limited region of the real line.

In many applications, the binomial function appears in the form $(u + v)^\alpha$ where $|v| < |u|$ and one is interested in the power series expansion in v/u. This is easily done:

$$(u+v)^\alpha = \left\{ u \left(1 + \frac{v}{u} \right) \right\}^\alpha = u^\alpha \left(1 + \frac{v}{u} \right)^\alpha$$

$$= u^\alpha + u^\alpha \sum_{n=1}^{\infty} \frac{\alpha(\alpha-1)\cdots(\alpha-n+1)}{n!} \left(\frac{v}{u} \right)^n \tag{10.18}$$

$$= u^\alpha + \sum_{n=1}^{\infty} \frac{\alpha(\alpha-1)\cdots(\alpha-n+1)}{n!} u^{\alpha-n} v^n, \qquad -|u| < v < |u|.$$

In practice, v is usually much smaller than u, and the requirement of convergence is overwhelmingly met.

The Hyperbolic Functions

The exponential function and the trigonometric functions have very similar power series: Except for (the crucial) coefficient $(-1)^k$, $\sin x$ appears to be the odd part of the expansion of e^x and $\cos x$ its even part. The $(-1)^k$ factor makes the trigonometric functions periodic. What if we take this factor away, and simply collect the even powers of e^x together and do the same to the odd powers? The resulting series will of course be (absolutely and uniformly) convergent because the exponential is so. So, let us introduce the following functions:

Maclaurin series of hyperbolic functions

$$\sinh x \equiv \sum_{k=0}^{\infty} \frac{x^{2k+1}}{(2k+1)!} = x + \frac{x^3}{3!} + \frac{x^5}{5!} + \cdots,$$

$$\cosh x \equiv \sum_{k=0}^{\infty} \frac{x^{2k}}{(2k)!} = 1 + \frac{x^2}{2!} + \frac{x^4}{4!} + \cdots, \tag{10.19}$$

---

[1] It is really defined for more than just positive values. For instance, if α is an integer, the function is defined for all values of x. For fractional powers such as $\alpha = 1/2$, $1 + x$ cannot be negative, so that we must restrict the values of x to $x > -1$.

$\sinh x$ (pronounced "sinch") is called the **hyperbolic sine** function. Similarly, $\cosh x$ (pronounced "kahsh") is called the **hyperbolic cosine** function. By their very definition, we have

$$e^x = \cosh x + \sinh x.$$

If we change x to $-x$, and note that $\sinh x$ is odd and $\cosh x$ is even, we can also write

$$e^{-x} = \cosh(-x) + \sinh(-x) = \cosh x - \sinh x.$$

Adding and subtracting the last two equations yields

$$\cosh x = \frac{e^x + e^{-x}}{2}, \qquad \sinh x = \frac{e^x - e^{-x}}{2}. \tag{10.20}$$

This is how the hyperbolic functions are usually defined. From these definitions, one can obtain a host of relations for the sinh and cosh that look similar to the relations satisfied by sine and cosine. For example, it is easy to show that

$$\cosh^2 x - \sinh^2 x = 1, \qquad \frac{d}{dx}\cosh x = \sinh x, \qquad \frac{d}{dx}\sinh x = \cosh x,$$

$$\cosh(x \pm y) = \cosh x \cosh y \pm \sinh x \sinh y, \tag{10.21}$$

$$\sinh(x \pm y) = \sinh x \cosh y \pm \cosh x \sinh y,$$

$$\cosh(2x) = \cosh^2 x + \sinh^2 x, \qquad \sinh(2x) = 2\sinh x \cosh x.$$

We give the derivation for the hyperbolic cosine of the sum, leaving the rest of them as problems for the reader. We start with the RHS:

$$\cosh x \cosh y + \sinh x \sinh y$$

$$= \left(\frac{e^x + e^{-x}}{2}\right)\left(\frac{e^y + e^{-y}}{2}\right) + \left(\frac{e^x - e^{-x}}{2}\right)\left(\frac{e^y - e^{-y}}{2}\right)$$

$$= \frac{(e^x + e^{-x})(e^y + e^{-y}) + (e^x - e^{-x})(e^y - e^{-y})}{4}$$

$$= \frac{e^{x+y} + e^{x-y} + e^{-x+y} + e^{-x-y} + e^{x+y} - e^{x-y} - e^{-x+y} + e^{-x-y}}{4}$$

$$= \frac{2e^{x+y} + 2e^{-x-y}}{4} = \frac{e^{x+y} + e^{-x-y}}{2} = \cosh(x+y).$$

We can also define the analogs of other trigonometric functions:

$$\tanh x \equiv \frac{\sinh x}{\cosh x} = \frac{e^x - e^{-x}}{e^x + e^{-x}}, \qquad \coth x \equiv \frac{\cosh x}{\sinh x} = \frac{e^x + e^{-x}}{e^x - e^{-x}}, \tag{10.22}$$

$$\operatorname{sech} x \equiv \frac{1}{\cosh x} = \frac{2}{e^x + e^{-x}}, \qquad \operatorname{cosech} x \equiv \frac{1}{\sinh x} = \frac{2}{e^x - e^{-x}}.$$

These functions have such properties as

$$\operatorname{sech}^2 x = 1 - \tanh^2 x, \qquad \operatorname{cosech}^2 x = \coth^2 x - 1,$$

and

$$\frac{d}{dx}\tanh x = \operatorname{sech}^2 x, \qquad \frac{d}{dx}\coth x = -\operatorname{cosech}^2 x.$$

The Logarithmic Function

Finally, we state the Maclaurin series for $\ln(1 + x)$, which occurs frequently in physics, and which the reader can verify:

Maclaurin series of logarithmic function

$$\ln(1 + x) = \sum_{n=1}^{\infty}(-1)^{n+1}\frac{x^n}{n}, \qquad -1 < x < 1. \qquad (10.23)$$

10.3   Helmholtz Coil

Power series are very useful tools for approximating functions, and the closer one gets to the point of expansion, the better the approximation. The essence of this approximation is replacing the infinite series with a finite sum, i.e., approximating the function with a polynomial.

In general, to get a very good approximation, one has to retain very large powers of the power series. So, the approximating polynomial will have to be of a high degree. However, suppose that a function $f(x)$ has the following expansion

$$f(x) = c_0 + c_1(x - a) + \cdots + c_m(x - a)^m + c_m(x - a)^{m+k} + \cdots,$$

where k is a fairly large number. Then the polynomial

$$p(x) = c_0 + c_1(x - a) + \cdots + c_m(x - a)^m$$

approximates the function very accurately because, as long as we are "close" enough to the point of expansion a, the next term in the series will not affect the polynomial much. In particular, if the series looks like

$$f(x) = c_0 + c_k(x - a)^k + \cdots, \qquad (10.24)$$

then the constant "polynomial" c_0 is an extremely good approximation to the function for values of x close to a.

The argument above can be used to design devices to produce physical quantities that are constant for a fairly large values of the variable on which the outcome of the device depends. A case in point is the **Helmholtz coil**, which is used frequently in laboratory situations in which homogeneous magnetic fields are desirable.

Figure 10.1 shows two loops of current-carrying wires of radii a and b separated by a distance L. We are interested in the z-component of the magnetic field midway between the two loops, which, to simplify expressions, we have chosen to be the origin. Example 4.1.4 gives the expression for the magnetic field of a loop at a point on its axis at a distance z from its center.

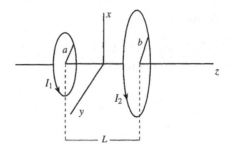

Figure 10.1: Two circular loops with different radii producing a magnetic field.

Let us denote the magnetic field of the loop of radius a by B_1 and that of the loop of radius b by B_2. Then Example 4.1.4 gives

$$B(z) \equiv B_1(z) + B_2(z) = \frac{2\pi k_m I_1 a^2}{[a^2 + (z + L/2)^2]^{3/2}} + \frac{2\pi k_m I_2 b^2}{[b^2 + (z - L/2)^2]^{3/2}}$$

$$= \frac{16\pi k_m I_1 a^2}{[4a^2 + (2z + L)^2]^{3/2}} + \frac{16\pi k_m I_2 b^2}{[4b^2 + (2z - L)^2]^{3/2}}. \tag{10.25}$$

We want to adjust the parameters of the two loops in such a way that the magnetic field at the origin is maximally homogeneous. This can be accomplished by setting as many derivatives of $B(z)$ equal to zero at the origin as possible, so that the Maclaurin expansion of $B(z)$ will have a maximum number of consecutive terms equal to zero, i.e., we will have an expression of the form (10.24).

The first derivative of $B(z)$ is

$$\frac{dB}{dz} = -\frac{96\pi k_m I_1 a^2 (2z + L)}{[4a^2 + (2z + L)^2]^{5/2}} - \frac{96\pi k_m I_2 b^2 (2z - L)}{[4b^2 + (2z - L)^2]^{5/2}}.$$

Setting this equal to zero at $z = 0$ gives

$$\frac{I_1 a^2}{(4a^2 + L^2)^{5/2}} = \frac{I_2 b^2}{(4b^2 + L^2)^{5/2}}. \tag{10.26}$$

The second derivative of $B(z)$ is

$$\frac{d^2 B}{dz^2} = -\frac{768\pi k_m I_1 a^2 [a^2 - (2z + L)^2]}{[4a^2 + (2z + L)^2]^{7/2}} - \frac{768\pi k_m I_2 b^2 [b^2 - (2z - L)^2]}{[4b^2 + (2z - L)^2]^{7/2}}.$$

Setting this equal to zero at $z = 0$ gives

$$\frac{I_1 a^2 (a^2 - L^2)}{(4a^2 + L^2)^{7/2}} + \frac{I_2 b^2 (b^2 - L^2)}{(4b^2 + L^2)^{7/2}} = 0. \tag{10.27}$$

Since both terms are positive, the only way that we can get zero in (10.27) is if each term on the LHS vanishes. It follows that $a = L = b$. Substituting this in Equation (10.26) gives $I_1 = I_2$ which we denote by I. Therefore, we can now write the magnetic field as

$$B(z) = 16\pi k_m I a^2 \left\{ \frac{1}{[4a^2 + (2z + a)^2]^{3/2}} + \frac{1}{[4a^2 + (2z - a)^2]^{3/2}} \right\}. \quad (10.28)$$

The reader may verify that not only are the first and the second derivatives of $B(z)$ of Equation (10.28) zero, but also its third derivative. In fact, we have

$$B(z) = \frac{32\pi k_m I}{5\sqrt{5}a} - \frac{4608\pi k_m I}{625\sqrt{5}a^5} z^4 + \cdots. \quad (10.29)$$

That only even powers appear in the expansion (10.29) could have been anticipated, because (10.28) is even in z as the reader may easily verify. It follows from Equation (10.29) that $B(z)$ should be fairly insensitive to the variation of z at points close to the origin. Physically, this means that the magnetic field is fairly homogeneous at the midpoint between the two loops as long as the loops are equal and separated by a distance equal to their common radius, and as long as they carry the same current. Figure 10.2 shows the plot of the magnetic field as a function of z. Note how flat the function is for even fairly large values of z.

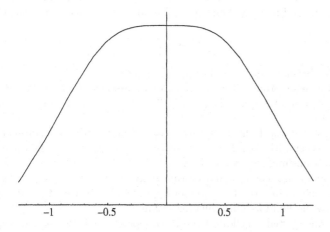

Figure 10.2: Magnetic field of a Helmholtz coil as a function of z. The horizontal axis is z in units of a.

Historical Notes

One of the problems faced by mathematicians of the late seventeenth and early eighteenth centuries was interpolation (the word was coined by Wallis) of tables of values. Greater accuracy of the interpolated values of the trigonometric, logarithmic, and nautical tables was necessary to keep pace with progress in navigation, astronomy, and geography. The common method of interpolation, whereby one takes the average of the two consecutive entries of a table, is called *linear interpolation* because it gives the exact result for a linear function. This gives a crude approximation for functions that are not linear, and mathematicians realized that a better method of interpolation was needed.

The general method which can give interpolations that are more and more accurate was given by **Gregory** and independently by **Newton**. Suppose $f(x)$ is a function whose values are given at a, $a + h$, $a + 2h$, $\ldots$, and we are interested in the value of the function at an x that lies between two table entries. The **Gregory–Newton formula** states that

$$f(a + r) = f(a) + \frac{r}{h}\Delta f(a) + \frac{\frac{r}{h}\left(\frac{r}{h} - 1\right)}{2!}\Delta^2 f(a) + \frac{\frac{r}{h}\left(\frac{r}{h} - 1\right)\left(\frac{r}{h} - 2\right)}{3!}\Delta^3 f(a) + \cdots,$$

where

$$\Delta f(a) = f(a + h) - f(a), \qquad \Delta^2 f(a) = \Delta f(a + h) - \Delta f(a),$$
$$\Delta^3 f(a) = \Delta^2 f(a + h) - \Delta^2 f(a), \quad \Delta^4 f(a) = \Delta^3 f(a + h) - \Delta^3 f(a), \ldots$$

To calculate f at any value y between the known values, one simply substitutes $y - a$ for r.

Brook Taylor's *Methodus incrementorum directa et inversa*, published in 1715, added to mathematics a new branch now called the *calculus of finite differences*, and he invented integration by parts. It also contained the celebrated formula known as Taylor's expansion, the importance of which remained unrecognized until 1772 when **Lagrange** proclaimed it the basic principle of the differential calculus.

Brook Taylor
1685–1731

To arrive at the series that bears his name, Taylor let h in the Gregory–Newton formula be Δx and took the limit of smaller and smaller Δx. Thus, the third term, for example, gave

$$\frac{r(r - \Delta x)}{2!}\frac{\Delta^2 f(a)}{\Delta x^2} \to \frac{r^2}{2!}f''(a)$$

which is the familiar third term in the Taylor series.

In 1708 Taylor produced a solution to the problem of the center of oscillation which, since it went unpublished until 1714, resulted in a priority dispute with *Johann Bernoulli*.

Taylor also devised the basic principles of perspective in *Linear Perspective* (1715). Together with *New Principles of Linear Perspective* the first general treatment of the vanishing points are given.

Taylor gives an account of an experiment to discover the law of magnetic attraction (1715) and an improved method for approximating the roots of an equation by giving a new method for computing logarithms (1717).

Taylor was elected a Fellow of the Royal Society in 1712 and was appointed in that year to the committee for adjudicating the claims of Newton and of Leibniz to have invented the calculus.

10.4   Indeterminate Forms and L'Hôpital's Rule

It is good practice to approximate functions with their power series representations, keeping as many terms as is necessary for a given accuracy. This practice is especially useful when encountering indeterminate expressions of the form $\frac{0}{0}$. Although L'Hôpital's rule (discussed below) can be used to find the ratio, on many occasions the substitution of the series leads directly to the answer, saving us the labor of multiple differentiation.

Example 10.4.1. Let us look at some examples of the ratios mentioned above. In all cases treated in this example, the substitution $x = 0$ gives $\frac{0}{0}$, which is indeterminate. Using the Maclaurin series (10.12) and (10.13), we get

$$\lim_{x \to 0} \frac{2e^x - 2 - 2x - x^2}{\sin x - x}$$

$$= \lim_{x \to 0} \frac{2(1 + x + x^2/2 + x^3/6 + x^4/24 + \cdots) - 2 - 2x - x^2}{x - x^3/6 + x^5/120 + \cdots - x}$$

$$= \lim_{x \to 0} \frac{x^3/3 + x^4/12 + \cdots}{-x^3/6 + x^5/120 - \cdots} = \lim_{x \to 0} \frac{1/3 + x/12 + \cdots}{-1/6 + x^2/120 - \cdots} = -2.$$

The series (10.14) and (10.23) can be used to evaluate the following limit:

$$\lim_{x \to 0} \frac{\ln(1 + x) - x}{\cos x - 1}$$

$$= \lim_{x \to 0} \frac{x - x^2/2 + x^3/3 - \cdots - x}{1 - x^2/2 + x^4/24 - \cdots - 1}$$

$$= \lim_{x \to 0} \frac{-x^2/2 + x^3/3 - \cdots}{-x^2/2 + x^4/24 - \cdots} = \lim_{x \to 0} \frac{-1/2 + x/3 - \cdots}{-1/2 + x^2/24 - \cdots} = 1.$$

With (10.12) and (10.15), we have

$$\lim_{x \to 0} \frac{\sqrt{1 + 2x} - x - 1}{e^{x^2} - 1}$$

$$= \lim_{x \to 0} \frac{1 + \frac{1}{2}(2x) + \frac{\frac{1}{2}(\frac{1}{2}-1)}{2!}(2x)^2 + \frac{\frac{1}{2}(\frac{1}{2}-1)(\frac{1}{2}-2)}{3!}(2x)^3 + \cdots - x - 1}{1 + x^2 + (x^2)^2/2! + \cdots - 1}$$

$$= \lim_{x \to 0} \frac{-x^2/2 + x^3/2 + \cdots}{x^2 + x^4/2 + \cdots} = \lim_{x \to 0} \frac{-1/2 + x/2 + \cdots}{1 + x^2/2 + \cdots} = -\frac{1}{2}. \quad \blacksquare$$

The method of expanding the numerator and denominator of a ratio as a Taylor series is extremely useful in applications in which mere substitution results in the indeterminate expression[2] of the form $\frac{0}{0}$. However, there are many other indeterminate forms that occur in applications. For example, a mere substitution of $x = 0$ in $(1+x)^{1/x}$ yields 1^∞ which is also indeterminate. Other examples of indeterminate expressions are $0 \times \infty$, $\frac{\infty}{\infty}$, 0^0, and ∞^0. Most of these expressions can be reduced to indeterminate ratios for which one can use **l'Hôpital's rule**:

l'Hôpital's rule

Box 10.4.1. (***L'Hôpital's Rule***). *If $f(a)/g(a)$ is indeterminate, then*

$$\lim_{x \to a} \frac{f(x)}{g(x)} = \lim_{x \to a} \frac{f'(x)}{g'(x)}, \tag{10.30}$$

where f' and g' are derivatives of f and g, respectively.

---

[2]An expression is indeterminate if it involves two parts each of which gives a result that is contradictory to the other. Thus the numerator of the ratio $\frac{0}{0}$ says that the ratio should be zero, while the denominator says that the ratio should be infinite.

In practice, one converts the indeterminate form into a ratio and differentiates the numerator and denominator as many times as necessary until one obtains a definite result or infinity. The following general rules can be of help:

- If $f(a) = 0$ and $g(a) = \infty$, then to find $\lim_{x \to a} f(x)g(x)$, rewrite the limit as

$$\lim_{x \to a} f(x)g(x) = \lim_{x \to a} \frac{f(x)}{\left[\dfrac{1}{g(x)}\right]} \quad \text{or} \quad \lim_{x \to a} f(x)g(x) = \lim_{x \to a} \frac{g(x)}{\left[\dfrac{1}{f(x)}\right]},$$

the first of which gives $\frac{0}{0}$ and the second $\frac{\infty}{\infty}$. In either case, one can apply L'Hôpital's rule.

- If $f(a) = 1$ and $g(a) = \infty$, first define $h(x) \equiv [f(x)]^{g(x)}$. Then to find

$$\lim_{x \to a} h(x) = \lim_{x \to a} [f(x)]^{g(x)},$$

take the natural logarithm of $h(x)$ and convert the result into the ratio

$$\lim_{x \to a} \ln[h(x)] = \lim_{x \to a} g(x) \ln[f(x)] = \lim_{x \to a} \frac{\ln[f(x)]}{\left[\dfrac{1}{g(x)}\right]}.$$

Then use Equation (10.30).

- If $f(a) = \infty$ (or $f(a) = 0$) and $g(a) = 0$, then to find

$$\lim_{x \to a} h(x) \equiv \lim_{x \to a} [f(x)]^{g(x)},$$

take the natural logarithm of $h(x)$ and convert the result into the ratio

$$\lim_{x \to a} \ln[h(x)] = \lim_{x \to a} g(x) \ln[f(x)] = \lim_{x \to a} \frac{\ln[f(x)]}{\left[\dfrac{1}{g(x)}\right]}.$$

Then use Equation (10.30).

Example 10.4.2. To find the $\lim_{x \to 0}(1+2x)^{1/x}$, we write $h(x) \equiv (1+2x)^{1/x}$ and note that

$$\lim_{x \to 0} \ln[h(x)] = \lim_{x \to 0} (1/x) \ln(1 + 2x) = \lim_{x \to 0} \frac{\ln(1 + 2x)}{x}$$

is indeterminate. Using Equation (10.30) yields

$$\lim_{x \to 0} \ln[h(x)] = \lim_{x \to 0} \frac{\ln(1 + 2x)}{x} = \lim_{x \to 0} \frac{\dfrac{2}{1 + 2x}}{1} = \lim_{x \to 0} \frac{2}{1 + 2x} = 2.$$

Therefore, $\lim_{x \to 0} h(x) = e^2$.

To find the $\lim_{x \to 0} x^x$, we write $h(x) \equiv x^x$ and note that

$$\lim_{x \to 0} \ln[h(x)] = \lim_{x \to 0} x \ln x = \lim_{x \to 0} \frac{\ln x}{1/x}$$

is indeterminate. Using Equation (10.30) yields

$$\lim_{x \to 0} \ln[h(x)] = \lim_{x \to 0} \frac{1/x}{-1/x^2} = \lim_{x \to 0} (-x) = 0.$$

Therefore, $\lim_{x \to 0} h(x) = e^0 = 1$. So, we have the interesting result $\lim_{x \to 0} x^x = 1$.
    The limit of $x^2/(1 - \cos x)$ as x goes to zero is obtained as follows:

$$\lim_{x \to 0} \frac{x^2}{1 - \cos x} = \lim_{x \to 0} \frac{2x}{\sin x} = \lim_{x \to 0} \frac{2}{\cos x} = 2.$$

Here we had to differentiate twice because the ratio of the first derivatives was also indeterminate.   ∎

It is instructive for the reader to verify all limits in Example 10.4.1 using L'Hôpital's rule to appreciate the ease of the Taylor expansion method.

10.5   Multipole Expansion

One extremely useful application of the power series representation of functions is in potential theory. The electrostatic or gravitational potential can be written as

$$\Phi(\mathbf{r}) = K \int_{\Omega} \frac{dQ(\mathbf{r}')}{|\mathbf{r} - \mathbf{r}'|}, \tag{10.31}$$

where K is k_e for electrostatics and $-G$ for gravity. Similarly, Q represents either electric charge or mass. In some applications, especially for electrostatic potential, the distance of the field point P from the origin is much larger than the distance of the source point P' from the origin. This means that $r >> r'$ and we can expand in the powers of the ratio r'/r which we denote by ϵ. The key to this expansion is a power series expansion of $1/|\mathbf{r} - \mathbf{r}'|$. First write

$$\frac{1}{|\mathbf{r} - \mathbf{r}'|} = \frac{1}{\sqrt{r^2 + r'^2 - 2\mathbf{r} \cdot \mathbf{r}'}} = \frac{1}{r\sqrt{1 + \epsilon^2 - 2\epsilon \hat{\mathbf{e}}_r \cdot \hat{\mathbf{e}}_{r'}}}$$
$$= \frac{1}{r} \left(1 + \epsilon^2 - 2\epsilon \hat{\mathbf{e}}_r \cdot \hat{\mathbf{e}}_{r'} \right)^{-1/2}.$$

Next use the binomial expansion (10.15) with $x = \epsilon^2 - 2\epsilon \hat{\mathbf{e}}_r \cdot \hat{\mathbf{e}}_{r'}$ and $\alpha = -\frac{1}{2}$. Up to second order in ϵ, this yields

$$\frac{1}{|\mathbf{r} - \mathbf{r}'|} = \frac{1}{r} \left\{ 1 - \tfrac{1}{2} \left(\epsilon^2 - 2\epsilon \hat{\mathbf{e}}_r \cdot \hat{\mathbf{e}}_{r'} \right) + \tfrac{3}{8} \left(\epsilon^2 - 2\epsilon \hat{\mathbf{e}}_r \cdot \hat{\mathbf{e}}_{r'} \right)^2 + \cdots \right\}$$
$$= \frac{1}{r} \left\{ 1 + \epsilon \hat{\mathbf{e}}_r \cdot \hat{\mathbf{e}}_{r'} + \epsilon^2 \left[-\tfrac{1}{2} + \tfrac{3}{2} (\hat{\mathbf{e}}_r \cdot \hat{\mathbf{e}}_{r'})^2 \right] + \cdots \right\}$$
$$= \frac{1}{r} + \frac{\hat{\mathbf{e}}_r \cdot \mathbf{r}'}{r^2} + \frac{r'^2}{r^3} \left[-\tfrac{1}{2} + \tfrac{3}{2} (\hat{\mathbf{e}}_r \cdot \hat{\mathbf{e}}_{r'})^2 \right] + \cdots. \tag{10.32}$$

Substituting this in Equation (10.31), we obtain

$$\Phi(\mathbf{r}) = \frac{K}{r}\int_\Omega dQ(\mathbf{r}') + \frac{K}{r^2}\hat{\mathbf{e}}_r \cdot \int_\Omega \mathbf{r}'\, dQ(\mathbf{r}')$$
$$+ \frac{K}{r^3}\int_\Omega r'^2 \left[-\tfrac{1}{2} + \tfrac{3}{2}(\hat{\mathbf{e}}_r \cdot \hat{\mathbf{e}}_{r'})^2\right] dQ(\mathbf{r}') + \cdots \qquad (10.33)$$
$$= \frac{KQ}{r} + \frac{K}{r^2}\hat{\mathbf{e}}_r \cdot \mathbf{p}_Q + \frac{K}{r^3}\int_\Omega r'^2 \left[-\tfrac{1}{2} + \tfrac{3}{2}(\hat{\mathbf{e}}_r \cdot \hat{\mathbf{e}}_{r'})^2\right] dQ(\mathbf{r}') + \cdots,$$

where

$$Q \equiv \int_\Omega dQ(\mathbf{r}')$$

electric dipole moment defined

is the total Q (charge, or mass)—also called the *zeroth Q moment*—and

$$\mathbf{p}_Q \equiv \int_\Omega \mathbf{r}'\, dQ(\mathbf{r}') \qquad (10.34)$$

is the *first Q moment*, which in the case of charge is also called the **electric dipole moment**. One can also define higher moments.

If the source of the potential is discrete, the integral in Equation (10.31) becomes a sum. The steps leading to (10.33) will not change except for switching all the integrals to summations. In particular, the dipole moment of N point sources $\{Q_k\}_{k=1}^N$, located at $\{\mathbf{r}_k\}_{k=1}^N$, turns out to be

$$\mathbf{p}_Q = \sum_{k=1}^N Q_k \mathbf{r}_k. \qquad (10.35)$$

For the special case of two electric charges $q_1 = +q$ and $q_2 = -q$, we obtain[3]

$$\mathbf{p} = q\mathbf{r}_1 - q\mathbf{r}_2 = q(\mathbf{r}_1 - \mathbf{r}_2). \qquad (10.36)$$

Thus, the dipole moment of a pair of equal charges of opposite sign is the magnitude of the charge times the displacement *vector* from the negative to the positive charge.

Example 10.5.1. Electric dipoles are fairly abundant in Nature. For example, an antenna is approximated as a dipole at distances far away from it; and in atomic transitions one uses the so-called **dipole approximation** to calculate the rate of transition and the lifetime of a state.

dipole approximation

Let us write the explicit form of the potential of a dipole, i.e., the second term on the RHS of Equation (10.33). In Cartesian coordinates, in which the dipole moment is in the z-direction (so that $\mathbf{p} = p\hat{\mathbf{e}}_z$), the potential can be written as

---

[3]It is customary to denote the electric dipole moment by $\mathbf{p}$ with no subscript.

$$\Phi_{\text{dip}}(x,y,z) = \frac{k_e}{r^2}\hat{\mathbf{e}}_r \cdot \mathbf{p} = \frac{k_e}{r^3}\mathbf{r} \cdot \mathbf{p} = \frac{k_e p z}{(x^2 + y^2 + z^2)^{3/2}}.$$

More important is the expression for potential in spherical coordinates:

electric potential
of a dipole

$$\Phi_{\text{dip}}(r,\theta,\varphi) = \frac{k_e p}{r^2}\overbrace{\hat{\mathbf{e}}_r \cdot \hat{\mathbf{e}}_z}^{\cos\theta} = \frac{k_e p}{r^2}\cos\theta. \tag{10.37}$$

The azimuthal symmetry (independence of φ) comes about because we chose $\mathbf{p}$ to lie along the z-axis. ∎

10.6   Fourier Series

Power series are special cases of the series of functions in which the nth function is $(x - a)^n$—or simply x^n—multiplied by a constant. These functions, simple and powerful as they are, cannot be used in all physical applications. More general functions are needed for many problems in theoretical physics.

The most widely used series of functions in applications are Fourier series in which the functions are sines and cosines. These are especially suitable for **periodic** functions which repeat themselves with a certain period. Suppose that a function $f(x)$ is defined in the interval (a,b). Can we write it as a series in sines and cosines, as we did in terms of orthogonal polynomials [see Theorem 7.5.2]? Let $L = b - a$ denote the length of the interval, and consider the functions

periodic functions

$$\sin\frac{2n\pi x}{L}, \quad \cos\frac{2n\pi x}{L}.$$

Let us try the series expansion

Fourier series
expansion

$$f(x) = a_0 + \sum_{n=1}^{\infty}\left(a_n\cos\frac{2n\pi x}{L} + b_n\sin\frac{2n\pi x}{L}\right), \tag{10.38}$$

where we have separated the $n = 0$ term. Now the sine and cosine terms have the following easily obtainable useful properties:

$$\int_a^b \sin\frac{2n\pi x}{L}\,dx = \int_a^b \cos\frac{2n\pi x}{L}\,dx = \int_a^b \sin\frac{2n\pi x}{L}\cos\frac{2m\pi x}{L}\,dx = 0,$$

$$\int_a^b \sin\frac{2n\pi x}{L}\sin\frac{2m\pi x}{L}\,dx = \begin{cases} 0 & \text{if } m \neq n, \\ L/2 & \text{if } m = n \neq 0, \end{cases} \tag{10.39}$$

$$\int_a^b \cos\frac{2n\pi x}{L}\cos\frac{2m\pi x}{L}\,dx = \begin{cases} 0 & \text{if } m \neq n, \\ L/2 & \text{if } m = n \neq 0. \end{cases}$$

These properties suggest a way of determining the coefficients of the series for a given function as in the case of orthogonal polynomials. If we integrate both sides of Equation (10.38) from a to b, we get[4]

$$\int_a^b f(x)\,dx = a_0 \int_a^b dx + \int_a^b \sum_{n=1}^{\infty} \left(a_n \cos \frac{2n\pi x}{L} + b_n \sin \frac{2n\pi x}{L} \right) dx$$

$$= (b-a)a_0 + \sum_{n=1}^{\infty} a_n \underbrace{\int_a^b \cos \frac{2n\pi x}{L}\,dx}_{=0} + \sum_{n=1}^{\infty} b_n \underbrace{\int_a^b \sin \frac{2n\pi x}{L}\,dx}_{=0}$$

or $\int_a^b f(x)\,dx = a_0 L$. This yields

$$a_0 = \frac{1}{L} \int_a^b f(x)\,dx. \tag{10.40}$$

Multiplying Equation (10.38) by $\cos(2m\pi x/L)$ and integrating both sides from a to b, we obtain

$$\int_a^b f(x) \cos \frac{2m\pi x}{L}\,dx$$

$$= a_0 \int_a^b \cos \frac{2m\pi x}{L}\,dx + \int_a^b \sum_{n=1}^{\infty} \left(a_n \cos \frac{2n\pi x}{L} + b_n \sin \frac{2n\pi x}{L} \right) \cos \frac{2m\pi x}{L}\,dx$$

$$= 0 + \sum_{n=1}^{\infty} a_n \int_a^b \cos \frac{2n\pi x}{L} \cos \frac{2m\pi x}{L}\,dx + \sum_{n=1}^{\infty} b_n \int_a^b \sin \frac{2n\pi x}{L} \cos \frac{2m\pi x}{L}\,dx$$

$$= a_m L/2,$$

where we used Equation (10.39). This yields

$$a_n = \frac{2}{L} \int_a^b f(x) \cos \frac{2n\pi x}{L}\,dx. \tag{10.41}$$

Similarly, multiplying both sides of Equation (10.38) by $\sin(2m\pi x/L)$ and integrating from a to b, yields

$$b_n = \frac{2}{L} \int_a^b f(x) \sin \frac{2n\pi x}{L}\,dx. \tag{10.42}$$

Equations (10.38), (10.40), (10.41), and (10.42) provide a procedure for representing a function $f(x)$ as a Fourier series. However, the RHS of Equation (10.38) is periodic. This means that for values of x outside the interval (a, b), $f(x)$ is also periodic. In fact, from Equation (10.38), we have

---

[4]Here we are assuming that the series converges uniformly so that we can switch the order of integration and summation. This assumption turns out to be correct, but we shall forego its (difficult) proof.

$$f(x + L) = a_0 + \sum_{n=1}^{\infty} \left\{ a_n \cos \frac{2n\pi(x + L)}{L} + b_n \sin \frac{2n\pi(x + L)}{L} \right\}$$

$$= a_0 + \sum_{n=1}^{\infty} \left\{ a_n \cos \left(\frac{2n\pi x}{L} + 2n\pi \right) + b_n \sin \left(\frac{2n\pi x}{L} + 2n\pi \right) \right\}$$

$$= a_0 + \sum_{n=1}^{\infty} \left(a_n \cos \frac{2n\pi x}{L} + b_n \sin \frac{2n\pi x}{L} \right) = f(x).$$

Thus, $f(x)$ repeats itself at the end of each interval of length L, i.e., it is periodic with period L. Fourier series is especially suited for representing such functions. In fact, any periodic function has a Fourier series expansion, and the simplicity of sine and cosine functions makes this expansion particularly useful in applications such as electrical engineering and acoustics where periodic functions in the form of waves and voltages are daily occurrences. Let us look at some examples.[5]

Example 10.6.1. In the study of electrical circuitry, periodic voltage signals of different shapes are encountered. An example is the so-called **square wave** of height V_0, and duration and "rest duration" T [see Figure 10.3(top)]. The potential as a function of time, $V(t)$, can be expanded as a Fourier series. The interval is $(0, 2T)$,   square wave because that is one whole cycle of potential variation. We thus write   potential

$$V(t) = a_0 + \sum_{n=1}^{\infty} \left(a_n \cos \frac{2n\pi t}{2T} + b_n \sin \frac{2n\pi t}{2T} \right) \tag{10.43}$$

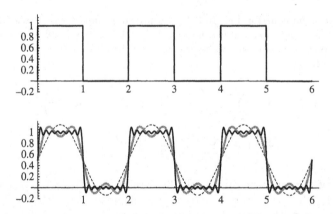

Figure 10.3: Top: The periodic square-wave potential with $V_0 = 1$ and $T = 2$. Bottom: Various approximations to the Fourier series of the square-wave potential. The dashed plot is that of the first term of the series, the thick gray plot keeps 3 terms, and the solid plot 15 terms.

---

[5]While Taylor series expansion demands that the function be (infinitely) differentiable, the orthogonal polynomial and Fourier series expansion require only **piecewise continuity**. This means that the function can have any (finite) number of discontinuities in the interval (a, b). Thus, the expanded function can not only be nondifferentiable, it can even be discontinuous.

with

$$a_0 = \frac{1}{2T} \int_0^{2T} V(t)\, dt,$$

$$a_n = \frac{2}{2T} \int_0^{2T} V(t) \cos \frac{2n\pi t}{2T}\, dt = \frac{1}{T} \int_0^{2T} V(t) \cos \frac{n\pi t}{T}\, dt, \qquad (10.44)$$

$$b_n = \frac{1}{T} \int_0^{2T} V(t) \sin \frac{n\pi t}{T}\, dt.$$

Substituting

$$V(t) = \begin{cases} V_0 & \text{if } 0 \le t \le T, \\ 0 & \text{if } T < t \le 2T, \end{cases}$$

in Equation (10.44), we obtain

$$a_0 = \frac{1}{2T} \int_0^T V_0\, dt = \tfrac{1}{2} V_0,$$

$$a_n = \frac{1}{T} \int_0^T V_0 \cos \frac{n\pi t}{T}\, dt = 0,$$

and

$$b_n = \frac{1}{T} \int_0^T V_0 \sin \frac{n\pi t}{T}\, dt = -\frac{V_0}{T} \frac{T}{n\pi} \cos \frac{n\pi t}{T} \Big|_0^T$$

$$= \frac{V_0}{n\pi}(1 - \cos n\pi) = \frac{V_0}{n\pi}\left[1 - (-1)^n\right].$$

Thus, there is no contribution from the cosine sum, and in the sine sum only the odd terms contribute ($b_n = 0$ if n is even). Therefore, let $n = 2k + 1$, where k now takes *all* values even and odd, and substitute all the above information in Equation (10.43), to obtain

$$V(t) = \tfrac{1}{2} V_0 + \sum_{k=0}^{\infty} \frac{V_0}{(2k+1)\pi}\left[1 - (-1)^{2k+1}\right] \sin \frac{(2k+1)\pi t}{T}$$

$$= \frac{V_0}{2}\left\{1 + \frac{4}{\pi} \sum_{k=0}^{\infty} \frac{\sin[(2k+1)\pi t/T]}{2k+1}\right\}.$$

The plots of the sum truncated at the first, third, and fifteenth terms are shown in Figure 10.3(bottom). Note how the Fourier approximation overshoots the value of the function at discontinuities. This is a general feature of all discontinuous functions and is called the **Gibb's phenomenon.**[6]

Gibb's phenomenon

sawtooth potential

Example 10.6.2. Another frequently used potential is the **sawtooth** potential. The interval is $(0, T)$ and the equation for the potential is

$$V(t) = V_0 \frac{t}{T} \qquad \text{for } 0 \le t < T.$$

[6]A discussion of Gibb's phenomenon can be found in Hassani, S. *Mathematical Physics: A Modern Introduction to Its Foundations*, Springer-Verlag, 1999, Chapter 8.

The coefficients of expansion can be obtained as usual:

$$a_0 = \frac{1}{T} \int_0^T V_0 \frac{t}{T} \, dt = \tfrac{1}{2} V_0,$$

$$a_n = \frac{2}{T} \int_0^T V_0 \frac{t}{T} \cos \frac{2n\pi t}{T} \, dt = \frac{2V_0}{T^2} \int_0^T t \cos \frac{2n\pi t}{T} \, dt$$

$$= \frac{2V_0}{T^2} \left\{ \frac{T}{2n\pi} t \sin \frac{2n\pi t}{T} \bigg|_0^T - \frac{T}{2n\pi} \int_0^T \sin \frac{2n\pi t}{T} \, dt \right\} = 0,$$

and

$$b_n = \frac{2}{T} \int_0^T V_0 \frac{t}{T} \sin \frac{2n\pi t}{T} \, dt = \frac{2V_0}{T^2} \int_0^T t \sin \frac{2n\pi t}{T} \, dt$$

$$= \frac{2V_0}{T^2} \left\{ -\frac{T}{2n\pi} t \cos \frac{2n\pi t}{T} \bigg|_0^T + \frac{T}{2n\pi} \int_0^T \cos \frac{2n\pi t}{T} \, dt \right\} = -\frac{V_0}{n\pi}.$$

Substituting the coefficients in the sum, we get

$$V(t) = \tfrac{1}{2} V_0 - \sum_{n=1}^{\infty} \frac{V_0}{n\pi} \sin \frac{2n\pi t}{T} = \frac{V_0}{2} \left\{ 1 - \frac{2}{\pi} \sum_{n=1}^{\infty} \frac{\sin(2n\pi t/T)}{n} \right\}.$$

The plot of the sawtooth wave as well as those of the sum truncated at the first, third, and fifteenth term are shown in Figure 10.4. ∎

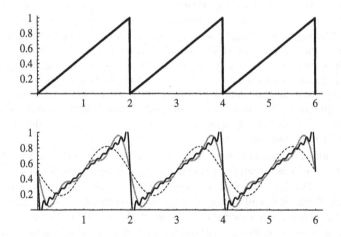

Figure 10.4: Top: The periodic sawtooth potential with $V_0 = 1$ and $T = 2$. Bottom: Various approximations to the Fourier series of the sawtooth potential. The dashed plot is that of the first term of the series, the thick gray plot keeps 3 terms, and the solid plot 15 terms.

Historical Notes

Although **Euler** made use of the trigonometric series as early as 1729, and d'Alembert considered the problem of the expansion of the reciprocal of the distance between two planets in a series of cosines of the multiples of the angle between the rays from the origin to the two planets, it was Fourier who gave a systematic account of the trigonometric series.

Joseph Fourier did very well as a young student of mathematics but had set his heart on becoming an army officer. Denied a commission because he was the son of a tailor, he went to a Benedictine school with the hope that he could continue studying mathematics at its seminary in Paris. The French Revolution changed those plans and set the stage for many of the personal circumstances of Fourier's later years, due in part to his courageous defense of some of its victims, an action that led to his arrest in 1794. He was released later that year, and he enrolled as a student in the Ecole Normale, which opened and closed within a year. His performance there, however, was enough to earn him a position as assistant lecturer (under **Lagrange** and **Monge**) in the *Ecole Polytechnique*. He was an excellent mathematical physicist, was a friend of **Napoleon**, and accompanied him in 1798 to Egypt, where Fourier held various diplomatic and administrative posts while also conducting research. Napoleon took note of his accomplishments and, on Fourier's return to France in 1801, appointed him prefect of the district of Isère, in southeastern France, and in this capacity built the first real road from Grenoble to Turin. He also befriended the boy **Champollion**, who later deciphered the *Rosetta stone* as the first long step toward understanding the hieroglyphic writing of the ancient Egyptians.

Like other scientists of his time, Fourier took up the flow of heat. The flow was of interest as a practical problem in the handling of metals in industry and as a scientific problem in attempts to determine the temperature at the interior of the Earth, the variation of that temperature with time, and other such questions. He submitted a basic paper on heat conduction to the Academy of Sciences of Paris in 1807. The paper was judged by **Lagrange**, **Laplace**, and **Legendre**, and was not published, mainly due to the objections of Lagrange, who had earlier rejected the use of trigonometric series. But the Academy did wish to encourage Fourier to develop his ideas, and so made the problem of the propagation of heat the subject of a grand prize to be awarded in 1812. Fourier submitted a revised paper in 1811, which was judged by the men already mentioned, and others. It won the prize but was criticized for its lack of rigor and so was not published at that time in the *Mémoires* of the Academy.

He developed a mastery of clear notation, some of which is still in use today. (The placement of the limits of integration near the top and bottom of the integral sign was introduced by Fourier.) It was also his habit to maintain close association between mathematical relations and physically measurable quantities, especially in limiting or asymptotic cases, even performing some of the experiments himself. He was one of the first to begin full incorporation of physical constants into his equations, and made considerable strides toward the modern ideas of units and dimensional analysis.

Fourier continued to work on the subject of heat and, in 1822, published one of the classics of mathematics, *Théorie Analytique de la Chaleur*, in which he made extensive use of the series that now bears his name and incorporated the first part of his 1811 paper practically without change. Two years later he became secretary of the Academy and was able to have his 1811 paper published in its original form in the *Mémoires*.

"The profound study of nature is the most fruitful source of mathematical discoveries."
Joseph Fourier

Joseph Fourier
1768–1830

10.7   Multivariable Taylor Series

The approximation to which we alluded at the beginning of this chapter is just as important when we are dealing with functions depending on several variables as those depending on a single variable. After all, most functions encountered in physics depend on space coordinates and time. We begin with two variables because the generalization to several variables will be trivial once we understand the two-variable case.

A direct—and obvious—generalization of the power series to the case of a function $f(u, v)$ of two variables about the point (u_0, v_0) gives

$$
\begin{aligned}
f(u,v) = {} & a_{00} + a_{10}(u-u_0) + a_{01}(v-v_0) + a_{20}(u-u_0)^2 \\
& + a_{02}(v-v_0)^2 + a_{11}(u-u_0)(v-v_0) + a_{30}(u-u_0)^3 \\
& + a_{21}(u-u_0)^2(v-v_0) + a_{12}(u-u_0)(v-v_0)^2 \\
& + a_{03}(v-v_0)^3 + \cdots .
\end{aligned}
\tag{10.45}
$$

The notation used above needs some explanation. All the a's are constants with two indices such that the first index indicates the power of $(u-u_0)$ and the second that of $(v-v_0)$. To obtain a Taylor series, we need to relate a's to derivatives of f. This is straightforward: To find a_{kj}, differentiate both sides of Equation (10.45) k times with respect to u and j times with respect to v and evaluate the result at (u_0, v_0). Thus, to evaluate a_{00}, we differentiate zero times with respect to u and zero times with respect to v and substitute u_0 for u and v_0 for v on both sides. We then obtain

$$
f(u_0, v_0) = a_{00} + 0 + 0 + \cdots + 0 + \cdots = a_{00}.
$$

By differentiating with respect to u and evaluating both sides at (u_0, v_0), we obtain

$$
\partial_1 f(u_0, v_0) = 0 + a_{10} + 0 + \cdots + 0 + \cdots = a_{10}.
$$

Similarly,

$$
\begin{aligned}
\partial_2 f(u_0, v_0) &= 0 + 0 + a_{01} + 0 + \cdots + 0 + \cdots = a_{01}, \\
\partial_1 \partial_1 f(u_0, v_0) &= \partial_1^2 f(u_0, v_0) = 2a_{20}, \\
\partial_2 \partial_2 f(u_0, v_0) &= \partial_2^2 f(u_0, v_0) = 2a_{02}, \\
\partial_2 \partial_1 f(u_0, v_0) &= a_{11}.
\end{aligned}
$$

We want to write Equation (10.45) in a succinct form to be able to extract a general formula for the coefficients. An inspection of that equation suggests that

$$
f(u,v) = \sum_{j=0}^{\infty} \sum_{k=0}^{\infty} a_{jk}(u-u_0)^j (v-v_0)^k.
$$

It is more useful to collect terms of equal total power together. Thus, writing $m = k + j$, and noting that j cannot be larger than m, we rewrite the above equation as

$$f(u, v) = \sum_{m=0}^{\infty} \sum_{j=0}^{m} a_{j,m-j}(u - u_0)^j (v - v_0)^{m-j}.$$

Let us introduce the notation $\partial_{k,n-k}$ for k differentiations with respect to the first variable, and $n - k$ differentiations with respect to the second:[7]

$$\partial_{k,n-k} f \equiv \frac{\partial^n f}{\partial u^k \partial v^{n-k}}$$

and apply it to both sides of the sum above. Evaluating the result at (u_0, v_0), we obtain

$$\partial_{k,n-k} f(u_0, v_0) = \sum_{m=0}^{\infty} \sum_{j=0}^{m} a_{j,m-j} \partial_{k,n-k} \left\{ (u - u_0)^j (v - v_0)^{m-j} \right\} \bigg|_{(u_0,v_0)}.$$

If $j < k$ or $m - j < n - k$ then the corresponding terms differentiate to zero. On the other hand, if $j > k$ or $m - j > n - k$ then some powers of $u - u_0$ or $v - v_0$ will survive and evaluation at (u_0, v_0) will also give zero. So, the only term in the sum that survives the differentiation is the term with $j = k$ and $m - j = n - k$ which gives $k!(n - k)!$. We thus have

Taylor series of a function of two variables

$$\partial_{k,n-k} f(u_0, v_0) = k!(n - k)! a_{k,n-k} \;\Rightarrow\; a_{k,n-k} = \frac{\partial_{k,n-k} f(u_0, v_0)}{k!(n - k)!},$$

and the Taylor series can finally be written as

$$f(u, v) = \sum_{n=0}^{\infty} \sum_{k=0}^{n} \frac{\partial_{k,n-k} f(u_0, v_0)}{k!(n - k)!} (u - u_0)^k (v - v_0)^{n-k}. \qquad (10.46)$$

Sometimes this is written in terms of increments to suggest approximation as in the single-variable case:

$$f(u + \Delta u, v + \Delta v) = \sum_{n=0}^{\infty} \sum_{k=0}^{n} \frac{\partial_{k,n-k} f(u, v)}{k!(n - k)!} (\Delta u)^k (\Delta v)^{n-k}, \qquad (10.47)$$

where we used (u, v) instead of (u_0, v_0). Once again, the first term in the expansion is $f(u, v)$ and the rest is a correction.

Taylor series of a function of three variables

The three-dimensional formula should now be easy to construct. We write this as[8]

$$f(u, v, w) = \sum_{n=0}^{\infty} \sum_{i+j+k=n} \frac{\partial^n_{ijk} f(u_0, v_0, w_0)}{i!j!k!} (u - u_0)^i (v - v_0)^j (w - w_0)^k. \qquad (10.48)$$

---

[7]This notation is not universal. Sometimes ∂^n_{kj} is used with the understanding that $k + j = n$.

[8]The symbol ∂^n_{ijk} represents the nth derivative with i differentiations with respect to the first variable, j differentiations with respect to the second variable, and k differentiations with respect to the third variable, such that $i + j + k = n$.

For a given value of n, suggested by the outer sum, the inner sum describes a procedure whereby all terms whose i, j, and k indices add up to n are grouped together. As a comparison, we also write Equation (10.46) in this notation:

$$f(u, v) = \sum_{n=0}^{\infty} \sum_{j+k=n} \frac{\partial_{jk}^{n} f(u_0, v_0)}{j! k!} (u - u_0)^j (v - v_0)^k. \qquad (10.49)$$

The three-dimensional Taylor series in terms of increments becomes

$$f(u + \Delta u, v + \Delta v, w + \Delta w)$$
$$= \sum_{n=0}^{\infty} \sum_{i+j+k=n} \frac{\partial_{ijk}^{n} f(u, v, w)}{i! j! k!} (\Delta u)^i (\Delta v)^j (\Delta w)^k, \qquad (10.50)$$

where again (u_0, v_0, w_0) has been replaced by (u, v, w).

Example 10.7.1. As an example we expand $e^x \sin y$ about the origin.[9] Using the notation in Equation (10.49), the coefficients, within a factor of $j! k!$, can be written as

$$\partial_{jk}^{n} (e^x \sin y) \Big|_{(0,0)} = \frac{\partial^n}{\partial x^j \partial y^k} (e^x \sin y) \Big|_{(0,0)} = \underbrace{\frac{\partial^j}{\partial x^j} (e^x) \Big|_{x=0}}_{=1} \frac{\partial^k}{\partial y^k} (\sin y) \Big|_{y=0}$$

$$= \frac{\partial^k}{\partial y^k} (\sin y) \Big|_{y=0} .$$

The first few terms of the Taylor expansion of this function can now be written down:
$$e^x \sin y = y + xy + \frac{x^2 y}{2} - \frac{y^3}{6} - \frac{xy^3}{6} + \frac{x^3 y}{6} + \cdots .$$

One could also obtain this result by multiplying the Taylor expansions of e^x and $\sin y$ term by term. ∎

10.8 Application to Differential Equations

One of the most powerful methods of solving an ordinary differential equation (ODE) is the power series method, and we shall use this method to solve some of the most recurring differential equations of mathematical physics in Chapters 25 through 27. Power series are uniformly and absolutely convergent, and can be differentiated term by term. This makes them a good candidate for representing the (unknown) solutions of differential equations. The relation among the derivatives, expressed in a differential equation, becomes a relation among coefficients of the power series, the so-called *recursion relation*, which is enough to determine all the relevant coefficients of the series, leaving only those coefficients which require initial conditions for their determination. The best way to understand the method is to look at an example.

---

[9]The use of x and y in place of u and v should not cause any confusion.

Example 10.8.1. The differential equation

$$\frac{dx}{dt} = bx$$

can be assumed to have a power series solution of the form

$$x(t) = \sum_{n=0}^{\infty} c_n t^n.$$

This power series will be uniformly and absolutely convergent for some interval on the real line, and as such, can be differentiated. Differentiating the foregoing equation and substituting the result in the differential equation, we get

$$\sum_{n=1}^{\infty} n c_n t^{n-1} = b \sum_{n=0}^{\infty} c_n t^n.$$

The essential property of power series is the equality of the corresponding coefficients when two such series are equal (see Theorem 10.1.4). Before using this property in the above equation, however, we need to reexpress the LHS so that the power of t is the same on both sides. We thus change the dummy index from n to $m = n - 1$, so that all n's are replaced by $m + 1$. We then get

$$LHS = \sum_{m+1=1}^{\infty} (m+1)c_{m+1} t^m = \sum_{m=0}^{\infty} (m+1)c_{m+1} t^m.$$

Since we are free to use any dummy index we please, let us change m to n so that we can compare the two sides of the equation. This gives

$$\sum_{n=0}^{\infty} (n+1)c_{n+1} t^n = \sum_{n=0}^{\infty} bc_n t^n \quad \Rightarrow \quad (n+1)c_{n+1} = bc_n. \qquad (10.51)$$

We can immediately test for the convergence of the series using the ratio test:

$$\lim_{n\to\infty} \left| \frac{c_{n+1} t^{n+1}}{c_n t^n} \right| = \lim_{n\to\infty} \left| \frac{tbc_n/(n+1)}{c_n} \right| = \lim_{n\to\infty} \left| \frac{bt}{n+1} \right| = 0$$

for all b and t. Thus, regardless of the value of b and t, the series converges.

We have established the convergence of the series representation of the solution of our differential equation. We now have to find the coefficients. This is done by rewriting Equation (10.51) as

$$c_{n+1} = \frac{b}{n+1} c_n \qquad (10.52)$$

recursion relation   which is called the **recursion relation** of the series. By iterating this relation we can obtain all the coefficients in terms of the first one as follows:

$$c_{n+1} = \frac{b}{n+1} c_n = \frac{b}{n+1} \left(\frac{b}{n} c_{n-1} \right) = \frac{b^2}{(n+1)n} c_{n-1}$$

$$= \frac{b^2}{(n+1)n} \left(\frac{b}{n-1} c_{n-2} \right) = \frac{b^3}{(n+1)n(n-1)} c_{n-2}.$$

Since we are interested in finding c_n, we can rewrite this equation as

$$c_n = \frac{b^3}{n(n-1)(n-2)}c_{n-3},$$

where we have lowered all n's on both sides by one unit. This relation can easily be generalized to an arbitrary positive integer j:

$$c_n = \frac{b^j}{n(n-1)\cdots(n-j+1)}c_{n-j}.$$

In particular, if we set $j = n$, we obtain

$$c_n = \frac{b^n}{n(n-1)\cdots 2\cdot 1}c_0 = \frac{b^n}{n!}c_0 \qquad (10.53)$$

which upon substitution in the original series, yields

$$x(t) = \sum_{n=0}^{\infty} c_0 \frac{b^n}{n!} t^n = c_0 \sum_{n=0}^{\infty} \frac{(bt)^n}{n!} = c_0 e^{bt},$$

where we have used Equation (10.12). The unknown c_0 is determined by the value of $x(t)$ at a given t, usually $t = 0$. ∎

There are of course much easier ways of solving the simple differential equation above, and the method used may appear to "kill a fly with a sledge-hammer." Nevertheless, it illustrates the almost mechanical way of obtaining the solution without resorting to any "tricks" used so often in arriving at the closed-form solutions of differential equations.

Example 10.8.2. Let us look at another familiar example. The motion of a mass m driven by a spring with spring constant k is governed by the differential equation

$$m\frac{d^2x}{dt^2} = -kx \quad \Rightarrow \quad \frac{d^2x}{dt^2} + \frac{k}{m}x = 0.$$

Once again we assume a solution of the form

$$x(t) = \sum_{n=0}^{\infty} a_n t^n = a_0 + a_1 t + a_2 t^2 + \cdots + a_n t^n + \cdots$$

and differentiate it twice to get

$$\frac{dx}{dt} = \sum_{n=1}^{\infty} n a_n t^{n-1} = a_1 + 2a_2 t + \cdots + n a_n t^{n-1} + \cdots,$$

$$\frac{d^2x}{dt^2} = \sum_{n=2}^{\infty} n(n-1) a_n t^{n-2} = 2a_2 + 3\cdot 2 a_3 t + \cdots + n(n-1) a_n t^{n-2} + \cdots.$$

Substitute $j = n - 2$ to bring the power of t into a form that can be compared with the RHS. This amounts to substituting $j + 2$ for all n's:

$$\frac{d^2x}{dt^2} = \sum_{j=0}^{\infty}(j+2)(j+1)a_{j+2}t^j = \sum_{n=0}^{\infty}(n+2)(n+1)a_{n+2}t^n.$$

In the last step we simply changed the dummy index. Substituting this and the series for $x(t)$ in the differential equation, we obtain

$$\sum_{n=0}^{\infty}(n+2)(n+1)a_{n+2}t^n + \frac{k}{m}\sum_{n=0}^{\infty}a_n t^n = 0$$

which gives the recursion relation

$$(n+2)(n+1)a_{n+2} + \frac{k}{m}a_n = 0 \Rightarrow a_{n+2} = -\frac{k/m}{(n+2)(n+1)}a_n. \qquad (10.54)$$

Application of the ratio test [as given by Equation (9.10) with $j = 2$] immediately yields that the series is convergent for all values of k/m and all values of t. If we lower the value of n by two units on both sides, we get

$$\begin{aligned}
a_n &= -\frac{k/m}{n(n-1)}a_{n-2} = -\frac{k/m}{n(n-1)}\left\{-\frac{k/m}{(n-2)(n-3)}a_{n-4}\right\} \\
&= \frac{(-k/m)^2}{n(n-1)(n-2)(n-3)}a_{n-4} \\
&= \frac{(-k/m)^2}{n(n-1)(n-2)(n-3)}\left\{-\frac{k/m}{(n-4)(n-5)}a_{n-6}\right\} \\
&= \frac{(-k/m)^3}{n(n-1)(n-2)(n-3)(n-4)(n-5)}a_{n-6} \\
&\quad\vdots \\
&= \frac{(-k/m)^i}{n(n-1)\cdots(n-2i+1)}a_{n-2i},
\end{aligned}$$

where i is some positive integer. Because of the form of this equation, we should consider two cases: For even n, we let $i = n/2$ or $n = 2i$ to obtain

$$a_{2i} = \frac{(-k/m)^i}{2i(2i-1)\cdots 2\cdot 1}a_0 = \frac{(-k/m)^i}{(2i)!}a_0$$

and for odd n we let $i = (n-1)/2$ or $n = 2i+1$ to get

$$a_{2i+1} = \frac{(-k/m)^i}{(2i+1)2i\cdots 2\cdot 1}a_1 = \frac{(-k/m)^i}{(2i+1)!}a_1.$$

Thus all even coefficients are given in terms of a_0, and all odd ones in terms of a_1. Absolute convergence of the series now allows us to rearrange terms and separate even and odd terms to write

$$\begin{aligned}
x(t) &= \sum_{n=even}^{\infty}a_n t^n + \sum_{n=odd}^{\infty}a_n t^n = \sum_{j=0}^{\infty}a_{2j}t^{2j} + \sum_{j=0}^{\infty}a_{2j+1}t^{2j+1} \\
&= \sum_{j=0}^{\infty}\frac{(-k/m)^j}{(2j)!}a_0 t^{2j} + \sum_{j=0}^{\infty}\frac{(-k/m)^j}{(2j+1)!}a_1 t^{2j+1} \\
&= a_0\sum_{j=0}^{\infty}\frac{(-1)^j}{(2j)!}\left(\sqrt{k/m}\,t\right)^{2j} + \frac{a_1}{\sqrt{k/m}}\sum_{j=0}^{\infty}\frac{(-1)^j}{(2j+1)!}\left(\sqrt{k/m}\,t\right)^{2j+1} \\
&= A\cos\left(\sqrt{k/m}\,t\right) + B\sin\left(\sqrt{k/m}\,t\right),
\end{aligned}$$

where $A = a_0$ and $B = a_1/\sqrt{k/m}$ are arbitrary constants to be determined by the initial conditions of the problem. The Maclaurin series for sine and cosine used above are given in Equations (10.13) and (10.14). ∎

The examples above, although illustrating the utility of the power series method of solving differential equations, should not give the impression that one needs no other methods. The closed-form solutions are sometimes essential for interpreting the physical properties of the system under consideration. For example, if the mass of the preceding example is in a fluid, so that a damping force retards the motion, the closed-form solution will turn out to be

$$x(t) = Ae^{-\gamma t}\cos(\omega t + \alpha), \qquad \omega \equiv \sqrt{\frac{k}{m}},$$

where γ is the *damping factor* and α is an arbitrary *phase*. Deciphering this closed form from its power series expansion, obtained by solving the differential equation by the series method, is next to impossible. The closed-form solution shows clearly, for instance, how the amplitude of the oscillation decreases with time, an information that may not be evident from the series solution of the problem. Nevertheless, on many occasions, a closed-form solution may not be available, in which case the power series solution will be the only alternative. In fact, many of the functions of mathematical physics were invented in the last century as the power series solutions of differential equations.

damping factor

10.9 Problems

10.1. Write the first five terms of the expansion of the binomial function (10.15) for (a) $\alpha = \frac{3}{2}$, (b) $\alpha = \frac{1}{3}$, and (c) $\alpha = \frac{3}{4}$.

10.2. Find the rational number of which each of the following decimal numbers is a representation:

(a) $0.5555\ldots$ (b) $0.676767\ldots$ (c) $0.123123\ldots$

(d) $1.1111\ldots$ (e) $2.727272\ldots$ (f) $1.108108\ldots$

10.3. Find the interval of convergence of the Maclaurin series for each of the familiar functions discussed in Section 10.2.

10.4. Using the series representation of the familiar functions evaluate the following series:

(a) $\sum_{k=1}^{\infty} \frac{(-1)^k x^{2k+1}}{2k}$.   (b) $\sum_{k=0}^{\infty} \frac{x^{2k+1}}{(2k)!}$.   (c) $\sum_{k=0}^{\infty} \frac{x^{k+1}}{(k+1)!}$.

(d) $\sum_{n=1}^{\infty} \frac{(-1)^{n-1} x^{3n-2}}{n3^n}$.   (e) $\sum_{n=0}^{\infty} \frac{(-1)^{n+2} x^{3n+1}}{3^{3n+1}(2n)!}$.   (f) $\sum_{m=0}^{\infty} \frac{x^{m+1}}{(2m+1)!}$.

10.5. Derive Equation (10.17).

10.6. Use the Maclaurin series to find the limits of the following ratios as $x \to 0$:

$$\frac{2\sqrt{1-x^2}+x^2-2}{2\cos x-2+x^2}, \qquad \frac{\sin x - \ln(1+x)}{e^x - x - \cos x}.$$

10.7. (a) Use the Maclaurin series expansion up to x^3 to find the following limit:

$$\lim_{x\to 0} \frac{2\sqrt[3]{1-6x}-2\cos x+4\sin x+7x^2}{\ln(1-x)+e^x-1}.$$

(b) Use the Maclaurin series expansion up to x^4 to find the following limit:

$$\lim_{x\to 0} \frac{e^x - \ln(1+x^2) - \cos x + \sin x - 2x}{2\sqrt{4+x^2}+\cos x - 5}.$$

10.8. In the special theory of relativity the energy E of a particle of mass m and speed v is given by

$$E = \frac{mc^2}{\sqrt{1-(v/c)^2}},$$

where c is the speed of light. Show that for ordinary speeds ($v << c$), one obtains the classical expression for the *kinetic energy*, defined to be E minus the *rest energy*.

10.9. The gravitational potential energy for a particle of mass m at a distance r from the center of a planet of radius R and mass M is given by

$$\Phi(r) = -\frac{GMm}{r} + C, \qquad r > R.$$

(a) Find C so that the potential at the surface of the planet is zero.
(b) Show that at a height $h << R$ above the surface of the planet, the potential energy can be written as mgh. Find g in terms of M and R and calculate the numerical value of g for the Earth, the Moon, and Jupiter. Look up the data you need in a table usually found in introductory physics or astronomy books.

10.10. Prove the hyperbolic identities of Equation (10.21).

10.11. Show that

$$\operatorname{sech}^2 x = 1 - \tanh^2 x, \qquad \operatorname{cosech}^2 x = \coth^2 x - 1,$$

and

$$\frac{d}{dx}\tanh x = \operatorname{sech}^2 x, \qquad \frac{d}{dx}\coth x = -\operatorname{cosech}^2 x.$$

10.12. Derive Equation (10.23).

10.13. Use L'Hôpital's rule to obtain the following limits:

(a) $\lim_{x \to 0} \frac{(2+x)\ln(1-x)}{(1-e^x)\cos x}$.

(b) $\lim_{x \to \infty} x \ln\left(\frac{x+1}{x-1}\right)$.

(c) $\lim_{x \to a} \frac{\sqrt[3]{x} - \sqrt[3]{a}}{x-a}$.

(d) $\lim_{x \to 0} \frac{xe^x}{1-e^x}$.

(e) $\lim_{x \to \frac{1}{2}\pi} (\tan x)^{\cos x}$.

(f) $\lim_{x \to 0} (\ln x)\tan x$.

10.14. Use L'Hôpital's rule to obtain the limits of Example 10.4.1.

10.15. Show that the following sequences converge and find their limits:

$$\frac{\ln n}{n^p}, \quad \frac{n^2}{2^n}, \quad n\ln\left(1+\frac{1}{n}\right), \quad P(n)\,e^{-n},$$

where p is a positive number and $P(n)$ is a polynomial in n.

10.16. The Yukawa potential of a charge distribution is given by

$$\Phi(\mathbf{r}) = \int_\Omega \frac{k_e e^{-\kappa|\mathbf{r}-\mathbf{r}'|}\, dq(\mathbf{r}')}{|\mathbf{r}-\mathbf{r}'|},$$

where κ is a constant. By expanding $|\mathbf{r}-\mathbf{r}'|$ up to the first order in r'/r, show that

$$\Phi(\mathbf{r}) \approx \frac{k_e Q e^{-\kappa r}}{r} + \frac{k_e(\kappa r + 1)e^{-\kappa r}}{r^2}\hat{\mathbf{e}}_r \cdot \mathbf{p},$$

where $\mathbf{p}$ is the dipole moment of the charge distribution.

10.17. A conic surface has an opening angle of 2α and a lateral length a as shown in Figure 10.5. It carries a uniform charge density σ.
(a) Show that the electrostatic potential Φ at a distance r from the vertex on the axis of the cone is

$$\Phi(r) = 2\pi k_e \sigma \sin\alpha\left(\sqrt{r^2 + a^2 - 2ar\cos\alpha} - r\right)$$
$$+ (2\pi k_e \sigma \sin\alpha\cos\alpha)r\ln\left|\frac{a - r\cos\alpha + \sqrt{r^2 + a^2 - 2ar\cos\alpha}}{r - r\cos\alpha}\right|.$$

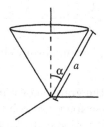

Figure 10.5: The cone of Problem 10.17.

(b) Now suppose that $r \gg a$, expand the square roots and the log up to the second power of the ratio a/r, and show that

$$\sqrt{r^2 + a^2 - 2ar\cos\alpha} \approx r - a\cos\alpha + \frac{a^2}{2r}\sin^2\alpha$$

and

$$\ln\left|a - r\cos\alpha + \sqrt{r^2 + a^2 - 2ar\cos\alpha}\right| \approx \ln|r - r\cos\alpha| + \frac{a}{r} + \frac{a^2}{2r^2}(1 + \cos\alpha).$$

(c) Put (a) and (b) together to show that the potential can be approximated by

$$\Phi(r) \approx \frac{\pi k_e \sigma a^2 \sin\alpha}{r}.$$

Write this expression in terms of the total charge in the cone. Do you get what you expect?

10.18. Recall from your introductory physics courses that the electric field at a distance ρ from a long uniformly charged rod has only a radial component which is given by $E = \lambda/2\pi\epsilon_0\rho$, where λ is the linear charge density. Show this result by setting $a = -L/2$ (why?) and taking the limit of infinite L in Equation (4.13).

10.19. After calculating the potentials of Problems 4.11 and 4.12 for finite L, find their limits when $L \to \infty$.

10.20. The potential of a certain charge distribution with total charge Q is given by

$$\Phi = \frac{k_e}{a_0} \int \left[\ln|\mathbf{r} - \mathbf{r}'| - \ln b\right] dq(\mathbf{r}'),$$

where k_e, a_0, and b are constants.
(a) Show that for $r' \ll r$, one can use the approximation

$$\ln|\mathbf{r} - \mathbf{r}'| \approx \ln r - \frac{r'}{r}\hat{\mathbf{e}}_r \cdot \hat{\mathbf{e}}_{r'}.$$

(b) Use (a) to show that the multipole expansion of Φ *only up to the dipole moment* is

$$\Phi \approx \frac{k_e Q}{a_0}\ln\frac{r}{b} - \frac{k_e}{a_0}\frac{\mathbf{p} \cdot \mathbf{r}}{r^2}.$$

10.21. Find the dipole moment of a uniformly charged sphere about its center.

10.22. A voltage is given by the graph shown in Figure 10.6.
(a) Write the function $V(t)$ describing the voltage for $0 \leq t \leq 2T$.
(b) If this voltage repeats itself periodically, find the Fourier series expansion of $V(t)$.

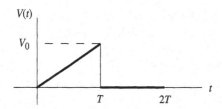

Figure 10.6: The voltage of Problem 10.22.

10.23. A periodic voltage with period $2T$ is given by

$$V(t) = \begin{cases} V_0 \cos(\pi t/T) & \text{if} \quad -T/2 \le t \le T/2, \\ 0 & \text{if} \quad T/2 \le |t| \le T. \end{cases}$$

(a) Sketch this function for the interval $-3T \le t \le 3T$.
(b) Find a_0 and a_1, the first two cosine coefficients of the Fourier series expansion of $V(t)$.
(c) Find a_n and all b_n, the sine coefficients.
(d) Write down the Fourier series of $V(t)$. Evaluate both sides at $t = 0$ to show that

$$\frac{\pi}{2} = 1 - 2 \sum_{n=1}^{\infty} \frac{(-1)^n}{4n^2 - 1}.$$

This is one of the many series representations of π.

10.24. An electric voltage $V(t)$ is given by

$$V(t) = V_0 \sin\left(\frac{\pi t}{2T}\right), \quad 0 \le t \le T$$

and repeats itself with period T.
(a) Sketch $V(t)$ for values of t from $t = 0$ to $t = 3T$.
(b) Find the Fourier series expansion of $V(t)$.

10.25. A periodic voltage is given by the formula

$$V(t) = \begin{cases} V_0 \sin(\pi t/2T) & \text{if} \quad 0 \le t \le T, \\ 0 & \text{if} \quad T \le t \le 2T. \end{cases}$$

(a) Sketch the voltage for the interval $(-4T, 4T)$.
(b) Find the Fourier series representation of this voltage.

10.26. A periodic voltage with period $4T$ is given by

$$V(t) = \begin{cases} V_0\left(1 - \dfrac{t^2}{T^2}\right) & \text{if} \quad -T \le t \le T \\ 0 & \text{if} \quad T \le |t| \le 2T. \end{cases}$$

(a) Sketch this function for the interval $-6T \leq t \leq 6T$.
(b) Find a_0, a_n, and b_n, the coefficients of the Fourier series expansion of $V(t)$.
(c) Write down the Fourier series of $V(t)$.
(d) Evaluate both sides at $t = T$. Do you obtain an identity? If not, what sort of relationship is obtained if we demand the equality of both sides?

10.27. Write out Equation (10.50) up to the second power in the Δ's.

10.28. Find the Taylor series expansion of $e^x \ln(1 + y)$ about $(0, 0)$.

10.29. (a) Find the multivariable Taylor series expansion of e^{xy} about $(0, 0)$.
(b) Now let $z = xy$, expand the function e^z, and substitute xy for z in the expansion. Show that the results of (a) and (b) agree.

10.30. Determine all the solutions of the differential equation

$$\frac{dx}{dt} + 2tx = 0$$

using infinite power series. From the power series solution guess the closed-form solution. Now suppose that $x(0) = 1$. What is the specific solution with this property?

10.31. Consider the differential equation

$$\frac{dx}{dt} + 3t^2 x = 0.$$

(a) Use a solution of the form $\sum_{n=0}^{\infty} a_n t^n$ and find a_1 and a_2.
(b) Find a recursion relation relating coefficients.
(c) From the recursion relation determine the radius of convergence of the infinite series.
(d) Find all coefficients in terms of only one.
(e) Guess the closed-form solution from the series. Now suppose that $x(0) = 2$. What is the specific solution with this property? What is the numerical value of $x(-2)$?

Chapter 11

Integrals and Series as Functions

The notion of a function as a mathematical entity has a long history as rich as the history of mathematics itself. With the invention of the coordinate plane in the seventeenth century, functions started to acquire graphical representations which, in turn, facilitated the connection between algebra and geometry. It was really calculus that triggered an explosion in function theory, and indeed, in all mathematics. With calculus came not only the concept of differentiation and integration, but also—in the hands of Newton and his contemporaries, as they were studying no smaller an object than the universe itself—that of differential equations. All these concepts, in particular integration and differential equation, had a dramatic influence on the notion of functions. The aim of this chapter is to give the reader a flavor of the variety of functions made possible by integration and differential equations.[1]

11.1 Integrals as Functions

Integrals are one of the most convenient media in which new functions can be defined. As we saw in Chapter 3, if the integrand or the limits of integration include parameters, those parameters can be treated as variables and the integral itself as a function of those parameters. In this section, we list some of the most important functions that are normally defined in terms of integrals.

---

[1] We shall not solve any differential equations in this chapter, but simply quote solutions to some of them in the form of power series. We shall come back to differential equations later in the book.

11.1.1   Gamma Function

Consider the integral

equation (11.1)
defines the
gamma function
evaluated at x.

$$\Gamma(x) \equiv \int_0^\infty t^{x-1}e^{-t}\,dt, \tag{11.1}$$

where x is a real number.[2] Integrate Equation (11.1) by parts with $u = t^{x-1}$ and $dv = e^{-t}\,dt$ to obtain

$$\Gamma(x) = \overbrace{t^{x-1}[-e^{-t}]}^{\equiv uv}\Big|_0^\infty + (x-1)\underbrace{\int_0^\infty \overbrace{t^{x-2}e^{-t}\,dt}^{\equiv -v\,du}}_{=\Gamma(x-1)}$$

$$\underbrace{}_{=0}$$

or

$$\Gamma(x) = (x-1)\Gamma(x-1). \tag{11.2}$$

In particular, if x is a positive integer n, then repeated use of Equation (11.2) gives

$$\Gamma(n) = (n-1)\Gamma(n-1) = (n-1)(n-2)\Gamma(n-2)$$
$$= (n-1)(n-2)\cdots 1 \cdot \Gamma(1) = (n-1)!,$$

where we used the fact that $\Gamma(1) = 1$ as the reader may easily verify using Equation (11.1). This equation is written as

for integers, the
gamma function
becomes a
factorial.

$$\Gamma(n+1) = n! \qquad \text{for positive integer } n. \tag{11.3}$$

Let us rewrite (11.2) as $\Gamma(x-1) = \Gamma(x)/(x-1)$. Then,

$$\lim_{x\to 1}\Gamma(x-1) = \lim_{x\to 1}\frac{\Gamma(x)}{x-1} \to \infty$$

because $\Gamma(1) = 1$. Thus, $\Gamma(0) = \infty$. Similarly,

$$\lim_{x\to 0}\Gamma(x-1) = \lim_{x\to 0}\frac{\Gamma(x)}{x-1} \to \frac{\Gamma(0)}{-1} \to \infty,$$

i.e., $\Gamma(-1) = \infty$. It is clear that $\Gamma(n) = \infty$ for any negative integer n or zero. It turns out that these are the only points at which $\Gamma(x)$ is not defined.

Definition 11.1.1. *The function defined by Equation (11.1) is called the* **gamma function**, *which, because it satisfies Equation (11.3), is the generalization of the factorials to noninteger values. We sometimes write*

$$\Gamma(x+1) = x! \qquad \text{for any real } x \tag{11.4}$$

and call Γ the **factorial function**. *The gamma function is defined for all values of its argument except zero and negative integers, for which the gamma function becomes infinite.*

---

[2]The most complete analytic discussion of $\Gamma(z)$ allows z to be complex and uses the full machinery of complex calculus. Here, we shall avoid such completeness and refer the reader to Hassani, S. *Mathematical Physics: A Modern Introduction to Its Foundations*, Springer-Verlag, 1999, where a full discussion of $\Gamma(z)$ can be found in Section 11.4.

It follows from Equation (11.2) that by repeatedly subtracting 1 from the argument of the gamma function, we can reduce the evaluation of $\Gamma(x)$ to the case where x lies between 0 and 1. Such an evaluation can be done numerically and the results tabulated.

the values of $\Gamma(x)$ for $0 < x \leq 1$ determine $\Gamma(x)$ for all x.

Example 11.1.1. In this example, we evaluate $\Gamma(\frac{1}{2})$. Equation (11.1) gives

$$\Gamma(\tfrac{1}{2}) = \int_0^\infty t^{-1/2} e^{-t}\, dt.$$

Change the variable of integration to $u = \sqrt{t}$ with $du = (1/2\sqrt{t})\, dt$. Then

$$\Gamma(\tfrac{1}{2}) = 2 \int_0^\infty e^{-u^2}\, du = 2\left(\tfrac{1}{2}\sqrt{\pi}\right) = \sqrt{\pi},$$

where we used the result of Example 3.3.1.

With $\Gamma(\frac{1}{2})$ at our disposal, we can evaluate the gamma function at any half-integer value by the remarks above. For example,

$$\Gamma(\tfrac{7}{2}) = \tfrac{5}{2}\Gamma(\tfrac{5}{2}) = (\tfrac{5}{2})(\tfrac{3}{2})\Gamma(\tfrac{3}{2}) = (\tfrac{5}{2})(\tfrac{3}{2})(\tfrac{1}{2})\Gamma(\tfrac{1}{2}) = \frac{15\sqrt{\pi}}{8}.$$

Similarly, with $\Gamma(\frac{1}{2}) = -\frac{1}{2}\Gamma(-\frac{1}{2})$, we obtain

$$\Gamma(-\tfrac{1}{2}) = -2\Gamma(\tfrac{1}{2}) = -2\sqrt{\pi}. \qquad \blacksquare$$

It is instructive to generalize the result of the example above and find a general formula for the gamma function of any half-integer. Such a formula is related to the notion of the double factorial:

double factorial

Definition 11.1.2. *The **double factorial** $(2n)!!$ [or $(2n-1)!!$] is defined as the product of all even (or odd) integers up to $2n$ (or $2n-1$).*

Problem 11.1 gives the detail of the derivation of the following formulas:

$$(2n)!! = 2^n n! = 2^n \Gamma(n+1), \qquad (2n-1)!! = \Gamma(n+\tfrac{1}{2})2^n \pi^{-1/2}. \qquad (11.5)$$

An extremely useful approximation to the gamma function is the so-called **Stirling approximation** which is valid for large arguments of the gamma function and which we present without derivation[3]

Stirling approximation

$$x! \equiv \Gamma(x+1) \approx \sqrt{2\pi}\, e^{-x} x^{x+1/2}. \qquad (11.6)$$

The Stirling formula works best when x is large. However, even for $x = 10$, it gives $\sqrt{2\pi}\, e^{-10} 10^{10.5} = 3598696$, which is surprisingly close to the exact value of $10! = 3628800$. For $x = 20$, the Stirling formula yields 2.42×10^{18} to three significant figures as opposed to the calculator result, which to the same number of significant figures is 2.43×10^{18}. For larger and larger values of x, the two results get closer and closer.

---

[3]For a derivation, see Hassani, S. *Mathematical Physics: A Modern Introduction to Its Foundations*, Springer-Verlag, 1999, Chapter 11.

11.1.2   The Beta Function

A function that sometimes shows up in applications is the **beta function**.
Consider

$$\Gamma(x)\Gamma(y) = \int_0^\infty t^{x-1}e^{-t}\,dt \int_0^\infty s^{y-1}e^{-s}\,ds = \int_0^\infty \int_0^\infty t^{x-1}s^{y-1}e^{-(t+s)}\,dt\,ds.$$

Introduce the new variable $u = t + s$ and use it to rewrite the s integral. Since
the lower limits of both s and t are 0, the lower limit of the u integral will
also be 0. Similarly, the upper limit of u will be infinity. However, since s
and t are positive and their sum is u, the upper limit of t cannot exceed u.
Therefore,

$$\Gamma(x)\Gamma(y) = \int_0^\infty du \int_0^u dt\, t^{x-1}(u-t)^{y-1}e^{-u}.$$

Now introduce another variable w by $t = uw$. Since in the t integration, u is
held constant, we have $dt = u\,dw$, and the limits of integration for w are 0
and 1. This will allow us to write

$$\Gamma(x)\Gamma(y) = \underbrace{\int_0^\infty du\, e^{-u}u^{x+y-1}}_{\equiv\Gamma(x+y)} \int_0^1 dw\, w^{x-1}(1-w)^{y-1}.$$

beta function
defined

The last integral defines the beta function. So,

$$B(x,y) \equiv \frac{\Gamma(x)\Gamma(y)}{\Gamma(x+y)} = \int_0^1 dt\, t^{x-1}(1-t)^{y-1}, \tag{11.7}$$

where we changed the (dummy) variable of integration from w to t.

We can find another representation of the beta function by substituting
$t = \sin^2\theta$. Then

$$dt = 2\sin\theta\cos\theta, \qquad 1 - t = 1 - \sin^2\theta = \cos^2\theta,$$

and the limits of integration become 0 and $\pi/2$. So,

$$B(x,y) = 2\int_0^{\pi/2} (\sin\theta)^{2x-1}(\cos\theta)^{2y-1}d\theta. \tag{11.8}$$

Historical Notes

Integration and differentiation and the whole machinery of calculus opened up entirely new ways of defining functions. Of these, one of the most important is the
gamma function, which arose from work on two problems, interpolation theory and
antidifferentiation. The problem of interpolation had been considered by **James Stirling** (1692–1770), **Daniel Bernoulli** (1700–1782), and **Christian Goldbach**. It was posed
to Euler and he announced his solution in a letter of October 13, 1729, to Goldbach.
A second letter, of January 8, 1730, brought in the integration problem.

The interpolation problem had to do with giving meaning to $n!$ for nonintegral values of n, and the integration problem was the evaluation of an integral already considered by Wallis, namely

$$\int_0^1 t^x (1-t)^y \, dt.$$

Euler showed that this integral led to our integral (11.1).

Leonhard Euler was Switzerland's foremost scientist and one of the three greatest mathematicians of modern times (**Gauss** and **Riemann** being the other two). He was perhaps the most prolific author of all time in any field. From 1727 to 1783 his writings poured out in a seemingly endless flood, constantly adding knowledge to every known branch of pure and applied mathematics, and also to many that were not known until he created them. He averaged about 800 printed pages a year throughout his long life, and yet he almost always had something worthwhile to say. The publication of his complete works was started in 1911, and the end is not in sight. This edition was planned to include 887 titles in 72 volumes, but since that time extensive new deposits of previously unknown manuscripts have been unearthed, and it is now estimated that more than 100 large volumes will be required for completion of the project. Euler evidently wrote mathematics with the ease and fluency of a skilled speaker discoursing on subjects with which he is intimately familiar. His writings are models of relaxed clarity. He never condensed, and he reveled in the rich abundance of his ideas and the vast scope of his interests. The French physicist **Arago**, in speaking of Euler's incomparable mathematical facility, remarked that "He calculated without apparent effort, as men breathe, or as eagles sustain themselves in the wind." He suffered total blindness during the last 17 years of his life, but with the aid of his powerful memory and fertile imagination, and with assistants to write his books and scientific papers from dictation, he actually increased his already prodigious output of work.

Leonhard Euler
1707–1783

Euler was a native of Basel and a student of *Johann Bernoulli* at the University, but he soon outstripped his teacher. He was also a man of broad culture, well versed in the classical languages and literatures (he knew the *Aeneid* by heart), many modern languages, physiology, medicine, botany, geography, and the entire body of physical science as it was known in his time. His personal life was as placid and uneventful as is possible for a man with 13 children.

Though he was not himself a teacher, Euler has had a deeper influence on the teaching of mathematics than any other person. This came about chiefly through his three great treatises: *Introductio in Analysin Infinitorum* (1748); *Institutiones Calculi Differentialis* (1755); and *Institutiones Calculi Integralis* (1768–1794). There is considerable truth in the old saying that all elementary and advanced calculus textbooks since 1748 are essentially copies of Euler or copies of copies of Euler. These works summed up and codified the discoveries of his predecessors, and are full of Euler's own ideas. He extended and perfected plane and solid analytic geometry, introduced the analytic approach to trigonometry, and was responsible for the modern treatment of the functions $\ln x$ and e^x. He created a consistent theory of logarithms of negative and imaginary numbers, and discovered that $\ln x$ has an infinite number of values. It was through his work that the symbols e, π, and $i = \sqrt{-1}$ became common currency for all mathematicians, and it was he who linked them together in the astonishing relation $e^{i\pi} = -1$. Among his other contributions to standard mathematical notation were $\sin x$, $\cos x$, the use of $f(x)$ for an unspecified function, and the use of $\sum$ for summation.

His work in all departments of analysis strongly influenced the further development of this subject through the next two centuries. He contributed many important ideas to differential equations, including substantial parts of the theory of second-order linear equations and the method of solution by power series. He gave the first systematic discussion of the calculus of variations, which he founded on his basic differential equation for a minimizing curve. He discovered the integral defining the gamma function and developed many of its applications and special properties. He also worked with **Fourier** series, encountered the **Bessel** functions in his study of the vibrations of a stretched circular membrane, and applied **Laplace** transforms to solve differential equations—all before Fourier, Bessel, and Laplace were born.

E. T. Bell, the well-known historian of mathematics, observed that "One of the most remarkable features of Euler's universal genius was its equal strength in both of the main currents of mathematics, the continuous and the discrete." In the realm of the discrete, he was one of the originators of number theory and made many far-reaching contributions to this subject throughout his life. In addition, the origins of topology—one of the dominant forces in modern mathematics—lie in his solution of the Königsberg bridge problem and his formula $V - E + F = 2$ connecting the numbers of vertices, edges, and faces of a simple polyhedron.

The distinction between pure and applied mathematics did not exist in Euler's day, and for him the entire physical universe was a convenient object whose diverse phenomena offered scope for his methods of analysis. The foundations of classical mechanics had been laid down by **Newton**, but Euler was the principal architect. In his treatise of 1736 he was the first to explicitly introduce the concept of a mass-point, or particle, and he was also the first to study the acceleration of a particle moving along any curve and to use the notion of a vector in connection with velocity and acceleration. His continued successes in mathematical physics were so numerous, and his influence was so pervasive, that most of his discoveries are not credited to him at all and are taken for granted in the physics community as part of the natural order of things. However, we do have Euler's angles for the rotation of a rigid body, and the all-important *Euler–Lagrange equation* of variational dynamics.

11.1.3   The Error Function

The **error function**, used extensively in statistics, is defined as

$$\operatorname{erf}(x) = \frac{1}{\sqrt{\pi}} \int_{-x}^{x} e^{-t^2} dt = \frac{2}{\sqrt{\pi}} \int_{0}^{x} e^{-t^2} dt \qquad (11.9)$$

and has the property that $\operatorname{erf}(\infty) = 1$. The error function $\operatorname{erf}(x)$ gives the area under the bell-shaped (normal) probability distribution located between $-x$ and $+x$.

11.1.4   Elliptic Functions

Recall from calculus[4] that the element of length of a curve parameterized by

$$x = f(t), \quad y = g(t), \quad z = h(t), \qquad t_1 \le t \le t_2,$$

---

[4]Or from our discussion of the parametric equation of curves in Chapter 4.

in Cartesian coordinates is

$$dl = \sqrt{dx^2 + dy^2 + dz^2} = \sqrt{[f'(t)]^2 + [g'(t)]^2 + [h'(t)]^2}\, dt,$$

where prime indicates the derivative. So, the length L of the curve connecting the initial point $(f(t_1), g(t_1), h(t_1))$ to the final point $(f(t_2), g(t_2), h(t_2))$ is

$$L = \int_{t_1}^{t_2} \sqrt{[f'(t)]^2 + [g'(t)]^2 + [h'(t)]^2}\, dt. \qquad (11.10)$$

The length of many curves, some very complicated-looking, can be found *analytically* using Equation (11.10). However, that of a simple curve such as an ellipse turns out to be impossible! Let us see what we get when we try to calculate the circumference of an ellipse. The parametric equation of an ellipse of respective semi-major and semi-minor axes a and b lying in the xy-plane is conveniently written as

there is no formula in closed form for the circumference of an ellipse!

$$x = a\sin t, \quad y = b\cos t, \quad z = 0, \qquad 0 \le t \le 2\pi. \qquad (11.11)$$

Substitution of these equations in (11.10) yields

$$L = \int_0^{2\pi} \sqrt{[a\cos t]^2 + [-b\sin t]^2 + [0]^2}\, dt = \int_0^{2\pi} \sqrt{a^2\cos^2 t + b^2\sin^2 t}\, dt$$

$$= \int_0^{2\pi} \sqrt{a^2(1 - \sin^2 t) + b^2\sin^2 t}\, dt = a\int_0^{2\pi} \sqrt{1 - k^2\sin^2 t}\, dt, \qquad (11.12)$$

where $k^2 = (a^2 - b^2)/a^2$. This innocent-looking integral does not succumb to *any* technique of integration. It was this resistance to analytical solution that prompted the nineteenth century mathematicians to study this and other related integrals as functions in their own right.

The **elliptic integral of the first kind** is defined as

elliptic integral of the first kind

$$F(\varphi, k) \equiv \int_0^{\varphi} \frac{dt}{\sqrt{1 - k^2\sin^2 t}} \qquad (11.13)$$

with F a function of two variables because the integral involves two parameters, one appearing in the integrand and the other appearing as a limit.

The **elliptic integral of the second kind** is defined as

elliptic integral of the second kind

$$E(\varphi, k) \equiv \int_0^{\varphi} \sqrt{1 - k^2\sin^2 t}\, dt. \qquad (11.14)$$

The elliptic integral of the second kind can be interpreted as the length of partial arcs of an ellipse. The circumference L of an ellipse with respective semi-major and semi-minor axes a and b is simply

$$L = aE(2\pi, k) \qquad \text{where} \quad k = \frac{\sqrt{a^2 - b^2}}{a}.$$

complete elliptic
integrals

It is common to define the **complete elliptic integral** of the first and second kinds:

$$K(k) \equiv F\left(\frac{\pi}{2}, k\right) = \int_0^{\pi/2} \frac{dt}{\sqrt{1 - k^2 \sin^2 t}},$$

$$E(k) \equiv E\left(\frac{\pi}{2}, k\right) = \int_0^{\pi/2} \sqrt{1 - k^2 \sin^2 t}\, dt. \qquad (11.15)$$

The reader may easily verify (Problem 11.10) that the total circumference of an ellipse can be given in terms of complete elliptic integrals.

The parameterization given in Equation (11.11) is that of a horizontal ellipse ($a > b$). However, one may wish to start with a *vertical* ellipse ($a < b$). Then, as the reader may verify, one ends up with an integral similar to (11.14), except that the coefficient of $\sin^2 t$ is $+k^2$. Would this be a *new* elliptic integral? Problem 11.9 shows that the new integral can be written as a sum of the existing elliptic integrals.

large-angle
pendulum and
elliptic integrals

Example 11.1.2. Elliptic integrals show up in areas of physics totally unrelated to the circumference of an ellipse. Consider a pendulum of mass m and length l displaced by an angle θ from its equilibrium position as shown in Figure 11.1. When the angle is θ, the velocity of the pendulum is $l\dot{\theta}$ and its height is h. Conservation of energy leads to

$$E = KE + PE = \tfrac{1}{2}m(l\dot{\theta})^2 + mgh = \tfrac{1}{2}ml^2\dot{\theta}^2 + mg(l - l\cos\theta),$$

where E is the total mechanical energy of the pendulum. If θ_m is the maximum angular displacement, then the total energy at this angle will be just the potential energy.[5] It then follows that

$$\tfrac{1}{2}m(l\dot{\theta})^2 + mgh = \tfrac{1}{2}ml^2\dot{\theta}^2 + mg(l - l\cos\theta) = mg(l - l\cos\theta_m),$$

or, after dividing both sides by ml,

$$\tfrac{1}{2}l\dot{\theta}^2 - g\cos\theta = -g\cos\theta_m. \qquad (11.16)$$

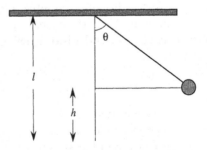

Figure 11.1: The pendulum displaced by an arbitrary angle θ.

---

[5]The KE is zero at θ_m because the pendulum comes to a momentary stop there.

The elementary treatment of the pendulum problem differentiates Equation (11.16) with respect to time, assumes that the maximum angle—and therefore any angle—is small, and approximates $\sin\theta$ with θ in radians. This leads to

$$l^2\dot{\theta}\ddot{\theta} + gl\dot{\theta}\sin\theta = 0 \quad \text{or} \quad \ddot{\theta} + \frac{g}{l}\sin\theta = 0 \quad \xrightarrow{\theta \to 0} \quad \ddot{\theta} + \frac{g}{l}\theta = 0,$$

which is the equation of a simple harmonic oscillator[6] with $\omega^2 = g/l$ or $T = 2\pi\sqrt{l/g}$. This is the famous result—known even to Galileo—that, for small angles, the period of oscillation is independent of the angle.

A more advanced treatment makes no approximation for the angle and simply integrates (11.16). Assuming that $\dot{\theta} > 0$, Equation (11.16) gives

$$\frac{d\theta}{dt} = \sqrt{\frac{2g}{l}}\sqrt{\cos\theta - \cos\theta_m} = 2\sqrt{\frac{g}{l}}\sqrt{\sin^2\left(\frac{\theta_m}{2}\right) - \sin^2\left(\frac{\theta}{2}\right)}, \quad (11.17)$$

where we used the trigonometric identity $\cos\theta = 1 - 2\sin^2(\theta/2)$. Introducing a new variable s given by

$$\sin\left(\frac{\theta}{2}\right) \equiv \sin\left(\frac{\theta_m}{2}\right)\sin s,$$

differentiating this equation with respect to t, and using Equation (11.17) yields

$$\frac{ds}{dt} = \sqrt{\frac{g}{l}}\sqrt{1 - \sin^2\left(\frac{\theta_m}{2}\right)\sin^2 s}. \quad (11.18)$$

This leads to

$$\sqrt{\frac{g}{l}}\,dt = \frac{ds}{\sqrt{1 - \sin^2(\theta_m/2)\sin^2 s}}$$

which can be integrated to yield

$$t = \sqrt{\frac{l}{g}}\int_0^s \frac{du}{\sqrt{1 - \sin^2(\theta_m/2)\sin^2 u}} \equiv \sqrt{\frac{l}{g}}F\left(s(\theta), \sin\frac{\theta_m}{2}\right), \quad (11.19)$$

where $s = \sin^{-1}[\sin(\theta/2)/\sin(\theta_m/2)]$, and we have assumed that at $t = 0$, the angle θ is zero and therefore $s = 0$ as well.

Of particular interest is the period of the oscillation which is four times the time it takes the pendulum to go from $\theta = 0$ to $\theta = \theta_m$. These values correspond to $s = 0$ and $s = \pi/2$. It follows that

period of a pendulum depends on the amplitude of oscillation.

$$T = 4\sqrt{\frac{l}{g}}\int_0^{\pi/2} \frac{du}{\sqrt{1 - \sin^2(\theta_m/2)\sin^2 u}}$$

$$\equiv 4\sqrt{\frac{l}{g}}F\left(\frac{\pi}{2}, \sin\frac{\theta_m}{2}\right) = 4\sqrt{\frac{l}{g}}K\left(\sin\frac{\theta_m}{2}\right). \quad (11.20)$$

---

[6]Recall that the equation of a simple harmonic oscillator (SHO)—such as a spring–mass system with mass m and spring constant k—is $m\ddot{x} + kx = 0$ or $\ddot{x} + (k/m)x = 0$. It is shown in elementary physics that the **angular frequency** of this SHO is $\omega = \sqrt{k/m}$. Thus, in any SHO equation in which the second derivative appears with no coefficient, the coefficient of the undifferentiated quantity is the square of the angular frequency.

This shows clearly that for large maximum angles, the period *does* depend on the amplitude. By expanding the integrand in a power series as developed in Chapter 10, one can obtain the deviation from constant period as powers of $\sin^2(\theta_m/2)$. We quote the result of such an expansion

$$T = 2\pi \sqrt{\frac{l}{g}} \left(1 + \frac{1}{4} \sin^2 \frac{\theta_m}{2} + \frac{9}{64} \sin^4 \frac{\theta_m}{2} + \cdots \right). \qquad (11.21)$$

The reader is urged to verify this result (see Problems 11.11 and 11.12).   ■

Historical Notes

The study of elliptical integrals can be said to have started in 1655 when Wallis began to study the arc length of an ellipse. In fact he considered the arc lengths of various cycloids and related these arc lengths to that of the ellipse. Both Wallis and Newton published an infinite series expansion for the arc length of the ellipse.

In 1679 *Jacob Bernoulli* attempted to find the arc length of a spiral and encountered an example of an elliptic integral. He made an important step in the theory of elliptic integrals in 1694. He examined the shape that an elastic rod will take if compressed at the ends. He showed that the curve could be expressed in terms of an integral, which was very similar to the one obtained by Wallis.

There is no doubt that Gauss obtained a number of key results in the theory of elliptic functions, because many of these were found after his death in papers he had never published. However, the acknowledged founders of the theory of elliptic functions were Abel and Jacobi.

Niels Henrik Abel was the son of a poor pastor. As a student in Christiania (Oslo), Norway, he had the luck to have Berndt Holmböe (1795–1850) as a teacher. Holmböe recognized Abel's genius and predicted when Abel was seventeen that he would become the greatest mathematician in the world. After studying at Christiania and at Copenhagen, Abel received a scholarship that permitted him to travel. In Paris, he was presented to Legendre, Laplace, and Cauchy, but they ignored him. Having exhausted his funds, he departed for Berlin and spent the years 1825–1827 with Crelle.

Niels Henrik Abel
1802–1829

He returned to Christiania so exhausted that he found it necessary, he wrote, to hold on to the gates of a church. To earn money he gave lessons to young students. He began to receive attention through his published works, and Crelle thought he might be able to secure him a professorship at the University of Berlin. But Abel became ill with tuberculosis and died in 1829 when he was only twenty-seven years old.

Abel knew of the work of Euler, Lagrange, and Legendre on elliptic integrals, and may have gotten ideas for his own work from the work of Gauss. Abel started to write papers in 1825. He presented his major paper to the Academy of Sciences in Paris in 1826. The paper was given to Cauchy to review it. But partly because of the length and the difficulty of the paper and partly to favor his own work, Cauchy laid it aside. After Abel's death, when his fame was established, the academy searched for the paper, found it, and published it in 1841.

The other discoverer of elliptic functions was **Carl Gustav Jacob Jacobi**. Unlike Abel, he lived a quiet life. Born in Potsdam to a Jewish family, he studied at the University of Berlin and in 1827 became a professor at Königsberg. In 1842 he had to give up his post because of ill health. He was given a pension by the Prussian government and retired to Berlin, where he died in 1851. His fame was great even during his lifetime, and his students spread his ideas to many centers.

Jacobi taught the subject of elliptic functions for many years. His approach became the model according to which the theory of functions itself was developed. He also worked in functional determinants (Jacobians), ordinary and partial differential equations, dynamics, celestial mechanics, and fluid dynamics.

Carl Gustav Jacobi
1804–1851

Jacobi's work on elliptic functions started in 1827 when he submitted a paper for publication without proof. Almost simultaneously, Abel wrote his research paper on elliptic functions. Both had arrived at the key idea of working with inverse functions of the elliptic integrals, an idea that Abel had had since 1823. Thereafter, they both published on the subject. But whereas Abel died in 1829, Jacobi lived to publish much more. In particular, his *Fundamenta Nova Theoriae Functionum Ellipticarum* of 1829 became a leading work on the subject.

11.2 Power Series as Functions

Differential equations have found their way into all areas of physics from the motion of planets around the Sun to standing waves on a rope or a drum, to electrical properties of conductors, and the behavior of electromagnetic fields and beyond. As is always the case, no mathematics can draw more attention than that which deals directly with Nature. The urgency of finding solutions to these differential equations prompted many mathematicians of the latter part of the eighteenth and the beginning of the nineteenth centuries to concentrate heavily on certain specific differential equations. It appeared that every differential equation dictated by Nature gave rise to a new function. The most common scheme for solving these differential equations was to assume a power series solution, substitute the assumed solution in the differential equation, and determine the (unknown) coefficients from the resulting equality of power series. We shall come back to this powerful method in Chapters 24 and 25 through 27. At this point, we want to simply give examples of solutions (functions) of certain differential equations that were discovered in the form of a power series.

Chapter 10 showed how *known functions* (such as trigonometric and logarithmic functions) can be *represented* as power series. These functions had been known prior to the popularity of infinite series, and the origin of their discovery lay in areas of mathematics outside calculus. One does not need a power series to calculate $\sin(35°)$; an appropriate right triangle and careful measurement of its sides and hypotenuse will do the job. The functions we are discussing here are *defined* in terms of power series and do not have independent existence. With some mathematical manipulation they may be written as a definite *integral*—which cannot be evaluated analytically. But that is just as abstract as an infinite series because in the latter case, the integrals become their definition.

11.2.1   Hypergeometric Functions

In their studies of second-order differential equations (DE), mathematicians, always in search of generalities, came up with the most general form of a second-order linear DE which appeared to encompass all known DEs of physical interest. This DE, called the **hypergeometric differential equation**, turned out to be[7]

$$x(1-x)y'' + [\gamma - (\alpha + \beta + 1)x]y' - \alpha\beta y = 0, \qquad (11.22)$$

hypergeometric differential equation

where α, β, and γ are constants.[8] The series solution of this DE, called the **hypergeometric function** can be written in terms of the gamma function as[9]

hypergeometric function

$$F(\alpha, \beta; \gamma; x) \equiv \frac{\Gamma(\gamma)}{\Gamma(\alpha)\Gamma(\beta)} \sum_{n=0}^{\infty} \frac{\Gamma(\alpha+n)\Gamma(\beta+n)}{\Gamma(\gamma+n)\Gamma(n+1)} x^n. \qquad (11.23)$$

From this series representation, we immediately note that the hypergeometric function is symmetric under interchange of α and β. Furthermore, if either α or β is a *negative integer*, say $-m$, then the denominator of the constant outside becomes infinite by Definition 11.1.1. However, the gamma function in the numerator of the first m terms of the sum will also be infinite. The cancellation of these infinities [see Problem 11.4(c)] gives a nonzero sum up to m, but the rest of the series will be zero. Therefore,

> **Box 11.2.1.** *The hypergeometric function is symmetric under interchange of α and β: $F(\alpha, \beta; \gamma; x) = F(\beta, \alpha; \gamma; x)$. Furthermore, $F(-m, \beta; \gamma; x)$ [and therefore $F(\alpha, -m; \gamma; x)$] is a polynomial if m is a positive integer.*

As mentioned before, many a time, the infinite series can be "integrated" and the resulting function written in terms of an integral. In this case, we start by multiplying and dividing the series of Equation (11.23) by $\Gamma(\gamma - \beta)$ to obtain

$$F(\alpha, \beta; \gamma; x) = \frac{\Gamma(\gamma)}{\Gamma(\alpha)\Gamma(\beta)\Gamma(\gamma-\beta)} \sum_{n=0}^{\infty} \frac{\Gamma(\alpha+n)}{\Gamma(n+1)} \underbrace{\frac{\Gamma(\gamma-\beta)\Gamma(\beta+n)}{\Gamma(\gamma+n)}}_{\equiv B(\gamma-\beta,\beta+n) \text{ by (11.7)}} x^n.$$

---

[7]For a comprehensive treatment of this differential equation, see Hassani, S. *Mathematical Physics: A Modern Introduction to Its Foundations*, Springer-Verlag, 1999, Chapter 14.

[8]Some authors use a, b, and c instead of α, β, and γ.

[9]Some authors use $_2F_1$ instead of F. Our use of F to represent both the elliptic integral of the first kind and the hypergeometric function should not cause any confusion because the two functions have different numbers of arguments (independent variables).

Now use $\Gamma(n+1) = n!$ and the integral representation of the beta function to get

$$F(\alpha, \beta; \gamma; x) = \frac{\Gamma(\gamma)}{\Gamma(\alpha)\Gamma(\beta)\Gamma(\gamma-\beta)} \sum_{n=0}^{\infty} \int_0^1 dt(1-t)^{\gamma-\beta-1} t^{\beta+n-1} \Gamma(\alpha+n) \frac{x^n}{n!}$$

$$= \frac{\Gamma(\gamma)}{\Gamma(\beta)\Gamma(\gamma-\beta)} \int_0^1 dt(1-t)^{\gamma-\beta-1} t^{\beta-1} \sum_{n=0}^{\infty} \frac{\Gamma(\alpha+n)}{\Gamma(\alpha)} \frac{(tx)^n}{n!}.$$

Using the result of Problem 11.4, we can now write

$$F(\alpha, \beta; \gamma; x) = \frac{\Gamma(\gamma)}{\Gamma(\beta)\Gamma(\gamma-\beta)} \int_0^1 dt(1-t)^{\gamma-\beta-1} t^{\beta-1}(1-tx)^{-\alpha}. \quad (11.24)$$

integral representation of the hypergeometric function

This is the integral representation of the hypergeometric function.

The generality of the hypergeometric DE results in the ability to express many functions—both elementary and the so-called special functions of mathematical physics—in terms of the hypergeometric function. For example, consider the complete elliptic integral of the second kind $E(k)$. The two factors of double factorials in both the numerator and denominator of its series expansion (see Problem 11.13), together with Equation (11.5) and the hypergeometric series (11.23), hint at the possibility of writing $E(k)$ as a hypergeometric function. This is indeed the case. Substituting (11.5) in the expansion of $E(k)$ as given in Problem 11.13 yields

$$E(k) = \frac{\pi}{2} \left\{ 1 - \sum_{n=1}^{\infty} \frac{\Gamma(n+\frac{1}{2})\Gamma(n+\frac{1}{2})\pi^{-1}}{\Gamma(n+1)\Gamma(n+1)} \frac{k^{2n}}{2(n-\frac{1}{2})} \right\}$$

$$= \frac{\pi}{2} - \frac{1}{4} \sum_{n=1}^{\infty} \frac{\Gamma(n+\frac{1}{2})\Gamma(n-\frac{1}{2})}{\Gamma(n+1)\Gamma(n+1)} (k^2)^n,$$

where we used $\Gamma(n+\frac{1}{2}) = (n-\frac{1}{2})\Gamma(n-\frac{1}{2})$. The sum starts with $n = 1$. To make it look like a hypergeometric series, we need to include the zero term as well. Adding and subtracting this term gives

$$E(k) = \frac{\pi}{2} - \frac{1}{4} \sum_{n=0}^{\infty} \frac{\Gamma(n+\frac{1}{2})\Gamma(n-\frac{1}{2})}{\Gamma(n+1)\Gamma(n+1)} (k^2)^n + \frac{1}{4} \left[\frac{\Gamma(\frac{1}{2})\Gamma(-\frac{1}{2})}{\Gamma(1)\Gamma(1)} (k^2)^0 \right]$$

$$= -\frac{1}{4} \sum_{n=0}^{\infty} \frac{\Gamma(n+\frac{1}{2})\Gamma(n-\frac{1}{2})}{\Gamma(n+1)\Gamma(n+1)} (k^2)^n$$

because $\Gamma(-\frac{1}{2}) = -2\Gamma(\frac{1}{2}) = -2\sqrt{\pi}$ by Example 11.1.1. We now note that except for a multiplicative constant, the sum is that of the hypergeometric function with $\alpha = \frac{1}{2} = -\beta$ and $\gamma = 1$. Inserting the multiplicative constant

$$\frac{\Gamma(1)}{\Gamma(\frac{1}{2})\Gamma(-\frac{1}{2})} = \frac{1}{(-2\pi)}$$

we obtain

$$E(k) = \frac{\pi}{4} F\left(\tfrac{1}{2}, -\tfrac{1}{2}; 1; k^2\right). \tag{11.25}$$

The reader may verify that

$$K(k) = \frac{\pi}{4} F\left(\tfrac{1}{2}, \tfrac{1}{2}; 1; k^2\right). \tag{11.26}$$

Historical Notes

Johann Carl Friedrich Gauss was the greatest of all mathematicians and perhaps the most richly gifted genius of whom there is any record. He was born in the city of Brunswick in northern Germany. His exceptional skill with numbers was clear at a very early age, and in later life he joked that he knew how to count before he could talk. It is said that Goethe wrote and directed little plays for a puppet theater when he was six and that Mozart composed his first childish minuets when he was five, but Gauss corrected an error in his father's payroll accounts at the age of three. At the age of seven, when he started elementary school, his teacher was amazed when Gauss summed the integers from 1 to 100 instantly by spotting that the sum was 50 pairs of numbers each pair summing to 101.

His long professional life is so filled with accomplishments that it is impossible to give a full account of them in the short space available here. All we can do is simply give a chronology of his almost uncountable discoveries.

Carl Friedrich
Gauss 1777–1855

1792–1794: Gauss reads the works of Newton, Euler, and Lagrange; discovers the prime number theorem (at the age of 14 or 15); invents the method of least squares; conceives the Gaussian law of distribution in the theory of probability.

1795: (only 18 years old!) Proves that a regular polygon with n sides is constructible (by ruler and compass) if and only if n is the product of a power of 2 and distinct prime numbers of the form $p_k = 2^{2^k} + 1$, and completely solves the 2000-year old problem of ruler-and-compass construction of regular polygons. He also discovers the law of quadratic reciprocity.

1799: Proves the **fundamental theorem of algebra** in his doctoral dissertation using the then-mysterious complex numbers with complete confidence.

1801: Gauss publishes his *Disquisitiones Arithmeticae* in which he creates the modern rigorous approach to mathematics; predicts the exact location of the asteroid Ceres.

1807: Becomes professor of astronomy and the director of the new observatory at Göttingen.

1809: Publishes his second book, *Theoria motus corporum coelestium*, a major two-volume treatise on the motion of celestial bodies and the bible of planetary astronomers for the next 100 years.

1812: Publishes *Disquisitiones generales circa seriem infinitam*, a rigorous treatment of infinite series, and introduces the **hypergeometric function** for the first time, for which he uses the notation $F(\alpha, \beta; \gamma; z)$; an essay on approximate integration.

1820–1830: Publishes over 70 papers, including *Disquisitiones generales circa superficies curvas*, in which he creates the intrinsic **differential geometry** of general curved surfaces, the forerunner of Riemannian geometry and the general theory of relativity.

From the 1830s on, Gauss was increasingly occupied with physics, and he enriched every branch of the subject he touched. In the theory of **surface tension**, he developed the fundamental idea of conservation of energy and solved the earliest problem in the **calculus of variations**. In **optics**, he introduced the concept of the focal length of a system of lenses. He virtually created the science of **geomagnetism**, and in collaboration with his friend and colleague Wilhelm Weber he invented the electromagnetic telegraph. In 1839 Gauss published his fundamental paper on the general theory of inverse square forces, which established **potential theory** as a coherent branch of mathematics and in which he established the **divergence theorem**.

Gauss had many opportunities to leave Göttingen, but he refused all offers and remained there for the rest of his life, living quietly and simply, traveling rarely, and working with immense energy on a wide variety of problems in mathematics and its applications. Apart from science and his family—he married twice and had six children, two of whom emigrated to America—his main interests were history and world literature, international politics, and public finance. He owned a large library of about 6000 volumes in many languages, including Greek, Latin, English, French, Russian, Danish, and of course German. His acuteness in handling his own financial affairs is shown by the fact that although he started with virtually nothing, he left an estate over a hundred times as great as his average annual income during the last half of his life.

The foregoing list is the published portion of Gauss's total achievement; the unpublished and private part is almost equally impressive. His scientific diary, a little booklet of 19 pages, discovered in 1898, extends from 1796 to 1814 and consists of 146 very concise statements of the results of his investigations, which often occupied him for weeks or months. These ideas were so abundant and so frequent that he physically did not have time to publish them. Some of the ideas recorded in this diary: **Cauchy Integral Formula**: Gauss discovers it in 1811, 16 years before Cauchy. **Non-Euclidean Geometry**: After failing to prove Euclid's fifth postulate at the age of 15, Gauss came to the conclusion that the Euclidean form of geometry cannot be the only one possible. **Elliptic Functions**: Gauss had found many of the results of Abel and Jacobi (the two main contributors to the subject) before these men were born. The facts became known partly through Jacobi himself. His attention was caught by a cryptic passage in the *Disquisitiones*, whose meaning can only be understood if one knows something about elliptic functions. He visited Gauss on several occasions to verify his suspicions and tell him about his own most recent discoveries, and each time Gauss pulled 30-year-old manuscripts out of his desk and showed Jacobi what Jacobi had just shown him. After a week's visit with Gauss in 1840, Jacobi wrote to his brother, "Mathematics would be in a very different position if practical astronomy had not diverted this colossal genius from his glorious career."

A possible explanation for not publishing such important ideas is suggested by his comments in a letter to Bolyai: "It is not knowledge but the act of learning, not possession but the act of getting there, which grants the greatest enjoyment. When I have clarified and exhausted a subject, then I turn away from it in order to go into darkness again." His was the temperament of an explorer who is reluctant to take the time to write an account of his last expedition when he could be starting another. As it was, Gauss wrote a great deal, but to have published every fundamental discovery he made in a form satisfactory to himself would have required several long lifetimes.

11.2.2 Confluent Hypergeometric Functions

The parameters α, β, and γ determine the behavior of the hypergeometric function completely. A great number of differential equations in mathematical physics correspond to the case where only two parameters are involved. The most effective way of accommodating this arises from the *confluence* $\beta \to \infty$. Let us see how this works.

Substitute $x = u/\beta$ in the hypergeometric DE using the—very simple—chain rule to transform the x-derivatives to the u-derivatives. This leads to the DE

$$\frac{u}{\beta}\left(1 - \frac{u}{\beta}\right)\beta^2 \frac{d^2 y}{du^2} + \left[\gamma - (\alpha + \beta + 1)\frac{u}{\beta}\right]\beta\frac{dy}{du} - \alpha\beta y = 0.$$

Dividing the entire equation by β, taking the limit $\beta \to \infty$—thus neglecting u/β and $1/\beta$—yields the so-called **confluent hypergeometric differential equation**:

$$xy'' + (\gamma - x)y' - \alpha y = 0, \tag{11.27}$$

confluent
hypergeometric
differential
equation

where we restored x as the independent variable.

The infinite series solution of this DE is called the **confluent hypergeometric function**. This solution, as well as its integral representation, can be obtained by taking the appropriate limit of the corresponding expression for the hypergeometric function. The limit of Equation (11.23) yields

confluent
hypergeometric
function

$$\Phi(\alpha; \gamma; x) \equiv \lim_{\beta \to \infty} F(\alpha, \beta; \gamma; x/\beta) = \frac{\Gamma(\gamma)}{\Gamma(\alpha)}\sum_{n=0}^{\infty}\frac{\Gamma(\alpha + n)}{\Gamma(\gamma + n)\Gamma(n + 1)}x^n, \tag{11.28}$$

where we used

$$\frac{\Gamma(\beta + n)}{\beta^n \Gamma(\beta)} = \frac{(\beta + n - 1)(\beta + n - 2)\cdots\beta\Gamma(\beta)}{\beta^n \Gamma(\beta)}$$

$$= \left(\frac{\beta + n - 1}{\beta}\right)\left(\frac{\beta + n - 2}{\beta}\right)\cdots\left(\frac{\beta}{\beta}\right) \xrightarrow{\beta \to \infty} 1.$$

Similarly, we have

$$\Phi(\alpha; \gamma; x) = \lim_{\beta \to \infty} F(\beta, \alpha; \gamma; x/\beta)$$

$$= \lim_{\beta \to \infty}\frac{\Gamma(\gamma)}{\Gamma(\alpha)\Gamma(\gamma - \alpha)}\int_0^1 dt\,(1 - t)^{\gamma - \alpha - 1}t^{\alpha - 1}\underbrace{\left(1 - \frac{tx}{\beta}\right)^{-\beta}}_{\to e^{tx}\ (\text{Prob. }11.3)},$$

where we have used the symmetry of the hypergeometric function under interchange of its first two parameters. It follows that the integral representation of the confluent hypergeometric function is

integral
representation of
the confluent
hypergeometric
function

$$\Phi(\alpha; \gamma; x) = \frac{\Gamma(\gamma)}{\Gamma(\alpha)\Gamma(\gamma - \alpha)}\int_0^1 dt\,(1 - t)^{\gamma - \alpha - 1}t^{\alpha - 1}e^{tx}. \tag{11.29}$$

We note that

$$\Phi(\alpha; \alpha; x) = \sum_{n=0}^{\infty} \frac{1}{\Gamma(n+1)} x^n = \sum_{n=0}^{\infty} \frac{x^n}{n!} = e^x$$

and Problem 11.20 shows that

$$\mathrm{erf}(x) = \frac{2x}{\sqrt{\pi}} \Phi(\tfrac{1}{2}; \tfrac{3}{2}; -x^2).$$

Many other functions encountered in mathematical physics can also be expressed in terms of confluent hypergeometric functions, and we shall point this out as we come across these functions in the sequel. We note in passing that, as in the case of hypergeometric function,

> **Box 11.2.2.** *If α happens to be a negative integer, then $\Phi(\alpha; \gamma; x)$ becomes a polynomial, i.e., the infinite series truncates.*

11.2.3   Bessel Functions

Bessel functions are arguably among the most utilized functions of mathematical physics. We shall come back to them when we consider solutions of Laplace's equation in cylindrical coordinates and discover their connection with other functions treated in this chapter. At this point, we simply introduce them as power series. The Bessel function $J_\nu(x)$ of order ν is a solution of the **Bessel differential equation**:

Bessel differential equation

$$x\frac{d^2 y}{dx^2} + \frac{dy}{dx} + \left(x - \frac{\nu^2}{x}\right) y = 0. \tag{11.30}$$

Chapter 27 shows how to obtain the power series expansion of $J_\nu(x)$:

$$J_\nu(x) = \left(\frac{x}{2}\right)^\nu \sum_{k=0}^{\infty} \frac{(-1)^k}{k!\,\Gamma(\nu+k+1)} \left(\frac{x}{2}\right)^{2k}. \tag{11.31}$$

The point to emphasize is that

> **Box 11.2.3.** *Bessel functions are always given in terms of their expansion in power series (or as an integral involving parameters). It is generally impossible to reduce Bessel functions to any functional combination of more elementary functions such as polynomials, or trigonometric and exponential functions.*

Properties and applications of Bessel functions are treated in some detail in Chapter 27.[10] However, some relations are elementary enough to be included here, as they also illustrate the use of summation symbols. First note that if ν is an integer $-m$, then

$$J_{-m}(x) = \left(\frac{x}{2}\right)^{-m} \sum_{k=0}^{\infty} \frac{(-1)^k}{k!\Gamma(-m+k+1)} \left(\frac{x}{2}\right)^{2k}$$

$$= \left(\frac{x}{2}\right)^{-m} \sum_{k=m}^{\infty} \frac{(-1)^k}{k!\Gamma(-m+k+1)} \left(\frac{x}{2}\right)^{2k}$$

because the first m terms of the first series have gamma functions in the denominator with negative integer (or zero) arguments. Now in the second series, replace k by $n = k - m$. This yields

$$J_{-m}(x) = \left(\frac{x}{2}\right)^{-m} \sum_{n=0}^{\infty} \frac{(-1)^{m+n}}{(m+n)!\Gamma(n+1)} \left(\frac{x}{2}\right)^{2m+2n} \tag{11.32}$$

$$= (-1)^m \left(\frac{x}{2}\right)^{m} \sum_{n=0}^{\infty} \frac{(-1)^{n}}{\Gamma(m+n+1)n!} \left(\frac{x}{2}\right)^{2n} = (-1)^m J_m(x),$$

where we used $\Gamma(j+1) = j!$ for positive integer j.

Example 11.2.1. Bessel functions of half-integer order are related to trigonometric functions. To see this, note that

$$J_{1/2} = \left(\frac{x}{2}\right)^{1/2} \sum_{k=0}^{\infty} \frac{(-1)^k}{k!\Gamma(k+\frac{3}{2})} \left(\frac{x}{2}\right)^{2k}$$

$$= \left(\frac{x}{2}\right)^{-1/2} \sum_{k=0}^{\infty} \frac{(-1)^k}{k!\Gamma(k+\frac{3}{2})2^{2k+1}} x^{2k+1}.$$

Now substitute for $\Gamma(k+\frac{3}{2})$ in terms of factorials as given in Problem 11.1 to obtain

$$J_{1/2} = \left(\frac{x}{2}\right)^{-1/2} \frac{1}{\sqrt{\pi}} \underbrace{\sum_{k=0}^{\infty} \frac{(-1)^k}{(2k+1)!} x^{2k+1}}_{=\sin x} = \left(\frac{2}{\pi x}\right)^{1/2} \sin x.$$

Similarly,

$$J_{-1/2} = \left(\frac{2}{\pi x}\right)^{1/2} \cos x$$

as the reader may verify.   ∎

---

[10]See also Hassani, S. *Mathematical Physics: A Modern Introduction to Its Foundations*, Springer-Verlag, 1999, Section 14.5.

Another formula of interest is a recursion relation connecting Bessel functions of different integer orders. Write $J_{m-1}(x)$ as

$$J_{m-1}(x) = \left(\frac{x}{2}\right)^{m-1} \sum_{k=0}^{\infty} \frac{(-1)^k}{k!\Gamma(m+k)} \left(\frac{x}{2}\right)^{2k} \qquad (11.33)$$

$$= \left(\frac{x}{2}\right)^{m-1} \left[\frac{1}{\Gamma(m)} + \sum_{k=1}^{\infty} \frac{(-1)^k}{k!\Gamma(m+k)} \left(\frac{x}{2}\right)^{2k}\right],$$

where we separated the $k = 0$ term from the rest of the sum. Similarly, write $J_{m+1}(x)$ as

$$J_{m+1}(x) = \left(\frac{x}{2}\right)^{m+1} \sum_{k=0}^{\infty} \frac{(-1)^k}{k!\Gamma(m+k+2)} \left(\frac{x}{2}\right)^{2k}$$

$$= \left(\frac{x}{2}\right)^{m+1} \sum_{j=1}^{\infty} \frac{(-1)^{j-1}}{(j-1)!\Gamma(m+j+1)} \left(\frac{x}{2}\right)^{2j-2} \qquad (11.34)$$

$$= -\left(\frac{x}{2}\right)^{m-1} \sum_{k=1}^{\infty} \frac{(-1)^k}{(k-1)!\Gamma(m+k+1)} \left(\frac{x}{2}\right)^{2k},$$

where in the second line, we substituted $j = k+1$ for k, and in the last line, we used $(-1)^{-1} = -1$, factored $(x/2)^{-2}$ out of the summation, and changed the dummy index back to k. Now add Equations (11.33) and (11.34) and use

$$\frac{1}{k!\Gamma(m+k)} - \frac{1}{(k-1)!\Gamma(m+k+1)} = \frac{m}{k!\Gamma(m+k+1)},$$

and $1/\Gamma(m) = m/\Gamma(m+1)$ to obtain

$$J_{m-1}(x) + J_{m+1}(x) = \left(\frac{x}{2}\right)^{m-1} \left[\frac{m}{\Gamma(m+1)} + \sum_{k=1}^{\infty} \frac{(-1)^k m}{k!\Gamma(m+k+1)} \left(\frac{x}{2}\right)^{2k}\right]$$

$$= m \left(\frac{x}{2}\right)^{-1} \underbrace{\left[\left(\frac{x}{2}\right)^m \sum_{k=0}^{\infty} \frac{(-1)^k}{k!\Gamma(m+k+1)} \left(\frac{x}{2}\right)^{2k}\right]}_{=J_m(x)}$$

or, finally,

$$J_{m-1}(x) + J_{m+1}(x) = \frac{2m}{x} J_m(x). \qquad (11.35)$$

The straightforward details are left as Problem 11.22. One can also show that

$$J_{m-1}(x) - J_{m+1}(x) = 2J'_m(x), \qquad (11.36)$$

where prime indicates differentiation. Equations (11.35) and (11.36) lead to

$$J_{m-1}(x) = \frac{m}{x} J_m(x) + J'_m(x),$$

$$J_{m+1}(x) = \frac{m}{x} J_m(x) - J'_m(x). \qquad (11.37)$$

These plus the results of Example 11.2.1 give all Bessel functions of half-integer order.

11.3 Problems

11.1. (a) Show that (see Definition 11.1.2 for the definition of the following notation):

$$(2n)!! = 2^n n! \quad \text{and} \quad (2n-1)!! = \frac{(2n)!}{2^n n!}.$$

Hint: For the second relation, supply the missing even factors in the "numerator" and the "denominator" of $(2n-1)!!$

(b) Using (a) and Example 11.1.1, show that

$$\Gamma(n + \tfrac{1}{2}) = \frac{(2n-1)!!}{2^n} \sqrt{\pi} = \frac{(2n)!}{2^{2n}n!} \sqrt{\pi}.$$

(c) Now use (b) to obtain the following result:

$$\Gamma(n + \tfrac{3}{2}) = \frac{(2n+1)!}{2^{2n+1}n!} \sqrt{\pi}.$$

11.2. Using the result of Problem 11.1, show that

$$(2n+m)! = \frac{1}{\sqrt{\pi}} 2^{2n+m} \Gamma\left(n + \frac{m}{2} + 1\right) \Gamma\left(n + \frac{m+1}{2}\right).$$

Hint: Consider the two cases of even m (with $m = 2k$) and odd m (with $m = 2k+1$) separately, and show at the end that both can be written as a single formula.

11.3. Using the result

$$\lim_{n \to \infty} \left(1 + \frac{1}{n}\right)^n = e$$

show that

$$\lim_{n \to \infty} \left(1 - \frac{t}{n}\right)^n = e^{-t}.$$

Hint: Let $n = -tm$.

11.4. (a) By using Equation (11.2) repeatedly, show that

$$\Gamma(a+n) = (a+n-1)(a+n-2)\cdots(a+n-k)\Gamma(a+n-k).$$

(b) Let $k = n$ in the above equation to show that

$$a(a+1)\cdots(a+n-1) = \frac{\Gamma(a+n)}{\Gamma(a)}.$$

(c) Using (b) show that

$$\alpha(\alpha-1)\cdots(\alpha-n+1) = (-1)^n \frac{\Gamma(n-\alpha)}{\Gamma(-\alpha)}.$$

11.5. Show that

$$\Gamma(x) = 2 \int_0^\infty e^{-t^2} t^{2x-1}\, dt$$

and

$$\Gamma(x) = \int_0^1 \left\{ \ln\left(\frac{1}{t}\right) \right\}^{x-1} dt.$$

11.6. Find the following integrals in terms of the gamma function:

(a) $\int_0^\infty t^{2x+1} e^{-at^2}\, dt.$ (b) $\int_0^\infty t^{2x} e^{-at^2}\, dt.$

11.7. Using *only its integral representation*, show that beta function is symmetric under interchange of its arguments.

11.8. Using the definition of the gamma function, show the justification for the frequently used equality $0! = 1$.

11.9. Show that

$$\int_0^\varphi \sqrt{1 + k^2 \sin^2 t}\, dt = \sqrt{1 + k^2}\left[E(k') - E\left(\frac{\pi}{2} - \varphi, k'\right)\right],$$

where $k' = k/\sqrt{1 + k^2}$. Hint: Change t to $s = \pi/2 - t$ and break up the interval of integration of the resulting integral into two.

11.10. Show that the circumference of an ellipse of respective semi-major and semi-minor axes a and b is $4aE(k)$ where $k = \sqrt{a^2 - b^2}/a$. Verify that you get the expected result when $a = b$.

11.11. (a) Expand the square roots in the definition of the elliptic integrals of the first and second kinds in powers of $k^2 \sin^2 t$, and keep the first three terms.
(b) Now integrate those terms to find an approximation to elliptic integrals for small k.
(c) Substitute $\pi/2$ for φ to obtain approximation for the complete elliptic integrals.

11.12. Use the result of Problem 11.11 to obtain Equation (11.21).

11.13. Use the integral

$$\int_0^{\pi/2} \sin^{2n} t\, dt = \frac{(2n-1)!!}{(2n)!!}\frac{\pi}{2}$$

to show that

$$E(k) = \frac{\pi}{2}\left\{ 1 - \sum_{n=1}^\infty \left[\frac{(2n-1)!!}{(2n)!!}\right]^2 \frac{k^{2n}}{2n-1} \right\},$$

$$K(k) = \frac{\pi}{2}\left\{ 1 + \sum_{n=1}^\infty \left[\frac{(2n-1)!!}{(2n)!!}\right]^2 k^{2n} \right\}.$$

11.14. Show that $E(0) = K(0) = \pi/2$, and that $E(1) = 1$, $K(1) = \infty$.

11.15. Use the ratio test on the hypergeometric series to determine its radius of convergence.

11.16. Verify that the complete elliptic integral of the first kind is related to the hypergeometric function as follows:

$$K(k) = \frac{\pi}{4}F\left(\tfrac{1}{2}, \tfrac{1}{2}; 1; k^2\right).$$

11.17. Show that $\ln(1 + x) = xF(1, 1; 2; -x)$.

11.18. Use the result of Problem 11.4 to express Equation (10.15) of Chapter 10 in terms of the gamma function; then show that

$$(1+x)^\alpha = \sum_{n=0}^{\infty} \frac{\Gamma(n-\alpha)}{\Gamma(-\alpha)\Gamma(n+1)}(-x)^n = F(-\alpha, \beta; \beta; -x)$$

for arbitrary β.

11.19. By using integral representations:
(a) Show that

$$B(a, b) = \frac{\Gamma(a)\Gamma(b+r)}{\Gamma(a+b+r)}F(a, r; a+b+r; 1),$$

where B is the beta function and r is any real number. Choose r appropriately and show that

$$B(a, b) = \frac{1}{a}F(a, 1-b; a+1; 1).$$

(b) Also prove that

$$F(\alpha, \beta; \gamma; 1) = \frac{\Gamma(\gamma)\Gamma(\gamma - \alpha - \beta)}{\Gamma(\gamma - \alpha)\Gamma(\gamma - \beta)}.$$

11.20. Expand the integrand of $\mathrm{erf}(x)$ in its Maclaurin series and use

$$2n + 1 = 2(n + \tfrac{1}{2}) = \frac{\Gamma(\tfrac{3}{2})}{\Gamma(\tfrac{1}{2})}$$

to show that

$$\mathrm{erf}(x) = \frac{2x}{\sqrt{\pi}}\Phi(\tfrac{1}{2}; \tfrac{3}{2}; -x^2).$$

11.21. Using the same procedure as in Example 11.2.1, show that

$$J_{-1/2} = \left(\frac{2}{\pi x}\right)^{1/2}\cos x.$$

11.22. Show that

$$\frac{1}{k!\Gamma(m+k)} - \frac{1}{(k-1)!\Gamma(m+k+1)} = \frac{m}{k!\Gamma(m+k+1)}$$

and use it to derive Equation (11.35).

11.23. Derive Equation (11.36).

11.24. Find $J_{3/2}(x)$ and $J_{-3/2}(x)$. Hint: Use Equation (11.37).

Part IV

Analysis of Vectors

Chapter 12

Vectors and Derivatives

One of the basic tools of physics is the calculus of vectors. A great variety of physical quantities are vectors which are functions of several variables such as space coordinates and time, and, as such, are good candidates for mathematical analysis. We have already encountered examples of such analyses in our treatment of the integration of vectors as in calculating electric, magnetic, and gravitational fields. However, vector analysis goes beyond simple vector integration. Vectors have a far richer structure than ordinary numbers, and, therefore, allow a much broader range of concepts.

Fundamental to the study of vector analysis is the notion of **field**, with which we have some familiarity based on our study of Chapters 1 and 4. Fields play a key role in many areas of physics: In the motion of fluids, in the conduction of heat, in electromagnetic theory, in gravitation, and so forth. All these situations involve a physical quantity that varies from point to point as well as from time to time,[1] i.e., it is a function of space coordinates and time. This physical quantity can be either a scalar, in which case we speak of a **scalar field**, or a vector, in which case we speak of a **vector field**. There are also **tensor fields**, which we shall discuss briefly in Chapter 17, and **spinor fields**, which are beyond the scope of this book.

scalar and vector fields

The temperature of the atmosphere is a scalar field because it is a function of space coordinates—equator versus the poles—and time (summer versus winter), and because temperature has no direction associated with it. On the other hand, wind velocity is a vector field because (a) it is a vector and (b) its magnitude and direction depend on space coordinates and time. In general, when we talk of a vector field, we are dealing with three functions of space and time, corresponding to the three components of the vector.

---

[1] In many instances fields are independent of time in which case we call them **static fields**.

12.1   Solid Angle

Before discussing the calculus of vectors, we want to introduce the concept of
a solid angle which is an important and recurrent concept in mathematical
physics, especially in the discussion of vector calculus.

12.1.1   Ordinary Angle Revisited

concept of angle
reexamined

We start with the concept of angle from a new perspective which easily gener-
alizes to solid angle. Consider a curve and a point P in a plane. The point P
is taken to lie off the curve [Figure 12.1(a)]. An arbitrary segment of the curve
defines an angle which is obtained by joining the two ends of the segment to
P. In particular, an element of length along the curve defines an infinitesimal
angle. We want to relate the length of this element to the size of its angle
measured in radians.

Connect P to the midpoint of the infinitesimal line element of length Δl,
and call the resulting vector $\mathbf{R}$ with the corresponding unit vector $\hat{\mathbf{e}}_R$ as shown
in Figure 12.1(a).[2] Let the angle between $\hat{\mathbf{e}}_R$ and the unit normal[3] to the
length element $\hat{\mathbf{e}}_n$ be α. As shown in the magnified diagram of Figure 12.1(b),
α is also the angle between the line element $\overline{QQ'}$ and the line segment obtained
by dropping a perpendicular $\overline{QH}$ onto the ray PQ'. It is clear from the
diagram that

$$\overline{QH} = \overline{QQ'}\cos\alpha \;\Rightarrow\; \overline{QH} = \Delta l\,\cos\alpha = \Delta l\,\hat{\mathbf{e}}_R\cdot\hat{\mathbf{e}}_n.$$

Now recall that the measure of an angle in radians is given by the ratio of
the length of the arc of a circle subtended by the angle to the radius of the
circle, and this measure is independent of the size of the circle chosen. To
find the measure of $\Delta\theta$ in radians, let us choose a circle of radius $R = |\mathbf{R}|$,

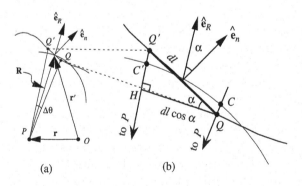

(a)   (b)

Figure 12.1: Defining angles as ratios of lengths.

---

[2]In actual calculations, it is convenient to denote the position vector of P by $\mathbf{r}$, say, and
that of the midpoint by $\mathbf{r}'$. Then $\mathbf{R} = \mathbf{r}' - \mathbf{r}$.

[3]There are two possible directions for this unit normal: one as shown in Figure 12.1,
and the other in the opposite direction. As long as we deal with open curves (no loops)
this arbitrariness persists.

the distance from P to the midpoint of the line element. The arc of this circle subtended by $\Delta\theta$ is CC', and the figure shows that the length of this arc is very nearly equal to $\overline{QH}$. One can think of CC' as the *projection of the line element* onto the circle. Thus,

$$\Delta\theta \approx \frac{\overline{QH}}{R} = \frac{\Delta l\,\hat{\mathbf{e}}_R \cdot \hat{\mathbf{e}}_n}{R}.$$

If we denote the location of P by $\mathbf{r}$ and that of Δl by $\mathbf{r}'$, then

$$\mathbf{R} = \mathbf{r}' - \mathbf{r}, \qquad \hat{\mathbf{e}}_R = \frac{\mathbf{r}' - \mathbf{r}}{|\mathbf{r}' - \mathbf{r}|},$$

and we obtain

$$\Delta\theta = \frac{\Delta l(\mathbf{r}')\,\hat{\mathbf{e}}_n \cdot (\mathbf{r}' - \mathbf{r})}{|\mathbf{r}' - \mathbf{r}|^2}, \tag{12.1}$$

where we have emphasized the dependence of Δl on $\mathbf{r}'$.

For a finite segment of the curve, we integrate to obtain the angle. This yields

angle as integral

$$\theta = \int_a^b \frac{dl\,\hat{\mathbf{e}}_R \cdot \hat{\mathbf{e}}_n}{R} = \int_a^b \frac{dl(\mathbf{r}')\,\hat{\mathbf{e}}_n \cdot (\mathbf{r}' - \mathbf{r})}{|\mathbf{r}' - \mathbf{r}|^2}, \tag{12.2}$$

where a and b are the beginning and the end of the finite segment. There is a way of calculating this finite angle which, although extremely simple-minded, is useful when we generalize to solid angle. Since the size of the circle used to measure the angle is irrelevant, let us choose a single fiducial circle of radius a centered at P (see Figure 12.2). Then, as we project elements of length from the curve, we obtain infinitesimal arcs of this circle with the property that

$$d\theta = \frac{dl\,\hat{\mathbf{e}}_R \cdot \hat{\mathbf{e}}_n}{R} = \frac{dl_c}{a},$$

where dl_c is the element of arc of the fiducial circle. From this equation, we obtain

$$\theta = \int_{a'}^{b'} \frac{dl_c}{a} = \frac{1}{a} \int_{a'}^{b'} dl_c = \frac{s}{a}, \tag{12.3}$$

where a' and b' are projections of a and b on the circle, and s is the length of the arc from a' to b'. This last relation is, of course, our starting point where we defined the measure of an angle in radians!

Of special interest is the case where the curve loops back on itself. For such a case, the direction of $\hat{\mathbf{e}}_n$ is predetermined by

Box 12.1.1. (*Convention*). *We agree that for angle calculations, the unit normal shall always point out of a closed loop.*

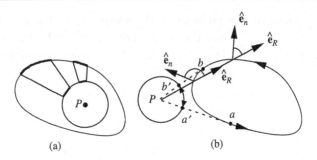

Figure 12.2: Total angle subtended by a closed curve about a point (a) inside and (b) outside.

If P happens to be inside the loop [Figure 12.2(a)], the *total* angle, corresponding to a complete traversal of the loop, is

$$\theta = \frac{s}{a} = \frac{2\pi a}{a} = 2\pi.$$

When P is *outside*, we get $\theta = 0$. This can be seen in Figure 12.2(b) where the projection of the closed curve covers *only a portion of the fiducial circle* and it does so twice, once with a positive sign—when $\hat{\mathbf{e}}_R$ and $\hat{\mathbf{e}}_n$ are separated by an acute angle—and once with a negative sign—when $\hat{\mathbf{e}}_R$ and $\hat{\mathbf{e}}_n$ are separated by an obtuse angle. Let us denote by θ_P^C the total angle subtended by the *closed* curve C about a point P and by U the region enclosed by C. Then, we have

total angle at a
point subtended
by a closed curve

$$\theta_P^C = \begin{cases} 2\pi & \text{if } P \text{ is in } U, \\ 0 & \text{if } P \text{ is not in } U. \end{cases} \tag{12.4}$$

Example 12.1.1. Point P is located outside a rectangle of sides $2a$ and $2b$ as shown in Figure 12.3. We want to verify Equation (12.4). The integration is naturally divided into four regions: right, top, left, and bottom. We shall do the

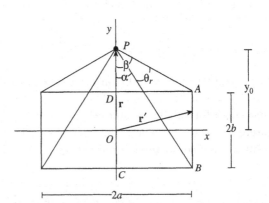

Figure 12.3: Total angle subtended by a rectangle about a point outside.

right-hand-side integration in detail, leaving the rest for the reader to verify. For the right side we have $\mathbf{r} = y_0\hat{\mathbf{e}}_y$, $\mathbf{r}' = a\hat{\mathbf{e}}_x + y'\hat{\mathbf{e}}_y$, and

$$dl = +dy', \ \mathbf{R} = \mathbf{r}' - \mathbf{r} = \langle a, y' - y_0 \rangle, \ \hat{\mathbf{e}}_n = \hat{\mathbf{e}}_x.$$

Therefore,

$$d\theta_r = \frac{dl\,\hat{\mathbf{e}}_R \cdot \hat{\mathbf{e}}_n}{R} = \frac{dy'\,\mathbf{R} \cdot \hat{\mathbf{e}}_x}{R^2} = \frac{a\,dy'}{a^2 + (y' - y_0)^2},$$

and the total integrated angle for the right side is

$$\theta_r = a\int_{-b}^{b} \frac{dy'}{a^2 + (y' - y_0)^2} = \overbrace{\tan^{-1}\left(\frac{y_0 + b}{a}\right)}^{=\angle CBP} - \overbrace{\tan^{-1}\left(\frac{y_0 - b}{a}\right)}^{=\angle DAP}$$

$$= \frac{\pi}{2} - \alpha - \left(\frac{\pi}{2} - \beta\right) = \beta - \alpha.$$

Similarly, one can easily show that $\theta_t = -2\beta$, $\theta_l = \beta - \alpha$, and $\theta_b = 2\alpha$, where t stands for "top," l for "left," and b for "bottom." The total subtended angle is, therefore zero, as expected. Note that only for the top side is the angle between $\hat{\mathbf{e}}_n$ and $\hat{\mathbf{e}}_R$ obtuse, and this fact results in the negative value for θ_t.   ∎

The purpose of the whole discussion of the ordinary angle in such a high-brow fashion and detail has been to lay the ground work for the introduction of the solid angle. As we shall see shortly, a good understanding of the new properties of the ordinary angle discussed above makes the transition to the solid angle almost trivial.

12.1.2   Solid Angle

We are now ready to generalize the notion of the angle to one dimension higher. Instead of a curve we have a surface, instead of a line element we have an area element, and instead of dividing by R we need to divide by R^2. This last requirement is necessary to render the "angle" dimensionless. Referring to Figure 12.4,

solid angle defined

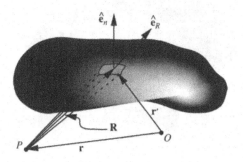

Figure 12.4: Solid angle as the ratio of area to distance *squared*.

Box 12.1.2. *We define the **solid angle** subtended by the element of area* Δa *as*

$$\Delta\Omega \approx \frac{\Delta a\,\hat{\mathbf{e}}_n \cdot \hat{\mathbf{e}}_R}{R^2} = \frac{\hat{\mathbf{e}}_R \cdot \Delta\mathbf{a}}{R^2} = \frac{(\mathbf{r}' - \mathbf{r}) \cdot \Delta\mathbf{a}}{|\mathbf{r}' - \mathbf{r}|^3},$$

where $\hat{\mathbf{e}}_n$ *is the unit normal to the surface and* $\Delta\mathbf{a} \equiv \hat{\mathbf{e}}_n \Delta a(\mathbf{r}')$.

The numerator is simply the projection of Δa onto a sphere of radius R as Figure 12.5 shows. This projection is obtained by the intersection of the fiducial sphere and the rays drawn from P to the boundary of Δa. As in the case of the angle, the choice of fiducial sphere is arbitrary. The integral form of the above equation is

$$\Omega = \iint_S \frac{\hat{\mathbf{e}}_R \cdot d\mathbf{a}}{R^2} = \iint_S \frac{\mathbf{R} \cdot d\mathbf{a}}{R^3} = \iint_S \frac{(\mathbf{r}' - \mathbf{r}) \cdot d\mathbf{a}(\mathbf{r}')}{|\mathbf{r}' - \mathbf{r}|^3}, \qquad (12.5)$$

where S is the surface subtended by the solid angle Ω.

Box 12.1.3. (***Convention***). *For any closed surface S, we take $\hat{\mathbf{e}}_n$ to be pointing outward.*

If we use a single fiducial sphere of radius b for all points of S, we obtain

$$\Omega = \iint_{S_b} \frac{da}{b^2} = \frac{1}{b^2} \iint_{S_b} da = \frac{A}{b^2}, \qquad (12.6)$$

where S_b is the projection of S onto the fiducial sphere and A its area. This equation is the analog of Equation (12.3) and can be used to define the measure

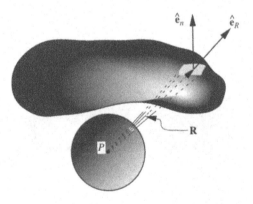

Figure 12.5: The relation between the $\hat{\mathbf{e}}_R \cdot \Delta\mathbf{a}$ and its projection on a fiducial sphere.

of solid angles. In particular, if the surface S is closed and P is inside, then A will be the total area of the fiducial sphere and we get $\Omega = 4\pi b^2/b^2 = 4\pi$. When P is outside, we get equal amounts of positive and negative contributions with the net result of zero.

total solid angle at a point subtended by a closed surface

Theorem 12.1.2. *Denote by Ω_P^S the total solid angle subtended by the* closed *surface S about a point P and by V the region enclosed by S. Then,*

$$\Omega_P^S = \begin{cases} 4\pi & \text{if } P \text{ is in } V, \\ 0 & \text{if } P \text{ is not in } V. \end{cases} \tag{12.7}$$

Example 12.1.3. As an example of the calculation of the solid angle, consider a square of side $2a$ with the point P located a distance z_0 from its center as shown in Figure 12.6. With $\mathbf{r} = \langle 0, 0, z_0 \rangle$ and $\mathbf{r}' = \langle x', y', 0 \rangle$, we have $\mathbf{R} = \mathbf{r}' - \mathbf{r} = \langle x', y', -z_0 \rangle$, and assuming that $\hat{\mathbf{e}}_n$ points in the negative z-direction,[4] we have

$$d\Omega = \frac{da\,\hat{\mathbf{e}}_n \cdot \hat{\mathbf{e}}_R}{R^2} = \frac{dx'\,dy'\,(-\hat{\mathbf{e}}_z) \cdot \mathbf{R}}{R^3} = \frac{z_0\,dx'\,dy'}{(x'^2 + y'^2 + z_0^2)^{3/2}}.$$

The solid angle is obtained by integrating this:

$$\Omega = z_0 \int_{-a}^{a} dx' \int_{-a}^{a} \frac{dy'}{(x'^2 + y'^2 + z_0^2)^{3/2}}$$

$$= 2az_0 \int_{-a}^{a} \frac{dx'}{\sqrt{x'^2 + a^2 + z_0^2}\,(x'^2 + z_0^2)} = 4\tan^{-1}\left(\frac{a^2}{z_0\sqrt{2a^2 + z_0^2}}\right).$$

An interesting special case is when $z_0 = a$. Then

$$\Omega = 4\tan^{-1}\left(\frac{a^2}{a\sqrt{3a^2}}\right) = 4\tan^{-1}\left(\frac{1}{\sqrt{3}}\right) = 4(\pi/6) = 2\pi/3.$$

The last result can also be derived in a simpler way. When $z_0 = a$, the point P will be at the center of a cube of side $2a$. Since the total solid angle subtended about P is 4π, and all six sides contribute equally, the solid angle subtended by one side is $4\pi/6$. ■

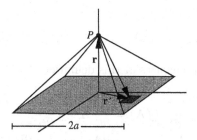

Figure 12.6: The solid angle subtended by a square of side $2a$.

---

[4]This assumption is not forced by any convention. It is chosen to make the final result positive.

Example 12.1.4. Let us replace the square of the last example with a circle of radius a. We can proceed along the same lines as before. However, in this particular case, we note that the solid angle is in the shape of a cone which is one of the primary surfaces of the spherical coordinate system. Placing the origin at P and projecting the area on a fiducial sphere, of radius b say, we may write

$$\Omega = \frac{A_b}{b^2} = \frac{2\pi b^2 (1 - \cos\alpha)}{b^2} = 2\pi(1 - \cos\alpha),$$

where $A_b \equiv 2\pi b^2 (1 - \cos\alpha)$ is the area of the projection of the circle on the fiducial sphere. The half-angle of the cone is denoted by α with

$$\tan\alpha = \frac{a}{z_0} \;\Rightarrow\; \cos\alpha = \frac{z_0}{\sqrt{a^2 + z_0^2}}.$$

The final result is

$$\Omega = 2\pi\left(1 - \frac{z_0}{\sqrt{a^2 + z_0^2}}\right). \tag{12.8}$$

It is instructive to obtain this result directly as in the previous example. ∎

12.2 Time Derivative of Vectors

Scalar and vector fields can be subjected to such analytic operations as differentiation and integration to obtain new scalar and vector fields. The derivative of a vector with respect to a variable (say time) in Cartesian coordinates amounts to differentiating each component:

$$\frac{\partial \mathbf{A}}{\partial t} = \frac{\partial A_x}{\partial t}\hat{\mathbf{e}}_x + \frac{\partial A_y}{\partial t}\hat{\mathbf{e}}_y + \frac{\partial A_z}{\partial t}\hat{\mathbf{e}}_z. \tag{12.9}$$

In other coordinate systems, one needs to differentiate the unit vectors as well.

In general, the derivative of a vector is defined in exactly the same manner as for ordinary functions. We have to keep in mind that a vector physical quantity, such as an electric field, is a function of space and time, i.e., its components are real-valued *functions* of space and time. So, consider a vector $\mathbf{A}$ which is a function of a number of independent variables $(t_1, t_2, \ldots, t_m)$. Then, we define the partial derivative as before:

$$\frac{\partial \mathbf{A}}{\partial t_k}(a_1, a_2, \ldots, a_n)$$
$$\equiv \lim_{\epsilon \to 0} \frac{\mathbf{A}(a_1, \ldots, a_k + \epsilon, \ldots, a_n) - \mathbf{A}(a_1, \ldots, a_k, \ldots, a_n)}{\epsilon}. \tag{12.10}$$

As immediate consequences of this definition, we list the following useful relations:

$$\frac{\partial}{\partial t_k}(f\mathbf{A}) = \frac{\partial f}{\partial t_k}\mathbf{A} + f\frac{\partial \mathbf{A}}{\partial t_k},$$
$$\frac{\partial}{\partial t_k}(\mathbf{A}\cdot\mathbf{B}) = \frac{\partial \mathbf{A}}{\partial t_k}\cdot\mathbf{B} + \mathbf{A}\cdot\frac{\partial \mathbf{B}}{\partial t_k}, \tag{12.11}$$
$$\frac{\partial}{\partial t_k}(\mathbf{A}\times\mathbf{B}) = \frac{\partial \mathbf{A}}{\partial t_k}\times\mathbf{B} + \mathbf{A}\times\frac{\partial \mathbf{B}}{\partial t_k}.$$

These relations can be used to calculate the derivatives of vectors when written in terms of unit vectors, keeping in mind that the derivative of a unit vector is not necessarily zero! Only Cartesian unit vectors are constant vectors, and for purposes of differentiation, it is convenient to write vectors in terms of these unit vectors, perform the derivative operation, and *then* substitute for $\hat{\mathbf{e}}_x$, $\hat{\mathbf{e}}_y$, and $\hat{\mathbf{e}}_z$ in terms of other—spherical or cylindrical—unit vectors.

only Cartesian unit vectors are constant.

Example 12.2.1. A vector whose magnitude is constant is always perpendicular to its derivative. This can be easily proved as follows:

$$\mathbf{A} \cdot \mathbf{A} = \text{const.} \quad \Rightarrow \quad \frac{\partial}{\partial t_k}(\mathbf{A} \cdot \mathbf{A}) = \frac{\partial}{\partial t_k}(\text{const.}) = 0.$$

On the other hand, the LHS can be evaluated using the second relation in Equation (12.11). This gives

$$\frac{\partial}{\partial t_k}(\mathbf{A} \cdot \mathbf{A}) = \frac{\partial \mathbf{A}}{\partial t_k} \cdot \mathbf{A} + \mathbf{A} \cdot \frac{\partial \mathbf{A}}{\partial t_k} = 2\mathbf{A} \cdot \frac{\partial \mathbf{A}}{\partial t_k}.$$

These two equations together imply that $\mathbf{A}$ and $(\partial/\partial t_k)(\mathbf{A})$ are perpendicular to one another. ■

An important consequence of the example above is that

Box 12.2.1. *A unit vector is always perpendicular to its derivative.*

Example 12.2.2. Newton's second law for a collection of particles leads directly to the corresponding law for rotational motion. Differentiating the total angular momentum

$$\mathbf{L} = \sum_{k=1}^{N} \mathbf{r}_k \times \mathbf{p}_k,$$

with respect to time and using the second law, $\mathbf{F}_k = d\mathbf{p}_k/dt$, for the kth particle, we get

$$\frac{d\mathbf{L}}{dt} = \sum_{k=1}^{N} \frac{d}{dt}(\mathbf{r}_k \times \mathbf{p}_k) = \sum_{k=1}^{N} (\dot{\mathbf{r}}_k \times \mathbf{p}_k + \mathbf{r}_k \times \dot{\mathbf{p}}_k) = \sum_{k=1}^{N} (0 + \mathbf{r}_k \times \mathbf{F}_k) \equiv \mathbf{T},$$

where an overdot indicates the derivative with respect to time and in the last line we used the definition of torque and the fact that velocity $\dot{\mathbf{r}}_k$ and momentum $\mathbf{p}_k$ have the same direction. ■

As a special case of the example above, we obtain the law of **angular momentum conservation**:

angular momentum conservation

Box 12.2.2. *When the total torque on a system of particles vanishes, the total angular momentum will be a constant of motion. This means that its components in a **Cartesian coordinate system** are constant.*

Since the unit vectors in other coordinate systems are not, in general, constant, a *constant* vector has *variable* components in these systems.

12.2.1   Equations of Motion in a Central Force Field

When one discusses the central-force problems in mechanics, for instance in the study of planetary motion, one uses spherical coordinates to locate the moving object. Thus, the position vector of the object, say a planet, is given in terms of spherical unit vectors. Newton's second law, on the other hand, requires a knowledge of the second time-derivative of the position vector.

In this subsection we find the second derivative of the position vector of a moving point particle P with respect to time in spherical coordinates. The coordinates (r, θ, φ) of P are clearly functions of time. First we calculate velocity and write it in terms of the spherical unit vectors

$$\mathbf{v} = \frac{d\mathbf{r}}{dt} = \frac{d}{dt}(\mathbf{r}) = \frac{d}{dt}(r\hat{\mathbf{e}}_r) = \hat{\mathbf{e}}_r \frac{dr}{dt} + r \frac{d\hat{\mathbf{e}}_r}{dt}.$$

We thus have to find the time-derivative of the unit vector $\hat{\mathbf{e}}_r$. The most straightforward way of taking such a derivative is to use the chain rule:

$$\frac{d\hat{\mathbf{e}}_r}{dt} = \frac{\partial \hat{\mathbf{e}}_r}{\partial r}\frac{dr}{dt} + \frac{\partial \hat{\mathbf{e}}_r}{\partial \theta}\frac{d\theta}{dt} + \frac{\partial \hat{\mathbf{e}}_r}{\partial \varphi}\frac{d\varphi}{dt} = \dot{\theta}\frac{\partial \hat{\mathbf{e}}_r}{\partial \theta} + \dot{\varphi}\frac{\partial \hat{\mathbf{e}}_r}{\partial \varphi},$$

where we have used the fact that the spherical unit vectors are independent of r [see Equation (1.39)]. We now evaluate the partial derivatives using (1.39) and noting that the Cartesian unit vectors are constant:

$$\frac{\partial \hat{\mathbf{e}}_r}{\partial \theta} = \hat{\mathbf{e}}_x \frac{\partial}{\partial \theta}(\sin\theta \cos\varphi) + \hat{\mathbf{e}}_y \frac{\partial}{\partial \theta}(\sin\theta \sin\varphi) + \hat{\mathbf{e}}_z \frac{\partial}{\partial \theta}(\cos\theta)$$

$$= \hat{\mathbf{e}}_x \cos\theta \cos\varphi + \hat{\mathbf{e}}_y \cos\theta \sin\varphi - \hat{\mathbf{e}}_z \sin\theta. \qquad (12.12)$$

We are interested in writing all vectors in terms of spherical coordinates. A straightforward way is to substitute for the above Cartesian unit vectors, their expressions in terms of spherical unit vectors. We can easily calculate such expressions using the method introduced at the end of Chapter 1. We leave the details for the reader and merely state the results:

$$\hat{\mathbf{e}}_x = \hat{\mathbf{e}}_r \sin\theta \cos\varphi + \hat{\mathbf{e}}_\theta \cos\theta \cos\varphi - \hat{\mathbf{e}}_\varphi \sin\varphi,$$
$$\hat{\mathbf{e}}_y = \hat{\mathbf{e}}_r \sin\theta \sin\varphi + \hat{\mathbf{e}}_\theta \cos\theta \sin\varphi + \hat{\mathbf{e}}_\varphi \cos\varphi, \qquad (12.13)$$
$$\hat{\mathbf{e}}_z = \hat{\mathbf{e}}_r \cos\theta - \hat{\mathbf{e}}_\theta \sin\theta.$$

Substituting these expressions in the previous equation, we get

$$\frac{\partial \hat{\mathbf{e}}_r}{\partial \theta} = (\hat{\mathbf{e}}_r \sin\theta \cos\varphi + \hat{\mathbf{e}}_\theta \cos\theta \cos\varphi - \hat{\mathbf{e}}_\varphi \sin\varphi)\cos\theta \cos\varphi$$

$$+ (\hat{\mathbf{e}}_r \sin\theta \sin\varphi + \hat{\mathbf{e}}_\theta \cos\theta \sin\varphi + \hat{\mathbf{e}}_\varphi \cos\varphi)\cos\theta \sin\varphi$$

$$- (\hat{\mathbf{e}}_r \cos\theta - \hat{\mathbf{e}}_\theta \sin\theta)\sin\theta,$$

which simplifies to

$$\frac{\partial \hat{\mathbf{e}}_r}{\partial \theta} = \hat{\mathbf{e}}_\theta. \qquad (12.14)$$

We could have immediately obtained this result by comparing Equation (12.12) with the expression for $\hat{\mathbf{e}}_\theta$ in Equation (1.39). The other partial derivative is obtained the same way:

$$
\begin{aligned}
\frac{\partial \hat{\mathbf{e}}_r}{\partial \varphi} &= \hat{\mathbf{e}}_x \frac{\partial}{\partial \varphi}(\sin\theta\cos\varphi) + \hat{\mathbf{e}}_y \frac{\partial}{\partial \varphi}(\sin\theta\sin\varphi) + \hat{\mathbf{e}}_z \frac{\partial}{\partial \varphi}(\cos\theta) \\
&= -\hat{\mathbf{e}}_x \sin\theta\sin\varphi + \hat{\mathbf{e}}_y \sin\theta\cos\varphi \\
&= -(\hat{\mathbf{e}}_r \sin\theta\cos\varphi + \hat{\mathbf{e}}_\theta \cos\theta\cos\varphi - \hat{\mathbf{e}}_\varphi \sin\varphi)\sin\theta\sin\varphi \\
&\quad + (\hat{\mathbf{e}}_r \sin\theta\sin\varphi + \hat{\mathbf{e}}_\theta \cos\theta\sin\varphi + \hat{\mathbf{e}}_\varphi \cos\varphi)\sin\theta\cos\varphi \\
&= \hat{\mathbf{e}}_\varphi \sin\theta.
\end{aligned}
\tag{12.15}
$$

Substituting this and Equation (12.14) in the expression for velocity, we obtain

$$
\mathbf{v} = \hat{\mathbf{e}}_r \dot{r} + r\left(\dot{\theta}\frac{\partial \hat{\mathbf{e}}_r}{\partial \theta} + \dot{\varphi}\frac{\partial \hat{\mathbf{e}}_r}{\partial \varphi}\right) = \hat{\mathbf{e}}_r \dot{r} + \hat{\mathbf{e}}_\theta r\dot{\theta} + \hat{\mathbf{e}}_\varphi r\dot{\varphi}\sin\theta.
\tag{12.16}
$$

components of velocity in spherical coordinates

To write the equations of motion, we need to calculate the acceleration which involves the differentiation of other unit vectors. The procedure outlined for $\hat{\mathbf{e}}_r$ can be used to obtain the partial derivatives of the other unit vectors. We collect the result of such calculations, including Equations (12.14) and (12.15) in the following:

$$
\begin{array}{lll}
\dfrac{\partial \hat{\mathbf{e}}_r}{\partial r} = 0, & \dfrac{\partial \hat{\mathbf{e}}_r}{\partial \theta} = \hat{\mathbf{e}}_\theta, & \dfrac{\partial \hat{\mathbf{e}}_r}{\partial \varphi} = \hat{\mathbf{e}}_\varphi \sin\theta, \\[2mm]
\dfrac{\partial \hat{\mathbf{e}}_\theta}{\partial r} = 0, & \dfrac{\partial \hat{\mathbf{e}}_\theta}{\partial \theta} = -\hat{\mathbf{e}}_r, & \dfrac{\partial \hat{\mathbf{e}}_\theta}{\partial \varphi} = \hat{\mathbf{e}}_\varphi \cos\theta, \\[2mm]
\dfrac{\partial \hat{\mathbf{e}}_\varphi}{\partial r} = 0, & \dfrac{\partial \hat{\mathbf{e}}_\varphi}{\partial \theta} = 0, & \dfrac{\partial \hat{\mathbf{e}}_\varphi}{\partial \varphi} = -\hat{\mathbf{e}}_r \sin\theta - \hat{\mathbf{e}}_\theta \cos\theta.
\end{array}
\tag{12.17}
$$

Similarly the time-derivatives of the unit vectors are given as follows:

$$
\begin{aligned}
\frac{d\hat{\mathbf{e}}_r}{dt} &= \dot{\theta}\hat{\mathbf{e}}_\theta + \dot{\varphi}\sin\theta\,\hat{\mathbf{e}}_\varphi, \\
\frac{d\hat{\mathbf{e}}_\theta}{dt} &= -\dot{\theta}\hat{\mathbf{e}}_r + \dot{\varphi}\cos\theta\,\hat{\mathbf{e}}_\varphi, \\
\frac{d\hat{\mathbf{e}}_\varphi}{dt} &= -\dot{\varphi}\sin\theta\,\hat{\mathbf{e}}_r - \dot{\varphi}\cos\theta\,\hat{\mathbf{e}}_\theta.
\end{aligned}
\tag{12.18}
$$

Differentiating Equation (12.16) with respect to t, inserting (12.18) in the result, and collecting the components, we get

$$
\begin{aligned}
\frac{d^2\mathbf{r}}{dt^2} = \frac{d\mathbf{v}}{dt} =\ & \hat{\mathbf{e}}_r\left(\ddot{r} - r\dot{\theta}^2 - r\dot{\varphi}^2\sin^2\theta\right) \\
&+ \hat{\mathbf{e}}_\theta\left(\dot{r}\dot{\theta} + \frac{d}{dt}(r\dot{\theta}) - r\dot{\varphi}^2\sin\theta\cos\theta\right) \\
&+ \hat{\mathbf{e}}_\varphi\left(\dot{r}\dot{\varphi}\sin\theta + r\dot{\theta}\dot{\varphi}\cos\theta + \frac{d}{dt}(r\dot{\varphi}\sin\theta)\right).
\end{aligned}
\tag{12.19}
$$

components of acceleration in spherical coordinates

One can use these expressions to write Newton's second law in spherical coordinates.

Now suppose that a particle (a planet) is under the influence of a **central force**, i.e., a force that always points toward, or away from, an origin (the Sun), and has a magnitude that is a function of the distance between the particle and the origin. This means that, in spherical coordinates, the force is of the form $\mathbf{F} = \hat{\mathbf{e}}_r F(r)$. The second law of motion now yields

$$m\frac{d^2\mathbf{r}}{dt^2} = \hat{\mathbf{e}}_r F(r) \quad \Rightarrow \quad \frac{d^2\mathbf{r}}{dt^2} = \hat{\mathbf{e}}_r \frac{F(r)}{m} \equiv \hat{\mathbf{e}}_r f(r)$$

central-force problem in spherical coordinates

which, together with Equation (12.19), gives

$$\ddot{r} - r\dot{\theta}^2 - r\dot{\varphi}^2 \sin^2\theta = f(r),$$

$$\dot{r}\dot{\theta} + \frac{d}{dt}(r\dot{\theta}) - r\dot{\varphi}^2 \sin\theta\cos\theta = 0, \qquad (12.20)$$

$$\dot{r}\dot{\varphi}\sin\theta + r\dot{\theta}\dot{\varphi}\cos\theta + \frac{d}{dt}(r\dot{\varphi}\sin\theta) = 0.$$

These equations are the starting point of the study of planetary motion. We shall not pursue their solution at this point, but consider some of their general properties, using angular momentum conservation. Since the force has only an $\hat{\mathbf{e}}_r$ component, its torque vanishes:

angular momentum is conserved in motions caused by central forces.

$$\mathbf{T} = \mathbf{r} \times \mathbf{F} = r\,\hat{\mathbf{e}}_r \times (F(r)\hat{\mathbf{e}}_r) = rF(r)\hat{\mathbf{e}}_r \times \hat{\mathbf{e}}_r = 0.$$

Therefore, by Box 12.2.2, the angular momentum of the particle relative to the origin is a constant vector. Equation (12.16) now yields

$$\begin{aligned}
\mathbf{L} = \mathbf{r} \times (m\mathbf{v}) &= mr\,\hat{\mathbf{e}}_r \times \left(\hat{\mathbf{e}}_r\dot{r} + \hat{\mathbf{e}}_\theta r\dot{\theta} + \hat{\mathbf{e}}_\varphi r\dot{\varphi}\sin\theta\right) \\
&= mr^2\,\hat{\mathbf{e}}_r \times (\hat{\mathbf{e}}_\theta\dot{\theta} + \hat{\mathbf{e}}_\varphi\dot{\varphi}\sin\theta) = mr^2(\hat{\mathbf{e}}_\varphi\dot{\theta} - \hat{\mathbf{e}}_\theta\dot{\varphi}\sin\theta) \\
&= mr^2\dot{\theta}(-\hat{\mathbf{e}}_x\sin\varphi + \hat{\mathbf{e}}_y\cos\varphi) \\
&\quad - mr^2\dot{\varphi}\sin\theta(\hat{\mathbf{e}}_x\cos\theta\cos\varphi + \hat{\mathbf{e}}_y\cos\theta\sin\varphi - \hat{\mathbf{e}}_z\sin\theta) \\
&= L_x\hat{\mathbf{e}}_x + L_y\hat{\mathbf{e}}_y + L_z\hat{\mathbf{e}}_z,
\end{aligned}$$

where L_x, L_y, and L_z are the constant *Cartesian* components of angular momentum and m is the mass of the particle. Equating the components of this vectorial relation gives

$$\begin{aligned}
L_x &= -mr^2(\dot{\theta}\sin\varphi + \dot{\varphi}\sin\theta\cos\theta\cos\varphi), \\
L_y &= mr^2(\dot{\theta}\cos\varphi - \dot{\varphi}\sin\theta\cos\theta\sin\varphi), \qquad (12.21) \\
L_z &= mr^2\dot{\varphi}\sin^2\theta.
\end{aligned}$$

The last equation gives

$$\dot{\varphi} = \frac{L_z}{mr^2\sin^2\theta}. \qquad (12.22)$$

From all of these relations, we obtain

$$L^2 = L_x^2 + L_y^2 + L_z^2 = m^2 r^4 \dot{\theta}^2 + \frac{L_z^2}{\sin^2 \theta}. \qquad (12.23)$$

Now suppose that we choose our coordinate axes so that *initially*, i.e., at $t = 0$, both the position and the velocity vectors of the particle lie in the xy-plane. Since $\mathbf{L}$ is perpendicular to both $\mathbf{r}$ and $\mathbf{v}$, it must be initially entirely in the z-direction. Conservation of angular momentum implies that $\mathbf{L}$ will always be in the z-direction. In particular, $L^2 = L_z^2$. Substituting this in Equation (12.23) yields

$$L^2 = m^2 r^4 \dot{\theta}^2 + \frac{L^2}{\sin^2 \theta} \;\Rightarrow\; 0 = m^2 r^4 \dot{\theta}^2 + \frac{L^2}{\sin^2 \theta} - L^2$$

or $0 = m^2 r^4 \dot{\theta}^2 + L^2 \cot^2 \theta$. Neither of the two terms on the RHS of this equation is negative. Thus, for their sum to be zero, each term must be zero. It follows that

$$m^2 r^4 \dot{\theta}^2 = 0 \;\Rightarrow\; \dot{\theta} = 0 \;\Rightarrow\; \theta = \text{const.},$$
$$L^2 \cot^2 \theta = 0 \;\Rightarrow\; \cot^2 \theta = 0 \;\Rightarrow\; \theta = \pi/2,$$

assuming that $r \neq 0$ and $L \neq 0$. These relations hold *for all times*. Thus, the particle is confined to a plane, our xy-plane, for eternity! This is why the planets do not wobble "up and down" out of their orbital planes.[5]

proof that planets move in a plane

If we substitute $\pi/2$ for θ and use (12.22) for $\dot{\varphi}$ in Equation (12.20), then the second and third relations are satisfied identically, and the first relation becomes

$$\ddot{r} - \frac{L^2}{m^2 r^3} = f(r) \qquad (12.24)$$

which is a single differential equation in *one variable*. The general problem of a particle's motion in three dimensions has reduced to a one-dimensional problem.

12.3   The Gradient

Analysis of vectors deals with the derivatives and integrals of vector fields. Because of its simplicity, we shall work in a Cartesian coordinate system at the beginning, and later generalize to other coordinates.

In many situations arising in physics, rates of change of certain scalar functions with distance are of importance. For instance, the way potential energy changes as we move in space is directly related to the force producing the potential energy. Similarly, the rate of change—derivative—of the electrostatic potential with respect to distance gives the electrostatic field. The concept of gradient makes precise the vague notion of a derivative with respect to distance.

---

[5] Actually, the planets, due to the influence of other planets, *do* wobble out of their orbits. But this is a very small effect.

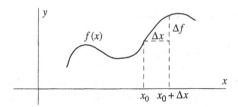

Figure 12.7: "Gradient" or differentiation with respect to distance in one dimension.

Let us analyze the notion of differentiation with respect to distance, start-

notion of ordinary derivative reexamined

ing with one variable. In Figure 12.7, a function $f(x)$ has an increment, Δf, corresponding to a change Δx in x. If Δx is small enough, we can write

$$\Delta f \approx \left(\frac{df}{dx}\right)_{x=x_0} \Delta x.$$

This shows that $(df/dx)_{x=x_0}$ is a measure of how fast the function f is changing at the point x_0.

With one variable, there is no ambiguity in defining the derivative, because there is only one line along which we can change x, the only coordinate. With two or more variables, the situation is completely different, as illustrated in Figure 12.8. A point $P_0 = (x_0, y_0)$ in the xy-plane is shown with the corresponding value of the function, $f(x_0, y_0) = z_0$. Out of the infinitude of points that are close to P_0 and cause a change in the function, only three are shown. These indicate how the change in $f(x, y)$ depends on the direction in which the neighboring point is located in relation to P_0. For example, if we move in the direction P_0P_1, there is very little change in $f(x, y)$, but if we move in the direction P_0P_2, we notice more change in the function, and if we move in the direction of P_0P_3, the change seems to be maximum. This maximum change, and the direction associated with it, is called the **gradient**.

notion of gradient analyzed

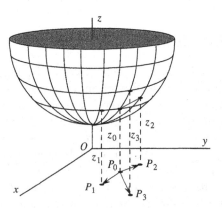

Figure 12.8: Gradient or differentiation with respect to distance is shown in two dimensions. The gradient is a vector *in the xy-plane*. Do not think of the surface as a variation in height! It could represent, for instance, the temperature at various points of the *xy*-plane.

Let us use $d\mathbf{r}$ to denote the infinitesimal displacement vector[6] connecting P_0 to a neighboring point *in the xy-plane*. If $f(x,y)$ is differentiable, Equation (2.12) gives

$$df = \left(\frac{\partial f}{\partial x}\right)_{P_0} dx + \left(\frac{\partial f}{\partial y}\right)_{P_0} dy,$$

where dx and dy are the components of the displacement from P_0 and df is (approximately) the change in f corresponding to the increments dx and dy. We can rewrite this equation as

$$df = (\boldsymbol{\nabla} f)_{P_0} \cdot d\mathbf{r} = |\boldsymbol{\nabla} f||d\mathbf{r}|\cos\theta, \qquad (12.25)$$

where, by definition,

gradient in two dimensions

$$(\boldsymbol{\nabla} f)_{P_0} \equiv \left\langle \frac{\partial f}{\partial x}, \frac{\partial f}{\partial y} \right\rangle_{P_0} \qquad (12.26)$$

is a vector in the xy-plane and θ is the angle between this vector and $d\mathbf{r}$. It is clear that df will be maximum when $\cos\theta = 1$, that is, when $d\mathbf{r}$ is in the direction of $\boldsymbol{\nabla} f$. We conclude, therefore, that $\boldsymbol{\nabla} f$ gives the direction along which f changes most rapidly. The vector in Equation (12.26) is the gradient of f at P_0.

The notion of gradient can be generalized to three variables although it is harder to visualize than the two-variable case. In three dimensions we deal with a function $f(x,y,z)$—which cannot be plotted as in Figure 12.8—and ask which $d\mathbf{r} = \langle dx, dy, dz \rangle$ maximizes the change in f. Once again, the three-dimensional version of Equation (12.25) shows that $d\mathbf{r}$ and

gradient in three dimensions

$$\boldsymbol{\nabla} f \equiv \left\langle \frac{\partial f}{\partial x}, \frac{\partial f}{\partial y}, \frac{\partial f}{\partial z} \right\rangle \qquad (12.27)$$

should be in the same direction for df to have a maximum.

Definition 12.3.1. *The **gradient** of a function $f(x,y,z)$ is defined as*

$$\boldsymbol{\nabla} f \equiv \hat{\mathbf{e}}_x \frac{\partial f}{\partial x} + \hat{\mathbf{e}}_y \frac{\partial f}{\partial y} + \hat{\mathbf{e}}_z \frac{\partial f}{\partial z}.$$

For the same small displacement $|\Delta\mathbf{r}|$, the change in f is maximum when $\Delta\mathbf{r}$ is in the direction of $\boldsymbol{\nabla} f$.

Example 12.3.1. As an example, let us find the gradient of the function

$$V(x,y,z) = f(r) = f\left(\sqrt{x^2 + y^2 + z^2}\right)$$

(which depends on r alone) at a point P with Cartesian coordinates (x,y,z). Using the chain rule, we have

---

[6]A better notation is $\Delta\mathbf{r}$. However, since there is no difference between differential and increment of an *independent* variable, and since eventually we will be interested in differentials, we use the latter notation.

$$\boldsymbol{\nabla} V = \hat{\mathbf{e}}_x \frac{\partial V}{\partial x} + \hat{\mathbf{e}}_y \frac{\partial V}{\partial y} + \hat{\mathbf{e}}_z \frac{\partial V}{\partial z} = \left\langle \frac{\partial V}{\partial x}, \frac{\partial V}{\partial y}, \frac{\partial V}{\partial z} \right\rangle$$

$$= \left\langle f'(r) \frac{\partial r}{\partial x}, f'(r) \frac{\partial r}{\partial y}, f'(r) \frac{\partial r}{\partial z} \right\rangle$$

$$= f'(r) \left\langle \frac{x}{r}, \frac{y}{r}, \frac{z}{r} \right\rangle = \frac{f'(r)}{r} \langle x, y, z \rangle = f'(r) \frac{\mathbf{r}}{r}.$$

The last equality shows that, for functions that depend on r alone, the gradient is proportional to the position vector of the point P, i.e., it is radial. ∎

Given a scalar function $f(x, y, z)$, we can consider surfaces on which this function maintains a constant value. If that constant value is C, the surface will be described by $f(x, y, z) = C$. One can, in principle, solve for z as a function of x and y to find the explicit dependence of the function. However, we are interested in the *implicit* dependence given above. Now consider two points P_1 and P_2 *on the surface* with coordinates (x, y, z) and $(x + \Delta x, y + \Delta y, z + \Delta z)$, respectively. We have

$$f(x, y, z) = f(x + \Delta x, y + \Delta y, z + \Delta z) \;\Rightarrow\; f(x, y, z) = f(x, y, z) + \Delta f$$

or $0 = \Delta f \approx \frac{\partial f}{\partial x} \Delta x + \frac{\partial f}{\partial y} \Delta y + \frac{\partial f}{\partial z} \Delta z$, if the increments of coordinates are small. This relation shows that $\boldsymbol{\nabla} f$ is perpendicular to the displacement from P_1 to P_2. The same argument applies to a curve $g(x, y) = C$; i.e., the two-dimensional gradient is perpendicular to the displacement from P_1 to P_2, both being points on the curve. Since P_1 and P_2 are completely arbitrary, we conclude that

Theorem 12.3.2. *The gradient $\boldsymbol{\nabla} f$ is perpendicular to all surfaces $f(x, y, z) = C$ for different C's. Similarly, $\boldsymbol{\nabla} g$ is perpendicular to all curves $g(x, y) = C$.*

electrostatic field is perpendicular to surfaces of conductors

For example, as we shall see later, the electrostatic field is the gradient of the electrostatic potential. Therefore, the electrostatic field is perpendicular to surfaces of constant potential such as conductors.

Example 12.3.3. The perpendicularity property of the gradient can be used to find the equation of the tangent plane to a surface $z = g(x, y)$ at a point P with coordinates (x_0, y_0, z_0). This surface can be written as

$$f(x, y, z) \equiv z - g(x, y) = 0.$$

Then, the normal to the surface at P—which is the same as the normal to the tangent plane at P—is the gradient of f at P:

$$(\boldsymbol{\nabla} f)_P = \left\langle \frac{\partial f}{\partial x}, \frac{\partial f}{\partial y}, \frac{\partial f}{\partial z} \right\rangle_P = \left\langle -\frac{\partial g}{\partial x}, -\frac{\partial g}{\partial y}, 1 \right\rangle_P.$$

derivation of the equation of a plane tangent to a surface

A point of the tangent plane at P is completely determined by the property that its displacement vector $\Delta \mathbf{r}$ from P should be perpendicular to the gradient at P (see Figure 12.9). If we denote the position vector of P by $\mathbf{r}_0$ and that of the point on the plane by $\mathbf{r} = \langle x, y, z \rangle$, then the equation of the tangent plane is given by

$$(\mathbf{r} - \mathbf{r}_0) \cdot (\boldsymbol{\nabla} f)_P = 0 \;\Rightarrow\; -(x - x_0) \left(\frac{\partial g}{\partial x} \right)_P - (y - y_0) \left(\frac{\partial g}{\partial y} \right)_P + (z - z_0) = 0$$

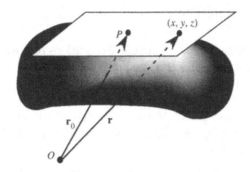

Figure 12.9: The plane tangent to the surface $z = g(x, y)$ at P.

or

$$z - z_0 = (x - x_0) \left(\frac{\partial g}{\partial x} \right)_P + (y - y_0) \left(\frac{\partial g}{\partial y} \right)_P \qquad \blacksquare$$

It is convenient to introduce a *differentiation operator* which we shall use
later.   the del operator

Definition 12.3.2. *The symbol $\boldsymbol{\nabla}$ can be thought of as a vector operator,
called **del** or **nabla**, whose components are $\partial/\partial x, \partial/\partial y$, and $\partial/\partial z$. Thus, we
can write*

$$\boldsymbol{\nabla} = \hat{\mathbf{e}}_x \frac{\partial}{\partial x} + \hat{\mathbf{e}}_y \frac{\partial}{\partial y} + \hat{\mathbf{e}}_z \frac{\partial}{\partial z}. \qquad (12.28)$$

*This vector operator $\boldsymbol{\nabla}$ operates on differentiable functions and produces vec-
tor fields.*

12.3.1 Gradient and Extremum Problems

The gradient is very nicely used to find the maxima and minima of functions
of several variables. A function $f(\mathbf{x})$ of n variables $\mathbf{x} = (x_1, x_2, \ldots, x_n)$ has a
local extremum (maximum or minimum) at a point $\mathbf{a}$ if its differential vanishes
at that point for arbitrary $d\mathbf{x}$:

$$df = \left. \frac{\partial f}{\partial x_1} \right|_{\mathbf{a}} dx_1 + \left. \frac{\partial f}{\partial x_2} \right|_{\mathbf{a}} dx_2 + \cdots + \left. \frac{\partial f}{\partial x_n} \right|_{\mathbf{a}} dx_n \equiv \big(\boldsymbol{\nabla} f(\mathbf{a}) \big) \cdot d\mathbf{x} = 0$$

where

$$\boldsymbol{\nabla} f \equiv \left\langle \frac{\partial f}{\partial x_1}, \frac{\partial f}{\partial x_2}, \ldots, \frac{\partial f}{\partial x_n} \right\rangle \qquad \text{and} \qquad d\mathbf{x} \equiv \langle dx_1, dx_2, \ldots, dx_n \rangle.$$

If the dot product of $\boldsymbol{\nabla} f(\mathbf{a})$ and $d\mathbf{x}$ is to vanish for arbitrary $d\mathbf{x}$, then $\boldsymbol{\nabla} f(\mathbf{a})$
must be zero. Thus for f to have an extremum at $\mathbf{a}$, we must have

$$\boldsymbol{\nabla} f(\mathbf{a}) = 0 \quad \text{or} \quad \left. \frac{\partial f}{\partial x_i} \right|_{\mathbf{a}} = 0, \quad i = 1, 2, \ldots, n. \qquad (12.29)$$

This is the generalization to n variables the familiar condition known from calculus.

In many situations, there are auxiliary conditions or **constraints** imposed on the independent variables. For example, let P_1, Q, and P_2 be three points in space, with P_1 and P_2 fixed but Q being allowed to move. Consider the path $P_1 Q P_2$ consisting of straight line segments $\overline{P_1 Q}$ and $\overline{Q P_2}$. What choice of Q gives the shortest path? If we denote the coordinates of Q by (x, y, z) and those of P_1 and P_2 with obvious subscripts, then we have to find the extremum of

$$f(x, y, z) = \sqrt{(x - x_1)^2 + (y - y_1)^2 + (z - z_1)^2}$$
$$+ \sqrt{(x - x_2)^2 + (y - y_2)^2 + (z - z_2)^2}.$$

So we set partial derivatives equal to zero and solve for (x, y, z). The answer, as expected, turns out to be the path for which Q lies on the line segment $\overline{P_1 P_2}$ between P_1 and P_2.

Now suppose we demand that Q lie on a sphere of radius a centered at the origin. Then the problem becomes extremizing $f(x, y, z)$ with the constraint condition that

$$g(x, y, z) \equiv x^2 + y^2 + z^2 - a^2 = 0.$$

To solve this problem, we *could* solve for one of the variables of the constraint equation in terms of the other two, substitute in $f(x, y, z)$, and solve the resulting two-variable problem. But there is a much more elegant way involving gradients, which we discuss now.

Suppose that we want to find the extremum of a function $f(\mathbf{x})$ of n variables $\mathbf{x} = (x_1, x_2, \ldots, x_n)$ subject to the condition that $\mathbf{x}$ must lie on the hypersurface $g(\mathbf{x}) = 0$. We cannot set ∇f equal to zero because $d\mathbf{x}$ is no longer arbitrary.

With constraint, $d\mathbf{x}$ is confined to the surface $g(\mathbf{x}) = 0$. Now, the only n-dimensional vector which has a vanishing dot product with *any* $d\mathbf{x}$ on the constrained surface is (a multiple of) the normal to the surface. Therefore, if $(\nabla f) \cdot d\mathbf{x}$ is to be zero for $d\mathbf{x}$ lying on the surface, then ∇f must be a multiple of the normal to the surface $g(\mathbf{x}) = 0$. But this normal is nothing but ∇g. Therefore, if f is to have an extremum subject to the constraint $g(\mathbf{x}) = 0$, then it must obey the following equation

$$\nabla f = -\lambda \nabla g \quad \text{or} \quad \nabla f + \lambda \nabla g = 0,$$

where λ is an arbitrary constant called the **Lagrange multiplier**. This equation shows that to find the extremum of the function f with constraint $g(\mathbf{x}) = 0$, one can define the function F of $n + 1$ variables

Lagrange
multipliers

$$F(x_1, x_2, \ldots, x_n; \lambda) \equiv f(x_1, x_2, \ldots, x_n) + \lambda g(x_1, x_2, \ldots, x_n),$$

and extremize it *without constraint*. Then we have

$$\frac{\partial F}{\partial x_i} = \frac{\partial f}{\partial x_i} + \lambda \frac{\partial g}{\partial x_i} = 0, \quad i = 1, 2, \ldots, n,$$

$$\frac{\partial F}{\partial \lambda} = g(x_1, x_2, \ldots, x_n) = 0. \tag{12.30}$$

The last equation is just the constraint condition, but it comes out conveniently as one of the extremal equations of F.

Example 12.3.4. A rectangular box is to be made out of a given amount A of material to have the largest volume. What dimensions should the box have? Here $f(x, y, z) = xyz$, the volume, and $g(x, y, z) = 2xy + 2xz + 2yz - A$. Setting the components of the gradient of

$$F(x, y, z; \lambda) = xyz + 2\lambda(xy + xz + yz - A/2)$$

equal to zero yields four equations

$$yz + 2\lambda(y + z) = 0,$$
$$xz + 2\lambda(x + z) = 0,$$
$$xy + 2\lambda(x + y) = 0,$$
$$2(xy + xz + yz) - A = 0.$$

Multiplying the first equation by x and the second equation by y and subtracting yields

$$2\lambda x(y + z) - 2\lambda y(x + z) = 0, \quad \text{or} \quad x = y.$$

Similarly, from the second and third equations we get $y = z$. So, the box should be a cube. The last equation then gives

$$6x^2 - A = 0, \quad \text{or} \quad x = y = z = \sqrt{\frac{A}{6}}.$$

Substituting this in any of the above equations involving λ yields $\lambda = -\frac{1}{4}\sqrt{A/6}$. ∎

The extremal problems may have several constraint equations such as

$$g_j(x_1, x_2, \ldots, x_n) \equiv g_j(\mathbf{x}) = 0, \quad j = 1, 2, \ldots, m. \tag{12.31}$$

We can "eliminate" the first constraint by replacing $f(x_1, x_2, \ldots, x_n)$ with

$$F_1(\mathbf{x}; \lambda_1) \equiv f(\mathbf{x}) + \lambda_1 g_1(\mathbf{x}),$$

where F_1 has only $m - 1$ constraint equations. Now eliminate the second constraint by defining

$$F_2(\mathbf{x}; \lambda_1, \lambda_2) \equiv F_1(\mathbf{x}; \lambda_1) + \lambda_2 g_2(\mathbf{x}) = f(\mathbf{x}) + \lambda_1 g_1(\mathbf{x}) + \lambda_2 g_2(\mathbf{x}).$$

Continuing, we can eliminate all constraints by defining

$$F(\mathbf{x}; \lambda_1, \lambda_2, \ldots, \lambda_m) \equiv f(\mathbf{x}) + \sum_{j=1}^{m} \lambda_j g_j(\mathbf{x}), \tag{12.32}$$

whose unconstrained extremization yields the extremal equations.

12.4   Problems

12.1. Find *directly* the solid angle subtended by a disk of radius a at a point P on its perpendicular axis located a distance b from the center.

12.2. A closed curve $\rho = 3a + a\cos\varphi$ in cylindrical coordinates bounds a region in the xy-plane. Find the solid angle subtended by this region at a point P on the z-axis a distance $2a$ above the xy-plane.

12.3. Derive Equation (12.11).

12.4. Show that when a moving particle is confined to a circle, its velocity is always perpendicular to its radius. If, furthermore, the *speed* of the particle is constant, then its acceleration is radial.

12.5. Derive Equations (12.17) and (12.18).

12.6. The vectors **a** and **b** are given by

$$\mathbf{a} = u\hat{\mathbf{e}}_x + v\hat{\mathbf{e}}_y, \qquad \mathbf{b} = v\hat{\mathbf{e}}_x - u\hat{\mathbf{e}}_y.$$

(a) Write $\hat{\mathbf{e}}_a$ and $\hat{\mathbf{e}}_b$ in terms of Cartesian unit vectors.
(b) Find the four vectors $\partial\hat{\mathbf{e}}_a/\partial u$, $\partial\hat{\mathbf{e}}_a/\partial v$, $\partial\hat{\mathbf{e}}_b/\partial u$, and $\partial\hat{\mathbf{e}}_b/\partial v$ in terms of Cartesian unit vectors.
(c) Express $\hat{\mathbf{e}}_x$ and $\hat{\mathbf{e}}_y$ in terms of $\hat{\mathbf{e}}_a$ and $\hat{\mathbf{e}}_b$.
(d) Express the four vectors $\partial\hat{\mathbf{e}}_a/\partial u$, $\partial\hat{\mathbf{e}}_a/\partial v$, $\partial\hat{\mathbf{e}}_b/\partial u$, and $\partial\hat{\mathbf{e}}_b/\partial v$ in terms of $\hat{\mathbf{e}}_a$ and $\hat{\mathbf{e}}_b$.
(e) If u and v are functions of time, find $d\hat{\mathbf{e}}_a/dt$ and $d\hat{\mathbf{e}}_b/dt$ in terms of $\hat{\mathbf{e}}_a$ and $\hat{\mathbf{e}}_b$.

12.7. Derive Equation (12.19).

12.8. Derive Equation (12.23).

12.9. Show that (12.22) and the assumption $\theta = \pi/2$ solve the last two equations of (12.20) and reduce the first one to (12.24).

12.10. (a) Obtain the time derivatives of the cylindrical unit vectors:

$$\frac{d\hat{\mathbf{e}}_\rho}{dt} = \dot{\varphi}\hat{\mathbf{e}}_\varphi, \qquad \frac{d\hat{\mathbf{e}}_\varphi}{dt} = -\dot{\varphi}\hat{\mathbf{e}}_\rho, \qquad \frac{d\hat{\mathbf{e}}_z}{dt} = 0.$$

(b) Use the result of (a) to show that if **A** is a vector written in terms of cylindrical unit vectors, then

$$\frac{d\mathbf{A}}{dt} = \left(\frac{dA_\rho}{dt} - A_\varphi\dot{\varphi}\right)\hat{\mathbf{e}}_\rho + \left(A_\rho\dot{\varphi} + \frac{dA_\varphi}{dt}\right)\hat{\mathbf{e}}_\varphi + \frac{dA_z}{dt}\hat{\mathbf{e}}_z.$$

12.11. A surface is given by $\dfrac{x^2}{a^2} + \dfrac{y^2}{4a^2} + \dfrac{z^2}{2a^2} = 1$. Find the unit normal to the surface and the equation of the tangent plane at $(a/2, a, a)$.

12.12. The potential of a certain charge distribution is given by

$$\Phi(x, y, z) = z^2 + \frac{y^2}{4} + \frac{x^2}{9}.$$

(a) Find the electric field $\mathbf{E} = -\nabla\Phi$ at $(3/\sqrt{2}, 1, 1/2)$ and show that it is normal to the surface

$$z^2 + \frac{y^2}{4} + \frac{x^2}{9} = 1.$$

(b) Show that the electric field is normal at *every point* of this surface.
(c) Show that the electric field is normal at *every point* of the surface obtained by replacing 1 on the RHS of the last equation by any arbitrary constant.

12.13. Show that $\nabla(fg) = (\nabla f)g + f(\nabla g)$ for any two (differentiable) functions f and g of (x, y, z).

12.14. Consider the plane $ax + by + cz = d$ and a point $P = (x_0, y_0, z_0)$ not lying in the plane. Use Lagrange multipliers to show that the parametric equation of the line passing through P that gives the minimum distance to the plane is

$$\mathbf{r} = \mathbf{r}_0 + t\mathbf{n}, \quad \text{where} \quad \mathbf{r} = \langle x, y, z \rangle, \quad \mathbf{r}_0 = \langle x_0, y_0, z_0 \rangle, \quad \mathbf{n} = \langle a, b, c \rangle.$$
$$(12.33)$$

From this deduce that the distance from P to the plane is

$$\frac{|d - ax_0 - by_0 - cz_0|}{\sqrt{a^2 + b^2 + c^2}}.$$

Hint: Take the dot product of (12.33) with $\mathbf{n}$ and use the fact that $\mathbf{n} \cdot \mathbf{r} = d$ when the tip of $\mathbf{r}$ is in the plane.

12.15. Consider the sphere $(x - a)^2 + (y - b)^2 + (z - c)^2 = d^2$ and a point $P = (x_0, y_0, z_0)$ not lying on the sphere. Use Lagrange multipliers to show that the shortest line segment connecting P to the sphere is that which extends through the center of the sphere.

12.16. For a vector $\mathbf{A}(\mathbf{r}, t)$ that is a function of position and time, show that

$$d\mathbf{A} = (d\mathbf{r} \cdot \nabla)\mathbf{A} + \frac{\partial \mathbf{A}}{\partial t} dt.$$

12.17. Find the gradient of

$$u(x, y, z, x', y', z') \equiv u(\mathbf{r} - \mathbf{r}') = |\mathbf{r} - \mathbf{r}'|^m,$$

first with respect to the components of $\mathbf{r}$ and then with respect to the components of $\mathbf{r}'$, and write the answer completely in terms of $\mathbf{r}$ and $\mathbf{r}'$. What is the answer when $m = -1$?

Chapter 13

Flux and Divergence

A vector field is a function with direction, and because of this directional property, many new kinds of differentiation and integration can be performed on it. For instance, a vector field can be made to pierce a surface or an element thereof, and as it pierces that surface its variation from point to point can be monitored. This leads to one kind of differentiation and integration which we discuss next. The integration leads to the concept of the flux of a vector field, and the associated differentiation to the notion of divergence.

13.1   Flux of a Vector Field

The paradigm of the concept of flux is that of the velocity field of a fluid (see Figure 13.1). A small ring of area Δa is situated in the flow. How much fluid is passing through the ring per unit time? It is clear that the answer depends on the density of the fluid,[1] the speed of the fluid, the size of the area Δa, and also on the relative orientation of the direction of the flow and the unit normal to the area, denoted by $\hat{\mathbf{e}}_n$. A little contemplation reveals that the amount of fluid of constant unit density passing through Δa is proportional to[2]

flux of flow velocity through a small area

$$\Delta \phi = \mathbf{v} \cdot \hat{\mathbf{e}}_n \Delta a \equiv \mathbf{v} \cdot \Delta \mathbf{a}, \qquad (13.1)$$

where $\Delta \phi$ is called the **flux of v** *through* Δa, and $\Delta \mathbf{a}$ is defined to be $\hat{\mathbf{e}}_n \Delta a$. If the ring is replaced by a large surface S then we have to divide the surface into small areas—not necessarily in the shape of a ring—and sum up the contribution of each area to the flux. In the limit of smaller and smaller areas and larger and larger numbers of such areas, we obtain an integral:

total flux of flow velocity through a large area

$$\phi = \lim_{\substack{\Delta a \to 0 \\ N \to \infty}} \sum_{i=1}^{N} \mathbf{v}_i \cdot \hat{\mathbf{e}}_{n_i} \Delta a_i \equiv \lim_{\substack{\Delta a \to 0 \\ N \to \infty}} \sum_{i=1}^{N} \mathbf{v}_i \cdot \Delta \mathbf{a}_i = \iint_S \mathbf{v} \cdot d\mathbf{a}, \qquad (13.2)$$

where ϕ is the total flux through S.

---

[1] For simplicity we assume that density is constant and we take it to be 1.

[2] We shall come back to a rigorous derivation of the flow of a substance through a small loop later (see the discussion after Theorem 13.2.2).

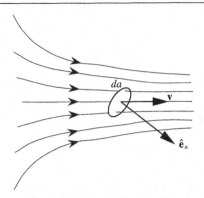

Figure 13.1: Flux of velocity vector through a small area Δa.

There is an arbitrariness in the direction of the unit vector normal to an element of area, because for any unit normal, there is another which points in the opposite direction. The flux for these two unit normals will have opposite signs. This may appear as if one could arbitrarily choose every one of the unit normals $\hat{\mathbf{e}}_{n_i}$ in the sum (13.2) to have either one of the two opposite orientations, leading to an arbitrary result for the integral. This is not the case, because the direction of the unit normal to an element of area is determined by the neighboring unit normals and the requirement of continuity. So, once the choice is made between the two possibilities of the unit normal for one element of area of the surface S, say the first one $\hat{\mathbf{e}}_{n_1}$, the second one can differ only slightly from $\hat{\mathbf{e}}_{n_1}$—in particular, it cannot be of opposite sign. The third one should point in almost the same direction as the second one, and so on. This requirement of continuity will uniquely determine the remaining unit normals. However, the initial choice remains arbitrary, and since the two orientations of the initial choice differ by a sign, the two total fluxes corresponding to these two orientations will also differ by a sign. We shall see shortly, however, that for *closed* surfaces, such an arbitrariness in sign can be overcome by convention.

the total flux can be determined only up to a sign.

The discussion above works for **orientable** surfaces. This means that on any closed loop entirely on the surface, the direction of a normal vector will not change when one displaces it on the loop continuously one complete orbit. It is clear that the lateral surface of a cylinder is orientable.

orientable surface

A cylinder is obtained by glueing the two edges of a rectangle. Now take the same rectangle and twist one of the (smaller) edges before glueing it to the opposite edge. The result—which the reader may want to construct—is a very famous mathematical surface called the **Möbius band**. A Möbius band is not orientable, because if one starts at the midpoint of the glued edges and moves perpendicular to it along the large circle (length of the original rectangle), then a unit normal displaced continuously and completely along the circle will be flipped.[3] In this book we shall never encounter nonorientable surfaces.

Möbius band

---

[3]The reader is urged to perform this surprising experiment using a (portion of a) toothpick as a unit normal.

Example 13.1.1. Consider the flow of a river and assume that the velocity of the water is given by

$$\mathbf{v} = v_0 \left(1 - \frac{4x^2}{w^2}\right) \hat{\mathbf{e}}_z,$$

where x is the distance from the midpoint of the river and w is the width of the river. Let us find the flux of the velocity, assuming that the cross section of the river is a rectangle with depth equal to h, as shown in Figure 13.2.

The normal to the area da is perpendicular to the xy-plane and is in the same direction as the velocity. Thus, we have $\mathbf{v} \cdot \mathbf{da} = v\,da = v\,dx\,dy$, and

$$\phi = \iint_S v\,dx\,dy = \int_{-h/2}^{h/2} dy \int_{-w/2}^{w/2} v_0 \left(1 - \frac{4x^2}{w^2}\right) dx$$

$$= hv_0 \int_{-w/2}^{w/2} \left(1 - \frac{4x^2}{w^2}\right) dx = hv_0 \left(w - \tfrac{1}{3}w\right) = \tfrac{2}{3} A v_0,$$

where S is the cross section of the river and A is its area. ∎

The concept of flux, although indicative of a flow, is not limited to the velocity vector field. We can *define* the flux of any vector field $\mathbf{A}$ in exactly the same way:

$$\phi = \iint_S \mathbf{A} \cdot \mathbf{da}. \tag{13.3}$$

flux can be defined not only for velocity, but for any vector field.

Whether such a definition is *useful* or not should be determined by experiment. It turns out that the flux of every physically relevant vector field is not only useful, but essential for the theoretical—as well as experimental—investigation of that field. For example, the flux of a gravitational field through a closed surface is related to the amount of mass in the volume enclosed in the surface. Similarly, the rate of change of the flux of a magnetic field through a surface gives the electric field produced at the boundary of the surface.

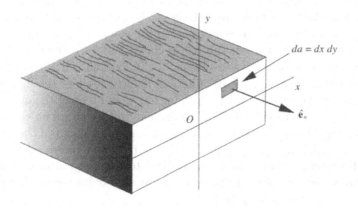

Figure 13.2: The river with its cross section.

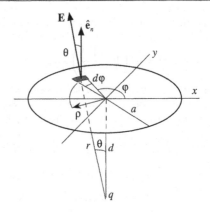

Figure 13.3: The flux of the electric field through a circle. The normal unit vector $\hat{e}_n$ could be chosen to be either up or down. We choose (quite arbitrarily) the up direction to make the flux positive for positive q.

Example 13.1.2. Consider the flux of the electric field of a point charge located at a distance d from the center of a circle of radius a as shown in Figure 13.3. The element of flux is given by

$$\mathbf{E} \cdot d\mathbf{a} = |\mathbf{E}| \cos \theta da = |\mathbf{E}| \cos \theta \rho d\rho d\varphi = \frac{kq}{r^2} \frac{d}{r} \rho d\rho d\varphi = \frac{kqd}{(d^2 + \rho^2)^{3/2}} \rho d\rho d\varphi,$$

where $\hat{e}_n$ is chosen to point up. The polar coordinates (ρ, φ) are used to specify a point in the plane of the circle at which point the element of area is $\rho \, d\rho \, d\varphi$. To find the total flux, we integrate the last expression above:

$$\phi = \iint\limits_{S} \frac{kqd}{(d^2 + \rho^2)^{3/2}} \rho \, d\rho \, d\varphi = kqd \int_0^{2\pi} d\varphi \int_0^a \frac{\rho \, d\rho}{(d^2 + \rho^2)^{3/2}}$$

$$= 2\pi kqd \left\{ -(d^2 + \rho^2)^{-1/2} \Big|_0^a \right\} = 2\pi kq \left(1 - \frac{d}{\sqrt{d^2 + a^2}} \right).$$

Note that since d represents a distance, as opposed to a coordinate, it is always positive and $d = \sqrt{d^2} = |d|$. ∎

It is often necessary to calculate the flux of a vector field through a *closed surface* bounding a volume. Intuitively, such a flux gives a measure of the strength of the source of the vector field in the volume. For instance, the flux of the velocity field of water through a closed surface bounding a fountain measures the rate of the water output of the fountain. If the surface does not enclose the fountain, the net flux will be zero because the flux through one "side" of the closed surface will be positive and that of the other "side" will be negative with the total flux vanishing. In the case of an electrostatic field, the flux through a closed surface measures the amount of charge in the volume bounded by that surface. The sign of the flux requires an *orientation* of the bounding surface which is equivalent to the assignment of a positive direction to the unit normal to the surface at each of its points. *We agree to adhere to the convention of Box 12.1.3.*[4]

> for a closed surface, one can uniquely determine the direction of normal at each point of the surface.

> out is positive!

---

[4]Only *orientable* surfaces can have a well defined orientation. Since we are excluding nonorientable surfaces from this book, all our surfaces respect Box 12.1.3.

Example 13.1.3. Let us consider the flux through a sphere of radius a centered at the origin of a vector field $\mathbf{A}$ given by $\mathbf{A} = kQr^m\hat{\mathbf{e}}_r$ with k a proportionality constant and Q the strength of the source. Assuming that the outward normal is considered positive (see Box 12.1.3) the total flux through the sphere is calculated as

$$\phi_Q = \iint\limits_S \mathbf{A} \cdot d\mathbf{a} = \iint\limits_S kQa^m\hat{\mathbf{e}}_r \cdot (\hat{\mathbf{e}}_r a^2 \sin\theta \, d\theta \, d\varphi)$$

$$= kQ \int_0^{2\pi} d\varphi \int_0^\pi a^m a^2 \sin\theta \, d\theta = 2\pi kQa^{m+2} \int_0^\pi \sin\theta \, d\theta = 4\pi kQa^{m+2}.$$

It is important to keep in mind that when calculating the flux of a vector field, one has to evaluate the field *at the surface*. That is why a appears in the integral rather than r. Notice how the flux depends on the radius of the sphere. If $m+2 > 0$, then the farther away one moves from the origin, the more total flux passes through the sphere. On the other hand, if $m+2 < 0$, although the size of the sphere increases, and therefore, more area is available for the field to cross, the field decreases too rapidly to give enough flux to the large sphere, so the flux decreases. The important case of $m = -2$ eliminates the dependence on a: The total flux through spheres of different sizes is constant. This last statement is a special case of the content of the celebrated **Gauss's law.** ∎

remember to evaluate the vector field at the surface when calculating its flux!

Historical Notes

Space vectors were conceived as three-dimensional generalizations of complex numbers. The primary candidates for such a generalization however turned out to be quaternions—discovered by Hamilton—which had four components. One could naturally divide a quaternion into its "scalar" component and its vector component, the latter itself consisting of three components. The product of two quaternions, being itself a quaternion, can also be divided into scalar and vector parts. It turns out that the scalar part of the product contains the dot product of the vector parts, and the vector part of the product contains the cross product of the vector parts. However, the full product contains some extra terms.

Physicists, on the other hand, were seeking a concept that was more closely associated with Cartesian coordinates than quaternions were. The first step in this direction was taken by James Clerk Maxwell. Maxwell singled out the scalar and the vector parts of Hamilton's quaternion and put the emphasis on these separate parts. In his celebrated *A Treatise on Electricity and Magnetism* (1873) he does speak of quaternions but treats the scalar and the vector parts separately.

Hamilton also developed a calculus of quaternions. In fact, the gradient operator introduced in Definition 12.3.2 and its name "nabla" were both Hamilton's invention.[5] Hamilton showed that if ∇ acts on the *vector part* $\mathbf{v}$ of a quaternion, the result will be a quaternion. Maxwell recognized the scalar part of this quaternion to be the divergence (to be discussed in the next section) of the vector $\mathbf{v}$, and the vector part to be the curl (to be discussed in the Section 14.2) of $\mathbf{v}$.

Maxwell often used quaternions as the basic mathematical entity or he at least made frequent reference to quaternions, perhaps to help his readers. Nevertheless, his work made it clear that vectors were the real tool for physical thinking and not just an abbreviated scheme of writing, as some mathematicians maintained. Thus

---

[5]He used the word "nabla" because ∇ looks like an ancient Hebrew instrument of that name.

by Maxwell's time a great deal of vector analysis was created by treating the scalar and vector parts of quaternions separately.

The formal break with quaternions and the inauguration of a new independent subject, vector analysis, was made independently by Josiah Willard Gibbs and Oliver Heaviside in the early 1880s.

13.1.1   Flux Through an Arbitrary Surface

It may be useful to have a general formula for calculating the flux through an arbitrary surface whose equation is given in parametric form in Cartesian coordinates. Let

$$x = f(u, v), \quad y = g(u, v), \quad z = h(u, v), \tag{13.4}$$

be the parametric equation of a surface. When v is held fixed and u is allowed to vary, a curve is traced on the surface whose infinitesimal displacement can be written as [see Equation (6.63)]

$$\vec{dl_1} = \hat{\mathbf{e}}_x \frac{\partial f}{\partial u} \, du + \hat{\mathbf{e}}_y \frac{\partial g}{\partial u} \, du + \hat{\mathbf{e}}_z \frac{\partial h}{\partial u} \, du.$$

Similarly infinitesimal displacement along curves of constant u is

$$\vec{dl_2} = \hat{\mathbf{e}}_x \frac{\partial f}{\partial v} \, dv + \hat{\mathbf{e}}_y \frac{\partial g}{\partial v} \, dv + \hat{\mathbf{e}}_z \frac{\partial h}{\partial v} \, dv.$$

The cross product of these two displacements is the element of area of the surface:

$$d\mathbf{a} = \vec{dl_1} \times \vec{dl_2} = \det \begin{pmatrix} \hat{\mathbf{e}}_x & \hat{\mathbf{e}}_y & \hat{\mathbf{e}}_z \\ \frac{\partial f}{\partial u} & \frac{\partial g}{\partial u} & \frac{\partial h}{\partial u} \\ \frac{\partial f}{\partial v} & \frac{\partial g}{\partial v} & \frac{\partial h}{\partial v} \end{pmatrix} du \, dv \equiv \det \begin{pmatrix} \hat{\mathbf{e}}_x & \hat{\mathbf{e}}_y & \hat{\mathbf{e}}_z \\ \frac{\partial x}{\partial u} & \frac{\partial y}{\partial u} & \frac{\partial z}{\partial u} \\ \frac{\partial x}{\partial v} & \frac{\partial y}{\partial v} & \frac{\partial z}{\partial v} \end{pmatrix} du \, dv.$$

Using this in (13.3) we get

$$\phi = \iint\limits_R \det \begin{pmatrix} A_x & A_y & A_z \\ \frac{\partial x}{\partial u} & \frac{\partial y}{\partial u} & \frac{\partial z}{\partial u} \\ \frac{\partial x}{\partial v} & \frac{\partial y}{\partial v} & \frac{\partial z}{\partial v} \end{pmatrix} du \, dv, \tag{13.5}$$

where A_x, A_y, and A_z are considered functions of u and v obtained by substituting (13.4) for their arguments. Equation (13.5) is an integral over a region R *in the uv-plane* determined by the range of the variables u and v sufficient to describe the surface S.

The special, but important case, of a surface given by $z = f(x, y)$ deserves special attention. In this case the parametrization is

$$x = u, \quad , y = v, \quad z = f(u, v)$$

and (13.5) yields

$$\phi = \iint\limits_{R} \det \begin{pmatrix} A_x & A_y & A_z \\ 1 & 0 & \frac{\partial z}{\partial u} \\ 0 & 1 & \frac{\partial z}{\partial v}, \end{pmatrix} du\, dv$$

or, writing (x, y) for (u, v)

$$\phi = \iint\limits_{R} \left(-A_x \frac{\partial z}{\partial x} - A_y \frac{\partial z}{\partial y} + A_z \right) dx\, dy, \tag{13.6}$$

where R is the projection of the surface S onto the xy-plane.

13.2   Flux Density = Divergence

The connection between flux and the strength of the source of a vector field was mentioned above. We now analyze this connection further. The variation in the strength of the source of a vector field is measured by the density of the source. For example, the variation in the strength—concentration—of the source of electrostatic (gravitational) field is measured by charge (mass) density. We expect this variation to influence the intensity of flux at various points in space.

13.2.1   Flux Density

Densities are physical quantities treated locally. A local consideration of flux, therefore, requires the introduction of the notion of **flux density**:

notion of flux density and divergence of a vector field introduced

> **Box 13.2.1.** *Take a small volume around a point P, evaluate the total flux of a vector field through the bounding surface of the volume, and divide the result by the volume to get the **flux density** or **divergence** of the vector field at P.*

We denote the flux density by ρ_ϕ for the moment. Later we shall introduce another notation which is more commonly used.

Let us quantify the discussion above for a vector field **A**. Consider a small *rectangular*[6] volume ΔV centered at P with coordinates (x, y, z). Let the sides of the box be $\Delta x, \Delta y,$ and Δz as in Figure 13.4. We are interested in

---

[6]The rectangular shape of the volume is not a restriction because it will be made smaller and smaller at the end. In such a limit, any volume can be built from—a large number of—these small rectangular boxes. Compare this with the rectangular strips used in calculating the area under a curve.

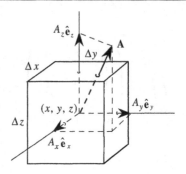

Figure 13.4: The flux of the vector field **A** through a closed infinitesimal rectangular surface.

the net outward[7] flux of the vector field, $\mathbf{A}(x, y, z)$. The six faces of the box are assumed to be so small that the angle between the normal to each face and the vector field **A** is constant over the area of the face. Since we are calculating the outward flux, we must assume that $\hat{\mathbf{e}}_n$ is always pointing out of the volume.

The total flux $\Delta\phi$ through the surface can be written as

$$\Delta\phi = (\Delta\phi_1 + \Delta\phi_2) + (\Delta\phi_3 + \Delta\phi_4) + (\Delta\phi_5 + \Delta\phi_6),$$

where each pair of parentheses indicates one coordinate axis. For instance, $\Delta\phi_1$ is the flux through the face having a normal component along the positive x-axis, $\Delta\phi_2$ is the flux through the face having a normal component along the negative x-axis, and so on. Let us first look at $\Delta\phi_1$, which can be written as

$$\Delta\phi_1 = \mathbf{A}_1 \cdot \hat{\mathbf{e}}_{n_1} \Delta a_1$$

or, since $\hat{\mathbf{e}}_{n_1}$ is the same as $\hat{\mathbf{e}}_x$,

$$\Delta\phi_1 = \mathbf{A}_1 \cdot \hat{\mathbf{e}}_x \Delta a_1 = A_{1x} \Delta a_1.$$

This requires some explanation. The subscript 1 in A_{1x} indicates the evaluation of the vector field at the midpoint[8] of the first face. The subscript x in A_{1x}, of course, means the x-component. So, A_{1x} means the x-component of **A** evaluated at the midpoint of the first face; Δa_1 is the area of face 1 which is simply $\Delta y\Delta z$ (see Figure 13.4). The center of the box—point P— has coordinates (x, y, z) by assumption. Thus, the midpoint of face 1 will have coordinates $(x + \Delta x/2, y, z)$. Therefore,

$$\Delta\phi_1 = A_x\left(x + \frac{\Delta x}{2}, y, z\right)\Delta y\Delta z. \tag{13.7}$$

---

[7]The choice of outward direction is dictated by Box 12.1.3.

[8]The restriction to midpoint is only for convenience. Since the area is small, any other point of the face can be used.

The flux density that we are evaluating will be the density at P. Thus, as a function of the three coordinates, the result will have to be given at the coordinates of P, namely at (x, y, z). This means that in Equation (13.7), all quantities must have (x, y, z) as their arguments. This suggests expanding the function on the RHS of Equation (13.7) as a Taylor series about the point (x, y, z). Recall from Chapter 10 that

$$f(x + \Delta x, y + \Delta y, z + \Delta z) = \sum_{n=0}^{\infty} \sum_{i+j+k=n} \frac{\partial_{ijk}^n f(x, y, z)}{i!j!k!} (\Delta x)^i (\Delta y)^j (\Delta z)^k.$$

We are interested only in the first power because the size of the box will eventually tend to zero. Therefore, we write this in the following abbreviated form:

$$f(x + \Delta x, y + \Delta y, z + \Delta z)$$
$$= f(x, y, z) + \Delta x \frac{\partial f}{\partial x} + \Delta y \frac{\partial f}{\partial y} + \Delta z \frac{\partial f}{\partial z} + \cdots, \qquad (13.8)$$

where it is understood that all derivatives are evaluated at (x, y, z). Applying this result to the function on the RHS of Equation (13.7), for which Δy and Δz are zero, yields

$$A_x \left(x + \frac{\Delta x}{2}, y, z \right) = A_x(x, y, z) + \frac{\Delta x}{2} \frac{\partial A_x}{\partial x} + 0 + 0 + \cdots$$

and

$$\Delta \phi_1 = \left\{ A_x(x, y, z) + \frac{\Delta x}{2} \frac{\partial A_x}{\partial x} \right\} \Delta y \Delta z + \cdots.$$

Similarly, for the second face we obtain

$$\Delta \phi_2 = \mathbf{A}_2 \cdot \hat{\mathbf{e}}_{n_2} \Delta a_2 = \mathbf{A}_2 \cdot (-\hat{\mathbf{e}}_x) \Delta a_2 = -A_{2x} \Delta y \Delta z$$
$$= -A_x \left(x - \frac{\Delta x}{2}, y, z \right) \Delta y \Delta z$$
$$= -\left\{ A_x(x, y, z) - \frac{\Delta x}{2} \frac{\partial A_x}{\partial x} + \cdots \right\} \Delta y \Delta z.$$

Adding the expressions for $\Delta \phi_1$ and $\Delta \phi_2$, we obtain

$$\Delta \phi_1 + \Delta \phi_2$$
$$= \left\{ A_x(x, y, z) + \frac{\Delta x}{2} \frac{\partial A_x}{\partial x} - A_x(x, y, z) + \frac{\Delta x}{2} \frac{\partial A_x}{\partial x} + \cdots \right\} \Delta y \Delta z$$

or

$$\Delta \phi_1 + \Delta \phi_2 = \frac{\partial A_x}{\partial x} \Delta x \Delta y \Delta z + \cdots = \frac{\partial A_x}{\partial x} \Delta V + \cdots.$$

The reader may check that

$$\Delta\phi_3 + \Delta\phi_4 = \frac{\partial A_y}{\partial y}\Delta V + \cdots,$$

$$\Delta\phi_5 + \Delta\phi_6 = \frac{\partial A_z}{\partial z}\Delta V + \cdots, \qquad (13.9)$$

so that the total flux through the small box is

$$\Delta\phi = \left(\frac{\partial A_x}{\partial x} + \frac{\partial A_y}{\partial y} + \frac{\partial A_z}{\partial z}\right)\Delta V + \cdots.$$

The flux density, or **divergence** as it is more often called, can now be obtained by dividing both sides by ΔV and taking the limit as $\Delta V \to 0$. Since all the terms represented by dots are of at least the fourth order, they vanish in the limit and we obtain

Theorem 13.2.1. *The relation between the flux density of a vector field and the derivatives of its components is*

$$\rho_\phi \equiv \operatorname{div} \boldsymbol{A} \equiv \boldsymbol{\nabla} \cdot \boldsymbol{A} = \lim_{\Delta V \to 0} \frac{\Delta\phi}{\Delta V} = \frac{\partial A_x}{\partial x} + \frac{\partial A_y}{\partial y} + \frac{\partial A_z}{\partial z}.$$

origin of the term   The term "divergence," whose abbreviation is used as a symbol of flux
"divergence"   density, is reminiscent of water flowing away from its source, a fountain. In this context, the flux density measures how quickly or intensely water "diverges" away from the fountain. The third notation $\boldsymbol{\nabla} \cdot \boldsymbol{A}$ combines the dot product in terms of components with the definition of $\boldsymbol{\nabla}$ as given in Equation (12.28).[9]

13.2.2   Divergence Theorem

The use of the word (volume) density for divergence suggests that the total flux through a (large) surface should be the (volume) integral of divergence. However, any calculation of flux—even locally—requires a surface, as we saw in the derivation of flux density. What are the "small" surfaces used in the calculation of flux density, and how is the large surface related to them? The answer to this question will come out of a treatment of an important theorem in vector calculus which we investigate now.

First consider two boxes with one face in common (Figure 13.5) and index quantities related to the volume on the left by a and those related to the one on the right by b. The total flux is, of course, the sum of the fluxes through all *six* faces of the *composite box*:

$$\Delta\phi = (\Delta\phi_1 + \Delta\phi_2) + (\Delta\phi_3 + \Delta\phi_4) + (\Delta\phi_5 + \Delta\phi_6),$$

---

[9]This notation is misleading because, as we shall see later, in non-Cartesian coordinate systems, the expression of divergence in terms of derivatives will not be equal to simply the dot product of $\boldsymbol{\nabla}$ with the vector field. One should really think of $\boldsymbol{\nabla} \cdot \boldsymbol{A}$ as a *symbol*, equivalent to ρ_ϕ or div $\boldsymbol{A}$ and not as an operation involving two vectors.

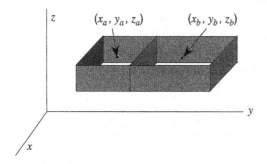

Figure 13.5: The common boundaries contribute no net flux.

where, as before, $\Delta\phi_1$ is the total flux through the face having a normal in the positive x-direction, and $\Delta\phi_2$ that through the face having a normal in the negative x-direction, and so on. It is evident from Figure 13.5 that

$$\Delta\phi_1 = \Delta\phi_{a_1} + \Delta\phi_{b_1},$$

where $\Delta\phi_{a_1}$ is the flux through the positive x face of box a and $\Delta\phi_{b_1}$ is the flux through the positive x face of box b. Using a similar notation, we can write

$$\Delta\phi_2 = \Delta\phi_{a_2} + \Delta\phi_{b_2},$$
$$\Delta\phi_5 + \Delta\phi_6 = \Delta\phi_{a_5} + \Delta\phi_{b_5} + \Delta\phi_{a_6} + \Delta\phi_{b_6}.$$

However, for the y faces we have $\Delta\phi_3 = \Delta\phi_{b_3}$ and $\Delta\phi_4 = \Delta\phi_{a_4}$, because the face of the composite box in the positive y-direction belongs to box b and that in the negative y-direction to box a. Now note that the outward flux through the left face of box b is the negative of the outward flux through the right face of box a; that is,

$$\Delta\phi_{b_4} = -\Delta\phi_{a_3} \;\Rightarrow\; \Delta\phi_{b_4} + \Delta\phi_{a_3} = 0.$$

Thus, we obtain

$$\Delta\phi_3 + \Delta\phi_4 = \Delta\phi_{b_3} + \Delta\phi_{a_4} = \Delta\phi_{a_3} + \Delta\phi_{b_3} + \Delta\phi_{a_4} + \Delta\phi_{b_4}.$$

Using all the above relations yields

$$\Delta\phi = (\Delta\phi_{a_1} + \Delta\phi_{a_2}) + (\Delta\phi_{a_3} + \Delta\phi_{a_4}) + (\Delta\phi_{a_5} + \Delta\phi_{a_6})$$
$$+ (\Delta\phi_{b_1} + \Delta\phi_{b_2}) + (\Delta\phi_{b_3} + \Delta\phi_{b_4}) + (\Delta\phi_{b_5} + \Delta\phi_{b_6})$$

or $\Delta\phi = \Delta\phi_a + \Delta\phi_b$, or $\Delta\phi = (\boldsymbol{\nabla} \cdot \mathbf{A})_a \Delta V_a + (\boldsymbol{\nabla} \cdot \mathbf{A})_b \Delta V_b$. These equations say that

Box 13.2.2. *The total flux through the outer surface of a composite box consisting of two adjacent boxes is equal to the sum of the total fluxes through the bounding surfaces of the two boxes, including the common boundary. Stated differently, in summing the total outward flux of adjacent boxes, the contributions of the common boundary cancel.*

It is now clear how to generalize to a large surface bounding a volume: Divide up the volume into N rectangular boxes and write $\phi \approx \sum_{i=1}^{N} (\boldsymbol{\nabla} \cdot \mathbf{A})_i \Delta V_i$. The LHS of this equation is the outward flux through the *bounding surface only*. Contributions from the sides of *all inner boxes* cancel out because each face of a typical inner box is shared by another box whose outward flux through that face is the negative of the outward flux of the original box. However, boxes at the boundary cannot find enough boxes to cancel *all* their flux contributions, leaving precisely the flux through the original surface. The use of the approximation sign here reflects the fact that N, although large, is not infinite, and that the boxes are not small enough. To attain equality we must make the boxes smaller and smaller and their number larger and larger, in which case we approach the integral:

$$\phi = \iiint\limits_{V} \boldsymbol{\nabla} \cdot \mathbf{A} \, dV. \tag{13.10}$$

the *very important* divergence theorem

Then, using Equation (13.2), we can state the important

Theorem 13.2.2. (*Divergence Theorem*). *The surface integral (flux) of any vector field* $\mathbf{A}$ *through a closed surface* S *bounding a volume* V *is equal to the volume integral of the divergence (or flux density) of* $\mathbf{A}$:

$$\iint\limits_{S} \mathbf{A} \cdot d\mathbf{a} = \iiint\limits_{V} \boldsymbol{\nabla} \cdot \mathbf{A} \, dV. \tag{13.11}$$

Let $\mathbf{A} = \mathbf{c}f$ where $\mathbf{c}$ is an arbitrary constant vector and f a function. Applying the divergence theorem to this $\mathbf{A}$ and using the readily verifiable identity $\boldsymbol{\nabla} \cdot (\mathbf{c}f) = \mathbf{c} \cdot \boldsymbol{\nabla} f$, we get

$$\iint\limits_{S} f\mathbf{c} \cdot d\mathbf{a} = \iiint\limits_{V} \mathbf{c} \cdot (\boldsymbol{\nabla} f) dV \quad \text{or} \quad \mathbf{c} \cdot \left(\iint\limits_{S} f d\mathbf{a} \right) = \mathbf{c} \cdot \left(\iiint\limits_{V} (\boldsymbol{\nabla} f) dV \right)$$

Since this holds for *any* $\mathbf{c}$, we must have

$$\iint\limits_{S} f d\mathbf{a} = \iiint\limits_{V} \boldsymbol{\nabla} f dV \tag{13.12}$$

Example 13.2.3. In this example we derive Gauss's law for fields which vary as the inverse of distance squared, specifically, gravitational and electrostatic fields. Let Q be a source point (a point charge or a point mass) located at P_0 with position vector $\mathbf{r}_0$ and S a closed surface bounding a volume V. Let $\mathbf{A}(\mathbf{r})$ denote the field produced by Q at the field point P with position vector $\mathbf{r}$ as shown in Figure 13.6(a). We know that

$$\mathbf{A}(\mathbf{r}) = \frac{KQ}{|\mathbf{r} - \mathbf{r}_0|^3} (\mathbf{r} - \mathbf{r}_0). \tag{13.13}$$

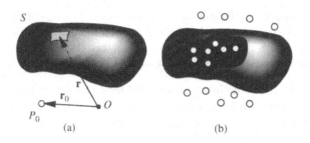

Figure 13.6: Derivation of Gauss's law for (a) a single point source, and (b) a number of point sources.

The flux of $\mathbf{A}$ through S can be written immediately:

$$\iint_S \mathbf{A} \cdot d\mathbf{a} = \iint_S \frac{KQ(\mathbf{r} - \mathbf{r}_0) \cdot d\mathbf{a}}{|\mathbf{r} - \mathbf{r}_0|^3}.$$

But the RHS is—apart from a constant—the solid angle subtended by S about P_0. Using Equation (12.7), we have

$$\iint_S \mathbf{A} \cdot d\mathbf{a} = \begin{cases} 4\pi KQ & \text{if } P_0 \text{ is in } V, \\ 0 & \text{if } P_0 \text{ is not in } V. \end{cases} \tag{13.14}$$

If there are N point sources $Q_1, Q_2, \ldots, Q_N$, then $\mathbf{A}$ will be the sum of individual contributions, and we have

$$\iint_S \mathbf{A} \cdot d\mathbf{a} = \iint_S \sum_{k=1}^{N} \mathbf{A}_k \cdot d\mathbf{a} = \sum_{k=1}^{N} \iint_S \frac{KQ_k(\mathbf{r}_k - \mathbf{r}_0) \cdot d\mathbf{a}}{|\mathbf{r}_k - \mathbf{r}_0|^3}$$

$$= K\sum_{k=1}^{N} Q_k \iint_S \frac{(\mathbf{r}_k - \mathbf{r}_0) \cdot d\mathbf{a}}{|\mathbf{r}_k - \mathbf{r}_0|^3} = K\sum_{k=1}^{N} Q_k \Omega_k,$$

where Ω_k is zero if Q_k is outside V, and 4π if it is inside [see Figure 13.6(b)]. Thus, only the sources enclosed in the volume will contribute to the sum and we have

$$\iint_S \mathbf{A} \cdot d\mathbf{a} = 4\pi KQ_{\text{enc}}, \tag{13.15}$$

global (integral) form of Gauss's law

where Q_{enc} is the amount of source enclosed in S.

For electrostatics, $K = k_e = 1/4\pi\epsilon_0$, $Q = q$, and $\mathbf{A} = \mathbf{E}$, so that

$$\iint_S \mathbf{E} \cdot d\mathbf{a} = q_{\text{enc}}/\epsilon_0. \tag{13.16}$$

For gravitation, $K = -G$, $Q = M$, and $\mathbf{A} = \mathbf{g}$, so that

$$\iint_S \mathbf{g} \cdot d\mathbf{a} = -4\pi GM_{\text{enc}}. \tag{13.17}$$

The minus sign appears in the gravitational case because of the permanent attraction of gravity. Gauss's law is very useful in calculating the fields of very symmetric source distributions, and it is put to good use in introductory electromagnetic discussions. The derivation above shows that it is just as useful in gravitational calculations. ∎

Equation (13.15) is the integral or global form of Gauss's law. We can also derive the differential or local form of Gauss's law by invoking the divergence theorem and assigning a volume density ρ_Q to Q_{enc}:

$$\text{LHS} = \iiint_V \boldsymbol{\nabla} \cdot \mathbf{A}\, dV, \qquad \text{RHS} = 4\pi K \iiint_V \rho_Q\, dV.$$

local (differential) form of Gauss's law

Since these relations are true for arbitrary V, we obtain

Theorem 13.2.4. (*Differential Form of Gauss's Law*). *If a point source produces a vector field* $\mathbf{A}$ *that obeys Equation (13.13), then for any volume distribution* ρ_Q *of the source we have* $\boldsymbol{\nabla} \cdot \mathbf{A} = 4\pi K \rho_Q$.

This can easily be specialized to the two cases of interest, electrostatics and gravity.

13.2.3 Continuity Equation

To improve our physical intuition of divergence, let us consider the flow of a fluid of density $\rho(x, y, z, t)$ and velocity $\mathbf{v}(x, y, z, t)$. The arguments to follow are more general. They can be applied to the flow (bulk motion) of many physical quantities such as charge, mass, energy, momentum, etc. All that needs to be done is to replace ρ—which is the mass density for the fluid flow—with the density of the physical quantity.

We are interested in *the amount of matter crossing a surface area* Δa *per unit time*. We denote this quantity momentarily by ΔM, and because of its importance and wide use in various areas of physics, we shall derive it in some detail. Take a small volume ΔV of the fluid in the shape of a slanted cylinder. The lateral side of this volume is chosen to be *instantaneously* in the same direction as the velocity $\mathbf{v}$ of the particles in the volume. For large volumes this may not be possible, because the macroscopic motion of particles is, in general, not smooth, with different parts having completely different velocities. However, if the volume ΔV (as well as the time interval of observation) is taken small enough, the variation in the velocity of the enclosed particles will be negligible. This situation is shown in Figure 13.7. The lateral length of the cylinder is $v\Delta t$ where Δt is the time it takes the particles inside to go from the base to the top, so that all particles inside will have crossed the top of the cylinder in this time interval. Thus, we have

$$\text{amount crossing top} = \text{amount in } \Delta V = \rho \Delta V.$$

But $\Delta V = (\mathbf{v}\Delta t) \cdot \Delta \mathbf{a} = \mathbf{v} \cdot \Delta \mathbf{a}\, \Delta t$, where the dot product has been used because the base and the top are not perpendicular to the lateral surface.

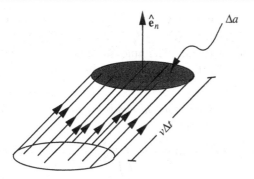

Figure 13.7: The flux through a small area is related to the current density.

Therefore,

$$\Delta M = \frac{\text{amount crossing top}}{\Delta t} = \frac{\rho \mathbf{v} \cdot \Delta \mathbf{a} \, \Delta t}{\Delta t} = (\rho \mathbf{v}) \cdot \Delta \mathbf{a}.$$

The RHS of this equation is the *flux* of the vector field $\rho \mathbf{v}$ which is called the mass **current density**, and usually denoted as $\mathbf{J}$. ◁ current density

As indicated earlier, this result is general and applies to any physical quantity in motion. We can therefore rewrite the equation in its most general form as

$$\Delta \phi_Q = (\rho_Q \mathbf{v}) \cdot \Delta \mathbf{a} \equiv \mathbf{J}_Q \cdot \Delta \mathbf{a}. \tag{13.18}$$

This is so important that we state it in words:

Box 13.2.3. *The amount of a flowing physical quantity Q crossing an area Δa per unit time is the flux $\mathbf{J}_Q \cdot \Delta \mathbf{a}$. The current density $\mathbf{J}_Q$ at each point is simply the product of volume density and velocity vector at that point.* ◁ relation between flux and current density

---

For a (large) surface S we need to integrate the above relation:

$$\phi_Q = \iint\limits_{S} (\rho_Q \mathbf{v}) \cdot d\mathbf{a} \equiv \iint\limits_{S} \mathbf{J}_Q \cdot d\mathbf{a} \tag{13.19}$$

and if S is closed, the divergence theorem gives

$$\phi_Q = \oiint\limits_{S} \mathbf{J}_Q \cdot d\mathbf{a} = \iiint\limits_{V} \nabla \cdot \mathbf{J}_Q \, dV. \tag{13.20}$$

Let Q, which may change with time, denote the total amount of physical quantity in the volume V. Then, clearly

$$Q(t) = \iiint\limits_{V} \rho_Q \, dV = \iiint\limits_{V} \rho_Q(\mathbf{r}, t) \, dV(\mathbf{r}),$$

where in the last integral we have emphasized the dependence of various quantities on location and time. Now, *if Q is a conserved quantity* such as energy, momentum, charge, or mass,[10] the amount of Q that crosses S *outward* (i.e., the flux through S) must precisely equal the rate of *depletion* of Q in the volume V.

global or integral
form of continuity
equation

Theorem 13.2.5. *In mathematical symbols, the conservation of a conserved physical quantity Q is written as*

$$\frac{dQ}{dt} = -\iint_S \mathbf{J}_Q \cdot d\mathbf{a}, \qquad (13.21)$$

which is the global or integral form of the **continuity equation**.

The minus sign ensures that *positive* flux gives rise to a *depletion*, and vice versa. The local or differential form of the continuity equation can be obtained as follows: The LHS of Equation (13.21) can be written as

$$\frac{dQ}{dt} = \frac{d}{dt} \iiint_V \rho_Q(\mathbf{r}, t)\, dV(\mathbf{r}) = \iiint_V \frac{\partial \rho_Q}{\partial t}(\mathbf{r}, t)\, dV(\mathbf{r}),$$

while the RHS, with the help of the divergence theorem, becomes

$$-\iint_S \mathbf{J}_Q \cdot d\mathbf{a} = -\iiint_V \boldsymbol{\nabla} \cdot \mathbf{J}_Q\, dV.$$

Together they give

$$\iiint_V \frac{\partial \rho_Q}{\partial t}\, dV = -\iiint_V \boldsymbol{\nabla} \cdot \mathbf{J}_Q\, dV$$

or

$$\iiint_V \left\{ \frac{\partial \rho_Q}{\partial t} + \boldsymbol{\nabla} \cdot \mathbf{J}_Q \right\}\, dV = 0.$$

This relation is true for all volumes V. In particular, we can make the volume as small as we please. Then, the integral will be approximately the integrand times the volume. Since the volume is nonzero (but small), the only way that the product can be zero is for the integrand to vanish.

local (differential)
form of continuity
equation

> **Box 13.2.4.** *The **differential form** of the continuity equation is*
>
> $$\frac{\partial \rho_Q}{\partial t} + \boldsymbol{\nabla} \cdot \mathbf{J}_Q = 0. \qquad (13.22)$$

---

[10]In the theory of relativity mass by itself is not a conserved quantity, but mass in combination with energy is.

Both integral and differential forms of the continuity equation have a wide range of applications in many areas of physics.

Equation (13.22) is sometimes written in terms of ρ_Q and the velocity. This is achieved by substituting $\rho_Q \mathbf{v}$ for $\mathbf{J}_Q$:

$$\frac{\partial \rho_Q}{\partial t} + \boldsymbol{\nabla} \cdot (\rho_Q \mathbf{v}) = 0$$

or

$$\frac{\partial \rho_Q}{\partial t} + (\boldsymbol{\nabla} \rho_Q) \cdot \mathbf{v} + \rho_Q \boldsymbol{\nabla} \cdot \mathbf{v} = 0.$$

However, using Cartesian coordinates, we write the sum of the first two terms as a total derivative:

$$\frac{\partial \rho_Q}{\partial t} + (\boldsymbol{\nabla} \rho_Q) \cdot \mathbf{v} = \frac{\partial \rho_Q}{\partial t} + \left\langle \frac{\partial \rho_Q}{\partial x}, \frac{\partial \rho_Q}{\partial y}, \frac{\partial \rho_Q}{\partial z} \right\rangle \cdot \left\langle \frac{dx}{dt}, \frac{dy}{dt}, \frac{dz}{dt} \right\rangle$$

$$= \underbrace{\frac{\partial \rho_Q}{\partial t} + \frac{\partial \rho_Q}{\partial x}\frac{dx}{dt} + \frac{\partial \rho_Q}{\partial y}\frac{dy}{dt} + \frac{\partial \rho_Q}{\partial z}\frac{dz}{dt}}_{\text{=total derivative=}d\rho_Q/dt} = \frac{d\rho_Q}{dt}.$$

Thus the continuity equation can also be written as

$$\frac{d\rho_Q}{dt} + \rho_Q \boldsymbol{\nabla} \cdot \mathbf{v} = 0. \tag{13.23}$$

Historical Notes

Aside from Maxwell, two names are associated with vector analysis (completely detached from their quaternionic ancestors): Willard Gibbs and Oliver Heaviside.

Josiah Willard Gibbs's father, also called Josiah Willard Gibbs, was professor of sacred literature at Yale University. In fact the Gibbs family originated in Warwickshire, England, and moved from there to Boston in 1658.

Gibbs was educated at the local Hopkins Grammar School where he was described as friendly but withdrawn. His total commitment to academic work together with rather delicate health meant that he was little involved with the social life of the school. In 1854 he entered Yale College where he won prizes for excellence in Latin and mathematics.

Remaining at Yale, Gibbs began to undertake research in engineering, writing a thesis in which he used geometrical methods to study the design of gears. When he was awarded a doctorate from Yale in 1863 it was the first doctorate of engineering to be conferred in the United States. After this he served as a tutor at Yale for three years, teaching Latin for the first two years and then Natural Philosophy in the third year. He was not short of money however since his father had died in 1861 and, since his mother had also died, Gibbs and his two sisters inherited a fair amount of money.

From 1866 to 1869 Gibbs studied in Europe. He went with his sisters and spent the winter of 1866–67 in Paris, followed by a year in Berlin and, finally spending 1868–69 in Heidelberg. In Heidelberg he was influenced by Kirchhoff and Helmholtz.

Gibbs returned to Yale in June 1869, where two years later he was appointed professor of mathematical physics. Rather surprisingly his appointment to the professorship at Yale came before he had published any work. Gibbs was actually

Josiah Willard
Gibbs 1839–1903

a physical chemist and his major publications were in chemical equilibrium and thermodynamics. From 1873 to 1878, he wrote several important papers on thermodynamics including the notion of what is now called the **Gibbs potential**.

Gibbs's work on vector analysis was in the form of printed notes for the use of his own students written in 1881 and 1884. It was not until 1901 that a properly published version appeared, prepared for publication by one of his students. Using ideas of **Grassmann**, a high school teacher who also worked on the generalization of complex numbers to three dimensions and invented what is now called *Grassmann algebra*, Gibbs produced a system much more easily applied to physics than that of **Hamilton**.

His work on statistical mechanics was also important, providing a mathematical framework for the earlier work of Maxwell on the same subject. In fact his last publication was *Elementary Principles in Statistical Mechanics*, which is a beautiful account putting statistical mechanics on a firm mathematical foundation.

Except for his early years and the three years in Europe, Gibbs spent his whole life living in the same house which his father had built only a short distance from the school Gibbs had attended, the college at which he had studied, and the university where he worked all his life.

Oliver Heaviside caught scarlet fever when he was a young child and this affected his hearing. This was to have a major effect on his life making his childhood unhappy, and his relations with other children difficult. However his school results were rather good and in 1865 he was placed fifth from 500 pupils.

Oliver Heaviside
1850–1925

Academic subjects seemed to hold little attraction for Heaviside, however, and at age 16 he left school. Perhaps he was more disillusioned with school than with learning since he continued to study after leaving school, in particular he learnt the Morse code, and studied electricity and foreign languages, in particular Danish and German. He was aiming at a career as a telegrapher and in this he was advised and helped by his uncle **Charles Wheatstone** (the piece of electrical apparatus the Wheatstone bridge is named after him).

In 1868 Heaviside went to Denmark and became a telegrapher. He progressed quickly in his profession and returned to England in 1871 to take up a post in Newcastle upon Tyne in the office of the Great Northern Telegraph Company which dealt with overseas traffic.

Heaviside became increasingly deaf but he worked on his own researches into electricity. While still working as chief operator in Newcastle he began to publish papers on electricity. One of these was of sufficient interest to **Maxwell** that he mentioned the results in the second edition of his *Treatise on Electricity and Magnetism*. Maxwell's treatise fascinated Heaviside and he gave up his job as a telegrapher and devoted his time to the study of the work. Although his interest and understanding of this work was deep, Heaviside was not interested in rigor. Nevertheless, he was able to develop important methods in vector analysis in his investigations.

His *operational calculus*, developed between 1880 and 1887, caused much controversy. Burnside rejected one of Heaviside's papers on the operational calculus, which he had submitted to the *Proceedings of the Royal Society*, on the grounds that it "contained errors of substance and had irredeemable inadequacies in proof." Tait championed quaternions against the vector methods of Heaviside and Gibbs and sent frequent letters to *Nature* attacking Heaviside's methods. Eventually, however, his work was recognized, and in 1891 he was elected a Fellow of the Royal Society. Whittaker rated Heaviside's operational calculus as one of the three most important discoveries of the late nineteenth Century.

Heaviside seemed to become more and more bitter as the years went by. In 1908 he moved to Torquay where he showed increasing evidence of a persecution complex. His neighbors related stories of Heaviside as a strange and embittered hermit who replaced his furniture with granite blocks which stood about in the bare rooms like the furnishings of some Neolithic giant. Through those fantastic rooms he wandered, growing dirtier and dirtier, with one exception: His nails were always exquisitely manicured, and painted a glistening cherry pink.

13.3 Problems

13.1. Using (13.6) find the flux of the vector field $\mathbf{A} = kx^2\hat{\mathbf{e}}_z$ through the portion of the sphere of radius a centered at the origin lying in the first octant of a Cartesian coordinate system.

13.2. Using (13.6) find the flux of the vector field $\mathbf{A} = y\hat{\mathbf{e}}_x + 3z\hat{\mathbf{e}}_y - 2x\hat{\mathbf{e}}_z$ through the portion of the plane $x + 2y - 3z = 5$ lying in the first octant of a Cartesian coordinate system.

13.3. A vector field is given by $\mathbf{A} = \mathbf{r}$. Using (13.6) find the flux of this vector field through the upper hemisphere centered at the origin. Verify your answer by calculating the flux using (the much easier) spherical coordinates.

13.4. Find the flux of the vector field $\mathbf{A} = x^2\hat{\mathbf{e}}_x + y^2\hat{\mathbf{e}}_y + z^2\hat{\mathbf{e}}_z$ through the portion of the plane $x + y + z = 1$ lying in the first octant of a Cartesian coordinate system.

13.5. Using (13.6), find the flux of the vector field $\mathbf{A} = k\mathbf{r}/r^3$ through the upper hemisphere centered at the origin. Verify your answer by calculating the flux using spherical coordinates.

13.6. Find the flux of the vector field $\mathbf{A} = y\hat{\mathbf{e}}_y + a\hat{\mathbf{e}}_z$ through the portion of the paraboloid $z = b^2 - x^2 - y^2$ above the xy-plane.

13.7. Derive Equation (13.9).

13.8. Find the flux of the vector

$$\mathbf{A} = \frac{6ka^2y}{\sqrt{x^2 + y^2 + a^2}}\hat{\mathbf{e}}_x + \frac{3ka^2z}{\sqrt{y^2 + z^2 + 4a^2}}\hat{\mathbf{e}}_y + \frac{2ka^2x}{\sqrt{x^2 + z^2 + 9a^2}}\hat{\mathbf{e}}_z$$

through the surface of the box shown in Figure 13.8:
(a) by integrating over the surface of the box; and
(b) by using the divergence theorem and integrating over the volume of the box.

13.9. The gravitational field of a certain mass distribution is given by

$$\mathbf{g}(x, y, z) = -kG\left\{(x^3y^2z^2)\hat{\mathbf{e}}_x + (x^2y^3z^2)\hat{\mathbf{e}}_y + (x^2y^2z^3)\hat{\mathbf{e}}_z\right\},$$

where k is a constant and G is the universal gravitational constant:
(a) Find the mass density of the source of this field.
(b) What is the total mass in a cube of side $2a$ centered about the origin?

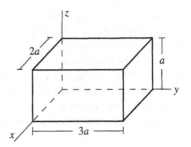

Figure 13.8: The box of Problem 13.8.

13.10. The gravitational field of a certain mass distribution in the first octant of a Cartesian coordinate system is given by

$$\mathbf{g}(x, y, z) = -\frac{GM}{a^3}\mathbf{r}e^{-(x+y+z)/a},$$

where $\mathbf{r}$ is the position vector, M and a are constants, and G is the universal gravitational constant.
(a) Find the mass density of the source of this field.
(b) What is the total mass in a cube of side a with one corner at the origin and sides parallel to the axes?

13.11. The electrostatic potential of a certain charge distribution in Cartesian coordinates is given by

$$\Phi(x, y, z) = \frac{V_0}{a^3}xyze^{-(x+y+z)/a},$$

where V_0 and a are constants.
(a) Find the electric field $\mathbf{E} = -\nabla\Phi$ of this potential.
(b) Calculate the charge density of the source of this field.
(c) What is the total charge in a cube of side a with one corner at the origin and sides parallel to the axes? Write your answer as a numerical multiple of $\epsilon_0 V_0 a$.

13.12. The electric field of a charge distribution is given by

$$\mathbf{E} = \frac{E_0}{a^4}xyze^{-(x+y+z)/a}\mathbf{r}.$$

(a) Write the Cartesian components of this electric field completely in Cartesian coordinates.
(b) Calculate the volume charge density giving rise to this field.
(c) Find the total charge in a cube of side a whose sides are parallel to the axes and one of whose corners is at the origin. Write your answer as a numerical multiple of $\epsilon_0 E_0 a^2$.

13.13. The velocity of a physical quantity Q is radial and given by $\mathbf{v} = k\mathbf{r}$ where k is a constant. Show that if the density ρ_Q is independent of position, then it is given by

$$\rho_Q(t) = \rho_{0Q}e^{-3kt}$$

where ρ_{0Q} is the initial density of Q.

Chapter 14

Line Integral and Curl

Last chapter introduced the concept of flux and the surface integral associated with it. Flux uses the directional property of a vector field to have it pierce an element of area. The directional property can also naturally assign a varying direction along a line. One can consider how a vector field changes direction as it moves along a curve in space. This change can also lead to a new kind of integration and differentiation of vector fields. The integration leads to the notion of a line integral and the associated differentiation to the concept of curl.

14.1 The Line Integral

The prime example of a line integral is the work done by a force. Consider the force field $\mathbf{F}(\mathbf{r})$ acting on an object and imagine the object being moved by a small displacement $\Delta\mathbf{r}$. Then the work done by the force in effecting this displacement is defined as

$$\Delta W = \mathbf{F}(\mathbf{r}) \cdot \Delta\mathbf{r},$$

where it is assumed that $\mathbf{F}(\mathbf{r})$ is (approximately) constant during the displacement.

To calculate the work for a finite displacement, such as the one shown in Figure 14.1, we break up the displacement into N small segments, calculate the work for each segment, and add all contributions to obtain $W \approx \sum_{i=1}^{N} \mathbf{F}(\mathbf{r}_i) \cdot \Delta\mathbf{r}_i$. The approximation sign can be removed by taking $\Delta\mathbf{r}_i$ as small as possible and N as large as possible. Then we have

line integral
defined

$$W = \int_{P_1}^{P_2} \mathbf{F}(\mathbf{r}) \cdot d\mathbf{r} \equiv \int_C \mathbf{F} \cdot d\mathbf{r}, \tag{14.1}$$

where C stands for the particular curve on which the force is displaced. This equation is, by definition, the **line integral** of the force field $\mathbf{F}$. In this particular case it is the work done by $\mathbf{F}$ in moving from P_1 to P_2. Of course, we can apply the line integral to any vector field, not just force. In electromagnetic

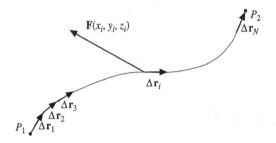

Figure 14.1: The line integral of a vector field **F** from P_1 to P_2.

theory, for example, the line integrals of the electric and magnetic fields play a central role.

The most general way to calculate a line integral is through parametric equation of the curve. Thus, if the Cartesian set of parametric equations of the curve is

$$x = f(t), \qquad y = g(t), \qquad z = h(t),$$

then the components of the vector field **A** will be functions of a single variable t obtained by substitution:

$$A_x(x, y, z) = A_x\big(f(t), g(t), h(t)\big) \equiv \mathcal{F}(t),$$
$$A_y(x, y, z) = A_y\big(f(t), g(t), h(t)\big) \equiv \mathcal{G}(t),$$
$$A_z(x, y, z) = A_z\big(f(t), g(t), h(t)\big) \equiv \mathcal{H}(t),$$

and the components of $d\mathbf{r}$ are

$$dx = f'(t)\, dt, \qquad dy = g'(t)\, dt, \qquad dz = h'(t)\, dt.$$

line integral in terms of the parametric equations of the curve

Then the line integral of **A** can be written as

$$\int_C \mathbf{A} \cdot d\mathbf{r} = \int_C (A_x\, dx + A_y\, dy + A_z\, dz)$$
$$= \int_a^b \big\{\mathcal{F}(t)f'(t) + \mathcal{G}(t)g'(t) + \mathcal{H}(t)h'(t)\big\}\, dt, \qquad (14.2)$$

where $t = a$ and $t = b$ designate the initial and final points of the curve, respectively. Other coordinate systems can be handled similarly. Instead of giving a general formula for these coordinate systems, we present an example using cylindrical coordinates.

Example 14.1.1. Consider the vector field given by

$$\mathbf{A} = c_1 z\varphi\hat{\mathbf{e}}_\rho + c_2 \rho z\hat{\mathbf{e}}_\varphi + c_3 \rho\varphi\hat{\mathbf{e}}_z,$$

where c_1, c_2, and c_3 are constants. We want to calculate the line integral of this field, starting at $z = 0$, along one turn of a uniformly wound helix of radius a whose

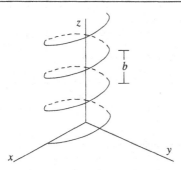

Figure 14.2: The helical path for calculating the line integral.

windings are separated by a constant value b (see Figure 14.2). The parametric equation of this helix in cylindrical coordinates is

$$\rho \equiv f(t) = a, \quad \varphi \equiv g(t) = t, \quad z \equiv h(t) = \frac{b}{2\pi}t.$$

Notice that as $\varphi = t$ changes by 2π, the height (i.e., z) changes by b as required. Substituting for the three coordinates in terms of t in the expression for $\mathbf{A}$, we obtain

$$\mathbf{A} \equiv \langle \mathcal{F}(t), \mathcal{G}(t), \mathcal{H}(t) \rangle = \left\langle c_1 \frac{b}{2\pi}t^2, c_2 a \frac{b}{2\pi}t, c_3 at \right\rangle.$$

Similarly,

$$d\mathbf{r} = \langle d\rho, \rho\, d\varphi, dz \rangle = \langle f'(t), f(t)g'(t), h'(t) \rangle dt = \left\langle 0, a, \frac{b}{2\pi} \right\rangle dt,$$

so that

$$\int_C \mathbf{A} \cdot d\mathbf{r} = \int_a^b \left\{ \mathcal{F}(t)f'(t) + \mathcal{G}(t)g'(t) + \mathcal{H}(t)h'(t) \right\} dt$$
$$= \int_0^{2\pi} \left\{ 0 + c_2 a^2 \frac{b}{2\pi}t + c_3 \frac{b}{2\pi}at \right\} dt = \pi ab(c_2 a + c_3). \qquad \blacksquare$$

Example 14.1.2. Consider the vector field $\mathbf{A} = K(xy^2\hat{\mathbf{e}}_x + x^2 y\hat{\mathbf{e}}_y)$. We want to evaluate the line integral of this field from the origin to the point (a, a) in the xy-plane along three different paths (i), (ii), and (iii), as shown in Figure 14.3. Since the vector field is independent of z and the paths are all in the xy-plane, we ignore z completely.

The first path is the straight line $y = x$. A convenient parameterization is $x = at$, $y = at$ with $0 \le t \le 1$. Along this path the components of $\mathbf{A}$ become

$$A_x = Kxy^2 = K(at)(at)^2 = Ka^3t^3, \qquad A_y = Kx^2 y = K(at)^2(at) = Ka^3t^3.$$

Furthermore, taking the differentials of x and y, we obtain $dx = a\, dt$ and $dy = a\, dt$. Thus,

$$\int_C \mathbf{A} \cdot d\mathbf{r} = \int_{(0,0)}^{(a,a)} (A_x dx + A_y dy) = K \int_0^1 \left[(a^3t^3)\, a\, dt + (a^3t^3)\, a\, dt \right]$$
$$= 2Ka^4 \int_0^1 t^3\, dt = \frac{Ka^4}{2}.$$

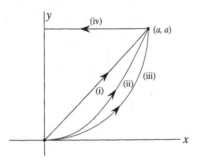

Figure 14.3: The three paths joining the origin to the point (a, a). Path (iv) is to illustrate the importance of parameterization.

Although parameterization is very useful, systematic, and highly recommended, it is not always necessary. We calculate the line integral along path (ii)—given by $y = x^2/a$—without using parameterization. All we have to notice is that all the y's are to be replaced by x^2/a [and therefore, dy by $(2x/a)\,dx$]. Thus,

$$A_x = Kxy^2 = Kx\left(\frac{x^2}{a}\right)^2 = K\frac{x^5}{a^2}, \qquad A_y = Kx^2 y = Kx^2\left(\frac{x^2}{a}\right) = K\frac{x^4}{a}.$$

The line integral can now be evaluated easily:

$$\int_{(0,0)}^{(a,a)} (A_x dx + A_y dy) = K\int_0^a \left[\left(\frac{x^5}{a^2}\right)dx + \left(\frac{x^4}{a}\right)\left(\frac{2x}{a}\,dx\right)\right]$$
$$= 3K\int_0^a \frac{x^5}{a^2}\,dx = \frac{Ka^4}{2}.$$

Finally, we calculate the line integral along the quarter of a circle. For this calculation, we return to the parameterization technique, because it eases the integration. A simple parameterization is

$$x = a - a\cos t, \quad y = a\sin t, \qquad 0 \le t \le \frac{\pi}{2},$$

with $dx = a\sin t\,dt$ and $dy = a\cos t\,dt$. This yields

$$A_x dx + A_y dy = K[(a - a\cos t)a^2 \sin^2 t]a\sin t\,dt + K[(a - a\cos t)^2 a\sin t]a\cos t\,dt$$
$$= Ka^4[(1 - \cos t)(1 - \cos^2 t) + (1 - \cos t)^2 \cos t]\sin t\,dt$$
$$= Ka^4(1 - 3\cos^2 t + 2\cos^3 t)\sin t\,dt.$$

This is now integrated to give the line integral:

$$\int_{(0,0)}^{(a,a)} (A_x dx + A_y dy) = Ka^4 \int_0^{\pi/2} (1 - 3\cos^2 t + 2\cos^3 t)\sin t\,dt$$
$$= Ka^4\left[-\cos t\Big|_0^{\pi/2} + \cos^3 t\Big|_0^{\pi/2} - \tfrac{1}{2}\cos^4 t\Big|_0^{\pi/2}\right] = \frac{Ka^4}{2}.$$

The fact that the three line integrals yield the same result may seem surprising. However, as we shall see shortly, it is a property shared by a special group of vector fields of which $\mathbf{A}$ is a member. ∎

Many a time parameterization makes life a lot easier! Suppose we want to calculate the line integral of a vector field along path (iv) of Figure 14.3. First let us attempt to calculate the line integral using the coordinates. Along path (iv) $d\mathbf{r} = -\hat{\mathbf{e}}_x\, dx$; so $\mathbf{A} \cdot d\mathbf{r} = -A_x\, dx$. Then

parameterization is essential for obtaining the correct sign for some line integrals!

$$\int_{(a,a)}^{(0,a)} \mathbf{A} \cdot d\mathbf{r} = -\int_a^0 A_x\, dx = \int_0^a A_x\, dx.$$

Thus, if $A_x > 0$ (try $A_x = x^2$), the integral will be positive. But this is wrong: A positive A_x should yield a negative $\mathbf{A} \cdot d\mathbf{r}$ because the two vectors are in opposite directions!

With parameterization, this problem is alleviated. A parameterization that represents path (iv) is

$$x = a(1-t), \quad y = a, \qquad 0 \le t \le 1.$$

Clearly, $t = 0$ corresponds to the beginning of path (iv) and $t = 1$ to its endpoint. The parameterization automatically gives $dx = -a\, dt$ and $dy = 0$. For instance, the vector field of Example 14.1.2 yields

$$\int_{(a,a)}^{(0,a)} \mathbf{A} \cdot d\mathbf{r} = \int_0^1 a(1-t)a^2(-a\, dt) = -a^4 \int_0^1 (1-t)\, dt = -\tfrac{1}{2}a^4.$$

This has the correct sign because A_x is positive and the direction of integration negative. The other method would have given a positive result!

14.2   Curl of a Vector Field and Stokes' Theorem

Line integrals around a closed path are of special interest. For example, if the velocity vector of a fluid has a nonzero integral around a closed path, the fluid must be turning around that path and a whirlpool must reside inside the closed path. It is remarkable that such a mundanely concrete idea can be applied verbatim to much more abstract and sophisticated concepts such as electromagnetic fields with proven success and relevance. Thus, for a vector field, $\mathbf{A}$, and a closed path, C, we denote the line integral as

$$\oint_C \mathbf{A} \cdot d\mathbf{r}$$

where the circle on the integral sign indicates that the path is closed and C denotes the particular path taken.

In our discussion of divergence and flux, we encountered Equation (13.11) where an integral (over volume V) was related to an integral over its boundary (surface S). This remarkable property has an analog in one lower dimension: Any closed curve bounds a surface inside it. Is it possible to connect the

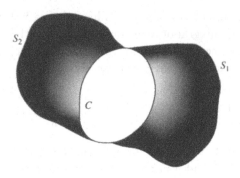

Figure 14.4: There is no "the" surface having C as its boundary. Both S_1 and S_2—as well as a multitude of others—are such surfaces.

line integral over the closed curve to a surface integral over the surface? The answer is yes, but we have to be careful here. What do we mean by "the" surface? A given closed curve may bound many different surfaces, as Figure 14.4 shows. It turns out that this freedom, which was absent in the divergence case,[1] is irrelevant and the relation holds for *any* surface whose boundary is the given curve.

Let us now develop the analog of the divergence theorem for closed line integrals. To begin, we consider a small closed rectangular path with a unit normal $\hat{\mathbf{e}}_n$, which is related to the direction of traversing the path by the right-hand rule (RHR):

Right-hand rule
(RHR) rules here!

> **Box 14.2.1. (*The Right-Hand Rule*).** *Curl the fingers of your right hand in the direction of integration along the curve, your thumb should then point in the direction of* $\hat{\mathbf{e}}_n$.

Without loss of generality we assume that the rectangle is parallel to the xy-plane with sides parallel to the x-axis and the y-axis and that $\hat{\mathbf{e}}_n$ is parallel to the z-axis (see Figure 14.5). The line integral can be written as

$$\oint_C \mathbf{A} \cdot d\mathbf{r} = \int_a^b \mathbf{A} \cdot d\mathbf{r} + \int_b^c \mathbf{A} \cdot d\mathbf{r} + \int_c^d \mathbf{A} \cdot d\mathbf{r} + \int_d^a \mathbf{A} \cdot d\mathbf{r}.$$

We do the first integral in detail; the rest are similar. Along ab the element of displacement $d\mathbf{r}$ is always in the positive x-direction and has magnitude dx,

---

[1] It should be clear that we cannot change the shape of the volume enclosed in S without changing S itself. This rigidity is due to the maximality of the dimension of the enclosed region: A volume is a three-dimensional object, and three is the maximum dimension we have. Theories with higher dimension than three *will* allow a deformability similar to the one discussed above.

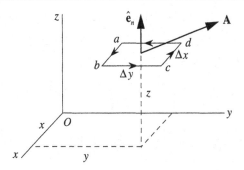

Figure 14.5: A closed rectangular path parallel to the xy-plane with center at (x, y, z).

so it can be written as $d\mathbf{r} = \hat{\mathbf{e}}_x \, dx$. Thus, the first integral on the RHS above becomes

$$\int_a^b \mathbf{A} \cdot d\mathbf{r} \equiv \int_a^b \mathbf{A}_1 \cdot d\mathbf{r}_1 = \int_a^b \mathbf{A}_1 \cdot (\hat{\mathbf{e}}_x \, dx) = \int_a^b A_{1x} \, dx,$$

where, as before, the subscript 1 indicates that we have to evaluate $\mathbf{A}$ at the midpoint of ab and the subscript x denotes the x-component. Now, since ab is small and *the angle between* $\mathbf{A}$ *and* $d\mathbf{r}$ *does not change appreciably on* ab,[2] we can approximate the integral with $A_{1x}\overline{ab}$ and write

$$\int_a^b \mathbf{A} \cdot d\mathbf{r} \approx A_{1x}\overline{ab} = A_{1x}\,\Delta x = A_x \underbrace{\left(x, y - \frac{\Delta y}{2}, z\right)}_{\substack{\text{coordinates of} \\ \text{midpoint of } \overline{ab}}} \Delta x$$

$$\approx \left\{ A_x(x, y, z) - \frac{\Delta y}{2}\frac{\partial A_x}{\partial y} \right\} \Delta x,$$

where in the last line we used the Taylor expansion of A_x. Similarly, we can write

$$\int_c^d \mathbf{A} \cdot d\mathbf{r} = \int_c^d \mathbf{A}_2 \cdot d\mathbf{r}_2 = \int_c^d \mathbf{A}_2 \cdot (-\hat{\mathbf{e}}_x \, dx) = -\int_c^d A_{2x} \, dx$$

$$\approx -A_{2x}\overline{cd} = -A_{2x}\,\Delta x = -A_x \underbrace{\left(x, y + \frac{\Delta y}{2}, z\right)}_{\substack{\text{coordinates of} \\ \text{midpoint of } \overline{cd}}} \Delta x$$

$$\approx -\left\{ A_x(x, y, z) + \frac{\Delta y}{2}\frac{\partial A_x}{\partial y} \right\} \Delta x.$$

Adding the contributions from sides ab and cd yields

$$\int_a^b \mathbf{A} \cdot d\mathbf{r} + \int_c^d \mathbf{A} \cdot d\mathbf{r} \approx -\frac{\partial A_x}{\partial y}\,\Delta x\,\Delta y.$$

---

[2]This condition is essential, because a rapidly changing angle implies a rapidly changing component A_{1x} which is not suitable for the approximation to follow.

The contributions from the other two sides of the rectangle can also be calculated:

$$\int_b^c \mathbf{A} \cdot d\mathbf{r} + \int_d^a \mathbf{A} \cdot d\mathbf{r} \approx A_{3y}\,\Delta y - A_{4y}\,\Delta y$$

$$= A_y\left(x + \frac{\Delta x}{2}, y, z\right)\Delta y - A_y\left(x - \frac{\Delta x}{2}, y, z\right)\Delta y$$

$$\approx \left\{A_y(x,y,z) + \frac{\Delta x}{2}\frac{\partial A_y}{\partial x}\right\}\Delta y - \left\{A_y(x,y,z) - \frac{\Delta x}{2}\frac{\partial A_y}{\partial x}\right\}\Delta y$$

$$= \frac{\partial A_y}{\partial x}\,\Delta x\,\Delta y.$$

The sum of these two equations gives the total contribution:

$$\oint_C \mathbf{A} \cdot d\,\mathbf{r} \approx \left(\frac{\partial A_y}{\partial x} - \frac{\partial A_x}{\partial y}\right)\Delta x\,\Delta y. \qquad (14.3)$$

Let us look at Equation (14.3) more closely. The expression in parentheses can be interpreted as the z-component of the cross product of the gradient operator ∇ with $\mathbf{A}$. In fact, using the mnemonic determinant form of the vector product, we can write

$$\nabla \times \mathbf{A} = \det\begin{pmatrix} \hat{\mathbf{e}}_x & \hat{\mathbf{e}}_y & \hat{\mathbf{e}}_z \\[2mm] \dfrac{\partial}{\partial x} & \dfrac{\partial}{\partial y} & \dfrac{\partial}{\partial z} \\[3mm] A_x & A_y & A_z \end{pmatrix}$$

$$= \left(\frac{\partial A_z}{\partial y} - \frac{\partial A_y}{\partial z}\right)\hat{\mathbf{e}}_x + \left(\frac{\partial A_x}{\partial z} - \frac{\partial A_z}{\partial x}\right)\hat{\mathbf{e}}_y + \left(\frac{\partial A_y}{\partial x} - \frac{\partial A_x}{\partial y}\right)\hat{\mathbf{e}}_z.$$

curl of a vector field defined

This cross product is called the **curl** of $\mathbf{A}$ and is an important quantity in vector analysis. We will look more closely at it later. At this point, however, we are interested only in its definition as applied in Equation (14.3). The RHS of that equation can be written as

$$\left(\frac{\partial A_y}{\partial x} - \frac{\partial A_x}{\partial y}\right)\Delta x\,\Delta y = (\nabla \times \mathbf{A})_z\,\Delta x\,\Delta y = (\nabla \times \mathbf{A})\cdot\hat{\mathbf{e}}_z\Delta a,$$

where $\Delta a = \Delta x\,\Delta y$ is the area of the rectangle. Noting that $\hat{\mathbf{e}}_z$ is in the direction normal to the area, we can replace it with $\hat{\mathbf{e}}_n$. Therefore, we can write Equation (14.3) as

$$\oint_C \mathbf{A} \cdot d\mathbf{r} \approx (\nabla \times \mathbf{A})\cdot\hat{\mathbf{e}}_n\Delta a = (\nabla \times \mathbf{A})\cdot\Delta\mathbf{a}. \qquad (14.4)$$

Equation (14.4) states that for a small rectangular path C the closed line integral is equal to the normal component of the curl of $\mathbf{A}$ evaluated at the center of the rectangle times the area of the rectangle. This statement *does not depend* on the choice of coordinate system. In fact, any rectangle (or any closed planar loop) defines a plane and we are at liberty to designate that plane the xy-plane. Thus, we can *define* the curl of a vector field this way:

Definition 14.2.1. *Given a small closed curve C, calculate the line integral of $\mathbf{A}$ around it and divide the result by the area enclosed by C. The component of the curl of $\mathbf{A}$ along the unit normal to the area is given by*

coordinate independent definition of curl

$$\text{Curl}\,\mathbf{A} \cdot \hat{\mathbf{e}}_n \equiv \boldsymbol{\nabla} \times \mathbf{A} \cdot \hat{\mathbf{e}}_n = \lim_{\Delta a \to 0} \frac{\oint_C \mathbf{A} \cdot d\mathbf{r}}{\Delta a}. \qquad (14.5)$$

The direction of $\hat{\mathbf{e}}_n$ is related to the sense of integration via the right-hand rule.

In Equation (14.5) we are assuming that the area is flat. This is always possible by taking the curve small enough. Definition 14.2.1 is completely independent of the coordinate system and we shall use it to derive expressions for the curl of vector fields in spherical and cylindrical coordinates as well. The reader should be aware that the notation $\boldsymbol{\nabla} \times \mathbf{A}$ is just that, a notation, and—except in Cartesian coordinates—should not be considered as a cross product.

What happens with a large closed path? Figure 14.6 shows a closed path C with an *arbitrary surface S*, whose boundary is the given curve. We divide S into small rectangular areas and assign a direction to their contours dictated by the direction of integration around C.[3] If we sum all the contributions from the small rectangular paths, we will be left with the integration around C because the contributions from the common sides of adjacent rectangles cancel.[4] This is because the sense of integration along their common side is

from small rectangles to large loops

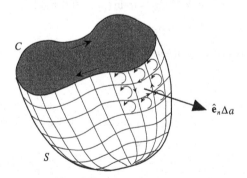

Figure 14.6: An arbitrary surface with the curve C as its boundary. The sum of the line integrals around the rectangular paths shown is equal to the line integral around C.

[3]The direction of the contour with one side on the curve C is determined by the direction of the integration of C. The direction of a distant contour is determined by working one's way to it one (small) rectangle at a time.

[4]This situation is completely analogous to the calculation of the total flux in the derivation of the divergence theorem.

opposite for two adjacent rectangles (see Figure 14.6). Thus, the macroscopic version of Equation (14.4) is

$$\oint_C \mathbf{A} \cdot d\mathbf{r} \approx \sum_{i=1}^{N} (\boldsymbol{\nabla} \times \mathbf{A})_i \cdot \hat{\mathbf{e}}_{n_i} \Delta a_i = \sum_{i=1}^{N} (\boldsymbol{\nabla} \times \mathbf{A})_i \cdot \Delta \mathbf{a}_i,$$

where $(\boldsymbol{\nabla} \times \mathbf{A})_i$ is the curl of $\mathbf{A}$ evaluated at the center of the ith rectangle, which has area Δa_i and normal $\hat{\mathbf{e}}_{n_i}$, and N is the number of rectangles on the surface S. If the areas become smaller and smaller as N gets larger and larger, we can replace the summation by an integral and obtain

the *most important* Stokes' theorem

Theorem 14.2.1. (*Stokes' Theorem*). *The line integral of a vector field* $\mathbf{A}$ *around a closed path* C *is equal to the surface integral of the curl of* $\mathbf{A}$ *on any surface whose only edge is* C. *In mathematical symbols, we have*

$$\oint_C \mathbf{A} \cdot d\mathbf{r} = \iint_S \boldsymbol{\nabla} \times \mathbf{A} \cdot d\mathbf{a}. \tag{14.6}$$

The direction of the normal to the infinitesimal area $d\mathbf{a}$ *of the surface* S *is related to the direction of integration around* C *by the right-hand rule.*

Example 14.2.2. In this example we apply the concepts of closed line integral and the Stokes' theorem to a concrete vector field. Consider the vector field

$$\mathbf{A} = K(x^2 y \hat{\mathbf{e}}_x + xy^2 \hat{\mathbf{e}}_y)$$

obtained from the vector field of Example 14.1.2 by switching the x- and y-components. We want to calculate the line integral around the two closed loops (the circle and the rectangle) of Figure 14.7 and verify the Stokes' theorem.

A convenient parameterization for the circle is

$$x = a \cos t, \quad y = a \sin t, \quad 0 \le t \le 2\pi,$$

with $dx = -a \sin t \, dt$ and $dy = a \cos t \, dt$. Thus,

$$\mathbf{A} \cdot d\mathbf{r} = K(a \cos t)^2 (a \sin t)(-a \sin t \, dt) + K(a \cos t)(a \sin t)^2 (a \cos t \, dt) = 0,$$

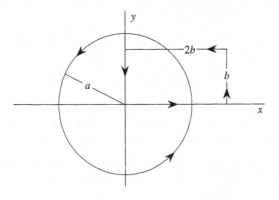

Figure 14.7: Two loops around which the vector field of Example 14.2.2 is calculated.

and the LHS of the Stokes' theorem is zero. For the RHS, we need the curl of the vector.

$$\nabla \times \mathbf{A} = K \begin{vmatrix} \hat{\mathbf{e}}_x & \hat{\mathbf{e}}_y & \hat{\mathbf{e}}_z \\ \frac{\partial}{\partial x} & \frac{\partial}{\partial y} & \frac{\partial}{\partial z} \\ x^2 y & xy^2 & 0 \end{vmatrix} = K(y^2 - x^2)\hat{\mathbf{e}}_z.$$

It is convenient to use cylindrical coordinates for integration over the area of the circle. Moreover, the right-hand rule determines the unit normal to the area of the circle to be $\hat{\mathbf{e}}_z$. Thus,

$$\iint_S \nabla \times \mathbf{A} \cdot d\mathbf{a} = K \int_0^a \int_0^{2\pi} (\rho^2 \sin^2 \varphi - \rho^2 \cos^2 \varphi)\rho \, d\rho \, d\varphi = 0$$

by the φ integration. Thus the two sides of the Stokes' theorem agree.

The two sides of the rectangular loop sitting on the axes will give zero because $\mathbf{A} = 0$ there. The contribution of the side parallel to the y-axis can be obtained by noting that $x = 2b$ and $dx = 0$, so that

$$\mathbf{A} \cdot d\mathbf{r} = A_x \, dx + A_y \, dy = 0 + 2bKy^2 \, dy$$

and

$$\int_{(2b,0)}^{(2b,b)} \mathbf{A} \cdot d\mathbf{r} = 2bK \int_0^b y^2 \, dy = \tfrac{2}{3} Kb^4.$$

To avoid ambiguity,[5] we employ parameterization for the last line integral. A convenient parametric equation would be

$$x = 2b(1 - t), \quad y = b, \qquad 0 \le t \le 1,$$

which gives $dx = -2b \, dt$, $dy = 0$, and for which the line integral yields

$$\int_{(2b,b)}^{(2b,0)} \mathbf{A} \cdot d\mathbf{r} = K \int_0^1 [2b(1-t)]^2 (b)(-2b \, dt) = -8b^4 K \int_0^1 (1-t)^2 \, dt = -\tfrac{8}{3} Kb^4.$$

So, the line integral for the entire loop (the LHS of the Stokes' theorem) is

$$\oint_C \mathbf{A} \cdot d\mathbf{r} = \tfrac{2}{3} Kb^4 - \tfrac{8}{3} Kb^4 = -2Kb^4.$$

We have already calculated the curl of $\mathbf{A}$. Thus, the RHS of the Stokes' theorem becomes

$$\iint_S \nabla \times \mathbf{A} \cdot d\mathbf{a} = K \iint_S (y^2 - x^2) \, dx \, dy$$

$$= K \underbrace{\int_0^{2b} dx \int_0^b y^2 \, dy}_{=2b(b^3/3)} - K \underbrace{\int_0^{2b} x^2 \, dx \int_0^b dy}_{(8b^3/3)b} = -2Kb^4$$

and the two sides agree. ∎

George Gabriel Stokes published papers on the motion of incompressible fluids in 1842–43 and on the friction of fluids in motion, and on the equilibrium and motion of elastic solids in 1845.

In 1849 Stokes was appointed Lucasian Professor of Mathematics at Cambridge, and in 1851 he was elected to the Royal Society and was secretary of the society from 1854 to 1884 when he was elected president.

He investigated the wave theory of light, named and explained the phenomenon of fluorescence in 1852, and in 1854 theorized an explanation of the Fraunhofer lines in the solar spectrum. He suggested these were caused by atoms in the outer layers of the Sun absorbing certain wavelengths. However, when Kirchhoff later published this explanation, Stokes disclaimed any prior discovery.

George Gabriel
Stokes 1819–1903

Stokes developed mathematical techniques for application to physical problems including the most important theorem which bears his name. He founded the science of geodesy, and greatly advanced the study of mathematical physics in England. His mathematical and physical papers were published in five volumes, the first three of which Stokes edited himself in 1880, 1883, and 1891. The last two were edited by Sir Joseph Larmor in 1887 and 1891.

14.3   Conservative Vector Fields

Of great importance are **conservative vector fields**, which are those vector fields that have vanishing line integrals around *every* closed path. An immediate result of this property is that

conservative
vector fields
defined

> **Box 14.3.1.** *The line integral of a conservative vector field between two arbitrary points in space is independent of the path taken.*

To see this, take any two points P_1 and P_2 connected by two different directed paths C_1 and C_2 as shown in Figure 14.8(a). The combination of C_1 and the negative of C_2 forms a closed loop [Figure 14.8(b)] for which we can write

$$\int_{C_1} \mathbf{A} \cdot d\mathbf{r} + \int_{-C_2} \mathbf{A} \cdot d\mathbf{r} = 0$$

because $\mathbf{A}$ is conservative by assumption. The second integral is the negative of the integral along C_2. Thus, the above equation is equivalent to

$$\int_{C_1} \mathbf{A} \cdot d\mathbf{r} - \int_{C_2} \mathbf{A} \cdot d\mathbf{r} = 0 \;\Rightarrow\; \int_{C_1} \mathbf{A} \cdot d\mathbf{r} = \int_{C_2} \mathbf{A} \cdot d\mathbf{r}$$

which proves the above claim.

Now take an arbitrary reference point P_0 and connect it via arbitrary paths to all points in space. At each point P with Cartesian coordinates (x, y, z), define the function $\Phi(x, y, z)$ by

$$\Phi(x, y, z) = -\int_{P_0}^{P} \mathbf{A} \cdot d\mathbf{r} \equiv -\int_{C} \mathbf{A} \cdot d\mathbf{r}, \tag{14.7}$$

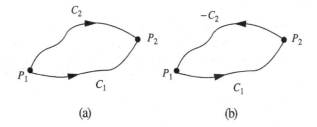

Figure 14.8: (a) Two paths from P_1 to P_2, and (b) the loop formed by them.

where C is any path from P_0 to P and the minus sign is introduced for historical reasons only. Φ is a well-defined function because its value does not depend on C and is called the **potential** associated with the vector field $\mathbf{A}$. We note that the potential at P_0 is zero. That is why P_0 is called the *potential reference point.*

the function Φ, so defined, has the mathematical property expected of a function, namely, that for every point P, the function has only one value that we may denote as $\Phi(P)$.

Now consider two arbitrary points P_1 and P_2, with Cartesian coordinates (x_1, y_1, z_1) and (x_2, y_2, z_2), connected by some path C. We can also connect these two points by a path that goes from P_1 to P_0 and then to P_2 (see Figure 14.9). Since $\mathbf{A}$ is conservative, we have

$$\int_{P_1}^{P_2} \mathbf{A} \cdot d\mathbf{r} = \int_{P_1}^{P_0} \mathbf{A} \cdot d\mathbf{r} + \int_{P_0}^{P_2} \mathbf{A} \cdot d\mathbf{r} = \Phi(x_1, y_1, z_1) - \Phi(x_2, y_2, z_2)$$

or

$$\Phi(x_2, y_2, z_2) - \Phi(x_1, y_1, z_1) = -\int_{P_1}^{P_2} \mathbf{A} \cdot d\mathbf{r}, \tag{14.8}$$

potential of a conservative vector field

which expresses the **potential difference** between the two points.

If P_1 and P_2 are displaced infinitesimally by $d\mathbf{r}$, then their infinitesimal potential difference will be

$$d\Phi = -\mathbf{A} \cdot d\mathbf{r}.$$

On the other hand, Φ, being a scalar differentiable function of x, y, and z, has infinitesimal increment

$$d\Phi = \frac{\partial \Phi}{\partial x} dx + \frac{\partial \Phi}{\partial y} dy + \frac{\partial \Phi}{\partial z} dz = (\boldsymbol{\nabla} \Phi) \cdot d\mathbf{r},$$

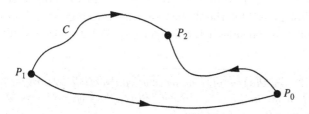

Figure 14.9: Any path C from P_1 to P_2 is equivalent to the path $P_1 \to P_0 \to P_2$.

so we have

$$-\mathbf{A} \cdot d\mathbf{r} = (\boldsymbol{\nabla}\Phi) \cdot d\mathbf{r}.$$

But this is true for an arbitrary $d\mathbf{r}$. Taking $d\mathbf{r}$ to be $\hat{\mathbf{e}}_x \, dx, \hat{\mathbf{e}}_y \, dy$, and $\hat{\mathbf{e}}_z \, dz$ in turn, we obtain the equality of the three components of $\boldsymbol{\nabla}\Phi$ and $-\mathbf{A}$. Therefore, we have

$$\mathbf{A} = -\boldsymbol{\nabla}\Phi, \qquad\qquad (14.9)$$

which states that

Theorem 14.3.1. *A conservative vector field can be written as the negative gradient of a potential function defined as*

$$\Phi(x,y,z) = -\int_{P_0}^{P} \mathbf{A} \cdot d\mathbf{r},$$

where (x,y,z) are the coordinates of P, and the integral is taken along any path connecting P_0 and P.

Another property of a conservative vector field can be obtained by rewriting Equation (14.4), which is true for an arbitrary infinitesimal closed path:

$$\oint_C \mathbf{A} \cdot d\mathbf{r} \approx (\boldsymbol{\nabla} \times \mathbf{A}) \cdot \hat{\mathbf{e}}_n \Delta a.$$

the curl of a conservative vector field is zero.

However, the LHS is zero because $\mathbf{A}$ is conservative. Thus we have

$$(\boldsymbol{\nabla} \times \mathbf{A}) \cdot \hat{\mathbf{e}}_n \Delta a = 0.$$

This is true for arbitrary Δa and $\hat{\mathbf{e}}_n$. Therefore, we have the important conclusion that $\boldsymbol{\nabla} \times \mathbf{A} = 0$ for a conservative vector field. It is important to note that although $\oint_C \mathbf{A} \cdot d\mathbf{r}$ is zero and C is small, we cannot deduce that $\mathbf{A} \cdot d\mathbf{r} = 0$ and, therefore, $\mathbf{A} = 0$. (Why?)

$\boldsymbol{\nabla} \times \mathbf{A} = 0$ does not necessarily imply that $\mathbf{A}$ is conservative!

A conservative vector field demands the vanishing of the curl. But is $\boldsymbol{\nabla} \times \mathbf{A} = 0$ sufficient for $\mathbf{A}$ to be conservative? The answer, in general, is no! (See Example 14.3.3 below.) If the vector field is well defined and well behaved (smoothly varying, differentiable, etc.) in a region of space U, then $\boldsymbol{\nabla} \times \mathbf{A} = 0$ in U implies that $\oint_C \mathbf{A} \cdot d\mathbf{r} = 0$ for all closed curves C *lying entirely in U*. In modern mathematical jargon such a region is said to be **contractible to zero**, which means that any closed curve in U can be contracted to a point (or "zero" closed curve) without encountering any singular point of the vector field (where it is not defined or well behaved). We state this result as follows:

Box 14.3.2. *Let the region U in space be contractible to zero for the vector field $\mathbf{A}$. Then for any closed curve C in U, the two relations $\boldsymbol{\nabla} \times \mathbf{A} = 0$ and $\oint_C \mathbf{A} \cdot d\mathbf{r} = 0$ are equivalent.*

Example 14.3.2. The line integral of the vector field of Example 14.1.2 was independent of the three paths examined there. Could it be that the vector field is conservative? The vector field is clearly well behaved everywhere. Therefore, the vanishing of its curl proves that it is conservative. But

$$\boldsymbol{\nabla} \times \mathbf{A} = K \begin{vmatrix} \hat{\mathbf{e}}_x & \hat{\mathbf{e}}_y & \hat{\mathbf{e}}_z \\ \dfrac{\partial}{\partial x} & \dfrac{\partial}{\partial y} & \dfrac{\partial}{\partial z} \\ xy^2 & x^2y & 0 \end{vmatrix} = (0)\hat{\mathbf{e}}_x + (0)\hat{\mathbf{e}}_y + (2xy - 2xy)\hat{\mathbf{e}}_z = 0.$$

So, $\mathbf{A}$ is indeed conservative.

Next we find the potential of $\mathbf{A}$ at a point (x_0, y_0) in the xy-plane.[6] Let the reference point be the origin. Since it does not matter what path we take, we choose a straight line joining the origin and (x_0, y_0). A convenient parametric equation is

$$x = x_0 t, \quad y = y_0 t, \quad 0 \le t \le 1,$$

which gives $dx = x_0\, dt$ and $dy = y_0\, dt$. We now have

$$\Phi(x_0, y_0) = -\int_{(0,0)}^{(x_0,y_0)} \mathbf{A} \cdot d\mathbf{r}$$

$$= -K \int_0^1 [(x_0 t)(y_0 t)^2 (x_0\, dt) + (x_0 t)^2 (y_0 t)(y_0\, dt)]$$

$$= -2Kx_0^2 y_0^2 \int_0^1 t^3 dt = -\tfrac{1}{2} K x_0^2 y_0^2.$$

We can now substitute (x, y) for (x_0, y_0) to obtain

$$\Phi(x, y) = -\tfrac{1}{2} K x^2 y^2.$$

The reader may verify that $\mathbf{A} = -\boldsymbol{\nabla}\Phi$.   ∎

It should be clear that $\boldsymbol{\nabla} \times \mathbf{A} \neq 0$ always implies that $\mathbf{A}$ is not conservative. However, $\boldsymbol{\nabla} \times \mathbf{A} = 0$ implies that $\mathbf{A}$ is conservative only if the region in question is contractible to zero.

Example 14.3.3. Consider the vector field

$$\mathbf{A} = \frac{ky}{x^2 + y^2}\hat{\mathbf{e}}_x - \frac{kx}{x^2 + y^2}\hat{\mathbf{e}}_y,$$

where k is a constant. Since the components of this vector are independent of z, the curl of the vector can have only a z-component:

$$\boldsymbol{\nabla} \times \mathbf{A} = \begin{vmatrix} \hat{\mathbf{e}}_x & \hat{\mathbf{e}}_y & \hat{\mathbf{e}}_z \\ \dfrac{\partial}{\partial x} & \dfrac{\partial}{\partial y} & \dfrac{\partial}{\partial z} \\ A_x & A_y & 0 \end{vmatrix} = \left(\frac{\partial A_y}{\partial x} - \frac{\partial A_x}{\partial y} \right)\hat{\mathbf{e}}_z.$$

---

[6]We completely ignore the z-coordinate because $\mathbf{A}$ has no component in that direction.

The reader may easily verify that

$$\frac{\partial A_y}{\partial x} = -\frac{k}{x^2 + y^2} + k\frac{2x^2}{(x^2 + y^2)^2}, \qquad \frac{\partial A_x}{\partial y} = \frac{k}{x^2 + y^2} + k\frac{2y^2}{(x^2 + y^2)^2},$$

so that

$$\frac{\partial A_y}{\partial x} - \frac{\partial A_x}{\partial y} = -\frac{2k}{x^2 + y^2} + k\frac{2(x^2 + y^2)}{(x^2 + y^2)^2} = 0$$

and $\nabla \times \mathbf{A} = 0$.

Now take a circle of radius a about the origin and calculate the line integral of $\mathbf{A}$ on this circle. For integration, use the parameterization

$$x = a\cos t, \quad y = a\sin t, \qquad 0 \le t \le 2\pi,$$

with $dx = -a\sin t\, dt$ and $dy = a\cos t\, dt$. Then

$$\mathbf{A} \cdot d\mathbf{r} = A_x dx + A_y dy = \frac{k(a\sin t)(-a\sin t\, dt)}{(a\cos t)^2 + (a\sin t)^2} - \frac{k(a\cos t)(a\cos t\, dt)}{(a\cos t)^2 + (a\sin t)^2} = -k\, dt$$

and, therefore

$$\oint_{\text{circ}} \mathbf{A} \cdot d\mathbf{r} = -k \int_0^{2\pi} dt = -2\pi k.$$

This is an example of a vector field whose curl vanishes but yields a nonzero result for a closed line integral. The reason is, of course, that the region inside the circle is *not* contractable to zero: At the origin the vector is infinite. ∎

If the vector field is conservative, in principle we can determine its potential either by direct antidifferentiation or by integration. The following example illustrates the former procedure.

Example 14.3.4. Consider the vector field

$$\mathbf{A} = (2xy + 3z^2)\hat{\mathbf{e}}_x + (x^2 + 4yz)\hat{\mathbf{e}}_y + (2y^2 + 6xz)\hat{\mathbf{e}}_z.$$

The reader may check that $\nabla \times \mathbf{A} = 0$. Thus, since $\mathbf{A}$ is well defined everywhere, it is conservative. To find its potential Φ, we note that

$$\frac{\partial \Phi}{\partial x} = -A_x = -2xy - 3z^2 \implies \Phi = -x^2 y - 3z^2 x + g(y, z),$$

where we have simply antidifferentiated A_x with respect to x—assuming that y and z are merely constants—and added a "constant" of integration: As far as x differentiation is concerned, any function of y and z is a constant. Now differentiate Φ obtained this way with respect to y and set it equal to $-A_y$:

$$-A_y = -(x^2 + 4zy) = \frac{\partial \Phi}{\partial y} = \frac{\partial}{\partial y}\left(-x^2 y - 3z^2 x + g(y, z)\right) = -x^2 + \frac{\partial g}{\partial y}.$$

This gives

$$\frac{\partial g}{\partial y} = -4yz \implies g(y, z) = -2y^2 z + h(z)$$

Note that our second "constant" of integration has no x-dependence because $g(y, z)$ does not depend on x. Substituting this back in the expression for Φ, we obtain

$$\Phi = -x^2 y - 3z^2 x + g(y, z) = -x^2 y - 3z^2 x - 2y^2 z + h(z).$$

Finally, differentiating this with respect to z and setting it equal to $-A_z$, we obtain

$$-A_z = -(2y^2 + 6xz) = \frac{\partial \Phi}{\partial z} = \frac{\partial}{\partial z}\left(-x^2y - 3z^2x - 2y^2z + h(z)\right) = -6xz - 2y^2 + \frac{dh}{dz}.$$

This gives

$$\frac{dh}{dz} = 0 \ \Rightarrow\ h(z) = \text{const.} \equiv C.$$

The final answer is therefore

$$\Phi(x, y, z) = -x^2y - 3z^2x - 2y^2z + C.$$

The arbitrary constant depends on the potential reference point, and is zero if we choose the origin as that point. It is easy to verify that $-\boldsymbol{\nabla}\Phi$ is indeed the vector field we started with. ∎

There are various vector identities which connect gradient, divergence, and curl. Most of these identities can be obtained by direct substitution. For example, by substituting the Cartesian components of $\mathbf{A} \times \mathbf{B}$ in the Cartesian expression for divergence, one can show that

$$\boldsymbol{\nabla} \cdot (\mathbf{A} \times \mathbf{B}) = \mathbf{B} \cdot \boldsymbol{\nabla} \times \mathbf{A} - \mathbf{A} \cdot \boldsymbol{\nabla} \times \mathbf{B}. \tag{14.10}$$

Similarly, one can show that

$$\boldsymbol{\nabla} \cdot (f\mathbf{A}) = \mathbf{A} \cdot \boldsymbol{\nabla}f + f\boldsymbol{\nabla} \cdot \mathbf{A},$$
$$\boldsymbol{\nabla} \times (f\mathbf{A}) = f\boldsymbol{\nabla} \times \mathbf{A} + (\boldsymbol{\nabla}f) \times \mathbf{A} \tag{14.11}$$
$$\mathbf{A} \times (\boldsymbol{\nabla} \times \mathbf{A}) = \tfrac{1}{2}\boldsymbol{\nabla}|\mathbf{A}|^2 - (\mathbf{A} \cdot \boldsymbol{\nabla})\mathbf{A}$$

We can use Equation (14.10) to derive an important vector integral relation akin to the divergence theorem. Let $\mathbf{B}$ be a *constant* vector. Then the second term on the RHS vanishes. Now apply the divergence theorem to the vector field $\mathbf{A} \times \mathbf{B}$:

$$\iint_S \mathbf{A} \times \mathbf{B} \cdot d\mathbf{a} = \iiint_V \boldsymbol{\nabla} \cdot (\mathbf{A} \times \mathbf{B})\, dV.$$

Using Equation (14.10), the RHS can be written as

$$\text{RHS} = \iiint_V \mathbf{B} \cdot \boldsymbol{\nabla} \times \mathbf{A}\, dV = \mathbf{B} \cdot \iiint_V \boldsymbol{\nabla} \times \mathbf{A}\, dV.$$

Moreover, the use of the cyclic property of the mixed triple product (see Problem 1.15) will enable us to write the LHS as

$$\text{LHS} = \iint_S (d\mathbf{a} \times \mathbf{A}) \cdot \mathbf{B} = \iint_S \mathbf{B} \cdot (d\mathbf{a} \times \mathbf{A}) = \mathbf{B} \cdot \iint_S d\mathbf{a} \times \mathbf{A}.$$

Equating the new versions of the two sides, we obtain

$$\mathbf{B} \cdot \iiint_V \boldsymbol{\nabla} \times \mathbf{A} \, dV = \mathbf{B} \cdot \iint_S d\mathbf{a} \times \mathbf{A}$$

or

$$\mathbf{B} \cdot \left(\iiint_V \boldsymbol{\nabla} \times \mathbf{A} \, dV - \iint_S d\mathbf{a} \times \mathbf{A} \right) = 0.$$

Since the last relation is true of arbitrary $\mathbf{B}$, the vector inside the parentheses must be zero. This gives the result we are after:

$$\iiint_V \boldsymbol{\nabla} \times \mathbf{A} \, dV = \iint_S d\mathbf{a} \times \mathbf{A}. \qquad (14.12)$$

14.4   Problems

14.1. Evaluate the line integral of

$$\mathbf{A}(x, y, z) = x^2 \hat{\mathbf{e}}_x + y^2 \hat{\mathbf{e}}_y - z^2 \hat{\mathbf{e}}_z$$

along the path given parametrically by

$$x = at^2, \quad y = bt, \quad z = c \sin(\pi t / 2)$$

from the origin to (a, b, c).

14.2. Evaluate the line integral of

$$\mathbf{A}(x, y, z) = x \hat{\mathbf{e}}_x + \frac{y^2}{b} \hat{\mathbf{e}}_y - \frac{z^2}{c} \hat{\mathbf{e}}_z$$

along the path given parametrically by

$$x = a \cos(\pi t / 2), \quad y = b \sin(\pi t / 2), \quad z = ct$$

from $(a, 0, 0)$ to $(0, b, c)$.

14.3. Evaluate the line integral of

$$\mathbf{A}(x, y) = x \hat{\mathbf{e}}_x + \frac{y^2}{b} \hat{\mathbf{e}}_y$$

along the *closed* ellipse given parametrically by

$$x = a \cos t, \quad y = b \sin t.$$

14.4. Show that $\boldsymbol{\nabla} \times (\mathbf{A} \times \mathbf{r}) = 2\mathbf{A}$.

14.5. Let

$$\mathbf{A}(x,y) = A_x(x,y)\hat{\mathbf{e}}_x + A_y(x,y)\hat{\mathbf{e}}_y$$
$$\mathbf{B}(x,y) = B_x(x,y)\hat{\mathbf{e}}_x + B_y(x,y)\hat{\mathbf{e}}_y$$

be vectors in two-dimensions.
(a) Apply the divergence theorem to $\mathbf{A}$ using a volume V enclosed by a cylinder whose bottom base is an arbitrary closed curve C in the xy-plane and whose top base is the same curve in a plane parallel to the xy-plane, and whose lateral side is parallel to the z-axis. Now conclude that

$$\oint_C (A_x dy - A_y dx) = \iint_R \left(\frac{\partial A_x}{\partial x} + \frac{\partial A_y}{\partial y} \right) dx\, dy$$

where R is the region enclosed by C in the xy-plane. This is the divergence theorem in two dimensions.
(b) Apply Stokes' theorem to $\mathbf{B}$ with C as above and S the region R defined above. Show that

$$\oint_C (B_x dx + B_y dy) = \iint_R \left(\frac{\partial B_y}{\partial x} - \frac{\partial B_x}{\partial y} \right) dx\, dy$$

This is the Stokes' theorem in two dimensions.
(c) Show that in two dimensions the Stokes' theorem and divergence theorem are the same.

14.6. Evaluate the line integral of

$$\mathbf{A}(x,y) = \left(x^2 + 3y\right)\hat{\mathbf{e}}_x + \left(y^2 + 2x\right)\hat{\mathbf{e}}_y$$

from the origin to the point $(1,2)$:
(a) along the straight line joining the two points; and
(b) along the parabola passing through the two points as well as the point $(-1,2)$.
(c) Is $\mathbf{A}$ conservative?

14.7. Is the vector field $\mathbf{A}(x,y) = xe^{x^2}\cos y\,\hat{\mathbf{e}}_x - \frac{1}{2}e^{x^2}\sin y\,\hat{\mathbf{e}}_y$ conservative? If so, find its potential.

14.8. A vector field is given by

$$\mathbf{A} = \frac{\Phi_0}{b^2}\left[y\left(1 + \frac{x}{b}\right)\hat{\mathbf{e}}_x + x\hat{\mathbf{e}}_y + \frac{xy}{b}\hat{\mathbf{e}}_z \right] e^{(x+z)/b},$$

where Φ_0 and b are constants.
(a) Determine whether or not $\mathbf{A}$ is conservative.
(b) Find the potential of $\mathbf{A}$ if it is conservative.

14.9. The components of a vector field are given by

$$A_x = V_0 k^3 yze^{k^2 xy}, \qquad A_y = V_0 k^3 xze^{k^2 xy} + V_0 k\sin ky, \qquad A_z = V_0 ke^{k^2 xy}.$$

(a) Determine whether $\mathbf{A}$ is conservative or not.
(b) If it is conservative, find its potential.

14.10. The Cartesian components of a vector are given by

$$A_x = 2axe^{kz}, \qquad A_y = 2aye^{kz}, \qquad A_z = ka(x^2 + y^2)e^{kz},$$

where a and k are constants.
(a) Test whether $\mathbf{A}$ is conservative or not.
(b) If $\mathbf{A}$ is conservative, find its potential.

14.11. Prove Equations (14.10) and (14.11).

14.12. Show that

$$\boldsymbol{\nabla}(\mathbf{A} \cdot \mathbf{B}) = (\mathbf{B} \cdot \boldsymbol{\nabla})\mathbf{A} + (\mathbf{A} \cdot \boldsymbol{\nabla})\mathbf{B} + \mathbf{B} \times (\boldsymbol{\nabla} \times \mathbf{A}) + \mathbf{A} \times (\boldsymbol{\nabla} \times \mathbf{B})$$

and that

$$\mathbf{A} \times (\boldsymbol{\nabla} \times \mathbf{B}) = \boldsymbol{\nabla}(\mathbf{A} \cdot \mathbf{B}) - (\mathbf{A} \cdot \boldsymbol{\nabla})\mathbf{B}$$

14.13. Verify the vector identity

$$\boldsymbol{\nabla} \times (\mathbf{A} \times \mathbf{B}) = (\mathbf{B} \cdot \boldsymbol{\nabla})\mathbf{A} - (\mathbf{A} \cdot \boldsymbol{\nabla})\mathbf{B} - \mathbf{B}(\boldsymbol{\nabla} \cdot \mathbf{A}) + \mathbf{A}(\boldsymbol{\nabla} \cdot \mathbf{B})$$

14.14. Verify that for constant $\mathbf{A}$ and $\mathbf{B}$

$$\boldsymbol{\nabla}[\mathbf{A} \cdot (\mathbf{B} \times \mathbf{r})] = \mathbf{A} \times \mathbf{B}$$

Chapter 15

Applied Vector Analysis

In the last three chapters, we introduced the operator ∇ and used it to make vectors out of scalars (gradient), scalars out of vectors (divergence), and new vector out of old vectors (curl). It is obvious that all these processes can be combined to form new scalars and vectors. For instance one can create a vector out of a scalar by the operation of gradient and use the resulting vector as an input for the operation of divergence. Since almost all equations of physics involve derivatives of at most second order, we shall confine our treatment to "double del operations" in this chapter.

15.1 Double Del Operations

We can make different combinations of the vector operator ∇ with itself. By direct differentiation we can easily verify that

$$\nabla \times (\nabla f) = 0. \tag{15.1}$$

Equation (14.9) states that a conservative vector field is the gradient of its potential. Equation (15.1) says, on the other hand, that if a field is the gradient of a function then it is conservative.[1] We can combine these two statements into one by saying that

> **Box 15.1.1.** *A vector field is conservative (i.e., its curl vanishes) if and only if it can be written as the gradient of a scalar function, in which case the scalar function is the field's potential.*

Example 15.1.1. The electrostatic and gravitational fields, which we denote generically by $\mathbf{A}$, are given by an equation of the form

$$\mathbf{A}(\mathbf{r}) = K \iint_{\Omega} \frac{dQ(\mathbf{r}')}{|\mathbf{r} - \mathbf{r}'|^3} (\mathbf{r} - \mathbf{r}').$$

---

[1] Assuming that the region in which the gradient of the function is defined is contractable to zero, i.e., the region has no point at which the gradient is infinite.

Furthermore, the reader may show that (see Problem 12.17)

$$\frac{\mathbf{r} - \mathbf{r}'}{|\mathbf{r} - \mathbf{r}'|^3} = -\boldsymbol{\nabla}\left(\frac{1}{|\mathbf{r} - \mathbf{r}'|}\right). \tag{15.2}$$

Substitution in the above integral then yields

$$\mathbf{A}(\mathbf{r}) = -K\int_{\Omega} dQ(\mathbf{r}')\boldsymbol{\nabla}\left(\frac{1}{|\mathbf{r} - \mathbf{r}'|}\right) = -\boldsymbol{\nabla}\left(K\int_{\Omega}\frac{dQ(\mathbf{r}')}{|\mathbf{r} - \mathbf{r}'|}\right)$$

$$= -\boldsymbol{\nabla}\Phi(\mathbf{r}), \tag{15.3}$$

where Φ, the potential of $\mathbf{A}$, is given by

$$\Phi(\mathbf{r}) \equiv K\int_{\Omega}\frac{dQ(\mathbf{r}')}{|\mathbf{r} - \mathbf{r}'|}. \tag{15.4}$$

Equation (15.3), in conjunction with Equation (15.1), automatically implies that both the electrostatic and gravitational fields are conservative. ∎

In a similar fashion, we can directly verify the following identity:

$$\boldsymbol{\nabla}\cdot(\boldsymbol{\nabla}\times\mathbf{A}) = 0. \tag{15.5}$$

Example 15.1.2. Magnetic fields can also be written in terms of the so-called vector potentials. To find the expression for the vector potential, we substitute Equation (15.2) in the magnetic field integral:

$$\mathbf{B} = \int_{\Omega}\frac{k_m dq(\mathbf{r}')\mathbf{v}(\mathbf{r}')\times(\mathbf{r} - \mathbf{r}')}{|\mathbf{r} - \mathbf{r}'|^3} = k_m\int_{\Omega} dq(\mathbf{r}')\mathbf{v}(\mathbf{r})\times\left\{-\boldsymbol{\nabla}\left(\frac{1}{|\mathbf{r} - \mathbf{r}'|}\right)\right\}.$$

We want to take the $\boldsymbol{\nabla}$ out of the integral. However, the cross product prevents a direct "pull out." So, we need to get around this by manipulating the integrand. Using the second relation in Equation (14.11), we can write

$$\boldsymbol{\nabla}\times\left(\frac{\mathbf{v}(\mathbf{r}')}{|\mathbf{r} - \mathbf{r}'|}\right) = \frac{1}{|\mathbf{r} - \mathbf{r}'|}\overbrace{\boldsymbol{\nabla}\times\mathbf{v}}^{=0} - \mathbf{v}\times\boldsymbol{\nabla}\left(\frac{1}{|\mathbf{r} - \mathbf{r}'|}\right)$$

$$= -\mathbf{v}(\mathbf{r}')\times\boldsymbol{\nabla}\left(\frac{1}{|\mathbf{r} - \mathbf{r}'|}\right).$$

We note that $\boldsymbol{\nabla}\times\mathbf{v} = 0$ because $\boldsymbol{\nabla}$ differentiates with respect to (x, y, z) of which $\mathbf{v}(\mathbf{r}')$ is independent. Substituting this last relation in the expression for $\mathbf{B}$, we obtain

$$\mathbf{B} = k_m\int_{\Omega} dq(\mathbf{r}')\boldsymbol{\nabla}\times\left(\frac{\mathbf{v}(\mathbf{r}')}{|\mathbf{r} - \mathbf{r}'|}\right) = \boldsymbol{\nabla}\times\left(k_m\int_{\Omega}\frac{dq(\mathbf{r}')\mathbf{v}(\mathbf{r}')}{|\mathbf{r} - \mathbf{r}'|}\right)$$

$$\equiv \boldsymbol{\nabla}\times\mathbf{A}, \tag{15.6}$$

where we have taken $\boldsymbol{\nabla}\times$ out of the integral since it differentiates with respect to the *parameters* of integration and Ω is assumed independent of (x, y, z). The **vector**

vector potential
defined

potential A is defined by the last line, which we rewrite as

$$\mathbf{A} = k_m\int_{\Omega}\frac{dq(\mathbf{r}')\mathbf{v}(\mathbf{r}')}{|\mathbf{r} - \mathbf{r}'|}. \tag{15.7}$$

If the charges are confined to one dimension, so that we have a current loop, then $dq(\mathbf{r}')\mathbf{v}(\mathbf{r}') = I\,d\mathbf{r}'$ and Equation (15.7) reduces to

$$\mathbf{A} = k_m I \oint \frac{d\mathbf{r}'}{|\mathbf{r} - \mathbf{r}'|}. \qquad (15.8)$$

An important consequence of Equations (15.6) and (15.5) is

$$\boldsymbol{\nabla} \cdot \mathbf{B} = 0. \qquad (15.9)$$

Since the divergence of a vector field is related to the density of its source, we conclude that *there are no magnetic charges.*

This statement is within the context of classical electromagnetic theory. Recently, with the advent of the unification of electromagnetic and weak nuclear interactions, there have been theoretical arguments for the existence of magnetic charges (or monopoles). However, although the theory predicts—very rare—occurrences of such monopoles, no experimental confirmation of their existence has been made. ∎

15.2 Magnetic Multipoles

The similarity between the vector potential [Equation (15.8)] and the electrostatic potential motivates the expansion of the former in terms of multipoles as was done in (10.33). We carry this expansion only up to the dipole term. Substituting Equation (10.32) in Equation (15.8), we obtain

$$\mathbf{A} = k_m I \oint \left(\frac{1}{r} + \frac{\hat{\mathbf{e}}_r \cdot \mathbf{r}'}{r^2} + \cdots \right) d\mathbf{r}' = \frac{k_m I}{r} \underbrace{\oint d\mathbf{r}'}_{=0} + \frac{k_m I}{r^2} \oint \hat{\mathbf{e}}_r \cdot \mathbf{r}' d\mathbf{r}'.$$

The reader can easily show that the first integral vanishes (Problem 15.5).

To facilitate calculating the second integral, choose Cartesian coordinates and orient your axes so that $\hat{\mathbf{e}}_r$ is in the x-direction. Denote the integral by $\mathbf{V}$. Then

$$\mathbf{V} = \oint \hat{\mathbf{e}}_r \cdot \mathbf{r}' d\mathbf{r}' = \oint \hat{\mathbf{e}}_x \cdot \mathbf{r}' d\mathbf{r}' = \oint x' d\mathbf{r}' = \oint x'(\hat{\mathbf{e}}_x \, dx' + \hat{\mathbf{e}}_y \, dy' + \hat{\mathbf{e}}_z \, dz').$$

We evaluate each component of $\mathbf{V}$ separately.

$$V_x = \oint x' dx' = \tfrac{1}{2} \oint d\left(x'^2\right) = \tfrac{1}{2} x'^2 \Big|_{\text{beginning}}^{\text{end}} = 0$$

because the beginning and end points of a loop coincide.

Now consider the identity

$$\oint (x' dy' + y' dx') = \oint d(x'y') = (x'y') \Big|_{\text{beginning}}^{\text{end}} = 0 \qquad (15.10)$$

with an analogous identity involving x' and z'. For the y-component of $\mathbf{V}$, we have

$$V_y = \oint x'\,dy' = \tfrac{1}{2}\oint x'\,dy' + \underbrace{\tfrac{1}{2}\oint x'\,dy' + \tfrac{1}{2}\oint y'\,dx' - \tfrac{1}{2}\oint y'\,dx'}_{\text{These add up to nothing!}}$$

$$= \tfrac{1}{2}\underbrace{\left(\oint x'\,dy' + \oint y'\,dx'\right)}_{=0 \text{ by Equation (15.10)}} + \tfrac{1}{2}\left(\oint x'\,dy' - \oint y'\,dx'\right)$$

$$= \tfrac{1}{2}\oint(x'\,dy' - y'\,dx') = \tfrac{1}{2}\oint(\mathbf{r}' \times d\mathbf{r}')_z = \tfrac{1}{2}\left(\oint \mathbf{r}' \times d\mathbf{r}'\right)\cdot\hat{\mathbf{e}}_z.$$

It follows that

$$A_y = \frac{k_m I}{r^2}V_y = \frac{k_m I}{2r^2}\left(\oint \mathbf{r}' \times d\mathbf{r}'\right)\cdot\hat{\mathbf{e}}_z \equiv \frac{k_m}{r^2}\boldsymbol{\mu}\cdot\hat{\mathbf{e}}_z,$$

magnetic dipole moment

where we have defined the **magnetic dipole moment** $\boldsymbol{\mu}$ as

$$\boldsymbol{\mu} \equiv \frac{I}{2}\oint \mathbf{r}' \times d\mathbf{r}'. \tag{15.11}$$

A similar calculation will yield

$$A_z = \frac{k_m I}{r^2}V_z = -\frac{k_m I}{2r^2}\left(\oint \mathbf{r}' \times d\mathbf{r}'\right)\cdot\hat{\mathbf{e}}_y \equiv -\frac{k_m}{r^2}\boldsymbol{\mu}\cdot\hat{\mathbf{e}}_y.$$

Therefore,

$$\mathbf{A} = A_x\hat{\mathbf{e}}_x + A_y\hat{\mathbf{e}}_y + A_z\hat{\mathbf{e}}_z = \frac{k_m}{r^2}\underbrace{\left(\hat{\mathbf{e}}_y\,\boldsymbol{\mu}\cdot\hat{\mathbf{e}}_z - \hat{\mathbf{e}}_z\,\boldsymbol{\mu}\cdot\hat{\mathbf{e}}_y\right)}_{=\boldsymbol{\mu}\times(\hat{\mathbf{e}}_y\times\hat{\mathbf{e}}_z)\text{ by bac cab rule}}.$$

Recalling that $\hat{\mathbf{e}}_y \times \hat{\mathbf{e}}_z = \hat{\mathbf{e}}_x$, and that by our choice of orientation of the axes $\hat{\mathbf{e}}_r = \hat{\mathbf{e}}_x$, we finally obtain

$$\mathbf{A} = \frac{k_m\boldsymbol{\mu}\times\hat{\mathbf{e}}_r}{r^2} = \frac{k_m\boldsymbol{\mu}\times\mathbf{r}}{r^3}. \tag{15.12}$$

There is a striking resemblance between the vector potential of a magnetic dipole [Equation (15.12)] and the scalar potential of an electric dipole [the second term in the last line of Equation (10.33)]: The *scalar* potential is given in terms of the scalar (dot) product of the electric dipole moment and the position vector, the *vector* potential is given in terms of the vector product of the magnetic dipole moment and the position vector.

magnetic dipole moment of a circular current loop

Example 15.2.1. Let us calculate the magnetic dipole moment of a circular current of radius a. Placing the circle in the xy-plane with its center at the origin, we have

$$\boldsymbol{\mu} = \frac{I}{2}\oint \mathbf{r}' \times d\mathbf{r}' = \frac{I}{2}\oint(a\hat{\mathbf{e}}_{\rho'}) \times (a\,d\varphi'\hat{\mathbf{e}}_{\varphi'}) = \frac{Ia^2}{2}\int_0^{2\pi}d\varphi'\hat{\mathbf{e}}_z = I\pi a^2\hat{\mathbf{e}}_z.$$

So, the magnitude of the magnetic dipole moment of a circular loop of current is the product of the current and the area of the loop. Its direction is related to the direction of the current by the right-hand rule. ∎

15.3 Laplacian

The divergence of the gradient is an important and frequently occurring operator called the **Laplacian**:

Laplacian of a function

$$\mathbf{\nabla} \cdot (\mathbf{\nabla} f) \equiv \nabla^2 f = \frac{\partial^2 f}{\partial x^2} + \frac{\partial^2 f}{\partial y^2} + \frac{\partial^2 f}{\partial z^2}. \tag{15.13}$$

Laplacian occurs throughout physics, in situations ranging from the waves on a drum to the diffusion of matter in space, the propagation of electromagnetic waves, and even the most basic behavior of matter on a subatomic scale, as governed by the Schrödinger equation of quantum mechanics.

Laplacian is found everywhere!

We discuss one situation in which the Laplacian occurs naturally. The result of the example above and Theorem 13.2.4 can be combined to obtain an important equation in electrostatics and gravity called the **Poisson equation**: $\mathbf{\nabla} \cdot (-\mathbf{\nabla}\Phi) = 4\pi K \rho_Q$, or

Poisson equation

$$\nabla^2 \Phi(\mathbf{r}) = -4\pi K \rho_Q(\mathbf{r}). \tag{15.14}$$

This is a partial differential equation whose solution determines the potential at various points in space.[2] In many situations the density in the region of interest is zero. Then the RHS vanishes and we obtain an important special case of the above equation called **Laplace's equation**:

Laplace's equation

$$\nabla^2 \Phi(\mathbf{r}) = 0. \tag{15.15}$$

Consider a fixed point P in space with Cartesian coordinates (x_0, y_0, z_0) and position vector $\mathbf{r}_0$. Take another (variable) point with Cartesian coordinates (x, y, z) and position vector $\mathbf{r}$. By direct differentiation, one can verify that

$$\mathbf{\nabla} \cdot \left(\frac{\mathbf{r} - \mathbf{r}_0}{|\mathbf{r} - \mathbf{r}_0|^3} \right) = 0$$

at all points of space except at $\mathbf{r} = \mathbf{r}_0$ for which the vector is not defined. Moreover, if S is any closed surface bounding a volume V, we have

$$\iint\limits_{S} \left(\frac{\mathbf{r} - \mathbf{r}_0}{|\mathbf{r} - \mathbf{r}_0|^3} \right) \cdot d\mathbf{a} \equiv \Omega_P^S = \begin{cases} 4\pi & \text{if } P \text{ is in } V, \\ 0 & \text{if } P \text{ is not in } V, \end{cases}$$

by Theorem 12.1.2. On the other hand, the divergence theorem relates the LHS of this equation with the volume integral of divergence. Thus,

$$\iiint\limits_{V} \mathbf{\nabla} \cdot \left(\frac{\mathbf{r} - \mathbf{r}_0}{|\mathbf{r} - \mathbf{r}_0|^3} \right) dV = \begin{cases} 4\pi & \text{if } P \text{ is in } V, \\ 0 & \text{if } P \text{ is not in } V. \end{cases} \tag{15.16}$$

---

[2]The reader should consider this, and any other differential equation, as a *local* equation, meaning that the derivatives on the LHS and the quantities on the RHS are to be evaluated *at the same point*.

This shows that $\mathbf{\nabla} \cdot [(\mathbf{r} - \mathbf{r}_0)/|\mathbf{r} - \mathbf{r}_0|^3]$ has the property that it is zero everywhere except at P, but whose volume integral is not zero. This is reminiscent of the three-dimensional Dirac delta function. In fact, it follows from Equation (15.16) that

$$\mathbf{\nabla} \cdot \left(\frac{\mathbf{r} - \mathbf{r}_0}{|\mathbf{r} - \mathbf{r}_0|^3} \right) = 4\pi\delta(\mathbf{r} - \mathbf{r}_0). \qquad (15.17)$$

relation between
Laplacian and
Dirac delta
function

Using Equation (15.2) and the definition of Laplacian, we also get

$$\nabla^2 \left(\frac{1}{|\mathbf{r} - \mathbf{r}_0|} \right) = -4\pi\delta(\mathbf{r} - \mathbf{r}_0). \qquad (15.18)$$

The last double-del operation we consider is

$$\mathbf{\nabla} \times (\mathbf{\nabla} \times \mathbf{A}) = \mathbf{\nabla}(\mathbf{\nabla} \cdot \mathbf{A}) - \nabla^2\mathbf{A} \qquad (15.19)$$

which holds only in Cartesian coordinates and can be verified component by component.

Example 15.3.1. Angular Momentum Operator   In quantum mechanics, the angular momentum $\mathbf{L} = \mathbf{r} \times \mathbf{p}$ becomes the differential operator $\mathbf{L} = -i\hbar\mathbf{r} \times \mathbf{\nabla}$, where $\hbar$ is the reduced Planck constant, which we set equal to 1 in the following discussion. The quantity $L^2 \equiv |\mathbf{L}|^2$ appears frequently in applications of quantum mechanics. It is therefore instructive to compute this quantity.

Since L^2 is a differential operator, we let it act on some function f and carry out the differentiation until we get a simple result. Since

$$L^2 = L_x^2 + L_y^2 + L_z^2,$$

we let each component act on f separately. First note that

$$
\begin{aligned}
L_x f &= -i\left(\mathbf{r} \times \mathbf{\nabla}f\right)_x = -i\left(y\frac{\partial f}{\partial z} - z\frac{\partial f}{\partial y} \right) \\
L_y f &= -i\left(\mathbf{r} \times \mathbf{\nabla}f\right)_y = -i\left(z\frac{\partial f}{\partial x} - x\frac{\partial f}{\partial z} \right) \qquad (15.20) \\
L_z f &= -i\left(\mathbf{r} \times \mathbf{\nabla}f\right)_z = -i\left(x\frac{\partial f}{\partial y} - y\frac{\partial f}{\partial x} \right)
\end{aligned}
$$

Therefore,

$$
\begin{aligned}
-L_x^2 f &= \left(y\frac{\partial}{\partial z} - z\frac{\partial}{\partial y} \right)\left(y\frac{\partial f}{\partial z} - z\frac{\partial f}{\partial y} \right) \\
&= y^2\frac{\partial^2 f}{\partial z^2} + z^2\frac{\partial^2 f}{\partial y^2} - y\frac{\partial f}{\partial y} - z\frac{\partial f}{\partial z} - 2yz\frac{\partial^2 f}{\partial y\partial z}
\end{aligned}
$$

. Similarly,

$$-L_y^2 f = x^2\frac{\partial^2 f}{\partial z^2} + z^2\frac{\partial^2 f}{\partial x^2} - x\frac{\partial f}{\partial x} - z\frac{\partial f}{\partial z} - 2xz\frac{\partial^2 f}{\partial x\partial z},$$

and

$$-L_z^2 f = x^2\frac{\partial^2 f}{\partial y^2} + y^2\frac{\partial^2 f}{\partial x^2} - x\frac{\partial f}{\partial x} - y\frac{\partial f}{\partial y} - 2xy\frac{\partial^2 f}{\partial x\partial y}.$$

Adding the three components and using a little algebra, we get

$$-L^2 f = r^2 \nabla^2 f - \left(x^2 \frac{\partial^2 f}{\partial x^2} + y^2 \frac{\partial^2 f}{\partial y^2} + z^2 \frac{\partial^2 f}{\partial z^2} \right)$$

$$- 2\mathbf{r} \cdot (\nabla f) - 2 \left(yz \frac{\partial^2 f}{\partial y \partial z} + xz \frac{\partial^2 f}{\partial x \partial z} + xy \frac{\partial^2 f}{\partial x \partial y} \right). \qquad (15.21)$$

Let A denote the sum of the two expressions in the large parentheses. We can write A in a compact form by expanding $(\mathbf{r} \cdot \nabla)(\mathbf{r} \cdot \nabla f)$:

$$(\mathbf{r} \cdot \nabla)^2 f \equiv (\mathbf{r} \cdot \nabla)(\mathbf{r} \cdot \nabla f) = \left(x \frac{\partial}{\partial x} + y \frac{\partial}{\partial y} + z \frac{\partial}{\partial z} \right) \left(x \frac{\partial f}{\partial x} + y \frac{\partial f}{\partial y} + z \frac{\partial f}{\partial z} \right)$$

$$= \underbrace{x \frac{\partial f}{\partial x} + x^2 \frac{\partial^2 f}{\partial x^2} + xy \frac{\partial^2 f}{\partial x \partial y} + xz \frac{\partial^2 f}{\partial x \partial z}}_{\text{comes from } x \text{ differentiation}} + \text{terms from } y \text{ and } z \text{ differentiation.}$$

Adding the terms from x, y, and z differentiations we obtain

$$(\mathbf{r} \cdot \nabla)^2 f = \mathbf{r} \cdot (\nabla f) + A \quad \text{or} \quad A = (\mathbf{r} \cdot \nabla)^2 f - \mathbf{r} \cdot (\nabla f).$$

Substituting this in (15.21) yields

$$L^2 f = -r^2 \nabla^2 f + \mathbf{r} \cdot (\nabla f) + (\mathbf{r} \cdot \nabla)^2 f. \qquad (15.22)$$

As a differential *operator*, L^2 is written as

$$L^2 = -r^2 \nabla^2 + \mathbf{r} \cdot \nabla + (\mathbf{r} \cdot \nabla)^2. \qquad (15.23)$$

We shall come back to this discussion in Chapter 17 to show how index manipulation eases the calculation (see Example 17.3.3). ∎

15.3.1   A Primer of Fluid Dynamics

We have already talked about the flow of a fluid in Section 13.2.3, where we derived the continuity equation, which states the conservation of mass in mathematical terms. We now want to take up the dynamics of a fluid, i.e., the motion of various parts of the fluid due to the forces acting on them.

Consider a volume V of the fluid bounded by a surface S. The pressure p exerted from outside at any point of S in the element of area $d\mathbf{a}$ is normal to S at that point and *pointing into the volume* V. Thus, the element of force due to pressure is $-p d\mathbf{a}$. If pressure is the only source of force on the volume V of the fluid, then the total force on V is

$$\mathbf{F} = -\iint_S p \, d\mathbf{a}.$$

Using Equation (13.12), we rewrite this as

$$\mathbf{F} = -\iint_S p \, d\mathbf{a} = -\iiint_V \nabla p \, dV.$$

This shows that $\boldsymbol{\nabla} p$ is a **force density**, whose volume integral gives the force. If the density of the fluid is ρ and the mass element dm in V has velocity $\mathbf{v}$, then the "mass time acceleration" is $dm\, d\mathbf{v}/dt = \rho\, dV\, (d\mathbf{v}/dt)$, and the total "mass time acceleration" is the volume integral of this quantity. If there are other forces acting on the fluid described by a force density $\mathbf{f}$, we can add it to the right-hand side. Thus, Newton's second law of motion gives

$$\iiint\limits_{V} \rho(d\mathbf{v}/dt)\, dV = -\iiint\limits_{V} \boldsymbol{\nabla} p\, dV + \iiint\limits_{V} \mathbf{f}\, dV,$$

and this holds for *any* volume V, in particular for an infinitesimal volume for which the integrals become the integrand. Hence, the second law of motion for the fluid is

$$\rho(d\mathbf{v}/dt) = -\boldsymbol{\nabla} p + \mathbf{f}. \tag{15.24}$$

The total time derivative of velocity is

$$\frac{d\mathbf{v}}{dt} = \frac{\partial \mathbf{v}}{\partial t} + \frac{\partial \mathbf{v}}{\partial x}\frac{dx}{dt} + \frac{\partial \mathbf{v}}{\partial y}\frac{dy}{dt} + \frac{\partial \mathbf{v}}{\partial z}\frac{dz}{dt} = \frac{\partial \mathbf{v}}{\partial t} + (\mathbf{v} \cdot \boldsymbol{\nabla})\mathbf{v}.$$

Euler's equation of fluid dynamics

Substituting this in (15.24) and dividing by ρ yields

$$\frac{\partial \mathbf{v}}{\partial t} + (\mathbf{v} \cdot \boldsymbol{\nabla})\mathbf{v} = \frac{-\boldsymbol{\nabla} p + \mathbf{f}}{\rho}. \tag{15.25}$$

This is **Euler's equation** and is one of the fundamental equations of fluid dynamics.

The force density $\mathbf{f}$ in Euler's equation is usually that of the gravitational force. Since the gravitational force on an element $\rho\, dV$ is $\mathbf{g}\rho\, dV$, where $\mathbf{g}$ is the gravitational acceleration (or field), the gravitational force density is $\rho\mathbf{g}$ and (15.25) becomes

$$\frac{\partial \mathbf{v}}{\partial t} + (\mathbf{v} \cdot \boldsymbol{\nabla})\mathbf{v} = \frac{-\boldsymbol{\nabla} p}{\rho} + \mathbf{g}. \tag{15.26}$$

Example 15.3.2. In hydrostatic situations with a uniform gravitational field the fluid is not moving and Equation (15.26) becomes

$$\boldsymbol{\nabla} p = \rho\mathbf{g},$$

and if $\mathbf{g}$ is in the negative z-direction, then

$$\frac{\partial p}{\partial x} = \frac{\partial p}{\partial y} = 0, \qquad \frac{\partial p}{\partial z} = -\rho g.$$

Thus the pressure is independent of x and y, and depends only on height z. We assume that the fluid (really the liquid) is incompressible, meaning that its density does not depend on the pressure. Then, integrating the z equation gives

$$p = -\rho g z + C.$$

If the liquid has a free surface at $z = h$ where the pressure is p_0, then $C = p_0 + \rho g h$, and

$$p = p_0 + \rho g(h - z). \qquad\blacksquare$$

Example 15.3.3. Stellar equilibrium A star is a large mass of fluid held together by gravitational attraction. If the star is in equilibrium, its fluid has no motion and (15.26) becomes

$$\nabla p = \rho \mathbf{g} \quad \text{or} \quad \nabla p = -\rho \nabla \Phi$$

where Φ is the gravitational potential. Dividing this equation by ρ, and taking the divergence of both sides, we obtain

$$\nabla \cdot \left(\frac{\nabla p}{\rho} \right) = -\nabla^2 \Phi \quad \text{or} \quad \nabla \cdot \left(\frac{\nabla p}{\rho} \right) = 4\pi G \rho$$

where we used the Poisson equation (15.14). For a spherically symmetric star, only the radial coordinate enters in the equation above, and borrowing from the next chapter the expressions (16.7) for gradient and (16.12) for divergence in spherical coordinates, the equation above takes the form

equation for stellar equilibrium

$$\frac{1}{r^2} \frac{d}{dr} \left(\frac{r^2}{\rho} \frac{dp}{dr} \right) = 4\pi G \rho$$

This is one of the fundamental equations of astrophysics. ∎

15.4 Maxwell's Equations

No treatment of vector analysis is complete without a discussion of Maxwell's equations. Electromagnetism was both the producer and the consumer of vector analysis. It started with the accidental discovery by Örsted in 1820 that an electric current produced a magnetic field. Subsequently, an intense search was undertaken by many physicists such as Ampère and Faraday to find a connection between electric and magnetic phenomena. By the mid-1800s, a fairly good theory of electromagnetism was attained which, in the contemporary language of vectors is translated in the following four equations:

the four equations that Maxwell inherited in integral form

$$(1) \iint_S \mathbf{E} \cdot d\mathbf{a} = \frac{Q}{\epsilon_0}; \qquad (2) \iint_S \mathbf{B} \cdot d\mathbf{a} = 0;$$

$$(3) \oint_C \mathbf{E} \cdot d\mathbf{r} = -\frac{d\phi_m}{dt}; \qquad (4) \oint_C \mathbf{B} \cdot d\mathbf{r} = \mu_0 I. \qquad (15.27)$$

The first integral, Gauss's law (or Coulomb's law in disguise), states that the electric flux through the closed surface S is essentially the total charge Q in the volume surrounded by S. The second integral says that the corresponding flux for a magnetic field is zero. The fact that this holds for an arbitrary surface implies that there are no magnetic charges. The third equation, **Faraday's law**, connects the electric field to the rate of change of magnetic flux ϕ_m. Finally, the last equation, **Ampère's law**, states that the source of the magnetic field is the electric current I. The constant ϵ_0 and μ_0 arise from a particular set of units used for charges and currents.

15.4.1   Maxwell's Contribution

Equations (15.27) can be cast in differential form as well. The differential form of the equations is important because it places particular emphasis on the fields which are the primary objects. The differential form of the equations above are:

the four equations
that Maxwell
inherited in
differential form

$$(1)\ \boldsymbol{\nabla} \cdot \mathbf{E} = \frac{\rho}{\epsilon_0}; \qquad\qquad (2)\ \boldsymbol{\nabla} \cdot \mathbf{B} = 0;$$

$$(3)\ \boldsymbol{\nabla} \times \mathbf{E} = -\frac{\partial \mathbf{B}}{\partial t}; \qquad (4)\ \boldsymbol{\nabla} \times \mathbf{B} = \mu_0 \mathbf{J}. \qquad (15.28)$$

We have already derived the first two equations in Theorem 13.2.4 and Equation (15.9). Here we derive the third equation and leave the derivation of the last equation—which is very similar to that of the third—to the reader. Stokes' theorem turns the LHS of the third equation of (15.27) into

$$LHS = \iint\limits_{S} \boldsymbol{\nabla} \times \mathbf{E} \cdot d\mathbf{a}.$$

The RHS is

$$-\frac{d\phi_m}{dt} = -\frac{d}{dt} \iint\limits_{S} \mathbf{B} \cdot d\mathbf{a} = \iint\limits_{S} \left(-\frac{\partial \mathbf{B}}{\partial t}\right) \cdot d\mathbf{a},$$

where we have assumed that the change in the flux comes about solely due to a change in the magnetic field. This makes it possible to push the time differentiation inside the integral, upon which it becomes a partial derivative because $\mathbf{B}$ is a function of position as well. Since the last two equations hold for arbitrary S, the integrands must be equal. This proves the third equation in (15.28).

Maxwell discovers
the inconsistency
of Equation
(15.28) with the
conservation of
electric charge,
and modifies the
last equation to
resolve the
inconsistency.

Maxwell inherited the four equations in (15.28), and started pondering about them in the 1860s. He noticed that while the second and third are consistent with other aspects of electromagnetism, the other two equations lead to a contradiction. Let us retrace his argument. By Equation (15.5), the divergence of the LHS of the last equation of (15.28) vanishes. Therefore, taking the divergence of both sides, we get $\boldsymbol{\nabla} \cdot \mathbf{J} = 0$. This contradicts the differential form of the continuity equation (13.22) for charges which expresses the conservation of electric charge. Because of the firm establishment of the charge conservation, Maxwell decided to try altering the four equations to make them compatible with charge conservation. The clue is in the first equation. If we differentiate that equation with respect to time, we obtain

$$\frac{\partial}{\partial t}\boldsymbol{\nabla} \cdot \mathbf{E} = \frac{1}{\epsilon_0}\frac{\partial \rho}{\partial t} \ \Rightarrow\ \boldsymbol{\nabla} \cdot \left(\frac{\partial \mathbf{E}}{\partial t}\right) = \frac{1}{\epsilon_0}\frac{\partial \rho}{\partial t} \ \Rightarrow\ \boldsymbol{\nabla} \cdot \left(\epsilon_0 \frac{\partial \mathbf{E}}{\partial t}\right) = \frac{\partial \rho}{\partial t}$$

This suggested to Maxwell that, if the four equations are to be consistent with charge conservation, the fourth equation had to be modified to include $\epsilon_0 \partial \mathbf{E}/\partial t$. With this modification, the four equations in (15.28) become

the four Maxwell
equations

(1) $\nabla \cdot \mathbf{E} = \dfrac{\rho}{\epsilon_0}$;   (2) $\nabla \cdot \mathbf{B} = 0$;

(3) $\nabla \times \mathbf{E} = -\dfrac{\partial \mathbf{B}}{\partial t}$;   (4) $\nabla \times \mathbf{B} = \mu_0 \mathbf{J} + \mu_0 \epsilon_0 \dfrac{\partial \mathbf{E}}{\partial t}$.   (15.29)

It was a great moment in the history of physics and mathematics when Maxwell, prompted solely by the forces of logic and pure deduction, introduced the second term in the last equation. Such moments were rare prior to Maxwell, and with the exception of **Copernicus**'s introduction of the heliocentric theory of the solar system and **Descartes**'s introduction of analytic geometry, deductive reasoning was the exception rather than the rule. Theories and laws were empirical (or inductive); they were introduced to fit the data and summarize, more or less directly, the numerous observations made. Maxwell broke this tradition and set the stage for deductive reasoning which, after a great deal of struggle to abandon the inductive tradition, became the norm for modern physics. *mathematics and the force of logic and human reasoning unravel one of the greatest secrets of Nature!*

Today, we aptly call all four equations in (15.29) **Maxwell's equations**, although his contribution to those equations was a "mere" introduction of the second term on the RHS of the last equation. However, no other "small" contribution has ever affected humankind so enormously. This very "small" contribution was responsible for Maxwell's prediction of the electromagnetic waves which were subsequently produced in the laboratory in 1887—only eight years after Maxwell's premature death—and put to technological use in 1901 in the form of the first radio. Today, Maxwell's equations are at the heart of every electronic device. Without them, our entire civilization, as we know it, would be nonexistent.

15.4.2 Electromagnetic Waves in Empty Space

Let us look at some of the implications of Maxwell equations. Taking the curl of the third Maxwell's equation and using (15.19) and the first and fourth equations of (15.29), we obtain for the LHS *from Maxwell's equations to wave equation*

$$LHS = \nabla \times (\nabla \times \mathbf{E}) = \nabla(\nabla \cdot \mathbf{E}) - \nabla^2 \mathbf{E} = \frac{1}{\epsilon_0}\nabla\rho - \nabla^2 \mathbf{E},$$

and for the RHS

$$RHS = -\nabla \times \left(\frac{\partial \mathbf{B}}{\partial t}\right) = -\frac{\partial}{\partial t}(\nabla \times \mathbf{B}) = -\frac{\partial}{\partial t}\left(\mu_0 \mathbf{J} + \mu_0 \epsilon_0 \frac{\partial \mathbf{E}}{\partial t}\right).$$

In particular, in free space, where $\rho = 0 = \mathbf{J}$, these equations give

$$\nabla^2 \mathbf{E} - \mu_0 \epsilon_0 \frac{\partial^2 \mathbf{E}}{\partial t^2} = 0.   \qquad (15.30)$$

This is a three-dimensional wave equation.[3] Recall that the inverse of the co-efficient of the second time derivative is the square of the speed of propagation of the wave. It follows that

$$v = \frac{1}{\sqrt{\mu_0\epsilon_0}} = \frac{1}{\sqrt{(4\pi \times 10^{-7})(8.854 \times 10^{-12})}} = 2.998 \times 10^8 \text{ m/s},$$

i.e., that the electric field propagates in empty space with the speed of light, c. The reader may check that the magnetic field also satisfies the same wave equation, and that it too propagates with the same speed. In fact, it can be shown that the so-called plane wave solutions of Maxwell's equations consist of an electric and a magnetic component which are coupled to one another and, therefore do not propagate independently (see Problem 15.9).

electromagnetic waves propagate at the speed of light.

Sometimes it is more convenient to work with potentials than the fields themselves. The vanishing of the divergence of magnetic fields suggests that $\mathbf{B} = \boldsymbol{\nabla} \times \mathbf{A}$ where $\mathbf{A}$ is the vector potential [see also Equation (15.6)]. The vector potential, as its scalar counterpart, has some degree of arbitrariness, because adding the gradient of an arbitrary function does not change its curl.

gauge transformation

This is an example of **gauge transformation** whereby a measurable physical quantity—the magnetic field, here—does not change when another (nonmeasurable) physical quantity is changed. Using this expression for $\mathbf{B}$ in the third Maxwell equation, we obtain

$$\boldsymbol{\nabla} \times \mathbf{E} = -\frac{\partial}{\partial t}(\boldsymbol{\nabla} \times \mathbf{A}) \;\Rightarrow\; \boldsymbol{\nabla} \times \left(\mathbf{E} + \frac{\partial \mathbf{A}}{\partial t}\right) = 0 \;\Rightarrow\; \mathbf{E} + \frac{\partial \mathbf{A}}{\partial t} = -\boldsymbol{\nabla}\Phi,$$

where we switched the order of differentiation with respect to position and time, and used the fact that if the curl of a vector vanishes, that vector is the gradient of a function (Box 15.1.1). We therefore write

$$\mathbf{E} = -\frac{\partial \mathbf{A}}{\partial t} - \boldsymbol{\nabla}\Phi \qquad \text{and} \qquad \mathbf{B} = \boldsymbol{\nabla} \times \mathbf{A}. \qquad (15.31)$$

Substituting these two expressions in the fourth Maxwell equation, we obtain

$$\boldsymbol{\nabla} \times (\boldsymbol{\nabla} \times \mathbf{A}) = \mu_0 \mathbf{J} + \frac{1}{c^2}\frac{\partial}{\partial t}\left(-\frac{\partial \mathbf{A}}{\partial t} - \boldsymbol{\nabla}\Phi\right).$$

Expanding the LHS using the double curl identity of Equation (15.19), and switching time and space partial derivatives yields

$$\boldsymbol{\nabla}\left(\boldsymbol{\nabla} \cdot \mathbf{A} + \frac{1}{c^2}\frac{\partial \Phi}{\partial t}\right) - \nabla^2 \mathbf{A} + \frac{1}{c^2}\frac{\partial^2 \mathbf{A}}{\partial t^2} = \mu_0 \mathbf{J}.$$

Because of the gauge freedom, we can choose $\mathbf{A}$ and Φ to satisfy

$$\boldsymbol{\nabla} \cdot \mathbf{A} + \frac{1}{c^2}\frac{\partial \Phi}{\partial t} = 0. \qquad (15.32)$$

---

[3]The reader may be familiar with the one-dimensional wave equation in which only one second partial derivative with respect to a single space coordinate appears.

This choice is called the **Lorentz gauge**, from which it follows that   Lorentz gauge

$$\nabla^2 \mathbf{A} - \frac{1}{c^2}\frac{\partial^2 \mathbf{A}}{\partial t^2} = -\mu_0 \mathbf{J}. \qquad (15.33)$$

Similarly, by taking the divergence of the first equation in (15.31) and using the first Maxwell equation and the Lorentz gauge, we obtain

$$\nabla^2 \Phi - \frac{1}{c^2}\frac{\partial^2 \Phi}{\partial t^2} = -\frac{\rho}{\epsilon_0}. \qquad (15.34)$$

Equations (15.32), (15.33), and (15.34) are the fundamental equations of electromagnetic theory. They not only give the solutions in empty space, where $\rho = \mathbf{J} = 0$, but also when the sources are not zero, i.e., when the mechanism of wave production becomes of interest, as in radiation and antenna theory.

Historical Notes

James Clerk Maxwell attended Edinburgh Academy where he had the nickname "Dafty." While still at school he had two papers published by the Royal Society of Edinburgh. Maxwell then went to Peterhouse, Cambridge, but moved to Trinity, where it was easier to obtain a fellowship. Maxwell graduated with a degree in mathematics from Trinity College in 1854.

He held chairs at Marischal College in Aberdeen (1856) and married the daughter of the Principal. However in 1860 Marischal College and King's College combined and Maxwell, as the junior of the department, had to seek another post. After failing to gain an appointment to a vacant chair at Edinburgh he was appointed to King's College in London (1860) and became the first Cavendish Professor of Physics at Cambridge in 1871.

James Clerk
Maxwell 1831–1879

Maxwell's first major contribution to science was a study of the planet Saturn's rings, and won him the Adams Prize at Cambridge. He showed that stability could be achieved only if the rings consisted of numerous small solid particles, an explanation now confirmed by the Voyager spacecraft.

Maxwell next considered the kinetic theory of gases. By treating gases statistically in 1866 he formulated, independently of **Ludwig Boltzmann**, the Maxwell–Boltzmann kinetic theory of gases. This theory showed that temperatures and heat involved only molecular movement.

This theory meant a change from a concept of certainty, heat viewed as flowing from hot to cold, to one of statistics, molecules at high temperature have only a high probability of moving toward those at low temperature. Maxwell's approach did not reject the earlier studies of thermodynamics but used a better theory of the basis to explain the observations and experiments.

Maxwell's most important achievement was his extension and mathematical formulation of Michael Faraday's theories of electricity and magnetic lines of force. His paper *On Faraday's lines of force* was read to the Cambridge Philosophical Society in two parts, 1855 and 1856. Maxwell showed that a few relatively simple mathematical equations could express the behavior of electric and magnetic fields and their interrelation.

The four partial differential equations, now known as **Maxwell's equations**, first appeared in fully developed form in *Treatise on Electricity and Magnetism* (1873). They are one of the great achievements of nineteenth-century mathematical physics. Solving these equations Maxwell predicted the existence of electromagnetic waves and the fact that these waves propagate at the speed of light (1862). He proposed that the phenomenon of light is therefore an electromagnetic phenomenon.

Maxwell left King's College, London, in the spring of 1865 and returned to his Scottish estate. He made periodic trips to Cambridge and, rather reluctantly, accepted an offer from Cambridge to be the first Cavendish Professor of Physics in 1871. He designed the Cavendish laboratory and helped set it up.

15.5 Problems

15.1. Show that the curl of the gradient of a function is always zero.

15.2. Show that the divergence of the curl of a vector is always zero.

15.3. Verify Equation (15.19) component by component.

15.4. Provide the details of Example 15.3.1:
(a) Compute the three components of **L** and verify Equation (15.20).
(b) Calculate $L_x^2 f$, $L_y^2 f$, $L_z^2 f$ and show that you obtain the expressions given in the example.
(c) Verify that $L^2 f$ is as given in Equation (15.21).
(d) Show that $A = (\mathbf{r} \cdot \nabla)^2 f - \mathbf{r} \cdot (\nabla f)$ and obtain (15.22). Here A is defined by the sum of the expressions in the two pairs of parentheses in Equation (15.21)

15.5. By taking each component of $d\mathbf{r}'$ separately in a convenient coordinate system show that its integral round any closed loop vanishes.

15.6. Recall that the total magnetic force on a current loop is given by

total magnetic force on a current loop in a constant magnetic field is zero.

$$\mathbf{F} = I \oint d\mathbf{r} \times \mathbf{B}.$$

Show that the total force on a current loop located in a *homogeneous* magnetic field is zero.

15.7. Derive the differential form of Maxwell's last equation from the corresponding integral form.

15.8. Starting with Maxwell's equations, show that the magnetic field satisfies the same wave equation as the electric field. In particular, that it, too, propagates with the same speed.

15.9. Consider $\mathbf{E} = \mathbf{E}_0 e^{i(\omega t - \mathbf{k} \cdot \mathbf{r})}$ and $\mathbf{B} = \mathbf{B}_0 e^{i(\omega t - \mathbf{k} \cdot \mathbf{r})}$, where $i = \sqrt{-1}$, $\mathbf{E}_0$, $\mathbf{B}_0$, k, and ω are constants. The **E** and the **B** so defined represent *plane waves* moving in the direction of the vector **k**.
(a) Show that they satisfy Maxwell's equations in free space if:

(1) $\mathbf{k} \cdot \mathbf{E}_0 = 0$;   (2) $\mathbf{k} \cdot \mathbf{B}_0 = 0$;

(3) $\mathbf{k} \times \mathbf{E}_0 = \omega \mathbf{B}_0$;   (4) $\mathbf{k} \times \mathbf{B}_0 = -\frac{\omega}{c^2} \mathbf{E}_0$.

(b) In particular, show that **k**, the propagation direction, and **E** and **B** form a mutually perpendicular set of vectors.

(c) By taking the cross product of **k** with an appropriate equation, show that $|\mathbf{k}| = \omega/c$.

15.10. Derive Equation (15.34).

Chapter 16

Curvilinear Vector Analysis

All the vector analytical quantities discussed in the previous chapters can also be calculated in other coordinate systems. The general procedure is to start with definitions of quantities in a coordinate-free way and substitute the known quantities in terms of the particular coordinates we are interested in and "read off" the vector analytic quantity. Instead of treating cylindrical and spherical coordinate systems separately, we lump them together and derive relations that hold not only in the three familiar coordinate systems, but also in all coordinate systems whose unit vectors form a set of right-handed mutually perpendicular vectors. Since the geometric definitions of all vector-analytic quantities involve elements of length, we start with the length elements.

16.1   Elements of Length

Consider **curvilinear coordinates**[1] (q_1, q_2, q_3) in which the *primary* line elements are given by

curvilinear coordinates

$$dl_1 = h_1(q_1, q_2, q_3)\, dq_1,\ dl_2 = h_2(q_1, q_2, q_3)\, dq_2,\ dl_3 = h_3(q_1, q_2, q_3)\, dq_3,$$

where h_1, h_2, and h_3 are some functions of coordinates. By examining the primary line elements in Cartesian, spherical, and cylindrical coordinates, we can come up with Table 16.1.

Denoting the unit vectors in curvilinear coordinate systems by $\hat{\mathbf{e}}_1$, $\hat{\mathbf{e}}_2$, and $\hat{\mathbf{e}}_3$, we can combine all the equations for the elements of length and write them as a single vector equation:

$$d\mathbf{r} = \vec{dl} = \hat{\mathbf{e}}_1 dl_1 + \hat{\mathbf{e}}_2 dl_2 + \hat{\mathbf{e}}_3 dl_3 = \hat{\mathbf{e}}_1 h_1 dq_1 + \hat{\mathbf{e}}_2 h_2 dq_2 + \hat{\mathbf{e}}_3 h_3 dq_3. \quad (16.1)$$

---

[1] As will be seen shortly, Cartesian coordinates are also included in such curvilinear coordinates. The former have lines (and planes) as their primary lengths and surfaces, thus the word "linear" in the name of the latter.

Curvilinear	Cartesian	Spherical	Cylindrical
q_1	x	r	ρ
q_2	y	θ	φ
q_3	z	φ	z
h_1	1	1	1
h_2	1	r	ρ
h_3	1	$r\sin\theta$	1

Table 16.1: The specifications of the three coordinate systems in terms of curvilinear coordinates.

This equation is useful in its own right. For example, we can obtain the curvilinear unit vectors as follows. Rewrite Equation (16.1) in terms of increments:

$$\Delta\mathbf{r} \approx \hat{\mathbf{e}}_1 h_1 \Delta q_1 + \hat{\mathbf{e}}_2 h_2 \Delta q_2 + \hat{\mathbf{e}}_3 h_3 \Delta q_3.$$

Keeping q_2 and q_3 constant (so that $\Delta q_2 = 0 = \Delta q_3$), divide both sides by Δq_1 to obtain

$$\frac{\Delta\mathbf{r}}{\Delta q_1} \approx \hat{\mathbf{e}}_1 h_1.$$

In the limit, the LHS becomes a partial derivative and we get

$$\hat{\mathbf{e}}_1 = \frac{1}{h_1}\frac{\partial\mathbf{r}}{\partial q_1}. \tag{16.2}$$

The other two unit vectors can be obtained similarly. We thus have

> **Box 16.1.1.** *The ith unit vector of a curvilinear coordinate system is given by*
> $$\hat{\mathbf{e}}_i = \frac{1}{h_i}\frac{\partial\mathbf{r}}{\partial q_i}, \quad i = 1, 2, 3. \tag{16.3}$$

This is a useful formula for obtaining the Cartesian components of curvilinear unit vectors, when the Cartesian components of the position vector are given in terms of curvilinear coordinates.

Example 16.1.1. As an illustration of the above procedure, we calculate the unit vectors in spherical coordinates. First we write

$$\mathbf{r} = x\hat{\mathbf{e}}_x + y\hat{\mathbf{e}}_y + z\hat{\mathbf{e}}_z = \hat{\mathbf{e}}_x r\sin\theta\cos\varphi + \hat{\mathbf{e}}_y r\sin\theta\sin\varphi + \hat{\mathbf{e}}_z r\cos\theta.$$

Now we differentiate with respect to r to get

$$\hat{\mathbf{e}}_1 \equiv \hat{\mathbf{e}}_r = \frac{\partial\mathbf{r}}{\partial r} = \hat{\mathbf{e}}_x \sin\theta\cos\varphi + \hat{\mathbf{e}}_y \sin\theta\sin\varphi + \hat{\mathbf{e}}_z \cos\theta.$$

Similarly,

$$\hat{\mathbf{e}}_2 \equiv \hat{\mathbf{e}}_\theta = \frac{1}{r}\frac{\partial \mathbf{r}}{\partial \theta} = \hat{\mathbf{e}}_x \cos\theta \cos\varphi + \hat{\mathbf{e}}_y \cos\theta \sin\varphi - \hat{\mathbf{e}}_z \sin\theta,$$

$$\hat{\mathbf{e}}_3 \equiv \hat{\mathbf{e}}_\varphi = \frac{1}{r\sin\theta}\frac{\partial \mathbf{r}}{\partial \varphi} = -\hat{\mathbf{e}}_x \sin\varphi + \hat{\mathbf{e}}_y \cos\varphi,$$

where we have used Table 16.1. These are the results we obtained in Chapter 1 from purely geometric arguments. ∎

We are now in a position to find the gradient, divergence, and curl of a vector field in general curvilinear coordinates. Once these are obtained, finding their specific forms in cylindrical and spherical coordinates entails simply substituting the appropriate expressions for q_1, q_2, and q_3 and h_1, h_2, and h_3.

16.2   The Gradient

The gradient is found by equating

$$df = \frac{\partial f}{\partial q_1}\,dq_1 + \frac{\partial f}{\partial q_2}\,dq_2 + \frac{\partial f}{\partial q_3}\,dq_3$$

to the differential of f in terms of the gradient:

$$df = \boldsymbol{\nabla} f \cdot d\mathbf{r} = (\boldsymbol{\nabla} f)_1 h_1\,dq_1 + (\boldsymbol{\nabla} f)_2 h_2\,dq_2 + (\boldsymbol{\nabla} f)_3 h_3\,dq_3.$$

The last two equations yield

$$(\boldsymbol{\nabla} f)_1 h_1 = \frac{\partial f}{\partial q_1}, \quad (\boldsymbol{\nabla} f)_2 h_2 = \frac{\partial f}{\partial q_2}, \quad (\boldsymbol{\nabla} f)_3 h_3 = \frac{\partial f}{\partial q_3},$$

which gives

gradient in curvilinear coordinates

> **Box 16.2.1.** *The gradient of a function f in a curvilinear coordinate system is given by*
>
> $$\boldsymbol{\nabla} f = \hat{\mathbf{e}}_1 \frac{1}{h_1}\frac{\partial f}{\partial q_1} + \hat{\mathbf{e}}_2 \frac{1}{h_2}\frac{\partial f}{\partial q_2} + \hat{\mathbf{e}}_3 \frac{1}{h_3}\frac{\partial f}{\partial q_3}. \tag{16.4}$$

This result, in conjunction with Table 16.1, agrees with the expression obtained for the gradient in the Cartesian coordinate system. In cylindrical coordinates, we obtain

$$\boldsymbol{\nabla} f = \hat{\mathbf{e}}_\rho \frac{\partial f}{\partial \rho} + \hat{\mathbf{e}}_\varphi \frac{1}{\rho}\frac{\partial f}{\partial \varphi} + \hat{\mathbf{e}}_z \frac{\partial f}{\partial z}, \tag{16.5}$$

so that the operator ∇ in cylindrical coordinates is given by

$$\nabla = \hat{e}_\rho \frac{\partial}{\partial \rho} + \hat{e}_\varphi \frac{1}{\rho} \frac{\partial}{\partial \varphi} + \hat{e}_z \frac{\partial}{\partial z}. \tag{16.6}$$

gradient of a
function in
spherical
coordinates

Similarly, in spherical coordinates, we get

$$\nabla f = \hat{e}_r \frac{\partial f}{\partial r} + \hat{e}_\theta \frac{1}{r} \frac{\partial f}{\partial \theta} + \hat{e}_\varphi \frac{1}{r \sin \theta} \frac{\partial f}{\partial \varphi} \tag{16.7}$$

with the operator ∇ given by

$$\nabla = \hat{e}_r \frac{\partial}{\partial r} + \hat{e}_\theta \frac{1}{r} \frac{\partial}{\partial \theta} + \hat{e}_\varphi \frac{1}{r \sin \theta} \frac{\partial}{\partial \varphi}. \tag{16.8}$$

Example 16.2.1. The electrostatic potential of an electric dipole was given in Example 10.5.1 in spherical coordinates. With the expression for the gradient given above, we can find the electric field $\mathbf{E} = -\nabla \Phi$ of a dipole in spherical coordinates:

$$E_r = -\frac{\partial \Phi_{\mathrm{dip}}}{\partial r} = -\frac{\partial}{\partial r} \left(\frac{k_e p \cos \theta}{r^2} \right) = \frac{2 k_e p \cos \theta}{r^3},$$

$$E_\theta = -\frac{1}{r} \frac{\partial \Phi_{\mathrm{dip}}}{\partial \theta} = -\frac{1}{r} \frac{\partial}{\partial \theta} \left(\frac{k_e p \cos \theta}{r^2} \right) = \frac{k_e p \sin \theta}{r^3},$$

$$E_\varphi = -\frac{1}{r \sin \theta} \frac{\partial \Phi_{\mathrm{dip}}}{\partial \varphi} = -\frac{1}{r \sin \theta} \frac{\partial}{\partial \varphi} \left(\frac{k_e p \cos \theta}{r^2} \right) = 0.$$

electric field of an
electric dipole

Summarizing, we have

$$\mathbf{E}_{\mathrm{dip}} = \frac{k_e p}{r^3} (2 \hat{e}_r \cos \theta + \hat{e}_\theta \sin \theta). \tag{16.9}$$

This is the characteristic field of a dipole.   ∎

Example 16.2.2. Just as electric charges can produce electric dipoles, electric currents can produce magnetic dipoles. We saw this in Subsection 15.2. In this example, we will calculate the magnetic *field* of a dipole directly. Consider the magnetic field of a circular loop of current as given in Equations (4.24) and (4.26). We change the coordinates of the field point P to spherical and assume that P is far away from the loop, i.e., that a is very small compared to r. Writing r^2 for $\rho^2 + z^2$ and $r \sin \theta$ for ρ, we expand the integrands of (4.24) and (4.26) in powers of a/r keeping only the first nonzero power. Thus,

$$\frac{1}{(r^2 + a^2 - 2ra \sin \theta \cos t)^{3/2}} = \frac{1}{r^3} \left[1 + \left(\frac{a}{r} \right)^2 - 2 \left(\frac{a}{r} \right) \sin \theta \cos t \right]^{-3/2}$$

$$= \frac{1}{r^3} \left[1 + 3 \left(\frac{a}{r} \right) \sin \theta \cos t \right] + \cdots,$$

$$\frac{r \sin \theta \cos t - a}{(r^2 + a^2 - 2ra \sin \theta \cos t)^{3/2}} = \frac{1}{r^2} \left(\sin \theta \cos t - \frac{a}{r} \right) \left[1 + \left(\frac{a}{r} \right)^2 - 2 \left(\frac{a}{r} \right) \sin \theta \cos t \right]^{-3/2}$$

$$= \frac{1}{r^2} \left(\sin \theta \cos t - \frac{a}{r} \right) \left[1 + 3 \left(\frac{a}{r} \right) \sin \theta \cos t \right] + \cdots$$

$$= \frac{1}{r^2} \left(\sin \theta \cos t - \frac{a}{r} + \frac{3a}{r} \sin^2 \theta \cos^2 t \right).$$

Substituting these in the integrals of (4.24) and (4.26) yields

$$B_\rho = \frac{k_m I a z}{r^3} \int_0^{2\pi} \cos t \left(1 + \frac{3a}{r} \sin\theta \cos t\right) dt = \frac{3 k_m I \pi a^2 \cos\theta \sin\theta}{r^3},$$

where we substituted $r\cos\theta$ for z. In an analogous way, we also obtain

$$B_z = -\frac{k_m I a}{r^2} \int_0^{2\pi} \left(\sin\theta\cos t - \frac{a}{r} + \frac{3a}{r}\sin^2\theta\cos^2 t\right) dt$$

$$= -\frac{k_m I a}{r^2} \left(-\frac{2\pi a}{r} + \frac{3a\pi}{r}\sin^2\theta\right).$$

We are interested in the spherical components of the magnetic field. To find these components, we first write

$$\mathbf{B} = B_\rho \hat{\mathbf{e}}_\rho + B_z \hat{\mathbf{e}}_z$$

and take the dot product with appropriate unit vectors:

$$B_r = \mathbf{B} \cdot \hat{\mathbf{e}}_r = B_\rho \hat{\mathbf{e}}_\rho \cdot \hat{\mathbf{e}}_r + B_z \hat{\mathbf{e}}_z \cdot \hat{\mathbf{e}}_r = B_\rho \sin\theta + B_z \cos\theta$$

$$= \frac{3 k_m I \pi a^2 \cos\theta\sin\theta}{r^3}\sin\theta + \frac{k_m I a}{r^2}\left(\frac{2\pi a}{r} - \frac{3a\pi}{r}\sin^2\theta\right)\cos\theta$$

$$= \frac{2 k_m I \pi a^2}{r^3}\cos\theta.$$

Similarly,

$$B_\theta = \mathbf{B} \cdot \hat{\mathbf{e}}_\theta = B_\rho \hat{\mathbf{e}}_\rho \cdot \hat{\mathbf{e}}_\theta + B_z \hat{\mathbf{e}}_z \cdot \hat{\mathbf{e}}_\theta = B_\rho \cos\theta - B_z \sin\theta$$

$$= \frac{3 k_m I \pi a^2 \cos\theta\sin\theta}{r^3}\cos\theta - \frac{k_m I a}{r^2}\left(\frac{2\pi a}{r} - \frac{3a\pi}{r}\sin^2\theta\right)\sin\theta$$

$$= \frac{k_m I \pi a^2}{r^3}\sin\theta.$$

Summarizing, we write

magnetic field of a
magnetic dipole

$$\mathbf{B} = \frac{k_m I \pi a^2}{r^3}(2\hat{\mathbf{e}}_r \cos\theta + \hat{\mathbf{e}}_\theta \sin\theta). \qquad (16.10)$$

This has a striking resemblance to Equation (16.9). In fact once we identify $I\pi a^2$ as the magnetic dipole of the loop, and change all magnetic labels to electric ones, we recover Equation (16.9). ∎

16.3   The Divergence

To find the divergence of a vector $\mathbf{A}$, we consider the volume element of Figure 16.1 and find the outward flux through the sides of the volume. For the front face we have

$$\Delta\phi_f = \mathbf{A}_f \cdot \hat{\mathbf{e}}_1 \Delta a_f,$$

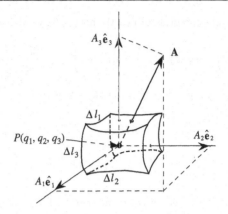

Figure 16.1: Point P and the surrounding volume element in curvilinear coordinates. Note that the midpoints of the front and back faces are $\Delta q_1/2$ away from P in the positive and negative $\hat{\mathbf{e}}_1$ directions, respectively. Similarly for the other four faces.

where $\mathbf{A}_f$ means the value of $\mathbf{A}$ at the center of the front face and Δa_f is the area of the front face. Following the arguments presented for the Cartesian case, we write

$$\Delta\phi_f \approx \mathbf{A}_f \cdot \hat{\mathbf{e}}_1 \Delta a_f = A_{1f}\Delta l_{2f}\Delta l_{3f}$$
$$= A_{1f}(h_2\Delta q_2)_f(h_3\Delta q_3)_f = A_{1f}h_{2f}h_{3f}\Delta q_2\Delta q_3$$

The subscript 1 in A_{1f}, for example, means component of $\mathbf{A}$ in the direction of the first coordinate. The subscript f implies evaluation—at the midpoint—on the front side whose second and third coordinates are the same as P, and whose first coordinate is $q_1 + \Delta q_1/2$. Thus, we have

$$\Delta\phi_f \approx A_1\left(q_1 + \frac{\Delta q_1}{2}, q_2, q_3\right) h_2\left(q_1 + \frac{\Delta q_1}{2}, q_2, q_3\right)$$
$$\times h_3\left(q_1 + \frac{\Delta q_1}{2}, q_2, q_3\right)\Delta q_2\Delta q_3$$

because, unlike the Cartesian case, h_1, h_2, and h_3 are functions of the coordinates. Using Taylor series expansion for the functions A_1, h_2, and h_3 yields

$$\Delta\phi_f \approx \left\{A_1(q_1, q_2, q_3) + \frac{\Delta q_1}{2}\frac{\partial A_1}{\partial q_1}\right\}\left\{h_2(q_1, q_2, q_3) + \frac{\Delta q_1}{2}\frac{\partial h_2}{\partial q_1}\right\}$$
$$\times \left\{h_3(q_1, q_2, q_3) + \frac{\Delta q_1}{2}\frac{\partial h_3}{\partial q_1}\right\}\Delta q_2\Delta q_3.$$

Multiplying out and keeping terms up to the third order (corresponding to the order of a volume element by which we shall divide shortly), we obtain

$$\Delta\phi_f \approx \left\{ A_1 h_2 h_3 + A_1 h_2 \frac{\partial h_3}{\partial q_1} + A_1 h_3 \frac{\partial h_2}{\partial q_1} + h_2 h_3 \frac{\partial A_1}{\partial q_1} \right\} \frac{\Delta q_1}{2} \Delta q_2 \Delta q_3$$

$$= \left\{ A_1 h_2 h_3 + \frac{\partial}{\partial q_1} (h_2 h_3 A_1) \right\} \frac{\Delta q_1}{2} \Delta q_2 \Delta q_3,$$

where we left out the explicit dependence of the functions on their independent coordinate variables. For the back face we have

$$\Delta\phi_b \approx \mathbf{A}_b \cdot (-\hat{\mathbf{e}}_1 \Delta a_b) = -A_{1b} \Delta l_{2b} \Delta l_{3b} = -A_{1b} (h_2 \Delta q_2)_b (h_3 \Delta q_3)_b$$

$$= -A_1 \left(q_1 - \frac{\Delta q_1}{2}, q_2, q_3 \right) h_2 \left(q_1 - \frac{\Delta q_1}{2}, q_2, q_3 \right)$$

$$\times h_3 \left(q_1 - \frac{\Delta q_1}{2}, q_2, q_3 \right) \Delta q_2 \Delta q_3.$$

Taylor expanding the three functions A_1, h_2, and h_3 as above, and multiplying out yields

$$\Delta\phi_b \approx -\left\{ A_1 h_2 h_3 - \frac{\partial}{\partial q_1} (h_2 h_3 A_1) \right\} \frac{\Delta q_1}{2} \Delta q_2 \Delta q_3.$$

Adding the front and back contributions, we obtain

$$\Delta\phi_1 \equiv \Delta\phi_f + \Delta\phi_b \approx \frac{\partial}{\partial q_1} (h_2 h_3 A_1) \, \Delta q_1 \Delta q_2 \Delta q_3.$$

Similarly, the fluxes through the faces perpendicular to $\hat{\mathbf{e}}_2$ and $\hat{\mathbf{e}}_3$ are

$$\Delta\phi_2 \approx \frac{\partial}{\partial q_2} (h_1 h_3 A_2) \, \Delta q_1 \Delta q_2 \Delta q_3,$$

$$\Delta\phi_3 \approx \frac{\partial}{\partial q_3} (h_1 h_2 A_3) \, \Delta q_1 \Delta q_2 \Delta q_3. \qquad (16.11)$$

Adding the three contributions and dividing by the volume

$$\Delta V = \Delta l_1 \Delta l_2 \Delta l_3 = h_1 h_2 h_3 \Delta q_1 \Delta q_2 \Delta q_3$$

and finally taking the limit of smaller and smaller volumes—which turns all approximations into equalities—we get

divergence in curvilinear coordinates

Theorem 16.3.1. *The divergence of a vector field* $\mathbf{A}$ *in a curvilinear coordinate system is given by*

$$\boldsymbol{\nabla} \cdot \mathbf{A} = \frac{1}{h_1 h_2 h_3} \left\{ \frac{\partial}{\partial q_1} (h_2 h_3 A_1) + \frac{\partial}{\partial q_2} (h_1 h_3 A_2) + \frac{\partial}{\partial q_3} (h_1 h_2 A_3) \right\}.$$

Now that we have a general formula for the divergence, we can use Table 16.1 to write the divergence in a specific coordinate system. For instance, substituting the entries of the second column gives the formula in Theorem 13.2.1, and the third column yields

divergence of a
vector field in
spherical
coordinates

$$\nabla \cdot \mathbf{A} = \frac{1}{r^2 \sin \theta} \left\{ \frac{\partial}{\partial r} \left(r^2 \sin \theta A_r \right) + \frac{\partial}{\partial \theta} \left(r \sin \theta A_\theta \right) + \frac{\partial}{\partial \varphi} \left(r A_\varphi \right) \right\}$$

$$= \frac{1}{r^2} \frac{\partial}{\partial r} \left(r^2 A_r \right) + \frac{1}{r \sin \theta} \left\{ \frac{\partial}{\partial \theta} \left(\sin \theta A_\theta \right) + \frac{\partial A_\varphi}{\partial \varphi} \right\}. \qquad (16.12)$$

To obtain the divergence in cylindrical coordinates, we use the last column and obtain

$$\nabla \cdot \mathbf{A} = \frac{1}{\rho} \left\{ \frac{\partial}{\partial \rho} \left(\rho A_\rho \right) + \frac{\partial}{\partial \varphi} \left(A_\varphi \right) + \frac{\partial}{\partial z} \left(\rho A_z \right) \right\}$$

$$= \frac{1}{\rho} \frac{\partial}{\partial \rho} \left(\rho A_\rho \right) + \frac{1}{\rho} \frac{\partial A_\varphi}{\partial \varphi} + \frac{\partial A_z}{\partial z}. \qquad (16.13)$$

Example 16.3.2. Consider the vector field defined by

$$\mathbf{A} = k r^\alpha \hat{\mathbf{e}}_r,$$

where k and α are constants. Let us verify the divergence theorem for a spherical surface of radius R (see Figure 16.2). The total flux is obtained by integrating over the surface of the sphere:

$$\phi = \iint_S \mathbf{A} \cdot d\mathbf{a} = \iint_S k R^\alpha \hat{\mathbf{e}}_r \cdot \hat{\mathbf{e}}_n R^2 \sin \theta \, d\theta \, d\varphi$$

$$= k R^{\alpha+2} \iint_S \sin \theta \, d\theta \, d\varphi = 4\pi k R^{\alpha+2}.$$

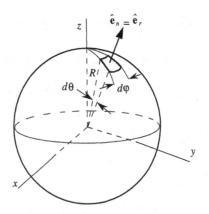

Figure 16.2: The element of area and its unit normal for a sphere.

On the other hand, using the expression for divergence in the spherical coordinate system and noting that $A_\theta = 0 = A_\varphi$, we obtain

$$\nabla \cdot \mathbf{A} = \frac{1}{r^2} \frac{\partial}{\partial r}(r^2 A_r) = \frac{1}{r^2} \frac{d}{dr}\left(kr^{\alpha+2}\right) = (\alpha + 2)kr^{\alpha-1},$$

where we have assumed that $\alpha \neq -2$. Therefore,

$$\iiint_V \nabla \cdot \mathbf{A}\, dV = \int_0^R (\alpha+2)kr^{\alpha-1}r^2\, dr \int_0^\pi \sin\theta\, d\theta \int_0^{2\pi} d\varphi = 4\pi k R^{\alpha+2}$$

which agrees with the surface integration.

For $\alpha = -2$ the divergence appears to vanish everywhere. However, a closer examination reveals that the statement is true only if $r \neq 0$. In fact, as we discussed before, the divergence of $\mathbf{A}$ is proportional to the Dirac delta function, $\delta(\mathbf{r})$ in this case [see Equation (15.2)]. ∎

16.4 The Curl

To calculate the curl, we choose a closed path perpendicular to one of the unit vectors, say $\hat{\mathbf{e}}_1$ and calculate the line integral of $\mathbf{A}$ around it. The situation is depicted in Figure 16.3. We calculate the contribution to the line integral from path (1) in detail and leave calculation of contributions from the remaining three paths to the reader. In all calculations, terms of higher order than the second will be omitted

$$\int_{(1)} \mathbf{A} \cdot d\mathbf{r} \approx \mathbf{A}_l \cdot \Delta\mathbf{r}_l = \mathbf{A}_l \cdot (-\hat{\mathbf{e}}_3 \Delta l_l) = -A_{3l}\Delta l_l = -A_{3l}h_{3l}\Delta q_3$$

$$= -A_3\left(q_1, q_2 - \frac{\Delta q_2}{2}, q_3\right)h_3\left(q_1, q_2 - \frac{\Delta q_2}{2}, q_3\right)\Delta q_3$$

$$= -\left\{A_3 - \frac{\Delta q_2}{2}\frac{\partial A_3}{\partial q_2}\right\}\left\{h_3 - \frac{\Delta q_2}{2}\frac{\partial h_3}{\partial q_2}\right\}\Delta q_3$$

$$\approx -A_3 h_3 \Delta q_3 + \frac{\partial}{\partial q_2}(h_3 A_3)\frac{\Delta q_2}{2}\Delta q_3.$$

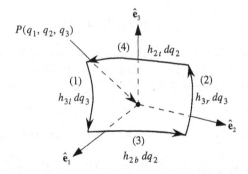

Figure 16.3: Path of integration for the first component of the curl of **A** in curvilinear coordinates.

Following similar steps, the reader may check that

$$\int_{(2)} \mathbf{A} \cdot d\mathbf{r} \approx A_3 h_3 \Delta q_3 + \frac{\partial}{\partial q_2}(h_3 A_3) \frac{\Delta q_2}{2} \Delta q_3,$$

$$\int_{(3)} \mathbf{A} \cdot d\mathbf{r} \approx A_2 h_2 \Delta q_2 - \frac{\partial}{\partial q_3}(h_2 A_2) \frac{\Delta q_3}{2} \Delta q_2, \qquad (16.14)$$

$$\int_{(4)} \mathbf{A} \cdot d\mathbf{r} \approx -A_2 h_2 \Delta q_2 - \frac{\partial}{\partial q_3}(h_2 A_2) \frac{\Delta q_3}{2} \Delta q_2.$$

Summing up all these contributions, we obtain

$$\oint \mathbf{A} \cdot d\mathbf{r} \approx \left\{ \frac{\partial}{\partial q_2}(h_3 A_3) - \frac{\partial}{\partial q_3}(h_2 A_2) \right\} \Delta q_2 \Delta q_3.$$

Dividing this by the area enclosed by the path

$$\Delta a = \Delta l_2 \Delta l_3 = h_2 h_3 \Delta q_2 \Delta q_3$$

we obtain the first component, the component along the unit normal to the area:

$$(\mathbf{\nabla} \times \mathbf{A})_1 = \frac{1}{h_2 h_3} \left\{ \frac{\partial}{\partial q_2}(h_3 A_3) - \frac{\partial}{\partial q_3}(h_2 A_2) \right\}.$$

Corresponding expressions for the other two components of the curl can be found by proceeding as above. We can put all of the components together in a mnemonic determinant form:

curl in curvilinear coordinates

Theorem 16.4.1. *The curl of a vector field* $\mathbf{A}$ *in a curvilinear coordinate system is given by*

$$\mathbf{\nabla} \times \mathbf{A} = \frac{1}{h_1 h_2 h_3} \begin{vmatrix} \hat{\mathbf{e}}_1 h_1 & \hat{\mathbf{e}}_2 h_2 & \hat{\mathbf{e}}_3 h_3 \\ \frac{\partial}{\partial q_1} & \frac{\partial}{\partial q_2} & \frac{\partial}{\partial q_3} \\ h_1 A_1 & h_2 A_2 & h_3 A_3 \end{vmatrix}. \qquad (16.15)$$

warning! $\mathbf{\nabla} \times \mathbf{A}$ **is** *not* **a cross product in general curvilinear coordinates!**

Note that $\mathbf{\nabla} \times \mathbf{A}$ is *not* a cross product (except in Cartesian coordinates), but a vector defined by the determinant on the RHS of (16.15).

If we substitute the appropriate values for h's and q's in spherical coordinates, we obtain

$$\mathbf{\nabla} \times \mathbf{A} = \frac{1}{r^2 \sin\theta} \begin{vmatrix} \hat{\mathbf{e}}_r & \hat{\mathbf{e}}_\theta r & \hat{\mathbf{e}}_\varphi r\sin\theta \\ \frac{\partial}{\partial r} & \frac{\partial}{\partial \theta} & \frac{\partial}{\partial \varphi} \\ A_r & rA_\theta & r\sin\theta A_\varphi \end{vmatrix}. \qquad (16.16)$$

In cylindrical coordinates we get

$$\nabla \times \mathbf{A} = \frac{1}{\rho} \begin{vmatrix} \hat{\mathbf{e}}_\rho & \hat{\mathbf{e}}_\varphi \rho & \hat{\mathbf{e}}_z \\ \dfrac{\partial}{\partial \rho} & \dfrac{\partial}{\partial \varphi} & \dfrac{\partial}{\partial z} \\ A_\rho & \rho A_\varphi & A_z \end{vmatrix}. \tag{16.17}$$

Example 16.4.2. We have already calculated the magnetic field of a dipole in Example 16.2.2. Here we want to obtain the same result using the vector potential of a dipole given in Equation (15.12). We take $\boldsymbol{\mu}$ to be along the z-axis. Then

$$\boldsymbol{\mu} = \mu \hat{\mathbf{e}}_z = \mu(\hat{\mathbf{e}}_r \cos\theta - \hat{\mathbf{e}}_\theta \sin\theta)$$

and

$$\boldsymbol{\mu} \times \hat{\mathbf{e}}_r = \mu(-\sin\theta \hat{\mathbf{e}}_\theta \times \hat{\mathbf{e}}_r) = \mu \sin\theta \hat{\mathbf{e}}_\varphi.$$

Therefore,

$$\mathbf{B} = \nabla \times \mathbf{A} = \nabla \times \left(\frac{k_m \boldsymbol{\mu} \times \hat{\mathbf{e}}_r}{r^2} \right) = \nabla \times \left(\frac{k_m \mu \sin\theta \hat{\mathbf{e}}_\varphi}{r^2} \right)$$

$$= \frac{k_m \mu}{r^2 \sin\theta} \begin{vmatrix} \hat{\mathbf{e}}_r & \hat{\mathbf{e}}_\theta r & \hat{\mathbf{e}}_\varphi r \sin\theta \\ \dfrac{\partial}{\partial r} & \dfrac{\partial}{\partial \theta} & \dfrac{\partial}{\partial \varphi} \\ 0 & 0 & r\sin\theta \dfrac{\sin\theta}{r^2} \end{vmatrix} = \frac{k_m \mu}{r^2 \sin\theta} \begin{vmatrix} \hat{\mathbf{e}}_r & \hat{\mathbf{e}}_\theta r & \hat{\mathbf{e}}_\varphi r \sin\theta \\ \dfrac{\partial}{\partial r} & \dfrac{\partial}{\partial \theta} & \dfrac{\partial}{\partial \varphi} \\ 0 & 0 & \dfrac{\sin^2\theta}{r} \end{vmatrix}$$

$$= \frac{k_m \mu}{r^2 \sin\theta} \left[\hat{\mathbf{e}}_r \left(\frac{2\sin\theta\cos\theta}{r} \right) - r\hat{\mathbf{e}}_\theta \left(-\frac{\sin^2\theta}{r^2} \right) \right] = \frac{k_m \mu}{r^3} (2\cos\theta \hat{\mathbf{e}}_r + \sin\theta \hat{\mathbf{e}}_\theta),$$

which is the expression obtained in Example 16.2.2. ∎

Example 16.4.3. Consider the vector field $\mathbf{B}$ described in cylindrical coordinates as

$$\mathbf{B} = \frac{k}{\rho} \hat{\mathbf{e}}_\varphi,$$

where k is a constant. The curl of $\mathbf{B}$ is easily found to be zero:

$$\nabla \times \mathbf{B} = \frac{1}{\rho} \begin{vmatrix} \hat{\mathbf{e}}_\rho & \hat{\mathbf{e}}_\varphi \rho & \hat{\mathbf{e}}_z \\ \dfrac{\partial}{\partial \rho} & \dfrac{\partial}{\partial \varphi} & \dfrac{\partial}{\partial z} \\ 0 & \rho(k/\rho) & 0 \end{vmatrix} = 0.$$

However, for any circle (of radius a, for example) centered at the origin and located in the xy-plane, we get[2]

$$\oint_C \mathbf{B} \cdot d\mathbf{r} = \int_0^{2\pi} \frac{k}{a} \hat{\mathbf{e}}_\varphi \cdot (\hat{\mathbf{e}}_\varphi a \, d\varphi) = 2\pi k \neq 0.$$

---

[2]See also Example 14.3.3 which discusses this same vector field in Cartesian coordinates.

The reason for this result is that the circle is *not* contractible to zero: At the origin—which is inside the circle and at which $\rho = 0$—**B** is not defined.

This vector field should look familiar. It is the magnetic field due to a long straight wire carrying a current along the z-axis. According to *Ampère's circuital law*, the line integral of **B** along any closed curve encircling the wire, such as the above circle, gives, up to a multiplicative constant, the current in the wire, and this current is not zero. ∎

central force fields are conservative

Example 16.4.4. A vector field that can be written as

$$\mathbf{F} = f(r)\mathbf{r},$$

where r is the displacement vector from the origin, is conservative. It is instructive to show this using both Cartesian and spherical coordinate systems.

First, in Cartesian coordinates

$$\mathbf{F} = xf(r)\hat{\mathbf{e}}_x + yf(r)\hat{\mathbf{e}}_y + zf(r)\hat{\mathbf{e}}_z$$

and the curl is

$$\boldsymbol{\nabla} \times \mathbf{F} = \begin{vmatrix} \hat{\mathbf{e}}_x & \hat{\mathbf{e}}_y & \hat{\mathbf{e}}_z \\ \dfrac{\partial}{\partial x} & \dfrac{\partial}{\partial y} & \dfrac{\partial}{\partial z} \\ xf & yf & zf \end{vmatrix} = \hat{\mathbf{e}}_x\left\{ \frac{\partial}{\partial y}(zf) - \frac{\partial}{\partial z}(yf) \right\}$$

$$+ \hat{\mathbf{e}}_y\left\{ \frac{\partial}{\partial z}(xf) - \frac{\partial}{\partial x}(zf) \right\} + \hat{\mathbf{e}}_z\left\{ \frac{\partial}{\partial x}(yf) - \frac{\partial}{\partial y}(xf) \right\}.$$

Concentrating on the x-component first and using the chain rule, we have

$$\frac{\partial}{\partial y}(zf) = z\frac{\partial f}{\partial y} = z\frac{df}{dr}\frac{\partial r}{\partial y} = zf'\frac{\partial r}{\partial y}.$$

But

$$\frac{\partial r}{\partial y} = \frac{\partial}{\partial y}\sqrt{x^2 + y^2 + z^2} = \frac{y}{r}.$$

Thus,

$$\frac{\partial}{\partial y}(zf) = yzf'.$$

Similarly,

$$\frac{\partial}{\partial z}(yf) = yzf'.$$

Therefore, the x-component of $\boldsymbol{\nabla} \times \mathbf{F}$ is zero. The y- and z-components can also be shown to be zero, and we get $\boldsymbol{\nabla} \times \mathbf{F} = 0$.

On the other hand, using spherical coordinates, we easily obtain

$$\boldsymbol{\nabla} \times \mathbf{F} = \frac{1}{r^2 \sin\theta}\begin{vmatrix} \hat{\mathbf{e}}_r & \hat{\mathbf{e}}_\theta r & \hat{\mathbf{e}}_\varphi r\sin\theta \\ \dfrac{\partial}{\partial r} & \dfrac{\partial}{\partial \theta} & \dfrac{\partial}{\partial \varphi} \\ rf(r) & 0 & 0 \end{vmatrix} = 0.$$

Obviously, the use of spherical coordinates simplifies the calculation considerably. ∎

The preceding example shows that

> **Box 16.4.1.** *Any* well-behaved *vector field whose magnitude is only a function of radial distance, r, and whose direction is along* **r** *is conservative. Such vector fields are generally known as* **central vector fields**.

16.4.1 The Laplacian

Combining divergence and the gradient gives the Laplacian. Using Equation (16.4) in Theorem 16.3.1, we get

Laplacian in curvilinear coordinates

Theorem 16.4.5. *The Laplacian of a function f is the divergence of gradient of f and—in a curvilinear coordinate system—is given by*

$$\nabla^2 f = \frac{1}{h_1 h_2 h_3} \left\{ \frac{\partial}{\partial q_1} \left(\frac{h_2 h_3}{h_1} \frac{\partial f}{\partial q_1} \right) + \frac{\partial}{\partial q_2} \left(\frac{h_1 h_3}{h_2} \frac{\partial f}{\partial q_2} \right) + \frac{\partial}{\partial q_3} \left(\frac{h_1 h_2}{h_3} \frac{\partial f}{\partial q_3} \right) \right\}.$$

For cylindrical coordinates the Laplacian is

$$\nabla^2 f = \frac{1}{\rho} \frac{\partial}{\partial \rho} \left(\rho \frac{\partial f}{\partial \rho} \right) + \frac{1}{\rho^2} \frac{\partial^2 f}{\partial \varphi^2} + \frac{\partial^2 f}{\partial z^2} \tag{16.18}$$

and for spherical coordinates it is

$$\nabla^2 f = \frac{1}{r^2} \frac{\partial}{\partial r} \left(r^2 \frac{\partial f}{\partial r} \right) + \frac{1}{r^2 \sin \theta} \left\{ \frac{\partial}{\partial \theta} \left(\sin \theta \frac{\partial f}{\partial \theta} \right) + \frac{1}{\sin \theta} \frac{\partial^2 f}{\partial \varphi^2} \right\}. \tag{16.19}$$

Equations (16.7) and (16.19) allow us to write the angular momentum differential operator derived in Example 15.3.1 in spherical coordinates, which is the most common way of writing it. We note that

$$\frac{\partial}{\partial r} \left(r^2 \frac{\partial f}{\partial r} \right) = 2r \frac{\partial f}{\partial r} + r^2 \frac{\partial^2 f}{\partial r^2},$$

and

$$\mathbf{r} \cdot (\nabla f) = r \frac{\partial f}{\partial r},$$

and

$$(\mathbf{r} \cdot \nabla)^2 f = r \frac{\partial}{\partial r} \left(\frac{\partial f}{\partial r} \right) = r \frac{\partial f}{\partial r} + r^2 \frac{\partial^2 f}{\partial r^2}.$$

Substituting these plus (16.19) in (15.22) yields

$$L^2 f = -\frac{1}{\sin \theta} \left\{ \frac{\partial}{\partial \theta} \left(\sin \theta \frac{\partial f}{\partial \theta} \right) + \frac{1}{\sin \theta} \frac{\partial^2 f}{\partial \varphi^2} \right\}. \tag{16.20}$$

Therefore, the angular momentum operator *depends only on angles* in spherical coordinates.

16.5 Problems

16.1. The divergence of a vector can be obtained in any coordinate system by brute force calculation. In this problem you are asked to find $\nabla \cdot \mathbf{A}$ in cylindrical coordinates.
(a) Express A_x in terms of cylindrical coordinates and components. Hint: Write $\mathbf{A}$ in cylindrical ccordinates and take the dot product with $\hat{\mathbf{e}}_x$ expressing everything in terms of cylindrical ccordinates.
(b) Use the chain rule

$$\frac{\partial A_x}{\partial x} = \frac{\partial A_x}{\partial \rho} \frac{\partial \rho}{\partial x} + \frac{\partial A_x}{\partial \varphi} \frac{\partial \varphi}{\partial x} + \frac{\partial A_x}{\partial z} \frac{\partial z}{\partial x}$$

where A_x is what you found in (a).
(c) Do the same with A_y and A_z, and add the three terms to obtain the divergence in cylindrical coordinates.

16.2. Find the divergence of a vector in spherical coordinates following the procedure outlined in Problem 16.1.

16.3. Find the gradient of a function in cylindrical and spherical coordinates following a procedure similar to the one outlined in Problem 16.1.

16.4. Find the curl of a vector in cylindrical and spherical coordinates following a procedure similar to the one outlined in Problem 16.1.

16.5. Start with the Laplacian in Cartesian coordinates.
(a) By using the chain rule and expressing the second derivatives in cylindrical coordinates, find the Laplacian in cylindrical coordinates.
(b) Do the same for spherical coordinates.

16.6. The **elliptic cylindrical coordinates** (u, θ, z) are given by

$$x = a \cosh u \cos \theta$$
$$y = a \sinh u \sin \theta$$
$$z = z$$

where a is a constant.
(a) What is the expression for the gradient of a function f in elliptic cylindrical coordinates?
(b) What is the expression for the divergence of a vector $\mathbf{A}$ in elliptic cylindrical coordinates?
(c) What is the expression for the curl of a vector $\mathbf{A}$ in elliptic cylindrical coordinates?
(d) What is the expression for the Laplacian of a function f in elliptic cylindrical coordinates?

16.7. The **prolate spheroidal coordinates** (u, θ, φ) are given by

$$x = a \sinh u \sin \theta \cos \varphi$$
$$y = a \sinh u \sin \theta \sin \varphi$$
$$z = a \cosh u \cos \theta$$

where a is a constant.

(a) What is the expression for the gradient of a function f in prolate spheroidal coordinates?

(b) What is the expression for the divergence of a vector $\mathbf{A}$ in prolate spheroidal coordinates?

(c) What is the expression for the curl of a vector $\mathbf{A}$ in prolate spheroidal coordinates?

(d) What is the expression for the Laplacian of a function f in prolate spheroidal coordinates?

16.8. The **toroidal coordinates** (u, θ, φ) are given by

$$x = \frac{a \sinh u \cos \varphi}{\cosh u - \cos \theta}$$
$$y = \frac{a \sinh u \sin \varphi}{\cosh \theta - \cos \theta}$$
$$z = \frac{a \sin u}{\cosh u - \cos \theta}$$

(a) What is the expression for the gradient of a function f in toroidal coordinates?

(b) What is the expression for the divergence of a vector $\mathbf{A}$ in toroidal coordinates?

(c) What is the expression for the curl of a vector $\mathbf{A}$ in toroidal coordinates?

(d) What is the expression for the Laplacian of a function f in toroidal coordinates?

16.9. The **paraboloidal coordinates** (u, v, φ) are given by

$$x = 2auv \cos \varphi$$
$$y = 2auv \sin \varphi$$
$$z = a(u^2 - v^2)$$

where a is a constant.

(a) What is the expression for the gradient of a function f in paraboloidal coordinates?

(b) What is the expression for the divergence of a vector $\mathbf{A}$ in paraboloidal coordinates?

(c) What is the expression for the curl of a vector $\mathbf{A}$ in paraboloidal coordinates?

(d) What is the expression for the Laplacian of a function f in paraboloidal coordinates?

16.10. The **three-dimensional bipolar coordinates** (u, θ, φ) are given by

$$x = \frac{a \sin\theta \cos\varphi}{\cosh u - \cos\theta}$$

$$y = \frac{a \sin\theta \sin\varphi}{\cosh u - \cos\theta}$$

$$z = \frac{a \sinh u}{\cosh u - \cos\theta}$$

(a) What is the expression for the gradient of a function f in three-dimensional bipolar coordinates?

(b) What is the expression for the divergence of a vector $\mathbf{A}$ in three-dimensional bipolar coordinates?

(c) What is the expression for the curl of a vector $\mathbf{A}$ in three-dimensional bipolar coordinates?

(d) What is the expression for the Laplacian of a function f in three-dimensional bipolar coordinates?

Chapter 17

Tensor Analysis

Our study of vectors in this part of the book has been limited to their analysis in specific coordinate systems, and although we touched on the general curvilinear coordinate system, our treatment aimed at orthogonal coordinates, and specifically at only three-dimensional spherical and cylindrical coordinate systems. Many situations in physics demand a three-fold generalization: non-orthogonal coordinate systems, higher-dimensional spaces, and objects, called tensors, whose components have more subscripts than one. This chapter is devoted to an analysis of tensors.

17.1   Vectors and Indices

Vector manipulations will be greatly simplified if equations are written in terms of a *general* component. How do we accomplish this? Start with a generic vector equation, which can be written as

$$\mathbf{U} = \mathbf{V},$$

where $\mathbf{U}$ and $\mathbf{V}$ are, in general, vector *expressions*. Examples of such an equation are

$$\mathbf{B} = \mathbf{\nabla} \times \mathbf{A}, \qquad \mathbf{E} = -\mathbf{\nabla}\Phi, \qquad \mathbf{A} = \int_a^b f(r)\hat{\mathbf{e}}_r \, dr.$$

You can also write each of these vector equations as three equations involving components. Thus, the foregoing generic equation becomes

$$U_x = V_x, \qquad U_y = V_y, \qquad U_z = V_z.$$

It is very helpful to convert letter indices into number indices. Let $x \to 1$, $y \to 2$, and $z \to 3$, and write[1]

$$U_1 = V_1, \qquad U_2 = V_2, \qquad U_3 = V_3.$$

---

[1] Note that the replacements here refer to *indices* not the Cartesian coordinates. The latter will have somewhat different symbols in the sequel.

These equations are abbreviated as

$$U_i = V_i, \qquad i = 1, 2, 3. \tag{17.1}$$

This is what we mean by an equation in terms of a general component: The index i refers to any one of the components of the vectors on either side of the equation. It is called a **free** index because it is free to take any one of the values between 1 and 3. An important property of a free index is that

free index defined

Box 17.1.1. *A **free index** appears once and only once on both sides of a vector equation.*

One can use any symbol to represent a free index, although the most common symbols used are i, j, k, l, m, and n. Thus, Equation (17.1) can be written in any one of the following alternative ways:

$$U_j = V_j, \qquad j = 1, 2, 3,$$
$$U_p = V_p, \qquad p = 1, 2, 3,$$
$$U_\heartsuit = V_\heartsuit, \qquad \heartsuit = 1, 2, 3.$$

indexed Cartesian coordinates

Of special interest are the components of the position vector $\mathbf{r}$. These are denoted by x_i rather than r_i. Thus, the vector relation $\mathbf{R} = \mathbf{r} - \mathbf{r}'$ is written as

$$X_j = x_j - x'_j, \qquad j = 1, 2, 3.$$

An abbreviation used for derivatives with respect to Cartesian coordinates (which coincide with the components of the position vector) is given as follows. First $\partial/\partial x$ is replaced by $\partial/\partial x_1$, and the latter by the much shorter notation, ∂_1. Similarly, $\partial/\partial y$ becomes ∂_2, and $\partial/\partial z$ becomes ∂_3. In particular, the general component of the gradient of a function f will be written as $\partial_k f$, $k =$

components of gradient

$1, 2, 3$.

All operations on vectors can be translated into the language of indexed relations. For example, $\mathbf{A} + \mathbf{B} = \mathbf{C}$ is equivalent to $A_k + B_k = C_k$, $k = 1, 2, 3$, and $\mathbf{A} = \alpha \mathbf{B}$ becomes $A_k = \alpha B_k$, $k = 1, 2, 3$, etc. The two operations of vector multiplication are a little more involved and we treat them separately in the following.

First let us consider the dot product. In terms of components, the dot product of $\mathbf{A}$ and $\mathbf{B}$ can be written as

$$\mathbf{A} \cdot \mathbf{B} = A_x B_x + A_y B_y + A_z B_z.$$

Converting to number indices, we get

$$\mathbf{A} \cdot \mathbf{B} = A_1 B_1 + A_2 B_2 + A_3 B_3 = \sum_{i=1}^{3} A_i B_i.$$

We now introduce a further simplification in notation due to Einstein, which gets rid of the clumsy summation sign:

Einstein summation convention

> **Box 17.1.2. (*Einstein Summation Convention*).** *Whenever an index is repeated, it is a dummy index and is summed from 1 to 3.*

Using this convention we write the dot product as

dot product

$$\mathbf{A} \cdot \mathbf{B} = A_i B_i. \tag{17.2}$$

No summation sign is needed as long as we remember that the *repeated* index i is summed over. Since the repeated index is a dummy index, we can change it to any other symbol. Thus,

$$\mathbf{A} \cdot \mathbf{B} = A_k B_k = A_j B_j = A_n B_n = A_\heartsuit B_\heartsuit = \cdots.$$

Example 17.1.1. In this example, we write some of the familiar vector relations in both vector form and component form:

$$\mathbf{E} = -\boldsymbol{\nabla}\Phi \qquad \Longleftrightarrow \qquad E_k = -\partial_k \Phi,$$

$$\boldsymbol{\nabla} \cdot \mathbf{A} \qquad \Longleftrightarrow \qquad \partial_j A_j,$$

$$\iint_S \mathbf{A} \cdot d\mathbf{a} = \iiint_V \boldsymbol{\nabla} \cdot \mathbf{A} \, dV \qquad \Longleftrightarrow \qquad \iint_S A_k \, da_k = \iiint_V \partial_j A_j \, dV,$$

$$\nabla^2 \Phi \qquad \Longleftrightarrow \qquad \partial_m \partial_m \Phi,$$

$$\boldsymbol{\nabla} \cdot (f\mathbf{A}) = \mathbf{A} \cdot \boldsymbol{\nabla} f + f \boldsymbol{\nabla} \cdot \mathbf{A} \qquad \Longleftrightarrow \qquad \partial_i (f A_i) = A_i \partial_i f + f \partial_i A_i.$$

The reader is urged to verify all these relations, remembering the Einstein summation convention. ∎

17.1.1 Transformation Properties of Vectors

Section 6.2.1 discussed the transformation of vectors, i.e., the way the components of a vector change when they are expressed in term of a new basis. To initiate the transformations relevant to the present chapter, let us begin with the position vector $\mathbf{r}$, which in one Cartesian coordinate system (with basis $\{\hat{\mathbf{e}}_1, \hat{\mathbf{e}}_2, \hat{\mathbf{e}}_3\}$) is represented by (x^1, x^2, x^3), and in another by $(\bar{x}^1, \bar{x}^2, \bar{x}^3)$. Here we are beginning to introduce new notation and terminology : instead of "vector space," we use "Cartesian coordinate system," and instead of subscripts, we use superscripts to label the coodinates.

coordinates with superscripts

Since both (x^1, x^2, x^3), and $(\bar{x}^1, \bar{x}^2, \bar{x}^3)$ are components of the same position vector, they are related via Equation (6.29):

$$\begin{aligned}
\bar{x}^1 &= a_{11}x^1 + a_{12}x^2 + a_{13}x^3, \\
\bar{x}^2 &= a_{21}x^1 + a_{22}x^2 + a_{23}x^3, \\
\bar{x}^3 &= a_{31}x^1 + a_{32}x^2 + a_{33}x^3.
\end{aligned} \tag{17.3}$$

In terms of a free index, we can rewrite this as

$$\bar{x}^i = a_{i1}x^1 + a_{i2}x^2 + a_{i3}x^3, \quad i = 1, 2, 3,$$

and using the summation notation

$$\bar{x}^i = \sum_{j=1}^{3} a_{ij}x^j, \quad i = 1, 2, 3.$$

Finally, using the Einstein summation convention and always keeping in mind that the free index i takes the values 1, 2, or 3, we come up with the following very succinct replacement for 17.3

$$\bar{x}^i = a_{ij}x^j. \tag{17.4}$$

Equations (17.3) and (17.4) are identical despite the enormous brevity of the latter.

As an application of the use of indices and summation convention, we conveniently express the rule of matrix multiplication, which we shall use frequently. Box 6.1.3 gives this rule. Let $C = AB$ be the product of A and B. Then the rule in Box 6.1.3 can be written as

$$c_{ij} = a_{ik}b_{kj}. \tag{17.5}$$

Notice that here we have two free indices i and j. The index k is being summed over on the right.

Of particular importance are transformations that leave the dot product intact. We called these transformations *orthogonal* (see Section 6.1.3). These orthogonal transformations satisfy Equation (6.20), which could be written in terms of indices. Noting that the ij-th element of the unit matrix is δ_{ij}, the familiar **Kronecker delta**, which as the reader may recall, is defined as

$$\delta_{ij} \equiv \begin{cases} 1 & \text{if } i = j, \\ 0 & \text{if } i \neq j, \end{cases} \tag{17.6}$$

we rewrite (6.20) as

$$\left(\tilde{A}\right)_{ik}(A)_{kj} = (1)_{ij} \quad \text{or} \quad a_{ki}a_{kj} = \delta_{ij}. \tag{17.7}$$

Kronecker delta in a sum   Now multiply both sides of (17.4) by a_{ik} and sum over i to get

$$a_{ik}\bar{x}^i = \underbrace{a_{ik}a_{ij}}_{=\delta_{kj}} x^j = x^k,$$

where in the last step we used the most important property of the Kronecker delta:

Thus the inverse of Equation (17.4) is

$$x^j = a_{ij}\bar{x}^i. \tag{17.8}$$

Note the difference in the position of the dummy index between this equation and (17.4).

Equations (17.4) and (17.8) give the transformation rules for the components of the position vector when one goes from one *Cartesian* coordinate system to another. It should be clear that the same transformation rules apply to the components of *any* vector, as long as one adheres to Cartesian coordinate systems. Thus if V_i and $\bar{V}_i$ represent the components of a vector **V** in two Cartesian coordinate systems, then

$$\bar{V}_i = a_{ij}V_j \qquad \text{and} \qquad V_j = a_{ij}\bar{V}_i. \tag{17.9}$$

In fact, it is customary to *define* vectors in terms of their transformation properties:

vectors defined in terms of their transformation property

Section 1.3 introduced the idea of expressing vectors in different coordinate systems, mainly Cartesian, cylindrical, and spherical. In all cases, care was taken to use *orthogonal* unit vectors. In fact, this has been the sole practice throughout the book so far, and for good reason: the dot product of two vectors—and hence length of a vector, defined as the square root of the dot product of the vector with itself—does not change when their components in one set of orthogonal unit vectors are written in terms of their components in another set of orthogonal unit vectors. This actually *defines* the orthogonal transformation of Section 6.1.3, and Equation (6.20) or (17.7) guarantees the invariance of the length of a vector.

Orthogonal transformations are not always the most suitable. As an example, consider a curve in space parametrized in a Cartesian coordinate system by $x^i = f_i(t)$, where $f_1(t)$, $f_2(t)$, and $f_3(t)$ are some smooth functions. The tangent to this curve—a vector—has components $\dot{x}^i \equiv dx^i/dt = f_i'(t)$. Now consider a new coordinate system, not necessarily Cartesian, given by

$$\bar{x}^i = g_i(x^1, x^2, x^3). \qquad (17.10)$$

The curve can be written in terms of the new coordinates by substituting $f_i(t)$ for each x^i:

$$\bar{x}^i = g_i(f_1(t), f_2(t), f_2(t)) \equiv h_i(t),$$

where the last identity defines the function $h_i(t)$. The components of the tangent to the curve in the new coordinate system are given by the chain rule:

$$\dot{\bar{x}}^i = h_i'(t) = \partial_1 g_i \frac{df_1}{dt} + \partial_2 g_i \frac{df_2}{dt} + \partial_3 g_i \frac{df_3}{dt} = \partial_1 g_i \dot{x}^1 + \partial_2 g_i \dot{x}^2 + \partial_3 g_i \dot{x}^3 = \partial_j g_i \dot{x}^j.$$

Recalling that $\partial_j g_i = \partial g_i / \partial x^j$ and that $g_i = \bar{x}^i$, this is usually written as

$$\dot{\bar{x}}^i = \frac{\partial \bar{x}^i}{\partial x^j} \dot{x}^j. \qquad (17.11)$$

It is instructive to see what happens if $\bar{x}^i$ is given by (17.4). In that case, we have

$$\frac{\partial \bar{x}^i}{\partial x^j} = \frac{\partial}{\partial x^j} \left(a_{ik} x^k \right) = a_{ik} \underbrace{\frac{\partial x^k}{\partial x^j}}_{=\delta_{kj}} = a_{ij}, \qquad (17.12)$$

where we have used an obvious property of partial derivative which is so useful that it is worth boxing it:

> **Box 17.1.5.** *If* $\{y_1, y_2, \ldots, y_m\}$ *are independent variables, then* $\partial y_i / \partial y_j = \delta_{ij}$.

Equation (17.12) shows that, when applied to *Cartesian* coordinate transformations, (17.11) is consistent with the definition of a vector as given in Box 17.1.4.

What about the inverse of (17.11)? Equation (17.10) can be treated as three equations in the three unknowns $\{x^1, x^2, x^3\}$. One can then solve these unknowns as functions of the independent variables $\{\bar{x}^1, \bar{x}^2, \bar{x}^3\}$. Whether or not one can actually solve (17.10) for $\{\bar{x}^i\}$ depends on the form of the functions $\{g_1, g_2, g_3\}$. If these functions satisfy certain (mild) mathematical properties, then Equation (17.10) is said to be **invertible** and each x^j can be written as a function of the independent variables $\{\bar{x}^i\}$. We assume that (17.10) is indeed invertible.

Treating x^j as dependent and $\{\bar{x}^i\}$ as independent variables, using the chain rule, and employing obvious notation, we can write

$$\dot{x}^j = \frac{dx^j}{dt} = \frac{\partial x^j}{\partial \bar{x}^1} \frac{d\bar{x}^1}{dt} + \frac{\partial x^j}{\partial \bar{x}^2} \frac{d\bar{x}^2}{dt} + \frac{\partial x^j}{\partial \bar{x}^3} \frac{d\bar{x}^3}{dt} = \frac{\partial x^j}{\partial \bar{x}^k} \frac{d\bar{x}^k}{dt} = \frac{\partial x^j}{\partial \bar{x}^k} \dot{\bar{x}}^k. \quad (17.13)$$

Is this consistent with Equation (17.11)? In other words, if we substitute $\dot{x}^j$ from this equation into the right-hand side of (17.11), do we get $\dot{\bar{x}}^i$? Let's try it!

$$RHS \text{ of } (17.11) = \frac{\partial \bar{x}^i}{\partial x^j}\dot{x}^j = \frac{\partial \bar{x}^i}{\partial x^j}\frac{\partial x^j}{\partial \bar{x}^k}\dot{\bar{x}}^k = \frac{\partial \bar{x}^i}{\partial \bar{x}^k}\dot{\bar{x}}^k = \delta_{ik}\dot{\bar{x}}^k = \dot{\bar{x}}^i,$$

where in the third equality we used the chain rule (2.16), in the fourth equality we used Box 17.1.5 as applied to the independent variables $\bar{x}^i$, and in the last equality we used Box 17.1.3. Thus, Equation (17.13) is indeed consistent with (17.11). It is tempting to call objects which transform according to (17.11) components of a vector. But before jumping to conclusions, let's look at another vector with which we are familiar.

17.1.2   Covariant and Contravariant Vectors

The gradient of a function was first defined in Section 12.3. It is a vector whose components are essentially derivatives of the function with respect to the coordinates. Because we are interested in the transformation properties of objects, we first have to clarify the notion of a *function*. A **scalar function** is a physical quantity, such as temperature, which takes on a single value at each point of space. Now, a point has an existence independent of any coordinate scalar function systems. Nevertheless, coordinates are useful for calculations. And if the point is described by (x^1, x^2, x^3) in a coordinate system, and ϕ denotes the scalar function, then we write $\phi(x^1, x^2, x^3)$ for the value of the scalar function at that point. The same point is described by $(\bar{x}^1, \bar{x}^2, \bar{x}^3)$ in another coordinate system, and the value of the scalar function in terms of the new coordinates is $\bar{\phi}(\bar{x}^1, \bar{x}^2, \bar{x}^3)$. It should be obvious that the form of the scalar function changes when one changes the coordinates. Thus the notation $\bar{\phi}$ instead of ϕ. Clearly,

$$\bar{\phi}(\bar{x}^1, \bar{x}^2, \bar{x}^3) = \phi(x^1, x^2, x^3). \tag{17.14}$$

Now differentiate both sides with respect to $\bar{x}^i$. The left side gives the ith component of the gradient of $\bar{\phi}$; and using the chain rule on the right side, we get

$$\frac{\partial \phi}{\partial \bar{x}^i} = \frac{\partial \phi}{\partial x^1}\frac{\partial x^1}{\partial \bar{x}^i} + \frac{\partial \phi}{\partial x^2}\frac{\partial x^2}{\partial \bar{x}^i} + \frac{\partial \phi}{\partial x^3}\frac{\partial x^3}{\partial \bar{x}^i} = \frac{\partial \phi}{\partial x^j}\frac{\partial x^j}{\partial \bar{x}^i}.$$

We thus obtain

$$\frac{\partial \bar{\phi}}{\partial \bar{x}^i} = \frac{\partial x^j}{\partial \bar{x}^i}\frac{\partial \phi}{\partial x^j}, \tag{17.15}$$

which is a different transformation than (17.11).

It appears that we have two kinds of vectors: those whose components transform according to (17.11) and those transforming according to (17.15). To further elucidate the discussion, let's look at the dot product. Let **A** and **B** be vectors which transform according to (17.11):

$$\bar{A}_i = \frac{\partial \bar{x}^i}{\partial x^j}A_j, \quad \bar{B}_i = \frac{\partial \bar{x}^i}{\partial x^k}B_k.$$

The dot product in the $\bar{x}$ coordinate system is $\bar{A}_i\bar{B}_i$ (sum over repeated indices understood!). Write this in terms of the x coordinates:

$$\bar{A}_i\bar{B}_i = \frac{\partial \bar{x}^i}{\partial x^j} A_j \frac{\partial \bar{x}^i}{\partial x^k} B_k = \frac{\partial \bar{x}^i}{\partial x^j} \frac{\partial \bar{x}^i}{\partial x^k} A_j B_k.$$

The right-hand side does not reduce to a dot product.

Now consider one vector $\mathbf{U}$ that transforms according to (17.11) and another $\mathbf{V}$ that transforms according to (17.15)

$$\bar{U}_i = \frac{\partial \bar{x}^i}{\partial x^j} U_j, \quad \bar{V}_i = \frac{\partial x^k}{\partial \bar{x}^i} V_k,$$

and take the dot product of these two vectors:

$$\bar{U}_i\bar{V}_i = \frac{\partial \bar{x}^i}{\partial x^j} U_j \frac{\partial x^k}{\partial \bar{x}^i} V_k = \underbrace{\frac{\partial x^k}{\partial \bar{x}^i}\frac{\partial \bar{x}^i}{\partial x^j}}_{\text{by the chain rule}} U_j V_k = \frac{\partial x^k}{\partial x^j} U_j V_k = \underbrace{\delta_{kj} U_j V_k}_{\text{by Box 17.1.3}} = U_j V_j.$$

$$(17.16)$$

This is the magic of a *general* coordinate transformation! Although the functions $\{g_1, g_2, g_3\}$ of (17.10) are *completely arbitrary* (except for invertibility), they respect the dot product, as long as one vector transforms according to (17.11) and the other according to (17.15).

So far we have been considering coordinates in a three-dimensional space. However, as this section's discussion easily points out, nothing prevents us from generalizing to n-dimensions: the only change we have to make is that the sums (and the repeated indices that imply them) should go from 1 to n. For example, (17.10) becomes

$$\bar{x}^i = g_i(x^1, x^2, \dots, x^n), \quad i = 1, 2, \dots, n. \qquad (17.17)$$

covariant and contravariant vectors

And this generalization is not purely academic, because, as we saw in Chapter 8, relativity demands a *four*-dimensional spacetime. Having this generalization in mind, we make the following definition of the two kinds of vector discussed above:

> **Box 17.1.6.** *The quantities* $\{A^1, A^2, \dots A^n\}$ *and* $\{B_1, B_2, \dots B_n\}$ *are said to constitute the components of a **contravariant** and a **covariant** vector, respectively, if, under a coordinate transformation (17.17) they transform according to*
>
> $$\bar{A}^i = \frac{\partial \bar{x}^i}{\partial x^j} A^j \quad \text{and} \quad \bar{B}_i = \frac{\partial x^j}{\partial \bar{x}^i} B_j. \qquad (17.18)$$

Note the placement of the indices on the two types of vector. Only when an "upper" index appears with a "lower" index in a sum is the result (the dot product) independent of the coordinate system used. Now the question arises: If one needs an upper and a lower index in the sum to get a quantity that is invariant, how does one define the length of a contravariant vector (which has only an upper index) or a covariant vector (which has only a lower index)? For this, we need to wait until we have introduced tensors and, in particular, the metric tensor.

17.2   From Vectors to Tensors

We have already discussed one kind of multiplication of vectors, the dot product [see Equation (17.2)]. Now we consider the cross product as a prototype of objects that have more than one index. The cross product of two vectors involves *different* components of those vectors (as opposed to the *same* components involved in the inner product). In terms of the index labels introduced above, this means that the cross product carries *two* indices. In fact, consider two (covariant) vectors A_i and B_j. The components of their cross product are of the form $A_i B_j - A_j B_i$. In another coordinate system related to the first by (17.10), the components are $\bar{A}_i \bar{B}_j - \bar{A}_j \bar{B}_i$. Using (17.18) in Box 17.1.6 for **A** and **B**, we get

$$\bar{A}_i \bar{B}_j = \frac{\partial x^k}{\partial \bar{x}^i} A_k \frac{\partial x^h}{\partial \bar{x}^j} B_h = \frac{\partial x^k}{\partial \bar{x}^i} \frac{\partial x^h}{\partial \bar{x}^j} A_k B_h,$$

and

$$\bar{A}_j \bar{B}_i = \frac{\partial x^k}{\partial \bar{x}^j} A_k \frac{\partial x^h}{\partial \bar{x}^i} B_h = \frac{\partial x^k}{\partial \bar{x}^j} \frac{\partial x^h}{\partial \bar{x}^i} A_k B_h = \frac{\partial x^h}{\partial \bar{x}^j} \frac{\partial x^k}{\partial \bar{x}^i} A_h B_k,$$

where in the last step we just changed the dummy indices [see Equation (9.4)]. Subtracting the last two equations, we get

$$\bar{A}_i \bar{B}_j - \bar{A}_j \bar{B}_i = \frac{\partial x^k}{\partial \bar{x}^i} \frac{\partial x^h}{\partial \bar{x}^j} (A_k B_h - A_h B_k).$$

Thus, if we define $C_{kh} \equiv A_k B_h - A_h B_k$ as the components of $\mathbf{A} \times \mathbf{B}$, the last equation gives their transformation property:

cross product as a two-indexed quantity

$$\bar{C}_{ij} = \frac{\partial x^k}{\partial \bar{x}^i} \frac{\partial x^h}{\partial \bar{x}^j} C_{kh}. \tag{17.19}$$

Cross products are special cases of a more general category of mathematical objects called **tensors** which carry multiple indices. Some of the indices may be upper, some lower. The most general tensor carries multiple upper and multiple lower indices.

Box 17.2.1. *A set of n^{r+s} quantities $T^{i_1\ldots i_r}_{j_1\ldots j_s}$ is said to constitute the components of a **tensor** $\mathbf{T}$ of type (r,s) if, under a coordinate transformation (17.17) they transform according to*

$$\bar{T}^{i_1\ldots i_r}_{j_1\ldots j_s} = \frac{\partial \bar{x}^{i_1}}{\partial x^{h_1}}\cdots\frac{\partial \bar{x}^{i_r}}{\partial x^{h_r}}\frac{\partial x^{k_1}}{\partial \bar{x}^{j_1}}\cdots\frac{\partial x^{k_s}}{\partial \bar{x}^{j_s}}T^{h_1\ldots h_r}_{k_1\ldots k_s} \qquad (17.20)$$

*$\{i_1\ldots i_r\}$ and $\{j_1\ldots j_s\}$ are called the **contravariant** and **covariant** indices, respectively. The **rank** of the tensor is defined as $r+s$.*

Note that for every index on the left there is an identical index on the right, and that only an upper index and its lower partner are repeated on the right. Here we are using the obvious convention that in the partial derivatives of the form $\partial x^k / \partial \bar{x}^j$ or $\partial \bar{x}^k / \partial x^j$, k is considered an upper index and j a lower one.

Example 17.2.1. When we introduced multipoles in Chapter 10, we were able to write the potential of a source distribution as an infinite sum of *moments of source* of higher and higher order. Although Cartesian coordinates are extremely clumsy for higher moments, the third moment can be handled neatly in Cartesian coordinates once we use the machinery of indices developed in this section.

Recall that the integrand of the third term in the expansion of potential is [see Equation (10.33)]

$$\text{Integrand} \equiv r'^2\left[-\frac{1}{2}+\frac{3}{2}(\hat{\mathbf{e}}_r\cdot\hat{\mathbf{e}}_{r'})^2\right] = -\frac{r'^2}{2}+\frac{3}{2}r'^2\left(\frac{\mathbf{r}\cdot\mathbf{r}'}{rr'}\right)^2.$$

Writing the position vectors in terms of their Cartesian components and rearranging terms yields

$$\begin{aligned}
\text{Integrand} &= \frac{3}{2}\frac{(xx'+yy'+zz')^2}{r^2}-\frac{r'^2}{2}\\
&= \frac{1}{2r^2}\left\{x^2(3x'^2-r'^2)+y^2(3y'^2-r'^2)+z^2(3z'^2-r'^2)\right.\\
&\qquad \left.+6xyx'y'+6xzx'z'+6yzy'z'\right\}.
\end{aligned} \qquad (17.21)$$

We want to express (17.21) in terms of indices. First let us concentrate on the terms involving x^2, y^2, and z^2. Since these *diagonal* terms involve $x^2 = x_1 x_1$, etc., it is natural to define a *two-indexed* quantity, say V'_{ij}, such that

$$\begin{aligned}
x^2(3x'^2 - r'^2) &\equiv x_1 x_1 V'_{11},\\
y^2(3y'^2 - r'^2) &\equiv x_2 x_2 V'_{22},\\
z^2(3z'^2 - r'^2) &\equiv x_3 x_3 V'_{33},
\end{aligned}$$

with

$$V'_{11}=3x'_1 x'_1 - r'^2,\quad V'_{22}=3x'_2 x'_2 - r'^2,\quad V'_{33}=3x'_3 x'_3 - r'^2.$$

Next, we note that the *off-diagonal* terms such as $6xyx'y'$ can be written as $6x_i x_j x'_i x'_j$ (no summation!). It appears as if we can write *all* terms in the last line of Equation (17.21) as $\sum_{i,j=1}^{3} x_i x_j V'_{ij}$ if we can define V'_{ij} properly. The off-diagonal sum suggests defining V'_{ij} as $V'_{ij} \equiv 3x'_i x'_j$. The reader may wonder why we did not include the factor of 6 in the definition. The reason is that when summed over indices, the symmetry of V'_{ij} under interchange of its indices automatically introduces a factor of 2. The problem with this definition is that when $i = j$, i.e., when evaluating the diagonal terms, the r'^2 term is absent. To remedy this, we change the definition to

$$V'_{ij} \equiv 3x'_i x'_j - r'^2 \delta_{ij}. \qquad (17.22)$$

Then, the Kronecker delta contributes *only to the diagonal terms* as it should. The reader is urged to show that

$$\text{Integrand} = \frac{1}{2r^2} \sum_{i,j=1}^{3} x_i x_j V'_{ij} = \frac{1}{2r^2} x_i x_j V'_{ij}, \qquad (17.23)$$

where in the last equality the summation convention is implied.

Now we substitute this in Equation (10.33) and denote the third term as $\Phi_3(\mathbf{r})$. This yields

$$\Phi_3(\mathbf{r}) = \frac{K}{r^5} x_i x_j \left[\frac{1}{2} \int_\Omega V'_{ij}\, dQ(\mathbf{r}') \right] \equiv \frac{K}{r^5} x_i x_j Q_{ij}. \qquad (17.24)$$

The last equation defines the components of the **quadrupole moment**:

quadrupole
moment defined

$$Q_{ij} = \frac{1}{2} \int_\Omega V'_{ij}\, dQ(\mathbf{r}') = \frac{1}{2} \int_\Omega (3x'_i x'_j - r'^2 \delta_{ij})\, dQ(\mathbf{r}'). \qquad (17.25)$$

One can use (17.25) to calculate the quadrupole moment of any source distribution. The quadrupole moment of electric charge distributions plays a significant role in nuclear physics. ∎

A scalar (function) is a tensor of type $(0,0)$; a contravariant vector is a tensor of type $(1,0)$; a covariant vector is a tensor of type $(0,1)$. Similarly, the cross product, the transformation of whose components is given in (17.19), is a tensor of type $(0,2)$. Of special interest is the zero tensor, which can be of any type. Box 17.2.1 shows clearly that

> **Box 17.2.2.** *If a tensor has zero components in one coordinate system, it has zero components in all coordinate systems.*

We have also encountered another two-indexed quantity, the Kronecker delta. Is it a tensor? If so, what type? We may think—since we have chosen both of its indices to be covariant—that it is of type $(0,2)$. However, that is not the case, for the following reason. Equation (17.6), which defines the Kronecker delta, must hold in all coordinate systems. If Kronecker delta were of type $(0,2)$, then it would transform according to

Kronecker delta
reindexed!

$$\bar{\delta}_{ij} = \frac{\partial x^k}{\partial \bar{x}^i} \frac{\partial x^h}{\partial \bar{x}^j} \delta_{kh} = \frac{\partial x^k}{\partial \bar{x}^i} \frac{\partial x^k}{\partial \bar{x}^j},$$

and the right-hand side does not satisfy Equation (17.6). For the same reason the Kronecker delta cannot be a tensor of type $(2,0)$. What if we define it to be a tensor of type $(1,1)$? Then

$$\bar{\delta}^i_j = \frac{\partial \bar{x}^i}{\partial x^k} \frac{\partial x^h}{\partial \bar{x}^j} \delta^k_h = \frac{\partial \bar{x}^i}{\partial x^k} \frac{\partial x^k}{\partial \bar{x}^j} = \frac{\partial \bar{x}^i}{\partial \bar{x}^j} = \delta^i_j.$$

This shows that the proper way of indexing the Kronecker delta is to give it one covariant and one contravariant index, i.e., to treat it as a tensor of type $(1,1)$.

Example 17.2.2. Chapter 8 introduced the idea of a four-vector, which is a vector with four components labeled $0,1,2,3$, with 0 being the time component and the rest the space components. It is common to label 4-vectors by Greek indices.

the dot product in relativity

For example, x^α represents the coordinates, $u^\alpha = dx^\alpha/d\tau$ represents the 4-velocity, $p^\alpha = mu^\alpha$ represents the 4-momentum, etc. The matrix η can be naturally assumed to be a tensor $\eta_{\alpha\beta}$, and the inner product of two 4-vectors a^α and b^β can be written as $\eta_{\alpha\beta}a^\alpha b^\beta$, with the summation over 0, 1, 2, 3 of a repeated index (one up, one down) understood. Because we have used i, j, k, etc., for the space part, we shall stick to this and write, for example $u^\alpha = (u^0, u^i)$, and

$$a^\alpha b_\alpha \equiv \sum_{i=0}^{3} a^\alpha b_\alpha = a^0 b_0 + a^i b_i \equiv a^0 b_0 + \sum_{i=1}^{3} a^i b_i. \qquad \blacksquare$$

The notation of the example above is very commonly used in relativity theory:

> **Box 17.2.3.** *Greek indices, representing the four-dimensional spacetime, run from 0 to 4, while Roman indices, representing the space part, run from 1 to 3.*

17.2.1 Algebraic Properties of Tensors

In our treatment of vectors, we saw that there were some formal operations which they obeyed. For instance, we could multiply a vector by a number, we could add two vectors, and we could multiply two vectors to get a third vector. Tensors also have some important properties which we summarize in the following.

Addition

If **T** and **S** are tensors of type (r,s), then their sum $\mathbf{U} = \mathbf{T} + \mathbf{S}$, defined componentwise as

$$U^{i_1\ldots i_r}_{j_1\ldots j_s} = T^{i_1\ldots i_r}_{j_1\ldots j_s} + S^{i_1\ldots i_r}_{j_1\ldots j_s},$$

is also a tensor of type (r,s). To show this, one simply has to demonstrate that $U^{i_1\ldots i_r}_{j_1\ldots j_s}$ transform according to (17.20) in Box 17.2.1.

Moreover, if we define $\mathbf{V} = \alpha\mathbf{T}$ componentwise as

$$V^{i_1\ldots i_r}_{j_1\ldots j_s} = \alpha T^{i_1\ldots i_r}_{j_1\ldots j_s},$$

where α is a real number, then **V** is also a tensor of type (r,s). The combination of these two operations makes the collection of tensors of type (r,s) a vector space.

Multiplication

If **T** is a tensor of type (r_1, s_1) and **S** is a tensor of type (r_2, s_2), then their **tensor product** $U = T \otimes S$, defined componentwise as

$$U^{i_1 \ldots i_{r_1 + r_2}}_{j_1 \ldots j_{s_1 + s_2}} = T^{i_1 \ldots i_{r_1}}_{j_1 \ldots j_{s_1}} S^{i_{r_1+1} \ldots i_{r_1+r_2}}_{j_{s_1+1} \ldots j_{s_1+s_2}} \qquad (17.26)$$

is a tensor of type $(r_1 + r_2, s_1 + s_2)$. For example, if **T** is a tensor of type $(2, 1)$ with components T^{ij}_k and **S** is a tensor of type $(0, 2)$ with components S_{lm}, then the components of their tensor product **U** are $U^{ij}_{klm} \equiv T^{ij}_k S_{lm}$, and they transform according to the following rule:

$$\overline{U}^{ij}_{klm} = \overline{T}^{ij}_k \overline{S}_{lm} = \frac{\partial \bar{x}^i}{\partial x^h} \frac{\partial \bar{x}^j}{\partial x^p} \frac{\partial x^q}{\partial \bar{x}^k} T^{hp}_q \frac{\partial x^r}{\partial \bar{x}^l} \frac{\partial x^s}{\partial \bar{x}^m} S_{rs}$$

$$= \frac{\partial \bar{x}^i}{\partial x^h} \frac{\partial \bar{x}^j}{\partial x^p} \frac{\partial x^q}{\partial \bar{x}^k} \frac{\partial x^r}{\partial \bar{x}^l} \frac{\partial x^s}{\partial \bar{x}^m} T^{hp}_q S_{rs} = \frac{\partial \bar{x}^i}{\partial x^h} \frac{\partial \bar{x}^j}{\partial x^p} \frac{\partial x^q}{\partial \bar{x}^k} \frac{\partial x^r}{\partial \bar{x}^l} \frac{\partial x^s}{\partial \bar{x}^m} U^{hp}_{qrs},$$

which shows that **U** is a tensor of rank $(2, 3)$.

Example 17.2.3. One can obtain a tensor of any type by multiplying contravariant and covariant vectors: take r contravariant vectors and s covariant vectors and multiply them to get a tensor of type (r, s). For example, if **A** is a contravariant vector with components A^i and **B** a covariant vector with components B_j, then $T^{ij} \equiv A^i A^j$ is a tensor of type $(2, 0)$, $S_{ijk} \equiv B_i B_j B_k$ is a tensor of type $(0, 3)$, and $U^{ij}_k \equiv A^i A^j B_k$ is a tensor of type $(2, 1)$. ∎

Contraction

Given a tensor of type (r, s), take a covariant index and set it equal to a contravariant index, i.e., sum over those two indices. The process is called **contraction** and the end result is a tensor of type $(r - 1, s - 1)$. For example, take the tensor of type $(2, 1)$ whose components are T^{ij}_k and set $k = j$. How do the components T^{ij}_j transform?

$$\overline{T}^{ij}_j = \frac{\partial \bar{x}^i}{\partial x^h} \frac{\partial \bar{x}^j}{\partial x^p} \frac{\partial x^q}{\partial \bar{x}^j} T^{hp}_q = \frac{\partial \bar{x}^i}{\partial x^h} \underbrace{\frac{\partial x^q}{\partial \bar{x}^j} \frac{\partial \bar{x}^j}{\partial x^p}}_{= \delta^q_p} T^{hp}_q = \frac{\partial \bar{x}^i}{\partial x^h} T^{hq}_q.$$

This shows that T^{ij}_j transform as components of a contravariant vector [see Equation (17.18)], i.e., a tensor of type $(1, 0)$.

Of special interest is a tensor of type $(1, 1)$. When you contract this tensor, you get a tensor of type $(0, 0)$, i.e., a scalar. For example, let **A** be a contravariant vector with components A^i and **B** a covariant vector with components B_j. Then $T^i_j \equiv A^i B_j$ is a tensor of type $(1, 1)$. When you contract it, you get $T^i_i \equiv A^i B_i$, which is the dot product of the two vectors, i.e., a scalar [see Equation (17.16)].

Symmetrization

symmetric and antisymmetric tensors

Some important tensors in physics have the property that when two of its indices are interchanged the tensor does not change or it changes sign. In the first case, we say that the tensor is **symmetric**, in the second case, **antisymmetric**. For example, if **T** is a tensor of type $(2, 0)$ and **U** of type $(0, 2)$, and if

$$T^{ij} = T^{ji} \qquad \text{and} \qquad U_{ij} = -U_{ji},$$

then **T** is symmetric and **U** is antisymmetric.

Given any tensor, one can always construct from it a tensor which is symmetric or antisymmetric in the interchange of any pair of its indices. In particular, if **T** is *any* tensor of type $(2, 0)$, then the tensors **S** and **A** with components

$$S^{ij} = \tfrac{1}{2}(T^{ij} + T^{ji}) \qquad \text{and} \qquad A^{ij} = \tfrac{1}{2}(T^{ij} - T^{ji})$$

are called the symmetric and antisymmetric parts of **T**, and

$$T^{ij} = \tfrac{1}{2}(T^{ij} + T^{ji}) + \tfrac{1}{2}(T^{ij} - T^{ji}) \equiv S^{ij} + A^{ij}. \tag{17.27}$$

The symmetric part S^{ij} is sometimes denoted by $T^{(ij)}$ and the antisymmetric part A^{ij} by $T^{[ij]}$.

17.2.2 Numerical Tensors

There are certain "constant" tensors which play important roles in tensor analysis. We have seen one such tensor already: the $(1, 1)$-type Kronecker delta. In fact, all the so-called *numerical tensors* are built form this funda-

generalized Kronecker delta

mental tensor. The **generalized Kronecker delta** $\delta^{i_1 \cdots i_r}_{j_1 \cdots j_r}$ is defined as

$$\delta^{i_1 \cdots i_r}_{j_1 \cdots j_r} = \det \begin{pmatrix} \delta^{i_1}_{j_1} & \delta^{i_1}_{j_2} & \cdots & \delta^{i_1}_{j_r} \\ \delta^{i_2}_{j_1} & \delta^{i_2}_{j_2} & \cdots & \delta^{i_2}_{j_r} \\ \vdots & \vdots & \vdots & \vdots \\ \delta^{i_r}_{j_1} & \delta^{i_r}_{j_2} & \cdots & \delta^{i_r}_{j_r} \end{pmatrix}. \tag{17.28}$$

The determinant of an $r \times r$ matrix is a sum of terms each consisting of the product of r matrix elements. In (17.28), each term is a product of r Kronecker deltas. Since the Kronecker delta is a $(1, 1)$-type tensor, each term, thus the determinant, and thus the generalized Kronecker delta, is an (r, r)-type tensor.

It is clear from (17.28) that the upper indices label the rows and the lower indices the columns of the matrix. Thus interchanging any two of the upper indices is equivalent to interchanging two rows of the matrix. This changes the sign of the determinant. Similarly for the interchange of two columns.

> **Box 17.2.4.** *The generalized Kronecker delta is a completely antisymmetric tensor in its upper and lower indices: interchanging any two of its upper indices or any two of its lower indices changes its sign.*

Example 17.2.4. In this example, we demonstrate a useful property of the generalized Kronecker delta. We illustrate the property for $r = 3$ and $n = 3$,[2] but the result can easily be generalized. Expand the determinant of δ^{ijk}_{lmp} about the last row starting from the right:

$$\delta^{ijk}_{lmp} = \det \begin{pmatrix} \delta^i_l & \delta^i_m & \delta^i_p \\ \delta^j_l & \delta^j_m & \delta^j_p \\ \delta^k_l & \delta^k_m & \delta^k_p \end{pmatrix} = \delta^k_p \det \begin{pmatrix} \delta^i_l & \delta^i_m \\ \delta^j_l & \delta^j_m \end{pmatrix} - \delta^k_m \det \begin{pmatrix} \delta^i_l & \delta^i_p \\ \delta^j_l & \delta^j_p \end{pmatrix} + \delta^k_l \det \begin{pmatrix} \delta^i_m & \delta^i_p \\ \delta^j_m & \delta^j_p \end{pmatrix}$$

$$= \delta^k_p \delta^{ij}_{lm} - \delta^k_m \delta^{ij}_{lp} + \delta^k_l \delta^{ij}_{mp}.$$

Now contract over the indices k and p to obtain

$$\delta^{ijk}_{lmk} = \delta^k_k \delta^{ij}_{lm} - \delta^k_m \delta^{ij}_{lk} + \delta^k_l \delta^{ij}_{mk} = 3\delta^{ij}_{lm} - \delta^{ij}_{lm} + \delta^{ij}_{ml} = 2\delta^{ij}_{lm} + \delta^{ij}_{ml} = \delta^{ij}_{lm} = \delta^i_l \delta^j_m - \delta^i_m \delta^j_l,$$

where in the next to the last step we used the antisymmetry of the generalized Kronecker delta. Note that because of the antisymmetry of the generalized Kronecker delta in both upper and lower indices, we can move both the upper and the lower last indices to the beginning:

$$\delta^{i_1 \cdots i_r}_{j_1 \cdots j_r} = \delta^{i_r i_1 \cdots i_{r-1}}_{j_r j_1 \cdots j_{r-1}}.$$

In particular,

$$\delta^{kij}_{klm} = \delta^{ijk}_{lmk} = \delta^i_l \delta^j_m - \delta^i_m \delta^j_l. \qquad \blacksquare$$

The procedure of the example above can be generalized to arbitrary r and n. Furthermore, one can contract over more than one pair of indices. The result is the following useful identity:

$$\delta^{i_1 \cdots i_s i_{s+1} \cdots i_r}_{j_1 \cdots j_s i_{s+1} \cdots i_r} = \frac{(n-s)!}{(n-r)!} \delta^{i_1 \cdots i_s}_{j_1 \cdots j_s}. \qquad (17.29)$$

From the generalized Kronecker delta two other important numerical tensors are built. These are called the **Levi-Civita symbols**. They are defined as follows:

$$\epsilon_{j_1 \cdots j_n} = \delta^{12 \cdots n}_{j_1 \cdots j_n} \qquad \text{and} \qquad \epsilon^{i_1 \cdots i_n} = \delta^{i_1 \cdots i_n}_{12 \cdots n}. \qquad (17.30)$$

Levi-Civita symbols

Note that both Levi-Civita symbols are antisymmetric in all their indices and will thus vanish if any two of their indices are equal. Moreover,

$$\epsilon_{12 \cdots n} = \delta^{12 \cdots n}_{12 \cdots n} = 1 \qquad \text{and} \qquad \epsilon^{12 \cdots n} = \delta^{12 \cdots n}_{12 \cdots n} = 1, \qquad (17.31)$$

so that we have

$$\epsilon_{i_1 \cdots i_n} = \epsilon^{i_1 \cdots i_n} = \begin{cases} +1 & \text{if } i_1 \cdots i_n \text{ is an even permutation of } 1,2,\ldots n, \\ -1 & \text{if } i_1 \cdots i_n \text{ is an odd permutation of } 1,2,\ldots n, \\ 0 & \text{otherwise.} \end{cases}$$

$$(17.32)$$

---

[2] Recall that n is the dimension of the space.

Now consider the quantity

$$A^{i_1 \cdots i_n}_{j_1 \cdots j_n} = \epsilon^{i_1 \cdots i_n} \epsilon_{j_1 \cdots j_n} - \delta^{i_1 \cdots i_n}_{j_1 \cdots j_n},$$

which is clearly antisymmetric in all its upper as well as lower indices. This means that the only nonzero elements of $A^{i_1 \cdots i_n}_{j_1 \cdots j_n}$ are those obtained from $A^{12 \cdots n}_{12 \cdots n}$. But this is zero by (17.31) and the definition of $A^{i_1 \cdots i_n}_{j_1 \cdots j_n}$. We have just shown the following important result

$$\epsilon^{i_1 \cdots i_n} \epsilon_{j_1 \cdots j_n} = \delta^{i_1 \cdots i_n}_{j_1 \cdots j_n}. \tag{17.33}$$

17.3 Metric Tensor

Let $\{x'^i\}$ denote a set of *Cartesian* coordinates, and $\{x^j\}$ some other coordinates of which $\{x'^i\}$ are functions. We then have

$$dx'^i = \frac{\partial x'^i}{\partial x^j} dx^j \quad \text{(sum over } j \text{ implied as usual).}$$

The element of length (squared)—which is customarily denoted by ds^2—in the Cartesian coordinate system is

$$ds^2 = (dx'^1)^2 + (dx'^2)^2 + \cdots + (dx'^n)^2 = \sum_{i=1}^{n} (dx'^i)^2.$$

In terms of the other coordinates, this can be written as

$$ds^2 = \sum_{i=1}^{n} (dx'^i)^2 = \sum_{i=1}^{n} dx'^i dx'^i$$

$$= \sum_{i=1}^{n} \left(\frac{\partial x'^i}{\partial x^j} dx^j \right) \left(\frac{\partial x'^i}{\partial x^k} dx^k \right)$$

$$= \left(\sum_{i=1}^{n} \frac{\partial x'^i}{\partial x^j} \frac{\partial x'^i}{\partial x^k} \right) dx^j dx^k.$$

The expression in parentheses on the last line, denoted by $g_{jk}(x)$, is a symmetric tensor of type $(0,2)$, which as indicated, is a function of the $\{x^j\}$:

$$g_{jk}(x) = \sum_{i=1}^{n} \frac{\partial x'^i}{\partial x^j} \frac{\partial x'^i}{\partial x^k}. \tag{17.34}$$

That $g_{jk}(x)$ is symmetric should be obvious. To show that it is a tensor of type $(0,2)$, let $\{\bar{x}^k\}$ be some new set of coordinates of which $\{x'^i\}$ are functions. We assume that all functional dependences are invertible. This means that $\{\bar{x}^k\}$ can be thought of as functions of $\{x'^i\}$, and through $\{x'^i\}$, as functions of $\{x^j\}$. In terms of the $\bar{x}$ variables,

$$\bar{g}_{jk}(\bar{x}) \equiv \sum_{i=1}^{n} \frac{\partial x'^i}{\partial \bar{x}^j} \frac{\partial x'^i}{\partial \bar{x}^k}.$$

Using the chain rule, this can be written as

$$\bar{g}_{jk}(\bar{x}) = \sum_{i=1}^{n} \frac{\partial x'^i}{\partial x^p} \frac{\partial x^p}{\partial \bar{x}^j} \frac{\partial x'^i}{\partial x^q} \frac{\partial x^q}{\partial \bar{x}^k} = \underbrace{\left(\sum_{i=1}^{n} \frac{\partial x'^i}{\partial x^p} \frac{\partial x'^i}{\partial x^q} \right)}_{=g_{pq}(x)} \frac{\partial x^p}{\partial \bar{x}^j} \frac{\partial x^q}{\partial \bar{x}^k} = \frac{\partial x^p}{\partial \bar{x}^j} \frac{\partial x^q}{\partial \bar{x}^k} g_{pq}(x),$$

which shows that g_{pq} transforms as a $(0,2)$-type tensor. In terms of this tensor, ds^2 is written as

$$ds^2 = \sum_{i=1}^{n} (dx'^i)^2 = g_{jk}(x) dx^j dx^k. \tag{17.35}$$

The matrix whose elements are g_{pq} is invertible. In fact, consider

$$h^{km}(x) \equiv \sum_{p=1}^{n} \frac{\partial x^k}{\partial x'^p} \frac{\partial x^m}{\partial x'^p},$$

which the reader can show to be a tensor of type $(2,0)$. Then

$$g_{jk}(x)h^{km}(x) = \left(\sum_{i=1}^{n} \frac{\partial x'^i}{\partial x^j} \frac{\partial x'^i}{\partial x^k} \right) \left(\sum_{p=1}^{n} \frac{\partial x^k}{\partial x'^p} \frac{\partial x^m}{\partial x'^p} \right)$$

$$= \sum_{i,p=1}^{n} \frac{\partial x'^i}{\partial x^j} \underbrace{\frac{\partial x'^i}{\partial x^k} \frac{\partial x^k}{\partial x'^p}}_{=\frac{\partial x'^i}{\partial x'^p}=\delta^i_p} \frac{\partial x^m}{\partial x'^p} = \sum_{i=1}^{n} \underbrace{\frac{\partial x^m}{\partial x'^i} \frac{\partial x'^i}{\partial x^j}}_{=\frac{\partial x^m}{\partial x^j}} = \delta^m_j,$$

where on the second line use was made of the chain rule and Box 17.1.5. This equation shows that the matrix whose elements are $h^{km}(x)$ is inverse to the matrix whose elements are $g_{jk}(x)$. It is common to use the same symbol for the inverse as for the original tensor. Thus, instead of $h^{km}(x)$, we use $g^{km}(x)$.

The $(0,2)$-tensor $g_{jk}(x)$ was defined in terms of the transformation rule between a Cartesian and a second coordinate system. It turns out that one can abstract the properties of $g_{jk}(x)$ and define the metric tensor:

> **Box 17.3.1.** *A **metric tensor g** with components g_{ij} is a symmetric type-$(0,2)$ tensor whose matrix has an inverse $\mathbf{g}^{-1}$ with components g^{km}. Every metric tensor defines a **geometry** in which the (square of the) element of length ds^2 is given by*
>
> $$ds^2 = g_{ij}(x)dx^i dx^j,$$
>
> *where $\{x^i\}$ are some appropriate coordinates in that geometry.*

geometry, manifold, and metric tensor

The word "geometry" in this Box is used rather loosely. A precise definition of "geometry" is beyond the scope of this book. Nevertheless, we mention that the notion of geometry starts with the concept of a **manifold**, which is a "space" that *locally* looks like a Euclidean space. For example, the surface of a sphere is a two-dimensional manifold, because a very small area of a sphere looks like a two-dimensional Euclidean space, i.e., a flat plane. Mathematicians study manifolds that have no metric tensors defined on them. However, in physics, almost all manifolds have a metric, and this metric defines the geometry of that manifold.

In our discussion of the inner product in Section 6.1.2, we also encountered the metric tensor, although we called it the metric matrix. There, we defined **Riemannian manifold** the notion of *positive definiteness*. In the context of the discussion here, this property becomes the cornerstone of a special kind of geometry: if ds^2 of Box 17.3.1 is always strictly greater than zero for nonzero dx^i and dx^j, then the manifold on which g_{ij} is defined is a called a **Riemannian manifold**. Relativity requires manifolds that are not Riemannian, i.e., for which ds^2 can be zero or negative.

Geometry is an intrinsic property of a space, while $g_{ij}(x)$ depends on the coordinates used. This is evident in Equation (17.35) where ds^2 is given in terms of Cartesian coordinates as well as the other general coordinates. Despite this coordinate dependence, the metric tensor does define the geometry of a manifold. In fact, there *are* some quantities obtained from the metric which characterize the *intrinsic* geometry of the manifold. We shall return to this discussion later.

Example 17.3.1. Let us find the metric tensor in spherical coordinates. Use spherical coordinate symbols as indices with r, θ, and φ as first, second, and third coordinates, respectively. Recalling that $x'^1 = x$, $x'^2 = y$, and $x'^3 = z$, with

$$x = r\sin\theta\cos\varphi, \quad y = r\sin\theta\sin\varphi, \quad z = r\cos\theta,$$

and using Equation (17.34), we get

$$g_{rr}(r,\theta,\varphi) = \left(\frac{\partial x}{\partial r}\right)^2 + \left(\frac{\partial y}{\partial r}\right)^2 + \left(\frac{\partial z}{\partial r}\right)^2 = (\sin\theta\cos\varphi)^2 + (\sin\theta\sin\varphi)^2 + (\cos\theta)^2 = 1$$

$$g_{r\theta}(r,\theta,\varphi) = \frac{\partial x}{\partial r}\frac{\partial x}{\partial \theta} + \frac{\partial y}{\partial r}\frac{\partial y}{\partial \theta} + \frac{\partial z}{\partial r}\frac{\partial z}{\partial \theta}$$

$$= (\sin\theta\cos\varphi)(r\cos\theta\cos\varphi) + (\sin\theta\sin\varphi)(r\cos\theta\sin\varphi) + (\cos\theta)(-r\sin\theta)$$

$$= r\sin\theta\cos\theta\cos^2\varphi + r\sin\theta\cos\theta\sin^2\varphi - r\cos\theta\sin\theta = 0.$$

Similarly, the reader can show that $g_{r\varphi} = 0$, and in fact all the off-diagonal elements vanish. On the other hand,

$$g_{\theta\theta}(r, \theta, \varphi) = \left(\frac{\partial x}{\partial \theta}\right)^2 + \left(\frac{\partial y}{\partial \theta}\right)^2 + \left(\frac{\partial z}{\partial \theta}\right)^2$$

$$= (r\cos\theta\cos\varphi)^2 + (r\cos\theta\sin\varphi)^2 + (-r\sin\theta)^2 = r^2$$

$$g_{\varphi\varphi}(r, \theta, \varphi) = \left(\frac{\partial x}{\partial \varphi}\right)^2 + \left(\frac{\partial y}{\partial \varphi}\right)^2 + \left(\frac{\partial z}{\partial \varphi}\right)^2$$

$$= (-r\sin\theta\sin\varphi)^2 + (r\sin\theta\cos\varphi)^2 = r^2\sin^2\theta.$$

Therefore,

$$ds^2 = (dr)^2 + r^2(d\theta)^2 + r^2\sin^2\theta(d\varphi)^2 = dr^2 + r^2 d\theta^2 + r^2\sin^2\theta d\varphi^2,$$

which agrees with Equation (2.25). Note how the parentheses have been removed from around the differentials. This is a very common (albeit inaccurate) practice.   ∎

17.3.1   Index Raising and Lowering

After Box 17.1.6, we mentioned that the length of a covariant or contravariant vector cannot be defined without a metric tensor. Now that we have a metric tensor, we define them. In fact, we can do better! We can define the *dot product* of any two vectors. If one vector is covariant and the other contravariant, their dot product is the usual one: the sum of the product of their components as shown in (17.16). If both vectors **A** and **B** are contravariant, define the dot product as

$$\mathbf{A} \cdot \mathbf{B} = g_{ij} A^i B^j, \qquad (17.36)$$

and if both vectors are covariant, define the dot product as

$$\mathbf{A} \cdot \mathbf{B} = g^{ij} A_i B_j. \qquad (17.37)$$

The reader can routinely show that $\bar{\mathbf{A}} \cdot \bar{\mathbf{B}} = \mathbf{A} \cdot \mathbf{B}$ in both cases.

Equations (17.36) and (17.37) have an interesting interpretation. Take the first equation and recall from Equation (17.26) that the product $g_{ij} A^k$ is a tensor of type $(1, 2)$. Contracting the indices i and k turns that into a tensor of type $(0, 1)$, i.e., a covariant vector, say **C** with components C_j. But now note that

$$\mathbf{C} \cdot \mathbf{A} = C_j A^j = g_{ij} A^i A j = \mathbf{A} \cdot \mathbf{A}.$$

It is therefore natural to denote $g_{ij} A^i$—which is equal to $g_{ji} A^i$ because of the symmetry of the metric tensor—by A_j. Thus, the metric tensor g_{ij} provides us with a way of changing contravariant vectors to covariant vectors, i.e., lowering their indices. Similar arguments show that the inverse of the metric tensor g^{ij} can be used to raise indices; and these two processes are consistent, in the sense that if we lower the index of a contravariant vector with g_{ij} and then raise the index of the resulting covariant vector with g^{ij}, we get the original contravariant vector. Here is a proof! Let $C^k = g^{kj} A_j$, where A_j is the covariant vector obtained from A^i. Then,

$$C^k = g^{kj} A_j = g^{kj} g_{ij} A^i = g^{kj} g_{ji} A^i = \delta_i^k A^i = A^k,$$

and the original contravariant vector is restored. The process of raising and lowering of indices works for arbitrary tensors:

> **Box 17.3.2.** *Any contravariant index i of a general tensor can be made into a covariant index j by multiplying the component that includes i by g_{ij}. Any covariant index i of a general tensor can be made into a contravariant index j by multiplying the component that includes i by g^{ij}.*

In Cartesian coordinates the (Euclidean) metric tensor is just the Kronecker delta. Therefore

$$A^j = g^{ij} A_i = \delta_j^i A_i = A_j, \quad \text{in Cartesian coordinates with Euclidean metric,} \tag{17.38}$$

and the distinction between covariant and contravariant vectors (and indices) disappears.

In special relativity and in Cartesian coordinates, the metric tensor is $\eta_{\alpha\beta}$, whose matrix is given in Equation (8.8). This tensor has components

$$\eta_{00} = 1, \ \eta_{11} = \eta_{22} = \eta_{33} = -1, \ \eta_{\alpha\beta} = 0 \text{ if } \alpha \neq \beta \quad \text{in special relativity.}$$

The inverse of $\eta_{\alpha\beta}$ is itself: $\eta^{\alpha\beta} = \eta_{\alpha\beta}$. In raising and lowering of an index, the time component does not change, while the space components change sign (see Box 17.2.3 for the meaning of Greek and Roman indices in relativity):

$$A^\alpha = \eta^{\alpha\beta} A_\beta \ \Rightarrow \ A^0 = A_0, A^i = -A_i \tag{17.39}$$

**components of
cross product**

Example 17.3.2. The Levi-Civita symbols are conveniently used to express the components of the cross product of two vectors in Cartesian coordinate systems. Since there is no difference between covariant and contravariant indices in Cartesian coordinate system, we use only covariant indices.

$$(\mathbf{A} \times \mathbf{B})_i = \epsilon_{ijk} A_j B_k, \qquad i = 1, 2, 3, \tag{17.40}$$

where a sum over j and k is understood. As a practice in index manipulation, the reader is urged to verify the above relation. The order of the two vectors on both sides of the equation is important!

Using Equation (17.40) and some properties of the Levi-Civita symbol, we can derive the *bac cab* rule:

$$\mathbf{A} \times (\mathbf{B} \times \mathbf{C}) = \mathbf{B}(\mathbf{A} \cdot \mathbf{C}) - \mathbf{C}(\mathbf{A} \cdot \mathbf{B}).$$

Start with a general component of the LHS and work through index manipulations until you reach the corresponding component of the RHS:

$$\{\mathbf{A} \times (\mathbf{B} \times \mathbf{C})\}_i = \epsilon_{ijk} A_j (\mathbf{B} \times \mathbf{C})_k = \epsilon_{ijk} A_j \epsilon_{kmn} B_m C_n$$
$$= \epsilon_{kij} \epsilon_{kmn} A_j B_m C_n = (\delta_{im}\delta_{jn} - \delta_{in}\delta_{jm}) A_j B_m C_n$$
$$= \delta_{im}\delta_{jn} A_j B_m C_n - \delta_{in}\delta_{jm} A_j B_m C_n = A_j B_i C_j - A_j B_j C_i$$
$$= B_i(A_j C_j) - C_i(A_j B_j) = B_i(\mathbf{A} \cdot \mathbf{C}) - C_i(\mathbf{A} \cdot \mathbf{B}).$$

On the second line we used (17.33) and the result obtained in Example 17.2.4. The last expression above is the ith component of the RHS of the bac cab rule. ∎

Example 17.3.3. Example 15.3.1 calculated the angular momentum differential operator using Cartesian coordinates. To illustrate the power of indices and the ease with which they allow some complex manipulations, we redo the calculation of Example 15.3.1 using indices.

We have $-L^2 f = (\mathbf{r} \times \boldsymbol{\nabla}) \cdot (\mathbf{r} \times \boldsymbol{\nabla})f$. Letting ∂_j stand for the partial derivative with respect to x_j, using Einstein summation convention, and recalling that no raising or lowering of indices is necessary for Euclidean space, we write

$$-L^2 f = (\mathbf{r} \times \boldsymbol{\nabla})_i (\mathbf{r} \times \boldsymbol{\nabla})_i f = (\epsilon_{ijk} x_j \partial_k)(\epsilon_{ilm} x_l \partial_m) f = \epsilon_{ijk}\epsilon_{ilm} x_j \partial_k (x_l \partial_m f),$$

where we used (17.40). Continuing, refer to (17.33) and write the above equation as

$$-L^2 f = (\delta_{jl}\delta_{km} - \delta_{jm}\delta_{kl}) x_j \partial_k (x_l \partial_m f) = x_j \partial_k (x_j \partial_k f) - x_j \partial_k (x_k \partial_j f)$$
$$= x_j \delta_{kj} \partial_k f + x_j x_j \partial_k \partial_k f - x_j \delta_{kk} \partial_j f - x_j x_k \partial_k \partial_j f \qquad (17.41)$$
$$= x_j \partial_j f + r^2 \nabla^2 f - 3 x_j \partial_j f - x_j x_k \partial_k \partial_j f = r^2 \nabla^2 f - 2(\mathbf{r} \cdot \boldsymbol{\nabla})f - x_j x_k \partial_k \partial_j f,$$

because $\partial_k x_j = \delta_{kj}$, $x_j x_j = r^2$, $\delta_{kk} = 3$, $x_j \partial_j = \mathbf{r} \cdot \boldsymbol{\nabla}$, and $\partial_k \partial_k = \nabla^2$. The last term in (17.41) above can be found from the following relation:

$$x_k \partial_k (x_j \partial_j f) = x_k \delta_{kj} \partial_j f + x_k x_j \partial_k \partial_j f = x_j \partial_j f + x_k x_j \partial_k \partial_j f,$$

or

$$x_k x_j \partial_k \partial_j f = (\mathbf{r} \cdot \boldsymbol{\nabla})^2 f - (\mathbf{r} \cdot \boldsymbol{\nabla})f.$$

Substituting in (17.41) yields Equation (15.22). Compare this derivation with the laborious calculation of Example 15.3.1! ∎

17.3.2   Tensors and Electrodynamics

Relativity was a logical outcome of the electromagnetic theory. It should therefore come as no surprise if the equations of electromagnetism found their most natural form in the language of relativity and tensors associated with it. In the discussion that follows, it is convenient and common practice to set the speed of light equal to 1; then since $c = 1/\sqrt{\epsilon_0 \mu_0}$, we have

$$c = 1, \quad \frac{1}{\epsilon_0} = \mu_0.$$

Consider the Lorentz force law

$$\mathbf{f} = q(\mathbf{E} + \mathbf{v} \times \mathbf{B}) \quad \text{or} \quad f_i = q(E_i + \epsilon_{ijk} v_j B_k), \qquad (17.42)$$

where as in Example 17.3.2, we used covariant indices for all tensors in the second equation. Since this is the fundamental force of electromagnetism, we expect it to have a natural expression in relativity.

As a starting point, we note that the magnetic part is of the form $v_j F_{ij}$, where $F_{ij} = \epsilon_{ijk} B_k$ is an antisymmetric tensor of rank two. The obvious generalization that might lead to a connection with relativity is to consider an expression of the form $u^\beta F_{\alpha\beta}$, where u^β is the velocity 4-vector and $F_{\alpha\beta}$ is an antisymmetric tensor of rank two which reduces to F_{ij} when both α and β are nonzero. Let us look at $u^\beta F_{\alpha\beta}$ when α is i:

$$u^\beta F_{i\beta} = u^0 F_{i0} + u^j F_{ij},$$

electromagnetic field tensor

where we used the convention of Example 17.2.2. Equation (8.21) now gives $u^0 = \gamma$, and $u^i = \gamma v^i$. Then the equation above gives

$$u^\beta F_{i\beta} = \gamma F_{i0} + \gamma v^j F_{ij} = \gamma \left(F_{i0} + v^j F_{ij} \right) = \gamma \left(F_{i0} + v_j \epsilon_{ijk} B_k \right),$$

where in the last step, we disregarded the difference between covariant and contravariant indices. Comparison with Equation (17.42) shows that it is natural to set $F_{i0} = E_i$. The second rank antisymmetric tensor $F_{\alpha\beta}$ is called the **electromagnetic field tensor**.

Maxwell's equations (15.29) take a specially simple form when written in terms of the electromagnetic field tensor. The first equation can be written as

$$\partial_i F_{i0} = \frac{\partial F_{i0}}{\partial x^i} = \frac{\rho}{\epsilon_0} = \mu_0 \rho. \tag{17.43}$$

The obvious generalization of the left-hand side to relativity is $\partial F_{\alpha\beta}/\partial x^\alpha$. But there is something wrong with this! Both α's are *lower* indices—recall that the *super*script of a coordinate in the denominator leads to a *sub*script—and you cannot sum over them. In the Euclidean case, this causes no problem because by (17.38), there is no difference between lower and upper indices and we can simply raise one of the i's. In relativity, however, there is a difference. So, we have to introduce the (inverse) η tensor. The left-hand side now becomes

$$\eta^{\alpha\nu} \frac{\partial F_{\alpha\beta}}{\partial x^\nu}.$$

Since β is a free index, we expect the right-hand side to have a free index as well. So, we write the generalization of Maxwell's first equation as

$$\eta^{\alpha\nu} \frac{\partial F_{\alpha\beta}}{\partial x^\nu} = \mu_0 V_\beta, \tag{17.44}$$

with V_β to be determined. For $\beta = 0$, we get

$$\eta^{\alpha\nu} \frac{\partial F_{\alpha 0}}{\partial x^\nu} = \mu_0 V_0, \quad \text{or} \quad \eta^{i\nu} \frac{\partial F_{i0}}{\partial x^\nu} = \mu_0 V_0, \quad \text{or} \quad -\frac{\partial F_{i0}}{\partial x^i} = \mu_0 V_0,$$

where we used the fact that $F_{\alpha\beta}$ is antisymmetric, so all its "diagonal" components are zero. We also used the fact that η is diagonal with the space elements being -1. Comparing with (17.43), we see that $V_0 = -\rho$.

Now let $\beta = i$ in (17.44). Then

$$\eta^{\alpha\nu}\frac{\partial F_{\alpha i}}{\partial x^\nu} = \mu_0 V_i, \quad \text{or} \quad \eta^{0\nu}\frac{\partial F_{0i}}{\partial x^\nu} + \eta^{j\nu}\frac{\partial F_{ji}}{\partial x^\nu} = \mu_0 V_i,$$

or

$$\frac{\partial F_{0i}}{\partial x^0} - \frac{\partial F_{ji}}{\partial x^j} = \mu_0 V_i, \quad \text{or} \quad -\frac{\partial E_i}{\partial t} + \epsilon_{ijk}\partial_j B_k = \mu_0 V_i.$$

This is the ith component of the vector equation

$$-\frac{\partial \mathbf{E}}{\partial t} + \nabla \times \mathbf{B} = \mu_0 \mathbf{V}.$$

Comparing this with the fourth Maxwell's equation, we identify $\mathbf{V}$ as $\mathbf{J}$. Thus, the first and fourth equations, the **inhomogeneous Maxwell's equations** are combined into

Maxwell's 1st and 4th equations and four-current

$$\eta^{\alpha\nu}\frac{\partial F_{\alpha\beta}}{\partial x^\nu} = \mu_0 J_\beta, \tag{17.45}$$

where $J_\beta = (-\rho, \mathbf{J})$ is the **4-current**. We leave it to the reader to verify that

Maxwell's 2nd and 3rd equations

$$\frac{\partial F_{\alpha\beta}}{\partial x^\nu} + \frac{\partial F_{\nu\alpha}}{\partial x^\beta} + \frac{\partial F_{\beta\nu}}{\partial x^\alpha} = 0 \tag{17.46}$$

combines the second and third equations, the **homogeneous Maxwell's equations**.

Equation (17.46) is satisfied if $F_{\alpha\beta} = \partial_\alpha A_\beta - \partial_\beta A_\alpha$ for *any* 4-vector A_α, as the reader can easily verify. For $\alpha = i$ and $\beta = 0$, this gives

$$F_{i0} = \partial_i A_0 - \partial_0 A_i, \quad \text{or} \quad E_i = \partial_i A_0 - \partial_0 A_i, \quad \text{or} \quad \mathbf{E} = \nabla A_0 - \frac{\partial \mathbf{A}}{\partial t}$$

Comparing this with (15.31) identifies A_0 with the negative of the scalar potential Φ and $\mathbf{A}$ with the vector potential. We can thus write

$$F_{\alpha\beta} = \partial_\alpha A_\beta - \partial_\beta A_\alpha, \quad A_\alpha = (-\Phi, \mathbf{A}). \tag{17.47}$$

Now that we have solved the homogeneous Maxwell's equations by introducing the 4-potential, we can insert the result in (17.45) to write the inhomogeneous Maxwell's equations in terms of the 4-potential as well. We then have

$$\eta^{\alpha\nu}\partial_\nu\left(\partial_\alpha A_\beta - \partial_\beta A_\alpha\right) = \mu_0 J_\beta,$$

or

$$\eta^{\alpha\nu}\partial_\nu\partial_\alpha A_\beta - \partial_\beta\left(\eta^{\alpha\nu}\partial_\nu A_\alpha\right) = \mu_0 J_\beta. \tag{17.48}$$

The expression in parentheses—when set equal to zero—gives the Lorentz gauge condition [see Equation (15.32)]. The remaining part of the equation gives the wave equation for $\mathbf{A}$ and Φ.

17.4   Differentiation of Tensors

Tensors represent many quantities, whose variation with coordinates (points in space) has physical significance. Therefore, the notion of a derivative of a tensor becomes important. Although we can always differentiate components of a tensor (they are just functions), the resulting derivative is not necessarily a tensor. To obtain a tensor, one needs to generalize the concept of the derivative, as we do in this section.

17.4.1   Covariant Differential and Affine Connection

Let us begin by noting that the differentials of coordinates form the components of a contravariant vector. In fact, when the new coordinates $\bar{x}^i$ are written as functions of the old coordinates x^j and one takes the differential of the new coordinates, one obtains

$$d\bar{x}^i = \frac{\partial \bar{x}^i}{\partial x^j} dx^j, \tag{17.49}$$

which is precisely the way a contravariant vector transforms. In fact, this is the archetypal example of a contravariant vector, and can be a guide in helping the reader remember the rule of transformation of the contravariant components of a tensor.

The differential of a scalar—a tensor of type $(0,0)$—is again a scalar, because

$$d\phi = \frac{\partial \phi}{\partial x^i} dx^i,$$

and the first term is the components of a covariant vector [see Equation (17.15)], and the second term the components of a covariant vector (as shown above).

Next take the differential of a contravariant vector A^i. How does it transform? By taking the differential of the transformation rule

$$\bar{A}^i = \frac{\partial \bar{x}^i}{\partial x^j} A^j, \tag{17.50}$$

one obtains

$$d\bar{A}^i = \frac{\partial \bar{x}^i}{\partial x^j} dA^j + d\left(\frac{\partial \bar{x}^i}{\partial x^j}\right) A^j = \frac{\partial \bar{x}^i}{\partial x^j} dA^j + \frac{\partial^2 \bar{x}^i}{\partial x^k \partial x^j} dx^k A^j. \tag{17.51}$$

If the second term on the right were absent, dA^j would transform as a contravariant vector. It turns out that one can add something to dA^j whose effect is to cancel the unwanted term.

Consider quantities Γ^j_{mp}, which transform according to

components of
affine connection

$$\bar{\Gamma}^j_{mp} = \frac{\partial \bar{x}^j}{\partial x^l} \frac{\partial x^h}{\partial \bar{x}^m} \frac{\partial x^k}{\partial \bar{x}^p} \Gamma^l_{hk} - \frac{\partial^2 \bar{x}^j}{\partial x^h \partial x^k} \frac{\partial x^h}{\partial \bar{x}^m} \frac{\partial x^k}{\partial \bar{x}^p}. \tag{17.52}$$

Any set of three-indexed symbols Γ^j_{mp} which transform according to this equation is said to constitute the components of an **affine connection**. An affine connection is not a tensor because of the second term on the right-hand side of (17.52). Since this term is the same for *all* affine connections, the *difference* between two affine connections is a tensor of type $(1, 2)$. If Γ^j_{mp} and Λ^j_{mp} are any two affine connections then

$$\overline{\Gamma}^j_{mp} - \overline{\Lambda}^j_{mp} = \frac{\partial \bar{x}^j}{\partial x^l} \frac{\partial x^h}{\partial \bar{x}^m} \frac{\partial x^k}{\partial \bar{x}^p} \left(\Gamma^l_{hk} - \Lambda^l_{hk} \right), \qquad (17.53)$$

showing that $\Gamma^l_{hk} - \Lambda^l_{hk}$ transform as components of a tensor of type $(1, 2)$. In particular, if $\Lambda^l_{hk} = \Gamma^l_{kh}$, then the difference $\Gamma^l_{hk} - \Gamma^l_{kh}$ is essentially the antisymmetric part of the affine connection Γ:

$$\Gamma^l_{hk} = \underbrace{\tfrac{1}{2} \left(\Gamma^l_{hk} + \Gamma^l_{kh} \right)}_{\text{symmetric part}} + \underbrace{\tfrac{1}{2} \left(\Gamma^l_{hk} - \Gamma^l_{kh} \right)}_{\text{antisymmetric part}} .$$

The antisymmetric part of an affine connection is called its **torsion tensor**. Clearly if it vanishes in one coordinate system then it vanishes in all coordinates (the zero tensor is zero in all coordinate systems). Thus, the torsion tensor of an affine connection is zero, if an only if the connection is *symmetric*.

torsion tensor

Lack of tensorial character of the affine connection is precisely what is needed to make dA^j, as well as dA_j a tensor:

> **Box 17.4.1.** *For any affine connection Γ^j_{kl}, the quantities DA^j and DA_j defined by*
>
> $$DA^j = dA^j + \Gamma^j_{kl} A^k dx^l \qquad and \qquad DA_j = dA_j - \Gamma^k_{jl} A_k dx^l$$
>
> *are, respectively, the components of a contravariant and a covariant vector. They are called the **covariant** or **absolute** differential of the vectors.*

We show that DA^j is a contravariant vector, leaving the proof of the second claim to the reader. In the bar coordinates, we have

$$D\bar{A}^j = d\bar{A}^j + \overline{\Gamma}^j_{kl} \bar{A}^k d\bar{x}^l.$$

Using Equations (17.49), (17.50), (17.51), and (17.52), we obtain

$$
D\bar{A}^j = \frac{\partial \bar{x}^j}{\partial x^k} dA^k + \frac{\partial^2 \bar{x}^j}{\partial x^k \partial x^l} dx^k A^l
$$
$$
+ \left(\frac{\partial \bar{x}^j}{\partial x^p} \frac{\partial x^q}{\partial \bar{x}^k} \frac{\partial x^r}{\partial \bar{x}^l} \Gamma^p_{qr} - \frac{\partial^2 \bar{x}^j}{\partial x^q \partial x^r} \frac{\partial x^q}{\partial \bar{x}^k} \frac{\partial x^r}{\partial \bar{x}^l} \right) \left(\frac{\partial \bar{x}^k}{\partial x^m} A^m \frac{\partial \bar{x}^l}{\partial x^s} dx^s \right)
$$
$$
= \frac{\partial \bar{x}^j}{\partial x^k} dA^k + \frac{\partial^2 \bar{x}^j}{\partial x^k \partial x^l} dx^k A^l
$$
$$
+ \frac{\partial \bar{x}^j}{\partial x^p} \underbrace{\frac{\partial x^q}{\partial \bar{x}^k} \frac{\partial \bar{x}^k}{\partial x^m}}_{=\delta^q_m} \underbrace{\frac{\partial x^r}{\partial \bar{x}^l} \frac{\partial \bar{x}^l}{\partial x^s}}_{=\delta^r_s} \Gamma^p_{qr} A^m dx^s - \frac{\partial^2 \bar{x}^j}{\partial x^q \partial x^r} \underbrace{\frac{\partial x^q}{\partial \bar{x}^k} \frac{\partial \bar{x}^k}{\partial x^m}}_{=\delta^q_m} \underbrace{\frac{\partial x^r}{\partial \bar{x}^l} \frac{\partial \bar{x}^l}{\partial x^s}}_{=\delta^r_s} A^m dx^s
$$
$$
= \frac{\partial \bar{x}^j}{\partial x^k} dA^k + \frac{\partial^2 \bar{x}^j}{\partial x^k \partial x^l} dx^k A^l + \frac{\partial \bar{x}^j}{\partial x^p} \Gamma^p_{qr} A^q dx^r - \frac{\partial^2 \bar{x}^j}{\partial x^q \partial x^r} A^q dx^r.
$$

The second term cancels the last term (remember that you can use any symbol for the dummy indices that are summed over). Therefore,

$$
D\bar{A}^j = \frac{\partial \bar{x}^j}{\partial x^k} dA^k + \frac{\partial \bar{x}^j}{\partial x^p} \Gamma^p_{qr} A^q dx^r = \frac{\partial \bar{x}^j}{\partial x^k} dA^k + \frac{\partial \bar{x}^j}{\partial x^k} \Gamma^k_{qr} A^q dx^r
$$
$$
= \frac{\partial \bar{x}^j}{\partial x^k} \left(dA^k + \Gamma^k_{qr} A^q dx^r \right) = \frac{\partial \bar{x}^j}{\partial x^k} DA^k,
$$

which is the transformation rule of a contravariant vector.

Absolute differential can be defined for any tensor. For a scalar ϕ, $D\phi = d\phi$. In the case of other tensors, for each contravariant index an affine connection term with a positive sign, and for each covariant index an affine connection term with a negative sign is introduced. For example, the covariant differential of T^{ij}_k is a tensor of type $(2,1)$ given by

$$
DT^{ij}_k = dT^{ij}_k + \left(\Gamma^i_{pq} T^{pj}_k + \Gamma^j_{pq} T^{ip}_k - \Gamma^p_{kq} T^{ij}_p \right) dx^q.
$$

Covariant differential has all the properties of ordinary differential when applied to tensors. For example, the covariant differential of the sum of two tensors of type (r,s) is a tensor of type (r,s), and $D(\alpha \mathbf{T}) = \alpha D\mathbf{T}$ for any constant α and any tensor $\mathbf{T}$. Covariant differential also obeys the Leibniz rule:

$$
D(\mathbf{T} \otimes \mathbf{S}) = D\mathbf{T} \otimes \mathbf{S} + \mathbf{T} \otimes D\mathbf{S}. \tag{17.54}
$$

17.4.2   Covariant Derivative

In the first equation of Box 17.4.1, write dA^j in terms of partial derivatives. Then, the equation becomes

$$
DA^j = \frac{\partial A^j}{\partial x^l} dx^l + \Gamma^j_{kl} A^k dx^l = \left(\frac{\partial A^j}{\partial x^l} + \Gamma^j_{kl} A^k. \right) dx^l
$$

Since the left-hand side and dx^l are contravariant vectors, we suspect that the expression in parentheses is a tensor of type $(1,1)$. This can in fact be shown directly. It is called the **covariant derivative** of A^j with respect to x^l and denoted by $A^j_{;l}$. Thus,

covariant derivative

$$A^j_{;l} \equiv \frac{\partial A^j}{\partial x^l} + \Gamma^j_{kl} A^k. \qquad (17.55)$$

This is the generalization of ordinary derivative to situations in which the affine connection is nonzero. Covariant derivative can similarly be defined for covariant vectors as well as arbitrary tensors. For example, the covariant derivative of T^{ij}_k is a tensor of type $(2,2)$ given by

$$T^{ij}_{k;q} = \frac{\partial T^{ij}_k}{\partial x^q} + \Gamma^i_{pq} T^{pj}_k + \Gamma^j_{pq} T^{ip}_k - \Gamma^p_{kq} T^{ij}_p.$$

Consider a curve in Euclidean space parametrized by t. Let $A^i(t)$ be the value of a vector field at a point on the curve. If $dA^i/dt = 0$, then the vector is constant along the curve, and we say that the vector is **parallel translated along the curve**. When the affine connection is nonzero, we divide both sides of the first equation in Box 17.4.1 by dt (which on the left we denote by Dt for aesthetic reasons), and say that a contravariant vector field is parallel translated along a curve if

parallel translation along a curve

$$\frac{DA^j}{Dt} = 0 \quad \text{or} \quad \frac{dA^j}{dt} + \Gamma^j_{kl} A^k \frac{dx^l}{dt} = 0, \qquad (17.56)$$

with a similar definition for a covariant vector field. Since A^j depends on t only through the coordinates, we use the chain rule $dA^j/dt = (\partial A^j/\partial x^l)dx^l/dt$ to rewrite the equation above as

$$\frac{DA^j}{Dt} = \left(\frac{\partial A^j}{\partial x^l} + \Gamma^i_{kj} A^k \right) \frac{dx^l}{dt} \equiv A^j_{;l} \frac{dx^l}{dt} \equiv A^j_{;l} \dot{x}^l = 0. \qquad (17.57)$$

A curve whose tangent vector is parallel translated along that curve is called a **geodesic**. The components of the vector tangent to a curve is $dx^i/dt \equiv \dot{x}^i$. If we substitute this in (17.56) we obtain the following second order differential equation called the **geodesic equation**:

geodesic and geodesic equation

$$\frac{D\dot{x}^j}{Dt} = 0, \quad \text{or} \quad \frac{d^2 x^j}{dt^2} + \Gamma^j_{kl} \frac{dx^k}{dt} \frac{dx^l}{dt} = 0, \quad \text{or} \quad \ddot{x}^j + \Gamma^j_{kl} \dot{x}^k \dot{x}^l = 0, \; (17.58)$$

where each super dot represents a differentiation with respect to t. Solving this differential equation yields the parametric equation of a geodesic.

17.4.3   Metric Connection

The affine connection, which is defined by its transformation property of (17.52) is completely arbitrary. One can define covariant differentials and covariant derivatives in terms of *any* set of quantities that transform according to Equation (17.52). With a metric tensor, however, one can define a unique *symmetric* (therefore, torsion-free) affine connection called **metric connection** given by

$$\Gamma^j_{kl} = \Gamma^j_{lk} = \frac{1}{2} g^{jm} \left(\frac{\partial g_{mk}}{\partial x^l} + \frac{\partial g_{ml}}{\partial x^k} - \frac{\partial g_{kl}}{\partial x^m} \right) \equiv g^{jm} \Gamma_{mkl}, \qquad (17.59)$$

where

$$\Gamma_{mkl} = \frac{1}{2} \left(\frac{\partial g_{mk}}{\partial x^l} + \frac{\partial g_{ml}}{\partial x^k} - \frac{\partial g_{kl}}{\partial x^m} \right), \qquad (17.60)$$

with all lower indices, is easier to remember. Note that it is the first index of Γ_{mkl} that is raised to give the components of the metric connection, and for this reason the metric connection is sometimes denoted by $\Gamma^j{}_{kl}$. The verification that (17.59) is indeed an affine connection—i.e., that it transforms according to (17.52)—is straightforward but tedious.

Example 17.4.1. If all components of a metric tensor are constant in some coordinate system, then all the components of the metric connection vanish. Note that this is true *only in that particular coordinate system*. Changing coordinates changes the affine connection, and in general, the components of a metric connection will not be zero even if they are zero in some coordinate system. If we use Cartesian coordinates, then the Euclidean metric is just the Kronecker delta. Therefore, all components of the metric connection are zero. Similarly, the metric of special relativity in Cartesian coordinates in $\eta_{\alpha\beta}$, whose components are either 0 or 1 or -1. Hence, all components of the metric connection of special relativity in Cartesian coordinates vanish. ∎

The metric connection has some special properties which are of physical importance. The first property which could be easily verified is that

$$g_{ij;k} = 0 \quad \text{or} \quad \frac{\partial g_{ij}}{\partial x^k} - \Gamma^p_{jk} g_{ip} - \Gamma^p_{ik} g_{pj} = 0. \qquad (17.61)$$

The second property is that between any two points passes a single geodesic of the metric connection, and this geodesic extremizes the distance between the two points. If the geometry is Riemannian (i.e., if the metric is positive definite) then the geodesic gives the *shortest* distance. In relativity, where the metric is not Riemannian, the geodesics give the *longest* distance.

Example 17.4.2. In this example, we find the geodesics of a sphere. The spherical angular coordinates θ and φ can be used on the surface of a sphere of radius a. From the element of length $ds^2 = a^2 d\theta^2 + a^2 \sin^2\theta d\varphi^2$ on this sphere, and using θ and φ to label components, we deduce that

$$g_{11} \equiv g_{\theta\theta} = a^2, \ g_{22} \equiv g_{\varphi\varphi} = a^2 \sin^2\theta, \ g_{12} \equiv g_{\theta\varphi} = g_{21} \equiv g_{\varphi\theta} = 0,$$

and similarly,

$$g^{11} \equiv g^{\theta\theta} = \frac{1}{a^2}, \ g^{22} \equiv g^{\varphi\varphi} = \frac{1}{a^2 \sin^2\theta}, \ g^{12} \equiv g^{\theta\varphi} = g^{21} \equiv g^{\varphi\theta} = 0.$$

Substituting these in (17.59), we can calculate the components of the affine connection. The nonzero components turn out to be

$$\Gamma^\theta_{\varphi\varphi} = -\sin\theta\cos\theta, \quad \Gamma^\varphi_{\theta\varphi} = \Gamma^\varphi_{\varphi\theta} = \cot\theta.$$

Using these in the geodesic equation (17.58), we obtain the following two differential equations:

$$\frac{d^2\theta}{dt^2} - \sin\theta\cos\theta\left(\frac{d\varphi}{dt}\right)^2 = 0,$$

$$\frac{d^2\varphi}{dt^2} + 2\cot\theta\frac{d\varphi}{dt}\frac{d\theta}{dt} = 0. \tag{17.62}$$

The second equation can be solved to give

$$\frac{d\varphi}{dt} = \frac{C}{\sin^2\theta} \Rightarrow d\varphi = \frac{C}{\sin^2\theta}dt, \tag{17.63}$$

where C is a constant of integration. Substituting this in the first equation of (17.62) gives

$$\frac{d^2\theta}{dt^2} - \frac{C^2\cos\theta}{\sin^3\theta} = 0. \tag{17.64}$$

To find the geodesic, it is more convenient to express θ as a function of φ. This means changing the independent variable in Equation (17.64) from t to φ. This is done formally by using the second equation of (17.63) to substitute for dt in (17.62). Thus, the first tem of (17.62) can be written as

$$\frac{d}{dt}\left(\frac{d\theta}{dt}\right) = \frac{Cd}{\sin^2\theta d\varphi}\left(\frac{Cd\theta}{\sin^2\theta d\varphi}\right) = \frac{C^2}{\sin^2\theta}\frac{d}{d\varphi}\left(\frac{1}{\sin^2\theta}\frac{d\theta}{d\varphi}\right).$$

Substituting this in (17.64) yields

$$\frac{d}{d\varphi}\left(\frac{1}{\sin^2\theta}\frac{d\theta}{d\varphi}\right) - \cot\theta = 0.$$

Differentiating the first term, we get

$$-2\frac{\cos\theta}{\sin^3\theta}\left(\frac{d\theta}{d\varphi}\right)^2 + \frac{1}{\sin^2\theta}\frac{d^2\theta}{d\varphi^2} - \cot\theta = 0,$$

which can be simplified to the following differential equation:

$$\sin\theta\frac{d^2\theta}{d\varphi^2} - 2\cos\theta\left(\frac{d\theta}{d\varphi}\right)^2 - \sin^2\theta\cos\theta = 0. \tag{17.65}$$

If we could solve this equation, we would find θ as a function of φ, and this should be the equation of a geodesic on a sphere. Instead, let us use our knowledge of the geodesics (curves giving the shortest distance) on a sphere, write it with θ as a function of φ and see if it satisfies (17.65). Our sphere is parametrized as

$$x = a\sin\theta\cos\varphi, \ y = a\sin\theta\sin\varphi, \ z = a\cos\theta.$$

The great circles—curves of shortest distance—are the intersection of a plane passing through the origin and the sphere. Such a plane has an equation of the form $Ax +$

$By + Cz = 0$. The intersection with the sphere is obtained by substituting for x, y, and z from the above equations:

$$Aa \sin \theta \cos \varphi + Ba \sin \theta \sin \varphi + Ca \cos \theta = 0.$$

Dividing by $Ca \sin \theta$ and redefining A to be $-A/C$ and B to be $-B/C$, we get

$$\cot \theta = A \cos \varphi + B \sin \varphi,$$

as the equation of geodesic on a sphere. It is straightforward to show that this equation indeed satisfies (17.65). ∎

17.5   Riemann Curvature Tensor

Consider a closed loop, such as a rectangle, on a flat surface. Start a vector at one point of the rectangle (the lower left corner) and carry it parallel to itself to the point diagonally opposite the initial point [Figure 17.1(a)]. In one case carry the vector to the right and then up. In the second case carry the vector up and then to the right. Compare the vector at the end of the two cases. They are equal. Do the same on a curved space such as the surface of a sphere. The two vectors at the end do not coincide [see Figure 17.1(b)]! The degree to which they are different is a measure of the curvature of the space.

Let us quantify the notion of the curvature. Suppose that the lower and upper curves of the "rectangle" are parametrized by t and the right and the left curves by s. Moving along a curve parametrized by t does not change s, and vice versa. Using a Taylor expansion, in which derivatives are replaced by covariant derivatives, parallel translate a contravariant vector A^j first to the right and then upward [see Figure 17.1(b) for clarification]. Assume that the lower left corner has (t, s) as the parameter values. As you move along the lower curve, the parameters change from (t, s) to $(t + \Delta t, s)$. So, to first order in Δt, we have

$$A^j(t + \Delta t, s) = A^j(t, s) + \frac{DA^j}{Dt} \Delta t = A^j(t, s) + A^j_{;l}(t, s) \frac{dx^l}{dt} \Delta t.$$

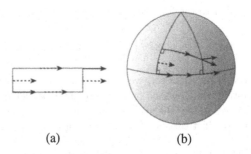

(a)   (b)

Figure 17.1: (a) In a flat space, the direction of the vector does not change when carried along two different paths. (b) In a curved space, the two vectors are different.

Now parallel translate this vector upward, the direction in which t is constant but s changes:

$$A^j(t + \Delta t, s + \Delta s) = A^j(t + \Delta t, s) + \frac{D}{Ds}(A^j(t + \Delta t, s))\Delta s$$

$$= A^j(t, s) + \frac{DA^j}{Dt}\Delta t + \frac{D}{Ds}\left(A^j(t, s) + A^j_{;l}(t, s)\frac{dx^l}{dt}\Delta t\right)\Delta s$$

$$= A^j(t, s) + \frac{DA^j}{Dt}\Delta t + \frac{DA^j}{Ds}\Delta s + \frac{D}{Ds}\left(A^j_{;l}(t, s)\frac{dx^l}{dt}\Delta t\right)\Delta s$$

$$= A^j(t, s) + \frac{DA^j}{Dt}\Delta t + \frac{DA^j}{Ds}\Delta s + A^j_{;l;m}(t, s)\frac{dx^m}{ds}\frac{dx^l}{dt}\Delta t \Delta s.$$

Since A^j is assumed to be parallel translated on both curves, $DA^j/Dt = 0 = DA^j/Ds$, and

$$A^j(t + \Delta t, s + \Delta s)_1 = A^j(t, s) + A^j_{;l;m}(t, s)\Delta x^l \Delta x^m,$$

where we used $\Delta x^l \approx (dx^l/dt)\Delta t$ and $\Delta x^m \approx (dx^m/ds)\Delta s$. The subscript 1 on the left hand side stands for the "first route." The "second route" is going up first and then to the right. It should be clear that the only difference in the final result is the interchange of l and m. We therefore have

$$A^j(t + \Delta t, s + \Delta s)_2 = A^j(t, s) + A^j_{;m;l}(t, s)\Delta x^l \Delta x^m.$$

Thus, using $A^j_{;lm}$ for the second covariant derivative, we have

$$A^j(t + \Delta t, s + \Delta s)_1 - A^j(t + \Delta t, s + \Delta s)_2 = \left(A^j_{;lm} - A^j_{;ml}\right)\Delta x^l \Delta x^m. \quad (17.66)$$

The difference in parentheses should be related to the curvature of the space (manifold) under consideration.

Finding this difference is straightforward. Using the rule of covariant differentiation for general tensors, we get

$$A^j_{;lm} = \frac{\partial A^j_{;l}}{\partial x^m} + \Gamma^j_{km}A^k_{;l} - \Gamma^p_{lm}A^j_{;p}$$

$$= \frac{\partial}{\partial x^m}\left(\frac{\partial A^j}{\partial x^l} + \Gamma^j_{kl}A^k\right) + \Gamma^j_{km}\left(\frac{\partial A^k}{\partial x^l} + \Gamma^k_{rl}A^r\right) - \Gamma^p_{lm}A^j_{;p}$$

$$= \frac{\partial^2 A^j}{\partial x^m \partial x^l} + \frac{\partial \Gamma^j_{kl}}{\partial x^m}A^k + \Gamma^j_{kl}\frac{\partial A^k}{\partial x^m} + \Gamma^j_{km}\frac{\partial A^k}{\partial x^l} + \Gamma^j_{km}\Gamma^k_{rl}A^r - \Gamma^p_{lm}A^j_{;p}.$$

In the last line switch l and m to get $A^j_{;ml}$:

$$A^j_{;ml} = \frac{\partial^2 A^j}{\partial x^l \partial x^m} + \frac{\partial \Gamma^j_{km}}{\partial x^l}A^k + \Gamma^j_{km}\frac{\partial A^k}{\partial x^l} + \Gamma^j_{kl}\frac{\partial A^k}{\partial x^m} + \Gamma^j_{kl}\Gamma^k_{rm}A^r - \Gamma^p_{ml}A^j_{;p}.$$

Subtracting, and changing the dummy indices when necessary, we obtain

$$A^j_{;lm} - A^j_{;ml} = \left(\frac{\partial \Gamma^j_{kl}}{\partial x^m} - \frac{\partial \Gamma^j_{km}}{\partial x^l} + \Gamma^j_{rm}\Gamma^r_{kl} - \Gamma^j_{rl}\Gamma^r_{km}\right)A^k - \left(\Gamma^p_{lm} - \Gamma^p_{ml}\right)A^j_{;p}.$$

$$(17.67)$$

It is straightforward but tedious to show that the expression in the first pair of parentheses transforms as a component of a tensor of type $(1,3)$. This tensor is denoted by R^{j}_{klm} and is called **Riemann curvature tensor**:

Riemann curvature tensor

$$R^{j}_{klm} = \frac{\partial \Gamma^{j}_{kl}}{\partial x^{m}} - \frac{\partial \Gamma^{j}_{km}}{\partial x^{l}} + \Gamma^{j}_{rm}\Gamma^{r}_{kl} - \Gamma^{j}_{rl}\Gamma^{r}_{km}. \qquad (17.68)$$

The expression in the second pair of parentheses in (17.67) is the torsion tensor introduced earlier [see Equation (17.53) and the discussion after it].

Example 17.5.1. Example 17.4.1 showed that the metric connection of Euclidean space and special relativistic spacetime *in Cartesian coordinates* are both zero. Equation (17.68) shows that for these spaces, the Riemannian curvature tensor expressed in Cartesian coordinates is zero. Since Riemannian curvature tensor is a *tensor*, it must be zero *in all coordinates*, as expressed in Box 17.2.2. Spaces that have zero Riemannian curvature tensor are called **flat**. We thus see that flatness is an *intrinsic* property of a space, independent of any coordinates used in that space. ∎

flat spaces (or manifolds)

The curvature tensor has some important properties which we state without proof. One property that is evident from (17.68) is

$$.R^{j}_{klm} = -R^{j}_{kml} \qquad (17.69)$$

The second property, which is true only if the torsion tensor vanishes, i.e., when the affine connection is symmetric, is

$$R^{j}_{klm} + R^{j}_{lmk} + R^{j}_{mkl} = 0. \qquad (17.70)$$

Bianchi identity

The third property, which involves the covariant derivative of the curvature tensor and is true only for torsion-free connections, is

$$R^{j}_{klm;i} + R^{j}_{kmi;l} + R^{j}_{kil;m} = 0. \qquad (17.71)$$

This is also called the **Bianchi identity**. The last property, which holds for Riemannian tensor of the metric connection, is that R^{j}_{klm} has $n^{2}(n^{2}-1)/12$ components.

Various other tensors can be obtained from the Riemann curvature tensor by contraction. For example, by contracting the contravariant index with the last covariant index one obtains the so-called **Ricci tensor**:

Ricci tensor and scalar curvature

$$R_{kl} = R^{j}_{klj} = R^{j}_{klj} = \frac{\partial \Gamma^{j}_{kl}}{\partial x^{j}} - \frac{\partial \Gamma^{j}_{kj}}{\partial x^{l}} + \Gamma^{j}_{rj}\Gamma^{r}_{kl} - \Gamma^{j}_{rl}\Gamma^{r}_{kj}, \qquad (17.72)$$

and by raising one of the Ricci tensor's indices and contracting, we obtain the **scalar curvature**:

$$R = R^{l}_{l} = g^{kl}R_{kl}. \qquad (17.73)$$

Einstein's general theory of relativity explains gravity as a manifestation of the curvature of spacetime. Since gravity is caused by mass, and since mass and energy are equivalent, the source of curvature is energy. Pursuing this idea, Einstein came up with an equation, the **Einstein equation**, that describes all (large scale) gravitational interactions. Defining the **Einstein curvature tensor** as

$$G_{ij} \equiv R_{ij} - \tfrac{1}{2}g_{ij}R, \qquad (17.74)$$

the Einstein equation is written as

$$G_{ij} = 8\pi G T_{ij}, \qquad (17.75)$$

where G is the universal gravitational constant and T_{ij} is the *energy momentum tensor*.

Einstein curvature tensor and Einstein equation

Example 17.5.2. For the sphere of Example 17.4.2, the Ricci curvature tensor can be written as

$$R_{kl} = \frac{\partial \Gamma^\theta_{kl}}{\partial \theta} - \frac{\partial \Gamma^\varphi_{k\varphi}}{\partial x^l} + \Gamma^\varphi_{\theta\varphi}\Gamma^\theta_{kl} - \Gamma^\theta_{\varphi l}\Gamma^\varphi_{k\theta} - \Gamma^\varphi_{\theta l}\Gamma^\theta_{k\varphi} - \Gamma^\varphi_{\varphi l}\Gamma^\varphi_{k\varphi}$$

Using this, it is easy to show that $R_{\theta\varphi} = 0 = R_{\varphi\theta}$, while

$$R_{\theta\theta} = 1, \quad R_{\varphi\varphi} = \sin^2\theta$$

Furthermore, since $g^{\theta\theta} = 1/a^2$ and $g^{\varphi\varphi} = 1/(a^2\sin^2\theta)$, the scalar curvature becomes

$$R = g^{ij}R_{ij} = g^{\theta\theta}R_{\theta\theta} + g^{\varphi\varphi}R_{\varphi\varphi} = \frac{2}{a^2}$$

showing that a sphere is a space of constant (and positive) curvature, as we expect. ∎

17.6 Problems

17.1. Write $\partial_i x_j$ in a form that includes the Kronecker delta. Now show that $\boldsymbol{\nabla} \cdot \mathbf{r} = 3$.

17.2. Recall that a homogeneous function f of n variable of degree q satisfies

$$qf(x_1, x_2, \ldots, x_n) = \sum_{i=1}^n x_i \partial_i f.$$

(a) Differentiate both sides with respect to x_j and show that

$$(q-1)\partial_j f(x_1, x_2, \ldots, x_n) = \sum_{i=1}^n x_i \partial_i \partial_j f.$$

(b) Multiply this equation by x_j and sum over j to obtain

$$q(q-1)f(x_1, x_2, \ldots, x_n) = \sum_{i,j=1}^n x_i x_j \partial_i \partial_j f.$$

17.3. Verify Equation (17.23).

17.4. Let the scalar function ϕ be given by $\phi(x, y, z) = x^2 + y^3 + z$ and

$$x = \sin \bar{x} + \cos \bar{y} + \bar{z}, \ y = \bar{x}\bar{y} + \bar{z}, \ z = \bar{x}^2.$$

What is the functional form of $\bar{\phi}$?

17.5. Show that the sum of two tensors of type (r, s) is a tensor of the same type.

17.6. Derive Equation (17.29). Show that $\delta^{12\cdots n}_{12\cdots n} = 1$.

17.7. Show that the inverse of a metric tensor given by

$$g^{km}(x) \equiv \sum_{p=1}^{n} \frac{\partial x^k}{\partial x'^p} \frac{\partial x^m}{\partial x'^p}$$

is a tensor of type $(2, 0)$. Here $\{x'^i\}$ are as defined in the beginning of Section 17.3.

17.8. Following Example 17.3.1, find the metric tensor for cylindrical coordinates.

17.9. Show that the dot products of Equations (17.36) and (17.37) do not change in a general coordinate transformation.

17.10. Verify Equation (17.40) component by component.

17.11. Using indices, show that the divergence of a curl and the curl of a gradient are both zero.

17.12. Using indices, prove the following "derivative" identities:

$$\boldsymbol{\nabla} \cdot (f\mathbf{A}) = (\boldsymbol{\nabla} f) \cdot \mathbf{A} + f\boldsymbol{\nabla} \cdot \mathbf{A},$$
$$\boldsymbol{\nabla} \times (f\mathbf{A}) = (\boldsymbol{\nabla} f) \times \mathbf{A} + f\boldsymbol{\nabla} \times \mathbf{A},$$
$$\boldsymbol{\nabla}(fg) = g\boldsymbol{\nabla} f + f\boldsymbol{\nabla} g.$$

17.13. Using indices, prove the **Green's identity**:

$$\boldsymbol{\nabla} \cdot (g\boldsymbol{\nabla} f - f\boldsymbol{\nabla} g) = g\nabla^2 f - f\nabla^2 g.$$

17.14. Prove the following vector identities using index notation for vectors:

$$\boldsymbol{\nabla} \cdot (\mathbf{A} \times \mathbf{B}) = \mathbf{B} \cdot \boldsymbol{\nabla} \times \mathbf{A} - \mathbf{A} \cdot \boldsymbol{\nabla} \times \mathbf{B},$$
$$\boldsymbol{\nabla} \times (\boldsymbol{\nabla} \times \mathbf{A}) = \boldsymbol{\nabla}(\boldsymbol{\nabla} \cdot \mathbf{A}) - \nabla^2 \mathbf{A}.$$

17.15. Show that the difference between any two affine connections is a tensor of type $(1, 2)$.

17.16. Verify that Equation (17.46) combines the second and third Maxwell's equations.

17.17. Verify that $F_{\alpha\beta} = \partial_\beta A_\alpha - \partial_\alpha A_\beta$ satisfies Equation (17.46).

17.18. Differentiate both sides of Equation (17.45) with respect to x^β and raise the index β to be able to sum over it; use the symmetry of second derivative and the antisymmetry of $F_{\alpha\beta}$ to show that the left-hand side is zero. On the right-hand side, you should have something like $\mu_0 \eta^{\beta\sigma} \partial_\sigma J_\beta$. Show that $\eta^{\beta\sigma} \partial_\sigma J_\beta = 0$ expresses charge conservation or continuity equation of Box 13.2.4.

17.19. With $c = 1$ and $\mu_0 = 1/\epsilon_0$, show that $\eta^{\alpha\nu} \partial_\nu A_\alpha = 0$ is the Lorentz gauge condition [Equation (15.32)]

$$\frac{\partial \Phi}{\partial t} + \boldsymbol{\nabla} \cdot \mathbf{A} = 0,$$

and that $\eta^{\alpha\nu} \partial_\nu \partial_\alpha A_\beta = \mu_0 J_\beta$ combines the two wave equations [Equations (15.33) and (15.34)]

$$\frac{\partial^2 \mathbf{A}}{\partial t^2} - \nabla^2 \mathbf{A} = \mu_0 \mathbf{J},$$

$$\frac{\partial^2 \Phi}{\partial t^2} - \nabla^2 \Phi = \mu_0 \rho.$$

17.20. Show that DA_j of Box 17.4.1 is a covariant vector.

17.21. Show that

$$\frac{\partial A^j}{\partial x^l} + \Gamma^j_{kl} A^k$$

is a tensor of type $(1, 1)$.

17.22. Show that Γ^j_{lk} given in Equation (17.59) is an affine connection, i.e., that it transforms according to Equation (17.52).

17.23. Show that the metric connection satisfies Equation (17.61).

17.24. (a) Find all the components of the affine metric connection on the surface of the sphere of Example 17.4.2.
(b) Derive Equation (17.62) from Equation (17.58).
(c) Show that (17.63) satisfies the second equation of (17.62).
(d) Show that $\cot \theta = A \cos \varphi + B \sin \varphi$ is a solution of (17.65).

17.25. Show that the Riemann curvature tensor of Equation (17.68) is a tensor of type $(1, 3)$.

17.26. Example 17.5.1 showed that the Riemannian curvature tensor of the Euclidean space, when expressed in Cartesian coordinates is zero. Since Riemannian curvature tensor is a tensor it should be zero when expressed in *any* coordinate system. Starting with the spherical components of the Euclidean metric obtained in Example 17.3.1, find the components of the metric connection in spherical coordinates. From these calculate the components of Riemannian curvature tensor and show that they all vanish.

17.27. Derive the expression for the Ricci curvature tensor of Example 17.5.2 and show that

$$R_{\theta\varphi} = 0 = R_{\varphi\theta}, \quad R_{\theta\theta} = 1, \quad R_{\varphi\varphi} = \sin^2\theta.$$

Part V

Complex Analysis

Chapter 18

Complex Arithmetic

Complex numbers were developed because there was a need to expand the notion of numbers to include solutions of algebraic equations whose prototype is $x^2 + 1 = 0$. Such developments are not atypical in the history of mathematics. The invention of irrational numbers occurred because of a need for a number that could solve an equation of the form $x^2 - 2 = 0$. Similarly, rational numbers were the offspring of the operations of multiplication and division and the quest for a number that gives, for example, 4 when multiplied by 3, or, equivalently, a number that solves the equation $3x - 4 = 0$.

There is a crucial difference between complex numbers and all the numbers mentioned above: All rational, irrational, and, in general, real numbers correspond to measurable physical quantities. However, there is no single measurable physical quantity that can be described by a complex number.

A natural question then is this: What need is there for complex numbers if no physical quantity can be measured in terms of them? The answer is that although no *single* physical quantity can be expressed in terms of complex numbers, *a pair* of physical quantities can be neatly described by a single complex number. For example, a wave with a given amplitude and phase can be concisely described by a complex number. Another, more fundamental, reason is that equations that describe the behavior of subatomic particles are *inherently* complex.

18.1 Cartesian Form of Complex Numbers

We demand a number system broad enough to include solutions to the equation

$$x^2 + 1 = 0 \qquad \text{or} \qquad x^2 = -1.$$

Clearly the solution(s) cannot be real because a real number raised to the second power gives a positive real number, and we want x^2 to be negative.

Cartesian form of
a complex number

So we broaden the concept of numbers by considering **complex numbers**. Such numbers are of the form

$$z = x + iy \quad \text{with} \quad i \equiv \sqrt{-1} \quad \text{and} \quad i^2 = -1. \quad (18.1)$$

It turns out that we don't need to introduce any other numbers to solve *all* algebraic equations—equations of the form $p(x) = 0$ with $p(x)$ a polynomial. In fact, the **fundamental theorem of algebra**, to which we shall return, states that all roots of any algebraic equation

$$a_n x^n + a_{n1-} x^{n-1} + \cdots + a_1 x + a_0 = 0$$

with arbitrary real or complex coefficients $a_0, a_1, \ldots, a_n$, are in the complex number system. In this sense, then, the complex number system is the most complete system.

complex plane,
real and imaginary
parts

A complex number can be conveniently represented as a point (or equivalently, as a vector) in the xy-plane, called the **complex plane**, as shown in Figure 18.1. In Equation (18.1), x is called the **real part** of z, written $\mathrm{Re}(z)$, and y is called the **imaginary part** of z, written $\mathrm{Im}(z)$. Similarly, the horizontal axis in Figure 18.1 is named the **real axis**, and the vertical axis is named the **imaginary axis**. The set of all complex numbers—or the set of points in the complex plane—is denoted by $\mathbb{C}$.

We can define various operations on $\mathbb{C}$ that are extensions of similar operations on the real number system, $\mathbb{R}$. The only proviso is that $i^2 = -1$, and that the final form of an equation must be written as Equation (18.1)—with real and imaginary parts. For instance, the sum of two complex numbers, $z_1 = x_1 + iy_1$ and $z_2 = x_2 + iy_2$, is

$$z_1 + z_2 = (x_1 + x_2) + i(y_1 + y_2).$$

This sum can be represented in the complex plane as the vector sum of z_1 and z_2, as shown in Figure 18.2. The product of z_1 and z_2 can also be obtained:

$$z_1 z_2 = (x_1 + iy_1)(x_2 + iy_2) = x_1 x_2 + x_1(iy_2) + iy_1 x_2 + iy_1(iy_2)$$
$$= x_1 x_2 + i(x_1 y_2 + y_1 x_2) - y_1 y_2 = x_1 x_2 - y_1 y_2 + i(x_1 y_2 + y_1 x_2).$$

Thus,

$$\mathrm{Re}(z_1 z_2) = x_1 x_2 - y_1 y_2,$$
$$\mathrm{Im}(z_1 z_2) = x_1 y_2 + x_2 y_1. \quad (18.2)$$

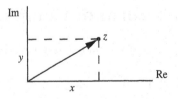

Figure 18.1: Complex numbers as points or vectors in a plane.

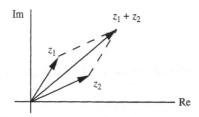

Figure 18.2: Addition of complex numbers as addition of vectors.

To obtain this equation, we have implicitly used the fact that two complex numbers are equal if and only if their real parts are equal and their imaginary parts are equal.

The factor i in z allows new operations for complex numbers that do not exist for real numbers. One such operation is **complex conjugation**. The *complex conjugate*, z^* or $\bar{z}$, of z is defined as

complex conjugation

$$z^* \equiv \bar{z} = (x + iy)^* = x - iy \qquad (18.3)$$

which is obtained from z by replacing i with $-i$. We note immediately that

$$zz^* = (x + iy)(x - iy) = x^2 + y^2 = z^*z$$

which is a positive real number. The positive square root of zz^* is called the **absolute value** of z and denoted by $|z|$. It is simply the length of the vector representing z in the xy-plane. Thus, we have

absolute value

$$|z| = \sqrt{zz^*} = \sqrt{z^*z} = \sqrt{x^2 + y^2} = \sqrt{(\mathrm{Re}(z))^2 + (\mathrm{Im}(z))^2}. \qquad (18.4)$$

We can also define the division of two complex numbers using complex conjugation.

> **Box 18.1.1.** *To find the real and imaginary parts of a quotient, multiply the numerator and denominator by the complex conjugate of the denominator.*

So, for the ratio of z_1/z_2, we get

$$\frac{z_1}{z_2} = \frac{z_1 z_2^*}{z_2 z_2^*} = \frac{(x_1 + iy_1)(x_2 - iy_2)}{|z_2|^2} = \frac{x_1 x_2 + y_1 y_2 + i(y_1 x_2 - x_1 y_2)}{|z_2|^2}$$

$$= \frac{x_1 x_2 + y_1 y_2}{|z_2|^2} + i\frac{y_1 x_2 - x_1 y_2}{|z_2|^2}.$$

Thus,

$$\mathrm{Re}\left(\frac{z_1}{z_2}\right) = \frac{x_1 x_2 + y_1 y_2}{x_2^2 + y_2^2} \qquad \text{and} \qquad \mathrm{Im}\left(\frac{z_1}{z_2}\right) = \frac{y_1 x_2 - x_1 y_2}{x_2^2 + y_2^2}. \qquad (18.5)$$

properties of
absolute value of
complex numbers

In particular,

$$\frac{1}{z} = \frac{z^*}{|z|^2} = \frac{x - iy}{x^2 + y^2} \qquad \text{and} \qquad \frac{1}{i} = -i.$$

Some useful properties of absolute values are as follows:

$$|z_1 z_2| = |z_1|\,|z_2|, \qquad \left|\frac{z_1}{z_2}\right| = \frac{|z_1|}{|z_2|},$$

$$\Big||z_1| - |z_2|\Big| \leq |z_1 + z_2| \leq |z_1| + |z_2|. \tag{18.6}$$

This last inequality is called the **triangle inequality** and it comes directly from the vector property of complex numbers. The right half of it can be generalized to more than two complex numbers:

$$\left|\sum_{k=1}^{n} z_k\right| \leq \sum_{k=1}^{n} |z_k|. \tag{18.7}$$

Example 18.1.1. Here we present some sample manipulations with complex numbers:

$$(1+i)^2 = (1)^2 + (i)^2 + 2i = 1 - 1 + 2i = 2i,$$

$$\frac{1}{1-i} - \frac{1}{1+i} = \frac{1 + i - (1 - i)}{(1-i)(1+i)} = \frac{2i}{|1+i|^2} = \frac{2i}{2} = i,$$

$$(1+i)^{-4} = \frac{1}{(1+i)^2(1+i)^2} = \frac{1}{(2i)(2i)} = \frac{1}{-4} = -\frac{1}{4},$$

$$\frac{2+i}{3-i} = \frac{(2+i)(3+i)}{|3-i|^2} = \frac{5+i5}{3^2+(-1)^2} = \frac{1}{2} + i\frac{1}{2},$$

$$\left|\frac{2i-1}{i-2}\right| = \frac{|-1+i2|}{|-2+i|} = \frac{\sqrt{(-1)^2+2^2}}{\sqrt{(-2)^2+1^2}} = 1.$$

The equation $|z - a| = b$, where a is a fixed complex number and b is real and positive, describes a circle of radius b with center at $a \equiv a_x + ia_y$. This is easily seen because

$$b^2 = |z - a|^2 = |(x + iy) - (a_x + ia_y)|^2$$
$$= |(x - a_x) + i(y - a_y)|^2 = (x - a_x)^2 + (y - a_y)^2.$$

We note that $|z - a|$ is the distance between the two complex numbers z and a. Therefore, $|z - a| = b$—with a a constant and z a variable—is the collection of all points z that are at a distance b from a.   ∎

properties of
complex
conjugation of
complex numbers

Complex conjugation satisfies some nice properties that we list below:

$$(z_1 + z_2)^* = z_1^* + z_2^*, \qquad (z_1 z_2)^* = z_1^* z_2^*, \qquad \left(\frac{z_1}{z_2}\right)^* = \frac{z_1^*}{z_2^*},$$

$$\text{Re}(z) = \tfrac{1}{2}(z + z^*), \qquad \text{Im}(z) = \frac{1}{2i}(z - z^*), \tag{18.8}$$

$$(z^*)^* = z, \qquad (z^n)^* = (z^*)^n.$$

The complex conjugate of a function of z is easily obtained by substituting z^* for z in that function.[1] This can be summarized as

$$(f(z))^* = f(z^*) \qquad (18.9)$$

which is equivalent to replacing every i with $-i$ in the expression for $f(z)$.

to find the complex conjugate of a function, change all its i's to $-i$.

Historical Notes

In the first half of the sixteenth century there was hardly any change from the attitude or spirit of Arabs, whose work had put practical arithmetical calculations in the forefront of mathematics, but merely an increase in the kind of activity Europeans had learned from Arabs. Moreover, the technological advances spurred by the Renaissance demanded further refinement in magnitudes such as trigonometric tables and astronomical observations.

By 1500 or so, zero was accepted as a number and irrational numbers were used more freely in calculations. However, the problem of whether irrationals were really numbers still troubled people. **Michael Stifel** (1486?–1567), the German mathematician, argued that

> Since, in proving geometrical figures, when rational numbers fail us irrational numbers ... prove exactly those things which rational numbers could not prove ... we are compelled to assert that they truly are numbers .... On the other hand, ... that cannot be called a true number which is of such a nature that it lacks precision [decimal representation].

He then argues that only whole numbers or fractions can be called true numbers, and since irrationals are neither, they are not real numbers. Even a century later, **Pascal**, **Barrow**, and **Newton** thought of irrational numbers as being understood in terms of geometric magnitude; they were mere symbols that had no existence independent of continuous geometrical magnitude.

Negative numbers were treated with equal suspicion by the sixteenth- and seventeenth-century mathematicians. They were considered "absurd." **Jerome Cardan** (1501–1576), the great Italian mathematician of the Renaissance, was willing to accept the negative numbers as roots of equations, but considered them as "fictitious," while he called the positive roots real. **François Vieta** (1540–1603), a lawyer by profession but recognized far more as the foremost mathematician of the sixteenth century, discarded negative numbers entirely. **Descartes** accepted them in part, but called negative roots of equations false, on the grounds that they represented numbers less than nothing.

An interesting argument against negative numbers was given by Antoine Arnauld (1612–1694), a theologian and mathematician who was a close friend of Pascal. Arnauld questioned the equality $-1 : 1 = 1 : (-1)$ because, he said, -1 is less than $+1$; hence, How could a smaller number be to a greater as a greater is to a smaller?

Without having fully overcome their difficulties with irrational and negative numbers, the Europeans were hit by another problem: the *complex numbers*! They obtained these new numbers by extending the arithmetic operation of square root

---

[1]This statement is not strictly true for *all* functions. However, only a mild restriction is to be imposed on them for the statement to be true. We shall not go into details of such restrictions because they require certain complex analytic tools which go beyond the scope of this book. See Hassani, S. *Mathematical Physics: A Modern Introduction to Its Foundations*, Springer-Verlag, 1999, Chapter 11 for details.

to whatever numbers appeared in solving quadratic equations. Thus Cardan sets up and solves the problem of dividing 10 into two parts whose product is 40. The equation is $x(10-x) = 40$, for which he obtains the roots $5 \pm \sqrt{-15}$ and then he says "Putting aside the mental torture involved," multiply these two roots and note that the product is $25 - (-15)$ or 40. He then states, "So progresses arithmetic subtlety the end of which, as is said, is as refined as it is useless."

Descartes also rejected complex roots and coined them "imaginary." Even Newton did not regard complex roots as significant, most likely because in his day they lacked physical meaning. The confusion surrounding complex numbers is illustrated by the oft-quoted statement by Leibniz, "The Divine Spirit found a sublime outlet in that wonder of analysis, that portent of the ideal world, that amphibian between being and not being, which we call the imaginary root of negative unity."

18.2   Polar Form of Complex Numbers

The introduction of polar coordinates in the complex plane makes available a powerful tool with which to facilitate complex manipulations. Figure 18.3 shows a complex number and its polar coordinates. In terms of these polar coordinates, z can be written as

polar
representation of a
complex number

$$z = x + iy = r\cos\theta + ir\sin\theta = r(\cos\theta + i\sin\theta). \qquad (18.10)$$

Assuming that series of complex numbers can be manipulated as those of real numbers,[2] we obtain the useful relation between imaginary exponentials and trigonometric functions.

In Chapter 10 we presented the Maclaurin series for the exponential and trigonometric functions. Let us assume that those functions are valid for complex numbers as well. Then, we have

a very important
relation

$$e^{i\theta} = \sum_{n=0}^{\infty} \frac{(i\theta)^n}{n!} = \sum_{n=\text{even}}^{\infty} \frac{(i\theta)^n}{n!} + \sum_{n=\text{odd}}^{\infty} \frac{(i\theta)^n}{n!} = \sum_{k=0}^{\infty} \frac{(i\theta)^{2k}}{(2k)!} + \sum_{k=0}^{\infty} \frac{(i\theta)^{2k+1}}{(2k+1)!}$$

$$= \sum_{k=0}^{\infty} (-1)^k \frac{\theta^{2k}}{(2k)!} + i \sum_{k=0}^{\infty} (-1)^k \frac{\theta^{2k+1}}{(2k+1)!} = \cos\theta + i\sin\theta \qquad (18.11)$$

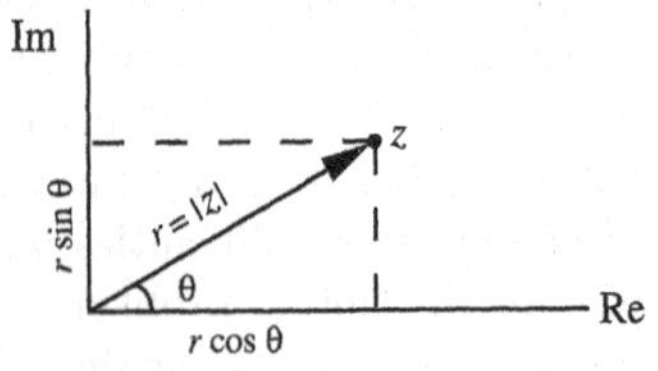

Figure 18.3: Complex numbers in polar coordinates.

[2]This assumption turns out to be correct. In particular, the power series expansion used in the following example plays a central role in complex analysis.

because $i^{2k} = (i^2)^k = (-1)^k$. This is probably the most important relation in complex number theory.

> **Box 18.2.1.** *The trigonometric and imaginary exponential functions are related by the **Euler equation**: $e^{i\theta} = \cos\theta + i\sin\theta$.*

The use of Equation (18.11) in (18.10) leads to another way of representing complex numbers:

$$z = re^{i\theta}, \qquad r = \sqrt{x^2 + y^2}, \qquad \theta = \tan^{-1}\left(\frac{y}{x}\right). \qquad (18.12)$$

Note that

> **Box 18.2.2.** *The angle θ is not uniquely determined: Any multiple of 2π can be added to it without affecting z.*

We can use Equation (18.12) together with $x = r\cos\theta$, $y = r\sin\theta$ to convert from Cartesian coordinates to polar coordinates, and vice versa. The coordinate θ is called the **argument** of z and written $\theta = \arg(z)$.

argument of a complex number

Example 18.2.1. Let us look at some numerical examples of polar-Cartesian conversion. In many cases, a diagram can be very helpful. For instance, take i whose real part is obviously zero and whose imaginary part is 1. If we were to use the formula, we would have $\tan\theta = 1/0$ which is not defined. However, Figure 18.4 shows that $z = i$ lies on the positive imaginary axis, and, thus, $\theta = \pi/2$. Since we can always add a multiple of 2π to the angle, we have

$$i = e^{i\pi/2 + i2n\pi}, \qquad n = 0, \pm1, \pm2, \ldots.$$

Similarly, the same figure makes it clear that

$$-i = e^{-i\pi/2 + i2n\pi} = e^{i3\pi/2 + i2n\pi}, \quad n = 0, \pm1, \pm2, \ldots.$$

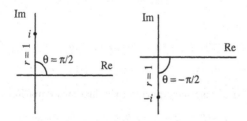

Figure 18.4: Cartesian and polar coordinates for i and $-i$.

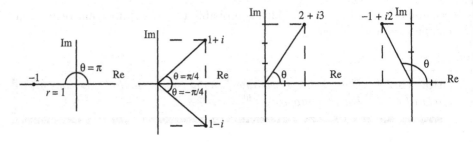

Figure 18.5: Cartesian and polar coordinates for some other complex numbers.

Referring to Figure 18.5, the reader may verify the following polar representations of complex numbers:

$$-1 = e^{i\pi + i2n\pi},$$

$$1 + i = \sqrt{2}\, e^{i\pi/4 + i2n\pi},$$

$$1 - i = \sqrt{2}\, e^{-i\pi/4 + i2n\pi} = \sqrt{2}\, e^{i7\pi/4 + i2n\pi},$$

$$2 + i3 = \sqrt{13}\, e^{i\tan^{-1}(3/2) + i2n\pi} = \sqrt{13}\, e^{i0.983 + i2n\pi},$$

$$-1 + i2 = \sqrt{5}\, e^{i\tan^{-1}(-2) + i2n\pi} = \sqrt{5}\, e^{i2.03 + i2n\pi}.$$

In all cases, n is an integer and angles are in radians. ∎

The complex conjugate of z in polar coordinates is

$$z^* = x - iy = r\cos\theta - ir\sin\theta = r\cos(-\theta) + ir\sin(-\theta) = re^{-i\theta}.$$

This equation confirms the earlier statement that complex conjugation is equivalent to replacing i with $-i$.

Generally speaking, polar coordinates are useful for operations of multiplication, division, and exponentiation, and Cartesian coordinates for addition and subtraction.

Example 18.2.2. We can use the polar representation of complex numbers to find some trigonometric identities. In all of the following, we set $r = 1$:

$$1 = e^{i\theta} e^{-i\theta} = (\cos\theta + i\sin\theta)(\cos\theta - i\sin\theta) = \cos^2\theta + \sin^2\theta.$$

Now consider the identity

$$e^{i(\theta_1 + \theta_2)} = \cos(\theta_1 + \theta_2) + i\sin(\theta_1 + \theta_2)$$

which can also be written as

$$e^{i(\theta_1 + \theta_2)} = e^{i\theta_1} e^{i\theta_2} = (\cos\theta_1 + i\sin\theta_1)(\cos\theta_2 + i\sin\theta_2)$$

$$= \cos\theta_1 \cos\theta_2 - \sin\theta_1 \sin\theta_2 + i(\sin\theta_1 \cos\theta_2 + \sin\theta_2 \cos\theta_1).$$

Equating the real and imaginary parts of the last two equations, we obtain

$$\cos(\theta_1 + \theta_2) = \cos\theta_1 \cos\theta_2 - \sin\theta_1 \sin\theta_2,$$

$$\sin(\theta_1 + \theta_2) = \sin\theta_1 \cos\theta_2 + \sin\theta_2 \cos\theta_1.$$

Similarly, equating the real and imaginary parts of

$$e^{i3\theta} = \cos 3\theta + i \sin 3\theta$$

and

$$e^{i3\theta} = \left(e^{i\theta}\right)^3 = (\cos\theta + i\sin\theta)^3 = \cos^3\theta + 3i\cos^2\theta\sin\theta - 3\sin^2\theta\cos\theta - i\sin^3\theta$$

gives the following trigonometric identity:

$$\cos 3\theta = 4\cos^3\theta - 3\cos\theta,$$
$$\sin 3\theta = 3\sin\theta - 4\sin^3\theta.$$ ∎

From

$$e^{in\theta} = \cos n\theta + i\sin n\theta \qquad \text{and} \qquad e^{in\theta} = \left(e^{i\theta}\right)^n = (\cos\theta + i\sin\theta)^n$$

we obtain the so-called **de Moivre theorem**:   de Moivre theorem

$$(\cos\theta + i\sin\theta)^n = \cos n\theta + i\sin n\theta. \qquad (18.13)$$

Equation (18.11) and its complex conjugate lead to the following useful
results: two important
                                                                    relations

$$\cos\theta = \tfrac{1}{2}\left(e^{i\theta} + e^{-i\theta}\right),$$
$$\sin\theta = \frac{1}{2i}\left(e^{i\theta} - e^{-i\theta}\right). \qquad (18.14)$$

As mentioned earlier, the exponential nature of polar coordinates makes
them especially useful in multiplication, division, and exponentiation. For
instance,

$$\frac{z_1}{z_2} = \frac{r_1 e^{i\theta_1}}{r_2 e^{i\theta_2}} = \frac{r_1}{r_2} e^{i(\theta_1 - \theta_2)},$$
$$z_1 z_2 = \left(r_1 e^{i\theta_1}\right)\left(r_2 e^{i\theta_2}\right) = r_1 r_2 e^{i(\theta_1 + \theta_2)}, \qquad (18.15)$$
$$\sqrt{z} = \sqrt{r e^{i\theta}} = \left(r e^{i\theta}\right)^{1/2} = r^{1/2}\left(e^{i\theta}\right)^{1/2} = \sqrt{r} e^{i\theta/2},$$

and so forth.

All of these relations have interesting geometric interpretations. For example, the second equation says that when you multiply a complex number z_1
by another complex number z_2, you dilate the magnitude of z_1 by a factor r_2
and increase its angle by θ_2. That is, multiplication involves both a dilation
and a rotation. In particular, if we multiply a complex number by $e^{i\omega t}$ where
t is time, we get a vector of constant length in the xy-plane that is rotating
with angular velocity ω.

Example 18.2.3. A *plane* wave is represented by a periodic function such as

$$A\cos(kx - \omega t) \qquad \text{or} \qquad B\sin(kx - \omega t).$$

On the other hand, sine and cosine are related by

$$\sin(kx - \omega t) = -\cos\left(kx - \omega t + \frac{\pi}{2}\right).$$

Therefore, one can concentrate solely on the cosine function with a phase angle added to its argument. Thus a typical periodic plane wave is represented as $A\cos(kx - \omega t + \alpha)$. To make connection with the material of this section, we note that

$$A\cos(kx - \omega t + \alpha) = A\ \mathrm{Re}\left(e^{i(kx - \omega t + \alpha)}\right) = \mathrm{Re}\left(Ae^{i(kx - \omega t + \alpha)}\right)$$

$$= \mathrm{Re}\left(Ae^{i\alpha}e^{i(kx - \omega t)}\right) = \mathrm{Re}\left(Ze^{i(kx - \omega t)}\right),$$

complex amplitude   where Z is a complex number—called **complex amplitude**—of magnitude A and argument α. It is therefore convenient to represent plane waves by the complex function $Ze^{i(kx - \omega t)}$ which includes the phase of the wave as the argument of Z. ∎

roots of complex numbers   Another interesting application of these ideas is finding roots of complex numbers. Suppose we are interested in all the nth roots of Z; i.e., all z's satisfying $z^n = Z$. To find the roots of a complex number Z, write it in polar form in the most general way:

$$Z = Re^{i\Theta + i2\pi k}, \qquad k = 0, \pm 1, \pm 2, \ldots,$$

Thus,

$$z^n = Re^{i\Theta + i2\pi k} \qquad \text{with}\quad k = 0, \pm 1, \pm 2, \ldots.$$

Taking the nth root of both sides, we obtain

$$z = Z^{1/n} = R^{1/n}e^{i\Theta/n + i2\pi k/n}, \qquad k = 0, \pm 1, \pm 2, \ldots,$$

and

Box 18.2.3. *The distinct nth roots $\{z_k\}$ of $Z = Re^{i\Theta}$ are*

$$z_k = R^{1/n}e^{i\Theta/n + i2\pi k/n}, \qquad k = 0, 1, 2, \ldots, n-1. \tag{18.16}$$

We see that the number of nth roots of a complex number is exactly n.

It is clear that z_k of Equation (18.16) repeats itself for $k \geq n$.

Example 18.2.4. Let us find the three cube roots of unity. With $n = 3$ and $Z = e^{i2\pi k}$, we have

$$z_k = e^{i2\pi k/3}, \quad k = 0, 1, 2,$$

or

$$z_0 = e^0 = 1,$$

$$z_1 = e^{i2\pi/3} = \cos\frac{2\pi}{3} + i\sin\frac{2\pi}{3} = -\frac{1}{2} + i\frac{\sqrt{3}}{2},$$

$$z_2 = e^{i4\pi/3} = \cos\frac{4\pi}{3} + i\sin\frac{4\pi}{3} = -\frac{1}{2} - i\frac{\sqrt{3}}{2}.$$

It is instructive to show directly that

$$\left(-\frac{1}{2}+i\frac{\sqrt{3}}{2}\right)^3 = 1 \quad \text{and} \quad \left(-\frac{1}{2}-i\frac{\sqrt{3}}{2}\right)^3 = 1.$$

Here are some more examples of finding roots:

$$\sqrt{1+i} = \left(\sqrt{2}\,e^{i\pi/4+i2n\pi}\right)^{1/2} = 2^{1/4}e^{i\pi/8+in\pi} \qquad n = 0, 1,$$

$$z_0 = 2^{1/4}e^{i\pi/8} = 2^{1/4}\left\{\cos\left(\frac{\pi}{8}\right)+i\sin\left(\frac{\pi}{8}\right)\right\} = 1.1 + i0.456,$$

$$z_1 = 2^{1/4}e^{i\pi/8+i\pi} = -2^{1/4}e^{i\pi/8} = -1.1 - i0.456.$$

The equation $z^3 = i$ has the roots

$$\sqrt[3]{i} = \left(e^{i\pi/2+i2n\pi}\right)^{1/3} = e^{i\pi/6+i2n\pi/3}, \qquad n = 0, 1, 2,$$

or

$$z_0 = e^{i\pi/6} = \cos\left(\frac{\pi}{6}\right) + i\sin\left(\frac{\pi}{6}\right) = \frac{\sqrt{3}}{2} + i\frac{1}{2},$$

$$z_1 = e^{i\pi/6+i2\pi/3} = \cos\left(\frac{5\pi}{6}\right) + i\sin\left(\frac{5\pi}{6}\right) = -\frac{\sqrt{3}}{2} + i\frac{1}{2},$$

$$z_2 = e^{i\pi/6+i4\pi/3} = \cos\left(\frac{3\pi}{2}\right) + i\sin\left(\frac{3\pi}{2}\right) = -i.$$

The reader is urged to show that $z_k^3 = i$ for $k = 0, 1, 2$.

Note how careful we were to include the factor of $e^{i2n\pi}$ when taking roots of complex numbers. ■

All nth roots of $Z = Re^{i\Theta}$ are equally spaced on a circle of radius $R^{1/n}$ in the complex plane. Figure 18.6 shows two circles on which the sixth and the eighth roots of unity are located.

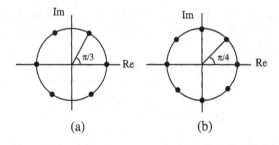

(a)   (b)

Figure 18.6: The (a) sixth and (b) eighth roots of unity.

Example 18.2.5. In certain applications of electromagnetic wave propagation (as in conductors) it becomes necessary to find an analytic expression for the Cartesian representation of the square root of a complex number. In this example, we derive such an expression.

Cartesian form of the square root of a complex number

We are trying to calculate the Cartesian representation of the square root of $z = x + iy$. First we express z in polar form; next we take its square root, and finally reexpress the result in Cartesian form. Thus,

$$z = re^{i(\theta+2n\pi)} \qquad \text{where} \quad r = \sqrt{x^2 + y^2}, \quad \tan\theta = \frac{y}{x}, \quad n = 0, \pm 1, \pm 2, \ldots.$$

Taking the square root of both sides yields

$$\sqrt{z} = z^{1/2} = r^{1/2}e^{i(\theta+2n\pi)/2} = (x^2+y^2)^{1/4}e^{i\theta/2+in\pi}$$

$$= \pm(x^2+y^2)^{1/4}e^{i\theta/2} = \pm(x^2+y^2)^{1/4}\left(\cos\frac{\theta}{2} + i\sin\frac{\theta}{2}\right)$$

because $e^{in\pi} = 1$ if n is even and $e^{in\pi} = -1$ if n is odd. All that is left now is to express the trigonometric functions in terms of x and y:

$$\cos\frac{\theta}{2} = \left[\tfrac{1}{2}(1+\cos\theta)\right]^{1/2} = \frac{1}{\sqrt{2}}\left(1 + \frac{1}{\sqrt{1+\tan^2\theta}}\right)^{1/2}$$

$$= \frac{1}{\sqrt{2}}\left(1 + \frac{1}{\sqrt{1+(y/x)^2}}\right)^{1/2} = \frac{1}{\sqrt{2}}\left(1 + \frac{|x|}{\sqrt{x^2+y^2}}\right)^{1/2}.$$

Similarly,

$$\sin\frac{\theta}{2} = \frac{1}{\sqrt{2}}\left(1 - \frac{|x|}{\sqrt{x^2+y^2}}\right)^{1/2}.$$

Collecting all these formulas together and simplifying, we obtain

$$\sqrt{x+iy} = \pm\frac{1}{\sqrt{2}}\left[\left(\sqrt{x^2+y^2}+|x|\right)^{1/2} + i\left(\sqrt{x^2+y^2}-|x|\right)^{1/2}\right]. \qquad (18.17)$$

The complexity of the expression for the square root rests on our insistence on an *analytic* form. The process of converting the Cartesian form of a complex *number* to polar, taking the square root, and converting the result back to Cartesian form is a far easier process than the one leading to Equation (18.17). ∎

18.3 Fourier Series Revisited

The connection between the trigonometric and exponential functions can be utilized to write the Fourier series expansion of periodic functions more succinctly. If we substitute

$$\cos\frac{2n\pi x}{L} = \frac{e^{2in\pi x/L} + e^{-2in\pi x/L}}{2},$$

$$\sin\frac{2n\pi x}{L} = \frac{e^{2in\pi x/L} - e^{-2in\pi x/L}}{2i},$$

in Equation (10.38) and collect the similar exponential terms, we obtain

$$f(x) = a_0 + \tfrac{1}{2}\sum_{n=1}^{\infty}\left[(a_n - ib_n)e^{2in\pi x/L} + (a_n + ib_n)e^{-2in\pi x/L}\right]$$

$$= a_0 + \tfrac{1}{2}\sum_{n=1}^{\infty}(a_n - ib_n)e^{2in\pi x/L} + \tfrac{1}{2}\sum_{n=1}^{\infty}(a_n + ib_n)e^{-2in\pi x/L}. \qquad (18.18)$$

In the second sum, let $n = -m$ to obtain

$$\text{2nd sum} = \tfrac{1}{2} \sum_{m=-1}^{-\infty} (a_{-m} + ib_{-m})\, e^{2im\pi x/L} = \tfrac{1}{2} \sum_{n=-1}^{-\infty} (a_{-n} + ib_{-n})\, e^{2in\pi x/L},$$

$$(18.19)$$

where in the last step, we switched the dummy index back to n. If we now introduce new coefficients A_n defined as

$$A_n = \begin{cases} \tfrac{1}{2}(a_n - ib_n) & \text{if} \quad 1 \le n \le \infty, \\ \tfrac{1}{2}(a_{-n} + ib_{-n}) & \text{if} \quad -\infty \le n \le -1, \\ a_0 & \text{if} \quad n = 0, \end{cases}$$

and use Equation (18.19) in (18.18), we obtain

$$f(x) = \sum_{n=-\infty}^{+\infty} A_n e^{2in\pi x/L} \qquad \text{where} \quad L = b - a, \qquad (18.20)$$

Fourier series in terms of complex exponentials

which is the equation we are after. To find A_n directly from this equation, multiply both sides by $e^{-2ik\pi x/L}$, integrate from a to b, and use the readily obtainable relation

$$\int_a^b e^{2i(n-k)\pi x/L} = \begin{cases} 0 & \text{if} \quad n \ne k \\ L & \text{if} \quad n = k \end{cases} = L\delta_{nk}, \qquad (18.21)$$

where δ_{nk} is the Kronecker delta. It follows that

$$A_k = \frac{1}{L} \int_a^b f(x) e^{-2ik\pi x/L} dx \quad \text{or} \quad A_n = \frac{1}{L} \int_a^b f(x) e^{-2in\pi x/L} dx. \quad (18.22)$$

It is customary to redefine the coefficients in the summation of Equation (18.20) in such a way that the summation giving $f(x)$ and the integral giving A_n are more symmetric, i.e., have the same constant in front of them. To this end, define $f_n \equiv \sqrt{L} A_n$. Then (18.20) and (18.22) become

$$f(x) = \frac{1}{\sqrt{L}} \sum_{n=-\infty}^{+\infty} f_n e^{2in\pi x/L}, \quad f_n = \frac{1}{\sqrt{L}} \int_a^b f(x) e^{-2in\pi x/L} dx \qquad (18.23)$$

Note that the coefficients f_n are complex; however, when $f(x)$ is a real function, the exponentials and their complex coefficients add up in such a way that the final result can be expressed as an infinite sum of trigonometric functions with *real* coefficients. In fact, we can show this generally using Equations (18.23). First, we note that, for real $f(x)$,

$$f_n^* = \frac{1}{\sqrt{L}} \int_a^b f(x) e^{+2in\pi x/L} dx = \frac{1}{\sqrt{L}} \int_a^b f(x) e^{-2i(-n)\pi x/L} dx = f_{-n}.$$

$$(18.24)$$

Next, we split the sum in (18.23) into positive integers, negative integers, and zero:

$$f(x) = \frac{1}{\sqrt{L}} \sum_{n=-\infty}^{-1} f_n e^{2in\pi x/L} + \frac{f_0}{\sqrt{L}} + \frac{1}{\sqrt{L}} \sum_{n=1}^{\infty} f_n e^{2in\pi x/L}. \qquad (18.25)$$

Changing the dummy index n to $-m$, the first sum can be rewritten as

$$\text{1st sum} = \frac{1}{\sqrt{L}} \sum_{-m=-\infty}^{-1} f_{-m} e^{-2im\pi x/L} = \frac{1}{\sqrt{L}} \sum_{m=1}^{\infty} f_{-m} e^{-2im\pi x/L}$$

$$= \frac{1}{\sqrt{L}} \sum_{m=1}^{\infty} f_m^* e^{-2im\pi x/L} = \frac{1}{\sqrt{L}} \sum_{n=1}^{\infty} f_n^* e^{-2in\pi x/L},$$

where we used Equation (18.24) and changed m back to n at the end. Substituting the last equation in (18.25) yields

$$f(x) = \frac{f_0}{\sqrt{L}} + \frac{1}{\sqrt{L}} \sum_{n=1}^{\infty} \left(f_n^* e^{-2in\pi x/L} + f_n e^{2in\pi x/L} \right)$$

$$= \frac{f_0}{\sqrt{L}} + \frac{2}{\sqrt{L}} \sum_{n=1}^{\infty} \text{Re} \left(f_n e^{2in\pi x/L} \right)$$

showing that $f(x)$ is indeed real. Equation (18.23) implies that f_0 is also real when $f(x)$ is. It is not hard to show that the expression in the parentheses of the first line is the sum of a sine and a cosine with real coefficients.

Example 18.3.1. Let us redo the square potential—whose Fourier series was calculated in Example 10.6.1—using exponentials. From Equation (18.23), for $n \neq 0$, we obtain

$$f_n = \frac{1}{\sqrt{2T}} \int_0^{2T} V(t) e^{-2in\pi t/(2T)} dt = \frac{1}{\sqrt{2T}} \int_0^T V_0 e^{-in\pi t/T} dt$$

$$= \frac{V_0}{\sqrt{2T}} \frac{T}{-in\pi} e^{-in\pi t/T} \Big|_0^T = \sqrt{\frac{T}{2}} \frac{V_0}{in\pi} [1 - (-1)^n]$$

because $e^{in\pi} = (e^{i\pi})^n = (-1)^n$. Similarly, $f_0 = V_0 \sqrt{T/2}$. We now substitute these in the Fourier series expansion

$$V(t) = \frac{1}{\sqrt{2T}} \sum_{n=-\infty}^{+\infty} f_n e^{2in\pi t/2T}$$

to get

$$V(t) = \frac{V_0}{2} + \sum_{n=-\infty}^{-1} \frac{V_0}{2in\pi} [1 - (-1)^n] e^{2in\pi t/2T} + \sum_{n=1}^{\infty} \frac{V_0}{2in\pi} [1 - (-1)^n] e^{2in\pi t/2T}.$$

If we change the dummy index of the first sum from n to $-m$, and back to n again, and put the two sums together, we obtain

$$V(t) = \frac{V_0}{2} + \sum_{n=1}^{\infty} \frac{V_0}{2in\pi}[1 - (-1)^n]\left(e^{in\pi t/T} - e^{-in\pi t/T}\right)$$

$$= \frac{V_0}{2} + \frac{2V_0}{\pi} \sum_{n=odd}^{\infty} \frac{1}{2in}\underbrace{\left(e^{in\pi t/T} - e^{-in\pi t/T}\right)}_{=2i\sin(n\pi t/T)}$$

$$= \frac{V_0}{2} + \frac{2V_0}{\pi} \sum_{k=0}^{\infty} \frac{\sin[(2k+1)\pi t/T]}{2k+1},$$

which is the expansion we obtained in Example 10.6.1 using trigonometric functions. ∎

18.4   A Representation of Delta Function

Consider the function $D_T(x - x_0)$ defined as

$$D_T(x - x_0) \equiv \frac{1}{2\pi} \int_{-T}^{T} e^{i(x-x_0)t} dt. \qquad (18.26)$$

The integral is easily evaluated, with the result

$$D_T(x - x_0) = \frac{1}{2\pi} \frac{e^{i(x-x_0)t}}{i(x-x_0)}\bigg|_{-T}^{T} = \frac{1}{\pi} \frac{\sin T(x - x_0)}{x - x_0}.$$

The graph of $D_T(x)$ as a function of x for various values of T is shown in Figure 18.7. Note that the width of the curve decreases as T increases. The area under the curve can be calculated:

$$\int_{-\infty}^{\infty} D_T(x - x_0)\, dx = \frac{1}{\pi} \int_{-\infty}^{\infty} \frac{\sin T(x - x_0)}{x - x_0}\, dx = \frac{1}{\pi} \underbrace{\int_{-\infty}^{\infty} \frac{\sin y}{y}\, dy}_{=\pi} = 1.$$

Figure 18.7 shows that $D_T(x - x_0)$ becomes more and more like the Dirac delta function as T gets larger and larger. In fact, we have

$$\delta(x - x_0) = \lim_{T\to\infty} D_T(x - x_0) = \lim_{T\to\infty} \frac{1}{\pi} \frac{\sin T(x - x_0)}{x - x_0}. \qquad (18.27)$$

To see this, we note that for any finite T we can write

$$D_T(x - x_0) = \frac{T}{\pi} \frac{\sin T(x - x_0)}{T(x - x_0)}.$$

Furthermore, for values of x that are very close to x_0,

$$T(x - x_0) \to 0 \quad \text{and} \quad \frac{\sin T(x - x_0)}{T(x - x_0)} \to 1.$$

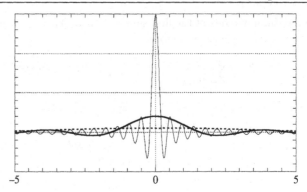

Figure 18.7: The function $\sin Tx/x$ also approaches the Dirac delta function as the width of the curve approaches zero. The value of T is 0.5 for the dashed curve, 2 for the heavy curve, and 15 for the light curve.

Thus, for such values of x and x_0, we have $D_T(x - x_0) \approx (T/\pi)$, which is large when T is large. This is as expected of a delta function: $\delta(0) = \infty$. On the other hand, the width of $D_T(x - x_0)$ around x_0 is given, roughly, by the distance between the points at which $D_T(x - x_0)$ drops to zero: $T(x - x_0) = \pm\pi$, or $x - x_0 = \pm\pi/T$. This width is roughly $\Delta x = 2\pi/T$, which goes to zero as T grows. Again, this is as expected of the delta function. Therefore, from (18.26) and (18.27), we have the following important representation of the Dirac delta function:

delta function as integral of imaginary exponential

$$\delta(x - x_0) = \frac{1}{2\pi} \int_{-\infty}^{\infty} e^{i(x-x_0)t} dt. \qquad (18.28)$$

Equation (18.28) can be generalized to higher dimensions, because (at least in Cartesian coordinates) the multi-dimensional Dirac delta function is just the product of the one-dimensional delta functions. Using the more common k instead of t as the variable of integration, the two-dimensional Dirac delta function can be represented as

$$\delta(\mathbf{r} - \mathbf{r}_0) = \frac{1}{(2\pi)^2} \int_{-\infty}^{\infty} \int_{-\infty}^{\infty} e^{i\mathbf{k}\cdot(\mathbf{r}-\mathbf{r}_0)} dk_x dk_y \equiv \frac{1}{(2\pi)^2} \iint_{\Omega_\infty} e^{i\mathbf{k}\cdot(\mathbf{r}-\mathbf{r}_0)} d^2k, \qquad (18.29)$$

where Ω_∞ means over all $k_x k_y$-plane and in the last integral we substituted d^2k for $dk_x dk_y$.

Similarly, the three dimensional Dirac delta function has the following representation:

$$\delta(\mathbf{r} - \mathbf{r}_0) = \frac{1}{(2\pi)^3} \iint_{\Omega_\infty} e^{i\mathbf{k}\cdot(\mathbf{r}-\mathbf{r}_0)} d^3k, \qquad (18.30)$$

where d^3k means a triple integral over k and Ω_∞ means over all k-space.

18.5   Problems

18.1. Find the real and imaginary parts of the following complex numbers:

(a) $(2 - i)(3 + 2i)$.   (b) $(2 - 3i)(1 + i)$.   (c) $(a - ib)(2a + 2ib)$.

(d) $\dfrac{i}{1 + i}$.   (e) $\dfrac{1 + i}{2 - i}$.   (f) $\dfrac{1 + 3i}{1 - 2i}$.

(g) $\dfrac{1 + 2i}{2 - 3i}$.   (h) $\dfrac{2}{1 - 3i}$.   (i) $\dfrac{1 - i}{1 + i}$.

(j) $\dfrac{5}{(1 - i)(2 - i)(3 - i)}$.   (k) $\dfrac{1 + 2i}{3 - 4i} + \dfrac{2 - i}{5i}$.

18.2. Convert the following complex numbers to polar form and find all cube roots of each:

(a) $2 - i$.   (b) $2 - 3i$.   (c) $3 - 2i$.   (d) i.

(e) $-i$.   (f) $\dfrac{i}{1 + i}$.   (g) $\dfrac{1 + i}{2 - i}$.   (h) $\dfrac{1 + 3i}{1 - 2i}$.

(i) $1 + i\sqrt{3}$.   (j) $\dfrac{2 + 3i}{3 - 4i}$.   (k) $27i$.   (l) -64.

(m) $2 - 5i$.   (n) $1 + i$.   (o) $1 - i$.   (p) $5 + 2i$.

18.3. Using polar coordinates, show that:

(a) $(-1 + i)^7 = -8(1 + i)$.   (b) $(1 + i\sqrt{3})^{-10} = 2^{-11}(-1 + i\sqrt{3})$.

18.4. Find the real and imaginary parts of the following:

(a) $(1 + i\sqrt{3})^3$.   (b) $(2 + i)^{53}$.   (c) $\sqrt[4]{i}$.   (d) $\sqrt[3]{1 + i\sqrt{3}}$.

(e) $(1 + i\sqrt{3})^{63}$.   (f) $\left(\dfrac{1 - i}{1 + i}\right)^{81}$.   (g) $\sqrt[6]{-i}$.   (h) $\sqrt[4]{-1}$.

(i) $\left(\dfrac{1 + i\sqrt{3}}{\sqrt{3} + i}\right)^{217}$.   (j) $(1 + i)^{22}$.   (k) $\sqrt[6]{1 - i}$.   (l) $(1 - i)^4$.

18.5. Find the Cartesian form of all complex numbers z which satisfy (a) $z^3 + 1 = 0$, and (b) $z^4 - 16i = 0$.

18.6. Find the absolute value of $\dfrac{3 + 4i}{3 - 4i}$ and $\dfrac{a + ib}{a - ib}$.

18.7. Derive the following trigonometric identities:

$$\cos 3\theta = 4\cos^3\theta - 3\cos\theta,$$
$$\sin 3\theta = 3\sin\theta - 4\sin^3\theta.$$

18.8. Show that Equation (18.11) leads to Equation (18.14).

18.9. Show that z is real if and only if $z = z^*$.

18.10. Show that $|\operatorname{Re}(z)| + |\operatorname{Im}(z)| \geq |z| \geq (|\operatorname{Re}(z)| + |\operatorname{Im}(z)|)/\sqrt{2}$.

18.11. Let $z_1 = x_1 + iy_1$ and $z_2 = x_2 + iy_2$ represent two planar vectors $\mathbf{z}_1$ and $\mathbf{z}_2$. Show that

$$z_1 z_2^* = \mathbf{z}_1 \cdot \mathbf{z}_2 - i\hat{\mathbf{e}}_z \cdot \mathbf{z}_1 \times \mathbf{z}_2.$$

18.12. Sketch the set of points determined by each of the following conditions:

(a) $|z - 2 + i| = 2$.   (b) $|z + 2i| \leq 4$.   (c) $|z + i| = |z - i|$.

(d) $\operatorname{Im}(z^* + i) = 2$.   (e) $2z + 3z^* = 1$.   (f) $z^2 + (z^*)^2 = 2$.

Hint: Find a relation between x and y.

18.13. Show that the equation of a circle of radius r centered at z_0 can be written as $|z|^2 - 2\operatorname{Re}(zz_0^*) = r^2 - |z_0|^2$.

18.14. Given that $z_1 z_2 \neq 0$, show that
(a) $\operatorname{Re}(z_1 z_2^*) = |z_1||z_2|$, and $|z_1 + z_2| = |z_1| + |z_2|$, if and only if $\arg(z_1) - \arg(z_2) = 2n\pi$, for $n = 0, \pm 1, \pm 2, \ldots$.
(b) What does the second equality mean geometrically?

18.15. Assume that $z \neq 1$ and $z^n = 1$. Show that $1 + z + z^2 + \cdots + z^{n-1} = 0$.

18.16. Substitute $x + iy$ for z in $z^2 + z + 1 = 0$ and solve the resulting equations for x and y. Compare these with the roots obtained by solving the equation in z directly.

18.17. Find the roots of $z^4 + 4 = 0$ and use them to factor $z^4 + 4$ into a product of quadratic polynomials with real coefficients. Hint: First factor $z^4 + 4$ into linear terms.

18.18. Evaluate the following roots and plot them on the complex plane:

(a) $\sqrt[5]{1+i}$.   (b) $\sqrt[4]{-1}$.   (c) $\sqrt[8]{1}$.   (d) $\sqrt[5]{-32}$.

(e) $\sqrt{3 + 4i}$.   (f) $\sqrt[3]{-1}$.   (g) $\sqrt[4]{-16i}$.   (h) $\sqrt[6]{-1}$.

18.19. Use binomial expansion to show directly that

$$\left(-\frac{1}{2} + i\frac{\sqrt{3}}{2}\right)^3 = 1 \qquad \text{and} \qquad \left(-\frac{1}{2} - i\frac{\sqrt{3}}{2}\right)^3 = 1.$$

18.20. Use $\int e^{ax} = e^{ax}/a$ to find the indefinite integral of $\sin^2 x$. Verify that the derivative of your answer is indeed $\sin^2 x$.

18.21. Use $\int e^{ax} = e^{ax}/a$ and $e^{ibx} = \cos(bx) + i\sin(bx)$ to verify the following relations by integrating a certain complex exponential:

$$\int e^{ax} \cos(bx)\, dx = \frac{e^{ax}}{a^2 + b^2}[a\cos(bx) + b\sin(bx)],$$

$$\int e^{ax} \sin(bx)\, dx = \frac{e^{ax}}{a^2 + b^2}[a\sin(bx) - b\cos(bx)],$$

where a and b are assumed to be real constants.

18.22. (a) Using $\sum_{k=1}^{N} r^k = (r^{N+1} - r)/(r-1)$, evaluate the sum $\sum_{k=1}^{N} e^{-i\beta k}$. In particular, show that

$$\sum_{k=1}^{N} e^{i(\alpha - \beta k)} = e^{i(\alpha - \beta)} \frac{e^{-i\beta N} - 1}{e^{-i\beta} - 1}.$$

(b) Now show that if $\beta = 2\pi/N$, then

$$\sum_{k=1}^{N} \cos(\alpha - \beta k) = 0 = \sum_{k=1}^{N} \sin(\alpha - \beta k).$$

18.23. Express $\cos 4\theta$ and $\sin 4\theta$ in terms of powers of $\cos\theta$ and $\sin\theta$.

18.24. Use mathematical induction to show the de Moivre theorem.

18.25. Using binomial expansion and the de Moivre theorem, show that

$$\cos n\theta = \sum_{m=0}^{[n/2]} (-1)^m \binom{n}{2m} \sin^{2m}\theta \cos^{n-2m}\theta,$$

$$\sin n\theta = \sum_{m=0}^{[n/2]} (-1)^m \binom{n}{2m+1} \sin^{2m+1}\theta \cos^{n-2m-1}\theta,$$

where $[x]$ stands for the greatest integer less than or equal to x.

18.26. Derive Equation (18.17) from the equations preceding it.

18.27. Find the following sums, where α and β are real:

(a) $\cos\alpha + \cos(\alpha + \beta) + \cos(\alpha + 2\beta) + \cdots + \cos(\alpha + n\beta)$.
(b) $\sin\alpha + \sin(\alpha + \beta) + \sin(\alpha + 2\beta) + \cdots + \sin(\alpha + n\beta)$.

Hint: Use the result of Problem 18.22.

18.28. Show that

$$\int_a^b e^{2i(n-k)\pi x/L} = \begin{cases} 0 & \text{if } n \neq k, \\ L & \text{if } n = k, \end{cases}$$

where $b = a + L$.

18.29. Use Equations (18.20) and (18.21) to obtain Equation (18.22)

18.30. Find the Fourier series expansion of Problem 10.22 using complex exponentials.

18.31. An electric voltage $V(t)$ is given by

$$V(t) = V_0 \sin\left(\frac{\pi t}{2T}\right), \quad 0 \leq t \leq T$$

and repeats itself with period T. Find the Fourier series expansion of $V(t)$ using complex exponential functions.

18.32. A periodic voltage is given by the formula

$$V(t) = \begin{cases} V_0 \sin(\pi t/2T) & \text{if} \quad 0 \le t \le T, \\ 0 & \text{if} \quad T \le t \le 2T, \end{cases}$$

in the interval $(0, 2T)$. Find the Fourier series representation of this voltage using complex exponential functions.

18.33. A periodic voltage with period $4T$ is given by

$$V(t) = \begin{cases} V_0 \left(1 - \dfrac{t^2}{T^2}\right) & \text{if} \quad -T \le t \le T, \\ 0 & \text{if} \quad T \le |t| \le 2T. \end{cases}$$

Write the Fourier series of $V(t)$ using complex exponential functions.

18.34. The function $f(x)$ is given by the integral

$$f(x) = \int_{-\infty}^{\infty} g(y) e^{ixy} \, dy.$$

Find $g(y)$ as an integral over $f(x)$. Hint: Multiply both sides of the equation by e^{-ixz} and integrate over x, changing the order of integration on the right-hand side and using (18.28).

Chapter 19

Complex Derivative and Integral

So far we have concerned ourselves with the *algebra* of the complex numbers. The subject of complex *analysis* is extremely rich and important. The scope and the level of this book does not allow a comprehensive treatment of complex analysis. Therefore, we shall briefly review some of the more elementary topics and encourage the reader to refer to more advanced books for a more comprehensive treatment. We start here, as is done in real analysis, with the notion of a function.

19.1 Complex Functions

A complex function $f(z)$ is a rule that associates one complex number to another. We write $f(z) = w$ where both z and w are complex numbers. The function f can be geometrically thought of as a correspondence between two complex planes, the z-plane and the w-plane. In the real case, this correspondence can be represented by a graph. It could also be represented by arrows from one real line (the x-axis) to another real line (the y-axis) joining a point of the first real line to the image point of the second real line. When the possibility of graph is available, the second representation of real functions appears prohibitively clumsy! For complex functions, no graph is available, because one cannot draw pictures in four dimensions![1] Therefore, the second alternative is our only choice. The w-plane has a real axis and an imaginary axis, which we can call u and v, respectively. Both u and v are real functions of the coordinates of z, i.e., x and y. Therefore, we may write

graph of a complex function is impossible to visualize because it lives in a four dimensional space.

$$f(z) = u(x, y) + iv(x, y). \tag{19.1}$$

---

[1]The "graph" of a complex function would be a collection of pairs $(z, f(z))$ just as the graph of a real function is a collection of pairs $(x, f(x))$. While in the latter case the graph can be drawn in the (x, y) plane, the former needs four dimensions because both z and $f(z)$ have two components each.

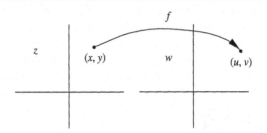

Figure 19.1: A map from the z-plane to the w-plane.

This equation gives a unique point (u, v) in the w-plane for each point (x, y) in the z-plane (see Figure 19.1). Under f, regions of the z-plane are mapped onto regions of the w-plane. For instance, a curve in the z-plane may be mapped into a curve in the w-plane.

Example 19.1.1. Let us investigate the behavior of some elementary complex functions. In particular, we shall look at the way a line $y = mx$ in the z-plane is mapped to lines and curves in the w-plane by the action of these functions.
(a) Let us start with the simple function $w = f(z) = z^2$. We have

$$w = (x + iy)^2 = x^2 - y^2 + 2ixy$$

with $u(x, y) = x^2 - y^2$ and $v(x, y) = 2xy$. For $y = mx$, these equations yield $u = (1 - m^2)x^2$ and $v = 2mx^2$. Eliminating x in these equations, we find $v = [2m/(1 - m^2)]u$. This is a line passing through the origin of the w-plane [see Figure 19.2(a)]. Note that the angle the line in the w-plane makes with its real axis is twice the angle the line in the z-plane makes with the x-axis.
(b) Now let us consider $w = f(z) = e^z = e^{x+iy}$, which gives $u(x, y) = e^x \cos y$ and $v(x, y) = e^x \sin y$. Substituting $y = mx$, we obtain $u = e^x \cos mx$ and $v = e^x \sin mx$. Unlike part (a), we cannot eliminate x to find v as an explicit function of u. Nevertheless, the last pair of equations are the *parametric equations* of a curve (with x as the parameter) which we can plot in a uv-plane as shown in Figure 19.2(b). ∎

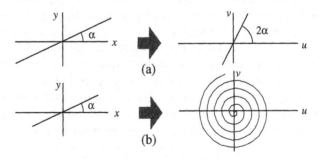

Figure 19.2: (a) The map z^2 takes a line with slope angle α and maps it onto a line with twice the angle in the w-plane. (b) The map e^z takes the same line and maps it onto a spiral in the w-plane.

19.1.1   Derivatives of Complex Functions

Limits of complex functions are defined in terms of absolute values. Thus, $\lim_{z \to a} f(z) = w_0$ means that, given any real number $\epsilon > 0$, we can find a corresponding real number $\delta > 0$ such that $|f(z) - w_0| < \epsilon$ whenever $|z - a| < \delta$. Similarly, we say that a function f is **continuous** at $z = a$ if $\lim_{z \to a} f(z) = f(a)$, or if there exist $\epsilon > 0$ and $\delta > 0$ such that $|f(z) - f(a)| < \epsilon$ whenever $|z - a| < \delta$.

The derivative of a complex function is defined as usual:

Definition 19.1.1. *Let $f(z)$ be a complex function. The derivative of f at z_0 is*

$$\left. \frac{df}{dz} \right|_{z_0} = \lim_{\Delta z \to 0} \frac{f(z_0 + \Delta z) - f(z_0)}{\Delta z}$$

provided the limit exists and is independent of Δz.

In this definition "independent of Δz" means independent of Δx and Δy (the components of Δz) and, therefore, independent of the direction of approach to z_0. The restrictions of this definition apply to the real case as well. For instance, the derivative of $f(x) = |x|$ at $x = 0$ does not exist because it approaches $+1$ from the right and -1 from the left.

It can easily be shown that all the formal rules of differentiation that apply to the real case also apply to the complex case. For example, if f and g are differentiable, then $f \pm g$, fg, and—as long as g is not zero—f/g are also differentiable, and their derivatives are given by the usual rules of differentiation.

> **Box 19.1.1.** *A function $f(z)$ is called **analytic** at z_0 if it is differentiable at z_0 and at all other points in some neighborhood of z_0. A point at which f is analytic is called a **regular point** of f. A point at which f is not analytic is called a **singular point** or a **singularity** of f. A function for which all points in $\mathbb{C}$ are regular is called an **entire function**.*

Example 19.1.2. Let us examine the derivative of $f(z) = x + 2iy$ at $z = 0$:

$$\left. \frac{df}{dz} \right|_{z=0} = \lim_{\Delta z \to 0} \frac{f(\Delta z) - f(0)}{\Delta z} = \lim_{\substack{\Delta x \to 0 \\ \Delta y \to 0}} \frac{\Delta x + 2i\Delta y}{\Delta x + i\Delta y}.$$

example illustrating path-dependence of derivative

In general, along a line that goes through the origin, $y = mx$, the limit yields

$$\left. \frac{df}{dz} \right|_{z=0} = \lim_{\Delta x \to 0} \frac{\Delta x + 2im\Delta x}{\Delta x + im\Delta x} = \frac{1 + 2im}{1 + im}.$$

This indicates that we get infinitely many values for the derivative depending on the value we assign to m—corresponding to different directions of approach to the origin. Thus, the derivative does not exist at $z = 0$. ∎

A question arises naturally at this point: Under what conditions does the limit in the definition of derivative exist? We will find the necessary and sufficient conditions for the existence of that limit. It is clear from the definition that differentiability puts a severe restriction on $f(z)$ because it requires the limit to be the same for *all paths* going through z_0, the point at which the derivative is being calculated. Another important point to keep in mind is that differentiability is a *local* property. To test whether or not a function $f(z)$ is differentiable at z_0, we move away from z_0 by a small amount Δz and check the existence of the limit in Definition 19.1.1.

For $f(z) = u(x, y) + iv(x, y)$, Definition 19.1.1 yields

$$\left.\frac{df}{dz}\right|_{z_0} = \lim_{\substack{\Delta x \to 0 \\ \Delta y \to 0}} \left\{ \frac{u(x_0 + \Delta x, y_0 + \Delta y) - u(x_0, y_0)}{\Delta x + i\Delta y} \right.$$

$$\left. + i \, \frac{v(x_0 + \Delta x, y_0 + \Delta y) - v(x_0, y_0)}{\Delta x + i\Delta y} \right\}.$$

If this limit is to exist for all paths, it must exist for the two particular paths on which $\Delta y = 0$ (parallel to the x-axis) and $\Delta x = 0$ (parallel to the y-axis). For the first path we get

$$\left.\frac{df}{dz}\right|_{z_0} = \lim_{\Delta x \to 0} \frac{u(x_0 + \Delta x, y_0) - u(x_0, y_0)}{\Delta x}$$

$$+ i \lim_{\Delta x \to 0} \frac{v(x_0 + \Delta x, y_0) - v(x_0, y_0)}{\Delta x} = \left.\frac{\partial u}{\partial x}\right|_{(x_0, y_0)} + i \left.\frac{\partial v}{\partial x}\right|_{(x_0, y_0)}.$$

For the second path ($\Delta x = 0$), we obtain

$$\left.\frac{df}{dz}\right|_{z_0} = \lim_{\Delta y \to 0} \frac{u(x_0, y_0 + \Delta y) - u(x_0, y_0)}{i\Delta y}$$

$$+ i \lim_{\Delta y \to 0} \frac{v(x_0, y_0 + \Delta y) - v(x_0, y_0)}{i\Delta y} = -i \left.\frac{\partial u}{\partial y}\right|_{(x_0, y_0)} + \left.\frac{\partial v}{\partial y}\right|_{(x_0, y_0)}.$$

If f is to be differentiable at z_0, the derivatives along the two paths must be equal. Equating the real and imaginary parts of both sides of this equation and ignoring the subscript z_0 (x_0, y_0, or z_0 is arbitrary), we obtain

$$\frac{\partial u}{\partial x} = \frac{\partial v}{\partial y} \quad \text{and} \quad \frac{\partial u}{\partial y} = -\frac{\partial v}{\partial x}. \tag{19.2}$$

These two conditions, which are necessary for the differentiability of f, are
Cauchy–Riemann   called the **Cauchy–Riemann (C–R) conditions**.
conditions   The arguments leading to Equation (19.2) imply that the derivative, if it exists, can be expressed as

$$\frac{df}{dz} = \frac{\partial u}{\partial x} + i\frac{\partial v}{\partial x} = \frac{\partial v}{\partial y} - i\frac{\partial u}{\partial y}. \tag{19.3}$$

The C–R conditions assure us that these two equations are equivalent.

Example 19.1.3. Let us examine the differentiability of some complex functions.
(a) We have already established that $f(z) = x + 2iy$ is not differentiable at $z = 0$.
We can now show that it is has no derivative at *any point* in the complex plane. This
is easily seen by noting that $u = x$ and $v = 2y$, and that $\partial u/\partial x = 1 \neq \partial v/\partial y = 2$,
and the first C–R condition is not satisfied. The second C–R condition is satisfied,
but that is not enough.
(b) Now consider $f(z) = x^2 - y^2 + 2ixy$ for which $u = x^2 - y^2$ and $v = 2xy$. The C–R
conditions become $\partial u/\partial x = 2x = \partial v/\partial y$ and $\partial u/\partial y = -2y = -\partial v/\partial x$. Thus, $f(z)$
may be differentiable. Recall that C–R conditions are only *necessary* conditions; we
do not know as yet if they are also sufficient.
(c) Let $u(x, y) = e^x \cos y$ and $v(x, y) = e^x \sin y$. Then $\partial u/\partial x = e^x \cos y = \partial v/\partial y$
and $\partial u/\partial y = -e^x \sin y = -\partial v/\partial x$ and the C–R conditions are satisfied. ■

The requirement of differentiability is very restrictive: the derivative must
exist along infinitely many paths. On the other hand, the C–R conditions
seem deceptively mild: they are derived for only two paths. Nevertheless,
the two paths are, in fact, true representatives of all paths; that is , the C–R
conditions are not only necessary, but also sufficient. This is the content of
the Cauchy–Riemann theorem which we state without proof:[2]

Theorem 19.1.4. (*Cauchy–Riemann Theorem*). *The function* $f(z) =$
$u(x, y) + iv(x, y)$ *is differentiable in a region of the complex plane if and only*
if the Cauchy–Riemann conditions

$$\frac{\partial u}{\partial x} = \frac{\partial v}{\partial y} \quad and \quad \frac{\partial u}{\partial y} = -\frac{\partial v}{\partial x}$$

are satisfied and all first partial derivatives of u and v are continuous in that
region. In that case

$$\frac{df}{dz} = \frac{\partial u}{\partial x} + i\frac{\partial v}{\partial x} = \frac{\partial v}{\partial y} - i\frac{\partial u}{\partial y}.$$

The C–R conditions readily lead to

$$\frac{\partial^2 u}{\partial x^2} + \frac{\partial^2 u}{\partial y^2} = 0, \qquad \frac{\partial^2 v}{\partial x^2} + \frac{\partial^2 v}{\partial y^2} = 0, \qquad (19.4)$$

i.e., both real and imaginary parts of an analytic function satisfy the two-
dimensional Laplace equation [Equations (15.13) and (15.15)]. Such functions
are called **harmonic functions**.

harmonic
functions defined

Example 19.1.5. Let us consider some examples of derivatives of complex functions.
(a) $f(z) = z$.

Here $u = x$ and $v = y$; the C–R conditions are easily shown to hold, and for

---

[2]For a simple proof, see Hassani, S. *Mathematical Physics: A Modern Introduction to
Its Foundations*, Springer-Verlag, 1999, Chapter 9.

any z, we have $df/dz = \partial u/\partial x + i\partial v/\partial x = 1$. Therefore, the derivative exists at all points of the complex plane, i.e., $f(z) = z$ is entire.

(b) $f(z) = z^2$.

Here $u = x^2 - y^2$ and $v = 2xy$; the C–R conditions hold, and for all points z of the complex plane, we have $df/dz = \partial u/\partial x + i\partial v/\partial x = 2x + i2y = 2z$. Therefore, $f(z)$ is differentiable at all points. So, $f(z) = z^2$ is also entire.

(c) $f(z) = z^n$ for $n \geq 1$.

We can use mathematical induction and the fact that the product of two entire functions is an entire function to show that $\dfrac{d}{dz}(z^n) = nz^{n-1}$.

(d) $f(z) = a_0 + a_1 z + \cdots + a_{n-1} z^{n-1} + a_n z^n$,

where a_i are arbitrary constants. That $f(z)$ is entire follows directly from (c) and the fact that the sum of two entire functions is entire.

(e) $f(z) = e^z$.

Here $u(x, y) = e^x \cos y$ and $v(x, y) = e^x \sin y$. Thus, $\partial u/\partial x = e^x \cos y = \partial v/\partial y$ and $\partial u/\partial y = -e^x \sin y = -\partial v/\partial x$ and the C–R conditions are satisfied at every point (x, y) of the xy-plane. Furthermore,

$$\frac{df}{dz} = \frac{\partial u}{\partial x} + i\frac{\partial v}{\partial x} = e^x \cos y + ie^x \sin y = e^x(\cos y + i \sin y) = e^x e^{iy} = e^{x+iy} = e^z$$

and e^z is entire as well.

(f) $f(z) = 1/z$.

The derivative can be found to be $f'(z) = -1/z^2$ which does not exist for $z = 0$. Thus, $z = 0$ is a singularity of $f(z)$. However, any other point is a regular point of f.

(g) $f(z) = 1/\sin z$.

This gives $df/dz = -\cos z/\sin^2 z$. Thus, f has (infinitely many) singular points at $z = \pm n\pi$ for $n = 0, 1, 2, \ldots$.   ∎

Example 19.1.5 shows that any polynomial in z, as well as the exponential function e^z is entire. Therefore, any product and/or sum of polynomials and e^z will also be entire. We can build other entire functions. For instance, e^{iz} and e^{-iz} are entire functions; therefore, the **complex trigonometric functions**, defined by

complex trigonometric functions

$$\sin z = \frac{e^{iz} - e^{-iz}}{2i} \qquad \text{and} \qquad \cos z = \frac{e^{iz} + e^{-iz}}{2} \tag{19.5}$$

are also entire functions. Problem 19.7 shows that $\sin z$ and $\cos z$ have only *real* zeros.

The **complex hyperbolic functions** can be defined similarly:

complex hyperbolic functions

$$\sinh z = \frac{e^z - e^{-z}}{2} \qquad \text{and} \qquad \cosh z = \frac{e^z + e^{-z}}{2}. \tag{19.6}$$

Although the sum and product of entire functions are entire, the ratio is not. For instance, if $f(z)$ and $g(z)$ are polynomials of degrees m and n, respectively, then for $n > 0$, the ratio $f(z)/g(z)$ is not entire, because at the zeros of $g(z)$—which always exist—the derivative is not defined.

The functions $u(x, y)$ and $v(x, y)$ of an analytic function have an interesting property which the following example investigates.

Example 19.1.6. The family of curves $u(x, y) = $ constant is perpendicular to the family of curves $v(x, y) = $ constant at each point of the complex plane where $f(z) = u + iv$ is analytic.

curves of constant u and v are perpendicular.

This can easily be seen by looking at the normal to the curves. The normal to the curve $u(x, y) = $ constant is simply $\nabla u = \langle \partial u/\partial x, \partial u/\partial y \rangle$ (see Theorem 12.3.2). Similarly, the normal to the curve $v(x, y) = $ constant is $\nabla v = \langle \partial v/\partial x, \partial v/\partial y \rangle$. Taking the dot product of these two normals, we obtain

$$(\nabla u) \cdot (\nabla v) = \frac{\partial u}{\partial x}\frac{\partial v}{\partial x} + \frac{\partial u}{\partial y}\frac{\partial v}{\partial y} = \frac{\partial u}{\partial x}\left(-\frac{\partial u}{\partial y}\right) + \frac{\partial u}{\partial y}\left(\frac{\partial u}{\partial x}\right) = 0$$

by the C–R conditions. ∎

Historical Notes

One can safely say that rigorous complex analysis was founded by a single man: Cauchy. **Augustin-Louis Cauchy** was one of the most influential French mathematicians of the nineteenth century. He began his career as a military engineer, but when his health broke down in 1813 he followed his natural inclination and devoted himself wholly to mathematics.

Augustin-Louis Cauchy 1789–1857

In mathematical productivity Cauchy was surpassed only by Euler, and his collected works fill 27 fat volumes. He made substantial contributions to number theory and determinants; is considered to be the originator of the theory of finite groups; and did extensive work in astronomy, mechanics, optics, and the theory of elasticity.

His greatest achievements, however, lay in the field of analysis. Together with his contemporaries Gauss and Abel, he was a pioneer in the rigorous treatment of limits, continuous functions, derivatives, integrals, and infinite series. Several of the basic tests for the convergence of series are associated with his name. He also provided the first existence proof for solutions of differential equations, gave the first proof of the convergence of a Taylor series, and was the first to feel the need for a careful study of the convergence behavior of Fourier series. However, his most important work was in the theory of functions of a complex variable, which in essence he created and which has continued to be one of the dominant branches of both pure and applied mathematics. In this field, Cauchy's integral theorem and Cauchy's integral formula are fundamental tools without which modern analysis could hardly exist.

Unfortunately, his personality did not harmonize with the fruitful power of his mind. He was an arrogant royalist in politics and a self-righteous, preaching, pious believer in religion—all this in an age of republican skepticism—and most of his fellow scientists disliked him and considered him a smug hypocrite. It might be fairer to put first things first and describe him as a great mathematician who happened also to be a sincere but narrow-minded bigot.

19.1.2   Integration of Complex Functions

We have thus far discussed the derivative of a complex function. The concept of integration is even more important because, as we shall see later, derivatives can be written in terms of integrals.

complex integrals
are
path-dependent.
The definite integral of a complex function is naively defined in analogy to that of a real function. However, a crucial difference exists: While in the real case, the limits of integration are real numbers and there is only one way to connect these two limits (along the real line), the limits of integration of a complex function are points in the complex plane and there are infinitely many ways to connect these two points. Thus, we speak of a definite integral of a complex function *along a path*. It follows that complex integrals are, in general, path-dependent.

$$\int_{\alpha_1}^{\alpha_2} f(z)\,dz = \lim_{\substack{N\to\infty \\ \Delta z_i \to 0}} \sum_{i=1}^{N} f(z_i)\Delta z_i, \tag{19.7}$$

where Δz_i is a small segment—situated at z_i—of the curve that connects the complex number α_1 to the complex number α_2 in the z-plane (see Figure 19.3). An immediate consequence of this equation is

$$\left| \int_{\alpha_1}^{\alpha_2} f(z)\,dz \right| = \lim_{\substack{N\to\infty \\ \Delta z_i \to 0}} \left| \sum_{i=1}^{N} f(z_i)\Delta z_i \right| \le \lim_{\substack{N\to\infty \\ \Delta z_i \to 0}} \sum_{i=1}^{N} |f(z_i)\Delta z_i|$$

$$= \lim_{\substack{N\to\infty \\ \Delta z_i \to 0}} \sum_{i=1}^{N} |f(z_i)|\,|\Delta z_i| = \int_{\alpha_1}^{\alpha_2} |f(z)|\,|dz|, \tag{19.8}$$

where we have used the triangle inequality as expressed in Equation (18.7).

Since there are infinitely many ways of connecting α_1 to α_2, there is no guarantee that Equation (19.7) has a unique value: It is possible to obtain different values for the integral of some functions for different paths. It may seem that we should avoid such functions and that they will have no use in physical applications. Quite to the contrary, most functions encountered, will not, in general, give the same result if we choose two completely arbitrary paths in the complex plane. In fact, it turns out that the only complex function that gives the same integral for any two arbitrary points connected by any two arbitrary paths is the constant function. Because of the importance of paths in complex integration, we need the following definition:

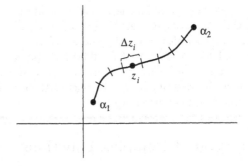

Figure 19.3: One of the infinitely many paths connecting two complex points α_1 and α_2.

> **Box 19.1.2.** *A **contour** is a collection of connected smooth arcs. When the beginning point of the first arc coincides with the end point of the last one, the contour is said to be a **simple closed contour** (or just closed contour).*

We encountered path-dependent integrals when we tried to evaluate the line integral of a vector field in Chapter 14. The same argument for path-independence can be used to prove (see Problem 19.21)

Theorem 19.1.7. (*Cauchy–Goursat Theorem*). *Let $f(z)$ be analytic on a simple closed contour C and at all points inside C. Then*

$$\oint_C f(z)\,dz = 0$$

Equivalently, $\int_{\alpha_1}^{\alpha_2} f(z)\,dz$ is independent of the smooth path connecting α_1 and α_2 as long as the path lies entirely in the region of analyticity of $f(z)$.

Example 19.1.8. We consider a few examples of definite integrals.
(a) Let us evaluate the integral $I_1 = \int_{\gamma_1} z\,dz$ where γ_1 is the straight line drawn from the origin to the point $(1,2)$ (see Figure 19.4). Along such a line $y = 2x$ and thus $\gamma_1(t) = t + 2it$ where $0 \le t \le 1$, and[3]

$$I_1 = \int_{\gamma_1} z\,dz = \int_0^1 (t + 2it)(dt + 2idt) = \int_0^1 (-3tdt + 4itdt) = -\tfrac{3}{2} + 2i.$$

For a different path γ_2, along which $y = 2x^2$, we get $\gamma_2(t) = t + 2it^2$ where $0 \le t \le 1$, and

$$I_1' = \int_{\gamma_2} z\,dz = \int_0^1 (t + 2it^2)(dt + 4itdt) = -\tfrac{3}{2} + 2i.$$

Therefore, $I_1 = I_1'$. This is what is expected from the Cauchy–Goursat theorem because the function $f(z) = z$ is analytic on the two paths and in the region bounded by them.

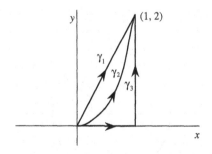

Figure 19.4: The three different paths of integration corresponding to the integrals I_1, I_1', I_2, and I_2'.

---

[3]We are using the parameterization $x = t$, $y = 2x = 2t$ for the curve.

(b) Now let us consider $I_2 = \int_{\gamma_1} z^2 dz$ with γ_1 as in part (a). Substituting for z in terms of t, we obtain

$$I_2 = \int_{\gamma_1} (t + 2it)^2 (dt + 2i dt) = (1 + 2i)^3 \int_0^1 t^2 dt = -\tfrac{11}{3} - \tfrac{2}{3} i.$$

Next we compare I_2 with $I_2' = \int_{\gamma_3} z^2 dz$ where γ_3 is as shown in Figure 19.4. This path can be described by

$$\gamma_3(t) = \begin{cases} t & \text{for} \quad 0 \le t \le 1, \\ 1 + i(t - 1) & \text{for} \quad 1 \le t \le 3. \end{cases}$$

Therefore,

$$I_2' = \int_0^1 t^2 dt + \int_1^3 [1 + i(t-1)]^2 (i\,dt) = \tfrac{1}{3} - 4 - \tfrac{2}{3} i = -\tfrac{11}{3} - \tfrac{2}{3} i,$$

which is identical to I_2, once again because the function is analytic on γ_1 and γ_3 as well as in the region bounded by them.

(c) An example of the case where equality for different paths is not attained is $I_3 = \int_{\gamma_4} dz/z$ where γ_4 is the upper semicircle of unit radius, as shown in Figure 19.5. A parametric equation for γ_4 can be given in terms of θ:

$$\gamma_4(\theta) = \cos\theta + i\sin\theta = e^{i\theta} \;\Rightarrow\; dz = ie^{i\theta} d\theta, \qquad 0 \le \theta \le \pi.$$

Thus, we obtain

$$I_3 = \int_0^\pi \frac{1}{e^{i\theta}} ie^{i\theta} d\theta = i\pi.$$

On the other hand,

$$I_3' = \int_{\gamma_4'} \frac{1}{z} dz = \int_{2\pi}^\pi \frac{1}{e^{i\theta}} ie^{i\theta} d\theta = -i\pi.$$

Here the two integrals are not equal. From γ_4 and γ_4' we can construct a counterclockwise simple closed contour C, along which the integral of $f(z) = 1/z$ becomes $\oint_C dz/z = I_3 - I_3' = 2i\pi$. That the integral is not zero is a consequence of the fact that $1/z$ is *not* analytic at all points of the region bounded by the closed contour C. ∎

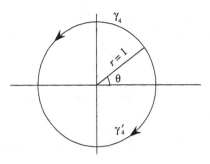

Figure 19.5: The two semicircular paths for calculating I_3 and I_3'.

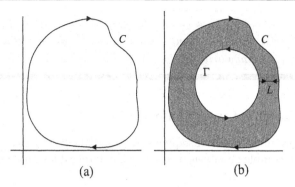

Figure 19.6: A contour of integration can be deformed into another contour. The second contour is usually taken to be a circle because of the ease of its corresponding integration. (a) shows the original contour, and (b) shows the two contours as well as the (shaded) region between them in which the function is analytic.

The Cauchy–Goursat theorem applies to more complicated regions. When a region contains points at which $f(z)$ is not analytic, those points can be avoided by redefining the region and the contour (Figure 19.6). Such a procedure requires a convention regarding the direction of "motion" along the contour. This convention is important enough to be stated separately.

convention for positive sense of integration around a closed contour

> **Box 19.1.3. (Convention).** *When integrating along a closed contour, we agree to traverse the contour in such a way that the region enclosed by the contour lies to our left. An integration that follows this convention is called integration in the **positive sense**. Integration performed in the opposite direction acquires a minus sign.*

Suppose that we want to evaluate the integral $\oint_C f(z)\,dz$ where C is some contour in the complex plane [see Figure 19.6(a)]. Let Γ be another—usually simpler, say a circle—contour which is either entirely inside or entirely outside C. Figure 19.6 illustrates the case where Γ is entirely inside C. We assume that Γ is such that $f(z)$ does not have any singularity in the region between C and Γ. By connecting the two contours with a line as shown in Figure 19.6(b), we construct a composite closed contour consisting of C, Γ, and twice the line segment L, once in the positive directions and once in the negative. Within this composite contour, the function $f(z)$ is analytic. Therefore, by the Cauchy–Goursat theorem, we have

$$-\oint_C f(z)\,dz + \oint_\Gamma f(z)\,dz + \int_L f(z)\,dz - \int_L f(z)\,dz = 0.$$

The negative sign for C is due to the convention above. It follows from this equation that the integral along C is the same as that along the circle Γ. This result can be interpreted by saying that

> **Box 19.1.4.** *We can always deform the contour of an integral in the complex plane into a simpler contour, as long as in the process of deformation we encounter no singularity of the function.*

19.1.3 Cauchy Integral Formula

One extremely important consequence of the Cauchy–Goursat theorem, the centerpiece of complex analysis, is the Cauchy integral formula which we state without proof.[4]

Theorem 19.1.9. *Let $f(z)$ be analytic on and within a simple closed contour C integrated in the positive sense. Let z_0 be any interior point of C. Then*

$$f(z_0) = \frac{1}{2\pi i} \oint_C \frac{f(z)}{z - z_0} dz.$$

*This is called the **Cauchy integral formula** (CIF).*

Example 19.1.10. We can use the CIF to evaluate the following integrals:

$$I_1 = \oint_{C_1} \frac{z^2 \, dz}{(z^2 + 3)^2 (z - i)}, \qquad I_2 = \oint_{C_2} \frac{(z^2 - 1) \, dz}{(z - \frac{1}{2})(z^2 - 4)^3},$$

$$I_3 = \oint_{C_3} \frac{e^{z/2} \, dz}{(z - i\pi)(z^2 - 20)^4},$$

where C_1, C_2, and C_3 are circles centered at the origin with radii $r_1 = \frac{3}{2}$, $r_2 = 1$, and $r_3 = 4$, respectively.

For I_1 we note that $f(z) = z^2/(z^2+3)^2$ is analytic within and on C_1, and $z_0 = i$ lies in the interior of C_1. Thus,

$$I_1 = \oint_{C_1} \frac{f(z)dz}{z - i} = 2\pi i f(i) = 2\pi i \frac{i^2}{(i^2 + 3)^2} = -i\frac{\pi}{2}.$$

Similarly, $f(z) = (z^2 - 1)/(z^2 - 4)^3$ for I_2 is analytic on and within C_2, and $z_0 = \frac{1}{2}$ is an interior point of C_2. Thus, the CIF gives

$$I_2 = \oint_{C_2} \frac{f(z)dz}{z - \frac{1}{2}} = 2\pi i f(\tfrac{1}{2}) = 2\pi i \frac{\frac{1}{4} - 1}{(\frac{1}{4} - 4)^3} = \frac{32\pi}{1125} i.$$

For the last integral, $f(z) = e^{z/2}/(z^2 - 20)^4$, and the interior point is $z_0 = i\pi$:

$$I_3 = \oint_{C_3} \frac{f(z)dz}{z - i\pi} = 2\pi i f(i\pi) = 2\pi i \frac{e^{i\pi/2}}{(-\pi^2 - 20)^4} = -\frac{2\pi}{(\pi^2 + 20)^4}. \qquad \blacksquare$$

why analytic functions "remote sense" their values at distant points

The CIF gives the value of an analytic function at every point inside a simple closed contour when it is given the value of the function only at points on the contour. It seems as though analytic functions have no freedom within a contour: They are not free to change inside a region once their value is fixed on the contour enclosing that region. There is an analogous situation in

---

[4] For a proof, see Hassani, S. *Mathematical Physics: A Modern Introduction to Its Foundations*, Springer-Verlag, 1999, Chapter 9.

certain areas of physics, for example, electrostatics: The specification of the potential at the boundaries, such as conductors, automatically determines it at any other point in the region of space bounded by the conductors. This is the content of the uniqueness theorem (to be discussed later in this book) used in electrostatic boundary-value problems. However, the electrostatic potential Φ is bound by another condition, Laplace's equation; and the *combination* of Laplace's equation and the boundary conditions furnishes the uniqueness of Φ.

It seems, on the other hand, as though the mere specification of an analytic function on a contour, without any other condition, is sufficient to determine the function's value at all points enclosed within that contour. This is not the case. An analytic function, by its very definition, satisfies another restrictive condition: Its real and imaginary parts separately satisfy Laplace's equation in two dimensions! [see Equation (19.4)]. Thus, it should come as no surprise that the value of an analytic function at a boundary (contour) determines the function at all points inside the boundary.

19.1.4 Derivatives as Integrals

The CIF is a very powerful tool for working with analytic functions. One of the applications of this formula is in evaluating the derivatives of such functions. It is convenient to change the dummy integration variable to ξ and write the CIF as

$$f(z) = \frac{1}{2\pi i} \oint_C \frac{f(\xi)\,d\xi}{\xi - z}, \tag{19.9}$$

where C is a simple closed contour in the ξ-plane and z is a point within C. By carrying the derivative inside the integral, we get

$$\frac{df}{dz} = \frac{1}{2\pi i} \frac{d}{dz} \oint_C \frac{f(\xi)\,d\xi}{\xi - z} = \frac{1}{2\pi i} \oint_C \frac{d}{dz}\left[\frac{f(\xi)\,d\xi}{\xi - z}\right] = \frac{1}{2\pi i} \oint_C \frac{f(\xi)\,d\xi}{(\xi - z)^2}.$$

By repeated differentiation, we can generalize this formula to the nth derivative, and obtain

Theorem 19.1.11. *The derivatives of all orders of an analytic function $f(z)$ exist in the domain of analyticity of the function and are themselves analytic in that domain. The nth derivative of $f(z)$ is given by*

$$f^{(n)}(z) = \frac{d^n f}{dz^n} = \frac{n!}{2\pi i} \oint_C \frac{f(\xi)\,d\xi}{(\xi - z)^{n+1}}. \tag{19.10}$$

Example 19.1.12. Let us apply Equation (19.10) directly to some simple functions. In all cases, we will assume that the contour is a circle of radius r centered at z.
(a) Let $f(z) = K$, a constant. Then, for $n = 1$ we have

$$\frac{df}{dz} = \frac{1}{2\pi i} \oint_C \frac{K\,d\xi}{(\xi - z)^2}.$$

Since ξ is always on the circle C centered at z, $\xi - z = re^{i\theta}$ and $d\xi = rie^{i\theta}d\theta$. So we have

$$\frac{df}{dz} = \frac{1}{2\pi i}\int_0^{2\pi}\frac{Kire^{i\theta}d\theta}{(re^{i\theta})^2} = \frac{K}{2\pi r}\int_0^{2\pi}e^{-i\theta}d\theta = 0.$$

That is, the derivative of a constant is zero.

(b) Given $f(z) = z$, its first derivative will be

$$\frac{df}{dz} = \frac{1}{2\pi i}\oint_C\frac{\xi\,d\xi}{(\xi - z)^2} = \frac{1}{2\pi i}\int_0^{2\pi}\frac{(z + re^{i\theta})ire^{i\theta}d\theta}{(re^{i\theta})^2}$$

$$= \frac{1}{2\pi}\left(\frac{z}{r}\int_0^{2\pi}e^{-i\theta}d\theta + \int_0^{2\pi}d\theta\right) = \frac{1}{2\pi}(0 + 2\pi) = 1.$$

(c) Given $f(z) = z^2$, for the first derivative Equation (19.10) yields

$$\frac{df}{dz} = \frac{1}{2\pi i}\oint_C\frac{\xi^2\,d\xi}{(\xi - z)^2} = \frac{1}{2\pi i}\int_0^{2\pi}\frac{(z + re^{i\theta})^2ire^{i\theta}d\theta}{(re^{i\theta})^2}$$

$$= \frac{1}{2\pi}\int_0^{2\pi}\left[z^2 + (re^{i\theta})^2 + 2zre^{i\theta}\right](re^{i\theta})^{-1}d\theta$$

$$= \frac{1}{2\pi}\left(\frac{z^2}{r}\int_0^{2\pi}e^{-i\theta}d\theta + r\int_0^{2\pi}e^{i\theta}d\theta + 2z\int_0^{2\pi}d\theta\right) = 2z.$$

It can be shown that, in general, $(d/dz)z^m = mz^{m-1}$. The proof is left as Problem 19.24. ∎

The CIF is a central formula in complex analysis. However, due to space limitations, we cannot explore its full capability here. Nevertheless, one of its applications is worth discussing at this point. Suppose that f is a *bounded* entire function and consider

$$\frac{df}{dz} = \frac{1}{2\pi i}\oint_C\frac{f(\xi)\,d\xi}{(\xi - z)^2}.$$

Since f is analytic everywhere in the complex plane, the closed contour C can be chosen to be a very large circle of radius R with center at z. Taking the absolute value of both sides yields

$$\left|\frac{df}{dz}\right| = \frac{1}{2\pi}\left|\int_0^{2\pi}\frac{f(z + Re^{i\theta})}{(Re^{i\theta})^2}iRe^{i\theta}d\theta\right|$$

$$\leq \frac{1}{2\pi}\int_0^{2\pi}\frac{|f(z + Re^{i\theta})|}{R}d\theta \leq \frac{1}{2\pi}\int_0^{2\pi}\frac{M}{R}d\theta = \frac{M}{R},$$

where we used Equation (19.8) and $|e^{i\theta}| = 1$. M is the maximum of the function in the complex plane.[5] Now as $R \to \infty$, the derivative goes to zero. The only function whose derivative is zero is the constant function. Thus

---

Box 19.1.5. *A bounded entire function is necessarily a constant.*

---

[5] M exists because f is assumed to be bounded.

There are many interesting and nontrivial *real* functions that are bounded and have derivatives (of all orders) on the entire real line. For instance, e^{-x^2} is such a function. No such freedom exists for *complex* analytic functions according to Box 19.1.5! Any nontrivial analytic function is either not bounded (goes to infinity somewhere on the complex plane) or not entire [it is not analytic at some point(s) of the complex plane].

A consequence of Proposition 19.1.5 is the **fundamental theorem of algebra** which states that any polynomial of degree $n \geq 1$ has n roots (some of which may be repeated). In other words, the polynomial

$$p(x) = a_0 + a_1 x + \cdots + a_n x^n \quad \text{for } n \geq 1$$

can be factored completely as $p(x) = c(x - z_1)(x - z_2)\ldots(x - z_n)$ where c is a constant and the z_i are, in general, complex numbers.

To see how Proposition 19.1.5 implies the fundamental theorem of algebra, we let $f(z) = 1/p(z)$ and assume the contrary, i.e., that $p(z)$ is never zero for any (finite) z. Then $f(z)$ is bounded and analytic for all z, and Proposition 19.1.5 says that $f(z)$ is a constant. This is obviously wrong. Thus, there must be at least one z, say $z = z_1$, for which $p(z)$ is zero. So, we can factor out $(z - z_1)$ from $p(z)$ and write $p(z) = (z - z_1)q(z)$ where $q(z)$ is of degree $n - 1$. Applying the above argument to $q(z)$, we have $p(z) = (z - z_1)(z - z_2)r(z)$ where $r(z)$ is of degree $n - 2$. Continuing in this way, we can factor $p(z)$ into linear factors. The last polynomial will be a constant (a polynomial of degree zero) which we have denoted as c.

(margin note: any nontrivial function is either unbounded or not entire.)

(margin note: fundamental theorem of algebra proved)

19.2 Problems

19.1. Show that $f(z) = z^2$ maps a line that makes an angle α with the real axis of the z-plane onto a line in the w-plane which makes an angle 2α with the real axis of the w-plane. Hint: Use the trigonometric identity $\tan 2\alpha = 2 \tan \alpha / (1 - \tan^2 \alpha)$.

19.2. Show that the function $w = 1/z$ maps the straight line $y = \frac{1}{2}$ in the z-plane onto a circle in the w-plane.

19.3. (a) Using the chain rule, find $\partial f / \partial z^*$ and $\partial f / \partial z$ in terms of partial derivatives with respect to x and y.
(b) Evaluate $\partial f / \partial z^*$ and $\partial f / \partial z$ assuming that the C–R conditions hold.

19.4. (a) Show that, when z is represented by polar coordinates, the C–R conditions on a function $f(z)$ are

$$\frac{\partial U}{\partial r} = \frac{1}{r}\frac{\partial V}{\partial \theta}, \qquad \frac{\partial U}{\partial \theta} = -r\frac{\partial V}{\partial r},$$

where U and V are the real and imaginary parts of $f(z)$ written in polar coordinates.

(b) Show that the derivative of f can be written as

$$\frac{df}{dz} = e^{-i\theta}\left(\frac{\partial U}{\partial r} + i\frac{\partial V}{\partial r}\right).$$

Hint: Start with the C–R conditions in Cartesian coordinates and apply the chain rule to them using $x = r\cos\theta$ and $y = r\sin\theta$.

19.5. Prove the following identities for differentiation by finding the real and imaginary parts of the function—$u(x,y)$ and $v(x,y)$—and differentiating them:

(a) $\dfrac{d}{dz}(f+g) = \dfrac{df}{dz} + \dfrac{dg}{dz}.$ (b) $\dfrac{d}{dz}(fg) = \dfrac{df}{dz}g + f\dfrac{dg}{dz}.$

(c) $\dfrac{d}{dz}\left(\dfrac{f}{g}\right) = \dfrac{f'(z)g(z) - g'(z)f(z)}{[g(z)]^2},$ where $g(z) \neq 0.$

19.6. Show that $d/dz(\ln z) = 1/z$. Hint: Find $u(x,y)$ and $v(x,y)$ for $\ln z$ using the exponential representation of z, then differentiate them.

19.7. Show that $\sin z$ and $\cos z$ have only real roots. Hint: Use definition of sine and cosine in terms of exponentials.

19.8. Use mathematical induction and the product rule for differentiation to show that $\dfrac{d}{dz}(z^n) = nz^{n-1}.$

19.9. Use Equations (19.5) and (19.6), to establish the following identities:

(a) $\mathrm{Re}(\sin z) = \sin x \cosh y,$ $\mathrm{Im}(\sin z) = \cos x \sinh y.$

(b) $\mathrm{Re}(\cos z) = \cos x \cosh y,$ $\mathrm{Im}(\cos z) = -\sin x \sinh y.$

(c) $\mathrm{Re}(\sinh z) = \sinh x \cos y,$ $\mathrm{Im}(\sinh z) = \cosh x \sin y.$

(d) $\mathrm{Re}(\cosh z) = \cosh x \cos y,$ $\mathrm{Im}(\cosh z) = \sinh x \sin y.$

(e) $|\sin z|^2 = \sin^2 x + \sinh^2 y,$ $|\cos z|^2 = \cos^2 x + \sinh^2 y.$

(f) $|\sinh z|^2 = \sinh^2 x + \sin^2 y,$ $|\cosh z|^2 = \sinh^2 x + \cos^2 y.$

19.10. Find all the zeros of $\sinh z$ and $\cosh z$.

19.11. Verify the following trigonometric identities:

(a) $\cos^2 z + \sin^2 z = 1.$

(b) $\cos(z_1 + z_2) = \cos z_1 \cos z_2 - \sin z_1 \sin z_2.$

(c) $\sin(z_1 + z_2) = \sin z_1 \cos z_2 + \cos z_1 \sin z_2.$

(d) $\sin\left(\dfrac{\pi}{2} - z\right) = \cos z,$ $\cos\left(\dfrac{\pi}{2} - z\right) = \sin z.$

(e) $\cos 2z = \cos^2 z - \sin^2 z,$ $\sin 2z = 2\sin z \cos z.$

(f) $\tan(z_1 + z_2) = \dfrac{\tan z_1 + \tan z_2}{1 - \tan z_1 \tan z_2}.$

19.12. Verify the following hyperbolic identities:

(a) $\cosh^2 z - \sinh^2 z = 1$.

(b) $\cosh(z_1 + z_2) = \cosh z_1 \cosh z_2 + \sinh z_1 \sinh z_2$.

(c) $\sinh(z_1 + z_2) = \sin z_1 \cosh z_2 + \cosh z_1 \sinh z_2$.

(d) $\cosh 2z = \cosh^2 z + \sinh^2 z$, $\quad \sinh 2z = 2 \sinh z \cosh z$.

(e) $\tanh(z_1 + z_2) = \dfrac{\tanh z_1 + \tanh z_2}{1 + \tanh z_1 \tanh z_2}$.

19.13. Show that

(a) $\tanh\left(\dfrac{z}{2}\right) = \dfrac{\sinh x + i \sin y}{\cosh x + \cos y}$. $\qquad$ (b) $\coth\left(\dfrac{z}{2}\right) = \dfrac{\sinh x - i \sin y}{\cosh x - \cos y}$.

19.14. Prove the following identities:

(a) $\cos^{-1} z = -i \ln(z \pm \sqrt{z^2 - 1})$. $\quad$ (b) $\sin^{-1} z = -i \ln[iz \pm \sqrt{1 - z^2}]$.

(c) $\tan^{-1} z = \dfrac{1}{2i} \ln\left(\dfrac{i - z}{i + z}\right)$. $\qquad$ (d) $\cosh^{-1} z = \ln(z \pm \sqrt{z^2 - 1})$.

(e) $\sinh^{-1} z = \ln(z \pm \sqrt{z^2 + 1})$. $\qquad$ (f) $\tanh^{-1} z = \frac{1}{2} \ln\left(\dfrac{1 + z}{1 - z}\right)$.

19.15. Prove that $\exp(z^*)$ is not analytic anywhere.

19.16. Show that $e^{iz} = \cos z + i \sin z$ for any z.

19.17. Show that both the real and imaginary parts of an analytic function are harmonic.

19.18. Show that each of the following functions—call each one $u(x, y)$— is harmonic, and find the function's harmonic partner, $v(x, y)$, such that $u(x, y) + iv(x, y)$ is analytic. Hint: Use C–R conditions.

(a) $x^3 - 3xy^2$. $\quad$ (b) $e^x \cos y$. $\qquad$ (c) $\dfrac{x}{x^2 + y^2}$ $\quad$ where $x^2 + y^2 \neq 0$.

(d) $e^{-2y} \cos 2x$. $\quad$ (e) $e^{y^2 - x^2} \cos 2xy$.

$\qquad$ (f) $e^x(x \cos y - y \sin y) + 2 \sinh y \sin x + x^3 - 3xy^2 + y$.

19.19. Describe the curve defined by each of the following equations:

(a) $z = 1 - it$, $\quad 0 \le t \le 2$. $\qquad\qquad$ (b) $z = t + it^2$, $\quad -\infty < t < \infty$.

(c) $z = a(\cos t + i \sin t)$ $\quad \dfrac{\pi}{2} \le t \le \dfrac{3\pi}{2}$. $\quad$ (d) $z = t + \dfrac{i}{t}$ $\quad -\infty < t < 0$.

19.20. Let $f(z) = w = u + iv$. Suppose that $\dfrac{\partial^2 \Phi}{\partial x^2} + \dfrac{\partial^2 \Phi}{\partial y^2} = 0$. Show that if f is analytic, then $\dfrac{\partial^2 \Phi}{\partial u^2} + \dfrac{\partial^2 \Phi}{\partial v^2} = 0$. That is, analytic functions map harmonic functions in the z-plane to harmonic functions in the w-plane.

19.21. (a) Show that $\int f(z)\,dz$ can be written as

$$\int \mathbf{A} \cdot d\mathbf{r} + i \int \mathbf{B} \cdot d\mathbf{r},$$

where $\mathbf{A} = \langle u, -v, 0 \rangle$, $\mathbf{B} = \langle v, u, 0 \rangle$, and $d\mathbf{r} = \langle dx, dy, 0 \rangle$.
(b) Show that both $\mathbf{A}$ and $\mathbf{B}$ have vanishing curls when f is analytic.
(c) Now use the Stokes' theorem to prove the Cauchy–Goursat theorem.

19.22. Find the value of the integral $\int_C [(z+2)/z]\,dz$, where C is: (a) the semicircle $z = 2e^{i\theta}$, for $0 \le \theta \le \pi$; (b) the semicircle $z = 2e^{i\theta}$, for $\pi \le \theta \le 2\pi$; and (c) the circle $z = 2e^{i\theta}$, for $-\pi \le \theta \le \pi$.

19.23. Evaluate the integral $\int_\gamma dz/(z - 1 - i)$ where γ is: (a) the line joining $z_1 = 2i$ and $z_2 = 3$; and (b) the path from z_1 to the origin and from there to z_2.

19.24. Use Equation (19.10) to show that $\dfrac{d}{dz}(z^m) = mz^{m-1}$. Hint: Use the binomial theorem.

19.25. Let C be the boundary of a square whose sides lie along the lines $x = \pm 3$ and $y = \pm 3$. For the positive sense of integration, evaluate each of the following integrals by using CIF or the derivative formula (19.10):

(a) $\displaystyle\oint_C \frac{e^{-z}}{z - i\pi/2}\,dz.$ (b) $\displaystyle\oint_C \frac{e^z}{z(z^2 + 10)}\,dz.$ (c) $\displaystyle\oint_C \frac{\cos z}{(z - \frac{\pi}{4})(z^2 - 10)}\,dz.$

(d) $\displaystyle\oint_C \frac{\sinh z}{z^4}\,dz.$ (e) $\displaystyle\oint_C \frac{\cosh z}{z^4}\,dz.$ (f) $\displaystyle\oint_C \frac{\cos z}{z^3}\,dz.$

(g) $\displaystyle\oint_C \frac{\cos z}{(z - i\pi/2)^2}\,dz.$ (h) $\displaystyle\oint_C \frac{e^z}{(z - i\pi)^2}\,dz.$ (i) $\displaystyle\oint_C \frac{\cos z}{z + i\pi}\,dz.$

(j) $\displaystyle\oint_C \frac{e^z}{z^2 - 5z + 4}\,dz.$ (k) $\displaystyle\oint_C \frac{\sinh z}{(z - i\pi/2)^2}\,dz.$ (l) $\displaystyle\oint_C \frac{\cosh z}{(z - \pi/2)^2}\,dz.$

(m) $\displaystyle\oint_C \frac{z^2}{(z - 2)(z^2 - 10)}\,dz.$

Chapter 20

Complex Series

As in the real case, representation of functions by infinite series of "simpler" functions is an endeavor worthy of our serious consideration. We start with an examination of the properties of sequences and series of complex numbers and derive series representations of some complex functions. Most of the discussion is a direct generalization of the results of the real series.

sequence, convergence to a limit, partial sums, and series

A sequence $\{z_k\}_{k=1}^{\infty}$ of complex numbers is said to converge to a limit z if $\lim_{k\to\infty} |z - z_k| = 0$. In other words, for each positive number ε there must exist an integer N such that $|z - z_k| < \varepsilon$ whenever $k > N$. The reader may show that the real (imaginary) part of the limit of a sequence of complex numbers is the limit of the real (imaginary) part of the sequence. Series can be converted into sequences by partial summation. For instance, to study the infinite series $\sum_{k=1}^{\infty} z_k$, we form the partial sums $Z_n \equiv \sum_{k=1}^{n} z_k$ and investigate the sequence $\{Z_n\}_{n=1}^{\infty}$. We thus say that the infinite series $\sum_{k=1}^{\infty} z_k$ converges to Z if $\lim_{n\to\infty} Z_n = Z$.

Example 20.0.1. A series that is used often in analysis is the geometric series $Z = \sum_{k=0}^{\infty} z^k$. Let us show that this series converges to $1/(1 - z)$ for $|z| < 1$. For a partial sum of n terms, we have

$$Z_n \equiv \sum_{k=0}^{n} z^k = 1 + z + z^2 + \cdots + z^n.$$

Multiply this by z and subtract the result from the Z_n sum to get (see also Example 9.3.3)

$$Z_n - zZ_n = 1 - z^{n+1} \quad \Rightarrow \quad Z_n = \frac{1 - z^{n+1}}{1 - z}.$$

We now show that Z_n converges to $Z = 1/(1 - z)$. We have

$$|Z - Z_n| = \left| \frac{1}{1 - z} - \frac{1 - z^{n+1}}{1 - z} \right| = \left| \frac{z^{n+1}}{1 - z} \right| = \frac{|z|^{n+1}}{|1 - z|}$$

and

$$\lim_{n\to\infty} |Z - Z_n| = \lim_{n\to\infty} \frac{|z|^{n+1}}{|1 - z|} = \frac{1}{|1 - z|} \lim_{n\to\infty} |z|^{n+1} = 0$$

for $|z| < 1$. Thus, $\sum_{k=0}^{\infty} z^k = 1/(1 - z)$ for $|z| < 1$. ∎

If the series $\sum_{k=0}^{\infty} z_k$ converges, both the real part, $\sum_{k=0}^{\infty} x_k$, and the imaginary part, $\sum_{k=0}^{\infty} y_k$, of the series also converge. From Chapter 9, we know that a necessary condition for the convergence of the real series $\sum_{k=0}^{\infty} x_k$ and $\sum_{k=0}^{\infty} y_k$ is that $x_k \to 0$ and $y_k \to 0$. Thus, a necessary condition for the convergence of the complex series is $\lim_{k \to \infty} z_k = 0$. The terms of such a series are, therefore, bounded. Thus, there exists a positive number M such that $|z_k| < M$ for all k.

absolute
convergence

A complex series is said to **converge absolutely**, if the *real* series

$$\sum_{k=0}^{\infty} |z_k| = \sum_{k=0}^{\infty} \sqrt{x_k^2 + y_k^2}$$

converges. Clearly, absolute convergence implies convergence.

20.1   Power Series

We now concentrate on the power series which, as in the real case, are infinite sums of powers of $(z - z_0)$. It turns out—as we shall see shortly—that for complex functions, the inclusion of negative powers is crucial.

power series

Theorem 20.1.1. *If the **power series** $\sum_{k=0}^{\infty} a_k(z - z_0)^k$ converges for z_1 (assumed to be different from z_0), then it converges absolutely for every value of z such that $|z - z_0| < |z_1 - z_0|$. Similarly if the power series $\sum_{k=0}^{\infty} b_k/(z - z_0)^k$ converges for $z_2 \neq z_0$, then it converges absolutely for every value of z such that $|z - z_0| > |z_2 - z_0|$.*

Proof. We prove the first part of the proposition; the second part is done similarly. Since the series converges for $z = z_1$, all the terms $|a_k(z_1 - z_0)^k|$ are smaller than a positive number M. We, therefore have

$$\sum_{k=0}^{\infty} |a_k(z - z_0)^k| = \sum_{k=0}^{\infty} \left| a_k(z_1 - z_0)^k \frac{(z - z_0)^k}{(z_1 - z_0)^k} \right|$$

$$= \sum_{k=0}^{\infty} |a_k(z_1 - z_0)^k| \left| \frac{z - z_0}{z_1 - z_0} \right|^k \leq \sum_{k=0}^{\infty} MB^k$$

$$= M \sum_{k=0}^{\infty} B^k = \frac{M}{1 - B},$$

where $B \equiv |(z - z_0)/(z_1 - z_0)|$ is a positive real number less than 1. Since the RHS is a finite (positive) number, the series of absolute values converges, and the proof is complete.   □

The essence of Theorem 20.1.1 is that if a power series—with positive powers—converges for a point at a distance r_1 from z_0, then it converges for *all interior* points of a circle of radius r_1 centered at z_0. Similarly, if a power series—with negative powers—converges for a point at a distance r_2 from z_0, then it converges for *all exterior* points of a circle of radius r_2 centered at z_0 (see Figure 20.1).

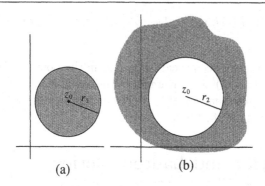

Figure 20.1: (a) Power series with positive exponents converge for the interior points of a circle. (b) Power series with negative exponents converge for the exterior points of a circle.

Box 20.1.1. *When constructing power series, positive powers are used for points inside a circle and negative powers for points outside it.*

The largest circle about z_0 such that the first power series of Theorem 20.1.1 converges is called the **circle of convergence** of the power series. It follows from Theorem 20.1.1 that the series cannot converge at *any* point outside the circle of convergence. (Why?)

circle of convergence

Let us consider the power series

$$S(z) \equiv \sum_{k=0}^{\infty} a_k(z - z_0)^k \qquad (20.1)$$

which we assume to be convergent at all points interior to a circle for which $|z - z_0| = r$. This implies that the sequence of partial sums $\{S_n(z)\}_{n=0}^{\infty}$ converges. Therefore, for any $\varepsilon > 0$, there exists an integer N_ε such that

$$|S(z) - S_n(z)| < \varepsilon \qquad \text{whenever} \quad n > N_\varepsilon.$$

In general, the integer N_ε may be dependent on z; that is, for different values of z, we may be forced to pick different N_ε's. When N_ε is independent of z, we say that the convergence is **uniform**. We state the following result without proof:

uniform convergence explained

Theorem 20.1.2. *The power series $S(z) = \sum_{n=0}^{\infty} a_n(z - z_0)^n$ is **uniformly** convergent for all points within its circle of convergence, and $S(z)$ is an analytic function of z there. Furthermore, such a series can be differentiated and integrated term by term:*

a power series is uniformly convergent and analytic; it can be differentiated and integrated term by term.

$$\frac{dS(z)}{dz} = \sum_{n=1}^{\infty} n a_n(z - z_0)^{n-1}, \qquad \int_\gamma S(z)\,dz = \sum_{n=0}^{\infty} a_n \int_\gamma (z - z_0)^n dz,$$

at each point z and each path γ located inside the circle of convergence of the power series.

By substituting the reciprocal of $(z - z_0)$ in the power series, we can show that if $\sum_{k=0}^{\infty} b_k/(z - z_0)^k$ is convergent in the annulus $r_2 < |z - z_0| < r_1$, then it is uniformly convergent for all z in that annulus, and the series represents a continuous function of z there.

20.2   Taylor and Laurent Series

Complex series, just as their real counterparts, find their most frequent utility in representing well-behaved functions. The following theorem, which we state without proof,[1] is essential in the application of complex analysis.

Theorem 20.2.1. *Let C_1 and C_2 be circles of radii r_1 and r_2, both centered at z_0 in the z-plane with $r_1 > r_2$. Let $f(z)$ be analytic on C_1 and C_2 and throughout S, the annular region between the two circles. Then, at each point z of S, $f(z)$ is* given *uniquely by the* **Laurent series**

$$f(z) = \sum_{n=-\infty}^{\infty} a_n (z - z_0)^n, \quad where \quad a_n = \frac{1}{2\pi i} \oint_C \frac{f(\xi)}{(\xi - z_0)^{n+1}} \, d\xi,$$

and C is any contour within S that encircles z_0. When $r_2 = 0$, the series is called **Taylor series**. *In that case $a_n = 0$ for negative n and $a_n = f^{(n)}(z_0)/n!$ for $n \geq 0$.*

We can see the reduction of the Laurent series to Taylor series as follows. The Laurent expansion is convergent as long as $r_2 < |z - z_0| < r_1$. In particular, if $r_2 = 0$, and if the function is analytic throughout the interior of the larger circle, then $f(\xi)/(\xi - z_0)^{n+1}$ will be analytic for negative integer n, and the integral will be zero by the Cauchy–Goursat theorem. Therefore, a_n will be zero for $n = -1, -2, \dots$. Thus, only positive powers of $(z - z_0)$ will be present in the series, and we obtain the Taylor series.

Maclaurin series   For $z_0 = 0$, the Taylor series reduces to the **Maclaurin series**:

$$f(z) = f(0) + f'(0)z + \cdots = \sum_{n=0}^{\infty} \frac{f^{(n)}(0)}{n!} z^n.$$

Box 19.1.4 tells us that we can enlarge C_1 and shrink C_2 until we encounter a point at which f is no longer analytic. Thus, we can include all the possible analytic points by enlarging C_1 and shrinking C_2.

Example 20.2.2. Let us expand some functions in terms of series. For entire functions there is no point in the entire complex plane at which they are not analytic.

---

[1] For a proof, see Hassani, S. *Mathematical Physics: A Modern Introduction to Its Foundations*, Springer-Verlag, 1999, Section 9.6.

Thus, only positive powers of $(z - z_0)$ will be present, and we will have a Taylor expansion that is valid for all values of z.

(a) We expand e^z around $z_0 = 0$. The nth derivative of e^z is e^z. Thus, $f^{(n)}(0) = 1$, and the Taylor (Maclaurin) expansion gives

$$e^z = \sum_{n=0}^{\infty} \frac{f^{(n)}(0)}{n!} z^n = \sum_{n=0}^{\infty} \frac{z^n}{n!}.$$

(b) The Maclaurin series for $\sin z$ is obtained by noting that

$$\frac{d^n}{dz^n} \sin z \bigg|_{z=0} = \begin{cases} 0 & \text{if } n \text{ is even,} \\ (-1)^{(n-1)/2} & \text{if } n \text{ is odd,} \end{cases}$$

and substituting this in the Maclaurin expansion:

$$\sin z = \sum_{n \text{ odd}} (-1)^{(n-1)/2} \frac{z^n}{n!} = \sum_{k=0}^{\infty} (-1)^k \frac{z^{2k+1}}{(2k+1)!}.$$

Similarly, we can obtain

$$\cos z = \sum_{k=0}^{\infty} (-1)^k \frac{z^{2k}}{(2k)!}, \qquad \sinh z = \sum_{k=0}^{\infty} \frac{z^{2k+1}}{(2k+1)!}, \qquad \cosh z = \sum_{k=0}^{\infty} \frac{z^{2k}}{(2k)!}.$$

It is seen that the series representation of all these functions is obtained by replacing the real variable x in their real series representation with a complex variable z.

(c) The function $1/(1+z)$ is not entire, so the region of its convergence is limited. Let us find the Maclaurin expansion of this function. Starting from the origin ($z_0 = 0$), the function is analytic within all circles of radii $r < 1$. At $r = 1$ we encounter a singularity, the point $z = -1$. Thus, the series converges for all points z for which $|z| < 1$.[2] For such points we have

$$f^{(n)}(0) = \frac{d^n}{dz^n}[(1+z)^{-1}] \bigg|_{z=0} = (-1)^n n!.$$

Thus,

$$\frac{1}{1+z} = \sum_{n=0}^{\infty} \frac{f^{(n)}(0)}{n!} z^n = \sum_{n=0}^{\infty} (-1)^n z^n. \qquad \blacksquare$$

The Taylor and Laurent series allow us to express an analytic function as a power series. For a Taylor series of $f(z)$ the expansion is routine because the coefficient of its nth term is simply $f^{(n)}(z_0)/n!$, where z_0 is the center of the circle of convergence. However, when a Laurent series is applicable in a given region of the complex plane, the nth coefficient is not, in general, easy to evaluate. Usually it can be found by inspection and certain manipulations of other known series. Then the uniqueness of Laurent series expansion assures us that the series so obtained is *the* unique Laurent series for the function in that region.[3]

there is only one Laurent series for a given function defined in a given region.

---

[2] As remarked before, the series diverges for *all* points outside the circle $|z| = 1$. This does not mean that the function cannot be represented by a series for points outside the circle. On the contrary, we shall see shortly that the Laurent series, with *negative* powers is designed precisely for such a purpose.

[3] See Hassani, S. *Mathematical Physics: A Modern Introduction to Its Foundations,* Springer-Verlag, 1999, p. 258.

As in the case of real series,

> **Box 20.2.1.** *We can add, subtract, and multiply convergent power series. Furthermore, if the denominator does not vanish in a neighborhood of a point z_0, then we can obtain the Laurent series of the ratio of two power series about z_0 by long division.*

Thus converging power series can be manipulated as though they were finite sums (polynomials). Such manipulations are extremely useful when dealing with Taylor and Laurent expansions in which the straightforward calculation of coefficients may be tedious. The following examples illustrate the power of infinite-series arithmetic. In these examples, the following equations are very useful:

$$\frac{1}{1-z} = \sum_{n=0}^{\infty} z^n, \quad \frac{1}{1+z} = \sum_{n=0}^{\infty} (-1)^n z^n, \quad |z| < 1. \tag{20.2}$$

Example 20.2.3. To expand the function $f(z) = \dfrac{2+3z}{z^2+z^3}$ in a Laurent series about $z = 0$, rewrite it as

$$f(z) = \frac{1}{z^2}\left(\frac{2+3z}{1+z}\right) = \frac{1}{z^2}\left(3 - \frac{1}{1+z}\right) = \frac{1}{z^2}\left(3 - \sum_{n=0}^{\infty}(-1)^n z^n\right)$$

$$= \frac{1}{z^2}(3 - 1 + z - z^2 + z^3 - \cdots) = \frac{2}{z^2} + \frac{1}{z} - 1 + z - z^2 + \cdots.$$

This series converges for $0 < |z| < 1$. We note that negative powers of z are also present. This is a reflection of the fact that the function is not analytic inside the entire circle $|z| = 1$; it diverges at $z = 0$. ∎

Example 20.2.4. The function $f(z) = z/[(z-1)(z-2)]$ has a Taylor expansion around the origin for $|z| < 1$. To find this expansion, we write[4]

$$f(z) = -\frac{1}{z-1} + \frac{2}{z-2} = \frac{1}{1-z} - \frac{1}{1-z/2}.$$

Expanding both fractions in geometric series (both $|z|$ and $|z/2|$ are less than 1), we obtain $f(z) = \sum_{n=0}^{\infty} z^n - \sum_{n=0}^{\infty} (z/2)^n$. Adding the two series yields

$$f(z) = \sum_{n=0}^{\infty} (1 - 2^{-n}) z^n \quad \text{for} \quad |z| < 1.$$

This is the unique Taylor expansion of $f(z)$ within the circle $|z| = 1$.

---

[4]We could, of course, evaluate the derivatives of all orders of the function at $z = 0$ and use the Maclaurin formula. However, the present method gives the same result much more quickly.

For the annular region $1 < |z| < 2$ we have a Laurent series. This can be seen by noting that

$$f(z) = \frac{1/z}{1/z - 1} - \frac{1}{1 - z/2} = -\frac{1}{z}\left(\frac{1}{1 - 1/z}\right) - \frac{1}{1 - z/2}.$$

Since both fractions on the RHS are analytic in the annular region ($|1/z| < 1$, $|z/2| < 1$), we get

$$f(z) = -\frac{1}{z}\sum_{n=0}^{\infty}\left(\frac{1}{z}\right)^n - \sum_{n=0}^{\infty}\left(\frac{z}{2}\right)^n = -\sum_{n=0}^{\infty}z^{-n-1} - \sum_{n=0}^{\infty}2^{-n}z^n$$

$$= -\sum_{n=-1}^{-\infty}z^n - \sum_{n=0}^{\infty}2^{-n}z^n = -\sum_{n=-\infty}^{\infty}a_n z^n,$$

where $a_n = -1$ for $n < 0$ and $a_n = -2^{-n}$ for $n \geq 0$. This is the unique Laurent expansion of $f(z)$ in the given region.

Finally, for $|z| > 2$ we have only negative powers of z. We obtain the expansion in this region by rewriting $f(z)$ as follows:

$$f(z) = -\frac{1/z}{1 - 1/z} + \frac{2/z}{1 - 2/z}.$$

Expanding the fractions yields

$$f(z) = -\sum_{n=0}^{\infty}z^{-n-1} + \sum_{n=0}^{\infty}2^{n+1}z^{-n-1} = \sum_{n=0}^{\infty}(2^{n+1} - 1)z^{-n-1}.$$

This is again the unique expansion of $f(z)$ in the region $|z| > 2$. ∎

The example above shows that a single function may have different series representations in different regions of the complex plane, each series having its own region of convergence.

Example 20.2.5. Define $f(z)$ as

$$f(z) = \begin{cases} (1 - \cos z)/z^2 & \text{for } z \neq 0, \\ \frac{1}{2} & \text{for } z = 0. \end{cases}$$

We can show that $f(z)$ is an entire function.

Since $1 - \cos z$ and z^2 are entire functions, their ratio is analytic everywhere except at the zeros of its denominator. The only such zero is $z = 0$. Thus, $f(z)$ is analytic everywhere except possibly at $z = 0$. To see the behavior of $f(z)$ at $z = 0$, we look at its Maclaurin series:

$$1 - \cos z = 1 - \sum_{n=0}^{\infty}(-1)^n\frac{z^{2n}}{(2n)!}$$

which implies that

$$\frac{1 - \cos z}{z^2} = \sum_{n=1}^{\infty}(-1)^{n+1}\frac{z^{2n-2}}{(2n)!} = \frac{1}{2} - \frac{z^2}{4!} + \frac{z^4}{6!} - \cdots.$$

The expansion on the RHS shows that the value of the series is $\frac{1}{2}$, which, by definition, is $f(0)$. Thus, the series converges for all z, and Box 20.1.2 says that $f(z)$ is entire. ∎

A Laurent series can give information about the integral of a function around a closed contour in whose interior the function may not be analytic. In fact, the coefficient of the first negative power in a Laurent series is given by

$$a_{-1} = \frac{1}{2\pi i} \oint_C f(\xi) \, d\xi. \qquad (20.3)$$

Thus,

> **Box 20.2.2.** *To find the integral of a (nonanalytic) function around a closed contour surrounding z_0, write the Laurent series for the function and read off a_{-1}, the coefficient of the $1/(z - z_0)$ term. The integral is $2\pi i a_{-1}$.*

Example 20.2.6. As an illustration of this idea, let us evaluate the integral $I = \oint_C dz/[z^2(z-2)]$, where C is a circle of radius 1 centered at the origin. The function is analytic in the annular region $0 < |z| < 2$. We can, therefore, expand it as a Laurent series about $z = 0$ in that region:

$$\frac{1}{z^2(z-2)} = -\frac{1}{2z^2}\left(\frac{1}{1-z/2}\right) = -\frac{1}{2z^2}\sum_{n=0}^{\infty}\left(\frac{z}{2}\right)^n$$

$$= -\frac{1}{2}\left(\frac{1}{z^2}\right) - \frac{1}{4}\left(\frac{1}{z}\right) - \frac{1}{8} - \cdots .$$

Thus, $a_{-1} = -\frac{1}{4}$, and $\oint_C dz/[z^2(z-2)] = 2\pi i a_{-1} = -i\pi/2$. Any other way of evaluating the integral is nontrivial. ∎

20.3   Problems

20.1. Expand $\sinh z$ in a Taylor series about the point $z = i\pi$.

20.2. Let C be the circle $|z - i| = 3$ integrated in the positive sense. Find the value of each of the following integrals using the CIF or the derivative formula (19.10):

(a) $\oint_C \dfrac{e^z}{z^2 + \pi^2}\, dz.$ (b) $\oint_C \dfrac{\sinh z}{(z^2 + \pi^2)^2}\, dz.$ (c) $\oint_C \dfrac{dz}{z^2 + 9}.$

(d) $\oint_C \dfrac{dz}{(z^2 + 9)^2}.$ (e) $\oint_C \dfrac{\cosh z}{(z^2 + \pi^2)^3}\, dz.$ (f) $\oint_C \dfrac{z^2 - 3z + 4}{z^2 - 4z + 3}\, dz.$

20.3. For $0 < r < 1$, show that

$$\sum_{k=0}^{\infty} r^k \cos k\theta = \frac{1 - r\cos\theta}{1 + r^2 - 2r\cos\theta} \quad \text{and} \quad \sum_{k=0}^{\infty} r^k \sin k\theta = \frac{r\sin\theta}{1 + r^2 - 2r\cos\theta}.$$

20.4. Find the Taylor expansion of $1/z^2$ for points inside the circle $|z - 2| < 2$.

20.5. Use mathematical induction to show that

$$\frac{d^n}{dz^n}(1 + z)^{-1}\bigg|_{z=0} = (-1)^n n!.$$

20.6. Find the (unique) Laurent expansion of each of the following functions in each of its regions of analyticity:

(a) $\dfrac{1}{(z - 2)(z - 3)}$. (b) $z \cos(z^2)$. (c) $\dfrac{1}{z^2(1 - z)}$. (d) $\dfrac{\sinh z - z}{z^4}$.

(e) $\dfrac{1}{(1 - z)^3}$. (f) $\dfrac{1}{z^2 - 1}$. (g) $\dfrac{z^2 - 4}{z^2 - 9}$. (h) $\dfrac{1}{(z^2 - 1)^2}$.

(i) $\dfrac{z}{z - 1}$.

20.7. Show that the following functions are entire:

(a) $f(z) = \begin{cases} \dfrac{e^{2z} - 1}{z^2} - \dfrac{2}{z} & \text{for } z \neq 0, \\ 2 & \text{for } z = 0. \end{cases}$ (b) $f(z) = \begin{cases} \dfrac{\sin z}{z} & \text{for } z \neq 0, \\ 1 & \text{for } z = 0. \end{cases}$

$$(c)\ f(z) = \begin{cases} \dfrac{\cos z}{z^2 - \pi^2/4} & \text{for } z \neq \pm\pi/2, \\ -1/\pi & \text{for } z = \pm\pi/2. \end{cases}$$

20.8. Obtain the first few nonzero terms of the Laurent-series expansion of each of the following functions about the origin by approximating the denominator by a polynomial and using the technique of long division of polynomials. Also find the integral of the function along a small simple closed contour encircling the origin.

(a) $\dfrac{1}{\sin z}$. (b) $\dfrac{1}{1 - \cos z}$. (c) $\dfrac{z}{1 - \cosh z}$. (d) $\dfrac{z^2}{z - \sin z}$.

(e) $\dfrac{1}{e^z - 1}$. (f) $\dfrac{1}{z^2 \sin z}$. (g) $\dfrac{z^4}{6z + z^3 - 6\sinh z}$.

20.9. Obtain the Laurent-series expansion of $f(z) = \sinh z/z^3$ about the origin.

Chapter 21

Calculus of Residues

One of the most powerful tools made available by complex analysis is the theory of residues, which makes possible the routine evaluation of certain *real* definite integrals that are impossible to calculate otherwise. Example 20.2.6 showed a situation in which an integral was related to expansion coefficients of Laurent series. Here we will develop a systematic way of evaluating both real and complex integrals using the same idea.

Recall that a singular point z_0 of $f(z)$ is a point at which f fails to be analytic. If, in addition, there is some neighborhood of z_0 in which f is analytic at every point (except, of course, at z_0 itself), then z_0 is called an **isolated singularity** of f. All singularities we have encountered so far have been isolated singularities. Although singularities that are not isolated also exist, we shall not discuss them in this book.

isolated singularity

21.1 The Residue

Let z_0 be an isolated singularity of f. Then there exists an $r > 0$ such that, within the "annular" region $0 < |z - z_0| < r$, the function f has the Laurent expansion[1]

$$f(z) = \sum_{n=-\infty}^{\infty} a_n(z - z_0)^n \equiv \sum_{n=0}^{\infty} a_n(z - z_0)^n + \frac{b_1}{z - z_0} + \frac{b_2}{(z - z_0)^2} + \cdots,$$

where

$$a_n = \frac{1}{2\pi i} \oint_C \frac{f(\xi)\, d\xi}{(\xi - z_0)^{n+1}} \qquad \text{and} \qquad b_n = \frac{1}{2\pi i} \oint_C f(\xi)(\xi - z_0)^{n-1}\, d\xi.$$

In particular,

$$b_1 = \frac{1}{2\pi i} \oint_C f(\xi)\, d\xi, \tag{21.1}$$

---

[1] We are using b_n for a_{-n}.

where C is any simple closed contour around z_0, traversed in the positive sense, on and interior to which f is analytic except at the point z_0 itself.

> **Box 21.1.1.** *The complex number b_1, which is $\frac{1}{2\pi i}$ times the integral of $f(z)$ along the contour, is called the **residue** of f at the isolated singular point z_0.*

It is important to note that the residue is independent of the contour C as long as z_0 is the only isolated singular point within C.

Example 21.1.1. We want to evaluate the integral $\oint_C \sin z \, dz/(z - \pi/2)^3$ where C is any simple closed contour having $z = \pi/2$ as an interior point.

To evaluate the integral we expand around $z = \pi/2$ and use Equation (21.1). We note that

$$\sin z = \cos\left(z - \frac{\pi}{2}\right) = \sum_{n=0}^{\infty}(-1)^n \frac{(z - \pi/2)^{2n}}{(2n)!} = 1 - \frac{(z - \pi/2)^2}{2} + \cdots$$

so

$$\frac{\sin z}{(z - \pi/2)^3} = \frac{1}{(z - \pi/2)^3} - \frac{1}{2}\left(\frac{1}{z - \pi/2}\right) + \cdots.$$

It follows that $b_1 = -\frac{1}{2}$; therefore, $\oint_C \sin z \, dz/(z - \pi/2)^3 = 2\pi i b_1 = -i\pi$. ∎

Example 21.1.2. The integral $\oint_C \cos z \, dz/z^2$, where C is the circle $|z| = 1$, is zero because

$$\frac{\cos z}{z^2} = \frac{1}{z^2}\sum_{n=0}^{\infty}(-1)^n \frac{z^{2n}}{(2n)!} = \frac{1}{z^2} - \frac{1}{2} + \frac{z^2}{4!} + \cdots$$

yields $b_1 = 0$ (no $1/z$ term in the Laurent expansion). Therefore, by Equation (21.1) the integral must vanish.

When C is the circle $|z| = 2$, $\oint_C e^z \, dz/(z - 1)^3 = i\pi e$ because

$$e^z = e e^{z-1} = e\sum_{n=0}^{\infty}\frac{(z - 1)^n}{n!} = e\left[1 + (z - 1) + \frac{(z - 1)^2}{2!} + \cdots\right]$$

and

$$\frac{e^z}{(z - 1)^3} = e\left[\frac{1}{(z - 1)^3} + \frac{1}{(z - 1)^2} + \frac{1}{2}\left(\frac{1}{z - 1}\right) + \cdots\right].$$

Thus, $b_1 = e/2$, and the integral is $2\pi i b_1 = i\pi e$. ∎

We use the notation $\mathrm{Res}[f(z_0)]$ to denote the residue of f at the isolated singular point z_0. Equation (21.1) can then be written as

$$\oint_C f(z) \, dz = 2\pi i \, \mathrm{Res}[f(z_0)].$$

What if there are several isolated singular points within the simple closed contour C? Let C_k be the positively traversed circle around z_k shown in Figure 21.1. Then the Cauchy–Goursat theorem yields

$$0 = \oint_{C'} f(z) \, dz = \oint_{\text{circles}} f(z) \, dz + \oint_{\substack{\text{parallel} \\ \text{lines}}} f(z) \, dz + \oint_C f(z) \, dz,$$

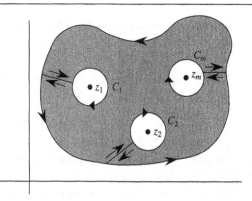

Figure 21.1: Singularities are avoided by going around them.

where C' is the union of all contours inside which union there are no singularities. The contributions of the parallel lines cancel out, and we obtain

$$\oint_C f(z)\,dz = -\sum_{k=1}^{m} \oint_{C_k} f(z)\,dz = \sum_{k=1}^{m} 2\pi i\,\mathrm{Res}[f(z_k)],$$

where in the last step the definition of residue at z_k has been used. The minus sign disappears in the final result because the sense of C_k, while positive for the shaded region of Figure 21.1, is negative for the interior of C_k because this interior is to our right as we traverse C_k in the direction indicated. We thus have

Theorem 21.1.3. (***The Residue Theorem***). *Let C be a positively integrated simple closed contour within and on which a function f is analytic except at a finite number of isolated singular points $z_1, z_2, \ldots, z_m$ interior to C. Then*

$$\oint_C f(z)\,dz = 2\pi i \sum_{k=1}^{m} \mathrm{Res}[f(z_k)]. \qquad (21.2)$$

Example 21.1.4. Let us evaluate the integral $\oint_C (2z-3)\,dz/[z(z-1)]$ where C is the circle $|z| = 2$. There are two isolated singularities in C, $z_1 = 0$ and $z_2 = 1$. To find $\mathrm{Res}[f(z_1)]$, we expand around the origin using Equation (20.2):

$$\frac{2z-3}{z(z-1)} = \frac{3}{z} - \frac{1}{z-1} = \frac{3}{z} + \frac{1}{1-z} = \frac{3}{z} + 1 + z + \cdots \qquad \text{for } |z| < 1.$$

This gives $\mathrm{Res}[f(z_1)] = 3$. Similarly, expanding around $z = 1$ gives

$$\frac{2z-3}{z(z-1)} = \frac{3}{(z-1)+1} - \frac{1}{z-1} = -\frac{1}{z-1} + 3\sum_{n=0}^{\infty} (-1)^n (z-1)^n$$

which yields $\mathrm{Res}[f(z_2)] = -1$. Thus,

$$\oint_C \frac{2z-3}{z(z-1)}\,dz = 2\pi i\{\mathrm{Res}[f(z_1)] + \mathrm{Res}[f(z_2)]\} = 2\pi i(3-1) = 4\pi i. \qquad \blacksquare$$

Let $f(z)$ have an isolated singularity at z_0. Then there exist a real number $r > 0$ and an annular region $0 < |z - z_0| < r$ such that f can be represented by the Laurent series

$$f(z) = \sum_{n=0}^{\infty} a_n (z - z_0)^n + \sum_{n=1}^{\infty} \frac{b_n}{(z - z_0)^n}. \qquad (21.3)$$

principal part of a function

The second sum in Equation (21.3), involving negative powers of $(z - z_0)$, is called the **principal part** of f at z_0. The principal part is used to classify isolated singularities. We consider two cases:

removable singular point

(a) If $b_n = 0$ for all $n \geq 1$, z_0 is called a **removable singular point** of f. In this case, the Laurent series contains only nonnegative powers of $(z - z_0)$, and setting $f(z_0) = a_0$ makes the function analytic at z_0. For example, the function $f(z) = (e^z - 1 - z)/z^2$, which is indeterminate at $z = 0$, becomes entire if we set $f(0) = 1/2$, because its Laurent series

$$f(z) = \frac{1}{2} + \frac{z}{3!} + \frac{z^2}{4!} + \cdots$$

has no negative power.

poles defined

(b) If $b_n = 0$ for all $n > m$ and $b_m \neq 0$, z_0 is called a **pole of order** m. In this case, the expansion takes the form

$$f(z) = \sum_{n=0}^{\infty} a_n (z - z_0)^n + \frac{b_1}{z - z_0} + \cdots + \frac{b_m}{(z - z_0)^m}$$

simple pole

for $0 < |z - z_0| < r$. In particular, if $m = 1$, z_0 is called a **simple pole**.

Example 21.1.5. Let us consider some examples of poles of various orders.
(a) The function $(z^2 - 3z + 5)/(z - 1)$ has a Laurent series around $z = 1$ containing only three terms: $(z^2 - 3z + 5)/(z - 1) = -1 + (z - 1) + 3/(z - 1)$. Thus, it has a simple pole at $z = 1$, with a residue of 3.
(b) The function $\sin z/z^6$ has a Laurent series

$$\frac{\sin z}{z^6} = \frac{1}{z^6} \sum_{n=0}^{\infty} (-1)^n \frac{z^{2n+1}}{(2n+1)!} = \frac{1}{z^5} - \frac{1}{6z^3} + \frac{1}{(5!)z} - \frac{z}{7!} + \cdots$$

about $z = 0$. The principal part has three terms. The pole, at $z = 0$, is of order 5, and the function has a residue of $1/120$ at $z = 0$.
(c) The function $(z^2 - 5z + 6)/(z - 2)$ has a removable singularity at $z = 2$, because

$$\frac{z^2 - 5z + 6}{z - 2} = \frac{(z - 2)(z - 3)}{z - 2} = z - 3 = -1 + (z - 2)$$

and $b_n = 0$ for all n.   ∎

The type of isolated singularity that is most important in applications is of the second type—poles. For a function that has a pole of order m at z_0, the calculation of residues is routine. Such a calculation, in turn, enables us to evaluate many integrals effortlessly. How do we calculate the residue of a function f having a pole of order m at z_0?

It is clear that if f has a pole of order m, then $g(z)$ defined by $g(z) \equiv (z - z_0)^m f(z)$ is analytic at z_0. Thus, for any simple closed contour C that contains z_0 but no other singular point of f, we have

$$\text{Res}[f(z_0)] = \frac{1}{2\pi i} \oint_C f(z)\, dz = \frac{1}{2\pi i} \oint_C \frac{g(z)\, dz}{(z - z_0)^m} = \frac{g^{(m-1)}(z_0)}{(m-1)!},$$

where we used Equation (19.10). In terms of f this yields[2]

$$\text{Res}[f(z_0)] = \frac{1}{(m-1)!} \lim_{z \to z_0} \frac{d^{m-1}}{dz^{m-1}}[(z - z_0)^m f(z)]. \qquad (21.4)$$

For the special, but important, case of a simple pole, we obtain

$$\text{Res}[f(z_0)] = \lim_{z \to z_0} [(z - z_0) f(z)]. \qquad (21.5)$$

The most widespread application of residues occurs in the evaluation of real definite integrals. It is possible to "complexify" certain real definite integrals and relate them to contour integrations in the complex plane. What is typically involved is the addition of a number of semicircles to the real integral such that it becomes a closed contour integral whose value can be determined by the residue theorem. One then takes the limit of the contour integral when the radii of the semicircles go to infinity or zero. In this limit the contributions from the semicircles should vanish for the method to work. In that case, one recovers the real integral. There are three types of integrals most commonly encountered. We discuss these separately below. In all cases we assume that the contribution of the semicircles will vanish in the limit.

application of the residue theorem in evaluating definite integrals

21.2   Integrals of Rational Functions

The first type of integral we can evaluate using the residue theorem is of the form

$$I_1 = \int_{-\infty}^{\infty} \frac{p(x)}{q(x)}\, dx,$$

where $p(x)$ and $q(x)$ are real polynomials, and $q(x) \neq 0$ for any real x. We can then write

$$I_1 = \lim_{R \to \infty} \int_{-R}^{R} \frac{p(x)}{q(x)}\, dx = \lim_{R \to \infty} \int_{C_x} \frac{p(z)}{q(z)}\, dz,$$

where C_x is the (open) contour lying on the real axis from $-R$ to $+R$. We now close that contour by adding to it the semicircle of radius R [see Figure 21.2(a)]. This will not affect the value of the integral because, by our

---

[2]The limit is taken because in many cases the mere substitution of z_0 may result in an indeterminate form.

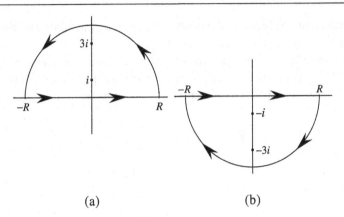

Figure 21.2: (a) The large semicircle is chosen in the UHP. (b) Note how the direction of contour integration is forced to be clockwise when the semicircle is chosen in the LHP.

assumption, the contribution of the integral of the semicircle tends to zero in the limit $R \to \infty$. We close the contour in the upper half-plane (UHP) if $q(z)$ has a zero there. We then get

$$I_1 = \lim_{R \to \infty} \oint_C \frac{p(z)}{q(z)} \, dz = 2\pi i \sum_{j=1}^{k} \text{Res} \left[\frac{p(z_j)}{q(z_j)} \right],$$

where C is the closed contour composed of the interval $(-R, R)$ and the semicircle C_R, and $\{z_j\}_{j=1}^{k}$ are the zeros of $q(z)$ in the UHP. We may instead close the contour in the lower half-plane (LHP), in which case

$$I_1 = -2\pi i \sum_{j=1}^{m} \text{Res} \left[\frac{p(z_j)}{q(z_j)} \right],$$

where $\{z_j\}_{j=1}^{m}$ are the zeros of $q(z)$ in the LHP. The minus sign indicates that in the LHP we (are forced to) integrate in the negative sense.

Example 21.2.1. Let us evaluate the integral $I = \int_0^\infty x^2 \, dx / [(x^2 + 1)(x^2 + 9)]$. Since the integrand is even, we can extend the interval of integration to all real numbers (and divide the result by 2). It is shown below that in the limit that the radius of the semicircle goes to infinity, the integral of that semicircle goes to zero. Therefore, we write the contour integral corresponding to I:

$$I = \frac{1}{2} \oint_C \frac{z^2 \, dz}{(z^2 + 1)(z^2 + 9)},$$

where C is as shown in Figure 21.2(a). Note that the contour is integrated in the positive sense. This is always true for the UHP. The singularities of the function in the UHP are the *simple* poles i and $3i$ corresponding to the simple zeros of the denominator. By (21.5), the residues at these poles are

$$\operatorname{Res}[f(i)] = \lim_{z \to i} \left[(z - i) \frac{z^2}{(z - i)(z + i)(z^2 + 9)} \right] = -\frac{1}{16i},$$

$$\operatorname{Res}[f(3i)] = \lim_{z \to 3i} \left[(z - 3i) \frac{z^2}{(z^2 + 1)(z - 3i)(z + 3i)} \right] = \frac{3}{16i}.$$

Thus, we obtain

$$I = \int_0^\infty \frac{x^2 \, dx}{(x^2 + 1)(x^2 + 9)} = \frac{1}{2} \oint_C \frac{z^2 \, dz}{(z^2 + 1)(z^2 + 9)} = \pi i \left(-\frac{1}{16i} + \frac{3}{16i} \right) = \frac{\pi}{8}.$$

It is instructive to obtain the same results using the LHP. In this case the contour is as shown in Figure 21.2(b). It is clear that the interior is to our right as we traverse the contour. So we have to introduce a minus sign for its integration. The singular points are at $z = -i$ and $z = -3i$. These are simple poles at which the residues of the function are

$$\operatorname{Res}[f(-i)] = \lim_{z \to -i} \left[(z + i) \frac{z^2}{(z - i)(z + i)(z^2 + 9)} \right] = \frac{1}{16i},$$

$$\operatorname{Res}[f(-3i)] = \lim_{z \to -3i} \left[(z + 3i) \frac{z^2}{(z^2 + 1)(z - 3i)(z + 3i)} \right] = -\frac{3}{16i}.$$

Therefore,

$$I = \int_0^\infty \frac{x^2 \, dx}{(x^2 + 1)(x^2 + 9)} = \frac{1}{2} \oint_C \frac{z^2 \, dz}{(z^2 + 1)(z^2 + 9)} = -\pi i \left(\frac{1}{16i} - \frac{3}{16i} \right) = \frac{\pi}{8}.$$

We now show that the integral of the large circle Γ tends to zero. On such a circle, $z = Re^{i\theta}$; therefore

$$\int_\Gamma \frac{z^2 \, dz}{(z^2 + 1)(z^2 + 9)} = \int_\Gamma \frac{R^2 e^{2i\theta} Re^{i\theta} \, d\theta}{(R^2 e^{2i\theta} + 1)(R^2 e^{2i\theta} + 9)}.$$

In the limit that $R \to \infty$, we can ignore the small numbers 1 and 9 in the denominator. Then the overall integral becomes $1/R$ times a finite integral over θ. It follows that as R tends to infinity, the contribution of the large circle indeed goes to zero. ∎

Example 21.2.2. Let us now consider a more complicated integral:

$$\int_{-\infty}^\infty \frac{x^2 \, dx}{(x^2 + 1)(x^2 + 4)^2}$$

which turns into $\oint_C z^2 \, dz / [(z^2 + 1)(z^2 + 4)^2]$. The poles in the UHP are at $z = i$ and $z = 2i$. The former is a simple pole, and the latter is a pole of order 2. Thus,

$$\operatorname{Res}[f(i)] = \lim_{z \to i} \left[(z - i) \frac{z^2}{(z - i)(z + i)(z^2 + 4)^2} \right] = -\frac{1}{18i},$$

$$\operatorname{Res}[f(2i)] = \frac{1}{(2 - 1)!} \lim_{z \to 2i} \frac{d}{dz} \left[(z - 2i)^2 \frac{z^2}{(z^2 + 1)(z + 2i)^2 (z - 2i)^2} \right]$$

$$= \lim_{z \to 2i} \frac{d}{dz} \left[\frac{z^2}{(z^2 + 1)(z + 2i)^2} \right] = \frac{5}{72i},$$

and

$$\int_{-\infty}^\infty \frac{x^2 \, dx}{(x^2 + 1)(x^2 + 4)^2} = 2\pi i \left(-\frac{1}{18i} + \frac{5}{72i} \right) = \frac{\pi}{36}.$$

Closing the contour in the LHP would yield the same result as the reader is urged to verify. ∎

21.3   Products of Rational and Trigonometric Functions

The second type of integral we can evaluate using the residue theorem is of the form

$$\int_{-\infty}^{\infty} \frac{p(x)}{q(x)} \cos ax \, dx \qquad \text{or} \qquad \int_{-\infty}^{\infty} \frac{p(x)}{q(x)} \sin ax \, dx,$$

where a is a real number, $p(x)$ and $q(x)$ are real polynomials in x, and $q(x)$ has no real zeros. These integrals are the real and imaginary parts of

$$I_2 = \int_{-\infty}^{\infty} \frac{p(x)}{q(x)} e^{iax} \, dx.$$

The presence of e^{iax} dictates the choice of the half-plane: If $a \geq 0$, we choose the UHP because

$$e^{iaz} = e^{ia(x+iy)} = e^{iax} e^{-ay} \qquad \text{where} \quad y > 0,$$

and the negative exponent ensures convergence for large R and y. For the same reason, we choose the LHP when $a \leq 0$. The following examples illustrate the procedure.

Example 21.3.1. Let us evaluate $\int_{-\infty}^{\infty} \cos ax \, dx/(x^2 + 1)^2$ where $a \neq 0$. This integral is the real part of the integral $I_2 = \int_{-\infty}^{\infty} e^{iax} \, dx/(x^2 + 1)^2$. When $a > 0$, we close in the UHP. Then we proceed as for integrals of rational functions. Thus, we have

$$I_2 = \oint_C \frac{e^{iaz}}{(z^2 + 1)^2} \, dz = 2\pi i \operatorname{Res}[f(i)] \qquad \text{for} \quad a > 0,$$

because there is only one singularity in the UHP at $z = i$ which is a pole of order 2. We next calculate the residue:

$$\operatorname{Res}[f(i)] = \lim_{z \to i} \frac{d}{dz} \left[(z - i)^2 \frac{e^{iaz}}{(z - i)^2 (z + i)^2} \right]$$

$$= \lim_{z \to i} \frac{d}{dz} \left[\frac{e^{iaz}}{(z + i)^2} \right] = \lim_{z \to i} \left[\frac{(z + i) i a e^{iaz} - 2 e^{iaz}}{(z + i)^3} \right] = \frac{e^{-a}}{4i}(1 + a).$$

Substituting this in the expression for I_2, we obtain $I_2 = (\pi/2) e^{-a}(1 + a)$ for $a > 0$.

When $a < 0$, we have to close the contour in the LHP, where the pole of order 2 is at $z = -i$ and the contour is taken clockwise. Thus, we get

$$I_2 = \oint_C \frac{e^{iaz}}{(z^2 + 1)^2} \, dz = -2\pi i \operatorname{Res}[f(-i)] \qquad \text{for } a < 0.$$

For the residue we obtain

$$\operatorname{Res}[f(-i)] = \lim_{z \to -i} \frac{d}{dz} \left[(z + i)^2 \frac{e^{iaz}}{(z - i)^2 (z + i)^2} \right] = -\frac{e^a}{4i}(1 - a)$$

and the expression for I_2 becomes $I_2 = (\pi/2) e^a (1 - a)$ for $a < 0$. We can combine the two results and write

$$\int_{-\infty}^{\infty} \frac{\cos ax}{(x^2 + 1)^2} \, dx = \operatorname{Re}(I_2) = I_2 = \frac{\pi}{2}(1 + |a|) e^{-|a|}. \qquad \blacksquare$$

Example 21.3.2. As another example, let us evaluate

$$\int_{-\infty}^{\infty} \frac{x \sin ax}{x^4 + 4} \, dx \qquad \text{where } a \neq 0.$$

This is the imaginary part of the integral $I_2 = \int_{-\infty}^{\infty} x e^{iax} \, dx/(x^4+4)$ which, in terms of z and for the closed contour in the UHP (when $a > 0$), becomes

$$I_2 = \oint_C \frac{z e^{iaz}}{z^4 + 4} \, dz = 2\pi i \sum_{j=1}^{m} \text{Res}[f(z_j)] \qquad \text{for } a > 0, \qquad (21.6)$$

where C is the large semicircle in the UHP. The singularities are determined by the zeros of the denominator: $z^4 + 4 = 0$ or $z = 1 \pm i, -1 \pm i$. Of these four simple poles only two, $1 + i$ and $-1 + i$, are in the UHP. We now calculate the residues:

$$\text{Res}[f(1+i)] = \lim_{z \to 1+i} (z - 1 - i) \frac{z e^{iaz}}{(z-1-i)(z-1+i)(z+1-i)(z+1+i)}$$

$$= \frac{(1+i)e^{ia(1+i)}}{(2i)(2)(2+2i)} = \frac{e^{ia}e^{-a}}{8i},$$

$$\text{Res}[f(-1+i)] = \lim_{z \to -1+i} (z + 1 - i) \frac{z e^{iaz}}{(z+1-i)(z+1+i)(z-1-i)(z-1+i)}$$

$$= \frac{(-1+i)e^{ia(-1+i)}}{(2i)(-2)(-2+2i)} = -\frac{e^{-ia}e^{-a}}{8i}.$$

Substituting in Equation (21.6), we obtain

$$I_2 = 2\pi i \frac{e^{-a}}{8i} (e^{ia} - e^{-ia}) = i \frac{\pi}{2} e^{-a} \sin a.$$

Thus,

$$\int_{-\infty}^{\infty} \frac{x \sin ax}{x^4 + 4} \, dx = \text{Im}(I_2) = \frac{\pi}{2} e^{-a} \sin a \qquad \text{for } a > 0. \qquad (21.7)$$

For $a < 0$, we could close the contour in the LHP. But there is an easier way of getting to the answer. We note that $-a > 0$, and Equation (21.7) yields

$$\int_{-\infty}^{\infty} \frac{x \sin ax}{x^4 + 4} \, dx = -\int_{-\infty}^{\infty} \frac{x \sin[(-a)x]}{x^4 + 4} \, dx = -\frac{\pi}{2} e^{-(-a)} \sin(-a) = \frac{\pi}{2} e^{a} \sin a.$$

We can collect the two cases in

$$\int_{-\infty}^{\infty} \frac{x \sin ax}{x^4 + 4} \, dx = \frac{\pi}{2} e^{-|a|} \sin a. \qquad \blacksquare$$

Example 21.3.3. The integral $\int_0^{\infty} \frac{\sin ax}{x} \, dx$ occurs frequently in physics. To evaluate it, first we assume that $a > 0$ and note that since the integrand is even, we can extend the lower limit of integration to $-\infty$ and write

$$\int_0^{\infty} \frac{\sin ax}{x} \, dx = \frac{1}{2} \int_{-\infty}^{\infty} \frac{\sin ax}{x} \, dx.$$

As in the previous examples, we are inclined to choose the contour C in the UHP. However, since C passes through the origin, this will not work because the origin is the pole of the integrand. So, let's avoid the origin by going around it on a small

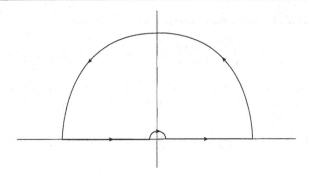

Figure 21.3: To avoid the origin move on an infinitesimal semicircle γ_ϵ of radius ϵ.

circle of radius ϵ as shown in Figure 21.3. This contour does not surround a pole. Therefore, we can write

$$0 = \oint_C \frac{e^{iaz}}{z}\, dz = \int_{-\infty}^{-\epsilon} \frac{e^{iax}}{x}\, dx + \int_{\gamma_\epsilon} \frac{e^{iaz}}{z}\, dz + \int_\epsilon^\infty \frac{e^{iax}}{x}\, dx$$

As $\epsilon \to 0$, the two integrals in x become a single integral over all real numbers. Thus, we get

$$\int_{-\infty}^\infty \frac{e^{iax}}{x}\, dx = -\lim_{\epsilon \to 0} \int_{\gamma_\epsilon} \frac{e^{iaz}}{z}\, dz$$

But on γ_ϵ, $z = \epsilon e^{i\theta}$. Thus

$$\lim_{\epsilon \to 0} \int_{\gamma_\epsilon} \frac{e^{iaz}}{z}\, dz = \lim_{\epsilon \to 0} \int_\pi^0 \frac{e^{ia\epsilon e^{i\theta}}}{\epsilon e^{i\theta}} i\epsilon e^{i\theta}\, d\theta = i \lim_{\epsilon \to 0} \int_\pi^0 e^{ia\epsilon e^{i\theta}}\, d\theta = -i\pi$$

and

$$\int_{-\infty}^\infty \frac{e^{iax}}{x}\, dx = i\pi.$$

Putting everything together, we obtain

$$\int_0^\infty \frac{\sin ax}{x}\, dx = \frac{1}{2} \int_{-\infty}^\infty \frac{\sin ax}{x}\, dx = \frac{1}{2} \operatorname{Im} \int_{-\infty}^\infty \frac{e^{iax}}{x}\, dx = \frac{1}{2} \operatorname{Im}(i\pi) = \frac{\pi}{2}$$

If $a < 0$, then $\sin ax = -\sin |a| x$ and we get the negative of the answer above. ∎

21.4   Functions of Trigonometric Functions

The third type of integral we can evaluate using the residue theorem involves only trigonometric functions and is typically of the form

$$\int_0^{2\pi} F(\sin \theta, \cos \theta)\, d\theta,$$

where F is some (typically rational) function[3] of its arguments. Since θ varies from 0 to 2π, we can consider it as the angle of a point z on the unit circle

---

[3]Recall that a rational function is, by definition, the ratio of two polynomials.

centered at the origin. Then $z = e^{i\theta}$ and $e^{-i\theta} = 1/z$, and we can substitute $\cos\theta = (z + 1/z)/2$, $\sin\theta = (z - 1/z)/(2i)$, and $d\theta = dz/(iz)$ in the original integral to obtain

$$\oint_C F\left(\frac{z - 1/z}{2i}, \frac{z + 1/z}{2}\right) \frac{dz}{iz}.$$

This integral can often be evaluated using the method of residues.

Example 21.4.1. Let us evaluate the integral $\int_0^{2\pi} d\theta/(1 + a\cos\theta)$ where $|a| < 1$. Substituting for $\cos\theta$ and $d\theta$ in terms of z, we obtain

$$\oint_C \frac{dz/iz}{1 + a[(z^2 + 1)/2z]} = \frac{2}{i}\oint_C \frac{dz}{2z + az^2 + a},$$

where C is the unit circle centered at the origin. The singularities of the integrand are the zeros of its denominator $2z + az^2 + a \equiv a(z - z_1)(z - z_2)$ with

$$z_1 = \frac{-1 + \sqrt{1 - a^2}}{a} \qquad \text{and} \qquad z_2 = \frac{-1 - \sqrt{1 - a^2}}{a}.$$

For $|a| < 1$ it is clear that z_2 will lie outside the unit circle C; therefore, it does not contribute to the integral. But z_1 lies inside, and we obtain

$$\oint_C \frac{dz}{2z + az^2 + a} = 2\pi i \operatorname{Res}[f(z_1)].$$

The residue of the simple pole at z_1 can be calculated:

$$\operatorname{Res}[f(z_1)] = \lim_{z \to z_1}(z - z_1)\frac{1}{a(z - z_1)(z - z_2)} = \frac{1}{a}\left(\frac{1}{z_1 - z_2}\right)$$

$$= \frac{1}{a}\left(\frac{a}{2\sqrt{1 - a^2}}\right) = \frac{1}{2\sqrt{1 - a^2}}.$$

It follows that

$$\int_0^{2\pi} \frac{d\theta}{1 + a\cos\theta} = \frac{2}{i}\oint_C \frac{dz}{2z + az^2 + a} = \frac{2}{i}2\pi i\left(\frac{1}{2\sqrt{1 - a^2}}\right) = \frac{2\pi}{\sqrt{1 - a^2}}. \qquad \blacksquare$$

Example 21.4.2. As another example, let us consider the integral

$$I = \int_0^{\pi} \frac{d\theta}{(a + \cos\theta)^2} \qquad \text{where} \quad a > 1.$$

Since $\cos\theta$ is an even function of θ, we may write

$$I = \frac{1}{2}\int_{-\pi}^{\pi} \frac{d\theta}{(a + \cos\theta)^2} \qquad \text{where} \quad a > 1.$$

This integration is over a complete cycle around the origin, and we can make the usual substitution:

$$I = \frac{1}{2}\oint_C \frac{dz/iz}{[a + (z^2 + 1)/2z]^2} = \frac{2}{i}\oint_C \frac{z\,dz}{(z^2 + 2az + 1)^2}.$$

The denominator has the roots $z_1 = -a + \sqrt{a^2 - 1}$ and $z_2 = -a - \sqrt{a^2 - 1}$ which are both of order 2. The second root is outside the unit circle because $a > 1$. The reader may verify that for all $a > 1$, z_1 is inside the unit circle. Since z_1 is a pole of order 2, we have

$$\text{Res}[f(z_1)] = \lim_{z \to z_1} \frac{d}{dz}\left[(z - z_1)^2 \frac{z}{(z - z_1)^2(z - z_2)^2}\right]$$

$$= \lim_{z \to z_1} \frac{d}{dz}\left[\frac{z}{(z - z_2)^2}\right] = \frac{1}{(z_1 - z_2)^2} - \frac{2z_1}{(z_1 - z_2)^3} = \frac{a}{4(a^2 - 1)^{3/2}}.$$

We thus obtain

$$I = \frac{2}{i}2\pi i \,\text{Res}[f(z_1)] = \frac{\pi a}{(a^2 - 1)^{3/2}}. \qquad\blacksquare$$

21.5   Problems

21.1. Evaluate each of the following integrals, for all of which C is the circle $|z| = 3$:

(a) $\oint_C \dfrac{4z - 3}{z(z - 2)}\,dz.$ (b) $\oint_C \dfrac{e^z}{z(z - i\pi)}\,dz.$ (c) $\oint_C \dfrac{\cos z}{z(z - \pi)}\,dz.$

(d) $\oint_C \dfrac{z^2 + 1}{z(z - 1)}\,dz.$ (e) $\oint_C \dfrac{\cosh z}{z^2 + \pi^2}\,dz.$ (f) $\oint_C \dfrac{1 - \cos z}{z^2}\,dz.$

(g) $\oint_C \dfrac{\sinh z}{z^4}\,dz.$ (h) $\oint_C z\cos\left(\dfrac{1}{z}\right)dz.$ (i) $\oint_C \dfrac{dz}{z^3(z + 5)}\,dz.$

(j) $\oint_C \tan z\,dz.$ (k) $\oint_C \dfrac{dz}{\sinh 2z}\,dz.$ (l) $\oint_C \dfrac{e^z}{z^2}\,dz.$

(m) $\oint_C \dfrac{dz}{z^2\sin z}\,dz.$ (n) $\oint_C \dfrac{e^z\,dz}{(z - 1)(z - 2)}.$

21.2. Find the residue of $f(z) = 1/\cos z$ at all its poles.

21.3. Evaluate the integral $\int_0^\infty dx/[(x^2 + 1)(x^2 + 4)]$ by closing the contour (a) in the UHP and (b) in the LHP.

21.4. Evaluate the following integrals in which a and b are nonzero real constants:

(a) $\displaystyle\int_0^\infty \dfrac{2x^2 + 1}{x^4 + 5x^2 + 6}\,dx.$ (b) $\displaystyle\int_0^\infty \dfrac{dx}{6x^4 + 5x^2 + 1}.$ (c) $\displaystyle\int_0^\infty \dfrac{dx}{x^4 + 1}.$

(d) $\displaystyle\int_0^\infty \dfrac{\cos x\,dx}{(x^2 + a^2)^2(x^2 + b^2)}.$ (e) $\displaystyle\int_0^\infty \dfrac{\cos ax}{(x^2 + b^2)^2}\,dx.$ (f) $\displaystyle\int_0^\infty \dfrac{dx}{(x^2 + 1)^2}.$

(g) $\displaystyle\int_0^\infty \dfrac{dx}{(x^2 + 1)^2(x^2 + 2)}.$ (h) $\displaystyle\int_0^\infty \dfrac{2x^2 - 1}{x^6 + 1}\,dx.$ (i) $\displaystyle\int_0^\infty \dfrac{x^2\,dx}{(x^2 + a^2)^2}.$

(j) $\displaystyle\int_{-\infty}^\infty \dfrac{x\,dx}{(x^2 + 4x + 13)^2}.$ (k) $\displaystyle\int_0^\infty \dfrac{x^3\sin ax}{x^6 + 1}\,dx.$ (l) $\displaystyle\int_0^\infty \dfrac{x^2 + 1}{x^2 + 4}\,dx.$

(m) $\displaystyle\int_{-\infty}^\infty \dfrac{x\cos x\,dx}{x^2 - 2x + 10}.$ (n) $\displaystyle\int_{-\infty}^\infty \dfrac{x\sin x\,dx}{x^2 - 2x + 10}.$ (o) $\displaystyle\int_0^\infty \dfrac{dx}{x^2 + 1}.$

(p) $\displaystyle\int_0^\infty \dfrac{x^2\,dx}{(x^2 + 4)^2(x^2 + 25)}.$ (q) $\displaystyle\int_0^\infty \dfrac{\cos ax}{x^2 + b^2}\,dx.$ (r) $\displaystyle\int_0^\infty \dfrac{dx}{(x^2 + 4)^2}.$

21.5. Evaluate each of the following integrals by turning them into contour integrals around the unit circle.

(a) $\displaystyle\int_0^{2\pi} \frac{d\theta}{5+4\sin\theta}$. 　　(b) $\displaystyle\int_0^{2\pi} \frac{d\theta}{a+\cos\theta}$ 　where $a > 1$.

(c) $\displaystyle\int_0^{2\pi} \frac{d\theta}{1+\sin^2\theta}$. 　　(d) $\displaystyle\int_0^{2\pi} \frac{d\theta}{(a+b\cos^2\theta)^2}$ 　where $a, b > 0$.

(e) $\displaystyle\int_0^{2\pi} \frac{\cos^2 3\theta}{5-4\cos 2\theta}\,d\theta$. 　(f) $\displaystyle\int_0^{\pi} \frac{d\phi}{1-2a\cos\phi+a^2}$ 　where $a \neq \pm 1$.

　　　　　　　　(g) $\displaystyle\int_0^{\pi} \frac{\cos^2 3\phi\,d\phi}{1-2a\cos\phi+a^2}$ 　where $a \neq \pm 1$.

(h) $\displaystyle\int_0^{\pi} \frac{\cos 2\phi\,d\phi}{1-2a\cos\phi+a^2}$ 　where $a \neq \pm 1$.

21.6. Use the method of residues to show that
$$\int_0^{\pi} \cos^{2n}\theta\,d\theta = \pi\frac{(2n)!}{2^{2n}(n!)^2}$$

21.7. Use the contour in Figure 21.4(a) to show that
$$\int_{-\infty}^{\infty} \frac{\sin x}{x}\,dx = \pi$$
by letting $X \to \infty$, $Y \to \infty$, and $\epsilon \to 0$.

21.8. Use the contour in Figure 21.4(b) to show that
$$\int_0^{\infty} \frac{1}{1+x^n}\,dx = \frac{\pi/n}{\sin(\pi/n)}$$
by letting $R \to \infty$.

21.9. Use the contour in Figure 21.4(c) to show that
$$\int_0^{\infty} \sin(x^2)\,dx = \int_0^{\infty} \cos(x^2)\,dx = \sqrt{\frac{\pi}{8}}$$
by letting $R \to \infty$.

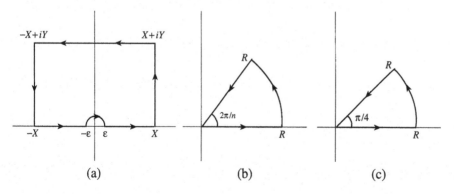

Figure 21.4: (a) The contour used for $\sin x/x$. (b) The contour used for $1/(1+x^n)$. (c) The contour used for $\sin(x^2)$.

Part VI

Differential Equations

Chapter 22

From PDEs to ODEs

Physics, as the most exact science, is characterized by its ability to make mathematical predictions. Predictions are based on two factors: the initial information (data), and the law governing the physical process. Knowing what the situation is *here* and *now* (initial data, initial conditions, boundary conditions) enables physics to predict what the situation will be *there* and *then*. This ability to predict is based on the intuitive belief that physical quantities, dependent on continuous parameters such as position and time, must be *continuous* functions of those parameters. Thus, knowledge of the values of those functions at one (initial) point and of how the functions change from one point to a neighboring point (given by the laws of physics) allows the values of the functions at the neighboring point to be predicted. Once the values of the functions are determined at the new point, their values can be predicted for its neighboring points, and the process can continue until a distant point is reached.

initial conditions are needed to predict the evolution of a physical system.

In mechanics, for example, knowledge of the force acting on a particle of mass m, located at $\mathbf{r}_0$ and moving with momentum $\mathbf{p}_0$ at time t_0, allows its momentum and position at a later time $t_0 + \Delta t$ to be predicted as follows. Because $d\mathbf{p}/dt = \mathbf{F}$ by Newton's second law of motion, we have

$$\Delta\mathbf{p} \approx \mathbf{F}(\mathbf{r}_0, \mathbf{p}_0, t_0)\Delta t$$

and

$$\mathbf{p}(t_0 + \Delta t) = \mathbf{p}_0 + \Delta\mathbf{p} \approx \mathbf{p}_0 + \mathbf{F}(\mathbf{r}_0, \mathbf{p}_0, t_0)\Delta t.$$

Similarly,

$$\mathbf{r}(t_0 + \Delta t) \approx \mathbf{r}_0 + \mathbf{v}_0 t \approx \mathbf{r}_0 + \frac{\mathbf{p}_0}{m}\Delta t.$$

The smaller Δt is, the better the prediction will be.

Newton's second law of motion,

$$\frac{d}{dt}\left(m\frac{d\mathbf{r}}{dt}\right) = \mathbf{F}(\mathbf{r}, d\mathbf{r}/dt, t)$$

ordinary
differential
equation (ODE)
is an example of an **ordinary differential equation** (ODE). A dependent variable $\mathbf{r}$ is determined from an equation involving a single independent variable t, the dependent variable $\mathbf{r}$, and its various derivatives.

In (point) particle mechanics there is only one independent variable, leading to ODEs. In other areas of physics, however, in which extended objects such as fields are studied, variations with respect to position are also present. Partial derivatives with respect to coordinate variables show up in the differ-

partial differential
equations (PDEs)
ential equations, which are therefore called **partial differential equations** (PDEs). For instance, in electrostatics, where time-independent scalar fields such as potentials, and vector fields such as electrostatic fields, are studied, the law is described by Poisson's equation, $\nabla^2 \Phi(\mathbf{r}) = -4\pi\rho(\mathbf{r})$, where Φ is the electrostatic potential and ρ is the volume charge density. Other PDEs occurring in mathematical physics include the heat equation, describing the transfer of heat, the wave equation, describing the propagation of various kinds of wave, and the Schrödinger equation, describing nonrelativistic quantum mechanical phenomena.

In fact, except for the laws of particle mechanics and electrical circuits, in which the only independent variable is time, almost all laws of physics are described by PDEs. We shall not study PDEs in their full generalities, but concentrate on the simplest ones encountered most frequently in ideal physical applications. The method of solution that works for all these equations is the **separation of variables**, whereby a PDE is turned into a number of ODEs.

Before embarking on the separation of variables, we need to formalize the discussion above. An ordinary or a partial DE will provide a unique solution to a physical problem only if the initial or the starting value of the solution

the meaning of
boundary
conditions (or
BCs) elaborated
is known. We refer to this as the **boundary conditions**, or BCs for short. For ODEs, boundary conditions amount to the specification of one or more properties of the solution at an initial time; that is why for ODEs, one speaks of **initial conditions**. BCs for PDEs involve specification of the solution on a surface (or a curve, if the PDE has only two variables).

22.1   Separation of Variables

We list here the PDEs encountered in undergraduate courses and initiate their transformation into ODEs. Let us start with the simplest PDE arising

Poisson equation
in electrostatic problems, the **Poisson equation**, derived in Chapter 15,

$$\nabla^2 \Phi(\mathbf{r}) = -4\pi\rho(\mathbf{r}). \qquad (22.1)$$

Laplace's equation
In vacuum, where $\rho(\mathbf{r}) = 0$, Equation (22.1) reduces to **Laplace's equation**,

$$\nabla^2 \Phi(\mathbf{r}) = 0. \qquad (22.2)$$

Many electrostatic problems involve conductors held at constant potentials and situated in a vacuum. In the space between such conducting surfaces, the electrostatic potential obeys Equation (22.2).

Next in complexity is the **heat equation**, whose most simplified version—
the one studied here—is

$$\frac{\partial T}{\partial t} = k^2 \nabla^2 T(\mathbf{r}, t), \qquad (22.3)$$

heat equation

where T is the temperature and k is a *real* constant characterizing the medium
in which heat is flowing.

Probably one of the most recurring PDEs encountered in mathematical
physics is the **wave equation**,

$$\nabla^2 \Psi - \frac{1}{c^2} \frac{\partial^2 \Psi}{\partial t^2} = 0. \qquad (22.4)$$

wave equation

This equation (or its simplification to lower dimensions) is applied to the
vibration of strings and drums, the propagation of sound in gases, solids, and
liquids, the propagation of disturbances in plasmas, and the propagation of
electromagnetic waves.

The **Schrödinger equation**, describing the nonrelativistic quantum phenomena, is

Schrödinger
equation

$$-\frac{\hbar^2}{2m} \nabla^2 \Psi + V(\mathbf{r}) \Psi = -i\hbar \frac{\partial \Psi}{\partial t}, \qquad (22.5)$$

where m is the mass of a subatomic particle, $\hbar$ is Planck's constant (divided by
2π), V is the potential energy of the particle, and $|\Psi(\mathbf{r}, t)|^2$ is the probability
density of finding the particle at $\mathbf{r}$ at time t.

Equations (22.3)–(22.5) have partial derivatives with respect to time. As
a first step toward solving these PDEs, let us separate the time variable. We
will denote the functions in all four equations by the generic symbol $\Psi(\mathbf{r}, t)$.

The separation of variables starts with separating the $\mathbf{r}$ and t dependence
into factors:[1]

$$\Psi(\mathbf{r}, t) \equiv R(\mathbf{r})T(t).$$

This factorization permits us to separate the two operations of space differentiation and time differentiation. As an illustration, we separate the time
and space dependence for the Schrödinger equation. The other equations are
done similarly. Substituting for Ψ, we get

time is separated
from space

$$-\frac{\hbar^2}{2m} \nabla^2 (RT) + V(\mathbf{r})(RT) = -i\hbar \frac{\partial}{\partial t}(RT),$$

or

$$-T \frac{\hbar^2}{2m} \nabla^2 R + V(\mathbf{r})(RT) = -iR\hbar \frac{dT}{dt},$$

where we have used ordinary derivatives for T because, by assumption, it is
a function of a single variable. Dividing both sides by RT yields

---

[1] Note that there is no a priori reason why the basic assumption underlying the separation
of variables is legitimate. After all, we cannot write $\sin(xt)$ as a product, $f(x)g(t)$. However,
in all cases of physical interest the separation of variables works.

$$-\frac{1}{R}\frac{\hbar^2}{2m}\nabla^2 R + V(\mathbf{r}) = -i\frac{1}{T}\hbar\frac{dT}{dt}. \qquad (22.6)$$

central argument
in separation of
variables

Now comes the crucial step in the process of the separation of variables. The LHS of Equation (22.6) is a function of *position alone*, and the RHS is a function of *time alone*. Since $\mathbf{r}$ and t are independent variables, the only way that (22.6) can hold is for *both sides to be constant*, say α:

$$-\frac{1}{R}\frac{\hbar^2}{2m}\nabla^2 R + V(\mathbf{r}) = \alpha \;\Rightarrow\; -\frac{\hbar^2}{2m}\nabla^2 R + V(\mathbf{r})R = \alpha R$$

and

$$-i\hbar\frac{1}{T}\frac{dT}{dt} = \alpha \;\Rightarrow\; \frac{dT}{dt} = \frac{i\alpha}{\hbar}T. \qquad (22.7)$$

We have reduced the original time-dependent Schrödinger equation, a PDE, to an ODE involving only time, and a PDE involving only the position variables. Most problems of elementary mathematical physics have the same property, i.e., they are completely equivalent to Equation (22.7) plus the equation before it, which we write generically as

$$\nabla^2 R + f(\mathbf{r})R = 0, \qquad (22.8)$$

where we have simplified the notation by including α in the function f.

The foregoing discussion is summarized in this statement:

> **Box 22.1.1.** *The time-dependent PDEs of mathematical physics can be reduced to an ODE in the time variable and the PDE given in Equation (22.8). For those PDEs involving second time derivatives, such as the wave equation, (22.7) will be a second-order ODE.*

With the exception of Poisson's equation, in all the foregoing equations the term on the RHS is zero. We will restrict ourselves to this so-called homogeneous case[2] and rewrite (22.8) as

$$\nabla^2\Psi(\mathbf{r}) + f(\mathbf{r})\Psi(\mathbf{r}) = 0. \qquad (22.9)$$

The rest of this section is devoted to the study of this equation in various coordinate systems.

22.2   Separation in Cartesian Coordinates

In Cartesian coordinates, Equation (22.9) becomes

$$\frac{\partial^2\Psi}{\partial x^2} + \frac{\partial^2\Psi}{\partial y^2} + \frac{\partial^2\Psi}{\partial z^2} + f(x,y,z)\Psi = 0.$$

---

[2]The most elegant way of solving inhomogeneous PDEs is the method of **Green's functions**, of which we shall have a brief discussion in Chapter 29. For a thorough discussion of Green's functions, see Hassani, S. *Mathematical Physics: A Modern Introduction to Its Foundations*, Springer-Verlag, 1999, Part VI.

As in the case of the separation of the time variable, we assume that we can separate the dependence on various coordinates and write

$$\Psi(x, y, z) = X(x)Y(y)Z(z).$$

Then the above PDE yields

$$YZ\frac{d^2X}{dx^2} + XZ\frac{d^2Y}{dy^2} + XY\frac{d^2Z}{dz^2} + f(x, y, z)XYZ = 0.$$

Dividing by XYZ gives

$$\frac{1}{X}\frac{d^2X}{dx^2} + \frac{1}{Y}\frac{d^2Y}{dy^2} + \frac{1}{Z}\frac{d^2Z}{dz^2} + f(x, y, z) = 0. \tag{22.10}$$

This equation is almost separated. The first term is a function of x alone, the second of y alone, and the third of z alone. However, the last term, in general, mixes the coordinates. The only way the separation can become complete is for the last term to be separated as well, that is, expressed as a sum of three functions, each depending on a single coordinate.[3] In such a special case we obtain

$$\frac{1}{X}\frac{d^2X}{dx^2} + \frac{1}{Y}\frac{d^2Y}{dy^2} + \frac{1}{Z}\frac{d^2Z}{dz^2} + f_1(x) + f_2(y) + f_3(z) = 0$$

or

$$\left[\frac{1}{X}\frac{d^2X}{dx^2} + f_1(x)\right] + \left[\frac{1}{Y}\frac{d^2Y}{dy^2} + f_2(y)\right] + \left[\frac{1}{Z}\frac{d^2Z}{dz^2} + f_3(z)\right] = 0.$$

The first term on the LHS depends on x alone, the second on y alone, and the third on z alone. Since the sum of these three terms is a constant (zero), independent of all variables, each term must be a constant. Denoting the constant corresponding to the ith term by $-\alpha_i$, we obtain

$$\frac{1}{X}\frac{d^2X}{dx^2} + f_1(x) = -\alpha_1, \quad \frac{1}{Y}\frac{d^2Y}{dy^2} + f_2(y) = -\alpha_2, \quad \frac{1}{Z}\frac{d^2Z}{dz^2} + f_3(z) = -\alpha_3,$$

which can be reexpressed as

$$\frac{d^2X}{dx^2} + [f_1(x) + \alpha_1]X = 0, \quad \frac{d^2Y}{dy^2} + [f_2(y) + \alpha_2]Y = 0,$$

$$\frac{d^2Z}{dz^2} + [f_3(z) + \alpha_3]Z = 0, \quad \alpha_1 + \alpha_2 + \alpha_3 = 0. \tag{22.11}$$

If $f(x, y, z)$ happens to be a constant C, then the first three terms of Equation (22.10) can be taken to be respectively $-\alpha_1$, $-\alpha_2$, and $-\alpha_3$, leading to

$$\frac{d^2X}{dx^2} + \alpha_1 X = 0, \quad \frac{d^2Y}{dy^2} + \alpha_2 Y = 0,$$

$$\frac{d^2Z}{dz^2} + \alpha_3 Z = 0, \quad \alpha_1 + \alpha_2 + \alpha_3 = C. \tag{22.12}$$

---

[3]This is where the limitation of the method of the separation of variables becomes evident. However, surprisingly, all physical applications, at our level of treatment, involve functions that are indeed separated.

These equations constitute the most general set of ODEs resulting from the separation of the PDE of Equation (22.9) in Cartesian coordinates.

Example 22.2.1. Let us consider a few cases for which (22.11) or (22.12) is applicable.

Laplace's equation
(a) In electrostatics, separation of Laplace's equation, for which $f(\mathbf{r}) = 0$, leads to these ODEs:

$$\frac{d^2 X}{dx^2} + \alpha_1 X = 0, \qquad \frac{d^2 Y}{dy^2} + \alpha_2 Y = 0, \qquad \frac{d^2 Z}{dz^2} - (\alpha_1 + \alpha_2)Z = 0.$$

The solutions to these equations are trigonometric or hyperbolic (exponential) functions, determined from the boundary conditions (conducting surfaces). The unsymmetrical treatment of the three coordinates—the plus sign in front of the first two constants and a minus sign in front of the third—is not dictated by the above equations. There is a freedom in the choice of sign in these equations. However, the boundary conditions will force the constants to adapt to values appropriate to the physical situation at hand.

(b) In quantum mechanics the time-independent Schrödinger equation for a free particle in three dimensions is

$$\nabla^2 \Psi + \frac{2mE}{\hbar^2} \Psi = 0.$$

Separation of variables yields the ODEs of Equation (22.12) with

$$\alpha_1 + \alpha_2 + \alpha_3 = \frac{2mE}{\hbar^2}.$$

After time is separated, the heat and wave equations also yield equations similar to (22.12).

(c) In quantum mechanics the time-independent Schrödinger equation for a three dimensional isotropic harmonic oscillator is

$$\nabla^2 \Psi - \left(\frac{m^2 \omega^2}{\hbar^2} r^2 - \frac{2mE}{\hbar^2} \right) \Psi = 0.$$

Thus,

$$f(\mathbf{r}) = -\frac{m^2 \omega^2}{\hbar^2} r^2 + \frac{2mE}{\hbar^2} = -\frac{m^2 \omega^2}{\hbar^2} (x^2 + y^2 + z^2) + \frac{2mE}{\hbar^2}.$$

Equation (22.11) then yields

$$\frac{d^2 X}{dx^2} - \frac{m^2 \omega^2}{\hbar^2} x^2 X + \alpha_1 X = 0,$$

$$\frac{d^2 Y}{dy^2} - \frac{m^2 \omega^2}{\hbar^2} y^2 Y + \alpha_2 Y = 0,$$

$$\frac{d^2 Z}{dz^2} - \frac{m^2 \omega^2}{\hbar^2} z^2 Z + \alpha_3 Z = 0,$$

with $\alpha_1 + \alpha_2 + \alpha_3 = 2mE/\hbar^2$.   ∎

22.3   Separation in Cylindrical Coordinates

Equation (22.9) takes the following form in cylindrical coordinates:[4]

$$\frac{1}{\rho}\frac{\partial}{\partial \rho}\left(\rho\frac{\partial \Psi}{\partial \rho}\right) + \frac{1}{\rho^2}\frac{\partial^2 \Psi}{\partial \varphi^2} + \frac{\partial^2 \Psi}{\partial z^2} + f(\rho,\varphi,z)\Psi = 0.$$

To separate the variables, we write $\Psi(\rho,\varphi,z) = R(\rho)S(\varphi)Z(z)$, substitute in the general equation, and divide both sides by RSZ to obtain

$$\frac{1}{R}\frac{1}{\rho}\frac{d}{d\rho}\left(\rho\frac{dR}{d\rho}\right) + \frac{1}{S}\frac{1}{\rho^2}\frac{d^2 S}{d\varphi^2} + \frac{1}{Z}\frac{d^2 Z}{dz^2} + f(\rho,\varphi,z) = 0.$$

We shall consider only the special (but important) case in which $f(\rho,\varphi,z)$ is a constant λ. In that case, the equation becomes

$$\underbrace{\left[\frac{1}{R}\frac{1}{\rho}\frac{d}{d\rho}\left(\rho\frac{dR}{d\rho}\right)\right] + \frac{1}{\rho^2}\left[\frac{1}{S}\frac{d^2 S}{d\varphi^2}\right]}_{\text{function of } \rho \text{ and } \varphi \text{ only}} + \underbrace{\left[\frac{1}{Z}\frac{d^2 Z}{dz^2}\right]}_{\text{fn. of } z} + \lambda = 0.$$

The sum of the first two terms is independent of z, so the third term must be as well. We thus get

$$\frac{1}{Z}\frac{d^2 Z}{dz^2} = \lambda_1$$

and

$$\left[\frac{1}{R}\frac{1}{\rho}\frac{d}{d\rho}\left(\rho\frac{dR}{d\rho}\right)\right] + \frac{1}{\rho^2}\left(\frac{1}{S}\frac{d^2 S}{d\varphi^2}\right) + \lambda_1 + \lambda = 0.$$

Multiplying this equation by ρ^2 yields

$$\underbrace{\left[\frac{\rho}{R}\frac{d}{d\rho}\left(\rho\frac{dR}{d\rho}\right) + (\lambda_1 + \lambda)\rho^2\right]}_{\text{function of } \rho \text{ only}} + \underbrace{\left(\frac{1}{S}\frac{d^2 S}{d\varphi^2}\right)}_{\text{fn. of } \varphi} = 0.$$

Since the first term is a function of ρ only and the second a function of φ only, both terms must be constants whose sum vanishes. Thus,

$$\frac{1}{S}\frac{d^2 S}{d\varphi^2} = \mu, \qquad \frac{\rho}{R}\frac{d}{d\rho}\left(\rho\frac{dR}{d\rho}\right) + (\lambda_1 + \lambda)\rho^2 + \mu = 0. \qquad (22.13)$$

Putting together all of the above, we conclude that when Equation (22.9) is separable in cylindrical coordinates and $f(\mathbf{r}) = \lambda$, it will separate into the following three ODEs:

$$\frac{d^2 Z}{dz^2} - \lambda_1 Z = 0, \qquad \frac{d^2 S}{d\varphi^2} - \mu S = 0,$$

$$\frac{d}{d\rho}\left(\rho\frac{dR}{d\rho}\right) + \left\{(\lambda_1 + \lambda)\rho + \left(\frac{\mu}{\rho}\right)\right\}R = 0, \qquad (22.14)$$

---

[4]See Chapter 16 for the expression of ∇^2 in spherical and cylindrical coordinate systems.

Bessel differential
equation

where in rewriting the second equation in (22.13), we multiplied both sides of the equation by R and divided it by ρ. The last equation of (22.14) is called the **Bessel differential equation**. This equation shows up in electrostatic and heat-transfer problems with cylindrical geometry and in problems involving two-dimensional wave propagation, as in drumheads.

Historical Notes

Jean Le Rond d'Alembert was the illegitimate son of a famous salon hostess of eighteenth-century Paris and a cavalry officer. Abandoned by his mother, d'Alembert was raised by a foster family and later educated by the arrangement of his father at a nearby church-sponsored school, in which he received instruction in the classics and above-average instruction in mathematics. After studying law and medicine, he finally chose to pursue a career in mathematics. In the 1740s he joined the ranks of the *philosophes*, a growing group of deistic and materialistic thinkers and writers who actively questioned the social and intellectual standards of the day. He traveled little (he left France only once, to visit the court of Frederick the Great), preferring instead the company of his friends in the salons, among whom he was well known for his wit and laughter.

Jean Le Rond
d'Alembert
1717–1783

d'Alembert turned his mathematical and philosophical talents to many of the outstanding scientific problems of the day, with mixed success. Perhaps his most famous scientific work, entitled *Traite de dynamique*, shows his appreciation that a revolution was taking place in the science of mechanics—the formalization of the principles stated by **Newton** into a rigorous mathematical framework. Later, d'Alembert produced a treatise on fluid mechanics, a paper dealing with vibrating strings, and a skillful treatment of celestial mechanics. d'Alembert is also credited with the use of the first partial differential equation as well as the first solution to such an equation using **separation of variables**.

Much of the work for which d'Alembert is remembered occurred outside mathematical physics. He was chosen as the science editor of the *Encyclopedie*, and his lengthy *Discours Preliminaire* in that volume is considered one of the defining documents of the Enlightenment. Other works included writings on law, religion, and music.

22.4   Separation in Spherical Coordinates

By far the most commonly used coordinate system in mathematical physics is the spherical coordinate system. This is because forces, potential energies, and most geometries encountered in Nature have a spherical symmetry. One of the consequences of this spherical symmetry is that the function $f(\mathbf{r})$ is a function of r and not of angles. We shall assume this to be true in this subsection.

In spherical coordinates, Equation (22.9) becomes [see Equation (16.19)]

$$\frac{1}{r^2}\frac{\partial}{\partial r}\left(r^2\frac{\partial\Psi}{\partial r}\right) + \frac{1}{r^2\sin\theta}\left\{\frac{\partial}{\partial\theta}\left(\sin\theta\frac{\partial\Psi}{\partial\theta}\right) + \frac{1}{\sin\theta}\frac{\partial^2\Psi}{\partial\varphi^2}\right\} + f(r)\Psi = 0. \quad (22.15)$$

To separate this equation means to write $\Psi(r, \theta, \varphi) = R(r)\Theta(\theta)\Phi(\varphi)$. If we substitute this in Equation (22.15) and note that each differentiation acts on only one of the three functions, we get

$$\Theta\Phi\frac{1}{r^2}\frac{d}{dr}\left(r^2\frac{dR}{dr}\right) + \frac{R}{r^2}\left[\frac{\Phi}{\sin\theta}\frac{d}{d\theta}\left(\sin\theta\frac{d\Theta}{d\theta}\right) + \frac{\Theta}{\sin^2\theta}\frac{d^2\Phi}{d\varphi^2}\right] + f(r)R\Theta\Phi = 0.$$

Now divide both sides by $R\Theta\Phi$ and multiply by r^2 to obtain

$$\underbrace{\frac{1}{R}\frac{d}{dr}\left(r^2\frac{dR}{dr}\right) + r^2 f(r)}_{\text{function of } r \text{ alone}} + \underbrace{\left[\frac{1}{\Theta\sin\theta}\frac{d}{d\theta}\left(\sin\theta\frac{d\Theta}{d\theta}\right) + \frac{1}{\Phi\sin^2\theta}\frac{d^2\Phi}{d\varphi^2}\right]}_{\text{function of } \theta \text{ and } \varphi \text{ only}} = 0.$$

Since each one of the two terms is a function of different variables, each must be a constant; and the two constants must add up to zero. Therefore, we have

$$\frac{1}{R}\frac{d}{dr}\left(r^2\frac{dR}{dr}\right) + r^2 f(r) = \alpha,$$

$$\frac{1}{\Theta\sin\theta}\frac{d}{d\theta}\left(\sin\theta\frac{d\Theta}{d\theta}\right) + \frac{1}{\Phi\sin^2\theta}\frac{d^2\Phi}{d\varphi^2} = -\alpha.$$

The second equation can be further separated. We add α to both sides and multiply the resulting equation by $\sin^2\theta$ to obtain

$$\underbrace{\frac{\sin\theta}{\Theta}\frac{d}{d\theta}\left(\sin\theta\frac{d\Theta}{d\theta}\right) + \alpha\sin^2\theta}_{\text{function of } \theta \text{ alone; set} = \beta} + \underbrace{\frac{1}{\Phi}\frac{d^2\Phi}{d\varphi^2}}_{=-\beta} = 0.$$

We have thus obtained three ODEs in three variables. We rewrite these ODEs in the following equations:

$$\frac{1}{r^2}\frac{d}{dr}\left(r^2\frac{dR}{dr}\right) + \left[f(r) - \frac{\alpha}{r^2}\right]R = 0,$$

$$\frac{1}{\sin\theta}\frac{d}{d\theta}\left(\sin\theta\frac{d\Theta}{d\theta}\right) + \left(\alpha - \frac{\beta}{\sin^2\theta}\right)\Theta = 0, \qquad (22.16)$$

$$\frac{d^2\Phi}{d\varphi^2} + \beta\Phi = 0.$$

radial, polar, and azimuthal equations

The first equation is called the **radial equation**, the second the **polar equation**, and the third the **azimuthal equation**. The radial equation can be further simplified by making the substitution $R = u/r$. This gives

$$\frac{d^2u}{dr^2} + \left[f(r) - \frac{\alpha}{r^2}\right]u = 0. \qquad (22.17)$$

Our task in this chapter was to separate the PDEs most frequently encountered in undergraduate mathematical physics into ODEs; and we have done this in the three coordinate systems regularly used in applications. We shall return to a thorough treatment of the ODEs so obtained later in the book, and in the process we shall be introduced to the so-called **special functions** that came into being in the nineteenth century as a result of the then newly discovered technique of the separation of variables.

22.5   Problems

22.1. Assume that two functions Φ_1 and Φ_2 satisfy the Poisson equation. Show that
(a) Φ defined by $\Phi = \Phi_1 - \Phi_2$ satisfies the Laplace's equation;
(b) $\nabla \cdot (\Phi \nabla \Phi) = |\nabla \Phi|^2$

22.2. Separate the solution of the heat equation (22.3): $T(\mathbf{r}, t) \equiv R(\mathbf{r})\tau(t)$, and show that
(a) the solution to the time equation is

$$\tau(t) = Ae^{-\alpha k^2 t},$$

(b) in which case, the space part must satisfy the following PDE:

$$\nabla^2 R + \alpha R = 0$$

22.3. Show that any function of the form $f(\mathbf{k} \cdot \mathbf{r} \pm \omega t)$ satisfies the wave equation (22.4) if $\omega = c|\mathbf{k}|$.

22.4. Separate the solution of the wave equation (22.4): $\Psi(\mathbf{r}, t) \equiv R(\mathbf{r})T(t)$, and show that
(a) the solution to the time equation is

$$T(t) = A \cos \omega t + B \sin \omega t,$$

(b) and the space part must satisfy the following PDE:

$$\nabla^2 R + k^2 R = 0$$

where $k = \omega/c$.

22.5. Provide the details of the derivation of Equation (22.16).

22.6. By substituting $R = u/r$ in the radial DE of spherical coordinates, show that it reduces to Equation (22.17).

Chapter 23

First-Order Differential Equations

The last chapter showed that all PDEs discussed there resulted in ODEs of *second order*, i.e., differential equations involving second derivatives. Thus, treating the first-order DEs (FODEs) may seem irrelevant. However, sometimes a second-order DE (SODE) may be expressed in terms of first derivatives. For example, take Newton's second law of motion along a straight line (free fall, say): $m\,d^2x/dt^2 = F$. If we write this in terms of velocity, we obtain $m\,dv/dt = F$, and if F is a function of v alone—as in a fall with air resistance—then we have a FODE. FODEs arise in other areas of physics beside mechanics. Therefore, it is worthwhile to study them here.

23.1 Normal Form of a FODE

The most general FODE is of the form $G(x, y, y') = 0$, where G is some function of three variables. We can find y' (the derivative of y) as a function of x and y if the function $G(x_1, x_2, x_3)$ is sufficiently well behaved. In that case, we have

$$y' \equiv \frac{dy}{dx} = F(x, y) \qquad (23.1)$$

the most general FODE in normal form

which is said to be a **normal** FODE.

Example 23.1.1. There are three special cases of Equation (23.1) that lead immediately to a solution.
(a) If $F(x, y)$ is independent of y, then $y' = g(x)$, and the most general solution can be written as $y = f(x) = C + \int_a^x g(t)\,dt$ where $C = f(a)$.
(b) If $F(x, y)$ is independent of x, then $dy/dx = h(y)$, and

$$\frac{dy}{h(y)} = dx \;\Rightarrow\; \underbrace{\int_C^y \frac{dt}{h(t)}}_{\equiv H(y)} - x + a = 0 \;\Rightarrow\; H(y) - x + a = 0$$

embodies a solution. That is, $H(y) = x - a$ can be solved for y in terms of x, say $y = f(x)$, and this y will be a solution of the DE. Note that $y|_{x=a} \equiv f(a) = C$.
(c) The third special case is really a generalization of the first two. If $F(x, y) = g(x)h(y)$, then $y' = g(x)h(y)$ or $dy/h(y) = g(x)\,dx$ and

$$\int_C^y \frac{dt}{h(t)} = \int_a^x g(t)\,dt \tag{23.2}$$

is an implicit solution.   ∎

The example above contains an information which is important enough to be "boxed."

> **Box 23.1.1.** *A differential equation is considered to be solved if its solution can be obtained by solving an algebraic equation involving integrals of known functions. Whether these integrals can be done in closed form or not is irrelevant.*

So, although we may not be able to actually perform the integration of (23.2), we consider the DE solved because, in principle, Equation (23.2) gives y as a (implicit) function of x.

As Example 23.1.1 shows, the solutions to a FODE are usually obtained in an implicit form, as a function u of two variables such that the solution y can be found by solving $u(x, y) = 0$ for y. Included in $u(x, y)$ is an arbitrary constant related to the initial conditions. The equation $u(x, y) = 0$ defines a curve in the xy-plane, which depends on the (hidden) constant in $u(x, y)$. Since different constants give rise to different curves, it is convenient to separate the constant and write $u(x, y) = C$. This leads to the concept of an integral of a differential equation.

integral of a normal FODE

Definition 23.1.1. *An **integral** of a normal FODE [Equation (23.1)] is a function of two variables $u(x, y)$ such that $u(x, f(x))$ is a constant for all possible values of x whenever $y = f(x)$ is a solution of the differential equation.*

an integral of a FODE is also called a constant of motion.

The integrals of differential equations are encountered often in physics. If x is replaced by t (time), then the differential equation describes the motion of a physical system, and a solution, $y = f(t)$, can be written implicitly as $u(t, y) = C$, where u is an integral of the differential equation. The equation $u(t, y) = C$ describes a curve in the ty-plane on which the value of the function $u(t, y)$ remains unchanged for all t. Thus, $u(t, y)$, the integral of the FODE, is also called a **constant of motion**.

Example 23.1.2. Consider a point particle moving under the influence of a force depending on position only. Denoting the position[1] by x and the velocity by v, we have, by Newton's second law, $m\,dv/dt = F(x)$. Using the chain rule, $dv/dt = (dv/dx)(dx/dt) = v\,dv/dx$, we obtain

$$mv\frac{dv}{dx} = F(x) \;\Rightarrow\; mv\,dv = F(x)\,dx, \tag{23.3}$$

---

[1]Here we are restricting the motion to one dimension.

which is easily integrated to

$$\tfrac{1}{2}mv^2 = \int F(x)\,dx + C \equiv -V(x) + C. \qquad (23.4)$$

The **potential energy** $V(x) = -\int F(x)\,dx$ has been introduced as an indefinite integral. We can write Equation (23.4) as

potential energy

$$\tfrac{1}{2}mv^2 + V(x) = C. \qquad (23.5)$$

Thus, the integral of Equation (23.3) is $u(x,v) = \tfrac{1}{2}mv^2 + V(x)$ which is the expression for the energy of the one-dimensional motion of a particle experiencing the potential $V(x)$. If v is a solution of Equation (23.3), then $u(x,v) = \text{constant}$. Since a solution of Equation (23.3) describes a *possible* motion of the particle, Equation (23.5) implies that the energy of a particle does not change in the course of its motion. This statement is the conservation of (mechanical) energy. ∎

23.2 Integrating Factors

Let D be a region in the xy-plane, and let $M(x,y)$ and $N(x,y)$ be continuous functions of x and y defined on D. The differential $M\,dx + N\,dy$ is **exact** if, for arbitrary points P_1 and P_2 of D, the line integral

exact differential

$$\int_{P_1}^{P_2} [M(x,y)\,dx + N(x,y)\,dy]$$

is independent of the path joining the two points. This condition is equivalent to saying that the line integral of the integrand around any closed loop in D vanishes. A necessary and sufficient condition for exactness is, therefore, that the curl of the vector $\mathbf{A} = \langle M, N, 0 \rangle$ be zero.[2] The vector $\mathbf{A}$ is then conservative, and we can define a (potential) function v such that $\mathbf{A} = \boldsymbol{\nabla} v = \langle \partial v / \partial x, \partial v / \partial y, 0 \rangle$, or

$$dv = \frac{\partial v}{\partial x}\,dx + \frac{\partial v}{\partial y}\,dy = M\,dx + N\,dy. \qquad (23.6)$$

Thus, $M\,dx + N\,dy$ is exact if and only if there exists a function $v(x,y)$ satisfying (23.6), in which case, $M = \partial v / \partial x$ and $N = \partial v / \partial y$.

Now consider all y's that satisfy $v(x,y) = C$ for some constant C. Then since $dC = 0$, we have

$$0 = dv = M\,dx + N\,dy.$$

It follows that $v(x,y) = C$ is an implicit solution of the differential equation. We therefore have

---

[2]The statement is true only if the region D does not contain any singularities of M or N. The region is then called **contractable to a point** (see Section 14.3).

Theorem 23.2.1. *If $M(x,y)\,dx + N(x,y)\,dy$ is an exact differential dv in a domain D of the xy-plane, then $v(x,y)$ is an integral of the DE*

$$M(x,y)\,dx + N(x,y)\,dy = 0$$

whose solutions are of the form $v(x,y) = C$.

We saw above that, for an exact differential, $M = \partial v/\partial x$ and $N = \partial v/\partial y$. A necessary consequence of this result is $\partial M/\partial y = \partial N/\partial x$. Could this relation be a sufficient condition as well? Consider the function $v(x,y)$ defined by

$$v(x,y) \equiv \int_a^x M(t,y)\,dt + \int_b^y N(a,t)\,dt,$$

and note that

$$
\begin{aligned}
dv &= \frac{\partial v}{\partial x}\,dx + \frac{\partial v}{\partial y}\,dy \\
&= \frac{\partial}{\partial x}\left[\int_a^x M(t,y)\,dt\right]dx + \left[\int_a^x \underbrace{\frac{\partial M}{\partial y}(t,y)}_{\partial N/\partial t}\,dt + \frac{\partial}{\partial y}\int_b^y N(a,t)\,dt\right]dy \\
&= M(x,y)\,dx + \underbrace{\left[N(t,y)\Big|_{t=a}^{t=x} + N(a,y)\right]}_{=N(x,y)}dy,
\end{aligned}
$$

and $v(x,y)$ indeed satisfies $dv = M\,dx + N\,dy$. It follows that (see Problem 23.1)

Theorem 23.2.2. *A necessary and sufficient condition for $M\,dx + N\,dy$ to be exact is $\partial M/\partial y = \partial N/\partial x$, in which case*

$$v(x,y) \equiv \int_a^x M(t,y)\,dt + \int_b^y N(a,t)\,dt$$

is the function such that $dv = M\,dx + N\,dy$.

Not very many FODEs are exact. However, there are many that can be turned into exact FODEs by multiplication by a suitable function. Such a function, if it exists, is called an **integrating factor**. Thus, if the differential $M(x,y)\,dx + N(x,y)\,dy$ is not exact, but

integrating factor

$$\mu(x,y)M(x,y)\,dx + \mu(x,y)N(x,y)\,dy = dv,$$

then $\mu(x,y)$ is an integrating factor for the differential equation

$$M(x,y)\,dx + N(x,y)\,dy = 0$$

whose solution is then $v(x,y) = C$. Integrating factors are not unique, as the following example illustrates.

Example 23.2.3. The differential $x\,dy - y\,dx$ is not exact. Let us see if we can find a function $\mu(x,y)$ such that $dv = \mu x\,dy - \mu y\,dx$, for some $v(x,y)$. We assume that the domain D of the xy-plane in which v is defined is contractable to a point. Then a necessary and sufficient condition for the equation above to hold is

illustration of nonuniqueness of integrating factor

$$\frac{\partial}{\partial x}(\mu x) = \frac{\partial}{\partial y}(-\mu y) \;\Rightarrow\; x\frac{\partial \mu}{\partial x} + y\frac{\partial \mu}{\partial y} + 2\mu = 0. \tag{23.7}$$

(a) Let us assume that μ is a function of x only. Then Equation (23.7) reduces to $x\,d\mu/dx = 2\mu$ or $\mu = C/x^2$ where $x \neq 0$. In this case we get

$$dv = C\left(\frac{1}{x}\,dy - \frac{y}{x^2}\,dx\right) = C\,d\left(\frac{y}{x}\right) \qquad \text{where} \quad x \neq 0.$$

Thus, as long as $x \neq 0$, any function C/x^2, with arbitrary C, is an integrating factor for $x\,dy - y\,dx = 0$. This integrating factor leads to the solution

$$v = \frac{Cy}{x} = \text{constant}. \tag{23.8}$$

In order to determine the constant, suppose that $y = m$ when $x = 1$. Then (23.8) determines the constant in terms of m:

$$\frac{Cm}{1} = \text{constant} \;\Rightarrow\; \text{constant} = Cm.$$

So, (23.8) becomes

$$\frac{Cy}{x} = Cm \;\Rightarrow\; y = mx.$$

(b) Now let us assume that μ is a function of y only. This leads to the integrating factor $\mu = C/y^2$ where $y \neq 0$. In this case $v = Cx/y$ is the integral of the DE, and a general solution is of the form $Cx/y = \text{constant}$. If we further impose the condition $y(1) = m$, we get $C/m = \text{constant}$. Equation (23.8) then yields

$$\frac{Cx}{y} = \frac{C}{m} \;\Rightarrow\; y = mx$$

as in (a).

(c) The reader may verify that

$$\mu = \frac{C}{x^2 + y^2} \qquad \text{where} \quad (x,y) \neq (0,0)$$

is also an integrating factor leading to the integral

$$v = \tan^{-1}\left(\frac{y}{x}\right) = \text{constant} \qquad \Rightarrow \qquad \frac{y}{x} = \tan(\text{constant}) \equiv C'.$$

Imposing $y(1) = m$ gives $C' = m$, so that $y = mx$ as before. ∎

The example above is a special case of the general fact that if a differential has one integrating factor, then it has an infinite number of them. Suppose that $\nu(x,y)$ is an integrating factor of $M\,dx + N\,dy$, i.e., $\nu M\,dx + \nu N\,dy$ is an exact differential, say du. Take *any* differentiable function $F(u)$. Then $\mu(x,y) \equiv \nu(x,y)F'(u)$ is also an integrating factor. In fact,

proof of nonuniqueness of integrating factor

$$\mu(M\,dx + N\,dy) = \nu F'(M\,dx + N\,dy) = \frac{dF}{du}\underbrace{(\nu M\,dx + \nu N\,dy)}_{=du} = dF.$$

23.3   First-Order Linear Differential Equations

order of a linear
DE

A *linear* DE is a sum of terms each of which is the product of a derivative of the dependent variable (say y) and a function of the independent variable (say x). The highest order of the derivative is called the **order of** the linear DE. The most general first-order *linear* differential equation (FOLDE) is

$$p_1(x)y' + p_0(x)y = q(x) \quad \Leftrightarrow \quad p_1\,dy + (p_0y - q)\,dx = 0. \qquad (23.9)$$

If this equation is to have a solution, then by the argument at the end of the last subsection, it must have at least one integrating factor. Let $\mu(x,y)$ be an integrating factor. Then there exists $v(x,y)$ such that

$$dv = \mu(p_0y - q)\,dx + \mu p_1\,dy = 0$$

The necessary and sufficient condition for this to hold is

$$\frac{\partial}{\partial y}[\mu(p_0y - q)] = \frac{\partial}{\partial x}(\mu p_1).$$

To simplify the problem, let us assume that μ is a function of x only (we are looking for any integrating factor, not the most general one). Then the above condition leads to the differential equation

$$\mu p_0 = \frac{d}{dx}(\mu p_1) = p_1\frac{d\mu}{dx} + \mu\frac{dp_1}{dx} \qquad (23.10)$$

or

$$p_1\frac{d\mu}{dx} = \mu\left(p_0 - \frac{dp_1}{dx}\right) \quad \Rightarrow \quad \frac{d\mu}{\mu} = \frac{p_0}{p_1}\,dx - \frac{dp_1}{p_1}.$$

Integrating both sides gives

$$\ln\mu = \int \frac{p_0}{p_1}\,dx - \ln p_1 + \ln C \quad \Rightarrow \quad \ln\left(\frac{\mu p_1}{C}\right) = \int \frac{p_0}{p_1}\,dx$$

or

$$\frac{\mu p_1}{C} = e^{\int p_0\,dx/p_1} \quad \Rightarrow \quad \mu = \frac{Ce^{\int p_0\,dx/p_1}}{p_1}.$$

Neglecting the unimportant constant of integration, we have found the integrating factor $\mu = \exp[(\int p_0\,dx/p_1)]/p_1$. Now multiply both sides of the original equation by μ to obtain

$$\mu p_1 y' + \mu p_0 y = \mu q. \qquad (23.11)$$

With the identity $\mu p_1 y' \equiv (\mu p_1 y)' - (\mu p_1)'y$ and the fact that $(\mu p_1)' = \mu p_0$ [the first equality of Equation (23.10)], Equation (23.11) becomes

$$\frac{d}{dx}(\mu p_1 y) = \mu q \quad \Rightarrow \quad \mu p_1 y = \int \mu(x)q(x)\,dx + C.$$

Therefore,

explicit solution of a general first-order linear differential equation

Theorem 23.3.1. *Any FOLDE of the form* $p_1(x)y' + p_0(x)y = q(x)$, *in which* p_0, p_1, *and* q *are continuous functions in some interval* (a, b), *has a general solution*

$$y = f(x) = \frac{1}{\mu(x)p_1(x)} \left[C + \int \mu(x)q(x)\, dx \right], \qquad (23.12)$$

where C *is an arbitrary constant, and*

$$\mu(x) = \frac{1}{p_1(x)} \exp\left[\int \frac{p_0(x)}{p_1(x)}\, dx \right]. \qquad (23.13)$$

Example 23.3.2. In an electric circuit with a resistance R and a capacitance C, Kirchhoff's law gives rise to the equation $R\, dQ/dt + Q/C = V(t)$, where $V(t)$ is the time-dependent voltage and Q is the (instantaneous) charge on the capacitor. This is a simple FOLDE with $p_1 = R$, $p_0 = 1/C$, and $q = V$. The integrating factor is

detailed treatment of an *RC* circuit

$$\mu(t) = \frac{1}{R} \exp\left[\int \frac{1}{RC}\, dt \right] = \frac{1}{R} e^{t/RC},$$

which yields

$$Q(t) = \frac{1}{\mu(t)p_1(t)} \left[B + \frac{1}{R} \int e^{t/RC} V(t)\, dt \right]$$

$$= B e^{-t/RC} + \frac{e^{-t/RC}}{R} \int e^{t/RC} V(t)\, dt.$$

Recall that an indefinite integral can be written as a definite integral whose upper limit is the independent variable—in which case we need to use a different symbol for the integration variable. For the arbitrary lower limit, choose zero. We then have

$$Q(t) = B e^{-t/RC} + \frac{e^{-t/RC}}{R} \int_0^t e^{s/RC} V(s)\, ds. \qquad (23.14)$$

Let $Q(0) \equiv Q_0$ be the initial charge. Then, substituting $t = 0$ in (23.14), we get $Q_0 = B$ and the charge at time t will be given by

$$Q(t) = Q_0 e^{-t/RC} + \frac{e^{-t/RC}}{R} \int_0^t e^{s/RC} V(s)\, ds. \qquad (23.15)$$

As a specific example, assume that the voltage is a constant V_0, as in the case of a battery. Then the charge on the capacitor as a function of time will be

$$Q(t) = Q_0 e^{-t/RC} + V_0 C(1 - e^{-t/RC}).$$

It is interesting to note that the final charge $Q(\infty)$ is $V_0 C$, independent of the initial charge. Intuitively, this is what we expect, of course, as the "capacity" of a capacitor to hold electric charge should not depend on its initial charge. ∎

Example 23.3.3. As a concrete illustration of the general formula derived in the previous example, we find the charge on a capacitor in an RC circuit when a voltage, $V(t) = V_0 \cos \omega t$, is applied to it for a period T and then removed. $V(t)$ can thus be written as

$$V(t) = \begin{cases} V_0 \cos \omega t & \text{if } t < T, \\ 0 & \text{if } t > T. \end{cases}$$

The general solution is given as Equation (23.15). We have to distinguish between two regions in time, $t < T$ and $t > T$.

(a) For $t < T$, we have (using a table of integrals)

$$Q(t) = Q_0 e^{-t/RC} + \frac{e^{-t/RC}}{R} \int_0^t e^{s/RC} V_0 \cos \omega s \, ds$$

$$= Q_0 e^{-t/RC} + \frac{V_0}{R} \frac{1}{(1/RC)^2 + \omega^2} \left(-\frac{1}{RC} e^{-t/RC} + \frac{\cos \omega t}{RC} + \omega \sin \omega t \right)$$

If $T \gg RC$, and we wait long enough,[3] i.e., $t \gg RC$, then only the oscillatory part survives due to the large negative exponents of the exponentials. Thus,

$$Q(t) \approx \frac{V_0}{R} \frac{1}{(1/RC)^2 + \omega^2} \left(\frac{\cos \omega t}{RC} + \omega \sin \omega t \right).$$

The charge $Q(t)$ oscillates with the same frequency as the driving voltage.

(b) For $t > T$, the integral goes up to T beyond which $V(t)$ is zero. Hence, we have

$$Q(t) = Q_0 e^{-t/RC} + \frac{e^{-t/RC}}{R} \int_0^T e^{s/RC} V_0 \cos \omega s \, ds$$

$$= Q_0 e^{-t/RC} + \frac{V_0/R}{(1/RC)^2 + \omega^2} \left[-\frac{e^{-t/RC}}{RC} + e^{(T-t)/RC} \left(\frac{\cos \omega T}{RC} + \omega \sin \omega T \right) \right].$$

We note that the oscillation has stopped (sine and cosine terms are merely constants now), and for $t - T \gg RC$, the charge on the capacitor becomes negligibly small: If there is no applied voltage, the capacitor will discharge. ∎

Although first-order *linear* DEs can always be solved—yielding solutions as given in Equation (23.12)—no general rule can be applied to solve a general FODE. Nevertheless, it can be shown that a solution of such a DE always exist, and, under some mild conditions, this solution is unique. Some special nonlinear FODEs can be solved using certain techniques some of which are described in the following examples as well as the problems at the end of the chapter.

falling object with air resistance

Example 23.3.4. In Problem 23.11 you are asked to find the velocity of a falling object when the air drag is proportional to velocity. This is a good approximation at low velocities for small objects; at higher speeds, and for larger objects, the drag force becomes proportional to higher powers of speed. Let us consider the case when the drag force is proportional to v^2. Then the second law of motion becomes

$$m\frac{dv}{dt} = mg - bv^2 \quad \Rightarrow \quad \frac{dv}{dt} = g - \gamma v^2, \qquad \gamma \equiv \frac{b}{m}.$$

---

[3] Of course, we still assume that $t < T$.

This equation can be written as

$$\frac{dv}{g - \gamma v^2} = dt \implies \frac{dv}{A^2 - v^2} = \gamma\, dt, \quad A^2 = \frac{g}{\gamma}. \tag{23.16}$$

Now we rewrite

$$\frac{1}{A^2 - v^2} = \frac{1}{2A}\left[\frac{1}{v + A} - \frac{1}{v - A}\right],$$

multiply both sides of Equation (23.16) by $2A$ and integrate to obtain

$$\ln|v + A| - \ln|v - A| = 2A\gamma t + \ln C,$$

where we have written the constant of integration as $\ln C$ for convenience. This equation can be rewritten as

$$\left|\frac{v + A}{v - A}\right| = Ce^{2A\gamma t}.$$

Suppose that at $t = 0$, the velocity of the falling object is v_0, then

$$\left|\frac{v_0 + A}{v_0 - A}\right| = C$$

and

$$\left|\frac{v + A}{v - A}\right| = \left|\frac{v_0 + A}{v_0 - A}\right| e^{2A\gamma t}.$$

Now note that $A > 0$, and $v > 0$ (if we take "down" to be the positive direction). Therefore, the last equation becomes

$$\frac{v + A}{|v - A|} = \frac{v_0 + A}{|v_0 - A|} e^{2A\gamma t}.$$

Suppose that $v_0 > A$; then we can remove the absolute value sign from the RHS, and since the two sides must agree at $t = 0$, we can remove the absolute value sign on the LHS as well. Similarly, if $v_0 < A$, then $v < A$ as well. It follows that

$$\frac{v + A}{v - A} = \frac{v_0 + A}{v_0 - A} e^{2A\gamma t} \implies (v + A)(v_0 - A) = (v - A)(v_0 + A)e^{2A\gamma t}.$$

Solving for v gives

$$\begin{aligned}
v &= A\frac{(v_0 + A)e^{2A\gamma t} + v_0 - A}{(v_0 + A)e^{2A\gamma t} - (v_0 - A)} \\
&= A\frac{v_0(e^{2A\gamma t} + 1) + A(e^{2A\gamma t} - 1)}{v_0(e^{2A\gamma t} - 1) + A(e^{2A\gamma t} + 1)} \\
&= A\frac{v_0\cosh(A\gamma t) + A\sinh(A\gamma t)}{v_0\sinh(A\gamma t) + A\cosh(A\gamma t)}.
\end{aligned} \tag{23.17}$$

It follows from Equation (23.17) that at $t = 0$, the velocity is v_0, as we expect. It also shows that, when $t \to \infty$, the velocity approaches $A = \sqrt{g/\gamma}$, the so-called **terminal velocity**. This is the velocity at which the gravitational force and the drag force become equal, causing the acceleration of the object to be zero. The terminal velocity can thus be obtained directly from the second law without solving the differential equation.

terminal velocity

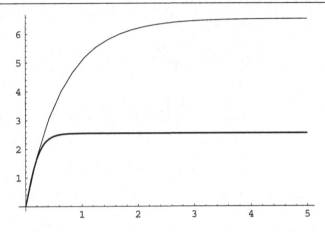

Figure 23.1: The achievement of terminal velocity for a drag force that is proportional to the square of speed (the heavy curve) is considerably faster than for a drag force that is linear in speed (the light curve) if γ has the same numerical value for both cases.

Figure 23.1 shows the plot of speed as a function of time for the two cases of the drag force being proportional to v and v^2 with the same proportionality constant. Because of the higher power of speed, the terminal velocity is achieved considerably more quickly for v^2 force than for v force. Furthermore, as the figure shows clearly, the terminal speed itself is much smaller in the former case. Since larger surfaces provide a v^2 drag force, parachutes that have very large surface are desirable. ■

Example 23.3.5. We consider here some other examples of (nonlinear) FODEs whose solutions are available:

Bernoulli's FODE

(a) **Bernoulli's FODE**: This equation is of the form $y' + p(x)y + q(x)y^n = 0$ where $n \neq 1$. This DE can be simplified if we substitute $y = u^r$ and choose r appropriately. In terms of u, the DE becomes

$$u' + \frac{p(x)}{r}u + \frac{q(x)}{r}u^{nr-r+1} = 0.$$

The simplest DE—whose solution could be found by a simple integration—would be obtained if the exponent of the last term could be set equal to 1. But this would require r to be zero, which is not acceptable. The next simplest DE results if we set the exponent equal to zero, i.e., if $r = 1/(1-n)$. Then the DE becomes

$$u' + (1-n)p(x)u + (1-n)q(x) = 0$$

which is a first-order *linear* DE whose solution we have already found.

homogeneous FODE

(b) **Homogeneous FODE**: This DE is of the form

$$\frac{dy}{dx} = w\left(\frac{y}{x}\right).$$

To find the solution, make the obvious substitution $u = y/x$, to obtain $y' = u + xu'$ and

$$u + xu' = w(u) \implies u' = \frac{w(u) - u}{x} \implies \frac{du}{w(u) - u} = \frac{dx}{x}$$

with the solution

$$\ln x = \int_c^u \frac{dt}{w(t) - t} \qquad \text{or} \qquad x = \exp\left[\int_c^{y/x} \frac{dt}{w(t) - t}\right],$$

where c is an arbitrary constant to be determined by the initial conditions. ■

23.4 Problems

23.1. Suppose that region D is contractible to zero. Using the equivalence of the vanishing of curl and vanishing of closed line integrals, show that $\partial M/\partial y = \partial N/\partial x$ is both necessary and sufficient condition for $M\,dx + N\,dy$ to be exact.

23.2. Verify that $\mu = C/(x^2 + y^2)$ is an integrating factor of $x\,dy - y\,dx$ which gives rise to

$$v = \tan^{-1}\left(\frac{y}{x}\right) = \text{constant} \qquad \Rightarrow \qquad \frac{y}{x} = C'$$

for a solution of $x\,dy - y\,dx$.

23.3. Find the general solution of Bernoulli's FODE

$$y' + p(x)y + q(x)y^n = 0 \qquad \text{where} \quad n \neq 1.$$

Hint: See Example 23.3.5.

23.4. Find a solution to the *linear fractional* DE

$$\frac{dy}{dx} = \frac{a_1 x + a_2 y}{b_1 x + b_2 y} \qquad \text{where} \quad a_1 b_2 \neq a_2 b_1.$$

Hint: Divide the numerator and denominator by x to obtain a homogeneous FODE.

23.5. Lagrange's FODE is $y - xp(y') - q(y') = 0$. Lagrange's FODE
(a) Let $y' = t$ and consider x as a function of t. Using the chain rule, find dx/dt in terms of dy/dt.
(b) Differentiate Lagrange's DE with respect to t. Use the result of this differentiation and that of (a) to arrive at $[t - p(t)]\dot{x} - \dot{p}x = \dot{q}$, where the dot indicates differentiation with respect to t.
(c) Find the (parametric) solution of the DE, considering two separate cases: $t = p(t)$ and $t \neq p(t)$.

23.6. Let $u(x, y) = C$ be a solution of the DE $M\,dx + N\,dy = 0$. Show that:
(a) $(\partial u/\partial x)/M = (\partial u/\partial y)/N$; and
(b) $\mu(x, y) \equiv (\partial u/\partial x)/M$ is an integrating factor for the DE.

23.7. Use direct differentiation to show that the function given in Equation (23.12) solves the FOLDE of Equation (23.9).

23.8. Analyze the capacitor's charge in an RC circuit in which a constant potential V_0 is applied for a time $T > 0$ and then disconnected. Consider the cases where $t < T$ and $t > T$.

23.9. Find all functions $f(x)$ whose definite integral from 0 to x equals the square of their reciprocal.

23.10. (a) Let $p_1 u' + p_0 u = 0$ be a homogeneous FOLDE in u. Solve it. (Note that it can easily be integrated.)
(b) Consider $p_1 y' + p_0 y = q$. Let $y = uv$, where u is as in (a), and obtain an equation for v. Solve this equation, and obtain a general solution for $p_1 y' + p_0 y = q$. This is the **method of variation of parameters**, which can also be used for second-order differential equations.

23.11. A falling body in air has a motion approximately described by the DE $m\, dv/dt = mg - bv$, where $v = dx/dt$ is the velocity of the body. Find this velocity as a function of time assuming that the object starts from rest.

23.12. Suppose that both the linear (av) and the quadratic (bv^2) terms are present in the fall of an object with air drag.
(a) Solve the DE and find the most general solution for the velocity as a function of time. Hint: Make the substitution $u = v + a/2b$.
(b) From this general solution, extract the solutions to the cases where only the linear and only the quadratic terms are present by taking the limits $b \to 0$ and $a \to 0$.

23.13. Take the limit of Equation (23.17) as $t \to \infty$ and show that it is equal to $\sqrt{g/\gamma}$.

Chapter 24

Second-Order Linear Differential Equations

The majority of problems encountered in physics lead to second order linear differential equations (SOLDEs) when the so-called nonlinear terms are approximated out. Thus, a general treatment of the properties and methods of obtaining solutions to SOLDEs is essential. In this section, we investigate their general properties, and leave methods of obtaining their solutions for the next section and later chapters.

The most general SOLDE is

$$p_2(x)\frac{d^2y}{dx^2} + p_1(x)\frac{dy}{dx} + p_0(x)y = p_3(x). \qquad (24.1)$$

Dividing by $p_2(x)$, and writing p for p_1/p_2, q for p_0/p_2, and r for p_3/p_2, reduces Equation (24.1) to the **normal form**,

normal form of a
SOLDE

$$\frac{d^2y}{dx^2} + p(x)\frac{dy}{dx} + q(x)y = r(x). \qquad (24.2)$$

Equation (24.2) is equivalent to (24.1) if $p_2(x) \neq 0$. The points at which $p_2(x)$ vanishes are called the **singular points** of the DE.

difference between
singular points of
linear and
nonlinear
differential
equations

There is a crucial difference between the singular points of linear DEs and those of nonlinear DEs. For a nonlinear DE such as $(x^2 - y)y' = x^2 + y^2$, the curve $y = x^2$ is the collection of singular points. This makes it impossible to construct solutions $y = f(x)$ that are defined on an interval $I = [a, b]$ of the x-axis because for any $a < x < b$, there is a $y = x^2$ for which the DE is undefined. On the other hand, linear DEs do not have this problem because the coefficients of the derivatives are functions of x only. Therefore, all the singular "curves" are vertical, and we can find intervals on the x-axis in which the DE is well behaved.

24.1 Linearity, Superposition, and Uniqueness

The FOLDE has only one solution; and we found this solution in closed form in Equation (23.12). The SOLDE may have (in fact, it does) more than one solution. Therefore, it is important to know how many solutions to expect for a SOLDE and what relation (if any) exists between these solutions.

We write Equation (24.1) as

$$\mathbf{L}[y] = p_3 \qquad \text{where} \qquad \mathbf{L} \equiv p_2 \frac{d^2}{dx^2} + p_1 \frac{d}{dx} + p_0. \tag{24.3}$$

It is clear that $\mathbf{L}$ is a *linear* operator[1] by which we mean that for constants α and β, $\mathbf{L}[\alpha y_1 + \beta y_2] = \alpha\mathbf{L}[y_1] + \beta\mathbf{L}[y_2]$. In particular, if y_1 and y_2 are two solutions of Equation (24.3), then

$$\mathbf{L}[y_1 - y_2] = \mathbf{L}[y_1] - \mathbf{L}[y_2] = p_3 - p_3 = 0.$$

homogeneous
SOLDE

That is, the difference between any two solutions of a SOLDE is a solution[2] of the **homogeneous equation** obtained by setting $p_3 = 0$. An immediate consequence of the linearity of $\mathbf{L}$ is that any linear combination of solutions of the *homogeneous* SOLDE (HSOLDE) is also a solution. This is called the **superposition principle**.

superposition
principle

We saw in the introduction to Chapter 22 that, based on physical intuition, we expect to be able to predict the behavior of a physical system if we know the DE obeyed by that system and equally importantly, the initial data. Physical intuition also tells us that if the initial conditions are changed by an infinitesimal amount, then the solutions will be changed infinitesimally. Thus, the solutions of linear DEs are said to be continuous functions of the initial conditions. Nonlinear DEs can have completely different solutions for two initial conditions that are infinitesimally close. Since initial conditions cannot be specified with mathematical precision in practice, nonlinear DEs lead to unpredictable solutions, or chaos. This subject has received much attention in recent years, and we shall present a brief discussion of chaos in Chapter 31.

By its very nature, a prediction is expected to be unique. This expectation for linear equations becomes—in the language of mathematics—an existence and a uniqueness theorem. First, we need the following[3]

Theorem 24.1.1. *The only solution $g(x)$ of the homogeneous equation $y'' + py' + qy = 0$, defined on the interval $[a, b]$, which satisfies $g(a) = 0 = g'(a)$, is the trivial solution, $g = 0$.*

Let f_1 and f_2 be two solutions of (24.2) satisfying the same initial conditions on the interval $[a, b]$. This means that $f_1(a) = f_2(a) = c$ and $f_1'(a) =$

---

[1]Recall from Chapter 7 that an operator is a correspondence on a vector space that takes one vector and gives another. A *linear* operator is an operator that satisfies Equation (7.3). The vector space on which $\mathbf{L}$ acts is the vector space of differentiable functions.

[2]This conclusion is not limited to the SOLDE; it holds for all linear DEs.

[3]For a proof, see Hassani, S. *Mathematical Physics: A Modern Introduction to Its Foundations*, Springer-Verlag, 1999, p. 354.

$f_2'(a) = c'$ for some given constants c and c'. Then it is readily seen that their difference, $g \equiv f_1 - f_2$, satisfies the homogeneous equation [with $r(x) = 0$]. The initial condition that $g(x)$ satisfies is clearly $g(a) = 0 = g'(a)$. By Theorem 24.1.1, $g = 0$ or $f_1 = f_2$. We have just shown

uniqueness of
solutions to
SOLDE

Theorem 24.1.2. (*Uniqueness Theorem*). *If p and q are continuous on $[a, b]$, then at most one solution of Equation (24.2) can satisfy a given set of initial conditions.*

The uniqueness theorem can be applied to any *homogeneous* SOLDE to find the latter's most general solution. In particular, let $f_1(x)$ and $f_2(x)$ be any two solutions of

$$y'' + p(x)y' + q(x)y = 0 \qquad (24.4)$$

defined on the interval $[a, b]$. Assume that the two vectors $\mathbf{v}_1 = (f_1(a), f_1'(a))$ and $\mathbf{v}_2 = (f_2(a), f_2'(a))$ are linearly independent.[4] Let $g(x)$ be another solution. The vector $(g(a), g'(a))$ can be written as a linear combination of $\mathbf{v}_1$ and $\mathbf{v}_2$, giving the two equations

$$g(a) = c_1 f_1(a) + c_2 f_2(a),$$
$$g'(a) = c_1 f_1'(a) + c_2 f_2'(a).$$

The function $u(x) \equiv g(x) - c_1 f_1(x) - c_2 f_2(x)$ satisfies the DE (24.4) and the initial conditions $u(a) = u'(a) = 0$. It follows from Theorem 24.1.1 that $u(x) = 0$ or $g(x) = c_1 f_1(x) + c_2 f_2(x)$. We have proved

basis of solutions

Theorem 24.1.3. *Let f_1 and f_2 be two solutions of the HSOLDE*

$$y'' + py' + qy = 0,$$

where p and q are continuous functions defined on the interval $[a, b]$. If $(f_1(a), f_1'(a))$ and $(f_2(a), f_2'(a))$ are linearly independent vectors, then every solution $g(x)$ of this HSOLDE is equal to some linear combination

$$g(x) = c_1 f_1(x) + c_2 f_2(x),$$

with constant coefficients c_1 and c_2. The functions f_1 and f_2 are called a ***basis of solutions*** *of the HSOLDE.*

The uniqueness theorem states that only one solution *can* exist for a SOLDE which satisfies a given set of initial conditions. Whether such a solution *does* exist is beyond the scope of the theorem. Under some mild assumptions, however, it can be shown that a solution does indeed exist. We shall not prove this *existence theorem* for a general SOLDE, but shall examine various techniques of obtaining solutions for specific SOLDEs in this and the next two chapters.

there is also an
existence theorem!

---

[4]If they are not, then one must choose a different initial point for the interval.

24.2   The Wronskian

To form a basis of solutions, f_1 and f_2 must be linearly independent. It is important to note that the linear dependence or independence of a number of functions, defined on the interval $[a, b]$, is a concept that must hold for all x in $[a, b]$. Thus, if

$$\alpha_1 f_1(x_0) + \alpha_2 f_2(x_0) + \cdots + \alpha_n f_n(x_0) = 0$$

for some $x_0 \in [a, b]$, it does not mean that the f's are linearly dependent. Linear dependence requires that the equality holds for all x in $[a, b]$.

Wronskian defined

The nature of the linear relation between f_1 and f_2 can be determined by their Wronskian.

Definition 24.2.1. *The **Wronskian** of any two differentiable functions $f_1(x)$ and $f_2(x)$ is defined to be*

$$W(f_1, f_2; x) = f_1(x)f_2'(x) - f_2(x)f_1'(x) = \det \begin{pmatrix} f_1(x) & f_1'(x) \\ f_2(x) & f_2'(x) \end{pmatrix}.$$

If we differentiate both sides of the definition of Wronskian and substitute from Equation (24.4), we obtain

$$\begin{aligned}
\frac{d}{dx} W(f_1, f_2; x) &= f_1' f_2' + f_1 f_2'' - f_2' f_1' - f_2 f_1'' \\
&= f_1(-pf_2' - qf_2) - f_2(-pf_1' - qf_1) \\
&= pf_1' f_2 - pf_1 f_2' = -p(x)W(f_1, f_2; x).
\end{aligned}$$

We can easily find a solution to this DE:

$$\frac{dW}{dx} = -pW \;\Rightarrow\; \frac{dW}{W} = -p\,dx \;\Rightarrow\; \ln W = -\int_c^x p(t)\,dt + \ln C,$$

where c is an *arbitrary* point in the interval $[a, b]$ and C is the constant of integration. In fact, it is readily seen that $C = W(c)$. We therefore have

$$W(f_1, f_2; x) = W(f_1, f_2; c)e^{-\int_c^x p(t)\,dt}. \tag{24.5}$$

Note that $W(f_1, f_2; x) = 0$ if and only if $W(f_1, f_2; c) = 0$, and that [because the exponential in (24.5) is positive] $W(f_1, f_2; x)$ and $W(f_1, f_2; c)$ have the same sign if they are not zero. This observation leads to

---

Box 24.2.1. *The Wronskian of any two solutions of Equation (24.4) does not change sign in the interval $[a, b]$. In particular, if the Wronskian vanishes at one point in $[a, b]$, it vanishes at all points in $[a, b]$.*

Let f_1 and f_2 be any two differentiable functions that are not necessarily solutions of any DE. If f_1 and f_2 are linearly dependent, then one is a multiple of the other, and the Wronskian is readily seen to vanish. Conversely, assume that the Wronskian is zero. Then $f_1(x)f_2'(x) - f_2(x)f_1'(x) = 0$. This gives

$$f_1 df_2 = f_2 df_1 \;\Rightarrow\; \frac{df_2}{f_2} = \frac{df_1}{f_1} \;\Rightarrow\; \ln f_2 = \ln f_1 + \ln C \;\Rightarrow\; f_2 = C f_1$$

and the two functions are linearly dependent. We have just shown that

differentiability is important in the statement of Box 24.2.2.

> **Box 24.2.2.** *Two differentiable functions, which are nonzero in the interval $[a, b]$, are linearly dependent if and only if their Wronskian vanishes.*

Example 24.2.1. Let $f_1(x) = x$ and $f_2(x) = |x|$ for $-1 \leq x \leq 1$. These two functions are linearly independent in the given interval, because $\alpha_1 x + \alpha_2 |x| = 0$ *for all* x if and only if $\alpha_1 = \alpha_2 = 0$. The Wronskian, on the other hand, vanishes for all $-1 \leq x \leq 1$:

$$
\begin{aligned}
W(f_1, f_2; x) &= x \frac{d|x|}{dx} - |x| \frac{dx}{dx} = x \frac{d|x|}{dx} - |x| \\
&= x \frac{d}{dx} \begin{cases} x & \text{if } x \geq 0 \\ -x & \text{if } x \leq 0 \end{cases} - \begin{cases} x & \text{if } x \geq 0 \\ -x & \text{if } x \leq 0 \end{cases} \\
&= \begin{cases} x - x = 0 & \text{if } x > 0, \\ -x - (-x) = 0 & \text{if } x < 0. \end{cases}
\end{aligned}
$$

This seems to be in contradiction to Box 24.2.2. It is not! Box 24.2.2 assumes that both functions are *differentiable* in their common interval of definition. However, $|x|$ is *not* differentiable at $x = 0$. ∎

24.3 A Second Solution to the HSOLDE

If we know one solution to Equation (24.4), we can use the Wronskian to obtain a second linearly independent solution. Let $W(x) \equiv W(f_1, f_2; x)$ be the Wronskian of the two solutions f_1 and f_2. Then, by definition and Equation (24.5), we have

$$f_1(x)f_2'(x) - f_2(x)f_1'(x) = W(x) = W(c)e^{-\int_c^x p(t)\,dt},$$

where c is an arbitrary point in the interval of interest. Given $f_1(x)$, this is a FOLDE in $f_2(x)$, which can be solved by the method of Subsection 23.3. In fact, $1/f_1^2(x)$ is an integrating factor, and dividing both sides by $f_1^2(x)$ gives

$$\frac{d}{dx}\left[\frac{f_2(x)}{f_1(x)}\right] = \frac{W(x)}{f_1^2(x)}$$

a second linearly
independent
solution can be
found from a
given solution

or

$$\frac{f_2(x)}{f_1(x)} = C + \int_\alpha^x \frac{W(s)}{f_1^2(s)}\, ds = C + \int_\alpha^x \frac{1}{f_1^2(s)} W(c) \exp\left[-\int_c^s p(t)\, dt\right] ds,$$

where C is an arbitrary constant of integration and α is a convenient point in the interval $[a, b]$. Thus,

second solution of
the HSOLDE
obtained from the
first

$$f_2(x) = f_1(x) \left\{ C + K \int_\alpha^x \frac{1}{f_1^2(s)} \exp\left[-\int_c^s p(t)\, dt\right] ds \right\}, \qquad (24.6)$$

where we substituted K for $W(c)$. We do not have to know $W(x)$ (this would require knowledge of f_2, which we are trying to calculate!) to obtain $K = W(c)$. In fact, it is a good exercise for the reader to show that f_2, as given by (24.6), indeed satisfies Equation (24.4) no matter what K is. Note also that $f_2(\alpha) = Cf_1(\alpha)$. Whenever possible—and convenient—it is customary to set $C = 0$ because its presence simply gives a term that is proportional to the known solution $f_1(x)$.

Example 24.3.1. (a) A solution to the SOLDE $y'' - k^2 y = 0$ is e^{kx}. To find a second solution, we let $C = 0$ and $K = 1$ in Equation (24.6). Since $p(x) = 0$, we have

$$f_2(x) = e^{kx} \left(0 + \int_\alpha^x \frac{ds}{e^{2ks}} \right) = -\frac{1}{2k} e^{-kx} + \frac{e^{-2k\alpha}}{2k} e^{kx}$$

which, ignoring the second term which is proportional to the first solution, leads directly to the choice of e^{-kx} as a second solution.

(b) The differential equation $y'' + k^2 y = 0$, which arises in mechanics in the study of the motion of a mass attached to the end of a spring, has $\sin kx$ as a solution. With $C = 0$, $c = \alpha = \pi/2k$, and $K = 1$, we get

$$f_2(x) = \sin kx \left(0 + \int_{\pi/2k}^x \frac{ds}{\sin^2 ks} \right) = -\sin kx \cot ks|_{\pi/2k}^x = -\cos kx.$$

Thus, $\sin kx$ and $\cos kx$ form a basis of solution, and a general solution is of the form

$$y(x) = A \cos kx + B \sin kx,$$

a result that should be familiar to the reader from introductory physics.

(c) For the solutions in part (a),

$$W(x) = \det \begin{pmatrix} e^{kx} & ke^{kx} \\ e^{-kx} & -ke^{-kx} \end{pmatrix} = -2k$$

and for those in part (b),

$$W(x) = \det \begin{pmatrix} \sin kx & k\cos kx \\ \cos kx & -k\sin kx \end{pmatrix} = -k.$$

Both Wronskians are constant. This is a special case of a result that holds for all DEs of the form $y'' + q(x)y = 0$. ∎

Most special functions used in mathematical physics are solutions of SOL-DEs. The behavior of these functions at certain special points is determined by the physics of the particular problem. In most situations physical expectation leads to a preference for one particular solution over the other. For example, although there are two linearly independent solutions to the **Legendre DE**,

$$\frac{d}{dx}\left[(1-x^2)\frac{dy}{dx}\right] + n(n+1)y = 0,$$

the solution that is most frequently encountered is a Legendre polynomial $P_n(x)$ discussed in Chapter 26. The other solution can be obtained by using Equation (24.6).

Legendre differential equation

24.4   The General Solution to an ISOLDE

We now determine the most general solution of an *inhomogeneous* SOLDE (ISOLDE). Let $g(x)$ be a particular solution of

$$\mathsf{L}[y] = y'' + py' + qy = r(x) \tag{24.7}$$

and let $h(x)$ be any other solution of this equation. Then $h(x) - g(x)$ satisfies Equation (24.4) and, by Theorem 24.1.3, can be written as a linear combination of a basis of solutions $f_1(x)$ and $f_2(x)$. It follows that

$$h(x) = c_1 f_1(x) + c_2 f_2(x) + g(x). \tag{24.8}$$

> **Box 24.4.1.** *If we have a particular solution of the ISOLDE of Equation (24.7) and two basis solutions of the HSOLDE, then the* most general *solution of (24.7) can be expressed as the sum of a linear combination of the two basis solutions and the particular solution.*

We know how to find a second solution to the HSOLDE once we know one solution. We now show that knowing one such solution will also allow us to find a particular solution to the ISOLDE. The method we use is called the **method of variation of constants**.

Let f_1 and f_2 be the two (known) solutions of the HSOLDE and $g(x)$ the sought-after solution to Equation (24.7). Write g as $g(x) = f_1(x)v(x)$ with v a function to be determined. Substitute this in (24.7) to get a SOLDE for $v(x)$:

method of variation of constants

with a solution of HSOLDE at our disposal, we can find a particular solution of an ISOLDE.

$$v'' + \left(p + \frac{2f_1'}{f_1}\right)v' = \frac{r}{f_1}.$$

This is a *first*-order linear DE in v' which has a solution of the form (see Problem 24.6)

$$v' = \frac{W(x)}{f_1^2(x)}\left[C + \int_a^x \frac{f_1(t)r(t)}{W(t)}\,dt\right],$$

where $W(x)$ is the (known) Wronskian of Equation (24.7). Substituting

$$\frac{W(x)}{f_1^2(x)} = \frac{f_1(x)f_2'(x) - f_2(x)f_1'(x)}{f_1^2(x)} = \frac{d}{dx}\left(\frac{f_2}{f_1}\right)$$

in the above expression for v' and setting $C = 0$ (we are interested in a *particular* solution), we get

$$\frac{dv}{dx} = \frac{d}{dx}\left(\frac{f_2}{f_1}\right)\int_a^x \frac{f_1(t)r(t)}{W(t)}\,dt$$

$$= \frac{d}{dx}\left[\frac{f_2(x)}{f_1(x)}\int_a^x \frac{f_1(t)r(t)}{W(t)}\,dt\right] - \frac{f_2(x)}{f_1(x)}\underbrace{\frac{d}{dx}\int_a^x \frac{f_1(t)r(t)}{W(t)}\,dt}_{=f_1(x)r(x)/W(x)}$$

and, by integration,

$$v(x) = \frac{f_2(x)}{f_1(x)}\int_a^x \frac{f_1(t)r(t)}{W(t)}\,dt - \int_a^x \frac{f_2(t)r(t)}{W(t)}\,dt,$$

where in the last integral, we used t as the variable of integration. This leads to the particular solution

$$g(x) = f_1(x)v(x) = f_2(x)\int_a^x \frac{f_1(t)r(t)}{W(t)}\,dt - f_1(x)\int_a^x \frac{f_2(t)r(t)}{W(t)}\,dt. \quad (24.9)$$

Note how symmetric f_1 and f_2 appear in the final result.

It thus follows that

> **Box 24.4.2.** *Given a single solution $f_1(x)$ of the homogeneous equation corresponding to an ISOLDE, one can use Equation (24.6) to find a second solution $f_2(x)$ of the homogeneous equation and Equation (24.9) to find a particular solution $g(x)$. The most general solution h, will then be*
>
> $$h(x) = c_1 f_1(x) + c_2 f_2(x) + g(x).$$

24.5   Sturm–Liouville Theory

We saw in Chapter 22 that the separation of PDEs normally results in expressions of the form

$$\mathsf{L}[u] + \lambda u = 0, \quad\text{or}\quad p_2(x)\frac{d^2 u}{dx^2} + p_1(x)\frac{du}{dx} + p_0(x)u + \lambda u = 0, \quad (24.10)$$

where u is a function of a single variable and λ is, a priori, an arbitrary constant. This is an eigenvalue equation for the operator L just as Equation (7.17) was an eigenvalue equation for the matrix T. In this section, we try to learn some properties of this eigenvalue problem, but first we need to understand the concept of the adjoint of a differential operator.

24.5.1   Adjoint Differential Operators

In our discussion of the eigenvalues and eigenvectors of matrices in Section 7.4, symmetric matrices seemed to be special (see Theorem 7.4.1). The analog of a symmetric matrix in the case of differential operators (DO) is a **self-adjoint** differential operator.

The HSOLDE

$$\mathsf{L}[y] \equiv p_2(x)y'' + p_1(x)y' + p_0(x)y = 0 \qquad (24.11)$$

is said to be **exact** if it can be written as

exact SOLDE

$$\mathsf{L}[y] = \frac{d}{dx}[A(x)y' + B(x)y]. \qquad (24.12)$$

An **integrating factor** for $\mathsf{L}[y]$ is a function $\mu(x)$ such that $\mu(x)\mathsf{L}[y]$ is exact. If an integrating factor exists, then Equation (24.11) reduces to

integrating factor for SOLDE

$$\frac{d}{dx}[A(x)y' + B(x)y] = 0 \ \Rightarrow \ A(x)y' + B(x)y = C,$$

a FOLDE with a constant inhomogeneous term whose solution is given in Theorem 23.3.1. Even the ISOLDE corresponding to Equation (24.11) can be solved, because

$$\mu(x)\mathsf{L}[y] = \mu(x)r(x) \ \Rightarrow \ \frac{d}{dx}[A(x)y' + B(x)y] = \mu(x)r(x)$$

$$\Rightarrow \ A(x)y' + B(x)y = \int_\alpha^x \mu(t)r(t)\,dt,$$

which is a general FOLDE. Thus, the existence of an integrating factor completely solves a SOLDE. It is therefore important to know whether or not a SOLDE admits an integrating factor.

If the SOLDE is exact, then (24.12) must equal (24.11), implying that $p_2 = A$, $p_1 = A' + B$, and $p_0 = B'$. It follows that $p_2'' = A''$, $p_1' = A'' + B'$, and $p_0 = B'$, which in turn give $p_2'' - p_1' + p_0 = 0$. Conversely if $p_2'' - p_1' + p_0 = 0$, then, substituting $p_0 = -p_2'' + p_1'$ in the LHS of Equation (24.11), we obtain

$$p_2 y'' + p_1 y' + p_0 y = p_2 y'' + p_1 y' + (-p_2'' + p_1')y$$
$$= p_2 y'' - p_2'' y + (p_1 y)' = (p_2 y' - p_2' y)' + (p_1 y)'$$
$$= \frac{d}{dx}(p_2 y' - p_2' y + p_1 y),$$

and the DE is exact. Therefore,

> **Box 24.5.1.** *The SOLDE of Equation (24.11) is exact if and only if* $p_2'' - p_1' + p_0 = 0$.

A general SOLDE is clearly not exact. Can we make it exact by multiplying it by an integrating factor as we did with a FOLDE? An immediate consequence of Box 24.5.1 is

> **Box 24.5.2.** *A function μ is an integrating factor of the SOLDE of Equation (24.11) if and only if it is a solution of the HSOLDE*
>
> $$\mathsf{M}[\mu] \equiv (p_2\mu)'' - (p_1\mu)' + p_0\mu = 0. \tag{24.13}$$

We can expand Equation (24.13) to obtain the equivalent equation

$$p_2\mu'' + (2p_2' - p_1)\mu' + (p_2'' - p_1' + p_0)\mu = 0. \tag{24.14}$$

adjoint of a second-order linear differential operator

The operator M given by

$$\mathsf{M} \equiv p_2\frac{d^2}{dx^2} + (2p_2' - p_1)\frac{d}{dx} + (p_2'' - p_1' + p_0) \tag{24.15}$$

is called the **adjoint** of the operator L and denoted by $\mathsf{M} \equiv \mathsf{L}^\dagger$. This is the equivalent of the transpose of a matrix T^t.

Box 24.5.2 confirms the existence of an integrating factor. However, the latter can be obtained only by solving Equation (24.14), which is at least as difficult as solving the original differential equation! In contrast, the integrating factor for a FOLDE can be obtained by a mere integration [see Equation (23.13)].

Although integrating factors for SOLDEs are not as useful as their counterparts for FOLDEs, they can facilitate the study of SOLDEs. Let us first note that the adjoint of the adjoint of a differential operator is the original operator: $(\mathsf{L}^\dagger)^\dagger = \mathsf{L}$ (see Problem 24.10). This suggests that if v is an integrating factor of $\mathsf{L}[u]$, then u will be an integrating factor of $\mathsf{M}[v] \equiv \mathsf{L}^\dagger[v]$. In particular, multiplying the first one by v and the second one by u and subtracting the results, we obtain [see Equations (24.11) and (24.13)] $v\mathsf{L}[u] - u\mathsf{M}[v] = (vp_2)u'' - u(p_2v)'' + (vp_1)u' + u(p_1v)'$, which can be simplified to

$$v\mathsf{L}[u] - u\mathsf{M}[v] = \frac{d}{dx}[p_2vu' - (p_2v)'u + p_1uv]. \tag{24.16}$$

Lagrange identities

Integrating this from a to b yields

$$\int_a^b \left(v\mathsf{L}[u] - u\mathsf{M}[v]\right)dx = \left[p_2vu' - (p_2v)'u + p_1uv\right]\Big|_a^b. \tag{24.17}$$

Equations (24.16) and (24.17) are called the **Lagrange identities**.

As in the case of matrices, a self-adjoint differential operator (corresponding to a symmetric matrix for which $\mathsf{T} = \mathsf{T}^t$) merits special consideration.

For $\mathbf{M}[v] \equiv \mathbf{L}^\dagger[v]$ to be equal to $\mathbf{L}[v]$, we must have [see Equations (24.11) and (24.14)] $2p_2' - p_1 = p_1$ and $p_2'' - p_1' + p_0 = p_0$. The first equation gives $p_2' = p_1$, which also solves the second equation. If this condition holds, then we can write Equation (24.11) as $\mathbf{L}[y] = p_2 y'' + p_2' y' + p_0 y$, or

$$\mathbf{L}[y] = \frac{d}{dx}\left[p_2(x)\frac{dy}{dx}\right] + p_0(x)y = 0.$$

Can we make all SOLDEs self-adjoint? Let us multiply both sides of Equation (24.11) by a function $w(x)$, to be determined later. We get the new DE

$$w(x)p_2(x)y'' + w(x)p_1(x)y' + w(x)p_0(x)y = 0,$$

which we desire to be self-adjoint. This will be accomplished if we choose $w(x)$ such that $wp_1 = (wp_2)'$, or $p_2 w' + w(p_2' - p_1) = 0$, which can be readily integrated to give

$$w(x) = \frac{1}{p_2}\exp\left[\int^x \frac{p_1(t)}{p_2(t)}\,dt\right].$$

We have just proved the following:

Theorem 24.5.1. *The SOLDE of Equation (24.11) is self-adjoint if and only if $p_2' = p_1$, in which case the DE has the form*

$$\frac{d}{dx}\left[p_2(x)\frac{dy}{dx}\right] + p_0(x)y = 0.$$

If it is not self-adjoint, it can be made so by multiplying it through by

$$w(x) = \frac{1}{p_2}\exp\left[\int^x \frac{p_1(t)}{p_2(t)}\,dt\right].$$

all SOLDEs can be made self-adjoint

Example 24.5.2. (a) The Legendre equation in normal form,

$$y'' - \frac{2x}{1-x^2}y' + \frac{\lambda}{1-x^2}y = 0,$$

is not self-adjoint. However, we get a self-adjoint version if we multiply through by $w(x) = 1 - x^2$:

$$(1-x^2)y'' - 2xy' + \lambda y = 0, \quad \text{or} \quad [(1-x^2)y']' + \lambda y = 0$$

(b) Similarly, the normal form of the Bessel equation

$$y'' + \frac{1}{x}y' + \left(1 - \frac{n^2}{x^2}\right)y = 0$$

is not self-adjoint, but multiplying through by $h(x) = x$ yields

$$\frac{d}{dx}\left(x\frac{dy}{dx}\right) + \left(x - \frac{n^2}{x}\right)y = 0,$$

which is clearly self-adjoint. ■

24.5.2   Sturm–Liouville System

Now that we know that every SOLDE can be made self-adjoint, let's apply the procedure to our starting DE (24.10). If we multiply that equation by the $w(x)$ of Theorem 24.5.1 it becomes self-adjoint, and can be written as

$$\frac{d}{dx}\left[p(x)\frac{du}{dx}\right] + [\lambda w(x) - q(x)]u = 0 \quad \text{or}$$

$$\mathbf{L}[u] = \frac{d}{dx}\left[p(x)\frac{du}{dx}\right] - q(x)u = -\lambda w(x)u \quad (24.18)$$

with $p(x) = w(x)p_2(x)$ and $q(x) = -p_0(x)w(x)$. Equation (24.18) is the standard form of the **Sturm-Liouville** (S-L) equation.

The appearance of w is the result of our desire to render the differential operator self-adjoint. It also appears in another context. Write the Lagrange identity (24.16) for a self-adjoint differential operator $\mathbf{L}$:

$$u\mathbf{L}[v] - v\mathbf{L}[u] = \frac{d}{dx}\{p(x)[u(x)v'(x) - v(x)u'(x)]\}. \quad (24.19)$$

If we specialize this identity to the S-L equation of (24.18) with $u = u_1$ corresponding to the eigenvalue λ_1 and $v = u_2$ corresponding to the eigenvalue λ_2, we obtain for the LHS

$$u_1\mathbf{L}[u_2] - u_2\mathbf{L}[u_1] = u_1(-\lambda_2 w u_2) + u_2(\lambda_1 w u_1) = (\lambda_1 - \lambda_2)w u_1 u_2.$$

Integrating both sides of (24.19) then yields

$$(\lambda_1 - \lambda_2)\int_a^b w u_1 u_2 dx = \{p(x)[u_1(x)u_2'(x) - u_2(x)u_1'(x)]\}_a^b. \quad (24.20)$$

A desired property of the solutions of a self-adjoint DE is their orthogonality when they belong to different eigenvalues. This property will be satisfied if we assume an inner product integral with weight function $w(x)$, and if the RHS of Equation (24.20) vanishes. There are various boundary conditions (BC) that fulfill the latter requirement. One such boundary conditions are **separated boundary conditions**:

$$\alpha_1 u(a) + \beta_1 u'(a) = 0,$$
$$\alpha_2 u(b) + \beta_2 u'(b) = 0, \quad (24.21)$$

Sturm–Liouville systems

where α_1, α_2, β_1, and β_2 are real constants. Another set of appropriate boundary conditions is the **periodic BC** given by

$$u(a) = u(b) \quad \text{and} \quad u'(a) = u'(b). \quad (24.22)$$

The collection of the DO and the boundary conditions is called a **Sturm–Liouville (S-L) system**.

Example 24.5.3. For fixed ν the DE

$$\frac{d^2u}{dr^2} + \frac{1}{r}\frac{du}{dr} + \left(k^2 - \frac{\nu^2}{r^2}\right)u = 0, \quad 0 \le r \le b \tag{24.23}$$

transforms into the **Bessel equation** $u'' + u'/x + (1 - \nu^2/x^2)u = 0$ if we make the substitution $kr = x$. Thus, the solution of the S-L equation (24.23) that is analytic at $r = 0$ and corresponds to the eigenvalue k^2 is $u_k(r) = J_\nu(kr)$—because Bessel functions $J_\nu(x)$ are entire functions. For two different eigenvalues, k_1^2 and k_2^2, the eigenfunctions are orthogonal if the boundary term of (24.20) corresponding to Equation (24.23) vanishes, that is, if

$$\{r[J_\nu(k_1r)J_\nu'(k_2r) - J_\nu(k_2r)J_\nu'(k_1r)]\}_0^b$$

vanishes, which will occur if and only if $J_\nu(k_1b)J_\nu'(k_2b) - J_\nu(k_2b)J_\nu'(k_1b) = 0$. A common choice is to take $J_\nu(k_1b) = 0 = J_\nu(k_2b)$, that is, to take both k_1b and k_2b as (different) roots of the Bessel function of order ν. We thus have $\int_0^b rJ_\nu(k_ir)J_\nu(k_jr)\,dr = 0$ if k_i and k_j are different roots of $J_\nu(kb) = 0$.

The **Legendre equation**

$$\frac{d}{dx}\left[(1 - x^2)\frac{du}{dx}\right] + \lambda u = 0, \qquad \text{where } -1 < x < 1,$$

is already self-adjoint. Thus, $w(x) = 1$, and $p(x) = 1 - x^2$. Solutions of this DE corresponding to $\lambda = n(n + 1)$ are the Legendre polynomials $P_n(x)$. The boundary term of (24.20) clearly vanishes at $a = -1$ and $b = +1$, and we obtain the orthogonality relation: $\int_{-1}^{+1} P_n(x)P_m(x)\,dx = 0$ if $m \ne n$.

The **Hermite equation** is

$$u'' - 2xu' + \lambda u = 0. \tag{24.24}$$

It is transformed into an S-L system if we multiply it by $w(x) = e^{-x^2}$. The resulting S-L equation is

$$\frac{d}{dx}\left[e^{-x^2}\frac{du}{dx}\right] + \lambda e^{-x^2}u = 0. \tag{24.25}$$

The function u is an eigenfunction of (24.25) corresponding to the eigenvalue λ if and only if it is a solution of (24.24). Solutions of this DE corresponding to $\lambda = 2n$ are the Hermite polynomials $H_n(x)$. The boundary term corresponding to the two eigenfunctions $u_1(x)$ and $u_2(x)$ having the respective eigenvalues λ_1 and $\lambda_2 \ne \lambda_1$ is

$$\{e^{-x^2}[u_1(x)u_2'(x) - u_2(x)u_1'(x)]\}_a^b.$$

This vanishes for arbitrary u_1 and u_2 if $a = -\infty$ and $b = +\infty$. We can therefore write $\int_{-\infty}^{+\infty} e^{-x^2}H_n(x)H_m(x)\,dx = 0$ if $m \ne n$. ∎

24.6   SOLDEs with Constant Coefficients

The SOLDEs with constant coefficients occur frequently and their solutions are easily accessible. In fact, we need not confine ourselves to the second order equations. The most general nth-order linear differential equation (NOLDE) with constant coefficients can be written as

$$\mathsf{L}[y] \equiv y^{(n)} + a_{n-1}y^{(n-1)} + \cdots + a_1y' + a_0y = r(x). \tag{24.26}$$

The corresponding *homogeneous* NOLDE (HNOLDE) is obtained by setting $r(x) = 0$.

24.6.1   The Homogeneous Case

The solution to the HNOLDE

$$\mathbf{L}[y] \equiv y^{(n)} + a_{n-1}y^{(n-1)} + \cdots + a_1 y' + a_0 y = 0 \qquad (24.27)$$

can be found by making the exponential substitution $y = e^{\lambda x}$, which results in the equation $\mathbf{L}[e^{\lambda x}] = (\lambda^n + a_{n-1}\lambda^{n-1} + \cdots + a_1\lambda + a_0)e^{\lambda x} = 0$. This equation will hold only if λ is a root of the **characteristic polynomial**

characteristic
polynomial of a
HNOLDE

$$p(\lambda) \equiv \lambda^n + a_{n-1}\lambda^{n-1} + \cdots + a_1\lambda + a_0$$

which, by the fundamental theorem of algebra, can be written as

$$p(\lambda) = (\lambda - \lambda_1)^{k_1}(\lambda - \lambda_2)^{k_2} \ldots (\lambda - \lambda_m)^{k_m}. \qquad (24.28)$$

The λ_i are the distinct roots of $p(\lambda)$ with λ_j having multiplicity k_j.

It is convenient to introduce $\mathbf{D} \equiv d/dx$ and define the **differential operator**

$$\mathbf{L} = p(\mathbf{D}) = \mathbf{D}^n + a_{n-1}\mathbf{D}^{n-1} + \cdots + a_1\mathbf{D} + a_0.$$

Since $\mathbf{D} - \mu$ and $\mathbf{D} - \lambda$ commute for arbitrary constants μ and λ, we can unambiguously factor out the above and obtain

$$\mathbf{L} = p(\mathbf{D}) = (\mathbf{D} - \lambda_1)^{k_1}(\mathbf{D} - \lambda_2)^{k_2} \ldots (\mathbf{D} - \lambda_m)^{k_m}. \qquad (24.29)$$

In preparation for finding the most general solution for Equation (24.27), we first note that

$$(\mathbf{D} - \lambda)e^{\lambda x} = \frac{d}{dx}e^{\lambda x} - \lambda e^{\lambda x} = 0 \qquad (24.30)$$

and

$$(\mathbf{D} - \lambda)(x^r e^{\lambda x}) = \frac{d}{dx}(x^r e^{\lambda x}) - \lambda x^r e^{\lambda x} = r x^{r-1} e^{\lambda x}.$$

If we apply $\mathbf{D} - \lambda$ twice, we get

$$(\mathbf{D} - \lambda)^2(x^r e^{\lambda x}) = (\mathbf{D} - \lambda)(r x^{r-1} e^{\lambda x}) = r(r-1)x^{r-2}e^{\lambda x}$$

and in general,

$$(\mathbf{D} - \lambda)^k(x^r e^{\lambda x}) = r(r-1)\ldots(r-k+1)x^{r-k}e^{\lambda x}$$

which, for $k = r$, gives

$$(\mathbf{D} - \lambda)^r(x^r e^{\lambda x}) = r!e^{\lambda x}.$$

If we apply $\mathbf{D} - \lambda$ one more time, we get zero by (24.30). Therefore,

$$(\mathbf{D} - \lambda)^k(x^r e^{\lambda x}) = 0 \qquad \text{if} \quad k > r. \qquad (24.31)$$

The set of functions

$$\{x^{r_1} e^{\lambda_1 x}\}_{r_1=0}^{k_1-1}, \ \{x^{r_2} e^{\lambda_2 x}\}_{r_2=0}^{k_2-1}, \ \ldots, \ \{x^{r_m} e^{\lambda_m x}\}_{r_m=0}^{k_m-1},$$

are all solutions of Equation (24.27). For example, an element of the first set yields

$$\mathbf{L}[x^{r_1}e^{\lambda_1 x}] = (\mathbf{D}-\lambda_1)^{k_1}(\mathbf{D}-\lambda_2)^{k_2}\ldots(\mathbf{D}-\lambda_m)^{k_m}(x^{r_1}e^{\lambda_1 x})$$
$$= (\mathbf{D}-\lambda_2)^{k_2}\ldots(\mathbf{D}-\lambda_m)^{k_m}\underbrace{(\mathbf{D}-\lambda_1)^{k_1}(x^{r_1}e^{\lambda_1 x})}_{=0\text{ because } k_1 > r_1} = 0.$$

If the root λ is complex and the coefficients of the DE are real, then the complex conjugate λ^* is also a root (see Problem 24.14). It follows that whenever $x^{r_j}e^{\lambda_j x}$ is a solution of the DE for complex λ_j, so is $x^{r_j}e^{\lambda_j^* x}$. Thus, writing $\lambda_j = \alpha_j + i\beta_j$ and using the linearity of $\mathbf{L}$, we conclude that

$$x^{r_j}e^{\alpha_j x}\cos\beta_j x \quad\text{and}\quad x^{r_j}e^{\alpha_j x}\sin\beta_j x, \quad\text{where}\quad r_j = 0,1,\ldots,k_j-1,$$

are all solutions of (24.27).

It is easily proved that the functions $x^{r_j}e^{\lambda_j x}$ are linearly independent (see Problem 24.13). Furthermore, $\sum_{j=1}^m k_j = n$ by Equation (24.28). Therefore, the set

$$\left\{x^{r_j}e^{\lambda_j x}\right\}, \quad\text{where}\quad r_j = 0,1\ldots,k_j-1 \text{ and } j = 1,2,\ldots,m,$$

contains exactly n elements. We have thus shown that there are at least n linearly independent solutions for the HNOLDE of Equation (24.27). In fact, it can be shown that there are exactly n linearly independent solutions.

Box 24.6.1. *Let $\lambda_1,\lambda_2,\ldots,\lambda_m$ be the roots of the characteristic polynomial of the real HNOLDE of Equation (24.27), and let the respective roots have multiplicities $k_1,k_2,\ldots,k_m$. Then the functions $x^{r_j}e^{\lambda_j x}$, where $r_j = 0,1\ldots,k_j-1$, are a basis of solutions of Equation (24.27).*

Example 24.6.1. An equation that is used in both mechanics and circuit theory is

$$\frac{d^2y}{dt^2} + a\frac{dy}{dt} + by = 0 \quad\text{with}\quad a,b > 0. \tag{24.32}$$

Its characteristic polynomial is $p(\lambda) = \lambda^2 + a\lambda + b$ which has the roots

$$\lambda_1 = \tfrac{1}{2}(-a + \sqrt{a^2-4b}) \quad\text{and}\quad \lambda_2 = \tfrac{1}{2}(-a - \sqrt{a^2-4b}).$$

We can distinguish three different possible motions depending on the relative sizes of a and b.

(a) $a^2 > 4b$ (**overdamped**): Here we have two distinct simple roots. The multiplicities are both one: $k_1 = k_2 = 1$ (see Box 24.6.1). Therefore, the power of t for both solutions is zero ($r_1 = r_2 = 0$). Let $\gamma \equiv \tfrac{1}{2}\sqrt{a^2-4b}$. Then the most general solution is

overdamped

$$y(t) = e^{-at/2}(c_1 e^{\gamma t} + c_2 e^{-\gamma t}).$$

Since $a > 2\gamma$, this solution starts at $y = c_1 + c_2$ at $t = 0$ and continuously decreases; so, as $t \to \infty$, $y(t) \to 0$.

critically damped   (b) $a^2 = 4b$ (**critically damped**): In this case we have one multiple root of order 2 ($k_1 = 2$); therefore, the power of x can be zero or 1 ($r_1 = 0, 1$). Thus, the general solution is

$$y(t) = c_1 t e^{-at/2} + c_0 e^{-at/2}.$$

This solution starts at $y(0) = c_0$ at $t = 0$, reaches a maximum (or minimum) at $t = 2/a - c_0/c_1$, and subsequently approaches zero asymptotically (see Problem 24.23).

underdamped   (c) $a^2 < 4b$ (**underdamped**): Once more, we have two distinct simple roots. The multiplicities are both one ($k_1 = k_2 = 1$); therefore, the power of x for both solutions is zero ($r_1 = r_2 = 0$). Let $\omega \equiv \frac{1}{2}\sqrt{4b - a^2}$. Then $\lambda_1 = -a/2 + i\omega$ and $\lambda_2 = \lambda_1^*$. The roots are complex, and the most general solution is thus of the form

$$y(t) = e^{-at/2}(c_1 \cos \omega t + c_2 \sin \omega t) = A e^{-at/2} \cos(\omega t + \alpha).$$

The solution is a harmonic variation with a decaying amplitude $A \exp(-at/2)$. Note that if $a = 0$, the amplitude does not decay. That is why a is called the **damping**
damping factor   **factor** (or the damping constant). All three cases are shown in Figure 24.1.

These equations describe either a mechanical system oscillating (with no external driving force) in a viscous (dissipative) fluid, or an electrical circuit consisting of a resistance R, an inductance L, and a capacitance C. For mechanical oscillators, $a = \beta/m$ and $b = k/m$, where β is the dissipative constant related to the drag force f_{drag} and the velocity v by $f_{\text{drag}} = \beta v$, and k is the spring constant (a measure of the stiffness of the spring).

For RLC circuits, $a = R/L$ and $b = 1/LC$. Thus, the damping factor depends on the relative magnitudes of R and L. On the other hand, the frequency

$$\omega \equiv \sqrt{b - \left(\frac{a}{2}\right)^2} = \sqrt{\frac{1}{LC} - \frac{R^2}{4L^2}}$$

depends on all three elements. In particular, for $R \geq 2\sqrt{L/C}$, the circuit does not oscillate.   ■

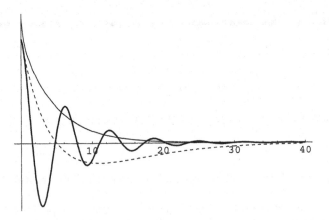

Figure 24.1: The solid thin curve shows the behavior of an overdamped oscillator. The critically damped case is the dashed curve, and the underdamped oscillator is the thick curve.

24.6.2 Central Force Problem

One of the nicest applications of the theory of DEs, and the one that initiated
the modern mathematical analysis, is the study of motion of a particle under
the influence of a central gravitational force. Surprisingly, such a motion can
be reduced to a one-dimensional problem, and eventually to a SOLDE with
constant coefficients as follows.

central force problem

Subsection 12.2.1 treated the equations of motion of a particle under the
influence of a central force. Conservation of angular momentum and the right
choice of the initial position and velocity (what amounted to setting $\mathbf{L} = L_z\hat{\mathbf{e}}_z = L\hat{\mathbf{e}}_z$) eliminated the polar angle θ by assigning it the value $\pi/2$. Thus
the particle is confined to the plane perpendicular to the angular momentum
vector, i.e., essentially the vector $\mathbf{r} \times \mathbf{v}$. The set of three complicated DEs
(12.20) reduces to a much simpler set consisting of (12.22) and (12.24) which
we rewrite here as

$$m\ddot{r} - \frac{L^2}{mr^3} = F(r), \qquad \dot{\varphi} = \frac{L}{mr^2}. \tag{24.33}$$

In principle, we can solve the first equation and find r as a function of t,
then substitute it in the second equation and integrate the result to find φ as
a function of time. However, it is more desirable to find r as a function of φ,
i.e., find the shape of the orbit of the moving particle.

In that spirit, we define a new dependent variable $u = 1/r$, and making
multiple use of the chain rule, we write the DEs with u as the dependent
variable and φ as the independent variable. We thus have

$$r = \frac{1}{u} \;\Rightarrow\; \dot{r} = -\frac{\dot{u}}{u^2} = -\frac{1}{u^2}\dot{\varphi}\frac{du}{d\varphi} = -r^2\dot{\varphi}\frac{du}{d\varphi} = -\frac{L}{m}\frac{du}{d\varphi},$$

$$\ddot{r} = -\frac{L}{m}\frac{d}{dt}\left(\frac{du}{d\varphi}\right) = -\frac{L}{m}\frac{d^2u}{d\varphi^2}\dot{\varphi} = -\frac{L}{m}\frac{d^2u}{d\varphi^2}\frac{L}{mr^2} = -\frac{L^2}{m^2}u^2\frac{d^2u}{d\varphi^2}.$$

Substituting for $\ddot{r}$ and r in terms of u and its derivative, Equation (24.33)
yields

$$\frac{d^2u}{d\varphi^2} + u = -\frac{m}{L^2u^2}F\left(\frac{1}{u}\right). \tag{24.34}$$

Historical Notes

Johannes Kepler (1571–1630) was a premature baby and a very delicate child
who was brought up by his grandparents. After elementary and secondary schooling,
Kepler entered Tübingen University to become a Protestant minister. At Tübingen
Kepler was taught astronomy by one of the leading astronomers of the day, Michael
Maestlin (1550–1631). The astronomy of the curriculum was, of course, geocen-
tric astronomy. At the end of his first year Kepler got 'A's for everything except
mathematics. Probably Maestlin was trying to tell him he could do better, because
Kepler was in fact one of the select pupils to whom he chose to teach more ad-
vanced astronomy by introducing them to the new, heliocentric cosmological system

Johannes Kepler
1571–1629

of Copernicus. It was from Maestlin that Kepler learned that the preface to Copernicus's book, explaining that this was 'only mathematics', was not by Copernicus. Kepler seems to have accepted almost instantly that the Copernican system was physically true, and from then on, astronomy and mathematics became his passion. Kepler also worked and wrote a book in optics, in which he used the idea of a 'ray of light' for the first time.

Kepler problem

For the **Kepler problem** this equation is easy to solve because[5]

$$F(r) = -\frac{K}{r^2} \;\Rightarrow\; F\left(\frac{1}{u}\right) = -Ku^2$$

and we have

$$\frac{d^2u}{d\varphi^2} + u = \frac{Km}{L^2}. \tag{24.35}$$

Let $v = u - Km/L^2$. Then Equation (24.35) becomes

$$\frac{d^2v}{d\varphi^2} + v = 0.$$

The characteristic polynomial of this equation is $\lambda^2 + 1$, whose roots are $\lambda = \pm i$. These simple roots give rise to the linearly independent solutions $v = \sin\varphi$ and $v = \cos\varphi$. The general solution can therefore be expressed as $v = C_1\cos\varphi + C_2\sin\varphi$ which, using Problem 24.22, can be rewritten as $v = A\cos(\varphi - \varphi_0)$. Therefore,

$$v = u - Km/L^2 = A\cos(\varphi - \varphi_0) \;\Rightarrow\; u = Km/L^2 + A\cos(\varphi - \varphi_0)$$

equation of the orbits in the Kepler problem

or

$$r = \frac{1}{(Km/L^2) + A\cos(\varphi - \varphi_0)}. \tag{24.36}$$

This is the equation of a conic section in plane polar coordinates (see Problem 24.15).

We now investigate the details of Equation (24.36). First we note that when $\varphi = \varphi_0$, r is either a maximum or a minimum depending on the sign of A. With an ellipse in mind, this corresponds to the (major) axis of the ellipse making an angle φ_0 with the x-axis. Thus setting $\varphi_0 = 0$ corresponds to choosing the axis of the conic section to be our x-axis. We adhere to this choice and write

$$r = \frac{1}{(Km/L^2) + A\cos\varphi}. \tag{24.37}$$

Next we want to determine the constant A in terms of the energy of the particle. The potential energy (PE) is clearly $-K/r$. So, let us concentrate

---

[5]Although the Kepler problem usually refers to the *gravitational* central force, we want to keep the discussion general enough so that electrostatic force is also included. Thus, K introduced below can be either GMm or $-k_e q_1 q_2$.

on the kinetic energy (KE). The velocity of the particle is given in Equation (12.16) with $\theta = \pi/2$ and $\dot{\theta} = 0$. Thus

$$KE = \tfrac{1}{2}mv^2 = \tfrac{1}{2}m(\dot{r}^2 + r^2\dot{\varphi}^2) = \tfrac{1}{2}m\dot{r}^2 + \frac{L^2}{2mr^2}, \qquad (24.38)$$

where we used the second equation in (24.33). The second term in (24.38) is sometimes called the **centrifugal potential** because (like a potential energy) it is a position-dependent energy that (like a centrifugal force) has resulted from a velocity-dependent term. Differentiating Equation (24.37) with respect to time gives

centrifugal potential

$$\dot{r} = \frac{A\dot{\varphi}\sin\varphi}{[(Km/L^2) + A\cos\varphi]^2} = Ar^2\dot{\varphi}\sin\varphi.$$

Squaring and using the second equation in (24.33), we obtain

$$\dot{r}^2 = A^2 r^4 \dot{\varphi}^2 \sin^2\varphi = \frac{L^2}{m^2}A^2\sin^2\varphi.$$

We can eliminate the sine term in favor of terms involving r by solving for $A\cos\varphi$ in (24.37):

$$A\cos\varphi = \frac{1}{r} - \frac{Km}{L^2} \;\Rightarrow\; A^2\sin^2\varphi = A^2 - \left(\frac{1}{r} - \frac{Km}{L^2}\right)^2.$$

It follows that

$$KE = \tfrac{1}{2}m\left[\frac{L^2 A^2}{m^2} - \frac{L^2}{m^2}\left(\frac{1}{r} - \frac{Km}{L^2}\right)^2\right] + \frac{L^2}{2mr^2}$$

$$= \frac{L^2 A^2}{2m} - \frac{K^2 m}{2L^2} + \frac{K}{r}$$

and

$$E = KE + PE = \frac{L^2 A^2}{2m} - \frac{K^2 m}{2L^2} + \frac{K}{r} - \frac{K}{r} = \frac{L^2 A^2}{2m} - \frac{K^2 m}{2L^2},$$

so that

$$A = \pm\sqrt{\frac{2mE}{L^2} + \frac{K^2 m^2}{L^4}}.$$

To avoid negative signs at later stages, we choose the negative sign now and finally write

$$r = \frac{L^2/(Km)}{1 - \sqrt{2EL^2/(K^2 m) + 1}\,\cos\varphi} = \frac{L^2/(Km)}{1 - e\cos\varphi}, \qquad (24.39)$$

where

$$e \equiv \sqrt{\frac{2EL^2}{K^2 m} + 1} \qquad (24.40)$$

is called the **eccentricity** of the conic section.

eccentricity of orbits

The eccentricity, which by its very definition is always positive, determines the shape of the orbit. Let us concentrate on the interesting case of elliptic orbits corresponding to $0 < e < 1$ indicating that the *total energy of the particle is negative*. Inspection of Problem 24.15 reveals that the semi-major and semi-minor axes of the ellipse are, respectively,

$$a^2 = \frac{L^4}{(1-e^2)^2 K^2 m^2} \quad \text{and} \quad b^2 = \frac{L^4}{(1-e^2)K^2 m^2}.$$

Substituting for e from Equation (24.40) and noting that $E < 0$, we obtain

$$a = -\frac{K}{2E} \Rightarrow E = -\frac{K}{2a} \quad \text{and} \quad b = \frac{L}{\sqrt{-2mE}} \Rightarrow L = \sqrt{\frac{mK}{a}}\, b.$$
$$(24.41)$$

The negativity of energy in an elliptic orbit is an indication of the stability of the orbit. The potential energy is negative and larger in absolute value than the kinetic energy. If the total energy is negative (and, of course, constant), the particle cannot move too far away from the center of attraction, because the magnitude of the PE may become too small to offset the positive KE. The binding energy   absolute value of this total negative energy is called the **binding energy**. For an ellipse this binding energy is $K/2a$.

Kepler's Laws

In 1609 Johannes Kepler, the German astronomer, after painstakingly analyzing the motion of Mars for many years announced what is now called Kepler's first law   **Kepler's first law** of planetary motion: The orbit of Mars is not a circle but an ellipse. In the context of a very resilient tradition—dating back to Pythagoras himself—in which circular orbits were given almost a divine status, this announcement was truly monumental. Kepler had a hunch that all planets obey this same law, but could not prove it. Equation (24.37) is the mathematical statement of Kepler's first law.

Kepler's second law   **Kepler's second law** of planetary motion states that *equal areas are swept out in equal times by the line joining the planet to the center of attraction (the Sun)*. In other words the rate of change of the area is a constant. This can be seen by referring to Figure 24.2 and noting that

$$\Delta A \approx \tfrac{1}{2} r \overline{AB} \approx \tfrac{1}{2} r (r \Delta\varphi) \Rightarrow \frac{\Delta A}{\Delta t} \approx \tfrac{1}{2} r^2 \frac{\Delta\varphi}{\Delta t} \rightarrow \frac{dA}{dt} = \tfrac{1}{2} r^2 \dot{\varphi}.$$

So, by the second equation in (24.33), $dA/dt = L/2m$ which is a constant.

After the first two laws, Kepler spent another 12 years searching for a "harmony" in the motion of planets. The imperfection he injected in the planetary motions by the assumption of elliptical orbits prompted him to seek for some sort of compensation. His third law was precisely that. He felt that this law, with its precise mathematical structure, gave sufficient harmony to the waltz of planets around the Sun to offset the imperfection of elliptical

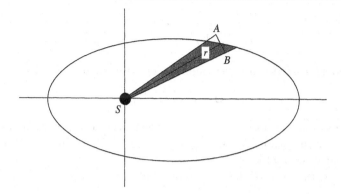

Figure 24.2: The shaded area is almost equal to the area of the triangle SAB.

orbits. **Kepler's third law** of planetary motion relates the period of each
planet to the length of its major axis. To derive it, we use Kepler's second   Kepler's third law
law:

$$T = \frac{\pi ab}{dA/dt} = \frac{\pi ab}{(L/2m)} = \frac{2\pi abm}{\sqrt{mK/ab}} = \frac{2\pi a^{3/2}m^{1/2}}{\sqrt{K}},$$

where we used Equation (24.41). For gravity, $K = GMm$, and squaring both
sides of the above equation gives

$$T^2 = \frac{4\pi^2 a^3}{GM}.$$

This is the mathematical statement of Kepler's third law.

24.6.3   The Inhomogeneous Case

When a driving force acts on a physical system, it will appear as the inhomogeneous term of the NOLDE. For the particular, but important, case in
which the inhomogeneous term is a product of polynomials and exponentials,
the solution can be found in closed form. This subsection shows how this is
done.

We assume that the inhomogeneous term in Equation (24.26) is of the
form $r(x) = \sum_k p_k(x)e^{\lambda_k x}$ where $p_k(x)$ are polynomials and λ_k are (complex)
constants. The most general solution of Equation (24.26) is a linear combi-
nation of a basis of solutions (as given in Box 24.6.1) of the homogeneous
NOLDE and a particular solution of the NOLDE. We need to find the latter.
Because $\mathbf{L}$ is a linear operator, it is clear that if y_1 is a particular solution
of $\mathbf{L}[y] = r_1(x)$ and y_2 that of $\mathbf{L}[y] = r_2(x)$, then $y_1 + y_2$ is a solution of
$\mathbf{L}[y] = r_1(x) + r_2(x)$. This suggests breaking up the inhomogeneous term into
smaller pieces. Thus, no generality is lost if we restrict $r(x)$ to be $p(x)e^{\lambda x}$
where $p(x)$ is a polynomial.

The reader may verify that, for any differentiable function f, we have

$$(\mathbf{D} - \lambda)[e^{\lambda x} f(x)] = e^{\lambda x} f'(x), \qquad (\mathbf{D} - \lambda)^2[e^{\lambda x} f(x)] = e^{\lambda x} f''(x),$$

and, in general,

$$(\mathbf{D} - \lambda)^k [e^{\lambda x} f(x)] = e^{\lambda x} \frac{d^k f}{dx^k}.$$

In particular, if $p(x)$ is a polynomial of degree n, then

$$(\mathbf{D} - \lambda)^k u = e^{\lambda x} p(x)$$

has a solution of the form $u = e^{\lambda x} q(x)$, where $q(x)$ is a polynomial of degree $n + k$ that is the primitive (indefinite integral) of $p(x)$ of order k [so that the kth derivative of $q(x)$ is $p(x)$].

If $\nu \neq \lambda$, then the reader may check that

$$(\mathbf{D} - \nu)[e^{\lambda x} f(x)] = e^{\lambda x} [(\lambda - \nu) f(x) + f'(x)]$$

and, therefore, $(\mathbf{D} - \nu)u = e^{\lambda x} p(x)$ has a solution of the form $u = e^{\lambda x} q(x)$, where $q(x)$ is a polynomial of degree k. Applying the last two equations repeatedly leads to

particular solution of nth order linear DE

> **Box 24.6.2.** *The NOLDE $\mathbf{L}[y] = e^{\lambda x} S(x)$, where $S(x)$ is a polynomial, has the particular solution $e^{\lambda x} q(x)$, where $q(x)$ is also a polynomial. The degree of $q(x)$ equals that of $S(x)$ unless $\lambda = \lambda_j$, a root of the characteristic polynomial of $\mathbf{L}$, in which case the degree of $q(x)$ exceeds that of $S(x)$ by k_j, the multiplicity of λ_j.*

Once we know the form of the particular solution of the NOLDE, we can find the coefficients in the polynomial of the solution by substituting in the NOLDE and matching the powers on both sides.

Example 24.6.2. We find the most general solutions of two differential equations subject to the boundary conditions $y(0) = 0$ and $y'(0) = 1$.
(a) The first DE we want to consider is

$$y'' + y = xe^x. \tag{24.42}$$

The characteristic polynomial is $\lambda^2 + 1$ whose roots are $\lambda_1 = i$ and $\lambda_2 = -i$. Thus, a basis of solutions is $\{\cos x, \sin x\}$. To find the particular solution we note that λ (the coefficient of x in the exponential part of the inhomogeneous term) is 1, which is neither of the roots λ_1 and λ_2. Thus, the particular solution is of the form $q(x)e^x$, where $q(x) = Ax + B$ is of degree 1 [same degree as that of $S(x) = x$]. We now substitute $u = (Ax + B)e^x$ in Equation (24.42) to obtain the relation

$$Axe^x + (2A + B)e^x + (Ax + B)e^x = xe^x.$$

Matching the coefficients, we have

$$2A = 1 \quad \text{and} \quad 2A + 2B = 0 \Rightarrow A = \tfrac{1}{2} = -B.$$

Thus, the most general solution is

$$y = c_1 \cos x + c_2 \sin x + \tfrac{1}{2}(x - 1)e^x.$$

Imposing the given boundary conditions yields $0 = y(0) = c_1 - \frac{1}{2}$ and $1 = y'(0) = c_2$. Thus,

$$y = \tfrac{1}{2} \cos x + \sin x + \tfrac{1}{2}(x - 1)e^x$$

is the unique solution.

(b) The next DE we want to consider is

$$y'' - y = xe^x. \tag{24.43}$$

Here $p(\lambda) = \lambda^2 - 1$, and the roots are $\lambda_1 = 1$ and $\lambda_2 = -1$. A basis of solutions is $\{e^x, e^{-x}\}$. To find a particular solution, we note that $S(x) = x$ and $\lambda = 1 = \lambda_1$. Box 24.6.2 then implies that $q(x)$ must be of degree 2 because λ_1 is a simple root, i.e., $k_1 = 1$. We, therefore, try

$$q(x) = Ax^2 + Bx + C \;\Rightarrow\; u = (Ax^2 + Bx + C)e^x.$$

Taking the derivatives and substituting in Equation (24.43) yields two equations,

$$4A = 1 \quad \text{and} \quad A + B = 0,$$

whose solution is $A = -B = 1/4$. Note that C is not determined, because Ce^x is a solution of the homogeneous DE corresponding to Equation (24.43), so when L is applied to u, it eliminates the term Ce^x. Another way of looking at the situation is to note that the most general solution to (24.43) is of the form

$$y = c_1 e^x + c_2 e^{-x} + (\tfrac{1}{4}x^2 - \tfrac{1}{4}x + C)e^x.$$

The term Ce^x could be absorbed in $c_1 e^x$. We, therefore, set $C = 0$, apply the boundary conditions, and find the unique solution

$$y = \tfrac{5}{4} \sinh x + \tfrac{1}{4}(x^2 - x)e^x. \qquad \blacksquare$$

The inhomogeneous DE (IDE) $\mathsf{L}[y] = r(x)$ can be thought of as a machine (or a black box) that produces a function $y(x)$ when a function $r(x)$ is fed into it. Such an interpretation is common in the study of electrical or acoustic filters. A signal, the function $r(x)$, is sent into the filter, and a second function, $y(x)$, is received as an output. In such a context, by far the most important input signal is a sinusoidal function of the general form $r(t) = A \cos(\omega t + \alpha)$, which, with $B = Ae^{i\alpha}$, can be written in complex notation as (see Example 18.2.3)

$$r(t) = \mathrm{Re}(R(t)) \quad \text{where} \quad R(t) \equiv Be^{i\omega t} = Ae^{i(\omega t + \alpha)},$$

where A, B, α, and ω, the angular frequency, are all constants, and t represents time (the independent variable). Assuming that $i\omega$ is not a root of $p(\lambda)$, the characteristic polynomial of L, Box 24.6.2 suggests a particular (complex) solution, $U = C(\omega)e^{i\omega t}$ where $C(\omega)$ is a (ω-dependent) constant. To determine it, we substitute U in $\mathsf{L}[U] = Be^{i\omega t}$:

$$\mathsf{L}[U] = \mathsf{L}[C(\omega)e^{i\omega t}] = C(\omega)\mathsf{L}[e^{i\omega t}] = C(\omega)p(i\omega)e^{i\omega t},$$

so that

$$\mathsf{L}[U] = Be^{i\omega t} \;\Rightarrow\; C(\omega)p(i\omega)e^{i\omega t} = Be^{i\omega t} \;\Rightarrow\; C(\omega) = \frac{B}{p(i\omega)}.$$

Writing the complex numbers in polar form

$$C(\omega) \equiv \rho(\omega)e^{i\gamma(\omega)}, \qquad B = Ae^{i\alpha}, \qquad p(i\omega) \equiv P(\omega)e^{i\theta(\omega)},$$

we obtain

$$\rho(\omega) = \frac{A}{P(\omega)} \qquad \text{and} \qquad \gamma(\omega) = \alpha - \theta(\omega).$$

The real solution, $u(t) = \text{Re}[U(t)]$, will then be

$$u(t) = \text{Re}[C(\omega)e^{i\omega t}] = \rho(\omega)\cos[\omega t + \gamma(\omega)]$$

$$= \frac{A}{P(\omega)}\cos[\omega t + \alpha - \theta(\omega)]. \qquad (24.44)$$

transfer function   The function $C(\omega)$ is called the **transfer function** associated with the linear operator **L**. Equation (24.44) shows that the output $u(t)$ has the same frequency as the input. It also indicates that the amplitude of $u(t)$ is frequency-dependent, making it possible to obtain large output amplitudes by varying the frequency until $P(\omega)$ is minimum. This is the phenomenon of resonance in AC circuits.

Example 24.6.3. Let us apply the analysis above to Example 24.6.1 and, for definiteness, take the underdamped case. In this case, $4b > a^2$; and $\omega_0 \equiv \sqrt{b}$ is called the **natural frequency** of the system. The characteristic polynomial is

natural frequency $p(\lambda) = \lambda^2 + a\lambda + b$. Thus,

$$p(i\omega) = -\omega^2 + i\omega a + b = (\omega_0^2 - \omega^2) + i\omega a$$

and

$$P(\omega) = \sqrt{(\omega_0^2 - \omega^2)^2 + \omega^2 a^2}, \qquad \theta(\omega) = \tan^{-1}\left(\frac{\omega a}{\omega_0^2 - \omega^2}\right).$$

gain function   The amplitude of the output signal, sometimes called the **gain function**, is

$$\rho(\omega) = \frac{A}{P(\omega)} = \frac{A}{\sqrt{(\omega_0^2 - \omega^2)^2 + \omega^2 a^2}}.$$

The minimum of the denominator occurs at $\omega = \omega_0$, that is, when the driving frequency equals the natural frequency. In such a situation we have $\rho(\omega) = A/(\omega_0 a)$, showing that the output signal will have a large amplitude when a, the damping coefficient, is small.

We have considered only the particular solution, $u(t)$, because the most general solution

$$y(t) = Ke^{-at/2}\cos(\omega_1 t + \beta) + u(t)$$

in which K and β are constants, eventually reduces to $u(t)$. The first term on the RHS, the **transient** term, decays to zero. The rate of this decay is determined by the **time constant** $2/a$, the time interval during which the amplitude of the transient term drops to $1/e$ of its initial value.   ■

transient term

time constant

The importance of the sinusoidal signal becomes clear when we recall that any periodic signal can be expanded in a Fourier series, $R(t) = \sum_{n=-\infty}^{\infty} b_n e^{in\omega t}$

where ω is the fundamental frequency. The linearity of $\mathbf{L}$ suggests the solution $u(t) = \mathrm{Re}[U(t)]$, where

$$U(t) = \sum_{n=-\infty}^{\infty} C_n(\omega)e^{in\omega t}.$$

Substituting in $\mathbf{L}[U] = R(t)$ gives

$$\sum_{n=-\infty}^{\infty} C_n(\omega)p(in\omega)e^{in\omega t} = \sum_{n=-\infty}^{\infty} b_n e^{in\omega t}.$$

Since the $e^{in\omega t}$ are orthonormal, we get $C_n(\omega) = b_n/[p(in\omega)]$, and

$$u(t) = \mathrm{Re}\left[\sum_{n=-\infty}^{\infty} \frac{b_n e^{in\omega t}}{p(in\omega)}\right].$$

Thus, $u(t)$ is also periodic and has the same fundamental frequency as $r(t)$.

24.7   Problems

24.1. Let f and g be two differentiable functions that are linearly dependent. Show that their Wronskian vanishes. (Note that f and g need not be solutions of a homogeneous SOLDE.)

24.2. Show that if (f_1, f_1') and (f_2, f_2') are linearly dependent at one point, then f_1 and f_2 are linearly dependent at all $a \le x \le b$. Here f_1 and f_2 are solutions of the DE of (24.4). Hint: Derive the identity

$$W(f_1, f_2; x_2) = W(f_1, f_2; x_1)\exp\left\{-\int_{x_1}^{x_2} p(t)\,dt\right\}.$$

24.3. Show by direct substitution that f_2 of Equation (24.6) indeed satisfies (24.4) no matter what K is.

24.4. Show that the solutions to the SOLDE $y'' + q(x)y = 0$ have a constant Wronskian.

24.5. Find a general integral formula for $G_n(x)$, the linearly independent "partner" of the Hermite polynomial $H_n(x)$ which satisfies the Hermite DE

$$y'' - 2xy' + 2ny = 0.$$

Specialize this to $n = 0, 1$. Is it possible to find $G_0(x)$ and $G_1(x)$ in terms of elementary functions?

24.6. Use Theorem 23.3.1 to construct

$$y = \frac{W(x)}{f_1^2(x)}\left[C + \int_a^x \frac{f_1(t)r(t)}{W(t)}\,dt\right],$$

a solution of

$$y' + \left(p + \frac{2f_1'}{f_1}\right)y = \frac{r}{f_1}.$$

24.7. Show that each pair of the following functions satisfy the DE next to it. Calculate the Wronskian, and give a solution satisfying the initial conditions $y(0) = 2$ and $y'(0) = 1$.

   (a) $\cos x$ and $\sin x$; $y'' + y = 0$.   (b) e^x and e^{3x}; $y'' + 4y' + 3y = 0$.

   (c) x and e^x; $y'' + \dfrac{x}{1-x}y' - \dfrac{1}{1-x}y = 0$.

24.8. For the HSOLDE $y'' + py' + qy = 0$, show that

$$p = -\frac{f_1 f_2'' - f_2 f_1''}{W(f_1, f_2)} \qquad \text{and} \qquad q = \frac{f_1' f_2'' - f_2' f_1''}{W(f_1, f_2)}.$$

Thus, knowing two solutions of an HSOLDE allows us to reconstruct the DE.

24.9. Show that the HSOLDE $y'' + py' + qy = 0$ can be cast in the form $u'' + S(x)u = 0$. Hint: Define $w(x)$ by $y = wu$, substitute in the DE, and demand that the coefficient of u' be zero to obtain

$$w(x) = C\exp\left[-\tfrac{1}{2}\int_\alpha^x p(t)\,dt\right].$$

Now show that the original DE can be written as $u'' + S(x)u = 0$ with

$$S(x) = q + p\frac{w'}{w} + \frac{w''}{w} = q - \tfrac{1}{4}p^2 - \tfrac{1}{2}p'.$$

24.10. Show that the adjoint of **M** given in Equation (24.14) is the original **L**.

24.11. Show that S-L equation (24.18) can be transformed into

$$\frac{d^2v}{dt^2} + [\lambda - Q(t)]v = 0,$$

by the so-called **Liouville substitution**, which changes both independent and dependent variables:

$$u(x) = v(t)[p(x)w(x)]^{-1/4}, \qquad t = \int_a^x \sqrt{\frac{w(s)}{p(s)}}\,ds.$$

Then

$$Q(t) = \frac{q(x(t))}{w(x(t))} + [p(x(t))w(x(t))]^{-1/4}\frac{d^2}{dt^2}[(pw)^{1/4}].$$

24.12. Show that

(a) the Liouville substitution (see Problem 24.11) transforms the Bessel DE $(xu')' + (k^2x - \nu^2/x)u = 0$ into

$$\frac{d^2v}{dt^2} + \left[k^2 - \frac{\nu^2 - 1/4}{t^2}\right]v = 0.$$

(b) Specialize to $\nu = \frac{1}{2}$ and show that

$$J_{1/2}(kt) = A\frac{\sin kt}{\sqrt{t}} + B\frac{\cos kt}{\sqrt{t}}.$$

(c) Use the fact that $J_\nu(x)$ is an analytic function of x to show that

$$J_{1/2}(kt) = A\frac{\sin kt}{\sqrt{t}}.$$

24.13. Show that the functions $x^r e^{\lambda x}$, where $r = 0, 1, 2, \ldots, k$, are linearly independent. Hint: Starting with $(\mathbf{D} - \lambda)^k$, apply powers of $\mathbf{D} - \lambda$ to a linear combination of $x^r e^{\lambda x}$ for all possible r's.

24.14. Suppose λ is a root of the polynomial

$$p(x) \equiv x^n + a_{n-1}x^{n-1} + \cdots + a_1 x + a_0,$$

where *all coefficients are real*. Show that λ^* is also a root of $p(x)$. Hint: Complex conjugate $p(\lambda) = 0$. Does the same result hold if the coefficients were complex?

24.15. Write Equation (24.39) in the more familiar Cartesian coordinates and show that $e = 0$ gives a circle, $0 < e < 1$ gives an ellipse, $e = 1$ gives a parabola, and $e > 1$ gives a hyperbola. Show that except for the case of a parabola, the Cartesian equation of the conic section is

$$\frac{(1-e^2)^2 K^2 m^2}{L^4}\left(x - \frac{L^2 e}{Km(1-e^2)}\right)^2 + \frac{(1-e^2)K^2 m^2}{L^4}y^2 = 1.$$

24.16. Derive all the formulas in Equation (24.41).

24.17. Find a basis of real solutions for each DE:

(a) $y'' + 5y' + 6 = 0.$ (b) $y''' + 6y'' + 12y' + 8y = 0.$

(c) $y^{(4)} = y.$ (d) $y^{(4)} = -y.$

24.18. Solve the following DEs subject to the given initial conditions.
(a) $y^{(4)} = y$, $y(0) = y'(0) = y'''(0) = 0, y''(0) = 1.$
(b) $y^{(4)} + y'' = 0$, $y(0) = y''(0) = y'''(0) = 0, y'(0) = 1.$
(c) $y^{(4)} = 0$, $y(0) = y'(0) = y''(0) = 0, y'''(0) = 2.$

24.19. Solve $y'' - 2y' + y = xe^x$ subject to the initial conditions $y(0) = 0, y'(0) = 1$.

24.20. Find the general solution of each equation:

(a) $y'' = xe^x$.

(b) $y'' - 4y' + 4y = x^2$.

(c) $y'' + y = \sin x \sin 2x$.

(d) $y'' - y = (1 + e^{-x})^2$.

(e) $y'' - y = e^x \sin 2x$.

(f) $y^{(6)} - y^{(4)} = x^2$.

(g) $y'' - 4y' + 4 = e^x + xe^{2x}$.

(h) $y'' + y = e^{2x}$.

24.21. Consider the Euler equation

$$x^n y^{(n)} + a_{n-1} x^{n-1} y^{(n-1)} + \cdots + a_1 x y' + a_0 y = r(x).$$

Substitute $x = e^t$ and show that such a substitution reduces this to a DE with constant coefficients. In particular, solve $x^2 y'' - 4xy' + 6y = x$.

24.22. Show that $v = C_1 \cos \theta + C_2 \sin \theta$ can be written as $v = A \cos(\theta - \theta_0)$. Find A and θ_0 in terms of C_1 and C_2.

24.23. (a) Show that the extremum (maximum or minimum) of the function

$$y(t) = c_1 t e^{-at/2} + c_0 e^{-at/2}$$

occurs at $t = 2/a - c_0/c_1$.
(b) Prove that if $c_1 > 0$, the extremum is maximum and if $c_1 < 0$, it is minimum.

24.24. Verify that, for any differentiable function f, we have

$$(\mathbf{D} - \lambda)[e^{\lambda x} f(x)] = e^{\lambda x} f'(x)$$

and if $\nu \neq \lambda$, then

$$(\mathbf{D} - \nu)[e^{\lambda x} f(x)] = e^{\lambda x}[(\lambda - \nu)f(x) + f'(x)].$$

24.25. Derive Equation (24.44).

Chapter 25

Laplace's Equation: Cartesian Coordinates

In Chapter 22 we discussed the technique of the separation of variables for the most important PDEs encountered in introductory physics and engineering courses. One such PDE deserving special attention is the Laplace equation

$$\nabla^2 \Phi = 0 \tag{25.1}$$

which shows up extensively in problems in electrostatics and steady-state heat conduction. The latter arises in situations in which the temperature does not change with time, so that the LHS of Equation (22.3) vanishes.

Aside from its significance in applications, Laplace's equation is important because its solution leads naturally to some of the most famous functions of mathematical physics. In fact, when separating this equation in various coordinate systems, one obtains not only such elementary functions as sines and cosines, but also the more advanced "special functions" such as Legendre polynomials and the Bessel functions. At the heart of such functions is the linearity of Laplace's equation which allows summing a (infinite) number of solutions to get a new solution. This leads naturally to solutions of Laplace's equation in terms of infinite series.

In a typical situation, Φ is given on some surfaces bounding a volume in space and its value is sought for all points in the volume. When the bounding surfaces are arbitrarily shaped, the solution can be found only by numerical techniques; but when they are primary surfaces of a coordinate system, then we can generally solve the problem by separating Laplace's equation in the appropriate coordinate system.

25.1   Uniqueness of Solutions

We shall see many examples of solutions to Laplace's equation in various coordinate systems in this and the following chapters. All of these solutions will be obtained in the form of infinite series. So, we know that solutions to Laplace's equation indeed exist. What we want to do in this section is to show that the solution which satisfies all the boundary conditions is unique. In other words, no matter how we find the solution, as long as it satisfies the boundary condition, it is *the* solution of Laplace's equation. In fact, we can be more general and prove the uniqueness for the *Poisson* equation $\nabla^2\Phi = \rho$.

Consider the volume V with some surfaces bounding it. Figure 25.1 shows two such volumes. Assume that two functions Φ_1 and Φ_2 satisfy the Poisson equation at every point of the volume, and that they both satisfy some other conditions related to the surfaces which we shall look into shortly. Let $\Phi = \Phi_1 - \Phi_2$ and note that Φ satisfies *Laplace's* equation because

$$\nabla^2\Phi = \nabla^2(\Phi_1 - \Phi_2) = \nabla^2\Phi_1 - \nabla^2\Phi_2 = \rho - \rho = 0.$$

For any function f, we have [see Equation (14.11)]

$$\nabla \cdot (f\nabla f) = \nabla f \cdot \nabla f + f\nabla^2 f = |\nabla f|^2 + f\nabla^2 f.$$

For Φ—since it satisfies Laplace's equation—we get

$$\nabla \cdot (\Phi\nabla\Phi) = \nabla\Phi \cdot \nabla\Phi + \Phi\underbrace{\nabla^2\Phi}_{=0} = |\nabla\Phi|^2.$$

Integrating both sides of the last equation over the volume V and using the divergence theorem on the LHS yields

$$\iint\limits_{S} (\Phi\nabla\Phi) \cdot \hat{\mathbf{e}}_n \, da = \iint\limits_{S} (\Phi_1 - \Phi_2)(\hat{\mathbf{e}}_n \cdot \nabla\Phi_1 - \hat{\mathbf{e}}_n \cdot \nabla\Phi_2) \, da$$

$$= \iiint\limits_{V} |\nabla\Phi|^2 dV. \tag{25.2}$$

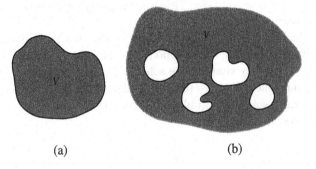

(a)   (b)

Figure 25.1: A volume (shaded region) with its bounding surface. (a) The volume is "inside" the bounding surface. (b) The volume is "outside" the bounding surface(s).

Now suppose:

- **Dirichlet boundary condition**: Φ_1 and Φ_2 take on the same value at every point of the bounding surface(s), i.e., $\Phi_1 - \Phi_2 = 0$ on S; or

  Dirichlet boundary condition

- **Neumann boundary condition**: the so-called **normal derivatives** $\hat{\mathbf{e}}_n \cdot \nabla\Phi_1$ and $\hat{\mathbf{e}}_n \cdot \nabla\Phi_2$ take on the same value at every point of the bounding surface(s), i.e., $\hat{\mathbf{e}}_n \cdot \nabla\Phi_1 - \hat{\mathbf{e}}_n \cdot \nabla\Phi_2 = 0$ on S.

  Neumann boundary condition; normal derivatives

Then, in either case, the first line of Equation (25.2) yields zero. Since the integrand of the RHS is never negative, the integrand must vanish.[1] It follows that

$$|\nabla\Phi|^2 = 0 \;\Rightarrow\; \nabla\Phi = 0 \;\Rightarrow\; \Phi = \text{constant} \;\Rightarrow\; \Phi_1 - \Phi_2 = \text{constant}$$

for all points in the volume V. Since $\Phi = 0$ on the bounding surface, the constant must be zero, i.e., $\Phi_1 = \Phi_2$ for all points in the volume V. We thus have

> **Box 25.1.1.** *Let V be a volume bounded by a (possibly disconnected) surface S. Then there exists a unique function which satisfies both Laplace's equation (or the Poisson equation) at every point of V and either Dirichlet or Neumann boundary conditions on S.*

Historical Notes

Pierre Simon de Laplace was a French mathematician and theoretical astronomer who was so famous in his own time that he was known as the **Newton** of France. His main interests throughout his life were celestial mechanics, the theory of probability, and personal advancement.

At the age of 24 he was already deeply engaged in the detailed application of Newton's law of gravitation to the solar system as a whole, in which the planets and their satellites are not governed by the Sun alone, but interact with one another in a bewildering variety of ways. Even Newton had been of the opinion that divine intervention would occasionally be needed to prevent this complex mechanism from degenerating into chaos. Laplace decided to seek reassurance elsewhere, and succeeded in proving that the ideal solar system of mathematics is a stable dynamical system that will endure unchanged for all time. This achievement was only one of the long series of triumphs recorded in his monumental treatise *Mécanique Céleste* (published in five volumes from 1799 to 1825), which summed up the work on gravitation of several generations of illustrious mathematicians. Unfortunately for his later reputation, he omitted all reference to the discoveries of his predecessors and contemporaries, and left it to be inferred that the ideas were entirely his own. Many anecdotes are associated with this work. One of the best known describes the occasion on which **Napoleon** tried to get a rise out of Laplace by protesting that

---

[1]The integral is the limit of a sum. If no term of this sum is negative, and the sum equals zero, then each term of the sum must be zero.

Pierre Simon de
Laplace 1749–1827

he had written a huge book on the system of the world without once mentioning God as the author of the universe. Laplace is supposed to have replied, "Sire, I had no need of that hypothesis." The principal legacy of the *Mécanique Céleste* to later generations lay in Laplace's wholesale development of potential theory, with its far-reaching implications for a dozen different branches of physical science ranging from gravitation and fluid mechanics to electromagnetism and atomic physics. Even though he lifted the idea of the potential from **Lagrange** without acknowledgment, he exploited it so extensively that ever since his time the fundamental equation of potential theory has been known as Laplace's equation.

After the French Revolution, Laplace's political talents and greed for position came to full flower. His compatriots speak ironically of his "suppleness" and "versatility" as a politician. What this really means is that each time there was a change of regime (and there were many), Laplace smoothly adapted himself by changing his principles—back and forth between fervent republicanism and fawning royalism—and each time he emerged with a better job and grander titles. He has been aptly compared with the apocryphal Vicar of Bray in English literature, who was twice a Catholic and twice a Protestant. The Vicar is said to have replied as follows to the charge of being a turncoat: "Not so, neither, for if I changed my religion, I am sure I kept true to my principle, which is to live and die the Vicar of Bray."

To balance his faults, Laplace was always generous in giving assistance and encouragement to younger scientists. From time to time he helped forward in their careers such men as the chemist **Gay–Lussac**, the traveler and naturalist *Humboldt*, the physicist **Poisson**, and—appropriately—the young **Cauchy**, who was destined to become one of the chief architects of nineteenth-century mathematics.

25.2   Cartesian Coordinates

The separation of Laplace's equation in Cartesian coordinates is obtained from Equation (22.12) by setting the constant C equal to zero.[2] This leads to the following three equations[3]

$$\frac{d^2 X}{dx^2} - \alpha_1 X = 0, \qquad \frac{d^2 Y}{dy^2} - \alpha_2 Y = 0, \qquad \frac{d^2 Z}{dz^2} + (\alpha_1 + \alpha_2)Z = 0, \quad (25.3)$$

where the α's could be any number (including zero and complex). The specific value that each α takes on depends on the boundary conditions (BCs). We consider bounding surfaces parallel to the planes of the Cartesian coordinates.

The most effective way of learning how to solve Laplace's equation is to go into the details of the solution of a number of specific examples. We do so in the following, hoping that the reader will examine these examples very carefully, taking note of steps taken with an eye on how each step would change in a different situation (different BCs, etc).

semi-infinite
electrically
conducting plates

Example 25.2.1. Two semi-infinite conducting plates starting on the y-axis and parallel to the x-axis are grounded (the potential Φ is zero on them) and separated by

---

[2]Recall from Subsection 22.2 that $\Phi(x, y, z) = X(x)Y(y)Z(z)$.

[3]We have changed the sign of the α's to illustrate how the boundary conditions force on us the correct functional form of X, Y, and Z.

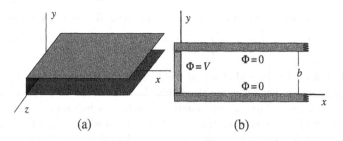

Figure 25.2: (a) The semi-infinite plates, and (b) the cross section of the two (grounded) plates and the strip maintained at potential V.

a distance b [Figure 25.2(a)]. Both plates extend from $-\infty$ to ∞ in the z-direction. A conducting strip of width b—also infinite in both directions of the z-axis—is located between the two plates and separated from them by an infinitesimal gap, so that the strip can be maintained at a different potential of $\Phi = V$. Figure 25.2(b) shows the cross section of the geometry of the problem. We want to find the potential in the region enclosed by the conductors.

The potential is independent of z because, as a small observer moves along the z-axis keeping the other two coordinates fixed, his detectors and instruments will not detect any change in the physics of the problem, because the physical environment of the detectors remains unchanged. So, $Z(z)$ is a constant which we absorb in $X(x)$ or $Y(y)$. Furthermore, substituting $Z = \text{const.}$ in the third equation of (25.3) yields $\alpha_1 + \alpha_2 = 0$.

Symmetry tells us that the potential is independent of z.

Thus the problem is reduced to finding $X(x)$ and $Y(y)$ which satisfy the differential equations of (25.3). First let us consider the Y equation. If $\alpha_2 = 0$, then the solution will be of the form

$$Y(y) = Ay + B.$$

The case of $\alpha_2 \neq 0$ is a SOLDE with constant coefficients whose most general solution can be written as

$$Y(y) = Ae^{\sqrt{\alpha_2}\,y} + Be^{-\sqrt{\alpha_2}\,y}. \tag{25.4}$$

The vanishing of Φ at $y = 0$ and $y = b$ means that

$$\Phi(x,0) = X(x)Y(0) = 0 \quad \text{for all } x \;\Rightarrow\; Y(0) = 0,$$
$$\Phi(x,b) = X(x)Y(b) = 0 \quad \text{for all } x \;\Rightarrow\; Y(b) = 0.$$

Therefore, for the case of $\alpha_2 = 0$, this implies

$$Y(0) = A \times 0 + B = 0 \;\Rightarrow\; B = 0,$$
$$Y(b) = Ab + B = Ab + 0 = 0 \;\Rightarrow\; A = 0.$$

Thus, if $\alpha_2 = 0$, we get $Y(y) = 0$ and $\Phi(x,y) = X(x)Y(y) = 0$ which is the trivial solution.

It follows that if we are interested in nontrivial solutions, we had better assume that $\alpha_2 \neq 0$. Then, Equation (25.4) gives

$$Y(0) = A + B = 0 \quad \text{and} \quad Y(b) = Ae^{\sqrt{\alpha_2}\,b} + Be^{-\sqrt{\alpha_2}\,b} = 0.$$

Multiplying the second equation by $e^{\sqrt{\alpha_2}\,b}$ and using $B = -A$, we obtain

$$A\left[e^{2\sqrt{\alpha_2}\,b} - 1\right] = 0 \;\Rightarrow\; A = 0 \;\text{ or }\; e^{2\sqrt{\alpha_2}\,b} = 1.$$

Boundary conditions force α_2 to be imaginary.

The first choice ($A = 0$) and $A = -B$ yields a trivial solution again. Therefore, we have to assume that the second choice holds. However, even with the second choice, if we restrict ourselves to the real numbers, the only solution of $e^{2\sqrt{\alpha_2}\,b} = 1$ would be $\alpha_2 = 0$ which is a contradiction because we are dealing precisely with the case of $\alpha_2 \neq 0$. It follows that $\sqrt{\alpha_2}$ must be a *complex number*. In fact, recalling that $e^{2in\pi} = 1$ for any integer n, we immediately get

$$2\sqrt{\alpha_2}\,b = 2in\pi \;\Rightarrow\; \sqrt{\alpha_2}\,b = in\pi \;\Rightarrow\; \alpha_2 = -\left(\frac{n\pi}{b}\right)^2, \qquad n = \pm 1, \pm 2, \ldots.$$

Note that $n = 0$ is excluded because this choice would make $\alpha_2 = 0$.

We now turn to the X equation. Since $\alpha_1 + \alpha_2 = 0$, we obtain

$$\alpha_1 = -\alpha_2 = \left(\frac{n\pi}{b}\right)^2, \qquad n = \pm 1, \pm 2, \ldots,$$

and

$$\frac{d^2 X}{dx^2} - \left(\frac{n\pi}{b}\right)^2 X = 0 \;\Rightarrow\; X(x) = Ce^{n\pi x/b} + De^{-n\pi x/b}.$$

To be physically meaningful, the potential must remain finite as $x \to +\infty$. It follows from the last equation that either n is negative and $D = 0$, or n is positive and $C = 0$. Either choice will lead to the same final result as the reader may verify. Choosing positive values of n with $C = 0$, the potential can be written as

$$\Phi_n(x, y) = ADe^{-n\pi x/b}\left[e^{in\pi y/b} - e^{-in\pi y/b}\right] = A_n e^{-n\pi x/b} \sin\left(\frac{n\pi y}{b}\right),$$

where we used $A = -B$ and introduced a new constant A_n. We also subscripted the potential because for every n, we get a different function for Φ. All such functions are solutions of Laplace's equation and therefore, so is their sum. In fact, it is only the sum that is general enough to result in the final solution. We thus write

$$\Phi(x, y) = \sum_{n=1}^{\infty} \Phi_n(x, y) = \sum_{n=1}^{\infty} A_n e^{-n\pi x/b} \sin\left(\frac{n\pi y}{b}\right). \tag{25.5}$$

This is a Fourier series in y with x dependent coefficients. The potential will be completely determined if the constants A_n can be determined. This is where the last unused information comes in: The potential at $x = 0$ is V. Substituting this information in Equation (25.5) yields

$$V = \Phi(0, y) = \sum_{n=1}^{\infty} A_n \sin\left(\frac{n\pi y}{b}\right)$$

from which A_n can be determined using the Fourier series techniques. We leave it for the reader to show that $A_n = 2V[1 - (-1)^n]/(n\pi)$ (see Problem 25.1), or

$$A_n = \begin{cases} \dfrac{4V}{n\pi} & \text{if } n \text{ is odd}, \\[2ex] 0 & \text{if } n \text{ is even}. \end{cases}$$

By writing $n = 2k + 1$ with $k = 0, 1, 2, \ldots$, the potential in the region of interest becomes

$$\Phi(x, y) = \frac{4V}{\pi} \sum_{k=0}^{\infty} \frac{e^{-(2k+1)\pi x/b}}{2k+1} \sin\left[\frac{(2k+1)\pi y}{b}\right] \qquad (25.6)$$

$$= \frac{4V}{\pi}\left[e^{-\pi x/b}\sin\frac{\pi y}{b} + \frac{e^{-3\pi x/b}}{3}\sin\frac{3\pi y}{b} + \frac{e^{-5\pi x/b}}{5}\sin\frac{5\pi y}{b} + \cdots\right].$$

Because of the exponential factor, the series converges very rapidly, and for large values of x the potential very quickly drops to zero. Figure 25.3 shows the potential function (in arbitrary units) as a function of x and y.   ∎

Example 25.2.1 illustrates the general feature of solving Laplace's equation by the separation of variables in Cartesian coordinates. This feature works in other coordinate systems as well. The separation of variables results in some ODEs which involve parameters (in the case above, the α's) to be determined by some of the BCs. All values of these parameters—which in all cases of interest to us will turn out to be integers—consistent with the used boundary conditions are allowed and must be taken into account, i.e., an infinite sum (with as yet undetermined coefficients) over such parameters is to be formed as the most general solution of Laplace's equation. By applying the remaining BCs, the undetermined coefficients can be evaluated, resulting in the unique solution appropriate for the geometry of the problem. If the geometry extends to infinity in a certain direction, then such an infinity is to be considered as a BC. It is extremely useful to take into account any symmetry of the problem as such symmetries will simplify the solution considerably. The symmetry in the z-direction of Example 25.2.1 saved us the trouble of solving one (out of three) complete ODE.

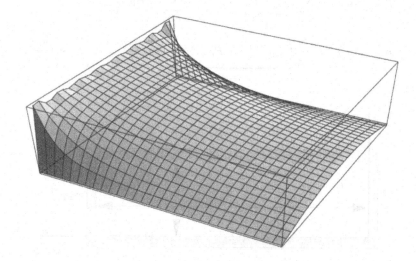

Figure 25.3: The potential function inside the semi-infinite box of Figure 25.2 when only 20 terms of the infinite series are kept. Note how quickly the potential drops to zero along the x-axis due to the exponential factor.

Example 25.2.2. Steady-state heat conduction problems also obey Laplace's equation. So, let us consider a rectangular medium infinite in the z direction enclosed by two pairs of parallel slabs of width a and b as shown in Figure 25.4. The temperatures of the slabs of width a—assumed parallel to the x-axis—are zero. The temperatures of the other two slabs are T_1 and T_2. We want to find the temperature at all points in the region enclosed after the equilibrium is reached.

As in Example 25.2.1, we can ignore the z-dependence and write $T(x,y) = X(x)Y(y)$ where X and Y satisfy Equation (25.3) with $\alpha_1 = -\alpha_2$. For exactly the same reason as in Example 25.2.1, α_2 cannot be zero and Y can only be of the form

$$Y(y) = A_n \sin \frac{n\pi y}{b}, \qquad n = 1, 2, \ldots,$$

where the subscript on A_n reminds us that different constants can be chosen to multiply different sine functions. The solution for X will, however, be different. We still have

$$X(x) = C_n e^{n\pi x/b} + D_n e^{-n\pi x/b}, \qquad n = 1, 2, \ldots,$$

but neither C_n nor D_n is zero this time. Multiplying the two functions and redefining the constants, we can write

$$T_n(x,y) = \left(A_n e^{n\pi x/b} + B_n e^{-n\pi x/b} \right) \sin \frac{n\pi y}{b}$$

and the most general infinite series solution becomes

$$T(x,y) = \sum_{n=1}^{\infty} \left(A_n e^{n\pi x/b} + B_n e^{-n\pi x/b} \right) \sin \frac{n\pi y}{b}. \qquad (25.7)$$

So far, we have used only two of the four BCs. The remaining two will determine the unknowns A_n and B_n. Substituting these BCs yields the following two equations:

$$T_1 = T(0,y) = \sum_{n=1}^{\infty} \underbrace{(A_n + B_n)}_{\equiv E_n} \sin \frac{n\pi y}{b} = \sum_{n=1}^{\infty} E_n \sin \frac{n\pi y}{b},$$

$$T_2 = T(a,y) = \sum_{n=1}^{\infty} \underbrace{\left(A_n e^{n\pi a/b} + B_n e^{-n\pi a/b} \right)}_{\equiv F_n} \sin \frac{n\pi y}{b} = \sum_{n=1}^{\infty} F_n \sin \frac{n\pi y}{b},$$

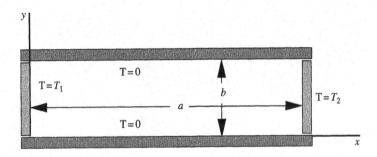

Figure 25.4: The cross section of the two pairs of parallel slabs maintained at different temperatures.

where we have redefined the constants multiplying the sine functions. As in Example 25.2.1 and Problem 25.1, we have

$$E_n = \frac{2T_1}{n\pi}[1 - (-1)^n], \qquad F_n = \frac{2T_2}{n\pi}[1 - (-1)^n].$$

These relations show that only odd terms of the infinite series are of relevance, and they are given by

$$E_{2k+1} \equiv A_{2k+1} + B_{2k+1} = \frac{4T_1}{\pi(2k + 1)}, \tag{25.8}$$

$$F_{2k+1} \equiv A_{2k+1}e^{(2k+1)\pi a/b} + B_{2k+1}e^{-(2k+1)\pi a/b} = \frac{4T_2}{\pi(2k + 1)}.$$

These are two equations in two unknowns which can be solved to get

$$A_{2k+1} = \frac{2(T_2 - T_1 e^{-(2k+1)\pi a/b})}{\pi(2k + 1)\sinh[(2k + 1)\pi a/b]},$$

$$B_{2k+1} = \frac{2(T_1 e^{(2k+1)\pi a/b} - T_2)}{\pi(2k + 1)\sinh[(2k + 1)\pi a/b]}.$$

Substituting in Equation (25.7)—with n replaced by $2k+1$—and rearranging terms yields

$$T(x,y) = \frac{4}{\pi}\sum_{k=0}^{\infty} \frac{T_1 \sinh\left[\frac{(2k+1)\pi(a-x)}{b}\right] + T_2 \sinh\left[\frac{(2k+1)\pi x}{b}\right]}{(2k + 1)\sinh\left[\frac{(2k+1)\pi a}{b}\right]} \sin\frac{(2k + 1)\pi y}{b}. \tag{25.9}$$

The reader is urged to verify that when $T_2 = 0$ and $a \to \infty$, we recover the result of Example 25.2.1—with V replaced by T_1—as we should. Figure 25.5 shows the potential function (in arbitrary units) as a function of x and y. ∎

The examples treated so far may give the impression that α_1 or α_2 is never zero. This has to do with the specific BCs imposed on Φ (or T). In

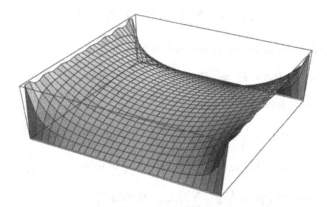

Figure 25.5: The potential function inside the box of Figure 25.4 for the special case of $a = b$ and $T_1 = T_2$ when only 20 terms of the infinite series are kept.

both examples, Y vanishes at both $y = 0$ and $y = b$. Such a BC excludes $\alpha_2 = 0$ because the corresponding Y, namely $Y = Ay + B$, cannot satisfy those conditions unless $Y = 0$ identically.

Example 25.2.3. To see how the $\alpha_1 = -\alpha_2 = 0$ terms can enter in the game, let us modify the temperatures of the plates and strips of Example 25.2.2 so that the bottom plate and the left strip are held at $T = 0$ while the top plate is held at T_1 and the right strip at T_2.

Let us write the most general solution of Laplace's equation obtained from separating variables *including* the $\alpha_1 = 0 = -\alpha_2$ term. Since the nonzero α_1 and α_2 are of opposite signs, one of them will be positive and will have real square roots and the other pure imaginary roots. Let us assume that α_1 is positive. Then X will be of exponential type and Y of imaginary exponential or trigonometric type. It follows that the most general solution of Laplace's equation can be written as

$$T(x,y) = (A_0 x + B_0)(C_0 y + D_0) \tag{25.10}$$
$$+ \sum_{\alpha}^{\infty} \left(A_\alpha e^{\sqrt{\alpha}\,x} + B_\alpha e^{-\sqrt{\alpha}\,x} \right) \left[C_\alpha \sin(\sqrt{\alpha}\,y) + D_\alpha \cos(\sqrt{\alpha}\,y) \right],$$

where we have used α for $\alpha_1 = -\alpha_2$. It is convenient to impose the y BCs first. So, since $T(x,0) = 0$, we have

$$0 = (A_0 x + B_0) D_0 + \sum_{\alpha}^{\infty} \left(A_\alpha e^{\sqrt{\alpha}\,x} + B_\alpha e^{-\sqrt{\alpha}\,x} \right) D_\alpha$$

which should hold for arbitrary values of x. This can happen only if $D_0 = D_\alpha = 0$. So, absorbing the multiplicative constant C_0 and C_α into the A's and B's, we get a new expression for the temperature:

$$T(x,y) = (A_0 x + B_0) y + \sum_{\alpha}^{\infty} \left(A_\alpha e^{\sqrt{\alpha}\,x} + B_\alpha e^{-\sqrt{\alpha}\,x} \right) \sin(\sqrt{\alpha}\,y).$$

The other y BC gives

$$T_1 = (A_0 x + B_0) b + \sum_{\alpha}^{\infty} \left(A_\alpha e^{\sqrt{\alpha}\,x} + B_\alpha e^{-\sqrt{\alpha}\,x} \right) \sin(\sqrt{\alpha}\,b).$$

The importance of $\alpha_1 = 0$ term is displayed here by the relation between B_0 and T_1.

For this to hold for arbitrary x, we need to have

$$A_0 = 0, \qquad B_0 b = T_1 \Rightarrow B_0 = \frac{T_1}{b}, \qquad \sin(\sqrt{\alpha}\,b) = 0 \Rightarrow \alpha = \left(\frac{n\pi}{b}\right)^2.$$

The temperature function reduces to

$$T(x,y) = \frac{T_1}{b} y + \sum_{n=1}^{\infty} \left(A_n e^{n\pi x/b} + B_n e^{-n\pi x/b} \right) \sin\frac{n\pi y}{b}.$$

We now impose the other two BCs. These will give us the following two equations:

$$0 = T(0,y) = \frac{T_1}{b} y + \sum_{n=1}^{\infty} (A_n + B_n) \sin\frac{n\pi y}{b},$$

$$T_2 = T(a,y) = \frac{T_1}{b} y + \sum_{n=1}^{\infty} \left(A_n e^{n\pi a/b} + B_n e^{-n\pi a/b} \right) \sin\frac{n\pi y}{b}.$$

Multiplying both sides of these equations by $\sin(m\pi y/b)$ and integrating from 0 to b yields the following two equations for A_m and B_m:

$$A_m + B_m = -\frac{2T_1}{b^2} \int_0^b y \sin\frac{m\pi y}{b}\, dy = \frac{2T_1}{m\pi}(-1)^m,$$

$$A_m e^{m\pi a/b} + B_m e^{-m\pi a/b} = \frac{2}{b} \int_0^b \left(T_2 - \frac{2T_1}{b}y\right) \sin\frac{m\pi y}{b}\, dy \qquad (25.11)$$

$$= \frac{2}{m\pi}\left[T_2 + (-1)^m(T_1 - T_2)\right].$$

These two equations can be solved to obtain the remaining unknown coefficients A_m and B_m. ∎

A couple of remarks are in order. The preceding example illustrated clearly the importance of the $\alpha_1 = 0$ term: Had we not included it in the expansion of T, we would not have obtained the answer. This is overlooked in most elementary treatments of Laplace's equation. It is worthwhile to emphasize this point.

> **Box 25.2.1.** *Always start with the most general solution of Laplace's equation, including the term corresponding to the case in which the constants of the separation of variables are zero, as given in Equation (25.10). Then apply the BCs, keeping in mind that there may be a preferred order for such an application.*

In Example 25.2.3, the order in which the y BCs were applied first was the preferred choice.

The second remark has to do with the choice of the functional form of X and Y. In Example 25.2.3, we chose X to be exponential and Y to be trigonometric. We could just as well have chosen Y to be exponential and X to be trigonometric. The *appearance* of the series would have changed, but the value of T at any point in the region of interest would have been the same for both series. This is due to the uniqueness of the solution of Laplace's equation.[4]

Example 25.2.4. The examples treated so far have been exclusively in two dimensions. We now consider a three-dimensional problem. Although this particular problem can be solved more quickly by relying on our intuition (as we did in Example 25.2.1, for example), we shall start from the most general solution, as prescribed by Box 25.2.1.

Suppose that the four lateral sides of widths a and b of a semi-infinite rectangular conducting tube are grounded and the closed base is held at potential V. The cross section of this tube is shown in Figure 25.4 where it is assumed that the tube starts at $z = 0$ and extends to infinity in the positive z-direction. We are interested in finding the potential inside this tube.

a three-dimensional example of the application of Laplace's equation

---

[4]The representation of the same function by different series should be familiar to the reader from calculus where $f(x)$ can be written as a Taylor expansion about *any point* in its domain of definition. Although such expansions look different, they all represent the same function.

We start with Equation (25.3) which holds for all solutions of Laplace's equation in Cartesian coordinates. There are four different cases to consider:

1. $\alpha_1 = 0 = \alpha_2$: In this case, $X(x)$ is of the generic form $Ax + B$, and with y or z replacing x, this is also the generic form of Y and Z. Let us denote these solutions as X_0, Y_0, and Z_0.

2. $\alpha_1 = 0$, $\alpha_2 \neq 0$: In this case, $X(x)$ is of the generic form $Ax + B$. But Y and Z are either exponential or trigonometric. Let us denote these solutions as $\mathfrak{X}_0$, Y_{α_2}, and Z_{α_2}.

3. $\alpha_1 \neq 0$, $\alpha_2 = 0$: In this case, $Y(y)$ is of the generic form $Ay + B$. But X and Z are either exponential or trigonometric. Let us denote these solutions as $\mathfrak{Y}_0$, X_{α_1}, and Z_{α_1}.

4. $\alpha_1 \neq 0$, $\alpha_2 \neq 0$: In this case X, Y, and Z are either exponential or trigonometric. Let us denote these solutions as $\mathfrak{X}_{\alpha_1}$, $\mathfrak{Y}_{\alpha_2}$, and $\mathfrak{Z}_{\alpha_1+\alpha_2}$.

The most general solution for the potential, encompassing all values of α_1 and α_2, is

$$\Phi(x,y,z) = X_0(x)Y_0(y)Z_0(z) + \mathfrak{X}_0(x)\sum_{\alpha_2} Y_{\alpha_2}(y)Z_{\alpha_2}(z) \qquad (25.12)$$

$$+ \mathfrak{Y}_0(y)\sum_{\alpha_1} X_{\alpha_1}(x)Z_{\alpha_1}(z) + \sum_{\alpha_1 \neq 0}\sum_{\alpha_2 \neq 0} \mathfrak{X}_{\alpha_1}(x)\mathfrak{Y}_{\alpha_2}(y)\mathfrak{Z}_{\alpha_1+\alpha_2}(z).$$

Boundary
conditions severely
restrict the terms
of the infinite
sums in (25.12).

We now apply the BCs. Since $\Phi(0,y,z) = \Phi(a,y,z) = 0$ for arbitrary y and z, and since each term in Equation (25.12) is independent of all others, we conclude that $X_0(0) = 0 = X_0(a)$ and $\mathfrak{X}_0(0) = 0 = \mathfrak{X}_0(a)$. It follows that A and B are both zero for X_0 and $\mathfrak{X}_0$. So, $X_0(x) = 0 = \mathfrak{X}_0(x)$. Similarly, $\mathfrak{Y}_0(y) = 0$, and Φ is reduced to the last term (the double sum) of (25.12). Furthermore, since both $\mathfrak{X}_{\alpha_1}$ and $\mathfrak{Y}_{\alpha_2}$ vanish at the two ends of their respective ranges, we expect them to be periodic, i.e., of trigonometric type. So, the most general solution is now

$$\Phi(x,y,z) = \sum_{\alpha_1,\alpha_2} [A_{\alpha_1}\cos(\sqrt{\alpha_1}\,x) + B_{\alpha_1}\sin(\sqrt{\alpha_1}\,x)]$$

$$\cdot [C_{\alpha_2}\cos(\sqrt{\alpha_2}\,y) + D_{\alpha_2}\sin(\sqrt{\alpha_2}\,y)]\,\mathfrak{Z}_{\alpha_1+\alpha_2}(z).$$

If this is to vanish at $x = 0$ for arbitrary y and z, then A_{α_1} must be zero; and if $\Phi(a,y,z) = 0$ for all y and z, then all coefficients of the product of the y and z functions in the sum must be zero. These coefficients—after setting A_{α_1} equal to zero—are of the form $\sin(\sqrt{\alpha_1}\,a)$. It follows that

$$\sqrt{\alpha_1}\,a = m\pi \;\Rightarrow\; \alpha_1 = \left(\frac{m\pi}{a}\right)^2, \qquad m = 1,2,\ldots,$$

where we have excluded the negative values of m as in the previous examples. An entirely analogous reasoning leads to $C_{\alpha_2} = 0$ and

$$\sqrt{\alpha_2}\,b = n\pi \;\Rightarrow\; \alpha_2 = \left(\frac{n\pi}{b}\right)^2, \qquad n = 1,2,\ldots.$$

The z-dependence is exponential, and since the potential cannot diverge at large values of z, the positive exponent will be absent. Absorbing all multiplicative constants into (a single doubly indexed) one, we can now write

$$\Phi(x,y,z) = \sum_{m,n=1}^{\infty} A_{mn}\sin\frac{m\pi x}{a}\sin\frac{n\pi y}{b}e^{-\pi\sqrt{m^2/a^2+n^2/b^2}\,z}. \qquad (25.13)$$

The unknown constants A_{mn} are determined by using the last BC:

$$V = \Phi(x, y, 0) = \sum_{m,n=1}^{\infty} A_{mn} \sin \frac{m\pi x}{a} \sin \frac{n\pi y}{b}. \tag{25.14}$$

This is a *double Fourier series*.

Theorem 25.2.5. *The coefficients of the* **double Fourier series** *(25.14) can be calculated by multiplying both sides by* $\sin \dfrac{j\pi x}{a} \sin \dfrac{k\pi y}{b}$ *and integrating the result from 0 to a in the x variable and from 0 to b in y:*

$$A_{jk} = \frac{4}{ab} \int_0^a \int_0^b \Phi(x, y, 0) \sin \frac{j\pi x}{a} \sin \frac{k\pi y}{b} dx\, dy$$

It now follows that

$$A_{jk} = \frac{4V}{ab} \int_0^a \int_0^b \sin \frac{j\pi x}{a} \sin \frac{k\pi y}{b} dx\, dy = \frac{4V}{ab} \int_0^a \sin \frac{j\pi x}{a} dx \int_0^b \sin \frac{k\pi y}{b} dy$$

$$= \frac{4V}{ab} \left(\frac{a}{\pi j}[1 - (-1)^j] \right) \left(\frac{b}{\pi k}[1 - (-1)^k] \right)$$

or

$$A_{jk} = \frac{4V}{\pi^2} \frac{1 - (-1)^j}{j} \frac{1 - (-1)^k}{k}.$$

It is clear that only the odd terms of the double sum will contribute. Thus, the final answer for the potential inside [Equation (25.13)] is

$$\Phi(x, y, z) = \frac{16V}{\pi^2} \sum_{m,n=1}^{\infty} \frac{\sin[(2m+1)\pi x/a]}{2m+1} \frac{\sin[(2n+1)\pi y/b]}{2n+1}$$
$$\cdot e^{-\pi\sqrt{(2m+1)^2/a^2 + (2n+1)^2/b^2}\, z}.$$

By its very construction, this function satisfies Laplace's equation as well as all the BCs. Therefore, by the uniqueness theorem it must represent the unique potential for the region of interest. ∎

25.3 Problems

25.1. Given that $V = \sum_{n=1}^{\infty} A_n \sin(n\pi y/b)$ where V is a constant in the interval $(0, b)$, show that $A_n = 2V[1 - (-1)^n]/(n\pi)$.

25.2. A long hollow cylinder with square cross section of side a has three sides grounded and the fourth side maintained at potential V_0 (see Figure 25.6). Find the potential at all points inside.

25.3. Example 25.2.1 treated the case in which the plate at $x = 0$ was held at the constant potential V. Now suppose that it is held at a potential that varies with y. Use Equation (25.5) to find the potential as a function of x and y when

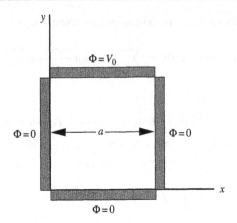

Figure 25.6: The cross section of the conducting cylinder extended along the z-axis.

(a) $\Phi(0, y) = \dfrac{V_0}{b^2} y(y - b)$

(b) $\Phi(0, y) = \dfrac{V_0}{b} y$

(c) $\Phi(0, y) = V_0 \sin \dfrac{\pi y}{b}$.

25.4. In Example 25.2.2, we assumed constant temperatures for the left and right plates. Now suppose that the top and bottom plates are as before, but the left plate is held at a varying temperature given by

$$T(0, y) = \frac{T_0}{b^2} y(y - b).$$

Use Equation (25.7) to find the temperature as a function of x and y when
(a) $T(a, y) = 0$;
(b) $T(a, y) = \dfrac{T_0}{b^2} y(y - b)$;
(c) $T(a, y) = T_0$;
(d) $T(a, y) = \dfrac{T_0}{b} y$;
(e) $T(a, y) = T_0 \sin \dfrac{\pi y}{b}$.

25.5. Suppose that the top and bottom plates of Example 25.2.2 are as before, but the left plate is held at a varying temperature given by

$$T(0, y) = T_0 \sin \frac{2\pi y}{b}.$$

Use Equation (25.7) to find the temperature as a function of x and y when
(a) $T(a, y) = 0$;
(b) $T(a, y) = \dfrac{T_0}{b^2} y(y - b)$;
(c) $T(a, y) = T_0$;

(d) $T(a, y) = \dfrac{T_0}{b} y$;

(e) $T(a, y) = T_0 \sin \dfrac{\pi y}{b}$.

25.6. Solve Equation (25.8) for A_{2k+1} and B_{2k+1} and substitute in (25.7) to obtain (25.9).

25.7. Verify that when $T_2 = 0$ and $a \to \infty$, Equation (25.9) approaches the result of Example 25.2.1—with V replaced by T_1.

25.8. Derive Equation (25.11). Assume that $T_1 = T_2 = T_0$ and solve for A_m and B_m.

25.9. Obtain the expression for A_{jk} in Theorem 25.2.5.

25.10. Find the potential inside a cube with sides of length a when the top side is held at a constant potential V_0 with all other sides grounded (zero potential).

25.11. Find the electrostatic potential inside a cube with sides of length a if all faces are grounded except the top, which is held at a potential given by:

(a) $\dfrac{V_0}{a} x, \quad 0 \le x \le a.$ (b) $\dfrac{V_0}{a} y, \quad 0 \le y \le a.$

(c) $\dfrac{V_0}{a^2} xy, \quad 0 \le x, y \le a.$ (d) $V_0 \sin\left(\dfrac{\pi}{a} x\right), \quad 0 \le x \le a.$

Chapter 26

Laplace's Equation: Spherical Coordinates

The separation of Laplace's equation in spherical coordinates is obtained from Equation (22.16) by substituting $f(r) = 0$. This will yield[1]

$$\frac{1}{r^2}\frac{d}{dr}\left(r^2\frac{dR}{dr}\right) - \frac{\alpha}{r^2}R = 0,$$

$$\frac{1}{\sin\theta}\frac{d}{d\theta}\left(\sin\theta\frac{d\Theta}{d\theta}\right) + \left(\alpha - \frac{\beta}{\sin^2\theta}\right)\Theta = 0, \qquad (26.1)$$

$$\frac{d^2 S}{d\varphi^2} + \beta S = 0.$$

We consider the case where S is the constant function.[2] This corresponds to problems with an **azimuthal symmetry**, i.e., problems for which it is a priori clear that the potential is independent of the azimuthal angle φ. For such situations, the third equation in (26.1) implies that $\beta = 0$ because S is a (nonzero) constant. The independent variables are reduced to two and, with $\Phi(r, \theta) = R(r)\Theta(\theta)$, the remaining ODEs simplify to

> azimuthal symmetry means independence from φ

$$\frac{1}{r^2}\frac{d}{dr}\left(r^2\frac{dR}{dr}\right) - \frac{\alpha}{r^2}R = 0,$$

$$\frac{1}{\sin\theta}\frac{d}{d\theta}\left(\sin\theta\frac{d\Theta}{d\theta}\right) + \alpha\Theta = 0. \qquad (26.2)$$

We shall now concentrate on the second equation and come back to the first after we have found solutions to the second.

---

[1] Here we have changed the symbol of the azimuthal function to S so that $\Phi(r, \theta, \varphi) = R(r)\Theta(\theta)S(\varphi)$.

[2] The case in which S is not constant—so that Φ depends on the azimuthal angle—is more complicated and will not be pursued here. Instead, the interested reader is referred to Hassani, S. *Mathematical Physics: A Modern Introduction to Its Foundations*, Springer-Verlag, 1999, Chapter 12 for details.

The appearance of $\sin\theta\,d\theta$ (the differential of $\cos\theta$) in the denominator suggests changing the independent variable from θ to $u \equiv \cos\theta$. For any function $f(\theta)$, the chain rule gives[3]

$$\frac{df}{du} = \frac{df}{d\theta}\frac{d\theta}{du} = \frac{df}{d\theta}\frac{1}{du/d\theta} = -\frac{1}{\sin\theta}\frac{df}{d\theta} \quad \text{or} \quad \frac{df}{d\theta} = -\sin\theta\frac{df}{du}, \quad (26.3)$$

which allows us to convert the derivative of a function with respect to u to the derivative with respect to θ and vice versa.

Introduce a new function $P(u)$ such that $P(u) \equiv \Theta(\theta)$. Using the chain rule, substituting in the second equation of (26.2), and writing $\sin^2\theta = 1 - u^2$, the DE becomes

$$-\frac{1}{\sin\theta}\frac{d}{d\theta}\left[(1-u^2)\frac{dP}{du}\right] + \alpha P = 0.$$

The term in the square brackets is a function of u. So, by Equation (26.3), we can convert the θ-derivative to a u-derivative and obtain

$$\frac{d}{du}\left[(1-u^2)\frac{dP}{du}\right] + \alpha P = 0, \quad (26.4)$$

which can also be written as

$$(1-u^2)\frac{d^2P}{du^2} - 2u\frac{dP}{du} + \alpha P = 0 \quad (26.5)$$

Legendre differential equation

or

$$\frac{d^2P}{du^2} - \frac{2u}{1-u^2}\frac{dP}{du} + \frac{\alpha}{1-u^2}P = 0. \quad (26.6)$$

Equation (26.4), or (26.5), or (26.6) is called the **Legendre equation**. We shall solve this DE using the so-called **Frobenius method** or the *method of undetermined coefficients*.

26.1　Frobenius Method

The basic assumption of the Frobenius method is that the solution of the DE can be represented by a power series. This is not a restrictive assumption because all functions encountered in physical applications can be written as power series as long as we are interested in their values lying in their interval of convergence. This interval may be very small or it may cover the entire real line.

A second order homogeneous linear DE can be written as

$$p_2(x)\frac{d^2y}{dx^2} + p_1(x)\frac{dy}{dx} + p_0(x)y = 0. \quad (26.7)$$

---

[3]Note that f can be considered a function of u as well as θ.

For almost all applications encountered in physics (certainly in *this* book), p_0, p_1, and p_2 are *polynomials*.[4] The first step in the implementation of the Frobenius method is to assume an infinite power series for y. It is common to choose the point of expansion to be $x = 0$. If $p_2(0) \neq 0$, only nonnegative powers of x need be considered.[5] If $p_2(0) = 0$, the DE loses its character of being "second order" and the solutions we are seeking may not be defined there. In such a case, we have two choices:

1. choose a different point of expansion $x_0 \neq 0$ so that $p_2(x_0) \neq 0$; or

2. allow nonpositive powers of x in the expansion of y.

The first choice is rarely used. It turns out that the most economic—but general—way of incorporating the second choice is to write the solution as

$$y = x^r \sum_{n=0}^{\infty} a_n x^n = \sum_{n=0}^{\infty} a_n x^{n+r} = a_0 x^r + a_1 x^{r+1} + a_2 x^{r+2} + a_3 x^{r+3} + \cdots,$$
(26.8)

where r is a *real number* (not necessarily a positive integer) to be determined by the DE.[6] It is customary to choose $a_0 = 1$ because any constant multiple of a solution is also a solution; so, if $a_0 \neq 1$, we simply multiply the series by $1/a_0$ to make it so.[7] Since a power series is uniformly convergent—within its radius of convergence—it can be differentiated term by term. So, we have

$$\frac{dy}{dx} = \sum_{n=0}^{\infty} a_n (n+r) x^{n+r-1} = r a_0 x^{r-1} + (r+1) a_1 x^r + \cdots,$$

$$\frac{d^2 y}{dx^2} = \sum_{n=0}^{\infty} a_n (n+r)(n+r-1) x^{n+r-1}$$
(26.9)

$$= r(r-1) a_0 x^{r-2} + (r+1) r a_1 x^{r-1} + (r+2)(r+1) a_2 x^r + \cdots.$$

We now substitute Equations (26.8) and (26.9) in the DE (26.7), multiply out the polynomials into the series, collect all distinct powers of x together, and set the coefficient of each term equal to zero. We thus obtain a set of equations whose solution determines r and the a_n's. The equation arising form the lowest power of x involves only r and is called the **indicial equation**. This is usually a quadratic equation in r which can be solved to obtain the

indicial equation

---

[4]The DE may not emerge in the form given here out of, say, the separation of variables, but can be cast in that form. The most complicated form of the coefficients of the derivatives in a DE are typically rational functions (ratios of two polynomials). Therefore, multiplying the DE by the product of all three denominators will cast the DE in the form given in (26.7).

[5]For a thorough discussion of the Frobenius method, including motivation and proofs for the claims cited here, consult Hassani, S. *Mathematical Physics: A Modern Introduction to Its Foundations*, Springer-Verlag, 1999, Chapter 14.

[6]As Problem 26.2 indicates, one can start with a solution of the form (26.8) even when $p_2(0) \neq 0$. The differential equation will then force r to be zero.

[7]The choice $a_0 = 1$ is convenient only when $p_2(0) = 0$. If $p_2(0) \neq 0$, we need not restrict a_0.

(two) possible values of r, each leading generally to a different solution. The other equations coming from higher powers of x give **recursion relations**, i.e., equations which give a_n in terms of a_{n-1} and a_{n-2}. By iterating this relation, one can obtain all a_n's in terms of only two which can be determined by the BCs. Let us summarize the procedure outlined above:

Theorem 26.1.1. (*Frobenius method*). *To solve the DE (26.7), assume a solution of the form (26.8). If $p_2(0) \neq 0$, choose $r = 0$, substitute y and its derivatives (26.9) in the DE, multiply out, collect all powers of x, and set their coefficients equal to zero. If $p_2(0) = 0$, let $a_0 = 1$ and solve the indicial equation to obtain r. Set the coefficients of all other powers of x equal to zero to find the recursion relation giving a_n in terms of a_{n-1} and a_{n-2}. Use this relation and the values of r obtained above to find all a_n's in terms of only two.*

26.2   Legendre Polynomials

We now apply the Frobenius method to the Legendre DE for which—using u as the independent variable—$p_2(u) = 1 - u^2$, $p_1(u) = -2u$, and $p_0(u) = \alpha$. Since $p_2(0) \neq 0$, we need not introduce an extra power of r for the series. Therefore, we may write

$$P(u) = \sum_{n=0}^{\infty} a_n u^n = a_0 + a_1 u + a_2 u^2 + a_3 u^3 + \cdots,$$

$$\frac{dP}{du} = \sum_{n=1}^{\infty} n a_n u^{n-1} = a_1 + 2a_2 u + 3a_3 u^2 + \cdots = \sum_{n=0}^{\infty} (n+1)a_{n+1} u^n,$$

$$\frac{d^2 P}{du^2} = \sum_{n=2}^{\infty} n(n-1) a_n u^{n-2} = 2a_2 + 6a_3 u + 12a_4 u^2 + \cdots$$

$$= \sum_{n=0}^{\infty} (n+1)(n+2) a_{n+2} u^n.$$

Multiplying each of the expressions above by its corresponding polynomial, we obtain

$$\alpha P(u) = \alpha a_0 + \alpha a_1 u + \alpha a_2 u^2 + \alpha a_3 u^3 + \cdots = \sum_{n=0}^{\infty} \alpha a_n u^n,$$

$$-2u\frac{dP}{du} = -2a_1 u - 4a_2 u^2 - 6a_3 u^3 + \cdots = -\sum_{n=0}^{\infty} 2(n+1)a_{n+1} u^{n+1},$$

$$(1 - u^2)\frac{d^2 P}{du^2} = 2a_2 + 6a_3 u + 12a_4 u^2 + 20a_5 u^3 + \cdots$$

$$- 2a_2 u^2 - 6a_3 u^3 - 12a_4 u^4 - 20a_5 u^5 + \cdots$$

$$= 2a_2 + 6a_3 u + (12a_4 - 2a_2)u^2 + (20a_5 - 6a_3)u^3 + \cdots.$$

We add these three series, noting that their sum must equal zero

$$0 = \alpha a_0 + \alpha a_1 u + \alpha a_2 u^2 + \alpha a_3 u^3 - 2a_1 u - 4a_2 u^2 - 6a_3 u^3 + 2a_2 + 6a_3 u$$
$$+ (12a_4 - 2a_2)u^2 + (20a_5 - 6a_3)u^3 + \cdots,$$
$$= (\alpha a_0 + +2a_2) + [(\alpha - 2)a_1 + 6a_3]u$$
$$+ [(\alpha - 6)a_2 + 12a_4]u^2 + [(\alpha - 12)a_3 + 20a_5]u^3 + \cdots.$$

The reader may note the pattern emerging in the expression for the coefficients. In fact, the coefficient of u^n can be written as $[\alpha - n(n+1)]a_n + (n+1)(n+2)a_{n+2}$. Setting this coefficient equal to zero, we obtain the recursion relation

$$a_{n+2} = \frac{n(n+1) - \alpha}{(n+1)(n+2)}a_n, \qquad n = 0, 1, 2, \ldots, \qquad (26.10)$$

recursion relation for the Legendre equation

which gives all a_n's in terms of a_0 and a_1.

Although writing out the series term by term is a sure way of arriving at each individual coefficient, and—by the discovery of a pattern—the recursion relation, manipulation with the summation symbols can also lead to the recursion relations without any expectation of pattern recognition. We go through the details of such a manipulation as a noteworthy exercise in working with the summation signs. The general procedure is to write all sums in such a way that the exponent of u agrees in all of them. To be specific, we write all sums over n so that the power of u is n. This may require redefining the summation index. So, the last term of the DE can be expressed as

How to get to the recursion relation (26.10) by manipulating summations.

$$\alpha P(u) = \sum_{n=0}^{\infty} \alpha a_n u^n. \qquad (26.11)$$

The term involving the first derivative is

$$-2u\frac{dP}{du} = -2u\sum_{n=1}^{\infty} na_n u^{n-1} = -\sum_{n=1}^{\infty} 2na_n u^n \qquad (26.12)$$

and the term involving the second derivative becomes

$$(1 - u^2)\frac{d^2 P}{du^2} = (1 - u^2)\sum_{n=2}^{\infty} n(n-1)a_n u^{n-2}$$
$$= \sum_{n=2}^{\infty} n(n-1)a_n u^{n-2} - \sum_{n=2}^{\infty} n(n-1)a_n u^n \qquad (26.13)$$
$$= \sum_{n=0}^{\infty} (n+2)(n+1)a_{n+2} u^n - \sum_{n=2}^{\infty} n(n-1)a_n u^n,$$

where in the first sum we replaced n with $n+2$ to change the power of u from $n-2$ to n.[8] The power of u in all the sums is now n.

---

[8] The phrase "in the first sum, we replaced n with $n+2$" is an abbreviation for a procedure whereby first a new dummy index m is defined by $m = n - 2$ (or $n = m + 2$), and then it is changed back to n.

The next step is to separate a sufficient number of terms of the "longer sums" so that all sums start with the same n corresponding to the shortest sum. In the case at hand, the shortest sum is the one that starts with $n = 2$. So, rewrite the sums in Equations (26.11), (26.12), and (26.13) as

$$\alpha P(u) = \alpha a_0 + \alpha a_1 u + \sum_{n=2}^{\infty} \alpha a_n u^n,$$

$$-2u\frac{dP}{du} = -2a_1 u - \sum_{n=2}^{\infty} 2n a_n u^n,$$

$$(1 - u^2)\frac{d^2 P}{du^2} = 2a_2 + 6a_3 u + \sum_{n=2}^{\infty} (n+2)(n+1)a_{n+2}u^n - \sum_{n=2}^{\infty} n(n-1)a_n u^n.$$

Adding these sums and noting that the LHS is zero gives

$$0 = \alpha a_0 + 2a_2 + (\alpha a_1 - 2a_1 + 6a_3)u$$
$$+ \sum_{n=2}^{\infty} \underbrace{\left[\alpha a_n - 2n a_n + (n+2)(n+1)a_{n+2} - n(n-1)a_n\right]}_{=[\alpha - n(n+1)]a_n + (n+2)(n+1)a_{n+2}} u^n.$$

By setting the coefficients of all powers of u equal to zero, we obtain

$$\alpha a_0 + 2a_2 = 0, \quad (\alpha - 2)a_1 + 6a_3 = 0,$$
$$[\alpha - n(n+1)]a_n + (n+2)(n+1)a_{n+2} = 0,$$

with the first two being special cases of the last one, which in turn happens to be the recursion relation (26.10).

Equation (26.10) is at the heart of the solution to the Legendre DE. It generates all the a_n's with even n from a_0, and all the odd a_n's from a_1. We derive a general formula for even a_n's, and leave the odd case to the reader. For $n = 0$, Equation (26.10) gives $a_2 = -(\alpha/2)a_0$, and for $n = 2$ we obtain

$$a_4 = \frac{2 \cdot 3 - \alpha}{4 \cdot 3}a_2 = \frac{2 \cdot 3 - \alpha}{4 \cdot 3}\left(-\frac{\alpha}{2}\right) = \frac{\alpha(\alpha - 2 \cdot 3)}{4!}a_0.$$

Similarly, for $n = 4$, we get

$$a_6 = \frac{4 \cdot 5 - \alpha}{6 \cdot 5}a_4 = -\frac{\alpha(\alpha - 2 \cdot 3)(\alpha - 4 \cdot 5)}{6!}a_0.$$

The reader may easily check that

$$a_8 = \frac{\alpha(\alpha - 2 \cdot 3)(\alpha - 4 \cdot 5)(\alpha - 6 \cdot 7)}{8!}a_0.$$

All these equations show a pattern that can be generalized to

$$a_{2n} = (-1)^n \frac{\alpha(\alpha - 2 \cdot 3)(\alpha - 4 \cdot 5) \cdots [\alpha - (2n - 2)(2n - 1)]}{(2n)!}a_0. \qquad (26.14)$$

Similarly, the odd terms can be calculated with the result

$$a_{2n+1} = (-1)^n \frac{(\alpha - 2)(\alpha - 3 \cdot 4)(\alpha - 5 \cdot 6) \cdots [\alpha - (2n-1)2n]}{(2n+1)!} a_1. \quad (26.15)$$

Inserting these coefficients in the series expansion of $P(u)$, we obtain

$$P(u) = a_0 \sum_{n=0}^{\infty} (-1)^n \frac{\alpha(\alpha - 2 \cdot 3)(\alpha - 4 \cdot 5) \cdots [\alpha - (2n-2)(2n-1)]}{(2n)!} u^{2n}$$

$$+ a_1 \sum_{n=0}^{\infty} (-1)^n \frac{(\alpha - 2)(\alpha - 3 \cdot 4)(\alpha - 5 \cdot 6) \cdots [\alpha - (2n-1)2n]}{(2n+1)!} u^{2n+1}.$$

$$(26.16)$$

If either of the series in Equation (26.16) is to have a physical utility, it must be convergent. The appearance of $(-1)^n$ may lead us to believe that the series is alternating. This is not true, because the terms involving α could be positive as well as negative. So, we cannot use the alternating series test. Let us use the ratio test. We apply this ratio test to the even series, the odd series calculation is identical. Calling the entire nth term of the series c_n, we have

The generalized ratio test shows that either of the series in Equation (26.16) diverges.

$$\lim_{n \to \infty} \left| \frac{c_{n+1}}{c_n} \right| = \lim_{n \to \infty} \left| \frac{a_{2n+2} u^{2n+2}}{a_{2n} u^{2n}} \right| = \lim_{n \to \infty} \left| \frac{2n(2n+1) - \alpha}{(2n+1)(2n+2)} \right| u^2 = u^2.$$

So, when $u^2 < 1$, the series converges. Recall that $u = \cos\theta$, and $\theta = 0, \pi$ are points of physical interest corresponding to $u = \pm 1$. Therefore, the series ought to converge there. In this case we cannot decide about the convergence of the series based on the ratio test. Let us apply the *generalized* ratio test. Then, for very large n, we have

$$\left. \left| \frac{c_{n+1}}{c_n} \right| \right|_{u^2=1} = \left| \frac{2n(2n+1) - \alpha}{(2n+1)(2n+2)} \right| = \left| \frac{n}{n+1} - \frac{\alpha}{(2n+1)(2n+2)} \right|$$

$$\approx 1 - \frac{1}{n+1} \approx 1 - \frac{1}{n},$$

and the generalized ratio test implies *divergence* for the series! This conclusion holds for both the even and odd series of (26.16).

There *is* a way of *making* the series convergent. Recall that the parameter α is completely arbitrary. In particular, we can—if it is helpful—put restrictions on it. Can we choose α in such a way that the series converges? We note that as long as the series is *infinite*, we have no luck because we get back to the generalized ratio test and divergence. However, if we choose α so that all a_n's after a certain finite number of terms vanish, then the series turns into a finite sum, and $P(u)$ becomes a polynomial for which the question of convergence is irrelevant. So, let us assume that a_0, a_2, and all the other a's up to a_{2k} can have nonzero values, but all the remaining coefficients are to

To make the series convergent, truncate it into a finite sum!

be zero. All we need to do is to choose α so that a_{2k+2} vanishes; then the recursion relation guarantees the vanishing of a_{2k+4}, a_{2k+6}, etc. Since

$$a_{2k+2} = \frac{2k(2k+1) - \alpha}{(2k+1)(2k+2)} a_{2k},$$

we must choose $\alpha = 2k(2k+1)$. A similar argument yields $\alpha = (2k-1)2k$ for the odd series.

Choosing α to turn one of the infinite sums into a finite polynomial is only a partial solution to the problem. Is it possible to choose α so that *both* the odd and the even series are truncated after a finite number of terms? Suppose we have chosen α to be $2k(2k+1)$, so that the even series has no term beyond the $(2k)$th term. The recursion relation for the odd series can be written as

$$a_{2n+1} = \frac{(2n-1)2n - 2k(2k+1)}{(2n+1)2n} a_{2n-1}.$$

Setting the numerator equal to zero gives a quadratic equation which can be solved for n to obtain

$$n = -k \qquad \text{or} \qquad n = k + \tfrac{1}{2}.$$

Neither of these is a *positive integer*! Thus, the value of α chosen to truncate the even series does not allow the truncation of the odd series. To avoid this dilemma, we resort to a choice of another arbitrary constant, a_1. By setting a_1 equal to zero, we completely avoid the odd series. Conversely, if $\alpha = (2k-1)2k$—chosen to truncate the odd series—then a_0 ought to be set equal to zero. By convention, a_0 and a_1 are determined so that $P(1) = 1$. Let us summarize our findings:

Theorem 26.2.1. *A solution to the Legendre DE (26.5) exists only if $\alpha = k(k+1)$ where k is a nonnegative integer. The corresponding solution is denoted by $P_k(u)$ and is a polynomial of degree k, called the kth **Legendre polynomial**, which has only even powers of u if k is even and odd powers of u if k is odd. By convention $P_k(1) = 1$ for all k.*

Thus, for each k we have a different solution, *and a different a_0 or a_1 to evaluate*. That is why it is more appropriate to write C_k for either a_0 or a_1. We can use either (26.14) and (26.15), or the recursion relation (26.10) to find the coefficients of each polynomial.

calculation of the
first five Legendre
polynomials using
the recursion
relation

Example 26.2.2. We calculate the Legendre polynomials up to order 4 using the recursion relation. P_0 is of degree zero, so it is a constant and $P_k(1) = 1$ forces it to be 1. So, $P_0(u) = 1$. Since, $P_1(u)$ is of degree 1 with no even "powers" of u, it can be only of the form $C_1 u$ where C_1 is a constant. But $P_1(1) = 1$; so $C_1 = 1$ and $P_1(u) = u$. For P_2, $\alpha = 2 \cdot 3 = 6$, and the recursion relation gives

$$a_2 = -\frac{\alpha}{2} a_0 = -3C_2 \;\Rightarrow\; P_2(u) = C_2 - 3C_2 u^2$$

because $P_2(u)$ has no u term. For $P_2(1)$ to be equal to 1, we must have $C_2 = -\frac{1}{2}$, so that $P_2(u) = \frac{1}{2}(3u^2 - 1)$. For P_3, $\alpha = 3 \cdot 4 = 12$, and the recursion relation gives

$$a_3 = \frac{2 - \alpha}{6}a_1 = \frac{2 - 12}{6}C_3 = -\tfrac{5}{3}C_3 \;\Rightarrow\; P_3(u) = C_3 u - \tfrac{5}{3}C_3 u^3$$

because $P_3(u)$ has no constant or u^2 term. For $P_3(1)$ to be equal to 1, we must have $C_3 = -\frac{3}{2}$, so that $P_3(u) = \frac{1}{2}(5u^3 - 3u)$. Finally, we calculate P_4 for which $\alpha = 4 \cdot 5 = 20$, and the recursion relations give

$$a_2 = -\tfrac{20}{2}a_0 = -10C_4 \qquad \text{and} \qquad a_4 = \frac{6 - 20}{12}a_2 = \tfrac{35}{3}C_4$$

and $P_4(u) = C_4 - 10C_4 u^2 - \frac{35}{3}C_4 u^4$. The condition $P_4(1) = 1$ gives $C_4 = 3/8$. Therefore,

$$P_4(u) = \tfrac{1}{8}(35u^4 - 30u^2 + 3).$$

Other Legendre polynomials can be obtained similarly. However, as we shall see shortly, there is a much easier way of calculating Legendre polynomials. ∎

With α determined to be of the form $k(k + 1)$, we can now calculate all coefficients of the Legendre polynomials. We start by rewriting the recursion relation (26.10) as

$$a_n = \frac{(n - 2)(n - 1) - k(k + 1)}{n(n - 1)}a_{n-2}$$

$$= -\frac{(k - n + 2)(k + n - 1)}{n(n - 1)}a_{n-2}, \quad n = 2, 3, \ldots, k. \tag{26.17}$$

Iterating this once, we obtain

$$a_n = (-1)^2 \frac{(k - n + 2)(k + n - 1)}{n(n - 1)} \frac{(k - n + 4)(k + n - 3)}{(n - 2)(n - 3)}a_{n-4}$$

$$= (-1)^2 \frac{[(k - n + 2)(k - n + 4)][(k + n - 1)(k + n - 3)]}{n(n - 1)(n - 2)(n - 3)}a_{n-4}.$$

By iterating a few times, the reader may check that

$$a_n = (-1)^m [(k - n + 2)(k - n + 4) \cdots (k - n + 2m)]$$

$$\cdot \frac{[(k + n - 1)(k + n - 3) \cdots (k + n - 2m + 1)]}{n(n - 1) \cdots (n - 2m + 1)}a_{n-2m}. \tag{26.18}$$

To proceed, we need to take the two cases of even and odd n separately. We treat the even case and leave the odd case as an exercise for the reader. Let us assume that $n = 2m$, then k must also be even[9] and the last equation above yields

$$a_{2m} = (-1)^m \frac{[(2j) \cdots (2j - 2m + 2)][(2j + 2m - 1) \cdots (2j + 1)]}{2m(2m - 1) \cdots 1}a_0$$

$$= (-1)^m \frac{[(2j)!!/(2j - 2m)!!][(2j + 2m - 1)!!/(2j - 1)!!]}{(2m)!}a_0,$$

---

[9]Recall from our discussion above that even Legendre polynomials correspond to even $\alpha = 2j(2j + 1)$.

where we set $k = 2j$. Using the relations (see Problem 11.1)

$$(2l - 1)!! = \frac{(2l)!}{2^l l!}, \qquad (2l)!! = 2^l l!,$$

we finally obtain

$$a_{2m} = \frac{(-1)^m j!}{2^j} \frac{(2j + 2m)!}{(j + m)!(j - m)!} \frac{1}{(2m)!} a_0$$

$$\equiv A_j (-1)^m \frac{(2j + 2m)!}{(j + m)!(j - m)!} \frac{1}{(2m)!}, \qquad (26.19)$$

where $A_j = a_0 (j!/2^j)$. The reader may check that

$$a_{2m+1} = B_j (-1)^m \frac{(2j + 2m + 2)!}{(j + m + 1)!(j - m)!} \frac{1}{(2m + 1)!} \qquad (26.20)$$

even Legendre polynomial

for some constant B_j. Therefore, the even Legendre polynomials will be given by

$$P_{2j}(x) = A_j \sum_{m=0}^{j} (-1)^m \frac{(2j + 2m)!}{(j + m)!(j - m)!} \frac{x^{2m}}{(2m)!} \qquad (26.21)$$

odd Legendre polynomial

and the odd polynomials by

$$P_{2j+1}(x) = B_j \sum_{m=0}^{j} (-1)^m \frac{(2j + 2m + 2)!}{(j + m + 1)!(j - m)!} \frac{x^{2m+1}}{(2m + 1)!}. \qquad (26.22)$$

We now introduce a new summation index $r = j - m$ in either sum and let $n = 2j$ in the even sum and $n = 2j + 1$ in the odd sum. Then both sums can be written simply as

$$P_n(x) = K_n \sum_{r=0}^{[n/2]} (-1)^r \frac{(2n - 2r)!}{(n - r)!r!} \frac{x^{n-2r}}{(n - 2r)!}, \qquad (26.23)$$

where $[a]$—for any real number a—denotes the largest integer less than or equal to a, and K_n is an arbitrary constant which, by convention, is taken to be $1/2^n$ so that $P_n(1) = 1$. This leads to

$$P_n(x) = \frac{1}{2^n} \sum_{r=0}^{[n/2]} (-1)^r \frac{(2n - 2r)!}{(n - r)!r!} \frac{x^{n-2r}}{(n - 2r)!}. \qquad (26.24)$$

Referring to the definition of the hypergeometric function (11.23) and Equation (26.24), the reader may verify that

$$P_{2n}(x) = (-1)^n \frac{(2n)!}{2^{2n}(n!)^2} F(-n, n + \tfrac{1}{2}; \tfrac{1}{2}; x^2) \qquad (26.25)$$

and

$$P_{2n+1}(x) = (-1)^n \frac{(2n+1)!}{2^{2n}(n!)^2} x F(-n, n + \tfrac{3}{2}; \tfrac{3}{2}; x^2). \qquad (26.26)$$

Historical Notes

Adrien-Marie Legendre came from a well-to-do Parisian family and received an excellent education in science and mathematics. His university work was advanced enough that his mentor used many of Legendre's essays in a treatise on mechanics. A man of modest fortune until the revolution, Legendre was able to devote himself to study and research without recourse to an academic position. In 1782 he won the prize of the Berlin Academy for calculating the trajectories of cannonballs taking air resistance into account. This essay brought him to the attention of **Lagrange** and helped pave the way to acceptance in French scientific circles, notably the Academy of Sciences, to which Legendre submitted numerous papers. In July 1784 he submitted a paper on planetary orbits that contained the now-famous *Legendre polynomials*, mentioning that Lagrange had been able to "present a more complete theory" in a recent paper by using Legendre's results. In the years that followed, Legendre concentrated his efforts on number theory, celestial mechanics, and the theory of elliptic functions. In addition, he was a prolific calculator, producing large tables of the values of special functions, and he also authored an elementary textbook that remained in use for many decades. In 1824 Legendre refused to vote for the government's candidate for the *Institut National*. Because of this, his pension was stopped and he died in poverty and in pain at the age of 80 after several years of failing health.

Adrien-Marie
Legendre
1752–1833

Legendre produced a large number of useful ideas but did not always develop them in the most rigorous manner, claiming to hold the priority for an idea if he had presented merely a reasonable argument for it. **Gauss**, with whom he had several quarrels over priority, considered rigorous proof the standard of ownership. To Legendre's credit, however, he was an enthusiastic supporter of his young rivals Abel and Jacobi and gave their work considerable attention in his writings.

Legendre also contributed to practical efforts in science and mathematics. He and two of his contemporaries were assigned in 1787 to a panel conducting geodetic work in cooperation with the observatories at Paris and Greenwich. Four years later the same panel members were appointed as the Academy's commissioners to undertake the measurements and calculations necessary to determine the length of the standard meter. Legendre's seemingly tireless skill at calculating produced large tables of the values of trigonometric and elliptic functions, logarithms, and solutions to various special equations.

26.3 Second Solution of the Legendre DE

Recall that any second order linear DE has two bases of solutions. We have so far found one solution of Legendre DE in the form of the Legendre polynomials. Once we have these solutions, we can obtain a second solution using Equation (24.6). To conform with Equation (24.6), we need to reexpress the Legendre DE as

$$\frac{d^2y}{dx^2} - \frac{2x}{1-x^2}\frac{dy}{dx} + \frac{n(n+1)}{1-x^2}y = 0.$$

This is an homogeneous second order linear DE with

$$p(x) = -\frac{2x}{1-x^2} \quad \text{and} \quad q(x) = \frac{n(n+1)}{1-x^2}.$$

Using $P_n(x)$ as our input, we can generate another set of solutions. Let $Q_n(x)$ stand for the linearly independent "partner" of $P_n(x)$. Then, setting $C = 0$ in Equation (24.6) yields[10]

$$Q_n(x) = KP_n(x) \int_\alpha^x \frac{1}{P_n^2(s)} \exp\left[\int_c^s \frac{2t}{1-t^2} dt\right] ds.$$

But

$$\int_c^s \frac{2t}{1-t^2} dt = -\ln|1-t^2|\Big|_c^s = -\ln\left|\frac{1-s^2}{1-c^2}\right| = \ln\left|\frac{1-c^2}{1-s^2}\right|$$

so that

$$\exp\left[\int_c^s \frac{2t}{1-t^2} dt\right] ds = \exp\left[\ln\left|\frac{1-c^2}{1-s^2}\right|\right] = \left|\frac{1-c^2}{1-s^2}\right| = \frac{|1-c^2|}{1-s^2},$$

because s, being the argument of a Legendre polynomial, is the cosine of an angle and therefore cannot exceed 1 so that $1 - s^2 \geq 0$. It now follows that

$$Q_n(x) = A_n P_n(x) \int_\alpha^x \frac{ds}{(1-s^2)P_n^2(s)}, \qquad (26.27)$$

Legendre functions of the second kind

where $A_n \equiv K|1-c^2|$ is an arbitrary constant determined by convention, and α is an arbitrary point in the interval $[-1, +1]$. The subscript for A_n indicates that the constant may be different for different n. These new solutions are called **Legendre functions of the second kind**. Note that, contrary to $P_n(x)$, $Q_n(x)$ is not well behaved at $x = \pm 1$ due to the presence of $1 - s^2$ in the denominator of the integrand of Equation (26.27). For this reason, we shall not use these second solutions in this book.

Example 26.3.1. Example 26.2.2 gives $P_0(x) = 1$. Therefore,

$$Q_0(x) = A_0 \int_\alpha^x \frac{ds}{1-s^2} = \frac{A_0}{2} \int_\alpha^x \left(\frac{1}{1+s} + \frac{1}{1-s}\right) ds$$

$$= A_0 \left[\frac{1}{2}\ln\left|\frac{1+x}{1-x}\right| - \frac{1}{2}\ln\left|\frac{1+\alpha}{1-\alpha}\right|\right].$$

The standard form of $Q_0(x)$ is obtained by setting $A_0 = 1$ and $\alpha = 0$:

$$Q_0(x) = \frac{1}{2}\ln\left|\frac{1+x}{1-x}\right| \qquad \text{for} \quad |x| < 1.$$

Similarly, since $P_1(x) = x$, we obtain

$$Q_1(x) = A_1 x \int_\alpha^x \frac{ds}{(1-s^2)s^2} = Ax + Bx\ln\left|\frac{1+x}{1-x}\right| + C \qquad \text{for} \quad |x| < 1,$$

---

[10]Since we are interested in a *different* second solution, we can ignore any *constant* multiple of the first solution that is added to the sought-after second solution.

where A, B, and C are constants, and to perform the integration, we used

$$\frac{1}{(1-s^2)s^2} = \frac{1}{s^2} + \frac{1}{2(1-s)} + \frac{1}{2(1+s)},$$

which renders the integral elementary. In the case of $Q_1(x)$, convention demands that $A = 0$, $B = \frac{1}{2}$, and $C = -1$. Thus,

$$Q_1(x) = \frac{1}{2}x \ln\left|\frac{1+x}{1-x}\right| - 1.$$

∎

26.4 Complete Solution

Having found the angular solution of Laplace's equation, we now tackle the radial part. With $\alpha = k(k+1)$, we can write the first equation in (26.2) as

$$r^2\frac{d^2R}{dr^2} + 2r\frac{dR}{dr} - k(k+1)R = 0. \tag{26.28}$$

Since $p_2(0) = 0$, we have to consider a solution of the form $R(r) = r^s \sum_{n=0}^{\infty} b_n r^n$. Differentiating this series and substituting it in Equation (26.28) gives

$$\sum_{n=0}^{\infty}[(n+s)(n+s+1) - k(k+1)]b_n r^{n+s} = 0$$

or

$$[(n+s)(n+s+1) - k(k+1)]b_n = 0 \quad \text{for} \quad n = 0, 1, 2, \ldots.$$

In particular, for $n = 0$, and assuming that $b_0 \neq 0$, we obtain the *indicial equation*

$$s(s+1) - k(k+1) = 0 \implies s = k \text{ or } s = -k-1.$$

an example of the indicial equation

For $s = k$, the equation for general nonzero n gives

$$[(n+k)(n+k+1) - k(k+1)]b_n = 0 \implies n(n+2k+1)b_n = 0.$$

Since neither n nor $n+2k+1$ is zero, we have to conclude that $b_n = 0$ for all $n \geq 1$. Thus, for $s = k$, we obtain the solution $R(r) = A_k r^k$ where A_k is an arbitrary constant (we called it b_0 before).

For $s = -k-1$, we have

$$[(n-k-1)(n-k) - k(k+1)]b_n = 0 \implies n(n-2k-1)b_n = 0$$

for which we can have either $n = 2k+1$ or $b_n = 0$. If $n = 2k+1$, then

$$R(r) = r^{-k-1}b_{2k+1}r^{2k+1} = b_{2k+1}r^k,$$

which is (a constant times) what we already have. So assume that $n \neq 2k+1$. Then $b_n = 0$ for all $n \geq 1$, and we obtain the solution $R(r) = B_k r^{-k-1}$ where

the most general
solution of the
spherical radial DE

B_k is another arbitrary constant. It follows that the most general solution of the radial DE is

$$R_k(r) \equiv A_k r^k + \frac{B_k}{r^{k+1}}, \qquad k = 0, 1, 2, \ldots.$$

We can now put the radial and the angular parts together:

Theorem 26.4.1. *To find the most general azimuthally symmetric solution of Laplace's equation in spherical coordinates, we multiply the radial solution and the angular solution (Legendre polynomial) for each k and sum over all possible values of k:*

$$\Phi(r, \theta) = \sum_{k=0}^{\infty} \left(A_k r^k + \frac{B_k}{r^{k+1}} \right) P_k(\cos \theta), \tag{26.29}$$

where we have substituted $\cos \theta$ *for* u.

From a known
solution of
Laplace's
equation, we find
a formula that
generates all
Legendre
polynomials.

Equation (26.29) gives the general solution of Laplace's equation, and we shall consider examples of how to use it to solve some representative problems, but first we will go backward: From a particular known solution of Laplace's equation, we want to find an important property of Legendre polynomials. Equation (15.18) shows that $1/|\mathbf{r} - \mathbf{r}_0|$ is a solution of Laplace's equation at all points of space except $\mathbf{r}_0$. In general, $|\mathbf{r} - \mathbf{r}_0|$ is not azimuthally symmetric. However, if we place $\mathbf{r}_0$ along the z-axis, the φ-dependence will disappear. In fact, with $\mathbf{r}_0 = a\hat{\mathbf{e}}_z$, we have

$$|\mathbf{r} - \mathbf{r}_0| = |\mathbf{r} - a\hat{\mathbf{e}}_z| = \sqrt{r^2 + a^2 - 2ar \cos \theta}.$$

According to (26.29) the solution $1/|\mathbf{r} - a\hat{\mathbf{e}}_z|$ can be written as a series:

$$\frac{1}{\sqrt{r^2 + a^2 - 2ar \cos \theta}} = \sum_{k=0}^{\infty} \left(A_k r^k + \frac{B_k}{r^{k+1}} \right) P_k(\cos \theta).$$

We are interested in the region of space inside the sphere of radius a. Since the origin is included in this region, no negative powers of r are allowed. Therefore, all coefficients of such powers must be zero, i.e., $B_k = 0$. To determine the other set of coefficients, evaluate both sides at $\theta = 0$ and use $P_k(1) = 1$. This gives

$$\frac{1}{\sqrt{r^2 + a^2 - 2ar}} = \frac{1}{|r - a|} = \frac{1}{a - r} = \sum_{k=0}^{\infty} A_k r^k.$$

Using the result of Example 9.3.3 and the fact that $r/a < 1$, the LHS can be expanded in powers of r/a:

$$\frac{1}{a - r} = \frac{1}{a(1 - r/a)} = \frac{1}{a} \sum_{k=0}^{\infty} \left(\frac{r}{a} \right)^k = \sum_{k=0}^{\infty} \frac{r^k}{a^{k+1}}.$$

Comparison of the last two equations gives $A_k = 1/a^{k+1}$. It follows that

$$\frac{1}{\sqrt{r^2 + a^2 - 2ar\cos\theta}} = \frac{1}{a}\sum_{k=0}^{\infty}\left(\frac{r}{a}\right)^k P_k(\cos\theta), \qquad r < a.$$

Introducing $t \equiv r/a$ and $u \equiv \cos\theta$ on both sides, we finally obtain the important relation

$$g(t, u) \equiv \frac{1}{\sqrt{1 + t^2 - 2tu}} = \sum_{k=0}^{\infty} t^k P_k(u). \tag{26.30}$$

The RHS can be considered as a Taylor (or Maclaurin) series in t for the function on the LHS.

Theorem 26.4.2. *The kth coefficient of the Maclaurin series expansion of* $g(t, u) \equiv 1/\sqrt{1 + t^2 - 2tu}$ *about* $t = 0$ *is* $P_k(u)$. *Specifically,*

$$P_k(u) = \frac{1}{k!}\frac{\partial^k}{\partial t^k}\frac{1}{\sqrt{1 + t^2 - 2tu}}\bigg|_{t=0}. \tag{26.31}$$

The function $g(t, u)$ *is called the* **generating function** *of the Legendre polynomials.*

Example 26.4.3. As an immediate application of the generating function to potential theory, consider the electrostatic or gravitational potential which can be written as

$$\Phi(\mathbf{r}) = K\iint_{\Omega}\frac{dQ(\mathbf{r}')}{|\mathbf{r} - \mathbf{r}'|}, \tag{26.32}$$

where K is k_e for electrostatics and $-G$ for gravity, and Q represents either electric charge or mass. Assuming that $r \gg r'$, we can expand in powers of the ratio r'/r which we denote by t. The key to this expansion is the following power series of $1/|\mathbf{r} - \mathbf{r}'|$:

$$\frac{1}{|\mathbf{r} - \mathbf{r}'|} = \frac{1}{\sqrt{r^2 + r'^2 - 2\mathbf{r}\cdot\mathbf{r}'}} = \frac{1}{r}\frac{1}{\sqrt{1 + t^2 - 2t\cos\gamma}} = \frac{1}{r}\sum_{k=0}^{\infty}t^k P_k(\cos\gamma),$$

where γ is the angle between $\mathbf{r}$ and $\mathbf{r}'$ and we used Equation (26.30). Substituting this expansion for $1/|\mathbf{r} - \mathbf{r}'|$ in (26.32), we obtain

Legendre polynomial and multipole expansion

$$\Phi(\mathbf{r}) = K\iint_{\Omega}\sum_{k=0}^{\infty}\frac{r'^k}{r^{k+1}}P_k(\cos\gamma)\,dQ(\mathbf{r}') = K\sum_{k=0}^{\infty}\frac{Q_k}{r^{k+1}}, \tag{26.33}$$

where we replaced t with r'/r and introduced Q_k, the so-called **k-th moment** of source (charge or mass), by[11]

$$Q_k \equiv \int_{\Omega}r'^k P_k(\cos\gamma)\,dQ(\mathbf{r}'). \tag{26.34}$$

---

[11] Do not confuse this Q_k with the second solution of Legendre DE introduced in Equation (26.27).

Recall that $\cos\gamma$ depends on θ and φ. Thus, once the integral over Ω is done, the result will depend on θ and φ as it should, because $\Phi(\mathbf{r})$ is, in general, dependent on these angles.

The moments Q_k are supposed to describe the intrinsic properties of charge (or mass) distributions and *should not depend on the observation point*—described, in part, by θ and φ. This is the reason that Cartesian coordinates are more useful—at this level of presentation—than spherical coordinates. In Cartesian coordinates, we can separate the primed from the unprimed coordinates (as we did in the definition of dipole in Chapter 10 and of quadrupole in Chapter 17), and define multipole moments entirely in terms of the density function of the distribution of the source. This does not mean, however, that a complete separation is impossible in spherical coordinates. In fact, there are techniques of performing such a separation—in terms of the so-called "spherical harmonics"—but they are much more complicated and beyond the scope of this book.[12]   ∎

26.5   Properties of Legendre Polynomials

From the Legendre DE, the generating function, and other formulas derived earlier, one can obtain a variety of relations connecting Legendre polynomials.

26.5.1   Parity

The easiest property to obtain is parity which is the content of the following formula:

$$P_k(-u) = (-1)^k P_k(u). \tag{26.35}$$

This is a direct consequence of the fact that $P_k(u)$ has only even powers of u if k is even, and odd powers if k is odd.

26.5.2   Recurrence Relation

Differentiate both sides of Equation (26.30) with respect to t to obtain

$$\frac{u-t}{(1+t^2-2tu)^{3/2}} = \sum_{k=1}^{\infty} kt^{k-1}P_k(u). \tag{26.36}$$

Rewrite the LHS as

$$\frac{u-t}{1+t^2-2tu}\frac{1}{\sqrt{1+t^2-2tu}} = \frac{u-t}{1+t^2-2tu}\sum_{k=0}^{\infty} t^k P_k(u), \tag{26.37}$$

where we used (26.30) for the term with the square root. Equating the RHS of (26.37) with the RHS of (26.36) and multiplying the result by $1+t^2-2tu$ yields

$$(t-u)\sum_{k=0}^{\infty} t^k P_k(u) + (1+t^2-2tu)\sum_{k=1}^{\infty} kt^{k-1}P_k(u) = 0$$

---

[12]See Hassani, S. *Mathematical Physics: A Modern Introduction to Its Foundations*, Springer-Verlag, 1999, Chapter 12 for a discussion of spherical harmonics.

or

$$\sum_{k=0}^{\infty} t^{k+1} P_k(u) - u \sum_{k=0}^{\infty} t^k P_k(u) + \sum_{k=1}^{\infty} k t^{k-1} P_k(u)$$

$$+ \sum_{k=1}^{\infty} k t^{k+1} P_k(u) - 2u \sum_{k=1}^{\infty} k t^k P_k(u) = 0.$$

All the coefficients of powers of t must vanish. To find these coefficients, change the dummy index in each sum so that all sums will have the same power of t. So, let $k = n - 1$ in the first and fourth sums, $k = n$ in the second and the last sums, and $k = n + 1$ in the third sum. Then the above equation can be written as

$$\sum_{n}^{\infty} [P_{n-1}(u) - u P_n(u) + (n+1) P_{n+1}(u)$$

$$+ (n-1) P_{n-1}(u) - 2un P_n(u)] t^n = 0,$$

where we have purposefully left out the lower limit of summation because different sums start at different initial values of n. Since a power series is zero only if all its coefficients are zero, we set the coefficients of the series above equal to zero to obtain

$$(2n+1) u P_n(u) = (n+1) P_{n+1}(u) + n P_{n-1}(u), \qquad n = 1, 2, 3, \ldots. \quad (26.38)$$

recurrence relation for Legendre polynomial

Using $P_0(u) = 1$ and $P_1(u) = u$, one can generate all Legendre polynomials from Equation (26.38).

Example 26.5.1. For $n = 1$, Equation (26.38) gives

$$3u P_1(u) = 2 P_2(u) + P_0(u) \Rightarrow 3u^2 = 2 P_2(u) + 1 \Rightarrow P_2(u) = \tfrac{1}{2}(3u^2 - 1)$$

For $n = 2$, Equation (26.38) gives

$$5u P_2(u) = 3 P_3(u) + 2 P_1(u) \Rightarrow \tfrac{5}{2} u(3u^2 - 1) = 3 P_3(u) + 2u \Rightarrow P_3(u) = \tfrac{1}{2}(5u^3 - 3u),$$

and so on. ∎

The recurrence relation can be used to obtain $P_n(0)$ which is a useful quantity. We quote the result and leave the details as an exercise for the reader. For odd n, we have $P_n(0) = 0$. The result for the even case is

$$P_{2n}(0) = (-1)^n \frac{(2n-1)!!}{(2n)!!} = (-1)^n \frac{(2n)!}{2^{2n} (n!)^2}. \quad (26.39)$$

Example 26.5.2. We can also obtain $P_n(0)$ by letting $u = 0$ in Equation (26.30):

$$(1 + t^2)^{-1/2} = \sum_{k=0}^{\infty} t^k P_k(0).$$

The binomial expansion of the LHS gives [see Equation (10.15)]

$$(1+t^2)^{-1/2} = 1 + \sum_{n=1}^{\infty} \frac{(-\frac{1}{2})(-\frac{1}{2}-1)\cdots(-\frac{1}{2}-n+1)}{n!} (t^2)^n$$

$$= 1 + \sum_{n=1}^{\infty} (-1)^n \frac{\frac{1}{2}(\frac{1}{2}+1)\cdots(n-\frac{1}{2})}{n!} t^{2n}$$

$$= 1 + \sum_{n=1}^{\infty} (-1)^n \frac{1 \cdot 3 \cdots (2n-1)}{2^n n!} t^{2n}.$$

Comparing this with the RHS of the first equation, we see that $P_n(0) = 0$ when n is odd and that

$$P_{2n}(0) = (-1)^n \frac{1 \cdot 3 \cdots (2n-1)}{2^n n!} = (-1)^n \frac{(2n-1)!!}{2^n n!}$$

which is the same as (26.39) because $2^n n! = (2n)!!$ by Problem 11.1.   ■

26.5.3   Orthogonality

The most useful property of the Legendre polynomials is their orthogonality. We have already seen in Chapters 6 and 7 how dot products can be defined for polynomials. We now show that Legendre polynomials of different orders are necessarily orthogonal once the dot product is defined in terms of suitable integrals (also see Example 24.5.3). Write the Legendre DE for P_n and P_m as

$$\frac{d}{du}[(1-u^2)P_n'(u)] + n(n+1)P_n(u) = 0,$$

$$\frac{d}{du}[(1-u^2)P_m'(u)] + m(m+1)P_m(u) = 0,$$

where the prime indicates derivative. Multiply both sides of the first equation by $P_m(u)$ and the second equation by $P_n(u)$ and integrate from -1 to $+1$:

$$\int_{-1}^{1} \frac{d}{du}[(1-u^2)P_n'(u)]P_m(u)\,du + n(n+1)\int_{-1}^{1} P_n(u)P_m(u)\,du = 0,$$

$$\int_{-1}^{1} \frac{d}{du}[(1-u^2)P_m'(u)]P_n(u)\,du + m(m+1)\int_{-1}^{1} P_m(u)P_n(u)\,du = 0. \quad (26.40)$$

Use integration by parts to write the first integral as

$$\int_{-1}^{1} \frac{d}{du}[(1-u^2)P_n'(u)]P_m(u)\,du = \underbrace{(1-u^2)P_n'(u)P_m(u)\big|_{-1}^{1}}_{=0 \text{ because of } (1-u^2)}$$

$$- \int_{-1}^{1} [(1-u^2)P_n'(u)]P_m'(u)\,du.$$

The first integral of the second line of Equation (26.40) gives exactly the same result. Therefore, if we subtract the two equations of (26.40), we obtain

$$[n(n+1) - m(m+1)] \int_{-1}^{1} P_n(u) P_m(u)\, du = 0.$$

It now follows that

Theorem 26.5.3. *If* $m \neq n$, *then* $\int_{-1}^{1} P_n(u) P_m(u)\, du = 0$, *i.e., if the inner product is defined as an integral from* -1 *to* $+1$, *then Legendre polynomials of different orders are orthogonal.*

We put this orthogonality relation to immediate use. Square both sides of Equation (26.30) *keeping in mind to introduce a new dummy index when multiplying the sums*, and integrate the result from -1 to $+1$:

$$\int_{-1}^{1} \frac{du}{1 + t^2 - 2tu} = \int_{-1}^{1} \left(\sum_{k=0}^{\infty} t^k P_k(u) \right) \left(\sum_{m=0}^{\infty} t^m P_m(u) \right) du. \qquad (26.41)$$

On the RHS, we switch the order of summation and integration:

$$RHS = \sum_{k=0}^{\infty} \sum_{m=0}^{\infty} t^{m+k} \underbrace{\int_{-1}^{1} P_k(u) P_m(u)\, du}_{=0 \text{ unless } m=k}.$$

As we perform the inner sum, by Theorem 26.5.3, all terms will vanish except one, i.e., only when $m = k$. So, the double sum reduces to a single sum

$$RHS = \sum_{k=0}^{\infty} t^{2k} \int_{-1}^{1} P_k^2(u)\, du.$$

The integral on the LHS of (26.41) can be done by substituting $y = 1 + t^2 - 2tu$ and $dy = -2t\, du$:

$$LHS = -\frac{1}{2t} \int_{(1+t)^2}^{(1-t)^2} \frac{dy}{y} = \frac{1}{t} [\ln(1+t) - \ln(1-t)].$$

The two natural log terms can be expanded using Equation (10.23). The reader may check that

$$\frac{1}{t} [\ln(1+t) - \ln(1-t)] = 2 \sum_{k=0}^{\infty} \frac{t^{2k}}{2k+1}. \qquad (26.42)$$

The fact that only even powers of t are present could have been anticipated because the function on the LHS of Equation (26.42) is even in t. Equating the RHS and the LHS of (26.41), we obtain

$$2 \sum_{k=0}^{\infty} \frac{t^{2k}}{2k+1} = \sum_{k=0}^{\infty} t^{2k} \int_{-1}^{1} P_k^2(u)\, du.$$

For these two power series in t to be equal, their coefficients must equal:

$$\int_{-1}^{1} P_k^2(u)\, du = \frac{2}{2k+1}.$$

Combining this with the orthogonality relation of Theorem 26.5.3 and using the Kronecker delta introduced in Equation (7.9), we have

$$\int_{-1}^{1} P_m(u)P_n(u)\, du = \frac{2}{2n+1}\delta_{mn}. \qquad (26.43)$$

26.5.4 Rodrigues Formula

We started our discussion of Legendre polynomials by representing them as infinite series and then truncating the series due to physical restrictions. We noted that the recursion relation obtained by the Frobenius method gave all the coefficients of the polynomials in terms of a_0 and a_1. Later, we found a "closed" expression for all Legendre polynomials in terms of derivatives of the generating functions which is a very useful function as the derivation of (26.43) demonstrated.

There is another "closed" expression of Legendre polynomials which we shall discuss now. This expression is called the **Rodrigues formula** and is given by[13]

$$P_n(x) = \frac{1}{2^n n!} \frac{d^n}{dx^n}\left[(x^2-1)^n\right]. \qquad (26.44)$$

To see that the RHS indeed gives the nth Legendre polynomial, we show that it satisfies the corresponding Legendre DE. The most elegant way to show this is to resort to complex analysis where derivatives are represented as integrals [see Equation (19.10)]. Thus, for $f(z) = (z^2-1)^n$, the Cauchy integral formula gives

$$(z^2-1)^n = \frac{1}{2\pi i}\oint_C \frac{(\xi^2-1)^n}{(\xi-z)}\, d\xi,$$

and Equations (19.10) and (26.44) yield

$$P_n(z) = \frac{1}{2^n n!}\frac{d^n}{dz^n}\left[(z^2-1)^n\right] = \frac{1}{2^n(2\pi i)}\oint_C \frac{(\xi^2-1)^n}{(\xi-z)^{n+1}}\, d\xi.$$

To find $P_n'(z)$ and $P_n''(z)$, we differentiate the integral, carrying the derivative inside and letting it differentiate the denominator:

$$\frac{dP}{dz} = \frac{1}{2^n(2\pi i)}\oint_C \frac{d}{dz}\left[\frac{(\xi^2-1)^n}{(\xi-z)^{n+1}}\right] d\xi = \frac{n+1}{2^n(2\pi i)}\oint_C \frac{(\xi^2-1)^n}{(\xi-z)^{n+2}}\, d\xi,$$

$$\frac{d^2P}{dz^2} = \frac{d}{dz}\left(\frac{dP}{dz}\right) = \frac{n+1}{2^n(2\pi i)}\oint_C \frac{d}{dz}\left[\frac{(\xi^2-1)^n}{(\xi-z)^{n+2}}\right] d\xi$$

$$= \frac{(n+1)(n+2)}{2^n(2\pi i)}\oint_C \frac{(\xi^2-1)^n}{(\xi-z)^{n+3}}\, d\xi.$$

---

[13]The fact that we are using x, rather than u, as the argument of the Legendre polynomial should not cause any confusion.

Substituting these expressions in the DE, the reader may check that

$$(1 - z^2)\frac{d^2 P}{dz^2} - 2z\frac{dP}{dz} + n(n+1)P$$

$$= \frac{n+1}{2^n(2\pi i)} \oint_C \frac{(\xi^2 - 1)^n[n\xi^2 - 2(n+1)\xi z + n + 2]}{(\xi - z)^{n+3}} \, d\xi.$$

The reader may also verify that

$$\frac{(\xi^2 - 1)^n[n\xi^2 - 2(n+1)\xi z + n + 2]}{(\xi - z)^{n+3}} = \frac{d}{d\xi}\left[\frac{(\xi^2 - 1)^{n+1}}{(\xi - z)^{n+2}}\right],$$

so that the integrand is the derivative of a function. Since the contour of integration is closed, the lower and upper limits of integration coincide and the integral vanishes. So, the Rodrigues formula indeed yields Legendre polynomials.

Example 26.5.4. As an illustration of the use of the Rodrigues formula, let us evaluate the integral

$$I = \int_{-1}^{1} x^k P_n(x) \, dx \qquad \text{for} \quad k \le n.$$

The procedure is to replace $P_n(x)$ by the RHS of Equation (26.44) and integrate by parts repeatedly. After one integration by parts, we get

$$I = \frac{1}{2^n n!} \int_{-1}^{1} \underbrace{x^k}_{u} \underbrace{\frac{d^n}{dx^n}\left[(x^2 - 1)^n\right] dx}_{dv}$$

$$= \frac{1}{2^n n!}\left\{ x^k \frac{d^{n-1}}{dx^{n-1}}\left[(x^2 - 1)^n\right]\Big|_{-1}^{1} - \int_{-1}^{1} kx^{k-1}\frac{d^{n-1}}{dx^{n-1}}\left[(x^2 - 1)^n\right] dx \right\}.$$

The first term on the RHS of the second line is zero because each differentiation reduces the power of $(x^2 - 1)^n$ by at most one unit. So after $n - 1$ differentiations, we get a sum of terms each having $(x^2 - 1)$ raised to various powers, *with the lowest power being one*. All these terms vanish at $x = 1$ as well as at $x = -1$. Continuing the integration by parts, we get

$$I = \underbrace{\frac{-kx^{k-1}}{2^n n!}\frac{d^{n-2}}{dx^{n-2}}\left[(x^2 - 1)^n\right]\Big|_{-1}^{1}}_{=0 \text{ for same reason as above}} + \frac{(-1)^2}{2^n n!}\int_{-1}^{1} k(k-1)x^{k-2}\frac{d^{n-2}}{dx^{n-2}}\left[(x^2 - 1)^n\right] dx.$$

After k integrations by part, we obtain

$$I = \frac{(-1)^k}{2^n n!}k!\int_{-1}^{1}\frac{d^{n-k}}{dx^{n-k}}\left[(x^2 - 1)^n\right] dx$$

because after k differentiations x^k yields $k!$. Now, if $k < n$, the integral vanishes for the same reason as above. If $n = k$, no differentiation will be left, and we have

$$I = \frac{(-1)^n}{2^n n!}n!\int_{-1}^{1}(x^2 - 1)^n \, dx.$$

Problem 26.14 shows how to evaluate the final integral and obtain

$$\int_{-1}^{1} (x^2 - 1)^n \, dx = (-1)^n 2^{2n+1} \frac{(n!)^2}{(2n+1)!}.$$

Therefore,

$$I = \frac{(-1)^n}{2^n} \int_{-1}^{1} (x^2 - 1)^n \, dx = \frac{2^{n+1}(n!)^2}{(2n+1)!}.$$

We summarize the above derivation

$$\int_{-1}^{1} x^k P_n(x) \, dx = \begin{cases} 0 & \text{if } k < n, \\ \dfrac{2^{n+1}(n!)^2}{(2n+1)!} & \text{if } k = n. \end{cases} \tag{26.45}$$

If instead of x^k we have a general polynomial of order k in x with $n > k$, the integral will still vanish. ∎

The result of the preceding example is summarized as

Box 26.5.1. *Any polynomial of degree less than n is orthogonal to P_n.*

26.6   Expansions in Legendre Polynomials

The orthogonality of Legendre polynomials—as the orthogonality of the Fourier trigonometric functions—makes them very useful for expansion of functions defined in the interval $(-1, +1)$. Let $f(x)$ be such a function. Then we write

$$f(x) = \sum_{n=0}^{\infty} c_n P_n(x) \tag{26.46}$$

and seek to find c_n. But c_n can be obtained by multiplying both sides of the series by $P_m(x)$ and integrating from -1 to $+1$. On the LHS, we get $\int_{-1}^{1} f(x) P_m(x) \, dx$, and on the RHS

$$\int_{-1}^{1} \left(\sum_{n=0}^{\infty} c_n P_n(x) \right) P_m(x) \, dx = \sum_{n=0}^{\infty} c_n \underbrace{\int_{-1}^{1} P_n(x) P_m(x) \, dx}_{[2/(2n+1)]\delta_{mn} \text{ by } (26.43)} = c_m \frac{2}{2m+1}.$$

Equating the RHS and the LHS, we obtain

$$c_m = \frac{2m+1}{2} \int_{-1}^{1} f(x) P_m(x) \, dx \quad \text{or} \quad c_n = \frac{2n+1}{2} \int_{-1}^{1} f(x) P_n(x). \tag{26.47}$$

Equations (26.46) and (26.47) give a procedure for expanding an arbitrary function defined in the interval $(-1, +1)$ in terms of Legendre polynomials.

If $f(x)$ happens to be a polynomial of degree k, then it can be written as a *finite sum* of Legendre polynomials of degree k and less. In fact, for $f(x) = x^k$, we have

$$c_n = \frac{2n+1}{2} \int_{-1}^{1} x^k P_n(x) dx = 0 \qquad \text{for} \quad n > k$$

by Box 26.5.1. Thus the coefficients in the sum (26.46) beyond k are all zero.

Example 26.6.1. We want to find the Legendre expansion of a function $f(x)$ defined as

$$f(x) = \begin{cases} V_0 & \text{if} \quad 0 < x \leq 1, \\ -V_0 & \text{if} \quad -1 \leq x < 0. \end{cases}$$

To find the coefficients of expansion, we use Equation (26.47):

$$\begin{aligned} c_n &= \frac{2n+1}{2} \int_{-1}^{1} f(x) P_n(x)\, dx \\ &= \frac{2n+1}{2} \int_{-1}^{0} \underbrace{f(x)}_{=-V_0} P_n(x)\, dx + \frac{2n+1}{2} \int_{0}^{1} \underbrace{f(x)}_{=+V_0} P_n(x)\, dx \qquad (26.48) \\ &= \frac{2n+1}{2} V_0 \left[-\int_{-1}^{0} P_n(x)\, dx + \int_{0}^{1} P_n(x)\, dx \right]. \end{aligned}$$

In the first integral of the last line, we make the substitution $x = -y$ so that

$$\int_{-1}^{0} P_n(x)\, dx = \int_{+1}^{0} P_n(-y)\,(-dy) = \int_{0}^{1} P_n(-y)\, dy = (-1)^n \int_{0}^{1} P_n(x)\, dx,$$

where we used (26.35) and, in the last equality, we changed the dummy variable of integration from y to x (Section 3.2). Inserting this in (26.48), we obtain

$$\begin{aligned} c_n &= \frac{2n+1}{2} V_0 [1 - (-1)^n] \int_{0}^{1} P_n(x)\, dx, \\ &= \frac{2n+1}{2} V_0 \begin{cases} 0 & \text{if} \quad n \text{ is even} \\ 2\int_{0}^{1} P_{2k+1}(x)\, dx & \text{if} \quad n = 2k+1 \end{cases} \end{aligned}$$

where we have written the odd n as $2k + 1$ for $k = 0, 1, \ldots$.

It remains to evaluate the integral of a Legendre polynomial of odd order in the interval $(0, 1)$. To this end, we use the Rodrigues formula:

$$\begin{aligned} \int_{0}^{1} P_{2k+1}(x)\, dx &= \frac{1}{2^{2k+1}(2k+1)!} \int_{0}^{1} \frac{d^{2k+1}}{dx^{2k+1}} \left[(x^2 - 1)^{2k+1} \right] dx \\ &= \frac{1}{2^{2k+1}(2k+1)!} \frac{d^{2k}}{dx^{2k}} \left[(x^2 - 1)^{2k+1} \right] \Big|_{0}^{1} \\ &= \frac{1}{2^{2k+1}(2k+1)!} \left\{ \frac{d^{2k}}{dx^{2k}} \left[(x^2 - 1)^{2k+1} \right] \Big|_{x=1} - \frac{d^{2k}}{dx^{2k}} \left[(x^2 - 1)^{2k+1} \right] \Big|_{x=0} \right\}. \end{aligned}$$

The first term gives zero because there is no sufficient number of differentiations to get rid of all factors of $(x^2 - 1)$. For the second term, we note that $(x^2 - 1)^{2k+1}$ is a polynomial in x whose derivatives of various orders consist of powers of x. These powers will give zero at $x = 0$ except for the constant term (of zeroth power). So, let us use binomial expansion for $(x^2 - 1)^{2k+1}$ which is equal to $-(1 - x^2)^{2k+1}$:

$$\frac{d^{2k}}{dx^{2k}}\left[(x^2 - 1)^{2k+1}\right]\Bigg|_{x=0} = -\frac{d^{2k}}{dx^{2k}}\left[\sum_{j=0}^{2k+1} \frac{(2k+1)!}{j!(2k+1-j)!}(-x^2)^j\right]\Bigg|_{x=0}$$

$$= -\sum_{j=0}^{2k+1} \frac{(2k+1)!}{j!(2k+1-j)!}(-1)^j \frac{d^{2k}}{dx^{2k}}\left(x^{2j}\right)\Bigg|_{x=0},$$

whose constant term is obtained when $k = j$, all the other terms of the sum will vanish either because of too many differentiations (when $j < k$, we end up differentiating constants) or too few differentiations (when $j > k$, a power of x will remain which evaluates to zero at $x = 0$). Therefore,

$$\frac{d^{2k}}{dx^{2k}}\left[(x^2 - 1)^{2k+1}\right]\Bigg|_{x=0} = -\frac{(2k+1)!}{k!(k+1)!}(-1)^k \frac{d^{2k}}{dx^{2k}}\left(x^{2k}\right)\Bigg|_{x=0}$$

$$= \frac{(2k+1)!}{k!(k+1)!}(-1)^{k+1}(2k)!$$

and

$$\int_0^1 P_{2k+1}(x)\,dx = -\frac{1}{2^{2k+1}(2k+1)!}\left[\frac{(2k+1)!}{k!(k+1)!}(-1)^{k+1}(2k)!\right] = \frac{(-1)^k(2k)!}{2^{2k+1}k!(k+1)!}. \tag{26.49}$$

Finally, we can write the coefficient c_{2k+1} as

$$c_{2k+1} = 2\frac{2(2k+1)+1}{2}V_0\int_0^1 P_{2k+1}(x)\,dx = \frac{(-1)^k(4k+3)(2k)!}{2^{2k+1}k!(k+1)!}V_0$$

with $c_n = 0$ for even n. The final expansion series can now be given:

$$f(x) = \begin{cases} V_0 & \text{if } 0 < x \le 1 \\ -V_0 & \text{if } -1 \le x < 0 \end{cases} = V_0\sum_{k=0}^{\infty} \frac{(-1)^k(4k+3)(2k)!}{2^{2k+1}k!(k+1)!}P_{2k+1}(x)$$

$$= V_0\left[\tfrac{3}{2}P_1(x) - \tfrac{7}{8}P_3(x) + \tfrac{11}{16}P_5(x) - \cdots\right]. \qquad \blacksquare$$

Example 26.6.2. We can easily obtain the Legendre expansion of the Dirac delta function. The expansion coefficients are given by

$$c_n = \frac{2n+1}{2}\int_{-1}^{1} f(x)P_n(x)\,dx = \frac{2n+1}{2}\int_{-1}^{1} \delta(x)P_n(x)\,dx = \frac{2n+1}{2}P_n(0).$$

From Equation (26.39) and the discussion preceding it we can find all values of $P_n(0)$. Substituting these values in the above equation, we conclude that $c_n = 0$ if n is odd, and

$$c_{2k} = \frac{4k+1}{2}\left[(-1)^k \frac{(2k)!}{2^{2k}(k!)^2}\right].$$

Legendre
expansion of the
Dirac delta
function
It now follows that

$$\delta(x) = \sum_{k=0}^{\infty}(-1)^k \frac{(4k+1)(2k)!}{2^{2k+1}(k!)^2}P_{2k}(x). \qquad \blacksquare$$

26.7 Physical Examples

The most common physical problems involving Laplace's equation are those from electrostatics in empty space, and steady-state heat transfer. In each case, a surface is held at some (not necessarily uniform) potential or temperature and the potential or temperature is sought in regions away from the surface. In the present context, these surfaces are typically (portions of) spheres.

Example 26.7.1. Two solid heat-conducting hemispheres of radius a, separated by a very small insulating gap, form a sphere. The two halves of the sphere are in contact—on the outside—with two (infinite) heat baths at temperatures T_0 and $-T_0$ [Figure 26.1(a)]. We want to find the temperature distribution $T(r, \theta, \varphi)$ inside the sphere. We choose a spherical coordinate system in which the origin coincides with the center of the sphere and the polar axis is perpendicular to the equatorial plane. The hemisphere with temperature T_0 is assumed to constitute the northern hemisphere.

two solid heat-conducting hemispheres held at temperatures T_0 and $-T_0$

Since the problem has azimuthal symmetry, T is independent of φ, and we can immediately write the general solution from Equation (26.29). However, since the origin is in the region of interest, we need to exclude all negative powers of r. This is accomplished by setting all the B coefficients equal to zero. Thus, we have

$$T(r, \theta) = \sum_{n=0}^{\infty} A_n r^n P_n(\cos \theta). \qquad (26.50)$$

It remains to calculate the constants A_n. This is done by noting that

$$T(a, \theta) = \begin{cases} T_0 & \text{if } 0 \leq \theta < \dfrac{\pi}{2}, \\ -T_0 & \text{if } \dfrac{\pi}{2} < \theta \leq \pi. \end{cases}$$

In terms of $u = \cos \theta$, this is written as

$$T(a, u) = \begin{cases} -T_0 & \text{if } -1 \leq u < 0, \\ T_0 & \text{if } 0 < u \leq 1. \end{cases}$$

Substituting this in Equation (26.50), we obtain

$$T(a, \theta) = \begin{cases} -T_0 & \text{if } -1 \leq u < 0 \\ T_0 & \text{if } 0 < u \leq 1 \end{cases} = \sum_{n=0}^{\infty} \underbrace{A_n a^n}_{\equiv c_n} P_n(u), \qquad (26.51)$$

which—except for using u instead of x—is entirely equivalent to the expansion of Example 26.6.1, where we found that even coefficients are absent and

$$c_{2k+1} \equiv A_{2k+1} a^{2k+1} = \frac{(-1)^k (4k+3)(2k)!}{2^{2k+1} k!(k+1)!} T_0.$$

Finding A_{2k+1} from this equation and inserting the result in (26.50) yields

$$T(r, \theta) = T_0 \sum_{k=0}^{\infty} \frac{(-1)^k (4k+3)(2k)!}{2^{2k+1} k!(k+1)!} \left(\frac{r}{a}\right)^{2k+1} P_{2k+1}(\cos \theta), \qquad (26.52)$$

where we have substituted $\cos \theta$ for u. ∎

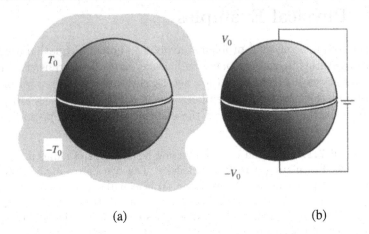

(a)   (b)

Figure 26.1: (a) Two heat-conducting hemispheres held at two different temperatures. (b) Two electrically conducting hemispheres held at two different potentials. The upper hemispheres have the polar angle range $0 \leq \theta < \pi/2$ or $0 < \cos\theta \leq 1$, and the lower hemispheres have the range $\pi/2 < \theta \leq \pi$ or $-1 \leq \cos\theta < 0$.

two electrically conducting hemispheres held at potentials V_0 and $-V_0$

Example 26.7.2. Consider two electrically conducting hemispheres of radius a separated by a small insulating gap at the equator. The upper hemisphere is held at potential V_0 and the lower one at $-V_0$ as shown in Figure 26.1(b). We want to find the potential at points outside the resulting sphere. Since the potential must vanish at infinity, we expect the first term in Equation (26.29) to be absent, i.e., $A_k = 0$. To find B_k, substitute a for r in (26.29), and let $\cos\theta \equiv u$. Then,

$$\Phi(a, u) = \sum_{k=0}^{\infty} \underbrace{\frac{B_k}{a^{k+1}}}_{\equiv c_k} P_k(u),$$

where

$$\Phi(a, u) = \begin{cases} -V_0 & \text{if } -1 < u < 0, \\ +V_0 & \text{if } 0 < u < 1. \end{cases}$$

The calculation of the coefficients is identical to that of Example 26.6.1. Thus, $c_k = 0$ for even k and

$$c_{2m+1} = \frac{B_{2m+1}}{a^{2m+2}} = (-1)^m \frac{(4m+3)(2m)!}{2^{2m+1}(m+1)!m!} V_0$$

or

$$B_{2m+1} = \frac{(-1)^m (4m+3)(2m)!}{2^{2m+1} m!(m+1)!} a^{2m+2} V_0.$$

Having found the coefficients, we can write the potential:

$$\Phi(r, \theta) = V_0 \sum_{m=0}^{\infty} (-1)^m \frac{(4m+3)(2m)!}{2^{2m+1} m!(m+1)!} \left(\frac{a}{r}\right)^{2m+2} P_{2m+1}(\cos\theta), \qquad (26.53)$$

where $\cos\theta$ has been restored. Equation (26.53) is the multipole expansion of the potential of the two hemispheres. It is interesting to note that the monopole term (the term with a single power of r in the denominator) is absent. It follows from Equation (10.33) that the total charge on the two spheres must be zero. This is consistent with the symmetry of the problem from which we expect equal surface charge densities of opposite signs on the two hemispheres.   ∎

Example 26.7.3. As yet another example of the solution of Laplace's equation in spherical coordinates, consider a grounded neutral conducting sphere of radius a placed in an originally uniform electric field E_0 which is assumed to be infinite in extent (see Figure 26.2). We want to find the electrostatic potential everywhere outside the sphere. Choosing the field to be in the positive z-direction and placing the center of the sphere at the origin, we will have a problem that is azimuthally symmetric. The general solution is therefore given by Equation (26.29). The boundaries outside the sphere consist of the sphere itself as well as infinity. The electric field at infinity is the original uniform field, because the field due to the charges induced on the sphere vanishes at infinity. The potential of this field (at infinity) can be deduced from[14]

(margin note:) conducting sphere in an originally uniform electric field

$$\mathbf{E} = E_0\hat{\mathbf{e}}_z = -\boldsymbol{\nabla}\Phi \;\Rightarrow\; E_0 = -\frac{\partial\Phi}{\partial z}, \quad \frac{\partial\Phi}{\partial x} = 0 = \frac{\partial\Phi}{\partial y}.$$

Thus, the potential at infinity is independent of x and y, and can be written as

$$\Phi(r,\theta) = -E_0 z = -E_0 r\cos\theta = -E_0 r P_1(\cos\theta) \qquad \text{for} \quad r \to \infty.$$

As $r \to \infty$, the B terms in Equation (26.29) will go to zero, and we must have the "limiting" equality

$$\sum_{k=0}^{\infty} A_k r^k P_k(u) \to -E_0 r P_1(u).$$

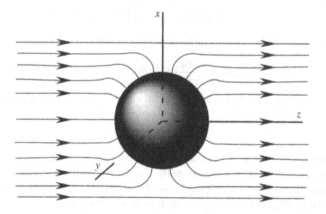

Figure 26.2: The electric field in the vicinity of a sphere placed in an external uniform field will change, but the field far away from the sphere will remain almost uniform.

---

[14] We could express the gradient in terms of spherical coordinates, but, as the reader will note, the initial manipulation is noticeably easier in the Cartesian coordinates.

The orthogonality of the Legendre polynomials requires the coefficients on both sides to be equal. This gives

$$A_k = 0 \quad \text{for } k = 0 \text{ and } k \geq 2, \quad A_1 = -E_0.$$

The B's are obtained by applying the boundary condition of the sphere itself, namely the fact that it is grounded. This means that $\Phi(a, \theta) = 0$, or

$$0 = A_1 a P_1(u) + \sum_{k=0}^{\infty} \frac{B_k}{a^{k+1}} P_k(u) = \frac{B_0}{a} + \left(\frac{B_1}{a^2} - E_0 a \right) P_1(u) + \sum_{k=2}^{\infty} \frac{B_k}{a^{k+1}} P_k(u).$$

Again orthogonality of the Legendre polynomials requires the coefficient of each polynomial to vanish. This yields

$$B_0 = 0, \quad B_1 = E_0 a^3, \quad \text{and} \quad B_k = 0 \quad \text{for } k \geq 2.$$

Inserting all these coefficients in Equation (26.29), we obtain

$$\Phi(r, \theta) = -E_0 \left(r - \frac{a^3}{r^2} \right) P_1(\cos \theta). \tag{26.54}$$

Because of the simplicity of the expression for potential, we can evaluate some other physical quantities of interest. For example, the electric field at all points in space is

$$\mathbf{E} = -\boldsymbol{\nabla} \Phi = -\hat{\mathbf{e}}_r \frac{\partial \Phi}{\partial r} - \hat{\mathbf{e}}_\theta \frac{1}{r} \frac{\partial \Phi}{\partial \theta}$$

or

$$E_r = -\frac{\partial \Phi}{\partial r} = E_0 \left(1 + 2 \frac{a^3}{r^3} \right) \cos \theta,$$

$$E_\theta = -\frac{1}{r} \frac{\partial \Phi}{\partial r} = -E_0 \left(1 - \frac{a^3}{r^3} \right) \sin \theta,$$

$$E_\varphi = 0.$$

This is the sum of the original uniform field

$$E_0 (\cos \theta \hat{\mathbf{e}}_r - \sin \theta \hat{\mathbf{e}}_\theta) = E_0 \hat{\mathbf{e}}_z$$

and the field due to the charges induced on the sphere

$$E_{\text{sph}} = \frac{a^3 E_0}{r^3} (2 \cos \theta \hat{\mathbf{e}}_r + \sin \theta \hat{\mathbf{e}}_\theta),$$

which (see Example 16.2.1) is the field of an electric dipole with dipole moment

$$p = \frac{a^3 E_0}{k_e} = 4\pi \epsilon_0 a^3 E_0.$$

It is interesting to note that at $r = a$, the only nonvanishing component of the field is E_r. This is consistent with the known fact that electrostatic fields are perpendicular to conducting surfaces. Furthermore, this perpendicular field is related to the surface charge density by $E = \sigma/\epsilon_0$. Therefore,

$$\sigma = \epsilon_0 E_r \big|_{r=a} = 3\epsilon_0 E_0 \cos \theta,$$

indicating an accumulation of positive charge on the "upper" (right in the figure) hemisphere and an identical distribution of negative charge on the "lower" (left in the figure) hemisphere. ∎

26.8 Problems

26.1. Show that by writing $P(u) \equiv \Theta(\theta)$—with $u = \cos\theta$—and using the chain rule, the second equation of (26.2) becomes

$$-\frac{1}{\sin\theta}\frac{d}{d\theta}\left[(1-u^2)\frac{dP}{du}\right] + \alpha P = 0.$$

26.2. Choose a solution of the form $u^r \sum_{n=0}^{\infty} a_n u^n$ for the Legendre DE, assume that a_0 and a_1 are both nonzero, and show that the only solution for r is $r = 0$.

26.3. Derive Equation (26.15).

26.4. Derive Equations (26.18), (26.19), and (26.20).

26.5. Derive Equations (26.21) and (26.22) and show that they can both be written as (26.23).

26.6. Show by mathematical induction (or otherwise) that Equation (26.24) satisfies $P_n(1) = 1$.

26.7. Show that Legendre polynomials and the hypergeometric function are related via (26.25) and (26.26).

26.8. Suppose that Q represents electric charge. Show that in (26.34) Q_0 is the total charge and Q_1 is the dot product of $\hat{\mathbf{e}}_r$ and the electric dipole moment.

26.9. (a) Change t to $-t$ and u to $-u$, and show that the generating function $g(t, u)$ of Legendre polynomials does not change.
(b) Now substitute $-t$ for t and $-u$ for u in Equation (26.30) and compare the resulting equation with (26.30) to derive the parity of Legendre polynomials.

26.10. (a) Show that $(1/t)[\ln(1+t) - \ln(1-t)]$ is an even function of t.
(b) Use the Maclaurin expansion of $\ln(1 \pm t)$ to derive the following series:

$$\frac{1}{t}[\ln(1+t) - \ln(1-t)] = 2\sum_{k=0}^{\infty} \frac{t^{2k}}{2k+1}.$$

26.11. (a) Show that $P_n(0) = 0$ if n is odd.
(b) Show that for $u = 0$, Equation (26.38) yields

$$P_{2n}(0) = -\frac{2n-1}{2n}P_{2n-2}(0).$$

(c) Iterate this relation and obtain

$$P_{2n}(0) = (-1)^n \frac{(2n-1)!!}{(2n)!!}P_0(0) = (-1)^n \frac{(2n-1)!!}{(2n)!!}.$$

Now use the result of Problem 11.1 to obtain the final form of (26.39).

26.12. Suppose $f(x) = \sum_{k=0}^{\infty} c_k P_k(x)$. Show that

$$\int_{-1}^{1} [f(x)]^2 \, dx = \sum_{m=0}^{\infty} \frac{2c_m^2}{2m+1}.$$

26.13. Show the following two equalities:

$$(1 - z^2)\frac{d^2 P}{dz^2} - 2z\frac{dP}{dz} + n(n+1)$$
$$= \frac{n+1}{2^n(2\pi i)} \oint_C \frac{(\xi^2 - 1)^n[n\xi^2 - 2(n+1)\xi z + n + 2]}{(\xi - z)^{n+3}} \, d\xi$$
$$= \frac{n+1}{2^n(2\pi i)} \oint_C \frac{d}{d\xi}\left[\frac{(\xi^2 - 1)^{n+1}}{(\xi - z)^{n+2}}\right] \, d\xi.$$

26.14. In the integral $\int_{-1}^{1}(x^2 - 1)^n \, dx$, let $u = (x^2 - 1)^n$ and $dv = dx$ and integrate by parts to show that

$$\int_{-1}^{1}(x^2 - 1)^n \, dx = -2n\int_{-1}^{1} x^2(x^2 - 1)^n \, dx.$$

Integrate by parts a few more times and show that

$$\int_{-1}^{1}(x^2 - 1)^n \, dx = (-2)^m\frac{n(n-1)\ldots(n-m+1)}{(2m-1)!!}\int_{-1}^{1} x^{2m}(x^2 - 1)^{n-m} \, dx.$$

Set $m = n$ and, using the result of Problem 11.1, obtain the following final result:

$$\int_{-1}^{1}(x^2 - 1)^n \, dx = (-1)^n 2^{2n+1}\frac{(n!)^2}{(2n+1)!}.$$

26.15. Use the procedures of Example 26.5.4 and the previous problem to show that for $m \geq n$:

$$\int_{-1}^{1} x^m P_n(x) \, dx = \begin{cases} 0 & \text{if } m \text{ and } n \text{ have opposite parities,} \\ \frac{2^{(m+n)/2+1} m!(\frac{m+n}{2})!}{(m-n)!(m+n+1)!} & \text{if } m \text{ and } n \text{ have the same parities,} \end{cases}$$

where having the same parity means being both even or both odd.

26.16. Show that $\int_0^1 P_{2k}(x) \, dx = 0$ if $k \geq 1$. Hint: Extend the interval of integration to $(-1, 1)$ and use the orthogonality of Legendre polynomials.

26.17. Find the Legendre expansion for the function $f(x) = |x|$ in the interval $(-1, +1)$. Hint: Break up the integrals into two pieces, employ the recurrence relation to express $xP_n(x)$ in terms of $P_{n-1}(x)$ and $P_{n+1}(x)$, and use the result of Example 26.6.1.

26.18. (a) Find the total charge on the upper and lower hemispheres and on the entire sphere of Example 26.7.3.
(b) Using $\mathbf{p} = \int \mathbf{r}' dq(\mathbf{r}')$, calculate the (induced) dipole moment of the sphere.

26.19. Suppose that the sphere of Example 26.7.3 is held at potential V_0.
(a) Find the potential $\Phi(r, \theta)$ and the electrostatic field at all points in space.
(b) Calculate the surface charge density on the sphere.
(c) Find the total charge on the upper and lower hemispheres and on the entire sphere.

26.20. Using the infinite series expansion, find the electrostatic potential both inside and outside a conducting sphere of radius a held at the constant potential V_0.

26.21. Find the electrostatic potential inside a sphere of radius a with an insulating small gap at the equator if the bottom hemisphere is grounded and the top hemisphere is maintained at a constant potential V_0.

26.22. A sphere of radius a is maintained at a temperature T_0. The sphere is inside a large heat-conducting mass. Find the expressions for the steady-state temperature distribution both inside and outside the sphere.

26.23. A ring of total charge q and radius a in the xy-plane with its center at the origin constitutes an azimuthally symmetric charge distribution whose potential is also azimuthally symmetric.
(a) Write the most general potential function valid for $r > a$.
(b) By direct integration show that

$$\Phi(r, \theta = 0) = \frac{1}{4\pi\epsilon_0} \int \frac{dq(\mathbf{r}')}{|\mathbf{r} - \mathbf{r}'|}\bigg|_{\theta=0} = \frac{q}{4\pi\epsilon_0} \frac{1}{\sqrt{r^2 + a^2}}.$$

(c) Expand this expression in powers of (a/r) and compare the result with the series in (a) to find the coefficients of Legendre expansion and show that

$$\Phi(r, \theta) = \frac{q}{4\pi\epsilon_0 r} \sum_{k=0}^{\infty} \frac{(-1)^k (2k)!}{2^{2k} (k!)^2} \left(\frac{a}{r}\right)^{2k} P_{2k}(\cos\theta).$$

(d) Find a similar expression for $\Phi(r, \theta)$ for $r < a$.

26.24. A conducting sphere of radius a is inside another conducting sphere of radius b. The inner sphere is held at potential V_1; the outer sphere at V_2. Find the potential inside the inner sphere, between the two spheres, and outside the outer sphere.

26.25. A conducting sphere of radius a is inside another conducting sphere of radius b which is composed of two hemispheres with an infinitesimal gap between them. The inner sphere is held at potential V_1. The upper half of the outer sphere is at potential $+V_2$ and its lower half at $-V_2$. Find the potential inside the inner sphere, between the two spheres, and outside the outer sphere.

26.26. A heat conducting sphere of radius a is composed of two hemispheres with an infinitesimal gap between them. The upper and lower halves of the sphere are in contact with heat baths of temperatures $+T_1$ and $-T_1$, respectively. The sphere is inside a second heat conducting sphere of radius b held at temperature T_2. Find the temperature inside the inner sphere, between the two spheres, and outside the outer sphere.

Chapter 27

Laplace's Equation: Cylindrical Coordinates

Before working specific examples of cylindrical geometry, let us consider a question that has more general implications. We saw in Chapter 22 that separation of variables led to ODEs in which certain constants appeared, and that different choices of signs for these constants can lead to a different functional form of the general solution. For example, an equation such as $d^2x/dt^2 - kx = 0$ can have exponential solutions if $k > 0$ and trigonometric solutions if $k < 0$. One cannot a priori assign a specific sign to k. Thus, the *general* form of the solution is indeterminate. However, once the *boundary conditions* are imposed, the unique solutions will emerge regardless of the initial functional form of the solutions. The following argument illustrates this point on the angular DE resulting from the separation of Laplace's equation in cylindrical coordinates.

27.1 The ODEs

The separation of variables $\Phi(\rho, \varphi, z) = R(\rho)S(\varphi)Z(z)$ for Laplace's equation $\nabla^2\Phi = 0$ yields the following three ODEs [see Equation (22.14) noting that $\lambda = 0$]. In what follows, we shall use λ for λ_1:

$$\frac{d}{d\rho}\left(\rho\frac{dR}{d\rho}\right) + \left(\lambda\rho + \frac{\mu}{\rho}\right)R = 0,$$

$$\frac{d^2S}{d\varphi^2} - \mu S = 0, \qquad \frac{d^2Z}{dz^2} - \lambda Z = 0. \qquad (27.1)$$

Let us concentrate on the second equation whose most general solution we
can write as

$$S(\varphi) = \begin{cases} Ae^{\sqrt{\mu}\,\varphi} + Be^{-\sqrt{\mu}\,\varphi} & \text{if} \quad \mu \neq 0, \\ C\varphi + D & \text{if} \quad \mu = 0. \end{cases} \tag{27.2}$$

No matter what type of boundary conditions are imposed on the potential Φ,
it must give the same value at φ and at $\varphi + 2\pi$ while keeping the other two
variables fixed.[1] This is because (ρ, φ, z) and $(\rho, \varphi + 2\pi, z)$ represent the same
physical point in space. It follows that

$$R(\rho)S(\varphi)Z(z) = R(\rho)S(\varphi + 2\pi)Z(z) \;\Rightarrow\; S(\varphi + 2\pi) = S(\varphi)$$

because the identity holds for all values of ρ and z. If the last relation is to
be true for the case of $\mu = 0$, we must have $C = 0$ and $S(\varphi) = D$. For $\mu \neq 0$,
Equation (27.2) yields

$$Ae^{\sqrt{\mu}\,(\varphi+2\pi)} + Be^{-\sqrt{\mu}\,(\varphi+2\pi)} = Ae^{\sqrt{\mu}\,\varphi} + Be^{-\sqrt{\mu}\,\varphi}$$

or

$$Ae^{\sqrt{\mu}\,\varphi}(e^{\sqrt{\mu}\,2\pi} - 1) + Be^{-\sqrt{\mu}\,\varphi}(e^{-\sqrt{\mu}\,2\pi} - 1) = 0.$$

This must hold for all φ. The only way that can happen (we want to keep A
and B nonzero) is to have

$$e^{\sqrt{\mu}\,2\pi} - 1 = 0 \qquad \text{and} \qquad e^{-\sqrt{\mu}\,2\pi} - 1 = 0$$

both of which are equivalent to $e^{\sqrt{\mu}\,2\pi} = 1$.[2] If we confine ourselves to real
μ, we get only trivial solutions. To avoid this, we have to have $\sqrt{\mu} = im$ for
$m = 0, \pm 1, \pm 2, \ldots$ or $\mu = -m^2$ for $m = 0, \pm 1, \pm 2, \ldots$. With this choice of μ,
the DE for $S(\varphi)$ becomes $S'' + m^2 S = 0$ whose general solution is a sum of
trigonometric functions. We summarize this finding:

Theorem 27.1.1. *For all physical problems for which the azimuthal angle
varies between 0 and 2π, one is forced to restrict the value of μ to the negative
of the square of an integer. The solution for the angular part then becomes*

$$S(\varphi) = A_m \cos m\varphi + B_m \sin m\varphi, \qquad m = 0, 1, 2, \ldots, \tag{27.3}$$

where A_m and B_m are constants that may differ for different m's.

The negative values of m will not give rise to any new solutions, so they
are not included in the range of m. The case of $\mu = 0$ need not be treated
separately, because the acceptable solution for this case is $S = D = $ const.,
which is what is obtained in (27.3) when $m = 0$.

---

[1]This argument is valid only for physical situations defined for the entire range of φ.
If the region of interest restricts φ to a subset of the interval $[0, 2\pi]$, the argument breaks
down.

[2]The second equation can be obtained by multiplying the first equation by $e^{-\sqrt{\mu}\,2\pi}$.

The DE for $Z(z)$ is independent of m and has an exponential solution if $\lambda > 0$ and a trigonometric solution if $\lambda < 0$. Assuming the former, and writing $\lambda \equiv l^2$, we have

$$Z(z) = Ae^{lz} + Be^{-lz}. \tag{27.4}$$

Least familiar is the radial DE which, in terms of $l = \sqrt{\lambda}$, can be rewritten as

$$\frac{d^2 R}{d\rho^2} + \frac{1}{\rho}\frac{dR}{d\rho} + \left(l^2 - \frac{m^2}{\rho^2}\right) R = 0. \tag{27.5}$$

Furthermore, if we define the variable $v = l\rho$, we can cast (27.5) in the form

$$\frac{d^2 R}{dv^2} + \frac{1}{v}\frac{dR}{dv} + \left(1 - \frac{m^2}{v^2}\right) R = 0. \tag{27.6}$$

Bessel differential equation

Equation (27.5), or (27.6), is one of the most famous DEs of mathematical physics called the **Bessel differential equation**. Our task for the remainder of this chapter is to find solutions of this DE and list some of their properties and examples of their usage.

Historical Notes

Friedrich Wilhelm Bessel showed no signs of unusual academic ability in school, although he did show a liking for mathematics and physics. He left school intending to become a merchant's apprentice, a desire that soon materialized with a seven-year unpaid apprenticeship with a large mercantile firm in Bremen. The young Bessel proved so adept at accounting and calculation that he was granted a small salary, with raises, after only the first year. An interest in foreign trade led Bessel to study geography and languages at night, astonishingly learning to read and write English in only three months. He also studied navigation in order to qualify as a cargo officer aboard ship, but his innate curiosity soon compelled him to investigate astronomy at a more fundamental level. Still serving his apprenticeship, Bessel learned to observe the positions of stars with sufficient accuracy to determine the longitude of Bremen, checking his results against professional astronomical journals. He then tackled the more formidable problem of determining the orbit of Halley's comet from published observations. After seeing the close agreement between Bessel's calculations and those of Halley, the German astronomer Olbers encouraged Bessel to improve his already impressive work with more observations. The improved calculations, an achievement tantamount to a modern doctoral dissertation, were published with Olbers's recommendation. Bessel later received appointments with increasing authority at observatories near Bremen and in Königsberg, the latter position being accompanied by a professorship. (The title of doctor, required for the professorship, was granted by the University of Göttingen on the recommendation of Gauss.)

Friedrich Wilhelm Bessel 1784–1846

Bessel proved himself an excellent observational astronomer. His careful measurements coupled with his mathematical aptitude allowed him to produce accurate positions for a number of previously mapped stars, taking account of instrumental effects, atmospheric refraction, and the position and motion of the observation site. In 1820 he determined the position of the vernal equinox accurate to 0.01 second, in agreement with modern values. His observation of the variation of the proper motion of the stars Sirius and Procyon led him to posit the existence of nearby, large, low-luminosity stars called dark companions. Between 1821 and 1833 he catalogued the positions of about 75,000 stars, publishing his measurements in detail. One of his most important contributions to astronomy was the determination of the distance to a star using parallax. This method uses triangulation, or the determination of the apparent positions of a distant object viewed from two points a known distance apart, in this case two diametrically opposed points of the Earth's orbit. The angle subtended by the baseline of the Earth's orbit, viewed from the star's perspective, is known as the star's parallax. Before Bessel's measurement, stars were assumed to be so distant that their parallaxes were too small to measure, and it was further assumed that bright stars (thought to be nearer) would have the largest parallax. Bessel correctly reasoned that stars with large proper motions were more likely to be nearby ones and selected such a star, 61 Cygni, for his historic measurement. His measured parallax for that star differs by less than 8% from the currently accepted value.

Given such an impressive record in astronomy, it seems only fitting that the famous functions that bear Bessel's name grew out of his investigations of perturbations in planetary systems. He showed that such perturbations could be divided into two effects and treated separately: the obvious direct attraction due to the perturbing planet and an indirect effect caused by the Sun's response to the perturber's force. The so-called Bessel functions then appear as coefficients in the series treatment of the indirect perturbation. Although special cases of Bessel functions were discovered by **Bernoulli**, **Euler**, and **Lagrange** the systematic treatment by Bessel clearly established his preeminence, a fitting tribute to the creator of the most famous functions in mathematical physics.

27.2   Solutions of the Bessel DE

The Frobenius method is an effective way of finding solutions for ODEs. We rewrite (27.6) by multiplying it by v^2 to turn all its coefficients into polynomials as suggested by Equation (26.7). This yields

$$v^2 \frac{d^2 R}{dv^2} + v \frac{dR}{dv} + \left(v^2 - m^2\right) R = 0. \tag{27.7}$$

Since v^2 vanishes at $v = 0$, we must assume a solution of the form

$$R(v) = v^s \sum_{k=0}^{\infty} c_k v^k = \sum_{k=0}^{\infty} c_k v^{k+s}$$

from which we obtain

$$v \frac{dR}{dv} = \sum_{k=0}^{\infty} c_k (k+s) v^{k+s},$$

$$v^2 \frac{d^2 R}{dv^2} = \sum_{k=0}^{\infty} c_k (k+s)(k+s-1) v^{k+s}.$$

Substituting these as well as $(v^2 - m^2) \sum_{k=0}^{\infty} c_k v^{k+s}$ in the DE yields

$$\sum_{k=0}^{\infty} c_k [\underbrace{k + s + (k + s)(k + s - 1)}_{=(k+s)^2} - m^2] v^{k+s} + \sum_{k=0}^{\infty} c_k v^{k+s+2} = 0.$$

To find the recursion relation, we need to have the same power of v in the sum. We do this by rewriting the first sum as

$$c_0 (s^2 - m^2) v^s + c_1 [(s+1)^2 - m^2] v^{s+1} + \sum_{k=2}^{\infty} c_k [(k+s)^2 - m^2] v^{k+s}$$

$$= c_0 (s^2 - m^2) v^s + c_1 [(s+1)^2 - m^2] v^{s+1}$$

$$+ \sum_{n=0}^{\infty} c_{n+2} [(n+2+s)^2 - m^2] v^{n+2+s},$$

where in the second line, we introduced $n = k - 2$. Since n is a dummy index, we can change it back to k. It then follows that

$$c_0 (s^2 - m^2) v^s + c_1 [(s+1)^2 - m^2] v^{s+1}$$

$$+ \sum_{k=0}^{\infty} \left\{ c_{k+2} [(k+2+s)^2 - m^2] + c_k \right\} v^{k+2+s} = 0.$$

Assuming that $c_0 \neq 0$ and setting the coefficients of all powers of v equal to zero, we get

$$s^2 = m^2, \qquad c_1 [(s+1)^2 - m^2] = 0,$$

$$c_{k+2} [(k+2+s)^2 - m^2] + c_k = 0.$$

The first equation gives $m = \pm s$. Inserting this in the second equation gives

$$c_1 (2s + 1) = 0 \ \Rightarrow \ c_1 = 0 \ \text{ or } \ s = -\tfrac{1}{2}.$$

The choice $s = -\tfrac{1}{2}$ gives $m = \mp \tfrac{1}{2}$ which is not acceptable,[3] as we decided that m is to be a positive integer. We therefore conclude that $s = \pm m$ and $c_1 = 0$. It follows from the recursion relation that all odd c's are zero. The Frobenius series will therefore look like

$$R(v) = v^{\pm m} \sum_{k=0}^{\infty} c_{2k} v^{2k}, \qquad \frac{c_{2k+2}}{c_{2k}} = -\frac{1}{(2k+2+s)^2 - m^2}. \qquad (27.8)$$

The ratio test for the convergence of series yields

$$\lim_{k \to \infty} \left| \frac{c_{2k+2} v^{2k+2}}{c_{2k} v^{2k}} \right| = \lim_{k \to \infty} \left| \frac{1}{(2k+2+s)^2 - m^2} \right| v^2 = 0,$$

---

[3] Actually, problems arising from other areas of physics beyond electrostatics and steady-state heat transfer allow noninteger values of m. However, we shall not deal with such problems here.

which indicates that

> **Box 27.2.1.** *The series of Equation (27.8) is convergent for all values of v.*

recursion relation
for Bessel
equation
We now use the recursion relation to obtain the coefficients of expansion. Rewrite the recursion relation as

$$c_{k+2} = -\frac{1}{(k+2+s)^2 - s^2}c_k = -\frac{1}{(k+2)(2s+k+2)}c_k,$$

where we substituted s^2 for m^2. This gives

$$c_2 = -\frac{1}{2(2s+2)}c_0,$$

$$c_4 = -\frac{1}{4(2s+4)}c_2 = (-1)^2\frac{1}{4(2s+4)}\frac{1}{2(2s+2)}c_0,$$

$$c_6 = -\frac{1}{6(2s+6)}c_4 = (-1)^3\frac{1}{6(2s+6)}\frac{1}{4(2s+4)}\frac{1}{2(2s+2)}c_0,$$

and, in general,

$$c_{2k} = (-1)^k\underbrace{\frac{1}{2k\cdot(2k-2)\ldots 2}}_{=2^k k!}\underbrace{\frac{1}{(2s+2k)[2s+(2k-2)]\ldots(2s+2)}}_{=2^k(s+k)(s+k-1)\ldots(s+1)}c_0.$$

Multiplying the numerator and denominator by $s!$, we obtain

$$c_{2k} = (-1)^k\frac{s!}{2^{2k}k!(s+k)!}c_0. \tag{27.9}$$

Substituting (27.9) in (27.8) yields

$$R(v) = c_0 s! v^s \sum_{k=0}^{\infty}\frac{(-1)^k}{2^{2k}k!(s+k)!}v^{2k} = c_0 s! 2^s\left(\frac{v}{2}\right)^s\sum_{k=0}^{\infty}\frac{(-1)^k}{k!(s+k)!}\left(\frac{v}{2}\right)^{2k},$$

where we substituted s for $\pm m$ in the exponent of v outside the summation. We also absorbed the powers of 2 in the denominator of the sum into the powers of v, and outside the sum, we multiplied and divided by 2^s. It is customary to choose the arbitrary constant c_0 to be equal to $1/(s!2^s)$. This leads to

> **Box 27.2.2.** *The Bessel function of order s is denoted by J_s and is given by the series*
>
> $$J_s(x) = \left(\frac{x}{2}\right)^s\sum_{k=0}^{\infty}\frac{(-1)^k}{k!(s+k)!}\left(\frac{x}{2}\right)^{2k} \tag{27.10}$$
>
> *which is convergent for all values of x.*

Although Equation (27.10) was derived assuming that m—and therefore s—was an integer, lifting this restriction will still yield a series which is convergent everywhere, and one can define Bessel functions whose orders are real or even complex numbers. The only difficulty is to correctly interpret $(s+n)!$ for non-integer s. But this is precisely what the gamma function was invented for (see Definition 11.1.1). Thus, we let Equation (27.10) stand for Bessel functions of *all* orders.

Equation (27.10) is valid not only for integer s, but also for real and even complex s.

27.3 Second Solution of the Bessel DE

As in the case of Legendre polynomials, we can obtain a second solution of the Bessel DE using Equation (24.6). For the Bessel DE, we have $p(x) = 1/x$. Using $J_m(x)$ as our input, we can generate another solution. With $C = 0$ in Equation (24.6), we obtain

$$Z_m(x) = K J_m(x) \int_\alpha^x \frac{1}{J_m^2(u)} \exp\left[-\int_c^u \frac{dt}{t}\right] du = A_m J_m(x) \int_\alpha^x \frac{du}{u J_m^2(u)},$$

where $A_m \equiv Kc$ and α are arbitrary constants determined by convention. Note that, contrary to $J_m(x)$, $Z_m(x)$ is not well behaved at $x = 0$ due to the presence of u in the denominator of the integrand.

Although the above procedure manufactures a second solution for the Bessel DE, it is not the customary procedure. It turns out that for *non-integer s*, the Bessel function $J_{-s}(x)$ is independent of $J_s(x)$ and can be used as a second solution.[4] However, a more common second solution is the linear combination

$$Y_s(x) = \frac{J_s(x)\cos s\pi - J_{-s}(x)}{\sin s\pi} \tag{27.11}$$

called the Bessel function of the second kind, or the **Neumann function**. For integer s the function is indeterminate because of Equation (11.32) and the identity $\cos n\pi = (-1)^n$. Therefore, we use l'Hôpital's rule and define

Bessel function of the second kind or Neumann function

$$Y_n(x) \equiv \lim_{s\to n} Y_s(x) = \lim_{s\to n} \frac{\frac{\partial}{\partial s}[J_s(x)\cos s\pi - J_{-s}(x)]}{\pi \cos n\pi}$$

$$= \frac{1}{\pi} \lim_{s\to n} \left[\frac{\partial J_s}{\partial s} - (-1)^n \frac{\partial J_{-s}}{\partial s}\right],$$

From (27.10) we obtain

$$\frac{\partial J_s}{\partial s} = J_s(x) \ln\left(\frac{x}{2}\right) - \left(\frac{x}{2}\right)^s \sum_{k=0}^\infty (-1)^k \frac{\Psi(s+k+1)}{k!\Gamma(s+k+1)} \left(\frac{x}{2}\right)^{2k},$$

where

$$\Psi(x) \equiv \frac{d}{dx} \ln[(x-1)!] = \frac{d}{dx} \ln \Gamma(x) = \frac{d\Gamma(x)/dx}{\Gamma(x)}.$$

---

[4]See Hassani, S. *Mathematical Physics: A Modern Introduction to Its Foundations*, Springer-Verlag, 1999, Chapter 14 for details.

Similarly,

$$\frac{\partial J_{-s}}{\partial s} = -J_{-s}(x)\ln\left(\frac{x}{2}\right) + \left(\frac{x}{2}\right)^{-s}\sum_{k=0}^{\infty}(-1)^k\frac{\Psi(-s+k+1)}{k!\Gamma(-s+k+1)}\left(\frac{x}{2}\right)^{2k}.$$

Substituting these expressions in the definition of $Y_n(x)$ and using $J_{-n}(x) = (-1)^n J_n(x)$ [Equation (11.32)], we obtain

$$Y_n(x) = \frac{2}{\pi}J_n(x)\ln\left(\frac{x}{2}\right) - \frac{1}{\pi}\left(\frac{x}{2}\right)^n\sum_{k=0}^{\infty}(-1)^k\frac{\Psi(n+k+1)}{k!\Gamma(n+k+1)}\left(\frac{x}{2}\right)^{2k}$$

$$-\frac{1}{\pi}(-1)^n\left(\frac{x}{2}\right)^{-n}\sum_{k=0}^{\infty}(-1)^k\frac{\Psi(k-n+1)}{k!\Gamma(k-n+1)}\left(\frac{x}{2}\right)^{2k}. \qquad (27.12)$$

It should be clear from (27.12) that the Neumann function $Y_s(x)$ is ill defined at $x = 0$, as expected of the second solution of the Bessel DE such as $Z_m(x)$ discussed above.

Since $Y_s(x)$ is linearly independent of $J_s(x)$ for any s, integer or noninteger, it is convenient to consider $\{J_s(x), Y_s(x)\}$ as a basis of solutions for the Bessel DE. In particular, the solution of the radial equation in cylindrical coordinates, i.e., the first equation in (27.1), becomes

$$R(\rho) = AJ_m(v) + BY_m(v) = AJ_m(l\rho) + BY_m(l\rho). \qquad (27.13)$$

27.4   Properties of the Bessel Functions

We have already considered some properties of the Bessel functions in Chapter 11. In this subsection, we quote those results and obtain other useful properties of the Bessel functions.

27.4.1   Negative Integer Order

Equation (11.32) gives a relation between a Bessel function of integer order and the Bessel function whose order is negative of the first one

$$J_{-m}(x) = (-1)^m J_m(x). \qquad (27.14)$$

27.4.2   Recurrence Relations

A number of recurrence relations involving Bessel functions of integer orders and their derivatives were derived in Chapter 11 which we reproduce here. The first one, involving no derivatives is

$$J_{m-1}(x) + J_{m+1}(x) = \frac{2m}{x}J_m(x). \qquad (27.15)$$

The second one, which includes derivatives of Bessel functions, is

$$J_{m-1}(x) - J_{m+1}(x) = 2J'_m(x). \qquad (27.16)$$

Combining these two equations, one obtains

$$J_{m-1}(x) = \frac{m}{x} J_m(x) + J'_m(x),$$

$$J_{m+1}(x) = \frac{m}{x} J_m(x) - J'_m(x). \qquad (27.17)$$

recurrence relations involving derivatives

We can use these equations to obtain new—and more useful—relations. For example, by differentiating $x^m J_m(x)$, we get

$$[x^m J_m(x)]' = m x^{m-1} J_m(x) + x^m J'_m(x)$$

$$= x^m \underbrace{\left[\frac{m}{x} J_m(x) + J'_m(x)\right]}_{=J_{m-1}(x) \text{ by } (27.17)} = x^m J_{m-1}(x).$$

Integrating (really, antidifferentiating) this equation yields

$$\int x^m J_{m-1}(x)\, dx = x^m J_m(x). \qquad (27.18)$$

Similarly, the reader may check that

$$\int x^{-m} J_{m+1}(x)\, dx = -x^{-m} J_m(x). \qquad (27.19)$$

27.4.3 Orthogonality

Bessel functions satisfy an orthogonality relation similar to that of the Legendre polynomials. However, unlike Legendre polynomials, the quantity that determines the orthogonality of different Bessel functions is not the order but a parameter in their argument (also see Example 24.5.3).

Consider two solutions of the Bessel DE corresponding to the same azimuthal parameter, but with different radial parameter. More specifically, let $f(\rho) = J_m(k\rho)$ and $g(\rho) = J_m(l\rho)$. Then

$$\frac{d^2 f}{d\rho^2} + \frac{1}{\rho}\frac{df}{d\rho} + \left(k^2 - \frac{m^2}{\rho^2}\right) f = 0,$$

$$\frac{d^2 g}{d\rho^2} + \frac{1}{\rho}\frac{dg}{d\rho} + \left(l^2 - \frac{m^2}{\rho^2}\right) g = 0.$$

The reader may check that by multiplying the first equation by ρg and the second equation by ρf and subtracting, one gets

$$\frac{d}{d\rho}[\rho(fg' - gf')] = (k^2 - l^2)\rho f g,$$

where the prime indicates differentiation with respect to ρ. Now integrate this equation with respect to ρ from some initial value (say a) to some final value (say b) to obtain

$$[\rho(fg' - gf')]_a^b = (k^2 - l^2) \int_a^b \rho f(\rho)g(\rho)\, d\rho.$$

In all physical applications a and b can be chosen to make the LHS vanish. Then, substituting for f and g in terms of Bessel functions, we get

$$(k^2 - l^2) \int_a^b \rho J_m(k\rho) J_m(l\rho)\, d\rho = 0.$$

It follows that *if* $k \neq l$, then the integral vanishes, i.e.,

$$\int_a^b \rho J_m(k\rho) J_m(l\rho)\, d\rho = 0 \qquad \text{if} \quad k \neq l. \tag{27.20}$$

This is the orthogonality relation for Bessel functions also derived in Example 24.5.3.

To complete the orthogonality relation, we must also address the case when $k = l$. This involves the evaluation of the integral $\int \rho J_m^2(k\rho)\, d\rho$, which, upon the change of variable $x \equiv k\rho$, reduces to $(\int x J_m^2(x)\, dx)/k^2$. By integration by parts, we have

$$I \equiv \int \underbrace{J_m^2(x)}_{u}\, \underbrace{x\, dx}_{dv} = \tfrac{1}{2}x^2 J_m^2(x) - \int J_m(x) J_m'(x) x^2\, dx.$$

In the last integral, substitute for $x^2 J_m(x)$ from the Bessel DE (27.6)—using x instead of v:

$$x^2 J_m(x) = m^2 J_m(x) - x J_m'(x) - x^2 J_m''(x).$$

Therefore,

$$I = \tfrac{1}{2}x^2 J_m^2(x) - \int J_m'(x)[m^2 J_m(x) \overbrace{-x J_m'(x) - x^2 J_m''(x)}^{=-(\frac{1}{2}x^2[J_m'(x)]^2)'}]\, dx$$

$$= \tfrac{1}{2}x^2 J_m^2(x) - m^2 \int \overbrace{J_m(x) J_m'(x)}^{=\frac{1}{2}[J_m^2(x)]'}\, dx + \tfrac{1}{2}\int \frac{d}{dx}\left(x^2[J_m'(x)]^2\right)\, dx$$

$$= \tfrac{1}{2}x^2 J_m^2(x) - \tfrac{1}{2}m^2 J_m^2(x) + \tfrac{1}{2}x^2[J_m'(x)]^2.$$

Returning back to ρ, we obtain the *indefinite* integral

$$\int \rho J_m^2(k\rho)\, d\rho = \frac{I}{k^2} = \frac{1}{2}\left(\rho^2 - \frac{m^2}{k^2}\right) J_m^2(k\rho) + \frac{1}{2}\rho^2[J_m'(k\rho)]^2. \tag{27.21}$$

In most applications, the lower limit of integration is zero and the upper limit is a positive number a. The RHS of (27.21) vanishes at the lower limit because of the following reason. The first term vanishes at $\rho = 0$ because $J_m(0) = 0$ for all $m > 0$ as is evident from the series expansion (27.10). For $m = 0$ (and $\rho = 0$), the parentheses in the first term of (27.21) vanishes. So, the first term is zero for all $m \geq 0$ at the lower limit of integration. The second term vanishes due to the presence of ρ^2. Thus, we obtain

$$\int_0^a \rho J_m^2(k\rho)\, d\rho = \frac{1}{2}\left(a^2 - \frac{m^2}{k^2}\right) J_m^2(ka) + \frac{1}{2}a^2[J_m'(ka)]^2 \qquad (27.22)$$

for all $m \geq 0$ and, by (27.14), also for all negative *integers*. As mentioned earlier, we shall confine our discussion to Bessel functions of integer orders. It is customary to simplify the RHS of (27.22) by choosing k in such a way that $J_m(ka) = 0$, i.e., that ka is a root of the Bessel function of order m. In general, there are infinitely many roots. So, let x_{mn} denote the nth root of $J_m(x)$. Then,

$$ka = x_{mn} \;\Rightarrow\; k = \frac{x_{mn}}{a}, \qquad n = 1, 2, \ldots,$$

and if we use Equation (27.17), we obtain

$$\int_0^a \rho J_m^2(x_{mn}\rho/a)\, d\rho = \tfrac{1}{2}a^2[J_{m+1}(x_{mn})]^2. \qquad (27.23)$$

Equations (27.20) and (27.23) can be combined into a single equation using the Kronecker delta:

orthogonality relations involving Bessel functions

Box 27.4.1. *The Bessel functions of integer order satisfy the orthogonality relations*

$$\int_0^a J_m(x_{mn}\rho/a) J_m(x_{mk}\rho/a)\rho\, d\rho = \tfrac{1}{2}a^2 J_{m+1}^2(x_{mn})\delta_{kn}, \qquad (27.24)$$

where $a > 0$ and x_{mn} is the nth root of $J_m(x)$.

27.4.4 Generating Function

Just as in the case of Legendre polynomials, Bessel functions of integer order have a generating function, i.e., there exists a function $g(x, t)$ such that

$$g(x, t) = \sum_{n=-\infty}^{\infty} t^n J_n(x). \qquad (27.25)$$

To find g, start with the recurrence relation

$$J_{m-1}(x) + J_{m+1}(x) = \frac{2m}{x} J_m(x),$$

multiply it by t^m, and sum over all m to obtain

$$\sum_{m=-\infty}^{\infty} t^m J_{m-1}(x) + \sum_{m=-\infty}^{\infty} t^m J_{m+1}(x) = \frac{2}{x} \sum_{m=-\infty}^{\infty} m t^m J_m(x). \qquad (27.26)$$

The first sum can be written as

$$\sum_{m=-\infty}^{\infty} t^m J_{m-1}(x) = t \sum_{m=-\infty}^{\infty} t^{m-1} J_{m-1}(x) = t \sum_{n=-\infty}^{\infty} t^n J_n(x) = t g(x,t),$$

where we substituted the dummy index $n = m - 1$ for m. Similarly

$$\sum_{m=-\infty}^{\infty} t^m J_{m+1}(x) = \frac{1}{t} \sum_{m=-\infty}^{\infty} t^{m+1} J_{m+1}(x) = \frac{1}{t} g(x,t)$$

and

$$\frac{2}{x} \sum_{m=-\infty}^{\infty} m t^m J_m(x) = \frac{2t}{x} \sum_{m=-\infty}^{\infty} m t^{m-1} J_m(x) = \frac{2t}{x} \frac{\partial g}{\partial t}.$$

It follows from Equation (27.26) that

$$\left(t + \frac{1}{t} \right) g(x,t) = \frac{2t}{x} \frac{\partial g}{\partial t},$$

or

$$\frac{x}{2} \left(1 + \frac{1}{t^2} \right) dt = \frac{dg}{g},$$

where x is assumed to be a constant because we have been differentiating with respect to t. Integrating both sides gives

$$\underbrace{\int \frac{x}{2} \left(1 + \frac{1}{t^2} \right) dt}_{= \frac{x}{2}\left(t - \frac{1}{t}\right)} = \ln g + \ln \phi(x),$$

where the last term is the "constant" of integration. Thus,

$$g(x,t) = \phi(x) \exp \left[\frac{x}{2} \left(t - \frac{1}{t} \right) \right].$$

To determine $\phi(x)$, we note that

$$g(x,t) = \phi(x) e^{xt/2} e^{-x/2t} = \phi(x) \sum_{n=0}^{\infty} \frac{(xt/2)^n}{n!} \sum_{m=0}^{\infty} \frac{(-x/2t)^m}{m!}$$

$$= \phi(x) \sum_{n,m=0}^{\infty} \frac{(-1)^m}{n! m!} \left(\frac{x}{2} \right)^{n+m} t^{n-m}.$$

In the last double sum, collect all terms whose power of t is zero, and call the sum S_0. This is obtained by setting $n = m$. Then,

$$S_0 = \phi(x) \sum_{n=0}^{\infty} \frac{(-1)^n}{n!n!} \left(\frac{x}{2}\right)^{2n} = \phi(x)J_0(x),$$

where we used Equation (27.10) with $s = 0$. But (27.25) shows that the collection of all terms whose power of t is zero is simply $J_0(x)$. Thus, $S_0 = J_0(x)$, and $\phi(x) = 1$. This leads to the final form of the Bessel generating function:

$$g(x,t) = \exp\left[\frac{x}{2}\left(t - \frac{1}{t}\right)\right] = \sum_{n=-\infty}^{\infty} t^n J_n(x). \tag{27.27}$$

generating function for Bessel functions

Example 27.4.1. The generating function for Bessel functions can be used to obtain a useful identity. First we note that

$$g(x + y, t) = g(x,t)g(y,t)$$

as the reader may easily verify. Expanding each side gives

$$\sum_{n=-\infty}^{\infty} t^n J_n(x+y) = \sum_{k=-\infty}^{\infty} t^k J_k(x) \sum_{m=-\infty}^{\infty} t^m J_m(y) = \sum_{k=-\infty}^{\infty} \sum_{m=-\infty}^{\infty} t^{k+m} J_k(x) J_m(y).$$

In the last double sum, let $n = k+m$, so that $k = n-m$. Since there is no limitation on the value of either of the dummy indices, the limits of the new indices n and m are still $-\infty$ and ∞. Therefore,

$$\sum_{n=-\infty}^{\infty} t^n J_n(x+y) = \sum_{n=-\infty}^{\infty} \sum_{m=-\infty}^{\infty} t^n J_{n-m}(x) J_m(y)$$

$$= \sum_{n=-\infty}^{\infty} t^n \left(\sum_{m=-\infty}^{\infty} J_{n-m}(x) J_m(y)\right).$$

Since each power of t should have the same coefficient on both sides, we obtain the so-called **addition theorem** for Bessel functions:

$$J_n(x+y) = \sum_{m=-\infty}^{\infty} J_{n-m}(x) J_m(y) = \sum_{m=-\infty}^{\infty} J_m(x) J_{n-m}(y), \tag{27.28}$$

addition theorem for Bessel functions

where the last equality follows from the symmetry of $J_n(x+y)$ under the exchange of x and y. ∎

The Bessel generating function can also lead to some very important identities. In Equation (27.27), let $t = e^{i\theta}$ and use (18.14) to obtain

$$e^{ix\sin\theta} = \sum_{n=-\infty}^{\infty} e^{in\theta} J_n(x). \tag{27.29}$$

This is a Fourier series expansion in θ—as given in (18.20)—whose coefficients are Bessel functions. To find these coefficients, we multiply both sides by $e^{-im\theta}$ and integrate from $-\pi$ to π [see also Equation (18.22)]. The LHS gives

$$LHS = \int_{-\pi}^{\pi} e^{ix\sin\theta} e^{-im\theta}\, d\theta = \int_{-\pi}^{\pi} e^{i(x\sin\theta - m\theta)}\, d\theta.$$

For the RHS, we obtain

$$\sum_{n=-\infty}^{\infty} \left[\int_{-\pi}^{\pi} e^{i(n-m)\theta} d\theta \right] J_n(x) = 2\pi J_m(x),$$

where we used the easily verifiable result [also see Equation (18.21)]:

$$\int_{-\pi}^{\pi} e^{i(n-m)\theta} d\theta = \begin{cases} 0 & \text{if } n \neq m \\ 2\pi & \text{if } n = m \end{cases} = 2\pi\delta_{mn}.$$

integral representation of the Bessel function

Equating the RHS and the LHS, we obtain

$$J_m(x) = \frac{1}{2\pi} \int_{-\pi}^{\pi} e^{i(x\sin\theta - m\theta)}\, d\theta. \tag{27.30}$$

The reader may check that this can be reduced to

$$J_m(x) = \frac{1}{\pi} \int_{0}^{\pi} \cos(x\sin\theta - m\theta)\, d\theta \tag{27.31}$$

Bessel's integral

which is called **Bessel's integral**.

Bessel functions can be written in terms of the confluent hypergeometric function. To see this, substitute $R(v) = v^\mu e^{-\eta v} f(v)$—with μ and η to be determined—in Equation (27.6) to obtain

$$\frac{d^2 f}{dv^2} + \left(\frac{2\mu+1}{v} - 2\eta \right) \frac{df}{dv} + \left[\frac{\mu^2 - m^2}{v^2} - \frac{\eta(2\mu+1)}{v} + \eta^2 + 1 \right] f = 0$$

which, if we set $\mu = m$ and $\eta = i$, reduces to

$$f'' + \left(\frac{2m+1}{v} - 2i \right) f' - \frac{(2m+1)i}{v} f = 0. \tag{27.32}$$

Making the further substitution $2iv = t$, and multiplying out by t, we obtain

$$t\frac{d^2 f}{dt^2} + (2m+1-t)\frac{df}{dt} - (m+\tfrac{1}{2})f = 0$$

relation between Bessel functions and confluent hypergeometric function

which is in the form of (11.27) with $\alpha = m + \frac{1}{2}$ and $\gamma = 2m+1$. Thus, Bessel functions $J_m(x)$ can be written as constant multiples of $x^m e^{-ix} \Phi(m+\frac{1}{2}, 2m+1; 2ix)$. In fact,

$$J_m(x) = \frac{1}{\Gamma(m+1)} \left(\frac{x}{2} \right)^m e^{-ix} \Phi(m+\tfrac{1}{2}, 2m+1; 2ix). \tag{27.33}$$

27.5   Expansions in Bessel Functions

The orthogonality of Bessel functions can be useful in expanding other functions in terms of them. The basic idea is similar to the expansion in Fourier series and Legendre polynomials. If a function $f(\rho)$ is defined in the interval $(0, a)$, then we may write

$$f(\rho) = \sum_{n=1}^{\infty} c_n J_m(x_{mn}\rho/a). \tag{27.34}$$

The coefficients can be found by multiplying both sides by $\rho J_m(x_{mk}\rho/a)$ and integrating from zero to a. The reader may verify that this yields

$$c_n = \frac{2}{a^2 J_{m+1}^2(x_{mn})} \int_0^a f(\rho) J_m(x_{mn}\rho/a)\rho \, d\rho. \tag{27.35}$$

Equations (27.34) and (27.35) are the analogues of Equations (10.38), (10.40), (10.41), and (10.42) for Fourier series, and Equations (26.46) and (26.47) for Legendre polynomials. Like those sets of equations, they can be used to expand functions in terms of Bessel functions of a specific order.

Example 27.5.1. The trigonometric functions can be expanded in Bessel functions with very little effort. In fact, Equation (27.29) leads immediately to

$$e^{ix} = \sum_{n=-\infty}^{\infty} i^n J_n(x)$$

or

$$\cos x + i \sin x = \sum_{k=-\infty}^{\infty} i^{2k} J_{2k}(x) + \sum_{k=-\infty}^{\infty} i^{2k+1} J_{2k+1}(x),$$

expansion of sine and cosine in Bessel functions

where we have separated the even and odd sums. The first sum is real and the second sum pure imaginary. Therefore,

$$\cos x = \sum_{k=-\infty}^{\infty} (-1)^k J_{2k}(x) = \sum_{k=-\infty}^{-1} (-1)^k J_{2k}(x) + J_0(x) + \sum_{k=1}^{\infty}(-1)^k J_{2k}(x).$$

The first sum can be written as

$$\sum_{k=-\infty}^{-1} (-1)^k J_{2k}(x) = \sum_{k=1}^{\infty}(-1)^{-k} J_{-2k}(x) = \sum_{k=1}^{\infty}(-1)^k(-1)^{2k} J_{2k}(x)$$

$$= \sum_{k=1}^{\infty}(-1)^k J_{2k}(x)$$

which is identical to the last sum. It follows that

$$\cos x = J_0(x) + 2 \sum_{k=1}^{\infty}(-1)^k J_{2k}(x). \tag{27.36}$$

Similarly,

$$\sin x = 2 \sum_{k=0}^{\infty}(-1)^k J_{2k+1}(x) \tag{27.37}$$

as the reader is urged to verify.   ∎

If we square Equation (27.34), multiply by ρ, and integrate from zero to a, we obtain

$$\int_0^a f^2(\rho)\rho\,d\rho = \sum_{n=1}^{\infty}\sum_{k=1}^{\infty} c_n c_k \underbrace{\int_0^a J_m(x_{mn}\rho/a)J_m(x_{mk}\rho/a)\rho\,d\rho}_{=\frac{1}{2}a^2 J_{m+1}^2(x_{mn})\delta_{kn} \text{ by (27.24)}}.$$

Parseval relation

This leads to the so-called **Parseval relation**:

$$\int_0^a f^2(\rho)\rho\,d\rho = \frac{1}{2}a^2 \sum_{n=1}^{\infty} c_n^2 J_{m+1}^2(x_{mn}) \tag{27.38}$$

for some m. This m can be chosen to make the integrations as simple as possible.

expansion of ρ^k in terms of Bessel functions

Example 27.5.2. Let us find the expansion of ρ^k in terms of Bessel functions. Equations (27.35) and (27.18) suggest expanding in terms of $J_k(x)$ because the integrals can be performed. Therefore, we write

$$\rho^k = \sum_{n=1}^{\infty} c_n J_k(x_{kn}\rho/a),$$

where

$$c_n = \frac{2}{a^2 J_{k+1}^2(x_{kn})}\int_0^a \rho^k J_k(x_{kn}\rho/a)\rho\,d\rho = \frac{2}{a^2 J_{k+1}^2(x_{kn})}\int_0^a \rho^{k+1} J_k(x_{kn}\rho/a)\,d\rho.$$

Introducing $y = x_{kn}\rho/a$ in the integral gives

$$c_n = \frac{2a^k}{x_{kn}^{k+2} J_{k+1}^2(x_{kn})}\int_0^{x_{kn}} y^{k+1} J_k(y)\,dy = \frac{2a^k}{x_{kn} J_{k+1}(x_{kn})},$$

where we used (27.18) with m replaced by $k+1$. Thus, we have

$$\rho^k = 2a^k \sum_{n=1}^{\infty} \frac{J_k(x_{kn}\rho/a)}{x_{kn} J_{k+1}(x_{kn})}. \qquad\blacksquare$$

27.6   Physical Examples

Our discussion of Laplace's equation has led us to believe that trigonometric functions and Legendre polynomials are, respectively, the "natural" functions of Cartesian and spherical geometries. It is of no surprise now to learn that Bessel functions are the natural functions of cylindrical geometry.

As in the case of Cartesian and spherical coordinates, unless the symmetry of the problem simplifies the situation, the separation of Laplace's equation results in two parameters leading to a double sum as in Example 25.2.4. The reason that we did not obtain double sums in spherical coordinates is that from the very beginning we assumed azimuthal symmetry. Thus, we expect

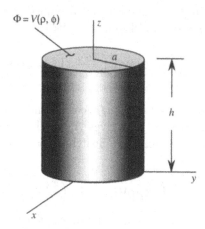

Figure 27.1: A conducting cylindrical can whose top has a potential given by $V(\rho, \theta)$ with the rest of the surface grounded.

a double summation in the most general solution of Laplace's equation in cylindrical geometries. One of these sums is over m which, as Equation (27.3) shows, appears in the argument of the sine and cosine functions. It also designates the order of the Bessel (or Neumann) function.

To understand the origin of the second summation, consider a cylindrical conducting can of radius a and height h (see Figure 27.1). Suppose that the potential at the top face varies as $V(\rho, \varphi)$ while the lateral surface and the bottom face are grounded. Let us find the electrostatic potential Φ at all points inside the can.

The general solution is a product of (27.3), (27.4), and (27.13):

$$\Phi(\rho, \varphi, z) = R(\rho)S(\varphi)Z(z).$$

Since $\Phi(\rho, \varphi, 0) = 0$ for arbitrary ρ and φ, we must have $Z(0) = 0$ yielding—to within a constant—$Z(z) = \sinh(lz)$.

Since $\Phi(0, \varphi, z)$ is finite, no Neumann function is allowed in the expansion, and, to within a constant, we have $R(\rho) = J_m(l\rho)$. Furthermore, since $\Phi(a, \varphi, z) = 0$ for arbitrary φ and z, we must have

$$R(a) = J_m(la) = 0 \implies la = x_{mn} \implies l = \frac{x_{mn}}{a}, \qquad n = 1, 2, \ldots,$$

where, as before, x_{mn} is the nth root of J_m.

We can now multiply R, S, and Z and sum over all possible values of m and n, keeping in mind that negative values of m give terms that are linearly dependent on the corresponding positive values. The result is the so-called **Fourier–Bessel** series:

Fourier–Bessel series

$$\Phi(\rho, \varphi, z) = \sum_{m=0}^{\infty} \sum_{n=1}^{\infty} J_m\left(\frac{x_{mn}}{a}\rho\right) \sinh\left(\frac{x_{mn}}{a}z\right) (A_{mn} \cos m\varphi + B_{mn} \sin m\varphi)$$

$$(27.39)$$

where A_{mn} and B_{mn} are constants to be determined by the remaining boundary condition which states that $\Phi(\rho, \varphi, h) = V(\rho, \varphi)$ or

$$V(\rho, \varphi) = \sum_{m=0}^{\infty} \sum_{n=1}^{\infty} J_m\left(\frac{x_{mn}}{a}\rho\right) \sinh\left(\frac{x_{mn}}{a}h\right) (A_{mn}\cos m\varphi + B_{mn}\sin m\varphi).$$

(27.40)

Multiplying both sides by $\rho J_m(x_{mk}a/\rho)\cos j\varphi$ and integrating from zero to 2π in φ, and from zero to a in ρ gives A_{jk}. Changing cosine to sine and following the same steps yields B_{jk}. Switching back to m and n, the reader may verify that

$$A_{mn} = \frac{2\int_0^{2\pi} d\varphi \int_0^a d\rho\, \rho V(\rho, \varphi) J_m\left(\frac{x_{mn}}{a}\rho\right)\cos m\varphi}{\pi a^2 J_{m+1}^2(x_{mn})\sinh(x_{mn}h/a)},$$

$$B_{mn} = \frac{2\int_0^{2\pi} d\varphi \int_0^a d\rho\, \rho V(\rho, \varphi) J_m\left(\frac{x_{mn}}{a}\rho\right)\sin m\varphi}{\pi a^2 J_{m+1}^2(x_{mn})\sinh(x_{mn}h/a)},$$

(27.41)

where we have used Equation (27.24).

The important case of azimuthal symmetry requires special consideration. In such a case, the potential of the top surface $V(\rho, \varphi)$ must be independent of φ. Furthermore, since $S(\varphi)$ is constant,[5] its derivative must vanish. Hence, the second equation in (27.1) yields $\mu = -m^2 = 0$. This zero value for m reduces the double summation of (27.39) to a single sum, and we get

$$\Phi(\rho, z) = \sum_{n=1}^{\infty} A_n J_0\left(\frac{x_{0n}}{a}\rho\right)\sinh\left(\frac{x_{0n}}{a}z\right).$$

(27.42)

The coefficients A_n can be obtained by setting $m = 0$ in the first equation of (27.41):

$$A_n = \frac{4}{a^2 J_1^2(x_{0n})\sinh(x_{0n}h/a)} \int_0^a \rho V(\rho) J_0\left(\frac{x_{0n}}{a}\rho\right) d\rho,$$

(27.43)

where $V(\rho)$ is the φ-independent potential of the top surface.

Example 27.6.1. Suppose that the top face of a conducting cylindrical can is held at the constant potential V_0 while the lateral surface and the bottom face are grounded. We want to find the electrostatic potential Φ at all points inside the can.

Since the potential of the top is independent of φ, azimuthal symmetry prevails, and Equation (27.43) gives

$$A_n = \frac{4V_0}{a^2 J_1^2(x_{0n})\sinh(x_{0n}h/a)} \int_0^a \rho J_0\left(\frac{x_{0n}}{a}\rho\right) d\rho = \frac{4V_0}{x_{0n}J_1(x_{0n})\sinh(x_{0n}h/a)},$$

where we used Equation (27.18). The detail of calculating the integral is left as Problem 27.15 for the reader. Therefore,

$$\Phi(\rho, z) = 4V_0 \sum_{n=1}^{\infty} \frac{J_0(x_{0n}\rho/a)\sinh(x_{0n}z/a)}{x_{0n}J_1(x_{0n})\sinh(x_{0n}h/a)}. \qquad \blacksquare$$

---

[5]$S(\varphi)$ must be a constant. Otherwise, the potential would depend on φ.

27.7 Problems

27.1. Derive (27.6) from the first equation of (27.1).

27.2. Show that *both* equations in (27.17) give

$$J_0'(x) = -J_1(x).$$

27.3. Show that

$$[x^{-m} J_m(x)]' = -x^{-m} J_{m+1}(x)$$

and derive Equation (27.19).

27.4. Obtain the following equation from the Bessel DE:

$$\frac{d}{d\rho}[\rho(fg' - gf')] = (k^2 - l^2)\rho fg,$$

where f and g are solutions of two Bessel DEs for which the "constants" of the DEs are k^2 and l^2, respectively.

27.5. (a) Show that for the Bessel generating function,

$$g(x+y,t) = g(x,t)g(y,t) \qquad \text{and} \qquad g(x,-t) = \frac{1}{g(x,t)}.$$

(b) Use the second relation to show that

$$\sum_{m=-\infty}^{\infty} J_{m-k}(x)J_m(x) = \delta_{0k} \equiv \begin{cases} 1 & \text{if} \quad k = 0, \\ 0 & \text{if} \quad k \neq 0. \end{cases}$$

Hint: Set the powers of t equal on both sides of $1 = g(x,t)g(x,-t)$.
(c) In particular,

$$1 = \sum_{m=-\infty}^{\infty} J_m^2(x) = J_0^2(x) + 2\sum_{m=1}^{\infty} J_m^2(x),$$

showing that $|J_0(x)| \leq 1$ and $|J_m(x)| \leq 1/\sqrt{2}$ for $m > 0$.

27.6. Derive Equation (27.31) from (27.30).

27.7. Use Equation (27.31) to show that $J_{-m} = (-1)^m J_m$.

27.8. Show that the substitution $R(v) = v^m e^{-iv} f(v)$ turns Equation (27.6) into (27.32).

27.9. Using the orthogonality of Bessel functions derive Equation (27.35) from (27.34).

27.10. Prove that

$$e^{ix\cos\theta} = \sum_{n=-\infty}^{\infty} i^n e^{in\theta} J_n(x).$$

27.11. Derive the expansion of the sine function in terms of Bessel functions. Hint: See Example 27.5.1.

27.12. The integral $\int_0^\infty e^{-ax} J_0(bx)\, dx$ may *look* intimidating, but leads to a very simple expression. To see this:
(a) Substitute for $J_0(bx)$ its series representation, and express the result of the integration in terms of the gamma function (a factorial, in this case).
(b) Use one of the results of Problem 11.1 to show that

$$\int_0^\infty e^{-ax} J_0(bx)\, dx = \frac{1}{a\sqrt{\pi}} \sum_{n=0}^\infty \frac{\Gamma(n+\frac{1}{2})}{n!} \left(-\frac{b^2}{a^2}\right)^n.$$

(c) Show that this result can also be expressed in terms of the hypergeometric function:

$$\int_0^\infty e^{-ax} J_0(bx)\, dx = \frac{1}{a} F\left(\frac{1}{2}, 1; 1; -\frac{b^2}{a^2}\right).$$

(d) Now use the result of Problem 11.4 to express the integral in a very simple form.

27.13. By writing the series representation of the Bessel function as in the previous problem, and using the result of Problem 11.2, show that for integer m

$$\int_0^\infty e^{-ax} J_m(bx)\, dx = \frac{1}{a\sqrt{\pi}} \left(\frac{b}{a}\right)^m \frac{\Gamma(m/2+1)\Gamma((m+1)/2)}{\Gamma(m+1)}$$
$$\cdot F\left(\frac{m}{2}+1, \frac{m+1}{2}; m+1; -\frac{b^2}{a^2}\right).$$

27.14. Multiply both sides of Equation (27.40) by $\rho J_m(x_{mk}a/\rho)\cos j\varphi$ and integrate appropriately to obtain A_{jk}. Switch cosine to sine and do the same to find B_{jk}.

27.15. Use Equation (27.18) to show that

$$\int_0^a \rho J_0\left(\frac{x_{0n}}{a}\rho\right) d\rho = \frac{a^2}{x_{0n}} J_1(x_{0n}).$$

27.16. Use the Parseval relation (27.38) for $f(\rho) = \rho^k$ to obtain

$$\sum_{n=1}^\infty \frac{1}{x_{mn}^2} = \frac{1}{4(m+1)}$$

for any m. Hint: See Example 27.5.2.

27.17. A long heat conducting cylinder of radius a is composed of two halves (with semicircular cross sections) with an infinitesimal gap between them. The upper and lower halves of the cylinder are in contact with heat baths of temperatures $+T_0$ and $-T_0$, respectively. Find the temperature both inside and outside the cylinder.

27.18. A long heat conducting cylinder of radius a is composed of two halves (with semicircular cross sections) with an infinitesimal gap between them. The upper and lower halves of the cylinder are in contact with heat baths of temperatures $+T_1$ and $-T_1$, respectively. The cylinder is inside a larger cylinder (and coaxial with it) held at temperature T_2. Find the temperature inside the inner cylinder, between the two cylinders, and outside the outer cylinder.

27.19. A long conducting cylinder of radius a is kept at potential V_1. The cylinder is inside a larger cylinder (and coaxial with it) held at potential V_2. Find the potential inside the inner cylinder, between the two cylinders, and outside the outer cylinder.

Chapter 28

Other PDEs
of Mathematical Physics

Chapters 25, 26, and 27 discussed one of the most important PDEs of mathematical physics, Laplace's equation. The techniques used in solving Laplace's equation apply to all PDEs encountered in introductory physics. Since we have already spent a considerable amount of time on these techniques, we shall simply provide some illustrative examples of solving other PDEs.

28.1 The Heat Equation

The heat equation, sometimes also called the **diffusion equation**, was introduced in Chapter 22 [see Equation (22.3)]. The separation of variables $T(t, \mathbf{r}) = g(t)R(\mathbf{r})$ yields

diffusion equation

$$\frac{\partial}{\partial t}[g(t)R(\mathbf{r})] = k^2 \nabla^2[g(t)R(\mathbf{r})] \;\Rightarrow\; R(\mathbf{r})\frac{dg}{dt} = k^2 g(t)\nabla^2 R.$$

Dividing both sides by $g(t)R(\mathbf{r})$, we obtain

$$\frac{1}{g}\frac{dg}{dt} = k^2 \underbrace{\frac{1}{R}\nabla^2 R}_{\equiv -\lambda}.$$

The LHS is a function of t, and the RHS a function of $\mathbf{r}$. The independence of these variables forces each side to be a constant. Calling this constant $-k^2\lambda$ for later convenience, we obtain an ODE in time and a PDE in the remaining variables:

$$\frac{dg}{dt} + k^2\lambda g = 0 \quad\text{and}\quad \nabla^2 R + \lambda R = 0. \tag{28.1}$$

The general solution of the first equation is

$$g(t) = Ae^{-k^2\lambda t} \tag{28.2}$$

λ of the heat
equation is always
positive

and that of the second equation can be obtained precisely by the methods of
the last chapter. We illustrate this by some examples, but first we need to
keep in mind that λ is to be assumed positive, otherwise the exponential in
Equation (28.2) will cause a growth of $g(t)$ (and, therefore, the temperature)
beyond bounds.

28.1.1   Heat-Conducting Rod

heat-conducting
rod

Let us consider a one-dimensional conducting rod with one end at the origin
$x = 0$ and the other at $x = b$. The two ends are held at $T = 0$. Initially,
at $t = 0$, we assume a temperature distribution on the rod given by some
function $f(x)$. We want to calculate the temperature at time t at any point
x on the rod.

Due to the one-dimensionality of the rod, the y- and z-dependence can be
ignored, and the Laplacian is reduced to a second derivative in x. Thus, the
second equation in (28.1) becomes

$$\frac{d^2 X}{dx^2} + \lambda X = 0, \tag{28.3}$$

where X is a function of x alone. The general solution of this equation is[1]

$$X(x) = B \cos(\sqrt{\lambda}\, x) + C \sin(\sqrt{\lambda}\, x).$$

Since the two ends of the rod are held at $T = 0$, we have the boundary
conditions $T(t, 0) = 0 = T(t, b)$, which imply that $X(0) = 0 = X(b)$. These
give $B = 0$ and[2]

$$\sin(\sqrt{\lambda}\, b) = 0 \;\Rightarrow\; \sqrt{\lambda}\, b = n\pi \qquad \text{for} \quad n = 1, 2, \ldots.$$

With a label n attached to λ, the solution, and the constant multiplying it,
we can now write

$$\lambda_n = \left(\frac{n\pi}{b}\right)^2 \qquad \text{and} \qquad X_n(x) = C_n \sin\left(\frac{n\pi}{b}x\right) \qquad \text{for} \quad n = 1, 2, \ldots.$$

The (subscripted) solution of the time equation is also simply obtained:

$$g_n(t) = A_n e^{-k^2 (n\pi/b)^2 t}.$$

This leads to a general solution of the form

$$T(t, x) = \sum_{n=1}^{\infty} B_n e^{-(n\pi k/b)^2 t} \sin\left(\frac{n\pi}{b}x\right), \tag{28.4}$$

where $B_n \equiv A_n C_n$. The initial condition $f(x) = T(0, x)$ yields

$$f(x) = \sum_{n=1}^{\infty} B_n \sin(n\pi x/b)$$

---

[1] The reader may check that the only solution for $\lambda = 0$ is the trivial solution.
[2] Consult Section 25.2.

which is a Fourier series from which we can calculate the coefficients

$$B_n = \frac{2}{b} \int_0^b \sin\left(\frac{n\pi}{b}x\right) f(x)\, dx.$$

Thus if we know the initial temperature distribution on the rod [the function $f(x)$], we can determine the temperature of the rod for all time. For instance, if the initial temperature distribution of the rod is uniform, say T_0, then

$$B_n = \frac{2T_0}{b} \int_0^b \sin\left(\frac{n\pi}{b}x\right) dx = \frac{2T_0}{b} \left[-\frac{b}{n\pi} \cos\left(\frac{n\pi}{b}x\right)\right]_0^b = \frac{2T_0}{n\pi}[1 - (-1)^n].$$

It follows that the odd n's survive, and if we set $n = 2m+1$, we obtain

$$B_{2m+1} = \frac{4T_0}{\pi(2m+1)}$$

and

$$T(t, x) = \frac{4T_0}{\pi} \sum_{m=0}^{\infty} \frac{e^{-[(2m+1)\pi k/b]^2 t}}{2m+1} \sin\left[\frac{(2m+1)\pi}{b}x\right].$$

This distribution of temperature for all time can be obtained numerically for any heat conductor whose k is known. Note that the exponential in the sum causes the temperature to drop to zero (the fixed temperature of its two end points) eventually. This conclusion is independent of the initial temperature distribution of the rod as Equation (28.4) indicates.

28.1.2   Heat Conduction in a Rectangular Plate

As a more complicated example involving a second spatial variable, consider a rectangular heat-conducting plate with sides of length a and b all held at $T = 0$. Assume that at time $t = 0$ the temperature has a distribution function $f(x, y)$. Let us find the variation of temperature for all points (x, y) at all times $t > 0$.

conduction of heat in a rectangular plate

The spatial part of the heat equation for this problem is

$$\frac{\partial^2 R}{\partial x^2} + \frac{\partial^2 R}{\partial y^2} + \lambda R = 0.$$

A separation of variables, $R(x, y) = X(x)Y(y)$, and its usual procedure leads to the following equation:

$$\underbrace{\frac{1}{X}\frac{d^2 X}{dx^2}}_{\equiv -\mu} + \underbrace{\frac{1}{Y}\frac{d^2 Y}{dy^2}}_{\equiv -\nu} + \lambda = 0.$$

This leads to the following two ODEs:

$$\frac{d^2 X}{dx^2} + \mu X = 0, \qquad \frac{d^2 Y}{dy^2} + \nu Y = 0, \qquad \lambda = \mu + \nu.$$

Due to the periodicity of the BCs, the general solutions of these equations are trigonometric functions. The four boundary conditions

$$T(0, y, t) = T(a, y, t) = T(x, 0, t) = T(x, b, t) = 0$$

determine the specific form of the solutions as well as the *indexed* constants of separation:

$$\mu_n = \left(\frac{n\pi}{a}\right)^2 \quad \text{and} \quad X_n(x) = A_n \sin\left(\frac{n\pi}{a}x\right) \quad \text{for} \quad n = 1, 2, \ldots,$$

$$\nu_m = \left(\frac{m\pi}{b}\right)^2 \quad \text{and} \quad Y_m(y) = B_m \sin\left(\frac{m\pi}{b}y\right) \quad \text{for} \quad m = 1, 2, \ldots.$$

So, λ becomes a double indexed quantity:

$$\lambda \equiv \lambda_{mn} = \mu_n + \nu_m = \left(\frac{n\pi}{a}\right)^2 + \left(\frac{m\pi}{b}\right)^2.$$

The solution to the g equation can be expressed as $g(t) = C_{mn}e^{-k^2\lambda_{mn}t}$. Putting everything together, we obtain

$$T(x, y, t) = \sum_{n=1}^{\infty}\sum_{m=1}^{\infty} A_{mn}e^{-k^2\lambda_{mn}t}\sin\left(\frac{n\pi}{a}x\right)\sin\left(\frac{m\pi}{b}y\right),$$

where $A_{mn} = A_n B_m C_{mn}$ is an arbitrary constant. To determine it, we impose the initial condition $T(x, y, 0) = f(x, y)$. This yields

$$f(x, y) = \sum_{n=1}^{\infty}\sum_{m=1}^{\infty} A_{mn}\sin\left(\frac{n\pi}{a}x\right)\sin\left(\frac{m\pi}{b}y\right)$$

from which we find the coefficients A_{mn} (see Theorem 25.2.5):

$$A_{mn} = \frac{4}{ab}\int_0^a dx \int_0^b dy f(x, y)\sin\left(\frac{n\pi}{a}x\right)\sin\left(\frac{m\pi}{b}y\right).$$

28.1.3 Heat Conduction in a Circular Plate

circular plate

In this example, we consider a circular plate of radius a whose rim is held at $T = 0$ and whose initial surface temperature is characterized by a function $f(\rho, \varphi)$. We are seeking the temperature distribution on the plate for all time. The spatial part of the heat equation in z-independent cylindrical coordinates,[3] appropriate for a circular plate, is

$$\frac{1}{\rho}\frac{\partial}{\partial\rho}\left(\rho\frac{\partial R}{\partial\rho}\right) + \frac{1}{\rho^2}\frac{\partial^2 R}{\partial\varphi^2} + \lambda R = 0$$

---

[3] See the discussion of Subsection 22.3.

which, after the separation of variables, $R(\rho, \varphi) = \mathcal{R}(\rho)S(\varphi)$, reduces to

$$S(\varphi) = A\cos m\varphi + B\sin m\varphi \quad \text{for} \quad m = 0, 1, 2, \ldots,$$

$$\frac{d^2\mathcal{R}}{d\rho^2} + \frac{1}{\rho}\frac{d\mathcal{R}}{d\rho} + \left(\lambda - \frac{m^2}{\rho^2}\right)\mathcal{R} = 0.$$

The solution of the last (Bessel) equation, which is well defined for $\rho = 0$ and vanishes at $\rho = a$, is

$$\mathcal{R}(\rho) = CJ_m\left(\frac{x_{mn}}{a}\rho\right) \quad \text{with} \quad \sqrt{\lambda} = \frac{x_{mn}}{a} \quad \text{and} \quad n = 1, 2, \ldots,$$

where, as usual, x_{mn} is the nth root of J_m. We see that λ is a double-indexed quantity. The time equation (28.2) has a solution of the form

$$g(t) = D_{mn}e^{-k^2\lambda_{mn}t} = D_{mn}e^{-k^2(x_{mn}^2/a^2)t}.$$

Multiplying the three solutions and summing over the two indices yields the most general solution

$$T(\rho, \varphi, t) = \sum_{m=0}^{\infty}\sum_{n=1}^{\infty} J_m\left(\frac{x_{mn}}{a}\rho\right)e^{-(kx_{mn}/a)^2t}(A_{mn}\cos m\varphi + B_{mn}\sin m\varphi).$$

The coefficients are determined from the initial condition

$$f(\rho, \varphi) = T(\rho, \varphi, 0) = \sum_{m=0}^{\infty}\sum_{n=1}^{\infty} J_m\left(\frac{x_{mn}}{a}\rho\right)(A_{mn}\cos m\varphi + B_{mn}\sin m\varphi).$$

Except for the hyperbolic sine term, this equation is identical to (27.40). Therefore, the coefficients are given by expressions similar to Equation (27.41). In the case at hand, we get

$$A_{mn} = \frac{2}{\pi a^2 J_{m+1}^2(x_{mn})}\int_0^{2\pi} d\varphi \int_0^a d\rho\, \rho f(\rho, \varphi) J_m\left(\frac{x_{mn}}{a}\rho\right)\cos m\varphi,$$

$$B_{mn} = \frac{2}{\pi a^2 J_{m+1}^2(x_{mn})}\int_0^{2\pi} d\varphi \int_0^a d\rho\, \rho f(\rho, \varphi) J_m\left(\frac{x_{mn}}{a}\rho\right)\sin m\varphi.$$

In particular, if the initial temperature distribution is independent of φ, then only the term with $m = 0$ contributes,[4] and we get

$$T(\rho, t) = \sum_{n=1}^{\infty} A_n J_0\left(\frac{x_{0n}}{a}\rho\right)e^{-(kx_{0n}/a)^2t}.$$

With $f(\rho)$ representing the φ-independent initial temperature distribution, the coefficient A_n is found to be

$$A_n = \frac{4}{a^2 J_1^2(x_{0n})}\int_0^a d\rho\, \rho f(\rho) J_0\left(\frac{x_{0n}}{a}\rho\right).$$

Note that the temperature distribution does not develop any φ dependence at later times.

---

[4]See the discussion after Equation (27.41).

28.2   The Schrödinger Equation

Chapter 22 separated the time part of the Schrödinger equation from its space part, and resulted in the following two equations:

$$\nabla^2 \psi + \frac{2m}{\hbar^2}[E - V(\mathbf{r})]\psi = 0 \quad \text{and} \quad \frac{dT}{dt} = \frac{iE}{\hbar}T, \quad (28.5)$$

where E, the energy of the quantum particle, is the constant of separation.[5] We have also used ψ instead of R, because the latter is usually reserved to denote a function of the radial variable r (or ρ) when separating the variables of the Laplacian in spherical (or cylindrical) coordinates.

The solution of the time part is easily obtained: It is simply

$$T(t) = Ae^{iEt/\hbar} = Ae^{i\omega t} \quad \text{where} \quad \omega \equiv \frac{E}{\hbar}. \quad (28.6)$$

time-independent
Schrödinger
equation

It is the solution of the first equation in (28.5), the **time-independent Schrödinger equation** that will take up most of our time in this section.

Historical Notes

Erwin Schrödinger was a student at Vienna from 1906 and taught there for ten years from 1910 to 1920 with a break for military service in World War I. While at Vienna he worked on radioactivity, proving the statistical nature of radioactive decay. He also made important contributions to the kinetic theory of solids, studying the dynamics of crystal lattices.

After leaving Vienna in 1920 he was appointed to a professorship in Jena, where he stayed for a short time. He then moved to Stuttgart, and later to Breslau before accepting the chair of theoretical physics at Zurich in late 1921. During these years of changing from one place to another, Schrödinger studied physiological optics, in particular the theory of color vision.

Zurich was to be the place where Schrödinger made his most important contributions. From 1921 he studied atomic structure. In 1924 he began to study quantum statistics soon after reading de Broglie's thesis which was to have a major influence on his thinking.

Schrödinger published very important work relating to wave mechanics and the general theory of relativity in a series of papers in 1926. Wave mechanics, proposed by Schrödinger in these papers, was the second formulation of quantum theory, the first being matrix mechanics due to Heisenberg. For this work Schrödinger was awarded the Nobel prize in 1933.

Erwin Schrödinger
1887–1961

Schrödinger went to Berlin in 1927 where he succeeded Planck as the chair of theoretical physics and he became a colleague of Einstein's.

Although he was a Catholic, Schrödinger decided in 1933 that he couldn't live in a country in which the persecution of Jews had become a national policy. He left, spending time in Britain where he was at the University of Oxford from 1933 until 1936. In 1936 he went to Austria and spent the years 1936–1938 in Graz. However, the advancing Nazi threat caught up with him again in Austria and he fled again, this time settling in Dublin, Ireland, in 1939.

---

[5]We used α in place of E in Chapter 22.

His study of Greek science and philosophy is summarized in *Nature and the Greeks* (1954) which he wrote while in Dublin. Another important book written during this period was *What Is Life* (1944) which led to progress in biology. He remained in Dublin until he retired in 1956 when he returned to Vienna.

During his last few years Schrödinger remained interested in mathematical physics and continued to work on general relativity, unified field theory, and meson physics.

28.2.1   Quantum Harmonic Oscillator

As an important example of the Schrödinger equation, we consider a particle in a one-dimensional harmonic oscillator potential.

quantum harmonic oscillator

The one-dimensional time-independent Schrödinger equation for a particle of mass μ in a potential $V(x)$ is

$$\frac{d^2\psi}{dx^2} + \frac{2\mu}{\hbar^2}[E - V(x)]\psi = 0,$$

where E is the total energy of the particle.

For a harmonic oscillator (with the "spring" constant k),

$$V(x) = \tfrac{1}{2}kx^2 \equiv \tfrac{1}{2}\mu\omega^2 x^2$$

and

$$\psi'' - \frac{\mu^2\omega^2}{\hbar^2}x^2\psi + \frac{2\mu}{\hbar^2}E\psi = 0, \qquad \omega \equiv \sqrt{\frac{k}{\mu}}.$$

To simplify the equation, we make the change of variables $x = (\sqrt{\hbar/\mu\omega})y$. The equation then becomes

$$\psi'' - y^2\psi + \frac{2E}{\hbar\omega}\psi = 0, \tag{28.7}$$

where the primes indicate differentiation with respect to y.

We could solve this DE by the Frobenius power series method. However, tradition suggests that we first look at the behavior of the solution at $y \to \infty$. In this limit, we can ignore the last term in (28.7), and the DE becomes

$$\psi'' - y^2\psi \approx 0$$

which can easily be shown to have (an approximate) solution of the form $e^{\pm y^2/2}$. Since the positive exponent diverges at infinity, we have to retain only the solution with negative exponent. Following the traditional steps, we consider a solution of the form $\psi(y) \equiv H(y)\exp(-y^2/2)$ in which the asymptotic function has been separated. Substitution of this separated form of ψ in (28.7) results in

$$H'' - 2yH' + \lambda H = 0 \quad \text{where} \quad \lambda = \frac{2E}{\hbar\omega} - 1. \tag{28.8}$$

Hermite
differential
equation

This is the **Hermite differential equation.**

To solve the Hermite DE by the Frobenius method, we assume an expansion of the form $H(y) = \sum_{n=0}^{\infty} c_n y^n$ with

$$H'(y) = \sum_{n=1}^{\infty} n c_n y^{n-1} = \sum_{n=0}^{\infty} (n+1) c_{n+1} y^n,$$

$$H''(y) = \sum_{n=1}^{\infty} n(n+1) c_{n+1} y^{n-1} = \sum_{n=0}^{\infty} (n+1)(n+2) c_{n+2} y^n,$$

where in the last step of each equation, we changed the dummy index to $m = n - 1$, and in the end, replaced m with n. Substituting in Equation (28.8) gives

$$\underbrace{\sum_{n=0}^{\infty} [(n+1)(n+2) c_{n+2} + \lambda c_n] y^n}_{\equiv S_1} - 2 \sum_{n=0}^{\infty} (n+1) c_{n+1} y^{n+1} = 0. \qquad (28.9)$$

Now separate the zeroth term of the first sum to obtain

$$S_1 = 2 c_2 + \lambda c_0 + \sum_{n=1}^{\infty} [(n+1)(n+2) c_{n+2} + \lambda c_n] y^n.$$

Changing the dummy index to $m = n - 1$ yields

$$S_1 = 2 c_2 + \lambda c_0 + \sum_{m=0}^{\infty} [(m+2)(m+3) c_{m+3} + \lambda c_{m+1}] y^{m+1}$$

whose dummy index can be switched back to n. Substitution of this last result in (28.9) now yields

$$2 c_2 + \lambda c_0 + \sum_{n=0}^{\infty} [(n+2)(n+3) c_{n+3} + \lambda c_{n+1} - 2(n+1) c_{n+1}] y^{n+1} = 0.$$

Setting the coefficients of powers of y equal to zero, we obtain

$$c_2 = -\frac{\lambda}{2} c_0,$$

$$c_{n+3} = \frac{2(n+1) - \lambda}{(n+2)(n+3)} c_{n+1} \qquad \text{for } n \geq 0,$$

recursion relation
for Hermite DE

or, replacing n with $n - 1$ and noting that the resulting recursion relation is true for $n = 0$ as well, we obtain

$$c_{n+2} = \frac{2n - \lambda}{(n+1)(n+2)} c_n, \qquad n \geq 0. \qquad (28.10)$$

Physics dictates
mathematics!

The ratio test yields easily that the series is convergent for all values of y. However, on physical grounds, i.e., the demand that $\lim_{x \to \infty} \psi(x) = 0$, the series must be truncated. Let us see why.

Construction of Hermite Polynomials

The fact that we are interested in the behavior of ψ (and therefore, H) as x (or y) goes to infinity permits us to concentrate on the very large powers of y in the series for $H(y)$. Hence, separating the even and odd parts of the series, we may write

$$H(y) = \sum_{k=0}^{\infty} c_{2k} y^{2k} + \sum_{k=0}^{\infty} c_{2k+1} y^{2k+1}$$

$$= P_{2M+1}(y) + \sum_{k=M+1}^{\infty} c_{2k} y^{2k} + \sum_{k=M+1}^{\infty} c_{2k+1} y^{2k+1} \qquad (28.11)$$

$$= P_{2M+1}(y) + \sum_{k=0}^{\infty} c_{2k+2M+2} y^{2k+2M+2} + \sum_{k=0}^{\infty} c_{2k+2M+3} y^{2k+2M+3},$$

where $P_{2M+1}(y)$ is a generic polynomial obtained by adding all the "small" powers of the series, and M is a very large number.[6] Now note that for very large n, the recursion relation yields

$$c_{n+2} \approx \frac{2n}{(n+1)(n+2)} c_n \approx \frac{2n}{(n)(n)} c_n \approx \frac{2}{n} c_n \Rightarrow c_n \approx \frac{2}{n-2} c_{n-2}.$$

A few iterations give

$$c_n \approx \frac{2^k}{(n-2)(n-4)\cdots(n-2k)} c_{n-2k}.$$

In particular,

$$c_{2k+N} \approx \frac{2^k}{(2k+N-2)(2k+N-4)\cdots(N)} c_N. \qquad (28.12)$$

To find the coefficients in Equation (28.11), first let $N = 2M + 2$ and obtain

$$c_{2k+2M+2} \approx \frac{2^k}{(2k+2M)(2k+2M-2)\cdots(2M+2)} c_{2M+2}$$

$$= \frac{2^k}{[2(k+M)][2(k+M-1)]\cdots[2(M+1)]} c_{2M+2} \qquad (28.13)$$

$$= \frac{1}{(k+M)(k+M-1)\cdots(M+1)} c_{2M+2} = \frac{M!}{(k+M)!} c_{2M+2}.$$

Similarly

$$c_{2k+2M+3} \approx \frac{2^k(2M+1)!!}{(2k+2M+1)!!} c_{2M+3}$$

$$= \frac{2^k[2(M+1)]!/[2^{M+1}(M+1)!]}{[2(k+M+1)]!/[2^{k+M+1}(k+M+1)!]} c_{2M+3}$$

$$= 2^{2k} \frac{[2(M+1)]!(k+M+1)!}{(M+1)![2(k+M+1)]!} c_{2M+3}, \qquad (28.14)$$

---

[6]In particular, M is very large compared to λ of Equation (28.10).

where we used the result of Problem 11.1. By using the Stirling approximation (11.6), the reader may verify that

$$c_{2k+2M+3} \approx \frac{(M+1)!}{(k+M+1)!} c_{2M+3}. \tag{28.15}$$

With the coefficients given in terms of two constants (c_{2M+2} and c_{2M+3}), Equation (28.11) becomes

$$H(y) = P_{2M+1}(y) + c_{2M+2}M!y^2 \sum_{k=0}^{\infty} \frac{y^{2k+2M}}{(k+M)!} + c_{2M+3}(M+1)!y \sum_{k=0}^{\infty} \frac{y^{2k+2M+2}}{(k+M+1)!}$$

$$= P_{2M+1}(y) + c_{2M+2}M!y^2 \sum_{j=M}^{\infty} \frac{y^{2j}}{j!} + c_{2M+3}(M+1)!y \sum_{j=M+1}^{\infty} \frac{y^{2j}}{j!}. \tag{28.16}$$

The first sum over j can be reexpressed as follows:

$$\sum_{j=M}^{\infty} \frac{y^{2j}}{j!} = \sum_{j=0}^{\infty} \frac{(y^2)^j}{j!} - \sum_{j=0}^{M-1} \frac{y^{2j}}{j!} = e^{y^2} - Q_{2M-2}(y),$$

where $Q_{2M-2}(y)$ is a polynomial of degree $2M-2$ in y. The second sum in (28.16) can be expressed similarly. Adding all the polynomials together, we finally get

$$H(y) \approx P_{2M+1}(y) + \underbrace{c_{2M+2}M!}_{\equiv \beta_M} y^2 e^{y^2} + \underbrace{c_{2M+3}(M+1)!}_{\equiv \alpha_M} y e^{y^2}$$

$$= P_{2M+1}(y) + (\alpha_M y + \beta_M y^2)e^{y^2}. \tag{28.17}$$

Let us now go back to $\psi(y)$ and note that

$$\psi(y) = H(y)e^{-y^2/2} \approx \underbrace{P_{2M+1}(y)e^{-y^2/2}}_{\to 0 \text{ as } y \to \infty} + \underbrace{(\alpha_M y + \beta_M y^2)e^{y^2/2}}_{\to \infty \text{ as } y \to \infty}$$

because any exponential decay outweighs any polynomial growth. It follows that, *if $H(y)$ is an infinite series, $\psi(y)$ will diverge at infinity.* From a physical standpoint this means that the quantum particle inside the harmonic oscillator potential well has an infinite probability of being found at infinity![7]

Truncation of the infinite series gives the quantization of harmonic oscillator energy levels.

To avoid this unrealistic conclusion, we have to reexamine $H(y)$. The case of Legendre polynomials tells us that the infinite series needs to be truncated. This will take place only if the numerator of the recursion relation vanishes for some n, i.e., if $\lambda = 2m$ for some integer m. An immediate consequence of such a truncation is the famous quantization of the harmonic oscillator energy:

$$2m = \lambda = \frac{2E}{\hbar\omega} - 1 \Rightarrow E = (m + \tfrac{1}{2})\hbar\omega.$$

Hermite polynomials

The polynomials obtained by truncating the infinite series are called the **Hermite polynomials.** We now construct them. With $\lambda = 2m$, the recursion relation (28.10) can be written as

---

[7]The **Copenhagen interpretation**, the only valid interpretation of quantum mechanics, states that $|\psi(x)|^2$ is the probability density for finding the particle at x.

$$c_n = \frac{2(n - m - 2)}{n(n - 1)} c_{n-2} = -\frac{2(m + 2 - n)}{n(n - 1)} c_{n-2}, \qquad m \geq n \geq 2.$$

The upper limit for n is due to the truncation mentioned above. After a few iteration, the pattern will emerge and the reader may verify that

$$c_n = (-1)^k \frac{2^k (m + 2 - n)(m + 4 - n) \cdots (m + 2k - n)}{n(n - 1) \cdots (n - 2k + 1)} c_{n-2k}. \qquad (28.18)$$

We need to consider the two cases of even and odd n separately. For $n = 2k$, we get

$$c_{2k} = (-1)^k \frac{2^k m(m - 2) \cdots (m + 2 - 2k)}{(2k)!} c_0. \qquad (28.19)$$

Now, since the numerator of (28.10) must vanish beyond some integer, and since $2n = 4k$, we must have $\lambda = 4j$ or $m = \lambda/2 = 2j$ for some integer j. Then, the reader may check that

$$c_{2k} = (-1)^k \frac{2^{2k} j!}{(2k)!(j - k)!} c_0 \qquad (28.20)$$

and

$$H_{2j}(y) = c_0^{(j)} \sum_{k=0}^{j} \frac{(-1)^k 2^{2k} j!}{(2k)!(j - k)!} y^{2k}, \qquad (28.21)$$

where we have given the constant a superscript to distinguish among the c_0's of different j's. The odd polynomials can be obtained similarly:

$$H_{2j+1}(y) = c_1^{(j)} \sum_{k=0}^{j} \frac{(-1)^k 2^{2k+1} j!}{(2k + 1)!(j - k)!} y^{2k+1}. \qquad (28.22)$$

The constants are determined by convention. To adhere to this convention, we define

$$c_0^{(j)} = \frac{(-1)^j (2j)!}{j!}, \qquad c_1^{(j)} = \frac{(-1)^j (2j + 1)!}{j!}.$$

The reader may check that, with these constants, the Hermite polynomials of all degrees (even or odd) can be concisely written as

$$H_n(y) = \sum_{r=0}^{[n/2]} \frac{(-1)^r n!}{(n - 2r)! r!} (2y)^{n-2r}, \qquad (28.23)$$

where $[a]$, for any real a, stands for the largest integer less than or equal to a.

Orthogonality of Hermite Polynomials

The Hermite polynomials satisfy an orthogonality relation resembling that of the Legendre polynomials. We can obtain this relation by multiplying the DE

weight function
for Hermite
polynomials

for $H_m(x)$ by $H_n(x)e^{-x^2}$, and the DE for $H_n(x)$ by $H_m(x)e^{-x^2}$ and subtracting. The factor e^{-x^2}, the so-called *weight function*, may appear artificial in this derivation, but an in-depth analysis of the classical orthogonal polynomials, of which Hermite and Legendre polynomials are examples, reveals that such weight functions are necessary. The reason we did not see such a factor in Legendre polynomials was that for them, the weight function is unity.[8] At any rate, the result of the above suggested calculation will be

$$(H_m'' H_n - 2x H_m' H_n - H_n'' H_m + 2x H_n' H_m)e^{-x^2} + (2m - 2n)H_m H_n e^{-x^2} = 0.$$

The reader may easily verify that the first term is the derivative of

$$(H_m' H_n - H_n' H_m)e^{-x^2},$$

so that

$$\frac{d}{dx}\left[(H_m' H_n - H_n' H_m)e^{-x^2}\right] + (2m - 2n)H_m H_n e^{-x^2} = 0$$

and if we integrate this over the entire real line, we obtain

$$(H_m' H_n - H_n' H_m)e^{-x^2}\Big|_{-\infty}^{\infty} + (2m - 2n)\int_{-\infty}^{\infty} H_m(x)H_n(x)e^{-x^2}\,dx = 0.$$

The first term vanishes because of the exponential factor. It now follows that if $m \neq n$, then

orthogonality of
Hermite
polynomials

$$\int_{-\infty}^{\infty} H_m(x)H_n(x)e^{-x^2}\,dx = 0. \tag{28.24}$$

Generating Function for Hermite Polynomials

We constructed the generating function for Legendre polynomials in Chapter 26. Here we want to do the same thing for Hermite polynomials. By definition, the generating function must have an expansion of the form

$$g(t,x) = \sum_{n=0}^{\infty} a_n t^n H_n(x),$$

where a_n is a constant to be determined. Differentiate both sides with respect to x assuming that t is a constant:

$$\frac{dg}{dx} = \sum_{n=1}^{\infty} a_n t^n H_n'(x).$$

The sum starts at 1 because $H_0'(x) = 0$. Use the result of Problem 28.7 to obtain

$$\frac{dg}{dx} = \sum_{n=1}^{\infty} a_n t^n (2n)H_{n-1}(x) = 2t\sum_{n=1}^{\infty} n a_n t^{n-1}H_{n-1}(x).$$

---

[8] We have no space to go into the details of the theory of classical orthogonal polynomials, but the interested reader can find a unified discussion of them in Hassani, S. *Mathematical Physics: A Modern Introduction to Its Foundations*, Springer-Verlag, 1999, Chapter 7.

Now choose the constant a_n in such a way that it satisfies the recursion relation $na_n = a_{n-1}$. It then follows that

$$\frac{dg}{dx} = 2t \sum_{n=1}^{\infty} a_{n-1} t^{n-1} H_{n-1}(x) = 2t \sum_{m=0}^{\infty} a_m t^m H_m(x) = 2tg.$$

Thus

$$\frac{dg}{g} = 2t\,dx \;\Rightarrow\; \ln g = 2tx + \ln C(t) \;\Rightarrow\; g(t,x) = C(t)e^{2tx},$$

where the "constant" of integration has been given the possibility of depending on the other variable, t. To find this constant of integration, we first determine a_n:

$$a_n = \frac{a_{n-1}}{n} = \frac{a_{n-2}}{n(n-1)} = \cdots = \frac{a_{n-k}}{n(n-1)\cdots(n-k+1)} \cdots = \frac{a_0}{n!}.$$

Using our results obtained so far, we get

$$C(t)e^{2tx} = a_0 \sum_{n=0}^{\infty} \frac{t^n}{n!} H_n(x) \to \sum_{n=0}^{\infty} \frac{t^n}{n!} H_n(x),$$

where in the last step we absorbed a_0 (really $1/a_0$) in the "constant" $C(t)$. To determine $C(t)$, evaluate both sides of the equation at $x = 0$ and use the result of Problem 28.8. This yields

$$C(t) = \sum_{n=0}^{\infty} \frac{t^n}{n!} H_n(0) = \sum_{k=0}^{\infty} \frac{t^{2k}}{(2k)!} \frac{(-1)^k (2k)!}{k!} = \sum_{k=0}^{\infty} \frac{(-t^2)^k}{k!} = e^{-t^2}.$$

It follows that

$$e^{2tx-t^2} = \sum_{n=0}^{\infty} \frac{t^n}{n!} H_n(x). \tag{28.25}$$

We now summarize what we have done:

generating function for Hermite polynomials

Box 28.2.1. *The nth coefficient of the Maclaurin series expansion of the generating function $g(t,x) \equiv e^{2tx-t^2}$ about $t = 0$ is $H_n(x)$. Specifically,*

$$H_n(x) = \frac{\partial^n}{\partial t^n} e^{2tx-t^2} \bigg|_{t=0}. \tag{28.26}$$

We can put the result above to immediate good use. Let us square both sides of Equation (28.25), multiply by e^{-x^2}, and integrate the result from $-\infty$

to $+\infty$. For the LHS, we have

$$LHS = \int_{-\infty}^{+\infty} e^{2(2tx-t^2)} e^{-x^2} dx = e^{-2t^2} \int_{-\infty}^{+\infty} e^{-x^2+4tx} dx$$

$$= e^{-2t^2} \int_{-\infty}^{+\infty} e^{-(x-2t)^2+4t^2} dx = e^{2t^2} \int_{-\infty}^{+\infty} e^{-(x-2t)^2} dx$$

$$= e^{2t^2} \int_{-\infty}^{+\infty} e^{-u^2} du = \sqrt{\pi} e^{2t^2} = \sqrt{\pi} \sum_{n=0}^{\infty} \frac{(2t^2)^n}{n!} = \sqrt{\pi} \sum_{n=0}^{\infty} \frac{2^n t^{2n}}{n!},$$

where we introduced $u = x - 2t$ for the integration variable, and used the result of Example 3.3.1.

To square the RHS, we need to write it as the product of two infinite sums *using two different dummy indices*! Therefore,

$$RHS = \int_{-\infty}^{+\infty} \left(\sum_{n=0}^{\infty} \frac{t^n}{n!} H_n(x) \right) \left(\sum_{m=0}^{\infty} \frac{t^m}{m!} H_m(x) \right) e^{-x^2} dx$$

$$= \sum_{m,n=0}^{\infty} \frac{t^{m+n}}{m!n!} \underbrace{\int_{-\infty}^{+\infty} H_m(x) H_n(x) e^{-x^2} dx}_{=0 \text{ unless } m = n \text{ by } (28.24)}$$

$$= \sum_{n=0}^{\infty} \frac{t^{2n}}{(n!)^2} \int_{-\infty}^{+\infty} [H_n(x)]^2 e^{-x^2} dx.$$

Comparing the LHS and the RHS, we conclude that

$$\int_{-\infty}^{+\infty} [H_n(x)]^2 e^{-x^2} dx = \sqrt{\pi}\, 2^n n!.$$

We can combine this result and Equation (28.24) and write

$$\int_{-\infty}^{+\infty} H_m(x) H_n(x) e^{-x^2} dx = \sqrt{\pi}\, 2^n n! \delta_{mn}, \qquad (28.27)$$

where δ_{mn} is the Kronecker delta which is 1 if $m = n$ and 0 if $m \neq n$.

Historical Notes

Charles Hermite
1822–1901

Charles Hermite, one of the most eminent French mathematicians of the nineteenth century, was particularly distinguished for the clean elegance and high artistic quality of his work. As a student, he courted disaster by neglecting his routine assigned work to study the classic masters of mathematics; and though he nearly failed his examinations, he became a first-rate creative mathematician while still in his early twenties. In 1870 he was appointed to a professorship at the Sorbonne, where he trained a whole generation of well-known French mathematicians, including *Picard*, *Borel*, and *Poincaré*.

The character of his mind is suggested by a remark of Poincaré: "Talk with M. Hermite. He never evokes a concrete image, yet you soon perceive that the most abstract entities are to him like living creatures." He disliked geometry, but was strongly attracted to number theory and analysis, and his favorite subject was elliptic functions, where these two fields touch in many remarkable ways. Earlier in the century the Norwegian genius Abel had proved that the general equation of the fifth degree cannot be solved by functions involving only rational operations and root extractions. One of Hermite's most surprising achievements (in 1858) was to show that this equation can be solved by elliptic functions.

His 1873 proof of the transcendence of e was another high point of his career.[9] If he had been willing to dig even deeper into this vein, he could probably have disposed of π as well, but apparently he had had enough of a good thing. As he wrote to a friend, "I shall risk nothing on an attempt to prove the transcendence of the number π. If others undertake this enterprise, no one will be happier than I at their success, but believe me, my dear friend, it will not fail to cost them some efforts." As it turned out, *Lindemann*'s proof nine years later rested on extending Hermite's method.

Several of his purely mathematical discoveries had unexpected applications many years later to mathematical physics. For example, the Hermitian forms and matrices that he invented in connection with certain problems of number theory turned out to be crucial for Heisenberg's 1925 formulation of quantum mechanics, and Hermite polynomials are useful in solving *Schrödinger*'s wave equation.

28.2.2 Quantum Particle in a Box

The behavior of an atomic particle of mass μ confined in a rectangular box with sides a, b, and c (an infinite three-dimensional potential well) is governed by the Schrödinger equation for a free particle, i.e., $V = 0$. With this assumption, the first equation of (28.5) becomes

quantum particle in a box

$$\nabla^2 \psi + \frac{2\mu E}{\hbar^2} \psi = 0.$$

A separation of variables, $\psi(x, y, z) = X(x)Y(y)Z(z)$, yields the ODEs:

$$\frac{d^2 X}{dx^2} + \lambda X = 0, \qquad \frac{d^2 Y}{dy^2} + \sigma Y = 0, \qquad \frac{d^2 Z}{dz^2} + \nu X = 0,$$

with $\lambda + \sigma + \nu = 2\mu E/\hbar^2$ (see Example 22.2.1).

These equations, together with the boundary conditions

$$
\begin{aligned}
\psi(0, y, z) = \psi(a, y, z) = 0 &\quad \Rightarrow \quad X(0) = 0 = X(a), \\
\psi(x, 0, z) = \psi(x, b, z) = 0 &\quad \Rightarrow \quad Y(0) = 0 = Y(b), \\
\psi(x, y, 0) = \psi(x, y, c) = 0 &\quad \Rightarrow \quad Z(0) = 0 = Z(c),
\end{aligned}
\qquad (28.28)
$$

---

[9]Transcendental numbers are those that are not roots of polynomials with integer coefficients.

lead to the following solutions:

$$X_n(x) = \sin\left(\frac{n\pi}{a}x\right), \qquad \lambda_n = \left(\frac{n\pi}{a}\right)^2 \qquad \text{for} \quad n = 1, 2, \ldots,$$

$$Y_m(y) = \sin\left(\frac{m\pi}{b}y\right), \qquad \sigma_m = \left(\frac{m\pi}{b}\right)^2 \qquad \text{for} \quad m = 1, 2, \ldots,$$

$$Z_l(z) = \sin\left(\frac{l\pi}{c}z\right), \qquad \nu_l = \left(\frac{l\pi}{c}\right)^2 \qquad \text{for} \quad l = 1, 2, \ldots,$$

where the multiplicative constants have been suppressed.

quantum
tunneling

The BCs in Equation (28.28) arise from the demand that the probability of finding the particle be continuous and that it be zero outside the box. This is not true for a particle inside a finite potential well, in which case the particle has a nonzero probability of "tunneling" out of the well.

The time equation has a solution of the form

$$T(t) = e^{-i\omega_{lmn}t} \qquad \text{where} \quad \omega_{lmn} = \frac{\hbar}{2\mu}\left[\left(\frac{n\pi}{a}\right)^2 + \left(\frac{m\pi}{b}\right)^2 + \left(\frac{l\pi}{c}\right)^2\right].$$

The solution of the Schrödinger equation that is consistent with the boundary conditions is, therefore,

$$\psi(x, y, z, t) = \sum_{l,m,n=1}^{\infty} A_{lmn} e^{-i\omega_{lmn}t} \sin\left(\frac{n\pi}{a}x\right) \sin\left(\frac{m\pi}{b}y\right) \sin\left(\frac{l\pi}{c}z\right).$$

The constants A_{lmn} are determined by the initial shape $\psi(x, y, z, 0)$ of the wave function. In fact, setting $t = 0$, multiplying by the product of the three sine functions in the three variables, and integrating over appropriate intervals for each coordinate, we obtain

$$A_{lmn} = \frac{8}{abc}\int_0^a dx \int_0^b dy \int_0^c dz\, \psi(x, y, z, 0) \sin\left(\frac{n\pi}{a}x\right) \sin\left(\frac{m\pi}{b}y\right) \sin\left(\frac{l\pi}{c}z\right).$$

The energy of the particle is

$$E = \hbar\omega_{lmn} = \frac{\hbar^2\pi^2}{2\mu}\left(\frac{n^2}{a^2} + \frac{m^2}{b^2} + \frac{l^2}{c^2}\right).$$

Each set of three positive integers (n, m, l) represents a **quantum state** of the particle. For a cube, $a = b = c \equiv L$, and the energy of the particle is

$$E = \frac{\hbar^2\pi^2}{2\mu L^2}(n^2 + m^2 + l^2) = \frac{\hbar^2\pi^2}{2\mu V^{2/3}}(n^2 + m^2 + l^2), \qquad (28.29)$$

where $V = L^3$ is the volume of the box. The ground state is $(1, 1, 1)$, has energy $E = 3\hbar^2\pi^2/2\mu V^{2/3}$, and is nondegenerate (only one state corresponds to this energy). However, the higher-level states are degenerate. For instance, the three distinct states $(1,1,2)$, $(1,2,1)$, and $(2,1,1)$ all correspond to the same energy, $E = 6\hbar^2\pi^2/2\mu V^{2/3}$. The degeneracy increases rapidly with larger values of n, m, and l.

Equation (28.29) can be written as

$$n^2 + m^2 + l^2 = R^2 \quad \text{where} \quad R^2 = \frac{2\mu E V^{2/3}}{\hbar^2\pi^2}.$$

This looks like the equation of a sphere in the nml-space. If R is large, the number of states contained within the sphere of radius R (the number of states with energy less than or equal to E) is simply the volume of the first octant[10] of the sphere. If N is the number of such states, we have

$$N = \frac{1}{8}\left(\frac{4\pi}{3}\right)R^3 = \frac{\pi}{6}\left(\frac{2\mu E V^{2/3}}{\hbar^2\pi^2}\right)^{3/2} = \frac{\pi}{6}\left(\frac{2\mu E}{\hbar^2\pi^2}\right)^{3/2}V.$$

Thus the **density of states** (the number of states per unit volume) is then   density of states

$$n = \frac{N}{V} = \frac{\pi}{6}\left(\frac{2\mu}{\hbar^2\pi^2}\right)^{3/2}E^{3/2}. \tag{28.30}$$

This is an important formula in solid-state physics, because the energy E is (with minor modifications required by spin) the **Fermi energy**. If the Fermi   Fermi energy
energy is denoted by E_f, Equation (28.30) gives $E_f = \alpha n^{2/3}$ where α is some constant.

28.2.3   Hydrogen Atom

When an electron moves around a nucleus containing Z protons, the potential energy of the system is $V(r) = -Ze^2/r$. In units in which $\hbar$ and the mass of the electron are set equal to unity, the time-independent Schrödinger equation of (28.5) gives

$$\nabla^2\Psi + \left(2E + \frac{2Ze^2}{r}\right)\Psi = 0.$$

The radial part of this equation is given by the first equation in (22.16) with $f(r) = 2E + 2Ze^2/r$. Defining $u = rR(r)$, we may write

$$\frac{d^2u}{dr^2} + \left(\lambda + \frac{a}{r} - \frac{\alpha}{r^2}\right)u = 0, \tag{28.31}$$

where $\lambda = 2E$ and $a = 2Ze^2$. This equation can be further simplified by defining $r \equiv kz$ (k is an arbitrary constant to be determined later):

$$\frac{d^2u}{dz^2} + \left(\lambda k^2 + \frac{ak}{z} - \frac{\alpha}{z^2}\right)u = 0.$$

---

[10]This is because n, m, and l are all positive.

Choosing $\lambda k^2 = -\frac{1}{4}$ and introducing $\beta \equiv a/(2\sqrt{-\lambda})$ yields

$$\frac{d^2 u}{dz^2} + \left(-\frac{1}{4} + \frac{\beta}{z} - \frac{\alpha}{z^2} \right) u = 0. \tag{28.32}$$

Let us examine the two limiting cases of $z \to \infty$ and $z \to 0$. For the first case, Equation (28.32) reduces to

$$\frac{d^2 u}{dz^2} - \tfrac{1}{4} u = 0 \;\Rightarrow\; u = e^{-z/2}.$$

For the second case the dominant term will be α/z^2 and the DE becomes

$$\frac{d^2 u}{dz^2} - \frac{\alpha}{z^2} u = 0$$

for which we try a solution of the form z^m with m to be determined by substitution:

$$\frac{d^2 u}{dz^2} = m(m-1)z^{m-2} \;\Rightarrow\; m(m-1)z^{m-2} - \frac{\alpha}{z^2} z^m = 0 \;\Rightarrow\; \alpha = m(m-1).$$

Recalling from Theorem 26.2.1 that $\alpha = l(l+1)$, we determine m to be $l+1$. Factoring out these two limits, we seek a solution for (28.32) of the form

$$u(z) = z^{l+1} e^{-z/2} f(z).$$

Substitution of this function in (28.32) gives a new DE:

$$f'' + \left[\frac{2(l+1)}{z} - 1 \right] f' - \frac{l+1-\beta}{z} f = 0. \tag{28.33}$$

Multiplying by z gives

$$z f'' + [2(l+1) - z] f' - (l+1-\beta) f = 0$$

which is a confluent hypergeometric DE [see Equation (11.27)]. Therefore, as the reader may verify, f is proportional to $\Phi(l+1-\beta, 2l+2; z)$. Thus, the solution of (28.31) can be written as

$$u(z) = C z^{l+1} e^{-z/2} \Phi(l+1-\beta, 2l+2; z).$$

Laguerre Polynomials

An argument similar to that used in the discussion of a quantum harmonic oscillator will reveal that the product $e^{-z/2} \Phi(l+1-\beta, 2l+2; z)$ will be infinite unless the power series representing Φ terminates (becomes a polynomial). This takes place only if (see Box 11.2.2)

$$l + 1 - \beta = -N \tag{28.34}$$

for some integer $N \geq 0$. In that case we obtain the **Laguerre polynomials**

$$L_N^j \equiv \frac{\Gamma(N+j+1)}{\Gamma(N+1)\Gamma(j+1)}\Phi(-N, j+1; z) \quad \text{where} \quad j = 2l+1, \quad (28.35)$$

Laguerre polynomials

where the factor in front of Φ is a standardization factor.

Condition (28.34) is the quantization rule for the energy levels of a hydrogen-like atom. Writing everything in terms of the original parameters, and redefining β as $\beta = N + l + 1 \equiv n$ to reflect its integer character, yields—after restoring all the μ's and the $\hbar$'s—the energy levels of a hydrogen-like atom:

Truncation of infinite series gives the quantization rule for the energy levels of the hydrogen atom.

$$E = -\frac{Z^2 \mu e^4}{2\hbar^2 n^2} = -Z^2 \left(\frac{\mu c^2}{2}\right)\alpha^2 \frac{1}{n^2},$$

where $\alpha = e^2/\hbar c = 1/137$ is the **fine structure constant**.

fine structure constant

The radial wave functions can now be written as

$$R_{n,l}(r) = \frac{u_{n,l}(r)}{r} = Cr^l e^{-Zr/na_0}\Phi\left(-n+l+1, 2l+2; \frac{2Zr}{na_0}\right),$$

where $a_0 = \hbar^2/me^2 = 0.529 \times 10^{-8}$ cm is the Bohr radius.

The explicit form of Laguerre polynomials can be obtained by substituting the truncated confluent hypergeometric series [see Equation (11.28)] in (28.35):

$$L_N^j(x) = \frac{\Gamma(N+j+1)}{\Gamma(N+1)\Gamma(j+1)}\frac{\Gamma(j+1)}{\Gamma(-N)}\sum_{k=0}^{N}\frac{\Gamma(-N+k)}{\Gamma(j+1+k)\Gamma(k+1)}x^k.$$

We now use the result of Problem 11.4 and write

$$\frac{\Gamma(k-N)}{\Gamma(-N)} = (-1)^k N(N-1)\cdots(N-k+1) = \frac{(-1)^k N!}{(N-k)!}.$$

It follows that

$$L_N^j(x) = \frac{\Gamma(N+j+1)}{\Gamma(N+1)}\sum_{k=0}^{N}\frac{(-1)^k N!}{(N-k)!}\frac{1}{\Gamma(j+1+k)\Gamma(k+1)}x^k.$$

Simplifying and writing all gamma functions in terms of factorials, we obtain the final form of the Laguerre polynomials:

$$L_N^j(x) = \sum_{k=0}^{N}\frac{(N+j)!(-1)^k}{(N-k)!k!(k+j)!}x^k. \quad (28.36)$$

The generating function of the Laguerre polynomials can be calculated using a procedure similar to the one used in the case of Hermite polynomials. We first write

$$g_j(t,x) = \sum_{n=0}^{\infty} a_n t^n L_n^j(x),$$

differentiate it with respect to x, and use the result of Problem 28.10 to obtain

$$\frac{dg_j}{dx} = -\sum_{n=0}^{\infty} a_n t^n L_{n-1}^{j+1}(x) = -t \sum_{n=0}^{\infty} a_n t^{n-1} L_{n-1}^{j+1}(x) = -t g_{j+1}, \qquad (28.37)$$

where we have taken $a_n = a_{n-1}$ as a natural choice whereby the last sum could be written in closed form. To find a solution of (28.37), we look at $g(t, 0)$. The recursion relation $a_n = a_{n-1}$ implies that all a_n are equal, and we set all of them equal to 1. Then

$$g_j(t, 0) = \sum_{n=0}^{\infty} t^n L_n^j(0) = \sum_{n=0}^{\infty} \frac{(n+j)!}{n! j!} t^n = (1-t)^{-j-1},$$

where we used the fact that the only contribution to $L_n^j(0)$ comes from the constant term in the polynomial [corresponding to $k = 0$ in (28.36)]. Furthermore, the last sum is the binomial series (10.15) with $x \to -t$ and $\alpha \to (-j - 1)$. This suggests defining a new function $g(t, x)$ via

$$g_j(t, x) = (1-t)^{-j-1} g(t, x).$$

Substitution of this in (28.37) gives

$$(1-t)^{-j-1} \frac{dg}{dx} = -t(1-t)^{-j-2} g \;\Rightarrow\; \frac{dg}{g} = \frac{-t}{1-t} dx \;\Rightarrow\; g = C(t) e^{-tx/(1-t)}$$

and

$$g_j(t, x) = (1-t)^{-j-1} C(t) e^{-tx/(1-t)} = \frac{C(t) e^{-tx/(1-t)}}{(1-t)^{j+1}}.$$

With the value of $g_j(t, 0)$ given, we determine $C(t)$ to be one.

Box 28.2.2. *The nth coefficient of the Maclaurin series expansion of the generating function $g_j(t, x) \equiv (1-t)^{-j-1} e^{-tx/(1-t)}$ about $t = 0$ is $L_n^j(x)$. Specifically,*

$$L_n^j(x) = \frac{1}{n!} \frac{\partial^n}{\partial t^n} \frac{e^{-tx/(1-t)}}{(1-t)^{j+1}} \bigg|_{t=0}. \qquad (28.38)$$

28.3   The Wave Equation

In the preceding sections the time variation has been given by a first derivative. Thus, as far as time is concerned, we have a FODE. It follows that the initial specification of the physical quantity of interest (temperature T or Schrödinger wave function ψ) is sufficient to determine the solution uniquely.

The wave equation

$$\nabla^2 \psi = \frac{1}{c^2} \frac{\partial^2 \psi}{\partial t^2} \qquad (28.39)$$

contains time derivatives of the second order, and, therefore, requires two arbitrary parameters in its general solution. To determine these, we expect two initial conditions. For example, if the wave is standing, as in a rope clamped at both ends, the initial shape of the rope is not sufficient to determine the wave function uniquely. One also needs to specify the initial (transverse) velocity of each point of the rope, i.e., the velocity profile on the rope.

Example 28.3.1. ONE-DIMENSIONAL WAVE   one-dimensional
The simplest kind of wave equation is that in one dimension, for example, a wave   wave
propagating on a rope. Such a wave equation can be written as

$$\frac{\partial^2 \psi}{\partial x^2} = \frac{1}{c^2}\frac{\partial^2 \psi}{\partial t^2},$$

where c is the speed of wave propagation. For a rope, this speed is related to the tension τ and the linear mass density ρ by $c = \sqrt{\tau/\rho}$.

Let us assume that the rope has length a and is fastened at both ends (located at $x = 0$ and $x = a$). This means that the "displacement" ψ is zero at $x = 0$ and at $x = a$.

A separation of variables, $\psi(x,t) = X(x)T(t)$, leads to two ODEs:

$$\frac{d^2 X}{dx^2} + \lambda X = 0, \qquad \frac{d^2 T}{dt^2} + c^2 \lambda T = 0. \tag{28.40}$$

The first equation and the spatial boundary conditions give rise to the solutions

$$X_n(x) = \sin\left(\frac{n\pi}{a}x\right), \qquad \lambda_n = \left(\frac{n\pi}{a}\right)^2 \qquad \text{for} \quad n = 1, 2, \ldots.$$

The second equation in (28.40) has a general solution of the form

$$T(t) = A_n \cos\omega_n t + B_n \sin\omega_n t,$$

where $\omega_n = cn\pi/a$ and A_n and B_n are arbitrary constants. The general solution is thus

$$\psi(x,t) = \sum_{n=1}^{\infty}(A_n \cos\omega_n t + B_n \sin\omega_n t)\sin\left(\frac{n\pi x}{a}\right). \tag{28.41}$$

Specification of the initial shape of the rope as $\psi(x,0) = f(x)$ gives a Fourier series,

$$f(x) = \sum_{n=1}^{\infty} A_n \sin\left(\frac{n\pi x}{a}\right)$$

from which we can determine A_n:

$$A_n = \frac{2}{a}\int_0^a f(x)\sin\left(\frac{n\pi x}{a}\right)\,dx.$$

What about B_n? Physically, the shape of the wave is not enough to define the problem uniquely. It is possible that the rope, while having the required initial shape, may be in an unspecified motion of some sort. Thus, we must also know the "velocity profile," which means specifying the function $\partial\psi/\partial t$ at $t = 0$. If it is given that $\partial\psi/\partial t|_{t=0} = g(x)$, then differentiating (28.41) with respect to t and evaluating both sides at $t = 0$ yields

$$g(x) = \sum_{n=1}^{\infty} \omega_n B_n \sin\left(\frac{n\pi x}{a}\right)$$

and B_n is also determined:

$$B_n = \frac{2}{a\omega_n} \int_0^a g(x) \sin\left(\frac{n\pi x}{a}\right) dx.$$

mode of oscillation

The frequency ω_n is referred to as that of the nth **mode of oscillation**. Thus a general solution is a linear superposition of infinitely many modes. In practice, it is possible to "excite" one mode or, with appropriate initial conditions, a finite number of modes. ∎

28.3.1   Guided Waves

Waveguides are hollow tubes (or tubes filled with some dielectric material) in which electromagnetic waves can propagate along an axis which we take to be the z-axis of either Cartesian or cylindrical coordinates. We assume that the dependence of the electric and magnetic fields on z and t is of the form $e^{i(\omega t - kz)}$ where ω and k are constants to be determined. We therefore write

$$\mathbf{E} = \mathbf{E}_0(x, y)e^{i(\omega t - kz)}, \qquad \mathbf{B} = \mathbf{B}_0(x, y)e^{i(\omega t - kz)}, \qquad (28.42)$$

for Cartesian coordinates. If cylindrical geometry is appropriate, then $\mathbf{E}_0$ and $\mathbf{B}_0$ will be functions of ρ and φ. Note that, in general, $\mathbf{E}_0$ and $\mathbf{B}_0$ have three components.

The electric and magnetic fields of (28.42) ought to satisfy the four Maxwell's equations. Let us assume that the waveguide is free of any charges or currents, so that Maxwell's equations for empty space are appropriate. Because of the nature of the dependence on z, it is useful to separate the **longitudinal** geometry—the geometry along z—from the **transverse** geometry—the geometry perpendicular to z. So, we write

longitudinal and transverse parts of guided waves.

$$\mathbf{E} = \mathbf{E}_t + \hat{\mathbf{e}}_z E_z = (\mathbf{E}_{0t} + \hat{\mathbf{e}}_z E_{0z})e^{i(\omega t - kz)},$$
$$\mathbf{B} = \mathbf{B}_t + \hat{\mathbf{e}}_z B_z = (\mathbf{B}_{0t} + \hat{\mathbf{e}}_z B_{0z})e^{i(\omega t - kz)}, \qquad (28.43)$$
$$\boldsymbol{\nabla} = \hat{\mathbf{e}}_x \frac{\partial}{\partial x} + \hat{\mathbf{e}}_y \frac{\partial}{\partial y} + \hat{\mathbf{e}}_z \frac{\partial}{\partial z} \equiv \boldsymbol{\nabla}_t + \hat{\mathbf{e}}_z \frac{\partial}{\partial z},$$

where the subscript t stands for transverse. With these assumptions, Maxwell's first equation becomes

$$0 = \boldsymbol{\nabla} \cdot \mathbf{E} = \left(\boldsymbol{\nabla}_t + \hat{\mathbf{e}}_z \frac{\partial}{\partial z}\right) \cdot \left[(\mathbf{E}_{0t} + \hat{\mathbf{e}}_z E_{0z})e^{i(\omega t - kz)}\right]$$
$$= (\boldsymbol{\nabla}_t \cdot \mathbf{E}_{0t})\, e^{i(\omega t - kz)} + (-ikE_{0z})\, e^{i(\omega t - kz)},$$

because $\boldsymbol{\nabla}_t \cdot (\hat{\mathbf{e}}_z E_{0z}) = 0$ and $\hat{\mathbf{e}}_z \cdot \mathbf{E}_{0t} = 0$. It follows that

$$\boldsymbol{\nabla}_t \cdot \mathbf{E}_{0t} = ikE_{0z}.$$

An analogous calculation gives a similar result for Maxwell's second equation. Putting these two equations together, we get

$$\boldsymbol{\nabla}_t \cdot \mathbf{E}_{0t} = ikE_{0z}, \qquad \boldsymbol{\nabla}_t \cdot \mathbf{B}_{0t} = ikB_{0z}. \tag{28.44}$$

The LHS of Maxwell's third equation gives

$$LHS = \boldsymbol{\nabla} \times \mathbf{E} = \left(\boldsymbol{\nabla}_t + \hat{\mathbf{e}}_z \frac{\partial}{\partial z}\right) \times \left[\mathbf{E}_0 e^{i(\omega t - kz)}\right]$$

$$= \boldsymbol{\nabla}_t \times \left[\mathbf{E}_0 e^{i(\omega t - kz)}\right] + \hat{\mathbf{e}}_z \times \left[-ik\mathbf{E}_0 e^{i(\omega t - kz)}\right]$$

$$= e^{i(\omega t - kz)} \left(\boldsymbol{\nabla}_t \times \mathbf{E}_0 - ik\hat{\mathbf{e}}_z \times \mathbf{E}_{0t}\right).$$

The RHS of the third equation is

$$-\frac{\partial \mathbf{B}}{\partial t} = -i\omega \mathbf{B}_0 e^{i(\omega t - kz)}.$$

Equating the two sides yields

$$-i\omega \mathbf{B}_0 = \boldsymbol{\nabla}_t \times \mathbf{E}_0 - ik\hat{\mathbf{e}}_z \times \mathbf{E}_{0t}. \tag{28.45}$$

The first term on the RHS can be written as

$$\boldsymbol{\nabla}_t \times \mathbf{E}_0 = \begin{vmatrix} \hat{\mathbf{e}}_x & \hat{\mathbf{e}}_y & \hat{\mathbf{e}}_z \\ \dfrac{\partial}{\partial x} & \dfrac{\partial}{\partial y} & 0 \\ E_{0x} & E_{0y} & E_{0z} \end{vmatrix} = \underbrace{\frac{\partial E_{0z}}{\partial y}\hat{\mathbf{e}}_x - \frac{\partial E_{0z}}{\partial x}\hat{\mathbf{e}}_y}_{\text{This is transverse.}} + \hat{\mathbf{e}}_z\left(\frac{\partial E_{0y}}{\partial x} - \frac{\partial E_{0x}}{\partial y}\right)$$

$$= \boldsymbol{\nabla} \times (E_{0z}\hat{\mathbf{e}}_z) + \hat{\mathbf{e}}_z\left(\frac{\partial E_{0y}}{\partial x} - \frac{\partial E_{0x}}{\partial y}\right).$$

The reader may easily check that the second line follows from the first. Equating the *transverse parts* of the two sides of Equation (28.45), we get

$$-i\omega \mathbf{B}_{0t} = \boldsymbol{\nabla} \times (E_{0z}\hat{\mathbf{e}}_z) - ik\hat{\mathbf{e}}_z \times \mathbf{E}_{0t}. \tag{28.46}$$

A similar calculation turns the fourth Maxwell equation into

$$i\frac{\omega}{c^2}\mathbf{E}_{0t} = \boldsymbol{\nabla} \times (B_{0z}\hat{\mathbf{e}}_z) - ik\hat{\mathbf{e}}_z \times \mathbf{B}_{0t}. \tag{28.47}$$

We now want to express the transverse components in terms of the longitudinal components. Multiply both sides of (28.47) by $-i\omega$ and substitute for $-i\omega\mathbf{B}_{0t}$ from (28.46):

$$\frac{\omega^2}{c^2}\mathbf{E}_{0t} = -i\omega\boldsymbol{\nabla} \times (B_{0z}\hat{\mathbf{e}}_z) - ik\hat{\mathbf{e}}_z \times [\boldsymbol{\nabla} \times (E_{0z}\hat{\mathbf{e}}_z) - ik\hat{\mathbf{e}}_z \times \mathbf{E}_{0t}]$$

$$= -i\omega\boldsymbol{\nabla} \times (B_{0z}\hat{\mathbf{e}}_z) - ik\hat{\mathbf{e}}_z \times [\boldsymbol{\nabla} \times (E_{0z}\hat{\mathbf{e}}_z)] - k^2\hat{\mathbf{e}}_z \times (\hat{\mathbf{e}}_z \times \mathbf{E}_{0t}).$$

Using the *bac cab* rule, the last term gives

$$\hat{\mathbf{e}}_z \times (\hat{\mathbf{e}}_z \times \mathbf{E}_{0t}) = \hat{\mathbf{e}}_z \underbrace{(\hat{\mathbf{e}}_z \cdot \mathbf{E}_{0t})}_{=0} - \mathbf{E}_{0t}(\hat{\mathbf{e}}_z \cdot \hat{\mathbf{e}}_z) = -\mathbf{E}_{0t}.$$

It now follows that

$$\left(\frac{\omega^2}{c^2} - k^2\right) \mathbf{E}_{0t} = -i\omega \boldsymbol{\nabla} \times (B_{0z}\hat{\mathbf{e}}_z) - ik\hat{\mathbf{e}}_z \times [\boldsymbol{\nabla} \times (E_{0z}\hat{\mathbf{e}}_z)]. \qquad (28.48)$$

The first term on the RHS can be simplified by using the second equation in (14.11):

$$\boldsymbol{\nabla} \times (B_{0z}\hat{\mathbf{e}}_z) = B_{0z} \underbrace{\boldsymbol{\nabla} \times \hat{\mathbf{e}}_z}_{=0} + (\boldsymbol{\nabla} B_{0z}) \times \hat{\mathbf{e}}_z = -\hat{\mathbf{e}}_z \times (\boldsymbol{\nabla}_t B_{0z})$$

because $\hat{\mathbf{e}}_z$ is a constant vector (both magnitude and direction), and B_{0z} is independent of z. The second term on the RHS of (28.48) can be simplified as follows:

$$\hat{\mathbf{e}}_z \times [\boldsymbol{\nabla} \times (E_{0z}\hat{\mathbf{e}}_z)] = \hat{\mathbf{e}}_z \times \left(\frac{\partial E_{0z}}{\partial y}\hat{\mathbf{e}}_x - \frac{\partial E_{0z}}{\partial x}\hat{\mathbf{e}}_y\right)$$

$$= \frac{\partial E_{0z}}{\partial y}\hat{\mathbf{e}}_y + \frac{\partial E_{0z}}{\partial x}\hat{\mathbf{e}}_x \equiv \boldsymbol{\nabla}_t E_{0z}.$$

Substituting these results in Equation (28.48) yields

$$\gamma^2 \mathbf{E}_{0t} = i\left[-k\boldsymbol{\nabla}_t E_{0z} + \omega\hat{\mathbf{e}}_z \times (\boldsymbol{\nabla}_t B_{0z})\right] \qquad \text{where} \quad \gamma^2 \equiv \frac{\omega^2}{c^2} - k^2.$$

A similar calculation gives an analogous result for the magnetic field. We assemble these two equations in

$$\gamma^2 \mathbf{E}_{0t} = i\left[-k\boldsymbol{\nabla}_t E_{0z} + \omega\hat{\mathbf{e}}_z \times (\boldsymbol{\nabla}_t B_{0z})\right], \qquad (28.49)$$

$$\gamma^2 \mathbf{B}_{0t} = i\left[-k\boldsymbol{\nabla}_t B_{0z} - \omega\hat{\mathbf{e}}_z \times (\boldsymbol{\nabla}_t E_{0z})\right], \qquad \gamma^2 \equiv \frac{\omega^2}{c^2} - k^2.$$

Although we derived (28.44) and (28.49) using Cartesian coordinates, the fact that the final result is written explicitly in terms of transverse and longitudinal parts—without reference to any coordinate system—implies that these equations are valid in cylindrical coordinates as well.

Three types of guided waves are usually studied.

transverse
magnetic (TM)
waves

1. Transverse magnetic (TM) waves have $B_z = 0$ everywhere. The boundary condition on $\mathbf{E}$ demands that E_z vanish at the conducting walls of the guide.

transverse electric
(TE) waves

2. Transverse electric (TE) waves have $E_z = 0$ everywhere. The boundary condition on $\mathbf{B}$ requires that the normal directional derivative

$$\frac{\partial B_z}{\partial n} \equiv \hat{\mathbf{e}}_n \cdot (\boldsymbol{\nabla} B_z)$$

vanish at the walls.

3. Transverse electromagnetic (TEM) waves have $B_z = 0 = E_z$. For a nontrivial solution, Equation (28.49) demands that $\gamma^2 = 0$. This form resembles a free wave with no boundaries.

transverse electromagnetic (TEM) waves

In the following, we consider some examples of the TM mode (see any book on electromagnetic theory for further details). The basic equations in this mode are

$$B_{0z} = 0, \quad \gamma^2 \mathbf{E}_{0t} = -ik\boldsymbol{\nabla}_t E_{0z}, \quad \gamma^2 \mathbf{B}_{0t} = -i\omega \hat{\mathbf{e}}_z \times (\boldsymbol{\nabla}_t E_{0z}).$$

Taking the dot product of $\boldsymbol{\nabla}_t$ with the middle equation and using the first equation in (28.44) yields

Basic equation for TM waves.

$$\boldsymbol{\nabla}_t^2 E_{0z} + \gamma^2 E_{0z} = 0. \tag{28.50}$$

This is the basic equation for TM waves propagating in a waveguide.

Example 28.3.2. RECTANGULAR WAVE GUIDES
For a wave guide with a rectangular cross section of sides a and b in the x and the y direction, respectively, (28.50) gives

rectangular wave guides

$$\frac{\partial^2 E_{0z}}{\partial x^2} + \frac{\partial^2 E_{0z}}{\partial y^2} + \gamma^2 E_{0z} = 0.$$

A separation of variables, $E_{0z}(x,y) = X(x)Y(y)$, leads to

$$\frac{d^2 X}{dx^2} + \lambda X = 0, \qquad X(0) = 0 = X(a),$$

$$\frac{d^2 Y}{dy^2} + \mu Y = 0, \qquad Y(0) = 0 = Y(b),$$

where $\gamma^2 = \lambda + \mu$. These equations have the solutions

$$X_n(x) = \sin\left(\frac{n\pi}{a}x\right), \qquad \lambda_n = \left(\frac{n\pi}{a}\right)^2 \quad \text{for} \quad n = 1, 2, \ldots,$$

$$Y_m(y) = \sin\left(\frac{m\pi}{b}y\right), \qquad \mu_m = \left(\frac{m\pi}{b}\right)^2 \quad \text{for} \quad m = 1, 2, \ldots.$$

The wave number is given by $k = \sqrt{(\omega/c)^2 - \gamma^2}$, or, introducing indexes for k,

$$k_{mn} = \sqrt{\frac{\omega^2}{c^2} - \left(\frac{n\pi}{a}\right)^2 - \left(\frac{m\pi}{b}\right)^2},$$

which has to be real if the wave is to propagate [an imaginary k leads to exponential decay or growth along the z-axis because of the exponential factor in (28.49)]. Thus, there is a cut-off frequency,

$$\omega_{mn} = c\sqrt{\left(\frac{n\pi}{a}\right)^2 + \left(\frac{m\pi}{b}\right)^2} \quad \text{for} \quad m, n \geq 1,$$

below which the wave cannot propagate through the wave guide. It follows that, for a TM wave, the lowest frequency that can propagate along a rectangular wave guide is $\omega_{11} = \pi c\sqrt{a^2 + b^2}/ab$.

The most general solution for E_z is, therefore,

$$E_z = \sum_{m,n=1}^{\infty} A_{mn} \sin\left(\frac{n\pi}{a}x\right) \sin\left(\frac{m\pi}{b}y\right) e^{i(\omega t \pm k_{mn}z)}.$$

The constants A_{mn} are arbitrary and can be determined from the initial shape of the wave, but that is not commonly done. Once E_z is found, the other components can be calculated using Equation (28.50). ∎

cylindrical wave
guide

Example 28.3.3. CYLINDRICAL WAVE GUIDE

For a TM wave propagating along the z-axis in a hollow circular conductor, we have [see Equation (28.50)]

$$\underbrace{\frac{1}{\rho}\frac{\partial}{\partial\rho}\left(\rho\frac{\partial E_{0z}}{\partial\rho}\right) + \frac{1}{\rho^2}\frac{\partial^2 E_{0z}}{\partial\varphi^2}}_{\equiv \boldsymbol{\nabla}_t^2 E_{0z}} + \gamma^2 E_{0z} = 0.$$

The separation $E_{0z} = R(\rho)S(\varphi)$ yields $S(\varphi) = A\cos m\varphi + B\sin m\varphi$ and the Bessel DE

$$\frac{d^2 R}{d\rho^2} + \frac{1}{\rho}\frac{dR}{d\rho} + \left(\gamma^2 - \frac{m^2}{\rho^2}\right)R = 0.$$

The solution to this equation, which is regular at $\rho = 0$ and vanishes at $\rho = a$, is

$$R(\rho) = C J_m\left(\frac{x_{mn}}{a}\rho\right) \qquad \text{and} \qquad \gamma = \frac{x_{mn}}{a}.$$

Recalling the definition of γ, we obtain

$$\frac{\omega^2}{c^2} - k^2 = \gamma^2 = \frac{x_{mn}^2}{a^2} \;\Rightarrow\; k = \sqrt{\frac{\omega^2}{c^2} - \frac{x_{mn}^2}{a^2}}.$$

This gives the cut-off frequency $\omega_{mn} = c x_{mn}/a$.

The solution for the azimuthally symmetric case ($m = 0$) is

$$E_z(\rho,\varphi,t) = \sum_{n=1}^{\infty} A_n J_0\left(\frac{x_{0n}}{a}\rho\right) e^{i(\omega t \pm k_n z)} \qquad \text{and} \qquad B_z = 0,$$

where $k_n = \sqrt{\omega^2/c^2 - x_{0n}^2/a^2}$. ∎

28.3.2   Vibrating Membrane

Waves on a circular drumhead are historically important because their investigation was one of the first instances in which Bessel functions appeared. The following example considers such waves.

For a circular membrane over a cylinder, the wave equation (28.39) in cylindrical coordinates becomes[11]

$$\frac{1}{\rho}\frac{\partial}{\partial\rho}\left(\rho\frac{\partial\psi}{\partial\rho}\right) + \frac{1}{\rho^2}\frac{\partial^2\psi}{\partial\varphi^2} = \frac{1}{c^2}\frac{\partial^2\psi}{\partial t^2}$$

---

[11] Assuming that the membrane is perpendicular to the z-axis, the wave amplitude will depend only on ρ and φ.

which, after separation of variables, reduces to

$$S(\varphi) = A\cos m\varphi + B\sin m\varphi \qquad \text{for} \quad m = 0, 1, 2, \ldots,$$
$$T(t) = C\cos \omega t + D\sin \omega t,$$
$$\frac{d^2 R}{d\rho^2} + \frac{1}{\rho}\frac{dR}{d\rho} + \left(\frac{\omega^2}{c^2} - \frac{m^2}{\rho^2}\right)R = 0.$$

The solution of the last (Bessel) equation, which is defined for $\rho = 0$ and vanishes at $\rho = a$, is

$$R(\rho) = E J_m\left(\frac{x_{mn}}{a}\rho\right) \qquad \text{where} \quad \frac{\omega}{c} = \frac{x_{mn}}{a} \quad \text{and} \quad n = 1, 2, \ldots.$$

This shows that only the frequencies $\omega_{mn} = (c/a)x_{mn}$ are excited.

If we assume an initial shape for the membrane, given by $f(\rho, \varphi)$, and an initial velocity of zero, then $D = 0$, and the general solution will be

$$\psi(\rho, \varphi, t) = \sum_{m=0}^{\infty}\sum_{n=1}^{\infty} J_m\left(\frac{x_{mn}}{a}\rho\right)\cos\left(\frac{cx_{mn}}{a}t\right)(A_{mn}\cos m\varphi + B_{mn}\sin m\varphi),$$

where

$$A_{mn} = \frac{2}{\pi a^2 J_{m+1}^2(x_{mn})}\int_0^{2\pi} d\varphi \int_0^a d\rho\, \rho f(\rho, \varphi) J_m\left(\frac{x_{mn}}{a}\rho\right)\cos m\varphi,$$

$$B_{mn} = \frac{2}{\pi a^2 J_{m+1}^2(x_{mn})}\int_0^{2\pi} d\varphi \int_0^a d\rho\, \rho f(\rho, \varphi) J_m\left(\frac{x_{mn}}{a}\rho\right)\sin m\varphi,$$

and the orthogonality of Bessel functions (27.24) has been used. In particular, if the initial displacement of the membrane is independent of φ, then only the term with $m = 0$ contributes, and we get

$$\psi(\rho, t) = \sum_{n=1}^{\infty} A_n J_0\left(\frac{x_{0n}}{a}\rho\right)\cos\left(\frac{cx_{0n}}{a}t\right),$$

where

$$A_n = \frac{4}{a^2 J_1^2(x_{0n})}\int_0^a d\rho\, \rho f(\rho) J_0\left(\frac{x_{0n}}{a}\rho\right).$$

Note that the wave does not develop any φ-dependence at later times.

28.4 Problems

28.1. Suppose $\lambda = 0$ in Equation (28.3). Show that $X(x) = 0$.

28.2. The two ends of a thin heat-conducting bar are held at $T = 0$. Initially, the first half of the bar is held at $T = T_0$, and the second half is held at $T = 0$. The lateral surface of the bar is then thermally insulated. Find the temperature distribution for all time.

28.3. The two ends of a thin heat-conducting bar of length b are held at $T = 0$. The bar is along the x-axis with one end at $x = 0$ and the other at $x = b$. The lateral surface of the bar is thermally insulated. Find the temperature distribution at all times if initially it is given by:

(a) $T(0, x) = \begin{cases} T_0 \text{ for the middle third of the bar,} \\ 0 \text{ for the other two-thirds.} \end{cases}$

(b) $T(0, x) = \begin{cases} 0 & \text{if } 0 \le x \le \dfrac{b}{3} \text{ or } \dfrac{2b}{3} \le x \le b, \\ T_0 \sin\left(\dfrac{3\pi}{b}x - \pi\right) & \text{if } \dfrac{b}{3} \le x \le \dfrac{2b}{3}. \end{cases}$

(c) $T(0, x) = T_0 \left| \dfrac{x}{b} - \dfrac{1}{2} \right| - \dfrac{T_0}{2}.$

(d) $T(0, x) = T_0 \sin\left(\dfrac{\pi}{b}x\right).$

28.4. Derive Equation (28.7) from the equation before it by changing variables.

28.5. Using the Stirling approximation (11.6), write all four factorials of Equation (28.14) as exponentials. Then simplify to arrive at Equation (28.15).

28.6. Derive Equations (28.20)–(28.23).

28.7. By differentiating both sides of Equation (28.23) with respect to y, and (slightly) manipulating the resulting sum, show that $H'_n(y) = 2nH_{n-1}(y)$.

28.8. Evaluate Equation (28.23) at $y = 0$ and note that only the last term survives. Now show that

$$H_n(0) = \begin{cases} 0 & \text{if } n \text{ is odd,} \\ \dfrac{(-1)^k (2k)!}{k!} & \text{if } n = 2k. \end{cases}$$

28.9. Use the substitution $u(z) = z^{l+1} e^{-z/2} f(z)$ in (28.32) to derive Equation (28.33).

28.10. Differentiate both sides of Equation (28.36) with respect to x and show that

$$\frac{d}{dx} L_N^j(x) = -L_{N-1}^{j+1}(x).$$

28.11. The two ends of a rope of length a are fixed. The midpoint of the rope is raised a distance $a/2$, measured perpendicular to the tense rope, and released from rest. What is the subsequent wave function?

28.12. A string of length a fastened at both ends has an initial velocity of zero and is given an initial displacement as shown in Figure 28.1. Find $\psi(x, t)$ in each case.

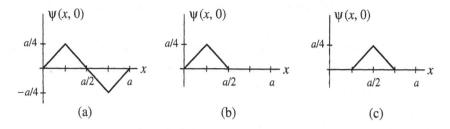

Figure 28.1: The initial shape of the waves.

28.13. Repeat Problem 28.12 assuming that the initial displacement is zero and the initial velocity distribution is given by each figure.

28.14. Repeat Problem 28.13 assuming that the initial velocity distribution is given by:

$$\text{(a)} \ g(x) = \begin{cases} v_0 & \text{if } 0 \le x \le \dfrac{a}{2}, \\ 0 & \text{if } \dfrac{a}{2} < x \le a. \end{cases}$$

$$\text{(b)} \ g(x) = \begin{cases} v_0 \sin \dfrac{2\pi x}{a} & \text{if } 0 \le x \le \dfrac{a}{2}, \\ 0 & \text{if } \dfrac{a}{2} < x \le a. \end{cases}$$

28.15. A wave guide consists of two coaxial cylinders of radii a and b $(b > a)$. Find the electric field for a TM mode propagating along the two cylinders in the region between them. Hint: Both linearly independent solutions of the Bessel DE are needed for the radial function.

Part VII

Special Topics

Chapter 29

Integral Transforms

Chapters 26 and 27 illustrated the Frobenius method of solving differential equations using power series, which gives a solution that converges within an interval of the real line. This chapter introduces another method of solving DEs, which uses **integral transforms**. The integral transform of a function v is another function u given by

$$u(x) = \int_a^b K(x,t)v(t)\,dt, \qquad (29.1)$$

where (a, b) is a convenient interval, and $K(x, t)$, called the **kernel** of the integral transform, is an appropriate function of two variables.

The idea behind using integral transform is to write the solution $u(x)$ of a DE in x in terms of an integral such as Equation (29.1) and choose v, the kernel, and the interval (a, b) in such a way as to render the DE more manageable. There are many kernels appropriate for specific DEs. However, two kernels are most widely used in physics, which lead to two important integral transforms, the Fourier transform and the Laplace transform.

kernel of integral transforms

Strategy for solving DEs using integral transforms

29.1   The Fourier Transform

Fourier transform has a kernel of the form $K(x, t) = e^{itx}$ and an interval $(-\infty, +\infty)$. Let us see how this comes about.

The Fourier series representation of a function $F(x)$ is valid for the entire real line as long as $F(x)$ is periodic. However, most functions encountered in physical applications are defined in some interval (a, b) without repetition beyond that interval. It would be useful if we could also expand such functions in some form of Fourier "series."

One way to do this is to start with the periodic series and then let the period go to infinity while extending the domain of the definition of the function. As a specific case, suppose we are interested in representing a function $f(x)$ that is defined only for the interval (a, b) and is assigned the value zero

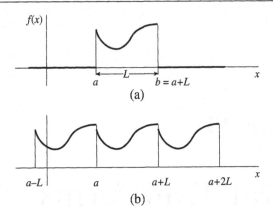

Figure 29.1: (a) The function we want to represent. (b) The Fourier series representation of the function.

everywhere else [see Figure 29.1(a)]. To begin with, we might try the Fourier series representation, but this will produce a repetition of our function. This situation is depicted in Figure 29.1(b).

Next we may try a function $f_\Lambda(x)$ defined in the interval $(a - \Lambda/2, b + \Lambda/2)$, where Λ is an arbitrary positive number:

$$f_\Lambda(x) = \begin{cases} 0 & \text{if} \quad a - \Lambda/2 < x < a, \\ f(x) & \text{if} \quad a < x < b, \\ 0 & \text{if} \quad b < x < b + \Lambda/2. \end{cases}$$

This function, which is depicted in Figure 29.2, has the Fourier series representation [see Equation (18.23)]

$$f_\Lambda(x) = \frac{1}{\sqrt{L + \Lambda}} \sum_{n=-\infty}^{\infty} f_{\Lambda,n} e^{2i\pi nx/(L+\Lambda)}, \tag{29.2}$$

where

$$f_{\Lambda,n} = \frac{1}{\sqrt{L + \Lambda}} \int_{a-\Lambda/2}^{b+\Lambda/2} e^{-2i\pi nx/(L+\Lambda)} f_\Lambda(x) \, dx. \tag{29.3}$$

We have managed to separate various copies of the original periodic function by Λ. It should be clear that if $\Lambda \to \infty$, we can completely isolate the function and stop the repetition. Let us investigate the behavior of Equations (29.2) and (29.3) as Λ grows without bound. First, we notice that the quantity k_n defined by $k_n \equiv 2n\pi/(L + \Lambda)$ and appearing in the exponent becomes almost continuous. In other words, as n changes by one unit, k_n changes only slightly. This suggests that the terms in the sum in Equation (29.2) can be lumped together in j intervals of width Δn_j, giving

$$f_\Lambda(x) \approx \sum_{j=-\infty}^{\infty} \frac{f_\Lambda(k_j)}{\sqrt{L + \Lambda}} e^{ik_j x} \Delta n_j,$$

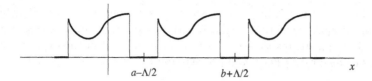

Figure 29.2: By introducing the parameter Λ, we have managed to separate the copies of the function.

where $k_j \equiv 2n_j\pi/(L+\Lambda)$, and $f_\Lambda(k_j) \equiv f_{\Lambda,n_j}$. Substituting $\Delta n_j = [(L+\Lambda)/2\pi]\Delta k_j$ in the above sum, we obtain

$$f_\Lambda(x) \approx \sum_{j=-\infty}^{\infty} \frac{f_\Lambda(k_j)}{\sqrt{L+\Lambda}} e^{ik_j x} \frac{L+\Lambda}{2\pi} \Delta k_j = \frac{1}{\sqrt{2\pi}} \sum_{j=-\infty}^{\infty} \tilde{f}_\Lambda(k_j) e^{ik_j x} \Delta k_j,$$

where we introduced $\tilde{f}_\Lambda(k_j)$ defined by $\tilde{f}_\Lambda(k_j) \equiv \sqrt{(L+\Lambda)/2\pi}\, f_\Lambda(k_j)$. It is now clear that the preceding sum approaches an integral in the limit that $\Lambda \to \infty$. In the same limit, $f_\Lambda(x) \to f(x)$, and we have

$$f(x) = \frac{1}{\sqrt{2\pi}} \int_{-\infty}^{\infty} \tilde{f}(k) e^{ikx} dk, \tag{29.4}$$

where

Fourier and inverse Fourier transforms

$$\tilde{f}(k) \equiv \lim_{\Lambda\to\infty} \tilde{f}_\Lambda(k_j) = \lim_{\Lambda\to\infty} \sqrt{\frac{L+\Lambda}{2\pi}}\, f_\Lambda(k_j)$$

$$= \lim_{\Lambda\to\infty} \sqrt{\frac{L+\Lambda}{2\pi}} \frac{1}{\sqrt{L+\Lambda}} \int_{a-\Lambda/2}^{b+\Lambda/2} e^{-ik_j x} f_\Lambda(x)\, dx,$$

or

$$\tilde{f}(k) = \frac{1}{\sqrt{2\pi}} \int_{-\infty}^{\infty} f(x) e^{-ikx} dx. \tag{29.5}$$

The function f in (29.4) is called the **Fourier transform** of $\tilde{f}$ and $\tilde{f}$ in (29.5) is called the **inverse Fourier transform** of f. Note that the difference between the two transforms is the sign of the exponential in the integrand.

Another notation that is commonly used for Fourier transform of a function f is $\mathcal{F}[f]$. The inverse Fourier transform of a function g is then denoted by $\mathcal{F}^{-1}[g]$. This means that $\mathcal{F}[f]$ is a *function* whose value at x is given by

$$\mathcal{F}[f](x) = \frac{1}{\sqrt{2\pi}} \int_{-\infty}^{\infty} f(k) e^{ikx} dk, \tag{29.6}$$

Similarly, $\mathcal{F}^{-1}[g]$ is a function whose value at k is given by

$$\mathcal{F}^{-1}[g](k) = \frac{1}{\sqrt{2\pi}} \int_{-\infty}^{\infty} g(x) e^{-ikx} dx, \tag{29.7}$$

Note that the use of k and x in these two equations is completely arbitrary. The only requirement is that the function and the variable in its argument on the left appear, respectively, in the integrand and in the exponent on the right. For example, (29.6) could be written as

$$\mathcal{F}[f](k) = \frac{1}{\sqrt{2\pi}} \int_{-\infty}^{\infty} f(x)e^{ikx}dx, \quad \text{or} \quad \mathcal{F}[h](t) = \frac{1}{\sqrt{2\pi}} \int_{-\infty}^{\infty} h(\omega)e^{i\omega t}d\omega,$$

and (29.7) as

$$\mathcal{F}^{-1}[g](x) = \frac{1}{\sqrt{2\pi}} \int_{-\infty}^{\infty} g(k)e^{-ikx}dk \quad \text{or} \quad \mathcal{F}^{-1}[f](y) = \frac{1}{\sqrt{2\pi}} \int_{-\infty}^{\infty} f(x)e^{-ixy}dx.$$

29.1.1   Properties of Fourier Transform

Equations (29.4) and (29.5) are reciprocals of one another. However, it is not obvious that they are consistent. In other words, if we substitute (29.4) in the RHS of (29.5), do we get an identity? Let's try this:

$$\tilde{f}(k) = \frac{1}{\sqrt{2\pi}} \int_{-\infty}^{\infty} dx\, e^{-ikx} \left[\frac{1}{\sqrt{2\pi}} \int_{-\infty}^{\infty} \tilde{f}(k')e^{ik'x}dk' \right]$$

$$= \frac{1}{2\pi} \int_{-\infty}^{\infty} dx \int_{-\infty}^{\infty} \tilde{f}(k')e^{i(k'-k)x}dk'.$$

We now change the order of the two integrations:

$$\tilde{f}(k) = \int_{-\infty}^{\infty} dk'\, \tilde{f}(k') \left[\frac{1}{2\pi} \int_{-\infty}^{\infty} dx\, e^{i(k'-k)x} \right].$$

But the expression in the square brackets is the Dirac delta function given by Equation (18.28). Thus, we have $\tilde{f}(k) = \int_{-\infty}^{\infty} dk'\, \tilde{f}(k')\delta(k'-k)$, which is an identity. In the $\mathcal{F}$ notation, this result can be written as

$$\mathcal{F}^{-1}\mathcal{F}[f] = \mathcal{F}\mathcal{F}^{-1}[f] = f, \tag{29.8}$$

for any function f. The second identity can be shown similarly. Another property enjoyed by the Fourier transform and its inverse in *linearity*. If a and b are constants and f and g functions, then

$$\mathcal{F}[af + bg] = a\mathcal{F}[f] + b\mathcal{F}[g], \quad \text{and} \quad \mathcal{F}^{-1}[af + bg] = a\mathcal{F}^{-1}[f] + b\mathcal{F}^{-1}[g]. \tag{29.9}$$

It is useful to generalize Fourier transform equations to more than one dimension. The generalization is straightforward:

$$\mathcal{F}[\tilde{f}](\mathbf{r}) \equiv f(\mathbf{r}) = \frac{1}{(2\pi)^{n/2}} \iint_{\Omega_{\infty}^k} d^n k\, e^{i\mathbf{k}\cdot\mathbf{r}} \tilde{f}(\mathbf{k}),$$

$$\mathcal{F}^{-1}[f](\mathbf{k}) \equiv \tilde{f}(\mathbf{k}) = \frac{1}{(2\pi)^{n/2}} \iint_{\Omega_{\infty}^x} d^n x\, f(\mathbf{r})e^{-i\mathbf{k}\cdot\mathbf{r}}. \tag{29.10}$$

where n is usually 2 or 3, Ω_{∞}^k is the entire k-space, and Ω_{∞}^x is the entire x-space.

29.1.2 Sine and Cosine Transforms

The complex exponential in the definition of Fourier transform or its inverse can be broken down into its trigonometric parts. Then for an even function, the cosine part contributes and for an odd function, the sine part contributes. In either case, the integration $\int_{-\infty}^{\infty}$ can be equated to $2\int_0^{\infty}$. This leads to the **sine transform** and **cosine transform** denoted by $\mathcal{F}_s[f]$ and $\mathcal{F}_c[f]$, respectively, for *any* function:

Sine and cosine transforms

$$\mathcal{F}_s[f](x) = \sqrt{\frac{2}{\pi}} \int_0^{\infty} f(k) \sin kx \, dk,$$

$$\mathcal{F}_c[f](x) = \sqrt{\frac{2}{\pi}} \int_0^{\infty} f(k) \cos kx \, dk. \qquad (29.11)$$

What is the inverse of a cosine transform? To find out, let $F(x)$ denote the left-hand side of the second equation in (29.11). Multiply both sides of the equation by $\cos k'x$—with $k' > 0$—and integrate over all positive values of x to get

$$\int_0^{\infty} F(x) \cos k'x \, dx = \sqrt{\frac{2}{\pi}} \int_0^{\infty} f(k) dk \int_0^{\infty} \cos kx \cos k'x \, dx. \qquad (29.12)$$

Writing the cosines in terms of exponential, the x integration on the right gives

$$\int_0^{\infty} \cos kx \cos k'x \, dx = \frac{1}{4} \int_0^{\infty} \left(e^{ikx} + e^{-ikx} \right) \left(e^{ik'x} + e^{-ik'x} \right) dx$$

$$= \frac{1}{4} \int_0^{\infty} \left[e^{ix(k+k')} + e^{-ix(k+k')} + e^{ix(k-k')} + e^{-ix(k-k')} \right] dx$$

$$= \frac{1}{4} \int_{-\infty}^{\infty} e^{ix(k+k')} dx + \frac{1}{4} \int_{-\infty}^{\infty} e^{ix(k-k')} dx$$

$$= \frac{\pi}{2} \left[\delta(k+k') + \delta(k-k') \right].$$

To go from the second to third line, we used $\int_0^{\infty} e^{-iax} dx = \int_{-\infty}^0 e^{iax} dx$, which the reader can easily verify; and to go from the third to the last line, we used Equation (18.28). Substituting the last result in (29.12), we obtain

Inverses of sine and cosine transforms

$$\int_0^{\infty} F(x) \cos k'x \, dx = \sqrt{\frac{\pi}{2}} \underbrace{\int_0^{\infty} f(k)\delta(k+k')dk}_{=0 \text{ (Reader, why?)}} + \sqrt{\frac{\pi}{2}} \int_0^{\infty} f(k)\delta(k-k')dk$$

$$= \sqrt{\frac{\pi}{2}} f(k'),$$

or

$$f(k') = \sqrt{\frac{2}{\pi}} \int_0^{\infty} F(x) \cos k'x \, dx.$$

This shows that the inverse of a cosine transform is another cosine transform. Similarly, one can show that the inverse of a sine transform is another sine transform. We shall not use sine or cosine transforms, as the Fourier transform, with the exponential in the integrand, is much more convenient.

29.1.3   Examples of Fourier Transform

Example 29.1.1. Let us evaluate the inverse Fourier transform of the function defined by

$$f(x) = \begin{cases} b & \text{if } |x| < a, \\ 0 & \text{if } |x| > a \end{cases}$$

(see Figure 29.3). From (29.5) and (29.7) we have

$$\mathcal{F}^{-1}[f](k) \equiv \tilde{f}(k) = \frac{1}{\sqrt{2\pi}} \int_{-\infty}^{\infty} f(x) e^{-ikx} dx = \frac{b}{\sqrt{2\pi}} \int_{-a}^{a} e^{-ikx} dx = \frac{2ab}{\sqrt{2\pi}} \left(\frac{\sin ka}{ka} \right),$$

which is the function encountered on page 491 and depicted in Figure 18.7.

This result deserves some detailed discussion. First, note that if $a \to \infty$, the function $f(x)$ becomes a constant function over the entire real line, and we get

$$\tilde{f}(k) = \frac{2b}{\sqrt{2\pi}} \lim_{a \to \infty} \frac{\sin ka}{k} = \frac{2b}{\sqrt{2\pi}} \pi \delta(k)$$

by Equation (18.27). This is the Fourier transform of an everywhere-constant function (see Problem 29.1). Next, let $b \to \infty$ and $a \to 0$ in such a way that $2ab$, which is the area under $f(x)$, is 1. Then $f(x)$ will approach the delta function, and $\tilde{f}(k)$ becomes

$$\tilde{f}(k) = \lim_{\substack{b \to \infty \\ a \to 0}} \frac{2ab}{\sqrt{2\pi}} \frac{\sin ka}{ka} = \frac{1}{\sqrt{2\pi}} \lim_{a \to 0} \frac{\sin ka}{ka} = \frac{1}{\sqrt{2\pi}}.$$

So the Fourier transform of the delta function is the constant $1/\sqrt{2\pi}$ as implied by (29.5).

Finally, we note that the width of $f(x)$ is $\Delta x = 2a$, and the width of $\tilde{f}(k)$ is roughly the distance, on the k-axis, between its first two roots, k_+ and k_-, on either side of $k = 0$: $\Delta k = k_+ - k_- = 2\pi/a$. Thus increasing the width of $f(x)$ results in a decrease in the width of $\tilde{f}(k)$. In other words, when the function is wide, its Fourier transform is narrow. In the limit of infinite width (a constant function), we get infinite sharpness (the delta function). The last two statements are very general. In fact, it can be shown that $\Delta x \Delta k \geq 1$ for any function $f(x)$. When both sides

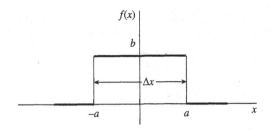

Figure 29.3: The square "bump" function.

of this inequality are multiplied by the (reduced) Planck constant $\hbar \equiv h/(2\pi)$, the result is the celebrated **Heisenberg uncertainty relation**:[1]

$$\Delta x \Delta p \geq \hbar,$$

Heisenberg uncertainty relation

where $p = \hbar k$ is the momentum of the particle.

Having obtained the transform of $f(x)$, we can write

$$f(x) = \frac{1}{\sqrt{2\pi}} \int_{-\infty}^{\infty} \frac{2b}{\sqrt{2\pi}} \frac{\sin ka}{k} e^{ikx} dk = \frac{b}{\pi} \int_{-\infty}^{\infty} \frac{\sin ka}{k} e^{ikx} dk.$$

Figure 29.4 shows the integral

$$\frac{b}{\pi} \int_{-K}^{K} \frac{\sin ka}{k} e^{ikx} dk$$

when $K = 10$, $K = 20$, and $K = 100$. It is seen that by making the limits of integration larger and larger, the graph approximates Figure 29.3 better and better. ∎

Example 29.1.2. Let us evaluate the Fourier transform of a Gaussian $g(x) = ae^{-bx^2}$ with $a, b > 0$:

$$\tilde{g}(k) = \frac{a}{\sqrt{2\pi}} \int_{-\infty}^{\infty} e^{-b(x^2 + ikx/b)} dx = \frac{ae^{-k^2/4b}}{\sqrt{2\pi}} \int_{-\infty}^{\infty} e^{-b(x + ik/2b)^2} dx.$$

To evaluate this integral rigorously, we would have to use the calculus of residues developed in Chapter 21. However, we can ignore the fact that the exponent is complex, substitute $y = x + ik/(2b)$, and write

$$\int_{-\infty}^{\infty} e^{-b[x + ik/(2b)]^2} dx = \int_{-\infty}^{\infty} e^{-by^2} dy = \sqrt{\frac{\pi}{b}}.$$

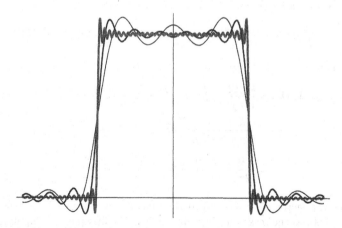

Figure 29.4: The thinnest plot represents $K = 10$; the next thinnest plot represents $K = 20$; and the thickest plot represents $K = 100$.

---

[1]In the context of the uncertainty relation, the width of the function—the so-called wave packet—measures the uncertainty in the position x of a quantum mechanical particle. Similarly, the width of the Fourier transform measures the uncertainty in k, which is related to momentum p via $p = \hbar k$.

Thus, we have $\tilde{g}(k) = \dfrac{a}{\sqrt{2b}}e^{-k^2/(4b)}$, which is also a Gaussian.

We note again that the width of $g(x)$, which is proportional to $1/\sqrt{b}$, is in inverse relation to the width of $\tilde{g}(k)$, which is proportional to $\sqrt{b}$. We thus have $\Delta x \Delta k \sim 1$. ∎

Example 29.1.3. In this example we evaluate the inverse Fourier transform of the Coulomb potential $V(r)$ of a point charge q at the origin: $V(r) = k_e q/r$. The inverse Fourier transform is important in scattering experiments with atoms, molecules, and solids. As we shall see in the following, the inverse Fourier transform of $V(r)$ is not **Yukawa potential** defined. However, if we work with the **Yukawa potential**,

$$V_\alpha(r) = \frac{k_e q e^{-\alpha r}}{r}, \qquad \alpha > 0,$$

the inverse Fourier transform will be well-defined, and we can take the limit $\alpha \to 0$ to recover the Coulomb potential. Thus, we seek the inverse Fourier transform of $V_\alpha(r)$.

We are working in three dimensions and therefore may write

$$\mathcal{F}^{-1}[V_\alpha](\mathbf{k}) \equiv \tilde{V}_\alpha(\mathbf{k}) = \frac{1}{(2\pi)^{3/2}} \int_{\Omega_\infty^x} d^3x\, e^{-i\mathbf{k}\cdot\mathbf{r}} \frac{k_e q e^{-\alpha r}}{r}.$$

It is clear from the presence of r that spherical coordinates are appropriate. We are free to pick any direction as the z-axis. A simplifying choice in this case is the direction of $\mathbf{k}$. So, we let $\mathbf{k} = |\mathbf{k}|\hat{\mathbf{e}}_z = k\hat{\mathbf{e}}_z$, or $\mathbf{k}\cdot\mathbf{r} = kr\cos\theta$, where θ is the polar angle in spherical coordinates. Now we have

$$\tilde{V}_\alpha(\mathbf{k}) = \frac{k_e q}{(2\pi)^{3/2}} \int_0^\infty r^2\, dr \int_0^\pi \sin\theta\, d\theta \int_0^{2\pi} d\varphi\, e^{-ikr\cos\theta} \frac{e^{-\alpha r}}{r}.$$

The φ integration is trivial and gives 2π. The θ integration simplifies if we make the substitution $u = \cos\theta$:

$$\int_0^\pi \sin\theta e^{-ikr\cos\theta}\, d\theta = \int_{-1}^1 e^{-ikru}\, du = \frac{1}{ikr}(e^{ikr} - e^{-ikr}).$$

We thus have

$$\tilde{V}_\alpha(\mathbf{k}) = \frac{k_e q (2\pi)}{(2\pi)^{3/2}} \int_0^\infty dr\, r^2 \frac{e^{-\alpha r}}{r} \frac{1}{ikr}(e^{ikr} - e^{-ikr})$$

$$= \frac{k_e q}{(2\pi)^{1/2}} \frac{1}{ik} \int_0^\infty dr \left[e^{(-\alpha+ik)r} - e^{-(\alpha+ik)r} \right]$$

$$= \frac{k_e q}{(2\pi)^{1/2}} \frac{1}{ik} \left(\frac{e^{(-\alpha+ik)r}}{-\alpha+ik} \bigg|_0^\infty + \frac{e^{-(\alpha+ik)r}}{\alpha+ik} \bigg|_0^\infty \right).$$

Note how the factor $e^{-\alpha r}$ has tamed the divergent behavior of the exponential at $r \to \infty$. This was the reason for introducing it in the first place. Simplifying the last expression yields $\tilde{V}_\alpha(\mathbf{k}) = (2k_e q/\sqrt{2\pi})(k^2+\alpha^2)^{-1}$. The parameter α is a measure of the range of the potential. It is clear that the larger α is, the smaller the range. In fact, it was in response to the short range of nuclear forces that Yukawa introduced α. For electromagnetism, where the range is infinite, α becomes zero and $V_\alpha(r)$ reduces to $V(r)$. Thus, the inverse Fourier transform of the Coulomb potential is

$$\tilde{V}_{\text{Coul}}(\mathbf{k}) = \frac{2k_e q}{\sqrt{2\pi}} \frac{1}{k^2}.$$

If a charge *distribution* is involved, the inverse Fourier transform will be interestingly different as the following example shows. ■

Example 29.1.4. The example above deals with the electrostatic potential of a point charge. Let us now consider the case where the charge is distributed over a finite volume. Then the potential is

$$V(\mathbf{r}) = \iiint \frac{k_e q \rho(\mathbf{r}')}{|\mathbf{r}' - \mathbf{r}|} d^3 x' \equiv k_e q \int \frac{\rho(\mathbf{r}')}{|\mathbf{r}' - \mathbf{r}|} d^3 x',$$

where $q\rho(\mathbf{r}')$ is the charge density at $\mathbf{r}'$, and we have used a single integral because $d^3 x'$ already indicates the number of integrations to be performed. Note that we have normalized $\rho(\mathbf{r}')$ so that its integral over the volume is 1. Figure 29.5 shows the geometry of the situation.

Making a change of variables, $\mathbf{R} \equiv \mathbf{r}' - \mathbf{r}$, or $\mathbf{r}' = \mathbf{R} + \mathbf{r}$, and $d^3 x' = d^3 X$, with $\mathbf{R} \equiv (X, Y, Z)$, we get

$$\mathcal{F}^{-1}[V](\mathbf{k}) \equiv \tilde{V}(\mathbf{k}) = \frac{1}{(2\pi)^{3/2}} \int d^3 x \, e^{-i\mathbf{k}\cdot\mathbf{r}} k_e q \int \frac{\rho(\mathbf{R}+\mathbf{r})}{R} d^3 X. \qquad (29.13)$$

To evaluate Equation (29.13), we substitute for $\rho(\mathbf{R} + \mathbf{r})$ in terms of its Fourier transform,

$$\rho(\mathbf{R}+\mathbf{r}) = \frac{1}{(2\pi)^{3/2}} \int d^3 k' \, \tilde{\rho}(\mathbf{k}') e^{i\mathbf{k}'\cdot(\mathbf{R}+\mathbf{r})}. \qquad (29.14)$$

Combining (29.13) and (29.14), we obtain

$$\tilde{V}(\mathbf{k}) = \frac{k_e q}{(2\pi)^3} \int d^3 x \, d^3 X \, d^3 k' \frac{e^{i\mathbf{k}'\cdot\mathbf{R}}}{R} \tilde{\rho}(\mathbf{k}') e^{i\mathbf{r}\cdot(\mathbf{k}'-\mathbf{k})}$$

$$= k_e q \int d^3 X \, d^3 k' \frac{e^{i\mathbf{k}'\cdot\mathbf{R}}}{R} \tilde{\rho}(\mathbf{k}') \underbrace{\left(\frac{1}{(2\pi)^3} \int d^3 x \, e^{i\mathbf{r}\cdot(\mathbf{k}'-\mathbf{k})} \right)}_{\delta(\mathbf{k}'-\mathbf{k}) \text{ by Equation (18.30)}}$$

$$= k_e q \tilde{\rho}(\mathbf{k}) \int d^3 X \, \frac{e^{i\mathbf{k}\cdot\mathbf{R}}}{R}. \qquad (29.15)$$

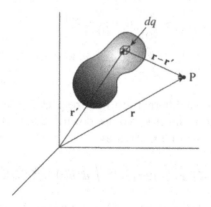

Figure 29.5: The inverse Fourier transform of the potential of a continuous charge distribution at P is calculated using this geometry.

What is nice about this result is that the contribution of the charge distribution, $\tilde{\rho}(\mathbf{k})$, has been completely factored out. The integral, aside from a constant and a change in the sign of $\mathbf{k}$, is simply the inverse Fourier transform of the Coulomb potential of a point charge obtained in the previous example. We can therefore write Equation (29.15) as

$$\tilde{V}(\mathbf{k}) = (2\pi)^{3/2}\tilde{\rho}(\mathbf{k})\tilde{V}_{\text{Coul}}(-\mathbf{k}) = \frac{4\pi k_e q\tilde{\rho}(\mathbf{k})}{|\mathbf{k}|^2}.$$

This equation is important in analyzing the structure of atomic particles. The inverse Fourier transform $\tilde{V}(\mathbf{k})$ is directly measurable in scattering experiments. In a typical experiment a (charged) target is probed with a charged point particle (electron). If the analysis of the scattering data shows a deviation from $1/k^2$ in the behavior of $\tilde{V}(\mathbf{k})$, then it can be concluded that the target particle has a charge distribution. More specifically, a plot of $k^2\tilde{V}(\mathbf{k})$ versus k gives the variation of **form factor** $\tilde{\rho}(\mathbf{k})$, the **form factor**, with k. If the resulting graph is a constant, then $\tilde{\rho}(\mathbf{k})$ is a constant, and the target is a point particle [$\tilde{\rho}(\mathbf{k})$ is a constant for point particles, where $\tilde{\rho}(\mathbf{r}') \propto \delta(\mathbf{r} - \mathbf{r}')$]. If there is any deviation from a constant function, $\tilde{\rho}(\mathbf{k})$ must have a dependence on k, and correspondingly, the target particle must have a charge distribution.

Fourier transform and the discovery of quarks   The above discussion, when generalized to four-dimensional relativistic space-time, was the basis for a strong argument in favor of the existence of point-like particles—quarks—inside a proton in 1968, when the results of the scattering of high-energy electrons off protons at the Stanford Linear Accelerator Center revealed deviation from a constant for the proton form factor. ∎

29.1.4   Application to Differential Equations

The Fourier transform is very useful for solving differential equations. This is because the derivative operator in $\mathbf{r}$ space turns into ordinary multiplication in $\mathbf{k}$ space. For example, if we differentiate $f(\mathbf{r})$ in Equation (29.10) with respect to x_j, we obtain

$$\frac{\partial}{\partial x_j} f(\mathbf{r}) = \frac{1}{(2\pi)^{n/2}} \iint_{\Omega_\infty^k} d^n k \frac{\partial}{\partial x_j} e^{i(k_1 x_1 + \cdots + k_j x_j + \cdots + k_n x_n)} \tilde{f}(\mathbf{k})$$

$$= \frac{1}{(2\pi)^{n/2}} \int_{\Omega_\infty^k} d^n k (ik_j) e^{i\mathbf{k}\cdot\mathbf{r}} \tilde{f}(\mathbf{k}). \tag{29.16}$$

That is, every time we differentiate with respect to any component of $\mathbf{r}$, the corresponding component of $\mathbf{k}$ "comes down." Thus, the n-dimensional gradient and Laplacian can be written as

$$\nabla f(\mathbf{r}) = (2\pi)^{-n/2} \int d^n k (i\mathbf{k}) e^{i\mathbf{k}\cdot\mathbf{r}} \tilde{f}(\mathbf{k})$$

$$\nabla^2 f(\mathbf{r}) = (2\pi)^{-n/2} \int d^n k (-k^2) e^{i\mathbf{k}\cdot\mathbf{r}} \tilde{f}(\mathbf{k}). \tag{29.17}$$

Let us illustrate the above points with a simple example. Consider the ordinary second-order differential equation

$$C_2 \frac{d^2 x}{dt^2} + C_1 \frac{dx}{dt} + C_0 x = f(t), \qquad (29.18)$$

where C_0, C_1, and C_2 are constants. We can "solve" this equation by simply substituting the following in it:

$$x(t) = \frac{1}{\sqrt{2\pi}} \int_{-\infty}^{\infty} d\omega \tilde{x}(\omega) e^{i\omega t}, \qquad \frac{dx}{dt} = \frac{1}{\sqrt{2\pi}} \int_{-\infty}^{\infty} d\omega \tilde{x}(\omega)(i\omega) e^{i\omega t},$$

$$\frac{d^2 x}{dt^2} = -\frac{1}{\sqrt{2\pi}} \int_{-\infty}^{\infty} d\omega \tilde{x}(\omega) \omega^2 e^{i\omega t}, \qquad f(t) = \frac{1}{\sqrt{2\pi}} \int_{-\infty}^{\infty} d\omega \tilde{f}(\omega) e^{i\omega t}.$$

This gives

$$\frac{1}{\sqrt{2\pi}} \int_{-\infty}^{\infty} d\omega \tilde{x}(\omega)(-C_2 \omega^2 + iC_1 \omega + C_0) e^{i\omega t} = \frac{1}{\sqrt{2\pi}} \int_{-\infty}^{\infty} d\omega \tilde{f}(\omega) e^{i\omega t}.$$

Equating the coefficients of $e^{i\omega t}$ on both sides, we obtain

$$\tilde{x}(\omega) = \frac{\tilde{f}(\omega)}{-C_2 \omega^2 + iC_1 \omega + C_0}. \qquad (29.19)$$

If we know $\tilde{f}(\omega)$ [which can be obtained from $f(t)$], we can calculate $x(t)$ by Fourier-transforming $\tilde{x}(\omega)$. The resulting integrals are not generally easy to evaluate. In some cases the methods of complex analysis may be helpful; in others numerical integration may be the last resort. However, the real power of the Fourier transform lies in the formal analysis of differential equations.

Example 29.1.5. A harmonically driven circuit consisting of an inductor L, a resistor R, and a capacitor C, obeys the following differential equation:

$$L \frac{d^2 Q}{dt^2} + R \frac{dQ}{dt} + \frac{Q}{C} = \mathcal{E} \cos \omega_0 t,$$

where Q is the charge on the capacitor. Except for the constants, this is identical to (29.18). The Fourier transform of cosine is a sum of two Dirac delta functions (see Problem 29.6). Substituting in Equation (29.19), we obtain

$$\tilde{Q}(\omega) = \mathcal{E} \sqrt{\frac{\pi}{2}} \frac{\delta(\omega - \omega_0) + \delta(\omega + \omega_0)}{-L\omega^2 + iR\omega + (1/C)}.$$

Therefore,

$$Q(t) = \frac{1}{\sqrt{2\pi}} \int_{-\infty}^{\infty} d\omega \tilde{Q}(\omega) e^{i\omega t} = \frac{\mathcal{E}}{2} \int_{-\infty}^{\infty} d\omega \frac{\delta(\omega - \omega_0) + \delta(\omega + \omega_0)}{-L\omega^2 + iR\omega + (1/C)} e^{i\omega t}$$

$$= \frac{\mathcal{E}}{2} \left(\frac{e^{i\omega_0 t}}{-L\omega_0^2 + iR\omega_0 + (1/C)} + \frac{e^{-i\omega_0 t}}{-L\omega_0^2 - iR\omega_0 + (1/C)} \right).$$

Noting that the second term in the outer parentheses is the complex conjugate of the first term, we obtain

$$Q(t) = \frac{\mathcal{E}}{2} 2 \operatorname{Re} \left(\frac{e^{i\omega_0 t}}{-L\omega_0^2 + iR\omega_0 + (1/C)} \right)$$

and using $\operatorname{Re}(z_1/z_2^*) = (x_1 x_2 + y_1 y_2)/|z_2|^2$, where x and y are the real and imaginary parts of a complex number z, we finally obtain

$$Q(t) = \mathcal{E} \frac{[(1/C) - L\omega_0^2] \cos \omega_0 t + R\omega_0 \sin \omega_0 t}{[-L\omega_0^2 + (1/C)]^2 + R^2 \omega_0^2}$$

for the charge on the capacitor and

$$I(t) = \frac{dQ}{dt} = \mathcal{E} \frac{-[(1/C) - L\omega_0^2]\omega_0 \sin \omega_0 t + R\omega_0^2 \cos \omega_0 t}{[-L\omega_0^2 + (1/C)]^2 + R^2 \omega_0^2}$$

for the current in the circuit. Note the occurrence of a resonance (large current) at the voltage source frequency of $\omega_0 = 1/\sqrt{LC}$. Note also that the $Q(t)$ obtained above is a *particular*, not the most general, solution of the differential equation (see Box 24.4.1). ∎

Example 29.1.6. The one-dimensional heat equation, the PDE governing the behavior of the temperature $T(x,t)$ along a rod, is

$$\frac{\partial T}{\partial t} = \kappa^2 \frac{\partial^2 T}{\partial x^2}, \tag{29.20}$$

where we have used κ [see Equation (22.3)] to leave k exclusively for Fourier transforms. Write $T(x,t)$ as a Fourier transform in the x variable

$$T(x,t) = \frac{1}{\sqrt{2\pi}} \int_{-\infty}^{\infty} \tilde{T}(k,t) e^{ikx} dk, \tag{29.21}$$

and substitute in (29.20) to obtain

$$\frac{1}{\sqrt{2\pi}} \int_{-\infty}^{\infty} \frac{\partial \tilde{T}}{\partial t} e^{ikx} dk = \frac{1}{\sqrt{2\pi}} \int_{-\infty}^{\infty} (-\kappa^2 k^2) \tilde{T}(k,t) e^{ikx} dk \quad \text{or} \quad \frac{\partial \tilde{T}}{\partial t} = -\kappa^2 k^2 \tilde{T}(k,t)$$

This is a first order ordinary differential equation which can be easily solved

$$\tilde{T}(k,t) = C(k) e^{-\kappa^2 k^2 t}, \tag{29.22}$$

where $C(k)$ is the constant of integration, which could depend on k. Now suppose that initially the temperature distribution on the rod is $T(x,0) = f(x)$, where $f(x)$ is a given known function. Then the last equation gives $\tilde{T}(k,0) = C(k)$, and (29.21) yields

$$f(x) = T(x,0) = \frac{1}{\sqrt{2\pi}} \int_{-\infty}^{\infty} \tilde{T}(k,0) e^{ikx} dk = \frac{1}{\sqrt{2\pi}} \int_{-\infty}^{\infty} C(k) e^{ikx} dk,$$

showing that $C(k)$ is the inverse Fourier transform of $f(x)$:

$$C(k) = \frac{1}{\sqrt{2\pi}} \int_{-\infty}^{\infty} f(x) e^{-ikx} dx.$$

Substituting this in (29.22) and the result in (29.21) yields

$$T(x,t) = \frac{1}{2\pi} \int_{-\infty}^{\infty} \left(\int_{-\infty}^{\infty} f(y)e^{-iky}dy \right) e^{-\kappa^2 k^2 t} e^{ikx} dk$$

$$= \frac{1}{2\pi} \int_{-\infty}^{\infty} f(y)\,dy \int_{-\infty}^{\infty} e^{-\kappa^2 k^2 t + ik(x-y)} dk. \qquad (29.23)$$

The inner integral can be done by completing the square in the exponent as in Example 29.1.2. The result is

$$\int_{-\infty}^{\infty} e^{-\kappa^2 k^2 t + ik(x-y)} dk = \sqrt{\pi} \frac{e^{-\frac{(x-y)^2}{4\kappa^2 t}}}{\kappa\sqrt{t}}.$$

Putting this in (29.23) and noting that $f(y) = T(y,0)$, we finally obtain

$$T(x,t) = \frac{1}{\sqrt{4\pi\kappa^2 t}} \int_{-\infty}^{\infty} T(y,0)e^{-\frac{(x-y)^2}{4\kappa^2 t}} dy. \qquad (29.24)$$

If we know the initial shape of the temperature distribution $T(y,0)$ on the rod, we can calculate the temperature at every point of the rod for any time. A simple example is if the temperature is infinitely hot at one point, say x_0 of the rod and zero every where else. Then

$$T(y,0) = T_0\delta(y - x_0),$$

and (29.24) yields

$$T(x,t) = \frac{1}{\sqrt{4\pi\kappa^2 t}} \int_{-\infty}^{\infty} T_0\delta(y - x_0)e^{-\frac{(x-y)^2}{4\kappa^2 t}} dy = \frac{T_0 e^{-\frac{(x-x_0)^2}{4\kappa^2 t}}}{\sqrt{4\pi\kappa^2 t}}. \qquad \blacksquare$$

29.2 Fourier Transform and Green's Functions

Suppose you are given a system of n linear equations in n unknowns and asked to slove them. An elegant approach would be to use matrices. So, let $\mathbf{L}$ be the matrix of coefficients, $\mathbf{y}$ the column vector of the n unknowns, and $\mathbf{f}$ the column vector of the constants appearing on the right-hand side of the system of equations. The matrix equation and the corresponding system of equations would look like the following:

$$\mathbf{Ly} = \mathbf{f}, \qquad \sum_{j=1}^{n} L_{ij}y_j = f_i, \quad i = 1, 2, \ldots, n. \qquad (29.25)$$

If $\mathbf{L}$ has an inverse $\mathbf{G}$, i.e., if there is a matrix $\mathbf{G}$ such that $\mathbf{LG} = \mathbf{1}$, then the solution to the above equation can formally be written as $\mathbf{y} = \mathbf{Gf}$, or in component form as

$$y_i = \sum_{j=1}^{n} G_{ij}f_j, \quad i = 1, 2, \ldots, n. \qquad (29.26)$$

Thus, the problem of solving the system of linear equations turns into the problem of finding the inverse of the coefficient matrix; and this is *independent of what* **f** *is*! Once I know the inverse of **L**, I can solve *any* system of linear equations, regardless of the constants on the right-hand side. Recalling that the elements of the unit matrix are just the correctly labeled Kronecker delta, the equation that **G** has to satisfy becomes

$$\mathbf{LG} = \mathbf{1}, \qquad \sum_{j=1}^{n} L_{ij} G_{jk} = \delta_{ik}, \quad i, k = 1, 2, \ldots, n. \tag{29.27}$$

From discrete matrices to continuous differential operators

Now think of a column vector **v** as a "machine:" feed the machine an *integer* between 1 and n, and it will give you a *real number*, i.e., the element of the column vector carrying the integer as an index. Similarly, think of a matrix **M** as another "machine" which gives you a *real number* if you feed it a *pair* of integers between 1 and n. Write this as

$$\mathbf{v}(i) = v_i, \qquad \text{and} \qquad \mathbf{M}(i, j) = M_{ij}, \quad i, j = 1, 2, \ldots, n. \tag{29.28}$$

Would it be beneficial to generalize the action of the machine to include all real numbers? A vector machine that feeds on real numbers is a *function*: feed a function a real number and it will spit out a real number. Replacing i with x, we have $\mathbf{v}(x) = v_x \equiv v(x)$, because v_x is not a common notation. Similarly, $\mathbf{M}(x, x') = M_{xx'} \equiv M(x, x')$. Furthermore, all summations have to be replaced by integrals. For example, the system of equations (29.25) becomes

$$\mathbf{Ly} = \mathbf{f} \qquad \int_a^b L(x, x') y(x') \, dx' = f(x),$$

where (a, b) is a convenient interval of the real line usually taken to be $(-\infty, \infty)$. What is the meaning of $L(x, x')$? It can be merely a function of two variables. But more interestingly, it can be a *differential operator*. However, a differential operator is a **local operator**, i.e., it is a linear combination of derivatives of various orders *at a single point*, say x. This requires the last integral above to collapse to x. The only way that can happen is if

$$L(x, x') = \delta(x - x') L(x) \equiv \delta(x - x') \mathbf{L}_x, \tag{29.29}$$

where $\mathbf{L}_x$ is by definition a differential operator in the variable x.

Now that we have a differential operator which is the generalization of a matrix, how do we find its inverse? In other words, how do we generalize Equation (29.27)? We suspect that the Kronecker delta turns into a Dirac delta function. With this suspicion, we generalize (29.27) to

$$\mathbf{LG} = \mathbf{1}, \qquad \int_a^b L(x, x') G(x', x_0) = \delta(x - x_0).$$

Differential equation for Green's function

Substituting (29.29) in the second equation yields

$$\int_a^b \delta(x - x') \mathbf{L}_x G(x', x_0) = \delta(x - x_0),$$

or

$$\mathbf{L}_x G(x, x_0) = \delta(x - x_0). \tag{29.30}$$

A function which satisfies this equation is called the **Green's function** for the differential operator $\mathbf{L}_x$. If we can find the Green's function for $\mathbf{L}_x$, then the solution to the differential equation $\mathbf{L}_x y(x) = f(x)$ can be written as the generalization of (29.26):

$$y(x) = \int_a^b G(x, x') f(x') \, dx'. \tag{29.31}$$

To show this, note that

$$\mathbf{L}_x y(x) = \mathbf{L}_x \int_a^b G(x, x') f(x') \, dx' = \int_a^b \mathbf{L}_x G(x, x') f(x') \, dx'$$

$$= \int_a^b \delta(x - x') f(x') \, dx' = f(x).$$

Green's functions are powerful tools for solving differential equations. Ordinary differential equations have ordinary derivatives and the differential operator involves a single variable. Partial differential equations correspond to differential operators involving several variables. If $\mathbf{x}$ denotes the collection of all these variables, then the differential operator can be denoted by $\mathbf{L}_\mathbf{x}$ and the Green's function by $G(\mathbf{x}, \mathbf{x}')$, which satisfies the partial differential equation

$$\mathbf{L}_\mathbf{x} G(\mathbf{x}, \mathbf{x}') = \delta(\mathbf{x} - \mathbf{x}'). \tag{29.32}$$

Since Fourier transform turns differentiation into multiplication, and the Dirac delta function has a very simple inverse Fourier transform, Green's function are very elegantly calculated via Fourier transform techniques. For example, if $\mathbf{L}_\mathbf{x}$ is a second order partial differential operator with constant coefficients in n variables, then Fourier transforming only the $\mathbf{x}$ variables and writing

$$G(\mathbf{x}, \mathbf{x}') = \frac{1}{(2\pi)^{n/2}} \int d^n k \, \tilde{G}(\mathbf{k}, \mathbf{x}') e^{i\mathbf{k}\cdot\mathbf{x}},$$

$$\delta(\mathbf{x} - \mathbf{x}') = \frac{1}{(2\pi)^n} \int d^n k \, e^{i\mathbf{k}\cdot(\mathbf{x}-\mathbf{x}')}, \tag{29.33}$$

the differential equation (29.32) becomes

$$\int d^n k \, \tilde{G}(\mathbf{k}, \mathbf{x}') \mathbf{L}_\mathbf{x} e^{i\mathbf{k}\cdot\mathbf{x}} = \frac{1}{(2\pi)^{n/2}} \int d^n k \, e^{i\mathbf{k}\cdot(\mathbf{x}-\mathbf{x}')} = \frac{1}{(2\pi)^{n/2}} \int d^n k \, e^{i\mathbf{k}\cdot\mathbf{x}} e^{-i\mathbf{k}\cdot\mathbf{x}'}.$$

When $\mathbf{L}_\mathbf{x}$ acts on the exponential, it produces a polynomial $p(k_j)$ of second degree in components k_j of $\mathbf{k}$. Therefore, equating the coefficient of $e^{i\mathbf{k}\cdot\mathbf{x}}$ on both sides, we obtain

Green's function in n dimensions

$$\tilde{G}(\mathbf{k}, \mathbf{x}') p(k_j) = \frac{1}{(2\pi)^{n/2}} e^{-i\mathbf{k}\cdot\mathbf{x}'} \quad \text{or} \quad \tilde{G}(\mathbf{k}, \mathbf{x}') = \frac{1}{(2\pi)^{n/2}} \frac{e^{-i\mathbf{k}\cdot\mathbf{x}'}}{p(k_j)}.$$

Substituting this in the first equation of (29.33), we get

$$G(\mathbf{x}, \mathbf{x}') = \frac{1}{(2\pi)^n} \int d^n k \, \frac{e^{i\mathbf{k}\cdot(\mathbf{x}-\mathbf{x}')}}{p(k_j)},$$

which shows that the Green's function is a function of the *difference* between its arguments. We therefore have

$$G(\mathbf{x} - \mathbf{x}') = \frac{1}{(2\pi)^n} \int d^n k \, \frac{e^{i\mathbf{k}\cdot(\mathbf{x}-\mathbf{x}')}}{p(k_j)}. \tag{29.34}$$

29.2.1   Green's Function for the Laplacian

Equation (29.17) tells us that $p(k_j) = -k^2$ for the Laplacian. Thus, with $n = 3$, (29.34) becomes

$$G(\mathbf{r} - \mathbf{r}') = -\frac{1}{(2\pi)^3} \int d^3 k \, \frac{e^{i\mathbf{k}\cdot(\mathbf{r}-\mathbf{r}')}}{k^2}. \tag{29.35}$$

To evaluate this integral, use spherical coordinates in the k-space, and choose the polar axis to be along the vector $\mathbf{r} - \mathbf{r}'$. Then, $d^3 k = k^2 \sin\theta \, dk d\theta d\varphi$ and (29.35) becomes

$$G(\mathbf{r} - \mathbf{r}') = -\frac{1}{(2\pi)^3} \int_0^\infty k^2 dk \int_0^\pi \sin\theta d\theta \int_0^{2\pi} d\varphi \, \frac{e^{ik|\mathbf{r}-\mathbf{r}'|\cos\theta}}{k^2}.$$

The φ integration gives 2π. For the θ integration, let $u = \cos\theta$. Then the integral becomes

$$G(\mathbf{r} - \mathbf{r}') = \frac{1}{(2\pi)^2} \int_0^\infty dk \int_1^{-1} du \, e^{ik|\mathbf{r}-\mathbf{r}'|u} = \frac{1}{(2\pi)^2} \int_0^\infty dk \, \frac{e^{ik|\mathbf{r}-\mathbf{r}'|u}}{ik|\mathbf{r}-\mathbf{r}'|} \Big|_1^{-1}$$

$$= \frac{1}{(2\pi)^2 |\mathbf{r}-\mathbf{r}'|} \int_0^\infty dk \, \frac{e^{-ik|\mathbf{r}-\mathbf{r}'|} - e^{ik|\mathbf{r}-\mathbf{r}'|}}{ik}$$

$$= -\frac{2}{(2\pi)^2 |\mathbf{r}-\mathbf{r}'|} \int_0^\infty dk \, \frac{\sin(k|\mathbf{r}-\mathbf{r}'|)}{k}.$$

Example 21.3.3 calculated the last integral and yielded $\pi/2$ for it. We thus obtain the important result that for the Laplacian, the Green's function is

$$G(\mathbf{r} - \mathbf{r}') = -\frac{1}{4\pi|\mathbf{r}-\mathbf{r}'|}. \tag{29.36}$$

From this and $\nabla^2 G(\mathbf{r} - \mathbf{r}') = \delta(\mathbf{r} - \mathbf{r}')$, we obtain another important result:

$$\nabla^2 \left(\frac{1}{|\mathbf{r}-\mathbf{r}'|} \right) = -4\pi\delta(\mathbf{r} - \mathbf{r}'). \tag{29.37}$$

With the Green's function of the Laplacian at our disposal, we can solve the **Poisson equation** $\nabla^2 \Phi(\mathbf{r}) = -4\pi k_e \rho(\mathbf{r})$ in electrostatics, using the three-dimensional version of Equation (29.31):

$$\Phi(\mathbf{r}) = -4\pi k_e \int d^3 x' G(\mathbf{r} - \mathbf{r}') \rho(\mathbf{r}') = k_e \int d^3 x' \frac{\rho(\mathbf{r}')}{|\mathbf{r} - \mathbf{r}'|},$$

Green's function solves Poisson equation

which is the electrostatic potential of a charge distribution described by the volume charge density $\rho(\mathbf{r})$.

29.2.2 Green's Function for the Heat Equation

The heat equation was given in (22.3), which, due to the special significance attached to the letter k in this chapter, we write as

$$\frac{\partial T}{\partial t} = \kappa^2 \nabla^2 T(\mathbf{r}), \quad \text{or} \quad \frac{\partial T}{\partial t} - \kappa^2 \nabla^2 T(\mathbf{r}) = 0. \tag{29.38}$$

This is a PDE in four variables. We let t be the "zeroth" coordinate, and $\mathbf{r}$ the remaining three. Similarly, the 4-dimensional k space consists of k_0 and $\mathbf{k}$. The polynomial $p(k_j)$ of Equation (29.34) is

$$p(k_j) = ik_0 + \kappa^2 \left(k_1^2 + k_2^2 + k_3^2\right) \equiv ik_0 + \kappa^2 k^2.$$

Hence, with $n = 4$, (29.34) gives

$$\begin{aligned}
G(\mathbf{x} - \mathbf{x}') &= \frac{1}{(2\pi)^4} \int d^4 k \frac{e^{ik_0(x_0 - x_0') + i\mathbf{k}\cdot(\mathbf{r} - \mathbf{r}')}}{ik_0 + \kappa^2 k^2} \\
&= \frac{1}{(2\pi)^4} \int d^3 k\, e^{i\mathbf{k}\cdot(\mathbf{r} - \mathbf{r}')} \int_{-\infty}^{\infty} dk_0 \frac{e^{ik_0(x_0 - x_0')}}{ik_0 + \kappa^2 k^2}.
\end{aligned} \tag{29.39}$$

Let's do the k_0 integration first. Multiply the numerator and denominator by $-i$ to change the denominator to $k_0 - i\kappa^2 k^2$; then use the calculus of residues and choose the contour in the UHP (assuming that $x_0 - x_0' > 0$). The only pole of the integrand is at $k_0 = i\kappa^2 k^2$. Thus,

$$\int_{-\infty}^{\infty} dk_0 \frac{e^{ik_0(x_0 - x_0')}}{ik_0 + \kappa^2 k^2} = -i \int_{-\infty}^{\infty} dk_0 \frac{e^{ik_0(x_0 - x_0')}}{k_0 - i\kappa^2 k^2} = -i\left(2\pi i e^{-\kappa^2 k^2 (x_0 - x_0')}\right).$$

Substituting this in (29.39) and using spherical coordinates in the 3-dimensional k-space with the polar axis along $\mathbf{r} - \mathbf{r}'$ yields

$$\begin{aligned}
G(\mathbf{x} - \mathbf{x}') &= \frac{1}{(2\pi)^3} \int d^3 k\, e^{i\mathbf{k}\cdot(\mathbf{r} - \mathbf{r}')} e^{-\kappa^2 k^2 (x_0 - x_0')} \\
&= \frac{1}{(2\pi)^3} \int_0^{\infty} k^2 e^{-\kappa^2 k^2 (x_0 - x_0')} dk \int_0^{\pi} \sin\theta d\theta \int_0^{2\pi} d\varphi\, e^{ik|\mathbf{r} - \mathbf{r}'|\cos\theta}.
\end{aligned}$$

The φ integration yields 2π, and as in the Laplacian case, the θ integration gives $2\sin(k|\mathbf{r} - \mathbf{r}'|)/(k|\mathbf{r} - \mathbf{r}'|)$; and since the resulting integrand of the k

integral is even, we can extend the lower limit of integration to $-\infty$ and introducing a factor of half. Thus, the equation above becomes

$$G(\mathbf{x} - \mathbf{x}') = \frac{2}{(2\pi)^2 |\mathbf{r} - \mathbf{r}'|} \int_0^\infty k e^{-\kappa^2 k^2 (x_0 - x_0')} \sin(k|\mathbf{r} - \mathbf{r}'|) dk$$

$$= \frac{1}{(2\pi)^2 |\mathbf{r} - \mathbf{r}'|} \int_{-\infty}^\infty k e^{-\kappa^2 k^2 (x_0 - x_0')} \sin(k|\mathbf{r} - \mathbf{r}'|) dk,$$

or, since sine is the imaginary part of complex exponential,

$$G(\mathbf{x} - \mathbf{x}') = \frac{1}{(2\pi)^2 |\mathbf{r} - \mathbf{r}'|} \operatorname{Im} \int_{-\infty}^\infty k e^{-\kappa^2 k^2 (x_0 - x_0')} e^{ik|\mathbf{r} - \mathbf{r}'|} dk$$

$$= \frac{1}{(2\pi)^2 |\mathbf{r} - \mathbf{r}'|} \operatorname{Im} \int_{-\infty}^\infty k e^{-\kappa^2 k^2 (x_0 - x_0') + ik|\mathbf{r} - \mathbf{r}'|} dk. \qquad (29.40)$$

Completing the square in the exponent, we have

$$-\kappa^2 k^2 (x_0 - x_0') + ik|\mathbf{r} - \mathbf{r}'| = -\kappa^2 (x_0 - x_0') \left(k - \frac{i|\mathbf{r} - \mathbf{r}'|}{2\kappa^2 (x_0 - x_0')} \right)^2 - \frac{|\mathbf{r} - \mathbf{r}'|^2}{4\kappa^2 (x_0 - x_0')}.$$

Call the imaginary number in the large parentheses $i\alpha$ and substitute the result in (29.40) to obtain

$$G(\mathbf{x} - \mathbf{x}') = \frac{e^{-\frac{|\mathbf{r} - \mathbf{r}'|^2}{4\kappa^2 (x_0 - x_0')}}}{(2\pi)^2 |\mathbf{r} - \mathbf{r}'|} \operatorname{Im} \int_{-\infty}^\infty k e^{-\kappa^2 (x_0 - x_0')(k - i\alpha)^2} dk. \qquad (29.41)$$

Change the variable of integration to $u = k - i\alpha$. Then the integral becomes

$$\int_{-\infty}^\infty (u + i\alpha) e^{-\kappa^2 (x_0 - x_0') u^2} du = i\alpha \int_{-\infty}^\infty e^{-\kappa^2 (x_0 - x_0') u^2} du = i\alpha \sqrt{\frac{\pi}{\kappa^2 (x_0 - x_0')}}.$$

The integral involving the u of $(u + i\alpha)$ vanishes because the integrand is odd. The Gaussian integral was evaluated in Example 3.3.1. Substituting this and the value of α in (29.41), we obtain

$$G(\mathbf{x} - \mathbf{x}') = \frac{e^{-\frac{|\mathbf{r} - \mathbf{r}'|^2}{4\kappa^2 (x_0 - x_0')}}}{(2\pi)^2 |\mathbf{r} - \mathbf{r}'|} \frac{|\mathbf{r} - \mathbf{r}'|}{2\kappa^2 (x_0 - x_0')} \sqrt{\frac{\pi}{\kappa^2 (x_0 - x_0')}},$$

or, recalling that $x_0 = t$ and assuming that $x_0' = t' = 0$, yields the final form of the Green's function for the heat equation:

Green's function
for heat equation

$$G(\mathbf{r} - \mathbf{r}'; t) = \frac{e^{-\frac{|\mathbf{r} - \mathbf{r}'|^2}{4\kappa^2 t}}}{(4\pi\kappa^2 t)^{3/2}}. \qquad (29.42)$$

29.2.3   Green's Function for the Wave Equation

The wave equation, which we write as

$$\frac{1}{c^2}\frac{\partial^2 \Psi}{\partial t^2} - \nabla^2 \Psi = 0, \tag{29.43}$$

with c the speed of the wave, is a PDE in 4 variables. As in the case of the heat equation, we let the fourth variable have 0 as subscript. Then

$$p(k_j) = -\frac{k_0^2}{c^2} + k_1^2 + k_2^2 + k_3^2 \equiv -\frac{k_0^2}{c^2} + k^2,$$

and the Green's function can be written as

$$G(\mathbf{x} - \mathbf{x}') = -\frac{1}{(2\pi)^4} \int d^4 k \frac{e^{ik_0(x_0 - x_0') + i\mathbf{k}\cdot(\mathbf{r} - \mathbf{r}')}}{k_0^2/c^2 - k^2}$$

$$= -\frac{c^2}{(2\pi)^4} \int d^3 k \, e^{i\mathbf{k}\cdot(\mathbf{r} - \mathbf{r}')} \int_{-\infty}^{\infty} dk_0 \frac{e^{ik_0 t}}{k_0^2 - c^2 k^2}, \tag{29.44}$$

where we substituted t for x_0 and assumed $x_0' = t' = 0$.

Let us concentrate on the k_0 integration and use the calculus of residues to calculate it. The integrand has two poles $k_0 = \pm ck$ on the real axis, and depending on how these poles are handled, different Green's functions are obtained. One way to handle the poles is to move them up slightly, i.e., give them an infinitesimal positive imaginary part. If $t > 0$, the contour of integration should be in the UHP with zero contribution from the large circle there. If $t < 0$, the contour of integration should be in the LHP for which the integral vanishes because there are no poles inside the contour. Thus, denoting the integrand by f, we have

$$\int_{-\infty}^{\infty} dk_0 \frac{e^{ik_0 t}}{k_0^2 - c^2 k^2} = 2\pi i \left[\mathrm{Res}(f(ck)) + \mathrm{Res}(f(-ck)) \right].$$

But

$$\mathrm{Res}(f(ck)) = \lim_{k_0 \to ck} \left\{ (k_0 - ck)\frac{e^{ik_0 t}}{k_0^2 - c^2 k^2} \right\} = \lim_{k_0 \to ck} \left\{ \frac{e^{ik_0 t}}{k_0 + ck} \right\} = \frac{e^{ickt}}{2ck}.$$

Similarly, $\mathrm{Res}(f(ck)) = -e^{-ickt}/2ck$, and the k_0 integral gives

$$\int_{-\infty}^{\infty} dk_0 \frac{e^{ik_0 t}}{k_0^2 - c^2 k^2} = 2\pi i \left(\frac{e^{ickt}}{2ck} - \frac{e^{-ickt}}{2ck} \right) = -2\pi \frac{\sin ckt}{ck}.$$

Substituting this in (29.44) yields

$$G(\mathbf{x} - \mathbf{x}') = \frac{c}{(2\pi)^3} \int d^3 k \, e^{i\mathbf{k}\cdot(\mathbf{r} - \mathbf{r}')} \frac{\sin ckt}{k},$$

which, through a by-now-familiar routine in k-space spherical integration yields

$$G(\mathbf{r} - \mathbf{r}'; t) = \frac{2c}{(2\pi)^2 |\mathbf{r} - \mathbf{r}'|} \int_0^\infty dk \sin(k|\mathbf{r} - \mathbf{r}'|) \sin ckt$$

$$= \frac{c}{(2\pi)^2 |\mathbf{r} - \mathbf{r}'|} \int_{-\infty}^\infty dk \frac{e^{ik|\mathbf{r}-\mathbf{r}'|} - e^{-ik|\mathbf{r}-\mathbf{r}'|}}{2i} \frac{e^{ickt} - e^{-ickt}}{2i}.$$

Multiply the exponentials and note that $\int_{-\infty}^\infty e^{-ix} dx = \int_{-\infty}^\infty e^{ix} dx$ to obtain

$$G(\mathbf{r} - \mathbf{r}'; t) = -\frac{c}{4(2\pi)^2 |\mathbf{r} - \mathbf{r}'|} 2 \int_{-\infty}^\infty dk \left[e^{ick(t+|\mathbf{r}-\mathbf{r}'|/c)} - e^{ick(t-|\mathbf{r}-\mathbf{r}'|/c)} \right]$$

$$= -\frac{1}{2(2\pi)^2 |\mathbf{r} - \mathbf{r}'|} \left[2\pi\delta(t + |\mathbf{r} - \mathbf{r}'|/c) - 2\pi\delta(t - |\mathbf{r} - \mathbf{r}'|/c) \right].$$

$$(29.45)$$

Retarded Green's function for wave equation

The first delta function vanishes because $t > 0$. Therefore, the final form of the Green's function for the wave equation is

$$G_{\text{ret}}(\mathbf{r} - \mathbf{r}'; t) = \frac{\delta(t - |\mathbf{r} - \mathbf{r}'|/c)}{4\pi |\mathbf{r} - \mathbf{r}'|}. \qquad (29.46)$$

The subscript "ret" on the Green's function stands for *retarded*. As the argument of the delta function implies, $G_{\text{ret}}(\mathbf{r} - \mathbf{r}'; t)$ is zero unless $t = |\mathbf{r} - \mathbf{r}'|/c$, i.e., unless the wave has had time to move from the source point $\mathbf{r}'$ to the observation point $\mathbf{r}$. The signal is "retarded" by this amount of time. Had we given the poles of the k_0 integral of (29.44) an infinitesimal negative imaginary part and chosen t to be negative, the first delta function of (29.45) would have

Advanced Green's function for wave equation

survived and we would have obtained the *advanced* Green's function:

$$G_{\text{adv}}(\mathbf{r} - \mathbf{r}'; t) = \frac{\delta(t + |\mathbf{r} - \mathbf{r}'|/c)}{4\pi |\mathbf{r} - \mathbf{r}'|}. \qquad (29.47)$$

29.3   The Laplace Transform

In the previous section, the power of the Fourier transform was illustrated by formalism and application. Fourier transform is by far the most important of all the transforms used in mathematical analysis. Another transform which is widely used in solving *ordinary* differential equations is the Laplace transform, the subject of this section.

Let $f(t)$ be a sufficiently well-behaved function. The **Laplace transform**

Laplace transform defined

of f is another function $\mathcal{L}[f]$ whose value at s is given by

$$\mathcal{L}[f](s) = \int_0^\infty e^{-st} f(t)\, dt. \qquad (29.48)$$

s could be complex, although it is usually taken to be real. To assure the convergence of the integral, s must have a positive real part. The left-hand side of (29.48) is usually denoted by $F(s)$. It is also common to write it (less precisely) as $\mathcal{L}[f(t)]$ with the letter s understood!

Example 29.3.1. The Laplace transform of the unit function—the function whose value everywhere is 1—evaluated at s can easily be shown to be $1/s$. The Laplace transform of $e^{i\omega t}$ can be readily calculated as well:

$$F(s) = \int_0^\infty e^{-st} e^{i\omega t}\, dt = \int_0^\infty e^{(-s+i\omega)t}\, dt = \frac{e^{(-s+i\omega)t}}{-s+i\omega}\bigg|_0^\infty = \frac{1}{s - i\omega}.$$

The Laplace transforms of $\sin \omega t$ and $\cos \omega t$ can now be evaluated:

$$\mathcal{L}[\cos \omega t] = \mathrm{Re}\left(\mathcal{L}[e^{i\omega t}]\right) = \mathrm{Re}\left(\frac{1}{s - i\omega}\right) = \mathrm{Re}\left(\frac{s + i\omega}{s^2 + \omega^2}\right) = \frac{s}{s^2 + \omega^2}, \quad (29.49)$$

and

$$\mathcal{L}[\sin \omega t] = \mathrm{Im}\left(\mathcal{L}[e^{i\omega t}]\right) = \mathrm{Im}\left(\frac{1}{s - i\omega}\right) = \mathrm{Im}\left(\frac{s + i\omega}{s^2 + \omega^2}\right) = \frac{\omega}{s^2 + \omega^2}. \quad (29.50)$$

The Laplace transform of the step function $\theta(t - a)$ is very useful in applications [see Section 5.1.3 for the definition of the step function].

$$\mathcal{L}[\theta(t - a)] = \int_0^\infty e^{-st}\theta(t - a)\, dt = \int_a^\infty e^{-st}\, dt = \frac{e^{-as}}{s}.$$

The lower limit of integration was changed because $\theta(t - a)$ is zero for $t < a$ (and it is equal to 1 for $t > a$).

Knowing $\mathcal{L}[1]$, we can find the Laplace transform of any power of t because

$$\mathcal{L}[t^n] = \int_0^\infty t^n e^{-st}\, dt = (-1)^n \frac{d^n}{ds^n}\int_0^\infty e^{-st}\, dt.$$

Since $\mathcal{L}[1](s) = 1/s$, we have

$$\mathcal{L}[t^n] = (-1)^n \frac{d^n}{ds^n}\left(\frac{1}{s}\right) = \frac{n!}{s^{n+1}}. \quad (29.51)$$

What if n in the above equation is not an integer? Let's evaluate $\mathcal{L}[t^\nu]$ directly.

$$\mathcal{L}[t^\nu] = \int_0^\infty t^\nu e^{-st}\, dt = \int_0^\infty \left(\frac{u}{s}\right)^\nu e^{-u}\frac{1}{s}\, du = \frac{1}{s^{\nu+1}}\int_0^\infty u^\nu e^{-u}\, du = \frac{\Gamma(\nu + 1)}{s^{\nu+1}}, \quad (29.52)$$

where Γ is the gamma function introduced in Section 11.1.1. Note that if $\nu = n$, we regain (29.51) because $\Gamma(n + 1) = n!$. ∎

29.3.1 Properties of Laplace Transform

In a typical application, one obtains the Laplace transform of a function from, say a differential equation, and inverts it to find the actual function. This is what was done in the case of the Fourier transform, and indeed in any other transform used. While the formula for inverting a Fourier transform [see

Equation (29.7)] is nice and symmetric, that for the Laplace transform is not as nice. Furthermore, Fourier transform adapts itself very naturally to partial differential equations as demonstrated in the previous section. However, the adaptation of Laplace transform to PDEs is not so natural. That is why the Fourier transform techniques are much more powerful—both formally and for calculations—than the Laplace transform.

Because of this drawback, one has to rely on some formal properties of the Laplace transform and its inverse—as well as a lot of examples—to be able to reconstruct the original function.

Linearity

One such property is the linearity of the Laplace transform and it inverse:

$$\mathcal{L}[af + bg] = a\mathcal{L}[f] + b\mathcal{L}[g], \quad \mathcal{L}^{-1}[af + bg] = a\mathcal{L}^{-1}[f] + b\mathcal{L}^{-1}[g]. \quad (29.53)$$

First shift property

Another is the **first shift property**. If $F(s)$ is the Laplace transform of $f(t)$, then $F(s - a)$ is the Laplace transform of $e^{at} f(t)$. This can easily be verified:

$$F(s-a) = \int_0^\infty e^{-(s-a)t} f(t)\, dt = \int_0^\infty e^{-st}\left(e^{at} f(t)\right) dt = \mathcal{L}[e^{at} f(t)]. \quad (29.54)$$

A more useful way of writing this equation is

$$\mathcal{L}^{-1}[F(s - a)] = e^{at}\mathcal{L}^{-1}[F(s)]. \quad (29.55)$$

Second shift property

The **second shift property** involves the step function:

$$\mathcal{L}[\theta(t - a)f(t - a)] = e^{-as}\mathcal{L}[f](s). \quad (29.56)$$

This is because

$$\mathcal{L}[\theta(t-a)f(t-a)] = \int_a^\infty e^{-st} f(t-a)\, dt = \int_0^\infty e^{-s(\tau+a)} f(\tau)\, d\tau = e^{-as}\mathcal{L}[f](s),$$

where in the second equality we changed the variable of integration to $\tau = t - a$. Denoting by $F(s)$ the Laplace transform of $f(t)$, we write (29.56) as

$$\mathcal{L}^{-1}[e^{-as}F(s)] = \theta(t - a)f(t - a) = \begin{cases} f(t - a) & \text{if } a > 0 \\ 0 & \text{if } a < 0. \end{cases} \quad (29.57)$$

Example 29.3.2. Since $\mathcal{L}[t^n] = n!/s^{n+1}$, using the first shift property, we get

$$\mathcal{L}[t^n e^{at}] = \frac{n!}{(s - a)^{n+1}}. \quad (29.58)$$

In particular, if $n = 0$, we have $\mathcal{L}[e^{at}] = 1/(s - a)$. From this, and the linearity property, we can find the Laplace transforms of the hyperbolic sine:

$$\mathcal{L}[\sinh \gamma t] = \mathcal{L}[\tfrac{1}{2}\left(e^{\gamma t} - e^{-\gamma t}\right)] = \frac{1}{2}\left(\mathcal{L}[e^{\gamma t}] - \mathcal{L}[e^{-\gamma t}]\right)$$

$$= \frac{1}{2}\left(\frac{1}{s - \gamma} - \frac{1}{s + \gamma}\right) = \frac{\gamma}{s^2 - \gamma^2} \qquad (29.59)$$

and hyperbolic cosine:

$$\mathcal{L}[\cosh \gamma t] = \mathcal{L}[\tfrac{1}{2}\left(e^{\gamma t} + e^{-\gamma t}\right)] = \frac{1}{2}\left(\mathcal{L}[e^{\gamma t}] + \mathcal{L}[e^{-\gamma t}]\right)$$

$$= \frac{1}{2}\left(\frac{1}{s - \gamma} + \frac{1}{s + \gamma}\right) = \frac{s}{s^2 - \gamma^2}. \qquad (29.60)$$

With our accumulated knowledge of the Laplace transform, let's see if we can find the inverse transform of $1/(s^2 + 2as + b^2)$. Complete the square in the denominator and consider three cases: $b > a$, $b < a$, and $b = 0$. First assume $b > a$ and define $\omega^2 = b^2 - a^2$. Then

$$\mathcal{L}^{-1}\left[\frac{1}{(s + a)^2 + b^2 - a^2}\right] = \frac{1}{\omega}\mathcal{L}^{-1}\left[\frac{\omega}{(s + a)^2 + \omega^2}\right] = \frac{e^{-at}}{\omega}\mathcal{L}^{-1}\left[\frac{\omega}{s^2 + \omega^2}\right] = \frac{e^{-at}}{\omega}\sin \omega t$$

where we used (29.55) and (29.50). Substituting for ω, we get

$$\mathcal{L}^{-1}\left[\frac{1}{s^2 + 2as + b^2}\right] = \frac{e^{-at}}{\sqrt{b^2 - a^2}}\sin(\sqrt{b^2 - a^2}\, t), \quad a < b.$$

For $b < a$ define $\gamma^2 = a^2 - b^2$. Then

$$\mathcal{L}^{-1}\left[\frac{1}{(s + a)^2 + b^2 - a^2}\right] = \frac{1}{\gamma}\mathcal{L}^{-1}\left[\frac{\gamma}{(s + a)^2 - \gamma^2}\right] = \frac{e^{-at}}{\gamma}\mathcal{L}^{-1}\left[\frac{\gamma}{s^2 - \gamma^2}\right] = \frac{e^{-at}}{\gamma}\sinh \gamma t$$

where we used (29.55) and (29.59). Substituting for γ, we get

$$\mathcal{L}^{-1}\left[\frac{1}{s^2 + 2as + b^2}\right] = \frac{e^{-at}}{\sqrt{a^2 - b^2}}\sinh(\sqrt{a^2 - b^2}\, t), \quad a > b.$$

If $b = a$, then the denominator is a complete square and

$$\mathcal{L}^{-1}\left[\frac{1}{(s + a)^2}\right] = e^{-at}t$$

by (29.58).

Similarly, we can show that

$$\mathcal{L}^{-1}\left[\frac{s}{s^2 + 2as + b^2}\right] = e^{-at}\cos(\sqrt{b^2 - a^2}\, t)$$
$$- \frac{a}{\sqrt{b^2 - a^2}}e^{-at}\sin(\sqrt{b^2 - a^2}\, t), b > a$$

$$\mathcal{L}^{-1}\left[\frac{s}{s^2 + 2as + b^2}\right] = e^{-at}\cosh(\sqrt{a^2 - b^2}\, t)$$
$$- \frac{a}{\sqrt{a^2 - b^2}}e^{-at}\sinh(\sqrt{a^2 - b^2}\, t), b < a$$

We shall use the formulas derived in this example in solving differential equations. ∎

Periodic functions

Although Fourier *series* are better suited for periodic functions, Laplace transform of periodic functions is also of interest. If $f(t)$ is periodic of period T, i.e., if $f(t+T) = f(t)$, then

$$\mathcal{L}[f(t)] = \underbrace{\int_0^T e^{-st} f(t)\, dt}_{\text{call this } F_1(s)} + \int_T^\infty e^{-st} f(t)\, dt = F_1(s) + \int_0^\infty e^{-s(u+T)} f(u+T)\, du$$

$$= F_1(s) + e^{-sT} \int_0^\infty e^{-su} f(u)\, du = F_1(s) + e^{-sT} \overbrace{\int_0^\infty e^{-st} f(t)\, dt}^{=\mathcal{L}[f(t)]}.$$

We thus have $\mathcal{L}[f(t)] = F_1(s) + e^{-sT}\mathcal{L}[f(t)]$, which upon solving for $\mathcal{L}[f(t)]$ yields

$$\mathcal{L}[f(t)] = \frac{1}{1 - e^{-sT}} F_1(s). \tag{29.61}$$

Example 29.3.3. The Laplace transform of the square wave function of Example 10.6.1 defined by

$$V(t) = \begin{cases} V_0 & \text{if } 0 \le t \le T, \\ 0 & \text{if } T < t \le 2T, \end{cases}$$

can be readily found. We simply note that the period is $2T$ and $F_1(s)$ is

$$F_1(s) = \int_0^{2T} e^{-st} V(t)\, dt = V_0 \int_T^{2T} e^{-st}\, dt = \frac{V_0}{s}(e^{-sT} - e^{-2sT}).$$

Substituting this in (29.61), we obtain

$$\mathcal{L}[V(t)] = \frac{1}{1 - e^{-2sT}} \left[\frac{V_0}{s}(e^{-sT} - e^{-2sT}) \right]$$

$$= \frac{V_0 e^{-sT}(1 - e^{-sT})}{s(1 - e^{-sT})(1 + e^{-sT})} = \frac{V_0 e^{-sT}}{s(1 + e^{-sT})} = \frac{V_0}{s(1 + e^{sT})}. \qquad \blacksquare$$

Convolution

The convolution of two functions is defined as

$$(f * g)(t) = \int_0^t f(u)g(t - u)\, du.$$

Let $v = t - u$ and change the variable of integration to v. Then

$$(f * g)(t) = \int_t^0 f(t - v)g(v)(-dv) = \int_0^t g(v)f(t - v)\, dv = (g * f)(t),$$

showing that convolution is commutative. Commutativity is only one of the following properties of convolution:

1. $c(f * g) = cf * g = f * cg$, c a constant;

2. $f * g = g * f$ (commutative property);

3. $f * (g * h) = (f * g) * h$ (associative property);

4. $f * (g + h) = f * g + f * h$ (distributive property).

The notion of convolution is useful for Laplace transform because one can show the following:

> **Box 29.3.1.** *The Laplace transform of the convolution of two functions is the product of the Laplace transforms of the two functions:*
>
> $$\mathcal{L}[f * g](s) = \mathcal{L}[f](s) \cdot \mathcal{L}[g](s) \quad \text{or} \quad (f * g)(t) = \mathcal{L}^{-1}[\mathcal{L}[f](s) \cdot \mathcal{L}[g](s)]$$

Example 29.3.4. Suppose we want to find the inverse transform of $s/(s^2 + a^2)^2$. We can write this as [see Equations (29.49) and (29.50)]

$$\frac{s}{(s^2 + a^2)^2} = \frac{1}{a} \frac{s}{s^2 + a^2} \cdot \frac{a}{s^2 + a^2} = \frac{1}{a} \mathcal{L}[\cos at] \cdot \mathcal{L}[\sin at].$$

Therefore,

$$\mathcal{L}^{-1}\left[\frac{s}{(s^2 + a^2)^2}\right] = \frac{1}{a} \cos at * \sin at = \frac{1}{a} \int_0^t \cos au \sin a(t - u)du = \frac{1}{2a} t \sin at.$$

Similarly

$$\mathcal{L}^{-1}\left[\frac{s^2}{(s^2 + a^2)^2}\right] = \mathcal{L}^{-1}[\mathcal{L}[\cos at] \cdot \mathcal{L}[\cos at]] = \cos at * \cos at$$

$$= \int_0^t \cos au \cos a(t - u)du = \frac{1}{2} t \cos at + \frac{1}{2a} \sin at. \qquad \blacksquare$$

29.3.2   Derivative and Integral of the Laplace Transform

By differentiating the integral of a Laplace transform, one can easily obtain the formulas

$$\frac{d^n}{ds^n} F(s) = \mathcal{L}[(-1)^n t^n f(t)] \quad \text{or} \quad \mathcal{L}^{-1}\left[F^{(n)}(s)\right] = (-1)^n t^n \mathcal{L}^{-1}[F(s)],$$
(29.62)

where $F^{(n)}$ denotes the nth derivative of F. For $n = 1$, this formula leads to the following useful relation:

$$\mathcal{L}^{-1}[F(s)] = -\frac{1}{t} \mathcal{L}^{-1}[F'(s)],$$
(29.63)

because sometimes it is easier to find the inverse Laplace transform of the derivative of a function than the function itself.

Example 29.3.5. It is not easy to find the inverse transform of $F(s) = \ln[(s + a)/(s + b)]$ directly. But the inverse transform of

$$F'(s) = \frac{d}{ds}[\ln(s + a) - \ln(s + b)] = \frac{1}{s + a} - \frac{1}{s + b}$$

is much easier to find. In fact,

$$\mathcal{L}^{-1}[F'(s)] = \mathcal{L}^{-1}\left[\frac{1}{s + a}\right] - \mathcal{L}^{-1}\left[\frac{1}{s + b}\right] = e^{-at} - e^{-bt},$$

by (29.58) with $n = 0$. Therefore, according to (29.63)

$$\mathcal{L}^{-1}\left[\ln \frac{s + a}{s + b}\right] = \frac{e^{-bt} - e^{-at}}{t}. \qquad \blacksquare$$

One can also find the primitive (antiderivative, indefinite integral) of $F(s)$. Recall that the indefinite integral of a function can be written as a definite integral with one of its limits being a variable [see Equation (3.18)]. Therefore, let's write the indefinite integral of $F(s)$ as $-\int_s^\infty F(u)\, du$. This integral can be easily evaluated:

$$\int_s^\infty F(u)\, du = \int_s^\infty du \int_0^\infty e^{-ut} f(t)\, dt = \int_0^\infty f(t)\, dt \int_s^\infty e^{-ut} du$$

$$= \int_0^\infty f(t)\, dt \left(-\frac{e^{-ut}}{t}\Big|_{u=s}^{u=\infty}\right) = \int_0^\infty e^{-st} \frac{f(t)}{t}\, dt.$$

This can be written as

$$\int_s^\infty \mathcal{L}[f](u)\, du = \mathcal{L}\left[\frac{f(t)}{t}\right]. \qquad (29.64)$$

Example 29.3.6. Let's use (29.64) to find the Laplace transform of $\sin \omega t/t$. From (29.50), we have

$$\mathcal{L}\left[\frac{\sin \omega t}{t}\right] = \int_s^\infty \frac{\omega}{u^2 + \omega^2}\, du = \tan^{-1}\left(\frac{u}{\omega}\right)\Big|_s^\infty = \frac{\pi}{2} - \tan^{-1}\left(\frac{s}{\omega}\right) = \tan^{-1}\left(\frac{\omega}{s}\right)$$

(see Problem 29.17 for the last equality). Similarly,

$$\mathcal{L}\left[\frac{\sinh \gamma t}{t}\right] = \int_s^\infty \frac{\gamma}{u^2 - \gamma^2}\, du = \frac{1}{2} \lim_{x \to \infty} \int_s^x \left(\frac{1}{u - \gamma} - \frac{1}{u + \gamma}\right) du$$

$$= \frac{1}{2} \lim_{x \to \infty} \left(\ln \frac{x - \gamma}{x + \gamma} - \ln \frac{s - \gamma}{s + \gamma}\right) = \frac{1}{2} \ln \frac{s + \gamma}{s - \gamma}. \qquad \blacksquare$$

29.3.3   Laplace Transform and Differential Equations

Certain differential equations with appropriate boundary conditions or initial values can be nicely solved by Laplace transform techniques. For the application of Laplace transform to differential equations, we need to know the transform of the derivative of a function. Using integration by parts, we have

$$\int_0^\infty e^{-st} f'(t)\, dt = e^{-st} f(t)\Big|_0^\infty + s \int_0^\infty e^{-st} f(t)\, dt = -f(0) + s\mathcal{L}[f](s).$$

Therefore,

$$\mathcal{L}[f'](s) = s\mathcal{L}[f](s) - f(0). \tag{29.65}$$

This can be iterated to give

$$\mathcal{L}[f''](s) = s\mathcal{L}[f'](s) - f'(0) = s[s\mathcal{L}[f](s) - f(0)] - f'(0),$$

or

$$\mathcal{L}[f''](s) = s^2\mathcal{L}[f](s) - sf(0) - f'(0). \tag{29.66}$$

We can continue iterating the formula, but since most differential equations encountered in applications are of second order, we stop at the second derivative.

To solve a differential equation, take the Laplace transform of both sides and use (29.65) and (29.66). Solve for $\mathcal{L}[f](s)$ and take the inverse transform to find the solution. Let's look at a specific example. Consider a mass m attached to a spring of spring constant k. The differential equation of motion of this system is

$$m\ddot{x} + kx = 0 \quad \text{or} \quad \ddot{x} + \omega_0^2 x = 0, \quad \omega_0 = \sqrt{\frac{k}{m}}.$$

Taking the Laplace transform of both sides gives

$$\mathcal{L}[\ddot{x}](s) + \omega_0^2 \mathcal{L}[x](s) = 0.$$

Using (29.66), this becomes

$$s^2\mathcal{L}[x](s) - sx(0) - \dot{x}(0) + \omega_0^2\mathcal{L}[x](s) = 0,$$

or, letting $x_0 = x(0)$ and $\dot{x}_0 = \dot{x}(0)$, we get

$$(s^2 + \omega_0^2)\mathcal{L}[x](s) = sx_0 + \dot{x}_0 \quad \text{or} \quad \mathcal{L}[x](s) = \frac{x_0 s + \dot{x}_0}{s^2 + \omega_0^2},$$

and from (29.49) and (29.50) we obtain

$$x(t) = x_0 \mathcal{L}^{-1}\left[\frac{s}{s^2 + \omega_0^2}\right] + \frac{\dot{x}_0}{\omega_0}\mathcal{L}^{-1}\left[\frac{\omega_0}{s^2 + \omega_0^2}\right]$$

$$= x_0 \cos\omega_0 t + \frac{\dot{x}_0}{\omega_0}\sin\omega_0 t.$$

Note how the initial values are automatically included in the solution.

A more general problem has a damping term as well as a driving force. This leads to a differential equation of the form

$$\ddot{x} + \gamma\dot{x} + \omega_0^2 x = f(t), \quad \omega_0 = \sqrt{\frac{k}{m}}. \tag{29.67}$$

To solve this, once again take the Laplace transform of both sides:

$$\mathcal{L}[\ddot{x}](s) + \gamma\mathcal{L}[\dot{x}](s) + \omega_0^2\mathcal{L}[x](s) = \mathcal{L}[f](s),$$

and use (29.65) and (29.66) to get

$$s^2 \mathcal{L}[x](s) - x_0 s - \dot{x}_0 + \gamma\{s\mathcal{L}[x](s) - x_0\} + \omega_0^2 \mathcal{L}[x](s) = \mathcal{L}[f](s),$$

where $x_0 = x(0)$ and $\dot{x}_0 = \dot{x}(0)$. Therefore,

$$(s^2 + \gamma s + \omega_0^2)\mathcal{L}[x](s) = \mathcal{L}[f](s) + x_0 s + \dot{x}_0 + \gamma x_0,$$

which yields

$$\mathcal{L}[x](s) = \frac{\mathcal{L}[f](s) + x_0 s + \dot{x}_0 + \gamma x_0}{s^2 + \gamma s + \omega_0^2}.$$

The solution can be obtained by inversion once we know $\mathcal{L}[f](s)$. Symbolically, we write

$$x(t) = \mathcal{L}^{-1}\left[\frac{\mathcal{L}[f](s)}{s^2 + \gamma s + \omega_0^2}\right] + x_0 \mathcal{L}^{-1}\left[\frac{s}{s^2 + \gamma s + \omega_0^2}\right]$$

$$+ (\dot{x}_0 + \gamma x_0)\mathcal{L}^{-1}\left[\frac{1}{s^2 + \gamma s + \omega_0^2}\right]. \tag{29.68}$$

We consider only the case of a damped harmonic *oscillator*, i.e., that $\omega_0 > \gamma/2$. The second and third inversions are given in Example 29.3.2 with $a = \gamma/2$ and $b = \omega_0$. Then, with $\Omega \equiv \sqrt{\omega_0^2 - (\gamma/2)^2}$, we have

$$\mathcal{L}^{-1}\left[\frac{s}{s^2 + \gamma s + \omega_0^2}\right] = e^{-\gamma t/2}\cos\Omega t - \frac{\gamma}{2\Omega}e^{-\gamma t/2}\sin\Omega t,$$

$$\mathcal{L}^{-1}\left[\frac{1}{s^2 + \gamma s + \omega_0^2}\right] = \frac{e^{-\gamma t/2}}{\Omega}\sin\Omega t. \tag{29.69}$$

Substituting these in (29.68), we obtain

$$x(t) = e^{-\gamma t/2}\left(x_0 \cos\Omega t + \frac{\dot{x}_0 + x_0\gamma/2}{\Omega}\sin\Omega t\right) + \mathcal{L}^{-1}\left[\frac{\mathcal{L}[f](s)}{s^2 + \gamma s + \omega_0^2}\right]. \tag{29.70}$$

Let us denote the last term of this equation by $\Phi(t)$ and evaluate the equation at $t = 0$ to obtain $x(0) = x_0 + \Phi(0)$ implying that $\Phi(0) = 0$. Similarly, differentiating the equation and evaluating the result at $t = 0$ yields $\dot{x}(0) = \dot{x}_0 + \dot{\Phi}(0)$ implying that $\dot{\Phi}(0) = 0$. This is an interesting result since $f(t)$ is quite arbitrary! The following example looks at a specific instance of this result.

Example 29.3.7. As an example of the general formula (29.70), let's consider a damped harmonic oscillator driven by a sinusoidal source $f(t) = A\sin\omega_0 t$ operating at the natural frequency of the oscillator as given in (29.67). Then by (29.50)

$$\mathcal{L}[f](s) = \mathcal{L}[A\sin\omega_0 t] = \frac{A\omega_0}{s^2 + \omega_0^2},$$

and the last term of (29.70) becomes

$$\mathcal{L}^{-1}\left[\frac{A\omega_0}{(s^2 + \omega_0^2)(s^2 + \gamma s + \omega_0^2)}\right].$$

Using partial fraction techniques, we can write this as

$$\frac{A\omega_0}{(s^2 + \omega_0^2)(s^2 + \gamma s + \omega_0^2)} = \frac{A}{\gamma\omega_0}\frac{s}{s^2 + \gamma s + \omega_0^2} + \frac{A}{\omega_0}\frac{1}{s^2 + \gamma s + \omega_0^2} - \frac{A}{\gamma\omega_0}\frac{s}{s^2 + \omega_0^2}.$$

Each term can now be inverted using the results we have obtained in several examples. Denoting the final result by $\Phi(t)$, we get

$$\Phi(t) \equiv \mathcal{L}^{-1}\left[\frac{\mathcal{L}[f](s)}{s^2 + \gamma s + \omega_0^2}\right] = -\frac{A}{\gamma\omega_0}\cos\omega_0 t + \frac{A}{\gamma\omega_0}e^{-\gamma t/2}\left(\cos\Omega t + \frac{\gamma}{2\Omega}\sin\Omega t\right).$$

Note that $\Phi(0) = 0$ as expected from the discussion above. Differentiating, we obtain

$$\dot{\Phi}(t) = \frac{A}{\gamma}\sin\omega_0 t - \frac{A}{2\omega_0}e^{-\gamma t/2}\left(\cos\Omega t + \frac{\gamma}{2\Omega}\sin\Omega t\right) + \frac{A}{\gamma\omega_0}e^{-\gamma t/2}\left(-\Omega\sin\Omega t + \frac{\gamma}{2}\cos\Omega t\right).$$

It is readily verified that $\dot{\Phi}(0) = 0$ as explained above. Substituting $\Phi(t)$ for the last term of (29.70) yields

$$x(t) = e^{-\gamma t/2}\left(x_0\cos\Omega t + \frac{\dot{x}_0 + x_0\gamma/2}{\Omega}\sin\Omega t\right)$$

$$- \frac{A}{\gamma\omega_0}\cos\omega_0 t + \frac{A}{\gamma\omega_0}e^{-\gamma t/2}\left(\cos\Omega t + \frac{\gamma}{2\Omega}\sin\Omega t\right). \tag{29.71}$$

After a long time (i.e., as $t \to \infty$), the terms containing an exponential—the so-called *transient terms*—will be negligible and $x(t) \to -\frac{A}{\gamma\omega_0}\cos\omega_0 t$ as expected from the elementary analysis of the same problem. ∎

We can understand this interesting behavior of $\Phi(t)$ in terms of the properties of convolution. Let $g(t)$ be the inverse transform of $1/(s^2 + \gamma s + \omega_0^2)$. Then invoking Box 29.3.1, the last term of (29.70) can be written as

$$\Phi(t) = \mathcal{L}^{-1}\left[\mathcal{L}[f](s) \cdot \mathcal{L}[g](s)\right] = (f * g)(t) = \int_0^t f(u)g(t-u)du,$$

whose derivative is (see Box 3.2.2)

$$\dot{\Phi}(t) = f(t)g(0) + \int_0^t f(u)\dot{g}(t-u)du.$$

It is now clear why $\Phi(0) = 0$. As for the derivative, we see that $\dot{\Phi}(0) = f(0)g(0)$. But $g(t)$ is given by (29.69) which is clearly 0 at $t = 0$.

29.3.4 Inverse of Laplace Transform

As mentioned earlier, the procedure for inverting a Laplace transform is important in solving differential equations, as the technique—like any other transform—yields the *transform* of the solution, and to get the solution, one has to invert that transform. So far, we have used various tricks and properties of the Laplace transform to get from $F(s) \equiv \mathcal{L}[f](s)$ to $f(t)$. Now, we provide a general formula that can always be used to yield the function. The procedure is the **Mellin inversion integral**:

$$f(t) = \frac{1}{2\pi i} \int_{\gamma-i\infty}^{\gamma+i\infty} F(s)e^{st}ds. \qquad (29.72)$$

The integration is along a line, called the **Bromwich contour**, parallel to the imaginary axis of the complex s plane. The real number γ is arbitrary as long as the integration line is to the right of all the singularities of $F(s)$. To find the actual value of the integral, one closes the contour with an infinite semicircle *to the left* of the line and uses the residue theorem.

To prove that the right-hand side of (29.72) is indeed $f(t)$, substitute the definition of $F(s)$,

$$F(s) = \int_0^\infty f(\tau)e^{-s\tau}d\tau,$$

in the integral and switch the order of integrations to get

$$\text{RHS} = \frac{1}{2\pi i} \int_0^\infty f(\tau)d\tau \underbrace{\int_{\gamma-i\infty}^{\gamma+i\infty} e^{s(t-\tau)}ds}_{\text{Denote this by } J}. \qquad (29.73)$$

Introduce a new variable of integration σ by $s = \gamma + i\sigma$ in the inner integral to get

$$J = \int_{-\infty}^\infty e^{(\gamma+i\sigma)(t-\tau)}id\sigma = ie^{\gamma(t-\tau)} \underbrace{\int_{-\infty}^\infty e^{i\sigma(t-\tau)}d\sigma}_{=2\pi\delta(t-\tau) \text{ by } (18.28)} = 2\pi i\delta(t-\tau).$$

The last step follows because $\delta(t-\tau) = 0$ unless $t = \tau$ in which case the exponent of the exponential is zero. Substituting this in (29.73) and noting that $\tau > 0$, we get RHS $= f(t)$.

To see why the integration line must lie to the right of all singularities, take the Laplace transform of both sides of (29.72):

$$\mathcal{L}[f(t)] = \frac{1}{2\pi i} \int_0^\infty e^{-st} \left(\int_{\gamma-i\infty}^{\gamma+i\infty} F(\sigma)e^{\sigma t}d\sigma \right) dt$$

$$= \frac{1}{2\pi i} \int_{\gamma-i\infty}^{\gamma+i\infty} F(\sigma)d\sigma \int_0^\infty e^{(\sigma-s)t}dt = -\frac{1}{2\pi i} \int_{\gamma-i\infty}^{\gamma+i\infty} \frac{F(\sigma)}{\sigma-s}d\sigma,$$

assuming that $\text{Re}(s) > \text{Re}(\sigma) = \gamma$. If $F(\sigma)$ is analytic to the right of the Bromwich contour, then closing the infinite semicircle on the right, there will be a single pole at $\sigma = s$ inside the closed contour, and the residue theorem gives the value of the integral as $-2\pi iF(s)$, with the negative sign coming from the clockwise integration. If any of the poles of F were on the right of the Bromwich contour we would not obtain $-2\pi iF(s)$ for the integration.

Example 29.3.8. Let us find the inverse Laplace transform of $F(s) = 1/(s^2 + \omega^2)$. This is given by

$$f(t) = \frac{1}{2\pi i} \int_{\gamma-i\infty}^{\gamma+i\infty} \frac{e^{st}}{s^2 + \omega^2} \, ds$$

where the contour of integration includes the infinite semicircle to the left. The poles of the integrand are at $\pm i\omega$, so as long as $\gamma > 0$, the contour encloses both poles. The residue theorem then yields

$$f(t) = \frac{1}{2\pi i} \left\{ 2\pi i \left[\text{Res} \left(\frac{e^{st}}{(s-i\omega)(s+i\omega)} \right) \Bigg|_{s=i\omega} + \text{Res} \left(\frac{e^{st}}{(s-i\omega)(s+i\omega)} \right) \Bigg|_{s=-i\omega} \right] \right\}$$

$$= \frac{e^{i\omega t}}{2i\omega} + \frac{e^{-i\omega t}}{-2i\omega} = \frac{1}{\omega} \sin \omega t$$

which is the expected result (see Example 29.3.1). ∎

We can similarly find the inverse Laplace transform of $F(s) = s/(s^2 + \omega^2)$:

$$f(t) = \frac{1}{2\pi i} \int_{\gamma-i\infty}^{\gamma+i\infty} \frac{se^{st}}{s^2 + \omega^2} \, ds$$

The contour of integration again includes the infinite semicircle to the left, and the poles of the integrand are at $\pm i\omega$, as above. The residue theorem now yields

$$f(t) = \frac{1}{2\pi i} \left\{ 2\pi i \left[\text{Res} \left(\frac{se^{st}}{(s-i\omega)(s+i\omega)} \right) \Bigg|_{s=i\omega} \right. \right.$$

$$\left. \left. + \text{Res} \left(\frac{se^{st}}{(s-i\omega)(s+i\omega)} \right) \Bigg|_{s=-i\omega} \right] \right\}$$

$$= \frac{i\omega e^{i\omega t}}{2i\omega} + \frac{-i\omega e^{-i\omega t}}{-2i\omega} = \cos \omega t$$

which is also treated in Example 29.3.1.

29.4 Problems

29.1. Find directly the Fourier transform of
(a) the constant function $f(x) = C$, and
(b) the Dirac delta function $\delta(x)$.

29.2. Show the second identity in (29.8).

29.3. Show that the inverse of a sine transform is another sine transform.

29.4. Show (29.9), the linearity property of Fourier transform and its inverse.

29.5. Suppose that $\tilde{f}(k)$ is the inverse Fourier transform of $f(x)$. Show that the inverse Fourier transform of $f(x + a)$ is $e^{iak}\tilde{f}(k)$.

29.6. Show that if $f(t) = \cos \omega_0 t$, then

$$\tilde{f}(\omega) = \sqrt{\frac{\pi}{2}} \left[\delta(\omega - \omega_0) + \delta(\omega + \omega_0) \right].$$

29.7. Show that
(a) $g(x)$ is real if and only if $\tilde{g}^*(k) = \tilde{g}(-k)$,
(a) $g(x)$ is imaginary if and only if $\tilde{g}^*(k) = -\tilde{g}(-k)$, and
(c) if $g(x)$ is even (odd), then $\tilde{g}(k)$ is also even (odd).

29.8. Evaluate the Fourier transform of

$$g(x) = \begin{cases} b - b|x|/a & \text{if } |x| < a, \\ 0 & \text{if } |x| > a. \end{cases}$$

29.9. Let

$$f(t) = \begin{cases} \sin \omega_0 t & \text{if } |t| < T, \\ 0 & \text{if } |t| > T. \end{cases}$$

Show that

$$\tilde{f}(\omega) = \frac{1}{\sqrt{2\pi}} \left\{ \frac{\sin[(\omega - \omega_0)T]}{\omega - \omega_0} - \frac{\sin[(\omega + \omega_0)T]}{\omega + \omega_0} \right\}.$$

Verify the uncertainty relation $\Delta\omega\Delta t \approx 4\pi$.

29.10. If $f(x) = g(x + a)$, show that $\tilde{f}(k) = e^{-iak}\tilde{g}(k)$.

29.11. For $a > 0$ find the Fourier transform of $f(x) = e^{-a|x|}$. Is $\tilde{f}(k)$ symmetric? Is it real? Verify the uncertainty relations.

29.12. The displacement of a damped harmonic oscillator is given by

$$f(t) = \begin{cases} Ae^{-\alpha t}e^{i\omega_0 t} & \text{if } t > 0, \\ 0 & \text{if } t < 0. \end{cases}$$

Find $\tilde{f}(\omega)$ and show that the frequency distribution $|\tilde{f}(\omega)|^2$ is given by

$$|\tilde{f}(\omega)|^2 = \frac{A^2}{2\pi} \frac{1}{(\omega - \omega_0)^2 + \alpha^2}.$$

convolution
theorem for
Fourier transform

29.13. Prove the **convolution theorem** for Fourier transform:

$$\int_{-\infty}^{\infty} f(x)g(y - x)\, dx = \int_{-\infty}^{\infty} \tilde{f}(k)\tilde{g}(k)e^{iky}\, dk.$$

What will this give when $y = 0$?

Parseval's relation

29.14. Prove **Parseval's relation** for Fourier transforms:

$$\int_{-\infty}^{\infty} f(x)g^*(x)\, dx = \int_{-\infty}^{\infty} \tilde{f}(k)\tilde{g}^*(k)\, dk.$$

29.15. Find the sine and cosine transform of e^{-ax}.

29.16. Following Example 29.1.6, substitute the Fourier Transform of the wave function $\Psi(x,t)$ in the one-dimensional wave equation

$$\frac{1}{c^2}\frac{\partial^2 \Psi}{\partial t^2} = \frac{\partial^2 \Psi}{\partial x^2},$$

and solve the differential equation in t to get

$$\tilde{\Psi}(k,t) = C(k)e^{\pm ickt}.$$

Assuming that the initial shape of the wave $\Psi(x,0)$ is given by a function $f(x)$, show that the solution $\Psi(x,t)$ can be written as

$$\Psi(x,t) = f(x \pm ct).$$

29.17. Show the relation used in Example 29.3.6:

$$\frac{\pi}{2} - \tan^{-1}\left(\frac{s}{\omega}\right) = \tan^{-1}\left(\frac{\omega}{s}\right).$$

Hint: Let x denote the left-hand side and $\alpha = \tan^{-1}(s/\omega)$. Take the tan of both sides of the definition of x and use $\cot \alpha = \tan(\pi/2 - \alpha) = 1/\tan \alpha$.

29.18. Let $f(t) = \sin \omega t$ be the periodic function of (29.61) and verify that the equation holds (for $T = 2\pi/\omega$). Do the same for $f(t) = \cos \omega t$.

29.19. Find the Laplace transform of the periodic sawtooth function with period T defined by

$$V(t) = V_0 \frac{t}{T} \qquad \text{for} \quad 0 \le t < T.$$

29.20. Find the Laplace transform of $2t + 4e^{2t} - 3\cos 3t$.

29.21. Compute $\mathcal{L}[\cosh^2 \gamma t]$ and $\mathcal{L}[\sinh^2 \gamma t]$.

29.22. Compute $\mathcal{L}[\cos^2 \omega t]$ and $\mathcal{L}[\sin^2 \omega t]$ directly from the definition of Laplace transform. Now show that

$$\mathcal{L}[\cos^2 \omega t] = \mathcal{L}[1] - \mathcal{L}[\sin^2 \omega t].$$

29.23. A function $N(t)$ is called a *null function* if

$$\int_0^t N(u)\,du = 0$$

for all $t > 0$. Show that $\mathcal{L}[N(t)] = 0$.

29.24. Compute $\mathcal{L}[e^{2t}\sin 3t]$, $\mathcal{L}[t^2 e^{-\gamma t}]$, $\mathcal{L}^{-1}[e^{-2s}/s^3]$, and

$$\mathcal{L}^{-1}\left[\frac{a}{s} - \frac{s}{s^2+1}e^{-bs}\right].$$

29.25. Find $\mathcal{L}[e^{3t}/\sqrt{t}]$, $\mathcal{L}[\sqrt{t}]$, and $\mathcal{L}^{-1}[e^{-2s}/\sqrt{s}]$.

29.26. (a) Show that

$$\frac{\partial}{\partial \nu} t^{\nu-1} = t^{\nu-1} \ln t.$$

(b) Now use (29.52) to prove that

$$\mathcal{L}[t^{\nu-1} \ln t] = \frac{\Gamma'(\nu) - \Gamma(\nu) \ln s}{s^\nu}.$$

29.27. Using Laplace transform, solve the following initial-value problems

(a) $\dfrac{d^2x}{dt^2} + 4x = \sin t, \qquad x(0) = 1, \ \dot{x}(0) = 0$

(b) $\dfrac{d^2x}{dt^2} - 2\dfrac{dx}{dt} - 3y = te^t, \qquad x(0) = 2, \ \dot{x}(0) = 1$

(c) $\dfrac{d^2x}{dt^2} + \dfrac{dx}{dt} = \theta(1 - t), \qquad x(0) = 1, \ \dot{x}(0) = -1,$ where θ is the step function.

(d) $\dfrac{d^2x}{dt^2} + x = \theta(\pi - t)\cos t, \qquad x(0) = 0, \ \dot{x}(0) = 0,$ where θ is the step function.

29.28. Using Laplace transform, solve the following boundary-value problems

(a) $\dfrac{d^2x}{dt^2} + \omega^2 x = \sin \omega t, \qquad x(0) = 1, \ x(\frac{\pi}{2\omega}) = \pi.$

(b) $\dfrac{d^2x}{dt^2} + \omega^2 x = t, \qquad x(0) = 1, \ \dot{x}(\frac{\pi}{\omega}) = -1.$

29.29. Find $\mathcal{L}^{-1}[\frac{1}{2s^2+2s+5}]$ and $\mathcal{L}^{-1}[\frac{1}{s^2-a^2}]$ using Mellin inversion integral (29.72).

Chapter 30

Calculus of Variations

In a typical multivariable extremum problem, you are given a function of n variables $f(x_1, x_2, \ldots, x_n)$ and asked to find those n values of the variables that maximize or minimize the function. The procedure is, of course, to set the partial derivative of the function with respect to each variable equal to zero and solve the resulting equations.

Geometrically, f is a function in an n-dimensional space, and the problem is to find the point in that space at which f has the highest (or lowest) value compared to the neighboring points. There is another geometric way of looking at the extremum problem. Think of $(x_1, x_2, \ldots, x_n)$ as a piecewise linear *path* in a two-dimensional coordinate system. The horizontal axis is restricted to the values $1, 2, \ldots, n$, and for each of these values i the value of the corresponding variable x_i determines one point with coordinates (i, x_i). Connecting the neighboring points by a straight line segment produces the path. Figure 30.1 shows a couple of such paths.

New way of looking at the multivariable extremum problem

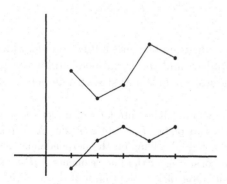

Figure 30.1: For each integer i between 1 and n, pick the real number x_i and draw a point with coordinates (i, x_i). Connect these points to form a path. Two such paths are shown for $n = 5$.

The extremum problem can now be stated in terms of paths: Find the path for which f has either the largest or the smallest value compared with its value at the *neighboring paths*. And to do so, we differentiate with respect to a point of the path. But let's be more general in anticipation of the problems typical of this chapter. Let x_α be a variable where α is not necessarily an integer between 1 and n. Differentiate the function with respect to x_α and set the result equal to zero:

$$\frac{\partial f}{\partial x_\alpha} = \sum_{i=1}^{n} \frac{\partial f}{\partial x_i} \frac{\partial x_i}{\partial x_\alpha} = \sum_{i=1}^{n} \frac{\partial f}{\partial x_i} \delta_{\alpha i} = 0. \tag{30.1}$$

If α is not equal to one of the integers between 1 and n, the sum vanishes identically, i.e., the left-hand side is identically zero because f is not a function of x_α. However, if α is one of the integers between 1 and n, (30.1) gives one of the equations to be solved for determining the extremizing path.

30.1   Variational Problem

Our treatment of the extremum problem above in terms of paths was motivated by situations in which variations of smooth paths are to be considered. A typical **variational problem** has a function whose value depends on the path, i.e., it takes a path and puts out a number. We say that it is a

Functional defined

functional, because its argument is a *function* rather than a set of numbers. If **L** is a functional and $x(t)$ represents a path in the tx-plane, then the value of the functional for this path is represented by $\mathbf{L}[x]$. The most common functional integrates a certain function of $x(t)$ and $\dot{x}(t)$ over some interval (a, b). If $L(x, \dot{x}, t)$ is such a function, then

$$\mathbf{L}[x] = \int_a^b L\big(x(t), \dot{x}(t), t\big)\, dt. \tag{30.2}$$

For every path, the integrand becomes a function of t which can be integrated to give a single number, and the variational problem asks for the path that yields either the largest or the smallest such number.

Example 30.1.1. Before delving into formalism, let's look at a very simple concrete example. Take two points $P_a = (a, y_a)$ and $P_b = (b, y_b)$ in the xy-plane. Consider points $P_Y = (\frac{a+b}{2}, Y)$ lying on the perpendicular bisector of the interval (a, b), and the path consisting of the line segments $\overline{P_a P_Y}$ and $\overline{P_Y P_b}$ as shown in Figure 30.2. For what value of Y is the length of this path minimum?

The length L of the path is given by

$$L = \int_a^b \sqrt{dx^2 + dy^2} = \int_a^b \sqrt{1 + \left(\frac{dy}{dx}\right)^2}\, dx. \tag{30.3}$$

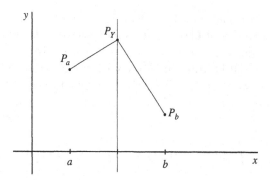

Figure 30.2: Here, a path consists of only two line segments. The middle point P_Y is constrained to move on the vertical line on which it is located to produce different paths.

The equation of the path can be shown to be

$$y(x) = \begin{cases} \dfrac{2(Y - y_a)}{b - a}x + \dfrac{(a + b)y_a - 2aY}{b - a} & \text{if } a < x < (a+b)/2, \\[3ex] \dfrac{2(y_b - Y)}{b - a}x + \dfrac{2bY - (a + b)y_a}{b - a} & \text{if } (a+b)/2 < x < b. \end{cases}$$

Substituting this in the integral gives

$$L = \int_a^{(a+b)/2} \sqrt{1 + \frac{4(Y - y_a)^2}{(b - a)^2}}\, dx + \int_{(a+b)/2}^b \sqrt{1 + \frac{4(y_b - Y)^2}{(b - a)^2}}\, dx$$

$$= \frac{1}{2}\left[\sqrt{(b - a)^2 + 4(Y - y_a)^2} + \sqrt{(b - a)^2 + 4(y_b - Y)^2}\right].$$

Differentiating with respect to Y and setting the result equal to zero leads to the following equation:

$$\frac{Y - y_a}{\sqrt{(b - a)^2 + 4(Y - y_a)^2}} = \frac{y_b - Y}{\sqrt{(b - a)^2 + 4(y_b - Y)^2}}.$$

Square both sides and simplify to get $Y - y_a = y_b - Y$, whose solution is $Y = (y_a + y_b)/2$, placing P_Y on the line joining P_a and P_b. Thus among all the paths $P_a P_Y P_b$ the shortest is the straight line joining P_a and P_b. ∎

30.1.1 Euler-Lagrange Equation

The preceding example showed that from among paths consisting of two specific straight line segments, the one whose middle point lies on the straight line joining the two end points gives the shortest length. What if the point P_Y is *not* on the perpendicular bisector of (a, b), or if the path has more than three points? There is a procedure which picks the minimizing path from among *all possible* paths. Let's discuss this procedure.

Functional
derivative
explained

Going back to Equation (30.2), we ask if there is a process whereby one can take the derivative of $\mathbf{L}[x]$, set it equal to zero, and solve for the desired path. Is there a derivative with respect to a path? To find out, let's see if we can generalize (30.1) from the discrete case of a path consisting of only n points to a *continuous* path. The derivative with respect to a path is called a **functional derivative** and δ is used instead of ∂ to symbolize it. So, let's write

$$\frac{\delta \mathbf{L}[x]}{\delta x(\tau)} = \frac{\delta}{\delta x(\tau)} \int_a^b L\big(x(t), \dot{x}(t), t\big)\, dt = \int_a^b \frac{\delta}{\delta x(\tau)} L\big(x(t), \dot{x}(t), t\big)\, dt. \quad (30.4)$$

In analogy with (30.1), and noting that L is to be considered as an ordinary *function* (not functional) of x, $\dot{x}$, and t, we have

$$\frac{\delta}{\delta x(\tau)} L\big(x(t), \dot{x}(t), t\big) = \frac{\partial L}{\partial x}\frac{\delta x(t)}{\delta x(\tau)} + \frac{\partial L}{\partial \dot{x}}\frac{\delta \dot{x}(t)}{\delta x(\tau)}, \quad (30.5)$$

because t is independent of $x(\tau)$. In the discrete case, we had $\frac{\partial x_i}{\partial x_\alpha} = \delta_{\alpha i}$. What is the generalization of the Kronecker delta to the continuous case? The Dirac delta function! This can be shown more rigorously, but the proof is outside the scope of this book. So, let's write the fundamental functional

A fundamental
functional
derivative

derivative:

$$\frac{\delta x(t)}{\delta x(\tau)} = \delta(t - \tau). \quad (30.6)$$

What about the functional derivative in the second term of (30.5)? Using the definition of the derivative and (30.6), we have

Another
fundamental
functional
derivative

$$\frac{\delta \dot{x}(t)}{\delta x(\tau)} = \frac{\delta}{\delta x(\tau)} \lim_{\epsilon \to 0} \frac{x(t+\epsilon) - x(t)}{\epsilon} = \lim_{\epsilon \to 0}\left[\frac{1}{\epsilon}\left(\frac{\delta x(t+\epsilon)}{\delta x(\tau)} - \frac{\delta x(t)}{\delta x(\tau)}\right)\right]$$
$$= \lim_{\epsilon \to 0}\left[\frac{1}{\epsilon}\big(\delta(t+\epsilon-\tau) - \delta(t-\tau)\big)\right] = \frac{d}{dt}\delta(t-\tau). \quad (30.7)$$

Putting (30.6) and (30.7) in (30.5) and the result in (30.4), we obtain

$$\frac{\delta \mathbf{L}[x]}{\delta x(\tau)} = \int_a^b \left[\frac{\partial L}{\partial x}\delta(t-\tau) + \frac{\partial L}{\partial \dot{x}}\frac{d}{dt}\delta(t-\tau)\right] dt = \frac{\partial L}{\partial x}(\tau) - \frac{d}{d\tau}\frac{\partial L}{\partial \dot{x}}(\tau), \quad (30.8)$$

where in the last step we used the properties of the Dirac delta function and its derivative as given in (5.10) and (5.11). We have assumed that τ lies in the interval (a, b).

Having found the functional derivative, we now equate it to zero and find the equation that determines the path—the function $x(t)$—which extremizes the functional. The equation is

Euler-Lagrange
equation

$$\frac{\partial L}{\partial x} - \frac{d}{dt}\frac{\partial L}{\partial \dot{x}} = 0, \quad (30.9)$$

and is called the **Euler-Lagrange equation**. It is at the heart of all variational problems. If we know the function L, we can differentiate it, substitute the derivatives in (30.9) and solve the resulting differential equation. We should emphasize that a path could be written as $y(x)$ or any other form, depending on the variables used in a particular problem.

Example 30.1.2. Shortest Length Example 30.1.1 looked at *very specific* paths connecting two points and found that the straight-line path minimizes the length. Is this true for *all* paths?

For any path $y(x)$, the length between (a, y_a) and (b, y_b) is given by the (30.3), where the independent variable is x and dependent variable is y. Thus, $L = \sqrt{1 + y'^2}$ and the Euler-Lagrange equation becomes

$$\frac{\partial L}{\partial y} - \frac{d}{dx}\frac{\partial L}{\partial y'} = 0 \quad \text{or} \quad \frac{d}{dx}\left(\frac{y'}{\sqrt{1 + y'^2}}\right) = 0. \qquad (30.10)$$

Differentiating the expression inside the parentheses yields

$$\frac{y''}{(1 + y'^2)^{3/2}} = 0, \quad \text{or} \quad y'' = 0, \quad \text{or} \quad y = cx + d,$$

where c and d are the constants of integration. This is the equation of a straight line. Thus out of all the possible paths between (a, y_a) and (b, y_b), the straight line gives the smallest length. Actually, we don't know if the straight line is the shortest or the longest distance. Euler-Lagrange equation, being the first derivative, is necessary, but not sufficient. As in calculus, to show minimality one has to look at the second derivatives. We shall do this later. ∎

30.1.2   Beltrami identity

Most variational problems have an L which is independent of t. In such a case, the Euler-Lagrange equation simplifies considerably. Consider the total derivative of L with respect to t:

$$\frac{dL}{dt} = \frac{\partial L}{\partial x}\dot{x} + \frac{\partial L}{\partial \dot{x}}\frac{d\dot{x}}{dt}.$$

Substitute for $\partial L/\partial x$ from Euler-Lagrange equation to obtain

$$\frac{dL}{dt} = \dot{x}\frac{d}{dt}\frac{\partial L}{\partial \dot{x}} + \frac{\partial L}{\partial \dot{x}}\frac{d\dot{x}}{dt} = \frac{d}{dt}\left(\dot{x}\frac{\partial L}{\partial \dot{x}}\right), \quad \text{or} \quad \frac{d}{dt}\left(L - \dot{x}\frac{\partial L}{\partial \dot{x}}\right) = 0.$$

This gives the **Beltrami identity**:

$$L - \dot{x}\frac{\partial L}{\partial \dot{x}} = C. \qquad (30.11)$$

Example 30.1.3. The Brachistochrone Problem A bead slides on frictionless bars of various shapes due to gravity. What shape gives the shortest time? This is the famous *brachistochrone problem* which started the calculus of variations. Specifically, consider various paths connecting $P_a = (x_a, y_a)$ and $P_b = (x_b, y_b)$ with $y_b < y_a$. A mass m starts from rest at P_a and moves on a frictionless path from P_a to P_b. Find the equation of the path that yields the shortest time.

For each element ds of the path, the time of travel is $dt = ds/v$, where v is the speed at ds. If ds is located at height y above the ground, then conservation of energy gives

$$mgy_a = \tfrac{1}{2}mv^2 + mgy \quad \text{or} \quad v = \sqrt{2g(y_a - y)}.$$

Therefore,

$$\mathbf{L}[y] = \int_{P_a}^{P_b} \frac{ds}{v} = \int_{P_a}^{P_b} \frac{\sqrt{dx^2 + dy^2}}{\sqrt{2g(y_a - y)}} = \int_{x_a}^{x_b} \sqrt{\frac{1 + y'^2}{2g(y_a - y)}} \, dx,$$

and $L(y, y') = \sqrt{(1 + y'^2)/[2g(y_a - y)]}$. Since L is independent of x, we can use the Beltrami identity:

$$\sqrt{\frac{1 + y'^2}{2g(y_a - y)}} - y' \frac{\partial}{\partial y'} \sqrt{\frac{1 + y'^2}{2g(y_a - y)}} = C, \quad \text{or}$$

$$\sqrt{1 + y'^2} - y' \frac{\partial}{\partial y'} \sqrt{1 + y'^2} = C\sqrt{2g(y_a - y)}.$$

Differentiating and simplifying the left-hand side gives

$$\frac{1}{\sqrt{1 + y'^2}} = C\sqrt{2g(y_a - y)}.$$

Square both sides, introduce a new constant, and solve for y' to get

$$\frac{dy}{dx} = \sqrt{\frac{k}{y_a - y} - 1}.$$

The substitution $u = k/(y_a - y)$ give $dy = (k/u^2)du$ and changes the differential equation to

$$\frac{k}{u^2} \frac{du}{dx} = \sqrt{u - 1} \quad \text{or} \quad \frac{du}{u^2 \sqrt{u - 1}} = \frac{1}{k} dx.$$

Integrating both sides—and using an integral table—yields

$$\frac{x}{k} = \frac{\sqrt{u - 1}}{u} + \tan^{-1}\left(\sqrt{u - 1}\right) + C.$$

As $y \to y_a$, $u \to \infty$ and $x \to x_a$. Therefore, $C = x_a/k - \pi/2$, and the solution becomes

$$\frac{x - x_a}{k} = \frac{\sqrt{u - 1}}{u} + \tan^{-1}\left(\sqrt{u - 1}\right) - \frac{\pi}{2}. \tag{30.12}$$

Let $\tan^{-1}\left(\sqrt{u - 1}\right) = \varphi$. Then $\sqrt{u - 1} = \tan \varphi$ and

$$u = 1 + \tan^2 \varphi = \sec^2 \varphi \quad \text{or} \quad y = y_a - k\cos^2 \varphi. \tag{30.13}$$

Substituting u in terms of φ in (30.12) yields

$$\frac{x - x_a}{k} = \sin \varphi \cos \varphi + \varphi - \frac{\pi}{2}.$$

Finally defining $\theta = 2\varphi - \pi$, this equation and (30.13) give x and y in terms of the parameter θ:

$$x - x_a = \frac{k}{2}(\theta - \sin \theta), \quad y - y_a = -\frac{k}{2}(1 - \cos \theta).$$

This is the parametric equation of a cycloid.   ■

Example 30.1.4. The Soap Film Problem   When a film of soap is stretched across a frame, the surface tension causes the area to be a minimum. The film of Figure 30.3 is an area of revolution with an element of area shown. This element of area is $2\pi y\sqrt{dx^2 + dy^2}$. Therefore, we have to extremize the functional

$$\mathsf{L}[y] = 2\pi \int_0^h y\sqrt{1 + y'^2}\, dx, \quad y(0) = a, \, y(h) = b.$$

Since $L(x, y, y') = y\sqrt{1 + y'^2}$ is independent of x, we can use the Beltrami identity and get

$$y\sqrt{1 + y'^2} - y'\frac{\partial}{\partial y'}\left(y\sqrt{1 + y'^2}\right) = C_1.$$

This yields

$$y = C_1\sqrt{1 + y'^2} \quad \text{or} \quad y' = \sqrt{(y/C_1)^2 - 1}.$$

Let $u = y/C_1$ to simplify this equation to

$$C_1 u' = \sqrt{(u^2 - 1)} \quad \text{or} \quad C_1\frac{du}{\sqrt{(u^2 - 1)}} = dx,$$

which can be easily integrated to give

$$x = C_1 \ln\left(u + \sqrt{u^2 - 1}\right) + C_2 \quad \text{or} \quad u + \sqrt{u^2 - 1} = e^{\frac{x - C_2}{C_1}} \equiv e^v,$$

where v is the exponent of the exponential. From this, we get

$$\sqrt{u^2 - 1} = e^v - u \quad \text{or} \quad u^2 - 1 = e^{2v} - 2ue^v + u^2 \quad \text{or} \quad u = \frac{e^v + e^{-v}}{2} = \cosh v.$$

Returning to y and x, we obtain

$$\frac{y}{C_1} = \cosh\left(\frac{x - C_2}{C_1}\right) \quad \text{or} \quad y = C_1 \cosh\left(\frac{x - C_2}{C_1}\right).$$

The constants C_1 and C_2 can be found by the conditions $y(0) = a$, $y(h) = b$.   ∎

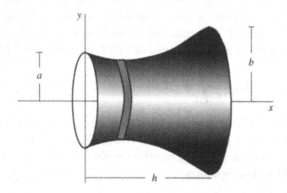

Figure 30.3: The soap film attaches itself to the two rings in such a way that the area obtained is minimum.

30.1.3 Several Dependent Variables

The path of (30.2) had only one dependent variable. One can consider paths in an m-dimensional space where L depends on $\{x_\alpha(t)\}_{\alpha=1}^{m}$ and their derivatives. Such a generalization is straightforward: In (30.4) instead of $x(\tau)$, we have $x_\alpha(\tau)$, which changes (30.5) to

$$\frac{\delta}{\delta x_\alpha(\tau)} L\big(\mathbf{x}(t), \dot{\mathbf{x}}(t)\big) = \sum_{\beta=1}^{m} \left[\frac{\partial L}{\partial x_\beta} \frac{\delta x_\beta(t)}{\delta x_\alpha(\tau)} + \frac{\partial L}{\partial \dot{x}_\beta} \frac{\delta \dot{x}_\beta(t)}{\delta x_\alpha(\tau)} \right],$$

where $\mathbf{x} = (x_1, x_2, \ldots, x_m)$. For this we need the equivalent of (30.6) and (30.7) which are easily shown to be

$$\frac{\delta x_\beta(t)}{\delta x_\alpha(\tau)} = \delta_{\alpha\beta}\delta(t-\tau), \quad \frac{\delta \dot{x}_\beta(t)}{\delta x_\alpha(\tau)} = \delta_{\alpha\beta}\frac{d}{dt}\delta(t-\tau). \tag{30.14}$$

Substituting this in the above sum yields

$$\frac{\delta}{\delta x_\alpha(\tau)} L\big(\mathbf{x}(t), \dot{\mathbf{x}}(t)\big) = \frac{\partial L}{\partial x_\alpha}\delta(t-\tau) + \frac{\partial L}{\partial \dot{x}_\alpha}\frac{d}{dt}\delta(t-\tau),$$

which replaces the x and $\dot{x}$ of (30.8) with x_α and $\dot{x}_\alpha$. We thus obtain the multivariable version of the Euler-Lagrange equation:

$$\frac{\partial L}{\partial x_\alpha} - \frac{d}{dt}\frac{\partial L}{\partial \dot{x}_\alpha} = 0, \quad \alpha = 1, 2, \ldots, m. \tag{30.15}$$

30.1.4 Several Independent Variables

Equation (30.15) is one generalization of the Euler-Lagrange equation. It still corresponds to a path, a (generally) curved *line*, albeit in a multi-dimensional space. There is another generalization which is also important: going from a path to a surface. In this case, our dependent variable is a function of several independent variables. So, consider a function ϕ of m variables which we collectively denote by $\mathbf{x}$, and instead of (30.2) consider the functional

$$\mathbf{L}[\phi] = \iint_\Omega d^m x \mathcal{L}(\phi; \phi_{,1}, \phi_{,2}, \ldots, \phi_{,m}; \mathbf{x}), \tag{30.16}$$

where $\phi_{,\alpha}$ denotes the derivative of ϕ with respect to x_α, and Ω is some region in the m-dimensional space. Note the change in notation: we use $\mathcal{L}$ instead of L when integration is over a multidimensional "volume." The variational derivative (30.5) now becomes

$$\frac{\delta}{\delta\phi(\mathbf{y})}\mathcal{L}\big(\phi; \phi_{,1}, \phi_{,2}, \ldots, \phi_{,m}; \mathbf{x}\big) = \frac{\partial\mathcal{L}}{\partial\phi}\frac{\delta\phi(\mathbf{x})}{\delta\phi(\mathbf{y})} + \sum_{\alpha=1}^{m}\frac{\partial\mathcal{L}}{\partial\phi_{,\alpha}}\frac{\delta\phi_{,\alpha}}{\delta\phi(\mathbf{y})}. \tag{30.17}$$

Furthermore, (30.6) and (30.7) generalize to

More fundamental
functional
derivatives

$$\frac{\delta\phi(\mathbf{x})}{\delta\phi(\mathbf{y})} = \delta(\mathbf{x}-\mathbf{y}), \quad \frac{\delta\phi_{,\alpha}(\mathbf{x})}{\delta\phi(\mathbf{y})} = \frac{\partial}{\partial x_\alpha}\delta(\mathbf{x}-\mathbf{y}). \tag{30.18}$$

Substituting these in the functional derivative of the integral (30.16) and setting the result equal to zero yields another Euler-Lagrange equation:

$$\frac{\partial\mathcal{L}}{\partial\phi} - \sum_{\alpha=1}^{m}\frac{\partial}{\partial x_\alpha}\frac{\partial\mathcal{L}}{\partial\phi_{,\alpha}} = 0. \tag{30.19}$$

Finally if we have several dependent variables $\{\phi^i\}_{i=1}^{N}$, collectively represented by $\mathbf{\Phi}$, and several independent variables $\{x_\alpha\}_{\alpha=1}^{m}$, collectively represented by $\mathbf{x}$, then the variational functional becomes

$$\mathbf{L}[\mathbf{\Phi}] = \int_\Omega d^m x \mathcal{L}(\mathbf{\Phi};\mathbf{\Phi}_{,1},\mathbf{\Phi}_{,2},\ldots,\mathbf{\Phi}_{,m};\mathbf{x}), \tag{30.20}$$

with the variational derivatives

$$\frac{\delta}{\delta\phi^i(\mathbf{y})}\mathcal{L}\big(\mathbf{\Phi};\mathbf{\Phi}_{,1},\mathbf{\Phi}_{,2},\ldots,\mathbf{\Phi}_{,m};\mathbf{x}\big) = \sum_{j=1}^{N}\frac{\partial\mathcal{L}}{\partial\phi^j}\frac{\delta\phi^j(\mathbf{x})}{\delta\phi^i(\mathbf{y})} + \sum_{j=1}^{N}\sum_{\alpha=1}^{m}\frac{\partial\mathcal{L}}{\partial\phi_{,\alpha}^j}\frac{\delta\phi_{,\alpha}^j}{\delta\phi^i(\mathbf{y})}, \tag{30.21}$$

and

... and more
fundamental
functional
derivatives

$$\frac{\delta\phi^j(\mathbf{x})}{\delta\phi^i(\mathbf{y})} = \delta_{ij}\delta(\mathbf{x}-\mathbf{y}), \quad \frac{\delta\phi_{,\alpha}^j(\mathbf{x})}{\delta\phi^i(\mathbf{y})} = \delta_{ij}\frac{\partial}{\partial x_\alpha}\delta(\mathbf{x}-\mathbf{y}). \tag{30.22}$$

Substitution of these in (30.20) leads to the Euler-Lagrange equations

$$\frac{\partial\mathcal{L}}{\partial\phi^i} - \sum_{\alpha=1}^{m}\frac{\partial}{\partial x_\alpha}\frac{\partial\mathcal{L}}{\partial\phi_{,\alpha}^i} = 0, \quad i=1,2,\ldots,N. \tag{30.23}$$

In many situations, the variational problem consists of various parts each having one or several dependent or independent variables.

30.1.5   Second Variation

Euler-Lagrange equation was obtained by setting the *first variational derivative* (30.8) equal to zero. As in the multivariable calculus, this only finds the *extremum*. And just as in the multivariable calculus, to see if we have a minimum or a maximum, we have to run the second derivative test.

The easiest way to apply the second derivative test in calculus is to consider the Taylor expansion of the function. And since we are interested in *local* minima and maxima, we ignore the third and higher orders in the Taylor expansion. Now recall from Section 10.7 that the Taylor series of a function

f of N independent variables $\left\{x^i\right\}_{i=1}^N \equiv \mathbf{x}$ up to the second order around $\mathbf{x}_0$ is

$$f(\mathbf{x}) = f(\mathbf{x}_0) + \sum_{i=1}^N (x_i - x_{0i}) f_{,i}(\mathbf{x}_0) + \frac{1}{2} \sum_{i,j=1}^N (x_i - x_{0i})(x_j - x_{0j}) f_{,ij}(\mathbf{x}_0), \quad (30.24)$$

where

$$f_{,i} \equiv \frac{\partial f}{\partial x_i} \qquad \text{and} \qquad f_{,ij} \equiv \frac{\partial^2 f}{\partial x_i \partial x_j}.$$

If $\mathbf{x}_0$ is an extremum of f, then $f_{,i}(\mathbf{x}_0) = 0$ and the above equation becomes

$$f(\mathbf{x}) - f(\mathbf{x}_0) = \frac{1}{2} \sum_{i,j=1}^N (x_i - x_{0i})(x_j - x_{0j}) f_{,ij}(\mathbf{x}_0) \equiv \delta_2 f(\mathbf{x}_0), \qquad (30.25)$$

where we introduced the abbreviation $\delta_2 f(\mathbf{x}_0)$—the second variation of f at $\mathbf{x}_0$—for the sum. The test for maximum or minimum of f can now be stated: If for *any* $\mathbf{x}$ that is close enough to $\mathbf{x}_0$, the second variation $\delta_2 f(\mathbf{x}_0)$ is positive, then $\mathbf{x}_0$ is a minimum point, and if $\delta_2 f(\mathbf{x}_0)$ is negative, then $\mathbf{x}_0$ is a maximum point.

The generalization to the variational problem follows from our usual passage from the discrete to the continuous. For the most general integral (30.20), the second variation is

$$\delta_2 \mathsf{L}[\mathbf{\Phi}_0] = \frac{1}{2} \sum_{i,j=1}^N \iint_\Omega d^m x \iint_\Omega d^m y \left(\phi^i(\mathbf{x}) - \phi_0^i(\mathbf{x}) \right) \left(\phi^i(\mathbf{y}) - \phi_0^i(\mathbf{y}) \right)$$
$$\frac{\delta^2 \mathsf{L}}{\delta \phi^i(\mathbf{x}) \delta \phi^j(\mathbf{y})} [\mathbf{\Phi}_0], \qquad (30.26)$$

where the last term means "find the second variational derivative and evaluate the result at the solution $\mathbf{\Phi}_0$ of the Euler-Lagrange equation." For a single dependent variable and several independent variables this becomes

$$\delta_2 \mathsf{L}[\phi_0] = \frac{1}{2} \iint_\Omega d^m x \iint_\Omega d^m y \left(\phi(\mathbf{x}) - \phi_0(\mathbf{x}) \right) \left(\phi(\mathbf{y}) - \phi_0(\mathbf{y}) \right) \frac{\delta^2 \mathsf{L}}{\delta \phi(\mathbf{x}) \delta \phi(\mathbf{y})} [\phi_0],$$
$$(30.27)$$

and for a single independent variable and several dependent variables we get

$$\delta_2 \mathsf{L}[\mathbf{x}_0] = \frac{1}{2} \sum_{i,j=1}^N \int_a^b dt \int_a^b d\tau \left(x_i(t) - x_{i0}(t) \right) \left(x_j(\tau) - x_{j0}(\tau) \right) \frac{\delta^2 \mathsf{L}}{\delta x_i(t) \delta x_j(\tau)} [\mathbf{x}_0],$$
$$(30.28)$$

and for the simplest case of a single independent variable and a single dependent variable (30.26) reduces to

$$\delta_2 \mathsf{L}[x_0] = \frac{1}{2} \int_a^b dt \int_a^b d\tau \left(x(t) - x_0(t) \right) \left(x(\tau) - x_0(\tau) \right) \frac{\delta^2 \mathsf{L}}{\delta x(t) \delta x(\tau)} [x_0]. \quad (30.29)$$

In the calculation of the second variation, we need to find the variational derivatives of second derivatives of dependent variables. It is not hard to show that

$$\frac{\delta\phi^j_{,\alpha\beta}(\mathbf{x})}{\delta\phi^i(\mathbf{y})} = \delta_{ij}\frac{\partial^2}{\partial x_\beta\partial x_\alpha}\delta(\mathbf{x}-\mathbf{y}).$$ (30.30)

Fundamental functional derivatives involving second partial derivatives

Example 30.1.5. The necessary condition for the straight line to be the shortest distance between two given points is that it satisfies the Euler-Lagrange equation (30.10). Example 30.1.2 showed that $y_0(x) = cx+d$ solves the Euler-Lagrange equation. To see if this is minimum or not, calculate the second variation (30.29). The first derivative is given by (30.8), which with the current symbols for independent and dependent variables, becomes

$$\frac{\delta L[y]}{\delta y(x)} = \frac{\partial L}{\partial y}(x) - \frac{d}{dx}\frac{\partial L}{\partial y'}(x) = -\frac{d}{dx}\left(\frac{y'}{\sqrt{1+y'^2}}\right) = -\frac{y''}{(1+y'^2)^{3/2}},$$

and

$$\frac{\delta^2 L[y]}{\delta y(x')\delta y(x)} = -\frac{\delta}{\delta y(x')}\left\{\frac{y''}{(1+y'^2)^{3/2}}\right\} = -\frac{\delta}{\delta y(x')}\left\{y''\left(1+y'^2\right)^{-3/2}\right\},$$

or

$$\frac{\delta^2 L[y]}{\delta y(x')\delta y(x)} = -\frac{\delta y''(x)}{\delta y(x')}\left(1+y'^2\right)^{-3/2} - y''\frac{\delta}{\delta y(x')}\left(1+y'^2\right)^{-3/2}.$$

Using (30.30), this yields

$$\frac{\delta^2 L[y]}{\delta y(x')\delta y(x)} = -\frac{\delta''(x-x')}{(1+y'^2)^{3/2}} + y''\frac{3}{2}(2y')\left(1+y'^2\right)^{-5/2}\frac{\delta y'(x)}{\delta y(x')}$$

$$= -\frac{\delta''(x-x')}{(1+y'^2)^{3/2}} + \frac{3y'y''\delta'(x-x')}{(1+y'^2)^{5/2}}.$$

Now we have to evaluate this at the solution $y_0(x)$ of the Euler-Lagrange equation for which $y_0' = c$ and $y_0'' = 0$. Thus,

$$\frac{\delta^2 L[y]}{\delta y(x')\delta y(x)}[y_0] = -\frac{\delta''(x-x')}{(1+c^2)^{3/2}}.$$

Substituting this in (30.29) and using the derivative property (5.12) of the Dirac delta function yields

$$\delta_2 L[y_0] = -\frac{1}{2(1+c^2)^{3/2}}\int_a^b dx\int_a^b dx'\left(y(x)-y_0(x)\right)\left(y(x')-y_0(x')\right)\delta''(x-x')$$

$$= -\frac{1}{2(1+c^2)^{3/2}}\int_a^b dx\left(y(x)-y_0(x)\right)\frac{d^2}{dx^2}\left(y(x)-y_0(x)\right).$$

The last integral can be integrated by parts to give

$$\underbrace{\left(y(x)-y_0(x)\right)\frac{d}{dx}\left(y(x)-y_0(x)\right)\Big|_a^b}_{=0 \text{ because } y(a)=y_0(a),\, y(b)=y_0(b)} - \int_a^b dx\left\{\frac{d}{dx}\left(y(x)-y_0(x)\right)\right\}^2.$$

Therefore,

$$\delta_2 \mathsf{L}[y_0] = \frac{1}{2(1+c^2)^{3/2}} \int_a^b dx \left\{ \frac{d}{dx} \big(y(x) - y_0(x) \big) \right\}^2 ,$$

which is a manifestly positive quantity for *any* $y(x)$. Hence, $y_0(x) = cx + d$ does indeed *minimize* the distance between any two given points. ∎

We should emphasize that although the calculation of the second variational derivative is rather straightforward, showing that the second variation $\delta_2 \mathsf{L}$—the integral of the second variational derivative as given in Equations (30.26) to (30.29)—is positive or negative is by no means trivial. Example 30.1.5 is one of those rare cases where the calculation of $\delta_2 \mathsf{L}$ is manageable.

30.1.6   Variational Problems with Constraints

The variational problems treated so far have been problems with boundary conditions, namely that all "paths," or extremal candidates, must go through the same boundary. In many applications, there are other auxiliary conditions or **constraints** that the extremal candidates must obey. A typical example is the problem of finding the closed curve of the largest area when the perimeter is a given fixed length. The most elegant way of treating the constrained variational problems is via Lagrange multipliers discussed in Section 12.3.1.

Isoperimetric problem

Suppose that we are looking for a curve that not only extremizes $\mathsf{L}[x]$ of (30.2), but also is such that another functional,

$$\mathsf{K}[x] = \int_a^b G\big(x(t), \dot{x}(t), t\big) \, dt, \tag{30.31}$$

takes a fixed value l. Such a problem is called **isoperimetric**. In exact analogy with the multivariable calculus, we form a new function $L + \lambda G$ and extremize that function. This means that we have to solve the Euler-Lagrange equation

$$\frac{\partial L}{\partial x} - \frac{d}{dt}\frac{\partial L}{\partial \dot{x}} + \lambda \left(\frac{\partial G}{\partial x} - \frac{d}{dt}\frac{\partial G}{\partial \dot{x}} \right) = 0. \tag{30.32}$$

Example 30.1.6. As an example of the isoperimetric variational problem, consider all curves of length l in the upper half plane passing through the points $(-a, 0)$ and $(a, 0)$. What is the equation of the curve that together with the interval $[-a, a]$ encloses the largest area? The sought-after function $y(x)$ must extremize

$$\mathsf{L}[y] = \int_{-a}^a y\,dx,$$

subject to the condition that

$$y(-a) = 0 = y(a), \qquad \mathsf{K}[y] = \int_{-a}^a \sqrt{1 + y'^2}\, dx = l.$$

Equation (30.32) with $L = y$ and $G = \sqrt{1 + y'^2}$ gives

$$1 + \lambda \frac{d}{dx} \frac{y'}{\sqrt{1 + y'^2}} = 0.$$

Integrating this yields

$$x + \lambda \frac{y'}{\sqrt{1+y'^2}} = C_1,$$

which can be solved for y' to give

$$y' = \pm \frac{C_1 - x}{\sqrt{\lambda^2 - (C_1 - x)^2}},$$

whose solution is

$$y = \pm \sqrt{\lambda^2 - (C_1 - x)^2} + C_2,$$

or

$$(x - C_1)^2 + (y - C_2)^2 = \lambda^2.$$

This is a circle of radius λ. The values of the three unknowns C_1, C_2, and λ are determined from the conditions

$$y(-a) = 0 = y(a), \qquad \mathbf{K}[y] = l. \qquad\qquad \blacksquare$$

There is another type of variational problem with constraint applicable to the case of one independent and several dependent variables, in which the constraint is given by an equation of the form

$$g\big(\mathbf{x}(t), \dot{\mathbf{x}}(t), t\big) = 0$$

Finite constraint problem

This is called the **finite constraint problem** and is similar to Equation (12.31) where the discrete index j has been replaced with the continuous index t. Thus, the Lagrange multipliers λ_j should be replaced with $\lambda(t)$ and the sum in (12.32) replaced with an integral over t, which is already present in the extremal problem. Therefore, the problem changes to finding the extremum of

$$\int_a^b \left\{ L\big(\mathbf{x}(t), \dot{\mathbf{x}}(t), t\big) + \lambda(t) g\big(\mathbf{x}(t), \dot{\mathbf{x}}(t), t\big) \right\} \, dt, \qquad (30.33)$$

and the Euler-Lagrange equation becomes

$$\frac{\partial L}{\partial x_i} - \frac{d}{dt}\frac{\partial L}{\partial \dot{x}_i} + \lambda\left(\frac{\partial g}{\partial x_i} - \frac{d}{dt}\frac{\partial g}{\partial \dot{x}_i} \right) - \frac{d\lambda}{dt}\frac{\partial g}{\partial \dot{x}_i} = 0, \quad i = 1, 2, \ldots, N. \quad (30.34)$$

If there are multiple constraint equations,

$$g_\alpha\big(\mathbf{x}(t), \dot{\mathbf{x}}(t), t\big) = 0, \quad \alpha = 1, 2, \ldots, m,$$

then there will be m Lagrange multipliers and a sum over α in (30.33),

$$\int_a^b \left\{ L\big(\mathbf{x}(t), \dot{\mathbf{x}}(t), t\big) + \sum_{\alpha=1}^m \lambda_\alpha(t) g_\alpha\big(\mathbf{x}(t), \dot{\mathbf{x}}(t), t\big) \right\} \, dt, \qquad (30.35)$$

leading to the following Euler-Lagrange equation:

$$\frac{\partial L}{\partial x_i} - \frac{d}{dt}\frac{\partial L}{\partial \dot{x}_i} + \sum_{\alpha=1}^m \left\{ \lambda_\alpha\left(\frac{\partial g_\alpha}{\partial x_i} - \frac{d}{dt}\frac{\partial g_\alpha}{\partial \dot{x}_i} \right) - \frac{d\lambda_\alpha}{dt}\frac{\partial g}{\partial \dot{x}_i} \right\} = 0, \quad i = 1, 2, \ldots, N.$$

$$(30.36)$$

Example 30.1.7. Among all curves lying on the sphere centered at the origin and of radius a and passing through two points (x_1, y_1, z_1) and (x_2, y_2, z_2), find the shortest one. This is a finite constraint problem with

$$\mathbf{L}[y, z] = \int_{x_1}^{x_2} \sqrt{1 + y'^2 + z'^2}\, dx$$

and

$$g(x, y, z) = x^2 + y^2 + z^2 - a^2.$$

The solution is the set of functions $\{y(x), z(x)\}$ which extremize the integral

$$\int_{x_1}^{x_2} \left\{ \sqrt{1 + y'^2 + z'^2} + \lambda(x)(x^2 + y^2 + z^2 - a^2) \right\} dx,$$

i.e., functions that satisfy the Euler-Lagrange equations

$$2y\lambda(x) - \frac{d}{dx} \frac{y'}{\sqrt{1 + y'^2 + z'^2}} = 0,$$

$$2z\lambda(x) - \frac{d}{dx} \frac{z'}{\sqrt{1 + y'^2 + z'^2}} = 0.$$

Solving these equations, we get the solutions in terms of four constants which can be determined from the boundary conditions

$$y(x_1) = y_1, \qquad y(x_2) = y_2,$$
$$z(x_1) = z_1, \qquad z(x_2) = z_2. \qquad \blacksquare$$

30.2   Lagrangian Dynamics

Variational calculus has become an indispensable tool in physics. Almost all (partial) differential equations of physics can be derived from some variational problem. Furthermore, symmetry considerations, which are the cornerstones of modern fundamental physics, find their natural settings in functionals. And a very elegant and powerful formulation of quantum mechanics done by Richard Feynman uses the variational techniques.

30.2.1   From Newton to Lagrange

For most conservative systems one can define functionals whose extremization leads to differential equations of motion of those systems. The second law of motion for a particle acted on by a conservative force can be written as

$$-\nabla\Phi = m\frac{d\mathbf{v}}{dt} \quad \text{or} \quad -\frac{\partial\Phi}{\partial x_i} = m\frac{d\dot{x}_i}{dt} \quad \text{or} \quad \frac{\partial}{\partial x_i}(-\Phi) - \frac{d}{dt}(m\dot{x}_i) = 0.$$
$$\tag{30.37}$$

This looks very much like (30.15)! Let's see if we can construct an L that leads to the equations of mechanics. Use x, y, and z for the moment with $n = 3$.

By equating the first term of (30.15) to the first term of the last equation in (30.37), we get

$$\frac{\partial L}{\partial x} = \frac{\partial}{\partial x}(-\Phi).$$

Antidifferentiation yields $L = -\Phi(x, y, z) + f(y, z, \dot{x}, \dot{y}, \dot{z})$, where f is the "constant" of integration. If the partials of L with respect to y and z are to be equal to the corresponding partials of $-\Phi$, then f cannot depend on y and z. So, f is a function of velocity components. If the second term of (30.37) is to equal the second term of (30.15), then

$$m\dot{x} = \frac{\partial L}{\partial \dot{x}} = \frac{\partial f}{\partial \dot{x}} \quad \text{or} \quad f(\dot{x}, \dot{y}, \dot{z}) = \tfrac{1}{2} m\dot{x}^2 + g(\dot{y}, \dot{z}),$$

where $g(\dot{y}, \dot{z})$ is the "constant" of this new integration. Applying the same argument to y and z, we conclude that f is just the kinetic energy. Therefore, we arrive at the important conclusion that for a single particle with position vector $\mathbf{r}$, the extremization of

$$L(\mathbf{r}, \dot{\mathbf{r}}, t) = -\Phi(\mathbf{r}) + \tfrac{1}{2} m \left|\dot{\mathbf{r}}\right|^2 = -\Phi(x, y, z) + \tfrac{1}{2} m \left(\dot{x}^2 + \dot{y}^2 + \dot{z}^2\right) \qquad (30.38)$$

gives the equation of motion of the particle. $L(\mathbf{r}, \dot{\mathbf{r}}, t)$ is called the **Lagrangian** of a single particle moving in potential Φ.

For N *non-interacting* particles in an *external* potential, the Lagrangian is the sum of the single-particle Lagrangians:

$$L = \sum_{i=1}^{N} L_i = \sum_{i=1}^{N} \left(-\Phi_i + \tfrac{1}{2} m_i \left|\dot{\mathbf{r}}_i\right|^2\right) = \sum_{i=1}^{N} \left[-\Phi_i + \tfrac{1}{2} m_i \left(\dot{x}_i^2 + \dot{y}_i^2 + \dot{z}_i^2\right)\right],$$

where $\Phi_i = \Phi(x_i, y_i, z_i)$. Note that this can be written as

$$L = KE - \Phi, \text{ where } KE = \sum_{i=1}^{N} \tfrac{1}{2} m_i \left(\dot{x}_i^2 + \dot{y}_i^2 + \dot{z}_i^2\right), \text{ and } \Phi = \sum_{i=1}^{N} \Phi_i.$$

$$(30.39)$$

If the particles are interacting, then Φ is no longer the sum of individual potentials, but a general function of all coordinates. It is therefore common to collect all the N triple coordinates into one big $3N$-component vector $\mathbf{q}$ and call it the **generalized coordinates** vector. Then the Lagrangian is written as

$$L(\mathbf{q}, \dot{\mathbf{q}}, t) = KE - \Phi = \sum_{i=1}^{3N} \tfrac{1}{2} \mu_i \dot{q}_i^2 - \Phi(q_1, q_2, \ldots, q_{3N}). \qquad (30.40)$$

We changed the mass to μ_i to avoid confusion with the m_i of the previous equation. For example, for three particles interacting gravitationally,

$$\Phi(q_1, q_2, \ldots, q_9) = \Phi(\mathbf{r}_1, \mathbf{r}_2, \mathbf{r}_3) = -\frac{Gm_1 m_2}{|\mathbf{r}_1 - \mathbf{r}_2|} - \frac{Gm_1 m_3}{|\mathbf{r}_1 - \mathbf{r}_3|} - \frac{Gm_2 m_3}{|\mathbf{r}_2 - \mathbf{r}_3|},$$

which can be written in terms of the q's, once the latter are defined in terms of the position vectors. Note that many of the μ_i's in (30.40) are equal. For instance, if $q_1 = x_1$, $q_2 = y_1$, and $q_3 = z_1$, then $\mu_1 = \mu_2 = \mu_3 = m_1$, etc.

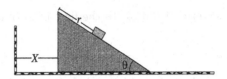

Figure 30.4: The inclined plane moves as m moves on it.

Example 30.2.1. A block of mass m slides on a frictionless inclined plane, which has mass M and moves on a frictionless horizontal surface as shown in Figure 30.4. The position of the incline is denoted by X and that of the block by r, or (x, y) with

$$x = X + r\cos\theta, \quad y = (l - r)\sin\theta,$$

where l is the length of the inclined plane. The kinetic energy of the system is

$$\begin{aligned}
KE &= \tfrac{1}{2}M\dot{X}^2 + \tfrac{1}{2}m\left(\dot{x}^2 + \dot{y}^2\right) \\
&= \tfrac{1}{2}M\dot{X}^2 + \tfrac{1}{2}m\left[\left(\dot{X} + \dot{r}\cos\theta\right)^2 + \dot{r}^2\sin^2\theta\right] \\
&= \tfrac{1}{2}M\dot{X}^2 + \tfrac{1}{2}m\left(\dot{X}^2 + \dot{r}^2 + 2\dot{X}\dot{r}\cos\theta\right),
\end{aligned}$$

and the potential energy

$$\Phi = mgy = mg(l - r)\sin\theta,$$

giving rise to the Lagrangian

$$L = \tfrac{1}{2}M\dot{X}^2 + \tfrac{1}{2}m\left(\dot{X}^2 + \dot{r}^2 + 2\dot{X}\dot{r}\cos\theta\right) - mg(l - r)\sin\theta.$$

The equations of motion

$$\frac{\partial L}{\partial X} - \frac{d}{dt}\left(\frac{\partial L}{\partial \dot{X}}\right) = 0, \quad \frac{\partial L}{\partial r} - \frac{d}{dt}\left(\frac{\partial L}{\partial \dot{r}}\right) = 0,$$

can now be calculated:

$$-M\ddot{X} - m\left(\ddot{X} + \ddot{r}\cos\theta\right) = 0, \quad mg\sin\theta - m\left(\ddot{r} + \ddot{X}\cos\theta\right) = 0.$$

Solving for the two accelerations, we get

$$\ddot{X} = \frac{-mg\sin\theta\cos\theta}{M + m\sin^2\theta}, \quad \ddot{r} = \frac{(m + M)g\sin\theta}{M + m\sin^2\theta}.$$

Note that for an infinitely heavy inclined plane, $\ddot{X} = 0$ and $\ddot{r} = g\sin\theta$, as expected. ∎

Historical Notes

was born Giuseppe Luigi Lagrangia but adopted the French version of his name. He was the eldest of eleven children, most of whom did not reach adulthood. His father destined him for the law—a profession that one of his brothers later pursued—and Lagrange offered no objections. But having begun the study of physics and geometry, he quickly became aware of his talents and henceforth devoted himself to the exact sciences. Attracted first by geometry, at the age of seventeen he turned to analysis, then a rapidly developing field.

In 1755, in a letter to the geometer Giulio da Fagnano, Lagrange speaks of one of Euler's papers published at Lausanne and Geneva in 1744. The same letter shows that as early as the end of 1754 Lagrange had found interesting results in this area, which was to become the *calculus of variations* (a term coined by Euler in 1766). In the same year, Lagrange sent Euler a summary, written in Latin, of the purely analytical method that he used for this type of problem. Euler replied to Lagrange that he was very interested in the technique. Lagrange's merit was likewise recognized in Turin; and he was named, by a royal decree, professor at the Royal Artillery School with an annual salary of 250 crowns—a sum never increased in all the years he remained in his native country. Many years later, in a letter to d´Alembert, Lagrange confirmed that this method of maxima and minima was the first fruit of his studies—he was only nineteen when he devised it—and that he regarded it as his best work in mathematics.

Joseph Louis
Lagrange
1736–1813

In 1756, in a letter to Euler that has been lost, Lagrange, applying the calculus of variations to mechanics, generalized Euler's earlier work on the trajectory described by a material point subject to the influence of central forces to an arbitrary system of bodies, and derived from it a procedure for solving all the problems of dynamics.

In 1757 some young Turin scientists, among them Lagrange, founded a scientific society that was the origin of the Royal Academy of Sciences of Turin. One of the main goals of this society was the publication of a miscellany in French and Latin, *Miscellanea Taurinensia ou Mélanges de Turin*, to which Lagrange contributed fundamentally. These contributions included works on the calculus of variations, probability, vibrating strings, and the principle of least action.

To enter a competition for a prize, in 1763 Lagrange sent to the Paris Academy of Sciences a memoir in which he provided a satisfactory explanation of the translational motion of the moon. In the meantime, the Marquis Caraccioli, ambassador from the kingdom of Naples to the court of Turin, was transferred by his government to London. He took along the young Lagrange, who until then seems never to have left the immediate vicinity of Turin. Lagrange was warmly received in Paris, where he had been preceded by his memoir on lunar libration. He may perhaps have been treated too well in the Paris scientific community, where austerity was not a leading virtue. Being of a delicate constitution, Lagrange fell ill and had to interrupt his trip. In the spring of 1765 Lagrange returned to Turin by way of Geneva.

In the autumn of 1765 d´Alembert, who was on excellent terms with Frederick II of Prussia, and familiar with Lagrange's work through *Mélanges de Turin*, suggested to Lagrange that he accept the vacant position in Berlin created by Euler's departure for St. Petersburg. It seems quite likely that Lagrange would gladly have remained in Turin had the court of Turin been willing to improve his material and scientific situation. On 26 April, d´Alembert transmitted to Lagrange the very precise and advantageous propositions of the king of Prussia. Lagrange accepted the proposals of the Prussian king and, not without difficulties, obtained his leave through the intercession of Frederick II with the king of Sardinia. Eleven months after his arrival in Berlin, Lagrange married his cousin Vittoria Conti who died in 1783 after a long illness. With the death of Frederick II in August 1786 he also lost his strongest support in Berlin. Advised of the situation, the princes of Italy zealously competed in attracting him to their courts. In the meantime the French government decided to bring Lagrange to Paris through an advantageous offer. Of all the candidates, Paris was victorious.

Lagrange left Berlin on 18 May 1787 to become *pensionnaire vétéran* of the Paris Academy of Sciences, of which he had been a foreign associate member since 1772. Warmly welcomed in Paris, he experienced a certain lassitude and did not immediately resume his research. Yet he astonished those around him by his extensive knowledge of metaphysics, history, religion, linguistics, medicine, and botany.

In 1792 Lagrange married the daughter of his colleague at the Academy, the astronomer Pierre Charles Le Monnier. This was a troubled period, about a year after the flight of the king and his arrest at Varennes. Nevertheless, on 3 June the royal family signed the marriage contract "as a sign of its agreement to the union." Lagrange had no children from this second marriage, which, like the first, was a happy one.

When the academy was suppressed in 1793, many noted scientists, including Lavoisier, Laplace, and Coulomb were purged from its membership; but Lagrange remained as its chairman. For the next ten years, Lagrange survived the turmoil of the aftermath of the French Revolution, but by March of 1813, he became seriously ill. He died on the morning of 11 April 1813, and three days later his body was carried to the Panthéon. The funeral oration was given by Laplace in the name of the Senate.

30.2.2 Lagrangian Densities

Particles are localized objects (indeed mathematical points), whose trajectories, determined by *ordinary* differential equations, describe curves in space. A Lagrangian of the form (30.40), with one independent variable (time), is therefore appropriate for particles.

Most of physical quantities, however, are not particles, but fields, which are not localized. In order to apply the variational techniques to fields, one has to consider a **Lagrangian density** $\mathcal{L}$, whose integral over some volume gives the Lagrangian, which can now be integrated over time as in (30.2). Thus, in field theories, the integration is over the 4-dimensional spacetime, a natural setting for relativity—which is very relevant because most field theories are relativistic—to operate. A physical field usually has several components, making Equation (30.23) relevant to the situation.

Electrodynamics Lagrangian

Section 17.3.2 derived the electromagnetic field tensor $F_{\alpha\beta}$ and wrote the four Maxwell's equations in terms of it. Since $F_{\alpha\beta}$ seems to be so fundamental, and the variational techniques seem to yield the (partial) differential equations of physics, there may be a chance that electrodynamics can be described by a Lagrangian density. In the language of tensors, a Lagrangian density is a scalar. Thus, we have to construct a scalar out of $F_{\alpha\beta}$. The simplest such scalar is $F^{\alpha\beta}F_{\alpha\beta}$. Equation (17.47) showed that the field tensor can be written as derivatives of the 4-potential A_α, which is therefore more "fundamental" than $F_{\alpha\beta}$. There is another 4-vector appearing in Maxwell's equations, namely the 4-current J_α. Thus, by taking the dot product $J^\alpha A_\alpha$, we form another scalar. We therefore write

$$\mathcal{L} = aF^{\alpha\beta}F_{\alpha\beta} + bJ^{\alpha}A_{\alpha} = a\eta^{\alpha\mu}\eta^{\beta\nu}F_{\mu\nu}F_{\alpha\beta} + bJ^{\alpha}A_{\alpha},$$

where a and b are to be determined later. Writing the field tensor in terms of the 4-potential, we get

$$\mathcal{L} = a\eta^{\alpha\mu}\eta^{\beta\nu}(\partial_{\mu}A_{\nu} - \partial_{\nu}A_{\mu})(\partial_{\alpha}A_{\beta} - \partial_{\beta}A_{\alpha}) + bJ^{\alpha}A_{\alpha}$$
$$\equiv a\eta^{\alpha\mu}\eta^{\beta\nu}(A_{\nu,\mu} - A_{\mu,\nu})(A_{\beta,\alpha} - A_{\alpha,\beta}) + bJ^{\alpha}A_{\alpha}. \qquad (30.41)$$

The Euler-Lagrange equation for A_{α} can be written as

$$\frac{\partial\mathcal{L}}{\partial A_{\sigma}} - \frac{\partial}{\partial x^{\rho}}\frac{\partial\mathcal{L}}{\partial A_{\sigma,\rho}} = 0. \qquad (30.42)$$

The first term is easy to calculate:

$$\frac{\partial\mathcal{L}}{\partial A_{\sigma}} = bJ^{\alpha}\frac{\partial A_{\alpha}}{\partial A_{\sigma}} = bJ^{\alpha}\delta^{\sigma}_{\alpha} = bJ^{\sigma}.$$

The second term is only slightly more complicated once we realize that

$$\frac{\partial A_{\beta,\alpha}}{\partial A_{\sigma,\rho}} = \delta^{\sigma}_{\beta}\delta^{\rho}_{\alpha}.$$

With this in mind, the second term of (30.42) can be shown to be

$$\frac{\partial}{\partial x^{\rho}}\frac{\partial\mathcal{L}}{\partial A_{\sigma,\rho}} = 4a\partial_{\rho}\left(\partial^{\rho}A^{\sigma} - \partial^{\sigma}A^{\rho}\right) = 4a\partial_{\rho}F^{\rho\sigma},$$

and (30.42) becomes

$$4a\partial_{\rho}F^{\rho\sigma} = bJ^{\sigma} \quad \text{or} \quad 4a\partial^{\rho}F_{\rho\sigma} = bJ_{\sigma}.$$

This becomes Maxwell's first and fourth equations combined [see Equation (17.45)] if $a = \frac{1}{4}$ and $b = \mu_{0}$. Thus the Lagrangian density for electrodynamics is

$$\mathcal{L} = \frac{1}{4}\eta^{\alpha\mu}\eta^{\beta\nu}(A_{\nu,\mu} - A_{\mu,\nu})(A_{\beta,\alpha} - A_{\alpha,\beta}) + \mu_{0}J^{\alpha}A_{\alpha}. \qquad (30.43)$$

This, like any other Lagrangian, can be multiplied by a constant without affecting the Euler-Lagrange equations.

Example 30.2.2. Charged Particle in EM Field Problem 30.18 shows that the Lagrangian density (30.43) can be written as

$$\mathcal{L} = \frac{1}{2}\left(|\mathbf{B}|^{2} - |\mathbf{E}|^{2}\right) + \mu_{0}\left(\rho\Phi - \mathbf{J}\cdot\mathbf{A}\right),$$

with the variational problem

$$\mathbf{L} = \int_{a}^{b}\left(\iiint_{\Omega}\mathcal{L}\,d^{3}x'\right)dt.$$

Now consider a single particle of charge q interacting with an electromagnetic field. For such a particle,

$$\rho = q\delta(\mathbf{r} - \mathbf{r}') \qquad \text{and} \qquad \mathbf{J} = \rho\mathbf{v} = q\mathbf{v}\delta(\mathbf{r} - \mathbf{r}'),$$

and $\mathbf{L}$ becomes

$$\mathbf{L} = \frac{1}{2}\int_a^b dt \int_\Omega \left(|\mathbf{B}|^2 - |\mathbf{E}|^2\right) d^3x' + \mu_0 \int_a^b dt \int_\Omega \left(q\Phi\delta(\mathbf{r} - \mathbf{r}') - q\mathbf{v}\cdot\mathbf{A}\delta(\mathbf{r} - \mathbf{r}')\right) d^3x'$$

$$= \frac{1}{2}\int_a^b dt \int_\Omega \left(|\mathbf{B}|^2 - |\mathbf{E}|^2\right) d^3x' + \mu_0 q \int_a^b dt \left\{\Phi(\mathbf{r}, t) - \mathbf{v}\cdot\mathbf{A}(\mathbf{r}, t)\right\}.$$

The particle also has kinetic energy, which needs to be added to this Lagrangian. When adding Lagrangians, one has to incorporate the freedom in multiplying Lagrangians by constants. In the case at hand, the kinetic energy of the particle should be added to the *negative* of the *scalar potential energy* (recall that $L = KE - \Phi$). To assure this, we have to divide the entire EM Lagrangian by $-1/\mu_0$ and add it to the kinetic energy of the particle. Hence the total Lagrangian becomes

$$\mathbf{L} = -\frac{1}{2\mu_0}\int_a^b dt \int_\Omega \left(|\mathbf{B}|^2 - |\mathbf{E}|^2\right) d^3x' + \int_a^b dt \left\{\tfrac{1}{2}m|\mathbf{v}|^2 - q\Phi(\mathbf{r}, t) + q\mathbf{v}\cdot\mathbf{A}(\mathbf{r}, t)\right\}.$$

Notice how the first integral is four-dimensional while the second integral is over a single variable.

We are interested in the motion of the particle. Therefore, the first integral is just a constant (independent of the coordinates and velocity components of the particle) and can be dropped. Thus, substituting $\dot{\mathbf{r}}$ for $\mathbf{v}$, we have

$$\mathbf{L}_{\text{part}} = \int_a^b dt \left\{\tfrac{1}{2}m|\dot{\mathbf{r}}|^2 - q\Phi(\mathbf{r}, t) + q\dot{\mathbf{r}}\cdot\mathbf{A}(\mathbf{r}, t)\right\},$$

Lagrangian of a charged particle in EM field

with the Lagrangian

$$L(\mathbf{r}, \dot{\mathbf{r}}, t) = \tfrac{1}{2}m|\dot{\mathbf{r}}|^2 - q\Phi(\mathbf{r}, t) + q\dot{\mathbf{r}}\cdot\mathbf{A}(\mathbf{r}, t). \qquad (30.44)$$

Let's look at the x-component of the motion:

$$\frac{\partial L}{\partial x} - \frac{d}{dt}\frac{\partial L}{\partial \dot{x}} = 0 \qquad \text{or} \qquad -q\frac{\partial \Phi}{\partial x} + q\dot{\mathbf{r}}\cdot\frac{\partial \mathbf{A}}{\partial x} - \frac{d}{dt}\left(m\dot{x} + qA_x\right) = 0,$$

or

$$m\ddot{x} + q\frac{\partial \Phi}{\partial x} + q\left(\frac{dA_x}{dt} - \dot{\mathbf{r}}\cdot\frac{\partial \mathbf{A}}{\partial x}\right) = 0. \qquad (30.45)$$

Now note that

$$\frac{dA_x}{dt} = \frac{\partial A_x}{\partial t} + \frac{\partial A_x}{\partial x}\dot{x} + \frac{\partial A_x}{\partial y}\dot{y} + \frac{\partial A_x}{\partial z}\dot{z},$$

and

$$\dot{\mathbf{r}}\cdot\frac{\partial \mathbf{A}}{\partial x} = \dot{x}\frac{\partial A_x}{\partial x} + \dot{y}\frac{\partial A_y}{\partial x} + \dot{z}\frac{\partial A_z}{\partial x}.$$

Putting these two equations in (30.45) and rearranging, we obtain

$$m\ddot{x} + q\underbrace{\left(\frac{\partial \Phi}{\partial x} + \frac{\partial A_x}{\partial t}\right)}_{=-E_x \text{ by } (15.31)} + q\dot{y}\underbrace{\left(\frac{\partial A_x}{\partial y} - \frac{\partial A_y}{\partial x}\right)}_{=-B_z \text{ by } (15.31)} + q\dot{z}\underbrace{\left(\frac{\partial A_x}{\partial z} - \frac{\partial A_z}{\partial x}\right)}_{=B_y \text{ by } (15.31)} = 0,$$

or

$$m\ddot{x} - qE_x - q\left(\dot{y}B_z - \dot{z}B_y\right) = 0. \qquad (30.46)$$

The expression in parentheses is just the x-component of $\mathbf{v} \times \mathbf{B}$. Thus, (30.46) is the x-component of the Lorentz force law, governing the motion of a charged particle in an electromagnetic field. ∎

Klein-Gordon Lagrangian

One of the first attempts at combining the special theory of relativity with quantum mechanics was made by Oskar Klein and Walter Gordon. In fact, Schrdinger himself started with the relativistic version of his equation, but abandoned it because of some difficulty he encountered when applying it to hydrogen atom. By the usual substitution

$$E \to i\hbar\frac{\partial}{\partial t}, \quad \mathbf{p} \to -i\hbar\boldsymbol{\nabla}$$

in the relativistic equation $E^2/c^2 - \mathbf{p} \cdot \mathbf{p} = m^2 c^2$, Klein and Gordon derived the equation that now bears their names:

$$\frac{1}{c^2}\frac{\partial^2 \phi}{\partial t^2} - \nabla^2\phi + \frac{m^2 c^2}{\hbar^2}\phi = 0,$$

which, in units $\hbar = 1 = c$, becomes

$$\frac{\partial^2 \phi}{\partial t^2} - \nabla^2\phi + m^2\phi = 0.$$

This equation can also be obtained from the Lagrangian density

$$\mathcal{L} = \eta^{\alpha\beta}\left(\partial_\alpha\phi\right)\left(\partial_\beta\phi\right) - \frac{m^2}{2}\phi^2, \qquad (30.47)$$

as the reader can verify.

30.3 Hamiltonian Dynamics

The Lagrangian formulation of mechanics treated in the previous section is a powerful tool for studying many different dynamical systems and fields. Furthermore, considerations of symmetry, an indispensable technique in the investigation of fundamental forces, is most adequately handled in the Lagrangian language. Once the Lagrangian is known, the Euler-Lagrange equations provide second-order differential equations to be solved under given boundary (or initial) conditions.

There is another formulation of mechanics, which instead of second-order differential equations, yields twice as many *first-order* DEs. It is called the **Hamiltonian formulation**. We describe only the case of several dependent and one independent variables, the other cases being very similar. Let us assume that our dynamical system has n generalized coordinates $\{q_i\}_{i=1}^{n}$ and a Lagrangian $L\left(\mathbf{q}, \dot{\mathbf{q}}, t\right)$. In the simplest case (30.40), $L = KE - \Phi$ where KE is a quadratic term in velocities alone and Φ dependent on the coordinates alone. In such a case,

$$\frac{\partial L}{\partial \dot{q}_j} = \frac{\partial KE}{\partial \dot{q}_j} = \mu_j \dot{q}_j,$$

which is the momentum associated with the jth generalized coordinate. It is therefore natural to generalize the concept of momentum as well, write

$$p_j \equiv \frac{\partial L\left(\mathbf{q}, \dot{\mathbf{q}}, t\right)}{\partial \dot{q}_j}, \tag{30.48}$$

and call p_j so defined the **generalized momentum** of the dynamical system.

The transition from Lagrangian to Hamiltonian formulation, from a strictly mathematical standpoint, is to go from the set of variables $(\mathbf{q}, \dot{\mathbf{q}}, t)$ to $(\mathbf{q}, \mathbf{p}, t)$. The procedure for making this transition is the *Legendre transformation* discussed in Section 2.2.2. To find the variables involved, consider the differential of the Lagrangian:

$$dL = \sum_{i=1}^{n} \left(\frac{\partial L}{\partial q_i} dq_i + \frac{\partial L}{\partial \dot{q}_i} d\dot{q}_i \right) + \frac{\partial L}{\partial t} dt,$$

and use (30.48) and the Euler-Laggrange equation to rewrite the above as

$$dL = \sum_{i=1}^{n} \left(\dot{p}_i \, dq_i + p_i \, d\dot{q}_i \right) + \frac{\partial L}{\partial t} dt.$$

If we want to switch the independent variable from $\dot{q}_i$ to p_i, then we have to define the Hamiltonian as

$$H\left(\mathbf{q}, \mathbf{p}, t\right) = \sum_{i=1}^{n} p_i \dot{q}_i - L\left(\mathbf{q}, \dot{\mathbf{q}}, t\right). \tag{30.49}$$

To verify this, we note that

$$dH = \sum_{i=1}^{n} \left(\dot{q}_i dp_i + p_i d\dot{q}_i \right) - dL = \sum_{i=1}^{n} \left(\dot{q}_i dp_i + p_i d\dot{q}_i \right) - \sum_{i=1}^{n} \left(\dot{p}_i \, dq_i + p_i \, d\dot{q}_i \right) - \frac{\partial L}{\partial t} dt.$$

Note that $p_i d\dot{q}_i$ terms cancel and we are left with

$$dH = \sum_{i=1}^{n} \left(\dot{q}_i dp_i - \dot{p}_i \, dq_i \right) - \frac{\partial L}{\partial t} dt.$$

On the other hand,

$$dH = \sum_{i=1}^{n} \left(\frac{\partial H}{\partial q_i} dq_i + \frac{\partial H}{\partial p_i} dp_i \right) + \frac{\partial H}{\partial t} dt.$$

Comparison of the last two equations gives

$$\dot{q}_i = \frac{\partial H}{\partial p_i}, \quad \dot{p}_i = -\frac{\partial H}{\partial q_i}, \quad -\frac{\partial L}{\partial t} = \frac{\partial H}{\partial t}, \quad i = 1, 2, \ldots, n, \qquad (30.50)$$

which are called **Hamilton** or **canonical equations**. Note that instead of n second-order DEs, we now have $2n$ first order DEs.

To discover the physical significance of the Hamiltonian, consider the familiar simple Lagrangian $L = KE - \Phi$, where KE is the usual kinetic energy term and Φ is the potential energy which is independent of velocities. Then, (30.48) yields $p_i = \mu_i \dot{q}_i$ and

Hamiltonian is the total energy.

$$H = \sum_{i=1}^{n} p_i \dot{q}_i - L = \underbrace{\sum_{i=1}^{n} \mu_i \dot{q}_i^2}_{=2KE} - KE + \Phi = KE + \Phi.$$

So H is the sum of the kinetic and potential energies, i.e., the total energy.

Example 30.3.1. Hamiltonian of a Charged Particle in EM Field The Lagrangian of a charged particle in an electromagnetic field is given by (30.44). Let's find the Hamiltonian of this system. First we need the generalized momentum (30.48):

$$p_i = \frac{\partial L}{\partial \dot{x}_i} = m\dot{x}_i + qA_i \quad \text{or} \quad \mathbf{p} = m\dot{\mathbf{r}} + q\mathbf{A}. \qquad (30.51)$$

This is an important equation in its own right. It says that the momentum of the system is not just that of the particle, but that it also includes a contribution from the EM field. In particular, that EM field *has momentum.*[1]

To find the Hamiltonian, compute $\dot{\mathbf{r}}$ from (30.51):

$$\dot{\mathbf{r}} = \frac{\mathbf{p} - q\mathbf{A}}{m},$$

and substitute in the definition of the Hamiltonian (30.49), where, in this case the sum is just the dot product:

$$H(\mathbf{r}, \mathbf{p}, t) = \mathbf{p} \cdot \left(\frac{\mathbf{p} - q\mathbf{A}}{m} \right) - \frac{1}{2}m \left| \frac{\mathbf{p} - q\mathbf{A}}{m} \right|^2 + q\Phi - q \left(\frac{\mathbf{p} - q\mathbf{A}}{m} \right) \cdot \mathbf{A}$$

$$= (\mathbf{p} - q\mathbf{A}) \cdot \left(\frac{\mathbf{p} - q\mathbf{A}}{m} \right) - \frac{1}{2} \frac{|\mathbf{p} - q\mathbf{A}|^2}{m} + q\Phi,$$

or

$$H(\mathbf{r}, \mathbf{p}, t) = \frac{|\mathbf{p} - q\mathbf{A}(\mathbf{r}, t)|^2}{2m} + q\Phi(\mathbf{r}, t). \qquad (30.52)$$

Thus, in the presence of an electromagnetic field, the Hamiltonian of a particle takes the same *form* as the total energy of a particle in a potential $q\Phi$, except that in the expression for the KE part, $\mathbf{p} - q\mathbf{A}$ replaces $\mathbf{p}$. Such a replacement is called the **minimal coupling** and plays a key role in the quantum mechanical treatment of charged particles interacting with EM fields. ∎

---

[1]This momentum is the source of *radiation pressure.*

30.4   Problems

30.1. Show that, in Example 30.1.6, $C_1 = 0$, $\lambda = \lambda_0$, and $C_2 = \sqrt{\lambda_0^2 - a^2}$, where λ_0 is the solution of the equation

$$\lambda \sin\left(\frac{l}{2\lambda}\right) = a.$$

30.2. Find the extremal of the functional

$$\mathsf{L}[x, y] = \int_0^{\pi/2} (\dot{x}^2 + \dot{y}^2 + 2xy)\, dt$$

subject to the boundary conditions

$$x(0) = 0, \qquad x(\pi/2) = 1, \qquad y(0) = 0, \qquad y(\pi/2) = 1.$$

30.3. Find the extremals of the following functionals:

(a) $\mathsf{L}[x, y] = \int_a^b (\dot{x}^2 + \dot{y}^2 + \dot{x}\dot{y})\, dt$,   (b) $\mathsf{L}[x, y] = \int_a^b (2xy - 2x^2 + \dot{x}^2 - \dot{y}^2)\, dt$.

30.4. Find the extremal of a functional of the form

$$\mathsf{L}[x, y] = \int_a^b L(\dot{x}, \dot{y})\, dt,$$

given that

$$\frac{\partial^2 L}{\partial \dot{x}^2} \frac{\partial^2 L}{\partial \dot{y}^2} - \left(\frac{\partial^2 L}{\partial \dot{x} \partial \dot{y}}\right)^2 \neq 0 \text{ for } a \leq x \leq b.$$

30.5. Find the extremal of the functional

$$\mathsf{L}[x] = \int_0^1 (\dot{x}^2 + t^2)\, dt,$$

subject to the boundary conditions

$$x(0) = 0, \qquad x(1) = 0, \qquad \int_0^1 x^2\, dt = 2.$$

30.6. Show that the extremization of (30.33) leads to the Euler-Lagrange equations (30.34).

30.7. Among all triangles with a given base line and a fixed perimeter, show that the isosceles triangle has the largest area.

30.8. An airplane with fixed air speed v_0 flies for a time T on a closed curve. The wind velocity $\mathbf{u}$ is constant in magnitude and direction and $|\mathbf{u}| < v_0$. What closed curve encloses the largest area?

30.9. Among all curves joining a given point $(0, b)$ on the y-axis to a point on the x-axis and enclosing a given area S together with the x-axis, find the curve which generates the least area when rotated about the x-axis.

30.10. An Atwood machine consists of two masses m_1 and m_2 connected by a light inextensible cord of length l which passes over a pulley whose radius is a and whose moment of inertia is I. Let x denote the distance of m_1 from the top of the pulley. Using Lagrangian methos, show that the acceleration of m_1 is

$$\ddot{x} = \frac{g(m_1 - m_2)}{m_1 + m_2 + I/a^2}.$$

30.11. Using polar coordinates, write the Lagrangian of a particle of mass m moving in a central force field with potential $\Phi(r)$. Show that the equations of motion are

$$m\ddot{r} = mr\dot{\theta}^2 - \frac{d\Phi}{dr}, \qquad \frac{d}{dt}(mr^2\dot{\theta}) = 0.$$

30.12. Using Lagrangian method, find the acceleration of a solid sphere rolling without sliding down an inclined plane having an angle θ with the horizontal.

30.13. Using Lagrangian method, find the acceleration of a solid sphere rolling without sliding down a movable wedge of mass M having an angle θ. The wedge moves on a frictionless horzontal surface.

30.14. Two blocks of equal mass m are connected by an inextensible cord whose linear mass density is μ. One block is placed on a smooth horizontal table, the other hangs over the edge of the table. What is the acceleration of the system? Use the Lagrangian method.

30.15. A simple pendulum of length l and mass m oscillates about its point of support which is attached to a block of mass M moving without friction along a horizontal line lying in the plane of the pendulum. Write the Lagrangian in terms of x, the position of M on the horizontal line, and θ, the angle l makes with the vertical. Find the equations of motion of m and M.

30.16. Find the equation of a curve describing the equilibrium position of a uniformly dense heavy flexible inextensible cord of length l fastened at its ends. Hint: The Lagrangian is just the potential energy written as an integral.

30.17. Show that the Lagrangian density (30.43) can be written as

$$\mathcal{L} = \frac{1}{2}\left(|\mathbf{B}|^2 - |\mathbf{E}|^2\right) + \mu_0\left(\rho\Phi - \mathbf{J}\cdot\mathbf{A}\right).$$

Hint: See Sections 17.3.1 and 17.3.2 and be careful about possible change of sign when raising or lowering indices.

30.18. Show that the Lagrangian density (30.47) leads to the Klein-Gordon equation.

Chapter 31

Nonlinear Dynamics and Chaos

A variety of techniques including the Frobenius method of infinite power series could solve almost all *linear* DEs of physical interest. However, some very fundamental questions such as the stability of the solar system led to DEs that were not linear, and for such DEs no analytic (including series representation) solution existed. In the 1890s, Henri Poincaré, the great French mathematician, took upon himself the task of gleaning as much information from the DEs describing the whole solar system as was possible. The result was the invention of one of the most powerful branches of mathematics (topology) and the realization that the *qualitative* analysis of (nonlinear) DEs could be very useful.

One of the discoveries made by Poincaré, which much later became the cornerstone of many developments, was that

Box 31.0.1. *Unlike the linear DEs, nonlinear DEs may be very sensitive to the initial conditions.*

In other words, if a nonlinear system starts from some initial conditions and develops into a certain final configuration, then starting it with slightly different initial conditions may cause the system to develop into a final configuration completely different from the first one. This is in complete contrast to the linear DEs where two nearby initial conditions lead to nearby final configurations.

In general, the initial conditions are not known with infinite accuracy. Therefore, the final states of a nonlinear dynamical system may exhibit an indeterministic behavior resulting from the initial (small) uncertainties. This is what has come to be known as **chaos**. The reader should note that the indeterminism discussed here has nothing to do with the quantum indeterminism.

chaos due to uncertainty in initial conditions

All equations here are completely deterministic. It is the divergence of the initially nearby—and completely deterministic—trajectories that results in unpredictable final states.

There are two general categories exhibiting chaotic behavior: systems obeying iterated maps and systems obeying DEs. We shall study the first category in some detail, and only outline some of the general features of the much more complicated category of systems obeying DEs.

31.1   Systems Obeying Iterated Maps

Consider the population of a species in consecutive years if the population is initially N_0. The simplest relation connecting N_1, the population after one year, to N_0 is

$$N_1 = \alpha N_0,$$

where α is a positive number depending on the environment in which the species lives. Under the most favorable conditions, α is a large number, indicating rapid growth of population. Under less favorable conditions, α will be small. And if the environment happens to be hostile, then α will be smaller than one, indicating a decline in population.

The above equation is unrealistic because we know that if $\alpha > 1$ and the population grows excessively, there will not be enough food to support the species. So, there must be a mechanism to suppress the growth. A more realistic equation should have a suppressive term which is small for small N_0 and grows for larger values of N_0. A possible term having such properties is one proportional to N_0^2. This leads to

$$N_1 = \alpha N_0 - \beta N_0^2 \qquad \text{where} \quad 0 < \beta \ll \alpha.$$

The minus sign causes the second term to decrease the population. Iterating this equation, we can find the population in the second, third, and subsequent years:

$$N_2 = \alpha N_1 - \beta N_1^2, \quad N_3 = \alpha N_2 - \beta N_2^2, \ldots,$$

and, in general,

$$N_{k+1} = \alpha N_k - \beta N_k^2. \tag{31.1}$$

It is customary to rewrite (31.1) in a slightly different form. First we note that since population cannot be negative, there exists a maximum number beyond which the population cannot grow. In order for N_{k+1} to be positive, we must have

$$\alpha N_k - \beta N_k^2 > 0 \ \Rightarrow \ N_k < \frac{\alpha}{\beta}$$

for all k. It follows that $N_{\text{max}} = \alpha/\beta$. Dividing (31.1) by N_{max} yields

$$x_{k+1} = \alpha x_k (1 - x_k), \tag{31.2}$$

where x_k is the *fraction* of the maximum population of the species after k years, and therefore, its value must lie between zero and one. Any equation of the form

$$x_{k+1} = f_\alpha(x_k), \tag{31.3}$$

where α is—as in the case of the logistic map—a control parameter, and in which a value of some (discrete) quantity at $k+1$ is given in terms of its value at k, is called an **iterated map**, and the function f_α is called the **iterated map function**. The particular function in (31.2) is called the **logistic map function**.

Starting from an initial value x_0, one can generate a sequence of x values by consecutively substituting in the RHS of (31.3). This sequence is called a *trajectory* or **orbit** of the iterated map.

iterated map, iterated map function, and logistic map function

31.1.1 Stable and Unstable Fixed Points

It is clear that the first few points of an orbit depend on the starting point. What may not be so clear is that, for a given α, the eventual behavior of the orbit is fairly insensitive to the starting point. There are, however, some starting points which are manifestly different from others. For example, in the logistic map, if $x_0 = 0$, no other point will be produced by iteration because $f_\alpha(0) = 0$ or $f_\alpha(x_0) = x_0$, and further application of f_α will not produce any new values of x. In general, a point x_α which has the property that

$$f_\alpha(x_\alpha) = x_\alpha \tag{31.4}$$

is called a **fixed point** of the iterated map associated with α. For the logistic map we have

fixed point of an iterated map

$$x_\alpha = \alpha x_\alpha(1 - x_\alpha) \;\Rightarrow\; x_\alpha(1 - \alpha + \alpha x_\alpha) = 0 \;\Rightarrow\; x_\alpha = 0,\, 1 - \frac{1}{\alpha}. \tag{31.5}$$

Since $0 \leq x_\alpha \leq 1$, there is only one fixed point (i.e., $x = 0$) for $\alpha \leq 1$, and two fixed points (i.e., $x = 0$ and $x = 1 - 1/\alpha$) for $\alpha > 1$.

What is the significance of fixed points? When $\alpha < 1$, Equation (31.2) shows—since both x_k and $1 - x_k$ are at most one—that the population keeps decreasing until it vanishes completely. And this is independent of the initial value of x. It is instructive to show this pictorially. Figure 31.1(a) shows the logistic map function with $\alpha = 0.5$. Start at any point x_0 on the horizontal axis; draw a vertical line to intersect the logistic map function at $f(x_0) \equiv x_1$; from the intersection draw a horizontal line to intersect the line $y = x$ at $y_1 = x_1$; draw a vertical line to intersect the logistic map function at $f(x_1) \equiv x_2$; continue to find x_3 and the rest of x's. The diagram shows that the x's are getting smaller and smaller.

graphical way of approaching a fixed point

What happens when $\alpha > 1$? Figure 31.1(b) shows the logistic map function with $\alpha = 2$. We note that the orbit is attracted to the fixed point at $x = 0.5$. We also note that the fixed point at $x = 0$ has now turned into a

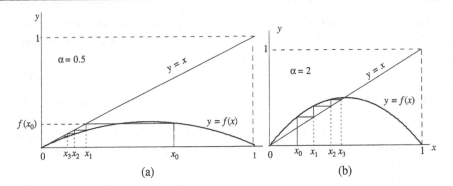

Figure 31.1: (a) Regardless of the value of x_0, the orbit always ends up at the origin when $\alpha < 1$. (b) Even for $\alpha > 1$, it appears that the orbit always ends up at some attractor regardless of the value of x_0. Note that now the origin has become a "repellor."

"repellor." We can treat the behavior of the logistic map at general fixed points analytically.

First let us consider a general (one-dimensional) iterated map as given by Equation (31.3). We are seeking the fixed points of (31.3). These points—commonly labeled by an asterisk—satisfy

$$x^* = f_\alpha(x^*),$$

i.e., they are intersections of the curves $y = x$ and $y = f_\alpha(x)$ in the xy-plane. An important property of fixed points is their **stability**—or whether they are attractors or repellors. To test this property, we Taylor-expand the iterated map function around x^*, keeping the first two terms:

stability of fixed points

$$x_{k+1} = f_\alpha(x_k) = f_\alpha(x^*) + \left.\frac{df_\alpha}{dx}\right|_{x^*}(x_k - x^*) = x^* + \left.\frac{df_\alpha}{dx}\right|_{x^*}(x_k - x^*)$$

or

$$\frac{|x_{k+1} - x^*|}{|x_k - x^*|} = \left|\frac{df_\alpha}{dx}\right|_{x^*} = |f_\alpha'(x^*)|.$$

So, x_{k+1} will be farther away from (or closer to) x^* than x_k if the absolute value of the derivative of the function is greater than one (or less than one).

analytic criterion for stability of a fixed point of an iterated map

> **Box 31.1.1.** *A fixed point x^* of an iterated map (31.3) is stable if $|f_\alpha'(x^*)| < 1$ and unstable if $|f_\alpha'(x^*)| > 1$.*

Example 31.1.1. For the logistic map, $f_\alpha(x) = \alpha x(1-x)$ so that $f_\alpha'(x) = \alpha - 2\alpha x$. The fixed points are $x_1^* = 0$ and $x_2^* = 1 - 1/\alpha$. Therefore,

$$f_\alpha'(x_1^*) = f_\alpha'(0) = \alpha \quad \text{and} \quad f_\alpha'(x_2^*) = f_\alpha'(1 - 1/\alpha) = 2 - \alpha. \tag{31.6}$$

It follows that the fixed point at $x = 0$ is stable (attractive) if $\alpha < 1$, while for this same value of α the fixed point x_2^* is unstable (repulsive). Thus, for $\alpha < 1$, all trajectories are attracted to the fixed point at $x = 0$.

Equation (31.6) also shows that for $1 < \alpha < 3$, the other fixed point becomes stable while the fixed point at the origin becomes unstable. This is also consistent with the behavior of the logistic map depicted in Figure 31.1(b). ∎

The criterion of Box 31.1.1 can also be stated graphically. Since 1 is the slope of the line $y = x$, and since a fixed point is an intersection of the two curves $y = x$ and $y = f_\alpha(x)$, the criterion of Box 31.1.1 is a comparison of the slope of the tangent to $y = f_\alpha(x)$ with the slope of $y = x$: A fixed point x^* of an iterated map (31.3) is stable, if the *acute* angle that the tangent line at $(x^*, f_\alpha(x^*))$ makes with the x-axis is smaller than the corresponding angle of the line $y = x$. If this angle is larger, then the fixed point is unstable. This is equivalent to the simpler statement:

graphical criterion for stability of a fixed point of an iterated map

> **Box 31.1.2.** *A fixed point x^* of an iterated map $f_\alpha(x)$ is stable (unstable) if immediately to the right of x^*, the curve $y = f_\alpha(x)$ lies below (above) the line $y = x$.*

31.1.2 Bifurcation

Although the logistic map has no *stable* fixed points beyond $x_\alpha = 1 - 1/\alpha$, we may ask whether there are points at which the iterated map is "semi-stable." What does this mean? Instead of demanding strict stability or instability, let us consider a case in which the map may *oscillate between two values*. This situation is neither completely stable nor completely unstable: Although the system moves away from the point in question, it does not leave it forever. Suppose that just above the largest value of a stable α, the system starts to oscillate between two values of x. This is an example of bifurcation:

bifurcation and period doubling

> **Box 31.1.3.** *When the development of a system splits into two regions as a parameter of the equations of motion of the system increases slightly, we say that a **bifurcation** has occurred and call the splitting of the trajectory a **period-doubling bifurcation**.*

Suppose that there are two "fixed" points x_1^* and x_2^* between which the function oscillates such as the two points illustrated in Figure 31.2(a). These fixed points must satisfy

$$x_2^* = f_\alpha(x_1^*), \qquad x_1^* = f_\alpha(x_2^*). \tag{31.7}$$

To gain further insight into the behavior of the logistic map, we introduce the so-called **second iterate** of f_α denoted by $f_\alpha^{[2]}$ and defined by

second iterate

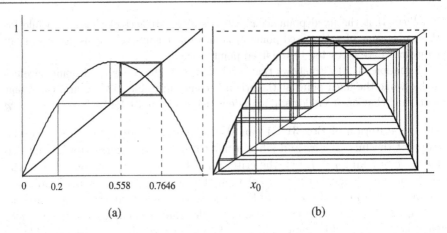

Figure 31.2: (a) For $\alpha = 3.1$, there are clearly two attractors located at $x = 0.5580$ and $x = 0.7646$. (b) For $\alpha = 3.99$, no attractor seems to exist because the iterations do not seem to converge in the diagram.

$$f_\alpha^{[2]}(x) \equiv f_\alpha(f_\alpha(x)). \tag{31.8}$$

From this definition, it is clear that *every fixed point of f_α is also a fixed point of $f_\alpha^{[2]}$*. However, the converse statement is not true. In fact, x_1^* and x_2^* defined in Equation (31.7) are fixed points of $f_\alpha^{[2]}$:

$$f_\alpha^{[2]}(x_1^*) = f_\alpha(f_\alpha(x_1^*)) = f_\alpha(x_2^*) = x_1^*,$$
$$f_\alpha^{[2]}(x_2^*) = f_\alpha(f_\alpha(x_2^*)) = f_\alpha(x_1^*) = x_2^*,$$

but not of f_α. It now follows that fixed points of $f_\alpha^{[2]}$ give information about period-doubling bifurcation.

For the logistic map, $f_\alpha^{[2]}$ can be found easily:

$$\begin{aligned} f_\alpha^{[2]}(x) &= f_\alpha(f_\alpha(x)) = \alpha f_\alpha(x)[1 - f_\alpha(x)] \\ &= \alpha[\alpha x(1-x)][1 - \alpha x(1-x)] = \alpha^2 x(1-x)(1 - \alpha x + \alpha x^2) \\ &= -\alpha^3 x^4 + 2\alpha^3 x^3 - (\alpha^2 + \alpha^3)x^2 + \alpha^2 x. \end{aligned} \tag{31.9}$$

The fixed points of $f_\alpha^{[2]}(x)$ are, therefore, determined by the equation

$$x = -\alpha^3 x^4 + 2\alpha^3 x^3 - (\alpha^2 + \alpha^3)x^2 + \alpha^2 x$$

which shows that there are, in general, four fixed points, one at $x = 0$, and three others satisfying the cubic equation

$$\alpha^3 x^3 - 2\alpha^3 x^2 + (\alpha^3 + \alpha^2)x - \alpha^2 + 1 = 0. \tag{31.10}$$

We can actually solve this equation because we know that one of its roots is $x_1(\alpha) = 1 - 1/\alpha$, a fixed point of $f_\alpha(x)$. The cubic polynomial in Equation

(31.10) can thus be factored out as $\alpha^3[x - x_1(\alpha)]$ times a quadratic polynomial whose roots give the remaining solutions to (31.10). The reader may verify that these roots are

$$x_2(\alpha) = \frac{1 + \alpha + \sqrt{\alpha^2 - 2\alpha - 3}}{2\alpha},$$

$$x_3(\alpha) = \frac{1 + \alpha - \sqrt{\alpha^2 - 2\alpha - 3}}{2\alpha}. \tag{31.11}$$

These two functions start out at the common value of $\frac{2}{3}$ when $\alpha = 3$. Then, as a function of α, $x_2(\alpha)$ monotonically increases and asymptotically approaches 1; $x_3(\alpha)$ monotonically decreases and asymptotically approaches 0.

We are interested in those values of α for which the fixed points are not completely unstable. In the present case, this means that the value of the iterated map must oscillate between only two values. This will happen only if the two points are *stable* fixed points of $f_\alpha^{[2]}$. Since by Box 31.1.1 stability imposes a condition on the derivative of the function, we need to look at the derivative of $f_\alpha^{[2]}$.

Using the chain rule, which in its most general form is

$$\frac{d}{dx}[g(h(x))] = g'(h(x))h'(x)$$

we obtain

$$\frac{d}{dx}f_\alpha^{[2]}(x) = [f_\alpha'(f_\alpha(x))]f_\alpha'(x). \tag{31.12}$$

In particular, if x happens to be a fixed point x^* of f_α, then

$$\left.\frac{df_\alpha^{[2]}}{dx}\right|_{x^*} = f_\alpha'(\underbrace{f_\alpha(x^*)}_{=x^*})f_\alpha'(x^*) = [f_\alpha'(x^*)]^2. \tag{31.13}$$

This shows that if x^* is a stable fixed point of f_α, then

$$|f_\alpha'(x^*)| < 1 \;\Rightarrow\; [f_\alpha'(x^*)]^2 < 1 \;\Rightarrow\; \left|\frac{d}{dx}f_\alpha^{[2]}(x^*)\right| < 1$$

and x^* is a *stable* fixed point of $f_\alpha^{[2]}$ as well. Furthermore, at the two fixed points of $f_\alpha^{[2]}$ discussed above, Equation (31.12) yields

$$\left.\frac{df_\alpha^{[2]}}{dx}\right|_{x_1^*} = f_\alpha'(\underbrace{f_\alpha(x_1^*)}_{x_2^*})f_\alpha'(x_1^*) = f_\alpha'(x_2^*)f_\alpha'(x_1^*),$$

$$\left.\frac{df_\alpha^{[2]}}{dx}\right|_{x_2^*} = f_\alpha'(\underbrace{f_\alpha(x_2^*)}_{x_1^*})f_\alpha'(x_2^*) = f_\alpha'(x_1^*)f_\alpha'(x_2^*). \tag{31.14}$$

It follows that $f_\alpha^{[2]}$ has the same derivatives at these two points.

*n*th iterate

The concept of iteration of f_α can be readily generalized. The ***n*th iterate** of f_α is

$$f_\alpha^{[n]} \equiv \underbrace{f(f(\ldots f(x)\ldots))}_{n \text{ times}}$$

and as in the case of $f_\alpha^{[2]}$, the fixed points of f_α are also fixed points of $f_\alpha^{[n]}$ and the *stable* fixed points of f_α are also stable fixed points of $f_\alpha^{[n]}$. The converse of neither of these statements is, in general, true.

The utility of the concept of the *n*th iterate comes in the analysis of the location of bifurcation points. To be specific, let us go back to the logistic map and Equation (31.6). The stable points, being characterized by the absolute value of the derivative of the map function, occur at $x = 0$ when $0 < \alpha < 1$ and at $x = 1 - 1/\alpha$ when $1 < \alpha < 3$. Within the α-range of stability, the derivative of f_α ranges between[1] -1 and $+1$, starting with $+1$ at $\alpha = 1$ and ending with -1 at $\alpha = 3$ [see the second equation in (31.6)]. Beyond this value of α—which is the parameter at which the 2-cycle fixed point occurs and which we now denote by α_1—f_α has no stable points. Equation (31.13), however, shows that the derivative of $f_\alpha^{[2]}$ is $+1$ there. This means that the derivative of $f_\alpha^{[2]}$ can decrease down to -1 as α increases beyond α_1. In fact, what happens as α increases past α_1 is precisely a repetition of what happened to f_α between $\alpha = 1$ and $\alpha = 3$: The derivative of $f_\alpha^{[2]}$ keeps decreasing until at a certain value of α denoted by α_2 a period-doubling bifurcation occurs

4-cycle fixed point

for $f_\alpha^{[2]}$. This corresponds to a **4-cycle** fixed point. Thus a 4-cycle fixed point x^*, as well as the corresponding value of α, is obtained by imposing the two requirements

$$f_{\alpha_2}^{[2]}(x^*) = x^* \quad \text{and} \quad \left.\frac{df_{\alpha_2}^{[2]}}{dx}\right|_{x^*} = -1. \tag{31.15}$$

This equation entails an important result. By Equation (31.14), the derivative of $f_{\alpha_2}^{[2]}$ at its two stable points are equal. Therefore, both stable points give rise to the same pair of equations (31.15). In particular,

> **Box 31.1.4.** *Any value of α_2 that gives a solution for the first fixed point must also give a solution for the second fixed point. In fact, we should expect two values of x^* for every α_2 that solves the pair of equations (31.15).*

For the logistic map, Equation (31.15) becomes [see (31.9)]

$$x^* = \alpha_2^2 x^*(1 - x^*)(1 - \alpha_2 x^* + \alpha_2 x^{*2}),$$
$$-1 = -4\alpha_2^3 x^{*3} + 6\alpha_2^3 x^{*2} - 2(\alpha_2^2 + \alpha_2^3)x^* + \alpha_2^2.$$

---

[1] The $x = 0$ is an exception because we are assuming that α is a positive quantity, therefore, $f'(0) = \alpha$ cannot be negative.

One can solve these equations and obtain eight possible pairs (x^*, α_2). The only two acceptable *real* pairs which have a value of α_2 larger than three are

$$\left(\frac{4 + \sqrt{6} \pm \sqrt{14 - 4\sqrt{6}}}{10}, 1 + \sqrt{6} \right) = (0.644949 \pm 0.204989, 3.44949). \quad (31.16)$$

In particular, $\alpha_2 = 1 + \sqrt{6} = 3.44949$ is the 4-cycle fixed point depicted in Figure 31.3 corresponding to the two x values of approximately 0.85 and 0.44.

The generalization to 2^n-cycles is now clear. One simply constructs the 2^nth iterate of f_α and solves the two equations

$$f_\alpha^{[2^n]}(x^*) = x^* \qquad \text{and} \qquad \left. \frac{df_\alpha^{[2^n]}}{dx} \right|_{x^*} = -1.$$

In practice, these are too complicated to solve analytically, but numerical methods are available for their solution. Each solution consists of a pair (x_n^*, α_n) where x_n^* is the 2^n-cycle fixed point and α_n is the corresponding control parameter. As in the case of $f_\alpha^{[2]}$, for each acceptable α_n, there are 2^n fixed points.

31.1.3   Onset of Chaos

Suppose we keep increasing α slowly. It may happen that at a certain value of α no finite set of "stable" points exists. A graphical analysis of this situation is depicted in Figure 31.2(b) showing that the behavior of the logistic map is chaotic. What is the relation between the value of α at which chaos sets in (which we denote by α_c) and α_n? Considering the chaotic behavior as

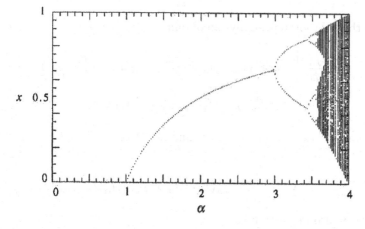

Figure 31.3: The bifurcation diagram for the logistic map. The behavior of the function is analytically very simple for $\alpha < 3$. For $\alpha > 3$, the behavior is more complicated. The 4-cycle fixed points are clearly shown to occur at $\alpha \approx 3.45$. From approximately 3.57 onward, chaotic behavior sets in.

one corresponding to an *infinite*-cycle fixed point, we conclude that, on a bifurcation diagram, chaos will occur at

$$\alpha_c = \alpha_\infty \equiv \lim_{n \to \infty} \alpha_n. \tag{31.17}$$

The limit in (31.17) is one way of characterizing chaos. A more direct way is to look at the trajectories. Two nearby points starting at x_0 and $x_0 + \epsilon$ will be separated after n iterations by a distance

$$d_n \equiv \left| f_\alpha^{[n]}(x_0 + \epsilon) - f_\alpha^{[n]}(x_0) \right|.$$

the n-th iterative Lyapunov exponent

If this separation grows exponentially, we have a chaotic behavior. We define $\lambda_{x_0}^{[n]}$, the nth **iterative Lyapunov exponent** at x_0, by

$$d_n \equiv d_0 e^{\lambda_{x_0}^{[n]} n} \equiv \epsilon e^{\lambda_{x_0}^{[n]} n}.$$

Then

$$\lambda_{x_0}^{[n]} = \frac{1}{n} \ln \left(\frac{d_n}{\epsilon} \right) = \frac{1}{n} \ln \left\{ \frac{\left| f_\alpha^{[n]}(x_0 + \epsilon) - f_\alpha^{[n]}(x_0) \right|}{\epsilon} \right\}.$$

As $\epsilon \to 0$, the RHS becomes the absolute value of the derivative of the n-th iterate at x_0. But by the chain rule

$$\left. \frac{df_\alpha^{[n]}}{dx} \right|_{x_0} = \left. \frac{d}{dx} \left[f_\alpha(f_\alpha^{[n-1]}(x)) \right] \right|_{x_0} = f_\alpha'(\underbrace{f_\alpha^{[n-1]}(x_0)}_{=x_{n-1}}) \left. \frac{df_\alpha^{[n-1]}}{dx} \right|_{x_0}$$

$$= f_\alpha'(x_{n-1}) \left. \frac{df_\alpha^{[n-1]}}{dx} \right|_{x_0}.$$

Using this relation repeatedly, we obtain

$$\left. \frac{df_\alpha^{[n]}}{dx} \right|_{x_0} = f_\alpha'(x_{n-1}) f_\alpha'(x_{n-2}) \dots f_\alpha'(x_{n-k}) \left. \frac{df_\alpha^{[n-k]}}{dx} \right|_{x_0}$$

$$= f_\alpha'(x_{n-1}) f_\alpha'(x_{n-2}) \dots f_\alpha'(x_0),$$

where in the last step we set $k = n$ and noted that $f_\alpha^{[0]} = f_\alpha$. It now follows that

$$\lambda_{x_0}^{[n]} = \frac{1}{n} \ln \left[|f_\alpha'(x_0)| \, |f_\alpha'(x_1)| \cdots |f_\alpha'(x_{n-1})| \right]$$

which can also be written as

$$\lambda_{x_0}^{[n]} = \frac{1}{n} \sum_{k=0}^{n-1} \ln |f_\alpha'(x_k)|. \tag{31.18}$$

It is common to define the **local Lyapunov exponent** as follows:

$$\lambda_{x_0} \equiv \lim_{n \to \infty} \lambda_{x_0}^{[n]} = \lim_{n \to \infty} \frac{1}{n} \sum_{k=0}^{n-1} \ln |f_\alpha'(x_k)| . \qquad (31.19)$$

To characterize the chaotic behavior of systems obeying iterated maps, one has to calculate λ_{x_0} for a sample of trajectory points and then take their average. The result is called the **Lyapunov exponent** for the system. It turns out that

> **Box 31.1.5.** *A necessary condition for a system obeying an iterated map function $f_\alpha(x)$ to be chaotic for α is for its Lyapunov exponent to be positive at α.*

31.2   Systems Obeying DEs

As a paradigm of a nonlinear dynamical system, we shall study the motion of a harmonically driven dissipative pendulum whose angle of oscillation is not necessarily small. The equation of motion of such a pendulum, coming directly from the second law of motion, is

$$m\frac{d^2x}{dt^2} = F_0 \cos(\Omega t) - b\frac{dx}{dt} - mg\sin\theta, \qquad (31.20)$$

where x is the length (as measured from the equilibrium position) of the arc of the circle on which mass m moves (see Figure 31.4).

The first term on the RHS of Equation (31.20) is the harmonic driving force with angular frequency Ω, the second is the dissipative (friction, drag, etc.) force, and the last is the gravitational force in the direction of motion.

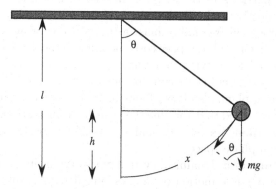

Figure 31.4: The displacement x and the gravitational force acting on the pendulum.

The minus signs appear because the corresponding forces oppose the motion. Since the pendulum is confined to a circle, x and θ are related via $x = l\theta$, and we obtain

$$ml\frac{d^2\theta}{dt^2} = F_0\cos(\Omega t) - bl\frac{d\theta}{dt} - mg\sin\theta.$$

Let us change t to $t = \tau\sqrt{l/g}$ where τ is a dimensionless parameter measuring time in units of $T/(2\pi)$ with T being the period of the small-angle pendulum.[2] Then, with the dimensionless constants

$$\gamma \equiv \frac{b}{m}\sqrt{\frac{l}{g}}, \qquad \phi_0 \equiv \frac{F_0}{mg}, \qquad \omega_D \equiv \Omega\sqrt{\frac{l}{g}},$$

the DE of motion becomes

$$\frac{d^2\theta}{d\tau^2} + \gamma\frac{d\theta}{d\tau} + \sin\theta = \phi_0\cos(\omega_D\tau).$$

It is customary to write this as

$$\ddot{\theta} + \gamma\dot{\theta} + \sin\theta = \phi_0\cos(\omega_D t), \tag{31.21}$$

where now t is the "dimensionless" time, and the dot indicates differentiation with respect to this t.

31.2.1   The Phase Space

The study of dynamical systems—i.e., systems obeying DEs—is considerably more complicated than systems obeying iterated maps. While in the latter case we were able to use a fair amount of analytical tools, the discussion of the former requires an enormous amount of numerical computation.

One of the devices that facilitates our understanding of dynamical systems is the **phase space diagram**. The phase space of a dynamical system is a Cartesian multidimensional space whose axes consist of positions and momenta of the particles in the system. Instead of momenta the velocities of particles are mostly used. Thus a single particle confined to one dimension (such as a particle in free fall, a mass attached to a spring, or a pendulum) has a two-dimensional phase space corresponding to the particle's position and speed. Two particles moving in a single dimension have a four-dimensional phase space corresponding to two positions and two speeds. A single particle moving in a plane also has a four-dimensional phase space because two coordinates are needed to determine the position of the particle, and two components to determine its velocity, and a system of N particles in space has a $6N$-dimensional phase space.

A **trajectory** of a dynamical system is a curve in its phase space corresponding to a possible motion of the system. If we can solve the equations

phase space diagram

phase space trajectory

---

[2]Recall that $T = 2\pi\sqrt{l/g}$. So $\tau = t/(T/2\pi)$ is indeed dimensionless.

of motion of a dynamical system, we can express all its position and velocity variables as a function of time, constituting a parametric equation of a curve in phase space. This curve is the trajectory of the dynamical system.

Let us go back to our pendulum, and consider the simplest situation in which there is no driving force, the dissipative effects are turned off, and the angle of oscillation is small. Then (31.21) reduces to $\ddot{\theta} + \theta = 0$, whose most general solution is $\theta = A\cos(t + \alpha)$ so that

$$x_1 \equiv \theta = A\cos(t + \alpha),$$

$$x_2 \equiv \omega \equiv \dot{\theta} \equiv \frac{d\theta}{dt} = -A\sin(t + \alpha). \tag{31.22}$$

This is a one-dimensional system (there is only one coordinate, θ) with a two-dimensional phase space. Equation (31.22) is the parametric equation of a circle of radius A in the x_1x_2-plane. Because A is arbitrary (it is, however, determined by initial conditions), there are (infinitely) many trajectories for this system, some of which are shown in Figure 31.5.

Let us now make the system a little more complicated by introducing a dissipative force, still keeping the angle small. The DE is now

$$\ddot{\theta} + \gamma\dot{\theta} + \theta = 0$$

and the general solution for the damped oscillatory case is

$$x_1 = \theta(t) = Ae^{-\gamma t/2}\cos(\omega_0 t + \alpha) \qquad \text{where} \quad \omega_0 \equiv \frac{\sqrt{4 - \gamma^2}}{2}$$

with

$$x_2 = \omega = \dot{\theta} = -Ae^{-\gamma t/2}\left\{\frac{\gamma}{2}\cos(\omega_0 t + \alpha) + \omega_0 \sin(\omega_0 t + \alpha)\right\}.$$

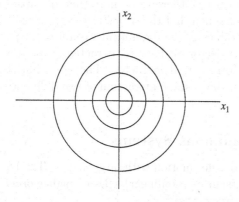

Figure 31.5: The phase space trajectories of a pendulum undergoing small-angle oscillations with no driving or dissipative forces. Different circles correspond to different initial conditions.

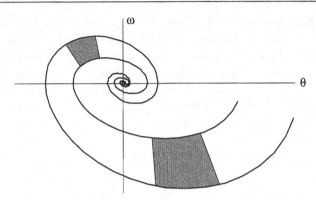

Figure 31.6: The phase space trajectories of a damped pendulum undergoing small-angle oscillations with no driving force. Different spirals correspond to different initial conditions. The larger shaded region, in time, shrinks to the smaller one.

The trajectories of this system are not as easily obtainable as the undamped linear oscillator discussed above. However, since the two coordinates of the phase space are given in terms of the parameter t, we can plot the trajectories. Two such trajectories for two different A's (but the same γ) are shown in Figure 31.6.

A new feature of this system is that regardless of where the trajectory starts at $t = 0$, it will terminate at the origin. The analytic reason for this is of course the exponential factor in front of both coordinates which will cause their decay to zero after a long time.[3] It seems that the origin "attracts" *all* trajectories, and for this reason is called an **attractor**.[4]

attractor

There are other kinds of attractors in nonlinear dynamics theory. For example, if trajectories approach an arc of a curve, or an area of a surface, then the curve or the area becomes the attractor. Furthermore, for a given value of the parameter, there may be more than one attractor for a given dynamical system (just as there were more than one fixed point for iterated maps); and it may happen that the trajectories approach these attractors only for certain initial values of the dynamical variables. The set of initial values corresponding to trajectories that are attracted to an attractor is called the

basin of attraction

basin of attraction for that attractor. The set of initial values that lie on the border between the basin of attraction of two different attractors is called

separatrix

a **separatrix**.

31.2.2   Autonomous Systems

Now we want to consider motion with large angles. The DE is then no longer linear. The discussion of (nonlinear) DEs of higher orders is facilitated by

---

[3] "Long" compared to $1/\gamma$.

[4] This is what we called a fixed point in our discussion of iterated maps. However, because of the existence of a variety of "fixed objects" for dynamical systems, it is more common to call these attractors.

treating derivatives as independent variables. The defining relations for these derivatives as well as the DE itself give a set of *first-order* DEs. For example, the third-order DE

$$\frac{d^3x}{dt^3} = x^4\frac{d^2x}{dt^2} + \sin(\omega t - kx)\frac{dx}{dt} + e^x\cos(\omega t)$$

can be turned into three first-order DEs by setting $\dot{x} \equiv x_1$ and $\ddot{x} \equiv x_2$. Then the DE splits into the following three first-order DEs:

$$\dot{x} = x_1, \qquad \dot{x}_1 = x_2,$$
$$\dot{x}_2 = x^4 x_2 + \sin(\omega t - kx)x_1 + e^x\cos(\omega t).$$

This is a set of three equations in the three unknowns x, x_1, and x_2, which, in principle, can be solved.

It is desirable to have a so-called **autonomous** system of first-order DEs. These are systems which have no explicit dependence on the independent variable (in our case, t). Our equations above clearly form a set of **nonautonomous** DEs. The nonautonomous systems can be reduced to autonomous ones by a straightforward trick: One simply calls t a new variable. More specifically, in the equation above, let $x_3 \equiv \omega t$. Then the nonautonomous equations above turn into the following autonomous system:

autonomous and nonautonomous dynamical systems

$$\dot{x} = x_1, \qquad \dot{x}_1 = x_2,$$
$$\dot{x}_2 = x^4 x_2 + \sin(x_3 - kx)x_1 + e^x\cos x_3,$$
$$\dot{x}_3 = \omega.$$

We have had to increase the dimension of our phase space by one, but in return, we have obtained an autonomous system of DEs.

Based on the prescription above, we turn the second-order DE of the driven pendulum into a set of first-order DEs. First we rewrite the DE describing a general pendulum [see Equation (31.21)] as

$$\ddot{\theta} + \gamma\dot{\theta} + \sin\theta = \phi_0\cos\alpha,$$

where α is simply $\omega_D t$. Then turn this equation into the following entirely equivalent set of three first-order DEs:

$$\dot{\theta} = \omega, \qquad \dot{\omega} = -\gamma\omega - \sin\theta + \phi_0\cos\alpha, \qquad \dot{\alpha} = \omega_D. \qquad (31.23)$$

The two-dimensional (θ, ω) phase space has turned into a three-dimensional (θ, ω, α) phase space. But the resulting system is autonomous.

Just as in the linear case, it is instructive to ignore the damping and driving forces first. We set γ and ϕ_0 equal to zero in Equation (31.23) and solve the set of DEs numerically.[5] For small angles, we expect a simple harmonic motion

---

[5]The solution can be given in terms of elliptic functions as discussed in Chapter 11.

(SHM). So, with $\theta(0) = \pi/10$ and $\omega(0) = 0$,[6] we obtain the plot on the left of Figure 31.7. This plot shows a simple trigonometric dependence of angle on time.

The initial angular displacement of the plot on the right of Figure 31.7 is approximately π radians corresponding to raising the mass of the pendulum all the way to the top.[7] The flattening of curves at the maxima and minima of the plot indicates that the pendulum almost stops once it reaches the top and momentarily remains motionless there. This is expected physically as $\theta(0) = \pi$ is a location of (unstable) equilibrium, i.e., with $\omega(0) = 0$, the pendulum can stay at the top forever. So, for $\theta(0) \approx \pi$, the pendulum is expected to stay at the top, not forever, but for a "long" time.

The phase space diagram of the pendulum can give us much information about its behavior. *With the initial angular velocity set at zero*, the pendulum will exhibit a periodic behavior represented by closed loops in the phase space. Figure 31.8 shows four such closed loops corresponding—from small to large

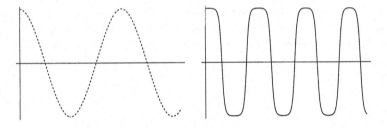

Figure 31.7: The undamped undriven pendulum shows an SHM for small initial angles (the plot on the left has a maximum angle of $\pi/10$). For large angles, the motion is periodic but not an SHM. The maximum angle of the plot on the right is slightly less than π.

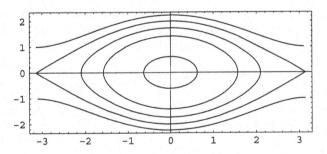

Figure 31.8: Phase space diagrams for a pendulum corresponding to different values of maximum displacement angles (horizontal axis). The inner diagrams correspond to smaller values; the outermost plot has a maximum angle of 179.98 degrees at which the angular speed is 1.

---

[6]It is important to keep $\omega(0)$ small, because a large initial angular velocity (even at a small initial angle) can cause the pendulum to reach very large angles!

[7]For this to be possible, clearly the mass should be attached to a rigid rod (not a string)!

loops—to the initial angular displacement of $\pi/5$, $\pi/2$, $2\pi/3$, and (almost) π. These loops represent *oscillations only*: the angular displacement is bounded between a minimum and a maximum value determined by $\theta(0)$. The closed loops are characterized by the fact that the angular speed vanishes at maximum (or minimum) θ, allowing the pendulum to start moving in the opposite direction.

The outermost curves result from $\theta(0) = -\pi$, $\omega(0) = 1$ (the upper curve), and $\theta(0) = \pi$, $\omega(0) = -1$ (the lower curve), and represent *rotations*. The angular displacement is unbounded: it keeps increasing for all times. Physically, this corresponds to forcing the pendulum to "go over the hill" at the top by providing it an initial angular velocity. If the pendulum is pushed over this hill once, it will continue doing it forever because there is no damping force. The rotations are characterized by a nonzero angular velocity at $\theta = \pm\pi$. This is clearly shown in Figure 31.8.

What happens when the damping force it turned on? We expect the trajectories to spiral into the origin of the phase space as in the case of the linear (small-angle) pendulum. Figure 31.9 shows two such trajectories corresponding to an initial displacement of just below π (on the right), and just above $-\pi$ (on the left). For both trajectories, the initial angular velocity is zero. It is intuitively obvious that regardless of the initial conditions, the pendulum will eventually come to a stop at $\theta = 0$ if there are no driving forces acting on it. So, Figure 31.9 is really representative of all dissipative motions of the pendulum.

The origin of the phase space is a fixed point of the pendulum dynamics. But it is not the only one. In general, any point in the phase space for which the time derivative of *all* coordinates of the trajectory are zero is a fixed point (see Problem 31.6). If we set all the functions on the RHS of Equation (31.23) equal to zero,[8] we obtain

$$\omega = 0 \qquad \text{and} \qquad \sin\theta = 0$$

corresponding to infinitely many fixed points at $(n\pi, 0)$ with n an integer. Points in the neighborhood of the origin, i.e., those lying in the basin of

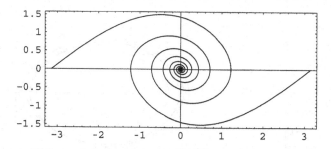

Figure 31.9: Phase space diagrams for a dissipative pendulum. Two trajectories starting at $\theta \approx -\pi$, $\omega = 0$ and $\theta \approx \pi$, $\omega = 0$ eventually end up at the origin.

---

[8]Still assuming no driving force.

attraction of the origin are attracted to the origin; the rest of the fixed points are repellors (or unstable) for such points.

The interesting motion of a pendulum begins when we turn on a driving force regardless of whether the dissipative effect is present or not. Nevertheless, let us place the pendulum in an environment in which $\gamma = 0.3$. Now drive this pendulum with a (harmonic) force of amplitude $\phi_0 = 0.5$ and angular frequency $\omega_D = 1$. A numerical solution of (31.23) will then give a result which has a transient motion lasting until $t \approx 32$. From $t = 32$ onward, the system traverses a closed orbit in the phase diagram as shown in Figure 31.10. This orbit is an *attractor* in the same sense as a point is an attractor for the logistic map and a dissipative nondriven pendulum. An attractor such as the

limit cycle   one exhibited in Figure 31.10 is called a **limit cycle**.

31.2.3   Onset of Chaos

As we increase the control parameter ϕ_0, the phase space trajectories go through a series of periodic limit cycles until they finally become completely aperiodic: chaos sets in. Figure 31.11 shows four trajectories whose common initial angular displacement θ_0, initial angular velocity ω_0, damping factor γ, and drive frequency ω_D are, respectively, π, 0, 0.5, and 2/3. The only (control) parameter that is changing is the amplitude of the driving force ϕ_0. This changes from 0.97 for the upper left to 1.2 for the lower right diagram.

A closer scrutiny of Figure 31.11—which we shall forego—will reveal that the chaotic behavior of the diagram at the lower right takes place after the pendulum goes through a bifurcation process as in the case of the logistic map. However, unlike the logistic map whose bifurcation stages were divided by fixed "points," the stages for the pendulum are characterized by limit cycles. In fact, the diagram at the upper left, corresponding to $\phi_0 = 0.97$, consists of two (very closely spaced) limit cycles. Bifurcations involving limit

Hopf bifurcation   cycles are called **Hopf bifurcation** after the mathematician E. Hopf who generalized the earlier results of Poincaré on such bifurcations to higher dimensions. The logistic map and the nonlinear pendulum have the following property in common: their "route to chaos" is via bifurcation. This is not

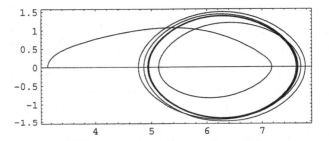

Figure 31.10: The moderately driven dissipative pendulum with $\gamma = 0.3$ and $\phi_0 = 0.5$. After a transient motion, the pendulum settles down into a closed trajectory.

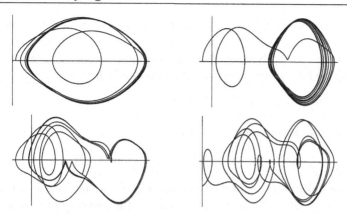

Figure 31.11: Four trajectories in the phase space of the damped driven pendulum. The only difference in the plots is the value of ϕ_0 which is 0.97 for the upper left, 1.1 for the upper right, 1.15 for the lower left, and 1.2 for the lower right diagrams.

true for all chaotic systems; there are other "routes to chaos," but we shall not investigate them here.

The main characteristic of chaos is the *exponential* divergence of neighboring trajectories. We have seen this behavior for the logistic map. A very nice illustration of this phenomenon for the nonlinear pendulum is depicted in Figure 31.12 where two nearby trajectories in the neighborhood of point $(-2, -2)$ are seen to diverge dramatically (in eight units of time).

The divergence of trajectories and the ensuing chaos has been termed the **butterfly effect** by Lorenz who, in the title of one of his talks, asked the   butterfly effect question: "Does the flap of a butterfly's wings in Brazil set off a tornado in Texas?" The point Lorenz is making in this statement is that if the atmosphere displays chaotic behavior (as a simple model proposed by Lorenz predicts), then a very small disturbance, such as the flapping of a butterfly's

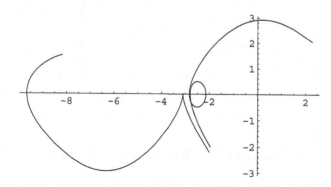

Figure 31.12: The projection onto the $\theta\omega$-plane of two trajectories starting at approximately the same point near $(-2, -2)$ diverge considerably after eight units of time. The loop does not contradict the DE uniqueness theorem!

wings, would make it impossible to predict the long-term behavior of the weather.

In general, for a dynamical system obeying an autonomous set of first-order DEs to be chaotic three requirements are to be met:

1. The trajectories must not intersect.

2. The trajectories must be bounded.

3. Nearby trajectories ought to diverge exponentially.

The first requirement is a direct consequence of the uniqueness theorem[9] for the solution of DEs: if two trajectories cross, the system will have a "choice" for its further development starting at the intersection point, and this is not allowed. The very notion of fixed point as well as the crisscrosses of Figures 31.11 and 31.12 may appear to violate the first property above. However, we have to remind ourselves that fixed points are (asymptotically) achieved after an *infinite* amount of time. As for the two figures, the reader recalls that all plots in those figures are *projections* of the three-dimensional trajectories onto the yz-plane. The three-dimensional trajectories never cross (see Figure 31.13).

The second requirement is important because unbounded regions of phase space correspond to infinities which are to be avoided. The third requirement is simply what defines chaos. It turns out that one- and two-dimensional phase spaces cannot accommodate all of these requirements. However, in three dimensions, one can "stretch" out the trajectories that want to loop in two dimensions as shown in Figure 31.13 where the loop of Figure 31.12 is seen to have been only a two-dimensional shadow! Thus,

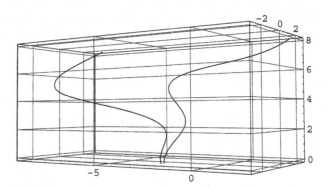

Figure 31.13: The two trajectories of Figure 31.12 shown in the full three-dimensional phase space.

---

[9]For our purposes this theorem states that if the dynamical variables and their first derivatives of a system are specified at some (initial) time, then the evolution of the system in time is uniquely determined. In the context of phase space this means that from any point in phase space only one trajectory can pass.

> **Box 31.2.1.** *A necessary condition for a system obeying autonomous DEs of first degree to be chaotic is to have a phase space that has at least three dimensions.*

The reader may wonder how a (one-dimensional) pendulum can satisfy the condition of Box 31.2.1. After all, the phase space of such a pendulum has only two dimensions. The answer lies in the fact that although a *driven* pendulum—with only θ and $\omega = \dot{\theta}$ regarded as independent variables—obeys DEs that are *not* autonomous, when time is turned into the third dimension of the phase space, a set of *three* autonomous DEs will result which allows chaotic behavior. This is in fact obvious from Equation (31.23) where α—the third dimension of the phase space—is seen to be essentially time in units of ω_D. A pendulum that is not driven does not exhibit chaotic behavior.

Only a *driven* pendulum (of large-angle oscillation) exhibits chaotic behavior.

31.3 Universality of Chaos

In the preceding sections, we examined two completely different systems displaying chaotic behavior. Although there are different "routes" to chaos, we shall concentrate only on the period-doubling route because it has been theoretically developed further than the other routes, and because it displays a universal character common to all such chaotic systems as discovered by one of the founders of the theory of chaos, Mitchell Feigenbaum.

31.3.1 Feigenbaum Numbers

In our theoretical investigation of the logistic map, we introduced the control parameters α_n at which the nth bifurcation takes place and for which there are a number of 2^n-cycle fixed points. It turns out that the ratio

$$\delta_n \equiv \frac{\alpha_n - \alpha_{n-1}}{\alpha_{n+1} - \alpha_n} \tag{31.24}$$

is almost the same for all large n, and that, in the limit as $n \to \infty$, it approaches a number δ^*, now called the **Feigenbaum delta**:

Feigenbaum delta

$$\delta^* \equiv \lim_{n \to \infty} \delta_n = \lim_{n \to \infty} \frac{\alpha_n - \alpha_{n-1}}{\alpha_{n+1} - \alpha_n} = 4.66920\ldots. \tag{31.25}$$

Feigenbaum looked at the same ratio for the so-called iterated sine function

$$x_{n+1} = \beta \sin(\pi x_n)$$

and found that exactly the same number was obtained in the limit. Later, he showed that δ^* is the same for *all* iterated map functions![10]

---

[10]This is not entirely true. The map functions should have a parabolic "shape" at their maximum. The logistic map and the sine function—as well as many other functions—have this property.

We can use δ^* to calculate approximations to α_n for large values of n, and, in particular, to find an approximate value for α_∞. First we note that, if we approximate δ_n with δ^*, then (31.24) yields

$$\alpha_{n+1} = \frac{\alpha_n - \alpha_{n-1}}{\delta_n} + \alpha_n \approx \frac{\alpha_n - \alpha_{n-1}}{\delta^*} + \alpha_n.$$

For example,

$$\alpha_3 \approx \frac{\alpha_2 - \alpha_1}{\delta^*} + \alpha_2, \qquad \alpha_4 \approx \frac{\alpha_3 - \alpha_2}{\delta^*} + \alpha_3,$$

or

$$\alpha_4 \approx \frac{(\alpha_2 - \alpha_1)/\delta^*}{\delta^*} + \alpha_3 = (\alpha_2 - \alpha_1)\left(\frac{1}{\delta^*} + \frac{1}{\delta^{*2}}\right) + \alpha_2.$$

We can easily generalize this to

$$\alpha_N \approx (\alpha_2 - \alpha_1)\left(\frac{1}{\delta^*} + \frac{1}{\delta^{*2}} + \cdots + \frac{1}{\delta^{*(N-1)}}\right) + \alpha_2. \qquad (31.26)$$

In the limit that $N \to \infty$, the sum becomes a geometric series which adds up to $1/(\delta^* - 1)$. So,

$$\alpha_\infty \approx \frac{\alpha_2 - \alpha_1}{\delta^* - 1} + \alpha_2. \qquad (31.27)$$

With $\alpha_1 = 3$ and $\alpha_2 = 1 + \sqrt{6}$, we obtain

$$\alpha_\infty \approx \frac{\sqrt{6} - 2}{3.66920} + 1 + \sqrt{6} = 3.572.$$

The actual value—obtained by more elaborate calculations—is $3.5699\ldots$.

Another quantity that seems to be universal is the ratio of the consecutive "bifurcation sizes." We mentioned earlier that there are several fixed points associated with the 2^n-cycle parameter α_n. At each stage of bifurcation, these fixed points come in pairs. For example, at $\alpha_2 \equiv 1+\sqrt{6}$, Equation (31.16) gives the two fixed points at $x = 0.849938$ and $x = 0.43996$. We define the "size" d_1 of the 4-cycle bifurcation as the (absolute value of the) difference between these x-values. In general, we define d_n, the size of the bifurcation pattern of period 2^n as the largest (in absolute value) of the differences between the two x's of each of the 2^n pairs of fixed points. On a bifurcation diagram, one would measure the vertical distance between the points where each curve of the diagram starts to branch out. If there are several such distances, one chooses the largest one. The second Feigenbaum number, the so-called **Feigenbaum alpha**, is then defined as

Feigenbaum alpha

$$\alpha^* \equiv \lim_{n \to \infty} \frac{d_n}{d_{n+1}} = 2.5029\ldots. \qquad (31.28)$$

Feigenbaum found that this number is obtained for the bifurcation pattern of all chaotic systems which reach chaos via bifurcation.

Aside from its universality as applied to *different* chaotic systems, this number suggests a general "size" scaling within the bifurcation pattern of a single system: For large enough values of n, the ratio of the size of each bifurcation is the same as the previous one. If we "blow up" the small bifurcations taking place for large values of n, they look almost identical to the ones occurring before them. This property is also called **self-similarity**.   self-similarity

31.3.2   Fractal Dimension

An elegant way of quantifying chaos is by examining the geometric properties of the trajectory of the chaotic system under study. Suppose we let the system run for a long time and suppose that it gravitates toward an attractor and remains there.[11] What is the "dimension" of the trajectory? The clarification of this question and the logic (as well as the application) of its answer is the subject of this subsection.

Intuitively, one assigns the dimension of 0 to points, 1 to curves, 2 to surfaces, 3 to volumes, and n to "solid" objects residing in spaces requiring n coordinates to describe their points. How can we go beyond intuition? We use the so-called **Hausdorff dimension**, whose calculation goes as follows.   Hausdorff
Try to cover the geometrical object by appropriate "boxes" of side length r.   dimension
Now count the number $N(r)$ of boxes required to contain all points of the geometric object. The Hausdorff dimension D is defined by

$$N(r) = \lim_{r \to 0} \left(k r^{-D} \right), \tag{31.29}$$

where k is an inessential proportionality constant which describes the shape of the "box." For example, as a box, we could use a "sphere" of radius r. Then the "volume" would be $2r$ for a line, πr^2 for a circle, and $\frac{4}{3}\pi r^3$ for a sphere. Thus, k is 2, or π or $\frac{4}{3}\pi$. If we choose "cubes," k will always be 1. Furthermore, by changing the unit of length, one can change k. Fortunately, as we shall see shortly, k will not enter the final definition of Hausdorff dimension.

Equation (31.29) can be solved for D,

$$D = \lim_{r \to 0} \left[-\frac{\ln N(r)}{\ln r} + \frac{\ln k}{\ln r} \right].$$

Now, we can see why k is not essential: As $r \to 0$ the denominator of the second term grows beyond bound. So,

$$D = - \lim_{r \to 0} \frac{\ln N(r)}{\ln r} \tag{31.30}$$

Let us test (31.30) on some familiar geometric objects. If the object is a single point on a line, then only one "box" is needed to cover it regardless of the size of the box. So, $N(r) = 1$ for all r, and Equation (31.30) gives $D = 0$.

---

[11]By an attractor, we mean any geometrical object on which the trajectory hovers. It can be a fixed point, a limit cycle, or some multidimensional object in the phase (hyper-)space.

In fact, the Hausdorff dimension of any finite number of points on a line is found to be zero. Similarly, the dimension of a finite number of points on a *surface* or in a *volume* is also zero.

If the object is a surface of area A, then we require A/r^2 boxes (squares) to cover the entire area. Thus,

$$D = -\lim_{r \to 0} \frac{\ln(A/r^2)}{\ln r} = -\lim_{r \to 0} \left(\frac{\ln A - 2\ln r}{\ln r} \right) = 2.$$

Similarly, the reader may check that the Hausdorff dimension for a curve is 1, and for a volume it is 3. So, the formula seems to be working for familiar geometric objects.

Cantor set

Example 31.3.1. A not-so-familiar geometric object is the **Cantor set**: Take the closed interval $[0, 1]$; remove its middle third; do the same with the remaining two segments; continue the process ad infinitum (Figure 31.14). What is left of the line segment is the Cantor set, named after the German mathematician whose work on set theory, controversial at the time, laid the foundation of modern formal mathematics. Figure 31.14 should convince the reader that after n steps, 2^n segments are left and that the length of each segment is $(1/3)^n$. Thus, denoting the size of the box after n steps by r_n, we have

$$r_n = (1/3)^n, \qquad N(r_n) = 2^n.$$

Therefore,

$$D = -\lim_{r \to 0} \frac{\ln N(r)}{\ln r} = -\lim_{n \to \infty} \frac{\ln N(r_n)}{\ln r_n} = -\lim_{n \to \infty} \frac{\ln(2^n)}{\ln[(1/3)^n]}$$

$$= -\lim_{n \to \infty} \frac{n\ln 2}{n\ln(1/3)} = -\frac{\ln 2}{\ln(1/3)} = \frac{\ln 2}{\ln 3} = 0.6309\ldots. \tag{31.31}$$

So, the Cantor set is more than just a set of points (dimension zero) and less than a line segment (dimension one). It is amusing to note—as the reader may verify—that the *length* of the Cantor set is zero! ∎

The Cantor set is only one example of geometrical objects whose dimensions are nonintegers:

fractal object or fractal

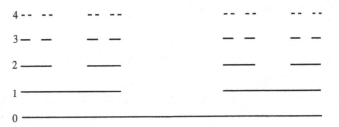

Figure 31.14: The Cantor set after one, two, three, and four "dissections."

> **Box 31.3.1.** *A geometrical object, whose Hausdorff dimension in not an integer, is called a **fractal object** or simply a **fractal**.*

Example 31.3.2. Another example of a fractal object is the so-called **Koch snowflake**. Start with an equilateral triangle of side L [Figure 31.15(a)]; remove the middle third of each side and replace it with two identical segments (a "wedge") to form a star [Figure 31.15(b)]. Do the same to the small segments so obtained [Figure 31.15(c)], and continue ad infinitum. The result is the Koch snowflake.

Koch snowflake

Let us find the Hausdorff dimension of the Koch snowflake.[12] It should be clear that the number of line segments on each side of the triangle at step n is 4^n so that $N(r_n) = 3 \times 4^n$, and the length r_n of each line segment is $L/3^n$. Therefore, the Hausdorff dimension of the Koch snowflake is

A finite area bounded by an infinite closed curve!

$$D = -\lim_{n \to \infty} \frac{\ln N(r_n)}{\ln r_n} = -\lim_{n \to \infty} \frac{\ln(3 \times 4^n)}{\ln(L/3^n)}$$

$$= -\lim_{n \to \infty} \frac{\ln 3 + n \ln 4}{\ln L - n \ln 3} = \frac{\ln 4}{\ln 3} = 1.2618595\ldots. \qquad (31.32)$$

The length of the perimeter of the snowflake is

$$\lim_{n \to \infty} N(r_n) r_n = \lim_{n \to \infty} (3 \times 4^n)\,(L/3^n) = 3L \lim_{n \to \infty} \left(\tfrac{4}{3}\right)^n \to \infty.$$

It is interesting to note that the *area* enclosed by the Koch snowflake is *finite* while its perimeter is *infinite*! ∎

The fractals discussed so far have the property which we called self-similarity. The present case is, however, a true (or regular) self-similarity because, as we scale the object, we obtain the *exact* replica of the original. In contrast, for

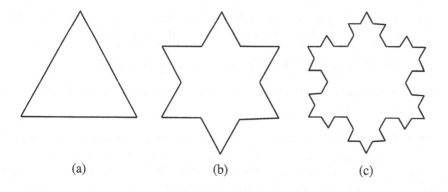

(a)   (b)   (c)

Figure 31.15: (a) Begin with an equilateral triangle. (b) Remove the middle third of each side and replace it with a "wedge" to form a star. (c) Remove the middle third of each new segment and replace them with "wedges." Continue ad infinitum to obtain the Koch snowflake.

---

[12]This is the dimension of the *perimeter*, not the area of the snowflake.

the logistic map, we obtained bifurcations which contained different scaling ratios: The ratio was α^* only for the largest bifurcation size at each stage. Comparison of the largest size with smaller sizes would not have yielded α^*. These (irregular) self-similarities occur frequently in chaotic systems and the determination of their Hausdorff dimension can give information about the long-term behavior of the dynamics of the system.

In the case of the logistic map, the Hausdorff dimension of the set of vertical (fixed) points on the bifurcation diagram—which is zero at all finite stages of bifurcation—will not be zero at α_∞. It has, in fact, been calculated to be $0.5388\ldots$. This is an example of attractors that have noninteger dimensions, i.e., they are neither points nor lines. If the attractor of a dissipative dynamical system has a fractal dimension, then we say that the system has a **strange attractor**. Strange attractors play a fundamental role in the theory of chaos.

strange attractor

31.4 Problems

31.1. Show that (31.1) leads to (31.2).

31.2. For the logistic map, assume that $1 < \alpha < 3$. Show that if $x_k > 1-1/\alpha$, then $x_{k+1} < x_k$, and if $x_k < 1 - 1/\alpha$, then $x_{k+1} > x_k$. Therefore, conclude that $x^* = 1 - 1/\alpha$ is a stable fixed point.

31.3. Write the cubic polynomial in Equation (31.10) as

$$\alpha^3 \left(x - 1 + \frac{1}{\alpha} \right) (x^2 + ax + b)$$

and determine a and b by expanding and comparing the result with (31.10). Now solve $x^2 + ax + b = 0$ to obtain $x_2(\alpha)$ and $x_3(\alpha)$ of (31.11).

31.4. Derive Equation (31.21) from the equation that precedes it.

31.5. Convince yourself that a system of N particles in space has a $6N$-dimensional phase space.

31.6. Consider a set of autonomous first-order DEs. Suppose that a point P of the phase space is a root of *all* functions on the RHS. By expanding each coordinate of a trajectory in a Maclaurin series in t and keeping only the first two terms, show that the trajectory does not move away from P. So, fixed points are determined by setting all functions on the RHS of an autonomous system equal to zero and solving for the coordinates.

31.7. Derive Equation (31.27) from (31.26).

31.8. Show that the dimension of a finite number of points on a *surface* or in a *volume* is zero.

31.9. Show that the Hausdorff dimension of any finite number of points is zero, of a curve is 1, and of a volume is 3.

31.10. Show that the Hausdorff dimension of the Cantor set is independent of the length of the original line segment.

31.11. Verify that the length of the Cantor set is zero.

Chapter 32

Probability Theory

Although probability theory did not flourish until after the Renaissance, and in particular in the 17th and 18th centuries, its roots go back to ancient history. Archaeological excavations reveal the presence of knuckle-bones (or astragali) in numbers far larger than any other kind of bones, indicating the possibility of the use of these bones in games. There is strong evidence that astragali were in use for board games at the time of The First Dynasty in Egypt (c.3500 B.C.). Other archaeological excavations, unearthing more recent periods, e.g. 1300 B.C. in Turkey, also reveal a definite connection between astragalus and recreation.

It seems that games of chance, such as the board game mentioned above, are, like counting, as old as civilization itself. Yet the science of counting, arithmetic, was already in an advanced stage of evolution when probability started to take root as a mathematical science in the 17th century. Why? Perhaps the reason is the crudeness with which "randomizers" such as dice—the artificial substitutes of astragali—were made for a long time. Abstraction requires perfection. Although the abstraction of counting from what was being counted took place naturally, the corresponding abstraction of randomness from what is random demanded an ideal device capable of producing random events, and a large number of experimental data for analysis, and this did not happen until well into the 17th century.

32.1 Basic Concepts

The reader no doubt has some familiarity with the notion of a **random event**. Any occurrence or experiment, whose outcome is uncertain is such an event. Flipping a coin, pulling a card out of a deck of cards, and throwing a die are all examples of experiments whose outcome are uncertain (if the coin, the deck of cards, and the die are all "unbiased"). The reader may also know intuitively that the chance of getting a head in the toss of a coin is 50% (or 1 out of 2, or 0.5); that the chance of getting a 3 in the throw of a die is 1

Random event: basis for probability

out of 6; and that the chance of getting a club in drawing a card is 1 out of 4. The aim of the theory of probability is to make precise these intuitive notions and to develop a mathematical procedure for answering questions related to random events. First we need to review some simple concepts from set theory.

32.1.1   A Set Theory Primer

Universal set

The most fundamental entity in any branch of mathematics is a **universal set**. It is the collection of all objects under consideration. For example, the universal set of plane geometry is a flat surface, and of solid geometry is the three-dimensional space. The universal set of calculus is the set of real numbers (or the real line), and the complex plane is the universal set of complex analysis. The generic universal set is denoted by S, but each specific universal set has its own symbol: $\mathbb{R}$ is the set of real numbers, $\mathbb{C}$ is the set of complex numbers, $\mathbb{Z}$ is the set of integers, and $\mathbb{N}$ is the set of natural numbers (nonnegative integers).

The simplest relation in set theory is that of belonging. We write $a \in S$ (and say "a belongs to S" or "a is in S" to express the fact that a is one of the objects in S. An object in S is called an **element** of S. A collection A of elements of S is called a **subset** of S, and we write $A \subset S$. In particular, $S \subset S$. Any subset can be considered as a set with its elements and subsets. Thus, $a \in A$ means that a is one of the elements of the subset A, $a \notin A$ means that a is not one of the elements A, and $B \subset A$ means that B consists of elements, all of which belong to A. Subsets are often specified either by enumeration or by some statement enclosed between a pair of curly brackets. For example,

Element and
subset of a set

$$\{0,1,2,3,\ldots\}, \quad \{2,4,6,\ldots\}, \quad \{2n+1 | n \in \mathbb{N}\},$$

$$\{(x,x) | x \in \mathbb{R}\}, \quad \left\{ -\frac{13.6}{n^2} \middle| n \in \mathbb{N}, n \neq 0 \right\}.$$

The first describes $\mathbb{N}$; the second, the set of even numbers; the third, the set of odd numbers; the fourth, the line $y = x$; and the fifth, the energy levels of the hydrogen atom in electron volt. Two subsets are equal if each is a subset of the other. In other words, if $A \subset B$ and $B \subset A$, then $A = B$. It is convenient to introduce the **empty** set, a subset $\emptyset$ of S, which has no element.

The subsets of a universal set have a rich mathematics which we can only briefly outline here. Given two sets[1] A and B, we can form another set, called the **union** of A and B and denoted by $A \cup B$, which consists of all elements belonging to either A or B or both. Thus,

$$A \cup B = \{x \in S | x \in A \text{ or } x \in B\}.$$

---

[1] It is very common to delete the prefix 'sub' and refer to subsets of a universal set as simply sets.

The **intersection** of A and B, denoted by $A \cap B$, consists of all elements that belong to both A and B:

Union, intersection, and complement

$$A \cap B = \{x \in S | x \in A \text{ and } x \in B\}.$$

The **complement** of a set A is the subset of S which contains all the elements of S which are not in A. Denoting this set by A^c, we have

$$A^c = \{x \in S | x \notin A\}.$$

The reader may easily verify that $S = A \cup A^c$ and $\emptyset = A \cap A^c$. When $A \cap B = \emptyset$, we say that A and B are **disjoint**.

Disjoint sets

The operations of union and intersection are commutative and associative:

$$A \cup B = B \cup A, \qquad\qquad A \cap B = B \cap A,$$
$$(A \cup B) \cup C = A \cup (B \cup C), \qquad (A \cap B) \cap C = A \cap (B \cap C).$$

Thus one can take the union and intersection of a number of sets without worrying about the order of the sets or where to put the parentheses. This makes it possible to introduce the following notations for the union and intersection of a family of sets:

$$\bigcup_{i=1}^{n} A_i \equiv A_1 \cup A_2 \cup \cdots \cup A_n,$$
$$\bigcap_{i=1}^{n} A_i \equiv A_1 \cap A_2 \cap \cdots \cap A_n. \tag{32.1}$$

We define the difference between two sets $A - B \equiv A \cap B^c$ as the collection of elements in A that are not in B. It is not hard to show that $A - B$, $B - A$, and $A \cap B$ are mutually disjoint. Furthermore,

$$A = (A - B) \cup (A \cap B),$$
$$B = (B - A) \cup (A \cap B), \tag{32.2}$$
$$A \cup B = (A - B) \cup (A \cap B) \cup (B - A).$$

Note that all sets on the right-hand side of each equation are mutually *disjoint*.

A useful way of picturing sets and operations on them is a **Venn diagram**. The universal set is depicted as a rectangle, and its subsets as circles in the rectangle. Figure 32.1 shows some examples of the use of Venn diagrams. Venn diagrams are intuitive representations of relations among sets. For example, the diagram on the right of Figure 32.1 shows clearly the equalities of Equation (32.2).

Venn diagrams

Using Venn diagrams, one can show that the operation of union distributes over intersection and vice versa:

$$A \cap (B \cup C) = (A \cap B) \cup (A \cap C),$$
$$A \cup (B \cap C) = (A \cup B) \cap (A \cup C), \tag{32.3}$$

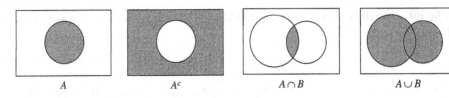

Figure 32.1: Venn diagrams of some sets. The grey region represents the set labeled at the bottom.

and more generally,

$$A \cap \left(\bigcup_{i=1}^{n} B_i \right) = \bigcup_{i=1}^{n} (A \cap B_i),$$

$$A \cup \left(\bigcap_{i=1}^{n} B_i \right) = \bigcap_{i=1}^{n} (A \cup B_i). \tag{32.4}$$

32.1.2   Sample Space and Probability

The underlying concept in probability theory is the **sample space**, which is the same as the universal set of the set theory and is also denoted by S. It is the collection of all possible outcomes of an experiment. For example, for the toss of a coin, $S = \{H, T\}$; for the toss of two coins, $S = \{HH, HT, TH, TT\}$; and for a die, $S = \{1, 2, 3, 4, 5, 6\}$. An **event** E is simply a subset of the sample space. Thus, the event $\{HT, TH\}$ is described as the outcome in the toss of two coins, in which one of the coins is head; and $\{2, 4, 6\}$ is the event that the roll of a die produces an even number. An event, therefore, can be **elementary** or **compound**, with the latter being a collection of the former.

Event: elementary and compound

We are now ready to define probability. Since the sample space—which is now also called the **probability space**—includes all possible events, its probability should be one, corresponding to absolute certainty. The probability of any event (any subset of the probability space) has to be a nonnegative number less than one. We may be tempted to say that the probability of the union of two events is the sum of their probabilities, but that would be wrong. For example, let $S = \{1, 2, 3, 4, 5, 6\}$ be the universal set of a die, and consider $E_1 = \{1, 3, 5\}$, the odd outcomes, and $E_2 = \{4, 5, 6\}$, all outcomes greater than 3. Intuitively, we know that the probability for each of these two events is $\frac{1}{2}$. But $E_1 \cup E_2 = \{1, 3, 4, 5, 6\}$, and if we were to add probabilities for the union, we would get that the probability of $\{1, 3, 4, 5, 6\}$ is one, which is clearly wrong. The reason for this is that we have actually double-counted $\{5\}$, the intersection of the two sets. Only if the two sets are disjoint, can we add the probabilities for the union. Now we define probability:

Probability space is sample space.

Box 32.1.1. $\mathcal{S}$ *is called a* **probability space** *if for each event $E \subset \mathcal{S}$ there is a number $P(E)$ satisfying the following conditions.*

1. $0 \leq P(E)$.

2. $P(\mathcal{S}) = 1$.

3. If E_1 and E_2 are disjoint *events, then $P(E_1 \cup E_2) = P(E_1) + P(E_2)$.*

Example 32.1.1. In this example we derive some relations involving probabilities.
(1) If $E_1 \subset E_2$, then $P(E_1) \leq P(E_2)$. To show this, use the first equation in (32.2) and the fact that $E_1 \cap E_2 = E_1$ to write

$$E_2 = (E_2 - E_1) \cup E_1 \quad \text{and} \quad P(E_2) = P(E_2 - E_1) + P(E_1).$$

Since $P(E_2 - E_1)$ is nonnegative, we get $P(E_2) \geq P(E_1)$.

(2) $P(E) \leq 1$ for every event E. This is a consequence of (1), because $E \subset \mathcal{S}$ and $P(\mathcal{S}) = 1$.

(3) For any two events E_1 and E_2,

$$P(E_1 \cup E_2) = P(E_1) + P(E_2) - P(E_1 \cap E_2). \tag{32.5}$$

Use Equation (32.2) to write

$$\begin{aligned}
P(E_1) &= P(E_1 - E_2) + P(E_1 \cap E_2), \\
P(E_2) &= P(E_2 - E_1) + P(E_1 \cap E_2), \\
P(E_1 \cup E_2) &= P(E_1 - E_2) + P(E_1 \cap E_2) + P(E_2 - E_1).
\end{aligned} \tag{32.6}$$

Substituting $P(E_1 - E_2)$ and $P(E_2 - E_1)$ of the first two equations in the last equation, we obtain the desired result.

Using E as E_1 and E^c as E_2, and noting that $E - E^c = E$ and $E \cap E^c = \emptyset$, the first (or second) equation in (32.6) gives $P(E) = P(E) + P(\emptyset)$, implying that
(4) $P(\emptyset) = 0$.

Using E as E_1 and E^c as E_2 again, and noting that $E - E^c = E$ and $E \cup E^c = \mathcal{S}$, the third equation in (32.6) and (4) give $P(\mathcal{S}) = P(E) + P(E^c)$, implying that
(5) $P(E^c) = 1 - P(E)$. ∎

Condition 3 of Box 32.1.1 can be generalized to the case of a collection of *mutually disjoint* sets $E_1, E_2, \ldots, E_m$:

$$P(E_1 \cup E_2 \cup \ldots \cup E_m) = P(E_1) + P(E_2) + \cdots + P(E_m) = \sum_{i=1}^{m} P(E_i). \tag{32.7}$$

A collection of mutually disjoint sets $E_1, E_2, \ldots, E_m$ with $\mathcal{S} = E_1 \cup E_2 \cup \ldots \cup E_m$ is called a **partition** of $\mathcal{S}$. Such a collection has the property that

partition of universal set

$$\sum_{i=1}^{m} P(E_i) = 1. \tag{32.8}$$

Up to now, we have not assigned any value to $P(E)$ for a given set E, and it cannot be done without some further assumptions concerning the physical properties of the probability space and the events that make it up. In fact, if $E_1, E_2, ..., E_m$ partition S, any set of nonnegative numbers p_1, p_2, ..., p_m adding up to 1 with $P(E_i) = p_i$ will satisfy the conditions of Box 32.1.1 and will turn S into a probability space. Physically, however, certain choices will not make sense. For instance, for $S = \{H, T\}$, the sample space of a single coin, one can set $P(H) = 0.75$ and $P(T) = 0.25$. However, this assignment is not very useful for ordinary coins, and in practice gives false results. For a probability space composed of elementary events, it is often natural to assign equal probability values to the elementary events. Thus if the E_i of Equation (32.8) are all elementary, then the natural assignment would be $P(E_i) = 1/m$ for $i = 1, 2, \ldots, m$. For a coin, $m = 2$ and $P(H) = P(T) = 0.5$ is a natural choice, while for a die $P(i) = 1/6$, and for a deck of cards, $P(E_i) = 1/52$.

32.1.3   Conditional and Marginal Probabilities

In many situations, the sample space is partitioned in two different ways. For example, a deck of cards can be partitioned either by 4 suits or by 13 values; the employees of a company can be partitioned by gender or by departments in which they work. Suppose $E_1, E_2, \ldots, E_m$ and $F_1, F_2, \ldots, F_n$ are two collections of events that partition S. It should be clear that $E_i \cap F_j$, $i = 1, 2, \ldots, m$; $j = 1, 2, \ldots, n$ is also a partition of S, and that

$$\bigcup_{j=1}^{n} (E_i \cap F_j) = E_i \qquad \text{and} \qquad \bigcup_{i=1}^{m} (E_i \cap F_j) = F_j. \qquad (32.9)$$

Since E_i, F_j, and $E_i \cap F_j$ are all partitions of S, we can define the probabilities $P(E_i)$, $P(F_j)$, and $P(E_i \cap F_j)$. Then, Equation (32.9) implies that

$$P(E_i) = \sum_{j=1}^{n} P(E_i \cap F_j) \qquad \text{and} \qquad P(F_j) = \sum_{i=1}^{m} P(E_i \cap F_j). \qquad (32.10)$$

Marginal and
conditional
probabilities

$P(E_i)$ and $P(F_j)$ are called **marginal probabilities**.
Associated with the marginal probability is the **conditional probability**. Suppose we know that E_i has occurred. What is the probability of F_j? For example, we draw a card from a deck of cards and somebody tells us that it is a heart. What is the probability that it is a jack? This *conditional probability* is denoted by $P(F_j|E_i)$ and is the *probability of F_j given that E_i has occurred*.

Example 32.1.2. The best way to understand marginal and conditional probabilities is to look at an example. Suppose that in a container, we have 100 marbles coming in three different sizes: small, medium, and large; and five different colors: white, black, red, green, and blue. Table 32.1 shows the distribution of the marbles according to color and size.
First note that from the very definition of probability, the chance of getting a medium red marble in a random drawing is 0.07, that of a large green marble is 0.03,

	White	Black	Red	Green	Blue	Total
Small	5	7	6	8	4	**30**
Medium	8	10	7	12	8	**45**
Large	9	5	4	3	4	**25**
Total	**22**	**22**	**17**	**23**	**16**	**100**

Table 32.1: The distribution of marbles according to size and color.

and there is a likelihood of 0.05 for drawing a small white marble. Similarly the probability that on a random drawing from the container, the ball is black is 0.22, and for the ball to be medium it is 0.45. This suggests the construction of another table, Table 32.2, which shows the distribution of the *probabilities* according color and size.

Each entry of the last row and last column of Table 32.2 is what we have called a marginal probability. The conditional probability that the marble is small *given that its color is white* is 5/22. This is because, by restricting the color to white, we limit the number of marbles to 22 rather than 100. Similarly, the probability that the marble is green *given that its size is medium* is 12/45; this also is a conditional probability. Conditional probabilities can be rewritten as ratios of probabilities. Thus, the probability that the marble is small given that its color is white is 0.05/0.22, and the second probability is 0.12/0.45. ∎

The results of the foregoing example can be easily generalized. Let $p_{ij} = P(E_i \cap F_j)$, construct a table with m rows and n columns, and fill the cells with the numbers p_{ij}. Add one more row for the totals with entries $P(F_1)$, $P(F_2)$, all the way to $P(F_n)$. Add one more column for the totals with entries $P(E_1)$, $P(E_2)$, all the way to $P(E_m)$. It should now be clear that $P(F_j|E_i)$, the probability of F_j given that E_i has occurred, is

$$P(F_j|E_i) = \frac{P(E_i \cap F_j)}{P(E_i)}. \tag{32.11}$$

Since any event and its complement partition the universal set, we can let $F_1 = A$ and $F_2 = A^c$ (only two F's), and write the equation above as

$$P(A|E_i) = \frac{P(E_i \cap A)}{P(E_i)}, \tag{32.12}$$

	White	Black	Red	Green	Blue	Total
Small	0.05	0.07	0.06	0.08	0.04	**0.3**
Medium	0.08	0.1	0.07	0.12	0.08	**0.45**
Large	0.09	0.05	0.04	0.03	0.04	**0.25**
Total	**0.22**	**0.22**	**0.17**	**0.23**	**0.16**	**1**

Table 32.2: The distribution of probabilities according to size and color.

or if we have two sets A and B and their complements as two partitions of S, then

$$P(A|B) = \frac{P(B \cap A)}{P(B)}, \tag{32.13}$$

and this is true for *any two* sets.

    If the probability $P(A|B)$ does not depend on the event B in any way, i.e., if $P(A|B) = P(A)$, then we say that the two events A and B are **statistically independent**. Equation (32.13) now yields

Statistically independent events

$$P(A) = \frac{P(B \cap A)}{P(B)} \quad \text{or} \quad P(A \cap B) = P(A)P(B), \tag{32.14}$$

and the second equation becomes the definition for two events to be statistically independent.

Difference between statistical independence and mutual exclusion

    It is important to differentiate between statistical independence and mutual exclusion. If two events are mutually exclusive then they have to be statistically *dependent* since the occurrence of one precludes the occurrence of the other. Similarly, Equation (32.14) shows that if $P(A) > 0$, $P(B) > 0$, and A and B are statistically independent, then $P(A \cap B) \neq 0$, implying that $A \cap B \neq \emptyset$ and, therefore, that A and B cannot be mutually exclusive.

    Equation (32.12) could be rewritten as

$$P(A \cap E_i) = P(E_i)P(A|E_i),$$

and since $A \cap E_i$ are mutually exclusive and their union is A, we have [see the second equation in (32.10) with $A = F_j$]

$$P(A) = \sum_{i=1}^{m} P(E_i)P(A|E_i). \tag{32.15}$$

Bayes' theorem

This is called **Bayes' theorem**.

Example 32.1.3. A selective four-year college admits mostly students whose ACT scores are 32 and higher, with a small number of admitted students whose scores are below 32. The college has a graduation rate of 97%. Of those who graduate, 98% have an ACT score of 32 and higher. Of those who drop out, 85% have an ACT score below 32. We want to calculate the probability of graduation for a student who has an ACT score below 32.

    Let E_1 and E_2 denote the events corresponding, respectively to graduating and dropping out. Let F_1 and F_2 denote the events corresponding to an ACT score of 32 or higher and lower than 32, respectively. We are after $P(E_1|F_2)$.

    Consider the following table, in which the most obvious probabilities are entered:

	F_1	F_2	Total
E_1			0.97
E_2			0.03
Total			1

Since we are given that $P(E_1) = 0.97$ and $P(F_1|E_1) = 0.98$, we can use Equation (32.11) to find the entry, p_{11}, in the first row and first column:

$$p_{11} = P(F_1 \cap E_1) = P(E_1 \cap F_1) = P(F_1|E_1)P(E_1) = 0.98 \times 0.97 = 0.9506.$$

The entry, p_{12}, in the first row and second column can now be calculated because the total is given as 0.97:

$$p_{12} = P(F_2 \cap E_1) = 0.97 - 0.9506 = 0.0194.$$

The table now looks like

	F_1	F_2	Total
E_1	0.9506	0.0194	**0.97**
E_2			**0.03**
Total			**1**

We are also given that $P(F_2|E_2) = 0.85$. So, using Equation (32.11) again, we can find p_{22}:

$$p_{22} = P(F_2 \cap E_2) = P(F_2|E_2)P(E_2) = 0.85 \times 0.03 = 0.0255.$$

The remaining entries are now trivial to calculate:

$$p_{21} = P(F_1 \cap E_2) = 0.03 - 0.0255 = 0.0045,$$
$$P(F_1) = 0.9506 + 0.0045 = 0.9551,$$
$$P(F_2) = 0.0194 + 0.0255 = 0.0449,$$

and the complete table becomes

	F_1	F_2	Total
E_1	0.9506	0.0194	**0.97**
E_2	0.0045	0.0255	**0.03**
Total	**0.9551**	**0.0449**	**1**

The desired probability is therefore,

$$P(E_1|F_2) = \frac{P(E_1 \cap F_2)}{P(F_2)} = \frac{0.0194}{0.0449} = 0.432.$$

So, there is almost a 43% chance for the graduation of a student whose ACT score is below 32. ∎

32.1.4 Average and Standard Deviation

When we are given a set of values—say the scores of students in a class—and asked to find the average, we add the values and divide by the total number of values. If $\{x_i\}_{i=1}^N$ is the set of values, then the average $\bar{x}$ is given by

$$\bar{x} = \frac{\sum_{i=1}^N x_i}{N} = \sum_{i=1}^N \left(x_i \frac{1}{N} \right).$$

This equation tacitly assumes that the probability is the same for all values and equal to $1/N$. If the probability depends on i, the definition of the average has to take this into account. Let p_i denote the probability for the occurrence of x_i, and change the notation for the average to the more common notation whereby *capital* letters are used inside angle brackets. Then the **average** or **mean** or **expectation value** of $\{x_i\}_{i=1}^{N}$ is defined as

Average, expectation value, mean

$$\langle X \rangle = \sum_{i=1}^{N} x_i p_i. \tag{32.16}$$

Standard deviation

Another quantity of interest is the **standard deviation**, which is a measure of how the values are spread from the mean. It is the average "distance" between $\bar{x}$ and x_i. The obvious choice $x_i - \bar{x}$ will have a zero average, because it is both positive and negative and the definition of $\bar{x}$ makes the positive and negative values cancel. To avoid this cancellation, one takes the *square* of these differences and then averages them. The **variance** σ^2 is defined by

Variance

$$\sigma^2 = \frac{\sum_{i=1}^{N}(x_i - \bar{x})^2}{N},$$

and the standard deviation by

$$\sigma = \sqrt{\frac{\sum_{i=1}^{N}(x_i - \bar{x})^2}{N}}. \tag{32.17}$$

When probability varies with x_i, the definition of the variance changes to

$$\sigma^2 = \sum_{i=1}^{N}(x_i - \langle X \rangle)^2 p_i. \tag{32.18}$$

In many situations one may be interested in the average of a quantity that depends on the random variable x_i. Thus, if $g(x_i)$ is such a quantity, one writes

$$\langle g(X) \rangle = \sum_{i=1}^{N} g(x_i) p_i. \tag{32.19}$$

In terms of such averages, one can show that

$$\sigma^2 = \langle X^2 \rangle - \langle X \rangle^2. \tag{32.20}$$

Moment generating function

Related to averages is the **moment generating function** defined by

$$\langle e^{tX} \rangle = \sum_{i=1}^{N} e^{tx_i} p_i. \tag{32.21}$$

The name comes from the fact that

$$\frac{d^k}{dt^k} \langle e^{tX} \rangle \bigg|_{t=0} = \langle X^k \rangle. \tag{32.22}$$

32.1.5   Counting: Permutations and Combinations

The probability space of many situations is discrete. In fact, one can say that *all* probability spaces are discrete, and only in the limit of large samples (atoms and molecules in thermodynamics, for example) can one approximate the random variable as a continuous variable. Therefore, it is important to find formulas that give the number of particular events of a universal set.

Suppose you have N **distinguishable** particles and you want to place them in M bins. There are two cases that are used in practice: each bin can hold as many particles as you place in it; or each bin can hold only one particle. For each case, we are interested in finding the number of distinct arrangements, or the number of **configurations**. Let this number be denoted by $\Omega(N, M)$.

If there is no restriction on the occupancy number, then you have M choices for the first particle, M choices for the second particle, etc. Therefore,

$$\Omega_{MB}(N, M) = M^N. \tag{32.23}$$

In statistical mechanics, this is called the **Maxwell-Boltzmann statistics**.

If the occupancy number is one, then you have M choices for the first particle, $M - 1$ choices for the second particle, etc. Therefore,

$$\Omega_p(N, M) = M(M - 1)(M - 2) \cdots (M - N + 1) = \frac{M!}{(M - N)!}, \quad M > N. \tag{32.24}$$

This is called the **permutation** of M objects taken N at a time. If $M = N$, then $\Omega(N, N) = N!$ is simply called the *permutation* of N objects.

The elementary constituents of nature are **indistinguishable** or **identical**. How does this affect the formulas above? Let's consider the single-occupancy case first because it is easier. Equation (32.24) is overcounting the arrangement by $N!$ because a permutation of the particles does not give any new arrangement. Therefore,

$$\Omega_b(N, M) = \frac{M!}{N!(M - N)!} \equiv \binom{M}{N}, \quad M > N. \tag{32.25}$$

In statistical mechanics, this is called the **Fermi-Dirac statistics**. It is also called the **combination** of M objects taken N at a time.

The multiple-occupancy case for indistinguishable particles is harder, but there is a trick that can make it easier to derive the formula. Figure 32.2(a) shows some bins with particles inside them. We can represent the arrangement by placing the particles outside and to the right of the bins and represent the bins by vertical lines as in Figure 32.2(b). Each vertical line has some particles to its right and left *except the bin on the extreme left*, which has particles only to its right. Since there is no limitation on the number of occupancy, the number of arrangements can be calculated by permuting *both* the lines and dots except the line on the extreme left. Since the dots are identical (as are the lines), the problem reduces to finding the number of permutations of

Maxwell-Boltzmann statistics

Permutation

Fermi-Dirac statistics

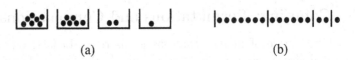

(a)   (b)

Figure 32.2: (a) The bins with particles inside them. (b) Bins are represented by vertical lines with the occupying particles to their right.

$N + M - 1$ objects N of which are identical and $M - 1$ of which are also identical, but different from the other N. Therefore,

$$\Omega_{BE}(N, M) = \frac{(N + M - 1)!}{N!(M - 1)!} = \binom{N + M - 1}{N}. \qquad (32.26)$$

Bose-Einstein
statistics

In statistical mechanics, this is called the **Bose-Einstein statistics**.

32.2   Binomial Probability Distribution

The Fermi-Dirac statistics is closely related to the so-called **binomial distribution**. Each of the M bins has two states: either it is occupied or empty. There are many situations where the binomial distribution applies. For example, in tossing n coins, each coin can be a head or a tail; a quantum mechanical spin-half particle can have its spin "up" or "down;" in a binary alloy system each site of the alloy can be occupied by an atom A or B.

Universality of
binomial
distribution

In fact, the binomial distribution is more general than this. In *any* trial, one can talk about success and failure, where success refers to one particular outcome (out of the many possible outcomes), and failure to the rest of the possible outcomes. Thus, if we are after a 6 in a toss of a die, then getting a 6 is a success, and getting 1, 2, 3, 4, or 5 is a failure.

Let p be the success probability, then the failure probability is $q = 1 - p$. What is the probability $P(m, n)$ that in n trials we have m successes? Because the events are statistically independent (what happens in each trial is independent of what has happened and what will happen), by (32.14), the probabilities multiply. Thus the probability of m successes and $n - m$ failures is $p^m q^{n-m}$; and since there are $\binom{n}{m}$ ways that this can happen in n trials, the probability $P(m, n)$ of m successes in n trials is

$$P(m, n) = \binom{n}{m} p^m q^{n-m} = \frac{n!}{m!(n - m)!} p^m q^{n-m}. \qquad (32.27)$$

Using the Stirling approximation $x! \approx \sqrt{2\pi}\, e^{-x} x^{x+1/2}$ of Equation (11.6), the reader can show that

$$P(m, n) \approx 2^n \sqrt{\frac{2}{n\pi}} e^{-(n-2m)^2/2n}\, p^m q^{n-m}, \qquad (32.28)$$

assuming that both m and $n - m$ are large, which is true in almost all cases of large systems.

The special case of $p = q = \frac{1}{2}$ is of importance:

$$P(m, n) = \frac{n!}{m!(n-m)!2^n} \approx \sqrt{\frac{2}{n\pi}} e^{-(n-2m)^2/2n}. \qquad (32.29)$$

Sometimes (32.28) is written in terms of the difference s between the number of successes and failures. This is conveniently equal to $2m - n$ which is the exponent of the exponential. We call s the **success excess**. Thus, (32.28) becomes

$$P(s, n) \approx \sqrt{\frac{2}{n\pi}} e^{-s^2/2n}, \qquad \Omega_b(s, n) = 2^n P(s, n) = 2^n \sqrt{\frac{2}{n\pi}} e^{-s^2/2n}, \qquad (32.30)$$

where Ω_b is the number of configurations now written in terms of s.

For the binomial distribution we can easily find the moment generating function. From its definition, we have

$$\langle e^{tX} \rangle = \sum_{x=0}^{n} e^{tx} P(x, n) = \sum_{x=0}^{n} e^{tx} \binom{n}{x} p^x q^{n-x}$$

$$= \sum_{x=0}^{n} \binom{n}{x} (pe^t)^x q^{n-x} = (pe^t + q)^n, \qquad (32.31)$$

the last equality following from the binomial theorem. Equation (32.31) allows us to easily calculate the average and variance for the binomial distribution. First note that

$$\frac{d}{dt}\langle e^{tX} \rangle = npe^t(pe^t + q)^{n-1}, \qquad \frac{d^2}{dt^2}\langle e^{tX} \rangle = npe^t(q + npe^t)(pe^t + q)^{n-2}.$$

Now evaluate these at $t = 0$—and note that $p + q = 1$—to obtain

$$\langle X \rangle = np, \quad \langle X^2 \rangle = n^2 p^2 + npq, \quad \sigma^2 = \langle X^2 \rangle - \langle X \rangle^2 = npq. \qquad (32.32)$$

Example 32.2.1. Assume that the probability at birth that the newborn is male (or female) is $\frac{1}{2}$. What is the probability that in a household of six, three are male? Blind intuition tells us that the probability is $\frac{1}{2}$; but that is wrong! Rephrasing the question to "What is the probability that in six trials we get three successes?" leads us to the binomial distribution and the following answer:

$$P(3, 6) = \frac{6!}{3!3!}\left(\frac{1}{2}\right)^3 \left(\frac{1}{2}\right)^{6-3} = \frac{6!}{3!3!}\left(\frac{1}{2}\right)^6 = 0.3125.$$

This result may be surprising, but even more surprising is the result we obtain if we ask the same question about a (small) school: "What is the probability that in a school with 200 pupils, 100 are male?"

$$P(100, 200) = \frac{200!}{100!100!}\left(\frac{1}{2}\right)^{100}\left(\frac{1}{2}\right)^{200-100} = \frac{200!}{100!100!}\left(\frac{1}{2}\right)^{200} = 0.056. \quad \blacksquare$$

The surprise encountered in the preceding example is due to the confusion caused by mixing the *expected* value with its probability. In a binomial distribution, the expected value (or the mean or average) $\langle X \rangle = np$, or $\langle X \rangle = n/2$ when $p = 0.5$, which is the answer we intuitively gave to the two questions in the example above. Since our surprise increases with n, let us investigate the behavior of the binomial distribution for values of m close to the mean for very large n.

For large n and m, we can use (32.29), from which we obtain

$$P(n/2, n) \approx \sqrt{\frac{2}{n\pi}}.$$

This shows that $P(n/2, n) \to 0$ as $n \to \infty$. Thus the probability of having $n/2$ successes in n trials becomes negligible as the number of trials increases. But this is the *maximum* probability! Therefore, any other probability goes to zero even faster. Where have all the probabilities gone?

Consider the graph of (32.29) for large n and plot it to a scale such that the peak of the maximum, although small, is conspicuous. Figure 32.3 shows such a graph for $n = 1000$. Note that the maximum probability has a value of only 0.025, and that the graph drops to a value that is indistinguishable from zero at about $m = 560$ on the right and $m = 440$ on the left. We can actually calculate the ratio r of the small probability at $m = 560$ to the maximum at $m = 500$ using (32.29):

$$r \equiv \frac{P(560, 1000)}{P(500, 1000)} = \frac{\sqrt{\frac{2}{1000\pi}} e^{-(-120)^2/2000}}{\sqrt{\frac{2}{1000\pi}}} = e^{-(-120)^2/2000} = 0.00075.$$

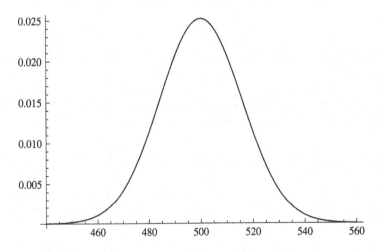

Figure 32.3: The plot of the binomial probability distribution for $n = 1000$.

This same ratio is obtained if we use $m = 440$, and we can therefore conclude that the nonzero probabilities are *essentially* concentrated between $m_- = 440$ and $m_+ = 560$ for $n = 1000$.

Now let's turn to a general n and find the corresponding values of m_- and m_+. These are the values of m at which the probability drops to 0.00075 of its maximum value. To find $m_\pm$, we have to solve the equation

$$r \equiv \frac{P(m, n)}{P(n/2, n)} = \frac{\sqrt{\frac{2}{n\pi}}e^{-(n-2m)^2/2n}}{\sqrt{\frac{2}{n\pi}}} = e^{-(n-2m)^2/2n} = 0.00075.$$

The answer is

$$m_\pm = \tfrac{1}{2}(n \pm 3.8\sqrt{n}), \tag{32.33}$$

as the reader may verify. So for a general n, a large fraction of the total probability is concentrated between $\tfrac{1}{2}(n - 3.8\sqrt{n})$ and $\tfrac{1}{2}(n + 3.8\sqrt{n})$. But how much? What is the probability that the number of successes lies between m_- and m_+?

To answer this question, we have to add all probabilities between m_- and m_+. For large numbers, one can replace the sum with an integral:

$$P(m_- \le m \le m_+, n) = \int_{m_-}^{m_+} \sqrt{\frac{2}{n\pi}}e^{-(n-2m)^2/2n}\,dm.$$

Define a new variable of integration x so that the exponent of the integrand becomes $-x^2$. This means

$$\frac{n - 2m}{\sqrt{2n}} = x, \quad \text{or} \quad m = \frac{n - \sqrt{2n}\,x}{2}, \quad dm = -\frac{\sqrt{2n}}{2}dx,$$

and the integral in terms of x becomes

$$\frac{1}{\sqrt{\pi}} \int_{-3.8/\sqrt{2}}^{3.8/\sqrt{2}} e^{-x^2}\,dx = 0.999855,$$

with the last result obtained by numerical integration. Therefore,

$$P(m_- \le m \le m_+, n) = 0.999855, \tag{32.34}$$

which is interestingly independent of n.

Let us investigate the meaning of this result. It says that for very large n, 99.99% of the time the successes lie between m_- and m_+, and the probability of *not* getting a success between m_- and m_+ is only 0.0145%. Note also that when n gets large, m_- and m_+ become very nearly equal to $n/2$. For example, if $n = 10^9$, then

$$\frac{n}{2} = 5 \times 10^8 \quad \text{and} \quad m_+ = \tfrac{1}{2}(10^9 + 3.8\sqrt{10^9}) = 5.001 \times 10^8.$$

So the probabilities are concentrated in a very narrow interval; i.e., the probability curve is extremely sharp.

Going back to the probability of the male gender, we note that in a very populous country such as China or India with approximately 10^9 inhabitants, although the probability that *exactly* half the population is male is extremely small, and of the order of

$$\sqrt{\frac{2}{n\pi}} = \sqrt{\frac{2}{10^9\pi}} = 0.000025,$$

the probability of the male population *deviating* too far from half is also small. So although the exact success (half male) is highly unlikely, a number of successes very close to exact is almost certain.

Example 32.2.2. An isolated spin-$\frac{1}{2}$ particle has equal probability of being in either spin-up or spin-down states. If there are n such particles, then the probability of m of them being in the up state is given by (32.29) or (32.30).

When a spin-$\frac{1}{2}$ particle with magnetic moment μ is placed in a magnetic field B, it has two possible states: in the direction of the field (called *up*) and opposite to it (called *down*). In the first case the energy of the particle is $-\mu B$ and in the second case $+\mu B$. The energy of the system is therefore determined by the success excess s, which in the present context is called the **spin excess**.

Now suppose that you have two systems that can exchange energy between themselves with the combined system isolated. This means that the total energy of the system is conserved. This energy is determined by the total spin excess s. Let $\Omega_{1b}(s_1, n_1)$ be the configuration number of the first system and $\Omega_{2b}(s_2, n_2)$ for the second. Let $\Omega_b(s, n)$ be the number of configurations for the combined system, where $n = n_1 + n_2$ and $s = s_1 + s_2$ is a constant. Since the total configuration number is the product of the configuration number of the components, we have

$$\Omega_b(s,n) = \Omega_{1b}(s_1, n_1)\Omega_{2b}(s - s_1, n_2) = C \exp\left(-\frac{s_1^2}{2n_1} - \frac{(s - s_1)^2}{2n_2}\right), \quad (32.35)$$

where C is independent of s_1.

What is the equilibrium state of the system? This corresponds to the most probable state of the combined system, i.e., the state that maximizes $\Omega_b(s, n)$. Instead of maximizing Ω_b, let's maximize its logarithm, which is

$$\ln \Omega_b = \ln C - \frac{s_1^2}{2n_1} - \frac{(s - s_1)^2}{2n_2}.$$

Differentiating with respect to s_1, we get

$$\frac{\partial \ln \Omega_b}{\partial s_1} = -\frac{s_1}{n_1} + \frac{s - s_1}{n_2}. \quad (32.36)$$

Note that the second derivative is

$$-\left(\frac{1}{n_1} + \frac{1}{n_2}\right),$$

which is negative, so the extremum is a maximum. Setting (32.36) equal to zero yields the most probable configuration:

Condition for thermal equilibrium

$$\frac{\hat{s}_1}{n_1} = \frac{s - \hat{s}_1}{n_2} = \frac{\hat{s}_2}{n_2} = \frac{s}{n},$$

where the last equality follows from the previous ones (see Problem 32.13) and a caret on a quantity indicates its value at maximum. If we substitute these in (32.35), we get the maximum number of configurations:

$$\Omega_b^{\mathrm{max}}(s,n) = C \exp\left(-\frac{\hat{s}_1^2}{2n_1} - \frac{\hat{s}_2^2}{2n_2}\right) = Ce^{-s^2/2n}. \tag{32.37}$$

The verification of the last equality is the subject of Problem 32.14.

Once in equilibrium, how likely is it for the system to move away from it? To investigate this, let s_1 and s_2 be slightly different from their equilibrium values

$$s_1 = \hat{s}_1 + \delta, \quad s_2 = s - s_1 = s - \hat{s}_1 - \delta = \hat{s}_2 - \delta.$$

Substituting these in (32.35) yields

$$C \exp\left(-\frac{\hat{s}_1^2 + 2\hat{s}_1\delta + \delta^2}{2n_1} - \frac{\hat{s}_2^2 - 2\hat{s}_2\delta + \delta^2}{2n_2}\right)$$

$$= \Omega_b^{\mathrm{max}}(s,n) \exp\left(-\frac{2\hat{s}_1\delta + \delta^2}{2n_1} + \frac{2\hat{s}_2\delta - \delta^2}{2n_2}\right).$$

But $\hat{s}_1/n_1 = \hat{s}_2/n_2$. Therefore,

$$\frac{\Omega_{1b}(\hat{s}_1 + \delta, n_1)\Omega_{2b}(\hat{s}_2 - \delta, n_2)}{\Omega_b^{\mathrm{max}}(s,n)} = \exp\left(-\frac{\delta^2}{2n_1} - \frac{\delta^2}{2n_2}\right). \tag{32.38}$$

As a realistic numerical example, let $n_1 = n_2 = 10^{23}$ and $\delta = 10^{13}$ so that the fractional deviation $\delta/n_1 = 10^{-10}$, a very small number. Then, the ratio in (32.38) is $e^{-1000} = 5 \times 10^{-435}$. The probability for fractional deviations *larger* that 10^{-10} is smaller than e^{-1000}. Assuming equal probability and adding the terms (about n_1 of them), the *upper bound* for the total probability becomes $n_1 e^{-1000}$ or 5×10^{-412} times the probability of finding the system in equilibrium.

How likely is it for the system to abandon its equilibrium state?

What is the meaning behind the statement "the probability to find the system with a fractional deviation larger than 10^{-10} is 5×10^{-412} of the probability of finding the system in equilibrium?" To have a reasonable chance of finding the system in such a deviated state, we have to sample 5×10^{412} similar systems. Even if we could sample at the rate of 10^{12} systems per second, we would have to sample for 5×10^{400} seconds, or over 10^{393} years, or 10^{383} times the age of the universe! Therefore, it is safe to say that deviations described above will *never* be observed. ∎

32.3 Poisson Distribution

Poisson processes are famous results in the probability theory. A Poisson process occurs in circumstances under which an event is repeated at a constant rate of probability. Suppose that dt is so small that the probability of the occurrence of two or more successes is negligible. Then the probability $P_1(dt)$ of one success in dt is νdt, where ν is a constant.

We are interested in $P_n(t)$, the probability of n successes in a time interval t. We can obtain a recursive differential equation involving $P_n(t)$ and $P_{n-1}(t)$, which we hope to solve to get $P_n(t)$. Consider $P_n(t+dt)$, the probability that n successes occur in time $t+dt$. This can be written as the sum of two disjoint probabilities, each consisting of the product of two independent probabilities: (a) n successes occur in time t and none in time dt, (b) $n-1$ successes occur in time t and one in time dt. In symbols,

$$P_n(t + dt) = P_n(t)P_0(dt) + P_{n-1}(t)P_1(dt).$$

But $P_1(dt) = \nu dt$ and $P_0(dt) = 1 - P_1(dt) = 1 - \nu dt$. Therefore,

$$P_n(t + dt) = P_n(t)(1 - \nu dt) + P_{n-1}(t)\nu dt.$$

Expanding the left-hand side as $P_n(t + dt) = P_n(t) + \frac{dP_n}{dt} dt$ and dividing both sides by dt, we obtain the desired recursive DE:

$$\frac{dP_n(t)}{dt} + \nu P_n(t) = \nu P_{n-1}(t). \tag{32.39}$$

For $n = 0$ the right-hand side is zero and the DE

$$\frac{dP_0(t)}{dt} + \nu P_0(t) = 0$$

has the solution $P_0(t) = Ae^{-\nu t}$. The fact that the probability of no success in zero time interval is 1 yields $A = 1$.

Equation (32.39) is a first order DE which can be solved. In fact, the solution is given in Theorem 23.3.1, where in the case at hand

$$\mu(t) = \exp\left[\int \nu dt\right] = e^{\nu t}$$

and

$$P_n(t) = \frac{1}{e^{\nu t}}\left[C + \nu \int_0^t e^{\nu t_1} P_{n-1}(t_1)\, dt_1\right].$$

We must have $P_n(0) = 0$ because there is no chance that n successes can be achieved in zero time interval. This sets $C = 0$, and we get the *integral recursion relation*:

$$P_n(t) = \nu e^{-\nu t} \int_0^t e^{\nu t_1} P_{n-1}(t_1)\, dt_1. \tag{32.40}$$

Substitute for $P_{n-1}(t_1)$ as an integral of P_{n-2} to get

$$P_n(t) = \nu e^{-\nu t} \int_0^t e^{\nu t_1} \left\{ \nu e^{-\nu t_1} \int_0^{t_1} e^{\nu t_2} P_{n-2}(t_2)\, dt_2 \right\} dt_1,$$

or

$$P_n(t) = \nu^2 e^{-\nu t} \int_0^t \int_0^{t_1} e^{\nu t_2} P_{n-2}(t_2)\, dt_2\, dt_1.$$

It should now be clear that if we repeat this k times, we get

$$P_n(t) = \nu^k e^{-\nu t} \int_0^t \int_0^{t_1} \cdots \int_0^{t_{k-2}} \int_0^{t_{k-1}} e^{\nu t_k} P_{n-k}(t_k) \, dt_k dt_{k-1} \cdots dt_1.$$

In particular when $k = n$,

$$P_n(t) = \nu^n e^{-\nu t} \int_0^t \int_0^{t_1} \cdots \int_0^{t_{n-2}} \int_0^{t_{n-1}} e^{\nu t_n} P_0(t_n) \, dt_n dt_{n-1} \cdots dt_1.$$

But $P_0(t) = e^{-\nu t}$ so $P_0(t_n) = e^{-\nu t_n}$, and the above equation becomes

$$P_n(t) = \nu^n e^{-\nu t} \int_0^t \int_0^{t_1} \cdots \int_0^{t_{n-2}} \int_0^{t_{n-1}} dt_n dt_{n-1} \cdots dt_1. \tag{32.41}$$

Starting with the innermost integral over t_n and integrating all the t's, the reader can show that the result will be $t^n/n!$. We thus finally obtain the **Poisson process**

$$P_n(t) = \frac{(\nu t)^n}{n!} e^{-\nu t}. \tag{32.42}$$

Poisson process is naturally a time-dependnent process and ν is the rate or the frequency of that process.

The discrete **Poisson distribution** $p(n)$ is defined by setting $\nu t = \lambda$ to obtain

$$p(n) = \frac{\lambda^n}{n!} e^{-\lambda}, \quad n = 0, 1, 2, \ldots, \infty. \tag{32.43}$$

Poisson probability distribution

The moment generating function is

$$\langle e^{tX} \rangle = \sum_{x=0}^{\infty} e^{tx} \frac{\lambda^x}{x!} e^{-\lambda} = e^{-\lambda} \sum_{x=0}^{\infty} \frac{(\lambda e^t)^x}{x!} = e^{\lambda(e^t - 1)}. \tag{32.44}$$

This gives

$$\frac{d}{dt} \langle e^{tX} \rangle = \lambda e^t e^{\lambda(e^t-1)}, \quad \frac{d^2}{dt^2} \langle e^{tX} \rangle = \lambda e^t e^{\lambda(e^t-1)} (\lambda e^t + 1).$$

Evaluating these at $t = 0$ yields

$$\langle X \rangle = \lambda, \quad \langle X^2 \rangle = \lambda(\lambda + 1), \quad \sigma^2 = \langle X^2 \rangle - \langle X \rangle^2 = \lambda. \tag{32.45}$$

Example 32.3.1. A city had 24 major fire accidents in a year. What is the probability that there will be (a) one major fire next month, (b) at least 5 major fires in the next 6 months?

Here ν, the frequency of fire is 24 per year or 2 per month. So for (a) we have $\lambda = 2 \times 1 = 2$ and

$$p(1) = \lambda e^{-\lambda} = 2e^{-2} = 0.27.$$

For (b), $\lambda = 2 \times 6 = 12$ and

$$p(n \geq 5) = 1 - p(0) - p(1) - p(2) - p(3) - p(4)$$

$$= 1 - e^{-12} - 12e^{-12} - \frac{12^2}{2!} e^{-12} - \frac{12^3}{3!} e^{-12} - \frac{12^4}{4!} e^{-12} = 0.992. \quad \blacksquare$$

Example 32.3.2. In a department store, 39 light bulbs burn out per year. The light bulbs are replaced from the store stock, which is replenished every week. What is the minimum number of light bulbs the stock should hold so that the store will have all its light bulbs working with a probability of at least 99%?

The frequency is $\nu = 39/52 = 0.75$ per week. Thus, $\lambda = 0.75 \times 1 = 0.75$. Let n stand for the number of bulbs burnt per week and m the number of bulbs in the stock. Then the store will have all its lights on as long as $n \leq m$. Therefore, we want $p(n \leq m) \geq 0.99$. This gives

$$p(n \leq m) = \sum_{n=0}^{m} \frac{\lambda^n}{n!} e^{-\lambda} \geq 0.99,$$

or

$$p(n \leq m) = \sum_{n=0}^{m} \frac{(0.75)^n}{n!} \geq 0.99 e^{0.75} = 2.096,$$

or

$$\left(1 + 0.75 + \frac{0.75^2}{2!} + \cdots + \frac{0.75^m}{m!}\right) \geq 2.096.$$

By trial and error, the reader can verify that $m = 3$. ∎

Poisson distribution is the limiting case of the binomial distribution when $n \to \infty$, $p \to 0$, and $\lambda = np$ is constant. To see this, expand $n!/(n-m)!$ using the Stirling approximation:

$$\frac{n!}{(n-m)!} \approx \frac{\sqrt{2\pi}\, e^{-n} n^{n+1/2}}{\sqrt{2\pi}\, e^{-n+m}(n-m)^{n-m+1/2}} = e^{-m} \left(\frac{n}{n-m}\right)^{n+1/2} (n-m)^m.$$

Now note that

$$\left(\frac{n}{n-m}\right)^{n+1/2} \approx \left(\frac{n}{n-m}\right)^n = \frac{1}{(1-m/n)^n} \to \frac{1}{e^{-m}} = e^m,$$

and $(n-m)^m \approx n^m$. Furthermore,

$$q^{n-m} = (1-p)^{n-m} = \left(1 - \frac{\lambda}{n}\right)^{n-m} \approx \left(1 - \frac{\lambda}{n}\right)^n \to e^{-\lambda}.$$

Substituting all this in (32.27) yields

$$P(m,n) \to n^m \frac{1}{m!} p^m e^{-\lambda} = \frac{(np)^m}{m!} e^{-\lambda} = \frac{\lambda^m}{m!} e^{-\lambda},$$

which is the Poisson distribution $p(m)$.

Example 32.3.3. A 3000-letter long message has been transmitted electronically with an error probability of 10^{-3}. What is the probability that there are at least two errors in the message?

This is a binomial distribution (error is success!) with small probability and large n. Therefore, we can use Poisson distribution (32.43) with $\lambda = np = 3000 \times 10^{-3} = 3$. Then

$$p(n \geq 2) = 1 - p(0) - p(1) = 1 - e^{-3} - 3e^{-3} = 0.8.$$

The probability that there is exactly one error in the message is

$$p(1) = 3e^{-3} = 0.149.$$ ∎

32.4   Continuous Random Variable

Most probability sample spaces are so large that approximating the discrete events with continuous variables becomes very useful and accurate. Take the case of the binomial distribution discussed above. We started with discrete counting, but when our sample grows to 10^{23}, not only does the discrete sum become unmanageable, it becomes unnecessary as well. This is also reflected in the replacement of the strictly discrete factorial with the more adaptable exponential function through the use of the Stirling approximation.

When continuous variables are used, probability is described by **probability density**. In the case of a single random variable x, the probability density $f(x)$ is used to give the probability that x lies in an interval of length dx:

Probability density

$$P(x - dx/2 < x < x + dx/2) = f(x)\,dx, \qquad \int_a^b f(x)\,dx = 1,$$

where (a, b) is the interval for which x is defined. This interval can be taken to be $(-\infty, \infty)$ by assigning zero probability density to points on the left of a and on the right of b. The integral describes the total probability, which is 1 as in the case of the discrete variable.

For more variables, the generalization is clear. If $\mathbf{x} = (x_1, x_2, \ldots x_m)$ and the probability density function is $f(\mathbf{x})$, then the probability that $\mathbf{x}$ is in an infinitesimal m-dimensional volume $d^m x$ is

$$P(\mathbf{x} \in d^m x) = f(\mathbf{x})\,d^m x, \qquad \int_\Omega f(\mathbf{x})\,d^m x = 1, \qquad (32.46)$$

where Ω is the region for which $f(\mathbf{x})$ is defined. If V is a subset of Ω, then

$$P(\mathbf{x} \in V) = \int_V f(\mathbf{x})\,d^m x$$

gives the probability that $\mathbf{x}$ lies in V.

For example, a quantum mechanical wave function $\Psi(\mathbf{r})$ gives rise to a density $f(\mathbf{r}) = |\Psi(\mathbf{r})|^2$, and all wave functions are normalized so that

$$\int_\Omega |\Psi(\mathbf{r})|^2\,d^3 x = 1 \quad \text{and} \quad P(\mathbf{r} \in V) = \int_V |\Psi(\mathbf{r})|^2\,d^3 x, \qquad (32.47)$$

where $\mathbf{r}$ is a set of convenient coordinates.

The average and variance is defined in exactly the same way. For example, the average of the ith component of $\mathbf{x}$, denoted by $\langle X_i \rangle$, is given by

$$\langle X_i \rangle = \int_\Omega x_i f(\mathbf{x})\,d^m x. \qquad (32.48)$$

Similarly

$$\sigma^2(X_i) = \int_\Omega (x_i - \langle X_i \rangle)^2 f(\mathbf{x})\,d^m x, \qquad (32.49)$$

and

$$\langle g(\mathbf{X}) \rangle = \int_\Omega g(\mathbf{x}) f(\mathbf{x}) \, d^m x. \tag{32.50}$$

Example 32.4.1. The ground state of the hydrogen atom is described in spherical coordinates by the wave function Ae^{-r/a_0}, where a_0 is the Bohr radius, A is a constant to be determined by the normalization (32.47), and r is the distance of the electron to the nucleus, which is placed at the origin. Using the volume element in spherical coordinates, we have

$$1 = A^2 \int_0^\infty \int_0^\pi \int_0^{2\pi} e^{-2r/a_0} r^2 \sin\theta \, dr \, d\theta d\varphi = 4\pi A^2 \underbrace{\int_0^\infty e^{-2r/a_0} r^2 \, dr}_{a_0^3/4} = \pi A^2 a_0^3,$$

giving $A = \sqrt{1/\pi a_0^3}$. Thus the *normalized* wave function for the ground state of the hydrogen atom is

$$\Psi(r,\theta,\varphi) = \sqrt{\frac{1}{\pi a_0^3}} \, e^{-r/a_0},$$

with $|\Psi(r,\theta,\varphi)|^2$ being the probability density of finding the electron at (r,θ,φ).

From this, we can calculate, for instance, the probability that the electron approaches the nucleus to within 10% of the Bohr radius. The second equation in (32.47) gives the answer where V is the volume of a sphere with radius $0.1a_0$. Therefore,

$$P(\mathbf{r} \in V) = \frac{1}{\pi a_0^3} \int_0^{0.1a_0} \int_0^\pi \int_0^{2\pi} e^{-2r/a_0} r^2 \sin\theta dr \, d\theta \, d\varphi$$

$$= \frac{4}{a_0^3} \int_0^{0.1a_0} e^{-2r/a_0} r^2 dr = \frac{1}{2} \int_0^{0.2} e^{-x} x^2 dx \approx 0.0013,$$

where we used the change of variables $r = a_0 x/2$ in the last integral to turn it into a numerical factor.

We can also calculate some averages. For instance, the average for the x coordinate of the electron is (remember that $x = r \sin\theta \cos\varphi$)

$$\langle X \rangle = \frac{1}{\pi a_0^3} \int_0^\infty \int_0^\pi \int_0^{2\pi} r \sin\theta \cos\varphi \, e^{-2r/a_0} r^2 \sin\theta dr \, d\theta \, d\varphi = 0.$$

The result of zero being due to the φ integration. Similarly, $\langle Y \rangle$ and $\langle Z \rangle$ also vanish. This null result should be expected because it is just as likely for the electron to have a positive x value as it is to have a negative value. On the other hand, r is always positive, and we expect its average value to be nonzero. In fact,

$$\langle R \rangle = \frac{1}{\pi a_0^3} \int_0^\infty \int_0^\pi \int_0^{2\pi} r \, e^{-2r/a_0} r^2 \sin\theta dr \, d\theta \, d\varphi = \frac{4}{a_0^3} \underbrace{\int_0^\infty e^{-2r/a_0} r^3 \, dr}_{=3a_0^4/8} = \tfrac{3}{2} a_0. \qquad \blacksquare$$

Independent
random variable

A random variable x_α is said to be **independent** of the rest of the variables if the probability density $f(\mathbf{x})$ factors out into

$$f(\mathbf{x}) = g(x_\alpha) h(x_1, x_2, \ldots, x_{\alpha-1}, x_{\alpha+1}, \ldots, x_m) \equiv g(x_\alpha) h_\alpha(\mathbf{x}),$$

where $h_\alpha(\mathbf{x})$ is a function of all x's except x_α. By multiplying it by a constant, we can always choose $g(x_\alpha)$ in such a way that

$$\int_{-\infty}^{\infty} g(x_\alpha)\, dx_\alpha = 1. \tag{32.51}$$

Then since

$$1 = \int_\Omega f(\mathbf{x})\, d^m x = \int_{-\infty}^{\infty} g(x_\alpha)\, dx_\alpha \int_{\Omega'} h_\alpha(\mathbf{x})\, d^{m-1}x,$$

where Ω' is the region of integration of the remaining variables, we also have

$$\int_{\Omega'} h_\alpha(\mathbf{x})\, d^{m-1}x = 1.$$

From this we conclude that the average of any function depending on x_α alone can be calculated using not the whole density $f(\mathbf{x})$, but $g(x_\alpha)$. In particular,

$$\langle X_\alpha \rangle = \int_{-\infty}^{\infty} x_\alpha g(x_\alpha)\, dx_\alpha, \qquad \sigma^2(X_\alpha) = \int_{-\infty}^{\infty} (x_\alpha - \langle X_\alpha \rangle)^2 g(x_\alpha)\, dx_\alpha.$$
$$\tag{32.52}$$

Define $\mathrm{cov}(X_\alpha, X_\beta)$, the **covariance** of x_α and x_β, for a general density function $f(\mathbf{x})$, by $\qquad$ Covariance defined

$$\mathrm{cov}(X_\alpha, X_\beta) \equiv \int_\Omega (x_\alpha - \langle X_\alpha \rangle)(x_\beta - \langle X_\beta \rangle) f(\mathbf{x})\, d^m x$$
$$\equiv \langle (X_\alpha - \langle X_\alpha \rangle)(X_\beta - \langle X_\beta \rangle) \rangle, \tag{32.53}$$

and note that by (32.49),

$$\mathrm{cov}(X_\alpha, X_\alpha) = \sigma^2(X_\alpha), \qquad \mathrm{cov}(X_\beta, X_\beta) = \sigma^2(X_\beta). \tag{32.54}$$

Now suppose that x_α is independent of the rest of the variables and $\beta \neq \alpha$. Then

$$\mathrm{cov}(X_\alpha, X_\beta) = \int_{-\infty}^{\infty} (x_\alpha - \langle X_\alpha \rangle) g(x_\alpha)\, dx_\alpha \int_{\Omega'} (x_\beta - \langle X_\beta \rangle) h_\alpha(\mathbf{x})\, d^{m-1}x = 0.$$

The result follows from the fact that the integration over Ω' is a constant and the integral over x_α can be done independently:

$$\underbrace{\int_{-\infty}^{\infty} (x_\alpha - \langle X_\alpha \rangle) g(x_\alpha)\, dx_\alpha = \int_{-\infty}^{\infty} x_\alpha g(x_\alpha)\, dx_\alpha}_{=\langle X_\alpha \rangle \text{ by } (32.52)} - \langle X_\alpha \rangle \underbrace{\int_{-\infty}^{\infty} g(x_\alpha)\, dx_\alpha}_{=1 \text{ by } (32.51)} = 0.$$

The preceding discussion shows that $\mathrm{cov}(X_\alpha, X_\beta)$ measures how much x_α is independent of the rest of the variables. If it is, then $\mathrm{cov}(X_\alpha, X_\beta) = 0$; if it is not, then $\mathrm{cov}(X_\alpha, X_\beta) \neq 0$. A quantity related to $\mathrm{cov}(X_\alpha, X_\beta)$, called **correlation**, is $\qquad$ Correlation defined

$$\text{cor}(X_\alpha, X_\beta) = \frac{\text{cov}(X_\alpha, X_\beta)}{\sigma(X_\alpha)\sigma(X_\beta)}. \tag{32.55}$$

The "strongest" correlation occurs when $\alpha = \beta$, in which case

$$\text{cor}(X_\alpha, X_\alpha) = \frac{\text{cov}(X_\alpha, X_\alpha)}{\sigma^2(X_\alpha)} = 1 \quad \text{by (32.54)}.$$

The "weakest" correlation occurs when $\alpha \neq \beta$ and x_α is independent of the rest of the variables, in which case $\text{cor}(X_\alpha, X_\beta) = 0$. Thus, $\text{cor}(X_\alpha, X_\beta)$ indeed measures how much x_α and x_β are correlated. Problem (32.22) shows that $|\text{cor}(X_\alpha, X_\beta)| \leq 1$.

32.4.1   Transformation of Variables

Sometimes it is necessary or convenient to change a given set of random variables to another set. Suppose that $\mathbf{x} = \{x_i\}_{i=1}^m$ is a set of variables, and $\mathbf{u} = \{u_i\}_{i=1}^m$ are new variables of which the x_i are functions. Given a density $f(\mathbf{x})$, the probability of finding $\mathbf{x}$ in an infinitesimal volume $d^m x$ is $f(\mathbf{x})d^m x$. What is the corresponding probability in terms of the u variables? What is the probability density $g(\mathbf{u})$ so that $g(\mathbf{u})d^m u$ is the probability that $\mathbf{u}$ lies in the infinitesimal volume $d^m u$? The answer is

$$g(\mathbf{u}) = f\big(x_1(\mathbf{u}), x_2(\mathbf{u}), \ldots, x_m(\mathbf{u})\big) J(\mathbf{x}, \mathbf{u}), \tag{32.56}$$

where $J(\mathbf{x}, \mathbf{u})$ is the Jacobian of the x-to-u transformation, whose special cases in two and three dimensions were given in (6.65) and (6.66). Equation (32.56) is obtained from $f(\mathbf{x})d^m x$ by writing x's in terms of u's, keeping in mind that $d^m x = J(\mathbf{x}, \mathbf{u})d^m u$.

In most cases, there are only two variables x and y, which are transformed into u and v. Then (32.56) yields

$$g(u, v) = f\big(x(u, v), y(u, v)\big) \begin{vmatrix} \frac{\partial x}{\partial u} & \frac{\partial y}{\partial u} \\ \frac{\partial x}{\partial v} & \frac{\partial y}{\partial v} \end{vmatrix}. \tag{32.57}$$

Example 32.4.2. The random variables x and y have the density function

$$f(x, y) = \begin{cases} c(x + y)e^{-x} & \text{if } 0 < x, 0 < y < 1; \\ 0 & \text{otherwise}, \end{cases} \tag{32.58}$$

where c is a positive constant. What is the density function $h(u)$ for the sum $u = x + y$?

As will become clear below, it is convenient to write $f(x, y)$ in terms of the θ function introduced in Section 5.1.3 Equation (5.18):

$$f(x, y) = c\theta(x)\theta(y)\theta(1 - y)(x + y)e^{-x}. \tag{32.59}$$

The reader is urged to verify that this is identical to (32.58). Let $x = v$. Then $u = x + y$ gives $y = u - v$, and the Jacobian for the transformation is

$$\begin{vmatrix} \dfrac{\partial x}{\partial u} & \dfrac{\partial y}{\partial u} \\[2mm] \dfrac{\partial x}{\partial v} & \dfrac{\partial y}{\partial v} \end{vmatrix} = \begin{vmatrix} 0 & 1 \\ 1 & -1 \end{vmatrix} = 1. \qquad (32.60)$$

Therefore, all is needed is to replace x and y in $f(x,y)$:

$$g(u,v) = c\theta(v)\theta(u-v)\theta(1-u+v)ue^{-v}.$$

This is the convenience we mentioned above: we don't have to worry about different cases corresponding to different limits of u and v; the θ function automatically takes care of that!

To find $h(u)$, we need—by definition—to integrate over all values of v. Because of the last θ factor in $g(u,v)$, we need to consider two cases: $0 < u < 1$ and $u > 1$. In the first case, $\theta(1-u+v) = 1$ because the first θ function requires v to be positive. Then the middle θ function sets the upper limit of v integration to u. Hence,

$$h(u) \equiv \int_{-\infty}^{\infty} g(u,v)\, dv = \int_0^u cue^{-v}\, dv = cu\left(1 - e^{-u}\right), \quad 0 < u < 1.$$

In the second case, $\theta(1-u+v)$ requires v to be grater than $u-1$, and the middle θ function still sets the upper limit of v integration to u. Therefore,

$$h(u) = \int_{u-1}^{u} cue^{-v}\, dv = -cu\, e^{-v}\Big|_{u-1}^{u} = cue^{-u}(e - 1), \quad u > 1.$$

The two cases can be combined using the θ function:

$$h(u) = cu\Big[\theta(u)\theta(1-u)\left(1 - e^{-u}\right) + \theta(u-1)e^{-u}(e-1)\Big]. \qquad \blacksquare$$

Suppose that x and y are *independent* random variables with the density function

$$f(x,y) = f_1(x)f_2(y).$$

We want to find the density function $h(u)$ of their sum $u = x + y$. Let $x = v$ and $y = u - v$, so that the sum is indeed $x + y$. The Jacobian of the transformation is 1 by (32.60). Therefore, by (32.57),

$$g(u,v) = f\big(x(u,v), y(u,v)\big) = f_1(v)f_2(u - v).$$

The density function of each variable is obtained by integrating over the other variable. Thus,

$$h(u) \equiv \int_{-\infty}^{\infty} g(u,v)\, dv = \int_{-\infty}^{\infty} f_1(v)f_2(u - v)\, dv. \qquad (32.61)$$

The reader may recall from our discussion of Laplace transform that h is the *convolution* of f_1 and f_2.

Example 32.4.3. Assume that x and y are independent variables with

$$f_1(x) = \frac{1}{\pi(x^2+1)}, \qquad -\infty < x < \infty;$$

$$f_2(y) = \frac{1}{\pi(y^2+1)}, \qquad -\infty < y < \infty.$$

Then their sum $u = x + y$ has the density function

$$h(u) = \frac{1}{\pi^2} \int_{-\infty}^{\infty} \frac{dv}{(v^2+1)[(u-v)^2+1]} = \frac{2}{\pi(u^2+4)}, \qquad -\infty < u < \infty$$

leaving the verification of the last integration for Problem 32.25.   ∎

32.4.2 Normal Distribution

One of the most frequently used probability distributions is **Gauss' normal distribution** given by

$$f(x) = \frac{1}{\sqrt{2\pi}\,\sigma} e^{-(x-m)^2/2\sigma^2}, \qquad -\infty < x < \infty. \tag{32.62}$$

It can be easily shown that $\langle X \rangle = m$ and, as the notation suggests, the variance is σ^2.

To find the probability that x lies in the interval (a, b), we have to integrate $f(x)$ from a to b:

$$p(a < x < b) = \frac{1}{\sqrt{2\pi}\,\sigma} \int_a^b e^{-(x-m)^2/2\sigma^2}\, dx.$$

Let $y = (x-m)/\sqrt{2}\,\sigma$ and substitute for x in terms of y. Then

$$p(a < x < b) = \frac{1}{\sqrt{\pi}} \int_{\frac{a-m}{\sqrt{2}\,\sigma}}^{\frac{b-m}{\sqrt{2}\,\sigma}} e^{-y^2}\, dy = \frac{1}{2}\left[\operatorname{erf}\left(\frac{b-m}{\sqrt{2}\,\sigma}\right) - \operatorname{erf}\left(\frac{a-m}{\sqrt{2}\,\sigma}\right)\right],$$

$$\tag{32.63}$$

where erf is the error function introduced in Equation (11.9). The error function has been tabulated precisely because of its relation to the normal distribution.

Suppose a and b are given in terms of their distance from the mean as a multiple of the standard deviation: $a = m + k_1\sigma$ and $b = m + k_2\sigma$, then we have the important relation

$$p(m + k_1\sigma < x < m + k_2\sigma) = \frac{1}{2}\left[\operatorname{erf}\left(k_2/\sqrt{2}\right) - \operatorname{erf}\left(k_1/\sqrt{2}\right)\right]. \tag{32.64}$$

In particular, if $k_1 = -k_2 \equiv -k$, then

$$p(m - k\sigma < x < m + k\sigma) = \frac{1}{2}\left[\operatorname{erf}\left(k/\sqrt{2}\right) - \operatorname{erf}\left(-k/\sqrt{2}\right)\right]. \tag{32.65}$$

For $k = 1, 2, 3$ this yields

$$p(m - \sigma < x < m + \sigma) = 0.6827$$
$$p(m - 2\sigma < x < m + 2\sigma) = 0.9545$$
$$p(m - 3\sigma < x < m + 3\sigma) = 0.9973.$$

Let x and y be random variables having the same normal distribution with mean m and variance σ^2

$$f_1(x) = \frac{1}{\sqrt{2\pi}\,\sigma} e^{-(x-m)^2/2\sigma^2}, \qquad f_2(y) = \frac{1}{\sqrt{2\pi}\,\sigma} e^{-(y-m)^2/2\sigma^2}.$$

We want to find the distribution of the sum $u = x + y$. This is a special case of (32.61). Therefore, we can immediately write

$$h(u) = \frac{1}{2\pi\sigma^2} \int_{-\infty}^{\infty} e^{-(v-m)^2/2\sigma^2} e^{-(u-v-m)^2/2\sigma^2}\, dv$$
$$= \frac{1}{2\pi\sigma^2} \int_{-\infty}^{\infty} \exp\left[-\frac{(v-m)^2 + (u-v-m)^2}{2\sigma^2}\right]\, dv.$$

The reader may verify that the exponent can be simplified to

$$-\frac{(v - u/2)^2 + (u - 2m)^2/4}{\sigma^2}.$$

Substituting back in the integrand gives

$$h(u) = \frac{e^{-(u-2m)^2/4\sigma^2}}{2\pi\sigma^2} \underbrace{\int_{-\infty}^{\infty} e^{-(v-u/2)^2/\sigma^2}\, dv}_{=\sqrt{\pi\sigma^2}} = \frac{e^{-(u-2m)^2/4\sigma^2}}{\sqrt{2\pi}\sqrt{2\sigma^2}}.$$

This shows that

> **Box 32.4.1.** *If the random variables x and y have the same normal distribution, then their sum has a normal distribution with twice the mean and twice the variance.*

Example 32.4.4. Let x and y be independent random variables whose density functions f_1 and f_2 are normal distribution with $m = 0$ and $\sigma = 1$. What is the density function $h(u)$ for the ratio x/y?

Let $x = v$, then $y = v/u$ makes u the sought-after ratio. The Jacobian of the transformation is

$$J = \begin{vmatrix} \frac{\partial x}{\partial u} & \frac{\partial y}{\partial u} \\ \frac{\partial x}{\partial v} & \frac{\partial y}{\partial v} \end{vmatrix} = \begin{vmatrix} 0 & -v/u^2 \\ 1 & 1/u \end{vmatrix} = \frac{|v|}{u^2}.$$

Then

$$g(u,v) = f_1(v)f_2\left(\frac{v}{u}\right)\frac{|v|}{u^2} = \frac{1}{2\pi}e^{-v^2/2}e^{-v^2/2u^2}\frac{|v|}{u^2}.$$

To find $h(u)$ we integrate $g(u,v)$ over all values of v, namely $-\infty$ to ∞. Since the integrand is even, we can integrate from 0 to ∞ and multiply the result by 2. Therefore,

$$h(u) = \frac{1}{\pi u^2}\int_0^\infty e^{-(\frac{1}{2}+\frac{1}{2u^2})v^2}v\,dv = \frac{1}{\pi(u^2+1)}.$$

The integration is straightforward as the reader can verify.   ∎

Equation (32.29), when written in the form (replacing m with x)

$$P(x,n) = \sqrt{\frac{2}{n\pi}}e^{-(x-n/2)^2/(n/2)}$$

displays the similarity of the binomial distribution and the normal distribution for the special case of $p = q = \frac{1}{2}$. The equation shows that the mean is $n/2$ and the variance $n/4$. We now generalize this to arbitrary p and q.

Using the Stirling approximation $x! \approx \sqrt{2\pi}\,e^{-x}x^{x+1/2}$ and replacing m with x, we write (32.27) as

$$P(x,n) \approx \frac{\sqrt{2\pi}\,e^{-n}n^{n+1/2}p^x q^{n-x}}{\sqrt{2\pi}\,e^{-x}x^{x+1/2}\sqrt{2\pi}\,e^{-n+x}(n-x)^{n-x+1/2}}$$

$$= \frac{n^{n+1/2}p^x q^{n-x}(n-x)^x}{\sqrt{2\pi}\,x^{x+1/2}(n-x)^{n+1/2}},$$

or, pulling out the power of $1/2$ and collecting all terms with equal powers together, we obtain

$$P(x,n) \approx \sqrt{\frac{n}{2\pi x(n-x)}}\left(\frac{nq}{n-x}\right)^n\left(\frac{p(n-x)}{xq}\right)^x.$$

In the approximation we are seeking, we assume that x is close to the mean np and write $x = np + \delta$ where δ is small compared to np. Then we get

$$P(x,n) \approx \sqrt{\frac{n}{2\pi(np+\delta)(n-np-\delta)}}\left(\frac{nq}{n-np-\delta}\right)^n\left(\frac{p(n-np-\delta)}{(np+\delta)q}\right)^{np+\delta}$$

$$= \sqrt{\frac{n}{2\pi(np+\delta)(nq-\delta)}}\left(\frac{nq}{nq-\delta}\right)^n\left(\frac{npq-p\delta}{npq+q\delta}\right)^{np+\delta},$$

or

$$P(x,n) \approx \sqrt{\frac{1}{2\pi npq}}\underbrace{\left(\frac{1}{1-\delta/nq}\right)^n\left(\frac{1-\delta/nq}{1+\delta/np}\right)^{np+\delta}}_{\equiv A}. \tag{32.66}$$

To proceed, we take the log of the term we have designated as A:

$$\ln A = -n\ln\left(1-\frac{\delta}{nq}\right) + (np+\delta)\left[\ln\left(1-\frac{\delta}{nq}\right) - \ln\left(1+\frac{\delta}{np}\right)\right]$$

$$= (-nq+\delta)\ln\left(1-\frac{\delta}{nq}\right) - (np+\delta)\ln\left(1+\frac{\delta}{np}\right).$$

Expanding the log terms up to the second order yields

$$\ln A \approx (-nq + \delta)\left(-\frac{\delta}{nq} - \frac{\delta^2}{2n^2q^2}\right) - (np + \delta)\left(\frac{\delta}{np} - \frac{\delta^2}{2n^2p^2}\right)$$

$$= -\frac{\delta^2}{2nq} - \frac{\delta^2}{2np} = -\frac{\delta^2}{2n}\left(\frac{1}{q} + \frac{1}{p}\right) = -\frac{\delta^2}{2npq} \Rightarrow A \approx e^{-\frac{\delta^2}{2npq}}.$$

Substituting this in (32.66) with $\delta = x - np$, we obtain

$$P(x, n) \approx \sqrt{\frac{1}{2\pi npq}}e^{-\frac{(x-np)^2}{2npq}},$$

which shows that $P(x, n)$ is a normal distribution with mean np and variance npq.

It can also be shown that the Poisson distribution (32.43) approaches the normal distribution when n and λ are both large and $\delta = n - \lambda$ is small compared to both:

$$p(n) \rightarrow \frac{1}{\sqrt{2\pi\lambda}}e^{-(n-\lambda)^2/2\lambda}.$$

We therefore have the **law of large numbers**:

law of large numbers

Box 32.4.2. *In the limit that the random variable and the mean go to infinity, both the binomial and Poisson distributions approach the Gauss' normal distribution.*

Normal distribution is a remarkable density function. We just saw that both binomial and Poisson distributions approach it in the limit of large n. But it goes beyond these two distributions. In fact it can be shown that a set of identically distributed random variables with an *arbitrary* distribution is approximately normally distributed if the number of components is large enough. This is the content of the **central limit theorem**, and the reason that normal distribution is the distribution of choice in many statistical applications.

central limit theorem

32.5 Problems

32.1. Using Venn diagrams, show that the operation of union distributes over intersection and vice versa:

$$A \cap (B \cup C) = (A \cap B) \cup (A \cap C),$$
$$A \cup (B \cap C) = (A \cup B) \cap (A \cup C).$$

32.2. Using Venn diagrams, show that

$$A \cap (B - C) = (A \cap B) - (A \cap C),$$
$$A - (B \cup C) = (A - B) - C.$$

32.3. Using Venn diagrams, show that

$$(A \cup B)^c = A^c \cap B^c \quad \text{and} \quad (A \cap B)^c = A^c \cup B^c.$$

32.4. Fill in the rest of the following table assuming that all probabilities p_{ij} are independent.

	F_1	F_2	Total
E_1			0.3
E_2			
Total	0.4		

32.5. Fill in the rest of the following table assuming that all probabilities p_{ij} are independent.

	F_1	F_2	F_3	Total
E_1				0.3
E_2				
E_3				0.5
Total	0.1		0.7	

32.6. Prove Equation (32.22).

32.7. What is the probability of obtaining 400 heads in 800 coin tosses? Of obtaining more than 500 heads? Of obtaining between 350 and 450 heads?

32.8. A graphic calculator is needed!   Plot the binomial distribution $P(m, n)$ as a function of m for $n = 50$ and $p = q = \frac{1}{2}$ using the exact formula (32.27).
(a) From the plot estimate m_-, the largest value on the left of maximum at which the probability is (almost) zero, and m_+, the smallest value on the right of maximum at which the probability is (almost) zero. Compare these values with (32.33).
(b) Sum the exact formula from m_- to m_+ to find the probability that the number of successes lies between m_- and m_+.
(c) Using the Stirling approximation (32.28) estimate the probability found in (b) and compare the two values.

32.9. Example 32.2.2 used the exponential approximation to the binomial distribution because the number of spins were assumed very large. Now assume two systems with $n_1 = 8$ and $n_2 = 12$, and the total energy being exchanged is represented by $s = 4$.
(a) Find $\hat{s}_1$ and $\hat{s}_2$, and show that $\hat{s}_1/n_1$ is (approximately) equal to $\hat{s}_2/n_2$ and s/n.
(b) Find the ratio of the probability that $s_1 = \hat{s}_1 - 1$ (and therefore, $s_2 = \hat{s}_2 + 1$) to the maximum probability. How does this compare with the same ratio found in Example 32.2.2?

32.10. Using the Stirling approximation $x! \approx \sqrt{2\pi}\, e^{-x} x^{x+1/2}$ of Equation (11.6), show that

$$\frac{n!}{m!(n-m)!} \approx 2^n \sqrt{\frac{2}{n\pi}} e^{-(n-2m)^2/2n},$$

assuming that both m and $n-m$ are large.

32.11. For the binomial distribution,
(a) show that

$$\sum_{m=0}^{n} P(m,n) = \sum_{m=0}^{n} \binom{n}{m} p^m q^{n-m} = 1,$$

(b) and that

$$\int_{-\infty}^{\infty} P(s,n)\, ds = \int_{-\infty}^{\infty} \sqrt{\frac{2}{n\pi}} e^{-s^2/2n}\, ds = 1.$$

32.12. Let $r \equiv \frac{P(m,n)}{P(n/2,n)}$. Solve for m and show that

$$m = \tfrac{1}{2}(n \pm \sqrt{n}\,\sqrt{-2\ln r}).$$

32.13. Show that if $a/b = c/d$ then $a/b = (a+c)/(b+d)$.

32.14. Derive (32.37) from (32.35).

32.15. Using the definitions of average and variance, show that $\sigma^2 = \langle X^2 \rangle - \langle X \rangle^2$.

32.16. Show directly that (32.43) satisfies $\sum_{n=0}^{\infty} p(n) = 1$.

32.17. A city had two earthquakes in a century. Find the probability that in this city, there will be one earthquake
(a) next year,
(b) in the next 50 years.
(c) What is the probability of three or more independent earthquakes in the same months?

32.18. The number of α particles emitted from a sample of a radioactive atom is counted every minute for 50 hours. The total count is 1500.
(a) What is ν for this Poisson distribution?
(b) What is the probability that in the next 6 minutes three α particles will be emitted?
(b) What is the probability that in the next 3 minutes at least four α particles will be emitted?

32.19. One of the first excited states of the H-atom has the wave function

$$\Psi(r,\theta,\varphi) = Are^{-r/2a_0} \cos\theta.$$

(a) Find A so that $\Psi(r, \theta, \varphi)$ is normalized to 1.

(b) Evaluate $\langle X \rangle$, $\langle Y \rangle$, and $\langle Z \rangle$. Are they all zero? Do you expect them to be?

(c) What is the expectation value of the distance of the electron from the nucleus for this state?

32.20. Suppose that ϕ is a function of x_α alone and x_α is independent of the rest of the variables of $f(\mathbf{x})$, the density function for a multidimensional probability space. Show that

$$\langle \phi(X_\alpha) \rangle = \int_\Omega \phi(x_\alpha) f(\mathbf{x}) \, d^m x = \int_{-\infty}^{\infty} \phi(x_\alpha) g(x_\alpha) \, dx_\alpha.$$

32.21. The uniform probability density function over (a, b) is

$$f(x) = \begin{cases} 1/(b-a) & \text{if } a < x < b; \\ 0 & \text{otherwise.} \end{cases}$$

What is the expectation value $\langle X \rangle$ for this distribution?

32.22. Consider the nonnegative function

$$x(t) = \langle [t(x_\alpha - \langle X_\alpha \rangle) + (x_\beta - \langle X_\beta \rangle)]^2 \rangle.$$

(a) Show that

$$x(t) = t^2 \sigma^2(X_\alpha) + 2t \, \mathrm{cov}(X_\alpha, X_\beta) + \sigma^2(X_\beta),$$

which is a parabola in the tx-plane.

(b) If the parabola is to be nonnegative, it should have at most one *real* root. Show that for this to happen, the following inequality must hold:

$$\mathrm{cov}^2(X_\alpha, X_\beta) \le \sigma^2(X_\alpha)\sigma^2(X_\beta).$$

32.23. Show that

$$\int_0^t \int_0^{t_1} \cdots \int_0^{t_{n-2}} \int_0^{t_{n-1}} dt_n \, dt_{n-1} \cdots dt_1 = \frac{t^n}{n!}.$$

32.24. Let c be a positive constant and

$$f(x, y) = \begin{cases} cx(x+y) & \text{if } 0 < x, \, 0 < y < 1; \\ 0 & \text{otherwise.} \end{cases}$$

Let $u = x + y$. Show that the density $h(u)$ for the variable u is

$$h(u) = \begin{cases} cu^3/2 & \text{if } 0 < u < 1; \\ cu(u - \frac{1}{2}) & \text{if } u > 1. \end{cases}$$

Show that this can be written as

$$h(u) = cu\left[\theta(u)\theta(1-u)u^2/2 + \theta(u-1)(u - \tfrac{1}{2})\right].$$

32.25. Using partial fractions write the integrand of Example 32.4.3 as

$$\frac{1}{(v^2+1)[(u-v)^2+1]} = \frac{av+b}{v^2+1} + \frac{cv+d}{(u-v)^2+1}.$$

Now write the right-hand side as a single fraction with the same denominator as the left-hand side. Set the coefficients of the powers of v in the numerator equal to zero, except the constant which must be equal to 1. Find a, b, c, and d. Show that the integral becomes

$$(b+d+cu) \int_{-\infty}^{\infty} \frac{dv}{v^2+1} = \frac{2}{u^2+4} \int_{-\infty}^{\infty} \frac{dv}{v^2+1}.$$

32.26. Using the Stirling approximation and a procedure similar to the one used for binomial distribution in the text, show that in the limit of large n and λ, the Poisson distribution of Equation (32.43) becomes the normal distribution.

32.27. Certain measurements are assumed to be normally distributed with 25 as the mean 25 and 0.5 as the standard deviation. What is the probability that a measurement lies between 23 and 27?

Bibliography

[1] Alligood, K.T., Sauer, T.D., and Yorke, J.A. *Chaos: An Introduction to Dynamical Systems*, Springer-Verlag, 1996.

[2] Auer, J.W. *Linear Algebra with Applications*, Prentice-Hall, 1991.

[3] Axler, S. *Linear Algebra Done Right*, Springer-Verlag, 1996.

[4] Baker, G.L. and Gollub, J.P. *Chaotic Dynamics: An Introduction*, Cambridge University Press, 1990.

[5] Birkhoff, G. and Rota, G.-C. *Ordinary Differential Equations*, 3rd ed., Wiley, 1978.

[6] Churchill, R. and Verhey, R. *Complex Variables and Applications*, 3rd ed., McGraw-Hill, 1974.

[7] Edwards, C.H., Penney, D.E. *Calculus and Analytic Geometry*, Prentice Hall, 1990.

[8] Finney, R.L., Thomas, G.B., Damana, F., and Waits, B.K. *Calculus*, Addison-Wesley, 1994.

[9] Friedberg, S., Insel, A., and Spence, L. *Linear Algebra*, Prentice-Hall, 1997.

[10] Gamow, G., *The Great Physicists: From Galileo to Einstein*, Dover, 1961.

[11] Gelfand, I. M. and Fomin, S. V., *Calculus of Variations*, translated and edited by Silverman, R. A., Dover, 1991.

[12] Goode, S.W. *An Introduction to Differential Equations and Linear Algebra*, Prentice-Hall, 1991.

[13] Halmos, P. *Finite Dimensional Vector Spaces*, 2nd ed., Van Nostrand, 1958.

[14] Hartle, J. *Gravity*, Addison Wesley, 2003.

[15] Hassani, S. *Mathematical Physics: A Modern Introduction to Its Foundations*, Springer-Verlag, 1999.

[16] Hilborn, R.C. *Chaos and Nonlinear Dynamics: An Introduction for Scientists and Engineers*, Oxford University Press, 1994.

[17] Holmgren, R.A. *A First Course in Discrete Dynamical Systems*, Springer-Verlag, 1996.

[18] Kaplan, W. *Advanced Calculus*, Addison-Wesley, 1991.

[19] Kline, M., *Mathematical Thought: From Ancient to Modern Times*, Vols. 1–3, Oxford University Press, 1972.

[20] Lang, S. *Calculus of Several Variables*, Springer-Verlag, 1988.

[21] Lang, S. *Complex Analysis*, 2nd ed., Springer-Verlag, 1985.

[22] Lorrain, P., Corson D., and Lorrain, F. *Electromagnetic Fields and Waves*, 3rd ed., W. H. Freeman, 1988.

[23] Lovelock, D. and Rund, H. *Tensors, Differential Forms, and Variational Principles*, Dover, 1989.

[24] Miller, R.K. *Introduction to Differential Equations*, Prentice-Hall, 1991.

[25] Pinsky, M.A. *Partial Differential Equations and Boundary-Value Problems with Applications*, McGraw-Hill, 1991.

[26] Rahman, M. *Ordinary Differential Equations* and *Partial Differential Equations*, Computational Mechanics Publication, 1991.

[27] Schiff, J. L. *The Laplace Transform*, Springer, 1991.

[28] Seydel, R. *From Equilibrium to Chaos: Practical Bifurcation and Stability Analysis*, Elsevier, 1988.

[29] Simmons, G. *Calculus Gems*, McGraw-Hill, 1992.

[30] Stewart, J. *Calculus*, 4th ed., Brooks/Cole, 1999.

Index

Abel, 331, 503, 675
Abel, Niels Henrik
    biography, 326
absolute differential, 463
Acceleration, 44
acceleration
    components
        spherical coordinates, 353
active transformation, 178
addition of velocities
    relativistic law of, 246
adjoint DO, *571–573*
advanced Green's function, 712
affine connection, *462–464*, 470
algebra
    fundamental theorem, 478, 511
amplitude
    complex, 486
angle
    as integral, 345
    solid, *344–350*
        total, 349
    total, 346
angular momentum, 28
    conservation, 351
        central force, 354
angular momentum operator, 412
    spherical coordinates, 435
antisymmetric tensor, 452
Arago, 321
Archimedes, 47
Archimedes, of Syracuse
    biography, 81
area element
    primary, 60
astragalus, 781
astrophysics, 415
    Poisson equation, 415
attractor, 766
    strange, 778

autonomous, 767
average, 790, 801
azimuth, 12
azimuthal
    angle, 12
    symmetry, 607
azimuthal equation, 549

Barrow, 87, 96, 97, 481
Barrow, Isaac
    biography, 47
basin of attraction, 766
basis, 16
    for plane, 175
    orthonormal, 186
    standard, 216
Bayes' theorem, 788
Beltrami identity, 731
Bernoulli, 272, 294, 320, 321, 326, 642
Bernoulli's FODE, 560
Bessel, 322
Bessel differential equation, 548, 641
    recursion relation, 644
Bessel equation
    Liouville substitution, 589
Bessel function, *333–335*, 644
    addition theorem, 651
    confluent hypergeometric, 652
    expansion in, *653–654*
        physical examples, *654–656*
    generating function, 651
    integral representation, 652
    Laplace's equation, *642–654*
    order
        negative integer, 646
    orthogonality relation, 649
    properties, *646–652*
    recurrence relation, 646
    second kind, 645
Bessel's integral, 652

Bessel, Friedrich Wilhelm
    biography, 641
beta function, *320*
Bezout, 210
Bianchi identity, 470
bifurcation, 757
    Hopf, 770
    period-doubling, 757
binding energy, 582
binomial probability distribution,
    792–797
binomial theorem, 265
biography
    Abel, Niels Henrik, 326
    Archimedes, of Syracuse, 81
    Barrow, Isaac, 47
    Bessel, Friedrich Wilhelm, 641
    Biot, Jean-Baptiste, 115
    Cauchy, Augustin-Louis, 503
    Cavalieri, Bonaventura, 90
    Cavendish, Henry, 23
    Cayley, Arthur, 192
    Coulomb, Charles, 23
    d'Alembert, Jean Le Rond, 548
    Descartes, Rene, 15
    Dirac, Paul Adrien Maurice, 151
    Euler, Leonhard, 321
    Fermat, Pierre de , 15
    Fourier, Joseph, 304
    Gauss, Johann Carl Friedrich,
        330
    Gibbs, Josiah Willard, 381
    Hamilton, William R., 10
    Heaviside, Oliver, 382
    Hermite, Charles, 674
    Jacobi, Carl Gustav Jacob, 326
    Kepler, Johannes, 579
    Laplace, Pierre Simon de, 593
    Legendre, Adrien-Marie, 617
    Leibniz, Gottfried Wilhelm, 103
    Maxwell, James Clerk, 419
    Newton, Isaac, 96
    Savart, Felix, 115
    Stokes, George Gabriel, 398
    Sylvester, James Joseph, 210
    Taylor, Brook, 294
    Wallis, John, 90
Biot, Jean-Baptiste
    biography, 115

Biot–Savart law, 30
    circuit, 111
    general, 110
bipolar coordinates, 73
    three-dimensional, 74, 438
Boltzmann, 419
Bose-Einstein statistics, 792
boundary condition, 542
    Dirichlet, 593
    Neumann, 593
boundary conditions
    periodic, 574
    separated, 574
brachistochrone, 731
Bromwich contour, 722
butterfly effect, 771

calculus
    fundamental theorem, 87
calculus of residues, *525–536*
canonical equations, 749
Cantor set, 776
Cardan, 481
Cartesian vector, 216
    component, 216
Cauchy, 279, 326, 331, 594
Cauchy criterion, 261
Cauchy integral formula, *508–509*
Cauchy, Augustin-Louis
    biography, 503
Cauchy–Goursat theorem, 505
Cauchy–Riemann conditions, 500
Cauchy–Riemann theorem, 501
Cavalieri, Bonaventura
    biography, 90
Cavendish, Henry
    biography, 23
Cayley, 192, 211
Cayley, Arthur
    biography, 192
center of mass, 21
central force, 354, *579–583*
    eccentricity, 581
central limit theorem, 809
centrifugal potential, 581
chain rule, *55–57*
Champollion, 304
chaos, 753
    theory
        systems obeying DE, *770–773*

systems obeying iterated map, 763
universality, *773–778*
coefficient, 173
cofactor, 205
collision
relativistic, *250–253*
column vector, 177
combination, 791
complement, 783
complex
conjugate, 479
function
analytic, 499
continuous, 499
regular point, 499
singular point, 499
singularity, 499
integral
positive sense, 507
number, 478
absolute value, 479
argument, 483
imaginary part, 478
real part, 478
plane, 478
complex amplitude, 486
complex function, *497–511*
derivative, *499–503*
derivative as integral, *509–511*
integration, *503–508*
complex number
Cartesian form, 478
Fourier series, 489
polar form, 482
roots, 486
complex numbers, *477–488*
Cartesian form, *477–481*
Fourier series, *488–491*
polar form, *482–488*
complex power series, 516
analyticity, 517
convergence circle, 517
differentiation, 517
integration, 517
uniform convergence, 517
complex series
absolute convergence, 516
component, 176
Compton wavelength, 253
conditional probability, *786–789*

conducting cylindrical can, 655
conductor
electrical, 594
heat, 598
confluent hypergeometric function, *332–333*
connection
affine, *462–464*, 470
metric, *465–468*
constraints, 360, 738
continuity equation, *378–381*
differential form, 380
integral form, 380
contour, 505
Bromwich, 722
simple closed, 505
contractible to zero, 400
contraction, 451
contravariant vector, *445–447*
convergence
test, *267–272*
convolution, 716
coordinate
generalized, 741
coordinate system, *11–15*
bipolar, 73
three-dimensional, 74, 438
Cartesian, 11, 12
cylindrical, 12
elliptic, 73, 213
elliptic cylindrical, 73, 213, 436
parabolic, 73
paraboloidal, 74, 437
polar, 11
prolate spheroidal, 74, 213, 437
spherical, 12
toroidal, 74, 213, 437
unit vector, *31–36*
vector, *16–31*
coordinate time, *239–240*
Copernicus, 97, 417, 580
correlation
probability, 803
cosine transform, 697
Coulomb, 744
Coulomb's law, 22, 24
Coulomb, Charles
biography, 23
covariance
in probability, 803

covariant derivative, *464–465*
covariant differential, *462–464*
covariant vector, *445–447*
Cramer, 210
Crelle, 326
cross product, 7, *28–31*
    as a tensor, 447
    Levi-Civita symbols, 458
    parallelepiped volume, 10
    parallelogram area, 9
curl
    curvilinear coordinates, *431–435*
    vector field, *391–398*
current density, 379
    and flux, 379
curvature, *468–471*
    scalar, 470
curve
    parametric equation, 61
    primary, 59
curvilinear
    vector analysis, *423–435*
curvilinear coordinates
    curl, *431–435*
    divergence, *427–431*
    gradient, *425–427*
    Laplacian, *435*
cycloid, 732

d'Alembert, 273, 303
d'Alembert, Jean Le Rond
    biography, 548
d'Alembert, 743
damping factor, 311
DE
    first-order, *551–561*
        integrating factor, *553–555*
        linear, *556–561*
    second-order, *563–570*
de Broglie, 666
de Moivre theorem, 485
del operator, 359
delta
    Kronecker, 442
delta function
    and Laplacian, 412
    cylindrical, 160
    derivative, 147, 159
    Legendre expansion, 630
    limit of sequence, 492

one-variable, *139–151*
point sources, 144
polar, 156
representation, *491–492*
spherical, 160
three-variable, *159–165*
two-dimensional, 155
two-variable, *154–159*
density, 45
    current, 379
    flux, *371–381*
    of states, 677
    probability, 801
density function
    surface, 154
derivative, *44–46*
    covariant, *464–465*
    functional, 730
    mixed, 52
    normal, 593
    partial, *47–59*
    time
        vector, *350–355*
    total, 86
Descartes, 46, 97, 103, 215, 417, 481, 482
Descartes, Rene
    biography, 15
determinant, *202–207, 222–227*
    parallelepiped volume, 10
differential, *53–54*
    absolute, 463
    covariant, *462–464*
    exact, 553
differential equation
    Bessel, 548, 641
        recursion relation, 644
        second solution, *645–646*
        solutions, *642–645*
    confluent hypergeometric, 332
    Hermite
        recursion relation, 668
    hypergeometric, 328
    Legendre, 608
        second solution, *617–619*
    order of, 556
    ordinary, 542
    partial, 542
    second-order linear
        adjoint, 572
        integrating factor, 571

differential operator, 217, 576
diffusion equation, 661
    time-dependent, 663
dimension, 176
    fractal, *775–778*
dipole
    approximation, 298
    magnetic, 410
dipole moment, 298
dipole potential, 299
Dirac, 26
    biography, 151
Dirac delta function
    in variational problems, 730
    step function, 153
Dirac, Paul Adrien Maurice
    biography, 151
disjoint sets, 783
distance
    spacetime, *240–242*
distribution, 146
    normal, *806–809*
        sum of two, 807
divergence, *371–381*
    curvilinear coordinates, *427–431*
    spherical coordinates, 430
    theorem, *374–378*
    vector field, 374
Doppler shift
    relativistic, 255
dot product, 5, 21
double del operation, *407–412*
double factorial, 319
dummy index, 262
dynamical system
    autonomous, 767
    nonautonomous, 767

eccentricity, 581
eigenvalue, 224
eigenvalue equation, 224
eigenvector, 224
Einstein, 215, 666
    summation convention, 441
Einstein curvature tensor, 471
Einstein equation, 471
electric field, 104
    point charge, 25
electrical conductor, 594
electrodynamics

Lagrangian density, 745
    tensors, *459–461*
element
    area, *59–68*
    Cartesian, *60–62*
    cylindrical, *65–68*
    length, *59–68*
    spherical, *62–64*
    volume, *59–68*
elliptic coordinates, 73, 213
elliptic cylindrical coordinates, 73, 213,
    436
elliptic functions, *322–326*
elliptic integral
    complete, 324
    first kind, 323
    second kind, 323
empty set, 782
energy
    relativistic, 249
    zero mass particle, 250
energy momentum tensor, 471
equation
    canonical, 749
    Klein-Gordon, 747
error function, *322*, 806
Euclid, 47, 80, 90
Euler, 272, 303, 326, 330, 503, 642, 743
Euler angles, 201
Euler equation, 483
Euler's equation, 414
Euler, Leonhard
    biography, 321
Euler-Lagrange equation, *729–731*,
    734–736, 738, 739
event, 784
    compound, 784
    elementary, 784
    random, 781
exact differential, 553
expectation value, 790
extremum problem, 727
    gradient, *359–361*

factorial
    double, 319
factorial function, 99, 318
Faraday, 26
Feigenbaum alpha, 774
Feigenbaum delta, 773

Feigenbaum numbers, *773–775*
Fermat, Pierre de
    biography, 15
Fermi energy, 677
Fermi-Dirac statistics, 791
Feynman, 26
field, *21–28*, 343
    electric, 104
    scalar, 343
    spinor, 343
    tensor, 343
    vector, 343
field point, 25, 78
fine structure constant, 679
finite constraint problem, 739
fixed point
    iterated map, 755
    stable, 756
flat space, 470
Florence Nightingale, 210
fluid dynamics, *413–415*
flux, *365–369*
    density, *371–381*
    vector field, *365–369*
FODE, *551–561*
    Bernoulli, 560
    homogeneous, 560
    integrating factor, *553–555*
    Lagrange, 561
    linear, *556–561*
    normal, 551
        integral of, 552
FOLDE, *556–561*
    explicit solution, 557
force
    central, 354
force density, 414
form factor, 702
four-acceleration, 248
four-momentum, *247–250*
four-vector, 243
four-velocity, *247–250*
Fourier, 115, 279, 322
Fourier series, *299–303*
    complex numbers, *488–491*
    to Fourier transform, *693–696*
Fourier transform, *693–712*
    and derivatives, *702–703*
    and quark model, 702
    application to DEs, *702–704*

convolution theorem, 724
Coulomb potential
    charge distribution, 701
    point charge, 700
    definition, 695
    examples, *698–702*
    Gaussian, 699
    Green's functions, *705–712*
    heat equation
        one-dimensional, 704
    higher dimensions, 696
    inverse, 695
    of delta function, 698
    properties, *696*
Fourier, Joseph
    biography, 304
Fourier-Bessel series, 655
fractal, 777
fractal dimension, *775–778*
free index, 440
frequency
    natural, 586
Frobenius method, *608–610*, 693
function
    analytic
        isolated singularity, 525
        principal part, 528
    antiderivative, 87
    as integral, *317–326*
    as power series, *327–335*
    Bessel, *333–335*, 644
        Laplace's equation, *642–654*
    beta, *320*
    complex, *497–511*
        derivative, *499–503*
        residue, 526
    complex hyperbolic, 502
    complex trigonometric, 502
    confluent hypergeometric, 332
    delta
        point sources, 144
    elliptic, *322–326*
    error, *322*
    even, 84
    factorial, 318
    gain, 586
    gamma, *318–319*
        Stirling approximation, 319
    harmonic, 501
    homogeneous, *57–59*

hypergeometric, *328–330*
  integral representation, 329
iterated map, 755
linear density, 143
logistic map, 755
odd, 84
periodic, 299
piecewise continuous, 82
primitive, 87
rational
  integral, *529–531*
sequence, *274–279*
series, *274–279*
special, 550
transfer, 586
functional, 728
functional derivative, 730
fundamental theorem of algebra, 478
fundamental theorem of calculus, 87

G-orthogonal, 187, 219
  matrix, 191, 222
    space, 200
  vector
    in space, 199
gain function, 586
Galileo, 26, 90, 97, 325
gamma function, *318–319*
  Stirling approximation, 319
gauge transformation, 418
Gauss, 279, 321, 326, 503, 617, 641
Gauss elimination, 231
Gauss's law, 369
  differential form, 378
  integral form, 377
Gauss, Johann Carl Friedrich
  biography, 330
Gaussian
  Fourier transform of, 699
Gay–Lussac, 594
generalized coordinates, 741
generalized momentum, 748
generating function
  Hermite polynomials, 673
geodesic, 465
  relativity, 466
  sphere, 466
geometric series, 271
geometry
  and metric tensor, 456

distance formula, 241
Gibb's phenomenon, 302
Gibbs, 370
Gibbs, Josiah Willard
  biography, 381
Goldbach, 320
gradient, *355–361*, 445
  components, 440
  curvilinear coordinates, *425–427*
  normal to surface, 358
  three dimensions, 357
  two dimensions, 357
Gram–Schmidt process, 221
  for space, 199
Grassmann, 382
Green, 210
Green's function
  advanced, 712
  differebtial eq. for, 707
  heat equation, *709–710*
  Laplacian, *708–709*
  Poisson equation, 709
  retarded, 712
  wave equation, *711–712*
Green's Functions, *705–712*
Gregory, 272, 294
guided wave, *682–686*
  TE, 684
  TEM, 685
  TM, 684

Halley, 641
Hamilton, 369, 382
Hamilton, William R.
  biography, 10
Hamiltonian, *747–749*
harmonic oscillator
  quantum, 667
    Hermite DE, 668
heat-conducting plate
  circular, *664–665*
  rectangular, *663–664*
heat-conducting rod, *662–663*
heat conductor, 598
heat equation, 543, *661–665*
  Green's function, *709–710*
  one-dimensional, 704
heat transfer
  time-dependent, 663

heat-conducting rod, 662
Heaviside, 370
Heaviside, Oliver
    biography, 382
Heisenberg, 26, 151, 675
Heisenberg uncertainty relation, 699
Helmholtz Coil, *291–293*
Helmholtz free energy, 54
Hermite DE
    recursion relation, 668
Hermite polynomial, 670
    orthogonality, 672
Hermite polynomials, 229, 575
    generating function, 673
Hermite, Charles
    biography, 674
HNOLDE, 575
    characteristic polynomial, 576
homogeneous
    function, 57
homogeneous function, *57–59*
homogeneous SOLDE
    exact, 571
Hooke, 97
Hopf bifurcation, 770
HSOLDE, 564
    second solution, 568
Huygens, 97, 103
hydrogen atom, *677–680*, 802
hyperbolic cosine, 290
hyperbolic sine, 290
hypergeometric function, *328–330*
    confluent, *332–333*
    integral representation, 329

identity matrix, 180
indeterminate form, *294–297*
index
    free, 440
indicial equation, 609
induction
    mathematical, *265–266*
inductive definition, 266
infinite series, *266–274*
inner product, *218–222*
    positive definite, 187
    Riemannian, 187
inner product matrix, 185
integral, 79
    as function, *317–326*

Bessel's, 652
    derivative of, *85–86*
    function of trigonometric, *534–536*
    indefinite, 87
    line, *387–391*
    Mellin inversion, 722
    rational function, *529–531*
    rational trigonometric, *532–534*
integral transform, 693
    kernel, 693
integrand, 80
integrating factor, *553–555*
integration, *77–80*
    application
        Cartesian coordinates, *104–107,
            112*, *115–117*
        cylindrical coordinates, *107–109,
            112–115*, *118–119*
        double integrals, *115–122*
        electricity, *104–109*
        general, *91–96*
        gravity, *104–109*
        magnetostatics, *109–115*
        mechanics, *101–103*
        single integral, *101–115*
        spherical coordinates, *120–122*
        triple integrals, *122–128*
    Cauchy integral formula, *508–509*
    change of dummy variable, 82
    complex function, *503–508*
    interchange of limits, 82
    linearity, 82
    parameter, 80
    partition of range, 82
    point, 79
    properties, *81–89*
    region of, 79
    small region, 83
    symmetric range, 84
    transformation of variable, 83
    variable, 80
intersection, 783
inverse
    matrix, 203, 207
    of a matrix, 180
inverse Fourier transform, 695
ionic crystal
    one-dimensional, 145
    potential energy, 164
    two-dimensional, 157

ISOLDE, 569
isoperimetric problem, 738
iterated map, *754–763*
    fixed point, 755
    orbit, 755

Jacobi, 211, 331
Jacobi, Carl Gustav Jacob
    biography, 326
Jacobian, *207–210*
    in probability, 804
Jacobian matrix, 208

Kaluza, 215
Kepler, 89, 97
Kepler's first law, 582
Kepler's second law, 582
Kepler's third law, 583
Kepler, Johannes
    biography, 579
kernel
    integral transform, 693
Klein-Gordon equation, 747
Koch snowflake, 777
Kronecker delta, 222, 442, 449, 489
    Euclidean metric, 466
    generalized, 452

Lagrange, 294, 304, 326, 330, 594,
      617, 642
    biography, 742
Lagrange identity, 572
Lagrange multiplier, 360, 738
Lagrangian, *740–745*
    interacting particles, 741
    Klein-Gordon, 747
    particle in EM field, 746
    single particle, 741
Lagrangian density, *744–745*
    electrodynamics, 745
Laguerre polynomials, 230, 679
Laplace, 115, 304, 322, 326, 744
Laplace transform, *712–723*
    and differential equations, *718–721*
    Bromwich contour, 722
    convolution, 716
    cosine, 713
    derivative, *717–718*
    first shift, 714
    gamma function, 713

imaginary exponential, 713
integral, *717–718*
inverse, *721–723*
linearity, 714
Mellin inversion integral, 722
periodic functions, 716
properties, *713–717*
second shift, 714
sine, 713
step function, 713
unit function, 713
Laplace's equation, 411, 542, 546
    Bessel functions, *642–654*
    Cartesian coordinates, *594–603*
    cylindrical coordinates, *639–656*
    Legendre polynomials, *610–617*
    radial equation, *619–622*
    solution
      uniqueness, 592
    spherical coordinates, *607–634*
    uniqueness of solution, *592–593*
Laplace, Pierre Simon de
    biography, 593
Laplacian, 411
    and Dirac delta function, 412
    curvilinear coordinates, *435*
    Green's function, *708–709*
Laurent series
    complex, *518–522*
Lavoisier, 744
law of addition of velocities, 237
law of large numbers, 809
law of motion
    relativistic, *253–254*
Legendre, 304, 326
Legendre equation, 575
    recursion relation, 611
Legendre functions
    second kind, 618
Legendre polynomial, 228, 614, 616
    expansion in, *628–630*
      physical examples, *631–634*
    generating function, 621
    Laplace's equation, *610–617*
    multipole expansion, 621
    orthogonality, 625
    parity, 622
    properties, *622–628*
    recurrence relation, 623
    Rodrigues formula, 626

Legendre polynomials, 229, 575
Legendre transformation, 54, 748
Legendre, Adrien-Marie
    biography, 617
Leibniz, 46, 87, 90, 97, 210, 272, 482
Leibniz, Gottfried Wilhelm
    biography, 103
length element
    primary, 59
Levi-Civita symbol, 453
Levi-Civita symbols
    cross product, 458
l'Hôpital's rule, *294–297*
limit cycle, 770
line integral, *387–391*
linear combination, 173
linear dependence, 174
linear equation, *230–234*
    compatible, 231
    echelon form, 232
    homogeneous, 234
    incompatible, 231
linear independence, 174
linear operator, 216
linear transformation, *216–218*
Liouville substitution, 588
logistic map, 755
    second iterate, 757
Lorentz gauge, 419
Lorentz transformation, *243–247*
    general, 244
    in 2 dimensions, 245
lowering indices, *457–459*
Lyapunov exponent, 763

Maclaurin, 210, 272
Maclaurin series, 287
Madelung constant, 165
magnetic charge, 409
magnetic dipole moment, 410
magnetic field
    moving charge, 30
magnetic force
    current loop, 420
    moving charge, 30
magnetic monopole, 409
manifold, 456, 469
map
    iterated, *754–763*
marginal probability, *786–789*

mathematical induction, *265–266*
matrix, 177
    G-orthogonal, 191, 222
        space, 200
    identity, 180
    inner product, 185
    inverse, 180, 203, 207
    Jacobian, 208
    metric, 185
    multiplication rule, 442
    orthogonal, 190
    symmetric, 182
    transformation
        in space, 195
    transpose, 181
    unit, 180
    zero, 180
Maxwell, 26, 369, 382
Maxwell's equations, *415–419*
    derivation of wave equation, 417
    relation to relativity, 237
Maxwell, James Clerk
    biography, 419
Maxwell-Boltzmann statistics, 791
mean, 790
Mellin inversion integral, 722
membrane, *686–687*
metric connection, *465–468*
    relativity, 466
metric matrix, 185
metric tensor, *454–461*
    definition, 456
    relativity, 458
minimal coupling, 749
Minkowski, 215
mode
    of oscillation, 682
Möbius band, 366
moment
    quadrupole, 449
moment generating function, 790
    binomial distribution, 793
    Poisson distribution, 799
moment of inertia, 122
momentum
    generalized, 748
    relativistic, 249
    zero mass particle, 250
Monge, 115, 304

motion
    constant of, 552
multipole expansion, *297–299*

Napoleon, 304, 593
natural frequency, 586
Neumann function, 645
Newton, 16, 26, 43, 46, 78, 87, 90, 103, 122, 272, 294, 317, 322, 326, 330, 481, 482, 548, 593
Newton, Isaac
    biography, 96
NOLDE, 575
nonautonomous, 767
normal distribution, *806–809*
    sum of two, 807
nth iterate, 760

ODE, 542
ODE and PDEs, *542–550*
Olbers, 641
operator
    angular momentum, 412
      spherical coordinates, 435
    del, 359
    differential, 576
      linear, 217
    linear, 216
orientable surface, 366
orthogonal
    matrix, 190
orthogonal polynomial
    standardization, 227
orthogonal polynomials, *227–230*
orthonormal
    basis, 186

parabolic coordinates, 73
paraboloidal coordinates, 74, 437
parallel translation, 465
Parseval relation, 654
Parseval's relation, 724
partial derivative, *47–59*
particle in a box, 675
Pascal, 15, 103, 481
passive transformation, 178
PDE, 542
    separation
      Cartesian coordinates, *544–546*
      cylindrical coordinates, *547–548*

      spherical coordinates, *548–550*
PDE and ODE, *542–550*
period-doubling, 757
periodic BC, 574
permutation, 791
phase space, *764–766*
    diagram, 764
    trajectory, 764
Planck, 666
plane
    basis, 175
Poincaré, 674
Poisson, 594
Poisson distribution, *797–800*
Poisson equation, 411, 542
    astrophysics, 415
    Green's function, 709
polar coordinates, 16
polar equation, 549
pole
    of order m, 528
    simple, 528
polynomial
    Hermite, 229, 670
    Laguerre, 230, 679
    Legendre, 228, 229, 614, 616
      Laplace's equation, *610–617*
    orthogonal, *227–230*
      standardization, 227
position vector, 19
potential, *21–28*, 399
    centrifugal, 581
    difference, 399
    of a dipole, 299
potential energy, 553
power series, *283–299*
    continuity, 285
    differential equations, 307
    differentiation, 285
    integration, 285
    operations, 520
    radius of convergence, 283
    zero, 285
pressure, 46
primary curve, 59
primary surface, 60
probability
    average, 790
    basic concepts, *781–792*
    binomial distribution, *792–797*

conditional , *786–789*
correlation, 803
covariance, 803
density, 801
expectation value, 790
independent random variable, 802
marginal , *786–789*
mean, 790
moment generating function, 790
Poisson distribution, *797–800*
sample space, *784–786*
set theory, *782–784*
standard deviation, 790
variance, 790
probability space, 784
prolate spheroidal coordinates, 74, 213, 437
proper time, *239–240*

quadrupole moment, 449
quantization
    hydrogen atom, 679
quantum harmonic oscillator, *667–674*
quantum mechanics
    angular momentum operator, 412
        spherical coordinates, 435
quantum particle
    in a box, *675–677*
quantum tunneling, 676
quaternions, 11

radial, 19
radial equation, 549
raising indices, *457–459*
random event, 781
random variable
    continuous, *801–809*
    independent, 802
    transformation, *804–806*
rate of change, 44
ratio test
    Waring, 273
recursion relation, 308, 610
relativistic collision, *250–253*
relativistic energy, 249
relativistic law of motion, *253–254*
relativistic momentum, 249
relativity
    geodesic, 466
    metric connection, 466

metric tensor, 458
principle, 238
special, 237
residue, 526
    calculus, *525–536*
residue theorem, 527
    definite integral
        rational function, 529
        rational trigonometric, 532
        trigonometric function, 534
retarded Green's function, 712
Ricci tensor, 470
Riemann, 321
Riemann curvature tensor, *468–471*
Riemann zeta function, 269
Riemannian manifold, 456
right-hand rule, 392
rigid transformation, 190
Rodrigues formula, 626
Rosetta stone, 304
row vector, 181

sample space, *784–786*
Savart, Felix
    biography, 115
scalar curvature, 470
scalar function, 445
Schrödinger, 675
    biography, 666
Schrödinger equation, 543, 546, *666–680*
    time-independent, 666
Schwarz inequality, 185, 220
Schwinger, 26
second iterate, 757
second variation, *735–738*
self-similarity, 775
separated boundary conditions, 574
separation of time, 543
separatrix, 766
sequence, *259–262*
    bounded, 261
    convergence, 260
        Cauchy criterion, 261
    divergence, 260
    functions, *274–279*
    limit, 260
    monotone decreasing, 261
    monotone increasing, 261
    partial sum, 259, 267
series, *266–274*

alternating
  test, 270
application to DE, *307–311*
complex, 518
  Laurent, *518–522*
  Taylor, *518–522*
convergence
  absolute, 268
  comparison test, 268
  conditional, 272
  generalized ratio test, 270
  integral test, 268
  n-th term test, 267
  ratio test, 269
convergent
  grouping, 273
  rearranging, 273
familiar functions, *287–291*
Fourier, *299–303*
  complex numbers, 489
Fourier–Bessel, 655
functions, *274–279*
  uniform convergence, 276
geometric, 271
harmonic
  order p, 269
Laurent
  complex, 518
Maclaurin, 287
  binomial function, 288
  complex, 518
  exponential function, 287
  hyperbolic function, 289
  logarithmic function, 291
  trigonometric function, 287
operations on, *273–274*
power, *283–299*
  differential equations, 307
Taylor, *286–287*
  complex, 518
  multivariable, *305–307*
uniform convergence
  differentiation, 278
  integration, 278
uniformly convergent, *277–279*
set theory, *782–784*
  complement, 783
  difference, 783
  disjoint sets, 783
  intersection, 783

union, 782
Venn diagrams, 783
sine transform, 697
soap film problem, 733
SOLDE, *563–570*
  basis of solutions, 565
  central force, 579
  constant coefficient, *575–587*
    homogeneous, *576–583*
    inhomogeneous, *583–587*
  homogeneous, 564
    second solution, *567–569*
  inhomogeneous
    general solution, *569–570*
  Kepler problem, 580
  linearity, *564–565*
  normal form, 563
  singular point, 563
  superposition, *564–565*
  superposition principle, 564
  uniqueness of solution, *564–565*
  uniqueness theorem, 565
  variation of constants, 569
  Wronskian, *566–567*
solid angle, *344–350*
  total, 349
source point, 25, 79
space
  dimension, 11
  flat, 470
  point, 11
  probability, 784
spacetime distance, *240–242*
  being zero, 242
span, 175
special functions, 550
standard basis, 216
standard deviation, 790
statistical independence, 788
statistics
  Bose-Einstein, 792
  Fermi-Dirac, 791
  Maxwell-Boltzmann, 791
stellar equilibrium, 415
step function, *152–153*
  Dirac delta function, 153
  Laplace transform, 713
Stifel, 481
Stirling, 320
Stirling approximation, 319, 792, 808

Stokes' theorem, *391–398*
Stokes, George Gabriel
    biography, 398
strange attractor, 778
Sturm–Liouville
    system, 574
Sturm-Liouville equation, 574
subset, 782
success excess, 793
summation, *262–266*
superposition principle, 25, 564
surface
    primary, 60
Sylvester, and Cayley, 192
Sylvester, James Joseph
    biography, 210
symmetric matrix, 182
symmetric tensor, 452

Taylor series, *286–287*
    complex, *518–522*
    multivariable, *305–307*
Taylor, Brook
    biography, 294
tensor, *447–454*
    addition, 450
    algebraic properties, *450–452*
    contraction, 451
    differentiation, *462–468*
    Einstein curvature, 471
    electrodynamics, *459–461*
    energy momentum, 471
    Levi-Civita symbols, 453
    metric, *454–461*
        definition, 456
        relativity, 458
    multiplication, 451
    numerical, *452–454*
    rank of, 448
    Ricci, 470
    Riemann curvature, *468–471*
    symmetrization, 452
    torsion, 463
terminal velocity, 559
theorem
    central limit, 809
time
    coordinate, *239–240*
    proper, *239–240*
time constant, 586

toroidal coordinates, 74, 213, 437
torque, 28
torsion tensor, 463
transfer function, 586
transform
    cosine, 697
    Fourier, *693–712*
        and quark model, 702
        application to DEs, *702–704*
        convolution theorem, 724
        examples, *698–702*
        Gaussian, 699
        Green's functions, *705–712*
        heat equation in 1D, 704
        inverse, 695
        of delta function, 698
        properties, *696*
    integral, 693
    Laplace, *712–723*
        and differential equations, *718–721*
        Bromwich contour, 722
        convolution, 716
        cosine, 713
        derivative, *717–718*
        first shift, 714
        gamma function, 713
        imaginary exponential, 713
        integral, *717–718*
        inverse, *721–723*
        linearity, 714
        Mellin inversion integral, 722
        periodic functions, 716
        properties, *713–717*
        second shift, 714
        sine, 713
        step function, 713
        unit function, 713
    sine, 697
transformation
    active, 178
    coordinate, 13
    differentiation, 197
    gauge, 418
    Legendre, 54, 748
    linear, 216
    Lorentz, *243–247*
    matrix
        in space, 195
    orthogonal, 442

passive, 178
    rigid, 190
transient term, 586
transpose
    of a matrix, 181
transposition, 181
    properties, 182
triangle inequality, 480
tunneling, 676

uncertainty relation, 699
uniform convergence
    Weierstrass M-test, 276
uniformly convergent series, *277–279*
union, 782
unit matrix, 180
unit vectors, 5
universal set, 782
    partition, 785

Van de Graff, 117
Vandermonde, 210
variable
    random
        continuous, *801–809*
        transformation, *804–806*
variance, 790, 801
variational problem, *728–740*
    constraints, *738–740*
    several dependent variables, 734
    several independent variables, 734
    soap film, 733
vector
    Cartesian
        component, 216
        n-dimensional, 216
    column, 177
    component, 176
    contravariant, *445–447*
    coordinate system, *16–31*
    covariant, *445–447*
    cross product, *7–10*
    field
        conservative, *398–404*
        curl, *391–398*
        flux, *365–369*
    G-orthogonal, 219
        in space, 199
    indices, *439–471*

inner product, *182–191*, *198–202*
    plane, *3–10*, *174–191*
    position, 19
    row, 181
    space, *3–10*, *192–207*
    time derivative, *350–355*
    transformation, *194–198*
    transformation of components, *176–
        182*
    transformation properties, *441–445*
    unit, 5
vector analysis
    curvilinear, *423–435*
vector field
    conservative
        curl, 400
    curl of, 394
    divergence, 374
vector potential, 408
vector space, 173, *215–227*
velocity, 44
    terminal, 559
Venn diagrams, 783
vibrating membrane, *686–687*
Vieta, 481

Wallis, 97, 293, 321, 326
Wallis, John
    biography, 90
wave equation, 543, *680–687*
    advanced Green's function, 712
    from Maxwell's equations, 417
    Green's function, *711–712*
    retarded Green's function, 712
wave guide, *682–686*
    cylindrical, 686
    longitudinal part, 682
    rectangular, 685
    transverse part, 682
weight function, 227
Wheatstone, 382
Wronskian, *566–567*

Yukawa potential, 700

zero mass, 250
zero matrix, 180
zero spacetime distance, 242
zeta function, 269

Printed in the United States of America

Printed in the United States
By Bookmasters